INTRODUCING HUMAN GEOGRAPHIES

Third edition

EDITED BY
PAUL CLOKE, PHILIP CRANG
AND MARK GOODWIN

LONDON AND NEW YORK

First edition published 1999 by Hodder Arnold

Second edition published 2005 by Hodder Arnold

Third edition published 2014 by Routledge
2 Park Square, Milton Park, Abingdon, Oxon OX14 4RN

and by Routledge
711 Third Avenue, New York, NY 10017

Routledge is an imprint of the Taylor & Francis Group, an informa business

British Library Cataloguing in Publication Data
A catalogue record for this book is available from the British Library

Library of Congress Cataloging in Publication Data
Introducing human geographies / [edited by] Paul Cloke, Philip Crang and Mark Goodwin. — Third edition.
pages cm
1. Human geography. I. Cloke, Paul J. editor of compilation.
GF41.I56 2013
304.2—dc23 2012046354

ISBN: 978-0-415-82663-1 (hbk)
ISBN: 978-1-4441-3535-0 (pbk)
ISBN: 978-0-203-52922-5 (ebk)

Typeset in AGaramond and Futura
by Keystroke, Station Road, Codsall, Wolverhampton

Printed and bound in India by Replika Press Pvt. Ltd.

INTRODUCING HUMAN GEOGRAPHIES

Introducing Human Geographies is the leading guide to Human Geography for undergraduate students, explaining new thinking on essential topics and discussing exciting developments in the field.

This new edition has been thoroughly revised and updated, and coverage is extended with new sections devoted to biogeographies, cartographies, mobilities, non-representational geographies, population geographies, public geographies and securities. Presented in three parts with 59 contributions written by expert international researchers, this text addresses the central ideas through which Human Geographers understand and shape their subject.

Part 1: Foundations engages students with key ideas that define Human Geography's subject matter and approaches, through critical analyses of dualisms such as local–global, society–space and human–nonhuman. *Part 2: Themes* explores Human Geography's main sub-disciplines, with sections devoted to biogeographies, cartographies, cultural geographies, development geographies, economic geographies, environmental geographies, historical geographies, political geographies, population geographies, social geographies, and urban and rural geographies. Finally, *Part 3: Horizons* assesses the latest research in innovative areas, from non-representational geographies to mobilities, securities and publics.

This comprehensive, stimulating and cutting-edge introduction to the field is richly illustrated throughout with full-colour figures, maps and photos. These are available to download on the companion website, located at www.routledge.com/cw/cloke.

Paul Cloke is Professor of Human Geography at the University of Exeter.

Philip Crang is Professor of Cultural Geography at Royal Holloway University of London.

Mark Goodwin is Professor of Human Geography and Deputy Vice-Chancellor at the University of Exeter.

Contents

List of contributors

Editors

Paul Cloke is Professor of Human Geography at the University of Exeter.

Philip Crang is Professor of Cultural Geography at Royal Holloway University of London.

Mark Goodwin is Professor of Human Geography and Deputy Vice-Chancellor at the University of Exeter.

Contributors

Peter Adey is Professor of Human Geography at Royal Holloway University of London.

Ben Anderson is Reader in Geography at Durham University.

Clive Barnett is Professor of Geography at the University of Exeter.

Katherine Brickell is Senior Lecturer in Human Geography at Royal Holloway University of London.

Harriet Bulkeley is Professor of Geography at Durham University.

David Conradson is Senior Lecturer in Human Geography at the University of Canterbury.

Jeremy Crampton is Associate Professor in Geography at the University of Kentucky.

Mike Crang is Professor of Human Geography at Durham University.

Tim Cresswell is Professor of Human Geography at Royal Holloway University of London.

Jessica Dempsey is Assistant Professor in Environmental Studies at the University of Victoria.

Klaus Dodds is Professor of Geopolitics at Royal Holloway University of London.

Felix Driver is Professor of Human Geography at Royal Holloway University of London.

Claire Dwyer is Senior Lecturer in Geography and Deputy Director of the Migration Research Unit at University College London.

Sally Eden is Reader in Environmental Issues and Geography at the University of Hull.

David Gilbert is Professor of Urban and Historical Geography at Royal Holloway University of London.

Pyrs Gruffudd is Senior Lecturer in Geography at Swansea University.

Muki Haklay is Professor of Geographical Information Science at University College London.

Sarah Hall is Associate Professor in Geography at the University of Nottingham.

Chris Hamnett is Professor of Human Geography at King's College London.

Stephen Hinchliffe is Professor of Human Geography at the University of Exeter.

Russell Hitchings is Lecturer in Human Geography at University College London.

Sarah L. Holloway is Professor of Human Geography at Loughborough University.

Peter Jackson is Professor of Human Geography at the University of Sheffield.

Nuala C. Johnson is Reader in Human Geography at Queen's University Belfast.

Andrew Jones is Professor of Economic Geography and Dean of the School of Arts and Social Sciences at City University London.

Scott Kirsch is Associate Professor in Geography at the University of North Carolina at Chapel Hill.

Rob Kitchin is Professor of Human Geography and Director of the National Institute of Regional and Spatial Analysis at the National University of Ireland, Maynooth.

Khalid Koser is Deputy Director and Academic Dean at the Geneva Centre for Security Policy.

Molly Kraft is a graduate student in geography at the University of British Columbia.

Peter Kraftl is Reader in Human Geography at the University of Leicester.

Lisa Law is Senior Lecturer in the School of Earth and Environmental Sciences at James Cook University.

Wen Lin is Lecturer in Human Geography at Newcastle University.

Jo Little is Professor of Gender and Geography at the University of Exeter.

Hayden Lorimer is Reader in Human Geography at the University of Glasgow.

Juliana Mansvelt is Senior Lecturer in Geography at Massey University.

Jon May is Professor of Geography at Queen Mary, University of London.

Claudio Minca is Professor and Head of the Cultural Geography chair group at Wageningen University.

David Murakami Wood is Associate Professor in Sociology and Geography and Canada Research Chair in Surveillance Studies at Queen's University.

Miles Ogborn is Professor of Geography at Queen Mary, University of London.

Kris Olds is Professor of Geography at the University of Wisconsin-Madison.

Hester Parr is Reader in Human Geography at the University of Glasgow.

Richard Phillips is Professor of Human Geography at the University of Sheffield.

Geraldine Pratt is Professor of Geography at the University of British Columbia.

Sarah A. Radcliffe is Professor of Latin American Geography and Fellow of Christ's College at the University of Cambridge.

Amanda Rogers is Lecturer in Human Geography at Swansea University.

Paul Routledge is Professor of Human Geography at the University of Glasgow.

Joanne P. Sharp is Professor of Geography at the University of Glasgow.

Juanita Sundberg is Associate Professor of Geography at the University of British Columbia.

Julia Verne is Lecturer in Human Geography at Goethe University Frankfurt am Main.

Lauren Wagner is Lecturer in Cultural Geography at Wageningen University.

Michael Watts is Professor of Geography at University of California, Berkeley.

Sarah Whatmore is Professor of Environment and Public Policy and Fellow of Keble College at the University of Oxford.

Mark Whitehead is Professor of Human Geography at Aberystwyth University.

Katie Willis is Professor of Human Geography at Royal Holloway University of London.

Keith Woodward is Assistant Professor of Geography at University of Wisconsin-Madison.

John Wylie is Professor of Cultural Geography at the University of Exeter.

Acknowledgements

Many deserve our thanks for their help with this third edition of *Introducing Human Geographies*. Once more we are indebted to the cohorts of first-year undergraduate students that we have taught over the many years that this book has been part of our professional lives: at Lampeter first, and then at Bristol, Aberystwyth, UCL, Royal Holloway University of London and Exeter. They have consistently shown enthusiasm for engaging with new, often complex ideas and materials, so long as that complexity is not used as an excuse for excluding all but a small clique from the conversation. We continue to strive to do that enthusiasm justice. We thank too our expert contributors, both those returning and those new to this third edition. It was a pleasure to edit their chapters and in the process to be reminded once again of the vibrant intellectual life across the range of Human Geography. This book began life overseen by Hodder Education; we thank in particular Bianca Knights and Beth Cleall for their Herculean efforts on the third edition. When Taylor & Francis acquired Hodder's Geography list, Andrew Mould and Faye Leerink at Routledge put in the hard yards to get us to publication. On a personal note, special thanks are due, as ever, to Viv, Liz and Will; Katharine, Esme and Evan; and Anne, Rosa and Sylvie.

Paul Cloke, Philip Crang and Mark Goodwin

Introducing Human Geographies: a guide

Introducing Human Geographies is a 'travel guide' into the academic subject of Human Geography and the things that it studies. Now in an updated and much extended third edition, the book is designed especially for students new to university degree courses. In guiding you through the subject, *Introducing Human Geographies* maps out the big, foundational ideas that have shaped the discipline past and present (in Part 1); explores key research themes being pursued in Human Geography's various sub-disciplines (in Part 2); and identifies some of the current research foci that are shaping the horizons of the subject (in Part 3).

Engaging with research literatures through academic journals and books is an important part of degree level study. The debates going on within them are exciting, challenging us not only to think about new subjects but also to think in new ways. However, it can take some time to get to grips with that published research. It is huge and diverse. It is dynamic, so that as a new student you can feel like you are coming into conversations halfway through, trying to figure out what people are talking about, why they are interested in it, and how come they are so animated about things. Academic publications are also by and large addressed to other researchers, deploying what may feel like rather arcane vocabularies as communicational shorthand. So not only has the conversation already begun, but it can also sound like it is in a foreign language. The ethos of *Introducing Human Geographies* is to make that cutting edge of contemporary Human Geography accessible; to map out key areas of study and debate; to guide you on forays into its somewhat daunting collections of ideas and interests; and to help you, as students of the subject, to participate in its conversations.

The Human Geography you will be introduced into here feels very different to some of the popular images of the subject. It is not a dry compendium of facts about the world, its countries, capital cities, and so on. Apologies in advance if this book is of limited help in getting the geography questions right in a quiz or television game show. Of course, knowing geographical facts and information is useful and important in all kinds of ways. But it is not enough. Human Geography today casts information in the service of two larger goals. On the one hand, it seeks out the realities of people's lives, places and environments in all their complexity. Geography is a subject that lives outside the classroom, the statistical dataset or the abstract model, gaining strength from its encounters with what (somewhat comically) we academics have a tendency to call 'the world out there'. On the other hand, Human Geographers are also acutely aware that this worldly reality is not easy to discern. The nature of the world is not laid out before our eyes, waiting for us to venture out blinking from the dark lecture theatre or library so that we can see it. 'Reality' only emerges through the carefully considered ways of thinking and investigating that we sometimes call 'theory'. As the contributions to this book show, Human Geography is characterized by a refusal to oppose 'reality' and 'theory', worldly engagement and contemplative, creative thought. Both are needed if we are to describe, explain, understand, question and maybe even improve the world's human geographies.

As you may have noticed (when putting your back out trying to pick it up . . .), this third edition of *Introducing Human Geographies* is a large book. It has a lot in it. Its contents are diverse. In this general introduction we therefore want to do three things. First, we focus on what unites this variety by addressing head on the question 'What is Human Geography?'. Second, we expand on the kinds of approaches and styles of thought that characterize Human Geography today across its range of substantive fields. Finally, we briefly map out the layout of the book itself, both in terms of structure and presentational style, offering some advice on how you might navigate around it.

What is Human Geography?

A common exercise for an initial Human Geography tutorial or seminar is a request to mine a week's news coverage and to come back with an example of something that seems to you to be 'human geography'. Have a go at doing this now. Think about the last week's news. Draw up a shortlist of two or three stories that strike you as the kinds of things that Human Geographers would study or that you think are 'human geography'. Then reflect on how you decided on these and what you thought was 'geographical' about them. What does your selection tell you about what human geography means to you?

The word 'geography' can be traced back to ancient Greece over 2,200 years ago. Specifically, it was Eratosthenes of Kyrene (ca. 288–205 BC), Librarian at Alexandria, who wrote the first scholarly treatise that established Geography as an intellectual field, the three-volume *Geographika* (Roller, 2010). In Greek, Geography means 'earth (*geo*) writing (*graphy*)'. Writing the earth was what Geographers did two millennia ago, and it still describes what Geographers do today. In all kinds of ways, it is a wonderful definition of the subject. It speaks to Geography as a fundamental intellectual endeavour concerned with understanding the world in which we live and upon which our lives depend. It expresses how Geography is all around us, a part of our everyday lives. It suggests that Geography is not confined to academic study but includes a host of more popular forms of knowledge through which we come to understand and describe our world. But it also raises questions, in particular about breadth and coherence. To return to that exercise of reviewing the week's news for examples of Human Geography, if what we were looking for were cases of 'earth writing' then an awful lot of stuff could fit that brief in some way. Most of the news is about things happening on the earth.

How do we deal with that breadth, with that seeming absence of specialization in Geography? We would suggest there are three sorts of responses: to recognize the underpinning intellectual commitments built into the very notion of geography; to embrace the diverse topics and events to which these relate; and to recognize the ways in which different areas of Geography are defined and organized. Let us take these in turn.

First, then, we need to think a little more directly about the 'geo' in geo-graphy, about what we mean by the *earth* in earth writing. This word is not just a general designation of everything around us but signals two interconnected cores to Human Geography's interests (Cosgrove, 1994): what we might call an 'earthiness' and a 'worldliness'; or, to use more current academic vocabulary, the relations between society and nature and between society and space (see Figure I). In terms of 'earthiness', the 'geo' in geography signifies 'the living planet Earth', the biophysical environments composed of land, sea, air, plants and animals that we live in and with. These are central concerns for Geographers. The relations between human beings and the 'nature' we are also part of have been a consistent preoccupation of Human Geography. There is a second meaning to 'geo' as well, though, that is equally central, one we use when we talk about 'the whole Earth' or 'the world'. Here, to write the earth means to explore its extents, to describe its areas, places and people, and to consider how and why these may have distinctive qualities. Human Geographers have long been fascinated with how various parts of the Earth's surface differ, with the relations between different areas, and with ways of knowing this (such as mapping and exploring). Geography endeavours to know the world and its varied features, both near to home and far away. The 'geo' in Geography designates this commitment to world knowledge.

The precise forms such concerns with 'earth' and 'world' have taken in Geography have varied over time of course, but both are central to the project of Human Geography today. Thus, Human Geographers lead debates over what are now often called the relations

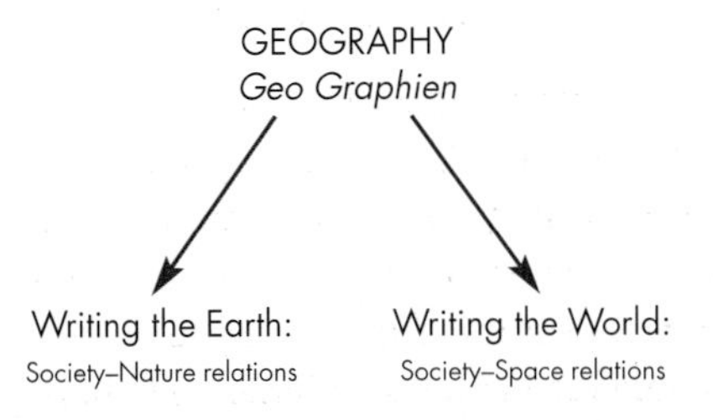

Figure I Human Geography: writing the earth and writing the world

between society and nature, on environmental understandings and values, on the causes and responses to environmental change. They do so at a variety of scales, from global concerns with climate change to local debates over particular environments and landscapes. Human Geographers are also concerned with how human lives, and our relations to nature, vary across the surface of the Earth. Everything happens somewhere and Human Geographers argue that this matters. A variety of central geographical notions reflect this: space, place, region, location, territory, distance, scale, for example, all try to express something about the 'where-ness' of things in the world. In contemporary parlance, Human Geographers emphasize the relations between society and space or what can be called **spatiality**. They argue both that human life is shaped by 'where it happens' and that 'where it happens' is socially shaped. The world and its differences are not innate; they are made. Human Geographers study that making.

Our argument, then, is that Human Geography today still lives up to the original meaning of its name, revolving around both 'writing the earth' (in contemporary academic parlance the relations between society and nature) and 'writing the world' (the relations between society and space). However, and this is the second point we want to develop, these core concerns are developed through a vast range of substantive topics. In this book you will find subject matters that range from the meanings of development and modernity to how we relate to plants when gardening, from the international financial system to tourism, from the post-9/11 'war on terror' to urban gentrification, from global climate change to shopping. And, for very good reasons, you'll also find a chapter largely devoted to a discussion of oven-ready chickens. It is quite common to have mixed feelings about

Figure II Which of these photographs of work look more like it should be in a Human Geography textbook to you? Why? Credit: (left) Jose Luis Palez, Inc./Corbis; (right) Brownie Harris/Corbis

this range. Many people choose to study Geography because of it, appreciating the wider understanding of human life such breadth seems to offer in comparison to many academic disciplines. In contrast, some react against it, worrying that Geographers seem to be 'jacks of all trades' and 'masters of none', complaining that Human Geography today seems to study things that 'aren't really Geography'.

In our view, the diversity of Human Geography is a strength not a weakness, for at least two reasons. First, it reflects how Geography existed well before, and exists well beyond, the kinds of specialization promoted by academic institutions over the last century or so (Bonnett, 2008). Geography is notable for how it challenges the divisions that have come to characterize academic organizations, spanning as it does the natural sciences, the social sciences and the arts and humanities. The world doesn't present itself to us in those categories and Geography resists being confined within them. As an academic discipline Geography has a healthy scepticism towards the disciplining of knowledge. Its diversity embodies that. Second, we would also encourage you to embrace the diversity of Human Geography in the spirit of being open to what might matter in the world. It is important that our thinking, and our academic disciplines, are not defined by inertia, pursuing topics simply because those are the subjects that we have traditionally pursued. Convention is not a good way to define and delimit what counts as Human Geography. You may find some of the subjects discussed in *Introducing Human Geographies* more familiar to you – for example, economic globalization – some less so – the idea of 'emotional geographies', perhaps – but all of them represent how Human Geography today is pursuing its tasks of 'writing the earth and the world'. Knowing the traditions of Human Geography is enormously valuable, but one of the crucial lessons we learn from that history is that what counts as Human Geography has always been subject both to change and to contestation (see Livingstone (1992) for an excellent, sustained analysis of this). For instance, shaped by the social worlds in which it was being produced, for much of its history Human Geography largely ignored over half the world's human beings. It reduced human to man. Even well into the latter half of the twentieth century,

Economic Geographers largely ignored the domestic work done by women at home; Development Geographers paid too little attention to the gendered nature of both development problems and practice; issues and understandings that were seen as feminine were routinely trivialized and cast as less worthy of academic attention. Human Geography was **masculinist** (Women and Geography Study Group, 1997; Domosh and Seager, 2001). Countering this involved introducing into Geography many novel topics and ideas. The issue for us, then, is not whether a topic is familiar as Geography but whether attending to it is part of 'writing the earth' in ways that have value.

Let's go back to the example of the oven-ready chicken that we mentioned earlier. We are assuming that if you told friends or family you were studying the Human Geography of chickens they might raise a quizzical eyebrow. But in fact, as Michael Watts explains in Chapter 27, the lives and deaths of chickens speak profoundly to how human societies today relate both to our living Earth and to the spatial organization of the world. An oven-ready chicken such as that pictured in Figure III embodies very particular ways for human beings to relate to nature, based on logics and practices of domestication, industrialized production, purposive modification and commodity consumption that reach well beyond this one member of the animal kingdom. The oven-ready chicken is also an embodiment of forms of spatiality that are very common in the world today. Different people and places are all connected together through the economic systems of the chicken world – the consumers eating it, the farmers raising it, the large companies controlling its production and its distribution, the scientists genetically modifying it – but at the same time these connections are forgotten or hidden through a distancing of places of chicken production and consumption – even avid meat eaters would be unlikely to want to see video footage of broiler production and death as they tuck into their roast bird. An oven-ready chicken presents us with the geography of the modern world on our plates. It is Geography.

Figure III Human Geography? Credit: British Chicken Information Service

Generally, then, *Introducing Human Geographies* presents a diverse and dynamic subject, and poses questions for you about what might count as valuable forms of geographical knowledge. There is, however, also a third response to the diversity inherent in Geography's intellectual remit: to organize it into various 'sub-disciplines' and research specialisms. The very idea of Human Geography already manifests this response, reflecting the widespread division between Physical Geography (placed in the natural sciences) and Human Geography (located in the social sciences and humanities). Contemporary research literatures and curricula take the process of dividing and specializing much further, organizing

Human Geography itself into the kind of sub-disciplines we present in Part 2 of this book (biogeographies, cartographies, cultural geographies, development geographies, historical geographies, political geographies and so on). These each possess their own research literatures (via their specialist journals) and, indeed, their own introductory textbooks. Quite often these sub-disciplinary designations form the basis of how Human Geography is taught within universities. Sub-disciplines are helpful in a number of ways. They map out the diversity of Geography into recognizable areas of work. They promote the development of expertise. They focus Geographers' engagements with other academic disciplines (Political Geographers engaging with Political Science and International Relations, Historical Geographers with History, and so on). But they can also be problematic. If one gets too hung up on sub-disciplines one can lose the anti-disciplinary holism that is one of the strengths of the subject. The much discussed 'divide' between Physical and Human Geography is a case in point. Furthermore, sub-disciplinary labels bear the imprint of university bureaucracy and job titling; we academics are very used to them but outside of universities they don't much help people relate to the Geography that we do. So, the useful foci provided by the various sub-divisions of Geography need to be accompanied by an ongoing commitment to seeing the distinctively geographical contribution that they make to understanding our worlds. At its best, Human Geography has a strong intellectual coherence, but applies it with an invigorating catholicism.

SUMMARY

- Geography means 'earth writing'. As a subject with that aim, Geography is notable for its wide-ranging concerns and interests.
- The first meaning of the 'geo' in Geography is 'the Earth'. The first of Human Geography's main intellectual contributions is to understand the relations between human beings and the natural world of which we are a part.
- The second meaning of the 'geo' in Geography is 'the world'. The second of Human Geography's main intellectual contributions is to understand the world both near and far. More abstractly, this means recognizing how all facets of human societies – the economic, the environmental, the political and so forth – are bound up with questions of 'spatiality'.
- These fundamental concerns of Human Geography are pursued across diverse and changing subject matters. We would encourage you to be open to that diversity and change; resist restricting your Human Geography to topics and approaches with which you are already familiar.

Approaching Human Geography today

Up to this point we have been outlining what Human Geography is about, emphasizing its foci on both 'the earth' (society–nature relations) and 'the world' (society–space relations). Now we turn to how Human Geographers approach these issues and the kinds of knowledge that they try to create. Our interest is not in well-defined schools of thought or even intellectual paradigms but in the looser sensibilities that shape how Human Geography is done today.

At the outset, it is important to note that the approaches of Human Geography have changed over time and differ from place to place. Human Geography in the 1920s or 1960s was different to Human Geography today. The approaches to Human Geography in Germany, Brazil or China are not identical to those in Britain. Even individual university departments can have distinctive research cultures. In fact the situation is more complex still; as you may find in your courses, at any one time and in any one place there are likely to be different kinds of Human Geography being done. There is not a single agreed view on what kinds of knowledge Human Geography should produce. *Introducing Human Geographies* contains some of that variety; it does not present a single version of the subject. But it does reflect and support some recurrent emphases that, in our view, characterize much of Human Geography today. We see these as commitments to five kinds of knowledge: *description*, *experience*, *interpretation*, *explanation* and *critique* (see Table I). Not all of these are equally endorsed by all Human Geographers, indeed they are often argued over; but they are commitments you will find frequently evidenced both in this book and in the course of your studies. Let us elaborate on each in turn.

First, then, Human Geography looks to describe the world. Sometimes dismissed with the epithet 'mere', in fact *description* has a very special value. Geographical description is not synonymous with dry compendia of information about a region or place. It involves attending to the world unusually carefully. The nature of that attention can vary. It might mean, for example, fashioning and mapping forms of statistical data (perhaps via a Geographical Information System (GIS)) that allow us to describe things that we can't fully see with our own eyes – spatial differences in wealth or access to services perhaps. It might involve tracing out the often hidden networks of connections linking people and places, as when Human Geographers 'follow' the things that people routinely consume (our food or clothes, for example) to see how they came to be, where they come from, and what kinds of trade govern their movements (e.g. Cook, 2004). Or it might mean being peculiarly observant in person. Think, for example, about how we normally move around the world, head often down, taking our surroundings somewhat for granted. Now contrast that to a more geographical engagement with place, perhaps a public square, where we look to document the details of the built environment, its history, the people who are present and absent, the kinds of action going on. Here, to describe a place geographically is to bear witness to its material textures and the forms of life that unfold through it. Our argument, then, is that Human Geography is an attentive discipline. It describes in order to reveal what we might otherwise overlook and to bring into focus what we might otherwise only vaguely perceive. It crafts ways of presenting the fruits of this attention, using forms of description that range from maps to statistics, prose, photography and film/video-making.

Second, Human Geography also commits to understand the world through *experience*. In part we see this in the discipline's commitment to fieldwork. Geography places a value on trying to understand issues not just from afar but through actually being there, in a place, amongst the action, conversing with people, getting a feel for things. The status of this kind of first-hand field knowledge is philosophically complex, but Human Geography tends to view understanding gained only from more 'remote' sorts of sensing with some suspicion. It is not a subject that is comfortable with being confined to the lab or library. Important here too are the people-centred approaches trumpeted initially under the label **humanistic geography** (for exemplary collections, see Ley and Samuels (1978) and Meinig (1979); for a more recent revisiting of such humanistic work see Holloway and Hubbard (2000)). Humanistic Geography emphasizes engaging with people's real lives, their values and beliefs, their daily preoccupations, their hopes and dreams, their loves and hates, what they think about things,

Type of knowledge	Approach	Illustrative examples
Description	Paying close attention to, and finding ways to represent, geographies that we normally struggle to perceive.	Statistical descriptions, GIS visualizations and maps; tracings of spatial networks and associations; detailed evocations of particular places.
Experience	Understanding geographies as part of human experience.	The emphasis placed on the experiential knowledge generated by fieldwork; humanistic concerns with understanding other people's diverse experiences of the world.
Interpretation	Recognizing and engaging with the meanings of the world's geographies.	Work focusing on geographical representations and on the discourses of which they are a part. Often associated with the so-called 'cultural turn'.
Explanation	Explaining why the world's geographies exhibit the forms and processes that they do.	Geographical explanations range from spatial science's search for spatial laws to (more commonly today) socio-spatial analyses of causal processes.
Critique	Rigorously evaluating and judging the world's geographies, as well as one's own and others' understandings of them.	Critique can be understood as a broad stance to geographical knowledge. It has also come to be associated with bodies of work that explicitly designate themselves as forms of 'critical geography'.

Table I Approaches to Human Geography today: a schema

the ways they feel about and sense their surroundings. Human Geographers are thus not only interested in experiencing places for themselves; they want to understand other people's geographical experiences and thoughts in all their variety.

A commitment to interpreting the meaningful nature of the world is apparent here too. Geographies are not just brute realities; it is fundamentally human to invest the world with meaning. We don't only sense the world, we make sense of it. Human Geography is concerned with *interpretation* insofar as it recognizes the importance of the meanings of things. Think, for example, about the interest Geography has in 'the Earth' and society–nature relations. The things we call 'natural', indeed the very notion of the 'natural', are deeply imbued with meanings. Reflect for a few seconds on geographical notions like 'wilderness' or 'rainforest' or 'the tropics'. These words are not narrowly factual; they come with a host of (often complex and even conflicting) meanings and connotations. The same is true of how we describe the world's different spaces. Consider what geographical designations such as 'urban', 'suburban' and 'rural' might mean to you and others; or the continents (Europe, Asia, Africa, Antarctica . . .); or a seemingly simple geographical label like 'The West' or 'The Western World'. All these terms are, to use a colloquialism, 'heavily loaded'. Human Geography's approaches here are informed by wider bodies of thought in the humanities on interpretation and meaning (with great names like 'hermeneutics', 'semiotics' and 'iconography'). They are also often identified with what has been called 'the cultural turn' taken within the discipline since the 1990s (Barnes and Duncan, 1992). Prominent is a focus on **representation**, with research teasing out the meanings given to geographies in forms both obviously imaginative (literature, the arts, film and television drama and so on) and less obviously so (maps, documentaries, news reports, policy documents, etc.). Interpreting these representations is important because they are not just an imaginative gloss that we humans add to our worlds, a subjective filter that obscures objective reality. Representations shape how we see things, think about them and act with and upon them. They partly make our worlds. They are part of reality. In academic terminology, by interpreting what things mean we engage with the **discourses** that produce the world as we know it. As an interpretive endeavour, Human Geography both looks to understand those discourses and to present other ways of ways of seeing, describing and acting upon our geographies.

So far we have outlined that when Human Geographers undertake their 'earth writing' (geo-graphy) they look to describe, experience and interpret. A fourth commitment has flickered in and out of these discussions: to *explanation*. Human Geography is not only concerned with what the geographies of the world are, but also with how they came to be. The nature of geographical explanation has varied over time and is subject to much debate. Divergent views are underpinned by different understandings of both the world 'out there' and the sorts of knowledge required to grasp it. For some, Human Geography should be a **spatial science**, formulating and testing theories of spatial organization, interaction and distribution in order to establish universal spatial laws about why geographical objects are located where they are and how they relate to each other. Emerging in the 1960s, spatial science distinguished itself from earlier regional geographies, criticizing them for being overly descriptive and lacking the explanatory power of scientific analysis. However, other approaches in Human Geography resist the equation of explanation with spatial science.

They are wary of its kind of 'social physics'. Historical Geographers, for instance, emphasize how forms of historical narrative can have explanatory power. To put it crudely, a historical approach explains the world today by understanding past events and processes. More generally, most Human Geography is wary of explaining things via reference only to spatial factors (what is called 'spatial reductionism'), emphasizing instead the two-way relations between 'space and society'. Aspects of society – the modern nation-state, for example, or the capitalist economy – are seen both to shape the nature of space and themselves to have spatial dimensions. Thus, there are no universal spatial laws that can explain our geographies; any explanation must recognize the socially produced nature of spatiality. There is also a concern about seeking universal laws as explanations; instead, a range of theories – most visibly represented by an approach known as 'critical realism' (Sayer, 2000) – have sought to understand causality in relation both to more abstract powers and more concrete, contingent, contextual factors. As you have probably gathered by now, it is hard to do justice to these sorts of complex debates in a brief introduction (sorry!). But, in essence, our view is that Human Geography today widely exhibits a commitment to explain the geographical phenomena it studies, but generally undertakes that explanation through nuanced accounts that weave together underlying tendencies/forces with more contextually specific factors.

Fifth, and finally, contemporary Human Geography is concerned with *critique*. It is easy to misunderstand this word. In everyday speech, when we say someone is being critical what we often mean is that they are being negative or finding fault. But that is not what we have in mind here. True critical thought is as much about seeing strengths as weaknesses. Critique, then, means exercising judgement. For Human Geography, a commitment to critique means that the subject not only describes, experiences, interprets and explains but also rigorously evaluates the world's geographies. A general consequence of this commitment is that the 'rightness' of our answers to geographical questions is not given. There is room for debate and argument. Critique is not just a matter of expressing one's opinion, but its reasoned judgement involves values, beliefs and perspectives. For all of us, as students of Human Geography, there are not often agreed correct answers that we simply have to remember. Doing Human Geography involves developing rigorous analyses of issues, evaluating both information and arguments, and thereby figuring out not only what the answers are but also what the most important questions might be. This means not taking things for granted, questioning the assumptions held by others and, crucially, ourselves. Critical thought – and this is a tricky balance – combines a determined, questioning scepticism with a profound openness to unfamiliar ideas and voices. It seeks to evaluate present and past conditions and to disclose future possibilities and alternatives.

More narrowly, these general critical attitudes have shaped distinctive bodies of philosophy, theory and practice that take them forward. Within Human Geography, 'critical geography' has emerged as a designation that folds in earlier appeals to radicality – as seen in the foundation in the 1970s of the 'radical journal of geography', *Antipode* – and the 'dissident geographies' of **feminist**, **Marxist** and **post-colonial** writers (Blunt and Wills, 2000). The words of *ACME*, an open access online journal of 'critical geography', give a sense of this; for this journal, 'analyses that are critical are understood to be part of the praxis of social and political change aimed at challenging, dismantling and transforming prevalent relations, systems and structures of exploitation, oppression, imperialism, neoliberalism, national aggression and environmental destruction' (ACME, 2012). A range of work discussed in this third edition of *Introducing Human Geographies* would fit that definition in some part, but critical thinking in the more general sense is not necessarily signed up to particular political colours. It spans, too, both cerebral philosophical thought and the kinds of work more directly invested in practical change. Critique, then, can be taken as a more general stance, committed to questioning, reasoned judgement, and a hopeful search for possible better futures. That stance can be usefully adopted within your own studies and writing of Human Geography.

Above, we have outlined various commitments that shape Human Geography today – to description, experience, interpretation, explanation and critique. Whilst keyed into wider debates over forms of knowledge and the interests they pursue, these five categories are, inevitably, something of a heuristic device. They are not exhaustive. They are also not mutually exclusive; many kinds of geographical description might also see themselves as interpreting and/or explaining and vice versa, for example. But, with those caveats, we believe that this schema conveys some of the principal rationales for why Human Geography undertakes its 'earth writing' and a sense of what you can achieve by studying it.

SUMMARY

- Human Geography undertakes its 'earth writing' for a number of reasons. It is helpful to reflect on these reasons as you develop your own geographical imagination.
- We have suggested five undertakings that shape Human Geography today. We termed these: description, experience, interpretation, explanation and critique.
- This is not an exhaustive list of all the rationales that underpin Human Geography but one, some or all of these commitments shape a great deal of the scholarship that you will be introduced to in this book.

Introducing Human Geographies: finding your way around

We have used the metaphor of a travel guide to describe this book. Guidebooks are not designed to be read from front to back in one go. They set scenes, provide contexts, and then as a reader we dip into them, dependent on our interests and our travel schedules. This new edition of *Introducing Human Geographies* is the same. It is designed to accompany and guide you as you find your way around Human Geography. Exactly how you read it, which parts you spend most time in and so on, will depend on your own intellectual itinerary and your programme of studies. The format we have created for the book, with a large number of comparatively short chapters organized into parts and sections, supports that kind of tailored reading 'on the go'.

Nonetheless, it may be helpful to explain the book's structure. The fifty-nine main chapters are organized into three parts – Foundations, Themes and Horizons. The nine chapters in Foundations (Part 1) give you the latest thinking on some of the 'big questions' that have long shaped the thinking of Human Geographers. An introduction to Part 1 says more about the individual chapters, but let us a say a little here about their remit. In setting out our foundations we have eschewed two common approaches: on the one hand, a narrative or episodic history of the subject; and on the other, abstract summaries of key theoretical approaches or '-isms'. (Often these are offered in combination; a chronology of different theoretical schools, dated on the basis of when they became influential within Human Geography.) There are excellent books that adopt variants of such approaches (for example Cresswell (2013), Livingstone (1992) and Nayak and Jeffrey (2011)) that we would encourage you to read, but for our purposes here we wanted to avoid a division of theoretical foundations from the geographies we live with every day. The foundations presented here therefore weave together conceptual ideas with examples and illustrations. Each chapter is framed around a binary relationship that frames both the topics Human Geography focuses on (its 'geo-') and how it thinks about them (the nature of its '-graphy'). Binaries are often central to how we think; critically engaging with them provides a powerful window on key elements of geographical thought (see also Cloke and Johnston (2005)). The chapters in Part 1 may not match with particular, substantive lectures in a taught course, and don't always exist as easily locatable debates in the discipline's journals. They crop up everywhere because in many ways they deal with some of the most important questions to think about as a new Human Geography student. They give you a sense of why Human Geographers pursue more specific studies in the way they do and introduce you to ideas and ways of thinking that you will be able to use across a range of substantive topics.

Those substantive areas of the subject are turned to directly in the second and largest part of the book, Themes. It has thirty-nine chapters, divided into eleven sections addressing major thematic 'sub-disciplines' of Human Geography in alphabetical order: biogeographies, cartographies, cultural geographies, development geographies, economic geographies, environmental geographies, historical geographies, political geographies, population geographies, social geographies, and urban and rural geographies. Each of these sections has its own brief editorial introduction, setting out both the sub-disciplinary field and how the following chapters engage with it. This part of the book provides you with thought-provoking arguments on the key issues currently being debated within sub-disciplines, as well as giving you a feel for the distinctive kind of Human Geography undertaken within each.

As we noted above, thematic sub-disciplines are one of the major ways in which teaching curricula are organized and research activity structured, to the extent that geographers are often labelled according to these specialisms (as economic geographers, political geographers, and so on). However, the world we live in is (unsurprisingly) resistant to these neat classifications. Economy and politics and culture and environment (and so on) all interweave with each other. You can't go out and find something that is purely 'economic' (or purely political, cultural or environmental).

In fact, a lot of the most innovative work in Human Geography goes on in the border zones between these sub-disciplinary territories. For these reasons, the final part of the book, Horizons, comprises eleven chapters organized around four contemporary research themes that do not fit neatly in any one sub-discipline. The four foci – non-representational geographies, mobilities, securities and publics – each have their own brief editorial introduction, contextualizing the chapters that follow. Each highlights current agendas in the discipline that are influencing debates in a number of its sub-disciplines.

Stylistically, while every chapter has its own authorial signature all the contributions combine discussions of challenging ideas and issues with accessible presentation. Unfamiliar academic terminology is kept to a minimum, but where central to an argument and not explained fully at the time it is marked in bold type and defined in the Glossary at the back of the book. Chapters include periodic summaries of key points, enabling you to pull out the central lines of argument. Potential discussion points are given at the end of chapters, offering options for group debates or individual essay plan development. Generally, *Introducing Human Geographies* aims to make you think and to challenge you intellectually, but to do that through being lively and engaging. Scholarly knowledge doesn't have to be dry and self-obsessed. Chapters are deliberately short and punchy, but there is guidance for how to develop and deepen your knowledge via suggested further readings included at the end of chapters and the section introductions.

The mention of further readings marks an appropriate place for us to stop introducing. Like any guidebook, the intention of *Introducing Human Geographies* is to take you around the subject so you can experience it for yourself. We rarely read guidebooks without travelling; the book is a companion on a journey not a destination in and of itself. Likewise, you shouldn't read this book without moving on from it to experience more directly the areas of research and debate it guides you towards. If it helps to mix metaphors, think of this book as an introduction agency, setting you up for a relationship with Human Geography. Studying a subject means getting to know it, figuring out what you like about it and what you don't, and maybe even falling in love with some of what it does. It also means 'asking it out'. Let Human Geography get to know you; introduce it to your life, your enthusiasms; liberate it from the library, lecture or textbook. Take some of the geographical ideas in this book to your favourite haunts and see what they make of each other. In other words, see what happens when not only are you introduced to Human Geography but Human Geography is introduced to you. Use this book as a guide both to reading Human Geography and to doing it yourself by thinking geographically. Join in the age-old endeavour of 'earth writing'.

DISCUSSION POINTS

1. Look at a newspaper from the last week. Identify three stories that seem to you to address Human Geography topics. Explain your choices and why you think they are 'geographical'.
2. What makes Human Geography a distinctive subject?

3. 'Human Geography is a down-to-earth subject, concerned with facts not theories.' Discuss this assertion.
4. Outline your understanding of Human Geography's commitment to one of the following: description, experience, interpretation, explanation, critique.

FURTHER READING

There are a number of other texts that fulfill different functions to this book, but offer valuable complementary overviews and resources that help introduce Human Geography. These include:

Bonnett, A. (2008) *What is Geography?* London: Sage.

In this book Alastair Bonnett develops his personal response to the question 'what is geography?'. His answer is thoughtful and thought-provoking, casting geography not as just another academic subject but as 'one of humanity's big ideas'. The book covers the two central foci identified in this chapter (what we called 'writing the earth and the world'); geographical interests in cities and mobilities; the doing of geography in forms of exploration, mapping, connection and engagement; and the institutionalization of geography within and beyond universities.

Cloke, P., Cook, I., Crang, P., Goodwin, M., Painter, J. and Philo, C. (2004) *Practising Human Geography*. London: Sage.

This book focuses on how research in Human Geography is done, covering both the production of geographical materials or 'data' and the production of varying kinds of geographical 'interpretations' of these data. This, and other books on geographical research methods, provide invaluable links between the kinds of materials introduced in this volume and the opportunities that exist for you to undertake your own geographical investigations in project work and independent dissertations.

Gregory, D., Johnston, R.J., Pratt, G., Watts, M.J. and Whatmore, S. (eds.) (2009) *The Dictionary of Human Geography (5th edn)*. Chichester: Wiley-Blackwell.

This dictionary has concise but comprehensive definitions and explanations relevant to almost every aspect of Human Geography. As a reference tool it is invaluable and has no better. Human Geography can be hard to engage with because of the density of its specialist terms. This is a book you will be able to use throughout your time studying Human Geography as you look to master that specialist vocabulary.

Kneale, P. (2011) *Study Skills for Geography, Earth and Environmental Science Students* (3rd edn). London: Hodder Education.

A guide to the study skills that Geography students need and use at university level. A very useful book.

Livingstone, D. (1992) *The Geographical Tradition*. Oxford: Blackwell.

A scholarly rendition of the history of Human Geography, a topic we pay comparatively little attention to in this book. Livingstone concentrates on the longer-term history of the subject rather than on its recent developments. Throughout, one gets fascinating insights into how the concerns of Human Geographers have run in parallel with wider social currents.

PART ONE

FOUNDATIONS

Introduction

Sometimes the start of Human Geography textbooks, and indeed courses, can be very daunting. This is because of the perception by some of the authors of the books and courses concerned that it is necessary to throw in a load of theoretical stuff at the beginning, before getting on with the more interesting stuff. While it may indeed be preferable that certain theoretical foundations are laid before dealing with systematic issues, the net result is likely to be that the reader/course-attender can either be bored to tears or bemused by the abstract nature of those foundations. Well, here's the bad news – we have also decided to begin this book with some theoretical dimensions. But, here's the good news – we utterly reject the false division between abstract theory and the substantive issues of everyday life. Indeed, we believe that our everyday lives are simply teeming with the kinds of issues and questions that are often pigeon-holed as theory. Much of the excitement and value in Human Geography lies in addressing these issues and questions by thinking through aspects of our own lives and of the world(s) in which we live.

As an illustration to get you thinking about Human Geography in terms of everyday life, here is a very short account of a typical journey to work for one of us – Paul Cloke. Neither the story nor the journey is in any way special; that is the point of narrating it. It could be any part of your everyday experience, whoever you are or wherever you live. What it does show is that different sets of Human Geography relationships crop up all over the place, and certainly not just in the abstract treatments of theory in books and lectures. So, imagine if you can a small hillside village in Devon, some 15 miles from the city of Exeter.

The alarm clock does its disturbing work and off we go. There's just the two of us now as our

Figure IV Bishopsteignton, south Devon. Source: Bishopsteignton Village Website

daughter lives in Horfield, Bristol and our son in Dalston, London. So we still get plenty of opportunities for city-time with them, but our home is distinctly rural. Throw open the curtains and there opening out before us is a familiar scene, described in a recent chapter on rural landscape:

> My gaze is drawn past landmark trees, across the tidal estuary of the River Teign, and up again to the valley side beyond. Rolling topography and ancient field enclosures – frequently re-patterned, re-coloured and re-lit with diurnal and seasonal change – are intersected by narrow lanes and straggly footpaths. The ebb and flow of the river continuously refresh the scene, imposing alternative senses of time on what can seem timelessly pastoral. A picture postcard? Yes, but so much more. This is where we walk our border collie, Ringo, where I ride my bike for exercise, where I am periodically enchanted by the affective capacity of bluebell woods, of the colour and texture of birch and rowan, of the persistence and beauty of goldfinch, blue tit and woodpecker, yet can remain relatively unaffected by the scenic presence of the view, or by the potential for hands-on

proximity with nature in the performance of gardening.

(Cloke, 2012: forthcoming)

Alongside its beautiful natural setting, though, rural life can be a place of tension and struggle. I struggle with political and religious conservatism; I struggle with social monoculturalism; I struggle with the vehemence of local opposition to new housing developments, especially from those who occupy the previous rounds of development. As I start the day, I reflect that this place that I call home is gazed on, lived in, performed and experienced in myriad different ways. The assemblage of human and non-human actors displayed out of that window is as diverse in its meaningful representation as it is it in its everyday life practices.

Away from the window, our home displays (consciously and unconsciously) all kinds of other geographies of connection. Paintings and photos provide constant reminders of other places that are precious to us – New Zealand, Khayelitsha in South Africa, Kenya – and there are other intentional reminders of ethical connections close to our heart, of fair trade, anti-slavery and anti-homelessness campaigns. And yes, glory of glories, the latest charitable craze from Cord and Tearfund of toilet twinning. Our toilet is proudly twinned with a latrine in Uganda, with a framed photo to mark the occasion. Charity must be regular . . . but there are uneasy relationships between private ethics and their public display. Of course we don't recognize other less progressive moral connections that will be evident to others in the exploitative relations entombed within our consumer goods, food miles, commuting and unsustainable lifestyles. No matter, to the background sound of music which can variously be drawn from Australia, Iceland and the USA as well as Britain, and with the foreground conversation of international news on the radio, we speed through breakfast. Food from all around the world, brought to us by multinational corporations via supermarkets. The global and the local come together at every turn.

Time marches on, so I begin my commute into Exeter. At first, my journey traverses farmland through tiny lanes, passing the local golf course; however early my commute, there is always earlier activity there. Then through Halden Forest to Telegraph Hill and on into the city. Halden represents a place of idyllic recreation and natural habitat for many, but its traverse is characterized by the modernism of a crowded dual-carriageway. In the winter its local microclimate renders it susceptible to heavy snowfall and ice which have in the past trapped unwary drivers for several hours. In many ways, then, natural, mechanical and human risk and hazard lie shallow beneath the surface of this kind of commuter journey. As I reach its outskirts, the city remains somewhat detached from the vantage point of the driver's seat, partly because of a necessary focus on traffic management and partly because the radio tends to fill in much of the 'thinking space' of the journey with national and international issues. Situated on the River Exe, Exeter snarls up at key bridging points during the rush hour, so along with many others I weave my way through the social geographies of housing estates and suburban lifescapes to avoid traffic on the way to the University. In so doing I bypass the centre of the city with its designer pubs and clubs, with Irishness here, and Walkabout there, interspersed with what are by now unremarkable Indian and Chinese restaurants. Designer-label beer and wine from all over the world is spilt here over designer T-shirts from all over the world. Where I used to work, in Bristol, the University was located in the heart of the city, and I would often encounter the heady contrast of financial centres and homelessness, side by side on my journey in. Exeter, however, is a campus university on the edge of the city, and these disturbing downtown hybridities are temporarily avoided by geography. Finally, it is up to Geography, passing through the multinational, but somehow overwhelmingly middle-class, throng of students on campus. Once inside my office, the first move is to fetch a cup of (fairly traded) coffee, switch on the PC and check my e-mails, hardly noticing the rows of shelves loaded with the production of particular knowledges about governments, policies, plans and politics, and how the lives of real people in real places intersect with so much in the geographical world.

There is so much else that I could (and perhaps should) have mentioned, but this much suffices to invite parallels with many of the themes covered in this opening section of the book. Philip Crang, in Chapter 1, discusses the relations between the global and the local, and the sights, sounds, histories and commodities of the global crop up time and again in the local story of my journey to work. Local places get their distinctive character from their past and present connections to the rest of the world, and therefore we need a global sense of the local. Conversely, global flows of information, ideas, money, people and things are routed into local geographies. We therefore also need a local sense of the global. Crang's core message is that ideas about global and local are not one-dimensional inputs to our Human geographical understanding. Rather local and global are interrelated and each helps to shape the other.

The same can be said for relations between society and space. In travelling from home through villages, suburbs and estates, my narratives are jam-packed with references to how and where different social groups live, work and take their leisure. In Chapter 2, Jo Little shows how spatial patterns can reflect social structures, and how spatial processes can be used as an index of social relations. My journey seems to traverse particular social areas, but she warns that social categories cannot be taken for granted. Such categories are constructed socially, politically, culturally, and are mediated by the organization of space; in other words, society and space are co-constructed. Moreover, we can no longer rely on two-dimensional maps of society and space. Beyond the obvious, there is complexity, ambiguity and multi-dimensional identity. Whether in rural communities, spaces of the night-time economy, or in the hopeful thirdspaces of liminality and change, society and space both shape each other, and are shaped by each other.

Just as local–global and society–space have seemed like binary terms but have been investigated by Human Geographers in terms of their co-dependence, so the relationship between human and non-human has also come under scrutiny. As Hayden Lorimer writes in Chapter 3, geographers have taken a strong interest in how humans understand and value the lives of other living creatures, not only in terms of issues around food and clothing, but also focusing on the companionship of pets (such as Ringo the border collie) and the lifeworlds of 'wild' animals (such as the deer that run free in Halden Forest). In so doing we have moved away from geographies that focus only on humans, and instead have emphasized the relations between humans and non-human beings, materials and ideas. One significant outcome of this shift has been an interest in the appropriate ethical responses that arise from these inter-relationships.

Part of the intellectual climate that has allowed Human Geographers to begin to deconstruct some of these key binary terms has arrived on the coat-tails of postmodernity.

Mark Goodwin's account in Chapter 4 of the shift from 'modern' to 'postmodern' charts the way in which wider society has moved away from the austere and geometrically planned patterns of life and thought under modernity into a more postmodern emphasis on diversity, plurality and playfulness. Tracing the outcomes of this shift in terms of architecture, cultural style and philosophical approach, Goodwin outlines a transformation in Human Geography by which many researchers have begun to reject any kind of search for universal truth, and instead have recognized that all knowledge is socially produced. As with other such categories of knowledge however, the boundaries between modern and postmodern are contested, and elements of each are visible in contemporary cultural and physical landscapes.

In Chapter 5, Paul Cloke explores the importance of 'self' and 'other' in these contestations over socially produced knowledge. Being reflexive about the self is a vital part of understanding how our knowledge of Human Geographies is situated. Our experience, politics, spirituality, identities, and so on, can add to our stories about the world, and denying their importance in search of 'objectivity' could well be dishonest. My journey to work will not be the same as yours, even if it follows much the same route. However, there is also a danger that we only see the world in terms of ourselves and those who are the same as us, thus creating categories of 'otherness' according to the essential characteristics of our selves. What escapes us are other 'others' – those whom we cannot categorize or pigeonhole; those who surprise us and cannot be accommodated in our organization of knowledge.

Gender is a fundamentally important dimension of how Human Geography can present understandings of how knowledge about the world (for example the domestic world of the household and the employment world of the academic workplace) is constructed. Geraldine Pratt and Molly Kraft, in Chapter 6, discuss how differences between masculine and feminine ways of bodily comportment lead to variations in self-perception and cognitive ability (especially spatial awareness). So the capacity to explore and know our environment can be conditioned, for example, by gendered (as well as racialized) geographies of fear and safety that characterize some local places. They argue that much of women's experience has long been ignored by Human Geographers, with the result that different types of masculinities have been formative in the production of geographical knowledge. It is therefore crucial that we seek to situate knowledge (see Chapter 5) so as both to acknowledge the validity of a range of perspectives, and to develop a commitment to communicate across different perspectives and types of knowledge. In the context of this chapter situating knowledge is important not least because gender itself is interwoven with other social identities that render it unstable over time and space.

From Chapter 7 onwards, this introductory section dealing with *Foundations* turns specifically to address the diversity in the ways in which Human Geography is studied and approached. In Chapter 7, David Gilbert notes the potentially confusing range of 'hard' and 'soft' approaches, ranging from scientific objectivity to having an opinion that counts. For example, my journey to work could have been portrayed in terms of time–space data and cartography rather than as a loose personal narrative. Alongside the continuing energetic focus on geographical information systems (GIS – see Chapter 14), Human Geography has over recent years mostly emphasized a *critical* social scientific approach to the subject, seeking to deal with issues of agency, meaningfulness, power and positionality. In parallel, however, there has also been a collaboration with the humanities – especially history, philosophy and literature – to investigate the importance of the creative imagination to places using analysis both of written texts (novels, travel writing and the like) and visual images (film, television, photography). Initially the focus here was on how these texts represented different places and people, but more recently Human Geographers have looked to the arts for inspiration about how we sense and move within the world in a more non-representational register (see Chapter 50).

The distinction between 'hard' and 'soft' approaches to Human Geography is often influential in how we develop conceptual frameworks, undertake research and interpret the findings. It is common to encounter significant divisions between quantitative and qualitative approaches – an expression of Human Geography identity according to methodological choice rather than substantive focus or theoretical viewpoint. Rob Kitchin in Chapter 8 explains that these methodological divisions of Human Geography identity are misleading; rather, it should be theory, philosophy and ideology that shape choices about methods. While quantitative research can produce explanation and prediction, and qualitative research can produce meaning and understanding, these methodological approaches should not be regarded as dualistic. Quantitative approaches fit the assumptions of some philosophies and qualitative approaches suit others. However, mixed methods including both quantification and qualification are often useful and even advisable.

A key part of the conceptual thinking through that Human Geographers have to engage in relates to how the truth of the world is variously represented in different circumstances. In Chapter 9, Mike Crang suggests that the relationship between representation and reality is more complex than a binary between truthful or deceptive depictions of places and people. Different people and organizations will understand different places and circumstances differently – living in a Devon village provokes many different portrayals, not all of which reflect idyllic rural life. It follows that all kinds of representations will contribute to ideas about and understandings of the world, and will help to shape the world rather than simply depict it. Human Geographers need to understand the selective angles from which representations are presented, not least because these representations are often subject to the power and control of the global political economy, which increasingly seems to trade on sign-values rather than 'truth'.

The nine chapters in this part of the book on 'Foundations', then, represent the very stuff of lively, interpretative, relevant and accessible Human Geographies. They help us to think through some of the recurring questions and issues involved in understanding the interconnections of people and places, and they help us to place ourselves in the picture as well. Far from being the 'boring theoretical stuff', they offer some keys with which to unlock thoughtful and nuanced accounts of the Human Geographies of everyday life. Enjoy!

CHAPTER 1
LOCAL–GLOBAL

Philip Crang

Introduction

My reasons for becoming a geographer were not particularly well considered or original, indeed they were pretty lame in some ways. Yet they still ring true to me today. I enjoyed Geography as a subject, and decided to do it for a degree at university, because through Geography I got to hear about, see pictures of, and maybe even go to a lot of different places. Geographers travel – both literally through an emphasis on fieldwork and various sorts of exploration, and more virtually in the form of slide shows and reportage. Why did I think that was a good thing? I valued the pleasures of getting to know particular, distinctive places, both familiar and unfamiliar. I enjoyed spending time in a place, getting a feel for it, finding out about it. A lot of my most powerful memories and attachments were with places of various sorts, from the house I grew up in, to the fields and moors I explored as a young runner, to the 'milk bar' where my grandmother took me for ice cream treats. But I also thought it was important to learn about areas of the world and people of which I would otherwise be largely ignorant. I was both moved and discomfited by how much of the world only came onto my TV screen when disasters struck, people died, and emergency problems needed responses. I knew my own life was parochial in the extreme, and while I enjoyed its confines, I also wanted to get beyond them.

Those feelings I had as a seventeen year old still animate my interest in Geography. I know much more about the subject now, but I still think my views then located something very close to its heart. They home in on a triumvirate of ideas that have long fostered Human Geography's understanding of itself as a distinctive intellectual endeavour. First, in the emphasis on the distinctive characters of particular places, they highlight the idea of the *local*. Second, bound up with a desire to broaden horizons and foster a greater 'world awareness' is the idea of the *global*. And third, central to this interest in both the local and global is an emphasis on *difference* (between places and people). This chapter examines the relations between these three ideas: the local, the global and difference. It will, I hope, give a sense of how productive they have been, and can still be, for geographers. However, it also argues for critical reflection. Notions of the local, the global and difference are not as simple and obvious as they might at first seem. It is important to think carefully about each of these ideas, and perhaps even more so about how they relate to each other. If we fail to do that, then we run the risk of unwittingly reproducing conventional arguments about our world's geographies, closing off other possible ways of thinking and acting. We may end up learning rather less about places, their particularities and their differences than we should as thoughtful 'travellers'.

The chapter starts by briefly outlining how and why ideas of the local and the global have been so important to Human Geography. I then set out three takes on local–global relations. I call these *mosaic*, *system* and *network*.

Local matters, global visions

Human Geography has long combined attention to local matters with some sort of global vision. To start with the local, it, and associated notions such as place and region, have long had a particular centrality in geographical imaginations. Many academic geographers have spent whole careers trying to document, understand and explain the individual 'personality' of an area (Dunbar, 1974; Gilbert, 1960). So, why is the local deemed so important to Human Geography's research and teaching? In his thoughtful book *The Betweeness of Place*, Nick Entrikin (1991) argued that geographers have been interested in the local for three interrelated reasons. First, they have emphasized the actually existing variations in economy, society and culture between places; or what Entrikin terms the 'empirical significance of place'. Despite the homogenizing ambitions attributed to the likes of McDonald's, everywhere is not the same. Landscapes vary. Life chances are materially affected by the lottery of location. Whether you happen to be born in Lagos or London or Los Angeles, or indeed in Compton or Beverly Hills, has an impact on the kind, and even length, of life you can expect. And location is not just something we encounter and deal with. It is part of us. Where we are is part of who we are. Most obviously, this is the case through the spatial partitioning of the world into nationalities, imaginative constructions that are part of our identities, so powerful as to get people to kill and die in their name (see Chapter 37). So, places and the differences between them can be seen to exist and have real effects.

But the local also matters in a second way. Spatial variations do not only exist. They are valued, or seen as a good thing, not least by Human Geographers. There is, then, what Entrikin calls a 'normative significance to place'. Sometimes this is expressed as a celebration of difference: whether out of a suspicion of the power of global, homogenizing forces ('the media', 'American multinationals', and so on); or out of a pleasure gleaned from experiencing variety and the unexpected. Sometimes the local is cherished for its communal forms of social organization, for embodying an ideal of small and democratic organizations (for a critical and suggestive review see Young, 1990). And sometimes this social idealization goes hand in hand with an environmental utopia of self-supporting, environmentally sustainable livelihoods (Schumacher, 1973), or at least an appeal to the local as a way of living more lightly on the planet, as when calls are made to reduce 'food miles' by 're-localizing' supply networks and supporting local producers. But whether culturally, socially or environmentally framed, in all such arguments the local does not just matter. It matters because it is in some way 'good'.

The third importance attached to the local within Human Geography, according to Entrikin, involves a concern with the impact of the local on the kinds of understanding or knowledges that geographers themselves produce; what he calls the '**epistemological**

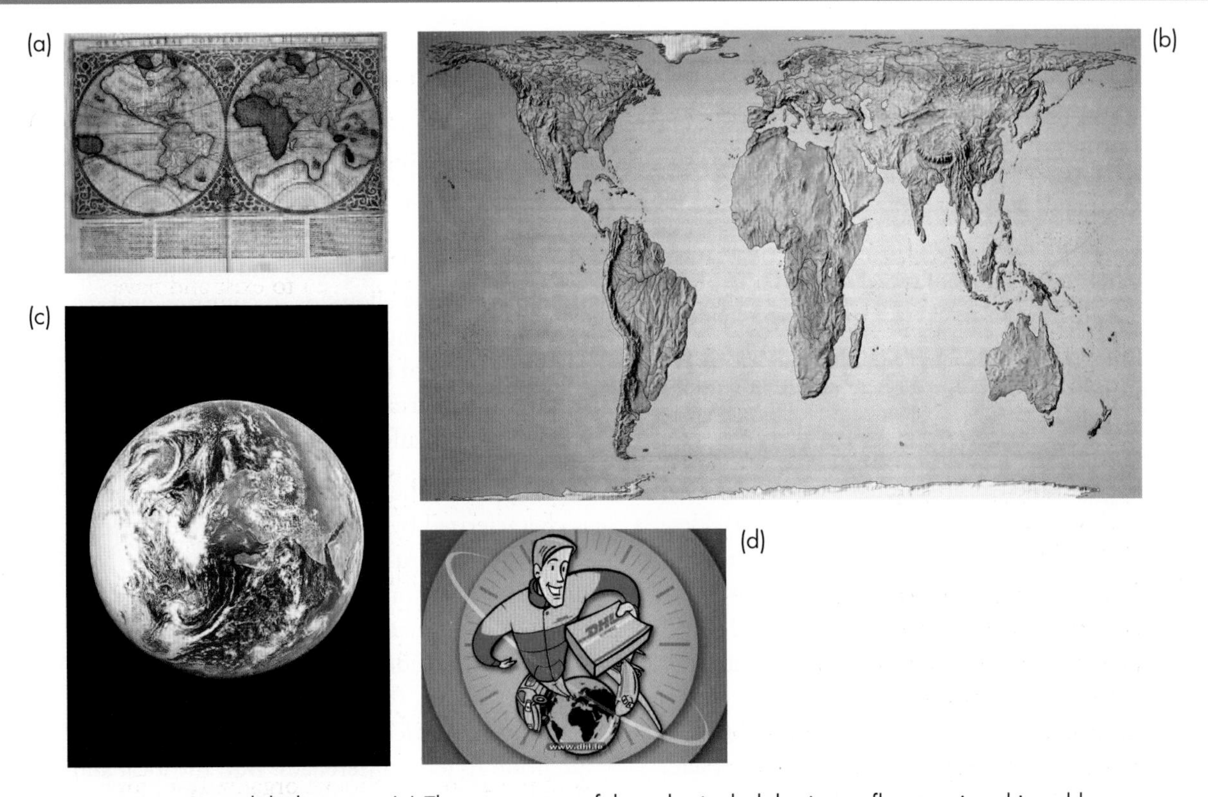

Figure 1.1 Four global visions. (a) The conversion of the spherical globe into a flat map is achieved here through a Mercator projection. Developed in the seventeenth century, the Mercator world map is ideal for exploration as a constant bearing appears as a straight line, but this is achieved by distorting sizes, which makes tropical regions look far smaller than they actually are. (b) The Peters projection, by contrast, is an equal area projection that distorts shape rather than size. First published in 1973, this projection was designed within development discourse to ensure the 'South' was given its proper global importance © Professor Arno Peters, Oxford Cartographers/Getty Images. (c) 'Spaceship earth' is an icon of contemporary environmentalism, portraying a living whole without apparent national boundaries or other political divisions. (d) The shrinking earth of 'globalization' and telecommunicational hype. Credit: (a) Royal Geographical Society, UK/www.bridgeman.co.uk; (b) © Oxford Cartographers and Huber Verlag; (c) NASA; (d) Courtesy of DHL

significance of place'. In part this involves a scepticism towards general theories that claim equal applicability everywhere. It also means a sensitivity to where knowledges come from (to their 'situatedness'). Geographers don't only know about localities, they produce local knowledges.

At the same time as having this local fixation, Human Geography is also determinedly global in its scope. Even as it values them, it also tries to break out of purely local knowledges through appeals to global awareness. Geographical interest in the global has been developed through a number of different emphases. Let me draw out four. Figure 1.1 displays a picture of the world that represents each.

First, we can identify a geographical concern with *exploration*, driven by a desire to 'know the world'. Exploration was central to geography's early history – such that geography's development as a science, from the sixteenth century onwards, went hand in glove with European explorations to the farthest corners of the earth (Driver, 2001; Livingstone, 1992; Stoddart, 1986). Today, exploration continues to excite popular cultures of geography, whether in forms of travel that offer experiences 'off the beaten track' (for more, see Chapter 53) or the mass-circulation *National Geographic*'s promotional claim to give American readers a 'window to the world of exotic peoples and places' (cited in Lutz and Collins, 1993: xi). Second, there is an emphasis on *development*, with its hope of 'improving the world'. Here, a world vision matters not only in order to rectify ignorance of the world's diversity, but also to explain and act against global inequalities between North and South. Third, there is global *environmentalism*, with its concern for 'saving the world' against planetary threats such as global warming or ozone depletion. Here, thinking globally is essential not only to recognize the scale of these problems but also to understand the true environmental impacts of our local actions (so, when I set the thermostat on my central heating I need to be aware of the impact of my domestic energy use on CO_2 emissions). Finally, there is a concern with global *compression* or the 'shrinking of the world' (see Harvey, 1989: 240–307). The emphasis here is on the increasingly dense interconnections between people and places on other sides of the world from each other, whether through telecommunications, global flows of money or migrations and other forms of travel. '**Globalization**' has become the most prevalent term to describe such compression (for a very good overview, see Murray, 2006). In a globalized world our local lives are led on a global scale. The food we eat, the clothes we wear, the television programmes we watch, the cars we drive or bicycles we ride, all these materials of our mundane, everyday lives come to us through enormously complex and globally extensive production and retail systems.

There are, then, many good reasons why Human Geography should not myopically focus on the local but also attend to the global: because global scale processes impact on, and result from, our local places and lives; because thinking globally allows us to compare, and even more usefully connect, our own lives and places to those of others; and because the global stands for important, 'big' issues and processes that we cannot afford to ignore.

My argument, then, is that Human Geography is characterized by a concern with *both* the local *and* the global. At times, these can be understood as competing scales of interest: as when calls are made for geographers to escape local trivia and address the really important global issues; or, conversely, when global accounts are criticized for not paying due attention to local differences. But, the local and the global can also be seen as two sides of the same coin. Travellers set out across the world to find new 'locals' to encounter and report back on. Environmentalists and multinational corporations both sloganize about 'thinking globally and acting locally'. So, how we understand and construct the global shapes our understanding of the local, and vice versa.

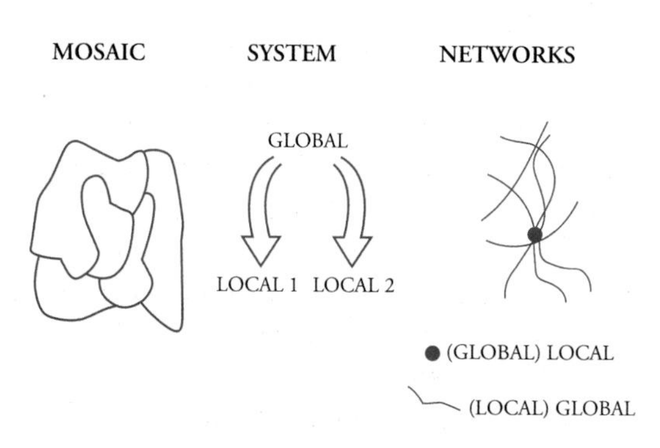

Figure 1.2 Figures of the local–global: mosaic, system and network

For the remainder of this chapter I want to turn more directly to these relations between the local and the global. They can, I want to suggest, be thought of in a number of different ways. To illustrate, I will review three schematic accounts of local–global relations: the world as *mosaic*, the world as *system*, and the world as *network* (see Figure 1.2).

SUMMARY

- Human Geography has fashioned itself as a distinctive intellectual endeavour through emphasizing its interest in *local* places and specificities, and showing how these matter empirically, normatively and epistemologically.
- Human Geography has also fashioned itself as a discipline through stressing various *global* concerns, for example with exploration, development, global environmental change and globalization.
- While the local and the global can be seen as alternative and competing scales of concern, we need to recognize that they are always constructed in relationship to each other.

Mosaic

One very popular way of thinking about Human Geographies is in terms of a mosaic. Here, the world is conceived as a collection of local peoples and places, each one being a piece in a broader global pattern. This way of seeing the world can be drawn out at a number of different scales, from neighbourhoods right up to whole continents. It is perhaps most obvious at the level of the nation-state. The whole idea of nationalities depends upon constructing distinctive pieces of an international mosaic; establishing borders and territories; and distinguishing between this country and that country, our people and those foreigners. Political maps of the world present this mosaic cartographically, national pieces set next to each other and the 'open' spaces of the sea. Mosaics are made too at the smaller scales of the city, in mappings of a patchwork of local areas, each characterized by different economies, residents and built environments. Think, for example, about how estate agents and others seeking value in the property market such as retail location analysts or gentrifiers, map out cities into areas, neighbourhoods and streets with supposedly distinctively different characters. (For more on the use of GIS, location decision making, and this sort of mapping, see Chapter 14.)

Move up to the supranational scale, and again mosaics are a common way of thinking about the world's geography. The global is divided up through reference to the compass points: South and North, East and West become designations of geopolitical and cultural entities when we refer to some people as, for example, being 'westerners'. (A classic account of this way of imagining the world is given in Edward Said's book *Orientalism*, which is about how 'the West' or Occident has defined itself through opposition to 'the East' or Orient. Said, 1995 [originally 1978].) At other times the mosaic pieces are defined in terms of latitude, as when the 'tropics' designate a part of the world with supposedly identifiable characteristics of 'tropicality' (Driver and Martins, 2005; Thompson, 2006). Or it may be distinctive continental economic, political, cultural units that are identified and contrasted in what Lewis and Wigen (1997) call a 'mythical metageography': Asia and Asians framed as different to Europe and Europeans, North America and North Americans as different to Africa and Africans, and so on.

These kinds of designations are commonplace, from geopolitical thought to popular culture. In many ways, the notion of the geographic mosaic has been so influential (see Gregory, 1994: 34–46) that it can be hard for us to see it as anything other than common sense, a description of an obvious reality. Tourism, the world's largest industry, feeds off and actively constructs such an understanding, as it showcases a world of different destinations that the holidaymaker can visit (see Chapter 53 for a fuller discussion). But the mosaic is only one possible way of framing local–global geographies and it is a very particular framing, with its own preoccupations and blind spots.

Three features are especially important. First, the mosaic puts an emphasis on boundaries and borders. Geographical difference is seen in terms of distinct areas that can have lines drawn around them. Second, these areas are understood in terms of their unique characters, personalities or traditions. That is, each piece of the mosaic is seen as having distinctive 'contents', whether that be its people, culture, economic activities and/or landscape, which cohere into some sort of unified geographical identity. Third, this means that any intrusions

Figure 1.3 Are global products a threat to local differences? Credit: Anders Ryman/Corbis

into a distinctive area tend to be seen as a threat to its unique character. For an example one could think of worries about how the global predominance of American popular culture, from fast food to TV programmes, is destroying local cultures and producing one Americanized global monoculture, where everybody, wherever they are, eats Big Macs, drinks Coca-Cola and watches American soaps (see Peet, 1989). Or one could think about claims that human migrations pose problems for the cultural integrity of receiving areas, overwhelming or in some way undermining indigenous culture unless immigrants are properly assimilated (see Chapter 41 for more on migration).

All these features of the mosaic model can be questioned evidentially. First, the world's differences do not fit into the frame of a geographical mosaic, no matter how many scales it is imagined at. The contents of any one area are never uniformly the same. To claim they are is to produce what statisticians call an 'ecological fallacy', applying the general, average qualities of an area to all its inhabitants. Second, one reason why difference refuses to be contained within the pieces of a mosaic is that the world does not stay still. If we think about the continental 'metageography' of people, then we know that Europeans haven't stayed in Europe, Africans haven't stayed in Africa, and so on (these population movements are sometimes called 'diasporas'; see Chapter 45). We know that our economies too are interlinked, with fluid forms of capital able to migrate around the world (see Chapter 28 on economic globalization). We cannot simply draw boundaries around local or national or continental economies. The world is not a fixed array of pieces; much of it is mobile, on the move.

Third, in analyzing the impacts of such mobilities, we cannot assume that the opening up of local places to global forces necessarily results in the destruction of difference. Instead, global forms are often 'indigenized' or 'localized' in different ways in different places. While living and researching in Trinidad the anthropologist Danny Miller was struck by the fact that he had to stop his research for an hour a day while everyone watched the daytime US soap *The Young and the Restless* (Miller, 1992). This might seem an obvious sign of homogenizing Americanization. In fact, Miller argues, 'paradoxically an imported soap opera has become a key instrument for forging a highly specific sense of Trinidadian culture' (1992: 165). In the extensive chat about this soap, what viewers identified was not an alien American world, to be aspired to or despised, but themes that resonated with deep existing structures of Trinidadian experience. In particular, viewers liked the way it dramatized what they called 'bacchanal', or the confusion and emergence of hidden truths through scandals, something also central to other Trinidadian cultural forms such as Carnival. So, this globally distributed American soap was not destructive of Trinidadian difference; as part of a dynamic local culture it helped to produce a distinctive Trinidadian sensibility. Elsewhere, Miller makes similar arguments about both older global imports – analysing how Coke became a 'black sweet drink from Trinidad' and part of the national drink of rum 'n' black – and newer global forms – reporting on Trinidadian uses of Facebook (Miller, 1998; 2011). To use a popular local expression, Miller suggests that all of these global products are not alien invaders but 'True True Trini', functioning as authentic forms of local cultural differentiation.

The problems with the figure of the mosaic are not only factual – they also stem from its political impulses and ramifications. To be fair, there are many positive elements to the notion of the geographical mosaic. Often underlying it is a desire both to recognize and respect differences; to appreciate, in both senses of the word, that everyone is not the same as you are, and that everywhere is not the same as here. But it is not enough just to appreciate difference. We have to think about how the idea of difference is being constructed and used. In the case of the mosaic, all too often either the impulse or the effect is defensive and exclusionary. Difference is locked into a geography of territories and borders. It is framed in terms of insiders and outsiders. The mosaic also depends on stereotyping. It understands and recognizes differences by simplifying them and their location. This way of seeing the world is not so much a description of it as a powerful way of claiming and attributing difference in spatial terms. It projects differences on to distant people and places in order to create some sense of unity 'at home'; 'they' and 'there' are different to 'us' and 'here'. It can legitimate claims for a place to belong to some and not to others. It entangles geography with a politics of 'purification', in which sameness should be here and difference should be there. 'Ethnic cleansing' would be an example of practices that have followed the mosaic and its logic of each different thing in its own different place to the most brutal conclusions.

The idea of a world made up of different geographical areas is commonplace and is likely, initially, to be seen as both obvious and non-contentious. However, while not without its merits – in particular its recognition of difference – the mosaic is but one way of thinking about local–global relations, and it can be deeply problematic because of *how* it recognizes difference. We need to think, then, about whether Human Geography can combine the local and the global in other ways too.

SUMMARY

- A very common way of imagining local–global relations is to envision a world of many different local places and peoples, each being a piece in a wider Human Geographic global mosaic.
- This constructs the local as a bounded area, made distinctive through the character of life and land within it. It also tends to construct global-scale processes as destructive to that local diversity.
- There are factual problems with this way of framing local–global relations. For example, local differences are not inevitably destroyed by global level processes; in fact they are often produced through them.
- There are also political dangers attached to it, in particular an impulse towards defensiveness and the exclusion of non-locals.
- The mosaic is only one way of imagining local–global relations, so rather than seeing it as a simple portrait of geographical reality the reasons for, and effects of, its use need to be analysed.

System

An alternative way of thinking about local–global relations is to see local differences as produced by a global system. That is, the differences between places are not seen as a consequence of their internal qualities but as a result of their location within the wider world. The mosaic of geographical difference is not innate but made systemically. We need to understand the processes and powers that make it. I have been intimating at this kind of argument already. We might, for example, argue that the very idea of a geographical mosaic is a framework that makes difference, forming the world through particular templates. However, perhaps the best examples of this argument come from within development studies and through attempts to understand the extreme differences that characterize our world.

One way of thinking about the differences and inequalities in wealth and life chances between different parts of the world would be to identify internal characteristics that explain them. So, we could say (and many do) that Europe and North America are so comparatively wealthy because of the economic innovation they have shown since the time of the Industrial Revolution or due to longer-term advantages conferred by temperate climates and the early adoption of agriculture. And then we might argue that the Philippines, say, are comparatively so poor because of their lack of natural resources, an inhospitable climate or some perceived deficiencies in their culture (e.g. endemic corruption or laziness). What this kind of explanation ignores, though, is the fact that Europe and the Philippines are not just separate places, they are places with long histories of interconnection through world political, economic and cultural systems. It is possible, then, that Europe and the Philippines are so different because of these relationships with each other rather than because of their internal qualities. To put it bluntly, maybe we need to think less about Europe and the Philippines separately, and rather more about whether Europe is rich precisely because the Philippines are poor. That is a very simplistic assertion but it has its virtues. It sensitizes us to the idea that there is a set of global relations between local places. In emphasizing how global relations actively produce differences between places it reorients our efforts away from just documenting diversity (Europe and the USA are like this, the Philippines are like that) and towards understanding the processes of that *differentiation*.

Central to such efforts of understanding how and why differences are produced at the global scale has been work focused on the **world-system**. Here the world is treated as a single economic and social entity. At the heart of its operations is the capitalist world economy. This is how Jim Blaut puts it, in arguing against the idea of a special European character that has led to its relative economic success:

> Capitalism arose as a world-scale process: as a world system. Capitalism became concentrated in Europe because colonialism gave Europeans the power both to develop their own society and to prevent development from occurring elsewhere. It is this dynamic of development and underdevelopment which mainly explains the modern world.
>
> (1993: 206)

A more concrete example may help to show the importance, and limits, of this systemic view of local–global relations. That example is the world coconut market as portrayed by James Boyce (1992).

Boyce notes two main things about the global coconut trade in the period 1960–85: first, 'the

Figure 1.4 Why do the Philippines produce coconuts? Credit: Getty Images

Philippines is king' with over 50 per cent of world exports; second, the Filipino producers of coconuts do not seem to be doing very well out of this dominant market position. Understanding either of these facts requires a global systemic focus. The prevalence of coconut production in the Philippines would have to be traced back to Spanish colonization (for example, a 1642 edict for all 'indios' to plant coconut trees to supply caulk and rigging for the colonizers' galleons), to demand in the nineteenth century from European and North American soap and margarine manufacturers, and to US colonial control and post-colonial patronage in the twentieth century (which led to preferential tariff rates for Filipino coconut products in the US market until 1974). It reflects, then, an emergent international system in which the Philippines was positioned, by external powers, as a supplier of an agricultural commodity, while those powers used that commodity for their own purposes (for their ships or their manufacturing industries). Low rewards for this agricultural production reflect declining global terms of trade, such that each barrel of coconut oil exported in 1985 would buy only half the imports it would have in 1962. The explanation for this decline is complex, but principally stems from the success of manufacturers of potential substitutes in the developed world – both ground nut oil producers and petro-chemical companies – at getting subsidies and protection from their governments, thereby depressing world prices for all traded fats and oils. That is, it is the political and economic power of developed-world producers and governments which means that the Filipino coconut industry gets an ever worse deal for its efforts. The world trading system not only differentiates through an international division of industries (you grow coconuts, we have petro-chemicals), it discriminates in relation to the value of these activities.

However, as well as stressing the global relations that have stimulated Filipino coconut production and worsened its terms of trade, Boyce's study also suggests some limits to purely global explanations. In particular, he stresses how the local trading relationships within the Philippines meant that while the majority of small growers reaped little reward, vast fortunes were made by a few powerful individuals. Under the guise of concern for small producers, the Marcos regime reorganized the industry to concentrate power in the hands of a single entity that controlled raw material purchases from farmers and marketing at home and overseas. This concentration was in turn used to reward a few close political associates, such as 'coconut king' Eduardo Cojuangco and the Defence Minister Juan Ponce Enrile, who siphoned off much of the dwindling national earnings from the coconut trade. Thus, existing inequalities in economic and political power within the Philippines allowed actions that made these inequalities greater still. Declining global terms of trade were experienced particularly severely and responded to in particularly unproductive ways, because of the distinctive (if not unique) political system in the Philippines. Local processes, as well as global processes, played their part in the impoverishment of coconut producers. Any attempt to rectify that impoverishment would have to deal with local and global trading relations and the political–economic structures of each.

The example of the Philippines and coconut trade illustrates how the differences between places cannot simply be understood through comparison. Differences are made through relations between places, as well as within them. In this section, I have used the notion of the 'system' to capture this emphasis on global interrelations, local agencies and their production of geographical differences. That kind of argument is well developed in accounts of the global differences associated with development and underdevelopment, but can be applied more generally too.

SUMMARY

- Differences between places are not just the result of their 'internal' characteristics. They are produced by systems of global relations between places.
- Human Geography should therefore do more than document diversity. It should investigate the *processes of differentiation* through which diversity and inequality are produced.
- These processes of differentiation operate at both global and local scales.

Networks

So far we have seen two differing ways to think about the relations between the global and local. In the model of the mosaic, the global is portrayed as a collection of smaller locals. In the model of the system, the global is portrayed as a set of relations through which local differences are produced, and the emphasis is less on collection and comparison than on connection. In this final section I want to take the idea of connection further. I want to suggest that we can see both the local and the global as made up of sets of connections and disconnections that we can call 'networks'. In consequence, we may need to view the local and the global not as different scales (small and large) but as two ways of approaching these networks, in which the local is global and the global is local.

Let's start by looking at the local (and its global character). In highly influential arguments, the British Geographer Doreen Massey coined the phrase 'a global sense of place' to reference how the distinctiveness of a particular place is not threatened by connections to the wider world but actually comes from them (Massey, 1991, 1994). Whether thinking about a metropolitan urban neighbourhood or a seemingly isolated rural village, Massey argued that localities gain their different, specific characters through distinctive historical and contemporary links to other places (see Case Study box). This also produces a more 'progressive' politics of place, in which the appreciation of local differences does not slip into a reactionary, defensive parochialism. Massey has since developed this argument within her book *For Space* (2005), a wider theorization of how Human Geography approaches core concepts such as space and place. Places for her are less 'things' (such as pieces in a geographical mosaic, to use my phrasing from this chapter) than they are 'ever-shifting constellations of trajectories' (Massey, 2005: 151). Places are less containers of different and distinctive contents than they are 'open and internally multiple'; less fixed, more of an 'event', a 'coming together of the previously unrelated' (Massey, 2005: 141, 138).

If we think of the world in terms of networks, then we see local places as gaining their different characters through their distinctive patterns of associations with other places. In turn, we begin to see how the global is less some neat, all-embracing system with a single logic, than a mass of globally extensive yet locally routed practices and technologies of connection. Not only do we need to globalize the local, but we also need to localize the global, understanding the global as something other than a single entity or system.

Writing in the context of debates over globalization, Arjun Appadurai (1990) provides a classic early intimation of such an approach. He argues that we can imagine the global as comprised of a range of interacting but distinctive '-scapes' or morphologies of flow and movement. 'Finanscapes' comprise global networks and flows of money (often in electronic and virtual forms, routed through the casino economies of major international financial centres in New York, Hong Kong, Tokyo and London). 'Ethnoscapes' are forged by global networks and flows of people (migrants, tourists, business travellers, even geographers), each with their own rather different patterns of movement. 'Mediascapes' are made up of communication technologies and product distributions. And so on. Many flows, many networks, often interconnected but possessing their own distinct geographies. So, for example, money moves across national borders with ease at the same time as the richest nations look to reinforce their disciplining of the movements of people cast as 'economic migrants'. Appadurai argues for breaking down

CASE STUDY

Doreen Massey on the global–local geographies of Kilburn, London and a Cambridgeshire village

Take a walk down Kilburn High Road, my local shopping centre. It is a pretty ordinary place, north-west of the centre of London. Under the railway bridge the newspaper stand sells papers from every county of what my neighbours, many of whom come from there, still often call the Irish Free State . . . Thread your way through the often stationary traffic . . . and there's a shop which as long as I can remember has displayed saris in the window . . . On the door a notice announces a forthcoming concert at Wembley Arena: Anand Miland presents Rekha, live, with Aamir Khan, Salman Khan, Jahi Chawla and Raveena Tandon . . . This is just the beginnings of a sketch from immediate impressions but a proper analysis could be done, of the links between Kilburn and the world . . . It is (or ought to be) impossible even to begin thinking about Kilburn High Road without bringing into play half the world and a considerable amount of British imperialist history.

(Massey 1991: 28)

Think of the [seemingly isolated] Cambridgeshire village. Quite apart from its more recent history, integrated into a rich agricultural trade, it stands in an area which in its ancient past has been invaded by Celts and Belgae, which was part of a Roman Empire which stretched from Hadrian's Wall to Carthage . . . The village church itself links this quiet place into a religion which had its birth in the Middle East, and arrived here via Rome.

(Massey 1995: 64)

the idea of the global into these different kinds of flows, and for seeing how those flows then come together (and sometimes clash) as the trajectories producing different places.

At least two influential broader approaches relate to such concerns for localizing the global and globalizing the local. The first has been called the 'new mobilities paradigm' (Sheller and Urry, 2006). In opposition to the static mapping of the world implied by the notion of a geographical mosaic, this approach emphasizes how the world is made through the interrelated mobilities of people, things and ideas. At its heart are explorations of the dialectics between fixity and fluidity. (Chapters 52–54 give much more information on this approach and examples of how it has been applied within Human Geography.) The second approach is most commonly called **Actor-Network Theory** (or ANT for short). As John Law puts it, ANT 'treat[s] everything in the social and natural worlds as a continuously generated effect of the webs of relations within which they are located' (Law, 2009: 141). Despite its name, ANT is less an explanatory theory than it is 'a toolkit for telling interesting stories about, and interfering in, those relations' and how they manage to 'assemble or don't' (Law, 2009: 141–2). ANT has been widely influential within Human Geography, but for our purposes here it is most important for how it emphasizes both 'localizing the global' and 'redistributing the local', to quote one of its principal advocates, Bruno Latour (2005: 173, 193). Contesting the idea that the global is a larger system within which local events must be contextualized, ANT proposes a much 'flatter' way of seeing the world, in which 'movements and displacements come first, places and shapes second' (Latour, 2005: 204). Rather than focusing on something like global capitalism, ANT advocates studying particular sites, such as Wall Street dealing rooms, tracing out how they become 'global' through the density and reach of the connections that they have to other sites. Like Massey, advocates of ANT also contest ideas of places being self-contained and bounded, instead looking to document how 'what is acting at the same moment in any one place is coming from many other places, many distant materials, many faraway actors' (Latour, 2005: 200). Both the local and the global are imagined as having 'networky shapes' comprising 'the intersections of many trails' (Latour, 2005: 204).

ANT has its own distinctive lexicon for describing these networky shapes and how they operate which, to be frank, can be slightly off-putting for the uninitiated. So, to illustrate the network approach I want to take a more accessible example: Ian Cook's work on the networked geographies of tropical fruits (2000; 2004). You are perhaps most likely to encounter the bananas and papaya he writes about on a supermarket aisle or in a fruit bowl in your house. Cook's interest is in part in the connections these fruit enact, linking as they do farm workers and farmers in the tropics (his own research focuses on the Caribbean), supermarket and fruit company technicians, managers and marketing people, and 'first world' consumers. When you eat a banana you become directly connected into a host of networks, obviously including those related to the production and retailing of the banana itself but also spreading out in multiple directions (the production of fertilizer for Caribbean farmers, the ships and planes that cross the Atlantic, the banking systems that allow payments to be made, the plastics in which the fruit are packaged, and so on). For Cook, these fruit represent geographies that cannot be contained in mosaic-like distinctions of here and there: both because of their individual travels from Caribbean farms to British or American mouths, and through their much longer implication in processes of botanical, economic and cultural exchange and their wider status as 'fruits of empire' (Walvin, 1996).

Figure 1.5 Not just a fruit bowl but networks of connections to many other places and actors. Credit: iStockphotos

But Cook is also interested in the disconnections these fruit networks enact. These fruit change status and meaning as they 'travel': changing from plants to be tended for a wage, to 'exotic' fruit, to domestic treats. Cognitive and emotional distances are made between the people and places that have these fruit in common. The Caribbean farm worker knows the papaya or banana eater only as 'the consumer' who dictates market pressures of demand. The banana or papaya eater knows the farm worker only as an invisible producer or as a vague stereotype, whether that be the smiling Caribbean labourer or the oppressed third world worker.

'Following' tropical fruit is just one way to access a world comprising multi-directional, multi-fibred networks, the geographies of which are not mappable on to neat territories or overarching systems. Faced with such networks, the task of Human Geography becomes not to produce knowledge of either the Caribbean or the UK, nor just to explain their differences through understanding global systems, but to explore the networks of connection and disconnection that bring these places and their differences into being.

SUMMARY

- Local places get their distinctive characters from their past and present links to the rest of the world. In consequence, we need a 'global sense of the local'.
- Global networks – with their flows of information, ideas, money, people and things – have locally routed geographies. In consequence, we need 'localized senses of the global'.
- Wider literatures on 'mobilities' and 'Actor-Network Theory' have informed recent attempts to map out 'networky' geographies that both 'localize the global' and 'redistribute the local'.

Conclusion

Human Geography is rightly interested in both the local – the specific place, with its distinctive qualities – and the global: the wider world, with its bigger picture. A crucial question that has always faced Human Geography is how to conceptualize the relations between these two. Three general arguments have informed the discussion here. First, that appeals to ideas of diversity – a global collection of many locals – may be problematic: factually, politically and conceptually. Second, that rather than diversity the conceptual keystone of geographical work in this area should be 'differentiation' – that is, an investigation of the ongoing productions of differences between peoples and places. Third, it is debatable whether these processes of differentiation accord to singular global logics (such as 'developed countries make other countries underdeveloped as part of their own development'). Rather, they may operate through the multiple networks that constitute both the local and the global. Tracing out these networks offers a particularly fruitful way of theorizing and studying local–global geographies.

DISCUSSION POINTS

1. Why might you, as a Human Geographer, be interested in 'the local' and 'the global'?
2. What are the strengths and weaknesses of seeing the world as made up of a mosaic of diverse places and peoples?

3. Why are some places, like the USA, so rich and other places, like the Philippines, so poor?
4. What does it mean to say the local and the global have 'networky shapes'?
5. Debate the relative merits of the 'mosaic', 'system' and 'network' models.

FURTHER READING

Blaut, J.M. (1993) The myth of the European miracle & after 1492. In: *The Colonizer's Model of the World.* New York: The Guilford Press, 50–151 and 179–213.

It is worth attempting a read of this for its powerful restatement of a world-systemic approach. It is particularly strong on debunking the idea that European 'development' stems from qualities internal to Europe itself.

Connell, J. and Gibson, C. (2003) *Sound Tracks. Popular Music, Identity and Place.* London: Routledge.

This is a book that surveys the geographies of popular music, exploring how they combine economic, cultural and political dynamics. I suggest it here because its approach is explicitly framed around seeing the geographies of music as *simultaneously* global and local, so it provides a great case study if you want to get a sense of how those dual emphases of Human Geography can be combined in practice. Chapter 1, called 'Into the music', sets out the book's approach in terms of a dialectic between fixity and fluidity.

Massey, D. (1994 [orig. 1991]) A global sense of place. In: *Space, Place and Gender.* Cambridge: Polity Press, 146–56.

Massey, D. (2005) The elusiveness of place. In: *For Space.* London: Sage, 130–42.

Doreen Massey is perhaps Human Geography's leading writer on issues of the local and the global. 'A global sense of place' is a classic essay. There are lots of ideas in it about globalizing the local and localizing the global and it is very accessible. The other extract suggested here is from Massey's later book *For Space*, and it provides a more explicitly conceptual elaboration of the thinking that underlies the notion of a global sense of place. They make a good pairing for getting into Massey's ideas.

Murray, W.E. (2006) *Geographies of Globalization.* Abingdon: Routledge.

An excellent textbook that both considers the implications of globalization for the practice of Human Geography and surveys the discipline's contributions to globalization debates.

CHAPTER 2
SOCIETY–SPACE

Jo Little

Introduction

At the heart of Human Geography lie questions about the relationship between social characteristics and places. How do the differences between groups and individuals within society map on to, reflect and reinforce spatial categories? Do the qualities of different places affect how society uses, enjoys and even accesses those places? And perhaps more challengingly – why do some social characteristics seem to have a much stronger relationship with place than others? Geographers have recognized the relevance of these questions across a range of scales from the global to the local and paying attention to them has shaped not only the content of what we study as Human Geographers but also the methodologies through which we conduct our research. While we may have long agreed that Human Geography is about this central relationship between space and place, *how* we have chosen to study it has been the subject of much more variation and debate.

In this chapter I will chart the development of Human Geographers' thinking on the relationship between society and space within the discipline, identifying three main 'phases' in the progression from spatial determination to the co-construction of people and place. The chapter will then go on to explore a series of examples in which we see clearly how space and society are inter-dependent and how the ways in which we think about and organize space are fundamental to the experiences of those who occupy, access and are excluded from certain spaces. The examples selected are not unique or even unusual. They speak of everyday situations and lives and are drawn from mainstream geographical topics, underlining the centrality of the relationship between society and space to the whole of Human Geography.

Three phases in the development of geographers' work on society and space

The three phases that will be discussed are spatial order and the mapping of social characteristics; society, space and power; and the **co-construction of society and space**.

Spatial order and the mapping of social characteristics

While both geographers and sociologists have questioned the ways in which space has influenced social processes since the latter part of the nineteenth century, more sustained attempts to conceptualize the interaction between society and space are traditionally seen as emerging in the 1960s in the form of urban ecology. A concern with measuring and predicting the spatial ordering of human behaviour dominated Human Geography in what became known as the quantitative revolution. Much has been written of the influence of this phase in the development of geography and of the emphasis placed on the identification of scientific laws to explain the spatial organization of human behaviour (see Johnston, 1991). Here it is important to appreciate the ways in which this scientific approach resulted in the classification of social characteristics and a belief that understanding socio-spatial relations emerged through the systematic and often very detailed mapping of key population variables.

Social geography was dominated by the idea of social segregation, showing, through the mapping of characteristics such as race, income, housing occupation and how different social groups were clustered in, for example, different residential areas (see Peach, 1975). This was a form of social area analysis which thrived on the development of computer mapping techniques and on the growing availability of forms of population data such as census and labour market statistics. This kind of geography became increasingly criticized, however, for being more concerned with the organization of social patterns than with their explanation and for a view of space that assumed neat, fixed and objective social ordering. In addition, it became clear that social area analysis only really included certain social characteristics (those easily mapped and traditionally seen as important), neglecting many that were less easy to study (e.g. sexuality).

Society, space and power

The key criticism levelled at social area analysis – that it failed to take account of power relations within society and of the ways in which such power relations underpinned the organization of space – became a central concern of Human Geography from the 1980s. Geographers turned to radical approaches, most notably Marxist approaches, to show how space was a product of social forces and to explain the processes whereby identity and difference was reflected in patterns of spatial inequality. This development in the conceptualization of socio-spatial relations was particularly significant in research on economic restructuring in the UK and the USA in the 1980s. It provided a new understanding of the spatial distribution of wealth and jobs across regions, countries and even globally, and of the social outcomes of the uneven development of resources. It also showed, as Massey (1994) asserted, the relevance of geography to political debate about inequality. Later, concerns grew about the assumed dominance of class within radical approaches to the study of society and space. Feminist geographers, in particular, argued that they were failing to recognize the differing experience of men and women and were thus blind to the gendered nature of the relationship between society and space (Bowlby *et al.*, 1989; McDowell, 1983). Such concerns gave rise to a number of geographical studies of the varying employment experiences of men and women within regions, communities and households (McDowell and Massey, 1984). Work showed how women were often disadvantaged within the labour market and, because of different roles and responsibilities, not able to make the same employment choices as men. Geographers argued, as a result of such work, that the spatial division of labour resulting from economic restructuring reflected not only class but other social characteristics such as gender and race and should be understood as an often complex interplay of social patterns.

As a result of such studies, recognition of the different ways in which social characteristics played out over space led to a richer and more nuanced geography. It was still a geography, however, in which places were seen to *reflect* the social characteristics of those who occupied them. Geographers had got much better at showing the subtleties and shifts in the relationship between people and places; they had demonstrated how different theoretical positions gave visibility to particular groups and highlighted different kinds of inequalities, yet space, even across these varying perspectives, remained effectively a container for social difference. While progress had certainly been made in moving away from a kind of spatial determinism in which spatial difference *caused* social inequality, there was still, at this time, little recognition of the interaction between the spatial and the social.

The co-construction of society and space

Developments in geography in the 1990s saw major shifts in the ways in which the relationship between society and space was understood. Engagement with postmodernism and the associated 'cultural turn' in geography were highly significant in challenging the conceptualization of both people and place in two key respects. The first was the new sensitivity to the variations between human beings in relation to characteristics such as gender, race, class, age, (dis)ability, etc., and to the differing experiences people have of space – what is termed '**spatial differentiation**'. The recognition of difference questioned the 'taken for granted' nature of social groupings and of peoples' varying experiences of broad social categories such as gender and age. Central to this recognition was an acceptance of categories as socially constructed and not fixed, and consequently open to contestation, resistance and negotiation. At this time a major area of social geographical research focusing on identity became firmly established, notably in respect to the marginalization of certain groups and individuals from particular spaces and places (this issue is developed further in Chapter 42).

The second area of work that emerged from the development of geographical thinking during the 1990s related to space and to its conceptualization as constructed. It became increasingly asserted that

> just as social identities [were] no longer regarded as fixed categories but . . . understood as multiple, contested and fluid, so too space [was] no longer understood as having particular fixed characteristics.
>
> (Valentine, 2001: 4)

Critically, space started to be understood not as a simple backdrop against which difference and inequality played out, but an active part of the construction of society and of the experiences of people within those places. Space, it was argued, could not be factored out (or in) to the operation of social relations and practices – it was a central part of how those relations were produced and reproduced. Geographers thus began to talk of society and space as mutually constituted and apparent in ways that were never fixed but always in the process of becoming.

These three phases in geographers' study of the relationship between society and space are summarised in Table 2.1 below.

Key phases in the conceptualization of socio-spatial relations	The scope and direction of research	Research content	Criticisms
Spatial order and the mapping of social characteristics	Social geography was dominated by the idea of social segregation, showing, through the mapping of characteristics such as race, income and housing occupation, how different social groups were clustered in, for example, different residential areas	Computer mapping techniques aided by the growing availability of forms of population data such as census and labour market statistics (see Johnston, 1991; Peach, 1975)	More concerned with the organization of social patterns than with their explanation. Belief that space was passive and assumed neat, fixed and objective social ordering
Society, space and power relations	Radical approaches, most notably Marxist, to show how space was a product of social forces and to explain the processes whereby identity and difference was reflected in patterns of spatial inequality	Research on economic restructuring and uneven development, often at the regional scale (see Massey, 1994)	Assumed dominance of class within radical approaches; failed to recognize gendered and racial characteristics (Bowlby *et al.*, 1989; McDowell and Massey, 1984)
The co-construction of society and space	A sensitivity to difference and hybrid identities and to the multiple, fluid and contested nature of both social characteristics and space	Poststructural and postmodern approaches and a focus on performance (see Panelli, 2004)	Mitigates against the recognition of broader patterns of disadvantage and may be difficult to mobilize politically

Table 2.1 Phases in the study of the relationship between society and space

SUMMARY

- Conventional approaches in the geographical study of the relationship between society and space were characterized by an initial concern to map the ways in which places were socially differentiated. Such mapping exercises were seen as useful to policy makers but provided only a very narrow view of people's experience of space and place.
- Geographers argued that understandings of society and space needed to take into account the inequalities and power relations reflected in social patterns and, in particular, the uneven development of the economy and unequal access to wealth.
- There was a recognition following the cultural turn in geography and the influence of postmodernism on the role of space in the construction of social difference. Geographers became interested in what was termed the co-construction of society and space.

Will we now discuss in more detail, and through the use of examples, how the relationship between society and space has come to be understood by geographers as *co-constructed*. That is, how space evolves to reflect and to shape the identities of those who use it and how the imagining of space in particular ways can act to exclude some and protect others.

Place and the social construction of space

Tim Cresswell (1996) used the notion of in place/out of place to explore how space becomes imbued with certain social and cultural values and assumptions. These values and assumptions drive ideas about which identities and behaviours we might deem to be appropriate and comfortable (in place) in those spaces and which we might see as inappropriate (out of place). These ideas may shift over time – they may, as we shall see later, be contested, but they are often powerful and hard to resist. They help to show how social and cultural characteristics are translated from society more broadly to the day-to-day experiences of particular people in particular places. The idea that space is an active agent in the ways in which social relations evolve and play out is now fundamental to geographical study. Not only does 'space matter' but indeed space is *part of* the very organization and operation of society. We can turn to research from almost every area of Human Geography to illustrate the relevance of the construction of space itself to our experience of place and performance of identity.

The rural community

The rural community provides a rich illustration of the ways in which our imagining and understanding of the spaces of the rural plays through the characteristics and organization of rural society and the day-to-day ways in which people live their lives. Geographers researching rural communities and lifestyles have recognized the power of taken-for-granted assumptions about rurality. In recent years, they have argued that any attempt to understand the nature of rural society must acknowledge and incorporate a set of timeless qualities associated with the countryside –

qualities such as the strength of the community, the slower pace of life and the closeness to nature. All of these 'rural imaginaries' will, it is asserted, underpin and inform the nature of both individual identity and the more general operation of the rural community.

The idea that rural social spaces are characterized by a more authentic and active sense of community is one such imaginary that has held an important place in academic writing and popular culture (Cloke, 2003). A uniquely rural way of 'doing' community is seen as so key to the formation and reproduction of rural society that it needs to be written into understandings of rural places and to the histories, attitudes and experiences of rural people. Constructions of rural community, as witnessed in many geographical studies (see Bell, M., 1994; Halfacree and Rivera, 2012) have proved a very strong 'pull factor' in people's decisions to migrate to rural areas in the UK and other Western countries and consequently highly relevant to the process of counter-urbanization (see Figure 2.1). Such strong expectations of community can prove a powerful force in mediating behaviour and identities of those living in rural areas – they may be more inclined to participate in community events to provide help and assistance to fellow villagers – and by doing so to fit in with the expectations of village life.

In my research on rural women in south-west England in the 1990s, I talked to many women who valued the 'sense of community' that existed in the village (Little and Austin, 1996). This community spirit, they believed, was not something they had experienced in previous (urban) places of residence. It ensured people 'looked out for one another' and that the elderly and vulnerable, in particular, were

Figure 2.1 A traditional view of 'rural life'. Credit: www.CartoonStock.com

Figure 2.2 Black and ethnic minority people are sometimes regarded as 'out of place' in rural and coastal spaces, even as visitors. Credit: Peter Lomas/Rex Features

not neglected. This is an example of the co-construction of society and space – the space of the rural community was socially constructed in line with past understandings and associated contemporary behaviour to be a place of friendship and cooperation. This, in itself, helped to shape behaviour and encourage villagers in caring and acts of mutual support. The women's identities as rural people, it seemed, had responded to the ways in which they felt village space to be constructed.

In a more negative reading of the notion of the rural community, other studies have drawn attention to those who do not fit (see Figure 2.2). They illustrate how the social construction of rurality in the UK in particular, is a very 'white' construction, appealing often to traditional ideas of Englishness (see Neal and Agyman, 2006). This construction sees black and ethnic minority rural residents (or would-be residents) and visitors as 'out of place' within such communities. This construction of rural community can support racist behaviour, as has been shown through studies of the experiences of black people and the attitudes of white residents (see Hubbard, 2004; Jay, 1992).

The night time economy

The urban **night time economy** provides us with another example of the relationship between space and society, which again shows the co-construction of place and identity as well as its fluid and contested nature. Geographers have been interested in the development of the entertainment industry, as an element of the night time economy, from a number of different angles and have looked at its role in regenerating flagging city centres and providing jobs where previous economies have declined (Bell, D., 2007; Chatterton and Hollands, 2002). They have also been concerned with the associated social changes to the city and how we use and feel about these spaces. This has, inevitably, included contemporary concerns about the behaviour of those using urban entertainment spaces and, in particular, how city centres have become constructed as exclusionary spaces, dominated by a very strong drinking culture (particularly in the UK) and in which only certain groups and identities are comfortable (Royal Geographical Society, 2012; Jayne *et al.*, 2006).

Displays of aggressive and drunken behaviour have been identified in both academic and popular reports as responsible for transforming the atmosphere of the city centre at night and making it, for some, a place of fear. The following passage from the *Observer* newspaper describes how the concentration of 'drinking spaces' has led to violence in a particular part of the centre of the Welsh city Cardiff.

> In the shadow of Cardiff's castle, dozens of bars and clubs have gained the Welsh capital its reputation as a party city. They line the roads through the city centre to St Mary Street, a pedestrianised zone lined with pubs of all types and chains. As an ambulance flashes by, three men are arguing loudly about where to go next. The level of aggression rises until one stomps off swearing loudly at the other two, who throw something at his back which smashes into the gutter.
>
> (McVeigh, *Observer*, 25 March 2012)

The expansion of vertical drinking spaces (as they have become known) in Cardiff and many other cities means that, for part of the day, these spaces are effectively 'no go areas' for anyone but those participating in the night time entertainment. Moreover, the normalization of aggressive and drunken behaviour is seen to reinforce the drinking culture. While clearly space does not cause violence, the enduring presence of aggression and the exclusion of other users of the space helps to cement the relationship between city centre space and drinking.

Some studies of this relationship between city centres and aggressive and alcohol-fuelled behaviour have argued that it has reinforced both spaces and associated identities as masculine (Figure 2.3).

The laddish culture generated in the drinking spaces (see Hubbard, 2009) – an element of which is the increasing presence of entertainment venues that objectify women (such as lap dancing and pole dancing clubs) – creates an atmosphere in which some women feel out of place, excluded and even fearful.

While the particular problem of today's city centre drinking spaces is relatively recent, the wider issue surrounding the ways in which some environments are experienced as dangerous or scary is not new. Over many years geographers have noted how public space is frequently viewed as dangerous or unwelcoming by women. Charting that research illustrates the different approaches that were introduced at the start of the chapter. So, during the 1980s the study of women's fear in public spaces was widely seen as a failure of planning and of the design of buildings and spaces – dangerous spaces were believed to be the outcome of a development process that ignored the particular needs of women and created bleak and functional public spaces (Little, 1994). Later, geographers argued that such ideas were overly environmentally determinist and through the adoption of more radical approaches sought to explain fear within public space in relation to broader understandings of women's feelings of vulnerability within patriarchal or male-dominated societies. More recently, however, with the interest in spaces as socially constructed, fear has been seen as the outcome

Figure 2.3 Evening drinking in city centres may often appear aggressive. Credit: Paul Panayiotou/Alamy

of socio-spatial relations and the performance of identity in place. This emphasis on the interdependence of identity and place has helped in understanding the more nuanced and complex relationship between the aggressive spaces of the night time economy and the performance of gender – particularly the different and very fluid circumstances under which both men and women experience such spaces as dangerous and threatening (Pain, 1997; Kern, 2005; Wesely and Gaarder, 2004).

Thirdspace

Throughout this chapter, examples from across Human Geography have been used to explore the interaction between society and space. They have shown how different ways of approaching the relationship between people and place can inform the understanding of this interaction and how geography has moved from simply mapping social characteristics in space to seeing space as bound up in how those characteristics are distributed and performed. We have seen that identities may become excluded from spaces to which they do not belong and also how space itself can take on particular qualities through the presence or absence of different identities. What is very important to stress about these socio-spatial interactions is their variability and fluidity. They are not fixed but made and re-made and while some relationships between society and space may be acknowledged, like the rural community, to be a product of historic associations, they are still constantly being negotiated and performed.

It is this negotiation that needs particular emphasis in this final section of the chapter.

While dominant constructions of space may be powerful, as the examples have shown, they may also be contested. Indeed, writing in the mid-1990s, Ed Soja (1996: 2) urged scholars to 'think differently' about the construction and lived experience of space. He argued that both practical and theoretical understandings of space and spatiality were in danger of being muddled by 'the baggage of tradition [and] by older definitions that no longer fit changing contexts'. Soja's concerns about the conceptualization of socio-spatial relations found momentum in a developing critique of geographers' thinking about space, and in particular, their use of dualisms such as inside/outside, home/work, belonging/excluded, white/black, public/private, etc. Such dualisms, it was argued, suggest the world can be understood as clear-cut, oppositional categories and that, used by geographers, these categories appeared to map onto space in straightforward and stable ways.

Challenges to the use of these dichotomies came in particular from post-colonial and feminist research in geography. Such research questioned the construction of knowledge, arguing the need to contest and destabilize the privileging of what were seen as western, masculine forms of knowledge, and to develop alternative approaches which recognized the varying and hybrid nature of identity and experience. Research demonstrated that traditional forms of identity were increasingly being reshaped in response to social, cultural and political change and that any attempt to understand people's lives needed to appreciate the complex and sometimes contradictory reworking of identity. Critically, the reworking of identity was seen to produce an alternative *spatiality*, 'a **thirdspace**' as Soja puts it, in which there was an opportunity to think and act politically and a responsibility to creatively re-think and re-theorize spatiality in conjunction with multiple forms of identity.

Soja (1996: 6) briefly states what thirdspace provides that takes us beyond other conceptualizations of space as follows:

> Thirdspace can be described as a creative recombination and extension, one that builds on a Firstspace perspective that is focused on the 'real' material world and a Secondspace perspective that interprets this reality through 'imagined' representations of spatiality.
>
> (Soja, 1996: 6)

In exploring the potential of thirdspace, Soja (1996) draws extensively on the work of bell hooks, the American academic and activist, and on her writing about the home and community as a space of nurture and resistance as well as oppression. hooks, Soja (1996: 13) suggests, is able to illustrate the 'radical openness' of thirdspace to the creation of alternative spatial imaginaries by those who wish to 'reclaim' spaces of oppression and make them into something else. Writing as a black, feminist activist, hooks (1990) talks in her book *Yearning* of the marginalization of African-American subjectivities and their place on the periphery of American political and intellectual life. According to hooks, this marginality can be used to provide a space, simultaneously material and symbolic, from which to challenge the dominant power of the mainstream and give voice to the ideas, beliefs and experiences of the silent 'other'.

The important thing for our discussion here is the notion of thirdspace as the spatialized expression of oppression and political action. It takes the experience of marginality and turns it into a space of resistance. Soja notes how the use of marginality in this sense as a form of resistance evokes the work of French philosopher Lefebvre, and transforms marginality into centrality. hooks, in her 'purposeful peripheralness' thus acquires a 'strategic positioning that disorders, disrupts and transgresses the center-periphery relationship itself' (Soja, 1996: 84).

For hooks, her activism and desire to counter the oppressive nature of the Black experience in the USA is a 'politics of location' which calls:

> those of us who would participate in the formation of counter-hegemonic cultural practice to identify the spaces where we begin the process of re-vision . . . For many of us, that moment requires pushing against oppressive boundaries set by race, sex and class domination . . . For me this space of radical openness is a margin – a profound edge.
> (hooks, 1990: 149, quoted in Soja, 1996: 85).

As part of the displacing of oppositional categories, thirdspace also challenges the division between academic theorizing and political action and does so through the use of multiple scales of analysis – from the global to the local and in between.

Thirdspace is a very useful way of highlighting the spatialization of resistance and of the breaking down of oppositional categories in geographical analysis. There are many examples we can draw on where space and place give expression to what may be seen as a challenge to conventional dualisms, enabling us to look beyond existing categories. Thirdspace recognizes not only the complex nature of identity but also the often contradictory ways subjectivities play out in space and time. It also allows us to think of the changing use of space and the ways in which space and identity may be co-constructed as temporary or transitory sites of resistance – for example, in a political march or rally or a protest camp.

Figure 2.4 Black women and spaces of resistance. Credit: Getty Images

In their edited book, *Writing Women and Space: Colonial and Postcolonial Geographies*, Alison Blunt and Gillian Rose (1994) introduce a collection of studies which illuminate the multiple and complex position of women in relation to the politics and processes of colonialization. The studies provide a very clear illustration of what might be termed thirdspace in contesting many of the dominant assumptions about not only the lives of the indigenous women but also the spaces of colonization. Blunt and Rose argue the need to deconstruct the binary opposition between colonizer and colonized and in re-thinking the varying subject positions of women and the fixity of constructions of otherness. The chapters in the book show the complicated relationship between gender, race and class and how the subject positions of women in post-colonial settings have been formed by the interaction of patriarchal and colonial discourses of difference. They call for a 're-mapping of colonization' to help understand the multiple subjectivities of the colonized and the colonial women together with the spaces in which they interact.

Blunt and Rose's book shows how thirdspace can contribute to the understanding of the relationship between space and subjectivity in the context of the gendered politics of post-colonialism. Many other examples have also made use of the concept in articulating a resistance to accepted binary subject positions and to the spatial politics of oppression. Take, for example, the occupation of certain spaces by gay, lesbian and bisexual people in gay pride marches (see Figure 2.5). Such marches reflect a desire to question and subvert the

Figure 2.5 Gay Pride march in New York. Credit: Getty Images

taken-for-granted heterosexual nature of public space. The march temporarily changes the relationship between society and space, creating 'gay space' – a thirdspace where behaviour and hybrid identities deemed unacceptable at other times are dominant. Returning to the issue of women's fear raised earlier, another example of thirdspace can be seen in the attempts by women to 'reclaim the night' by refusing to be fearful, and contesting the dominant assumptions of masculinity and violence that surround public space at night (Koskela, 1997). Space and place are sexualized in that they reflect an acceptance of or hostility towards particular spatial norms and identities – what is known as the sexualization of space.

SUMMARY

- In the past, Human Geography conceptualized space as a backdrop for social relations and was concerned with first, the mapping of spatial inequalities and, second, the articulation of the broad power relations through which those inequalities developed.
- Through time geographers have become more aware of the constructed nature of both identity and space, and have recognized that socio-spatial relations are negotiated, created and reinforced through everyday performance.
- It is clear that the co-construction of identity and space means that some identities are accepted, where others may be seen as out of place. Notions of belonging, community and exclusion are all central to the understanding of the relationship between society and space.
- The exclusion of some identities means that different, hybrid identities may emerge to destabilize and contest the dominant patterns of belonging in space and ensure the relationship of society with space is never fixed but always ongoing and in the process of being renegotiated.

Conclusion

The study of the relationship between society and space by geographers has developed over time through the different phases outlined in the chapter. This development has been presented, perhaps rather misleadingly in the chapter, as a rather neat sequence – primarily to assist understanding. In reality, however, it has not been a case of one approach replacing another but rather there has been a shift over time from geographical studies that sought to map spatial characteristics and the differentiation of social and economic characteristics to studies that represented social constructions of space and, finally, studies that challenged dominant socio-spatial constructions and focused on space as a form of resistance. Some would suggest that such a development has allowed geographers to think of space in a different way. We have moved from thinking of space as simply containing or reflecting social difference to being a part of how that difference is constructed, performed and contested. Space has moved from being passive to more active in the production of social change and in the experience of place. Understanding space as part of the process of reproducing and resisting social change allows us to think in much more hopeful and positive ways about the strategic role of space in resisting oppression and celebrating diversity and opportunity.

DISCUSSION POINTS

1. Why did geographers become dissatisfied with the mapping of spatial patterns and how did they seek to address the limitations of such approaches?
2. How have different approaches to understanding social spatial relations been reflected in understandings of spaces as frightening or 'scary'?
3. What is meant by the term thirdspace and why might we associate this concept with feelings of hope ?

FURTHER READING

Bell, M.M. (2004) *Childerley: Nature and Morality in a Country Village*. Chicago: University of Chicago Press.

This is a book about a rural community, and provides a detailed examination of how constructions of rural space are formulated and contested by those living and visiting the countryside. It is helpful in illustrating the idea of the co-construction of space and society as discussed in the chapter.

Cresswell, T. (2004) *Place: A Short Introduction*. Oxford: Blackwell.

As Tim Cresswell writes in his book, Place is a form of space – it is space 'invested with meaning'. The book thus takes the ideas surrounding the relationship between society and space as discussed in this chapter and applies them to the notion of 'place'.

Holloway, L. and Hubbard, P. (2001) *People and Place: The Extraordinary Geographies of Everyday Life*. Edinburgh: Prentice Hall.

A series of very accessible chapters that talk about both the way geographers have understood space and also how different identities have experienced everyday space at the local level.

Johnston, L. and Longhurst, R. (2010) *Space, Place and Sex*. Plymouth: Rowman and Littlefield.

This book explores different spaces of sexual identity, the assumptions and challenges that surround the relationship between sex and space.

Soja, E. (1996) *Thirdspace*. Oxford: Blackwell.

A challenging book but one which provides a critical discussion of the concept of thirdspace as first used by geographers. It situates geographers' use of thirdspace within intellectual and empirical discussions.

CHAPTER 3
HUMAN–NON-HUMAN

Hayden Lorimer

Introduction: all the stuff that we stuff in our mouths

As sure as eggs are eggs, you'll have eaten a meal in the last few hours. Right? Thought so. And which one was it? Breakfast, lunch or dinner? If you can stomach it, then pause for a minute and just remember what you chewed up and swallowed. Whether it was a smorgasbord or a simple snack, review all that matter journeying through your system, the stuff processing its way down, channelled through digestive tract and gut into bowel, washed over and worked on by gastric juices. And while you go about the work of mental regurgitation, you might place your hands over the swell of your stomach – go on now, all the way around – listening out for the embarrassingly loud noises that your organs have a habit of making as they go about doing the necessaries. Mine just gurgled while I wrote these words. Makes you think, doesn't it? The experience gets odder still if, only momentarily, you re-cast universal human actions – like eating – as being, well, a little bit out of the ordinary (Bennett, 2007). There's no room for the squeamish here. Keep the metabolic experiment going by revisiting in more detail, *exactly*, what was on that plate. What did you scoop from inside the foil casing or plastic wrap? If the waste bin is nearby, be brave, flip it open and pick out the discarded food packaging. Take a close look at the ingredients listed. How much does the information provided *really* tell you about source, sort or standard? What geographies does it disclose? Or discretely pull a veil over? Elspeth Probyn (2011) offers a few quick pointers for the more inquisitive geographer-eater:

> The global food crisis has brought a renewed public and academic attention to questions of what we eat, where it comes from, how much it costs, and whether it is sustainable . . .
>
> Coinciding, in ways that are more than coincidental, with a growing awareness and at times panic about global warming and climate change, people are becoming attuned to how what we always deemed as edible (corn, soya) are being turned into non-edible things like bio-fuels. And, as one of the most virulent forms of globalization, there is a seemingly endless circulation of food scares about things we had thought were edible – chickens that carry flu, cows that turn mad, eggs that are bad.
>
> (Probyn, 2011: 33)

Half a century ago, Claude Levi-Strauss, a French anthropologist, put all that in a nutshell, declaring that 'food is good to think with' (1969).

Now take a deep breath and back we go to the belly of the beast. You might well have eaten some choice cuts (or, some not-so-select bits) from part of another animal's body. Perhaps you enjoyed the differing tastes of more than one kind of flesh (quite possibly without even knowing it). Prior to that, you may have sliced or diced the meat – there's just a slim chance you even gutted or filleted a carcass, or plucked it clean – as part of the preparations for cooking. Or, maybe you didn't eat a scrap of flesh, fowl or fish. If so, then this could just be a fairly arbitrary occurrence, explained by your failure to get the shopping done yesterday or the fact that the fridge or cupboard is looking a bit bare right now. Perhaps your religious beliefs or your family upbringing mean that you consider some meat types as unfit for human consumption (and, by contrast, others as palatable because they come from animals slaughtered in proper observance of recognized custom). Or, it could be that the absence of products derived from animals' bodies in your diet is because you've actually made a moral choice, at some stage earlier in your life, to consciously limit the range of foodstuffs that you consume. Like it or not, that decision places you in a minority and confers a badge of identity (vegetarian, pescetarian or vegan). Depending on which of these terms fits best, then what you just ate may have contained a mycoprotein meat substitute product (such as Quorn). Possibly this is because you find you still have to suppress strong carnivorous urges for a certain taste, tang and texture. For many among us, the first bite taken from a bacon roll is hard to beat, whatever the time of day. And even having commited to a meat-free lifestyle, it's not always possible to be entirely sure. Unless you have been an extra-specially careful consumer, there might be rendered animal tissue in the beauty products you apply to your body or face, or in those sweets you sometimes treat yourself to between meals, or the shoes that protect the soles of your feet. And what about me? Seeing as I've been doing all the quizzing so far, it's reasonable to expect an answer. I'm one of those fish-eating sort-of-vegetarians. By some sorts of judgement, that stance makes me a contradiction in terms.

So, you might reasonably ask, what's the purpose of all this prying into personal habits and mealtime preferences? The answer is simple enough, and it is central to the material conditions of our existence here on Earth. Attitudes to meat (and an extraordinary range of animal by-products) tell us a great deal about how we humans understand and value the lives of other living creatures, or 'non-human animals', to adopt a semi-technical term. When we're not processing bits of them on the inside, we're wearing bits of their bodies on the outside. It's a scale of intimacy and sort of immediacy that's all too easily, and comfortably, forgotten. For much of modern life, and for many millions of the world's population, it's just seemed better that way. The attitudes we hold about the lives and the deaths of millions of non-human animals, specifically reared to be eaten, are for most of the time, kept in a mental 'black-box', along with a range of beliefs, judgements, imaginings, tastes, morals and ethics that inform our sense of place as humans amid a greater planetary ecology of relations (Foer, 2009; Baggini, 2005). There are powerful ideas bundled up in there and emotions that can pack a punch. And, when it comes to the central concern of this chapter, the edibility of non-human animals might be tellingly illustrative, but that doesn't even cover the half of it.

Non-human relations and non-human agency

As geographers today are coming to realize, it is important to trace the different forms of relation and connection that exist between the human and what is referred to increasingly as 'the non-human'. Such a project has the potential to radically alter the way that you configure the everyday world around you, in ways reaching far beyond the bounds of our introductory exercise about personal patterns of consumption, and that demand new mind-maps to navigate by. Rethinking relations has real kinds of analytical and material consequence, redrawing what we understand as the very constitution, and the basic boundaries, of a world of 'humans', 'non-humans' and a great medley of other 'things' that make up the material culture of everyday life. If these words already begin to read like a significant challenge to generally accepted values – by decentring our separate condition as human beings – then that is no accident. The direction of travel in current geographical thought is away from the cherished idea of sovereign species (or what I'll later on refer to as 'human exceptionalism' and 'ontological separatism') and towards one of a world populated by post-human entities, like 'hybrids', 'cyborgs' and 'monsters' (Whatmore, 2002; Davies, 2003). Why is that? Well, in truth, these days it's not easy to say where the human ends and the non-human begins. As scientific visionaries plan possible futures for radically different kinds of life on Earth, with new biotech redesigns based on genetic sequencing, and contemplate grand plans for the geo-engineering of earth and atmospheric systems, our long-standing appreciation of organisms and physical phenomena as things with an individual existence, identifiably separate and sealed, is being buffeted about. The future is no longer the heady stuff of science fiction (as it undoubtedly was for your parents' generation). A tumbledown world is widely predicted and projected, rather than a perfect one. Elements of it have even arrived early. And what's still to come does not promise to be a simple matter.

The trends in post-human geographical thought that this chapter explores are ones placing in question generally accepted orthodoxies about **ontology**, that is to say the very nature and reality of existence. According to post-human principles, rather than bodies or matter being conceived in terms of fixed, essential states, their properties are instead to be understood as vital and always in flux. To acknowledge this restlessness and vitality is also to accept that non-human entities have real and significant **agency**. Whether we like it or not, other living things and artificially intelligent systems have the potential to act up, to bite back, to spread virally or with volatility, to undermine the great certainties of human will, to evade or corrupt original designs, or simply to move beyond our full control.

Such observations about ontology and agency might be rooted in questions of existential philosophy, but their implications are deeply political. Depending on your view, they hold significant promise or pose real threats. They re-make us as 'human becomings', rather than human beings. They open up new horizons, where genomic data (like DNA profiles) and ID biometrics might be the sort of evidence used to tell us why we are the way we are. They also topple – or just gently nudge – us from an elevated and exceptional position, based on an age-old assumption about the all-powerful dominion that humans hold over nature. These are really big considerations to take on. Should some reassurance be necessary as we delve deeper into the world of human and non-human relations, it might prove helpful if I elaborate on a new typology of non-humans and hybrid entities. We can

Figure 3.1 An illustrated page from a bestiary: an ancient kind of book containing descriptions in text and image of all sorts of animals, real and imaginary, monstrous and fabulous. Creatures were not described in scientific terms, but rather in humorous or imaginative ways, sometimes with moral judgements cast on aspects of non-human behaviour. Credit: British Library

figure this as something akin to a contemporary version of the medieval bestiary (see Figure 3.1) or compendium of living creatures and fabled beings.

For a start, this will mean reconfiguring some standard disciplinary labels and accepted classificatory terms – those first enshrined in school classrooms and still in common usage in university lecture theatres. So, what happens if we expand the domain of our given subject area, contemporary Human Geography, so that it becomes a 'more-than-human geography' (Braun, 2005)? That would be a version of the subject with its parameters stretched to better accommodate the great tangle of spatial and temporal relations we humans have with other kinds of living organism, materials, objects and a host of other 'things' besides, taking place in a vast array of settings. What would the 'more-than . . .' version of Human Geography include? Just to begin, numbered here would be the entire animal kingdom, all fauna inclusive of birds, fish and insects. Every kind of flora too: plants and trees, mosses and algal blooms, fruit and vegetables. But then what of living things that are less easily mapped upon, or tracked across a landscape, operating at a micro or molecular scale, perhaps internal to bodies? They need accounting for too. So the geography unfolding is one also inclusive of the bacteria and the bacillus, the germ and the genetically modified life form. After all, in the twenty-first century, the pervasive presence of biotechnology has begun to normalize to a degree where public attitudes seem increasingly tolerant (or unquestioning) about the most fundamental kinds of change. Biotech innovations range from crops of Canola, engineered to be tolerant of herbicides and pesticides, to 'Enviropig™' (see Figure 3.2), an enhanced line of livestock with a capability to digest plant phosphorous more efficiently and less toxically.

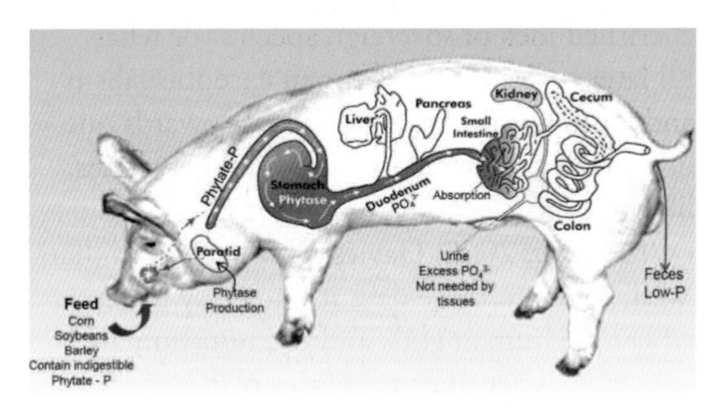

Figure 3.2 Enviropig™ is a genetically enhanced strain of swine, derived from the Yorkshire pig. Biotechnological intervention has altered the creature's digestive process so that the manure produced from cereal grain consumption has lower phosphorous content. The manure can then be spread on farmland with less environmental risk of phosphorous leaching into freshwater ponds, streams and rivers, leading to reductions in water quality or fish populations. Credit: © 2012 University of Guelph – All Rights Reserved

And there's more space required yet. The category of the 'more-than-human' must also encompass non-human beings that, while not being fully sentient or necessarily organic in origin, are nonetheless active and dynamic forms that significantly shape the conditions of contemporary living. **Non-human agency** flourishes, in everything from the yeasts that rise in bread to fruit crops that fall from trees, in the tides generating supplies of renewable energy to the pulses that transmit electrical power to industrial destinations and domestic households (Bennett, 2010). The more-than-human realm must also take in genetic data and chemical compounds, the classic experimental apparatus of pipettes and petri dishes, through to laser-guided neuroscience. Witness the fact that some geographers now research in the company of scientists who are operating on the frontiers of experimental biomedical science where trans-species mutants proliferate (Davis, in press), or developing programmes for animal conservation through species 'back-breeding' (Lorimer and Driesson, in press). Still more capacity is necessary for entities from this 'more-than . . .' world. It must take in the host of machines, mechanical and digital (from robot milking machines for dairy cows to unmanned military drones). It must encompass software environments (from android apps to iClouds and online social networks) now so very deeply programmed into the fabrics and rhythms of life as to sometimes seem inseparable from the very core of existence. As I sit here typing, the smart device nestled snug in my pocket gently vibrated, as if to verify the modern maxim that 'you're never alone with a phone'.

In certain instances, transplanted or implanted biomedical technologies operate internal to bodies (maintaining stabilities in heart-rate or mood), or they can work as sensory fixes and anatomical add-ons, augmenting the capacities and competencies of naturally evolved human form. To illustrate one such socio-technical advance, consider the case of Oscar Pistorius, South African track athlete, gold-medallist paralympian and first ever paralympic athlete to qualify for the Olympic Games (see Figure 3.3). A double amputee, dubbed 'The Bladerunner' and 'the fastest man on no legs', Pistorius runs using Cheetah Flex-Foot carbon fibre transtibial artificial limbs, fitted below

Figure 3.3 Oscar Pistorius is an athlete with a difference, who poses interesting questions for the worlds of sport, law and ethics. Hi-tech prosthetics, fitted to his kneecaps, enable him to compete at the highest levels of international track competition with able-bodied athletes. In 2012, his personal best for running 400m stands at 45.07 seconds. Credit: Getty Images

the kneecap. In the process, he troubles internationally accepted rules, set to ensure that competition takes place on a 'level playing field'. Sporting arbitrators and authorities are anxious that the pioneering design of his prosthetics could actually place Pistorius at a freakish advantage when running against able-bodied athletes. Perhaps the Bladerunner is blazing a trail for other 'cyborg-athletes' who yet might perform superhuman feats, rewrite the record books and win the human race. Hybrid anatomical designs are not always so glamorous or so quick to grab headlines. It is now standard practice for dentists to offer a patient the option of a cow bone implant as part of a surgical procedure for fusing a replacement denture to a remaining tooth root. Whether cosmetic, commercial or biomedical, this is only one in an increasingly diverse range of treatments and interventions available that depend on trans-species fusions to rebuild, regenerate and replace parts of bodies.

A point could be reached where, quite properly, questions need to be asked about the possible extent of this experiment in relational thinking. What is to be left out from the category of non-human beings? Is it a case of everything *and* the kitchen sink? Where do these complex material assemblages of related stuff stop? Is it actually possible to differentiate between sorts of being? Surely, of necessity, there must be some spatial and temporal limits established, otherwise wouldn't everything end up being connected to everything else? Where and how to place spatial limits around the interactions of human and non-human is an important consideration for geographers, and some of the more conceptually driven thinking that can enable this to happen is introduced in the next section.

SUMMARY

- 'More-than-human geographies' is a label that invites a full disciplinary rethink about the hybrid forms that life seems increasingly to take, and about how multivariate entities are related to one another.
- The agency of non-human animals, objects and entities is a phenomenon being treated seriously by geographers, based on a growing recognition that we humans are not in sole control of social situations.
- Thinking about how environments and situations happen in relational terms can be enlightening, but simultaneously it is worth wondering about the spatial and social reach of these relations.

Into the mangle of post-humanism

In recent years, a variety of big ideas has been utilized by geographers who are thinking about diverse assemblies of human and non-human entities. Arguably, it is along the threshold of human–non-human existence that some of geography's most inventive thinking is happening. Some of this originality draws on the discipline's own intellectual heritage, though it also reflects a lively traffic in ideas with other subject areas, like anthropology, sociology and philosophy. To begin with, encountering these exchanges can be a fairly daunting business, partly because it means becoming reasonably literate in unfamiliar sorts of language. This section will begin that task by mapping out three key conceptual influences and highlighting some of the new geographies that are being produced as a result.

The emergence of 'new animal geographies' (Emel and Wolch, 1998; Philo and Wilbert, 2000) is a good place to turn to first. This field of study reminded many geographers of the need for a far greater understanding of the spacing of the lives of non-human animals in Human Geographies, and encouraged a subtler appreciation of the placing of the lives of humans in animal geographies. The subjects of these relations range all the way from wolves (Brownlow, 2000) to foxes (Woods, 2000), pet dogs (Howell, 2000) to feral cats (Griffiths *et al.,* 2000). Sometimes concerned with animals as symbolic representations and sometimes as substantial lively things, this work ensured that – whatever the nature of the relation encountered – matters of social power and moral–ethical concern were kept to the fore. Such studies of inter-species relations have since extended to include other kinds of non-human agencies and biotechnical assemblages. The 'hybrid geographies' written about by Lewis Holloway (2007; 2009) and Carol Morris (2009) are those employed in the commercial breeding of farm livestock. They show how the genetic revolution has created entirely new spaces and scales of knowledge, meaning that cows are very differently understood as creatures. They consider such developments as a powerful expression of 'biopower', a concept that originates with Michel Foucault, a French philosopher, relocated, so as to include animal lives. Biopower captures the human will to regulate conditions of living and the nature of life itself.

Second, geographers have learned some important lessons from social anthropology about the ways that the lifeworlds of humans and non-humans are enmeshed or co-constituted. The 'relational ecology' of Tim Ingold (2000; 2011) has had a telling effect. Drawing on ethnographic observations of the lifestyles of non-Western, indigenous peoples, Ingold explains the cosmological beliefs that inform systems of environmental perception in these worlds. Here, animals, birds, trees, rivers, weather *and* humans are all 'persons', who might come to share in each other's wisdoms, sometimes even shifting identities and bodies as they do so. This makes indigenous knowledge systems and languages about the skies, sea and land among the world's most ancient, but also shows how well attuned indigenous ideas are to prevailing environmental theories concerned with ontological hybridity and fluidity. 'Old ways' that originate in the lived experiences and practical skills required for the upkeep of extended communities of humans and non-humans serve to remind us that not everything is new under the sun. Geographers have taken these ideas on different travels, exploring nearby worlds and familiar landscapes, the kinds found in fruit orchards (Jones and Cloke, 2002) and among herd animals (Lorimer, 2006), showing the meshing together of agencies into local swirls of life.

Finally, when it comes to challenging the principles of 'human exceptionalism' and 'ontological separatism', Bruno Latour, a French sociologist, has his own ideas. Like Ingold's, these imports have been highly influential in geography. Latour offers up a powerful argument about the state of humankind: 'we have never been modern' (1993). By this provocation, Latour means to expose something he believes has been hiding in plain sight for centuries. Namely, that the model separating 'culture' and 'nature' is a false intellectual construct. Instead of imposing this model to claim power and making distinctions, we need to apprehend worlds as they actually are. That is, as a succession of fabricated environments, comprising human and non-human beings, involved in spatially distributed interactions, normally through socially equivalent conditions. Latour's particular way of thinking places greatest emphasis on the practical relations existing between actors and agents and intermediary objects and technologies. It is attentive to a social realm made up of networks, circulations and translations. It has been applied, rather like a helpful tool or template, to all manner of socio-technical spaces and ordinary situations. He once used it to explain the operations of an entire public transportation system, affording agency to its constituent parts (Latour, 1996). Some geographers have taken up his toolkit, using it to explain the ways that water vole conservation happens (Hinchliffe *et al.,* 2005), how the hunting of foxes has been represented in the British countryside (Woods, 2000) and how elephants were hunted in the British Empire (Lorimer and Whatmore, 2009). As well as scrambling nature and culture, Latourian thinking can break down all manner of other dualisms: organic/inorganic, inside/outside, architectural/environmental, biological/artificial. What results from this melding together of social, natural and technical environments? Bruno Latour has likened what results to a 'parliament of things'. In so doing, he aims to provide the practical impetus for a new social ideal, where emerging sciences and technologies can be subject to public scrutiny, and as a consequence become more transparent.

In different ways, the proliferation of big ideas has helped more-than-human geographers to grapple with very tricky questions concerning the extent to which it is ever possible to claim to fully know **animality**, and how to write about the agency of non-human entities. To different degrees, this work still struggles with concerns raised about anthropomorphism, or what is called 'x-morphism' in the case of other objects (Laurier and Philo, 1999). Anxieties also remain about what ultimately is bound to remain unknowable, since for all the inter-species affinities that are detected, there is still a 'beastliness of being animal' that must also be respected. In the following section, I want to turn the focus of attention to an alternative kind of experiment told as a true animal story that will provide some imaginative resources to work through these ethical and moral conundrums.

'One pig': the animal–art–agriculture–advocacy assemblage

Having decided to buy himself a pig, Matthew Herbert took a trip to market. Well, sort of. Truth be told, this most traditional kind of transaction took place by more modern means. Herbert's own record label 'Accidental' (a micro-operation run out of the second floor of Unit 11, Block A, Greenwich Quarry, London) agreed to pay £100 to take legal ownership of one pig, selected from a litter of eleven piglets born to a sow on a family-run piggery in Kent, England. You can see the invoice (No. 1421) (see Figure 3.4), processed in March 2009. Over the next 20 months, Herbert amassed an archive of field and farm recordings that track the lifecycle of the pig, all the way from birth to slaughter to plate. As his purchase steadily put on the pork, Herbert kept a blog, documenting its development and describing its health and general welfare. In the process, one pig morphed into 'One Pig': an unlikely centrefold star in an experiment combining recorded music and live performance, art and appetite, animal-rights activism and animal husbandry, travelling all the way from farm to fork, and then to places beyond. Clearly, the life and the

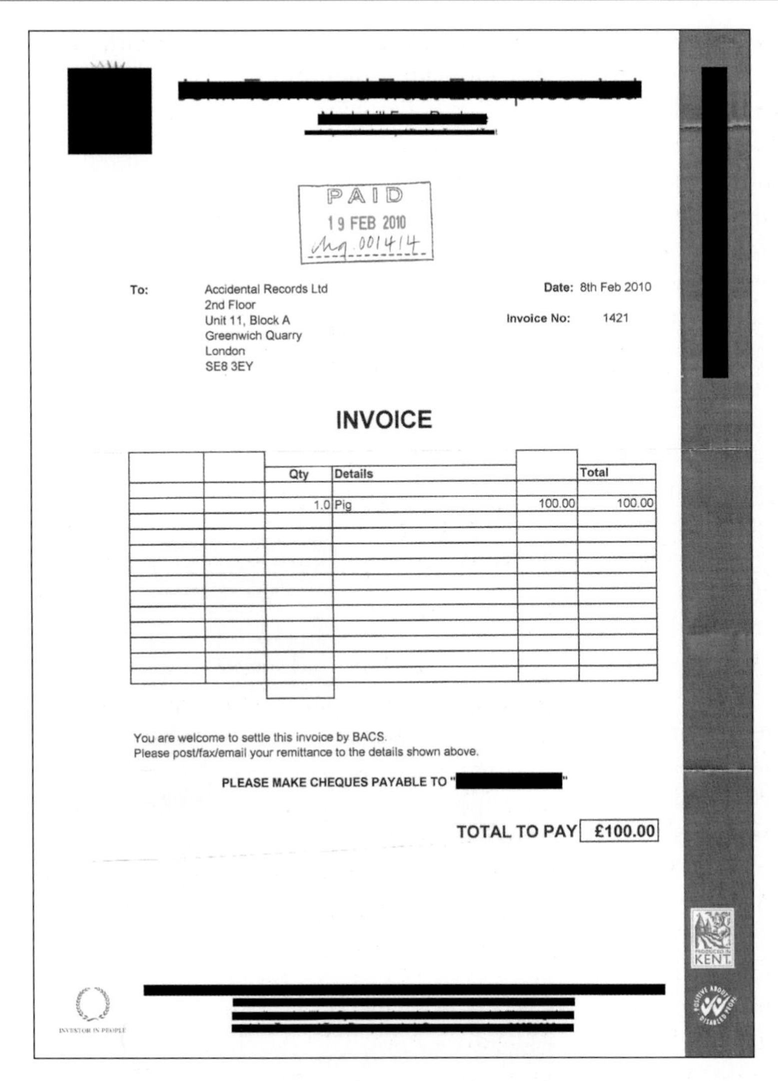

PAID
19 FEB 2010
Mg 001414

To: Accidental Records Ltd
2nd Floor
Unit 11, Block A
Greenwich Quarry
London
SE8 3EY

Date: 8th Feb 2010

Invoice No: 1421

INVOICE

Qty	Details		Total
1.0	Pig	100.00	100.00

You are welcome to settle this invoice by BACS.
Please post/fax/email your remittance to the details shown above.

PLEASE MAKE CHEQUES PAYABLE TO " "

TOTAL TO PAY £100.00

Figure 3.4 Just as there are many ways 'to skin a cat', there are lots of methods for purchasing a pig. Here is documentary evidence of Matthew Herbert becoming the proud owner of 'One Pig'. Credit: Matthew Herbert/Solar Management Ltd

death of this pig is going to require a little more explaining . . .

The first thing to clear up about Matthew Herbert is that he is no pig farmer. Nor is he a geographer. He is a musician – critically acclaimed among the cognoscenti of a genre known as 'electronica' (see Figures 3.5 and 3.6) – and he's a meat-eater too. Herbert has a history of audio experimentation, especially when it comes to sourcing the sounds that are used in his recordings. Some years ago, he wrote an influential artists' manifesto entitled 'Personal Contract for the Composition of Music' (PCCOM), which appeals for recording artists to avoid using drum machines and pre-existing samples in their work. Herbert suggests that sounds should come live-from-life. Sticking to his principles, he uses everyday life in the twenty-first century as a sound palette: 'With the invention of the sampler, I can now explicitly root my work in the literal, critical present. I can describe the real in the frame of the imaginary' (www.matthewherbert.com/biography).

This creative vision for truth-telling has shaped his recording practice ever since. On *The Mechanics of Destruction*, Herbert samples McDonald's products and Gap merchandise as a protest against corporate globalism. *Plat Du Jour*, another activist album, contains one track of compressed sounds that retell 'The Truncated Life of a Modern Industrialised Chicken'. The album *One Pig* pushed the political–ethical project of human–non-human relations further still. It is a musical portrait of an animal bred, ultimately, for human consumption. In the process, Herbert attempts various things: to make more meaningful the lives of animals reared for meat; to force us to ask critical questions about the global meat industry; and to make more transparent the direct consequences of our appetites and actions. So is it a recording to relish? You can decide for yourself by listening at: www.matthewherbert.com (or search for 'One Pig' on YouTube). I should be honest. Some tracks don't necessarily make for the easiest of listens. It ranks as the kind of thing music critics tend to describe as 'challenging'. Spliced through the crunchy beats and melodic blips are classic farmyard sounds (hay shuffling under foot and trotter), darker echoing squeals, distorted grunts, a chorus of competing snorts and oinks, mechanical clunks and clangs, a medley of manipulated scraping, sucking, sipping and slurping sounds, a knife being sharpened (then cleaned, perhaps?), a hacksaw cutting (for legal reasons, Herbert was not able to record actual sounds from the pig's slaughter), human voices, kitchen clatter, chomping and chewing noises and the appreciative kind of exhalation ('aaaahhhhhhh') made by a happy eater whose taste buds are being given an extra special treat. The butchering of One Pig was announced plainly on Herbert's blog: 'Wednesday, February 10, 2010 at 11:18PM. The pig is now dead.' Track by track, the affect of listening can be comfortably appealing, sometimes funny, but also disorienting and disturbing.

The afterlife of this individual pig keeps on happening, taking diverse forms of cultural expression and occupying unexpected spaces. The album artwork (see Figure 3.7) displays a range of lavish dishes cooked with the pig's meat ('ballotine of pork shoulder with tomato jelly', 'five-spice braised pig's head with borage and organic summer vegetables') and a selection of by-products derived from body parts (pig trotter candelabra, pig fat candles, a pigskin drum). The drum was one of the percussion

Figure 3.5/Figure 3.6 Matthew Herbert keeping company with pigs during field recordings at a piggery. Such intimacy in relations between humans and livestock is not new; in the pre-modern world, a swineherd was a person who looked after pigs. Credit: Matthew Herbert/Solar Management Ltd

instruments used for a string of live dates during recent European and UK tours. In these performances, Herbert's quintet appeared on stage dressed in traditional butcher's outfits (shirt-and-tie, knee-length white coat). He described the shows as an alternative kind of remembrance service, the music built from memories of the ghost-pig, and backed by slide projections from its former life (a bit like a farm-family album). Digital and material technologies were crucial to the spectacle of pig re-presentation. Performances centred on the 'StyHarp', a custom-built instrument formed of glowing red wires that mark the perimeter walls of an otherwise invisible pigpen. It is played by plucking and pulling, actions that activate a series of sound modules. Part way through the set, a chef joined the musicians on stage. The sizzling sound of pork frying was amplified. The smell of meat cooking filled the venue.

Figure 3.7 Some of the culinary products of One Pig's life and Herbert's artistic labours. Credit: Matthew Herbert/Solar Management Ltd

Audience participation was encouraged. As a finale, taste samples were dished out to the most curious on the dance floor, an act rendering One Pig as an assemblage of animal-art-agriculture-advocacy.

SUMMARY

- A range of contrasting theories and concepts currently inform geographical thought about how it is possible to reconfigure relations, persons and entities as 'post-human'.
- When it comes to a world of hybrids, capitalism, compassion and creativity are all motors of invention, but the formation of new entities can occur as an accidental event too.
- It is possible to identify possibilities and problems in the post-human condition, though it seems highly unlikely that a universal moral–ethical judgement can be cast. Instead there will be local geographies of reaction, ranging from opposition to enchantment.

Conclusion: Between creaturely wonder and animal welfare

It is very difficult to say with any degree of moral certainty if One Pig was luckier than the anonymous millions that are reared annually in agro-industry and then processed through the global meat industry. Quite possibly it was, in its short life. But for all the creaturely wonder engendered, ultimately this single animal befell the same fate. Arguably, the creation of ambiguous feeling is precisely the point of Herbert's more-than-human project.

It seems only proper that this chapter finds a way to end by returning to where it began, with gut feelings about non-human animals . . . as foodstuffs (Highmore, 2011). At present, commercially reared pigs really are big business on the world stage, in spite of recent food scares (Mizelle, 2011; Law and Mol, 2008). Pork is a key foodstuff catering to growing appetites and shifting dietary patterns among the expanding populations of Asia. Recently, Tulip (the UK's largest maker of Danepak Bacon and Spam) signed a £50 million pork export deal with China, the world's biggest pork market. The *Guardian* newspaper reported that:

> Much of the exported pork will be offal, tripe, trotters, ears and other parts of the so-called 'fifth quarter' – the parts even meat-eating Brits tend to turn their nose up at, but the Chinese savour.
>
> [The *Guardian*, 2012]

To me, this seems like capitalism in its purest form. As a globalized rationale for slaughterhouse efficiency the visceral concept of the 'fifth quarter' recalls the earliest days of the pork industry when American meat packers boasted about how they had found a use for 'everything but the squeal'. As well as pork, UK exports of live breeding pigs to China are being stepped up. Up to 900 at a time travel east by jumbo jet, on a non-stop twelve-hour flight, then 'once breeding herds have been established, farmers send bottles of semen to keep the production line going, supplemented by a new batch of sows every year' (Kollewe, 2012). What we have here are sites, flows and things (in the shape of farming practices, health protocols, live animals, trade emissaries, rendered meat, food safety officers and bodily fluids) *all* on the move around the globe. For Emma Roe, this is a situation raising critical issues around standards of animal welfare and about the material realities of agri-industrial production. Her unflinching account of a pig 'slaughter event' contains genuinely discomforting details:

> The pig carcass is put in hot water at 60 degrees Celsius to loosen its hairs. The pig is wet and slippery when it comes out of the hot bath. Some smaller pig carcasses slip onto the floor and are dragged back again onto the table. Then the carcass is put in a big tumbler to dry and 'rub off' as many hairs as possible. A pig has edible skin, so the hairs are meticulously removed, and as little water as possible is used to clean the meat (should any faecal matter slip out of the rectum as the whole of the digestive system is removed).
>
> (Roe, 2010: 271)

The graphic journey that Roe takes her readers on, along the brimming gutters of the food factory, was one once undertaken by Upton Sinclair (1906) in his novel *The Jungle*, and is today rehearsed by celebrity chefs and television film crews.

We might feel like we know that script. But what stuff really matters here? It seems that even at the same time as the categories of human and non-human blur (or all but dissolve), a sense for what is humane and inhumane must be protected. As sentient human-animals we feel the suffering of others. That offers us very good grounds for establishing a new relational ethics that can encompass more of life that is more-than-human. We must also acknowledge that these relational ethics will shift in shape and expression, according to the multiple beings enrolled into their constitution and the locally global spaces in which they keep on taking place.

DISCUSSION POINTS

1. What parts of the post-human condition are you comfortable about, and what bits make you most concerned? Ask yourself why.
2. What happens to your normal daily round of work and leisure activities if you try to rethink them as assemblages of hybrid entities, enrolled together by relations and connections?
3. Discuss whether the concept of more-than-human geographies might have significant implications for *physical geography*.
4. Discuss how the experience of listening to Matthew Herbert's 'One Pig' recordings made you feel.
5. Read through a daily newspaper (online or hard copy) and try to identify an article or feature that is concerned with something that poses a challenge to the idea of sealed-in human and non-human entities.

FURTHER READING

Davies, G. (2003) 'A geography of monsters?' In: *Geoforum* 34(4) 409–12.

A short commentary piece that throws open the possibility of thinking about the world as populated by monsters, old and new.

Lorimer, H. (2006) 'Herding memories of humans and animals'. In: *Environment and Planning D: Society and Space* 24(4) 497–518.

A paper that explores how it is possible to entwine the life stories of the humans and animals that make up a herd and the place of indigenous knowledge systems in this social arrangement.

Roe, E. (2010) Ethics and the non-human: the matterings of animal sentience in the meat industry. In: B. Anderson and P. Harrison (eds.) *Taking-place: Non-Representational Theories and Geography*. London: Ashgate, 261–83.

A book chapter offering a detailed and insightful consideration of the ethics of the meat industry. It draws on fieldwork findings from inside the slaughterhouse and leaves very little to the imagination.

Whatmore, S. (2002) *Hybrid Geographies: Natures, Culture, Spaces*. London: Sage.

An excellent, radical and influential book that explains the theoretical conditions for thinking of geographies as hybrid.

WEBSITE

www.matthewherbert.com

As well as the full album version, One Pig's life recordings have been compressed into a three-minute-long montage track, available on digital format. Alternatively, courtesy of Micachu, you can listen to an EP of dancefloor-friendly remixes of original One Pig tracks.

CHAPTER 4

MODERN–POSTMODERN

Mark Goodwin

Introduction

The term **modern** has been used for many centuries to distinguish a new social order from previous ones, and ideas of the modern are most commonly defined through their opposition to the old and the traditional. This 'oppositional definition' has taken many forms. In post-Roman Europe the term *modernus* was used to distinguish a Christian present from a pagan past (Johnston *et al.,* 2000), while in the late seventeenth century the quarrel between the 'Ancients' and the 'Moderns' spilled out from a debate over literature to embrace ideas of religion and social issues, causing the term 'modern' to enter widespread public usage for the first time. Towards the end of the eighteenth century, the term modern acquired another meaning, this time denoting a qualitative and not just a chronological difference from pervious eras. To live in a modern age denoted not just newness but also progress and betterment. Linked to the **Enlightenment** search for rational scientific thought, the idea began to emerge that humans could change history for the better, and that progress could be controlled and ordered – rather than history being done to people in a manner that was preordained

Throughout the nineteenth and twentieth centuries this notion of **modernity** as progress held sway (see also Chapter 32). Rapid changes in economy, technology, culture and society meant that, in Europe at least, each generation could claim to be qualitatively different from previous ones. Stephen Kern, for instance, summarizes the changes that were taking place at the end of the nineteenth century:

> From around 1880 to the outbreak of World War I a series of sweeping changes in technology and culture created distinctive new modes of thinking about and experiencing time and space. Technological innovations including the telephone, wireless, telegraph, x-ray, cinema, bicycle, automobile and airplane established the material foundation for this reorientation; independent cultural developments such as the stream of consciousness novel, psychoanalysis, Cubism, and the theory of relativity shaped consciousness directly. The result was a transformation of the dimensions of life and thought.
>
> (Kern, 1983: 1–2)

Our experiences of these 'transformations in life and thought' became labelled as modernity (Berman, 1982). Their artistic, cultural and aesthetic expression was called **modernism**. So, while ideas of being modern can be traced back several centuries, notions of modernity and modernism coalesced around a very particular twentieth-century experience – one to be especially found in the emerging cosmopolitan urban centres of Berlin, Paris and New York. The impacts of modernity and modernism spread out from these cultural heartlands to influence us all. Even at the beginning of the twenty-first century, the built environment that most of us inhabit has largely been shaped by modernism. The houses we live in, the offices and factories we work in, the chairs we sit on and the tables we sit at, and the graphic design we see around us – on shop fronts and in newspapers and magazines – have all been created by the aesthetics and ideology of modernist design.

Towards the end of the twentieth century, however, modernism was challenged by a new movement which significantly did not label itself as another stage in modernity. Instead it self-consciously proclaimed itself to be **postmodern** – to be different from, and moving beyond modernity. In the arts and literature, in philosophy and in the social sciences, **postmodernism** and **postmodernity** began to flourish. As the geographer Michael Dear put it in 1994 'Postmodernity is everywhere, from literature, design and philosophy, to MTV, ice cream and underwear' (1994: 3).

What I want to do in the rest of this chapter is to trace the continuities and discontinuities between the modern and the postmodern and to sketch how geography and geographers have been influenced by, and in turn influenced, both movements.

CASE STUDY

Marhall Berman's description of Modernity

There is a mode of vital experience – experience of space and time, of the self and others, of life's possibilities and perils – that is shared by men and women all over the world today. I will call this body of experience 'modernity'. To be modern is to find ourselves in an environment that promises us adventure, power, joy, growth, transformation of ourselves and the world – and, at the same time, that threatens to destroy everything we have, everything we know, everything we are. Modern environments and experiences cut across all boundaries of geography and ethnicity, of class and nationality, of religion and ideology: in this sense, modernity can be said to unite all mankind. But it is a paradoxical unity, a unity of disunity: it pours us all into a maelstrom of perpetual disintegration and renewal, of struggle and contradiction, of ambiguity and anguish. To be modern is to be part of a universe in which, as Marx said, 'all that is solid melts into air'.

(From Berman, 1982: 15)

Modernism and post-modernism: continuities and discontinuities

Dear (1994: 3-4) identifies three components of postmodernism and postmodernity – style, epoch and method. We will use this classification to trace and analyse the shift from the modern to the postmodern.

Style

While we can trace the shift from a modern to a postmodern style across art, literature and music, architecture has become paradigmatic for discussing such a shift. Indeed, it has often been used as the starting point for discussions of postmodernity more generally, perhaps because it provides a very visible and public presence of changes in style. It also provides an immediate link to the concerns of geographers interested in the changing form and function of the built environment. The modern movement certainly stamped its authority on the architecture of the age: at the core of the movement lay the idea that the world had to be completely rethought and that following the carnage of the First World War, a more rational and enlightened society could be built – both socially and architecturally. The result was a set of sweeping changes in urban design, both in

Figure 4.1 Villa La Roche, designed by Le Corbusier.

terms of planning whole neighbourhoods and designing individual buildings. The latter came first, with initial appearances of architectural modernism being confined to small-scale infill buildings. The famous Villa La Roche, for instance, designed in 1925 by Le Corbusier, perhaps the most famous of all modernist architects, for a Swiss banker and art collector, lies at the end of a cul-de-sac in the Parisian district of Auteuil, still surrounded by nineteenth-century housing (see Figure 4.1).

In the 1930s, inspired by the famous Bauhaus movement, this modern style of architecture began to be used to design and construct housing estates, office blocks and whole communities. It reached its zenith with the large-scale urban renewal schemes of the 1950s and 1960s which can be found in almost every major city in the western world. The watch words of this urban design were rationality, order and efficiency, and the result was a technocratic and industrialized ordering of public space. Figure 4.2 contrasts an early Le Corbusier vision for the complete reordering of Paris (never built of course!) with the layout of Stuyvesant Town in New York, a private housing community which was built immediately after the Second World War. The universalism of modern architecture means that the same forms can be found across the globe, the result of new construction techniques and the mass use of 'new' materials such as glass, steel and concrete. Somewhat ironically, by the end of the 1960s, modernist architecture came to be seen as drab, functional and commonplace and had lost its early rationale as a revolutionary opposition to the traditional forms of what the modern movement perceived as the reactionary nineteenth century.

Reaction to modernist architecture formed part of the anti-modernist movements which developed towards the end of the 1960s. These eventually crystallized around the emergence of a new postmodern style. In opposition to the austerity and formalism of modern architecture, postmodernism developed as a more playful alternative, emphasizing pastiche and collage. Rather than emphasizing the universalism of functional modernism, postmodern architecture was centred around vernacular and traditional styles, often rooted in regional traditions, with the result that diversity and pluralism replaced uniformity. An early example of such architecture was provided by the AT&T building in New York (now the Sony Building), designed by Philip Johnson

Figure 4.2 (a) Le Corbusier's dream for Paris in the 1920s; (b) the achieved design for Stuyvesant Town, New York. Credit: (a) © FLC/ADAGP, Paris and DACS, London 2013; (b) Alec Jordan (Creative Commons)

and completed in 1984 (see Figure 4.3). Here the playfulness of postmodernism is celebrated by a pediment at the top resembling Chippendale furniture and an arched entry some seven stories high. As construction techniques developed, architects began to create even more distinct building forms. Figure 4.4 shows the Disney Concert Hall in Los Angeles, designed by Frank Gehry; the contrast with the traditional modernist towers in the background is stark.

Figure 4.3 AT&T building in New York, an early example of postmodern architecture, with more austere and functional modernist blocks adjacent and behind. Credit: Getty Images

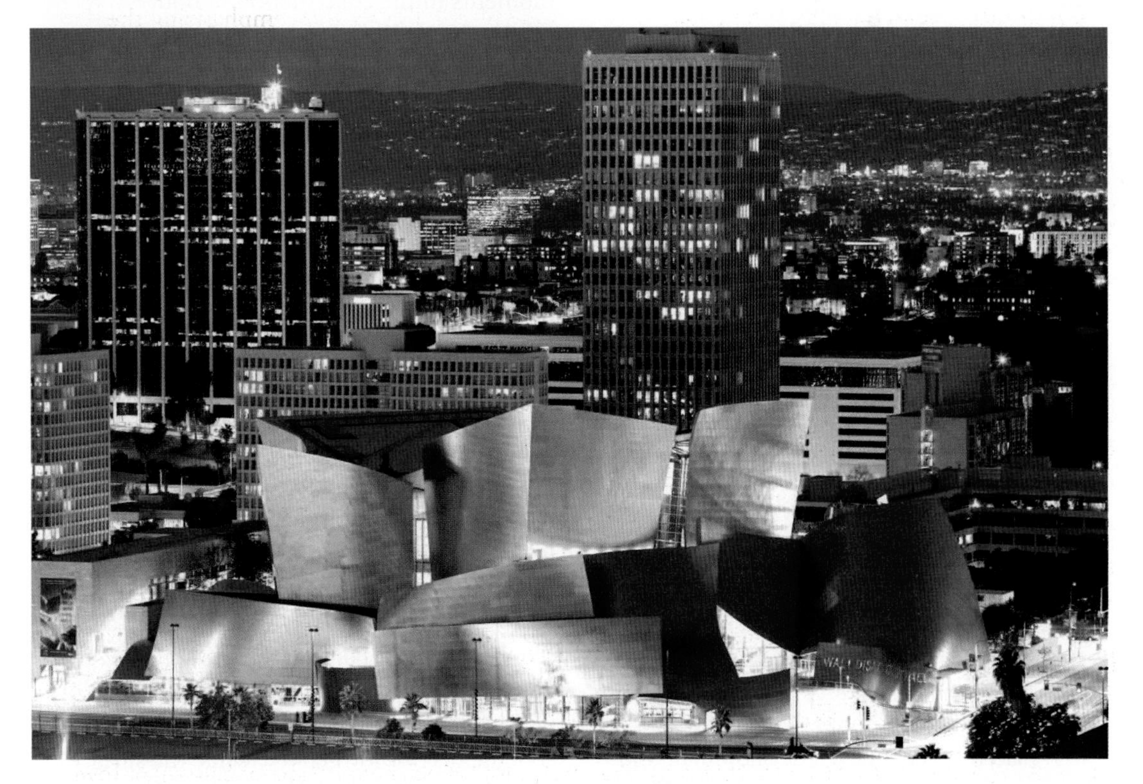

Figure 4.4 The Disney Concert Hall in Los Angeles in the foreground, with more traditional and functional modernist office towers in the background. Credit: Getty Images

SUMMARY

- Modernism developed in opposition to the perceived tradition and conservatism of nineteenth-century art, literature and architecture.
- Modernism developed to encompass a global style, emphasizing universality, functionalism and order.
- Postmodernism developed in opposition to the perceived austerity of modernism and emphasized diversity, playfulness and plurality.

Epoch

The surface differences between the architectural styles of modernism and postmodernism are clear to see. They are reinforced by other cultural differences in literary styles and in photography and design (Harvey, 1989; Huyssen, 1984). The question is whether these are enough to signal a radical break with past social and cultural trends and thereby form a distinct epoch following that of modernity. One key problem for those attempting to chart such a break is the difficulty of theorizing contemporary trends. As Dear puts it:

> any landscape is simultaneously composed of obsolete, current, and emergent artifacts: but how do we begin to codify and understand this variety? And at what point is the accumulated evidence sufficient to announce a radical break with the past? The idea that we are living in 'new times' is seductive, but there are no clear answers to these questions.
> (Dear, 1994: 3)

Geographers have played a key role in the search for such answers. In particular, two extremely influential books, both published in 1989, placed space, uneven development and urbanism at the centre of these debates. The first, David Harvey's *Condition of Postmodernity* became a bestseller across the humanities and social sciences. The second, Ed Soja's *Postmodern Geographies* consciously set out to reassert the role of space in social theory more generally. Both emphasized the role of geography and uneven development as active moments in the construction of both modernity and postmodernity, rather than as simple reflections of them. Harvey had the more critical take on ideas of postmodernity as a distinct new epoch. His understanding was traced through the four parts of the book. Part one examines the major themes in the transition from modernism to postmodernism across elements as diverse as urban planning, painting, literature and music. But Harvey differs from many academics that have identified similar changes by relating them to underlying shifts in the capitalist economy. The second part continues this theme, by setting out and analysing a political–economic transition away from **Fordism** and situating the rise of postmodern representations within this transition. Here again, the cultural practices of postmodernity are related to underlying economic shifts. For Harvey, the shift towards flexible accumulation helps to explain why postmodernity often appears as fleeting and ephemeral. But for Harvey, while the surface appearances of capitalism may have changed, its underlying logic, which leads to a constant set of crises, remains the same. In Part three of the book, Harvey examines different conceptions of space and time under modernity and postmodernity. Again, he tends to highlight the similarities between the two rather than the differences by emphasizing that both modern and postmodern periods are dominated by what Harvey calls '**time–space compression**'. This is not just about ever-quicker forms of global communication and travel, but also refers to the ever-accelerated rate at which capital has to turnover to make a profit. For Harvey this leads to an accentuation of volatility and ephemerality in fashion, consumer goods and production techniques as well as in ideas, ideologies and values. In this sense, Berman's idea that modernity is characterized by all that is solid meting into air (see Case Study on p. 52) is now transferred to postmodernity. In Part four Harvey merges political, economic and cultural analyses to produce an understanding which emphasizes the continuing internal contradictions of capitalism, rather than the categorical distinction between modernism and postmodernism (see Table 4.1).

Soja stresses the differences of postmodernism from modernism, but also refuses to see it as a distinct epoch, preferring instead to characterize it 'as another deep and broad restructuring of modernity, rather than a complete break' (1989: 5). However, Soja uses the move towards postmodernity to emphasize the role of space in social thought. This is not just about empirically investigating the new spaces and architectural forms of postmodernity, but of building a spatial awareness into the very foundations of social theory. According to Soja (1989: 31), the traditions of the two dominant social theories within modernism – Positivism and Marxism – caused the 'virtual annihilation of space by time

CASE STUDY

Ed Soja's description of postmodernity

With exquisite irony, contemporary Los Angeles has come to resemble more than ever before a gigantic agglomeration of theme parks, a lifespace comprised of Disneyworlds. It is a realm divided into showcases of global village cultures and mimetic American landscapes, all-embracing shopping malls and crafty Main Streets, corporation-sponsored magic kingdoms, high-technology-based experimental prototype communities of tomorrow, attractively packaged places for rest and recreation, all cleverly hiding the buzzing workstations and labour processes which hold it together. The experience of living here can be extremely diverting and exceptionally enjoyable, especially for those who can afford to remain inside long enough to establish their own modes of transit and places to rest. And of course, the enterprise has been enormously profitable over the years. After all, it was built on what began as relatively cheap land, has been sustained by a constantly replenishing army of even cheaper imported labour, is filled with the most modern technological gadgetry, enjoys extraordinary levels of protection and surveillance, and runs under the smooth aggression of the most efficient management systems, almost always capable of delivering what is promised just in time to be useful.

(Soja, 1989: 246)

Fordist modernity	Flexible postmodernity
economies of scale/master code/ hierarchy homogeneity/detail division of labour	economies of scope/idiolect/anarchy diversity/social division of labour
paranoia/alienation/symptom public housing/monopoly capital	schizophrenia/decentering/desire homelessness/entrepreneurialism
purpose/design/mastery/determinancy production capital/universalism	play/chance/exhaustion/indeterminancy fictitious capital/localism
state power/trade unions state welfarism/metropolis	financial power/individualism neo-conservatism/counterurbanization
ethics/money commodity God the Father/materiality	aesthetics/moneys of account The Holy Ghost/immateriality
production/originality/authority blue collar/avant-gardism interest group politics/semantics	reproduction/pastiche/eclecticism white collar/commercialism charismatic politics/rhetoric
centralization/totalization synthesis/collective bargaining	decentralization/deconstruction antithesis/local contracts
operational management/master code phallic/single task/origin	strategic management/idiolect androgynous/multiple tasks/trace
metatheory/narrative/depth mass production/class politics technical-scientific rationality	language games/image/surface small-batch production/social movements/pluralistic otherness
utopia/redemptive art/concentration specialized work/collective consumption	heterotopias/spectacle/dispersal flexible worker/symbolic capital
function/representation/signified industry/protestant work ethic mechanical reproduction	fiction/self-reference/signifier services/temporary contract electronic reproduction
becoming/epistemology/regulation urban renewal/relative space	being/ontology/deregulation urban revitalization/place
state interventionism/industrialization internationalism/permanence/time	laissez-faire/deindustrialization geopolitics/ephemerality/space

Table 4.1 David Harvey's characterization of modernity and postmodernity – using both political–economic and cultural–ideological relations. (Source: D. Harvey, *The Condition of Postmodernity*, 1989: 340)

in critical social thought', squeezing geography out of the picture. But if modernist social theory emphasized time, progress and historical development, for Soja, postmodernity entails an analysis of space and uneven development. Like Harvey, Soja makes this manoeuvre by emphasizing the uneven urban and regional development entailed in new forms of economic production. He illustrates this in the book by analysing what he calls 'the dynamics of capitalist spatialisation' (1989: 191) and uses Los Angeles as an exemplary case study of 'postmodern urbanism'. But he does so by building a new type of urban analysis, after noting how language and description tends to be linear and sequential, making it difficult to envisage the simultaneity of past and present that is inherent in all landscapes. In his final chapter on Los Angeles he offers a view from above, taking the reader on an imaginary cruise around the circumference of the sixty-mile circle which encompasses the built-up area of LA. The Case Study on p. 57 provides an example of how Soja renders the experience of postmodernity especially visible.

SUMMARY

- Geographers such as Harvey and Soja have played a key role in seeking to understand the transition from the modern to the postmodern.
- Both authors conclude that postmodernity is an extension rather than a replacement of modernism.
- Both insist on joining political and economic analysis to an understanding of cultural and ideological shifts.
- Both use the debate about whether postmodernity is a distinct epoch to reassert the importance of space in social theory.

Method

The third component of modernity and postmodernity identified by Dear (1994: 3-4) is that of method. Here, Dear is identifying different ways of viewing and understanding the world. As we noted earlier, modernism grew out of the enlightenment search for order, rationality and science. Within this search was a concern for uncovering the universal laws which underpinned both the physical and the social worlds. In the physical world, such universality is more straightforward, and grand narratives built around a universal understanding of gravity or evolution or relativity or nuclear fusion, are commonplace – if occasionally contested. In matters of the social world, things are not so straightforward. The grand social theories of modernism, however, were conducted as if they were. Positivism and Marxism took on the character of 'metanarratives', used by their adherents to explain all kinds of social and economic behaviour. For the positivists, rational man, acting to optimize his own individual interests, provided the foundation for economics and economic development. For Marxists, class struggle provided the motor for history and historical development. The developing discipline of geography was also enrolled into this search for universal laws. For some, like Soja, it responded too enthusiastically, resulting in what he termed 'Modern Geography's fixation on empirical appearances and involuted description' (1989: 51).

Indeed, by the 1960s Human Geography had embraced a so-called 'scientific' approach, which had as its rationale the search for universally applicable laws of human behaviour. People were reduced to little more than dots on a map or integers in an equation and were all assumed to operate according to the same general laws – indeed, it was the very search for these controlling laws that drove this entire approach. This kind of reasoning dominated Human Geography in the 1960s and most of the 1970s and generated the search for law-like statements of order and regularity that could be applied to spatial patterns and processes. Hence the succession of models that appeared in geography over this period, for instance, Christaller's model of settlement hierarchy, Alonso's land use model, Zipf's rank size rule of urban populations and Weber's model of industrial location. All were an attempt to use law-like statements in order to explain and predict the spatial outcomes of human activity.

One such model that Human Geographers used to explain patterns of flow between two or more centres was the so-called **gravity model**. This proposed that we can estimate the spatial interaction between two regions by multiplying together the mass of the two (equated conveniently with population size) and dividing it by some function of the distance separating them. The model was used to 'explain' all kinds of flows, from those of migration to passenger traffic, telephone conversations and commodity flows. Noticeable by their absence are any references to the actual motivations for the behaviour of the individuals who are migrating or commuting or speaking to each other on the phone or purchasing the commodities. The freedom to choose one's behaviour is given no space whatsoever and people's actions are assumed to conform to a general pattern, which is itself based on a model derived from a crude analogy with Newton's law of universal gravitation developed in 1687. Thus what was originally conceived as a way of accounting for the behaviour of distant bodies in the universe was being used to explain a whole host of social, economic and cultural activities by reference to the two variables of population and distance. These, and the relation between them, were felt to govern, or control, the rate and nature of population movement.

Postmodernity consciously rejected the search for universal truths and instead emphasized that all knowledge is socially produced by those with particular positions and particular interests. This led to a strategy of deconstruction, 'a mode of critical interpretation that seeks to demonstrate how the (multiple) positioning of an author (or reader) in terms of class, culture, race, gender, etc., has influenced the writing (and reading) of a text' (Johnston *et al.,* 2000: 621). The outcome was a destabilization of meaning, which in turn cast doubt on the authority of those who claimed to be privileged interpreters. Local knowledges were prioritized alongside scientific ones, and postmodernism sought to undermine the modernist belief that theory can mirror and explain reality. As Dear puts it, 'more than most, therefore, postmodernists, learn to contextualise, to tolerate relativism and to be conscious always of difference' (1994: 4). However, critics of postmodernism seized on such relativism to argue that this amounts to a kind of 'anything goes' academia, where every single viewpoint is equally valid. Geographers were again at the forefront of these debates as they sought to understand how the meanings and interpretations of all kinds of texts – from books, to maps, to landscapes – were socially derived and mediated. In the end though it is worth remembering a warning from Dear, that 'in our shifting world, postmodern thought has not removed the necessity for political and moral judgements: what it has done is to question the basis for such judgements' (1994: 4).

SUMMARY

- Modernism and postmodernism contain their own ways of viewing and understanding the world.
- Modernism has tended to search for universal laws, emphasizing rationality and order.
- Postmodernism has tended to emphasize the relative and socially situated basis of all knowledge claims.

Conclusions

Geography and geographers have inevitably been heavily influenced by the social and cultural movements we know as modernism and postmodernism, both in terms of what they have studied and how they have studied it. Geographers have also played leading roles in the interpretation of modernity and postmodernity and especially in analysing whether we have passed from one to the other. Putting these two elements together, we can now see the way in which an explicit concern with modernity and postmodernity revolutionized geography in the 1980s and 1990s. It caused geographers to ask all kinds of questions: about the relationship between the past and the present; about the relationship between society and space (see Chapter 2); about the role of space in social theory and social change; about diversity and difference, and how they should be rendered visible (see Chapters 5 and 6); and about how we represent and understand different types of meanings and interpretations (see Chapter 9). In many ways then, debates around notions of the modern and the postmodern presaged a flowering of geographical enquiry and a serious reintegration of Human Geography into broader debates within social science and philosophy. The encounter has undoubtedly changed Human Geography, but Human Geographers have also changed our understandings of what it means to be modern or postmodern.

DISCUSSION POINTS

1. Choose a geography that is well known to you – for instance, your journey to university, the place where you live, places you have worked – and see if you can find elements of modern or postmodern architecture in the built environment.

2. Examine a contentious social issue and see how many different interpretations of it you can establish. Think about why these different voices are there and analyse which ones are most powerful.
3. Using particular examples, explore the ways in which cultural and economic change are linked.
4. Do you think modernity has come to an end?

FURTHER READING

Berman, M. (2010) *All That is Solid Melts into Air*. London: Verso.

A new edition of a wonderful book which examines the experience of modernity by charting the impact of modernism on art, literature and architecture.

Harvey, D. (1989) *The Condition of Postmodernity*. London: Blackwell.

Soja, E. (1989) *Postmodern Geographies*. London: Verso.

The two books which opened up geography's engagement with debates around modernity and postmodernity – and which brought a geographical sensitivity to subsequent debates on these themes.

Le Corbusier (1927) *Towards a New Architecture*. London: The Architectural Press.

Try to find a copy in a library of this English translation of Le Corbusier's 1923 classic French text *Vers une Architecture*. There is also a new 2008 edition (published by Frances Lincoln Ltd, London) with an introduction which sets the book in its original context. Read this to understand why modernism gained such a hold for those seeking to build a new world after the First World War by sweeping away the disorder and chaos of the old one.

Jencks, C. (2007) *Critical Modernism – Where is Post Modernism Going?* London: Wiley Academy.

A book by one of the doyens of the postmodern movement which provides an overview of both postmodernism and its relationship with modernism.

CHAPTER 5
SELF–OTHER

Paul Cloke

Introduction: self-centred geographies?

Some people say that you should not judge a book by its cover. However, it is often interesting to pause and reflect on why books, organizations or in this case subjects such as geography are represented by particular 'cover' images. Figure 5.1 shows the cover of the 1994 Annual Report of the Royal Geographical Society, which is the organization representing academic and non-academic geographers in Britain. The image was designed to show geography in a positive light, as a subject that causes adventurous individuals to embark on exciting expeditions of learning in which they can discover the secrets of far-flung places and understand the lives of exotically different people. It is the 'us here' subjecting the 'them there' to serious geographical scrutiny.

This image, however, unintentionally poses other questions about 'us' and 'them'. The 'us' might suggest that Human Geographers can somehow be categorized as a homogeneous group of people, studying our geography in a somewhat standardized way – a bizarre supposition on a number of counts, not least the 'maleness' of the encounter that is represented. The 'them' seems to have been selected on the grounds of their exotic difference to us. They, too, are in danger of being stereotyped. The strangeness of the place along with differences in skin colour, language, dress and 'culture' seem to be sufficient to mark out an appropriately 'other' subject of study. 'Us' encountering 'them' is on our terms. Exotic difference is defined by our mapping out of people and places in the world, and our assumptions about what is, and what is not, a normal view of life.

Perhaps these questions read too much from one particular image, especially since the RGS/IBG has subsequently sought to rectify in its output any previous perceptions of social or cultural insensitivity. However, these questions do reflect some of the most important themes to have arisen in Human Geography over recent years. The first is a highlighting and questioning of the geographical self. Not so many years ago, Human Geographers were taught to be objective in their studies, so that anyone else tackling the same subject would come up with the same results. They were, in effect, being positioned as some kind of scientific automaton whose background, identity, experience, personality and worldview needed to be subjugated to the need for objectivity. The 'I' was personal pronoun *non grata* when it came to doing geography. However, the *self* does matter, and does

Figure 5.1 Annual Report of the Royal Geographical Society, 1994. Credit: Royal Geographical Society (with IBG)

influence the geography we practise. We do have different place- and people-experiences, different political and spiritual worldviews, different aspects to our identity and nature, and all of these factors will influence how we see the world, why our geographical imaginations are fired up by particular issues and, ultimately, what and how we choose to study.

The danger of *not* acknowledging and reflecting on the self is not only that we can unknowingly buy into other people's orthodoxies, but also that we can assume that everyone sees the same world as we do. We can, thereby, impose our 'sameness' on to others. The second set of questions, then, concerns recognition of how we deal with 'others'. It is extraordinarily difficult sometimes to do anything but see things from our own perspective, however hard we try to escape from our self-centred geographies. Yet as soon as we move beyond the samenesses of self, we immediately begin to stylize and stereotype the differences of 'the other'. This has been the subject of much recent questioning across the range of human sciences, including Human Geography, under the rubric of debates on '**Otherness**' and 'Othering'. How do we think about people who are not like us without 'othering' them, without prioritizing the self and at best offering benign tolerance to others (Shurmer-Smith and Hannam, 1994)? Do we categorize others in order to control them – socially, culturally, politically, economically, spatially (Leeuw *et al.*, 2011)? Do we equate difference with abnormality or deviance?

Dealing with otherness and difference is therefore fraught with difficulty, as these are by no means neutral categories, and any critical assumptions about them being 'obvious' or even 'threatening' require very considerable reflection. For example, we need to challenge any assumptions that appropriate dealings with others are somehow automatically transacted through our citizenship – both in terms of our status as 'citizens' of Human Geography, which is somehow already sufficiently attuned to issues of otherness, and in terms of our state-citizenship through which it might be thought that welfare and aid functions already take care of the need to deal with others. The French anthropologist Marc Augé has suggested that we need to adopt a two-pronged approach to understanding otherness. First, we should seek *a sense for the other*. In the same way that we have a sense of direction, or family, or rhythm, he argues that we have a sense of otherness, and he sees this sense both disappearing and becoming more acute. It is being lost as our tolerance for others – for difference – disappears. Yet it is becoming more acute as that very intolerance itself creates and structures othernesses such as nationalism, regionalism and 'ethnic cleansing', which involve 'a kind of uncontrolled heating up of the processes that generate otherness' (Augé, 1998: xv). Second, we should seek *a sense of the other*, or a sense of what has meaning for others; that which they elaborate upon. This involves listening to other voices and looking through other windows on to the world so as to understand some of the social meanings that are instituted among and lived out by people within particular social or identity groups. This combination of an intellectual understanding of the other and an emotional, connected and committed sense of appreciation for the other can perhaps best be summarized in terms of attempting to achieve solidarity with the other by participation and involvement in their worlds. Rather than converting 'them' into 'our' world, such solidarity involves a conversion of ourselves for the other, hence as I wrote a decade ago:

> I believe that any re-radicalized geography will be measured to some extent by the degree to which radical and critical geographers achieve a going beyond the self

> in order to find a sense for the other in practices of conversion for the other.
> (Cloke, 2004: 101)

As the remainder of this chapter suggests, the attempt to inculcate the curricula and research of Human Geography with senses of the other, and with reflections on the self, has proved to be a complex and politicized process. Perhaps this reflects less the novelty of the ideas being worked with than the way they speak to and critique an absolutely central concern of Human Geography: developing knowledge of people and places beyond those one already knows. This chapter argues that this critique is worthwhile, and therefore discusses some of the delights, as well as difficulties, of bringing explicit reflections of self and other into our Human Geographies.

Self-reflections

In many ways, '**reflexivity**' has become one of the most significant passwords in Human Geography over recent years. To reflect on the self in relation to space and society has been seen as a key with which to open up new kinds of Human Geographies that relate to individuals more closely, and that individuals can relate to more closely. In particular, reflexivity has been used by **feminist** and **post-colonial** geographers in their respective political projects to persuade Human Geographers to reflect something other than male, white orthodoxies. A poem by Clare Madge (see Case Study box opposite) urges geography to connect 'in here' rather than 'out there' by becoming a subject 'on my terms and in my terms'.

Her frustration with the subject is echoed in the book *Feminist Geographies* (Women and Geography Study Group, 1997) where the writing (usually by men) in geography is critiqued, but the problems of proposing alternative forms of writing (usually by women) are starkly acknowledged:

> Much academic writing . . . is characterized by a dispassionate, distant, disembodied narrative voice, one which is devoid of emotion and dislocated from the personal. In contrast to this, writing which is personal, emotional, angry or explicitly embodied is implicitly (and often explicitly) portrayed as its antithesis: something which (maybe) has a place in the world of fiction and/or creative writing, but which, quite definitely, is out of place in the academic world . . . to be masculine often means not to be emotional or passionate, not to be explicit about your values, your background, your own felt experiences. Feminist academics wishing to challenge those exclusions from the written voice of Geography find themselves in a dilemma, however, for if academic masculinity is dispassionately rational and neutral, writing which is overly emotional or explicitly coming from a particular personalized position is often dismissed as irrational, as too emotional, as too personal – as too feminine, in other words. Thus feminists who want to assert the importance of the emotional in their work, or feminists who want to acknowledge the personal particularities of their analysis, run the risk of being read as incapable of rational writing, of merely being emotional women whose work cannot be universally relevant.
> Women and Geography Study Group, 1997: 23)

It has therefore been important for Human Geographers not only to *theorize* the self in new ways, but also to position the self appropriately in the *practising* of Human Geography, such that knowledge is situated in the conscious and subconscious subjectivities of both the author/researcher and the subjects of writing and research. In terms of new ways of

CASE STUDY

Clare Madge: An Ode to Geography

Geography,
What are you?
What makes you?
Whose knowledge do you represent?
Whose 'reality' do you reflect?
Geography,
You are not just space 'out there'
To be explored, mined, colonised.
You are also space 'in here'
The space within and between
That binds and defines and differentiates us as people.
Geography,
I want you to become a subject
On my terms and in my terms,
Delighting and exploring
The subtleties and inconsistencies
Of the world in which we live.
The world of pale moonlight and swaying trees in a bluebell wood.
The world of sand and bone and purple terror.
The world of bright lights flying past factory, iron and engine.
The world of jasmine scents and delicate breeze.
The world of subversion, ambiguity and resistance.
The world of head proud, shoulders defiant under the gaze of cold eyes laying bare the insecurity underlying prejudice.
The world of music, laughter and light,
Of torment and exploding violence
Of tar and steel strewn with hate
While the moon gently observes and heals.
Geography, could you be my world?
Will you ever have the words, concepts and theories
To encapsulate
The precarious, exhilarating, exquisite, unequal world in which we live?
I believe so.
By looking within and without, upside down and inside out,
Come alive geography, come alive!

(Women and Geography Study Group, 1997)

theorizing the self, Steve Pile and Nigel Thrift (1995) discuss four interconnected ideas that map out the territory of the human subject:

1. *The body*: orders our access to and mobility in spaces and places; interfaces with technology and machinery; encapsulates our experiences of the world around us; harbours unconscious desires, vulnerabilities, alienations and fragmented aspects of self, as well as expressions of sexuality and gender; and is a site of cultural consumption where choices of food and clothing and jewellery, for example, will inscribe meanings about the person.
2. *The self*: can be understood in a variety of ways, ranging from a personal identity formed by an ongoing series of experiences and relationships, but where there is no distinctive characteristic in these experiences and relationships to suppose an inner, fixed personality, to a personal identity in which self-awareness serves to characterize each experience as belonging to a distinct self.
3. *The person*: a description of the cultural framework of the self and allows for different selves in different frameworks. For example, if you were born and brought up in Rwanda, Albania or Cuba, your person would reflect the cultural frameworks of life in those places.
4. *Identity*: where the person is located within social structures with which they identify. Traditionally this would have been seen to involve rigid structures such as class and family, but more recently identities have tended to be constructed reflexively and therefore often flexibly leading to new identity issues – for example, focusing on alternative sexualities, ethnicities or resistance to local change.

The subject is therefore 'in some ways detachable, reversible and changeable', while in other ways it is 'fixed, solid and dependable'. It is certainly 'located in, with and by power, knowledge and social relationships' (1995: 12).

Some of these theoretical distinctions may at first be difficult to grasp, but they do serve to emphasize just how difficult it actually is to be reflexive about the self in our Human Geography. To what extent is it possible to know and to reflect on our selves, to appreciate fully how, precisely, the self is responsible for how we think, how our imaginations are prompted, how we interpret places, people and events, and so on? How much more difficult is it to understand the selves of others whom we might wish to study? These practices, which I have identified earlier as being important political and personal projects in Human Geography, are perhaps more difficult than they first appear. The multidimensionality of the body, the relationally dependent and often subconscious nature of the self, the culturally framed (and therefore flexible) person and the changeable and overlapping influences of identity render reflexivity a most complex, and some would say impossible, task. Indeed a whole new angle on Human Geography – **non-representational theory** (see Thrift, 2007; Anderson and Harrison, 2010a) has sprung up in which the focus has switched specifically to non-reflexive accounts of human being and becoming in which the instinctive, habitual and performative are emphasized.

Nevertheless, the breaking down of detached and personally irrelevant orthodoxies in Human Geography has remained a task that many continue to consider sufficiently worthwhile to warrant attempts to bring reflexivity into a prominent position in the practice of Human Geography. Three interconnected and often overlapping strategies are briefly outlined here. First, a strategy of **positionality** can be identified in which 'telling where you are coming from' can be employed tactically as a contextualization of the interpretations that are to follow (see Browne *et al.*, 2010). Sometimes this involves the identification of key political aspects of the self, for example, a feminist positioning, which will self-evidently influence what occurs subsequently and which provides us with new positions from which to speak. On other occasions, particular spatial or social experiences will be described that are used to claim expertise or insight into particular situations. Take, for example, George Carney's autobiographical preface to *Baseball, Barns and Bluegrass*, his book on the geography of American folklife. Here, he describes his childhood in the foothills of the Ozarks, and how the folk knowledge accumulated during that time has been translated into a scholarly pursuit of cultural tactics of American folklife more generally. Not only does his folk heritage equip him for this work, but it also punctuates what he writes and how he writes it.

Second, a more radical strategy of **autoethnography** can be pursued involving different kinds of self-narrative that use the perspective of self-involvement to produce wider understandings of different kinds of social contexts. Autoethnography represents one significant way of both presenting the 'self to self', and presenting the 'self to others', and takes a number of forms (see Butz and Besio, 2009), including:

- analysing our own biographies in order to interrogate and illuminate wider phenomena
- reflective narratives on empirical or ethnographic encounters
- responses from subaltern groups to their representation by others
- 'indigenous' ethnographies from members of subordinated groups who take on academic positions from which to speak
- other forms of 'insider' research, often involving participant observation, witnessing and testimony (see Kindon, 2010; Dewsbury, 2009).

Autoethnography opens up intriguing possibilities for studying, for example, our gender, race/ethnicity, sexuality, sense of place, and also our work, leisure, tourism and other activity geographies through the medium of our personal involvement. At the same time it is important to recognize the challenges inherent in these approaches, including: the difficulties of knowing the self well enough, especially given its dynamic and multi-faced nature; the risk of self-obsession and consequent failure to communicate with others; the dangers inherent in presenting what are perhaps sensitive self-narratives to others who may be in positions of power and/or judgement; and the problematic task of shaping our self-representations to different audiences. Butz and Besio (2009) argue that these challenges resonate with the need for what they call *autoethnographic sensibility*, involving efforts to:

- perceive ourselves as an inevitable part of what is being researched and signified
- understand research subjects as autoethnographers in their own right
- interconnect self-narratives with the contexts, narratives and voices of others in any particular network of social relations.

Even with autoethnography, then, there are strong arguments for including 'other' voices in our own stories.

A third strategy therefore is to acknowledge *intertextuality* in our practice of Human Geography, by finding ways of recognizing the significance of our selves as important influences that shape our geographies, while at the same time seeking to listen to other voices. The texts that result from such encounters are complex dialogues. The Human Geographer will shape the conversation, both by the

CASE STUDY

George Carney's autobiographical preface

The first eighteen years of my life were spent on a 320-acre farm in Deer Creek Township, Henry County, Missouri, some six miles south of Calhoun (population 350), ten miles northwest of Tightwad (population 50), and five miles west of Thrush (population 4). My parents, Josh and Aubertine, inherited the acreage and farmstead buildings from my grandpa and grandma Carney, who retired and moved to Calhoun. The eighty acres to the north of the farmstead consisted of hardwood timber (walnut, hickory, and oak), Minor Creek, which flowed in an easterly direction as a tributary to Tebo Creek, and some patches of grazing land. The remaining 240 acres, south of the farmstead, were relatively rich farmland where my Dad planted and harvested a variety of crops ranging from corn and soybeans to alfalfa and oats. Classified as a diversified farmer, he also raised beef and dairy cattle, hogs, sheep, and chickens. Thus, my roots lay in a rural, agrarian way of life in the foothills of the Ozarks.

My early years fit the description that is often used to define the *folk* – a rural people who live a simple way of life, largely unaffected by changes in society, and who retain traditional customs and beliefs developed within a strong family structure. I was experiencing the *folklife* of the Ozarks. Folklife includes objects that we can see and touch (tangible items), such as food (Mom's home-made yeast rolls) and buildings (Dad's smokehouse). It also consists of other traditions that we cannot see or touch (intangible elements), such as beliefs and customs (Grandpa Whitlow's chaw of tobacco poultice used to ease the pain of a honeybee sting). Both aspects of folklife, often referred to as material and nonmaterial culture, are learned orally as they are passed down from one generation to the next – such as Grandpa Carney teaching me to use a broad axe – or they may be learned from a friend or neighbor – for example, Everett Monday, a neighbor, instructing me on the techniques of playing a harmonica.

Through this oral process, I learned many of the traditional ways from the folk who surrounded my everyday life – parents, relatives, friends, neighbors, teachers, preachers, and merchants. The most vivid memories associated with my early life among the Ozark folk are the six folklife traits selected for this anthology – architecture, food and drink, religion, music, sports, and medicine.

Since leaving the Ozarks for the Oklahoma plains some thirty-five years ago, I have developed a greater awareness and deeper appreciation for American folklife and all its spatial manifestations. My teaching and research interests have been strongly influenced by those folk experiences of yesteryear. Students in my introductory culture geography classes are annually given a heavy dose of lectures and slides on the folklife traits covered in this reader. My research has increasingly focused on two of these traits – music and architecture. Clearly, my roots have made a lasting impression – one that I have converted into a scholarly pursuit.

(Carney, 1998: xv–xxii)

individuality of their own subject-experience and by the questions that are asked of the 'other'. In turn, other individuals will have different, changing and even competing experiences and will represent themselves differently to different people. The 'results' of the encounter will usually be 'interpreted' by the Human Geographer in the light of their self-positioning. This may involve a process of 'finding new places to speak from', and bringing them into the conversation, or it may involve a tactic of 'letting people speak for themselves' and seeking for a plurality of voices (a 'polyphony') to emerge. Interpretations are then usually written down, often using quoted extracts of other voices, but almost always with the author in control, exerting power over what is included and excluded, what is contextualized and how, and what storylines are used to shape the narrative of the 'findings'. In all these processes and practices, the need to recognize the interconnections between the powerful self and the 'subjected to' other is paramount.

The increasing use of ethnographic strategies and qualitative methods in Human Geography (see Cloke *et al.*, 2004; Crang and Cook, 2007) has certainly helped to provide research practices with which we can be more reflexive about our selves, and the relationships between our selves and others. In the end, however, we have to realize just how 'easy' it can become to think and write about ourselves, and how difficult it is to know enough about our selves to be reflexive in our geographies. Delvings into **psychoanalysis** (Sibley, 1995) have begun to help our understandings here but there still seems to be an inbuilt desire to empower the self over the other, however much a many-voiced, polyphonic geography is being aimed at. In the more general context of the problems in the world, such preoccupations with the self might be regarded as inappropriate, if not positively dangerous!

SUMMARY

- Reflexivity – reflecting on the self in relation to society and space – is an essential process in recognizing how our individualities contribute to all aspects of our practice of Human Geography. It also gives us grounds on which to challenge seemingly 'orthodox' geographies and to make our Human Geography more relevant to us and to others.
- The difficulties involved in understanding the self are often underestimated. The human subject is a complex mix of body, self, person and identity, and for some, spirit and soul will also be important considerations.
- There is an interconnected range of strategies by which the self can consciously be included in the practice of Human Geography.
- The dangers of exaggerating the self in our reading, thinking, researching and writing about Human Geography are very real and can divert us from important issues relating to others.

Sensing the other

John Paul Jones (2010, 43) argues that 'a central problem in social geography is how to sort out relations between identity, on the one hand, and space on the other, particularly in terms of how their interplay affects the well-being of people and the prospects of the places they inhabit and move through.' It is important, therefore, to take serious notice of different kinds of people who are situated in different kinds of spaces and places, and who experience, mould and negotiate these spaces and places in a different way to ourselves. This interest in the differences of the other has implications for the ways in which we conceptualize and practise our Human Geographies, and also for the ways in which these geographies are politicized. Dealing with the 'other' is of course linked to dealing with the 'self'. To reiterate, the arrogance of the self is often manifest in an assumption that others must see the world in the same way as we do. Alternatively, we will often place ourselves in the centre of some 'mainstream' identity that is defined not only around our self-characteristics but also in opposition to others who are not the same as us. Think, for example, about the way white people often assume that only 'non-white' people have an ethnicity and find their own whiteness unremarkable. As Chris Philo has suggested, then, we are often 'locked into the thought-prison of "the same"' (1997: 22), which makes it impossible for us to appreciate the workings of the other. Indeed we will often seek either to *incorporate* the other into our sameness, or to *exclude* the other from our sameness, in order to cope with the threat that difference seems to present to the perceived mainstream nature of our identity (see Sibley, 1995). Both incorporation and exclusion are highly political acts that trap the other in the logic of the same.

The interest in recognizing 'other' Human Geographies focuses attention not only on that which is remote to us, but also should make us rethink what is close to home. Two examples serve here to highlight some of the principal themes in the recognition of otherness in proximal and remote situations. The first relates to the neglect of 'other' geographies close to home and focuses on rural geographies (see Chapter 48), although the principles involved relate to a wide range of Human Geography contexts. Philo's (1992) review of 'other' rural geographies emphasized that most accounts of rural life have viewed the mainstream interconnections between culture and rurality through the lens of typically white, male, middle-class narratives:

> there remains a danger of portraying British rural people . . . as all being 'Mr Averages', as being men in employment, earning enough to live, white and probably English, straight and somehow without sexuality, able in body and sound in mind, and devoid of any other quirks of (say) religious belief or political affiliation.
>
> (Philo, 1992: 200)

Such a list is important in its highlighting of neglect for others, but also runs the risk of immediately producing a formulaic view of what is other. Thus, we can recognize that individuals and groups of people can be marginalized from a sense of belonging to, and in, the rural on the grounds of their gender, age, class, sexuality, disability, and so on. However, as David Bell and Gill Valentine (1995b) remind us, the mere listing of socio-cultural variables represents neither a commitment to deal seriously with the issues involved nor a complete sense of the *range* of other geographies. Indeed, our very recognition of *these others* serves to 'other' *different others* and exclude them from view.

A specific illustration within this rural context is offered in Figure 5.2, which presents a well-known self-portrait by the photographer Ingrid Pollard (see Kinsman, 1995). Her autobiographical notes suggest that the photograph is a self-aware comment on race, representation and the British landscape. She sets herself in the countryside, and through juxtaposing her identity as a 'black photographer' with the cultural construction of landscape and rurality as an idyll-ized space of white heartland, she graphically expresses a sense of her own unease, dread, non-belonging – of other. The black presence in 'our' green and pleasant land says much about whiteness = sameness in this content. However, as the Women and Geography Study Group (1997) point out, the otherness in this representation is by no means a unidimensional matter of race. They suggest that:

> Pollard is claiming a different position from which to look at and enjoy English landscapes (albeit an uneasy pleasure); a right to be there and a right to be represented and make representations. She challenges, disrupts and complicates the notion of a generalisable set of shared ideas about England and the implicitly white and masculinised position from which it is usually viewed.
>
> (1997: 185–6)

Ingrid Pollard the 'black *woman* photographer', then, exposes another critical edge of otherness in this content and clearly the multidimensional nature of identity is by no means exhausted by these labels. In our seemingly known worlds, therefore, we make assumptions about the nature of people and places; about who belongs where, and who doesn't fit into the sameness of our mainstream; about who, what, where and when is other.

The second illustration is even better known within Human Geography, having achieved

Figure 5.2 'Pastoral Interlude' (1988) '. . . it's as if the Black experience is only ever lived within the urban environment. I thought I liked the Lake District; where I wandered lonely as a Black face in a sea of white. A visit to the countryside is always accompanied by a feeling of unease; dread . . . feeling I don't belong. Walks through leafy glades with a baseball bat by my side . . .' Credit: Ingrid Pollard. Courtesy of the artist.

Figure 5.3 A guard with a zither player in an interior, by Ludwig Deutsch (1855–1935). The illustration was used on the cover of Edward Said's *Orientalism*, 1995. Credit: Christie's Images/The Bridgeman Art Library

almost cult status in attempts to formulate post-colonial approaches to the subject. Edward Said is Professor of English and Comparative Literature at Columbia University in the USA. He is a Palestinian, born in Jerusalem and educated in Egypt and America, who is most famous for his analysis of the way the West imagines the Orient or East (including the Arabic Middle East) as different to itself (for a review of these and other 'imaginative geographies' see Chapter 16). In his classic book *Orientalism* (1978; 1995) Said traces how the Arab world has come to be imagined, represented and constructed in terms of its otherness to Europe:

> the French and the British – less so the Germans, Russians, Spanish, Portuguese, Italians and Swiss – have had a long tradition of what I shall be calling Orientalism, a way of coming to terms with the Orient that is based on the Orient's special place in European Western experience. The Orient is not only adjacent to Europe; it is also the place of Europe's greatest and richest and oldest colonies, the source of its civilization and languages, its cultural contestant, and one of its deepest and most recurring images of the other.
>
> (1995: 1)

Representations of the romantic, mystical Orient, he argues, act as a container for western desires and fantasies that cannot be accommodated within the boundaries of what is normal in the West. Yet at the same time, representations of the cruel, detached and money-grabbing nature of the Oriental Arab serve to underline the assumed hegemony of the West over political–economic and socio-cultural norms:

> Arabs, for example, are thought of as camel-riding, terroristic, hook-nosed, venal lechers whose undeserved wealth is an affront to real civilisation. Always there lurks the assumption that although the Western consumer belongs to a numerical minority, he is entitled either to own or to expend (or both) the majority of the world's resources . . . a white middle-class westerner believes it his human prerogative not only to manage the non-white world but also to own it, just because by definition 'it' is not quite as human as 'we' are.
>
> (Said, 1995: 108)

Through the process of Orientalism, the societies and cultures concerned are marginalized, devalued and insulted, while the imperialism and moral superiority of the West are legitimized. Said's contestation of the othering of Orientalism points the way for wide-ranging inquiry by Human Geographers into how different people and places are similarly othered. It also shows us that at the heart of what we take to be familiar, natural, at home, actually lurk all kinds of relations and positionings to that which is unfamiliar, strange and uncanny (Bernstein, 1992).

From these illustrations it becomes clear that whether otherness is close to home or positioned in some far-off exotic space, it is often difficult to detach ourselves, both conceptually and empirically from a frame of study that validates the self, the same and the familiar as waymarkers for the understanding of others. Two sets of issues arise from this conclusion. First, there is a need to think through much more deeply about what constitutes otherness in Human Geographical study, otherwise our main contribution may only be to further emphasize the othernesses that are *reinforced by* such study. At one level, this requires a grasp of the multidimensional nature of identity. As Mike Crang (1998a) puts it:

> very few people are the 'same' as others – everyone is different in some respects. The most we could say is that certain groups share certain things in common, so who is counted as part of a group or excluded from it will depend on which things are chosen as being significant . . . Belonging in a group depends on which of all the possible characteristics are chosen as 'defining' membership. The characteristics that have been treated as definitive vary over space and time with significant political consequences attached to deciding what defines belonging.
> (1998a: 60)

We need to recognize, therefore, that 'same' and 'other' identities are:

- Contingent – in that differences which define them are a part of an open and ongoing series of social processes.
- Differentiated – in that individuals and groups of people will occupy positions along many separate lines of difference at the same time
- Relational – in that the social construction of difference is always in terms of the presence of some opposing movement.

(Jones and Moss, 1995)

Even with greater sensitivity for other identities, we are usually still trapped in a concern for what Marcus Doel (1994) calls 'the Other of the Same' – that is, we translate othernesses into our language, our conceptual frameworks, our categories of thought, and thereby effectively obscure the other with the familiarities of the samenesses of our self. The real difficulty, then, is to find ways of accessing 'the Other of the Other' – that is, the unfamiliar, unexpected, unexplainable other that defies our predictive, analytical and interpretative powers and our socio-cultural positionings.

The second set of issues relates to the methods we employ in order to encounter 'others'. As with our self-reflections, the increasing use of **ethnographic** and qualitative methods is important to this project. However, researching the other through ethnography takes a long time. Drawing lessons from anthropology, we would have to conclude that to carry out appropriate studies of unknown peoples and worlds can take several years. Consider, for example, the account of French anthropologist Pierre Clastres (1998), who spent two years with so-called 'savage' tribes of Indians in Paraguay in the 1960s. He acknowledges that even 'being there' with his research subjects did not break down the very considerable barriers of communication and cross-referenced understanding, until circumstances changed many months into his research. Even over this timescale it proved difficult to form a bridge between himself (and here we might wonder whether his concept of 'savages' got in the way of effective communication) and the mythologies, embodiments and social practices that lay at the heart of the very existence of the Guayaki Indians (see Case Study box).

CASE STUDY

Pierre Clastres: Chronicle of the Guayaki Indians

They really were savages, especially the *Iroiangi*. They had only been in contact with the white man's world for a few months, and that contact had for the most part been limited to dealings with one Paraguayan. What made them seem like savages? It was not the strangeness of their appearance – their nudity, the length of their hair, their necklaces of teeth – nor the chanting of the men at night, for I was charmed by all this; it was just what I had come for. What made them seem like savages was the difficulty I had in getting through to them: my timid and undoubtedly naive efforts to bridge the enormous gap I felt to exist between us were met by the Atchei with total, discouraging indifference, which made it seem impossible for us ever to understand one another. For example, I offered a machete to a man sitting under his shelter of palm leaves sharpening an arrow. He hardly raised his eyes; he took it calmly without showing the least surprise, examined the blade, felt the edge, which was rather dull since the tool was brand-new, and then laid it down beside him and went on with his work. There were other Indians nearby; no one said a word. Disappointed, almost irritated, I went away, and only then did I hear some brief murmuring: no doubt they were commenting on the present. It would certainly have been presumptuous of me to expect a bow in exchange, the recitation of a myth, or status as a relative! Several times I tried out the little Guayaki I knew on the Iroiangi. I had noticed that, although their language was the same as that of the Atchei Gatu, they spoke it differently: their delivery seemed much faster, and their consonants tended to disappear in the flow of the vowels, so that I could not recognize even the words I knew – I therefore did not understand much of what they said.

Figure 5.4 Jyvukugi, chief of the Atchei Gatu. Credit: Pierre Clastres

But it also seemed to me that they were intentionally disagreeable. For example, I asked a young man a question that I knew was not indiscreet, since the Atchei Gatu had already answered it freely: *'Ava ro nde apa?* Who is your father?' He looked at me. He could not have been amazed by the absurdity of the question, and he must have understood me (I had been careful to articulate clearly and slowly). He simply looked at me with a slightly bored expression and did not answer. I wanted to be sure I had pronounced everything correctly. I ran off to look for an Atchei Gatu and asked him to repeat the question; he formulated it exactly the way I had a few minutes earlier, and yet the *Iroiangi* answered him. What could I do? Then I remembered what Alfred Métraux had said to me not long before: 'For us to be able to study a primitive society, it must already be starting to disintegrate.'

I was faced with a society that was still green, so to speak, at least in the case of the *Iroiangi*, even though circumstances had obliged the tribe to live in a 'Western' area (but in some sense, wasn't their recent move to Arroyo Moroti more a result of a voluntary collective decision than a reaction to intolerable outside pressure?). Hardly touched, hardly contaminated by the breezes of our civilization – which were fatal for them – the Atchei could keep the freshness and tranquillity of their life in the forest intact: this freedom was temporary and doomed not to last much longer, but it was quite sufficient for the moment; it had not been damaged, and so the Atchei's culture would not insidiously and rapidly decompose. The society of the Atchei *Iroiangi* was so healthy that it could not enter into a dialogue with me, with another world. And for this reason the Atchei accepted gifts that they had not asked for and rejected my attempts at conversation because they were strong enough not to need it: we would begin to talk only when they became sick.

Old Paivagi died in June 1963; he certainly believed that he had no more reason to remain in the world of the living. In any case, he was the oldest of the Atchei Gatu, and because of his age (he must have been over seventy) I was often eager to ask him about the past. He was usually quite willing to engage in these conversations but only for short periods, after which he would grow tired and shut himself up in his thoughts again. One evening when he was getting ready to go to sleep beside his fire, I went and sat down next to him. Evidently he did not welcome my visit at all, because he murmured softly and unanswerably: 'Cho ro tuja praru. Nde ro mita kyri wyte. I am a weak old man. You are still a soft head, you are still a baby.' He had said enough; I left Paivagi to poke his fire and went back to my own, somewhat upset, as one always is when faced with the truth.

This was what made the Atchei savages: their savagery was formed of silence; it was a distressing sign of their last freedom, and I too wanted to deprive them of it. I had to bargain with death; with patience and cunning, using a little bribery (offers of presents and food, all sorts of friendly gestures, and gentle, even unctuous language), I had to break through the Strangers' passive resistance, interfere with their freedom, and make them talk. It took me about five months to do it, with the help of the Atchei Gatu.

We need to acknowledge just how difficult it is to form a bridge between ourselves and the complicated essential existences of others, whether far off or close to home. It can be argued that the pressure to publish in the contemporary academy has run the risk of too many 'quickie' ethnographies of othered subjects. As with the Guayaki, an appreciation of the other geographies and experiences of, say, homeless people in a city like Bristol require long-term commitment rather than brief encounters. Only by reconceptualizing otherness, and reviewing the quality of our encounters with it, are Human Geographers likely to become any more attuned to a sense for the other and a sense of the other as suggested by Augé at the beginning of this chapter.

SUMMARY

- Sensing the other is inextricably linked with understanding the self. By assuming that others are somehow the same as us, we can be locked into the 'thought prison' of the same, which makes it impossible to sense the other appropriately.
- Geographies of other people and places can be close to home or in far-off exotic worlds. In either case, Human Geographers should see themselves as observers who are situated *within* the objects and worlds of their observation.
- At the heart of what we take to be familiar, natural and belonging lurk all kinds of relations and positionings with that other that is unfamiliar, strange and uncanny.
- There is a need to think through much more deeply what constitutes otherness in Human Geography. It is usually very difficult to bridge over between self and other.
- There is also a need to avoid methodological shortcuts in encounters with others.

Conclusion

This discussion of the interconnections of self and other raises a number of important issues about our Human Geographies. First, there is the risk that in acknowledging our selves in our work, we become too self-centred and too little concerned with political and other priorities in the world around us. Second, there is the potential for losing our sense of otherness. Third, there is the conceptual and methodological complexity involved in encountering the other of the same, let alone the other of the other. Finally, there a concern over the way in which we can sometimes privilege certain kinds of otherness without giving due attention to the need for sustained, empathetic and contextualized research under appropriate ethical conditions. There can be a tendency to 'flit in and flit out' of intellectually groovy subjects, with the danger that research becomes mere tourism or voyeurism of the subjects concerned.

When we have negotiated these tricky questions, there is one further important issue of self–other interrelations to resolve. In the words of Derek Gregory, 'By what right and on whose authority does one claim to speak for those "others"? On whose terms is a space

Figure 5.5
The power to exclude when engaging in touristic or voyeuristic geographies. Credit: Mikkel Ostergaard/ Panos Pictures

created in which "they" are called upon to speak? How are they (and we) interpellated?' (1994: 205).

In seeking to encounter the stories of other people and worlds, is it inevitable that we become mere tourists, burdened by the authority of our selves and the power of our authorship? Or are there ways in which we can be sufficiently sensitive about the positionality and intertextuality of our authorship that we can legitimately seek to understand and write about the stories of others, without polluting them with our voyeuristic or touristic tendencies, the exclusionary power of which are so graphically illustrated in Figure 5.5? I believe that in this we can learn much from Gregory's emphatic and optimistic answer:

> Most of us have not been very good at listening to others and learning from them, but the present challenge is surely to find ways of comprehending those other worlds – including our relations with them and our responsibilities toward them – without being invasive, colonizing and violent . . . we need to learn how to reach beyond particularities, to speak of larger questions without diminishing the significance of the places and the people to which they are accountable. In so doing, in enlarging and examining our geographical imaginations, we might come to realise not only that our lives are 'radically entwined with the lives of distant strangers' but also that we bear a continuing and unavoidable responsibility for their needs in times of distress more.
> (Gregory, 1994: 205)

In this agenda lies a pathway towards more sensitive and meaningful engagements of self and other in Human Geography.

DISCUSSION POINTS

1. What aspects of your self are significant in shaping *your* Human Geography? How do you know?
2. To what extent does non-representational theory involve a complete rethink of the self?
3. How is it possible for researchers to represent the other when they are 'so thoroughly saturated with the ideological baggage of their own culture' (Ley and Mountz, 2001)?
4. To what extent is the distinction between *the* self and *the* other crude and oversimplified, given that identity is 'always stitched together imperfectly, and therefore able to join with another to see together without claiming to be another' (Haraway, 1996)?
5. What evidence do you see in contemporary Human Geography of an emotional, connected and committed sense of the other?

FURTHER READING

Butz, D. and Besio, K. (2009) Autoethnography. *Geography Compass*, 3, 1660–74

Ellis, C. and Bochner, A. (2000) Autoethnography, personal narrative, reflexivity: researcher as subject. In: N. Denzin and Y. Lincoln (eds.), *Handbook of Qualitative Research*, London: Sage.

Pile, S. and Thrift, N. (eds.) (1995) *Mapping the Subject: Geographies of Cultural Transformation*. London: Routledge.

These offer comprehensive discussions of subjectivity and self and ethnography and self.

Cloke, P. (2004) Deliver us from evil? Prospects for living ethically and acting politically in Human Geography. In: P. Cloke, P. Crang and M. Goodwin (eds.), *Envisioning Human Geographies*. London: Arnold, 210–28.

Ruddick, S. (2004) Activist geographies: building possible worlds. In: P. Cloke, P. Crang and M. Goodwin (eds.), *Envisioning Human Geographies*. London: Arnold, 229–41.

Smith, D. (2004) Morality ethics and social justice. In: P. Cloke, P. Crang and M. Goodwin (eds.), *Envisioning Human Geographies*. London: Arnold, 195–209.

These present a series of manifestos on how to develop ethical Human Geographies.

Cloke, P., Cook, I., Crang, P., Goodwin, M., Painter, J. and Philo, C. (2004) *Practising Human Geography*. London: Sage.

Crang, M. and Cook, I. (2007) *Doing Ethnographies*. London: Sage.

Further commentary on ethnography and qualitative methodologies in Human Geography.

Crang, P. (1994) It's showtime: on the workplace geographies of display in a restaurant in southeast England. *Environment and Planning D. Society and Space* 12, 675–704.

An excellent case study of the conceptual and practical outworking of self–other in ethnography.

Cloke, P. and Little, J. (eds.) (1997) *Contested Countryside Cultures: Otherness, Marginality and Rurality*. London: Routledge.

Said, E. (1995) Orientalism. London: Penguin (reprint with Afterword).

For a continuation of illustrations of rural otherness and Orientalism, respectively.

Cloke, P. (2004) Exploring boundaries of professional/personal practice and action: being and becoming in Khayelitsha township, Cape Town. In: D. Fuller and R. Kitchin (eds.), *Radical Theory/ Critical Praxis: Making a Difference Beyond the Academy*. Praxis (e) Press, 92–102.

For a simple illustration of self–other issues in the context of an academic and community work in a South African township.

CHAPTER 6
MASCULINITY–FEMININITY

Geraldine Pratt and Molly Kraft

Introduction

Geraldine: Come into my office and, please, do sit down. If you take a peek around, you can learn quite a bit about me. For instance, I have a child and a man in my life. The walls of my colleagues' offices are equally or possibly even more chatty about their family relations, especially their roles as fathers (see Figure 6.1).

Molly: Come into mine and you will learn – just as I have in writing this – that I unconsciously quiet certain parts of my sexuality. Take for example my loud appreciation of the **heteronormative** family that raised me along with the active silencing of my queerness and current relationships. This might come from my vulnerability as a graduate student, not yet established in the academy, but it also highlights norms of visibility.

We invite you to consider the many ways that heterosexual masculinities and femininities permeate your everyday life as a geography student and the geographical knowledges that you learn. It is likely that you come to this text as a gendered and sexed subject – just as we do – often unselfconsciously living out particular masculine and feminine norms. Geographical knowledges – both everyday and scholarly ones – are gendered, traditionally as masculine. But we now have many ideas of how to disrupt or unsettle this gendering of knowledge: by asking different questions, using different methods and telling or writing different stories. We want to explore in particular the tactics of situating knowledge, exploring linkages across spheres and scales of life that typically are conceived as separate, and queering the binaries that structure much of our social life.

Embodied geographies and knowledge acquisition

Take a look at how you are sitting at the moment. In 1979, a German photographer, Marianne Wex, published hundreds of photographs of men and women to demonstrate feminine and masculine modes of bodily comportment. Standing, sitting or lying on the beach, women tend to hold their arms close to their sides and their legs locked together, while men often sprawl and take up more space. Undergraduate students like to argue that Wex's photographs are dated,

Figure 6.1 Family relations come to work: (a) Geraldine Pratt's desk; (b) Derek Gregory's message board; (c) Trevor Barnes' office wall; (d) Molly Kraft's desk. Credit: Geraldine Pratt/Molly Kraft

but more recent photographs by artist-photographer Amelia Butler suggest otherwise. Her photographs (Figure 6.2), taken at IKEA in Vancouver, Canada in 2003, show customers who agreed to pose in a display room of their choice. Consider how you interpret the bodies that do not conform to feminine and masculine norms.

These bodily practices may seem trivial but it is arguable that they inhabit our psyches and shape cognitive abilities. Iris Young (1990c) has argued that masculine and feminine modes of bodily comportment reflect the fact that boys are freer to simply be in space as subjects rather than objects of visual inspection. Boys then tend to see themselves as 'creating' space and spatial relationships rather than being positioned in space. Young argued that these ways of moving the body affect cognitive ability and self-perceptions of competence. Indeed, it was documented in Britain in the late 1970s that from age eight (middle-class) parents tend to allow boys to explore more independently at much greater distances from home (Matthews, 1987). It is immensely suggestive that it is from age nine that boys begin to perform better on tests that measure spatial aptitudes. (These are paper and pencil tests that measure various spatial aptitudes by asking, for example, what a two-dimensional representation of a three-dimensional figure would look like if it were rotated by 90 degrees.) Spatial aptitude is a cognitive ability for which there is some of the most persistent and robust evidence of gender difference (Lawton, 2010) and some forms of spatial aptitude are prerequisite for math competency. To trace one consequence of this: in 2010 fewer than fifteen per cent of faculty in math-intensive fields in US universities were women (Ceci and Williams, 2011).

Boys' relative freedom to explore no doubt reflect their parents' fears about girls' vulnerability in public space and there is considerable evidence that these gendered geographies of fear persist with age, and that women tend both to be more fearful and vulnerable to extreme physical violence in isolated public spaces. Such a generalization requires several qualifications, however. Kristen Day (2001) argues that generalizations about women's fear in public places is as much a

Figure 6.2 Feminine and masculine bodily comportment on display at IKEA. Credit: Amelia Butler

construction of masculinity as it is of women's actual experiences. Whereas young men interviewed in Irvine, California revealed that they are fearful in a few selective places, they thought young women were vulnerable everywhere and persistently in need of their (masculine) protection. Other research suggests that parental norms may be changing due to increased fears of 'stranger danger'. Valentine (1997) has found that middle-class British mothers (especially South Asian mothers) are now equally, if not more, protective of their sons (who they regard as less mature and somewhat 'dizzier' than their daughters and more vulnerable to random violence in public spaces). This not only calls attention to the fact that gender is lived differently within different racialized groups, but suggests that, as parenting norms change, gendered differences in both freedom to explore independently and spatial aptitudes may also disappear.

There is evidence that some spatial aptitudes can be developed through video games (Terlecki and Newcombe, 2005) and there is a gendered geographical story to tell about this as well. McNamee (1998) found that middle-class British boys tend to have better access to computers than do girls, and this results from gendered assumptions and territorial control within the home. In households in which there is only one computer, it is often housed in the boy's bedroom and female siblings must gain entry to this room in order to access computer technology (see box). This is a new set of spatial relations, then, that may reinforce gendered cognitive development.

CASE STUDY

SM (Interviewer)]: You've got a sister, Phil. You said that your Nintendo was shared with her – do you ever fight about going on it?

Phil: Well, it's always kept in my room and there's a lot of arguments because I won't let her on it because it's in my room. I won't let her in my room.

SM: Did you say your sister was older or younger?

Phil: Younger.

SM: How much younger?

Phil: Ermm, she's eleven.

SM: So you won't let her in your room, you won't let her use it?

Phil: Yeah. She starts getting real annoyed and that and starts saying 'Oh well that's it now. Next time I'm gonna trash your room' and all this lot.

SM: What does your mum say?

Phil: She tells us both to pack it in. She usually blames it on me sister.

Source: McNamee, 1998, 197–8

SUMMARY

- There are masculine and feminine ways of bodily comportment and these are thought to affect self-perception and cognitive abilities, especially spatial abilities.
- Spatial mobility is conditioned by gendered (and racialized) perceptions of safety, and this affects our capacity to explore and know our environment.
- Environmental experience and organization of domestic space affect spatial aptitudes with ongoing consequences for women's presence in math-intensive professions.

Masculinist geographical knowledges

Assuming that you already embody geographical knowledges, you likely also encounter gendered geographies in your everyday life, and may, unfortunately, be exposed to more as a student through the geographical knowledges that you are taught.

Gendered geographies of the everyday

Popular geographical imaginations reflect and structure our social worlds, and they draw upon and organize existing gender relations, sometimes brutally so. Trying to understand why two white middle-class nineteen-year-old male university students were given light sentences for their brutal murder of an Aboriginal woman in Regina, Saskatchewan in Canada, for instance, Sherene Razack (2000: 196) reasons that the judge's decision rested on spatial assumptions: the murdered woman lived in a poor, racialized and gendered space of prostitution 'where violence is innate' while the men lived middle-class lives 'far removed from spaces of violence'. Plainly put: 'She was of the space where murders happen; they were not'.

Melissa Wright (2011) makes a similar analysis of the wasting of women's bodies in Cuidad Juarez, Mexico, where some government officials have rendered unremarkable the murdering and dumping of the bodies of dozens of young women and girl factory workers through a spatial argument. They do this by intimating that 'public women' working outside of the home invite violence and suggesting that families can best protect their daughters by keeping them at home. These are extreme cases, but we ask you to think about how everyday gendered assumptions about public and private or other spaces permeate your own life.

Masculinist scholarly knowledges

It is likely that some of what you are taught in your geography courses will be riddled with such assumptions. The argument about the **masculinism** of scholarly geographical knowledge has been made in different ways. The first is nicely summarized by the title of an article written by Jan Monk and Susan Hanson in 1982: 'On not excluding half of the human in Human Geography'. They argued that geographers have ignored vast areas of social life, ones that are understood to be feminine.

Geographers have found some questions more compelling than others (for example, the journey to work as compared to the journey to childcare) and certain activities are both undervalued and effectively invisible. Writing almost thirty years later in 2008, Monk and Hanson maintain the enduring urgency of their message – for the discipline as a whole and not just a subgroup of dedicated feminist geographers.

Women's labour in the home is a prime example. In most countries it is virtually impossible to obtain comprehensive information about it. Every country in the world keeps a national account of economic activity and productivity, but much of women's daily work – household work, volunteer activities, child- and eldercare, subsistence activities, bartering – lies outside these measurements of the economy (Domosh and Seager, 2001). In 1996 the Canadian government was one of the first and only national governments to collect information in the national census on unpaid work. Analysis of the 2006 census (the last year in which this data was and will be collected) documented the vast quantity of this work and the fact that women do the bulk of it: of the 25 billion hours of unpaid work done in Canada in 2006, worth an estimated $319 billion in the money economy (or 41 per cent of GDP), two-thirds was done by women (www.unpac.ca/economy/unpaidwork.html#3). The racialization of so much care work likely figures into its marginalization in everyday life and scholarly accounts (Pratt, 2012).

A second observation about the masculinism of knowledge is that when social life is recorded, it is often done in gendered terms, in subtle and unacknowledged ways. Gibson-Graham (1996), for example, argues that popular and scholarly understandings of globalization are organized through masculine representations of capitalism and a metaphor of penetration and rape of feminized, vulnerable local economies. She argues that workers have absorbed the politics of fear in ways analogous to rape victims. Following feminist efforts to re-script women's vulnerability to rape in ways other than pure victim-hood, Gibson-Graham considers ways of rhetorically diminishing the perceived power of multi-national corporations by exploring their vulnerabilities as masculine bodies. We are asked to envision seminal fluid as leaky, often misdirected and wasted. Violating the norms of heterosexuality (within which the male body is seen as having well defined borders and (literally) as an agent rather than recipient of penetration), Gibson-Graham represents the male body as penetrable. By analogy, we are also asked to consider that money (capitalism's semen) might also misfire and that non-capitalist enterprises have the capacity to penetrate and reshape capitalism. '**Queering**' globalization in this way, Gibson-Graham claims, liberates alternative ways of imagining globalization and creates opportunities for non-capitalist economic and social forms.

Third, Gillian Rose (1993a) has argued that Human Geography is masculinist in a more general **epistemological** sense. She identifies two traditions of masculinist geographical knowledge production, which produce and relate to constructions of femininity in specific but differing ways. She argues that masculinity and femininity are not simply social constructions but a self-reinforcing binary constructed within a masculinist frame of reference. Masculinity defines itself in opposition to what is conceived as feminine: the subjective, emotional, embodied. Social-scientific masculine knowledge represses femininity and is conceived as rational, neutral, universal, exhaustive and the product of a detached, objective observer. Aesthetic masculinity, which she attributes to **humanistic geography**, includes the emotional and non-rational but continues to operate within the dualism of masculine and feminine because it feminizes nature, landscapes and place and posits its authority to know these places in universalizing ways.

SUMMARY

- Much of women's experience has been ignored by geographers. Government agencies fail to collect data on many aspects of social life and the racialization of much of women's work renders it doubly invisible.
- Nonetheless, social life is often interpreted through a gendered lens.
- In the production of geographical knowledge, different types of masculinities, which produce and play upon the masculine–feminine duality in varying ways, structure epistemology.

Beyond the duality of masculine–feminine

On the cover of the book in which Gillian Rose developed her critique of the masculinisms underpinning geographical knowledge is a photograph by the artist, Barbara Kruger. Framing the image of a woman's face is the text: 'We won't play nature to your culture'. This is a refusal to participate in the dualisms (i.e., nature–culture, feminine–masculine) that structure masculinist knowledges. How might we refuse these dualisms, once the masculinist limits of existing geographical knowledge have been identified?

Situated knowledges

One strategy has been to rethink the concept of scientific objectivity and refuse the masculinist dualisms (objectivity–subjectivity, mind–body, scientifically detached–politically engaged) on which it has been based. Donna Haraway (1991) has likened the traditional, detached scientific account to a 'God trick' because it is located both nowhere and everywhere. In her view, such accounts are dangerous because they overgeneralize one way of looking at the world, and colonize other knowledges and perspectives through powerful claims to objectivity. She argues that all accounts of the world, including scientific ones, come from a social location. Of necessity scientists view the world through enabling technologies (whether these be questionnaires, maps, computer modelling, to name just few). Responsible objective descriptions of the world are situated, partial ones, that attempt to reveal the technologies upon which they are built. This is what Haraway calls 'accountable positioning'.

There are three implications that follow from the notion of situated knowledge. First, it dissolves an unfortunate dualism between quantitative and qualitative and scientific and non-scientific modes of research. Haraway wanted to contain the claims of scientists through the notion of situated knowledge, but she also made it clear that scientists have developed very powerful and useful technologies for seeing, and that the stakes and resources of science are too high *not* to participate in it. This is a pressing concern for geographers: reflecting on her time as editor of the journal *Gender, Place and Culture*, Linda Peake (2008) noted that over 95 per cent of article submissions used qualitative methods. In her view, this signalled a dangerous trend towards a generation of feminist geographers that has no inclination or practical skills to engage with quantitative methods or with scientific knowledge more generally. However, there are inspiring exceptions, for instance, Mei-Po Kwan (2002). Echoing Haraway, one goal of (the relatively few) feminist quantitative geographers has been to explore both the strengths and limits of quantitative methods and the powerful visualization techniques associated with them. To see what this looks like, you might visit http://li-gis.cancer.gov/. This is a website that evolved out of the activism of women on Long Island, New York, who were concerned about the high rate of breast cancer among themselves, friends and family, and possible environmental causes for it. Their efforts led to a $27 million geographical information system to evaluate exposure to a large number of environmental contaminants as well as ongoing research projects linking environmental hazards, public health and mapping (Cromley and McLafferty, 2011). There is a clear and detailed statement on the website of the quality of the quantified data through which the maps have been created (Figure 6.3). This is an explicit effort at 'accountable positioning'.

Second, situating scientific claims opens space for other types of knowledge, constructed from different vantage points, using other technologies for seeing. These might include indigenous or community-based knowledges or simply different research methodologies. One of the most exciting contributions of feminist geographers has been their enthusiasm for methodological experimentation with, for instance, ethnography, participatory research, video, theatre and different modes of media and writing (e.g. Moss, 2001). Figure 6.4 shows a scene from a testimonial play that I (Pratt) developed in collaboration with the Philippine

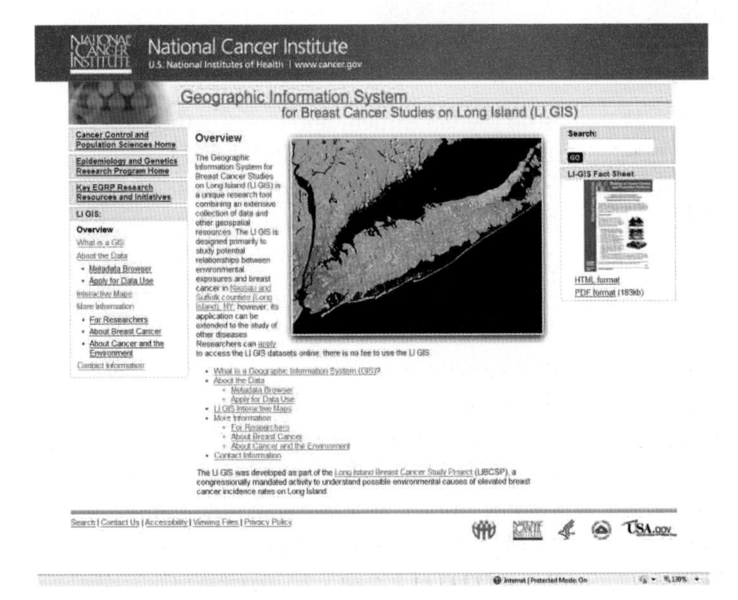

Figure 6.3 Accountable positioning: revealing the limitations of quantified data. Credit: www.healthgis-li.com/

Figure 6.4 Turning research into testimonial theatre: a play performed in Nanay featuring a Filipino migrant worker. Credit: top photo, Charles Venzon; bottom photo, Caleb Johnston

Women Centre (PWC) and Vancouver theatre artists, directly from research that the PWC and I have done with Filipino migrant domestic workers in Canada (Johnston and Pratt, 2010). The play is a series of monologues constructed verbatim from interview transcripts and is one way of bringing scholarly research to a wide public audience. As another example, Gill Valentine's (1998) account of being stalked and receiving obscene letters presents an early and striking presentation of the power of experimental methods. Her writing is autobiographical and deeply personal and is both an intervention to stop the harassment and a fascinating description of the ways that sexual harassment reorders the meaning of places such as the home. She describes how her new home, previously a source of pride and pleasure, became a site of great anguish. Her writing blurs all sorts of distinctions in productive ways: emotional involvement and detached assessment; scholarly reflection and activist intervention; autobiography and social geography.

Third, and most importantly, the significance of specifying scientific claims and proliferating other ways of seeing is to generate conversations across partial knowledges so that we might develop less singular, less dangerous, more complex, multidimensional and reliable understandings of the world. Situated knowledges open opportunities to learn from other perspectives and ways of knowing, and to engage in processes of translation across non-reducible knowledges. This process of translation is a laborious, halting one and it demands a rich geographical imagination that is both knowledgeable and sensitive to the particularity of context and culture (Katz, 2001), as well as close attention to the ways that masculinist (and colonial and imperialist geopolitical) positions can creep back.

Sunera Thobani (2007), for instance, has criticized prominent western feminist critiques of the US-led war on terror, arguing that they reinscribe conventional notions of whiteness by refusing to critically engage with the political demands of those who challenge the West. But there are also examples of western feminists attempting to rethink their perspectives through conversations with women's organizations elsewhere. Drawing upon the experiences of non-governmental organizations (NGOs) working with migrant domestic workers in Southeast Asia, Aihwa Ong (2006) invites western feminists to rethink their preoccupations with liberal rights, including rights to citizenship. NGOs argue that few migrant domestic workers in Southeast Asia are interested in gaining citizenship in the country in which they labour as domestic workers. What they want is access to employment (that is, unrestricted mobility across national borders) and moral legitimacy once employed as domestic workers. In line with the perspectives of these NGOs, Ong argues that moral legitimacy is more effectively achieved through the discourse of the at-risk body of the female mother, life-giver and life-nurturer than through a language of abstract legal rights. A discourse of the suffering maternal body is more likely to appeal to employers' Asian family values, she argues, and to elicit their sympathy for domestic workers' bodily needs; that is, their need for a day of rest, reduced over-time hours and adequate health care. Redeploying this seemingly paternalistic valuing of the maternal body is a profoundly challenging perspective within western feminism and offers an important moment to reflect on the possibilities and difficulties of translation across world views.

SUMMARY

- One way of disrupting the dichotomy between subjective and objective knowledge is to reconceptualize the meaning of objectivity. Situated knowledge has been an important concept for doing this.
- Situated knowledge is the idea that all knowledge claims – even scientific ones – are partial and located from within a particular perspective. Objectivity emerges through elaborating and understanding of this partiality.
- This opens opportunities to acknowledge the validity of other perspectives and knowledges. Situated knowledge requires a commitment to communicate across different perspectives and types of knowledge.

Exploring linkages and disrupting binaries

Another strategy for disrupting the duality of masculinity–femininity is to question the classification of social life into private and public spheres and to explore linkages across different 'spheres' of life and scales of analysis. We have already noticed that certain types of work – mostly non-waged work of **social reproduction** done in and around the home – are considered to lie outside of the formal economy, and most governments keep no record of them. It is not only women's work that disappears within the domestic sphere: many issues and relations that are fundamentally important to the equality and material and psychological well-being of women (and other marginalized groups) have been designated as private and non-political. Much feminist analysis and activism has been dedicated to reframing and politicizing issues that have been enclaved within the private sphere, for example, childcare, domestic violence and sexual freedom.

The other side of this is to demonstrate how what is conceived as private (and feminine) is ever present in other realms of life. Reconsider the family pictures on our office desks (Figure 6.1). So too, women's continuing overwhelming responsibility for domestic work and care affects where they work, whether they work full-time or part-time, and their long-term career aspirations and progress (Milligan and Miles, 2010). Kay Anderson and Susan Smith (2001) ask us to consider how emotions permeate all aspects of public life, whether in the workplace, the housing market or urban planning decisions. Recognizing this emotionality is an intensely political act. By feminizing emotions and relegating them to private sphere, we sustain the myth that economic matters and government policies are purely rational and set the stage for a broader neo-liberal agenda: 'the logic of efficiency *depends on* the silencing of emotions' (Anderson and Smith, 2001: 8). Understanding the public nature of emotions, how they are suppressed, manipulated and harnessed to both maintain and transform social relations is therefore an important focus of feminist inquiry (Sharp, 2009).

Another aspect of disrupting conventional gendered classifications of social life involves rethinking assumptions about the scale at which particular social processes occur. Scale itself is gendered: the body, home and neighbourhood tend to be feminized, and 'large' scale processes such as the economy, government and geopolitical security are typically viewed as masculine. These 'scalar narratives' are gendered social constructions (Marston, 2000). Not only does business and state administration involve more than rational calculation, it is accomplished at a range of scales, including the body. Alison Mountz (2010) describes how the administration of Canadian federal immigration law is embodied at a variety of scales, including very intimate ones. For instance, Canadian immigration officers interpret the criminality of refugee claimants by reading their bodies. In the case of smuggled migrants from Fujian, China apprehended off the coast of British Columbia in 1999, immigration officers made a close study of their body tattoos as one means of differentiating 'snakeheads' (smugglers) from those who were being smuggled. Also challenging conventional narratives about scale, in this case of nation and war, Tamar Mayer (2004) argues that the war in former Yugoslavia was the paradigm for a new warfare because the rape of tens of thousands of women became, not 'collateral damage', but a key means through which the war was conducted. War was conducted in the most intimate of ways as women were systematically raped, first in their homes, then in public spaces, then in detention camps, then in special rape camps and eventually in brothels.

SUMMARY

- The division between public and private is a gendered classification, and what is deemed private is also often classified as feminine and emotional. The private and emotional tend to be depoliticized. Feminist analysis shows the effects of this classificatory distinction, and demonstrates the links between public and private and the blurring of emotional and rational registers of experience.
- Geographical scale is also a gendered social construction. New analytical possibilities are opened by 're-scaling' our analyses (or telling other 'scalar narratives').

Dismantling gender

Finally, the duality of masculine–feminine is exhausted by proliferating the category of gender – beyond the number two. This has been accomplished in different ways. One is to document the existence of many different masculinities and femininities, many different ways of 'doing' and experiencing gender. In her study of the financial district in London, England, McDowell (1997) identifies two very different masculinities at work: the youthful virility of the 'Big swinging dick' on the trading floor and the sober, rational patriarch of corporate finance. Alison Blunt (1994) has described how Mary Kingsley, an upper-middle-class British woman who travelled to West Africa in the 1890s, was able to transgress the norms of white bourgeois femininity while travelling in Africa. Femininity could thus be lived differently in different places and geographical mobility (and colonial privilege) were key to getting some distance from Victorian norms of femininity.

Dismantling gender norms also means becoming attuned to their instability. Transgender (or trans) signals a slippage between a person's assigned gender and that which they identify with or perform (Doan, 2010). For many, the either/or of the traditional roles have become tiresome and often oppressive as neither accurately describes their lived realities. Or sometimes, both do. Gender includes a 'splendid array' of different experiences (Doan, 2010: 638) but the heteronormativity of space often requires 'choosing' when this is not possible or desired (for example, at public toilets). One way of challenging this has been to think about 'queer time' and 'queer space' (Halberstam, 2005; Oswin, 2008). The term queer and its associated politics have come to signify both a self-description for those not aligned with gender norms as well as a way of opening 'new life narratives and alternative relations to time and space' (Halberstam, 2005: 2).

Another strategy for dismantling gender comes from recognizing that gender intersects with other social identities, including race, sexuality, disability and class. There are two different approaches to thinking about this. One is to emphasize the way that intersectionality creates interlocking matrixes of oppression. Feminized racialization of Filipino men, for instance, has been key to their deployment as cooks and cleaners within the United States military (Lee, 2012). Racialization, gendering and sexualization can feed into each other or interlock to reinforce marginalization or privilege. One implication of this is that racialized women and men do not share the same gendered experiences as those (non) racialized as white, and geographical scholarship that ignores these differences or generalizes white persons' experiences as the norm can do the work of maintaining race and 'first world' privilege (Mohanty, 2005). As another approach to intersectionality, Gill Valentine (2007) suggests that it can reveal the complexity and fluidity of identities as well as how they are made and unmade in our everyday lives. An example of this was observed in a focus group of Asian-Canadian youths in Vancouver, Canada in 1997. One young woman described embodying different femininities and racial identities as she moved between languages and different spaces of the city. Her performances were not freely chosen and she expressed anxiety about mis-performing her identities in various places in the city (see box). Sexuality also has been a rich site for proliferating and troubling gender dualisms (see Chapter 43). Dismantling genders does more than make us think about gender in new ways; it profoundly unsettles a core dichotomy that structures our subjectivity, our patterns of thought (indeed, this chapter!) in profound and far reaching ways.

CASE STUDY

When I speak Chinese, I feel very formal. I feel like I should be sitting with my back straight. It seems that when I speak Chinese, I am trying to fit the stereotype of what I am supposed to be in Chinese culture . . . My voice when I speak English isn't extremely high pitched. It's quite low. But when I speak Chinese my voice goes up. It just shoots up! . . . [Which language I speak] depends on where I am because I don't want to be seen as an outsider. There's no way I'm going to go [to a downtown shopping area] where there are lots of people speaking English, there's no way I'm going to speak Chinese. I don't want to get looks [and verbal insults.] But if I'm in Richmond [a suburb with majority Chinese-Canadian residents] or in Chinatown, there's no way I'm going to speak English because I'll get that same kind of negative thing coming at me: 'Oh, look at her. She's got yellow skin but she can't speak in Chinese like a good Chinese girl.'

(Pratt, 2004: 134–6)

SUMMARY

- One strategy to move beyond gender dualism is to demonstrate the gender itself is plural, interwoven with other social identities, and unstable in itself and across time and place.

Conclusion

Gender dualism structures our subjectivities as teachers and students, our everyday lives, and the geographical knowledges that we teach and learn. A sustained critique of the masculinism of geographical knowledge over the last 35 years has led to many strategies to free our thinking from the limits imposed by this dualistic gender hierarchy. Many of these strategies are inherently geographical. They involve locating and contextualizing one's knowledge claims in particular places and times, moving across and making connections between spaces in our conceptualizations and dismantling gender classifications by proliferating masculinities and femininities and exploring their instability and particularity for specific groups in particular time and places.

Human Geography has not only benefited from feminist analyses and critiques of gender dualisms; a spatial imagination is central to both understanding and challenging gendered, sexualized, racialized and other social categories and norms.

DISCUSSION POINTS

1. Are you equally feminine (or masculine) in different places? Do you live different femininities (or masculinities) in different places, in relation to different people?
2. Does Kristen Day's argument that young men assume and even promote women's fear as a way of experiencing and performing their masculinity ring true for you?
3. Thinking about Tamar Mayer's discussion of rape as a means of conducting war, can you think of other or more recent examples where gender and sexuality are mobilized as tactics of war?
4. It is much easier to talk about translating across different knowledges in the abstract than in practice. Feminists have had notoriously difficult times talking across issues such as female circumcision or the rights of Islamic girls and women to wear the veil (or, more recently in France, head scarves in secular public schools). Discussion usually founders on the difficulties of reconciling universal notion of women's equality (which some see as essentially western) with the desire to respect cultural differences. Consider one such controversial issue, by first elaborating the arguments for and against, and then thinking about how you might translate across these conflicting positions. How might a geographical education and attention to geographical specificity help in this process of translation? Thinking back to Chandra Mohanty's argument, how might the privilege of a white western perspective structure these debates and how might we escape this?

FURTHER READING

Haraway, D. (1991) Situated knowledges: the science question in feminism as a site of discourse on the privilege of partial perspective. In: D. Haraway (ed.) *Simians, Cyborgs, and Women: The Reinvention of Nature*. London and New York: Routledge.

This essay is not easy and it assumes some knowledge of key philosophical debates around objectivity and relativism. But it is a classic and has had a profound effect within Human Geography since the early 1990s. Don't expect to understand it fully the first time that you read it but enjoy the writing for its great wit, passion and energy.

Moss, P. (ed.) (2002) *Feminist Geography in Practice: Research and Methods*. Oxford: Blackwell.

This is an edited volume that introduces undergraduate students to a range of methodologies and feminist research strategies.

Rose, G. (1993) *Feminism and Geography: The Limits of Geographical Knowledge*. Minneapolis: University of Minnesota Press.

Rose elaborates a critique of geographical knowledge as masculinist (as well as white, bourgeois and heterosexist) through case studies of time–geography, humanistic geography and traditions of studying 'the landscape' in geography.

Seager, J. (2009) *The State of Women in the World Atlas* (4th edn). New York: Penguin.

This atlas demonstrates the power of mapping, for seeing patterns and generating questions. A wide range of themes and statistics are mapped – from women's formal political participation and representation to migration patterns of domestic workers.

Seager, J. and Nelson, L. (eds.) (2004) *Companion to Feminist Geography*. Oxford: Blackwell.

A large compendium of articles, this provides a good introduction to a full range of topics.

CHAPTER 7
SCIENCE–ART

David Gilbert

Introduction: the indiscipline of Human Geography

Each year, alongside the more formal teaching that I do in lectures and seminars, I work with a small group of first years. We meet weekly to discuss essays and assignments and other issues arising from the courses that they are doing. In the last session before Christmas, we look back on the first term's work. Often this discussion concentrates on practicalities, perhaps about the differences between learning at school and university, or about moving to a British university from abroad, or about the challenges of balancing study and other commitments. But we also talk about their early impressions of 'doing geography'. Two reactions are very common: reactions that are two sides to the same coin. Some students enthuse about the diversity of what they have been doing, and about the range of subjects and approaches addressed in the first few weeks. This variety (as Philip Crang points out in Chapter 1) is often a strong motivation for taking a course in geography. But for others, it's a source of anxiety. As one (rather insightful) student put it, 'other people here at university are in disciplines, but we seem to be in an indiscipline'.

When we talk in our group we often identify a number of different dimensions to this 'indiscipline'. The first and most obvious is about the huge range of topics studied. Our students take broad introductory modules in both Human and Physical Geography in their first term. At 10 o'clock in the morning they may be learning about what the crisis in the global economy tells us about the connections between places, and an hour later the focus has shifted dramatically to the analysis of ice cores as evidence of long-term environmental change. Even within the confines of the Human Geography course, students have to consider a dizzying range of topics and case studies, from the lasting effects of the slave trade, through Nike's global sub-contracting arrangements, to the significance of sexual identity for urban change in Manchester, and the travel writings of Alexander von Humboldt or Bill Bryson. This diversity may just be a matter of chaotic course design (I'm responsible for this module), but I suspect that something similar happens in many introductory courses to Human Geography.

However, the discussion can go on to identify other significant differences within geography. Drawing on their experiences in classes in research methods, as well as the lecture series, some students talk about differences in

Figure 7.1 Images from undergraduate prospectuses showing the wide variety of practices associated the discipline of geography. Different ways of 'doing geography' are associated with different ways of thinking about the world. Credit: (a) and (b) Jenny Kynaston, Department of Geography, Royal Holloway University of London; (c) Michelle Stewart and Nathan Stewart; (d) Chris Perkins

approach, as well as content. They point not just to the differences in what geographers study, but also differences in the ways that they undertake research. At the most obvious level, this is about the practices of doing geography. After they've been on their first year field course, students are even more likely to comment on this – different kinds of geographer have very different ways of finding out about the world. This may involve the measurement of physical features (perhaps the slope profile of a beach or the size of particles on a river bed) as a way testing the validity of a particular model of the processes creating a particular landform. Or it may involve undertaking questionnaires or interviews with local people, to be analysed afterwards using either **quantitative** or **qualitative methods**. Students are often surprised to be asked to read novels, look at paintings or watch films set in the place that they are studying. This surprise is heightened when it is suggested that this work is more than just a way of getting a background feel for the place, but a significant form of geographical research with particular methods and techniques of analysis and interpretation. (You can read more about these in Chapter 9 Representation–reality and Chapter 16

Imaginative geographies). Closely linked to these differences in approach are differences in the technology and equipment used to do geography and the wide variety of sites that geography takes place in. This includes not just the range of places that we call 'the field' (which might be a glacier face far from the nearest human habitation or a housing project in the heart of a giant city), but also laboratories, libraries and archives (Crang, 1998b; Livingstone 2000, 2001; DeLyser and Starrs, 2001).

The diversity of geography is also reflected in the ways that it is written. Alongside textbooks like this one, this is the time when many of our students are taking their first look at academic journals, now usually online. While all of the journals they read have the same general aim of presenting new research and ideas to the academic community, the ways in which they do this are often very different. This isn't just a difference between those journals that specialize in physical geography and those that publish the work of Human Geographers. There are very marked differences in style and structure between say, *Regional Studies* and the *Journal of Historical Geography*, or even between articles in a single issue of general journals like *Transactions of the Institute of British Geographers* or *Annals of the Association of American Geographers*. Of course, students also experience this diversity actively, when they have to write and are expected to follow particular conventions of structure and writing style.

A final set of differences often comes out in these discussions with my students. Sometimes this can get a bit personal – students will talk about how different aspects of geography seem to attract different kinds of people. Very often this is done by reference to other disciplines. Some geographers are described as being like 'hard scientists', while others are seen as having much more in common with academics teaching in sociology, history, literature or cultural studies. Students point to differences in teaching styles, in the use of language, in attitudes, even in ways of dressing. Students, too, are often tempted to 'take sides', and many rapidly identify themselves as a certain kind of geographer. In many ways this is what higher education should be about. In a good geography degree programme there is an opportunity not just to specialize in certain aspects of the discipline, but also to develop a commitment to particular ways of approaching the study of the world – ways of organizing evidence, solving problems, developing theories or searching for meanings. However, all too often this development is accompanied by a growing lack of understanding of alternative approaches. It's not unusual to hear geographers who regard themselves as doing 'science' contrasting their work with supposedly 'softer' or more 'arty-farty' approaches. Conversely, other geographers will oppose the subtlety and sophistication of their work to the philistinism, one-dimensionality and political naivety of 'nerdy' scientists.

Such stand-offs are not, of course, limited to the discipline of geography. Indeed we can find earlier examples of remarkably similar kinds of posturing and name-calling from as long ago as the seventeenth century, from those promoting ways of approaching the study of the world that came to be known as the sciences and the arts or humanities. Writing in the 1950s, the English novelist and commentator C.P. Snow famously pointed to this much more general division in academic life, marked by differences between 'literary intellectuals at one pole – at the other scientists, and as the most representative, the physical scientists'. Snow suggested that there was 'a gulf of mutual incomprehension' between the two groups and that 'their attitudes are so different that, even on the level of emotion they can't find much common ground' (Snow, 1959: 4).

Snow was writing about what he called the problem of the 'two cultures' in the very specific context of mid-twentieth-century Britain and its universities (see Gould, 2004: 69.) The intervening years have seen dramatic changes in universities and intellectual life more generally, most obviously in the great expansion of the social sciences, with subjects like Sociology, Anthropology and Politics, characterized by a distinctive third kind of culture. Nonetheless, his remarks draw our attention to a couple of features that are particularly important in understanding the diversity of Human Geography, which seems to encompass all three intellectual cultures and more. First, he points to the ways in which academic work has a 'culture' and always takes place in particular social contexts (and indeed, as we pointed out earlier, in particular geographical contexts too). When we discuss different approaches to Human Geography we cannot talk about these in abstract – we are always talking about the work of particular groups of people, with particular interests, social networks and social characteristics. We need to think about geographical knowledge being socially and geographically situated. Second, Snow's comments draw attention to 'mutual incomprehension' and the possibility of hostility and even conflict between different approaches. In celebrating the diversity of Human Geography we shouldn't underplay the extent to which the discipline is contested – particularly in the kinds of claims about knowledge made by different approaches.

SUMMARY

- The disciplines of geography in general and Human Geography in particular are characterized by marked diversity. This diversity is expressed in the focus, methods, practices, equipment and geographical sites associated with different approaches. Different approaches to geography are also marked by very different forms of writing.
- Thinking about the diversity of Human Geography also draws our attention towards contestation and conflict between different perspectives, and between different groups of geographers.

Histories of Human Geography

To be confronted with this 'indiscipline' at the start of a programme in geography can be more than a little daunting. This feeling isn't restricted to new undergraduates, as Noel Castree and Thomas MacMillan have pointed out, commenting on the remarkable dynamism of Human Geography in recent times. As they put it:

> Intellectual change, it seems, is now the only disciplinary constant. Some have experienced the procession of new 'isms', 'ologies', and 'turns' in Human Geography as a threat to subject identity. Others are no doubt exhausted by the ceaseless profusion of theories, methodologies, and data sources.
>
> (Castree and MacMillan, 2004: 469)

The most common way of making sense of this diversity has been to examine the history of the discipline, looking at the development of major approaches, and the competition between them. The history of the discipline of geography can help us to understand its present characteristics in a number of ways. First, looking at the past shows that geography has always been characterized by diversity – as Mike Heffernan argues, 'geography, whether defined as a university discipline, a school subject, or a forum for wider debate, has always existed in a state of uncertainty and flux' (2003: 19). There was no golden age when geography had a single stable set of questions and methods, although there have been plenty of attempts to impose a unified vision on the discipline. This historical emphasis moves us away from trying to frame an overall, everlasting definition of what geography is, towards thinking about a changing tradition (Livingstone, 1992). The introduction to this book suggests that we might answer the question 'What is Human Geography?' by reference to very general kinds of intellectual contribution – particularly in understanding the relations between human beings and the natural world, and in understanding questions of space and place in human life. Viewed historically, these questions can be regarded as being at the core of the geographical tradition, but we can also see how the balance between them has shifted, and how they have been framed in dramatically different ways by different groups of geographers at different times and in different places.

A second benefit of a historical approach is that it draws our attention towards the situated nature of Human Geography. Put simply, changes in the discipline need to be understood in the context of changes in the world beyond (although this isn't to make the stronger claim that such disciplinary developments are completely determined by external events.) At one level this is unsurprising, particularly in respect of a rapidly changing human world. To take one very obvious example from the past 25 years, the development of the internet, mobile telecommunications and other forms of technology has prompted geographers not only to study in detail the ways that these are used, but also to think in new ways about the nature of space and place.

But we also need to think about the situatedness of Human Geography as more than just a passive response to new developments in human life, by focusing on how the discipline has been actively bound up with those developments. A very significant example of this concerns the emergence of a distinctive academic discipline in the late nineteenth century, with departments and chairs (i.e. professorships) of geography established in universities in Germany, France and Britain. This incorporation of geography into the university system came at a time of dramatic European colonial expansion, particularly in what became known as the 'scramble for Africa'. (Between 1870 and 1914 the main European powers divided up political rule of almost all of the African continent between them.) Late nineteenth-century European geography must be understood as a science of empire, valued above all for its power to map and categorize newly colonized lands and peoples (see Chapter 16; Driver, 1992; Bell *et al.*, 1994; Godlewska and Smith, 1994).

This example of 'imperial geography' may seem to belong to a distant past. Certainly the specific ways that geographers of the late nineteenth and early twentieth centuries framed and answered questions about human–nature relations or issues of spatial order and organization – often by searching for climatic, environmental or racial 'explanations' for the supposed superiority of European 'civilization' – now seem abhorrent. But we are in a discipline that is very clearly **post-colonial**, not just in its interest in the historical geographies of imperialism and their lasting consequences for the modern world, but also in its own long-term development as an intellectual tradition. (Indeed, in the geography of Geography, while what we've identified here as core questions for Human Geography are to be found in some guise in almost all cultures, not all academic cultures have a separate academic discipline called Geography or Human Geography. A world map of current university geography departments – as opposed to cartography and other technical disciplines – would have some distinct similarities with the map of the British Empire shown in Figure 16.4.) While we may share none of the views of the imperial geographers, we can trace connections back to them through generations of academics and students in particular universities. There have been constant disputes about what and how geography should study, and struggles about who should be doing that study (as Gillian Rose, 1993a, in her work on the **masculinist** history of geography and many others have argued, academic disciplines have a long history of elitism and exclusion), but new generations and new approaches have usually had a vested interest in continuing and developing the geographical tradition itself. Many have sought to radically transform Human Geography while relatively few have sought to do away with it completely.

Taking a historical view also leads us to question simple notions that the discipline 'progresses'. There is a common assumption made by many new students that today's Human Geography has developed bit by bit from earlier versions of the discipline through the addition of 'new' knowledge (gained by empirical research) and by the development of 'better' theories (produced from new evidence, or possibly just by academics thinking that bit harder than their predecessors). Looking at the history of Human Geography, and indeed at the history of most other academic disciplines, indicates that change is much more episodic than this, often involving periodic radical changes in their central interests and assumptions. One significant way of thinking about this has drawn on Thomas Kuhn's notions of **paradigms** and paradigmatic change (Kuhn, 1962) (see box).

Kuhn's model of scientific change was severely criticized by many philosophers and historians of science, and he made substantial adjustments to his ideas (Lakatos and Musgrave, 1970; Kuhn, 1977). Kuhn also doubted that his ideas had much applicability to the social sciences, which he saw as characterized by much less coherence and agreement about central assumptions. The history of Human Geography doesn't fit the strict model of paradigm change very well at all, and it is extremely hard to identify anything that even closely resembles a Kuhnian paradigm in contemporary Human Geography. Nonetheless, Kuhn's terminology is still used regularly to identify general approaches that are broader than specific theories and that are shared by significant sections of a discipline or, as is often the case in the social sciences and humanities, by groups of academics across different disciplines. (See Dixon and Jones, 2004, for a review of developments in Human Geography that use Kuhn's terminology in this looser way.) Thinking about paradigms and paradigm shifts directs our attention to communities of Human Geographers committed to very different ways of doing geography, and to the ways that advocates of a new paradigm will often promote their approach by a critique (i.e. by systematic criticisms) of an existing approach.

To see the extent to which central ideas can shift in Human Geography we can compare this book with one of its predecessors. In 1972, Peter Haggett published the first edition of

CASE STUDY

Paradigm shifts

Thomas Kuhn's *The Structure of Scientific Revolutions* was first published in 1962, and despite much criticism, its detailed arguments have been an important reference point for the discussion of change in academic disciplines ever since. Kuhn's work applies specifically to change within scientific disciplines (rather than disciplines in the social sciences or humanities) and was a sociological study, providing a descriptive model of the behaviour of disciplinary communities of scientists. In its broadest sense a paradigm is a stable consensus about the aims, shared assumptions and practices of a particular disciplinary community. Kuhn argued that most scientific activity works within a stable consensus ('normal science') and does not challenge these fundamental assumptions. Eventually normal science becomes disrupted by anomalies or issues that cannot be explained within the existing framework and new innovative work may trigger a 'paradigm shift' to a new set of shared assumptions and practices. Perhaps the best known example of this kind of change was seen in physics, where the set of stable assumptions commonly described as Newtonian Physics, after Sir Isaac Newton, was undermined by anomalies discovered in the late nineteenth century. These led to a period of 'extraordinary research' associated with Einstein's theory of relativity and the development of quantum mechanics, and a paradigm shift to a new set of broadly held assumptions in the disciplinary community. The term paradigm has entered popular usage in a looser sense since Kuhn's initial formulation, and is regularly used to refer to general approaches, theoretical frameworks and methodologies held by significant groups within disciplines and, particularly in the social sciences and humanities, across disciplines. In this far looser sense the history of Human Geography can be said to have seen significant 'paradigm shifts', and today's Human Geography can be said to be 'multi-paradigmatic'.

Geography: A Modern Synthesis. In many ways this was an equivalent book to this one, designed to give an introductory overview of the discipline and to work as a bridge between study at school and university. Like this book, *A Modern Synthesis* was organized around the very general questions of human–nature relations and the significance of space in human life. Like this book, Haggett recognized the diversity of geography, but as the title of the book suggests, his response was very different. *Introducing Human Geographies* celebrates not just the range of topics covered by Human Geography, but also the range of approaches to those subjects – in this sense, it can be seen to be an introduction to a multi-paradigmatic discipline (hence the reference to 'geographies' in its title.) By contrast, *A Modern Synthesis* sought to find a single central approach to Human Geography that could be used to integrate the discipline (again, reflected in its title). This is unsurprising as *A Modern Synthesis* was published towards the end of perhaps the most self-consciously and deliberately paradigmatic period in Human Geography's history (Haggett, 1972: 16). Peter Haggett was one of the leading advocates of what was known as **spatial science**, arguing that Geography should be concerned with formulating and testing theories of spatial organization using methods that looked explicitly towards those of the natural sciences. It is some measure of the extent of change over 40 years that there is little direct application in this volume of the approaches that Haggett heralded as the core of a new stable paradigm in 1972 (see box).

CASE STUDY

Introducing Human Geographies, 1972-style

Just like this volume, Peter Haggett's *Geography: A Modern Synthesis* was intended as an introductory text, and opened by drawing attention to the diversity of geography:

> Starting a course in a new subject at college is like driving into an unfamiliar city. We see the sprawling new suburbs, the bustling new freeways, the pockets of decay, but find it hard to get an overall impression of its structure or know where we are. Geography is a Los Angeles among academic cities in that it sprawls over a very large area, it merges with its neighbors, and we have a hard time finding its central business district.
>
> (Haggett, 1972: xix)

Figure 7.2 Peter Haggett's *Geography: A Modern Synthesis* (cover of third edition from 1977) New York: Harper & Rowe. Credit: TBC

But rather than promote a diversity of ways of approaching geography, Haggett argued for a new 'supermodel' or paradigm that could integrate or synthesize the discipline into a more coherent whole:

> Modern geography is settling down again . . . The last 10 years have involved a massive swing toward more analytical work in which mathematical models . . . play a dominant role.
>
> (Haggett, 1972: 16)

However, far from settling down, the next few years saw the development of radically new ways of approaching Human Geography that provided a fundamental critique of Haggett's suggested synthesis.

SUMMARY

- Looking at the history of geography in general and Human Geography in particular can help us to understand its variety and fragmentation, particularly through an emphasis on the historical and geographically situated nature of geographical knowledges. It also helps us to understand geography as a changing and contested tradition, rather than a fixed way of seeing the world.
- While Human Geography cannot be said to develop paradigmatically in the strict sense suggested by Kuhn, a looser usage of the term 'paradigm' draws our attention to the existence of communities of Human Geographers who share and develop common sets of fundamental assumptions and practices. It also draws our attention to the episodic nature of change in the discipline and to the ways that proponents of a new paradigm will often promote their approach through critique of the fundamental assumptions of existing paradigms.

Human Geography: science or social science?

In this more general sense, Human Geography has been influenced by many paradigms over the past 30 years. A non-exhaustive list, in roughly chronological order, includes: **spatial science**, **humanistic geography**, **Marxism**, **structuration theory**, **feminism**, the **linguistic (or cultural) turn**, post-structuralism, **post-modernism**, **post-colonialism**, **non-representational theory** and **Actor-Network Theory**. (You can start to see why Castree and MacMillan (2004) talked of exhaustion!) Very brief definitions of these terms are to be found in the Glossary of this book, and the assumptions and perspectives of many of these inform the discussions of individual chapters. The list also gives some indication of the way in which Human Geography has regularly been influenced by broader intellectual movements originating outside of the discipline (although some geographers have made very significant contributions to these movements).

By way of an introduction to some very general differences between these paradigms, it is useful to think about Human Geography's position in relation to the three broad intellectual cultures identified at the opening of this chapter. C.P. Snow drew attention to general cultural differences between these approaches, but we can go beyond this to identify different ways in which they construct knowledge about the world. **Epistemology** is a technical term used in philosophy to refer to theories of knowledge. The sciences, social sciences and humanities are marked by distinctively different epistemologies: by different ideas about their relationships with the 'real world', what it is possible to know about it, and how it is possible to express that knowledge. In very general terms we can say that the sciences work from a basic assumption that the natural world is *ordered* and that the task of science is to uncover its fundamental regularities – often in the form of statements or theories about basic laws. Science involves the use of systematic methods of investigation that are *objective*, to assemble evidence that can be used to support or challenge theories. Science's claim to objectivity is very important. Scientific enquiry should be *replicable* – that is to say, that if the methods and techniques are applied properly and consistently, then any scientist undertaking the same work will reproduce the same results. Although science often seeks to construct general theories or laws, its conclusions are also always somewhat *tentative* – new evidence or arguments can lead to theories being revised or even rejected (Giddens, 1993: 20).

Many studies of the history and social contexts of disciplines like physics, chemistry or biology have emphasized the complexity of the ways they work in practice, and the twentieth century saw the development of ideas like relativity and quantum mechanics that questioned simple assumptions of an ordered natural world. In Physical Geography, too, there has been perceptive comment about the difference between any general model of scientific epistemology and the practice of a field science where it is often impossible to reproduce the exact conditions of any measurement or observation (see, for example, Rhoads and Thorn, 1996). Nonetheless, for many there is still something very seductive about the idea that the study of human society can be modelled on the approaches of the natural sciences, and that the social sciences can produce the 'same kind of precise, well-founded knowledge that natural scientists have developed in respect of the physical world' (Giddens, 1993: 20). Because of geography's long history as a discipline that has combined study of the physical and human worlds and its emphasis on the relationship between them, there has been additional pressure to approach Human Geography as a form of science (the paradigm of **environmental determinism**, significant in geography in the early twentieth century, can be seen as a different expression of this tendency). The epistemology of orthodox scientific method was adopted wholeheartedly by spatial science, so influential in Human Geography in the 1960s and 1970s. Spatial science developed a methodology based upon quantitative measurement, model building and hypothesis testing in a search for general scientific laws of spatial patterning and processes.

The claim that the human world can be studied using the methods of natural science is sometimes described as positivism. It is ironic that just at the time that Human Geography was turning to the positivism of spatial science, positivist assumptions were being subjected to sustained critique in disciplines like sociology and anthropology. By the early 1970s, this critique of positivism had been extended to Human Geography. At the heart of this critique was an argument that there were important and necessary differences between the sciences and social sciences. There were a number of dimensions to the challenge to positivism in Human Geography, which we can summarize as being about issues of *human agency*, *language and meaning*, *power* and *positionality* (see box). These are core issues for what we can generally describe as 'critical social science' or 'critical social theory'. While the paradigms that have influenced Human Geography over the past 30 years have approached these issues in very different ways, it is possible to identify a very fundamental shift in the ways that Human Geographers have come to understand the nature of their discipline. As Ron Johnston puts it, 'before the 1970s, very few Human Geographers identified their discipline as a social science: two decades later, most did' (Johnston, 2003: 51).

One key difference between the epistemologies of the natural and social sciences involves the idea of *human agency*, as Mark Goodwin discusses in Chapter 4. Human beings are not like atoms, molecules or grains of sand, and

CASE STUDY

A lasting tradition of spatial science?

While a majority of Human Geographers would endorse a critique of positivism in terms of agency, meaningfulness, power and positionality, a significant community within the discipline has continued to develop specialist techniques in spatial modelling and analysis that draw directly upon parallel work in the natural sciences. This has been particularly associated with the development and use of geographical information systems (GIS) and other uses of information technology to simulate, analyse and visualize geographical data. For example, the Centre for Advanced Spatial Analysis at University College London continues to work with methods and an epistemology that have recognizable continuities with the spatial science of the 1960s and 1970s, albeit revolutionized by the use of advanced digital technologies (see www.bartlett.ucl.ac.uk/casa).

The rejection of positivism was often accompanied by a knee-jerk reaction against both quantitative data and cartography, but these are not necessarily linked to a particular approach or epistemology. The geographer Danny Dorling has shown the power of numbers and maps as an important element of a critical social science, producing images that make us see the human world differently and question existing social orders (see www.sasi.group.shef.ac.uk/ and www.worldmapper.org).

studying them as if they were unthinking objects takes away what is distinctively human about them. Unlike objects in the physical world, conscious human beings have awareness of themselves and do not always act in the same way in a particular situation. This is a key issue for the social sciences. The idea of agency highlights human decision-making and creativity, but draws particular attention to epistemological issues associated with *language and meaning*. Put simply, in studying the social world we deal with objects and activities that have meaning for people. A social scientist has to pay very careful attention to the terms that people use to describe, interpret and understand their lives. Seen in this way, even apparently obvious terms used in Human Geography, like 'city', 'urban' or 'rural', become problematic. We are in a difficult position if we choose to define a city in terms of some measurable objective criterion (such as size of population) and then fail to grasp the richness and diversity of ways that people use the term in their lives to make sense of particular places and spaces. This emphasis on language and meaning is sometimes associated with the term 'hermeneutics'. While a scientific approach can be said to seek *explanation* through causal laws, a hermeneutic approach seeks *understanding* of self-aware human beings. The study of language and meaning raises difficult issues about how we can claim to understand completely what others mean by particular terms, and about the balance between individual creativity and wider social rules in the use and development of language. Different approaches to these questions (such as humanism and post-structuralism) have led to very different kinds of approaches to Human Geography.

Another distinctive issue for the epistemology of the social sciences is raised by the importance of *power* in human societies. Again, the contrast with the study of the natural world is significant. Human societies are marked by inequalities of political and economic power, and different people and groups within human society have different *interests* in particular forms of social organization. While, for example, we can talk of, say, more powerful electro-magnetic forces, it makes no sense to think of electrons or atoms having an interest in (or benefiting from) the way that the natural world is organized. Critical social sciences are concerned with uncovering orderings of power in human societies; critical Human Geographers have argued that such orderings are both expressed in their spatial arrangements, and that those spatial arrangements actively contribute to making and maintaining structured inequalities (see Jo Little's discussion of these issues in Chapter 2). Very often, critical social science works through questioning simple, superficial notions of a natural order in human societies – for example, that it is 'natural' to find cities divided between rich and poor, or that it is 'natural' for women and men to have very different employment prospects and career histories (this type of argument is sometimes referred to as 'naturalization'). Instead, critical social science asks questions about how a particular form of social ordering developed, in whose interests it works and how it is sustained. One important feature of critical social science is that its epistemology sees connections between knowledge and power, criticizing other approaches for serving dominant interests.

This was an important dimension of the critique of spatial science and positivist assumptions that took place in Human Geography from the 1970s. Radical geographers, particularly those influenced by **Marxist** perspectives, argued that the models of spatial science were merely descriptions of spatial patterns, rather than examinations of fundamental social processes, and that the search for explanation in terms of abstract laws of spatial organization worked to naturalize the social, political and economic ordering of capitalism. (Indeed, Marxists argued that division of academic work into distinctive disciplines like geography, biology or sociology was a powerful way of fragmenting knowledge and of obscuring the fundamental inequalities of contemporary society.) Marxist geography gave precedence to the organization of capitalist society and particularly to class divisions in its analyses. Other forms of critical Human Geography (notably those influenced by some **feminist** and post-colonial perspectives) have rejected this reduction of all inequalities of power to a single dimension, but also work to reveal obscured or naturalized social and spatial orderings. Implicit within all approaches in critical social science, and explicit in most, is the idea that such critical knowledge has an emancipatory role; unlike the archetype of the disinterested, objective scientist, the critical social scientist seeks to change the world that they are studying.

A final feature of social science epistemology acknowledges that the social scientist is also part of the world that they are studying. This has a number of implications. On the one hand, it means that the social scientist has a great deal of almost taken-for-granted knowledge about the ways that the human world works, and how others are likely to think and feel about it. It is this kind of knowledge that makes any attempt to understand others' lives possible. On the other hand, this raises important issues about the extent to which the social scientist is able to speak for others and how their *positionality* shapes their knowledge of the world (see Paul Cloke's discussion in

Chapter 5). Again, different paradigmatic approaches to Human Geography have approached this issue in different ways.

Attention to these issues of agency, meaningfulness, power and positionality has been a feature of the various paradigms that have influenced Human Geography over the past 30 years, during its period as a critical social science. We can think of each of the paradigmatic approaches in the list at the start of this section as a particular response to these epistemological challenges and the fragmented or multi-paradigmatic character of contemporary Human Geography as the result of contestation between these approaches.

SUMMARY

- Human Geography has been influenced by approaches with a range of epistemologies, that is to say, claims about what it is possible to know about the world and how it is possible to express that knowledge. We can draw a broad distinction between the epistemologies of the natural and social sciences.
- For the past 30 years a majority of Human Geographers have regarded their discipline as a critical social science and have rejected positivism (or the direct use of the epistemologies of the natural sciences in the study of human society). Debates about Human Geography's epistemology have been shaped around issues of agency, meaningfulness, power and positionality.
- A significant minority of Human Geographers, particularly those closely involved in the development of GIS, have maintained a research interest in spatial modelling and analysis, using epistemologies and methodologies that retain positivistic elements.

The art of Human Geography

Many new students of Human Geography are surprised at the strength of its links with the humanities, the second of C.P. Snow's 'two cultures'. However, Human Geography has always had some connections with disciplines usually described as humanities, such as history, philosophy and the study of literature. In particular, the sub-discipline of Historical Geography has seen important debates about its epistemologies and methods that have positioned it in relation to the approaches of both history and geography. More generally, throughout the history of their discipline, geographers have pointed to the importance of the creative imagination in the ways that we respond to places. Writing in 1962, H.C. Darby, one of the most influential geographers of his time, argued that geography needed to be both a science and an art to provide a fully meaningful description of places. He quoted these words of Margaret Anderson to promote his ideas: 'no deadly accurate, purely technical description can bring vividly to life a mountain, a great river, or even a climate, can make it our own to love and remember, as an imaginative description by a great writer can do' (Anderson, quoted in Darby, 1962: 3). You may have read a novel set in a place that has also been the subject of a case study in a textbook or academic journal. In conventional terms the novel is fictional while the case study is factual, but it is worth reflecting instead on the different kinds of knowledge and understanding of that place that you have taken from each of them.

What has happened in recent times is that Human Geography's engagement with the humanities has become both deeper and wider. This is in part because of a more general blurring of the distinctions between the humanities and the social sciences. This has sometimes been called the 'cultural turn', a broadly based fusion between critical social science and the traditional focus of the humanities on human creativity and the interpretation of texts, visual imagery, music and other cultural phenomena (Blunt, 2003: 73).

Human Geographers working within this cultural turn have drawn upon ideas of agency, meaningfulness, power and positionality to analyse written texts, such as novels and travel writing, and visual images, particularly landscape art but also other forms of visual representation such as photography, film and television (see Chapters 16, 18 and 19). The 1990s were marked by a concern for the power of **representations** of the world and its geographies. Cultural geographers have used approaches such as **discourse** analysis and **iconography** (both distinctive features of the broader cultural turn in the social sciences) to ask questions about the creation of texts and images, about their formal structure and content, and about the ways that they are received and understood by their wider audiences. These interpretations have often highlighted the ways that texts and images address what we have identified as central themes for Human Geography – the ways that different cultures shape their ideas about the relationships between humans and the natural world, and the significance of space and place in human relations. One important feature of Human Geography's recent engagement with the humanities has been the increasingly explicit way that researchers in other disciplines, such as literary studies or art history, have also explored the significance of these geographical questions.

We can go one step further in thinking about Human Geography as an 'art' as well as a science or social science. An important feature of the humanities is the emphasis that they place not just on the critical analysis of texts, images and other cultural products, but also on active creativity itself. Put simply, the humanities are about 'doing' as well as about commentating and criticizing. Recent years have seen collaborations between Human Geographers and practitioners in the visual and performing arts (see Hawkins, 2011). Commenting on one early example of such collaboration at Royal Holloway, University of London, the geographers Catherine Nash and Felix Driver and the artist Kathy Prendergast pointed to how they disrupt the expected distinctions between the rigorous analytical epistemologies of the academics and the 'imaginative indeterminacy' that supposedly characterizes artistic creativity. Rather than artists simply illustrating the work of the geographers, or the work of geographers providing a starting point (or inspiration) for artists, 'the categories of the "artist" and "researcher/scientist" begin to shift, realign, dissolve and sometimes re-crystallise' (Driver *et al.,* 2002: 8). What comes out of such collaborations is a recognition that there are unacknowledged creative and aesthetic dimensions to most forms of academic work, that escape standard accounts of the epistemologies of science or social science.

CASE STUDY

Between Love and Paradise

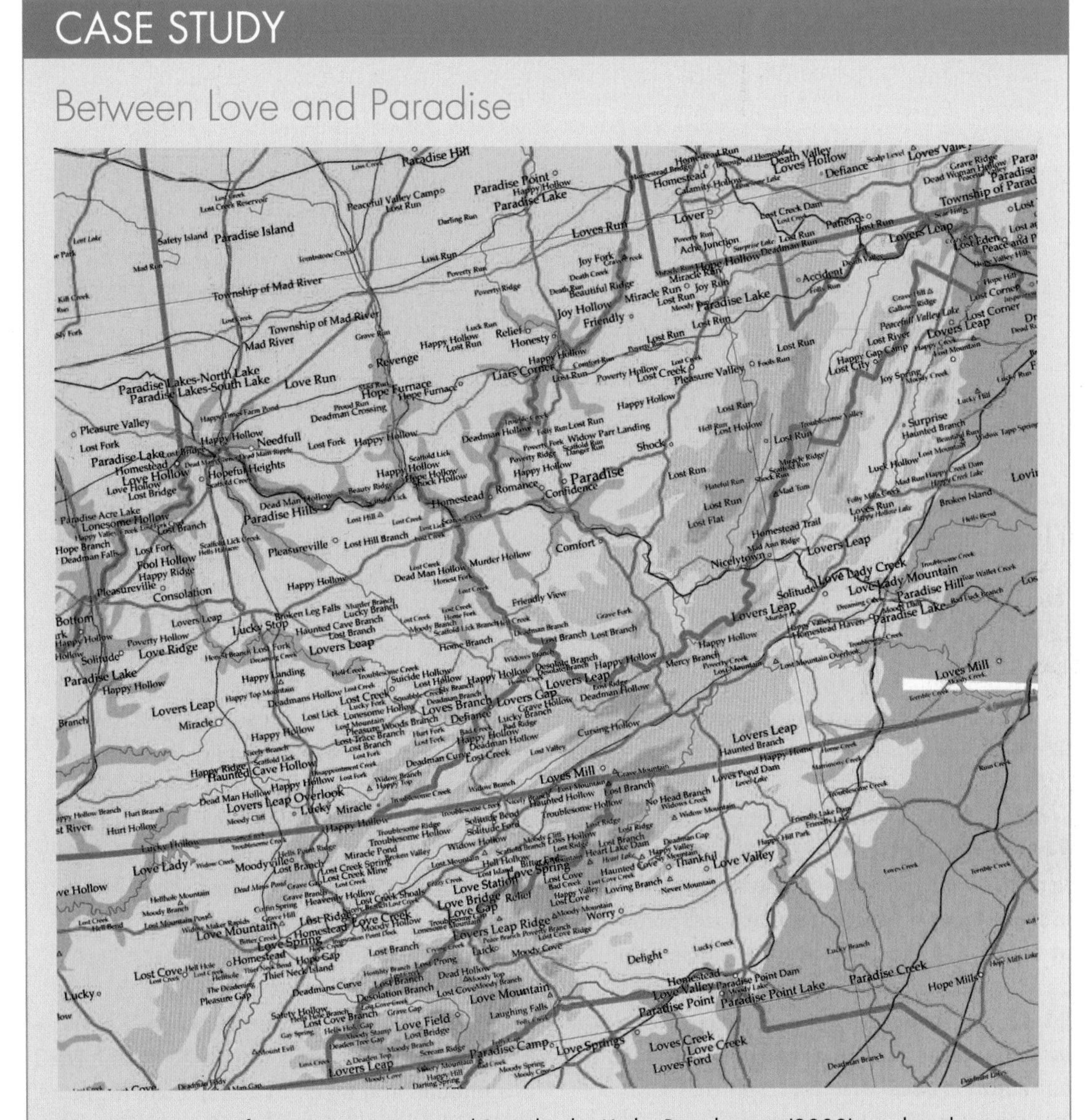

Figure 7.3 Extract from *Between Love and Paradise* by Kathy Prendergast (2002) produced as part of a project bringing together artists and academic geographers. Credit: Kathy Prendergast

Between Love and Paradise is a work produced by the visual artist Kathy Prendergast, during her time as a research fellow in the Department of Geography at Royal Holloway University of London. It is part of a longer series of works called 'mapping emotion', but also draws upon her close collaboration with the cultural geographer, Catherine Nash. Aided by sophisticated cartographic software, *Between Love and Paradise* creates what at first sight is a standard topographic map of the USA. Closer inspection reveals that this is a mapping of emotional place names, used to open up questions about place and identity. In her accompanying commentary, Catherine Nash points to the ambivalence of place names:

> I am interested in the ability of place names to suggest partial narratives of settlement, displacement, migration, possession, loss and authority. I like their taken for granted nature and their burden of meaning. There is something neat, contained and sensible about their reference to location, but there is also something that is elusive and infinite about them. . . . To say they have poetics and politics only begins to trace their diverse registers of meaning.
> (Prendergast and Nash, 2002: 48)

Between Love and Paradise works to stimulate creative speculation about the places it maps. We look at the words on the map and start to fill in stories about hopes fulfilled and hopes unfulfilled for the people that named these places.

This emphasis on 'doing' geography (or what is sometimes described as practice) chimes closely with one of the newer paradigms (again, used in the looser sense suggested earlier in the chapter) adopted by some Human Geographers, **non-representational theory** (see Chapter 50). Put very simply, this approach stresses that we should be as concerned with how we sense and move within the world, as with how the world is represented in art, or interpreted and analysed. This is a move towards a geography that incorporates all of the human senses (including touch, smell, and taste as well as vision) and that regards human beings' relationships with the world both as **embodied** and as working through our emotions and affective responses (see Chapter 49). This approach has also looked towards the arts for inspiration and collaboration. Excellent recent examples of this are the Carrlands and Warplands projects – collaborations between the performance artist Mike Pearson, the musical composer John Hardy and the sound-artist and engineer Hugh Fowler, working with academic geographers from the University of Nottingham. These works are examples of what Hayden Lorimer has usefully described as a 'more-than-representational' approach (Lorimer, 2005). They are 'site-specific', drawing intensely upon the distinctive experiences of a bleak and seemingly empty corner of Lincolnshire in Eastern England and combine words taken from different kinds of writing about these places with music and sounds recorded in those places. Ideally the works should be downloaded as audiofiles and taken to 'the site where you become both listener and participant' (www.carrlands.org.uk).

SUMMARY

- The past 20 years have seen a growing interest in Human Geography's relationship with disciplines and approaches in the humanities. This has been part of a broader engagement (known as the 'cultural turn') between critical social theory and the traditional interests of the humanities in human creativity and the interpretation of texts, images and other cultural products.
- Geographers are also developing interests in the practices associated with the humanities, particularly through collaborations with artists and performers. This is an expression of a broader interest in the significance of embodiment and senses other than just the visual.

Conclusion

Human Geography, then, is a discipline that spans the epistemologies of the sciences, social sciences and humanities, which we can only make sense of as a contested tradition, subjected to episodic changes associated with new approaches (or in a loose sense, new paradigms). As Driver, *et al.* suggest, 'there can be as much difficulty communicating across widely varying research areas, methodologies and theoretical approaches within geography and amongst geographers as across disciplines' (2002: 8). In many ways this is more of an issue for new students bewildered by the range and diversity of what is on offer than for those of us working as academics. We work in academic systems that encourage us to specialize, and that can bring 'star' status to those most closely associated with the development of new perspectives: put bluntly, we have a vested interest in emphasizing the distinctiveness and originality of our own work. Most Human Geographers also have close links with researchers outside the discipline who share their interests and approaches (for example, I'm as likely to present my own work at conferences in cultural studies or modern history as in geography). If there is one thing guaranteed to unite most Human Geographers (albeit in opposition), it is the suggestion that it is possible to find some overarching single epistemological approach to the discipline. Appealing as it may be to someone starting out in geography or Human Geography, this is no longer a discipline that lends itself to a 'modern synthesis'.

There are, I think, a couple of responses to this fragmentation that I'd recommend to a new Human Geographer. The first goes beyond simple enthusiasm about the range of courses on offer, to stress that Human Geography offers the chance not just to study different topics, but to think in different ways about the world. While the diversity of geography can encourage a clash of (intellectual) cultures – that sterile division between 'arty-farties' and 'nerds' – it can also, much more positively, encourage critical reflection on the strengths and limits of different approaches. In the terminology used in this chapter, this discipline gives an unparalleled chance to experience different epistemologies in practice. My second recommendation stems from Bruno Latour's observations about the limitations of conventional academic disciplines and ways of thinking. (See the discussion of the different ways of thinking about the relationships between the human and non-human world in Chapter 3.) Latour suggests that the most significant issues of the contemporary world are **hybrids**, which escape and cross the expertise and conventional epistemological frameworks of academic disciplines. (We might add that they certainly escape and cross the various specialisms and approaches within Human Geography.) Latour asks us to think about the limits to specialism, by following the journeys made by objects in the world (an AIDS virus, a chlorofluorocarbon molecule, a rainforest tree) – see box. This strikes me as a profoundly geographical observation, not just in the way that it invites us to trace the paths of such objects through time and space, but also to look for where and how they appear (and disappear) in seemingly self-contained approaches to economic geography, climate change, geopolitics, development, cultural

CASE STUDY

Bruno Latour: Indisciplined thinking

On page four of my daily newspaper, I learn that the measurements taken above the Antarctic are not good this year: the hole in the ozone layer is growing ominously larger. Reading on, I turn from upper-atmosphere chemists to Chief Executive Officers of Atochem and Monsanto, companies that are modifying their assembly lines in order to replace the innocent chlorofluorocarbons, accused of crimes against the ecosphere. A few paragraphs later, I come across heads of state of major industrialized countries who are getting involved with chemistry, refrigerators, aerosols and inert gases. But at the end of the article, I discover that the meteorologists don't agree with the chemists; they're talking about cyclical fluctuations unrelated to human activity. So now the industrialists don't know what to do. The heads of state are also holding back. Should we wait? Is it already too late? Toward the bottom of the page, Third World countries and ecologists add their grain of salt and talk about international treaties, moratoriums, the rights of future generations and the right to development.

The same article mixes together chemical reactions and political reactions. A single thread links the most esoteric sciences and the most sordid politics, the most distant sky and some factory in the Lyon suburbs, dangers on a global scale and the impending local elections or the next board meeting. The horizons, the stakes, the time frames, the actors — none of these is commensurable, yet there they are, caught up in the same story . . .

The smallest AIDS virus takes you from sex to the unconscious, then to Africa, tissue cultures, DNA and San Francisco, but the analysts, thinkers, journalists and decision-makers will slice the delicate network traced by the virus for you into tidy compartments were you will find only science, only economy, only social phenomena, only local news, only sentiment, only sex. Press the most innocent aerosol button and you'll be heading for the Antarctic, and from there to the University of California at Irvine, the mountain ranges of Lyon, the chemistry of insert gases, and then maybe to the United Nations, but this fragile thread will be broken into as many segments as there are pure disciplines.

(Latour, 1993: 1)

geography and so on. What courses in geography offer (often as much by accident as design) is a chance to trace and follow these journeys, and to test the limits of conventional ways of thinking.

DISCUSSION POINTS

1. Identify three pieces of geographical writing (perhaps by looking at articles from different journals) that you would consider to be written from within a scientific, social scientific and a humanities tradition. For each, provide a short description of the ways that they approach their subjects: that is, instead of summarizing their contents, try to distinguish between them in terms of how they organize their material, the ways that they relate theories to evidence and the kind of language that they use to make their case.
2. If you have access to a library or an online journal service, look briefly at the contents pages and abstracts of all Human Geography articles for 1970, 1980, 1990, 2000 and 2010 in these journals: *Area; Transactions of the Institute of British Geographers; Annals of the Association of American Geographers*. Does your survey provide any evidence for paradigm shifts in the approaches, theories and methods of Human Geography?
3. Look again at the text extract by Bruno Latour (see box). In this passage Latour uses the examples of the AIDS virus and CFCs to show the limits of conventional disciplines. Identify another example of something in the world that escapes or crosses disciplinary boundaries. Draw up a list of the different ways that it is possible to have knowledge of this object (for example, how it might be understood and interpreted by a biologist, a sociologist, a historian and an artist.) What do you think a Human Geographer can bring to the study of this object?

FURTHER READING

Holloway, S, Price, S. and Valentine, G. (2008) *Key Concepts in Geography* (2nd edn.). London: Sage.

The best next step in thinking about these issues is to look at the four essays on the disciplinary nature of geography in this book: Mike Heffernan looks at the long-term development of the discipline of geography; Keith Richards, Ron Johnston and Alison Blunt provide excellent introductory accounts of geography's relationship with the sciences, social sciences and humanities.

Nayak, A. and Jeffrey, A. (2011) *Geographical Thought: An Introduction to Ideas in Human Geography*. London: Pearson

This is an excellent text for thinking about the main approaches to Human Geography and is particularly good at showing the continuing relevance of different perspectives for the discipline today. It has a broadly chronological structure, covering Human Geography's relationship with empire, its ambitions to become a spatial science, and the turns towards Marxism, feminism and various forms of cultural theory. The chapters on post-colonialism, critical geopolitics and 'Emotions, Embodiment and Lived Geographies' are particularly good as a kind of road map of key approaches in contemporary Human Geography.

Johnston, R. and Sidaway, J. (2011) *Geography and Geographers: Anglo-American Human Geography Since 1945* (7th edn.). London: Arnold.

The most comprehensive account of modern changes in the discipline of Human Geography. (As the title suggests, the account is primarily limited to change in the UK and USA.) It has a useful discussion of the nature of academic disciplines and the extent to which it is useful to draw upon the notion of paradigms in interpreting change in Human Geography. Johnston wrote the first edition of the book in the 1970s and has described contemporary Human Geography as an 'abundance of turbulence'. You can get some measure of that turbulence and a rough-and-ready sense of the direction of travel in Human Geography by comparing the changing contents pages of the seven editions of this book (from 1979, 1983, 1987, 1991, 1997, 2004 and 2011).

Hawkins, H. (2011) Dialogues and doings: Sketching the relationships between Geography and art. In: *Geography Compass, 5/7*, 464–78.

A useful summary of the developing relationship between geographers and artists that stresses new approaches to landscapes and the political significance of art in urban change.

Driver, F., Nash, C., Prendergast, K. and Swenson, P. (2002) *Landing: Eight Collaborative Projects Between Artists and Geographers*. Egham: Royal Holloway, University of London.

Reflections on the relationship between art and geography associated with one of the pioneering collaborative projects from the early 2000s. Examples of the connections between Human Geography and practitioners in the visual and performing arts can be found in the regular 'Cultural geographies in practice' section of the journal *Cultural Geographies*.

WEBSITES

www.bartlett.ucl.ac.uk/casa

Centre for Advanced Spatial Analysis at University College London

www.sasi.group.shef.ac.uk/

Social and Spatial Inequalities group at The University of Sheffield.

www.worldmapper.org

A collection of world maps.

www.carrlands.org.uk

The Carrlands Project: collaborations between a performance artist, musical composer and sound-artist/engineer working with academic geographers from the University of Nottingham.

www.nottingham.ac.uk/landscape/index.aspx

The Warplands project: part of a major exploration of the arts of landscape and environment funded by the UK's Arts and Humanities Research Council and directed by the cultural geographer Stephen Daniels.

CHAPTER 8
EXPLANATION–UNDERSTANDING

Rob Kitchin

Introduction

Undertaking research is the key means through which we find out about and assess the world around us. Research can take many different forms, and there are numerous textbooks that detail the mechanics and techniques for how effective and valid research should be undertaken. This chapter, rather than focusing on the nuts and bolts of conducting research – how to go about generating and analysing data in practice – examines instead how research, and by implication all academic work, is conceptually framed. Conducting research, it is important to note, is not simply a set of standard practices that, if implemented correctly, will provide a valid understanding or explanation of the world; it consists of more than a set of techniques that are mechanically applied by rule and rote, selected for general utility, convenience and expediency. Instead, empirical research practices – generating, analysing, interpreting, writing and so on – are contextually embedded within a philosophically informed framing that works across four inter-related domains: ontology (what one can know about the world), epistemology (how one can know it), methodology (how one can measure it) and ideology (what one does with the knowledge produced) (Hubbard *et al.,* 2002).

Taken together, **ontology**, **epistemology**, methodology and **ideology** provide the parameters of an academic world view and the means through which to conceptually frame how one approaches and undertakes a piece of research and how one interprets and makes sense of the research findings. They shape the kinds of questions one might legitimately ask, how those questions are operationalized and how data are gathered, analysed and interpreted. In addition, they frame positions and debates about issues such as research ethics, positionality, validity and the politics of research. Unfortunately, when discussing the conceptual bases, practices and practicalities of conducting research this wider philosophical framing often gets reduced to the level of methodology and, in particular, to the kind of data that is generated and analysed. Indeed, it is not uncommon to hear some researchers describe themselves as either performing **quantitative** or **qualitative** research, or even being a quantitative or qualitative geographer (see DeLyser *et al.,* 2010; Fotheringham *et al.,* 2000). In other words, they define themselves by the methodology they use, rather than by the focus of their research (e.g. a political, economic, cultural geographer) or by their wider theoretical viewpoint (e.g. a positivist, feminist, Marxist, poststructuralist geographer). This division between quantitative and qualitative research maps onto the notion that quantitative methods seek *explanation* (where quantitative data analysis explains causal relationships between variables) and qualitative methods create *understanding* (where qualitative data analysis produces insight and reveals meaning).

Such a description or identity is misleading because it suggests that the kinds of questions researchers ask, how they choose to try and answer them and their identity as a researcher, can simply be reduced to the kind of data they generate and how they analyse them – that one can meaningfully define themselves as a quantitative or qualitative geographer. This is not the case. Methodology is how research is operationalized; how we seek to ask and answer questions. The methodology devised, and the specific methods used, are shaped by our world view as to what is the most appropriate and valid way to make sense of the world. In other words, the methodology stems from the ontological, epistemological and ideological tenets of a researcher's world view, not vice versa. For example, a central tenet of **positivism**, an approach that seeks to apply the scientific approach used in the natural sciences to the social world, is that science should not seek to answer metaphysical questions because they are empirically unknowable and unverifiable (see Kitchin, 2006). Such a question would be: is there a god? This is a question of faith and cannot be definitely proven through scientific measurement. Positivists would also be wary of subjective information such as opinions, attitudes, values, ethics, principles and beliefs, again because they are difficult to analytically measure and verify, rather preferring to focus on observed behaviour to explain actions. Feminists, on the other hand, who are interested in how power is mobilised and circulates within society, would have no difficulties in dealing with such information, arguing that it can be validly and rigorously examined (Seager and Nelson, 2004). They would also be interested in the power dynamics within the research process itself between researcher and researched and seek to find methods that are sensitive to imbalances in power such as participatory approaches or which openly acknowledge the **positionality** and situatedness of the researcher (how they are approaching the research foci theoretically, politically, ethically) (Kindon *et al.*, 2007; Rose, 1997). In other words, to characterize research as either being quantitative or qualitative in nature – to be about either explanation or understanding – then is somewhat deceptive.

SUMMARY

- Conducting research is more than the rote application of standard methodological practices.
- Research is contextually embedded within a philosophically informed framing.
- Ontology, epistemology, methodology and ideology provide the parameters of philosophical thought and shape choices over the methodology and techniques adopted.

The qualitative/ quantitative divide

In broad terms, quantitative data consist of numeric information. The information gathered is either extensive and relates to physical properties of phenomena (such as length, height, distance, weight, area, volume, etc.) or representative and relates to non-physical characteristics of phenomena (such as social class, educational attainment, social deprivation, quality of life rankings, etc.). In geography, a combination of extensive and representative quantitative information can be gathered to explain social and economic issues. These data can be nominal (placed into categories), ordinal (ranked in relation to each other), interval (measured on a fixed, continuous scale) or ratio (measured on a scale with a fixed zero origin) in character (Kitchin and Tate, 1999). These forms of data can be analysed using descriptive or inferential statistics or be used as the inputs to predictive and simulation models. Descriptive statistics provide summary overviews of the trends within a data set and include techniques such as standard deviations, graphs, histograms, pie charts, maps and so on. Inferential statistics seek to determine statistically whether there are relationships and patterns within the data, whether the data differs significantly from other groups, or to make wider inferences about a larger population based on the sample. Quantitative data in the social sciences is mostly generated through surveys and questionnaires, much of which is derived through instruments such as censuses, household surveys, passenger surveys, political polling, etc., or is generated from large databases such as those held by government departments and health and financial institutions. More recently it can be generated from sensor and scanner technologies.

Qualitative data is non-numeric information. It can consist of text, images and sounds, including literature, diaries, policy documents, interview transcripts, photographs, art, video, movies, and music (Hay, 2010). While these data can be converted into quantitative data through classification, much of the richness of the material is lost through such a translation process. Qualitative data analysis then generally seeks to work with the original materials, seeking to tease out and build up meaning and understanding, using analytical techniques such as content analysis and deconstruction. Qualitative data in the social sciences is often generated through interviews, focus groups, observation, ethnography and participatory methods, or is accessed through archive collections, and generally consists of case studies focused on particular individuals, communities and places.

There is little doubt then that quantitative and qualitative data are different in nature: one being numeric, the other non-numeric. They are also generally understood to differ in terms of parameters of data generation, so that quantitative data is gathered by prescription, has large sample sizes, concentrates on incidence and frequency and focuses on populations, while qualitative data is gathered personally, has small sample sizes, concentrates on concepts and categories and focuses on individuals (see Table 8.1). The fundamental issues, however, are whether the type of data used in a study defines the approach taken, whether the two kinds of data are used in mutually exclusive ways and the ways in which the two broad data types are used to create and reproduce cleavages in geographical praxis.

As noted above, some geographers define themselves as quantitative or qualitative researchers, which suggests some kind of mutually exclusive relationship – you either

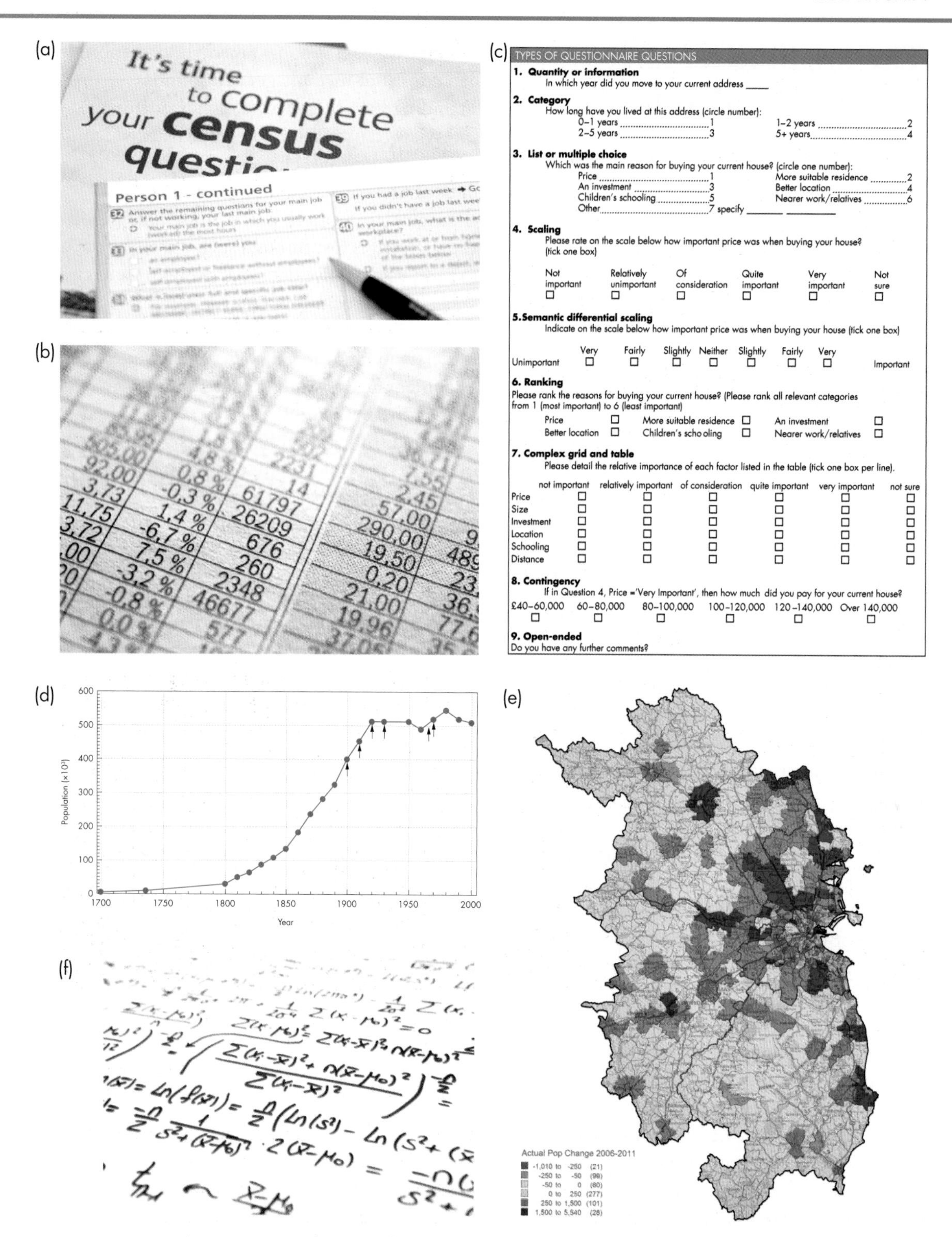

(c) TYPES OF QUESTIONNAIRE QUESTIONS

1. Quantity or information
In which year did you move to your current address ____

2. Category
How long have you lived at this address (circle number):
0–1 years ....1 1–2 years 2
2–5 years ....3 5+ years 4

3. List or multiple choice
Which was the main reason for buying your current house? (circle one number):
Price ....1 More suitable residence 2
An investment ....3 Better location 4
Children's schooling ....5 Nearer work/relatives 6
Other ....7 specify ____ ____

4. Scaling
Please rate on the scale below how important price was when buying your house? (tick one box)

Not important	Relatively unimportant	Of consideration	Quite important	Very important	Not sure
☐	☐	☐	☐	☐	☐

5. Semantic differential scaling
Indicate on the scale below how important price was when buying your house (tick one box)

	Very	Fairly	Slightly	Neither	Slightly	Fairly	Very	
Unimportant	☐	☐	☐	☐	☐	☐	☐	Important

6. Ranking
Please rank the reasons for buying your current house? (Please rank all relevant categories from 1 (most important) to 6 (least important)
Price ☐ More suitable residence ☐ An investment ☐
Better location ☐ Children's schooling ☐ Nearer work/relatives ☐

7. Complex grid and table
Please detail the relative importance of each factor listed in the table (tick one box per line).

	not important	relatively important	of consideration	quite important	very important	not sure
Price	☐	☐	☐	☐	☐	☐
Size	☐	☐	☐	☐	☐	☐
Investment	☐	☐	☐	☐	☐	☐
Location	☐	☐	☐	☐	☐	☐
Schooling	☐	☐	☐	☐	☐	☐
Distance	☐	☐	☐	☐	☐	☐

8. Contingency
If in Question 4, Price ='Very Important', then how much did you pay for your current house?

£40–60,000	60–80,000	80–100,000	100–120,000	120–140,000	Over 140,000
☐	☐	☐	☐	☐	☐

9. Open-ended
Do you have any further comments?

Figure 8.1 (a)–(f) Quantitative data. Credit: (a)–(c) iStockphoto; (d) JeremyA (Wikimedia Creative Commons); (e) Rob Kitchin; (f) iStockphoto

Figure 8.2 Qualitative data. Credit: (a) Christine Leerink; (b) iStockphoto; (c) Jacqueline Veissid/Getty Images; (d) Runner 1616 (Wikimedia Creative Commons); (e) iStockphoto

work with quantitative or qualitative data, but not both. Indeed, the quantitative/qualitative distinction has become shorthand for a whole set of binaries that appear to operate with respect to research praxis as set out in Table 8.1. These binaries concern a whole set of inter-related issues which frame how academic research is conceptually and practically located from the type of data, sample size, form of data generation, focus, scope, objective,

Quantitative	Qualitative
Data are numbers	Data are words, pictures and sounds
Data gathered by technology or prescription	Data gathered personally
Large sample sizes	Small sample sizes
Incidence and frequency	Concepts and categories
Populations	Individuals
Explanation and prediction	Meaning and understanding
Scientific	Humanistic
Nomothetic	Idiographic
Realistic	Idealistic
Deductive	Inductive
Objective	Subjective
Functionalist	Interpretative
Generalization	Extrapolation
God's eye view	Situated
Inquiry from the outside	Inquiry from the inside
Subjects/objects	Participants
Artificial	Natural
Macro	Micro
Generality	Specificity
Society	Self

Table 8.1 The qualitative/quantitative divide. Adapted from Kitchin and Tate (1999)

philosophical framing and rationale, to method and technique.

In Table 8.1 the distinction between qualitative and quantitative research is mapped onto a conceptual framing focused on methodology and methodological considerations. In very broad terms, the distinction between quantitative and qualitative research is seen as one of *explanation and prediction* versus *meaning and understanding*. This divide is captured by a whole series of interlinked binaries that are worth elaborating a little in order to make it clear how the distinction is being cleaved:

- The scientific versus humanistic divide concerns a distinction between an approach that is specifically designed to capture the complexities of people and society and one that seeks to take the procedures and practices of natural science and apply them to society. This is captured somewhat in the divide between an idiographic and nomothetic approach (Schaefer, 1953). An idiographic approach focuses on the specificity and uniqueness of individuals and places whereas a nomothetic approach seeks to determine generalizable laws to explain phenomenon. The quantitative revolution in Geography, for example, is often framed as an attempt to shift Geography from a discipline interested in understanding unique regions, landscapes and cultures, to one that discovered spatial laws that held across places and people (Johnston and Sidaway, 2012).
- Idealistic versus realistic approaches is a divide between research that seeks to understand and take account of the metaphysical, spiritual and non-materialistic aspects of life and approaches that concentrate on empirically measurable facts only (Kitchin and Tate, 1999).
- Qualitative studies often adopt an inductive approach to interpretation and theory building; that is, they generate the data and then use these to build a theory as to what is observed. Quantitative studies, in contrast, often use a deductive approach by constructing a theory and then testing whether that theory has any validity by examining the veracity of hypotheses.
- Subjective versus objective is framed as the difference between a point of view (such as values, opinions, beliefs) and observable, measurable facts, and between the researcher shaping the research process and a neutral, value-free analysis and interpretation (Rose, 1997). Qualitative data generation, because of its open-ended format and the kinds of questions asked, is thought by many to produce subjective data, but also to be open to subjectivity on behalf of the researcher (that is, qualitative methods are more open to the interpretation of the researcher). On the other hand, quantitative methods, it is argued, generate and analyse factual information and the use of statistical techniques (both descriptive and inferential) provides answers free of researcher bias and is therefore more objective.
- Related to this, situated knowledge is that which recognizes the experience, context and positionality of the researcher in investigating a topic, whereas a 'God's eye view' is the idea that we can stand outside of our personal history, beliefs and experiences when conducting research and interpreting findings – that we can rise above ourselves and see the world for what it is, free of any influences.
- Inquiry from the inside suggests that it is possible to become close to a group of people and to see the world from their perspective; that it is possible to conduct studies with and for people, rather than an inquiry from the outside that is detached, disembodied and is a study of a group. One consequence is that the people investigated within qualitative studies are often thought of as participants, whereas in quantitative studies they are viewed as subjects or objects. A critique of the latter is that people are effectively reduced to an essence that denies their complex messiness; they are simply a number in the analysis, not a person (hence the critique that quantitative geography is 'peopleless') (Hubbard *et al.*, 2002).
- Extrapolation versus generalization is related to sample sizes and the representativeness of

the data. Qualitative studies often have quite small sample sizes, so drawing conclusions from them with respect to large populations involves extrapolation. With quantitative data it is often aggregated for analysis, so it tends to generalize individual data by hiding and reducing data variability.

SUMMARY

- Quantitative data is numeric and can be analysed using descriptive or inferential statistics; qualitative data is non-numeric and is generally analysed discursively.
- A whole series of binaries relating to philosophical framing and research practices are often mapped onto the supposed quantitative/qualitative divide.
- Quantitative research is seen as producing explanation and prediction, whereas quantitative research produces meaning and understanding.

A false dualism

Discussed in this way, it is easy to see how the cleavage between research that generates and analyses quantitative and qualitative has been wedged apart into two seemingly mutually exclusive camps, one centred on providing explanation, the other on creating understanding. It is important to note, however, that this cleavage represents something of a false dualism and the binaries set out in Table 8.1 hide a huge amount of messiness. Even at the level of data, the divide is artificial in the sense that qualitative data can be made quantitative through codification, classification and digital rendering, and quantitative data can be described textually as narrative and visually as images, graphs and maps. And if the characteristics of qualitative data can be made to match those ascribed to quantitative data generation and vice versa, then it is relatively easy to see that the distinction between the other binaries related to data generation can also become blurred and indistinct.

While qualitative data are often gathered personally through interviews, focus groups, and ethnographic fieldwork, it can equally be derived from national archives, large photo and film libraries and technically produced documents such as parliamentary minutes, where the researcher had no personal involvement in generating or classifying the data or applying its metadata. Likewise, quantitative data can be generated through interviews, diaries, ethnographic research and so on, and not necessarily through a highly prescriptive technique (for instance, it might be derived from open-ended interviews where information is subsequently categorized). Equally, qualitative studies can have very large sample sizes, deal with populations such as cultural and linguistic groups or communities and deal with issues of incidence and frequency, as with health geography concerning illnesses and treatments or space–time diaries. Quantitative-based projects can have quite small sample sizes, deal with categories and concepts and do modelling around individual lives.

More broadly, as discussed in the introduction, it is the philosophical tenets of a worldview that shape research praxis, not data types. Geography as a discipline is theoretically plural and diverse with respect to the ontological, epistemological and ideological positions adopted. Indeed, there are numerous conceptual approaches practised within Geography and across the social sciences (Johnston and Sidaway, 2012). The extent to which qualitative and quantitative methodology is used in a mutually exclusive manner varies by philosophical approach, with some conceptual framings favouring an interpretive emphasis that seeks to generate meaning and understanding, others favouring a functionalist approach that aims to explain and predict, and a few that seek to do both, using both quantitative and qualitative methods in combination with each other.

Sometimes there can even be disagreements within a family of related approaches. For example, some feminists reject quantitative approaches as being masculinist and reductionist, producing peopleless geographies that not only fail to take account of the politics and power that shape everyday life but reproduce such relations, whereas others argue that quantitative data and techniques can be used effectively and sensitively to illustrate the effects of patriarchy on men and women's lives. This internal conflict between researchers who share the same ideological goals – to dismantle patriarchal social relations to create a more just society – was aired in the 'Should Women Count?' debate (see *Professional Geographer* 47(4): 426-466). This debate consisted of five papers that each explored the extent to which quantitative methods can be used to undertake research that adheres to feminist ideals and principles; that conform to feminist ways of seeing, doing and knowing the world. The argument forwarded across the papers is that women can and should count (in quantitative method terms), but feminists should be aware and open about how the approach they have taken has shaped the questions asked and how they have been answered.

To illustrate the relationship between conceptual thought and methodology further, Table 8.2 outlines how eight different approaches used within Geography would generally conceptually frame and investigate issues of poverty (Kitchin and Tate, 1999). What the table makes clear is that each philosophical position approaches poverty quite differently and this has a profound effect on the types of question they ask and how they seek to answer them. In other words, in order to operationalize poverty research, each philosophy requires certain kinds of data to be generated and for that data to be analysed and interpreted in a particular way in order to comply with the ontological and epistemological principles of that approach. For example, empiricism relies on the weight of facts, positivism statistically tests the relationship between variables, phenomenology tries to reconstruct the life world of people who are poor, Marxism seeks to uncover the capitalist structures that shape life chances, and so on. It should be noted that while each approach pushes a researcher towards a particular kind of data, what is most important is that the data is analysed in such a way that it does not break the ontological and epistemological assumptions of that philosophy. For example, a positivist can use qualitative data in their analysis, but only if it relates to non-metaphysical matters and is analysed scientifically through statistics. In general, this means avoiding metaphysical questions (that is, questions that cannot be empirically measured and verified) and converting qualitative data into quantitative data through classification and codification.

Philosophy	How poverty is researched	Main methodologies
Empiricism	Facts and statements about poverty are collected and presented for interpretation by the reader (e.g. indices of poverty – social welfare status, housing tenure). Data are understood to 'speak for themselves'	Presentation of experienced facts and statements; descriptive statistics
Positivism	Poverty is explained through testing a hypothesis by collecting and scientifically testing data related to poverty (e.g. statistically testing whether poverty is a function of educational attainment)	Surveys, questionnaires, secondary analysis of other quantitative data sets Would rarely use qualitative methods/data
Phenomenology	To understand poverty it is necessary to reconstruct the lifeworld of people who are poor (e.g. we need to try and see the world through the eyes of poor people). This might be attempted by talking to them about their life experiences	In-depth interviews; ethnography Would rarely use quantitative methods/data
Existentialism	Poverty is understood by trying to gain insight into how people who are poor come to know, ascribe meaning and interact with the world (e.g. interviewing poor people about how they decide how much money they spend on different things)	In-depth interviews; ethnography; participant observation Would rarely use quantitative methods/data
Pragmatism	Poverty is understood by observing how individuals in society interact to produce conditions which sustain destitution (e.g. examining whether poor people remain poor because they live in a cycle of crime, under-education, low self-esteem)	Ethnography; participant observation
Marxism	Poverty is explained through the examination of how society is structured for the purposes of capital accumulation (e.g. we need to examine how the interests of capital are served by retaining unskilled, low wage jobs rather than distributing fully corporate profit)	Observation; quantitative analysis of secondary sources; deconstruction of policy documents; interviews

Philosophy	How poverty is researched	Main methodologies
Post-structuralism	Poverty is understood through an examination and deconstruction of complex (and often contradictory and paradoxical) exclusionary practices of society, as expressed through discursive and material practices (e.g. deconstructing cultural norms, myths and practices that reproduce exclusionary processes which seek to marginalize poor people from material wealth)	Observation; deconstruction of documents and practices Would critique and deconstruct quantitative data/methods but would rarely use them
Feminism	Poverty is understood by examining the ways in which power works to create and reproduce certain social and spatial relations (e.g. examining the unequal access to work and wealth between men and women and the role of patriarchy in reproducing such relations)	Interviews; focus groups; ethnography; participatory methods; surveys; questionnaires; analysis of secondary data sets

Table 8.2 Different approaches used to conceptually frame and investigate issues of poverty. Adapted from Kitchin and Tate (1999)

Deciding how one researches and makes sense of poverty then – and indeed any other issue or phenomenon – is far from simply choosing whether one uses qualitative or quantitative methods of data generation and analysis. Instead, a much more fundamental thinking needs to occur regarding how one thinks the world works and how best to formulate meaningful questions and answer to those questions. For some researchers this means pursuing approaches that prioritize description, interpretation and meaning (such as empiricism, phenomenology, post-structuralism) and for others adopting approaches that seek explanation and prediction (such as positivism and Marxism).

In many instances, as detailed in Table 8.2, it might mean generating and analysing both quantitative and qualitative data in order to try and produce both explanation and meaning. A mixed methods approach, combining quantitative and qualitative methods, is a perfectly legitimate way of generating and analysing data within some philosophical approaches. For example, in the 'Should women count?' debate, one of the principle conclusions of the participants was that a mixed methods approach provides a middle way forward that can remain true to the principles and values of feminism.

Such a mixed method approach is also common in Marxist, realist and empiricist approaches. In the latter case, data are seen to speak for themselves and knowledge about the world conveyed through the weight of evidence. Here, by combining insights into quantitative data with that of qualitative data it is hoped that both a broad and detailed understanding of an issue can be conveyed. So, for example, if one is interested in issues of migration one might start by conducting a broad-based analysis of patterns of migration to provide a generalized understanding of migratory flows. This might involve an analysis of the census or undertaking a large scale survey of migrants. The next step might be to interview in depth a smaller sample of migrants in order to try and gain a deeper understanding of the reasons for and experiences of migration. In other words, the quantitative data is used to frame and contextualize the qualitative element of the research. The process can also work in reverse. So, for example, in-depth interviews are undertaken to determine the main reasons for and experiences of migration. The findings are then used to design a larger survey which is distributed to a much larger sampler. In both cases, quantitative and qualitative methods work in concert with each other to enhance insights. In the former case, the broad picture is used to help frame and deepen understanding. In the latter case, in-depth understanding provides the basis for trying to establish wider explanation.

SUMMARY

- The quantitative/qualitative divide is somewhat of a false dualism.
- The philosophical world-view shapes the methodology and form of data generated, not vice versa.
- Quantitative approaches fit the assumptions of some philosophies while qualitative approaches more suit others and a mixed methods approach, utilising quantitative and qualitative approaches, is possible.

Conclusion

Conceptually, research is sometimes framed as producing either explanation or understanding, and these are mutually exclusive. This division is mapped onto a crude division between quantitative and qualitative data and methods, and a range of associated binaries such as scientific/humanistic, nomothetic/ idiographic, deductive/inductive, objective/ subjective, functionalist/interpretative, generalization/ extrapolation and so on. The argument in this chapter has been that binaries are somewhat false and that the methodology and purpose of research is not defined by data type and method, but rather by a wider philosophical framing with respect to ontology, epistemology and ideology.

While some philosophical approaches largely foreclose the use of quantitative or qualitative methods, others permit the use of both, and it is possible to conduct research that seeks both understanding and explanation of a phenomena. What is important then as a researcher is to develop a coherent worldview and a sense of one's ontological, epistemological and ideological positioning and to use this to frame the questions one can legitimately and validly ask, how one asks them and for what purpose. This is no simple task and involves serious engagement and reflection upon philosophical thought. This tends to be an evolving process as ideas are mulled over, teased out and tested. It is, however, a vital process for determining how one's research is practised and defended.

DISCUSSION POINTS

1. Do you consider yourself a qualitative geographer, a quantitative geographer, neither or both? On what basis do your rationalize your choice?
2. What philosophy underpins your worldview and how does this shape your methodological approach to research?
3. In what ways do quantitative and qualitative data and methods of analysis differ from each other? What do those differences mean with respect to what a research study might discover?
4. To what extent can qualitative and quantitative methods be used in conjunction with each other?

FURTHER READING

Aitken, S. and Valentine, G. (2006) *Approaches to Human Geography*. London: Sage.

A general introduction to different philosophical approaches used in geographical enquiry.

Kitchin, R. and Tate, N. (1999) *Conducting Research in Human Geography*. Harlow: Pearson.

A general introduction to the theory and practice of conducting research in Human Geography.

Multi-method research in population geography. *Professional Geographer* 51(1), 40–89.

A collection of five papers that examine the use of mixed and multiple methods in migration and population geography.

Should women count? *Professional Geographer* 47(4), 426–66.

A collection of five papers that debate feminist theory and the use of quantitative methods.

CHAPTER 9
REPRESENTATION–REALITY

Mike Crang

Introduction

Geographers spend a lot of time making representations of reality. They do so using both words and images – be they pictures, charts, graphs or even maps. Representing is a profoundly geographical act – it is making presenting something from elsewhere to an audience. It is also something that has been regarded with philosophical suspicion for centuries. The classical philosopher Plato, for instance, provided a critique in his analogy of the cave – where a group of prisoners were chained for all their lives facing the wall of a dark cave with a great fire behind them and all they could see were the shadows cast on it as people passed in front of the fire. For these prisoners, the shadows or representations become the real world, and they mistook representation for reality. If they were brought out into the sunlight they would be baffled by the colours – taking them as 'unreal'. Philosophers have gone on to argue then that we are all precisely confined to a 'prison chamber' of representation – be it images or language. Representation is then often seen as obscuring or failing to capture the reality of the world. Representation is thus often seen as a problem. The best form of representation reflects exactly what is out there, like a mirror. This then is a correspondence model of truth – that good representation reflects the world. Even when we are talking about textual representations, this visual sense of 'mirroring' and picturing comes through very strongly as underpinning our sense of both what is true and how truthful representation works. The philosopher Martin Heidegger suggested that a crucial shift in how western people experienced the world was when it became conceived as a picture. The world became seen as separate and detached from the viewer. Up until then, Heidegger argued, people had seen themselves as part of the world. This change can be linked to the rise of new techniques for producing images, such as the camera obscura. At its simplest, this is a darkened room with a hole in one wall, while the wall opposite forms a screen on which an image of the outside world appears – like a large pin-hole camera. Observers could draw directly from life. The truthfulness of these images could be assessed by their direct correspondence to the outside world. It produced this image through the seclusion and detachment of the observer, separated from the world. This way of producing images thus became a model of truth that saw a world that could be known and represented through a detached observer. In this model transparency is the best way to represent the world and anything less gets in the way and misleads people. Nor is this idea of truthful knowledge confined to geography: Clifford Geertz comments of scientific writing in general 'that "symbolic" opposes to "real" as fanciful to sober, figurative to literal, obscure to plain, aesthetic to practical, mystical to mundane, and decorative to substantial' (in Baker 1993: 10). Image is taken to imply the opposite of real, in a series of binary pairs where two terms are opposed and we are trained through years of education to value the second terms on Geertz's list.

This chapter will suggest that the relationship of representation and reality is rather more complex than this – where one way of viewing creates a truthful depiction and others are deceptive. Although we might look at how images refract, reflect and alter our knowledge of the world and we know that images can be deliberately promoted, massaged or altered to achieve desired ends, they all go into forming the ideas and understandings of the world. It is on this basis that people make choices and act. Representations have impacts on the world in terms of how they shape action by people – they are performative, not just reflective of reality; they shape the world as well as depict it. There are geographies of images in terms of what areas they do, or do not, show and how they move through society. Moreover, Geographers produce images of the world, so we need to re-evaluate what role images play in geographical knowledge. Changes in how the world is seen can tell us something about those doing the seeing. Geographers have studied '**mental maps**' to see not just how these diverge from 'truthful' and detached representations (birds-eye view maps of layouts) but also how they show the local 'reality.' They show how their response to the city is shaped by social status, by race and fear. Rather than contrasting the 'perception' of risk to its 'real' statistical likelihood, or tourist images as glamorizing real places, or facades and regenerated areas as images concealing real economic processes or literature as a 'subjective' representation of a region, we might look at how each of those ways of representing constructs different 'truths' (plural) about a place. They accentuate some factors and hide others – just as a detached representation does.

Figure 9.1 *The Geographer* by Jan Vermeer. Credit: akg-images

Situating observers: partial perception and objective depiction

People do not take in the whole world as they go about their business. Everyone selects and filters what they see and what they make of it. We might look at this through three ideas: biology, position (both physical and social) and cultural frames.

Biology

The first idea highlights that human senses connect us to the world in particular ways. With the philosopher of science Donna Haraway, I would suggest walking a dog shows how specific is human experience. The dog's world is visually poor but full of odours carrying information. Our dependence on the visual, our sense of scale and our sense of location, are all based around the human body. Perception does not start as a free-floating moment, but is grounded in frailties and adaptations of the body. We are disposed to make sense of the world and make order from the sensory stimuli our body gets – think of the parlour games of optical illusions that play on that tendency (see Figure 9.2). Not that this is bad – just think how useful it is that we can 'see' three dimensions on a flat piece of paper.

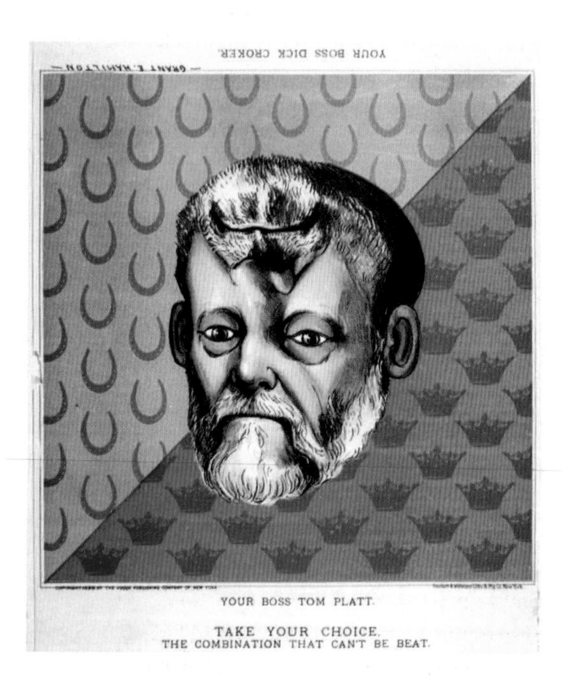

Figure 9.2 Things are not always as they appear. Credit: Bettman/Corbis

Position

Our orientation also plays an important part in ordering experience. The world comes to us in terms of high and low, near and far, present and absent. We understand ourselves spatially as we recall the world in mental maps to situate ourselves. These are egocentric maps, where our life and experiences are centre stage, and the world fades off into the distance around us. We do not know every part of the planet equally, nor our country, our city or even our neighbourhood. We have areas of concern or interest and different ways of understanding these (see Figure 9.3). We are inescapably immersed in the world, so our immediate surroundings are grasped as left and right, while more distant places are subjects of abstract images. These images may be our own experiences as they are remembered or they may be memories of images produced by others – blurring the boundaries of sensed and imagined worlds. These remembered spaces are not based on the geometrical spaces of latitude and longitude; they are shaped by our experiences, travels and the tasks we have at hand (see Figure 9.4). Our perception of the world is spatial in the way we define objects of interest as a foreground set off against a background and in relation to our viewing position. Images create a relationship between three terms: the perceiving subject, the viewed object and the relationship between them.

Cultural frames

Our representations of the world are not simply our own but are derived from social sources.

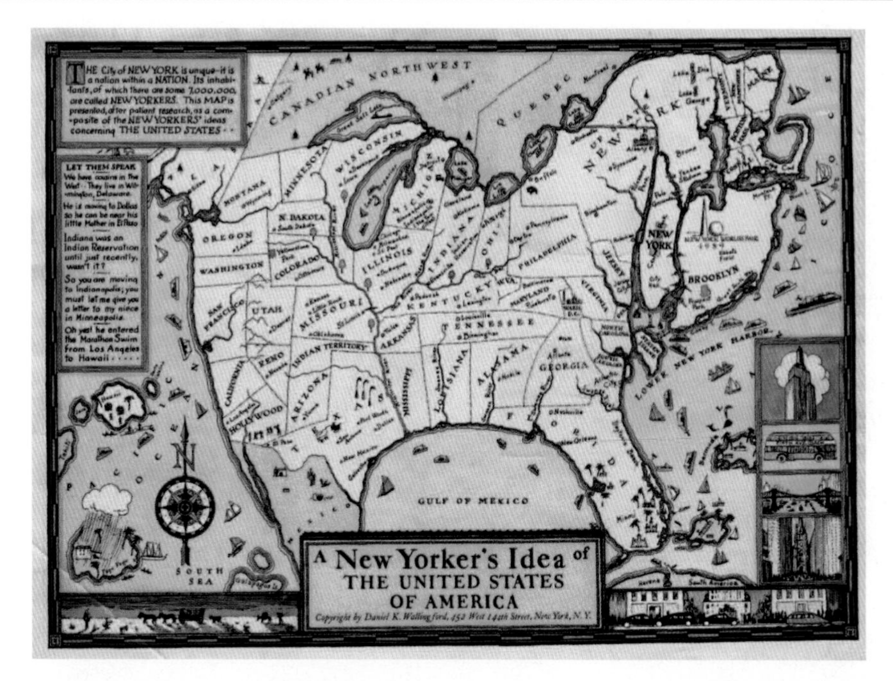

Figure 9.3 *A New Yorker's Idea of the United States of America* by Daniel K. Wallingford, c. 1939. Credit: George Glazer Gallery

Different cultures have different ways of seeing the world and representing it. For example, twisted and notched sticks formed maps for travelling in the Arctic for the Inuit. Even colours can shift between cultures, so that we read Homer's 'wine dark sea', because he did not clearly distinguish blue and green (Eco, 1985: 159). Colours are described by reference to other objects ('red like fire', 'white as snow', 'yellow as cornfields'). Our descriptions refer to other things, not direct experience that reflect our own collective cultures and individual experiences, and are shaped by our specific viewing positions. All representations imply a 'positioned spectator'. Everything is seen and pictured from somewhere. Therefore, vision and representation are always geographically embedded, creating differing epistemological structures (Rogoff, 2000: 11). The viewing position affects the knowledge created and can be characteristic of an era, a culture or social position or all of these.

The detached viewing point that offers perfect vision as if from nowhere has thus been called a 'god-trick' by Donna Haraway. One example is the viewpoint of maps, which appeals to the sense of the detached observer outside of what is depicted seeing a true representation. However, although apparently corresponding to reality, maps can also have hidden implications of who is looking at whom from where. Thus, British students are familiar with a map centred on Europe and the Atlantic, not one centred on the Americas or on the southern hemisphere (Figure 9.4) What implicit assumptions are there in the familiar Mercator map about who is of central importance and who is not? The projection downplays the size of places near the equator and exaggerates those near the poles, making the 'West' disproportionately large. An alternative projection – the Peter's Projection – equalizes for area but is less useful for navigation (Figure 9.5). Each emphasizes certain features of the world and not others.

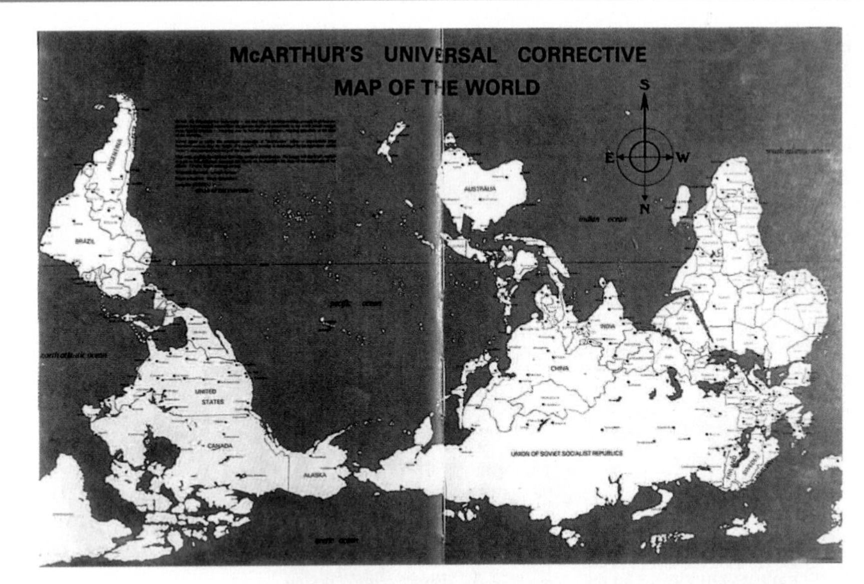

Figure 9.4 McArthur's corrective map of the world (1979) by Stuart McArthur. Credit: Stuart McArthur

Figure 9.5 Peter's Projection world map. Credit: Oxford Cartographers and Huber Verlag/Getty Images

SUMMARY

- The perceptual world may be more 'real' and truthful to our experiences, and thus as important in orchestrating actions, as any scientifically defined reality. As the world and our experience of it changes, the images that are produced may also change.
- Geographers need to be aware of 'from what angle', literally and metaphorically, people perceive the world. Particular ways of creating images imply relationships between the viewer and the world.
- The detached point of observation from which a coherent picture can be made may depend on relationships of power and control, more than its 'realistic' depiction of the world.

Making pictures

The way some representations deny that the spectator is embedded has been seen as a key weakness in accounts that depend upon representational models of the truth. For instance, the landscape and world has often been turned into a spectacle or exhibition during the nineteenth century. Landscape painting is a specific relationship of the viewer to the world, looking on the scene for pleasure, as an aesthetic object, not as an immersive workaday environment. Landscapes reflect how such paintings were often commissioned by owners to show their estate. A famous example is Gainsborough's *Mr & Mrs Andrews* (Figure 9.6), which depicts Mr Andrews standing by his seated wife in the left-hand corner and overlooking the house and estate. The composition is of Mr Andrews showing off his belongings – with his wife included in that category. Thus it is not only the content that tells us about rural life; the composition of these images reveals a class-based and gendered way of looking at the world. This image relates to scenic country parks created by excluding the rural poor, removing traditional rights and privatizing the land. This was a period where villages and public rights of way were moved to create views for the owners of country houses and where new laws criminalized intrusions. Many rural people had traditionally hunted game, which was redefined as poaching. At the start of the nineteenth century, over a quarter of all prosecutions in England and Wales were for breaches of these 'game laws'. The image of beautiful and charming rural scenery reflects real struggles over access to the land. The detachment from the scene is also linked to patriarchal power and a masculine viewing position. Landscapes and women tend to be the objects of a male viewer separated from what is depicted, tending to link the natural and the feminine (Rose, 1997). The art critic John Berger (1972) suggests that, in most art, men act or are viewing subjects while women appear as objects of that view. The viewing subject in this type of painting echoes the positioning in colonial photography, tending to be invisible, as though all-seeing yet not physically present, and meanwhile it turns the people depicted into distanced objects of knowledge (Gregory, 2003).

The form of representation thus objectifies what it depicts, and also gives one version as

Figure 9.6 *Mr and Mrs Andrews* by Gainsborough. Credit: National Gallery, London, UK/The Bridgeman Art Library

'facts' that are closed and complete. So during the nineteenth century, along with the expansion of western empires that dominated and colonized the world, there was a rise in great exhibitions that showed the conquered lands, animals and people to viewers in western countries. Products, buildings and scenes from around the world were imported to or recreated in specially built parks in western cities, often populated with 'native' performers. This was laid out as a spectacle of visual consumption for the western populace. This form of representation then did try and equate sight with truth, as when the 1893 World's Columbian Exposition in Chicago could have the motto *To See is to Know*. This is a 'scopic regime', or way of ordering how we see, that equates sight with fact and visibility with knowledge. It also positions the observer as a knowing subject, separate from objects of knowledge which are disciplined into a spatial layout. Objective evidence was thus created by objectifying the world (and its inhabitants) depicted. The form of representation, its epistemology, is connected to a politics that produces 'others' as passive objects of knowledge. The overall form encodes a cartographic imagination of cultures in the world – that is they are made visualizable as locatable, bounded objects (what in Chapter 3 is called the 'mosaic' view of global diversity).

This envisioning of cultures is not only confined to views of exotic others, but in the same period rises in terms of views of rural folk within the west. At the start of the twentieth century, the Nobel Prize winning author Selma Lagerlof published *Nils Holgersson's Wonderful Journey through Sweden* (1906) as an educational device for children to learn about their nation. It seats the eponymous and magically shrunken hero upon a Goose's back so he can behold each region in turn. The visual grasp of the diversity of Swedish regions was thus activated in the service of a nationalizing project where the old sense of 'lanskap' as province and working environment is replaced by the view of landscape from above. Later nineteenth century painters like George Clausen focused intensively on a rural culture they saw as disappearing. He exhibited his painting *Haying* in 1882 in London, showing two young women who are seemingly effortlessly turning forkfuls of mown grass, standing beside the blanket on which are the remains of their picnic lunch. The middle class, urban public the public swooned at this rural idyll. And yet records of agriculture at the time, in the place he painted, show haymaking was a highly skilled, almost entirely male task marked out by a period of fighting connected with heavy drinking. The same year his *Winter Work*, with bleak wintry tones and male and female workers bent over labouring amidst the viscous mud, instead echoed fears about the conditions for women workers and was dismissed in *The Times* as 'really too ugly' (Robins, 2002: 13, 18). If Gainsborough's *Mr and Mrs Andrews* showed a detachment from labour and lived engagement with the landscape depicted for the owners, we also have to think of the viewer being set up as a distanced observer.

Not all images rely on the detached vantage points found in classic landscape painting, and we can usefully consider why they may not. In art, the rise of cubism and surrealism, after the First World War, did away with realistic perspective and notions of correspondence to seen reality. What does the emergence of these images tell us about the world? Stephen Kern (1983) argues that they express changing experiences of space and time. This was a period that saw the climax of rapid changes in transport (the expansion of railways, steam ships and mass transit systems) and communication technologies (the telegraph, telephone, then radio). People, goods and information were circulating more rapidly, in greater numbers and over greater distances than ever before, offering no stable vantage point at which the whole picture came together or from where it could all be controlled. The great cities, modern communications and transport created a fragmented experience, not a coherent whole. The world could no longer be depicted in the same way.

So at this moment what we find emerging are a range of aesthetic movements and practices seeking to refashion how we might see the world, how images might portray the new experiences encountered. They tend to suggest the loss of a stable viewing point and coherent, singular perspectival view on the world – expressing the instability and fluidity found in the world. All these changes in images suggest novel ways of organizing and understanding the world and the shifts going on in society at large. Henri Lefebvre (1991: 25) suggested a connection between changes in how the world was experienced in different epochs and how it was represented: 'The fact is that around 1910 a certain space was shattered. It was the space of common sense, of knowledge, of social practice, of political power . . . the space, too, of classical perspective and geometry, developed from the Renaissance onwards'. The engine of this were technologies that changed how the observer was situated and especially how representations circulated.

Geographies of circulation: representation as movement

Representation has implied taking an image about and of one place and showing it to people in another. In an era of satellite pictures beamed around the world, of global news corporations and media events, the conventional geography of national broadcasters breaks up, and we need to think about the circulation of images (Christophers, 2009). Images circulate over increased distances at an increasing rate and in vast numbers. Issues of power and control are highlighted when we remember that technologies of vision have been an important part of military development over the last century.

Control of vision has been a vital stake in military power – who can see and not be seen. Reconnaissance flights, high-altitude spy planes, night sights, spy satellites . . . all create new images. Our now taken for granted satellite imagery is a result of technologies developed for intelligence gathering in the Cold War. Both Gulf Wars have seen western forces deploying advanced image-producing technology. One issue has been control of pictures: the US military blamed pictures from the Vietnam conflict for undermining support, so reporters have been tightly controlled and 'embedded' in military units. The military also provided pictures from cameras in 'smart bombs' and war planes ready for broadcast.

Our viewing position is from one very particular angle. Unsurprisingly that viewpoint emphasizes successful missions and not the carpet bombing of conscripted troops in the 1991 campaign, nor the literally uncounted casualties in 2002. It shows the local inhabitants as the potential threats felt by western forces, not the western forces as invaders into the worlds of the locals which might be the view from the other equally located perspective among the locals. It is revealing that the military commanders in the 1990s had to remind viewers that these pictures were not 'video games' but involved real people. The images were similar, though, with pilots only seeing their enemy via screens and describing missions in arcade game terms. With the rising use of drones in Afghanistan and Iraq to conduct a remotely controlled war, the military were using pilots based in Nevada sitting at screens and using joysticks actually modelled on video game interfaces. It becomes less clear, then, where a division of representation and reality might be.

If we turn from the representation of violence to that of pleasure we find some surprising continuities of processes of observation. Some of the most ubiquitous images are tourist snapshots. Hundreds of billions are taken every year. What is made visible by these images and

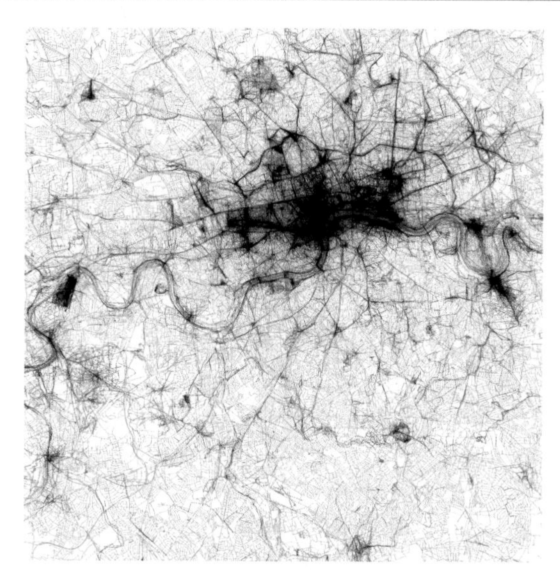

Figure 9.7 A map of all the photographs taken on 1 January. Credit: triposo.com/Eric Fischer (Creative Commons)

how is the world represented? Taking those pictures posted and circulated through photo sharing sites like Flickr it is possible to map this (see Figure 9.7). If we drill down we find dense clusters around the great cities and the key sights to see – so in London it is the Tower, not Romford, the Taj Mahal in Agra, and so on – tailing off into darkened hinterlands, in the slums of Mumbai, into Harlem, former mining villages around Durham, the poor rural interiors of Africa. There are more photos taken of Paris in a day than in Africa in a year. So we have a geography of what is regarded as photogenic and who has the means to represent it.

To see what is deemed photogenic we can look at the representations of peoples and places in brochures and on postcards. These emphasize the good points of a place: the weather is sunny, the scenery picturesque and the beaches clean. We can think of the types of place these pictures market and depict, so we might think of how they show a world where ethnic stereotypes pervade. Thus, there are Amerindians in tribal clothing, next to totem poles; exotic cultures of Asia represented by attractive women – stereotypes of all sorts (Dann, 1996; Desmond, 1999; Edwards, 1996; Waitt and Head, 2002). We know these images will be selective in what they portray. They illustrate and help shape the desires of tourists – what they want to see – and thus play a vital role in shaping flows of tourists, even if 'inaccurate' (Bhattacharyya, 1997; Gilbert, 1999; Koshar, 1998; McGregor, 2000). These images do not just reflect reality but actually shape it.

Representations making new realities

An example of how images shape reality comes from tourism, when places are altered to conform to expected images. More subtly, the experience of these places is shaped through images. So, in Don DeLillo's novel *White Noise*, the narrator visits the most photographed barn in America:

> We counted five signs before we reached the site. There were forty cars and a tour bus in the makeshift lot. We walked along a cowpath to the slightly elevated spot set aside for viewing and photographing. All the people had cameras; some had tripods, telephoto lenses, filter kits. A man in a booth sold postcards and slides – pictures of the barn taken from the elevated spot. We stood near a grove of trees and watched the photographers. Murray maintained a prolonged silence, occasionally scrawling some notes in a little book.
>
> 'No one sees the barn,' he said finally. 'Once you've seen the signs about the barn, it becomes impossible to see the barn . . . We're not here to capture an image, we're here to maintain one . . . They are taking pictures of taking pictures . . . We can't get outside the aura. We're part of the aura.'
>
> (Cited in Frow, 1991: 126; and Nye, 1991)

The image-worthy element of the barn is not its inherent qualities but that it appears in so many images. If you think this is a far-fetched fictional example, we can look at Myrtos beach on the Greek island of Kefalonia (Figure 9.8). This is one of the most-often pictured beaches in tourist brochures, with its stunning white sand pictured from above, and where there is now a large viewing platform for visitors to take similar pictures – often without actually visiting the beach itself (Figure 9.9). We can see this taken a step further where not only not visiting, but the actual itself, is surpassed. For instance, we have all seen so many images of wild animals filmed with advanced equipment that actual encounters can be disappointing. Umberto Eco has called this '**hyper-reality**', where the copies are more important and realistic than their originals. Going one step further, images become **simulacra** – that is, copies for which there is no original. They become entirely self-sustaining without referring to any exterior reality. Jean Baudrillard argues that this has been a long-running trend where, gradually, images have come to stand for and then replace things, calling the whole category of reality into question. Examples might include themed shopping malls that have facades conjuring up images of Parisian cafés that have never existed, or films like *The Lord of the Rings*, where tours of Aotorea/New Zealand now invite you to walk through Moria. The real landscape is imagined in terms of a film about a book that invented a fictional world. Other films may connect mythic or real events to places. The film *Troy* sparked a surge of interest and motivated a stream of visitors – but they did not travel to the site of historic Troy in Anatolia. They travelled to the island of Malta

Figure 9.8 View of Myrtos beach, Kefalonia: 'the most photographed beach in the world'. Credit: Mike Crang

Figure 9.9 Tourists taking pictures of Myrtos beach. Credit: Mike Crang

Figure 9.10 World weatherman

where the film was shot. The island of Kefalonia too has been made famous by a novel (*Captain Corelli's Mandolin*) that became a film actually shot on the island, depicting the beautiful scenery in lavish detail. When people visit the island, which island are they visiting – the one represented in the book, the film or the brochures? And when they take their pictures what are they representing? Very often their representations are trying to confirm that they saw what everyone is expecting to see and to picture. Several studies suggest that 'pre-existing stereotypes are not dismantled by actual experiences, but instead serve as standards against which the visited culture is evaluated' (Andsager and Drzewiecka, 2002: 403).

One example of how powerful ways of representing the world can be, not just in shaping actions but in how they are understood, is the Ethiopian famine of 1984. Here striking news footage of starving people in relief camps produced an enormous charitable effort, global fundraising events and a massive relief effort. The images also suggested helpless victims, fuelling stereotypes about Africa, and said nothing about the way southern Ethiopia remained a food exporter. The cause of the disaster was simplistically presented as a lack of rain and consequent decline in food availability (Figure 9.10). These images can be criticized for inaccurately depicting the causes, yet we have to acknowledge their enormous real-world impacts.

SUMMARY

- Images selectively portray peoples and places. We need to consider who in the current global economy decides what is shown and controls how it is circulated.
- How peoples are represented and whether they represent themselves can be the subject of political struggles.

- Images do not just reflect reality but shape actions, experiences and beliefs. Intuitively we think reality comes first and images second. However, the relationship can be more circular. It is possible to suggest that, in some cases, this could be a closed circle of image referring to image without needing to refer to an external reality.

Conclusion: the truth is out there?

These issues not only affect people 'out there', but also 'academic' representations. The representations produced by geographers are not exempt from the sorts of issues outlined above. We spend most of our time working with metaphors, images, models and so forth. How do we assume they relate to reality? Traditionally, geographers have seen themselves as making progressively more accurate representations of the world. We often have an implicit assumption that academic accounts should be judged upon and strive for correspondence to reality. Yet a perfect correspondence would lead to some bizarre conclusions. The novelist Jorge Luis Borges discussed the perfect map, at a one-to-one scale, which was thus as large as the territory it depicted. More imaginatively, even if you managed to shrink this map, if it shows all the activities and features in the territory, then somewhere on it there would have to be an image of you holding it. And on that image there would have to be an even smaller image of you holding it. This is known as a problem of 'infinite regress'. It may be, then, that trying to find some underlying reality behind the images is the wrong approach. It reminds me of a story about a young pupil who approached a Lama and asked,

> 'Oh wise one, what is the world on which we dwell?'
>
> The Lama paused, then replied, 'The world is a great disk with seven mountains and nine seas surrounded by the outer ocean.'
>
> The pupil thought, then asked, 'Oh wise one, on what does the disk of the world sit?'
>
> 'It sits on the backs of four celestial elephants, and it shudders as they shift its weight on their shoulders.'
>
> The pupil thought for a minute in silence.
>
> 'Oh wise one, on what do the four celestial elephants stand?'
>
> 'They stand, facing out towards the stars, upon the vast carapace of the world turtle.'
>
> The pupil hesitated.
>
> 'Oh wise one . . .'
>
> 'Forget it,' said the lama, 'after that it's turtles all the way down.'

A silly example, but we have to face the possibility that we may only be able to understand reality through representations. We need to think, then, not of how they may distort reality, but what effects and meanings they have for their beholders. In this way we need to see representations as actively creating the world rather than simply transmitting a prior reality. And representation then not as a matter of fixed images of the world but as a set of practices in which we are all engaged.

DISCUSSION POINTS

1. If representation is not simply about reflecting reality, nor distorting an underlying pre-representational truth, it cannot be assessed by how closely representations correspond to the 'world outside'. How then do we discriminate among different representations to assess their 'truth value'?
2. How do different viewpoints relate to different truths? How are those viewpoints valued and how do they relate to who is representing whom, and how they are portrayed?
3. How does control of representations create outcomes and effects in the world? How is such control evident and enacted?
4. Are there systematic differences between different processes of representation – textual, performance, oral, pictorial, photographic? How do they interrelate with each other and non-discursive practices?
5. What happens if we say that we can never know the world outside of representation? Does this mean we cannot think about physical processes? Does it focus us too much on processes of meaning and away from embodied experience? If we look to the non-representational, non-discursive do we risk seeing representation as 'inferior'?
6. How do we represent the unspoken, the unseen and unwritten? Are there things that are unsayable and unseeable? How would we represent them?

FURTHER READING

Berger, J. (1972) *Ways of Seeing*. London: BBC Books.

An old book mostly on photography, but a classic and trenchant account that is accessibly written.

Daniels, S. (1993) *Fields of Vision: Landscape Imagery and National Identity in England and the US*. Cambridge: Polity Press.

A synthetic argument linking how place has been represented in literature and art to national identity.

Rose, G. (2007) *Visual Methodologies: An Introduction to the Interpretation of Visual Materials*. London: Sage.

A great guide to working with images and different approaches to seeing how representations work.

Schwartz, J. and Ryan, J. (eds.) (2003) *Picturing Place: Photography and the Geographical Imagination*. London: I.B. Tauris.

A selection of historical essays that track the production and effects of photographic representations in geography.

Wood, D. (1992) *The Power of Maps*. New York: Guilford Press.

A really good introduction to how apparently factual cartographic representation can be freighted with hidden assumptions.

PART TWO

THEMES

SECTION ONE

BIOGEOGRAPHIES

Introduction

As an academic discipline, Geography is distinctive for how it spans the natural and social sciences, addressing issues both of nature and society. This section on 'biogeographies' sets out responses to that span from the perspective of contemporary thought in Human Geography. It acts as a counterpoint to the later discussion of 'environmental geographies' (Chapters 29–31), where environmentalism, sustainability and the Human Geographies of climate change are directly considered.

In some ways, the title of this section could be seen as contentious. There is a well-established body of biogeographical work within Geography that records geographic differences in vegetation and wildlife, and which explains these in relation to both past and present dynamics. Distinct sub-fields exist within that framing of biogeography: for example, zoogeography studies past and present distributions of animals, while ecological biogeography focuses on the relations between organisms and the environment. Whole textbooks are devoted to such work from the perspective of Physical Geography and Biology (see, for example, MacDonald (2003)). In this book, however, we do not directly consider these fields, because of their conventional location within Physical Geography curricula. Rather, this section explores how and why Human Geographers are increasingly problematizing that very division between Physical and Human Geographies, and in so doing making their Human Geographies more biogeographical. The chapters in this section thus come at the issue of biogeographies from Human Geography perspectives.

In the opening chapter Sarah Whatmore sets the scene by considering Human Geography's intellectual approaches to 'nature' and the natural world. Generally she is in favour of geographers challenging what she calls the 'culture–nature binary' through which both worldly phenomena and our knowledge of them are divided up into separate realms of nature and culture. She is looking to promote forms of work that transgress the divisions institutionalized within Geography between the study of nature (via the natural science of Physical Geography) and the study of culture or society (via the social science of Human Geography). In outline she distinguishes two broad ways in which Human Geographers have tackled questions of nature: the first concerned with the social construction of nature; the second with what she terms 'more-than-human geographies'.

In the first of these approaches, the emphasis is placed on how nature itself is 'socially constructed'. One strand of such work focuses on this social construction in a more material sense, exploring how 'nature' rarely exists in some pure, original, God-given state but rather is socially produced within human economies and forms of production. Picture a field of wheat, for example. This is not purely natural but part of an agro-industrial system of food production and retail. It is not a 'first nature', existing outside of human societies, but a 'second nature', produced by them. Another strand of Human Geography takes this focus on the social construction of nature in another direction, emphasizing how humans represent and understand nature. Nature becomes part of culture through the ways in which we imagine it and give it meaning. Think of that field of wheat again. For some it symbolizes human achievement in food production, agricultural progress and modernization; for others it might mean a problematic industrialization of our relations to the land and environment, and becomes a symbol of human domination over and disregard for the natural world of which we are a part. A field of wheat thus becomes part of social contests over what sort of society, and social nature, people see as best.

Sarah recognizes the value of seeing how nature is both materially and imaginatively produced within 'culture', rather than being some pure and separate realm. But she also raises concerns. In particular she is worried that in emphasizing the social construction of nature, Human Geographers have presented 'marvellous worlds of exclusively human achievement' in which the energies of the earth, plants and animals get ignored. This is particularly odd when perhaps the most pressing concerns spanning the nature–culture binary (from climate change to food scares or bio-security) highlight the limits to human control and agency. Sarah therefore signals an alternative approach, one she helped to pioneer, that emphasizes what have been called 'more-than-human geographies'. Here the focus is on according due attention not only to humans but also to non-humans (animals, plants, physical forces and forms, machines, and so on). Rather than looking to divide up nature and culture, this way of thinking emphasizes how what we see in the world are diverse human and non-human entities relating together, indeed sometimes even fusing (think, for example, of transgenic organisms or the developments in life sciences of artificial intelligence and bionic enhancements).

In Chapter 11, Russell Hitchings develops this account of 'more-than-human geographies' through a more detailed account of recent research in Human Geography on animals and plants. His account starts from where Sarah Whatmore's ends. Russell seeks to avoid the culture–nature binary, as Sarah termed it, by focusing instead on much more mundane relations that people have with animals and plants in their everyday lives. He identifies emergent schools of work that are 'focused on specific animals and plants' and that ask 'how people live with them and how they live with us . . . [and] how we might feasibly co-exist in better ways'. In surveying what has become known as the field of 'animal geographies', Russell maps out how an early emphasis on how human societies organize and understand animals has more recently been complemented by greater attention to the role animals themselves play in these relations. Work on 'plant geographies' is more diffuse, but Russell draws out some principal substantive foci including work on human relations with trees and on planted spaces, such as allotments and private gardens. Running through such work, he argues, are three main sets of questions: how do humans shape plant life and how do plants shape human subjectivities (read the chapter if you want to find out about what Paul Robbins calls 'lawn people'); what are the ethics of our relations with plants and what role do particular plants play in provoking wider ethical relations to nature (again, read the chapter if want to learn where the phrase 'tree-hugger' comes from); and what can plants do for humans, in helping us to make happier, more sustainable societies?

If Russell Hitchings' focus is on the ethics of everyday relations with animals and plants, in Chapter 12 Juanita Sundberg and Jessica Dempsey turn to a more explicitly politicized area of research known as 'political ecology'. This is a significant and increasingly well-established area of Human Geography. Juanita and Jessica define political ecology as a body of scholarship that challenges dominant framings of pressing concerns like deforestation, famine and climate change, as well as a way of thinking that can be applied more broadly. For them, political ecology is a 'stance', a way of thinking and researching that sees ecology and politics as fundamentally linked. Political ecology also has a normative quality, in so far as it aims generally to promote better ways for humans to live in the world, and more specifically to recognize and support marginalized people and communities. Juanita and Jessica spend some time outlining how this political ecology stance can be undertaken in practice, i.e. how one goes about doing political ecology. They then illustrate this, and the importance of political ecology, through exploring contemporary concerns with biofuels. In response to concerns over global climate change as well as energy security, the production of fuels from renewable biological resources (biofuels) has often been cast as some sort of panacea. However, others have gone so far as to label them 'a crime against humanity', not least because of their impact on food production and food security. Juanita and Jessica illustrate how political ecologists navigate such a contentious issue, paying attention both to the economic power relations bound up in biofuel production (who will benefit from them and how) and to the ways in which ecological knowledge is politicized in such a context (for example, a crucial issue is how and by whom some lands are designated as 'marginal' and hence ripe for biofuel cultivation; often those who live, work and grow food on such lands are given little say in the matter).

FURTHER READING

Castree, N. (2005) *Nature*. Routledge: Abingdon and New York.

In this book Noel Castree surveys how (Anglophonic) Geographers past and present have variously understood nature, drawing out the implications of their varying ideas. Its focus, then, is on the ways in which we can think about nature, and its argument is that these ideas of nature have real material importance and consequences.

Castree, N. (2013) *Making Sense of Nature*. Routledge: Abingdon and New York.

In this newer book, Noel Castree sets out to 'denaturalize' nature, to show how it is socially produced at the same time as questioning a simple division between nature and culture. The book pays particular attention to how we understand nature through the representations produced by various sorts of 'experts' and 'media'. In comparison to his earlier book, *Nature*, listed above, this book's remit extends beyond Geography and explores the wider production and circulation of ideas. The book aims to address a wide readership.

Hinchliffe, S. (2007) *Geographies of Nature. Societies, Environments, Ecologies*. London: Sage.

This book sets out a 'more-than-human geographies' approach to nature, framed in terms of moving beyond an emphasis solely on the social construction or cultural representations of nature. It is not always the easiest read, perhaps, but in Part I it sets out its overall approach and argument, and in Part II illustrates how and why geographies of nature matter through addressing areas of public and policy debate such as biosecurity, conservation and sustainability.

MacDonald, G.M. (2003) *Biogeography. Space, Time and Life*. New York: John Wiley and Sons.

This textbook surveys the field of Biogeography as traditionally defined and as developed as a field emanating more from the natural science perspectives of Physical Geography.

Robbins, P. (2007) *Lawn People. How Grasses, Weeds and Chemicals Make Us Who We Are.* Philadelphia, PA: Temple University Press.

A fascinating study that applies political ecological approaches and contemporary thinking on 'more-than-human geographies' to the question of middle class American suburbia and its turf grass lawns. Well written and readable.

Whatmore, S. (2002) *Hybrid Geographies. Natures, Cultures, Spaces.* London: Sage.

This book sets out at greater length the argument for not viewing nature and culture as antitheses, but as intimately linked. The case studies considered extend those discussed in Sarah's chapter here, focusing on food, wildlife and the management of wilderness spaces. This book was in many respects foundational in establishing wider moves towards developing 'more-than-human geographies'.

CHAPTER 10

NATURE AND HUMAN GEOGRAPHY

Sarah Whatmore

Introduction: nature and culture

Has anyone noticed how many television wildlife programmes seem to be scheduled around mealtimes? It's a mundane coincidence that illustrates just one of the ways in which we confront the tricky borders between culture and nature in our everyday lives. The feeding habits of the creatures on display and the food on the viewer's fork collide momentarily in millions of homes. In that moment, the cordon separating the things we call 'natural' from those we call 'cultural' loses its grip. Which is on the screen and which on the plate? At first glance, the big cat tearing into the flesh of its prey seems to embody nature at its most elemental – a world apart. But look again. This vision of nature 'red in tooth and claw' has been carefully framed by the hidden crews and technologies of film-making. They, in turn, are shaped by the conventions of science and television, which establish our expectations of how a particular type of animal should eat and which aspects of feeding make good viewing. The meal in front of us, on the other hand, is more obviously of human making. But on closer inspection we cannot fail to be reminded that, however *haut* the cuisine or industrial the ingredients, we share the metabolic urges of our animal kin. Culture and nature, it seems, are not so easy to pin down.

Geography asserts itself as a subject uniquely concerned with this interface between human culture and natural environment. While the overt sexism of exploring 'Man's role in changing the face of the earth' (Thomas *et al.*, 1956) may have become outmoded, this classic description of the geographical project has lost none of its appeal (see, for example, Simmons, 1996). But it has also become shorthand for one of the underlying difficulties with the way the discipline is organized. The assumption that everything we encounter in the world already belongs either to 'culture' or to 'nature' has become entrenched in the division between 'Human' and 'Physical' Geography, and reinforced by the faltering conversation between them. As a result, even as geographers set about trafficking between culture and nature, a fundamental asymmetry in the treatment of the things assigned to these categories has been smuggled into the enterprise. Geography, like history, becomes the story of exclusively human activity and invention played out over, and through, an inert bedrock of matter and objects made up of everything else.

This division of the world into two all-encompassing and mutually exclusive kinds of things, the so-called culture–nature *binary*, casts a long shadow over the way we imagine and inhabit it. It has not always been so and does not hold universal sway today. Rather, it can be traced to the European **Enlightenment** which, beginning in the fifteenth and sixteenth centuries, came to embrace the world through networks of commerce, empire and science. The geographical tradition of exploration and expedition played an important role in extending and mapping these networks and has left us with a thoroughly *modern* sense of nature as the world that lies beyond their reach (Livingstone, 1992). From this European vantage point, nature comes to be associated with the places most remote from where 'we' are – like jungles and wildernesses or, more recently, nature reserves and national parks. The trouble is that, at the start of the twenty-first century, 'we' seem to be everywhere – from the hole in the ozone layer to the cloning of sheep. Where is this pristine nature to be found now? In this climate it is not surprising that geographers, like the rest of us, are having problems holding the line between the natural and the cultural.

This chapter examines some of the ways in which contemporary Human Geography handles the relationship between nature and culture. The opening themes of food and wildlife are used to illustrate various approaches and interpretations, and to show the difference they make to the ways in which we understand and act in the world. The chapter focuses on two well-established kinds of account, which explore different aspects of the ways in which human societies have refashioned natural environments over time. It concludes by looking at the growing dissatisfaction with such accounts and their assumption that we can best make sense of the world by first setting ourselves apart from everything else in it.

Social constructions of nature

We begin, then, with established efforts by Human Geographers to make sense of the ways in which the ideas, activities and devices of human societies reshape the natural world. To put it another way, Human Geographers have treated nature first and foremost as a **social construction** although, as we shall see, they disagree over what this means. Two different, but in some ways complementary, traditions of academic work have been particularly influential over the last 25 years or so. The first is the **Marxist** tradition, which has been concerned with the material transformation of nature as it is put to a variety of human uses under different conditions of production. The second is Cultural Geography, which has focused on the changing idea of nature, what it means to different societies and how they go about representing it in words and images.

Producing nature

Writing at the height of the industrial revolution in Europe in the mid-nineteenth century, Karl Marx observed the ways in which plants and animals were being physically transformed by farmers using careful selection and breeding methods to produce commercially more valuable crops and livestock (Marx, 1976 [1867]). The lesson he drew from this observation was that with the rise of industrial capitalism, those things that we are accustomed to think of as natural were increasingly becoming refashioned as the products of human labour. This apparently contradictory idea of 'the production of nature' has become a central theme for Human Geographers.

Noel Castree (1995) has identified three reasons for its geographical importance. First, to acknowledge that nature is produced undermines the familiar, but misleading, idea that it is something fixed and unchanging. Instead we are forced to look at the specific ways in which human societies have interacted with natural environments in different times and places – from hunter-gatherer to post-industrial societies, from economies based on slave labour to those based on wage labour, for example. Second, it captures the double-edged sense in which the process of producing goods for human use and exchange simultaneously transforms the physical fabric of the natural world *and* people's relationship to it. For those of us whose idea of provisioning is stacking our shopping trolleys at the supermarket, it is difficult to imagine the intimate bonds that characterize societies in which the medicinal properties of plants or the seasonal habits of animals are part of everyday knowledge and practice. Third, it alerts us to the way in which capitalist production, in particular, seems to stop at nothing in its quest for profitability, turning landscapes, bodies and, these days, even the molecular structure of cells into marketable commodities.

Neil Smith's book *Uneven Development*, first published in 1983 and in a revised edition in 1990, has been one of the most influential elaborations of this analytical approach in contemporary Human Geography. Capitalism, he argues, for the first time in history puts human society in the driving seat, replacing God as the creative force fashioning the natural world. We can get more of the flavour of the argument from his own words:

> In its constant drive to accumulate larger and larger quantities of social wealth under its control, capital transforms the shape of the entire world. No God-given stone is left unturned, no original relation with nature is unaltered, no living thing unaffected. Uneven development is the concrete process and pattern of the production of nature under capitalism. With the development of capitalism, human society has put itself at the centre of nature.
>
> (Smith, 1990: xiv)

This revolutionary social capacity to produce nature is termed *second nature* by Smith and other Marxist geographers, in order to distinguish it from nature in its 'God-given' or 'original' state, so-called *first nature*. These terms have an explicit historical dimension, marking off modern, or more particularly capitalist, societies from all those that have gone before in terms of their relationship with the natural world. In the same vein, a further transition is deemed to be going on today as we move towards **postmodern** social forms, accompanied by a *third nature* of computer-simulated and televisual landscapes and creatures.

The transition between first, second and third nature also has significant geographical dimensions, which are illustrated through the example of the potato in the three-part sequence of Figure 10.1. The potato arrived as an exotic curiosity from 'the New World' among the booty of fifteenth- and sixteenth-century explorers like Christopher Columbus and Walter Raleigh. Since then, the humble spud has become a staple of northern European diets and, in the guise of the McDonald's french fry, of a global fast-food cuisine. The image in Figure 10.1(a) dates from around 1600 and comes from Guaman Poma's encyclopaedic survey of the ancient Inca state of Tahuantinsuyu (in modern-day Peru) for the

(a)

(b)

(c)

Figure 10.1 (a) First nature: seventeenth-century Spanish illustration of the Inca state of Tahuantinsuyu – potatoes being dug up. (b) Second nature: industrial potato cultivation in the UK, 2000. (c) Third nature: a genetically engineered potato. Credit: (a) in public domain; (b) Richard Morrell/Corbis; (c) Pugh/*The Times*/www.cartoonstock.com

King of Spain. It shows potatoes being harvested and removed for storage in the month of June. It is the kind of image that from our own time seems to capture just what is meant by first nature – plants (and animals) in their 'original' state, remaining essentially unchanged by their encounter with a 'pre-modern' society. Figure 10.1(b) is a photograph of potato harvesting in Brittany in the 1980s. The large, featureless field, the monotony of the crop and the presence of the tractor tell us that this is an industrial agricultural landscape – a readily recognizable picture of second nature, wearing its human fabrication on its sleeve. The third image, Figure 10.1(c), is a cartoon illustrating some of the popular anxieties associated with the current transition to a third nature. Here, not only has the location and landscape of potato growing become a human artefact but the genetic structure of the potato plant itself has been mapped and engineered – this one can speak!

SUMMARY

- Nature is socially constructed in the sense that it is transformed through the labour process, and fashioned by the technologies and values of human production.
- From this perspective, nature–society relations are seen to have changed progressively over time from first (original) nature, to second (industrial) nature to today's third (virtual) nature.

Representing nature

A rather different interpretation of what is meant by the social construction of nature is that associated with the cultural tradition of Human Geography. In this geographical enterprise the natural world is understood to be shaped as powerfully by the human imagination as by any physical manipulation. This is because 'nature' does not come with handy labels naming its parts or making sense of itself, like a plant from the garden centre. Such naming and sense-making are the attributes of human cultures. The importance of this approach is that it forces us to recognize that our relationship with those aspects of the world we call natural are unavoidably filtered through the categories, technologies and conventions of human **representation** in particular times and places. As Alex Wilson, a Canadian landscape architect puts it:

> Our experience of the natural world – whether touring the Canadian Rockies, watching an animal show on TV, or working in our own gardens – is always mediated. It is always shaped by rhetorical constructs like photography, industry, advertising, and aesthetics, as well as by institutions like religion, tourism and education.
>
> (Wilson, 1992: 12)

For Cultural Geographers, then, nature itself is first and foremost a category of the human imagination, and therefore best treated as a part of culture.

This can be a rather unnerving starting point for those who look to nature as the reassuring bedrock of a 'real' world that stubs your toe when you trip over it, regardless of any attempt to 'imagine' it otherwise. And one could be forgiven for not taking it very seriously if Cultural Geographers were arguing that nature is 'just a figment of our imaginations'. But, of course, they are not. What they are saying is that the relationship between the 'real' and the 'imagined' is no less slippery than that between nature and culture (see also Chapters 9 and 16). These arguments are brought to a head in the concept of landscape (see Chapter 18). In everyday speech, landscape refers both to physical places in which we encounter the natural world and to artistic representations of such encounters and places (as in 'landscape painting' or 'landscape photography'). Cultural Geography builds on these ambiguities to direct attention to the ways in which the relationship between the two – the 'real' and the 'represented' landscapes of nature – is far from straightforward.

In their influential book *The Iconography of Landscape* (1988), Stephen Daniels and Denis Cosgrove suggest that landscape is a way of seeing the world which can take a variety of forms:

> in paint on canvas, in writing on paper, in earth, stone, water and vegetation on the

> ground. A landscape park is more palpable but no more real, nor less imaginary, than a landscape painting or poem.
>
> (Daniels and Cosgrove, 1988: 1)

Whatever their form, these 'ways of seeing' the natural world share three common principles. The first of these is that the representation of nature is not a neutral process that simply produces a mirror image of a fixed external reality, like a photocopy. Rather, it is instrumental in constituting our sense of what the natural world is like. This is easy to accept for paintings or literature, where we make allowance for 'artistic licence' in terms of the artist's vision and the technical qualities and stylistic conventions of their chosen medium – oils or poetry, say. But it holds equally well for representational genres such as natural history programme-making, or the geographical art of map-making, in which the nuts and bolts of the process of representation are less readily apparent or more actively hidden from view.

It follows that the second principle of landscape is not to take representations of the natural world at face value, however much they seem or claim to be 'true to life'. The 'real' and the 'represented' cannot be so surely distinguished or firmly held apart in the practical business of 'seeing the world'. The work of the imagination has begun before a single brush stroke has been made on the canvas or a single frame of documentary footage has been shot. What, for example, has brought the artist to this particular spot and made it a worthy subject for painting? As much as anything else, it is an established repertoire of cultural reference points for interpreting the natural world that repeat and ricochet off one another down the ages, like the biblical imagery of the 'wilderness' or the 'ark'. Likewise, representations of the natural world shift effortlessly from being understood as depictions of what it *is* like, to being used as blueprints of what it *should be* like in the guise, for example, of management plans for the conservation or restoration of historic landscapes.

The third principle of Cultural Geography's approach to landscape is that there are many incompatible ways of seeing the same natural phenomenon, event or environment. For example, the drawings and accounts of eighteenth- and nineteenth-century European colonists depicted Australia as a 'waste and barbarous' land. Yet this representation could hardly have been more at odds with the 'dreamtime' landscapes of its Aboriginal peoples whose communal stories and dances teem with plant and animal life. Such irreconcilable landscapes underline the importance of carefully situating different representations of the natural world, including our own, in the historical and social contexts that make them meaningful. They also alert us to the highly political nature of the representational process. British colonization and settlement of Australia were justified by treating it as an 'empty' continent, as a wasteland parts of which could be recovered and civilized. The prior claims and land rights of its Aboriginal inhabitants, their ways of seeing landscape, went unrecognized in the country's constitution and legal process until a historic ruling in the Australian High Court in 1992.

We can illustrate these points by returning to the case of natural history programmes that opened this chapter. Television wildlife documentaries are widely taken to 'tell it like it is', providing us with a direct lens on to nature's creatures and landscapes. But as David Attenborough, one of the world's leading wildlife film-makers, made clear in an interview with geographers in the early 1980s, 'there is precious little that is natural . . . in any film'. He goes on to explain why.

> You distort speed if you want to show things like plants growing, or look in detail at the

Figure 10.2 Taking a close-up of a lion. Credit: Michele Westmorland/Getty Images

way an animal moves. You distort light levels. You distort distribution, in the sense that you see dozens of different species in a jungle within a few minutes, so that the places seem to be teeming with life. You distort size by using close-up lenses. And you distort sound.

(Burgess and Unwin, 1984: 103)

Figure 10.2 illustrates this more complex relationship between the 'real' and 'represented' landscapes of wildlife. On the left, we catch a rare glimpse of the people and equipment that mediate between an animal and its celluloid image. It gives us some sense of the discomfort and risk that film-makers face to 'get the shot'. But look closely at the lion. She is within spitting distance of the cameraman yet completely uninterested in him and his paraphernalia. The reason is that this photograph was taken in Serengeti National Park, which has become a favoured location for filming African wildlife precisely because, after decades of intensive management and tourism, the animals have become habituated to human presence. Careful editing will be needed to make this animal and place live up to the standards of a documentary 'wildlife' landscape. Figure 10.3 adds a further twist. It shows the

Figure 10.3 The Monarch butterfly's journey to Lake Windermere. Credit: Jonathan Anstee/ *The Independent on Sunday*

BBC Natural History Unit being caught out by a national newspaper over its filming of the Monarch butterfly for the series *Incredible Journeys*. In order to save money, the programme-makers decided not to travel to its native region in the Great Lakes of North America but to film captive-bred Monarchs which they had released into the comparable scenery of the English Lake District. Again, the film satisfies our sense of the wild aesthetically, but this time the representational process has left unknown consequences in its wake for both the butterflies and the regional ecology into which they were let loose.

SUMMARY

- Nature is socially constructed in the sense that it is shaped as powerfully by the human imagination as by any physical manipulation. Our relationships with nature are unavoidably filtered through the categories and conventions of human representation.
- From this perspective, the landscapes of nature are understood as 'ways of seeing' the world in which the 'real' and the 'imagined' are intricately interwoven.

Enlivening the geographical landscape: towards more-than-human geographies

In this final part of the chapter I want to turn to a third, more recent approach to nature within Human Geography. Whether the emphasis has been on its material transformation or on its changing meaning, in the approaches outlined above Human Geographers have treated the natural world primarily as an object fashioned by the imperatives of human societies in particular times and places. Each perspective illuminates different aspects of the convoluted relationship between the things of human making (culture) and those that are not of our making (nature). But in different ways the creative energies of the earth itself, in rivers, soils, weather and oceans, and of the living plants and creatures assigned to 'nature', are eclipsed in both accounts. In their eagerness to stress the capitalist capacity for producing nature, the Marxist tradition, for example, too readily overlooks the active role of these entities and energies in making the geographies we inhabit. Likewise, the Cultural Geography argument that our relationship to the natural world is always culturally mediated has tended to fix attention on the powers of the human imagination, ignoring the multitude of other lives and capacities bound up in the fashioning of landscapes.

In these marvellous worlds of exclusively human achievement, nature appears destined to be relentlessly and comprehensively colonized by culture. Human Geography's long march from environmental determinism (i.e. the deeply problematic idea that human cultures are determined by the natural environments in which they are located) to social constructionism (the idea that even nature is not innate, but produced and imagined) seems to have brought us, as the environmentalist Bill McKibben puts it, to 'the end of nature' (1990). Whatever their

Figure 10.4 The cows come home. Credit: Tony Reeve

differences, both accounts of this triumph of human culture over the matter of nature are grounded in the assumption that the collective 'us' of human society is somehow removed from the rest of the world. Only by first placing it at a distance can human society be (re)connected to everything else on such asymmetrical terms as those between producer and product (in the Marxist account), or viewer and view (in Cultural Geography accounts). These are geographies whose only subjects, or active inhabitants, are people, while everything consigned to nature becomes so much putty in our hands.

Such **humanist** geographies do not square with the anguish and infrastructure of environmental concern that characterizes the twenty-first century. In unimaginable and unforeseen ways the forcefulness of all manner of 'non-humans' has come to make itself felt in our social lives. From climate change to 'mad cow disease' there is a growing sense that our actions, and indifference to their consequences, are returning to haunt us. The popular face of this growing sensibility is illustrated by the image in Figure 10.4, which was circulating as a postcard at the Edinburgh Fringe Festival in 1995. Likewise, the pets and viruses, plants and wildlife that share the most urbanized of living spaces, and the peoples who over centuries have inhabited the deserts, jungles and swamplands where 'we' have seen only nature, make it apparent that 'the whole idea of nature as something separate from human experience is a lie' (Wilson, 1992: 13).

Over the past few years, then, there has been mounting unease about the ways in which Geography has built this binary division between nature and culture into its descriptions and explanations of the changing world. This unease stems from several different concerns, not least the crippling effect this polarization had on the contribution that Geography can make to informing more sustainable living practices (Adams, 1996). In Human Geography, it centres on a growing recognition of the intricate and dynamic ways in which people, technologies, organisms and geophysical processes are woven together in the making of spaces and places. This recognition is central to what have come to be characterized as 'more-than-human' styles of geographical work (Whatmore, 2002). Three of the most important currents in the emergence of these 'more-than-human' geographies give a flavour of work ongoing.

The first of these currents is concerned with showing that the idea of nature as a pristine space 'outside society' is a historical fallacy. This idea is so pervasive today that it is difficult for many of us to recognize it as a particular and contestable way of seeing the world. But, for example, Historical Geographers are helping to expose the ways in which the presence of native peoples was actively erased from the landscapes that came to be seen as wildernesses in colonial European eyes, and that are now revered by many environmentalists as remnants of 'pristine' nature (Cronon, 1995) (see Chapter 33). Try looking again at the image of 'first nature' potatoes in Figure 1.1a, in this light.

The second current extends this historical repudiation of the separation of human society and the natural world by paying close attention to the mixed-up, mobile lives of people, plants and animals in our own everyday lives (see Chapter 11). The place of animals had largely fallen off, or, more accurately, between the agendas of contemporary Human and Physical Geography. But a new focus on 'animal geographies' is emerging that seeks to demonstrate the ways in which they are caught up in all manner of social networks from the wildlife safari to the city zoo, the international pet trade to factory farming, which disconcert our assumptions about their 'natural' place in the world (Philo and Wilbert, 2002).

Finally, and most provocatively, a third current of work against the grain of the nature–culture binary is trying to come to terms with the ways in which the seemingly hard-and-fast categories of human, animal and machine are becoming blurred. This blurring is achieved by technologies like genetic engineering and artificial intelligence, which are seen to recombine the qualities associated with these categories in new forms, such as transgenic organisms, bionic enhancements and the like (Hinchliffe and Bingham, 2008). Here, the body is emerging as an important new site for geographical research in ways that force us to rethink the 'human' in Human Geography as much as the 'nature' of the world out there (Whatmore, 1999). Returning again to the example of food, Figure 10.5 illustrates the ways in which body spaces are being reorganized at the most intimate of scales. Where does nature end and culture start for this cow?

Human Geography has come a long way from defining itself as the study of 'man's [sic] role in changing the face of the earth'. The geographies

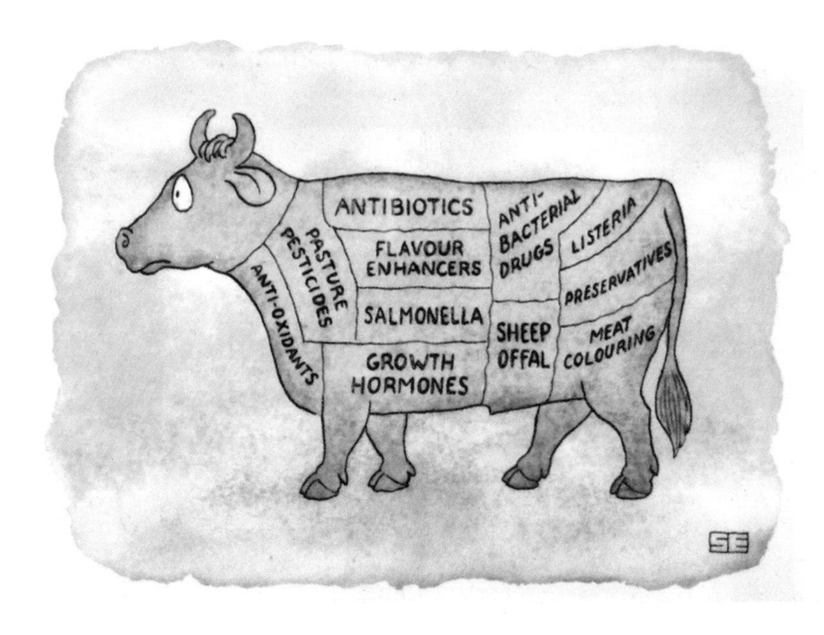

Figure 10.5 Mapping hybrid body spaces. Credit: Stan Eales

now emerging challenge us to look again at how and where we locate nature, how and where we draw the line between nature and culture, and to recognize that this densely and diversely inhabited planet is a much more unruly place than these categories admit.

DISCUSSION POINTS

1. In what senses has nature been understood as 'socially constructed'?
2. On what grounds have social constructionist accounts of the natural world been critiqued in recent geographical work?
3. Discuss some of the ways in which 'non-humans' are now being incorporated into the fabric of 'human' Geography.

FURTHER READING

Anderson, K. (1997) A walk on the wildside. In: *Progress in Human Geography* 21(4), 63–85.

A thorough review of geographical work on domestication as one of the longest-running processes connecting human, animal and plant life worlds in new ways and helping to put the novelty of genetic engineering into historical perspective.

Cronon, W. (1995) The trouble with wilderness: or getting back to the wrong nature. In: W. Cronon (ed.) *Uncommon Ground: Towards Reinventing Nature*. New York: W.W. Norton, 69–90.

A very readable piece by an eminent US environmental historian, which reflects on both the intellectual and political problems of treating nature as 'outside' culture.

Hinchliffe, S. and Bingham, N. (2008) Securing life: the emerging practices of biosecurity. In: *Environment and Planning A*, 40(7), 1534–51.

A difficult piece that summarizes recent work on the spaces and politics of the micro-biological engineering of animal bodies in industries like food and pharmaceuticals, and attempts to 'manage' the potency of bacterial and viral agents as vectors of disease in human societies.

McKibben, B. (1990) *The End of Nature*. Harmondsworth: Penguin.

This popular bestseller argues that nature in its 'true' sense as a 'separate realm' has been eradicated by the relentless industrialization of human society. It is a passionate example of an environmental politics premised on maintaining the distinction between nature and culture. It is short and worth reading in its entirety, but the basic case is set out in Part 1.

Wilson, A. (1992) *The Culture of Nature*. London: Routledge.

Written by a Canadian landscape designer, this book is a visual and literary feast, concerned with the numerous ways in which nature and culture shape each other in post-war North American landscapes. The introduction and chapters on nature films (Chapter 4) and nature parks (Chapter 7) are particularly good.

CHAPTER 11
ANIMALS AND PLANTS

Russell Hitchings

Introduction

My chapter title may strike you as rather banal. Most of us feel we have a pretty good grasp of what can be called 'animal' and what constitutes a 'plant'. Yet questioning these basic terms has been exactly what certain geographers have recently sought to do. Motivated by some of the arguments introduced earlier in this book by Hayden Lorimer (Chapter 3) and Sarah Whatmore (Chapter 10) about the development of 'more-than-human geographies', their aim has been to sidestep big concepts like 'nature' or 'the environment' in order to better understand how humans are part of life on earth. Rather than thinking generally about human relations to the natural world, these geographers have effectively gone 'back to basics' by looking closely at some of the components that were previously placed (rather unhelpfully in their view) within the concept of nature. Focused on specific animals and plants, they ask how people live with them and how they live with us, and what this reveals about how we might feasibly co-exist in better ways. This chapter aims to give you a sense of why this body of research has taken this particular path and, by looking at animals and plants in turn, to show how it can be both useful and intellectually exciting.

The nuts and bolts of nature

It is fair to say that one of the defining characteristics of Geography as an academic discipline has been a willingness to think holistically. If we agree that Geography is, at least partly, about examining different places, we can see why this is so. When we arrive anywhere new, it is immediately evident that what we are presented with – the place we are encountering – is the outcome of many social and environmental factors. To understand that place fully, all of these need to be considered. So, while more broadly it makes sense to divide up the task of producing knowledge, so that 'social scientists' tend to know more about people and 'natural scientists' understand physical processes, geographers have always been quite keen on putting these different insights together. Geography as a discipline spans the natural sciences, the social sciences and the arts and humanities. In that context, it is not surprising that Human Geographers, even though they are supposed to be studying people, have also been keen to pay attention to the natural environment and how various groups relate to it.

In Chapter 10, Sarah Whatmore outlined some of the main contemporary approaches through which this has been done. She argued that recently a number of geographers, among whom I would include myself, have sought to recast the idea of nature in our research and to avoid its framing within a nature–culture binary. Indeed such work, I would argue, is highly suspicious of the whole idea of 'nature'. This is partly because nature is such a powerful concept with all kinds of assumed meanings. Think about how when we say something is 'natural' it suddenly becomes quite hard to argue against it. It is also because as soon as something is labelled as belonging to nature it becomes imbued with certain qualities that might not necessarily apply. We might like to think we appreciate nature when out for a walk in the woods, for example, but, were we to be stung by a bee, we might find ourselves appreciating it less. With this example in mind, the contention of researchers in this subfield has been that it is not at all clear that the various phenomena we often find ourselves describing as part of nature have that much in common at all. Is a snake really the same as snow or a tree comparable to a toad? Maybe we would do better to avoid the idea of nature altogether and instead start afresh with the various phenomena that were previously subsumed within this category?

Such an endeavour has been bolstered by arguments that geographers have taken on board from other subjects that span binaries of nature and culture, the human and the physical. In science and technology studies, for example, some influential thinkers, associated with what has become known as Actant-Network Theory (ANT), have argued we should look at how people interact with all manner of 'nonhumans' to show how 'social life' is not really always all about people, but also in large part about how people live with technologies and (crucially for this chapter) animals and plants (Latour, 1993). Indeed, rather than presuming we know how our interactions with them are likely to work, we should recognise the **agency** of all things in terms of their ability to influence us. So, instead of assuming people 'naturally' share certain characteristics as a result of being human and other entities 'naturally' share certain characteristics as a result of nominally belonging to nature, we might prefer to approach the situation agnostically by exploring how different humans and nonhumans interact in particular instances. Maybe we should even do away with the fundamental distinction between society and environment altogether, because it stands in the way of a fuller appreciation of how people live with the nuts and bolts of the various phenomena that were hitherto trapped inside an unhelpful notion of nature.

And so my chapter title no longer seems so simple. Certainly the 'naturalized' ideas we may have about what qualities animals and plants possess has now become something worth investigating, rather than something we can easily assume. By evaluating these two categories in turn, I now want to give a sense of how geographical work on these topics has progressed and why examining them has proved attractive.

SUMMARY

- Geographers seek to understand the world holistically, rather than confining themselves to established academic divisions such as the social and the natural. In consequence, Human Geographers have long been interested in human relations to the natural world.

- The idea of studying 'animals' and 'plants' has gained currency in Human Geography recently because it helps us sidestep the problems associated with the term 'nature' and allows us to examine the various relations people have with particular creatures and forms of life.

Animals, agency and Angelica the octopus

We all have quite an ambivalent relationship with animals when we think about it. Many of us keep pets, lavishing attention on them and treating them almost as if they were people. We feel some obligation to meet their needs and treat them, at least in some way, as extra members of the family (Power, 2008). Yet, at the same time, many of us eat other animals, without ever really considering in depth their killing or our consumption of their flesh though, in the back of our minds, we may know there are ethical concerns about the conditions in which farm animals are kept and indeed over meat consumption *per se*. Somehow the situation has been arranged such that particular animals become friends, others pets, and yet others merely a source of protein (Holloway, 2001; see also Tuan, 2004). The concept of **relational identity** has been important for geographers trying to understand such ambivalent relationships with animals. It suggests that our relationships with, and beliefs about, different animals are the outcome of the ways in which we encounter them in particular places, and not the product of any pre-given natural order of things. Consider how pigs are supposed to be particularly intelligent; certainly cleverer than the goldfish. Yet many people care for goldfish and many people eat pigs. Why? Well, it is probably largely because our choice is less about actively deciding which animal is most likely to suffer through its farming and much more about how we have, for various reasons of interest to geographers, become inclined to live with particular animals in particular ways. In the UK, goldfish and guinea pigs and dogs have become pets not food; pigs have become farm animals not pets or wildlife.

This interest in our relationships with animals is relatively new. Previously, animals, as a topic of concern, featured comparatively infrequently in Human Geography. The main way in which they appeared was as part of the narrating of human social evolution, in a focus on how, when and where particular animals (and plants) became domesticated as part of agricultural or other human practices. In turn, those historical geographies of domestication were seen as offering material evidence for how cultures and societies changed and spread over time and space (Sauer, 1952; for an overview see Emel *et al.*, 2002). It was in the 1990s that geographers, inspired by various strands of cultural theory and ethical writing, started to consider more directly how our relationships with animals were reflective of particular cultural attitudes and ideas. A good example is provided by a historical study of Adelaide Zoo, where Kay Anderson (1995) explores how the animals found there were used, in effect, to tell visitors stories about society. More specifically, if we looked closely at the way the Zoo was organized, she argues, we would see how it reinforced particular ideas about the benefits of colonialism, its power to assemble a menagerie of animals for people to admire, and how this thereby legitimized the whole colonial project by making certain people seem superior and certain styles of government appropriate.

This links to a wider theme within the research on animal geographies: what the examination of *where* certain animals are placed says about how people understand themselves (see also Chapter 17 for further discussion). Consider the idea of animals in the city. As Chris Philo describes, partly because urbanizing societies in nineteenth century Euro-America liked to think of the city as a place of 'civilized' living, away from the 'primitive' rural world, livestock were increasingly removed from the city and placed in rural spaces (Philo, 1995). Farming became no less important to urban lives, but became distanced from city people, its animals becoming 'a shadow population' to city dwellers (Wolch *et al.,* 1995). The later rise of factory farms did not disrupt the placement. These factories – emblematic of industrial capitalism and its processes of commodification (see Chapter 27) – remained in rural space rather than, say, on industrial estates hidden from urban view. Contemporary 'city farms' are the exception that proves the rule. In these places, charities hope to re-establish a connection between city people and livestock (see Figure 11.1). Yet it can still feel strange to be confronted by sheep grazing alongside an inner city railway. The sheep may seem happy, but somehow this is not where they 'belong'. By exploring these kinds of issues, a number of 'animal geographers' sought to expose how the places where animals were put (both practically, regarding physical location, and mentally, regarding cultural understanding) said something revealing about how people sustained particular beliefs about themselves.

Figure 11.1 City farms: the idea of 'urban livestock' can sometimes feel unsettling because this is not really where we feel they 'belong'. Credit: Isabelle Plasschaert/Getty Images

But what if animals escape the places in which they have been put, both physically and mentally? At the Berlin Zoo in 2004 a 'spectacled bear', that was previously contemplated and cooed at from a reassuringly safe distance, suddenly became a threat when it staged a bold escape by swimming across a moat and surfacing near a children's park. In the English town of Malmesbury in 1998, two Tamworth breed pigs escaped from their lorry at the abattoir where they were due to be slaughtered. A media frenzy ensued. Dubbed the Tamworth Two, and individually Butch and Sundance (after the fugitive outlaws famously portrayed on film by Paul Newman and Robert Redford), these sibling pigs were lauded for their escape. They became not potential pork but identifiable characters. Bought by a national newspaper in order to save their lives, once recaptured they were housed in an animal sanctuary, their deaths over a decade later also making the national news.

Such events are suggestive of how we are not always so in control of animals as we might otherwise like to think. We cannot simply place them. What animals decide to do clearly has at least some impact on how we think of them.

Indeed, some would say the whole idea of humans being 'in charge' belongs to a somewhat arrogant belief about **human exceptionalism**, within which people have come to assume they are logically dominant over other creatures. In questioning it, geographers have increasingly attended to how animals display their agency. This extended recognition of agency beyond humans alone is partly motivated by an ethical impulse: to recognize animals as sentient creatures or, as Whatmore and Thorne (2000) evocatively put it, 'strange persons' whom we might want to resist categorizing at the start and rather examine according to how they do things to us, and what we do to them.

The ethical impulse in animal geographies introduces new questions about how the agencies of humans and animals relate. How might we live together with animals most effectively, and in ways that are sensitive both to their needs and ours? How, for example, might cities be better organized so that animals and humans can co-exist most happily together there (Wolch *et al.*, 1995)? Because they take pride in being sensitive to context and to how various factors play out in different places, geographers are in a very good position to answer such questions. They have, for instance, critically explored what is allowed into our 'circle of concern' (Murdoch, 2003) with regard to the particular creatures different groups are willing to care for, and how this may be encouraged (see also Lynn, 1998). Building on the argument that animals display different qualities (and accordingly receive a different reception from people), Jamie Lorimer (2007) has sought to identify what exactly makes us like particular animals. This has very real implications in terms of conservation, because people are more likely to donate money to safeguard those animals with more 'charisma'. Think of how elephants are enlisted to encourage us to care about conservation projects that are themselves more broadly concerned with ecosystems rather than just the elephants (Figure 11.2). Elephants appeal, partly because they seem so wise and considered and they evidently seem to love their young. In the UK context, the same kind of argument could be made about donkey sanctuaries and the volume of donations that they receive. Donkeys are funded at levels above many other species and that is probably something to do with personal memories of riding donkeys on the beach and how donkeys seem loveably doleful. How animals look, how they behave and how this makes an emotional impact upon us has profound impacts upon the ways we feel it appropriate to handle them and, in effect, whether we are bothered to care.

Animals, then, are not ethically engaged as part of a general relationship with nature. Their specific species characteristics shape our relationships with them. Even this argument, however, can be pushed further as we look to disaggregate the idea of nature. It still works with an idea of species – a sense that particular categories of animal share certain attributes by virtue of their evolutionary history and a

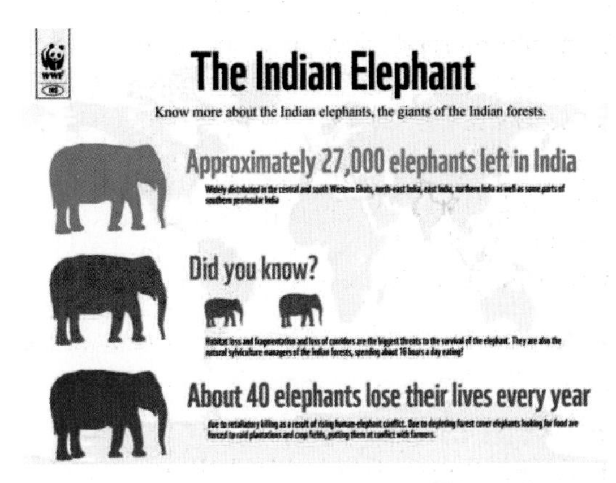

Figure 11.2 Elephants as 'flagship' conservation species. But what is it about them that makes us like them so much? Credit: WWF International

gradual journey towards slotting into specific ecological 'niches'. Of course, in many ways, they do – that is exactly why the idea of 'species' is a cornerstone to how biologists work. Yet, if we really want to push the argument for thinking geographically about how relations with animals are the product of what happens in specific contexts, we might want to consider individual biography as much as evolutionary history. In other words, we might want to see how specific people live with individual animals. People often say their pets have their own personalities. So should we really treat all members of a given dog species, for example, the same? We don't do that with people. This is an argument pursued by Chris Bear (2011) with reference to an octopus living in a public aquarium in Yorkshire. They called her 'Angelica' because she reminded the staff there of a spoilt character from a cartoon show called Rugrats who had lots of toys to play with but was seldom satisfied. Was Angelica really the same as all the others in her species? She had had quite a different life in what, with a refreshing honesty, we call 'captivity', after all. Or had her experiences in the aquarium made her something different?

Dealing with animals 'relationally' can represent quite a challenge to biologists and other academics who prefer to think of all members of a species as similar. Yet for a group that has come to call themselves 'animal geographers', this is exactly the point. Indeed, by examining the varied ways in which we live with animals in different times and places, they have usefully examined the ethics involved and how we might live with animals better.

SUMMARY

- A field of 'Animal Geography' has emerged that focuses on examining the various relations people have with particular creatures.
- Initially this was largely with regard to how societies organized animals, what this told us about ourselves as people and how our modes of living with animals revealed particular cultural traits, values or anxieties.
- More recently, geographers have started to think more 'relationally' by examining how animals physically do things and how people receive these capacities for action, or 'agency'.
- This approach has been used, in particular, to question how ethics are 'practised', in terms of how specific relations with animals are sustained in different contexts, how they came about, and whether they are right.

Researching a living landscape

Like animals, vegetation can be thought about in different ways. On looking at a domestic garden, for example, we might find ourselves admiring the gardener's ingenuity and how they managed to create a pleasing combination of colours and effects that says something about their personal style. Yet, if we looked again, we might think about how that same garden represents a struggle to survive on the part of various plants who must work hard to get the nutrition and light they need. They must also persuade the gardener they are attractive enough to remain there. It is by trying to reconcile these different ways of understanding how specific environments come to be arranged – one more human focused, one more focused on other biological agents – that another group of Human Geographers have studied how people relate to vegetation or, put more simply, plants and trees.

In many ways, this account parallels that of animals since, at the start of this enterprise, the people were cast as relatively dominant. In earlier work, when Human Geographers concerned themselves with how societies related to trees and plants, this was largely done with reference to how different human groups organized the physical world around them to meet their requirements and tastes and to create what was termed a cultural landscape (see also Chapter 18). Several studies have very effectively shown how particular landscapes – from historic stately gardens to suburban lawns – can be examined to reveal the cultural mores at the time of their production (see Cosgrove and Daniels, 1988 for a classic collection analysing the cultures of landscape; on lawns see Robbins, 2007).

Yet others have since questioned whether we should always give so much weight to the visual and the cultural when thinking about vegetation in Geography. Are plants and trees always just the passive paint that people easily deploy on the canvases of their cultural landscapes? For Tim Cresswell (2003), this is a central problem with the very idea of 'landscape' which, by definition, prioritizes the visual. For him, this concept centres unhelpfully on more nominally 'academic' forms of humanities analysis that leave us with much less sense of how landscapes are practically lived with. In short, he believes the idea of landscape 'obliterates practice', by encouraging us to contemplate landscapes in removed ways that leave us with little sense of how people engage with their environments. Though Cresswell places his focus on human practice, his argument also opens up the question of how component parts to landscapes, such as trees and plants, might do things to us by virtue of being 'alive'.

Of course, the tendency to overlook plant agencies is partly a reflection of how successful societies have been in making landscapes seem like the outcome of a controlling human culture alone. Think of how city parks can be so neatly organized that they allow us to do any number of activities on them without ever really thinking about the grass. Think of football pitches that are now so well managed by their ground staff that they almost look computer generated when we see them on television (Figure 11.3). Yet, were we to look more closely at how people experience plants and trees, we might uncover new insights. In this respect, Human Geographers have been inspired by anthropologists who framed landscape in less visual and distanced terms, perhaps because they study cultural groups who seemed much less removed from their vegetation and much more reliant on skilled interactions with it in order to survive.
If we think about tribespeople subsisting by foraging in the Amazon, for example (see Rival, 1998), claiming landscapes are 'cultural' (in the sense of being predominantly the product of what people do and how they want to arrange them) seems much less appropriate. Human Geographers interested in plants also have long traditions of bio-geographical and ecological thinking upon which to draw. Approaching the subject from more natural science perspectives initially, and then pursuing these into questions

Figure 11.3 Though we may seldom think about it, the modern football pitch is dependent on the properties and behaviour of grass. Credit: Worakit Sirijinda/ Shutterstock

of social and environmental policy, this work has also highlighted the practical character to human–plant relations, for example through notions of 'ecosystem services' – the various benefits for humans from ecosystem resources and processes (Fisher *et al.,* 2009).

Studies of plants have taken this more practical turn for two main reasons. First was the ethics of our relations to other living things, as with the work on animal geographies described earlier. Take, for example, writings on trees. In the UK work has considered how people become attached to particular trees and how that could be encouraged as a means of promoting a broader environmental sensibility (Jones and Cloke, 2002). When authorities propose to chop down trees because their presence has become an obstacle to development, people often resist, revealing themselves as quite keen on them. These trees then operate as talismans for how people should, more generally, live with the environment respectfully. Perhaps trees have this talismanic quality because, like the elephants discussed above, they are bigger than us and seem to represent an enduring wisdom we benefit from having around. Such ethical ideas were also tapped into by spokespeople for the 'Chipko' or 'tree hugger' social movement, within which international support was sought for a struggle against corporate logging in India (Figure 11.4). This was done partly by suggesting the villagers involved had such a close relation with their trees that they thought of them as allies in processes of mutual support, rather than as mere resources to be used (Shiva, 1991). Though some of these arguments may have misrepresented the villagers involved, part of the power of the campaign related to how it tapped into deeper cultural ideas about the right way of co-existing with trees.

A second motivation for focusing on our relations with plants supplements this ethical orientation. Its emphasis is on the practical benefits of being respectful to plants, of how this is of value to us. The slow growth of plants can help relax us, getting us to engage, at least briefly, with rhythms that stand outside modern, rushed lifestyles. This is, after all, part of the pleasure of activities like gardening which are now, for example, sometimes encouraged by the state as a means of coping with mental health problems (Parr, 2007). Particular kinds of places can thus be understood in terms of their relationships with plants and the social benefits of these. Allotments, for instance, have been cast as places of practice that foster values of sustainability, care, sociality, refuge, and productive making (Crouch and Ward, 1997). Often under threat from land uses of apparently more economic benefit (at least for some) (DeSilvey, 2003), allotments enact a somewhat counter-cultural form of urban life through the relations between humans, plants, earth and air that they perform. For Crouch and Ward, allotments may not be grand landscapes that we gaze at in awe, but they are important places, allowing the 'making [of] a knowledge of the world almost incidentally . . . by moving around the place, and getting to know it with others' (1997: vii-viii). Allotments sit within wider debates over the relations between the politics of gardening and the spaces that constitute it. Along with so-called 'guerrilla gardening' (tending to plants outside of private property or defined gardens, for example in verges or disused plots within urban space), allotments are contemporary forms of a longer tradition of 'radical gardening' (McKay, 2011). The attention paid to them is indicative of how exploring our intimate relations with living vegetation is a means of finding effective ways of promoting happier, more sustainable societies.

Figure 11.4 In the Indian 'Chipko' or 'tree hugging' movement, for some commentators at least, the villagers involved showed us an ethical way of relating to trees that modern societies had somehow forgotten. Credit: Waging Nonviolence.org

One of the main spaces in which these issues have been explored is the domestic garden. Paul Robbins, for example, considers the suburban American garden and especially its emphasis on 'perfect' turf grass lawns (Robbins, 2007). On the one hand, Robbins argues, turf grass is very much the product of human beings, both imaginatively (as a model of what constitutes a garden that is by no means timeless or placeless, but related to particular suburban cultures) and practically (in the work put into mowing, weeding, seeding, watering, spraying and treating). It is also a product with severe ecological consequences, especially in relation to chemical usage. But, on the other hand, the lawn is also a plant-form, a 'Plant Geography' if you will, that acts upon humans, changing what and who we are. As Robbins puts it,

> lawns make "turf grass subjects", they turn us into a "lawn person", their green

Figure 11.5 Domestic gardens can be thought of in different ways. Who really made the landscape shown? Was it the plants or was it people? Credit: Artens/Shutterstock

> monocultural aesthetics *requiring* us to behave in certain ways, their seasonal and daily rhythms . . . becom[ing] . . . those of suburban communities and subjects
> (Robbins, 2007: xviii).

In Australia vast amounts of water are used to safeguard the wellbeing of plants that might not otherwise thrive in domestic gardens. The preferred size of gardens can also be complicit in potentially unsustainable processes of urban sprawl. On the other hand, argue Lesley Head and Pat Muir, the Australian suburban garden is also a vital space that 'ruptures . . . the more separationist views of nature', getting people near 'nature' through engagements with plants and animals (especially birds) (Head and Muir, 2006: 522). The suburban garden here is a 'paradoxical space', characterized both by the policing of order and a control over plants, and by the potential for alternative relations to emerge (Longhurst, 2006; see also Barker, 2008). In the patio gardens of barrio dwellers in Managua, Nicaragua, Laura Shillington finds relations to plants that make these urban environments habitable and dwellings homely (2008).

In my own work, I sought to understand whether people's relations with living plants have been changing within the domestic gardens of London (Hitchings, 2007). In many cases they were, because modern consumers did not have the patience or confidence to either wait for, or grow, their plants. Take the example of the 'plant guarantee'. In some ways, this feature, provided by many garden centres at the time of the study, makes sense because it allays the fears of buyers who want to purchase a dependable 'product'. Yet, on the other hand, surely a plant is alive and its success in the garden is therefore dependent on the environmental conditions of the garden and whether a potential gardener is capable of caring for it? If this is so, can it really be guaranteed? Indeed, surely doing so undermines the whole pleasure of learning to grow? By examining such contradictions in how societies now handle plants, geographers interested in gardens have examined what this means for future sustainability and well-being. Domestic gardens, in particular, have provided some of the most fertile ground for this kind of enquiry, presumably because of the particularly close relationship between some of the people and plants found within them.

In any case, an interest in the 'liveliness' of plants and trees, and how people deal with this, means geographers have something distinct and valuable to add to wider debates that often draw on comparatively removed discussions of green space provision or landscape planning. Because these bigger concepts can sometimes downplay the personal experiences associated with the sites involved, a contextually sensitive approach to 'plant encounters' fills an important gap. So, whenever you come across a tree or plant, it is worth thinking about how it came to be there, how people interact with it, and how they are likely to be thinking about it as a consequence of that interaction. There is good reason for undertaking this exercise since, though it might initially feel strange, doing so has the potential to reveal insights regarding how societies live with vegetation and what this tells us about how best to deliver on various wider public objectives.

SUMMARY

- For similar reasons to the animal geographers, a number of other Human Geographers have directed their attention to how different groups of people handle trees and plants.
- This was partly associated with a movement away from a geographical preoccupation with cultural landscapes that stood in the way of a fuller appreciation of how these landscapes were lived with.
- The motivations for focusing on our relations with plants were partly about exploring the ethics of living with the 'nonhuman' world, but also about seeing what this approach suggests for making better, more sustainable societies.
- Particular foci have included plant types, such as trees, and plant(ed) spaces, such as allotments and domestic gardens.

Conclusion

I began this chapter by saying the research topics I would describe might initially feel a bit banal. Certainly you can feel a little self-conscious when telling others you are researching what people feel about flowers or how they handle fish. Surely a 'proper' academic should be researching weightier matters and bigger issues? Yet hopefully you are now (at least, partly) persuaded by the value of this branch of Human Geography. Not only does it allow us to get beyond thorny, and often quite limiting, debates about what is 'culture' and what is 'nature' but, once we are beyond them, a whole new world of exciting phenomena opens up. And it is crammed full of potential research topics by virtue of the fact that we are allowing ourselves to see all the different creatures and interactions we could potentially study. Driven by the imperatives detailed here, geographers seem set to say much more about how we live with plants and animals, what this says about both us and them, and how we could harness this knowledge to help these relations take the most positive future paths.

DISCUSSION POINTS

1. Think of the last time you interacted with an animal. How did you feel about it? To what extent were these feelings a consequence of the attributes of that animal? Might you have interacted with it differently in another context? Why?
2. Where do particular plants or animals 'belong'? Can you think of places where certain plants or animals really shouldn't go and how is it that you came to feel this way?
3. How are the 'green spaces' near your home organized? Who 'uses' them and what is it about the space that draws them there? Do people notice the plants involved and how might this vegetation be managed better in view of this?

FURTHER READING

Bear, C. (2011) Being Angelica? Exploring individual animal geographies. In: *Area* 43(3), 297–304.

An interesting paper that evaluates some of the main currents in Animal Geography and feeds into them through a novel case study of how one specific animal is handled.

Philo, C. and Wilbert, C. (eds.) (2000) *Animal Spaces, Beastly Places.* London: Routledge.

Something of an animal geography 'classic', this book sets out some of the main arguments sustaining this subfield and contains a wealth of historical and contemporary case studies.

Head, L. and Atchison, C. (2009) Cultural ecology – emerging human plant geographies. In: *Progress in Human Geography* 33(2), 236–45.

A relatively detailed overview of the 'state of the art' in plant geographies, evaluating why and how this work is being undertaken.

Jones, O. and Cloke, P. (2002) *Tree Cultures: The Place of Trees and Trees in Their Place.* Oxford: Berg.

Though conceptually challenging, this book demonstrates, through interlinked case studies, how trees could be researched in a geographically sensitive way and what such an approach can reveal.

CHAPTER 12
POLITICAL ECOLOGY

Juanita Sundberg and Jessica Dempsey

Political ecology as stance

Political ecology is a dynamic body of scholarship that makes important contributions to the discipline of Geography and the wider social sciences by challenging dominant framings of pressing concerns like deforestation, famine and climate change.

In this chapter, we define political ecology as a political stance towards the world and, therefore, research. This stance has a number of implications for the practice of political ecology. First, in highlighting the 'political', political ecology emphasizes the practices and processes through which power is negotiated and wielded, and the importance of ecologies to these (Paulson and Gezon, 2004: 28). Political ecology sees the two parts of its name as fundamentally connected; it is underpinned by the supposition that, as David Harvey (1996: 174) so neatly puts it, 'all socio-political projects are seen as ecological and vice versa.' From this perspective, the world around us is always already political; no ecologies may be understood outside of politics. At the same time, our politics are always intensely material; no political projects may be fully understood without analysing their ecologies. While power relations infuse all socio-ecological activities, political ecologists have consistently demonstrated that the effects of such activities are unevenly distributed. Political ecological research shows these uneven distributions across space and time, identifying which actors (including non-humans) are made more or less vulnerable, or, as Donna Haraway (2008) puts it, more or less 'killable'. So, political ecology attends to issues of life (and death). The concern is with who or what lives and dies, and also – critically – *who* benefits from this living and dying?

Second, political ecologists tend to take a normative stance, meaning they embrace the idea that there are better ways of living together that are also less coercive and less damaging. Research in political ecology aims to contribute to more equitable relations between humans and non-humans that also empower marginalized groups. The influential political ecologist Piers Blaikie established the importance of normative research that prioritizes the needs of marginalized peoples and invites them to participate in producing knowledge relevant to their daily lives (Forsyth, 2008). The agenda-setting volume, *Liberation Ecologies* (first published in 1996) furthered Blaikie's concerns by pointing to the emancipatory potential of social movements (and political ecological research) that highlight the connections between social justice and ecology (Peet and Watts, 2004).

Third, political ecology is a stance in the sense that, like any research, it comes from somewhere and therefore offers a partial view. Not only are political ecologists embedded in particular bodies of scholarship, and physically located in a geopolitical world characterized by uneven development, they also are marked corporeally by power relations organized around the constructs of race, gender, class and sexuality. Donna Haraway (1991b) links **positionality** to power and knowledge; in other words, one's geographic and embodied position informs one's analytical categories, research questions, as well as the collection and interpretation of data. As an example of how geopolitical position matters, Joel Wainwright (2008) invites North American and European political ecologists to think more critically about the very geopolitical categories – like 'Third World'/'First World' or 'undeveloped'/ 'developed' – used to describe the places where political ecologists do fieldwork. Political ecology is renowned for its diverse range of research locations and commitment to people in places classified as 'Third World'. However, taking these categories for granted, Wainwright argues, risks naturalizing them as static, bounded regions with particular sociopolitical and ecological traits, available for political ecologists to enter, observe and map (see also Sundberg, 2005). Arguably, such imperial gestures of possession are at odds with the normative positions espoused by political ecologists.

For another example of how positionality matters, we point to gender. In the 1990s, feminist political ecologists disrupted conventional framings of land managers as male to show how gender matters to understanding local resource use. Dianne Rocheleau (1995) along with colleagues (Rocheleau *et al.*, 1995) demonstrated that women use and manage resources differently from men and therefore have distinct environmental knowledges. As she and others argued, neglecting such differences may result in dispossession, as women's spaces and labour are subordinated to men's when land reform or development projects frame men as *the* primary environmental actor (Carney, 1996; Schroeder, 1999; Rocheleau *et al.*, 1996). In sum, who we are and where we stand has profound implications for the knowledge we produce. Hence, situating one's own stance is good academic practice and advances the normative dimensions of political ecological research.

Finally, in defining political ecology as a political stance, we are suggesting that the focus of study need not be strictly 'environmental'. In other words, a political ecological stance is valuable for studying issues in a wide range of sites and spaces, reflecting how politics and ecologies are more generally intertwined beyond issues that we might think of as being about 'the environment'. For instance, Sundberg (2011) elaborates a political ecological approach to studying the United States' boundary enforcement operations in the US–Mexico borderlands. In the late 1990s, the US government implemented new border security policies to push unauthorized border traffic away from urban centres into outlying areas, where, officials calculated, 'natural barriers such as rivers, mountains, and the harsh terrain of the desert' would serve as 'deterrents to illegal entry' (US Government Accountability Office, 2001: 24). As a result, undocumented migrants began traversing borderland environments, over 40 per cent of which are designated as protected areas by the federal government (e.g. national parks and national wildlife refuges). Pushed into this harsh terrain, over 6000 people have died attempting to enter the USA. As Sundberg

(2011) demonstrates, non-humans such as the Sonoran Desert, Tamaulipan Thornscrub and ocelots (a small feline) inflect, disrupt and obstruct boundary enforcement, forcing state actors to call for more funding, infrastructure and boots on the ground. Ultimately, Sundberg's political ecological approach demonstrates that non-humans are constitutive of border politics and, as such, compel alternate explanations for the escalation of US enforcement strategies at its southern border.

Having set out how political ecology is a stance characterized by concerns for power, positionality and the intertwining of politics and ecologies, the rest of this chapter considers how this stance is elaborated in practice. First, we address more generally how political ecology is done, how it approaches research topics and what kinds of methods it might involve. Second, we focus in on a particular example, outlining a potential political ecological approach to the pressing issue of biofuels.

SUMMARY

- Political ecology is known for challenging dominant framings of pressing concerns like deforestation, climate change, food production and famine.
- Political ecology is a dynamic body of scholarship that adopts a political stance, emphasizing questions of power and positionality.
- Political ecology recognizes a general intertwining of politics and ecology, such that political ecological approaches are valuable in understanding a wide range of issues and spaces that we might not think of as 'environmental'.

Doing political ecology

How do political ecologists do their research? As is true for all scholars, political ecologists develop methodologies to guide the research process. Methodological considerations include:

- thinking about the research questions (e.g. what do I want to know?)
- the assumptions underlying one's categories (e.g. am I assuming land managers are male?)
- the kinds of information needed to answer the questions (e.g. remote sensing data, oral histories)
- the best methods for producing this information (e.g. vegetation inventories, interviews)
- where the research should be conducted and at what scale (e.g. in the lab, with land managers in the field).

All of these considerations are political, both in the sense that each choice we make will shape the knowledge produced and because in many ways our choices are inextricably rooted in our experiences as geopolitical, embodied beings. A great strength of political ecology has been its attention to the politics of methodology. For instance, feminist political ecologists have pioneered the development of collaborative methodologies to understand the categories and distinctions used by local people themselves (Fortmann, 1996; Rocheleau, 1995; Sundberg, 2004). Their aim is to avoid assuming and imposing academic categories derived from different places and times that may erase local knowledges and ways of being. Methods like participatory mapping, which in Fortmann's (1996) case were produced on the ground using sticks and stones to represent community resources, are effective ways of producing scholarly knowledge that acknowledges and learns from other knowledges (Haraway, 1991b; Nightingale, 2003; Rose, 1997b).

Political ecologists are keen to understand how power relations configure specific socio-environmental projects. To demonstrate how such questions of power factor into methodology, Paul Robbins (2004, 2012) compares political ecology to *apolitical ecology*. An apolitical ecology typically focuses on the observable, proximate causes of a given environmental issue or conflict, as in the behaviour of local actors involved in, say, grazing or farming. In so doing, apolitical ecological explanations ignore the power relations that manifest in, for example, particular policies (free trade agreements, land tenure laws) that inform the decisions of local actors.

You may be familiar with a prominent *apolitical* explanation for environmental problems: overpopulation. To illustrate how overpopulation works as an *apolitical* explanation, Robbins (2004: 3–4) provides the example of habitat decline and wildlife losses in Kenya. An apolitical ecological approach would highlight rising populations and the expansion of farms into previously forested areas, thereby pointing the finger at local people for causing the decline of giraffes, gazelles and buffalo. However, population growth is not the only factor to consider when examining this issue. Drawing from Homewood *et al.* (2001), Robbins shows how Kenya experiences greater wildlife decline than its neighbour, Tanzania, due to increased private landholdings and more intensive cropping for cereal grains that enter the global food chain. These factors result from decisions made by politicians in the Kenyan government and multilateral institutions like the World Bank. In short, focusing on the actions of local actors to explain declining wildlife reveals only a small part of the story and in many cases, blames those most vulnerable to policies enacted without their participation.

While an apolitical ecology ignores power relations, policy structures and the market economy, a political ecological approach includes them. This means that political ecology necessitates methodologies to examine phenomena that may be difficult to fully observe – colonialism, globalization, racism, sexism – but which leave their marks on bodies, landscapes and soils. Hence, political ecologists tend to employ a mix of methods to trace the connections between ecological changes and sociopolitical dynamics (Rocheleau, 1995; Bassett and Zuéli, 2000). For example, in her study of a community forestry programme in Nepal, Andrea Nightingale (2003) combined aerial photo interpretation with ecological oral histories to analyse the effectiveness and sustainability of community management of forest resources. These two methods are likely to produce very different kinds of knowledge. With aerial photographs, the researcher aims to observe and interpret things that are visible to the eye: land forms and vegetation. The knowledge produced relies upon an interpretation of observable phenomena and is therefore considered objective. In contrast, the purpose of ecological oral histories is to analyse landscape change by asking local people to narrate their experiences over time. The knowledge produced will likely shed light on observable and unobservable phenomena as perceived by embodied actors, many of whom may not be considered legitimate knowledge producers by state institutions because they are poor, illiterate and seen to have an interest in the forest. Nightingale (2003) treated the results of both aerial photo interpretation and oral histories as partial and placed them into conversation to produce new insights about the pace and location of forest regeneration as well as how and why local people claimed the programme a success. In so doing, she also framed local people as legitimate producers of environmental knowledge.

SUMMARY

- Political ecologists build their political stance into their research methodologies, studying things differently to 'apolitical ecology'. For example, political ecology involves recognizing the wider factors that create environmental problems rather than focusing solely on local people or ecologies as the problem.
- Political ecology also involves working collaboratively with local people, recognizing them as legitimate producers of environmental knowledge.
- Political ecology often combines research methods cast as scientific and objective – such as aerial photography – with research methods from the humanities seen as subjective, such as oral history.

The political ecology of biofuels

In this section we elaborate a contemporary example – that of biofuels – to clarify what political ecology does: the kinds of questions asked as well as the dilemmas political ecologists face in deciding how to approach and represent their research sites.

Global climate change is *the* socio-environmental issue of our time. It is also deeply political. The historical and contemporary unevenness/inequities of greenhouse gas (GHG) production is stark. (For an excellent interactive map showing these inequities see www.guardian.co.uk/environment/interactive/2011/dec/08/carbon-emissions-global-climate-talks.) A key issue facing inhabitants of the Global North with energy-intensive lives is how to supply fuel for cars, heat houses and power mobile phones, computers, tablets, etc. Biofuels are an energy source held up as part of the answer. Biofuels are liquid fuels that are directly derived from renewable biological resources, especially from purpose-grown energy crops. They include bioethanols, which are alcohol-based fuels made by fermenting the sugar components of plant materials and made mostly from sugar and starch crops (such as wheat, corn, sugar beets, sugar cane) and biodiesels, which are made from vegetable oils, animal fats and recycled greases (and thus associated with crops such as soy, rapeseed, jatropha, mahua, mustard, flax, sunflower, palm oil, hemp, field pennycress, pongamia pinnata and algae). Global biofuel production has boomed in the last few years, with global output of bioethanol quadrupling between 2000 and 2009, and biodiesel increasing ten-fold in the same time period. As Figure 12.1 demonstrates, it is poised to at least double over the next decade (Pike Research, 2011). A recent report released by international institutions like the Food and Agriculture Organization (FAO), the World Bank and the World Trade Organization (WTO) says this growth has largely been driven by government policies of countries in the Global North (FAO *et al.*, 2011: 26). In the USA, Canada and the European Union, for example, the development of biofuels has been supported through subsidies, tax exemptions or mandatory blending in gasoline or diesel (see the series of reports by the Global Subsidies Initiative called *Biofuels: At what cost*, available at www.iisd.org/gsi/biofuel-subsidies/biofuels-what-cost). In regions like the EU, it is unlikely that targets for increased use of biofuels for blending may be met through domestic production. As a result, biofuel expansion is also growing in many Southern countries such as Malaysia, Brazil and Tanzania, not only to feed domestic energy supplies but also for export to the Global North.

However, in the last few years the perception of biofuels has shifted from a panacea for a range of problems – climate change, energy insecurity and underdevelopment – to a 'crime against humanity,' in the words of the UN Special Rapporteur on the Right to Food (Ziegler, 2007). Biofuel production is linked to biodiversity loss (UNEP-WCMC, 2009) and tropical deforestation (Lapola *et al.*, 2010). In addition, the carbon emissions savings from biofuels are under debate. One study calculates that the conversion (direct and indirect) of rainforests, peatlands, savannahs and grasslands creates a kind of 'biofuel carbon debt' by releasing 17 to 420 times more carbon than conventional (hydrocarbon-based) fuels (Fargione *et al.*, 2008). A more recent study focused on Brazil found that it would take 250 years to repay the carbon debt from biofuel expansion due to indirect land use changes (Lapola *et al.*, 2010). Another major issue –

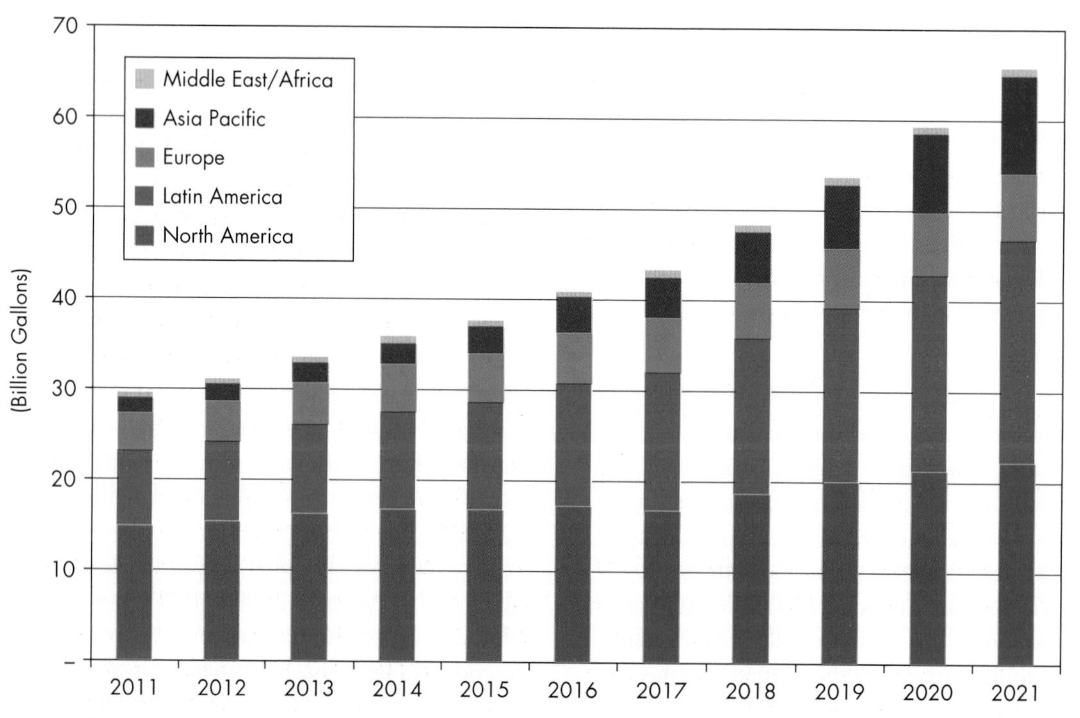

Figure 12.1 Projected growth in biofuels by region 2011–21. Credit: Pike Research (2011)

and why the UN Rapporteur on the Right to Food made the above statement – is that biofuels compete with arable land for food production and cause increases in food prices (FAO *et al.*, 2011). For this reason, the FAO, WTO and other institutions have called for an end to governmental subsidies for biofuels (FAO *et al.*, 2011). Moreover, increasing evidence suggests that 'dispossession' is rising with biofuel expansion. Some wealthy countries are now acquiring tracts of agricultural land in the Global South, especially in Africa, to grow biofuels and food for their own national consumption. For example, in the Tana Delta region of Kenya, villagers have been evicted to make way for a sugar cane plantation for biofuel production. Companies such as Canada's Bedford Biofuels and the UK's G4 Industries are buying up land through long-term leases (McVeigh, 2011).

Despite these varied problems, a recent market report predicts that the biofuel market will double over the next decade from 82.7 US$ billion in 2011 to 185.3 US$ billion in 2021 (Pike Research in Clean Technica, 2011) due to higher oil prices, new policies, feedstocks and advanced technologies. How is it, then, that biofuels continue to expand in the face of the widely recognized problems noted above? What are the consequences of biofuel production? How is power wielded and negotiated with respect to biofuel production, the biofuel market and biofuel policy? These are the kinds of questions that might animate a political ecologist in approaching the issue. In turn, a political ecology stance suggests a number of avenues for enquiry.

One option is to start by analysing the **political economy** of global energy. This may be pursued at various scales. A political ecologist might begin to investigate the principal actors and institutions promoting biofuels and the specific mechanisms through which this is accomplished. As with hydrocarbon oil and gas production, there is profit to be made in biofuels. In the words of one commentator, 'Whoever produces abundant biofuels could end up making more than just big bucks – they will make history . . .The companies, the countries, that succeed in this will be the economic winners of the next age to the same extent that the oil rich nations are today' (Craig Venter, quoted in ETC Group, 2010). Following the money will lead to investment structures and state development initiatives and subsidies that make biofuel expansion a profitable enterprise. Practically, such research might then involve interviewing key decision makers within the institutions driving the expansion of biofuels to understand the explanations put forth to rationalize increased production. It might also involve analysing the regulatory contexts governing biofuels. What specific legal mechanisms are created to develop this market? Who has helped to shape these and who has not?

While the political economy of energy is crucial to understanding the global biofuel boom, another equally important scale of analysis are the constellations of power and knowledge that establish *truths* about biofuels. In other words, we might pursue research on how powerful institutions and corporations produce knowledge that comes to dominate the public script, thereby marginalizing other perspectives. Like many other issues that span the categories of nature and society, biofuels are not only implicated in the politics of economic power, but also in a politics of knowledge. For instance, biofuel expansion is currently driven by the argument that if production is directed at lands that are otherwise 'not useable' then many of the problems chronicled above (such as competition with food) will be avoided. These arguments come from powerful and influential sources, including academic institutions. For example, a 2011 study claimed that biofuel crops cultivated on 'marginal land' would produce up to half of the world's current fuel consumption without affecting food crops or pastureland (Cai *et al.,* 2011). The study team defined marginal land as land with low productivity that has been abandoned or degraded or that is of low quality for agriculture (hence crop land, pasture land and forests were excluded from the analysis). A political ecologist might seek to analyse how this particular fact was produced. This question is crucial, given that the 2011 study was partially funded by the Energy Biosciences Institute (EBI), a public–private partnership between several large universities in the USA and BP, one of the world's largest energy companies. Again, interviews with key actors would be an important research method here, along with analysis of relevant documents pertaining to biofuel production in the Global South.

Pursuing this angle further, we might ask how the category 'marginal lands' is produced. Who determines what counts as 'marginal land' and how? And with what methods? What do these methods render (in)visible? These questions are critical, for one study of 17 bioenergy feasibility studies found that 'land reported to be degraded is often the base of subsistence for the rural population' (Berndes *et al.,* 2003). As Jonathan Davies, global coordinator for the World Initiative for Sustainable Pastoralism, says: 'These marginal lands do not exist on the scale people think. In Africa, the lands in question are actively managed by pastoralists, hunter-gatherers and dry land farmers' (cited in Gaia Foundation *et al.,* 2008: 3). In other words, people in one part of the world are using so-called marginal lands to support their families, while people in another part of the world slate these same lands for biofuel development on the basis of data from remote sensing maps. What counts as marginal land for some is the source of livelihood and subsistence for others. That difference is about a politics of knowledge that political ecologists look to contest, thereby improving the science on the subject. Researchers sitting in a US university examining remotely sensed data might have a very different view of 'marginal' land than a political ecologist working alongside rice growers on that land.

Thus, as well as examining the global economic, scientific and policy networks that shape biofuel production, political ecologists may also look to research this issue at a completely different scale of analysis: that of the land managers whose livelihoods are most directly affected by the logics of capitalist accumulation and knowledge production outlined above. As noted, numerous **non-governmental organizations** (NGOs) challenge the notion of abandoned lands awaiting biofuel production. 'What governments or corporations often call "marginal" lands are in fact lands that have been under communal or traditional customary use for generations, and are not privately owned, or under intensive agricultural production. The lives of the peoples living on these lands are all too often ignored' (Gaia Foundation *et al.,* 2008: 1). Who are these ignored communities? In what ways are they managing lands for subsistence and through what institutional arrangements? And, what narrative strategies are used to render their livelihood strategies unproductive in the context of modern economic models?

One approach to such questions might be to work collaboratively with an NGO like the African Biodiversity Network, which is involved in investigating the policies or legislation at various scales that classifies and governs marginal lands. Another approach is to work directly with affected communities.

A common method of approaching communities is through an existing NGO already working alongside them. For instance, the World Initiative for Sustainable Pastoralism, based in Nairobi, Kenya emerged to support pastoral peoples' livelihoods in dry land areas, which are at the fore of resisting biofuel expansion. Given that many affected communities are identified as small-scale farmers, pastoralists and indigenous peoples, appropriate and respectful research protocols would need to be elaborated to contact such communities and to establish agreed upon and mutually beneficial research relationships. Developing such relationships and forms of collaboration is complex and difficult, but seeing the issue from such a perspective would offer important insights and contest the dominant politics of knowledge through which biofuels are debated and biofuel policy made.

For instance, political ecologists working in collaboration with local people to understand the implications of biofuel-driven conversion of lands are in an ideal position to investigate the specific mechanisms through which communities are dispossessed of communally managed lands. For instance, how specifically did the Tanzanian government forcibly displace 1000 rice farmers in the Usangu Plains from their lands to make way for a large sugarcane plantation for bioethanol production (African Biodiversity Network, 2007: 13)? What legal mechanisms were used? And what are the implications for the community, now that the company has denied them access to the river? Moreover, what are the short- and long-term effects on the region's ecology? What lessons might this case have for resisting or dealing with actual and potential dispossessions elsewhere?

SUMMARY

- The growth in biofuel production, and evidence of its problematic implications, is an exemplary issue that calls for a political ecology stance.
- Political ecology emphasizes the politics of both biofuel economies and biofuel knowledge.
- A particular issue highlighted by political ecology is the trend towards locating biofuel production on so-called 'marginal lands' in the Global South. Political ecology approaches involve both critical analysis of how such marginal lands are defined and mapped and collaborative work with the people currently using them for their own subsistence.

Conclusion

Enacting a political ecology stance and, more specifically, applying it to research on a topic like biofuels poses challenges and choices. In concluding this chapter, we draw out two such challenges in particular, relating each back to the case of biofuels for the purposes of illustration. The first challenge relates to scale.

As we have seen, in order to understand the political ecologies of biofuels one has to trace out wider factors related to money, knowledge and power. Indeed, for Robbins (2004, 2012), attending to such wider forces is central to distinguishing political ecology from what he calls 'apolitical ecology'. The powers of money and knowledge seem to operate at a scale far removed from the ordinary people affected by biofuels, whether rice farmers in Tanzania or teenagers driving gas guzzlers in Texas. On the other hand, political ecologists are also committed to working locally with people affected by an issue like biofuels, taking seriously the knowledge that they have on the subject and making visible their experiences. One way to deal with this dual focus on both wider causes and local knowledge/practices is to resist framing the former – the 'big' forces at play – as abstract global processes, reduced solely to labels such as capitalist accumulation, globalization or modernization. Instead, the challenge is to analyse how specific processes materialize in multiple places, for example, the sites where embodied actors elaborate, negotiate or resist biofuel production (or knowledge thereof). Rather than framing the global and local in opposition, many political ecologists argue that such places are simultaneously global, national and local (see also Chapter 1).

Decisions about scale also relate to a second challenge, namely choosing which actors are privileged within political ecological analyses. In the biofuels story, the actors include multinational companies, governments and big academic research institutions as well as NGOs, national bureaucrats, entrepreneurs, families and ecologies. Focusing attention on one set of actors over others is not necessarily right or wrong, but such decisions will significantly shape the analysis at each step of the way. For instance, a focus on the global arrangements that lead to the dispossession of rice farmers in Africa may have the effect of portraying local rice farmers as helpless in the face of global forces. Conversely, attention to the efforts of displaced farmers to feed their families by harvesting resources within a national park without examining how and why they were displaced may end up portraying such farmers as illegal poachers and as 'the problem'. Many political ecologists therefore choose to study across a variety of scales and actors and employ mixed methods to produce a holistic portrayal of a given situation.

In addressing such challenges, political ecologists are committed to tracing the operation of power in the production of particular socionatural projects. Rather than understanding ecologies as natural processes that sit outside of social life, they research across nature–culture binaries. They aim to disentangle how ecological truths are produced and to understand how inequalities in human relations with ecologies may be resisted or improved. Ascertaining how particular socionatural arrangements come into being allows political ecologists to rob them of their 'naturalness' – one of their primary sources of power – and to reveal their potential for reconfiguration.

DISCUSSION POINTS

1. Discuss David Harvey's suggestion that 'all socio-political projects are . . . ecological and vice versa' (1996: 174).
2. On what basis might we distinguish between 'apolitical ecology' and 'political ecology' (Robbins, 2004, 2012)?
3. What research methods do political ecologists use, and why are they often 'mixed'?
4. Outline the implications of a political ecology stance for analyses of contemporary trends in biofuel development.

FURTHER READING

Nightingale, A. (2003) 'A Feminist in the Forest: Situated Knowledges and Mixing Methods in Natural Resource Management'. In: *ACME: An International E-Journal for Critical Geographies* 2(1), 77–90.

As noted in the main text of the chapter, this article provides an excellent analysis of mixed methods in relation to feminist political ecology.

Peet, R. and Watts, M. (eds.) (2004) *Liberation ecologies. Environment, development and social movements* (2nd edn.). London and New York: Routledge.

This is the second edition of a classic volume (first published in 1996) that framed political ecology as concerned with struggles over resources and livelihoods and argued for listening to the social movements concerned with such issues on the ground. The substantive focus is on cases from the Global South and the politics enacted at the intersection of environment and development.

Robbins, P. (2012) *Political ecology: a critical introduction* (2nd edn.). Oxford and Malden MA: Wiley-Blackwell.

Updating a first edition published in 2004, this is an excellent synthesis of the field, focused on contemporary approaches and concerns within political ecology. It approaches political ecology as 'a field that seeks to unravel the political forces at work in environmental access, management and transformation' (p.3), thereby contributing both to environmental science and practical debates over environmental sustainability and justice.

Sundberg, J. (2011) 'Diabolic Caminos in the Desert and Cat Fights on the Rio: A Posthumanist Political Ecology of Boundary Enforcement in the United States–Mexico Borderlands'. In: *Annals of the Association of American Geographers* 101(2), 318–36.

As briefly discussed in the chapter, this paper is illustrative of how political ecology approaches offer important insights to issues and topics beyond those defined as 'environmental', in this case the policing of the border between the USA and Mexico.

WEBSITE

www.iisd.org/gsi/biofuel-subsidies/biofuels-what-cost

If you want to read more about the substantive case of biofuels, then an excellent resource is the series of reports by the Global Subsidies Initiative called *Biofuels: At what cost?*, available at the website address above. The reports critically consider the subsidies being offered for the production of biofuels in a range of countries.

SECTION TWO

CARTOGRAPHIES

Introduction

Ask most people what they associate with Geography and before too long they are likely to mention maps. Maps, in the public mind, are a way of knowing and presenting the world 'geographically'. Many academic Human Geographers have been more wary of that association. Tellingly, if you browse the back issues of a range of Human Geography research journals you won't find that many maps in some (indeed most) of them. You will see an awful lot of writing in various sorts of academic prose; perhaps quotations from people interviewed as part of the research; some diagrams; photos, maybe, or tables and graphs of data; but, on average, probably less than one map for every article you scan. One commentator in Urban Geography went so far as to wonder if Geographers suffered from 'mapphobia' (Wheeler, 1998). Why is that so? And why, then, are we devoting a section of this book to the subject of cartographies?

To simplify somewhat, we can identify two reasons for a past marginalization of cartography within the discipline. First, many were concerned that maps struggled to present the kind of Human Geographies they were studying. Maps seemed more suited to presenting factual information than, for example, the complex spatial and social relations that shape the economic geographies of the world today or the rich experiences that people have of place and environments. Second, such concerns intensified as mapping was increasingly subsumed within the development of geographical information systems (GIS). The digitization of data and the digital production of maps have been hugely significant for the practice of cartography. GIS has also been very important within the recent development of Human Geography but, at least in its early development in the 1980s and 1990s, this was somewhat opposed to a range of other contemporaneous developments in geographical thought. GIS in particular, and mapping more generally, became understood by some as a technical, applied branch of the subject, committed to particular understandings of geographical science, and most often deployed in the service of government and commercial projects.

We want to challenge this neglect of maps in Human Geography. It is certainly true, in our view, that Human Geography is about much more than maps. But, equally, to cast maps and cartography as a narrow, separate, technical area of the discipline is grossly mistaken. Maps are central to past, present and future forms of geographical knowledge. Maps matter. As objects, they are fascinating cultural artefacts, representing and shaping the ways in which we know our worlds and their geographies. As practice, map-making offers the opportunity to engage publics 'geo-visually' rather than solely through the written word. Maps and map-making deserve careful, critical, imaginative analysis. Helpfully, there are substantial bodies of scholarship and research that do precisely that.

On the one hand, for example, work on the history of cartography has shown how maps are social products, exploring how they came to be made, what work they did, and how they were used in practice. Historically maps have performed religious cosmologies, claims for private property, formations of national territory, imperial imaginations, geopolitical propaganda, among other things. On the other hand, research in GIS and digital map-making has been energized by the possibilities for making maps in new ways for new purposes, intervening in a digital culture in which GIS technologies have become an everyday part of people's geographical experiences. Increasingly, these two bodies of work, despite very different intellectual lineages, are speaking to each other within a wider field of critical map studies. Cartography has become one of the most vibrant areas of thought in Human Geography today. The three chapters in this section seek to explain and express that vibrancy.

Taking his title from a seminal book by Denis Wood (more details of which are given in the suggestions for further reading below), Jeremy Crampton opens with a discussion of 'the power of maps'. In the first part of the chapter he argues that maps do not simply present the world but politically intervene into it. Careful analysis of maps, and the social and technical details of their making, gives us insight into the nature of that intervention. Jeremy illustrates this with reference to one historic map of the British Empire, one much more recent map (about social unrest and 'riots' in the UK) and through reflection on debates over world map projections. In the second part of the chapter, he considers how changes in the economies, technologies and cultures of map-making associated with what is sometimes called 'the new spatial geoweb' are delivering a new and different cartographic landscape. 'Bottom-up' mapping, rather than mapping only by powerful institutions, has become the norm. Maps, Jeremy suggests, are no longer simply the tools of professional geographers but, to quote him, 'part of our geographically lived experience'. Maps both represent our worlds and are part of our immersion within them.

In Chapter 14 Muki Haklay focuses on the place of Geographical Information Systems (GIS) within contemporary mapping. A GIS involves the representation of geographies in digital computers. As Muki explains, GIS is now a widespread and varied form of mapping, both within the academy and beyond. In the chapter, he speaks to that variety by considering the use of GIS both within practices such as location planning, where it is underpinned by the intellectual paradigm of spatial science and quantitative data, and within emergent fields of 'critical' and 'qualitative GIS', where GIS could be focused on representing the experiences of marginalized groups of people, for example. Generally, Muki argues against the equation of GIS with only one sort of Human Geography, showing how it can be used as a technology within various kinds of research. More specifically, his account shows how current work is pursuing those options through careful consideration of both the wider issues of power and representation present in mapping and the detailed, technical and scientific challenges within GIS development.

In Chapter 15 Wen Lin extends this discussion of GIS, and of the sorts of bottom-up mapping flagged earlier by Jeremy Crampton, through a focus on what she calls 'counter-cartographies'. Three interrelated schools of work are outlined and exemplified: Participatory GIS (PGIS), a label most often applied to a range of 'counter-mapping' projects working with local and often indigenous communities in the Global South; Public Participation GIS (PPGIS), a label applied to the use of GIS to involve often excluded publics in planning practice within the developed world; and forms of mapping undertaken by artists within projects that seek to subvert dominant ways of seeing places, environments and landscapes, again often through collaboration with wider publics. Wen's analysis points to the strengths of this kind of work, both in opening up the power of maps and in enacting forms of more participatory, collaborative research in Human Geography. She also, however, shows how work in these areas is not naively enthusiastic, but reflects critically on the kinds of 'empowerment' such counter-cartographies perform.

Generally, cartography is a particularly exciting area of interest at the moment. This is not only true within academic Human Geography but also in wider public culture. The last decade has seen a proliferation of exhibitions, books and television programmes devoted to the subject of maps. The further reading (and viewing) suggested below aims to give you a feel for that.

FURTHER READING

A number of journals present research on mapping and cartography. *Cartographica* publishes research on a wide range of issues concerned with cartography, geovisualization and GIS; *The Cartographic Journal* has a similar remit. *Cartographic Perspectives* is an open access journal that publishes work on both the history of cartography and contemporary developments in mapping. As its subtitle explains, *Imago Mundi: The International Journal for the History of Cartography* publishes scholarly studies exploring the history of maps and mapping. The *Journal of Maps* is an online journal that publishes new maps and spatial diagrams. Finally, the regular progress reports on Cartography in the journal *Progress in Human Geography* offer overviews of the issues and debates shaping the field, albeit often in quite a compressed form.

Cosgrove, D. (ed.) (1999) *Mappings*. London: Reaktion.

A collection of critical interpretive essays on the politics and cultures of mapping, past and present.

Crampton, J. (2010) *Mapping. A Critical Introduction to Cartography and GIS*. Oxford: Wiley-Blackwell.

An important book that brings together what had been often oddly disconnected debates over 'critical geography' and developments in cartography and GIS.

Dodge M., Kitchin, R. and Perkins, C. (eds.) (2009) *Re-thinking Maps: New Frontiers in Cartographic Theory*. Abingdon: Routledge.

This collection of essays assesses the diverse forms that mapping now takes and comments on new thinking with respect to cartography.

Dodge, M., Kitchin, R. and Perkins, C. (eds.) (2011) *The Map Reader: Theories of Mapping Practice and Cartographic Representation*. Oxford: Wiley-Blackwell.

A large collection of key writings that reflect on maps and mapping practice. Themes covered include general approaches to cartography, the relations between maps and power, the design and aesthetics of maps, the reading and use of maps and technology and mapping.

Garfield, S. (2012) *On the Map: Why the World Looks the Way It Does*. London: Profile Books.

An engaging, journalistic rendition of the ideas generated by more scholarly work on the history and cultures of cartography.

Wood, D., with Fels, J. and Krygier, J. (2010) *Rethinking the Power of Maps*. New York: Guilford Press.

This is a much revised edition of Denis Wood's classic book, *The Power of Maps*, first published in 1992. In that first book, Wood argued that maps are not objective and impartial but powerful forms of persuasion that present particular points of view and represent particular interests. In this later edition he revisits that argument, adding in additional discussion of the contemporary ways in which maps are being used in more radical and creative ways and unbound from the interests of state or industrially sponsored cartography.

WEBSITES

Two BBC documentary series in 2011 focused on maps:

Maps: Power, Plunder and Possession: **www.bbc.co.uk/programmes/b00s5m7w**

The Beauty of Maps: **www.bbc.co.uk/programmes/b00s2w83**

In both cases clips from the series are available for viewing from those websites or via YouTube.

A British Library exhibition in 2010 called 'Magnificent Maps: Power, Propaganda and Art' still has an interactive website at: **www.bl.uk/magnificentmaps/index.html**

You can explore particular maps from the exhibition and view curatorial commentary on them. You can also listen to podcasts of discussions on topics ranging from historical work on the social production and display of maps, to geopolitics and maps, to new open source mapping technologies and Google maps.

CHAPTER 13

THE POWER OF MAPS

Jeremy Crampton

Introduction

In physics and engineering, '**power**' is defined as 'a measure of doing work'. More generally, we usually think of power as the capacity to do something. In the phrase 'the power of maps' we can see right away that maps are *active and not passive* – they are a capacity to do things. Not only are maps involved in achieving a goal, they also define or identify that goal: they *frame the narrative*.

In this chapter we shall ask what it is that maps have a capacity to do. We shall see that the answer is a political one, understanding politics here to be the issue of deciding what is a problem and of deciding how to solve it. (Some authors call this 'problematizing' but I shall avoid that rather ugly word here.) In doing so, maps (and their creators and readers) actively assess and create new knowledge.

This means that maps are political interventions in the world. Why? When you decide that something – but not something else – is a problem, you are making a political decision. After all, the root word in 'politics' is Greek for 'city' [*polis*]. Something that is political is therefore a question of how to run the city (or the place, more generally). Power and politics then (in the sense used here) go together.

SUMMARY

- The power of maps refers to the way maps are used to interpret and affect the world.
- Maps actively construct new knowledges.
- All maps are political interventions.

How maps are powerful

This is easy to understand in say, a propaganda map or the 'subversive cartographies' of the 1960s Situationists or the famous 1929 surrealist map of the world (Figure 13.1). In this map the author (thought to be Paul Éluard (1895–1952)) is obviously playing on notions of territorial sovereignty by deliberately misplacing or omitting countries (including the USA). A meandering Equator seems to get lost in the Pacific islands. Easter Island (Isle de Paques) is almost as big as South America. Only two cities are named: Paris and Constantinople (Istanbul). The whole thing is a visual pun on map projections and their distortions of lived space.

Figure 13.1 Surrealist map of the world, possibly by Paul Éluard, 1929. Source: Paul Éluard, *Variétés*

Figure 13.2 'Mapping the riots with poverty'. Red indicates areas with high social deprivation and blue indicates areas with low. Credit: *Guardian Newspapers*

But let's say we have a more everyday map. Figure 13.2 shows one from *The Guardian* newspaper following the riots in the UK in the summer of 2011. In this map, the newspaper has created a map mashup by combining Google locator pins with a layer shaded to represent social deprivation, which is a measure of poverty. (Compare this to the Charles Booth maps of poverty in Chapter 14.) How is this an example of the power of maps? Here are four reasons. A map may use one or more of these.

Framing the narrative

As we examine Figure 13.2 the first thing to notice is that the argument is already being framed for us. It is being framed as one of a purported relationship between rioting and poverty (or lack of poverty). The map has a **narrative**. This does not mean the data are wrong or poorly chosen. We can still frame the argument when the data are bang up to date. The terms of debate that this map establishes is whether there is a relationship between rioting and social deprivation. We can visually scan the map to decide for ourselves, or we can perform sophisticated geostatistical analyses. But in either case we're still working within the framework of the basic question. So the map is not just a description of the way the world is, but also a prescription for what might be done about the problem as identified.

Location of incidents vs. location of perpetrators (cause and effect)

Another way Figure 13.2 exercises power is that it maps the location of the riots. Fair enough, you might say. But if we're looking for a geographical pattern, a relationship between poverty and riot action, does it necessarily follow that people will riot in their own neighbourhood? Or, might they riot in other neighbourhoods, perhaps looting in shopping centres or along the high street a few blocks away? We don't know. The map is silent on that issue. It exercises power by allowing discourse of one thing, but silencing (or marginalizing) other discourses (see Case Study box). This is an example of cause and effect. We are required to reason about cause and effect in the terms provided, even if there might be other sorts of cause and effect. An example of another sort of cause and effect is quite geographical. Is the effect caused by the local conditions or rather by what are called 'spatial externalities' (that is, factors from outside the local area)? What is the proper scale of analysis here? Can we carve out a square of London here, or do we need to bring in other places and even networks of relationships?

A good example of externalities is provided by the concept of poverty itself. In Figure 13.2, poverty is indicated by area, for where people live. We often have to map data this way because of the way it is collected in the census. Typically, a census form is delivered to a household (there are household address databases that the government maintains, although these might be incomplete and not every citizen has an address). However, let's assume that these data are correct and up to date. If we're interested in what *causes* poverty though, this map is very unhelpful indeed. Remember, data on social deprivation are used not just to see where it is, but also in order to direct efforts to get rid of it (funding services and assessing need). But the causes of poverty might very well be low wages at the place of work, low-value skill sets of the residents or government policies. In that case, a map of poverty would not concentrate on where people live, but would show places of employment, poor schools and the offices of the County Council or Westminster!

One of the first people to think about causes of poverty and how we map and talk about it is the Development Geographer Lakshman Yapa. Yapa has worked closely with community groups in Philadelphia, USA. Yapa found that the discourse on poverty, and on a larger scale the development of countries, was always already framed in certain ways, primarily as lack of financial resources (low income). Additionally, a norm was invoked, that of the well-off suburbs:

> If one-fourth of people live in the suburbs, three-fourths cannot . . . the sheer quantity of resources needed per family to sustain a suburban lifestyle automatically disallows a majority of people from also adopting it. We must begin to see that suburban life is not the only model of good living if we are to have the power to solve the problem of scarcity . . .
> (Kusch, 2000: 3)

KEY CONCEPT

Marginalization

Marginalization and silencing (i.e. what is not on the map) are as important a component of mapping as what is included. On one level this is a factor of scale and generalization, and will involve explicit and implicit choices by the map-maker or what is afforded by their mapping environment. Many GISs, for example, do not offer a direct way to make a form of map known as a dasymetric map, even though it is very good at representing certain kinds of data (see also **representation**). On another, data that could appear on the map is excluded because it does not comport with the map's narrative. For example, Christopher Columbus silenced existing indigenous place names from his maps of the Caribbean because he wanted to replace them with Christian names. 'America' itself is also not an indigenous name, but comes from the Italian explorer Amerigo Vespucci's first name.

It was as though there was only one way to be well-resourced: to have what people in the suburbs had.

One particularly pressing problem is that of 'food deserts', where healthy nutritious food is made expensive and hard to get, while fast food is made inexpensive and highly visible. Many inner city areas are food deserts. If it is a question of feeding your family while living in poverty (as measured by income), it is very challenging while working two jobs and taking long time-wasting rides on public transport. Yapa has helped identify urban co-ops and food gardens as viable alternatives – community resources not included in a simple income-based measure of poverty. (*The Guardian* uses the UK Government's Indices of Deprivation, which use multiple measures but are heavily weighted by income and employment.)

The lesson here then is to look behind the map and the way that it frames cause and effect. This of course is not always easy, but almost certainly will involve the next question.

What counts as data?

This is a map of riots. But what is a 'riot'? How many people need to be involved? What must they do for it to be a riot and who gets to say? Governments have often labelled people protesting peacefully as rioters, so these are important questions. Luckily *The Guardian* map is quite transparent. It presents a Google fusion table (online spreadsheet) of all the incidents they call riots and that appear on the map. Examining these gives a clearer picture of the wide range of activities deemed riots. Here are some examples – are all these riots, in your mind?

'Gang of youths gather', 9 August 2011: Cabot Circus, Bristol.

'Bottles thrown at police', 9 August 2011: Gloucester Road, Bristol.

'Former Oasis singer Liam Gallagher's shop looted', 9 August 2011: Pretty Green, Manchester.

'Approximately 50 youths congregated and caused damage to property', 7 August 2011: Oxford Circus, London.

'Dozens of rioters, many wearing hoods and scarves, ransacked the shop. A small number of riot police left the scene when they came under light bombardment from projectiles', 8 August 2011: Curry's, Clapham Junction.

What role does symbology and map design play?

Map symbology is a very important part of the power of maps. Sometimes this is very explicit and almost impossible to miss. An example of this is shown in Figure 13.3 (see also Chapter 16 for a further discussion of this map). This is a famous map from 1886 called 'Imperial Federation – Map of the World Showing the Extent of the British Empire'. It shows the empire in its classic pink or red colour. But perhaps what is most remarkable about the map are the marginal illustrations. These act to support and highlight the positive virtues of the British Empire. In the centre is a woman dressed with a Greek helmet and holding a triton (these evoke the Greek warrior goddess Athena). Her shield has the union flag of the United Kingdom and she's literally sitting on top of the world (you can see Atlas balancing it on his head).

The woman is Britannia, the embodiment of Britain. Surrounding her, and turned in obedience towards her, are the fruits and products of the Empire. There are the fishes of the sea, two native women positioned almost as

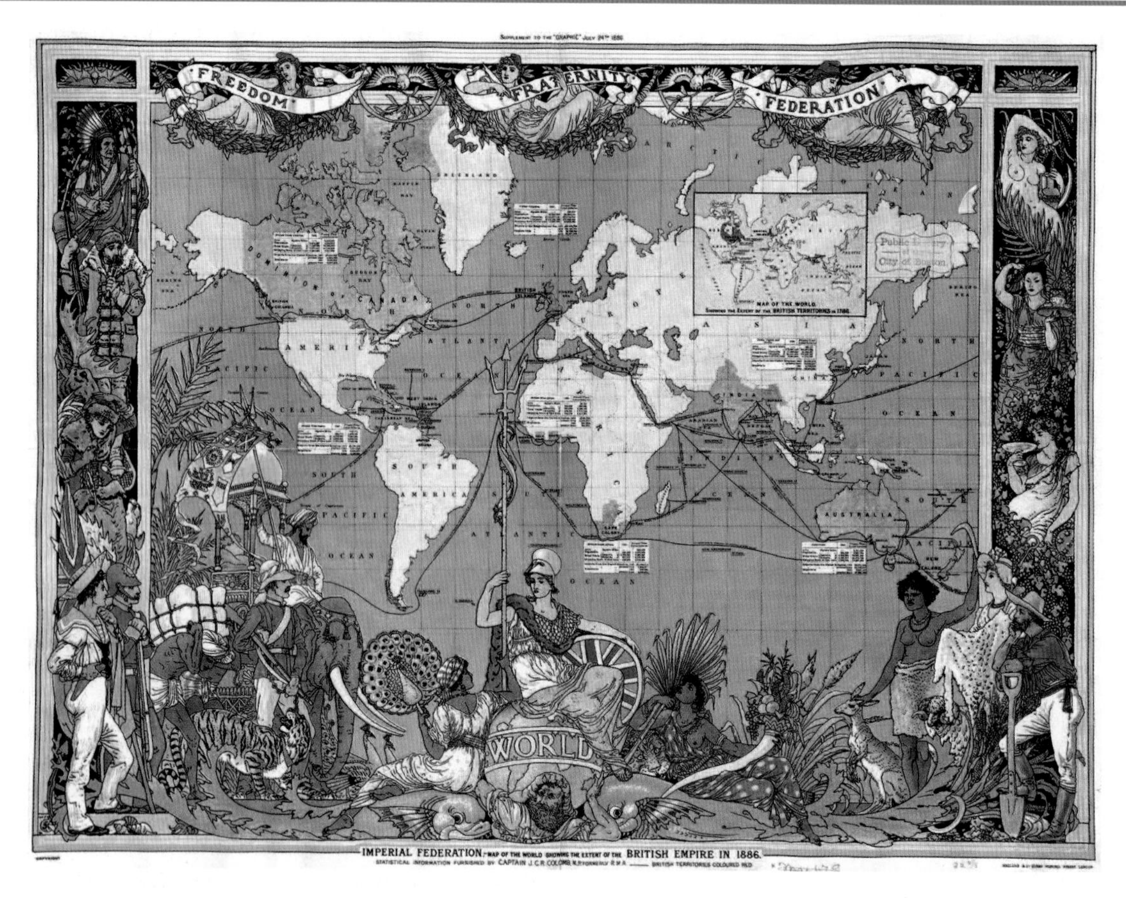

Figure 13.3 'Imperial Federation Map Showing the Extent of the British Empire in 1886', *The Graphic* Supplement, 24 July 1886. Credit: in public domain

ladies in waiting, one holding a cornucopia (and leaning on a lion's head), the other a fancy peacock fan. Peoples dressed in local costume adorn the map, as if coming to pay homage to her. The East is symbolically eroticized, with a sensually reclining naked woman holding a flask of massage oils (compare the white settler women or Britannia herself who appear fully clothed.) Even the wild beasts are tamed, as if in awe of her majesty.

The idea here is *dominion*, that is, control and possession. Unmistakably there is a reference to *Genesis*, where man is given the power to rule 'over the fish of the sea and the birds of the air, over the livestock, over all the earth' (Gen. 1).

The message is clear: Britain has a divine right to rule over the earth, to establish its empire. And, as it does, it brings prosperity (the cornucopia), peace and enlightenment everywhere it reaches. (This peace will be brought by the measured use of force if necessary; note that the soldier has slung his rifle at the ready and the sailor rests his hand on the hilt of his gun.)

None of this is said directly, but it's fairly explicit. In any case, we know who produced the map, a man called Walter Crane (Biltcliffe, 2005). Interestingly, Crane was no nativist conservative, but rather a Victorian illustrator and member of the Arts and Crafts movement.

As a socialist he sought to achieve a utopia that valued the dignity of the worker. (Too small to see here, Crane labelled Atlas with the words 'Human Labour'; that is, that all this wealth and enlightenment is done with and through the working class and don't you forget it!) The map was produced totally within the mainstream of opinion and was exhibited at the Colonial and Indian Exhibition in November 1886 (Biltcliffe, 2005).

Could we produce a map like this today? Well, not literally, since the British Empire has ended. (What empires exist today – the twentieth century was called the American century, but what will we call the twenty-first?) But Figure 13.4 shows a contemporary example that in fact evoked more controversy than did the Imperial Federation map in its day. It first came to prominence in 1974 and was developed by Arno Peters, a German historian.

Peters' map is much more restrained than the pre-Raphaelite-inspired Imperial Federation map (later versions were published by the development aid magazine *New Internationalist*). But is it any less political? At first glance it appears to be a simple map projection, showing the continents in realistic colours. Like any map projection, it takes the three-dimensional sphere of the earth and flattens it out (as in *King Lear*: 'strike flat the thick rotundity o' the world!'). It falls into a class of major projections known as cylindrical projections, of which the most famous is the Mercator projection (1569). So how could this map be so controversial as to attract nearly 100 critical articles published in the cartographic literature and evoke a resolution by seven North American cartographic organizations condemning rectangular world maps (including the Peters map)?

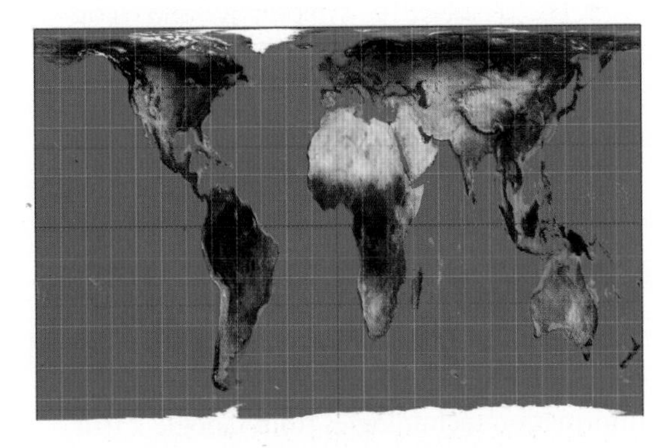

Figure 13.4 Peters' projection or world map. Credit: Strebe (Wikimedia Creative Commons)

True, the continents do look a little odd. Peters' most prominent critic, the American cartographer Arthur Robinson, once memorably described it as looking like long wet ragged underwear, left to dry from the North Pole! In fact, the odd look comes from a specific goal that motivated Peters: to make an equal-area world map. (There are plenty of odder-looking projections, but they are usually reserved for specialized uses.) As a child, Peters' family had been visited by the American civil rights leader, William Pickens. Pickens, whose parents were liberated slaves, was a graduate of Yale and vice-President of Morgan College in Baltimore. He had been involved in the National Association for the Advancement for Colored People since its inception in 1910 and was probably America's best known black man and later president of the NAACP. A fluent speaker of German, Pickens travelled to Germany at least four times (Avery, 1989). He expressed much admiration for the resistance of ordinary people to the increasingly authoritarian rule of the Nazi party. Pickens inscribed a copy of his autobiographical book *Bursting Bonds* (Pickens, 1991(1923)) to Lucy Peters, Arno's mother (ODT Inc., 2008).

Thus inspired, Peters (whose son was a cartographer) started to notice that the Mercator map failed badly at representing the world in its true relative size. (The Mercator retains good shape, known to cartographers as conformality.) This eventually resulted in his equal-area projection. For Peters, the Mercator map had negative political effects, and he argued that it was essentially racist:

> This geographical view of the world is designed to eternalize the personal overestimation of the white man and in particular the European while keeping colored peoples conscious of their impotence . . .
>
> [The Mercator] map is an expression of the epoch of the Europeanisation of the world, the age in which the white man ruled the world, the epoch of the colonial exploitation of the world by a minority of well-armed, technically superior, ruthless white master races . . .
>
> (Peters, 1974)

So if Peters' map was political, it was being offered as an antidote as it were to the Mercator. (Plus, for Peters, the Mercator symbolized the square TV-watching passivity of modern society.)

Peters' map became very popular, much to the dismay of the cartographic profession. The latter were already well aware of the limitations of the Mercator, and indeed that there were a number of alternatives (Robinson himself managed to get the American Geographical Society to adopt his projection for about ten years as its official world base map). Cartographers were also incensed by Peters' inadvisable claims that his map was the only one needed from now on and leapt gleefully on some small technical errors that Peters had made in earlier versions of the map. Today, if you ask a cartographer (and many geographers) what they think of Peters . . . well, just try it for yourself!

What is interesting here though is that cartographers and Peters basically talked past each other. For Peters and all those who bought his map, including the United Nations and hundreds of NGOs, the point of the map was its political effect: equal representation for all peoples. No longer would Africa, a continent of over 50 countries, appear smaller than its true relative size. Europe's tiny territorial extent would also be correctly shown and not inflated as on the Mercator. For cartographers, the point was all about Peters' incorrect cartographic statements and irritating claims of originality (they dug up a previous similar map from 1855 constructed by an obscure minister in Scotland named James Gall and pointedly called the map the 'Gall-Peters projection'). For Peters, the map was a political act. For cartographers, it was a cartographic act.

How did it come to be then that the discipline of cartography itself refused the political content of mapping? That is, that they refused the power of maps? To answer this, we need to look at the new power of maps.

SUMMARY

The power of maps is a result of:

- Framing the argument or narrative;
- Making a cause and effect claim;
- Defining what counts as data; and
- Using effective symbology and map design.

The new power of maps

Today, as Haklay describes in his chapter on 'Geographical Information Systems', you can interact with a multitude of geographical information technologies from Google Earth to your broadband smartphone (Chapter 14). Many of these applications and services

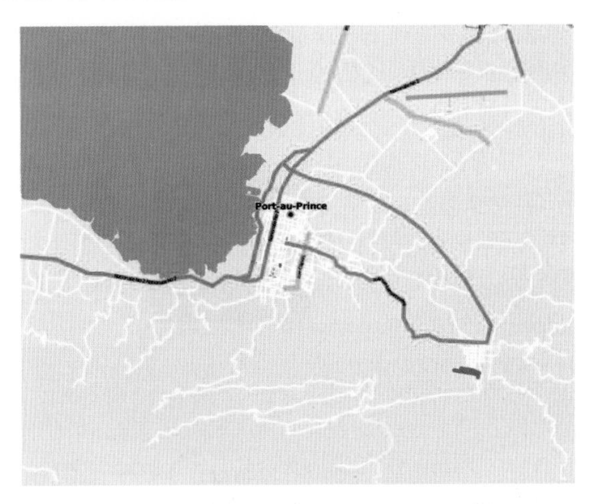

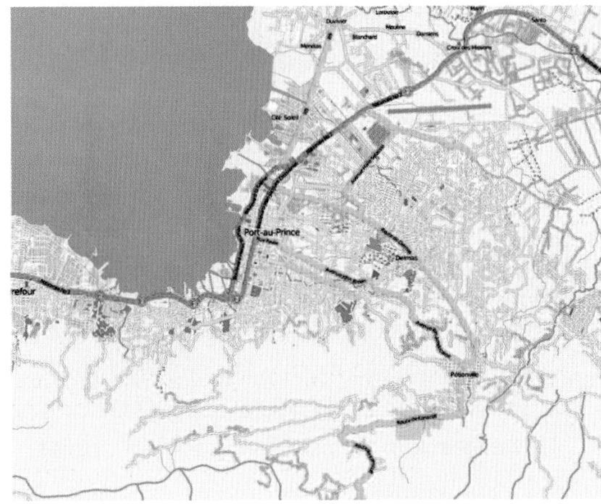

Figure 13.5 (a) Before and (b) after views of Port-au-Prince, Haiti in OpenStreetMap. Credit: (a) and (b) Mikel Maron (Creative Commons)

themselves draw on a mixture of data sources. Take, for example, geotagging pictures in Twitter and the project known as OpenStreetMap (OSM). Some sources are traditional and official. Others, like OSM, rely on the contributions of thousands of volunteers worldwide. Following the earthquakes in Haiti for example, OSM contributors got a working map of the affected areas prepared well before official mapping agencies could react (Figure 13.5).

Although we may be familiar with these sorts of services and mapping capabilities, they represent a recent turn in the history of mapping. Google Maps and Google Earth, for example, date from only 2005. Here's how it happened, featuring one of the most influential cartographers you've never heard of.

In late 2004, Paul Rademacher was looking for somewhere to live in San Francisco using Craig's List, the classified housing list. Actually Rademacher is not really a cartographer; he was an animator for DreamWorks, the studio behind films such as *Shrek*. The problem was that the piles of printouts and maps were hard to use. Rademacher wondered if 'it would be better to have one map with all the listings on it' (Ratliff, 2007: 157). Then, on 8 February 2005, Google Maps went online. It provided a sliding (or 'slippy') base map of the roads, rivers and town names. It was great if you wanted to look somewhere up, but it was not meant to be more than a digital version of existing locational maps. Rademacher then took a brilliant unprecedented step. He 'reverse engineered' the JavaScript for Google Maps so that it contained Craig's List apartments and houses for sale. This was the world's first 'map mashup'. A mashup is a website or program that combines two or more different sources of content into one tailor-made experience (Butler, 2006; Miller, 2006).

Google's response to this development was the key to their now pre-eminent position in mapping. Rather than prosecuting Rademacher or fixing the 'hacked' map, they actually opened up their database to more mashups. Their insight was that by embracing the 'open-source' model of knowledge they could both be successful in getting their maps used and provide a societal good. And it worked. Google Maps are everywhere, including your phone. It worked for Rademacher, too – he was hired by

KEY CONCEPT

Free and open-source software (FOSS)

Open source software is 'configured fundamentally around the right to distribute, not the right to exclude' (Weber, 2004: 16). In the open source community 'free' refers to freedom, rather than free-of-charge (i.e *libre*, not *gratis*). These freedoms include:

1. The freedom to run (use) the program for any purpose.
2. The freedom to modify the program (at the code level if necessary).
3. The freedom to distribute copies (sell or donate).
4. The freedom to distribute modified copies.

(Stallman, 1999: 56)

An extension of this idea that seeks to balance author rights versus these ideal goals is © Creative Commons, founded in 2001.

There is also an Open Source Geospatial Foundation for geographical data (OSGEO).

Google to work on Maps and Google Earth. So perhaps he is a cartographer after all!

So now we have thousands and thousands of map mashups, OpenStreetMap is mapping the world without copyright restrictions and databases are opening up. You can even be involved in public participation GIS by contributing to ongoing scientific projects or by developing your own (see Chapter 15 for further discussion).

Another emerging practice is citizen politics. Many people are so turned off by politics that voting rates are in decline, with the consequence that their voice is not heard and they are disenfranchised from decision-making and resource allocation. Some states in the USA are addressing this by providing websites where citizens can participate. During 2011 and 2012 for example, many US states went through a decennial process of redrawing their voting districts (a process called redistricting). Usually this is done in secret by elite groups of law makers. Perhaps inevitably, these districts favour incumbents by drawing lines around voters likely to re-elect them. But voters should choose politicians, not the other way round! Using online websites and web mapping services (WMS) you can now draw your own districts and make sure they have the right number of people in them (by law in the USA, districts should be equally sized). This will not solve political disenfranchisement, but it is an indication of shifting power dynamics, with maps playing a central role.

All of this, however, is a direct challenge to traditional ways of providing maps and geographical knowledge. It is bottom-up, whereas traditionally it was top-down. Many people celebrate this development as the new power of maps. Perhaps today, the Peters world map would be just another user-contributed map mashup? On the other hand, it disrupts and is in tension with established norms and as such, gives rise to questions about the information being provided. By definition it is not authoritative, but rather asserted. Can you trust it? Is it deliberately misleading or biased? In response, some geospatial organizations have developed qualifying certificates based on experience, for example URISA GIS Professional Certificate (GISP).

KEY ISSUE

A challenge to the discipline?

Cartography and Geography are *disciplines*. This can refer to the traditional academic sense of the major or subject that you study, but also to the fact that they discipline (control) a field of knowledge. There are norms and standards. The new power of maps refers to challenges to these norms. For example, that only governments and institutions can produce reliable geospatial data. There are many other names used to describe this new power of maps, including the geospatial web, neogeography, new spatial media and **volunteered geographic information** (VGI).

SUMMARY

- The new power of maps refers to the development of bottom-up mapping and GIS capabilities.
- These include map mashups, citizen participation in science and official map-making, and an emphasis on volunteered geographic information (VGI).
- Some celebrate these bottom-up forms of cartography for putting the power of maps in the hands of people rather than just official map-making institutions. Some are concerned about the quality of geospatial data that they may use.

Conclusion

Maps are powerful ways of describing and intervening in the world. Although they have traditionally been associated with government and institutional projects (e.g. colonialism, territorial governance; see Chapter 33 for Richard Phillips' analysis of maps in relation to colonialism and post-colonialism), in the early twenty-first century that situation is radically changing. Mapping and GIS capabilities are available to increasingly more groups (although this does not mean the end of the digital divide). The new spatial geoweb and VGI provide bottom-up products and services that are deeply imbricated in daily lives and mobilities (see also Chapter 54). What this means is that maps and mappings are more than just tools for geographers (although they remain that) – they are becoming part of our geographically lived experience.

DISCUSSION POINTS

1. Power is both an abstraction and real social practice. What are some everyday ways in which mappings have been used to exercise power?
2. Is the exercise of power necessarily negative or can it also be productive?
3. Have you observed maps and/or geospatial data being used in mobile situations? What were they? Which ones worked or didn't work, and why?

FURTHER READING

Hannah, M. (2000) *Governmentality and the Mastery of Territory in Nineteenth-Century America.* Cambridge: Cambridge University Press.

Hannah draws lightly on Foucault to examine one of the great mapping projects of the late nineteenth century. Has surprising relevance to questions of information, surveillance and identity today.

Harley, J.B. (2001) *The New Nature of Maps: Essays in the History of Cartography.* Baltimore, MD.: Johns Hopkins University Press.

A collection of essays by one of the first people responsible for critically re-evaluating maps and the production of knowledge. Readable, with an infectious, enquiring spirit.

Krygier, J. and Wood, D. (2011) *Making Maps: A Visual Guide to Map Design for GIS* (2nd edn.). New York: Guilford Press.

A practical and visually rich guide to making your maps more powerful. Wood also wrote the classic book *The Power of Maps* (1991), which was in turn reworked in a much developed second edition called *Rethinking the Power of Maps* (2010).

CHAPTER 14 GEOGRAPHICAL INFORMATION SYSTEMS

Muki Haklay

Introduction

It is morning, and I am heading to a meeting at the offices of Google in London. Before leaving home, I log in to Transport for London's 'Journey Planner' website to find out the best way to arrive there by public transport. I do that by entering my home postcode and the office's address. Within seconds I receive instructions of which bus to take and details of possible delays on the way due to road works. On my way to the bus, I notice that a coffee chain has just opened a new shop on our local high street. After the meeting, in which we consider how Google Earth, a 'virtual globe' that allows 'flying' to any spot on earth and viewing information about it, can be used in teaching geographical concepts, I use my phone to register the location with a taxi company, which then sends a taxi to pick me up. In the taxi, the driver checks his Personal Navigation Device (PND or Satellite Navigation/SatNav in the UK), which relies on the satellites of the Global Positioning System (GPS) to identify our location, while receiving updates over the mobile phone network that show the current level of congestion on our route.

Almost without noticing, I was interacting directly with, or influenced by, five different geographical information technologies, most of which can be included under the banner of geographical information systems or GIS. GIS emerged from early experiments with computers about a half century ago, and they are now shaping and influencing many aspects of Human Geography. For example, in the opening vignette, it is easy to overlook the mention of the coffee shop. In all likelihood, the coffee shop chain made the decision to open a shop in my neighbourhood following geographical analysis by a GIS at their head office. This type of influence on reality is more hidden than the direct influence of the SatNav but is no less profound.

GIS have multiple incarnations within Human Geography – as a tool in the analysis and visualization of public and private datasets in the daily operation of government and business; as part of the toolset of spatial science and the focus of Geographical Information Science (GIScience) as it was defined by Goodchild (1992); as the focus of attention of studies that explore its influence on society;

Figure 14.1 Google Earth. Credit: Google Earth

or as a tool that is used for resistance and empowerment of marginalized groups in society (Longley *et al.*, 2010). At the same time, the technology on which it is based puts limits on what can be represented and analysed with a GIS. Even more than the technology, the concepts and methods for processing information allow for some things to be done with GIS easily, while making other tasks very difficult. Interestingly, many academic disciplines such as Sociology, Public Health, Anthropology or History are happy to adapt GIS to their needs and epistemological frameworks, while some Human Geographers are reluctant to use it (see Chapter 8) – precisely because they are acutely aware of the disciplinary baggage that GIS carries with it everywhere it goes. Unfortunately, the assumption that GIS is a locked and immutable technology is a simplistic and, ironically reductionist view, that fails to notice the potential of using a technology that should be 'owned' by geographers.

This chapter is not aimed at focusing on the technical description of GIS and its applications – for that there are plenty of sources, including the popular *Geographic Information Systems and Science* (Longley *et al.*, 2010). Instead, the focus is on the relationship of GIS with Human Geographies. As a scientific tool, GIS is used by both Physical and Human Geographers, and techniques move between the two areas. Thus, the analysis method Kriging, originally developed for estimating the distributions of gold deposits in mines, once computerized has been applied widely to socio-economic data sets. However, while Physical Geography relies on the scientific method, the role of scientific analysis in Human Geography is contested (see Chapter 7) and, as a result, so is the way that GIS is perceived.

Too often, the role of GIS within Human Geography is assumed to be only as a scientific tool that enforces positivist-like **epistemology** on its use. This might be true about its origin and was clearly articulated during debates in the early 1990s about its role in geography – of which the most cited are Taylor (1990), Openshaw (1991), Taylor and Overton (1991) and Openshaw (1992). However, by the mid-1990s the terms of the debate changed, with the geographers who were developing GIS and getting used to their own identification as geographical information scientists starting to engage with their critics (see Schuurman, 2000, for a comprehensive overview). As a result, geographers started to pay attention to 'GIS and society' issues, which today is a flourishing and active area of research (see Nyerges *et al.*, 2011). Therefore, to understand the role of GIS in Human Geographies today, it is vital to explore these multiple interpretations.

We start with a short explanation of the nature of GIS. Following this, we will look at two aspects of GIS. Since, historically, GIS was associated and linked to **spatial science**, we will start with an example of a location decision, similar to the one that resulted in the coffee shop in my neighbourhood. In the second part we will look at what is termed 'critical GIS' and 'qualitative GIS', to demonstrate how the technology can escape the narrow confines that critics sometimes assume it to be limited by.

The nature of geographical information systems

So what is a geographical information system? Most common definitions (and there are many) define it as a computer system and a set of organizational and personal practices that allow the collection, organization, analysis and visualization of geographic information in a digital form. Although definitions such as this one are a useful short description of what we call GIS, it is important to understand the tool in more practical terms.

First and foremost, GIS are about representation of geographical realities in digital computers. Their origins are in the 1960s, at about the time spatial science emerged with the suggestion that Human Geography should become similar to the physical sciences through the extensive use of empirical evidence and the development of mathematical models (see Chapter 8). At that time, several research centres around the world started using computers for geographical research, most notably the Harvard Laboratory for Computer Graphics and Spatial Analysis and the Experimental Cartography Unit in the Royal College of Art in London. This early work was mostly concerned with proving that computers are capable of analysing geographical data and producing useful results. Early researchers had a vision of computers as machines that are capable of analysing vast amounts of data – for example, in 1963, Malcolm Pivar and his colleagues, while demonstrating a rudimentary display of oceanographic information on an early computer screen, suggested that:

> Only a high-speed computer has the capacity and speed to follow the quickly shifting demands and questions of a human mind exploring a large field of numbers. The ideal computer-compiled oceanographic atlas will be immediately responsive to any demand of the user, and will provide the precise detailed information requested without any extraneous information. The user will be able to interrogate the display to evoke further information; it will help him track down errors and will offer alternative forms of presentation.
>
> (Pivar, 1963: 396)

The reality was very different. In 1995 one of the pioneers of GIS, Ian McHarg, reflected in unsavoury language that the quality of output from early GIS was: 'Absolutely terrible. I mean there wasn't a left-handed . . . technician who couldn't do better than the best computer' (GIS World, 1995). Yet, because of their belief in the ability of computers and the goal of providing geography with empirical models, they continued to develop the technology until it became usable.

One of the core problems they faced was the need to store representations of the world in digital computers. To do that, the information had to be converted to numbers – in the same way that later on music, images and videos were encoded as numbers so they could be transmitted and played on digital devices. To achieve the encoding of information about the world as numbers, geographers used cartographic representations that allow the use of a pair of coordinates to describe the location of a shop or a long list of coordinates to describe a road. As the ability of computers progressed, GIS became capable of manipulating large data sets and processing complex geographical models.

To make GIS operational, these systems are based on knowledge from many areas of academic research – surveying and geodesy (the science of measuring the earth) to collect data and turn it into numerical representations; software engineering to create the software; database engineering to store large quantities of data and efficiently retrieve it; statistics (and especially spatial statistics) to develop new methods of analysis; computer graphics to visualize the information; cartography to produce useful and meaningful maps; and, of course, geography for the concepts and theories that are tested and represented by these systems.

Not surprisingly, with such a mixture of backgrounds and epistemologies, the end product is never complete or without problems. For most of their existence, GIS were notoriously difficult to use (Traynor and Williams, 1995; Haklay, 2010). Only since the late 1990s, with the introduction of graphical user interfaces that actually allow users to view the information they are manipulating and visually explore the data as Pivar had suggested almost four decades earlier, have GIS become more readily available to a wider group of people. With the introduction of specialized applications such as SatNav devices or web maps, the ability to use GIS became commonplace.

SUMMARY

- Geographical information systems are computerized systems that hold digital representations of the world, with numbers that represent points, lines, areas and continuous surfaces. To implement them, their creators had to overcome complex technical and conceptual challenges.
- The processing of this information is achieved using knowledge from multiple areas and the results are always a compromise between epistemologies, ontologies, practicality and efficiency. As a result, similarly named operations will produce different results in different systems.

Using GIS within the spatial science paradigm: location planning

So far, in describing what GIS are we have ignored the type of information that is stored on the system – and, indeed, there isn't a specialized GIS for handling socio-economic data as opposed to handling physical geography data. Nonetheless, socio-economic applications of GIS have received specific attention over the years (e.g. Martin, 1996; Harris *et al.*, 2005) and we now turn to an example of these applications to understand how GIS is used in the scientific interpretation of Human Geography. The example relates to location planning; the wider point is to illustrate how GIS has been used within what is termed **spatial science.**

The scientific interpretation of Human Geography – spatial science – is based on a method of collecting verifiable facts about human activities in a systematic and comprehensive way, analysing them using quantitative analysis techniques and testing the results with statistical tests which verify that the observed pattern is unlikely to be there by chance. Most importantly, establishing theories and then testing them with models that are based on empirical observations allows for the prediction of future events, in addition to an accurate and truthful description of the present.

In the UK, an example of a large-scale empirical data collection exercise is the population census that runs every ten years and surveys the full population of the United Kingdom, asking a range of questions, from the number of people living in a household to the location of the place of work. Each census form can be associated with an output area – an area that contains about 125 households and has been created through sophisticated GIS processing (Martin, 2004). The census is used by the government and other public bodies to understand the changing characteristics of the population – such as changes in family size or planning how many school and university places will be required in the future.

However, there is another use to this data source. It is possible to link it to other sources of information, such as credit card rating, home ownership and patterns of spending. This is because each record is linked to a postcode, which is also common to the census. Using statistical techniques for identifying areas with similar properties in the statistical sense of similarity (clustering), it is possible to group areas together. Using such a technique, the more than two million postcode areas of the UK, each representing about 25 households, can be grouped into 150 or so types. Next, each group is named using characteristics that are the most common to its members. For example, a group that is mostly concentrated in an urban area and contains young professionals who spend a large proportion of their income on consumer electronics can be called 'technology savvy professionals'.

The process that we just described is the creation of geodemographic classification (Harris *et al.*, 2005) – see Figure 14.2. These classifications are created by commercial producers (one of the most famous is Mosaic UK) and they are used for direct marketing, but also for location decisions. Using such a classification, the coffee shop chain identifies locations across the country in which the people who have visited their shops live. They can find out about who is visiting their shops from loyalty cards and also from analysis of which shops are doing especially well, and use GIS and geodemographic classification to see the type of people that live nearby.

(a)

Figure 14.2 (a) Geodemographic classification groups the population into 'types' that have certain characteristics in common; (b) Geodemographic classification then maps where different groups of the population live, allowing firms to see where would be most profitable to target. Credit: (a) Experian Ltd; (b) Alex Singleton

(b)

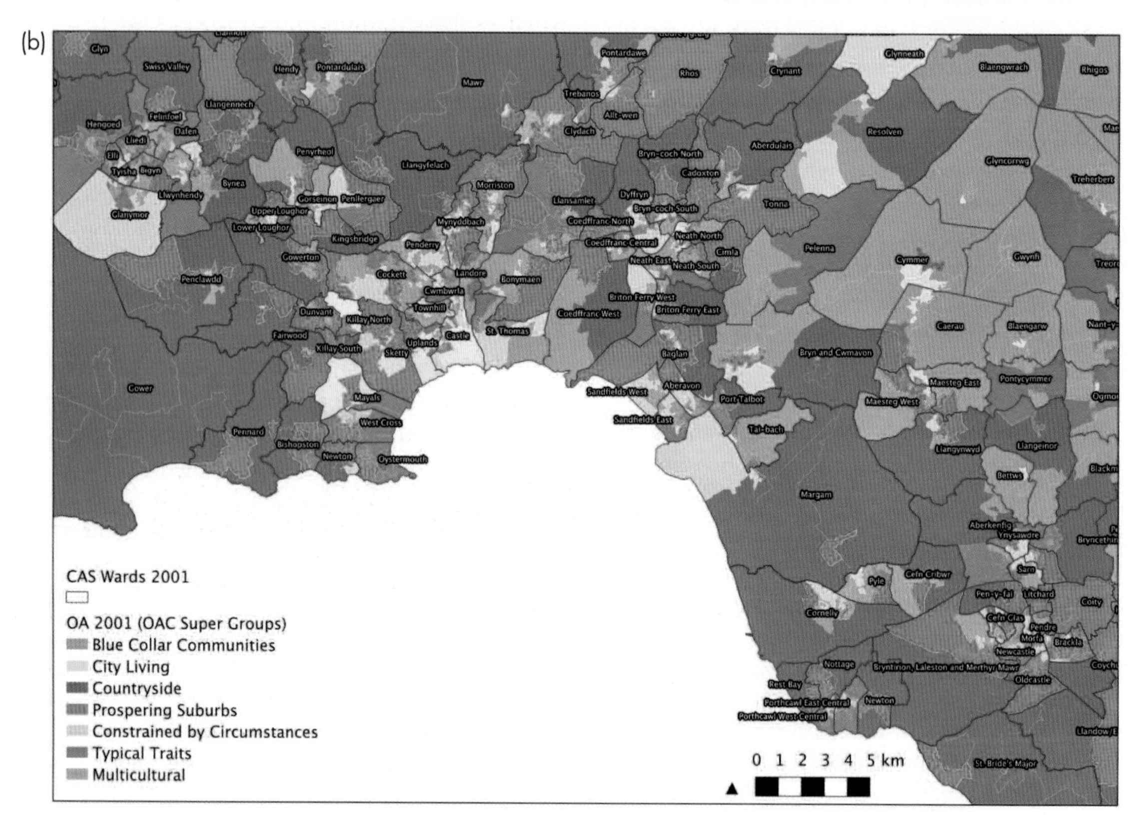

Location analysis takes into account other aspects, too. For the coffee shop chain, this might mean using a database of property prices and rental costs. They then see which locations are most profitable in terms of the income from a shop, calculated using a mathematical model that may include the size of the shop, the place it is located and the likely number of daily clients, etc. It must be that my local high street met these criteria, and this is why I've noticed the appearance of the new coffee shop. For a more detailed description of similar processes, see Chapter 2 in Longley *et al.* (2010).

This may seem to be a rational, objective, scientific and accurate process. However, we can notice many issues. First, throughout the process we are dealing with 'ecological fallacies' – the assumption that, because statistically the area shows certain characteristics, this is true for everyone that lives in it. Second, the processes of selecting the number of groups, deciding on which datasets to include to create the geodemographic classification and naming the groups are all subjective, and they can be very different between analysts. There is no single perfect analysis and each has a level of uncertainty. Most geographers who use these techniques will be aware of the inaccuracies and complexities of using them in Human Geography. The coffee shop chain managers also understand that the model does not describe universal rules that will work every time. Yet, too often, the model and the analysis are represented as accurate and authoritative descriptions of places rather than as complex, constructed, uncertain (at least to some degree) and subjective (at least in some respects). This should be avoided and, indeed, these limitations are discussed widely in the GIS literature.

Quantitative analysis in Human Geography is increasing. The growing availability of 'digital trails' – the signal from mobile phones, credit card transactions and the like – makes it very tempting to assume that it is possible to find out from this information many details about human geographies by analysing them quantitatively. However, care needs to be taken. We cannot ignore, for instance, the socio-economic distribution of the devices that leave the trails and the implications for whose human geographies get analysed and whose get ignored – for example, illegal immigrants tend not to have access to credit and therefore credit cards. Even more importantly, while these trails tell us 'what' happened or at least provide a description of what happened, they cannot tell us 'why' and provide detailed explanation of the causality that led to the observed phenomena.

SUMMARY

- The scientific view of Human Geography is based on empirical data collection, its analysis and the development of mathematical models. GIS facilitate and support these.
- The patterns identified through statistical analysis are repeatable and can be observed in other places and times, yet they are rarely universal and correct everywhere. Nonetheless, the ability to measure phenomena quantitatively allows for confirmation that an observation or hypothesis is correct and can be demonstrated using information that was collected in a systematic and dispassionate way as far as possible.
- Examples of the use of GIS in this way include processes such as geodemographic classification and location analysis. GIS science understands such processes as complex, rather than the production of simple truths about places.
- The growing availability of digital trails to people's lives offers many opportunities for using GIS. However, care has to be taken with regard to both the partiality of such trails and their ability to explain the phenomena they describe.

Using GIS in critical and cultural studies: critical and qualitative GIS

Over the past twenty years, two important changes happened that have challenged the association of GIS as a tool that is only suitable for spatial science. First, it was demonstrated that GIS can be used within studies that question social structures and power relationships, thus GIS can be part of 'critical' scholarship. Second, with the increased integration of multimedia abilities with GIS, the area of 'qualitative GIS' emerged, in which GIS are used as a container of qualitative information and as a tool for its analysis. I will now discuss these two trends in turn.

Geographical analysis and cartographic visualization are very effective and powerful in communicating social conditions. At the end of the nineteenth century, Charles Booth, a wealthy merchant who was not satisfied with a declaration that in London 25 per cent of the population lived in abject poverty, set out a study that surveyed the city systematically and demonstrated that the real level of poverty was even higher and therefore required policy action. Famously, part of the outputs from his investigation were maps that showed the location of deprived communities in London.

Today, GIS allow for such an analysis to be done more easily and rapidly. For example, Danny Dorling has published extensively on the basis of such geographical analyses, highlighting social and spatial inequalities at national and international levels (e.g. Dorling, 2010). An example of one of his global maps is provided in Figure 14.4 and clearly demonstrates the unfair distribution of votes in the International Monetary Fund, which follows the pre-modern principle where only men with property had a vote. This map was produced at the request of the African Nation Group when negotiations over vote reallocation were carried out in 2006. Another example is the work of Mei-Po Kwan (2002a, 2002b) on the **feminist** interpretation of GIS, visualizing the specific way in which the world is experienced from a woman's perspective.

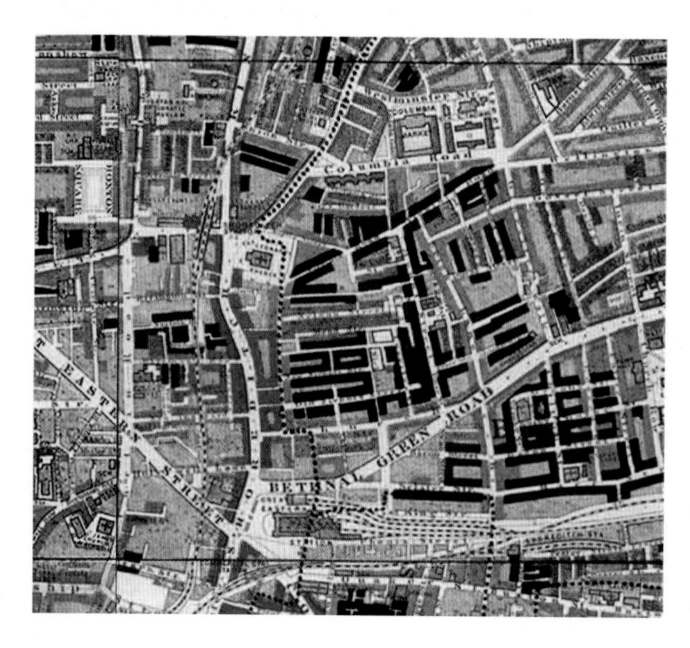

Figure 14.3 A section of Booth's map of poverty in London. Credit: www.icaci.org

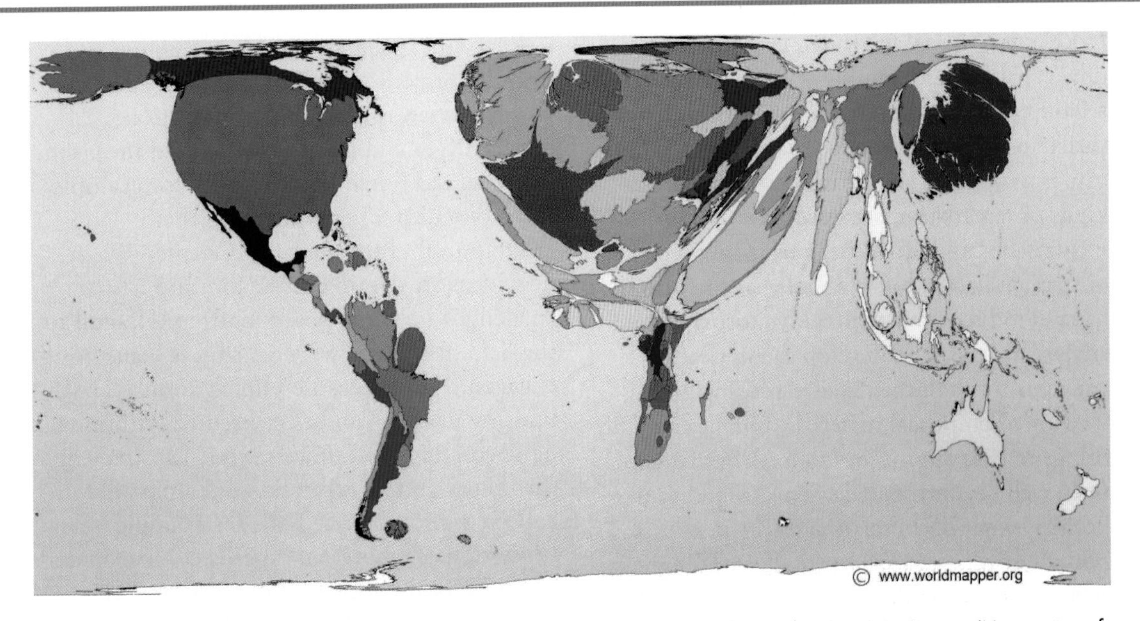

Figure 14.4 Distribution of power in the International Monetary Fund. Credit: © SASI Group (University of Sheffield) and Mark Newman (University of Michigan)

In these analyses, the GIS software remains the same, but the focus turns from the spatial sciences to supporting hypotheses and visualizing ways of thinking in alternative social theories. In Kwan's case, she has led a host of projects that have looked to move beyond the use of official data sets organized via official areal units, instead deploying in-depth empirical research to focus on individual experiences and how these are shaped by multiple axes of social differentiation, such as gender, race, age, class and religion (see also Chapter 43). For instance, working with Jae Yong Lee in a study focused on Korean residents of Columbus, Ohio, a recent project considered the issue of socio-spatial isolation and the role for the geovisualization capabilities of GIS in analysing and presenting the study's data on activity patterns (including interactive 3-D visualizations of activity paths in time and space and activity density surfaces mapped across the city) (Lee and Kwan, 2011). This development of 'critical GIS' (O'Sullivan, 2006), as this area came to be known, also includes participatory applications of GIS, where mapping activities are used to empower marginalized communities. Chapter 15 considers that issue more directly.

Yet the origin of GIS as a mostly scientific tool posed a challenge for critical GIS in relation to the materials or data that it could operate with. Within the social scientific and **humanistic** interpretations of Human Geography, words, images and personal descriptions of place and space are important parts of the data being collected during research activities and in presenting the results. Quantification can be problematic in handling such materials, losing much of what they are able to tell us about the world and people's experiences of it. However, as GIS and computing evolved, it became possible to store text, imagery, video and audio in a GIS, and use the tool to organize this information and analyse it. This form of GIS became known as Qualitative GIS (Cope and Elwood, 2009). This burgeoning field recognizes GIS as a powerful form of representation and mapping, but one that is now capable of containing within it and relating multiple representational forms from a variety of sources (see also Chapter 13). GIS is cast as a tool that can be used within various kinds of research; it is argued that GIS is not inextricably wedded to the analytical precepts of spatial science, but is a tool that can also be used in inductive, interpretative forms of data analysis that look to develop theoretical concepts through the detailed reading and viewing of empirical materials. Qualitative GIS brings to bear approaches to its data that come from well beyond spatial science: whether that be analysing GIS in relation to the poetics and politics of **representation** (understanding it as a means through which we can picture our world in various ways) or framing GIS as producing geo-visualizations that can **affect** viewers and provoke emotional responses (an emphasis related to thinking in **non-representational theory**; see Chapter 50) (Aitken and Crane, 2009).

Critical and Qualitative GIS are interesting from both methodological and theoretical points of view. Through the interaction between GIS and theories being used in specific branches of Human Geography – for example, the exploration of emotions and their geographical expressions (see Chapter 49) – new ideas are coming to the fore and new challenges are being posed, both to GIS and to the field of enquiry with which it is being engaged. This is because GIS, as a tool, is both limiting in what can be represented within its data structures and processes but also challenges the theory and its expression in empirically collected information. Take, for example, the expression of emotional experience, as recorded during a local walk with interviewees that reflect on their memories and feelings. If the participant talks about emotions that are sparked by a local site, but the emotion or memory is about somewhere else, where and how should it be presented on a GIS map, if at all? This is a common and typical challenge in this area of activity.

SUMMARY

- 'Critical GIS' uses GIS within studies that question social structures and power relationships. It uses a variety of approaches to the ability of GIS to represent geographical information, from the scientific to the humanistic.
- 'Qualitative GIS' recognizes the growing ability of GIS to store multimedia, in the form of text, image, video and audio, and how this has made the GIS a suitable tool for organizing and analysing qualitative data with reference to a geographical location.
- The use of GIS within the critical and qualitative frameworks is not without challenges, including the appropriate way to represent the information, visualize it and understand its meaning.

Conclusions

GIS and geographical information technologies are becoming part of everyday life. SatNav devices, maps on mobile phones and GIS analysis are influencing the way the world is perceived, ordered and operates. An important literature on GIS and society critically reflects on this influence. There is also a body of research that falls under the area of GIScience. This focuses on a wide range of issues that relate to the development of the tools themselves and understanding their use, as well as fundamental issues such as the human understanding of space. GIS is thus an important and vibrant area of geographical thought and practice.

This chapter demonstrated that GIS can be used within different frameworks of understanding in Human Geography, from the scientific, to the critical, to the humanistic. It is an effective tool to collect, organize and visualize information that is linked to a location. The growing availability of digital location information about human activities means that the need and opportunity to use GIS to make sense of human geographies will continue to increase. It is therefore important to understand what is possible and impossible to achieve with GIS, while also continuing to consider how to adapt them to the scientific, social scientific and humanistic world views that are part of Human Geography.

DISCUSSION POINTS

1. Is GIS a tool that can be used flexibly within any point of view in Geography, or do its origins and internal mechanisms limit it for use only within a specific way of understanding the world?
2. What part of the GIS is more significant for Human Geographers – the large set of analysis functions, the ability to store very large data sets or the ability to visualize them in maps, graphs and other forms of geographical visualization?
3. What are the advantages and limitations of using GIS as a community empowerment tool?

FURTHER READING

Longley, P.A., Goodchild, M.F., Maguire, D.J. and Rhind, D.W. (2010) *Geographic Information Systems and Science* (3rd edn.) Chichester: Wiley.

A very popular textbook on GIS and GIScience, which adopts the scientific view of GIS but also describes the challenges of meaningful geographical analysis as well as organisational and business issues.

Pickles, J. (ed.) (1995) *Ground Truth: The Social Implications of Geographic Information Systems.* New York: Guilford Press.

A collection of papers from the early phase of the GIS and society debate within Geography, providing a range of views on some of the conceptual challenges of using GIS in Human Geography.

Nyerges, T., Couclelis, H. and McMaster, R. (eds.) (2011) *The SAGE Handbook of GIS and Society*. London: Sage.

A comprehensive collection that surveys the landscape of participatory and critical GIS, while also including examples of the scientific framing of GIS and society issues.

Cope, M. and Elwood, S. (eds.) (2009) *Qualitative GIS: A Mixed Methods Approach*. London: Sage.

A collection of case studies and examples of using GIS to analyse and visualize qualitative information, which also considers the integration of qualitative GIS within critical social theory frameworks.

WEBSITES

www.esri.com

The website of the leading vendor of GIS, which is also committed to the use of the tool as an implementation of geographical thinking in the world through the use of such systems.

www.google.co.uk/intl/en_uk/earth/index.html

The website of the Google Earth application that provides a virtual globe application and allows the exploration of information across the world.

www.openstreetmap.org

The OpenStreetMap is a project aimed at creating a free and comprehensive map of the world, in the same way that Wikipedia is aimed at creating a free encyclopaedia. It allows any person to join in and create their own geographic information.

www.ppgis.net

A comprehensive collection of information about participatory GIS, which includes an active forum in which academics and professionals share experience and studies.

CHAPTER 15 COUNTER-CARTOGRAPHIES

Wen Lin

Introduction

Maps are powerful. 'The power of maps to include and exclude peoples and territories' has long been the subject of enquiry in social sciences, including Human Geography (Cooke, 2003: 265; Harley, 1989; see Chapter 13). Traditionally, map creation has been in the hands of experts (Crampton and Krygier, 2006). However, the past two decades have witnessed an explosion of mapping initiatives conducted by non-experts including community groups and local people in many different domains throughout the world (Poole, 1995; Craig *et al.,* 2002). Counter-cartographies here refer to this broad range of mapping practices by less privileged groups and individuals, which look to counter and overcome predominant power hierarchies and to further progressive goals (Hodgson and Schroeder, 2002; Harris and Hazen, 2006).

A number of labels have been used with respect to such counter-cartographies, in a variety of contexts. The phrase 'counter-mapping' was first coined by Nancy Peluso (1995) in her examination of forest resource mapping in Indonesia and the efforts by marginalized groups to contest state maps that had long undermined their interests in these resources. Such counter-mapping has, dependent on context, also been referred to as participatory mapping, indigenous mapping, community-based mapping and ethnocartography (IFAD, 2009). When it involves the use of geographic information systems (GIS), this kind of work is often termed Participatory GIS (PGIS). Counter-mapping can also be employed to describe less privileged groups' usage of mapping technologies and GIS in more developed countries. Here, the terminology and acronym has differed slightly, and it is often labelled as Public Participation GIS (PPGIS), reflecting an emphasis on getting publics involved in established planning and policy-making processes, especially in the contexts of urban revitalization and environmental management (Sieber, 2006). There is also a growing trend among artists to use mapping to make a political statement (Hurrell *et al.,* 2010), producing what have been described as 'subversive cartographies' (Crampton and Krygier, 2006) – different and new ways of seeing and mapping places, environments and the world. While some of this 'map art' is very much driven by the creative imagination of the artist, mapping has also been used by artists as a way to involve publics more collaboratively within the production of art works.

This chapter covers some of this range of counter-cartographies. Though there are differences among these initiatives in their objectives, methods, applications and users, I argue that key in the practices of counter-cartography are the ways in which unequal power relations might be questioned, transformed or influenced through the process of mapping by community groups (IFAD, 2009; Sieber, 2006). In the following section, I will discuss the major theoretical ideas underpinning counter-cartographies before moving on to discuss some examples and applications, drawn from the fields of PGIS, PPGIS and arts projects. Lastly, the final section concludes this chapter with some critical reflections on counter-cartographies. Counter-cartographies can be an important means to counter dominant power, but their production is complex and their effects are highly contingent. Therefore, it is important to recognize these possible tensions and reflect critically on the processes and outcomes of counter-cartographies.

SUMMARY

- Counter-cartographies refer to a broad range of mapping practices by less privileged groups and individuals that aim to counter predominant power relations and to empower marginalized communities and individuals.
- A range of terms and acronyms designate these mapping practices and literatures about them. Foci for this chapter will include: Participatory mapping/GIS (PGIS) with indigenous communities; Public Participation mapping/GIS (PPGIS) with groups often excluded from consultation, policy and planning processes; and the 'subversive cartography' of arts practice.

Theoretical background

While mapping practices by indigenous and local communities are not new (Chambers, 2006; Crampton and Krygier, 2006), there are two important contexts of change that underline the emergence and explosion of counter-cartographies since the 1990s: first, the advent of participatory research methodologies within development practice, planning practice and arts practice; and second, the growing accessibility of mapping technologies (Wainwright and Bryan, 2009). Let me briefly discuss these trends in turn.

Since the 1980s, there has been an emphasis on local communities and grassroots engagement in development initiatives, through the employment of various participatory approaches to collect, analyse and communicate community information (IFAD, 2009). In these development initiatives, **non-governmental organizations (NGO)** and **community-based organizations (CBO)** are considered to play an important role in promoting participation and community empowerment among less privileged groups. There is a growing family of approaches and methods, such as Rapid Rural Appraisal and Participatory Rural Appraisal, to enable people to share their local knowledge among themselves and with outsiders in development interventions (Chambers, 1994; Corbett *et al.*, 2006). A wide variety of mapping approaches are used, including ground, paper and GIS maps. Participatory GIS approaches merge Participatory Learning and Action methods with GIS technologies (Harris and Weiner, 1998). Such efforts provide new political possibilities through which struggles over resources and development are linked to fundamental questions of culture, identity and power (Hodgson and Schroeder, 2002).

This paradigm shift to participatory methods was influenced by Paulo Freire's theory of critical pedagogy (1972), which suggests that 'poor and exploited people can and should be enabled to conduct their own analysis of their reality' (Chambers, 1994: 954). The critical pedagogy approach believes that 'human liberation from existing exploitation and historical prejudice can be achieved through education' (Corbett and Keller, 2005: 96). The idea is that by learning new skills of reading and writing the poor would obtain 'critical consciousness', referring to a clearer realization of the injustices of their own situation. This notion of **empowerment** through education and skills enhancement underlines many participatory mapping approaches and initiatives, within and beyond those explicitly framed in terms of 'development' (Corbett and Keller, 2005: 96). Freire's thinking has been important too within theories and practices of planning in many parts of the world, and more generally the notion of public participation in decision-making processes has been appealed to in a range of spheres, from urban regeneration to climate change response. Parallel debates have also occurred within the art world as, for example, the notion of 'public art' has shifted from simply being about the location of an already produced work to being about some public involvement in the process of making art works in relation to specific places and communities. Across these fields, then, participatory forms of research and knowledge making have been emphasized, and counter-cartographies are a crucial part of this.

The wide spread of counter-cartographies is also associated with the emergence and, more recently, democratization of new mapping technologies such as GIS, GPS (Global Positioning System) and remotely sensed images (Crampton, 2009; Elwood, 2010). In Chapters 13 and 14 Jeremy Crampton and Muki Haklay set out these developments in more depth. Generally, a body of work on cartography had emerged in the 1980s that viewed mapping as a political act; that is, as a process that embodies power, rather than merely reflecting or mirroring realities and facts (Crampton, 2001; Crampton and Krygier, 2006). As GIS entered many fields in society in the 1990s, critical concerns were raised regarding its social implications (Pickles, 1995; Sheppard, 1995; Curry, 1995). GIS (as a form of mapping) was criticized for the kinds of knowledge it could deal with and promote, tending, it was argued, to serve a narrowly instrumental application of geographical information rather than facilitating a geographical knowledge that could better understand the world in which we live or engage critically with it. The high costs and technical complexity of GIS were also seen as posing significant barriers for its use, positioning GIS as a tool of powerful institutions like corporations or parts of the state (Pickles, 1995). However, more recent work has focused on 'rethinking the power of maps' (Wood *et al.*, 2010), arguing that the politics of mapping is not fixed but can vary, including as the political economy of map making changes (see Chapter 13 for more on this). More specifically, within GIS, schools of work labelled as 'critical GIS' and 'qualitative GIS' have focused on how GIS can be integrated with commitments to social critique and progressive politics and to forms of knowledge that are more interpretative in nature (see Chapter 14). As such, there have been a lot of discussions and debates about the possibilities of GIS in empowering community in the fields of PGIS and PPGIS (Elwood, 2004; Corbett and Keller, 2005; Obermeyer, 1998). Also emerging is an emphasis on increasing access to mapping technologies, thereby empowering marginalized social groups. During the past two decades, these areas of enquiry have evolved and expanded considerably, informed by diverse theoretical frameworks derived from social and political theory, science and technology studies, organizational theory and feminist theory (Obermeyer, 1998; Craig *et al.*, 2002; Sieber, 2006; Elwood, 2009a).

For instance, a rich body of work has examined factors influencing inequitable access to GIS and other mapping technologies (Barndt, 1998; Sawicki and Craig, 1996; Laituri, 2003). Because of resource constraints for sufficient support of software and staff, grassroots groups and community organizations often rely on external expertise and organizations, such as non-governmental organizations and university-community partnerships for GIS data and technology. These practices are highly contingent, shaped by institutional culture, local political context and broader political economy (Sieber, 1997; Ghose, 2007). More recently, the internet has been increasingly utilized as a platform to provide spatial data and representation, in order to overcome the time and geography barriers for disadvantaged social groups and individuals (Kingston *et al.*, 1999; Carver, 2001). The combination of mapping practices with the emergence of Web 2.0 technologies has been recognized as an important new form of spatial data provision and productive for the generation of counter-cartographies and participatory mapping (Goodchild, 2007; Sui, 2008). For example, Google Maps 'mashups' allow users to integrate information from other sources with the base map provided by Google Maps, a process with greater user-friendliness and a relatively low level of technology requirement from users (Miller, 2006; Tulloch, 2008). Open source mapping codes and platforms (e.g. OpenStreetMap) have also been utilized to elicit publics' inputs for mapping data about resource management and planning (Hall *et al.*, 2010). Wider trends to the 'wikification' of GIS have been highlighted (Sui, 2008), as **volunteered geographic information (VGI)** rather than official information is used to provide data. Pursuing these questions of access further into the issues of what kind of knowledge maps and GIS produce, scholars have sought to investigate the process of spatial knowledge production and ways of incorporating local knowledge and indigenous knowledge into GIS. These efforts include community organizations' active reworking of the meaning of mapping through traditional GIS (e.g. Elwood, 2009b), the creative engagement of visualization and multimedia representations (e.g. Harris and Weiner, 1998) and efforts at rewriting GIS software so that it can embody multiple forms of spatial data (e.g. Sieber, 2004). For example, in their research in South Africa, Harris and Weiner (1998) attempted to incorporate various forms of local knowledge through a multimedia GIS, which embeds narratives, pictures and videos within a GIS representation.

While recognizing the positive potential of such developments, the literature on counter-cartographies is not uncritically promotional. It also attends to the problems that can be encountered both in practice and in the conceptualization of what these participatory forms of mapping can achieve. For example, it is cautioned that not all local knowledge should be reduced to fit current GIS standards (Sieber, 2006; Rundstrom, 1995). There have also been discussions on conceptualizing the **empowerment** process in participatory mapping practices (Kyem, 2001; Elwood, 2002). Elwood (2002) provides a particularly insightful multi-dimensional analysis of empowerment in the context of GIS usage in community planning. Her case is the use of GIS by the Powderhorn Park Neighborhood Association (PPNA), a community-based organization (CBO) within an inner city area of Minneapolis, USA. Conceptually, within her framework she understands the empowerment attributed to counter-cartographies as containing three dimensions: distributive changes (such as greater access to services or a greater opportunity to participate in a political process); procedural changes (such that the views of citizens or community groups are given greater legitimacy within decision-making processes); and capacity building ('an expansion in the ability of citizens or communities to take action on their own behalf' (Elwood 2002: 909)). Among these three dimensions, distributive change is the least sustainable and capacity building is the most sustainable. Moreover, in considering the case of the PPNA, it was apparent that forms of cartography and GIS might empower in one of these dimensions, but disempower in another: they might empower one group of people and disempower another. The empowerment offered by counter-cartography is not a simple outcome, but a complex and often contradictory process (see also Harris and Weiner, 1998). I will return to such critical considerations in the concluding part of this chapter.

SUMMARY

- Counter-cartographies have been influenced by a range of theoretical themes and technology development. In particular, the emergence of participatory research methods and a critical engagement with the increasing availability of mapping technologies play an important role.
- The theorization of counter-cartographies has not been purely promotional, but has also carefully reflected upon what it might mean for them to be 'empowering'.

Counter-cartographies in practice

In this section, I want to provide more of a feel for counter-cartographies in practice. As noted earlier, counter-cartography practices are highly diverse, ranging from the field of land and natural resource management to urban neighbourhood revitalization, and having spread worldwide over the past two decades (Nietschmann, 1995; Poole, 1995; Peluso, 1995; Sieber, 2006). For the sake of simplicity, this section therefore addresses counter-cartographies in three exemplary groups: PGIS, PPGIS and artistic 'subversive cartographies', while recognizing that the lines among these categories are often blurred.

PGIS practices are often addressed through a lens of **colonization**, dispossession and conflicting interests, especially in relation to the place of indigenous communities within the mapping of territories governed by **neo-colonial** institutions or post-colonial nation-states. The objectives of PGIS vary significantly but include, without being limited to, the following: '1) gaining recognition of land rights; 2) demarcation of traditional territories; 3) protection of demarcated lands; 4) gathering and guarding traditional knowledge; 5) management of traditional lands and resources; and 6) community awareness and cohesion, mobilization and conflict resolution' (Hodgson and Schroeder, 2002: 80; see also Harris and Hazen, 2006). Within these objectives, using mapping to make claims to land rights by indigenous communities is one of the classic forms of counter-mapping practices. As stated by Barney Nietschmann (1995: 37), 'more indigenous territory can be reclaimed and defended by maps than by guns'. Community groups use a variety of approaches including basic sketch mapping, transect mapping, participatory three-dimensional models, digital mapping tools such as GIS, satellite images and global positioning systems to elicit and represent their spatial knowledge (Hodgson and Schroeder, 2002; Poole, 1995).

PPGIS practices usually refer to the broadening of mapping technologies to facilitate public participation by disadvantaged groups in industrialized nations. These practices may also be led by government agencies through a top-down approach to elicit input from community groups for development or planning initiatives (Sieber, 2006). PPGIS practices involve a wide range of technological interventions to facilitate public participation, including expanding the range of forms of data representation (such as audio, pictures and texts) in GIS visualization and incorporating multiple mediums, such as web-based mapping platforms, to provide more in-depth and non-Cartesian spatial knowledge. These efforts are related to a new strand of work called 'qualitative GIS', which aims to incorporate qualitative data to accommodate the needs of community groups and to challenge hegemonic forms of data representation (Sieber, 2004; Kwan, 2002; Cope and Elwood, 2009) (see also Chapter 14). PPGIS has also involved establishing a range of partnerships and networks including academic actors and non-governmental groups, in order to address the resource constraints of grassroots and marginalized groups in PPGIS practices (Elwood and Leitner, 2003; Ghose, 2005).

Another field of counter-cartographies is map experimentation by the artistic community. Introducing an edited collection of the journal *Cartographic Perspectives* on the relations of art and mapping, Cosgrove assessed that 'maps and mapping have proved astonishingly fertile material for artistic expression and intervention' (2006: 4). Published collections have presented visually arresting arrays of this 'map art' (Harmon, 2004, 2010). Pinder (2007) in turn positions artistic practice as one of the ways in which cartography has become 'unbound' from powerful institutions and their particular world views. Sometimes labelled as 'subversive cartographies' (Crampton and Krygier, 2006; Pinder, 1996), artistic forms of mapping look to see both the world and its maps in new ways by appropriating cartographic methods and conventions and deploying them differently. In the words of Catherine D'Ignazio (2009), a.k.a. the artist kanarinka, artists are 'symbol saboteurs' who often engage in mapping to challenge the status quo. 'The map is increasingly used in contemporary art as a political tool for commentary and/or intervention' (Watson, 2009: 297). Sometimes these mappings are clearly generated by the creative imagination of the artist. At other times, though, instead of being a sole creator-genius, 'the new artist is a conduit, at times a facilitator of events or environments that seek to engage with new audiences, employing terms familiar to most users' (Watson, 2009: 300). In the art world too, then, counter-cartographies often operate in tandem with commitments to participatory practice.

CASE STUDY

'Through The Eyes of Hunter-gatherers' (Rambaldi *et al.*, 2007)

This 2006–2008 project is part of the 'Strengthening the East African Regional Mapping and Information Systems Network'. In this project, the Ogiek indigenous people from the village of Nessuit, Nakuru District, Kenya, employed a 'participatory three-dimensional modeling' (P3DM) exercise to assess the vulnerability of their situation and land and natural resource loss.

The project was implemented by the NGO Environmental Research Mapping and Information Systems in Africa (ERMIS-Africa) and technically and financially supported by the Technical Centre for Agricultural and Rural Cooperation (CTA) and the Indigenous Peoples of Africa Coordinating Committee (IPACC). A key component in this participatory mapping process was to produce a map legend (see Figure 15.1). This collaboratively-produced map legend allows a representation of local spatial knowledge that may contrast with official maps. It is also crucial because it provides an important graphic vocabulary that is understood by all parties involved, and through which dialogue and negotiation using maps can be carried out (Rambaldi *et al*,. 2007: 113).

Figure 15.1 Map legend produced by the Ogiek indigenous people through community participation. Source: Rambaldi *et al.* (2007)

CASE STUDY

'Prospects of Qualitative GIS at the Intersection of Youth Development and Participatory Urban Planning' (Dennis, 2006)

This article reports on a participatory planning project in a community engaged in a ten-year revitalization effort, the South Allison Hill community of Harrisburg, Pennsylvania, USA. In this project, facilitated by a partnership between the author and several youth-serving organizations, local youths collected qualitative data on their neighbourhood through a variety of techniques including photography, drawings and narratives. For example, the youths were asked to think about particular places in their neighbourhood that they liked (and wanted to protect or preserve) and places that they did not like (and wanted to change) (Dennis, 2006: 2045). The qualitative data generated were incorporated in public forums and discussed along with other quantitative data on the neighbourhood, working towards a qualitative GIS. This study shows that the youths felt a sense of empowerment by contributing their data and viewpoints in the neighbourhood revitalization process. Giving some sense of the local knowledges that this mapping process was able to tease out and represent, Figure 15.2 is a sketch map of a 'bad' intersection drawn by a group of boys. Its 'badness' related to drug dealing undertaken there. Note how only two of the four street corners were considered 'bad' and how the existing vacant lots and alleyways provided shortcuts through the blocks in order to avoid the bad corners. This understanding of neighbourhood space was not previously apparent to planners or indeed many adults within the community.

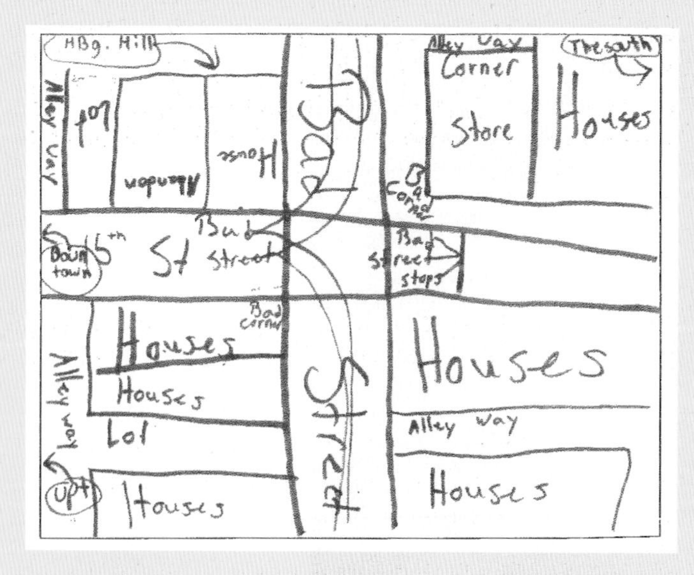

Figure 15.2 A sketch map by neighborhood youth on a 'bad' intersection. Source: Dennis (2006)/www.pion.co.uk and www.envplan.com

CASE STUDY

'By Day, But Not By Night': Counter-mapping Platt Fields Park (Hurrell *et al.*, 2010)

This project was carried out by a group of women students from the University of Manchester, UK. It aims to look at women students' experiences of exclusion from 'public space' through fear and to challenge the dominant perception of 'public space' as equally accessible to all. The focus was Platt Fields Park in the Rusholme area of Manchester, an area perceived as dangerous at night by/for women students. A combination of ethnographic and open-ended questions was used to collect data. In particular, one route through the park, which women would not take at night, was used for mapping. The map was then painted on the neck, shoulders and lower back of the body to emphasize the femininity of the image. The UK Ordnance Survey (OS) was used for the map base. This OS base map represents a masculine ideal of a 'scientific' and 'neutral' mapping service, while also making a social claim that space is 'public' despite the reality that access to this public space is structured by gendered geographies of fear and violence (Figure 15.3). In this way, the map intended to deliver a political message to challenge both the notion that space is simply public and the kind of cartography that represents it as such (Hurrell *et al.*, 2010: 20).

Figure 15.3 **The final map of 'By day, but not by night'. Source: Hurrell *et al.* (2010). Reproduced by kind permission of the Society of Cartographers**

SUMMARY

- Practices of counter-cartography are highly diverse. Examples range from mappings of indigenous community lands, to forms of qualitative GIS that look to engage marginalized groups with neighbourhood planning issues and politics, to artists' imaginative uses of cartography's visual languages and mapping practices.
- The shared theme across this diversity is the effort of addressing and questioning unequal power relations through mapping.

Conclusion

Mapping is intrinsically a political act (Peluso, 1995). Counter-cartographies, in a broad sense, are mapping practices by local communities, non-governmental organizations, universities and other actors that seek to contest dominant power relations and to increase the empowerment of marginalized groups. The widespread adoption of such practices with a diversity of objectives has been nourished by both the development of participatory methods and the increasing availability and affordability of mapping technologies. These mapping practices utilize a wide range of tools, ranging from sketch maps to computerized mapping tools such as GIS. Counter-cartographies differ from traditional cartography with respect to the agency of mapping, the process by which the mapping is carried out and its subsequent usage.

While counter-cartographies may be an important means to map against power, there have been some critiques regarding the effects. It is with these critical reflections that I want to end. I will highlight three. First, the access to these technologies for marginalized groups remains a challenge, and new forms of exclusion may be imposed within the social groups that employ them. The recent emergence of Web 2.0 technologies may provide another important source of spatial narratives, with its greater user-friendliness and relatively lower technical barriers; nonetheless, even here there is exclusion of possible participants without internet access, a significant issue in many parts of the world (Elwood, 2010; Crampton, 2009).

Second, the employment of the language of mapping, which tends to be based on Western traditions, may result in counter-cartographies having to inhabit uneasily some of the forms of geographical knowledge that they look to contest. For instance, one recurring issue in indigenous mapping projects is the loss of indigenous geographic knowledge in the process of translating fluid and flexible boundaries of land and resource use, generally recognized by indigenous peoples, to Western cartographic representations that favour fixed boundaries (Johnson *et al.,* 2006; Rundstrom, 1995). Some scholars have called for a 'two-pronged' approach for indigenous communities where indigenous groups become literate in modern cartographic technologies, while developing a critical consciousness of problems that accompany their use (Johnson *et al.,* 2006: 83).

Third, one should not idealize the effects of counter-cartographies. For example, while these maps may contribute to community cohesion, they can also result in conflicts and disagreement within or among communities (IFAD, 2009; Kyem, 2004). In mapping out community land relations, the knowledge and property claims of women, minorities and other vulnerable and disenfranchised groups may be overlooked (Hodgson and Schroder, 2002). There are also concerns about the exploitation of local people following the documenting of sensitive information (such as valuable resources or archaeological sites) through community mapping (IFAD, 2009). It is further pointed out that there are political dilemmas embedded in many counter-cartography practices. For example, there may be conflicts inherent in conservation efforts that involve territorialization, privatization, integration and indigenization. As such, 'community-level' political engagement needs to combine 'mapping efforts with broader legal and political strategies' (Hodgson and Schroeder, 2002: 82). Nonetheless, while counter-cartographies may not be able simply to reverse dominant social relations and unequal power structures, they provide important possibilities of reworking them.

SUMMARY

- While counter-cartographies can be an important means to counter dominant power, they are highly contingent and may result in new conflicts and exclusions. Therefore, it is important to recognize these possible tensions and reflect critically on the processes and outcomes of counter-cartographies.

DISCUSSION POINTS

1. Discuss the role of counter-cartographies in different contexts, such as indigenous mapping, urban neighbourhood revitalization and artist engagement. What are the benefits and pitfalls of such practices?
2. What are the impacts of the internet, including the recent emergence of Web 2.0 technologies, on counter-cartographies?
3. Can you think of some conditions and possibilities that you might engage with counter-cartographies?

FURTHER READING

Brosius, P., Tsing, A.H. and Zerner, C. (eds.) (2005) *Communities and Conservation: Histories and Politics of Community-based Natural Resource Management*. Lanham, MD: Altamira Press.

This volume on community-based natural resource management discusses important concepts of community, territory, locality and conservation. Four chapters focus on community-based mapping in conservation, providing intriguing and insightful studies of mapping against power in various sites.

Craig, W.J., Harris, T.M. and Weiner, D. (eds.) (2002) *Community Participation and Geographical Information Systems*. London: Taylor and Francis.

This volume provides a rich range of case studies of participatory mapping practices.

D'Ignazio, C. (2009) Art and cartography. In: R. Kitchin and N. Thrift (eds.) *The International Encyclopedia of Human Geography*. Oxford: Elsevier, 190–206.

This short review summarizes past and present artistic interests in cartography. The author is also known as kanarinka, an artist and 'experimental geographer' whose own practice has engaged with forms of cartography.

Schuurman, N. (1999) Critical GIS: theorizing an emerging discipline. In: *Cartographica* 36(4), 1–108.

This monograph, where the term of 'critical GIS' is coined, takes the early critiques of GIS and argues for alternative models of examining GIS and engaging with the technology.

WEBSITES

www.nativemaps.org/

The Aboriginal Mapping Network website provides mapping resources related to participatory mapping and aboriginal mapping, and forums for discussions from interested participants.

www.ppgis.net/

The PPGIS network website is a rich web resource for PGIS, PPGIS and community mapping practices and research.

SECTION THREE

CULTURAL GEOGRAPHIES

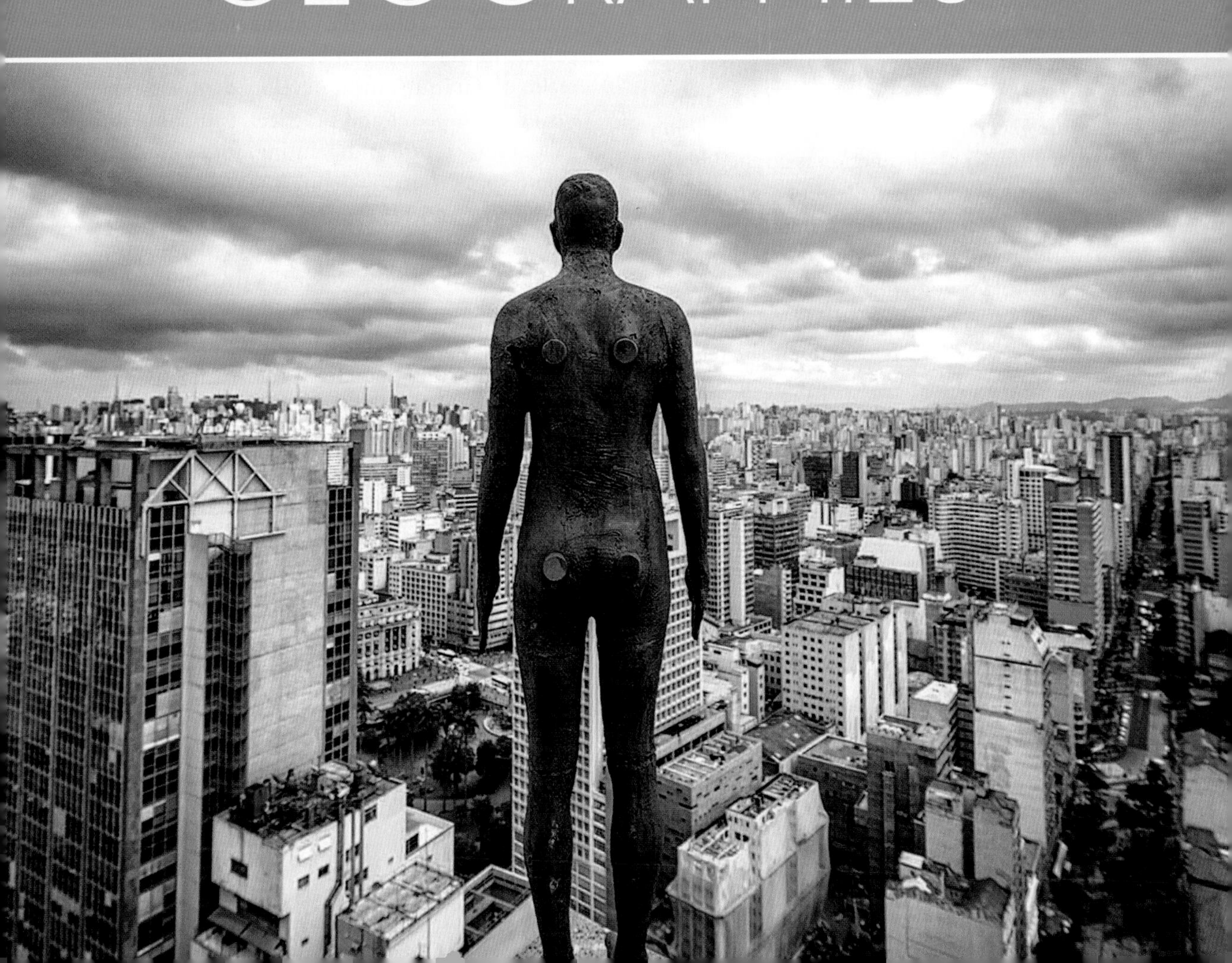

Defining the cultural in cultural geography

In approaching the sub-discipline of Cultural Geography, it makes sense to start by reflecting on what the 'cultural' in its name means. Raymond Williams famously suggested that 'culture is one of the two or three most complicated words in the English language' (1983: 87). To tackle its complexity, Williams outlined the history of the word and its usage. Broadly, we can identify three sets of current meanings of 'culture' from his account; these in turn speak to central concerns of Cultural Geography.

First, the oldest usage of the word is as a noun of process, referring to practices of 'cultivation'. Nowadays, this is a sense of the word that we deploy most directly when talking about a 'bacterial culture' or 'agri*culture*', but those somewhat specialist usages point to a much more general framing of 'culture' in relation to 'nature'. This distinction is applied both in accounts of the external world – such that agriculture is about cultivating the natural world, transforming it into domesticated crops and animals, humanizing it in various ways – and in understandings of human subjectivity – where we routinely talk about the relative impacts of nature and culture (usually referenced by its more alliterative synonym 'nurture') in shaping our own characteristics. In both cases, culture speaks to the relations between human beings and the biophysical worlds in which we exist. In Cultural Geography this meaning of culture is reflected in still vibrant traditions of work concerned with the relations between 'land and life', to use a phrase from one of the founding figures of the sub-discipline, Carl Sauer (Leighly, 1963). Traditions of scholarship on landscape are particularly pertinent here and we will say more about these below. Such concerns are also developed in geographical research focused on the question of 'nature', considered in this book in the earlier section on 'biogeographies' (Chapters 10–12).

A second meaning of 'culture' shares the focus on 'cultivation', but takes us in a rather different direction focused on human subjects and minds. This is a sense of the term that has become institutionalized in the organization of governments (where one gets Ministries or Departments of Culture), in descriptions and measurements of the economy (where one gets references to 'the cultural sector' or 'the cultural industries'), and in something as mundane as broadsheet newspapers (where there may be a 'Culture' section or supplement). In all these cases, 'culture' has come to refer to the imaginative activities that make people 'cultivated' or 'cultured'. On the one hand, culture in this sense can be cast as a recognized realm of 'Culture', most developed in particular institutions, places and practices. Museums and galleries, literature and painting, for example, would speak to many people's senses of what counts as 'Culture'. On the other hand, culture can also be seen as a human universal, as central to what makes humanity distinctive from other animals. From rock paintings to rock music, 'culture' designates the acts of imagination and meaning making that characterize humanity. This tension between the specificity and universality of culture plays out in recurrent debates over what forms of human creativity and imagination are of value and deserve recognition as 'Culture'.

Cultural Geography has a general interest in how we incorporate the world into our human imaginations and create our 'maps of meaning', to use the title of a seminal text (Jackson, 1989). More specifically, these understandings of culture are reflected in bodies of work that study a range of cultural forms, both the more recognized – novels, poetry, films, paintings, theatre, museums – and the less so – popular music, video games, comic strips, advertising, for example. Key interests include both the geographies imagined and performed through these creative forms, and how the cultural practices and institutions that produce them are organized spatially.

A third sense of the word 'culture' extends this interest by thinking about how there may be different maps of meaning associated with different cultural identities and values. It is the sense of culture we evoke when we talk about British, American or Singaporean cultures; middle-class, working-class or youth cultures; or indeed the culture of the 'Swinging Sixties' or fin-de-siècle Paris. Here culture operates as a noun of configuration, referring to distinct ways of life. Significantly, in this usage the singular noun culture is pluralized, with multiple 'cultures' identified. Indeed the notion of culture in this sense has become strongly attached to ideas of difference, referring to what distinguishes different peoples, places and times. In Cultural Geography, such senses of culture have been of longstanding interest. However, earlier concerns with trying to map out the spatial extents and diffusions of different cultures have given way to more direct and critical interrogations of how, and with what consequences, cultural differences and values are produced, used and changing. Culture now is approached more as an ongoing process that operates through geographical entities such as place, space and environment, and less as a 'thing' that can be called upon to explain some state of affairs. Cultures are framed not as static entities that exist above social action; rather, they are understood as suites of more or less habitual activities and practices that are pursued, regulated, and in various ways rub up against each other. Culture this raises issues of 'cultural politics'.

Introducing cultural geographies

In tracing out these three common meanings of the word 'culture' we can also see, then, some of the broad concerns that animate the sub-discipline of Cultural Geography: material interactions with the world; imagination, meaning and the arts; difference and identity. The chapters in this section of the book introduce you to these concerns and to how different senses of the word culture come together in geographical thought by focusing on four key concepts within Cultural Geography.

In the first, Felix Driver introduces the idea of 'imaginative geographies'. Originally advanced in the 1970s by the Palestinian literary and political critic Edward Said, and very influential in Cultural Geography since, the concept of 'imaginative geographies' posits that imaginations of other people and places help to fashion our own senses of identity. Felix outlines how this can be seen in imaginations of the world and its cultures. He looks critically at representations that distinguish between a civilized West/Europe and more 'primitive' cultures in the rest of the world. Paying attention to such representations, he argues, is not to be distracted from the geographies of the real world; imaginative geographies really matter. Images and words are important both for how we understand the world and how we act in it. Images are not the opposite of reality; they *shape* realities. In consequence, Felix also makes an argument to take imagery seriously as a form of geographical knowledge. Focusing on maps, he shows how close visual analysis can draw out their representations of peoples and places, of cultures and their differences, sometimes unearthing surprising complexity to the geographies they imagine. In discussing exploration, he shows how important visual images were to the activity, and illustrates how geography's historic visual archive (exploration's paintings, photographs, films) can be re-exhibited today to present previously hidden aspects. Geographers, he argues, need to take seriously both how they look at and how they make visual imagery. Overall, in attending to 'imaginative geographies', Cultural Geography shows how we make the world meaningful through practices of representation and argues that this shapes how we see different cultures and places.

In the second chapter, Tim Cresswell focuses on 'place'. Place is a central term in Human Geography more generally, but Cultural Geography brings particular approaches to its conceptualization. Tim understands places as ways of seeing and framing the world according to what and who is said to be 'in' and 'out of place'. Places, then, are ways of imagining the world through categorization. Our place categories may often seem mundane, for example distinguishing between 'rural' and 'urban' places or 'public' and 'private' places, but Tim argues we need to consider them critically. Our ideas of place help to define what is 'right' in a particular context; they put things and people in their place; they map out moral geographies of conduct and judgement. Places, he says, are 'ideological', laden with meanings that create or reinforce relations of power. Tim argues that for the most part we take place categorization for granted but occasionally, especially when that categorical order is 'transgressed' by something being out of place, it becomes a matter of reflection and contention. In the chapter he offers two sets of examples to illustrate this argument. First, he considers how ideas of 'rural' and 'urban' place enact ideas of identity and of our relations to nature. Second, he discusses how protest movements can involve transgressing the established order of place, focusing on the Plaza de Mayo Madres in Argentina and the more recent Occupy movement in the USA and UK. Overall, he argues that Cultural Geographers understand places critically, as ways of imagining the world that both shape our routine conduct of everyday life and are open to challenge through practices of transgression.

Our third chapter in this section focuses on 'landscape' and brings together the preceding chapters' foci on imaginative geographies and on the politics of place. Its author, John Wylie, notes at the outset how landscape has been a central concern of Cultural Geographers for nearly a century now. Schematically, four broad emphases on landscape can be drawn out. The first of these frames landscapes as material entities that one can identify in the world out there, comprised of hills, fields, roads, woods, buildings and so on. Since the seminal work of Carl Sauer in the early twentieth century, Cultural Geographers have approached these as 'cultural landscapes', simultaneously 'natural' and 'cultural', formed in interactions between physical environments and the cultural groups that interact with them. The second approach emphasizes how landscapes are not only 'out there' but 'in our imaginations', not only material but also meaningful. Coming to the fore since the 1980s, one such perspective treats landscapes as 'ways of seeing'. Most obviously, we think of landscapes like this when we talk of 'landscape painting' or 'landscape photography'. These representations of landscape give us visions of the world shaped by particular values and ways of thinking as well as particular visual conventions. They involve both artistic portrayals of landscapes (in paintings, photos, films, poems and so on) and material interventions in the world to produce landscapes that look how we want them to (such as landscape gardening or landscape architecture).

A third emphasis centres the relations between landscape and power. These relations are especially clear when we look to Germanic and Scandinavian traditions of thought; the German word 'landschaft', for example, references a political territory defined by culture and customary law. But Cultural Geographers have argued more generally that landscapes are political. For instance, far from being a purely aesthetic form, traditions of landscape painting are politically implicated, whether that be through their early association with claims to landownership and private property (it was landowners who often commissioned landscape paintings, after all) or ongoing implications with projects of nationalism (landscape painting is notable for its 'national' schools presenting visions of their nations, from John Constable and JMW Turner in Britain, to Thomas Cole and Edwin Church in the USA and the 'Group of Seven' in Canada).

A fourth approach, returning to the fore over the last decade or so, frames landscapes as speaking to how we inhabit the world through our bodies and senses. Neither external realities nor ways of seeing, here landscapes are indicative of how people and their environments mesh together in practices of 'dwelling'. Landscapes are not the scenery inthe background of life's theatre but part of a dynamic action of 'landscaping' through which our experiences of the world and our senses of self are co-produced. Studies of this sort within Cultural Geography have included both more mundane ways in which people inhabit landscapes – walking, cycling, driving or sitting looking out a rail carriage or airplane window, for instance – and what are to varying extents more specialist cultures that engage us with the land – the examples of touristic sightseeing, gardening, caving and exploration are suggestive of the range of topics covered.

In Chapter 18, John Wylie's general argument is that while it can be helpful to distinguish such approaches to landscape conceptually, in practice Cultural Geographers should bring them together rather than oppose them. He does this by taking two examples, each initiated by a particular landscape image. The first of these, an 'ideal' picturesque landscape created for the cover of a UK government report on ecosystem assessment, speaks to how landscapes can 'naturalize' (i.e. make seemingly natural) the cultural politics of national identity. The second image is an unnerving photograph of the London landscape taken by the 'urban explorer' (and Cultural Geographer) Bradley Garrett while documenting an unauthorized ascent to the top of the Shard, Europe's tallest building. For John, this image illustrates how ways of seeing landscape co-exist with ways of doing landscape (the view and the climb are fused in the image), as well as highlighting the politics of landscape and place through urban exploration's emphasis on transgression and trespass. Overall, John's narrative presents landscape as a still vibrant area of concern for Cultural Geography that brings together issues of imagination, inhabitation and power.

The final chapter in the section, by Philip Crang, considers 'material geographies'. If Felix Driver's opening discussion of 'imaginative geographies' introduced us to how Cultural Geographers take imagery and meaning seriously, then Philip's account closes this section by showing how material objects are also central to Cultural Geography. His argument is that material 'stuff' plays an important role in cultural processes of making distinctive ways of life and giving meaning to the world, its places and people. The examples discussed range from stuff one might expect geographers to study – for example, houses as a material part of 'cultural landscapes' – to things that may be less expected – the iPod, for example, considered as a technology for personal environmental management. Cultural Geographers have long been interested in the material aspects of culture. Traversing that intellectual history, Philip sets out four ways in which material things can be seen as having cultural geographies: as evidence or indicators of cultural difference and distribution; as materials involved in the reproduction of distinctive cultural groups and places; as tools and technologies used in the fashioning of everyday spaces; and as forms of cross-cultural 'traffic' that can be traced out through following things as they move around the world. The general flow of argument is away from viewing the geography of things as limited to their distribution across space, and towards cultural geographic analyses that get inside things, their meanings and uses, teasing out their role in the production of the places and spaces that people imagine and inhabit.

FURTHER READING

One of the ways to move beyond the introductory discussions in this section is to familiarize yourself with the research articles and essays published in the key journals of Cultural Geography. *Cultural Geographies* (founded in 1994 under the title *Ecumene* and renamed in 2002) is the leading international journal focused on the field. *Social and Cultural Geography* (founded in 2000) represents how research in Cultural Geography has come into close association with research in Social Geography, through shared interests in issues of identity, social difference and exclusion, for example. If you want more compressed summaries of developments, the regular 'progress reports' on Cultural Geography in the journal *Progress in Human Geography* are also an excellent resource, though they can be quite dense.

Anderson, J. (2010) *Understanding Cultural Geography: Places and Traces*. Abingdon: Routledge.

In this comparatively recent textbook on the field of Cultural Geography, Jon Anderson focuses on what he terms 'a culturally geographical approach to place'. The book's starting point is the importance of context and place for how Human Geographers think about the world. It argues that places are formed by cultural 'traces', that is marks, residues or remnants of cultural life. It brings together the foci of this section – place, landscape, imaginative and material geographies – in the development of that argument.

Anderson, K., Domosh, M., Pile, S. and Thrift, N.J. (eds.) (2003) *Handbook of Cultural Geography*. London: Sage.

This big, sprawling book collects together state-of-the-art thinking on Cultural Geography. Its range includes themes directly addressed in this section, such as landscape, place and our geographical imaginations and knowledge, and the intersections of Cultural Geography with other sub-disciplines, such as Social, Economic and Political Geography. The pitch can be uneven – sometimes explaining issues carefully, sometimes assuming a lot of knowledge – but this is a useful resource, especially for later on in your studies.

Crang, M. (1998) *Cultural Geography*. London: Routledge.

Though now a little dated, this is short, punchy and suggestive. It is written specifically for first or second year undergraduates.

Foote, K.E., Hugill, P.J., Mathewson, K. and Smith, J.M. (eds.) (1994) *Re-reading Cultural Geography*. Austin: University of Texas Press.

A voluminous collection of classic papers and newly commissioned pieces, representing an account of Cultural Geography's interests, approaches and history from a distinctively American perspective.

Jackson, P. (1989) *Maps of Meaning*. London: Routledge.

This book codified the re-emergence of British Cultural Geography in the late 1980s (labelled as a 'New' Cultural Geography), connecting the sub-discipline to wider intellectual currents in so-called cultural studies as well to developments in Social Geography. Inevitably a little dated now, this is still an inspiring piece of work that rightly has become a standard reference.

Mitchell, D. (2000) *Cultural Geography: A Critical Introduction*. Oxford: Blackwell.

A combination of textbook and manifesio, this book is much longer than Crang's and consequently simplifies much less and has a more explicit authorial agenda. It emphasizes how issues of culture are bound up with issues of power.

CHAPTER 16

IMAGINATIVE GEOGRAPHIES

Felix Driver

Introduction

Geography is a subject that has always had a reputation for being down to earth. After all, its focus is on the real world, the shape of its landscapes and the pattern of their use by human beings. This sense of the rootedness of the discipline in the material world is often associated with an image of the geographer as an active inquirer, engaged with the world rather than distanced from it. Or, to put it bluntly, many people are convinced that you simply can't do geography without getting your boots muddy. This sense of geography as an active pursuit was once embodied in the figure of the intrepid explorer, determined to seek out the truth in the field with his or her own eyes rather than rely on the speculations of 'armchair' geographers. Today, relatively few geographers consider themselves explorers in quite this sense; indeed, one (Lowenthal, 1997) has gone as far as to describe himself as an 'extrepid implorer'! Nonetheless, a sense of engagement with the world, and a more general commitment to addressing real-world problems – such as geopolitical conflict, climate change, poverty and injustice – remains a strong feature of a modern geographical education. This is one of the main reasons why we say that geography matters.

The purpose of this chapter is to consider how people imagine the geography of cultures and places, and why this matters too. The inclusion of such a topic in a geography book might seem a luxury, given the emphasis on practical fieldwork and worldly relevance that has had such an influence on our discipline. Why focus on 'imaginative' geographies when there are so many real-world problems to deal with? Can't we leave that to other disciplines, concerned with fictions rather than facts, with subjective impressions rather than objective realities? The argument of this chapter is that, far from being a diversion, the study of imaginative geographies is one of the reasons why geography remains relevant to the troubled world of the twenty-first century.

Imaginative geographies and why they matter

In this chapter, the term **imaginative geographies** refers to more than the subjective perceptions of individuals. While every human being is unique, and each of us experiences the world in a particular way, the images we live by are at the same time inherently social. Think of the words we use or the pictures we draw: these depend on shared systems of communication, codes or languages, which depend on a wider community whose spatial extent usually extends far beyond our actual experience. While Western writers have long understood 'imagination' in subjective terms, associating it with creative license or individual genius, there is no reason why we cannot think of imaginations in other ways. Ever since the origins of modern anthropology in the eighteenth century, cultures have been understood in terms of shared beliefs and practices; and at least since the birth of modern psychology in the late nineteenth century, we have come to recognize that these may be rooted in the unconscious realm as well as in the realms of thought and action. Of course, there are many different ways of conceiving these ways of thinking or patterns of behaviour; the fundamental point here is that they are more than the work of individual minds. In other words, imaginations are social as well as individual.

Turning to the content of these imaginative geographies, let us start with the question of identity, the sense we have of our place in the world. I have a birth certificate that tells me where I came from, and a passport that states my nationality. I may also identify myself as belonging to a certain region, generation, class, gender and ethnic group. But how do these define my sense of identity? It very much depends how I imagine them to be related to each other; on the extent to which I belong to a number of different communities; and, not least, on how I am identified by others. These senses of identification – subjective, intersubjective or imposed – may well change over time. Moreover, the answer to questions about personal identity depends very much on the circumstance in which they are asked: at a border post, in a bar, on a train, at home, in a questionnaire or a census. In other words, identities are complicated things: they are shaped not just by our physical characteristics or our social positions, but also by images – those we ourselves compose to make sense of the world, those of others and those in the society in which we live. In this chapter, I shall consider only a few of the many ways in which cultures and places are imagined: the more general point is that these imaginative geographies help to shape our sense not only of the meaning of cultures, but of our most intimate sense of our selves (Valentine, 1999; May, 1996).

Imaginative geographies make a difference, that is to say, they are real. Take passports for example: these are documents made up of images – print, photographs, impressions, stamps, digital codes, biometric data – that together compose one very material form of identification. That document shapes our lives by making certain things like residence, nationality and mobility available to us on certain terms, and it simultaneously restricts our access to these things elsewhere in the world. Obviously this is a very particular example of the effects printed images may have: the visual codes that constitute a passport are enormously powerful, marking the extent to which the definition of identity in the modern world is bound up with the power of states. Let us then take another, less obvious, example: that of images of childhood. How do these images shape the geography of the world we inhabit? The answer to such a question is inevitably complex, especially given the wide variety of ways in which childhood is and has been imagined. In Europe, for example, images of childhood have undergone considerable change, notably in the nineteenth century, the era in which mass schooling began. Many of our ideas about what it means to be, to look like or to behave as a child have their origins in this period, which also saw the gradual exclusion of young children from the world of paid work, the emergence of child protection movements and the development of new conceptions of juvenile delinquency closely associated with modern urbanization. The point here is that these changes were in some measure imaginative – they required new visions of childhood – and all of them had practical consequences for the geography of children's lives. The same might be said for other aspects of identity, such as our conceptions of masculinity and femininity, or able-bodied and disabled bodies, or whiteness and blackness, or madness and sanity. Images, too, have real effects.

SUMMARY

- The study of imaginative geographies takes images seriously: it treats words and pictures as both objects of study in their own right and as evidence for understanding the ways in which identities are constructed.
- Human Geographers are concerned with the realms of the imagination, not as a contrast to, or an escape from, the real world 'out there', but because they help to make sense of, and shape, that world.

CASE STUDY

Geography and the visual image: embodied knowledge

Visual images have long been considered important sources of geographical information. The patient, almost forensic, skills long employed by geographers in the interpretation of maps and aerial photographs, for example, continue to play a role in our discipline today, alongside the use of high-precision microscopes and computer-aided analysis of remotely sensed data. But visual images are not simply frozen data banks, reflections of patterns ready to be brought to life: they also re-present information in particular ways that, themselves, are far from neutral or self-evident. Geographers must therefore be interested in the nature of forms and codes of **representation**, how they have evolved over time and how they may be related to the circulation of imaginative geographies of culture, landscape and identity (Harley, 2001; Rose, 2011; Schwartz, 1996).

This emphasis on the significance of representation, particularly within the field of Cultural Geography, has recently been challenged. Some critics argue that the emphasis on the ways people represent the world has diverted attention from questions of practice and habit: it would be better to start, they suggest, with the question of what people actually *do* in the world, than with how they think about or represent what they do. Such arguments actually have a long history within social sciences stretching back at least two centuries. Moreover, the debate over what is now called '**non-representational theory**' is not about the importance of images or imaginative geographies in themselves: it is about how worlds are actually made, and where best to begin the process of interpretation (Thrift, 1999).

The process of making imaginative geographies involves a variety of embodied practices and knowledges: it is not simply a product of conscious thought. Take, for example, the striking landscape sketch in Figure 16.1, a drawing from the logbook of a young midshipman on a British naval vessel in the spectacular harbour of Rio de Janeiro in 1817, en route for a tour of duty in Australia (discussed in Driver and Martins, 2002). This depicts the well-known Sugar Loaf Mountain, with nearby coastal features and fortifications. Such views formed part of an established way of seeing, a common visual code through which landscapes could be recognized and mapped from the sea. In interpreting this image it is necessary to consider the precise forms and techniques through which this knowledge took shape; the networks through which these skills were learned and applied; and the material, institutional and social contexts in which they acquired value. Coastal survey, like mapping in general, was an embodied labour, requiring particular kinds of skill and discipline. Such images as these are not simply projections of the mind's eye, but are also embodied and laborious engagements with the world.

Figure 16.1 John Septimus Roe, *Views of the Harbour of Rio de Janeiro, June 1817*. Credit: State Library of Western Australia/301A/1

Imagining cultures

How are cultures 'imagined'? Consider the picture in Figure 16.2. It formed the frontispiece of a *Concise History of the World*, a forerunner of today's multimedia encyclopaedias, published in 1935 by Associated Newspapers. Such reference works aimed to provide within a single printed work – now within the space of a compact disc – the complete story of humanity, from the very beginnings to the present day. The story of human development was often conceived as a kind of family tree, quite literally in this case. Here, the narrative of human history is presented through a procession of emblematic figures, each standing for a distinct civilization or culture, culminating at the foot of the page with an image of the modern family, recognizable to British readers in the 1930s, complete with domesticated pet.

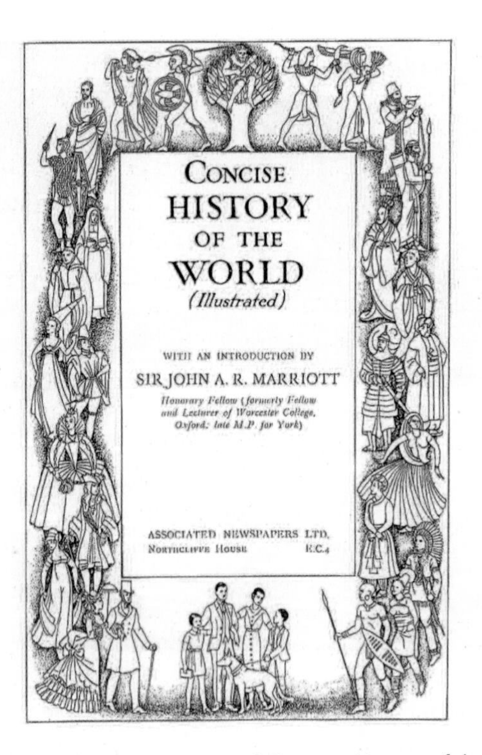

Figure 16.2 Frontispiece of *Concise History of the World (Illustrated)*. Credit: Associated Newspapers, 1935

Let us examine this curious image a little more closely. Its picture of world history as a sequence of stages through historical time owes something to the impact of evolutionary thinking on popular publishing. H.G. Wells' famous book, *The Outline of History* (1920), for example, had set the pattern for a large number of similar works, intended for a wide audience. Promising nothing less than an epic history of the world, Figure 16.2 presents world history as a stately procession or pageant. Significantly, a fundamental parting of the ways appears early in the story. The figures on the left-hand side correspond to a sequence of civilizations that would have been recognized by early twentieth-century readers: through the Persians, Greeks and Romans, to the medieval, the renaissance, the Victorian, and into the twentieth century. Each of the figures is represented in subtly different ways, their posture and dress signifying the characteristics of their time. There is a distinct sense of progression, too, from the era of martial prowess, through courtly ritual, to the modern world of the nuclear family. This is a story of progress by domestication: an evolutionary tale in which, ultimately, the values of the modern win out not only over the primitive, but perhaps even over nature itself. Yet on the other side of the image, nothing like the same logic applies. Here we find a motley collection of cultures, including the Ancient Egyptian, Phoenician, Chinese, Japanese, Indian and Native American. There is no sense of progression in this sequence, and as if to emphasize this, a thinly clad African couple is placed at the foot of the chain, the man armed with spear and shield. The overall message is plain: human history is divided in two. The West is the domain of progress, where history has evolved in successive stages towards the modern family form; the rest of humanity, meanwhile, is pictured more as a spectacle than a pageant. Far than being genuine subjects, these peoples are to be kept in their place: in imaginative terms, these are what one anthropologist called 'the people without history' (Wolf, 1982).

Looked at in this way, this image can be connected to wider traditions of thought about cultural history – and geography. For example, it could be interpreted as a popular version of more scholarly traditions of 'Orientalism', as analysed by Edward Said in an influential book published in 1978, in which the term 'imaginative geographies' was first coined. Said argued that non-western cultures in general (and those of the 'Orient' in particular) have often been represented by western commentators as being static, exotic and backward. He also maintained that these images have played an important part in the historical construction of a contrasting image of Europe – and the West in general – as dynamic and progressive. For Said, such imaginative geographies – based on a binary opposition between the West and the Rest – have played an important role in the history of global power relationships during the last two centuries. Whatever the merits of this thesis, which has been hotly debated (Gregory, 1995; MacKenzie, 1995), it does seem that something like this pattern is pictured in Figure 16.2. Said's argument raises fundamental questions about our understandings of global history and culture. In place of an imaginative geography based on a binary opposition between the West and the Rest, it encourages us to re-think global history as a more complex pattern of interconnected histories and intersecting geographies.

Said's work emphasized the role of ideas of the **Other** in the construction of imaginative geographies of culture; that is to say, he was concerned with the ways in which European history has often been conceived on the basis of an essential opposition between the civilized European and the uncivilized, or yet-to-be-civilized, non-European (Hall, 1992b). While it would be simplistic to suggest that this was the *only* way in which Europeans have imagined other cultures, it is remarkable how common and enduring such stereotypes have been. Look, for example, at the images in Figure 16.3, which appeared in a book entitled *Life in the Southern Isles*, published by a missionary (and Fellow of the Royal Geographical Society) in 1876. They show two contrasting visions of a Pacific Island village scene, supposed to be 'before' and 'after' conversion to Christianity. The first depicts the islanders engaged in a ceremonial dance, wearing grass skirts and shells, abandoned to a hedonistic life; in the second, they have been transformed into a sedate community, domesticated and cultivated, reminiscent of an idealized eighteenth-century English landscape. Such 'before and after' images were a staple part of missionary writings during this period and were intended to promote the missionary cause in Europe. As historians have shown, they often had little to do with the realities on the ground (Thomas, 2010). Yet these images mattered, not simply because they reflected the missionary imagination, but also because they were framed with a particular audience in mind – the mission supporters and sponsors back in England. Looking more closely at the second image in Figure 16.3, for example, it is clear that the 'heathen' natives have not simply been converted to Christianity and commerce: they and their landscape have been Anglicized in ways which would have been recognizable to readers. An imagined transformation – the promised impact of Europeans on Pacific islanders – has been effected through an act of geographical inversion: imaginatively, at least, the island has become England.

A VILLAGE IN PUKAPUKA, UNDER HEATHENISM

THE SAME VILLAGE, UNDER CHRISTIANITY.

Figure 16.3 An imaginative geography of missionary work: a village in Pukapuka under (a) Heathenism and (b) Christianity. Credit: (a) and (b) W. Gill, *Life in the Southern Isles*, 1876

SUMMARY

- Images of culture frequently draw upon shared collective imaginations about cultural identity and difference. This provided the starting point for Edward Said's influential book, *Orientalism*.
- Such images transmit messages about the global geography of cultures to a wider audience, in school texts, popular books, film or television, for example.
- These images can be regarded as 'real', not because they reproduce the world accurately, but because they reflect and sustain people's imagination of that world, and in turn, help to influence the worlds we now inhabit.

Mapping cultures

Studies of the imaginative geographies of empire have often focused on the idea of mapping in general, and the practice of cartography in particular: in both its iconography and its concrete effects, the map has often been interpreted as one of empire's most powerful imaginative tools (Harley, 2001) (see also Chapter 13). A century ago, maps of the British empire, with their scattered patches of imperial pink, were not only found in atlases: they adorned many classroom walls and were widely reproduced in newspapers, commercial advertising and official publications. Often represented in the form of a global map on the Mercator projection, centred on the Greenwich meridian, the effect was to foreground the global reach of empire while obscuring the actual territorial fragility of rule from a small island.

Figure 16.4 shows a famous example of such an imperial map, originally produced as a supplement to *The Graphic* (an illustrated paper) in 1886, which has been widely reproduced in paper and electronic form. Today it still features as an illustration in many history books and educational web resources, and may be purchased as a framed print from the Victoria and Albert Museum. Its exemplary status as an icon of empire rests on its striking combination of two somewhat different imperial registers: the infrastructural (represented on the map by the statistics of trade and by lines connecting major ports of call) and the fantastical (notably in the use of human bodies, flora and fauna around its crowded margins to denote whole continents, races and landscapes).

Figure 16.4, with its combination of graphic and pictorial elements, positively overflows with symbolism, the interpretation of which requires close consideration of both content and context (see Driver, 2010). One important clue is provided by the banners being unfurled on its upper frame, celebrating 'freedom', 'fraternity' and 'federation'. The last term – federation – refers to a movement within the British Empire for the political reorganization of imperial government, associated with an organization (the Imperial Federation League) which sponsored the production of the map. The second clue is provided by the date, July 1886, which connects the image to a climactic moment of imperial image-making associated with the hugely popular Colonial and Indian

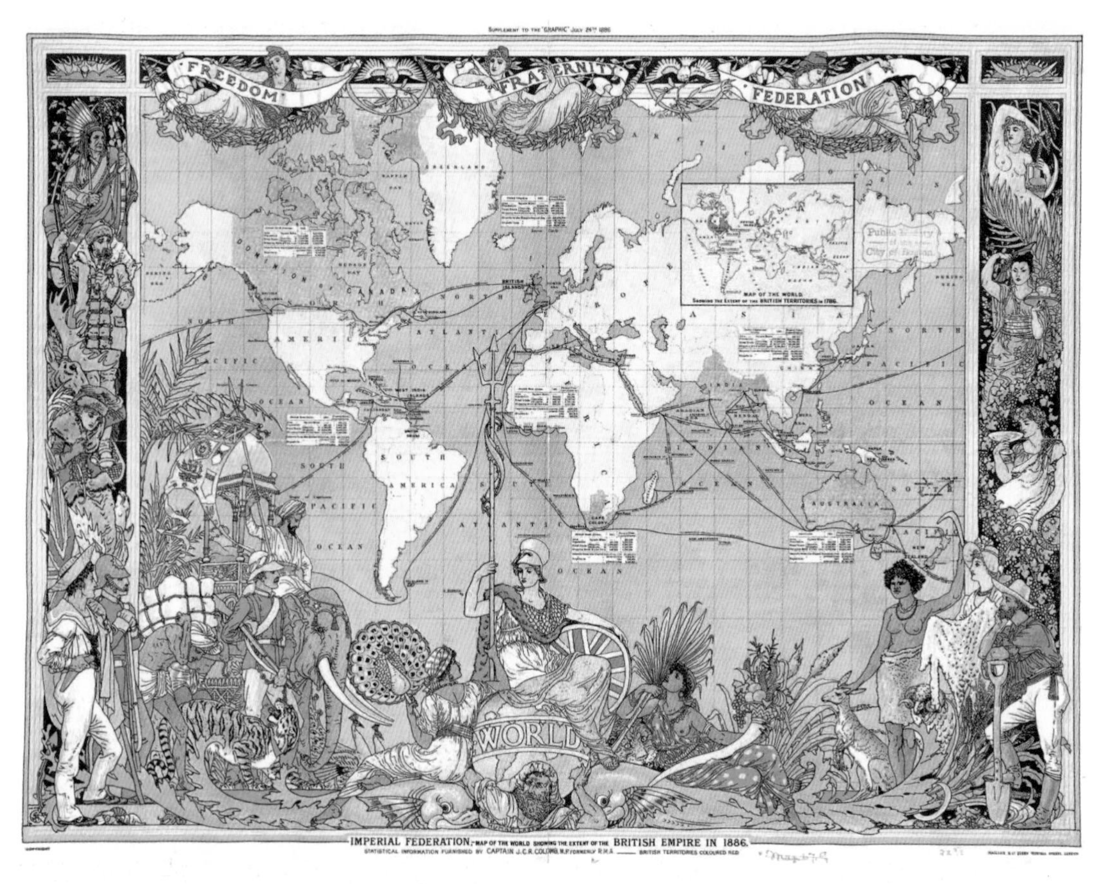

Figure 16.4 'Imperial Federation Map Showing the Extent of the British Empire in 1886', *The Graphic*, Supplement, 24 July 1886. Credit: in public domain

Exhibition in London. Further clues are suggested by some of the details of the iconography – including, for example, the juxtaposition of luxuriously dressed figures (such as the woman in furs, representing Canada) with others pictured struggling amidst plenty (such as the half-clothed Asian porter, bent double under the weight of his burden). You might also wonder about the tributes to the virtues of work scattered throughout the image, notably in the words 'human labour' discreetly inscribed on a banner worn by the god Atlas as he supports Britannia's world. Or you might be puzzled by the headwear of the classically draped female figures floating above the scene, especially as they bear a distinct resemblance to the red Phrygian cap worn by liberated slaves in ancient Rome. This cap was adopted as a revolutionary symbol of liberty across Europe and North America, and was widely used as an anti-colonial icon. Indeed, it still features on both the flag of the Haitian Republic and in the US Army seal.

Once one looks for them, there are enough clues in this image – the caps of liberty, the banners to freedom, the highly gendered pastoral iconography and the screen-like framework surrounding the map (creating a visual effect not unlike that of book illustration or stage design) – to suggest a rather unexpected source for this most imperial of

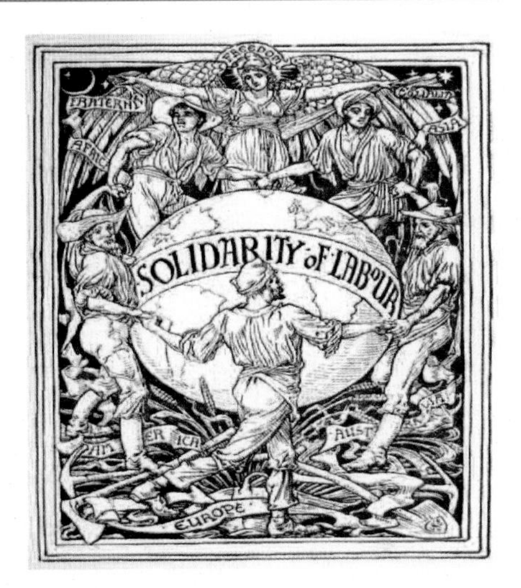

Figure 16.5 'Labour's May Day, Dedicated to the Workers of the World', by Walter Crane, frontispiece to W. Morris, *News from Nowhere*, 1890. Credit: in public domain

maps. In fact, the graphic was the work of a well-known socialist designer and illustrator Walter Crane, whose distinctive monogram combining the sketch of a crane with his initials appears in its very bottom left hand corner. The link to Crane, only recently discovered by a Cultural Geographer (Biltcliffe, 2005), raises interesting questions about the connections between the iconography of empire and of socialism in this period, especially in the context of contemporary ideas concerning colonial settlement, race and labour. It is intriguing, for example, to draw comparisons as well as contrasts between the iconography of Crane's imperial map and that of his frontispiece for the American edition of William Morris' influential socialist work, *News from Nowhere*, published just four years later (Figure 16.5). Here, the solidarity of labour is enacted through the linked arms of a transcontinental yet all-male brotherhood, while the angel of liberty looks down, unfurling her banners of fraternity and equality.

SUMMARY

- Empires, like nations, may be understood as imagined communities, constructed in part through the mobilization of images.
- Maps provide powerful ways of representing the relationships between places and between peoples; in the age of empire, maps were important means of diffusing ideas of imperial identity and difference.
- The detailed study of the production and use of maps may reveal divergent and sometimes unexpected aspects of the history of imaginative geographies, running against the grain of popular interpretation.

CASE STUDY

Geographers and the visual image: from theory to practice

Geographers' interest in the imaginative realm raises questions about our capacity not merely to interpret the world as we find it, but also to intervene in it. That requires further reflection on the ways in which images of peoples and places are made and the role they play in communication. It also means taking seriously the idea that in our professional and pedagogic practice as geographers, we are already engaged in the production of imaginative geographies: the making of geographical knowledge, whether in the classical age of exploration or in the world of digital media, has always been about the mobilization of images. In this context, it is notable that geographers are increasingly involved in creative collaborations with visual artists and others for whom the project of an imaginative geography offers the potential for inspiration as well as critique (Driver *et al.*, 2002; Hawkins, 2010).

The potential of visual media to challenge taken-for-granted assumptions provided the starting point for a recent public exhibition held at the Royal Geographical Society (with IBG), entitled *Hidden Histories of Exploration* (Driver and Jones, 2009) (see Figure 16.6). The aim of the exhibition was to highlight the contributions of the many people who made exploration possible but are rarely centre stage. In place of the usual emphasis on the heroic explorer, the exhibition drew attention to the role of locals and intermediaries whose work made voyages of exploration possible: guides and interpreters, pilots and porters, interpreters and field assistants. It did so by presenting various exemplary stories using materials from the unique historical collections of the RGS-IBG – including manuscripts, sketches, maps, paintings, prints, photographs and film – in an exhibition space designed to draw visitors into an alternative narrative.

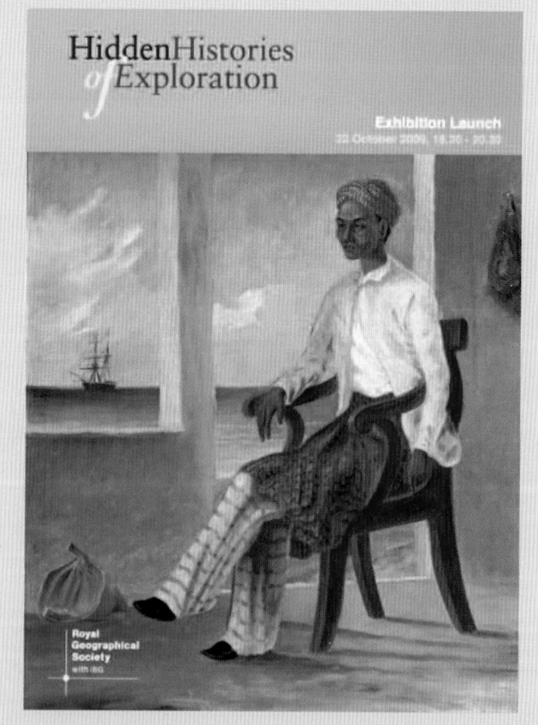

Figure 16.6 Exhibition poster for 'Hidden Histories of Exploration'. Credit: Royal Geographical Society (with IBG)

Making visible the agency of those usually hidden from history is, in part, a design challenge. In this exhibition, a variety of spatial strategies helped to dramatize the overall narrative. The first involved role reversal: the portraits and stories of the 'local assistants' (often leaders in their own right) were given

Figure 16.7 Dr Rita Gardner, Director of the Royal Geographical Society (with IBG), opens the 'Hidden Histories of Exploration' exhibition in October 2009. The 1936 Sherpa portraits are on the wall behind. Credit: Philip Hatfield with permission of Dr Rita Gardner, RGS

prominence, while those of European explorers were subservient to the main narrative. If this approach tended simply to reverse the terms of the 'heroic' model, a second strategy – involving the juxtaposition of images of European explorers alongside their collaborators – served to raise questions about the collective nature of expedition work. A key section of the exhibition, entitled 'uneasy partnerships', was inspired by the idea that the geographical knowledge resulting from expeditions was actually co-produced. Finally, through the selective re-scaling of archival images, the exhibition designer achieved perhaps the most dramatic effect of all. By its magnified reproduction on the wall of the gallery, an archival sheet of photographs of Sherpa porters recruited for a 1936 Everest expedition was transformed into a series of individualized portraits which attracted much visitor comment. By such means, subjects normally hidden from view were brought powerfully into presence.

Conclusion

Images of the kind we have considered in this chapter help to compose imaginative geographies: representations of place, space and landscape that structure people's understandings of the world, and in turn help to shape their actions. They have a real existence and real effects; in other words, they matter. In some circumstances, particular expressions of an imaginative geography may be judged false or partial, but that is not to say that such images are inconsequential. Whether or not we regard them as essentially true or false, imaginative geographies have significant implications for the way in which people behave. And they are certainly not simply things of the past.

The resurgence of new forms of Orientalism in representations of recent military conflict is a reminder of the continuing significance of imaginative geographies. Prior to the American invasion of Taliban-ruled Afghanistan in 2001, one military historian claimed to identify an essential difference between 'western' and 'Oriental' approaches to warfare: 'Westerners fight face to face, in stand-up battle . . . Orientals, by contrast, shrink from pitched battle' (Keegan, cited in Driver, 2003). In view of the nature of modern military technology, which enables unimaginable slaughter to take place without the inconvenience of face-to-face confrontation, these remarks are cruelly ironic. Expressing doubts as to whether the conflict could best be understood in religious terms, the same commentator reverted to what, if anything, was a still cruder thesis: 'This war belongs within the much larger spectrum of a far older conflict between settled, creative productive Westerners and predatory, destructive Orientals'. As Derek Gregory argues in his book *The Colonial Present* (2004), such claims replicate the discursive structures of Orientalism, reducing complex political situations to an essential opposition between the civilized West and its barbarian Others. The US-led 'war on terror' has raised far-reaching questions about the role of such imaginative geographies in the pursuit of geopolitical power.

Imaginative geographies, however, are not simply the instruments of states and elites. They also continue to circulate in a wide variety of public and commercial realms, for example, through television, film, newspapers and magazines, school textbooks, advertising, computer games or the web. Their influence extends beyond texts and images to material landscapes, too, thus housing estates, suburban homes, shopping centres, nature reserves, tourist destinations and theme parks may provide fertile sites for the study of imaginative geographies in the twenty-first century. What, for example, should we make of the Japanese fascination for reproductions of 'exotic' cultural landscapes in their own country? In Maruyama, a major transnational corporation was commissioned by the city authorities to construct a 'Shakespeare Country Park', a concrete abstraction of Elizabethan culture, complete with townhouse, inn, theatre, garden and village green. It would be easy to interpret this as simply an exercise in consumerism, a Disneyfied myth of 'Merrie England'. Yet the attention to detail manifested in the architecture and design of the site, together with the distinctive tradition of recreational pilgrimage within Japanese culture, suggests a more complex story (Chaplin, 1997).

The case of the contemporary theme park raises some wider methodological issues: how should we go about interpreting the meaning and effects of imaginative geographies? This is no easy task. Indeed, if it were, there would be no need for you to read any further! To interpret a landscape, a text or an image, it is clearly necessary to enter the world of those who created these artefacts and to understand their ideas, values and relationships. Once we recognize that the process of creation involves more than a single intention and usually requires the participation of many different people, the matter begins to look much more complex. And that is not all: a theme park, just like a map, is likely to convey quite different meanings to different people, and they too will carry with them a wide variety of experiences, preconceptions and desires. Interpreting imaginative geographies, then, turns out to require us to think not only about their content, but about their form, and not only about what they say, but how they have said it and to whom. While we may find it useful to think of imaginative geographies as 'ways of seeing', we must not forget that they are also sites of interaction and intervention. The challenge for the geographer is not to make imaginative geographies more real, but rather to make them more truly human.

DISCUSSION POINTS

1. Why should geographers concern themselves with the realms of the imagination?
2. What are imaginative geographies, and how are they generated?
3. How far does Edward Said's *Orientalism* provide a model for the understanding of cultural geographies of identity in general?
4. What role should imaginative geographies play in the making of geographical knowledge in the twenty-first century?

FURTHER READING

Cosgrove, D. (2008) *Geography and Vision: Seeing, Imagining and Representing the World.* London: I.B. Tauris.

An evocative set of essays on geographical visions of the world by an influential Cultural Geographer.

Driver, F. (2010) In search of the imperial map: Walter Crane and the image of empire. In: *History Workshop Journal* 69, 146–57.

A case study of an iconic image of empire, highlighting the importance of both close reading and contextualization.

Driver, F. (2013) Hidden histories made visible? Reflections on a geographical exhibition. In: *Transactions of the Institute of British Geographers*, 38, 420–35.

This paper considers the potential of public exhibitions to challenge long taken-for-granted assumptions about the history of exploration and geography, focusing especially on the role of design strategies in shaping exhibition space.

Gregory, D. (2004) *The Colonial Present: Afghanistan, Palestine, Iraq*. Oxford: Blackwell.

A passionate critique of the 'war on terror', which traces the imaginative geographies of military intervention in the Middle East into the colonial present.

Lutz, C. and Collins, J. (1993) *Reading 'National Geographic'*. Chicago: University of Chicago Press.

An influential study of the world's most popular geographical magazine, focusing on the photographic representation of place, race and gender.

Said, E. (1978) *Orientalism*. Harmondsworth: Penguin.

The classic work on the imaginative geography of the Orient as represented in scholarly and political writings in the West.

Schwartz, J. and Ryan, J. (2003) *Picturing Place: Photography and the Geographical Imagination*. London: I.B. Tauris.

An important collection of studies of the role of photography in representing imaginative geographies of place, nation and empire.

WEBSITES

http://hiddenhistories.rgs.org/

Hidden Histories of Exploration is an accessible website containing numerous image galleries, exhibition panels, research resources, weblinks and guides to further study.

CHAPTER 17
PLACE

Tim Cresswell

Place, order and categorization

Place is one of the central terms in Human Geography. It is a term that eludes easy definition and has been used in a number of disparate ways throughout geography's history (see Entrikin, 1991; Massey, 1993; Sack, 1997; Tuan, 1977). Place has been used as an alternative to 'location'. While location refers to position within a framework of abstract space, often indicated by 'objective' markers such as degrees of longitude and latitude, or distance from another location, place has come to refer to a mixture of 'objective' and 'subjective' facets including location but adding other, more subtle, attributes of the world we inhabit. John Agnew (1987) has argued, for instance, that place consists of:

- location: a point in space with specific relations to other points in space
- locale: the broader context (both built and social) for social relations
- sense of place: subjective feelings associated with a place.

Sense of place refers to the subjective feelings evoked by a place for both insiders (people who live there) and outsiders (people who visit or imagine that place). We can see, then, that place is a much richer idea than its precursor, location. It is not surprising, therefore, that geographers have studied place and places in a number of ways. While the importance of meaning to the definition of place has made it central to the concerns of Cultural Geographers, it has appeared throughout the sub-disciplines of Human Geography. Political Geographers, for instance, have looked at the construction of particular kinds of politics influenced by specific local places such as the north of Italy or Cornwall (Agnew, 2002; MacLeod and Jones, 2001). Economic Geographers, meanwhile, have enquired into why and how specific places become sites of clusters of particular kinds of economic activity. Place's central role in Human Geography means that it transcends sub-disciplines. In this chapter I focus on places from the perspective of Cultural Geography, more particularly understanding places as ways of seeing and framing the world according to what and who is said to belong where. This approach is only one possible way of thinking about place.

Place is more than an academic term – it is a word we frequently use in our everyday lives. Some of the ways we use it point to the richness of place as an idea. Here are some examples:

- He knew his place.
- She was put in her place.
- Everything in its place.

Figure 17.1 A woman's place is in the home? A stereotypical image of a post-war US housewife. Credit: Bettman/Corbis

Terms such as these point towards the social and cultural significance of place. In each of these phrases the word place suggests simultaneously a geographical location and a position on a social hierarchy. Think, for example, of a dinner table, either at home or in a more formal setting such as an annual dinner of an organization. Everything from the flowers to the position of the cutlery to the seating arrangement is in some way related to place in the social sense. Old fashioned notions of the patriarch sitting at the head of the table live on in households today and are formalized in the formal business dinner with its 'head table' and peripheral space for secretaries and janitors. A place for everything and everything in its place.

The human mind makes sense of the world by dividing it up into categories. As the examples above reveal, place and space are fundamental forms of categorization. Philosophers (most famously Kant) have insisted that the two basic dimensions of life are space and time, which form the basis for all other forms of categorization. Indeed, our conceptions of space and time are so fundamental they appear to pre-exist our conception and **representation** of them – that is to say they appear as nature. When we say that something is natural we are saying that it is not social – it 'just is' and is therefore unchangeable. This makes categories of space and time potent **ideological** weapons. They are ideological because they are laden with meanings that tend to create and reinforce relations of domination and subordination. As the French theorist Pierre Bourdieu (1984; 1990) has claimed, categorization schemes that remain unarticulated (seemingly as nature) inculcate adherence to the established order of things. This is the case because categorizations in space and time, for the most part, are not recognized discursively (we do not speak about them, write about them or even think about them) but practically (we act upon them). As we cannot possibly think about everything we do throughout our lives, the vast majority of our actions are fairly uncritical acts that conform to the expectations of those around us. We do what is defined as 'normal'. In the remainder of this chapter I will examine the way in which place acts as a category that serves to reproduce this normality and the existing 'order of things'. In addition, we will see how challenges to the taken-for-granted relations between place and actions can trouble the existing order of things, bringing it into view and questioning it.

SUMMARY

- Place is a complicated term that refers to both objective location and subjective meanings attached to it.
- Place is simultaneously geographical and social.
- Place is 'ideological', in the sense that it embodies taken-for-granted categories and orders. This is apparent in the judgements we make about things being in and out of place.

Rural and urban

Most places can be categorized in various ways and these categories can be challenged by people and other things being 'out of place'. One of the main ways in which places are categorized is as 'urban' or 'rural', and these kinds of categories bring expectations with them (see for example Chapter 48 by Paul Cloke on rurality). In the UK in the late 1990s, a group called the Countryside Alliance (self-designated as 'the voice of countryside') organized large marches on and in London by British people claiming to represent the 'rural way of life'. A significant part of their complaint is that present attitudes to rural pursuits are driven by urban perceptions of the rural. This was brought to a head by legislation (now passed as The Hunting Act 2004; still campaigned against) banning the practice of hunting animals with dogs, in particular fox hunting. Thus, the claim is made that fox hunting is an age-old rural tradition under threat from urban do-gooders who do not understand the intricacies of country life. Indeed, there is a long tradition of people and actions labelled 'urban' being described as out of place in the country, just as there are things associated with the rural that are labelled out of place in the city.

Prostitution in the country

A television advertisement for Hovis bread in the late 1970s and early 1980s featured a little boy struggling up a steep hill with his bicycle, complete with freshly baked wholesome brown bread and Dvorak's 'New World Symphony' playing in the background (see Figure 17.2). The place was more than a backdrop in this advertisement – it was a central actor that suggested a simpler and more wholesome and honest life of rural innocence. In the advertisement we are led to believe by the voiceover that the town is in the north of England but, as it turns out, the location was Shaftesbury in the southern county of Dorset. The street up the hill in the middle of Shaftesbury was already a tourist attraction before the Hovis advertisement appeared on television. Since then, thousands of tourists every year have climbed the hill or simply taken a picture of it. When people see it they recognize it as the site where the little boy pushed his bicycle up the hill. It has become a place where a certain view of rural and rustic Englishness is reproduced.

Figure 17.2 Shaftesbury, Dorset, as featured in an advertisement for Hovis bread. Credit: The Advertising Archives

In the autumn of 2002 a policeman noticed a website called 'Complete Excellence', which advertised the sexual services of a Shaftesbury resident who was described, in bucolic terms, as an 'English rose' who lived in a 'beautiful home deep in the West of England'. Potential clients were offered an overnight stay followed by 'sensuous morning tea in bed'. The police raided the Shaftesbury home and arrested Michael Chubb, the husband of 'English rose' Jilly Bywater, for running a brothel. The arrest of pimps is not usually front-page news in Britain but the arrest of Chubb and the revelations about Bywater's activities made all the newspapers. The press expressed surprise at the existence of prostitution in rural Britain. In one article which appeared in *The Guardian* the Shaftesbury area is referred to as both 'Hovisland' and 'Thomas Hardy country':

> But this 'English Rose' is not advertising in the red light district of London, Manchester or Edinburgh. She lives 'in a discreet cottage in the Shaftesbury area of Dorset' . . . Thomas Hardy country is not the place you'd expect to find a hooker, but Rosie is no normal lady of the night . . .
>
> Almost everyone in Shaftesbury and the surrounding villages is now talking about how all is not wholesome in Hovisland. 'What did she do for £500 a night? The ladies in my bridge circle want to know', says Janet Brady, 48, her dog, Max, by her feet as she sips coffee in the Rose Café. This part of Englande is the land of the tea shoppe.
>
> (Source: *The Guardian*, 28 July 2003: 6)

Thus, the disclosure that Jilly Bywater (like several other rural prostitutes) had resorted to prostitution following the failure of farming in the area in order to pay for her daughter's private boarding school fees is framed within implicit understandings of what it means to be a rural place in England. People have 'bridge circles' and England is spelled Englande. As *The Guardian* points out, prostitution would not be surprising if it occurred in 'London, Manchester or Edinburgh' but when it occurs at the top of Gold Hill, where the Hovis advert was filmed, it becomes national news. Shaftesbury is not simply any place in the English countryside but, thanks to Hovis, a place that has become an icon of the countryside and an imagined way of life. Such is the power of place and the rural–urban distinction in framing expectations about behaviour.

Animals in the city

The division between urban and rural mirrors the more fundamental division between culture and nature (see Chapter 10). Cities are supposed to be places of culture and society while rural landscapes have often been thought of as natural landscapes. Particular problems arise, therefore, when 'nature' appears in the city. Animals, for instance, are subjected to many of the place-based forms of control that marginalized social groups experience. Animals, like people, have their place. This is not usually understood as being the city (see Chapter 11 for further discussion of these 'animal geographies'). Of course, many cities have wildlife that inhabit them, often on the peri-urban or suburban fringe. In London the suburbs are frequently inhabited by more foxes than you are ever likely to see in the countryside thanks to the plentiful supply of leftovers among the garbage. In North America we might think of the incursions of cougar or coyote. But encounters with animals in the city are often framed in terms of their 'out of placeness'.

Let us focus on an example, the appearance of possums in the homes of suburban Australians. Emma Power, in research on Australian homeowners, has shown how the possum frequently crosses over the border that separates the 'place of animals' from 'human place'. This transgression of animal/human borders creates anxiety in many homeowners who think of their home as their own special place and are reluctant to share it with uninvited guests. You are probably familiar with how we often frame unwanted animal co-habitees as 'pests' (in the same way that we frame unwanted plant co-habitees in our domestic gardens as 'weeds'). In this case, the possums live in all the invisible and often unreachable spaces that are part of every home. These include the ducts, the cavities inside walls and the spaces inside roofs. The homeowners that Power spoke to sometimes enjoyed the company of wildlife in their suburban homes but mostly found their presence (often experienced as sound and smell) unsettling. Having possums in their personal places detracted from their sense of home as theirs. Many thought that possums did not belong in an urban environment.

Participants' encounters with possums were framed by discussions about whether possums belonged, or did not belong, in the urban and urban bushland environments that participants lived in. Participants' beliefs about whether possums belonged in these environments shaped their own sense of homeyness in these spaces and in the house as home. Although urban environments have traditionally been framed as human-only spaces, participants expressed a more complex conception of possum-belonging that drew on narratives of nativeness, invasion, colonialism and contemporary urban development (Power, 2009: 38)

The contradictory sense of possums both belonging and not belonging in urban space was also manifested in the participants sometimes claiming that the presence of possums in and around their homes led to a feeling of co-habitation and 'homeyness'. The presence of possums also increased their sense of a shared human-nonhuman space.

Possums, perceived as wild, native animals, extended the home into nature and brought nature into the home to make the house more homely. These experiences conflict with views of home as a human space and illustrate some ways that wildness and homeyness can come together within the house-as-home. They extend discussion of designed, integrated living spaces that blur inside and outside, because here blurring was a product of uninvited ruptures in home. It went beyond human design to instead reflect the particular impact of possum-agency on home and homemaking. (Power, 2009: 45)

The relationship between homeowners and possums revealed by Power's research is a complicated one. For the most part, the homeowners believed that the possums belonged elsewhere – that this was not their place. On the other hand there was also a more subtle recognition that the suburban areas that they were living in were built over the place of the possum and that, therefore, the possum had a place in and around their homes. Here, binaries of the urban and rural, the social and the wild, the human and the natural, were destabilized. Thus the process of categorizing things in relation to place – here possum and suburban homes – does not always result in a clear categorical verdict (in or out of place). It can be ambivalent. Its importance is in the ideas and feelings called upon and provoked, which together form judgements about places and the wider order of things.

Figure 17.3 An urban possum. Credit: Getty Images

So while prostitutes have been seen as the urban in the rural, the possums in an Australian suburb were, for the most part seen as the rural (wild even) in the urban. In these moments, one sees taken for granted categorizations of rural and urban brought to the fore, provoking responses that sought to define the 'proper' activities for such places and expel the intruding transgressors. However, one can also see responses where the presence of something out of place provokes a questioning of categorical distinctions such as rural and urban. Both responses reveal how place is implicated in the construction of particular kinds of morality, in framing what is right and wrong, and on what account. Place, then, is fundamentally about the production of 'moral geographies'.

SUMMARY

- The categories of urban and rural are signified by particular places and particular actions.
- While prostitution has often been associated with urban space, it becomes newsworthy when it is discovered in a rural area. Rural places have been associated with wholesome innocence, and the location of prostitution in a rural town confounds normal expectations of the links between commercial sex and place.
- Certain kinds of animals associated with the rural and the wild share urban and suburban places with humans. They are frequently experienced and represented as 'out of place'. However, they can also provoke more ambivalent reactions, in which the distinctions between rural and urban are disturbed.

Place, public space and the politics of protest

Neither prostitutes in a rural village nor possums in suburban homes are deliberately out to challenge the meanings given to place or the practices associated with them. Sometimes, however, being 'out of place' is used deliberately as a form of protest or resistance to accepted forms of power. This is particularly true of the ways in which public space is used. Public space is often simultaneously the site of the assertion of power and ideology by the **nation-state**, corporations and local governments, and of counter-ideological practices. Just as power is spatialized as place is given meaning, resistance can take the form of spatialized disobedience.

A case in point is the actions of the Plaza de Mayo Madres (sometimes called the Mothers of the Disappeared) in Buenos Aires, Argentina, during the late 1970s and early 1980s. More recently, we have seen the highly visible protests of the Occupy movement in public spaces across the world. The application of meaning to space (the creation of place) is a supremely political process that tends to inscribe a particular idea of order on the lives of the people who inhabit (but do not build) that space (Duncan, 1990). The creation of public places such as streets, parks and public squares are often acts of ordering of the first magnitude. These spaces are constructed in order to convey the legitimate order to citizens. However, in consequence such public spaces are also potential stages for resistance through the 'transgression' of who and what is expected within them.

Mothers of the Disappeared

Jennifer Schirmer recounts some of the history of how gender is produced and reproduced in western spaces. She tells the story of the French eighteenth-century assertion of freedom for men to speak, move and think, which arose at the same time as the banishment of women's public speech and political life. Women in public were 'out of place'. The best kind of woman, the most virtuous woman, was one who knew her place and did not speak out of turn:

> Boundaries between the public and private, the political and social, the productive and reproductive, and justice and family were established, and justified by women's absence in the first and presence in the second.
>
> (Schirmer, 1994: 188)

Schirmer goes on to tell the story of women who have taken it upon themselves to transgress the boundary of public and private space in order to make political points (in places supposedly non-political). Thus, the Plaza de Mayo Madres inserted their bodies into the public space of Plaza de Mayo in order to protest against the 'disappearance' of family members (see Figure 17.4). During the period 1976–82 the military junta of General Videla

Figure 17.4 Protest by the Plaza de Mayo Madres (the Mothers of the Disappeared) in Buenos Aires, 1985. More than 10,000 attended. Credit: Micheline Pelletier/Sygma/Corbis

fought against a so-called International Conspiracy of Subversion that was said to be against all the (western) values that Argentina stood for. This included getting rid of all those who were out of place in Argentina – the 'alien bodies' who sought to subvert the state. Over 30,000 people became targets for official and unofficial state security forces to abduct, torture and disappear. The victims were erased from public consciousness. In protest, the Plaza de Mayo Madres began to circle the main square in Buenos Aires in 1977.

The square itself, like many grand squares around the world, had been built to symbolize elements of official history and ideology. In this case it was a symbol of the Inquisition, in addition to contemporary ideas of governance. Every Thursday at 3.30 p.m. the Plaza de Mayo Madres would walk arm in arm, their heads covered, demanding the return of their disappeared relatives and the punishment of the people responsible. The presence of these women in public space is an immediate transgression of two place-based categorizations – the association of public space with masculinity and the association of such places with an absence of politics. In addition to these transgressions, however, the women used many of the symbols of motherhood and domesticity to make their case.

> The Plaza de Mayo is flanked by monumental buildings that are incongruous with the private lives of domesticity: the presidential palace (the Casa Rosada), which was used by the juntas; the cathedral; and the Ministry of Social Welfare. This site of masculinist power is demystified by older women, humbly circling the plaza wearing on their heads diapers first, and later white headscarves, embroidered with the names of their disappeared son or daughter or husband, together with worn photographs of their loved ones pinned on their breasts or placed on large placards at marches and demonstrations.
>
> (Schirmer, 1994: 203–4)

Here the Madres are confusing the relationship between place and meaning in complicated ways, both transgressing the expectations of **gender** and politics, public and private space, and reaffirming (strategically) the 'normal' and 'proper' association between femininity and the nexus of the family and the domestic.

Occupy movement

The power of place in political protest has arisen recently in reactions to the Occupy movement that has spread across global cities in order to protest the workings of contemporary capitalism. Occupy started as 'Occupy Wall Street' in Zuccotti Park in Manhattan, New York on 17 September 2011. Within a month it had spread to over 100 countries. The Occupy movement seeks to shine a spotlight on the extreme levels of inequality produced by a capitalist economy. It took some of its inspiration from the way citizens took over public spaces in North African and Middle Eastern countries during the Arab Spring in order to affect change in the political cultures of those places. The most prominent example of this was the crowd that gathered in Tahrir Square in Cairo. Typically the Occupy style of protest is to take over a public space and create a semi-permanent place of protest consisting of tents, complete with libraries, childcare facilities, universities and places to eat. The high visibility of these camps and their reliance on a geography of place was noted by a *New York Times* commentator:

> We tend to underestimate the political power of physical places. Then Tahrir Square comes along. Now it's Zuccotti Park, until four weeks ago an utterly obscure city-block-size downtown plaza with a few

Figure 17.5 Occupy London protesters camped outside St Paul's Cathedral, London. Credit: Getty Images

> trees and concrete benches, around the corner from ground zero and two blocks north of Wall Street on Broadway. A few hundred people with ponchos and sleeping bags have put it on the map.
>
> Kent State, Tiananmen Square, the Berlin Wall: we clearly use locales, edifices, architecture to house our memories and political energy. Politics troubles our consciences. But places haunt our imaginations.
>
> (Michael Kimmerman: 'In Protest, The Power of Place', *New York Times*, 15 October 2011

In his article, Kimmerman reflects on the relationship between the virtual world of social media and the concrete world of urban places where protest so often takes place. A lot has been said about the role of Facebook and other apps in the creation of sudden, uncontrollable gatherings, but less has been said about the visibility that place brings to forms of popular protest. In London, the Occupy movement camped outside St Paul's Cathedral in the City of London – the financial heart of the metropolis (see Figure 17.5). The significance of the place (like the significance of the Plaza de Mayo) is integral to the visibility of the protest.

Local road signs were relabelled 'Tahrir Square' in recognition of the way in which the meaning of a place can travel and influence the meaning of other places, elsewhere. The administration of St Paul's had to more or less publicly debate the merits of the protesters' cause while considering whether or not to go to Court in order to have the protesters removed. Occupy relies on the power of place to attract attention and to lodge itself in people's memories. As Kimmerman commented in the context of New York:

> Not since 9/11 have so many people been asking 'Have you been there?' 'Have you seen it?' about any place in Manhattan. The occupation of the virtual world along with Zuccotti Park is of course jointly propelling the Occupy Wall Street movement now, and neither would be so effective minus the other.
>
> (Michael Kimmerman: 'In Protest, The Power of Place', *New York Times*, 15 October 2011

Another element of place that arose in response to the Occupy movement's use of it was the kind of place that the camps became. It was suggested by one writer in the online commentary of the British newspaper *The Guardian* that the Occupy Wall Street group had replicated the sense of place of an American small town:

> The vision the American Occupy movement's camps in New York, Oakland, Los Angeles, Atlanta, Chicago, Boston, Washington, DC and a hundred smaller cities and hamlets most resembled was our inner picture of the American small town. But then I didn't know it myself until we lost it. Every small town has its post office, its town library and its diner, as the places where its citizens meet informally. Occupy Wall Street had its communications centre, its people's library and its kitchen. The way that they were laid out in the space of Zuccotti Park, at least, seemed like a scale model of the casual way small towns are set up. In between, people lived in their little humble homes. These were the sleeping bags, and only much later the tents, which you politely stepped around, and which filled the spaces where gatherings occurred.
>
> (Mark Greif: 'Occupy London: What did the St Paul's camp represent?' *The Guardian*, 29 February 2012)

In the various Occupy camps, then, we see place acting in two main ways: one is the contrast between the camps of the protestors as places and the places they were located in or adjacent to (the financial and administrative centres of major cities); the other was the nature of the camps themselves as semi-permanent places that suggested a different way of doing things. As the *New York Times* commentator suggested, it seems likely that place plays a major role in the power and symbolism of these inventive protest movements.

SUMMARY

- The division of the world into private and public places denotes particular relationships between geography, social group and behaviour. Public places are often seen and experienced as masculine spaces.
- Place can be used very effectively in forms of popular protest. Public spaces are often used by those seeking to challenge established ways of doing things in ways that challenge the meanings of places and create new senses of place instead.

Conclusion

In the introduction to this chapter I discussed the way in which the association between place as meaningful location and categories of people and actions is often invisible because it is so deeply engrained. Knowing one's place thus seems 'natural' or 'inevitable'. In the examples that followed, the relationship between place and categories became most apparent when the relationships were *transgressed* – when people (or animals or things) were said to be 'out of place'. Thus beliefs about what happens and belongs in the countryside are underlined when something (such as prostitution) is described as 'out of place' there. The order that is constructed by and through place is not inevitable and is often transgressed. It is in these moments of transgression that a great deal is said (in the media, by politicians and figures of authority) about what and who belongs where. Transgression, then, is a key concept for geographers who want to describe, explain and critically question the construction of 'normality' through the creation and maintenance of particular types of place.

DISCUSSION POINTS

1. How is place different from location?
2. In addition to distinctions between urban and rural places, what other kinds of binary distinctions characterize our thinking about place?
3. In what ways can places be given meaning?
4. How do pre-existing images of place affect our judgement of events located in them?
5. How might place be thought of as ideological?
6. How might places be used to resist dominant forms of power? Can you think of other protests that have created and/or subverted place as one of their tactics?

FURTHER READING

Cresswell, T. (1996) *In Place/Out of Place: Geography, Ideology and Transgression*. Minneapolis: University of Minnesota Press.

In this book I expand upon the connections between place, meaning and power through an examination of events and people labelled out of place. It includes discussions of graffiti, travellers and Greenham Common peace protesters.

Cresswell, T. (2004) *Place: A Short Introduction*. Oxford: Blackwell.

This book surveys in greater depth the variety of ways place has been used in geography and beyond. It explores the development of the concept and provides concrete examples of the role of place in a variety of social and cultural processes.

Power, E. (2009) Border-processes and homemaking: encounters with possums in suburban Australian homes. In: *Cultural Geographies* 16(1), 29–54.

Schirmer, J. (1994) The claiming of space and the body politic within national-security states. In: J. Boyarin (ed.) *Remapping Memory: The Politics of Timespace*. Minneapolis: University of Minnesota Press, 185–220.

Some of the case studies presented in this chapter are discussed in more depth in the above texts.

Sibley, D. (1995) *Geographies of Exclusion: Society and Difference in the West.* London: Routledge.

David Sibley develops a similar theme in this excellent book. His focus includes a consideration of psychoanalytic theory and he provides numerous examples of how space and place are implicated in the exclusion of people and actions defined as 'Other'.

WEBSITES

http://societyandspace.com/2011/11/18/forum-on-the-occupy-movement/

The online Forum discussion organized by the journal *Environment and Planning D: Society and Space* offers a series of varied responses to the geographies of the Occupy movement.

CHAPTER 18
LANDSCAPE

John Wylie

Introduction

I have a problem here. Let me spell it out as follows:

- A landscape is a particular type of painting.
- The landscape is whatever you see in front of you at any given moment (especially if it's a view of fields, hills etc.).
- A landscape is an area of the earth's surface that has been shaped, gardened – *landscaped* – so as to be pleasing to the eye; made to look neat and trim.
- A landscape doesn't have to be spectacular or scenic, or even 'rural'. The gardens, streets and houses of the suburbs are a landscape, and so is the city centre – the urban landscape.
- Landscape is a page format.
- A landscape is a certain size of area, bigger than a place or a location, but smaller than a country.
- The landscape is whatever's outside – the outdoor world.
- The surface of the earth is a mosaic or jigsaw of different landscapes, some 'natural', some the product of interactions between people and their environment.
- The landscape isn't just the view, something 'out there' which you appreciate visually from a distance – it's the world all around you, the world you live in and are part of, the world you inhabit with all of your senses.

I hope you see my problem? Landscape can mean many different things, and at least some of these are likely to be already familiar. If I had a more specialist term to write about – spatiality, say, or colonialism or globalization – I could more easily start from scratch and say to the reader, in this chapter I will explain what this term means. But landscape is a word in everyday use, in multiple ways. I very much doubt if any reader of this chapter, someone just starting to study university-level geography, could come to it completely free of preconceived ideas of what a landscape is. So, landscape arrives already gilded – or perhaps tarnished – by a wide range of associations. Certainly in my experience students will commonly offer definitions such as those above when prompted to do so at the start of a set of classes on landscape. Associations with countryside, scenery and painting are usually prominent.

Cultural geographies of landscape

The problem I've highlighted might solve itself if I could say, at this point, please forget all those pre-existing definitions; when *Cultural Geographers* talk about landscape they mean something different altogether. But that wouldn't be true exactly. Cultural Geographers approach landscape in diverse ways and 'landscape' has been a key context and reference point for Cultural Geography for nearly a hundred years now. For example, for much of the twentieth century Cultural Geographers defined landscape as the outcome of how different human cultures interacted with, and were influenced by, natural environmental conditions – by soil, terrain, vegetation and climate. Landscape, in part, was understood as the imprint of human culture upon the earth; as a 'cultural landscape' (Sauer, 1925). This chimes quite closely with one of my initial bullet points above – the idea that the surface of the earth presents a mosaic of different types of landscape, reflecting both the so-called 'natural environment' and human cultural diversity, in terms of differences in agriculture, architecture, belief systems and so on.

My focus in this chapter will be upon more recent cultural geographical understandings of landscape – but here, too, everyday meanings and associations creep back in. One influential and productive line of inquiry argues that landscape is a particular *way of seeing* the world (for pioneering examples, see Cosgrove, 1985; Cosgrove and Daniels, 1988). In other words, landscape is a certain format or framework for visualizing and depicting the world – a standard visual technology, almost like a map or telescope or microscope. As a way of seeing, a visualization, landscape takes the form of paintings and photographs. Thus, *these* are the landscapes to be studied by geographers. This understanding might not be the approach that most people recalling their schooldays associate with gography, but to say that a landscape is a type of painting is commonplace.

Another strand of current thinking argues that landscape needs to be understood as a term which connotes how humans (principally) inhabit and dwell in the world. Partly this involves thinking about landscape as the accretion, over time and space, of local customs, knowledges, traditions and laws (see Olwig, 2002). It is also partly about understanding that human landscapes and 'lifeworlds' emerge and are sustained through everyday practices of inhabitation – through moving, interacting, working, playing, remembering and dreaming (see Seamon and Mugerauer, 1985; Ingold, 2000). To link back to my initial bullet points, the landscape is therefore the world we live in – the world we're immersed in through our senses. In this conception, a landscape is a lived, evolving web of pathways and dwelling places.

So, the definitions of landscape that Cultural Geographers mostly work with today *do* still connect with 'everyday' understandings. However, two important differences must be noted. First, Cultural Geographers consistently stress the role of *power* in the making and perception of landscape. Who owns and controls the landscape? Whose purposes does it serve? Who has the power to influence and direct how we see landscape? How do landscapes reflect and reinforce power relations between different groups in society? And how can we as geographers make these processes more visible? Second, Cultural Geographers – and others – have come to understand landscape in terms of *performance*. Landscape has often been conceived as a rather static background phenomena, or to adopt a theatrical metaphor, as the relatively inert scenery against which the actors perform. Recent work, though, emphasizes landscape as both scene *and* performance, and thus advances a perhaps more dynamic account of landscape-as-performance

– both in terms of everyday actions and artistic performances.

That concludes my introduction to landscape. In what follows, I will work in detail through two landscape 'examples' (two images) to flesh out the points I've made so far and more fully introduce contemporary cultural geographies of landscape. An alternative structure for the chapter would be to recount the story of how the study of landscape in Cultural Geography has evolved and changed over time. But this would involve compressing and vacuum-packing a mass of often technical arguments into a short space. In addition there are already several sources available which tell such a 'story' of landscape – including pieces by myself (Wylie, 2007, 2010). I want to try something fresher and more lightly referenced here. Having spent the last dozen or so years studying and writing about landscape, I find myself routinely on the lookout, so to speak, for new forms and examples, and this is primarily how I have come across the two I'll discuss here. They are: a) the front cover image of a scientific report; and b) an opportunistic but still hi-spec photograph, taken from an unusual vantage point. They are quite different from each other, in terms of composition, purpose and audience, and this in turn will allow me to explore different issues and perspectives about landscape.

SUMMARY

- Cultural Geographers have approached landscape in various ways.
- A common approach for much of the twentieth century emphasized the notion of the 'cultural landscape' – the human imprint upon the natural world.
- More recently, landscape has been understood both as a particular 'way of seeing' and as designating how people 'inhabit' and 'dwell' in the world.
- These approaches to landscape are currently developed with particular emphases placed on issues of power and performance.

The UK National Ecosystem Assessment

Images of landscape, painted or photographic, may express not just a sense of locality, but of nationality too. This link between landscapes and national identity has long been a key topic of study for geographers and others (see notably Daniels, 1993; Schama, 1995; Matless, 1998). It is crucial to note that the focus of such work has *not* been upon which landscapes 'best' or most 'accurately' depict a given nation. Instead, the critical focus is upon how certain types of landscape imagery communicate and reinforce certain senses of nationhood, and often elite visions of the nation. So, this imaging of the nation via landscape is not an innocent or consequence-free process. Instead it is a process in which *particular* national ideas and values are reinforced and elevated over others.

In order to critically understand how certain national values are expressed via landscape, we

Synthesis of the Key Findings

Figure 18.1 Front cover of the UK National Ecosystem Assessment Report, 2011. Credit: UK National Ecosystem Assessment

need to keep two sets of questions in mind. First, what is included in the image and what is excluded? Who is there, and who is not? What belongs and is at home – and what apparently does not belong? Second, how is the natural world being 'nationalized' in such imagery? Landscape art often pictures relationships between culture and nature (see also Chapter 10). In this specific case, we can argue, landscape involves the creation and depiction of a 'national nature'. So, how are nature, topography, flora and fauna enrolled by landscapes into the project of expressing a sense of national identity?

Figure 18.1 is the front cover of a recently published UK government document – the UK National Ecosystem Assessment (UKNEA) (2011). A lengthy and detailed report compiled by a large number of academics, experts, and stakeholders, the UKNEA aims to provide an audit of the health of the UK's ecosystems and the 'services' they provide to humans. It is therefore a kind of 'state of the nation' report, an assessment of how the UK's ecosystems currently stand and of what the future might hold. In this context, the landscape imagery found in the report has a particular, national resonance.

The front cover we see here, and the report as a whole, was designed by a company called NatureBureau™. It is not, obviously, a picture of a real scene. Rather, it is a stylized ideogram intended to depict, in some way, the key motifs of the UK landscape and ecosystems addressed by the UKNEA report. It is, however, very much a *landscape*, in the popular sense of presenting a scenic view of, for the most part, countryside. In this way the image works by drawing upon a visual landscape 'format' that readers in the UK will very likely be familiar with, to the point of almost not noticing it. So, what image of the UK does this landscape depict?

First of all, the UK is pictured here as a *rural* nation. Despite the fact that the UK is actually one of the world's most urbanized countries, at some point a decision has been taken to place the UKNEA Report within this rather traditional, rural landscape frame. This landscape therefore suggests that the heart of the UK remains in the countryside. There *is* a city pictured here, as you can see, and industry too, as well as just one little car, on an otherwise empty road. But these features take their place snugly within the landscape; instead of dominating or clashing, they are placed unthreateningly in the comfortable middle distances. City, industry and transport are pictured at home, so to speak, within this rural landscape. In this way, we can see that the image reproduces and perpetuates a characteristically rural landscape ideology when it comes to senses of national identity in the UK. What difference would it have made if the UKNEA frontispiece had depicted a busy city scene? To whom does a rural vision of the UK appeal – whose values does it endorse and confirm?

In addition to rurality, another national landscape framing at work here is the *coast*. Clearly, this image has been designed to include as many key 'ecosystem' elements as possible, so we see the city and the country, industry and agriculture, work and leisure, humans and non-humans. In this context, the inclusion of the coastline – and of what looks like a working fishing boat – might simply be taken as a visual acknowledgement of the role of marine ecosystems, but the symbolic placing of the coastline right at the visual centre of this landscape demands that we pay further attention here. The coast is an important signifier of national identity in the UK; beyond being a physical feature it appeals to a sense of being an 'island nation' – an independent and autonomous place, a place made distinctive and strong *through* its 'islandness'. The popular success of the BBC2/Open University television series *Coast* is another example of how such a sense of islandness persists in the UK, and is perpetuated by various media.

What about the people of the UK? Crucially, here we have a *family*. Placed in the foreground of the image, in conformity with the visual conventions of centuries worth of landscape painting, we see a man, woman and child – a recognizable family unit. Just as we see stock representatives of 'nature' here (birds, insects), so we are presented with what is almost a caricature of 'Mr and Mrs Average UK'. Their ethnicity is indeterminate, but they are clearly a heterosexual family. And this family has chosen to go for a walk in the countryside, to interact healthily with the ecosystems. They are not obese. The parents (if we assume this role) stick responsibly to the path, while the child strays playfully onto the grass, although this is an acceptable transgression – interacting with 'nature' is good for children, the UKNEA claims. More widely, this family unit embodies the nation just as the whole landscape does. Like the countryside, the nuclear family denotes continuity, solidity and domesticity. As citizens, they inherit the national landscape and assume some measure of stewardship of it. They might well be family members of the National Trust. The child denotes the future of the nation – that for whose sake the UK's ecosystems must be sustained and preserved. This family, in sum, adds visual focus, coherence and meaning to the landscape, and help to make it very much a picture of the nation.

A 'naturalized' family unit takes centre stage in this image, but other emblems of nature are also placed carefully in the landscape. A bird sails through a summer sky – a symbol here of nature 'free' and unmanaged, though still, by implication, in need of recognition and protection. Elsewhere, a lonely cow faces a fence even it could hop over, and represents perhaps the one troubling feature of this otherwise highly composed scene. Cows and sheep are stereotypical elements of traditional pastoral British landscape art. But, after BSE or 'mad cow disease' and the foot-and-mouth epidemic of 2001, the cow in the British landscape today connotes danger and disaster as much as placid harmony. It conjures a sense of humans and nature worryingly out of kilter, of a once supposedly 'green and pleasant land' despoiled by the 'unnatural' practices of intensive agriculture, animal transportation, global commodity markets and so on. However, if the cow is an uneasy presence here – placed awkwardly on the surface of the landscape, like 'Fuzzy felt', rather than nestled within it – then some reassurance can be found elsewhere, especially perhaps on the right-hand edge of the frame, where a tree, an oak tree maybe, confidently stands. Of course, the oak is one of the most enduring symbols of 'English nature' and its presence here acts as a kind of visual reassurance – a sort of promise that, however the UK landscape might develop and change, some elements of familiar character will endure.

Lastly we should ask: where is this landscape? It seeks to visually represent and express the UK as a nation. It communicates themes core to a certain elite discourse of UK nationhood: rurality, tradition, islandness, family. However, because it is in landscape format, it has to represent somewhere in particular. As we've seen, this is a stylized landscape, designed to include a series of key elements, but despite this, the landscape presented still has a particular topographic quality. It recalls most specifically the landscape of southern England and of coastal counties such as Devon, Dorset or Sussex, for example. The landscape is not just rural – it is a quite specific rural vision.

It is not the highlands of Scotland or the mountains, uplands and moors of the northern and western UK generally. Nor is it the flatlands of eastern England or much of the Midlands. Instead, the varied landscapes of the UK – of England, Scotland, Wales and Northern Ireland – are represented here from a specifically English, southern and rural point of view. *This* type of landscape is the one chosen to symbolize the nation as a whole.

The choice might seem unremarkable and unproblematic. The landscape might even strike the casual viewer as a typical or normal British scene, but this is only because of a centuries-old ideology through which the rural southern 'shires' of England have been claimed to be the heart of, or the most essential expression of, the UK national landscape. This pastoral vision of British national identity stands in opposition to that of the cities and the uplands (see Cosgrove, 1993). Intentionally or not, in choosing such a landscape for the front cover of a UK-wide official report, the UKNEA buys into and reinforces a vision of rural southern England as the essential UK landscape – the one best-suited to symbolizing the nation.

SUMMARY

- Landscapes and national identities are often linked. National values are expressed via landscape.
- Cultural Geographers critically analyse these national landscape expressions. They ask questions about who and what is included in them. They consider how forms of 'nature' are enrolled into national 'culture' through landscape.
- In the example shown in Figure 18.1, the landscape picture on the front cover of the 2011 UK National Ecosystem Assessment Report, we see the reproduction of some common ways of constructing the UK's national landscape.

'Don't trip'

Figure 18.2, '*Don't trip*', shows us another landscape, but one very different to that considered in the previous section. Here, in contrast to the conventional perspectives and comfortable, perhaps even safe scenery offered to the viewer in the UKNEA example, we find ourselves occupying an extreme, vertiginous viewpoint. We are looking down, in fact, from the very top of the Shard, London's (and Europe's) tallest building, completed in spring 2012.

To make sense of this landscape, we first need to know something about its origins. The image was posted on the Place Hacking website on 7 April 2012, but the photo itself was taken several months earlier. The photographer is in fact the figure we see in the image – this is therefore a self-portrait, among other things. This figure is Bradley Garrett, the author/editor/owner of Place Hacking, and at that time a cultural and urban geographer from Royal Holloway, University of London. Garrett studies, and practices, contemporary 'urban exploration' – the exploration, by small and

Figure 18.2 'Don't trip'. Credit: Bradley L. Garrett/www.placehacking.co.uk

often anonymous groups of people, of derelict, forgotten, hidden and forbidden city spaces, such as abandoned underground tunnels, inaccessible rooftops or securitized building sites (Garrett, 2010). The ascent to the top of the Shard is a good example of the ethos and agendas of urban exploration. The aim is to subvert the usual ordering and policing of the urban landscape, to get past security, to access unusual or extreme locations, and then to document and evidence the event in published photographs.

What we see documented in this landscape, therefore, is a deliberate act of *trespass*. The ascent of the Shard might seem simply a prank or publicity stunt, but a long tradition links landscape and trespass in the UK, insofar as landscape has historically been deeply implicated in issues of ownership, property and rights of access. The best known example of this is the Kinder Scout mass trespass, in the Derbyshire Peak District in 1932, when a group of ramblers, mostly working class and from the industrial cities of northern England, trespassed upon the estates of the Duke of Devonshire, leading to a pitched battle with gamekeepers and subsequent arrests and convictions. Despite this apparent defeat, however, the Kinder trespass is today seen as a victory, insofar as it became pivotal to the later establishment of legal rights of access to countryside in the UK, as most recently extended and expressed in the 2000 Countryside and Rights of Way Act[1].

Both the Kinder trespass, and the ascent of the Shard pictured in *Don't trip,* may therefore be understood as *forms of landscaping,* or direct action, which deliberately set out to contest and unsettle official spatial zonings, demarcations and claims to ownership (see also Chapter 17 on the relationships between place, order and transgression). What we see here is landscape expressed openly in terms of conflict and contestation. I am stressing this point because it connects with my introductory argument that Cultural Geographers commonly and critically understand landscapes in terms of power relations. In my first example of the UKNEA Report, it took a critical reading to draw this out and make visible the ways in which power was at work, as the landscape implicitly reinforced certain senses of what is acceptable, what is 'normal' and so on. But in Figure 18.2, the landscape is explicitly configured as an act of opposition and subversion. This cuts usefully against a tendency to consider landscape as inherently conformist and conservative. As Tim Cresswell (2003: 269) writes, the whole notion of landscape can seem to be 'too much about the already-accomplished'; the word itself 'altogether too quaint'. A first key message, though, of *Don't trip*, is that it need not always be so.

If *Don't trip* enables us to conceive landscape in terms of trespass, rather than encouraging us to conform to the country (or city) code, then more widely a second message here is that landscape is something kinetic and dynamic. Again, this is to cut against a common preconception. Because a 'landscape' is often understood as something relatively fixed and stable – or as a view from a fixed point – it can have a static, motionless connotation. Like a map or architectural plan – visual forms to which landscape art is in fact related – a classical landscape painting imposes a static grid upon the world; the better, it seems, to produce a picture that people will naively accept as 'objective' and 'real'. Putting this in terms of the human body and its moods, landscape is perhaps more usually commonly associated with calm contemplation, rather than involved or intense action. Even here, with *Don't trip*, what do we see? A certain moment, frozen.

Much recent work by Cultural Geographers and others, however, has explored an alternative perspective, one in which landscape is understood as mobile, dynamic and performative. If you read the paragraph above and thought, no, sometimes when I'm walking/cycling/driving/canoeing/skateboarding, etc., the landscape *does* seem intense and alive and I'm fully involved with it, then you may already have an intuitive sense of what this perspective might mean. To go back to the introduction again, the task is to set aside any idea of landscape as simply the stage upon which a performance is enacted and think instead of landscape-as-performance. The ascent of the Shard documented in *Don't trip* certainly helps convey a sense of how landscape is articulated by, or *comes alive through*, moving, living bodies. The landscape is energized through, and in actuality made to exist by, the ascent itself. Rather than being the static backdrop to action, the landscape here is realized in and through action. Other examples develop the point. Think of *parkour*, the practice of 'urban free running', in which participants seek to jump across rooftops and use walls, ramps, stairwells and so on in ways more dynamic and expressive than that intended (see Saville, 2008). In doing so, they make the landscape anew; they *make another landscape*.

The 'sense' of landscape here is one of engagement, involvement and immersion – a landscape sensibility in which the whole body, and not just the eyes, participates. It is important to note that this understanding of landscape in terms of our mobile and sensory inhabitation of the world is *not* one that only applies to extreme or unusual situations and events. Rather, the argument is that such situations help bring to the fore the fact that humans – *all* human bodies, male or female, young or old, able-bodied or otherwise, of all ethnicities and creeds – *always* find themselves embedded within ongoing contexts of engagement with landscape and with other bodies. This insight – that the term landscape names the world in which we move and dwell – shapes many contemporary cultural geographies of landscape.

I've got this far without making what might seem to be the most obvious point about *Don't trip*. It works as an image and makes an impact precisely because the perspective is so vertiginous and alarming. You take a first, quick and casual look and then you look again. *How high up is he?* Another, closer look – and perhaps now you begin to feel a shiver of dread, literally feel it, in your hands (holding on tight), as you imagine yourself in the situation of the figure in the image. But then you can relax, I imagine, because in all likelihood you're looking and reading from somewhere fairly safe and secure – a library, a bedroom.

There is a name for this odd mixture of fear and fascination, this 'pleasurable dread' as it is sometimes called – it's called the *sublime*. If notions of sublimity, of vast, wild and potentially dangerous beauty, have traditionally been associated with 'natural' landscapes – with mountain ranges, polar ice caps, towering cliffs, raging seas – then today we can also clearly see a contemporary urban and technological sublime at work. This surfaces in our awe (or revulsion) before the New York or Dubai skylines, in the popularity of superhero movies and apocalyptic sci-fi scenarios and in the devastated urban landscapes of many war-based video console games. This, we can argue, is the tradition that *Don't trip* fits into and draws upon for its appeal. Just as the UKNEA image drew on pastoral landscape traditions which idealize certain kinds of rurality, so here the sublime provides the aesthetic framework – the context according to which we comprehend the landscape as striking, beautiful and so on. Here, we look downward into an abyssal, sublime depth, but also outwards, out and over a glittering urban surface to a far horizon. The human figure in the foreground supplies both drama and scale, but it's very much a picture of the city too, in all its awful nighttime glory.

But before we get too carried away, one final, crucial point must be made. The contemporary London skyline, with its Gherkin and Shard, may have some striking visual appeal, but this is also of course a landscape founded on, and expressive of, inequalities at many levels. With skyscrapers erected on the back of property speculation and ballooning land and rental values, it reflects the current dominance of finance and banking sectors in many western economies and societies, and the grotesque disparities in wealth and power that consequently arise. It is also undeniably a phallic landscape, symbolic of a certain conception of masculine might and potency. As Feminist Geographers have noted and discussed, the urban landscape, with its skyscrapers and statues, is often coded as a masculine space and is more widely reflective of longstanding patriarchal discourses, though which differences of power between genders and ages and social classes are maintained (Bondi, 1992; Rose, 1993). The ascent of the Shard in *Don't trip* is perhaps a commentary on the absurdly inflated sense of entitlement that the rich and wealthy in society possess today, as well as being an exercise in landscape subversion. Does it also – or should it also – ironically comment upon itself? An heroic 'first ascent' by a band of brothers, with a gung-ho motto: 'explore everything'?

SUMMARY

- Landscape is political not only because it can reinforce what is understood to be normal and acceptable but also because acts of 'landscaping', such as trespass, can contest these norms.
- Landscape is increasingly approached not only as something that is understood to be fixed and stable but also as mobile, dynamic and performative, formed in the relations of living bodies to their environments.
- The interpretation of urban exploration developed here illustrates how we can usefully roll together different understandings of landscape: as a way of dwelling and a way of seeing; as a powerful expression of inequalities and as a resistant practice.

Conclusion

My aim in this chapter has been twofold. First, to outline some of the major lines of inquiry being pursued today by cultural geographies of landscape and second, to illustrate and contextualize these by presenting them through two extended landscape examples. In concluding I also want to do two things.

Inevitably in a chapter of this length, much has to be left out and it is important to at least highlight some missing elements. I will focus here upon the role of time and memory in relation to landscape. I will then briefly return to some of the more conceptual issues flagged up in the introduction, regarding how Cultural Geographers understand and use the term landscape.

Through the UKNEA frontispiece and '*Don't trip*' I've explored the key issues which currently cluster around landscape. Issues such as how entrenched systems of identity and power are enacted and reproduced through landscape and how landscape, as an influential and widespread form of visualizing and picturing, acts to frame and direct our understanding of what is 'natural', what is 'beautiful' and what is 'normal'. A further issue that has been considered is how landscape may also be approached and understood in terms of dwelling and inhabitation – in terms of the moving, sensing and performing body.

Both of the landscape examples I have worked through are *contemporary*. Although I have sought to show how both draw significantly upon long established discourses, for instance visual and aesthetic traditions like the 'pastoral' and the 'sublime', I am still left with a slight worry – a feeling that I've underplayed here the extent to which many geographers emphasize the *historicity* and *temporality* of landscape. For an archaeologist or geologist, a landscape is physically composed of time – layers of time stacked successively on top of each other. Equally, geographers often stress the role of time and memory in the cultural composition of landscape. For example, landscapes of heritage and collective memory play an important role in the performance of identity, especially national identity, and these also require a critical reading alert to issues of power and authority (e.g. see Tolia-Kelly's (2011) work on Hadrian's Wall; to continue with the example of urban exploration, see Garrett (2011); see also Chapter 32). The temporality of landscape – the fact that it has been shaped and created, often over millennia, by the interacting of human and non-human forces – also needs to be critically stressed for a different reason. Because a landscape can often appear to be, or masquerade as, a slice of a fixed, pre-given 'nature', whose human histories are invisible, or at least not readily apparent. In other words, a landscape can be a forgetting, a sort of vanishing or absenting, and one task for Cultural Geographers is to critically remember, to conjure up the ghosts and traces of those who have seemingly vanished into the landscape (Wylie, 2009). Lastly, issues of time and memory are central to an understanding of landscape as dwelling-in-the-world. In my examples, I've stressed, as others have, the role of the moving, sensing body in performing landscape. However, thinking of landscape as dwelling – as a gradual and always evolving web of dwellings and pathways in which land and human life are very closely intertwined – must involve paying attention also to how, sometimes, landscapes and lifeworlds can grow and decay over long durations.

This takes me back to my introduction, where I noted that two major approaches in Cultural Geography understand landscape as either a way of seeing the world or as dwelling in the world. A few years ago it might have been common to see these approaches as opposed to each other, as indeed something of an either/or choice – a choice between, for instance, a critical visual analysis *of* landscape and an embodied engagement *with* landscape. To adopt my own phrase (Wylie, 2007), there are certainly creative tensions here. But perhaps these tensions have indeed proved to be creative and productive? Certainly at present in Cultural Geography, in the UK especially, there is a range of landscape writing which rolls together the key issues I have sought to associate with landscape in this chapter: visuality, power, identity, memory, embodiment and performance. These are the issues I would stress as central for anyone seeking better and deeper understanding of landscape.

DISCUSSION POINTS

1. What were your understandings of 'landscape' before you read this chapter? How have they featured, or not, within the discussion here?
2. What does it mean to understand 'landscape as a way of seeing'?
3. What does it mean to understand 'landscape as dwelling'?
4. How does power have a role in the making and perception of landscape?
5. Take a particular landscape and consider how it rolls together issues of visuality, power, identity, memory, embodiment and performance.

FURTHER READING

Cosgrove, D. (1985) Prospect, perspective and the evolution of the landscape idea. In: *Transactions of the Institute of British Geographers*, New Series 10(1), 45–62.

A classic essay on 'landscape as a way of seeing'.

DeLue, R. and Elkins, J. (eds.) (2008) *Landscape Theory*. Abingdon: Routledge.

A more advanced and conceptual but still accessible text on landscape. Comprises numerous short essays on landscape, plus a full transcription of a seminar discussion on landscape by a group of leading experts.

Ingold, T. (2000) The temporality of landscape. In: *The Perception of the Environment*. London: Routledge.

A classic essay on 'landscape as dwelling'.

Rose, G. (1993) Looking at landscape: the uneasy pleasures of power. In: *Feminism and Geography*. Cambridge: Polity Press, 86–112.

A classic essay/chapter on landscape and gender relations in geography and beyond.

Thompson, I. (2009) *Rethinking Landscape: A Critical Reader*. Abingdon: Routledge.

This is a compendium of numerous classic essays on landscape, with commentaries from the author/editor, Ian Thompson. A very valuable resource for more in depth landscape study.

Winchester, H., Kong, L. and Dunn, K. (2003) *Landscapes: Ways of Imagining the World*. Harlow: Pearson.

An introductory text on landscape, which is also a more general introduction to Cultural Geography, with a notable focus on Australasian and SE Asian contexts.

Wylie, J. (2007) *Landscape: Key Ideas in Geography*. Abingdon: Routledge.

My own textbook on landscape is a detailed study of how and why Cultural Geographers and others have defined landscape in different ways (i.e. as a way of seeing, as dwelling, etc.). It also gives numerous grounded examples of landscape research.

NOTE

1. The 80th anniversary of the Kinder trespass led to a number of commentaries and commemorations. See: www.kindertrespass.com for more information.

CHAPTER 19
MATERIAL GEOGRAPHIES

Philip Crang

Introduction: the stuff of geography

Cultural Geographers have long been interested in things, objects and materials – various kinds of 'stuff' in other words. In this chapter we will see how and why they have studied, among other things, buildings, fences, trees, iPods, knitting, toys and Italian motor scooters. All of these things are treated as forms of 'material culture', a designation that emphasizes their role in cultural processes of making distinctive ways of life and giving meaning to the world, its places and people.

As it developed in the first half of the twentieth century, Cultural Geography was devoted to work on material culture, documenting and explaining the spatial distributions of all manner of things from house types to foods, domesticated plants and agricultural tools (for example, I spent much of my undergraduate studies reading up on plough and digging-stick distributions in Latin America). This interest never really went away but, with the emergence of a so-called 'New Cultural Geography' in the 1980s and 1990s, it was downplayed. There was a concern that focusing on such 'stuff' meant neglecting more important questions of **cultural politics**. An interest in material culture was seen to 'fetishize' things, making them the centre of our attention rather than the social, political and economic relations that brought them into being.

However, in the last decade there have been moves to 're-materialize' Cultural Geography (Jackson, 2000), shaped by various bodies of theoretical and substantive work. First, the emergence of a multi-disciplinary field of 'material culture studies' has countered criticisms of object fetishism, arguing that: a) rather than dismissing things as trivial ephemera we should recognize how their triviality and taken-for-grantedness makes them crucial **ideological** mechanisms; and b) rather than material forms simply being reflections of underlying social relations they in fact help to make our social worlds (Miller, 2010). These insights have been pursued in studies ranging across both 'cultures of consumption' – how we use things made by others – and 'cultures of making' – how we craft things by working with objects and materials. Second, there has been a wider interest in developing 'more-than-human-geographies'. In Chapters 10 and 11 respectively, Sarah Whatmore and Russell Hitchings trace this out by considering new approaches to 'nature', plants and animals. That concern for other non-human forms of life has been paralleled by a resurgent interest in non-living stuff too with, for example,

'**Actor-Network Theory**' (ANT) arguing that objects, technologies and materials are fundamental to how we relate to each other and the world (Latour, 2005). Third, this 'relational materialism' has been joined by philosophical ruminations on materiality and the material forces that flux across human and non-human bodies, what Anderson and Wylie call a broadly 'affective materialism' (Anderson and Wylie, 2009; Bennett, 2010). Less abstractly, the anthropologist Tim Ingold has influenced a range of work that focuses on embodied responses to the lively world of materials, in practices as diverse as walking, carpentry and being exposed to the weather (Ingold, 2011; see especially Part I).

This chapter responds to Cultural Geography's changing interests in material culture by setting out four approaches to material geographies. First, we look at scholarship in Cultural Geography that treats material culture as an *inscription* or indicator of cultures and their spatial extents and influences. Our empirical focus will be pioneer colonization in North America. Second, we look at work that sees material culture as involved in processes of social and cultural *reproduction*, thereby making the cultural politics of material forms much more evident. Here our principal substantive focus will be on one elite residential neighbourhood of the city of Vancouver in British Columbia, Canada: Shaughnessy Heights. Third, we turn to research that examines material culture as part of everyday *practices*, illustrated through the examples of iPods, toys and skateboarding. Here the focus is on what people do with things and on the powers that those things have to shape our everyday worlds. Finally, we look at material culture as a realm of cross-cultural *traffic*, highlighting work that has 'followed' things as they move around the world through the example of the 'Italian' motor scooter.

The stuff of cultural inscription

Most summers, myself, my partner and our two children squeeze ourselves into our VW campervan ('Rosie') and head through the Channel Tunnel from England for a short summer holiday on 'the continent'. We don't ever get very far. It's an old van and it doesn't go very fast. But as we potter along through France, Belgium or the Netherlands, it's become a family tradition that I inflict a bit of Cultural Geography on everyone. I point out how things look different. The road signs, the houses, the fields – they are not the same as those in our home area of southern England. Noticing and explaining such things, I tell them, is a long tradition within Cultural Geography.

Figure 19.1 Rosie, the campervan. Credit: Philip Crang

This is particularly true in the USA. Developing Carl Sauer's (1925) interest in the **cultural landscape** – the ways in which people shape natural landscapes into culturally distinctive forms – a fascinating body of work emerged that examined the regionalization of various material culture forms. A favourite of mine is Fred B. Kniffen's writings which focus on American vernacular or folk housing, as well as bridges, barns and fences (Kniffen, 1965, 1990). In this kind of work, emphasis is placed on observational and archival fieldwork that together can document and explain the morphology of visible material culture forms. With a particular focus on 'ordinary landscapes' (Meinig, 1979), the broad argument is that the 'built' components of those landscapes can be taken as 'un-witting autobiography' (Lewis, 1979: 12), an imprint of the cultural influences apparent in any place.

An example may help elucidate this approach. Terry Jordan and Matti Kaups were interested in the historical geography of European settlement and 'pioneering' in North America (Jordan and Kaups, 1989). Specifically, their focus was the 'backwoods' pioneers of the eighteenth century, 'a group of highly successful forest colonizers who formed the vanguard of European settlement in eastern North America' (Jordan, 1994: 215), moving European settlement westwards from the Delaware Valley far faster than the 'Yankees' in the North or the Planters in the South. Jordan and Kaups view the success of these colonizers as stemming from a congruence between their ways of life (especially their ways of relating to land and nature) and the requirements of the pioneer environments they lived in:

> . . . living in isolated log cabins; throwing up zigzag fences around their small ax- and fire-cleared fields of corn; skilfully hunting deer and bears in the woods; gathering berries, herbs, wild fruits, and honey; and herding small semiferal droves of pigs and cattle, these pioneers were ideally suited to the pioneer life.
>
> (Jordan, 1994: 215)

More specifically, they identify one short-lived (1638–55) seventeenth-century colony – New Sweden in the lower Delaware Valley, and its population of ethnic Finns from the Savo-Karelian districts of Finland – as having had a vital role to play in developing this successful pioneer culture. The evidence for this surprising influence is detected in American frontier material culture itself. Styles of log carpentry, house roof construction, house plans and fence types are all seen to bear the imprint of Savo-Karelian culture from northern Europe. The Karelian weighted board roof, the open-passage dog-trot house plan, the zig-zag or 'worm' fence; these specific material forms are witness, say Jordan and Kaups, to the influence of New Sweden, being found only in northern Europe and the American backwoods frontier (see Figure 19.2).

Their account is, of course, more subtle and complex than this brief summary allows. For example, they also attend to the cross-cultural contact between European pioneers and indigenous Amerindians. But, for our purposes, the two central tenets of this approach to material culture and its geographies are established. First, material forms can be seen as the unwitting imprint or inscription of distinctive cultures and peoples. Second, by tracing out the spatial distribution of material forms we can therefore map out the geographies of cultures, in terms of their origins, their patterns of diffusion and their regionalization.

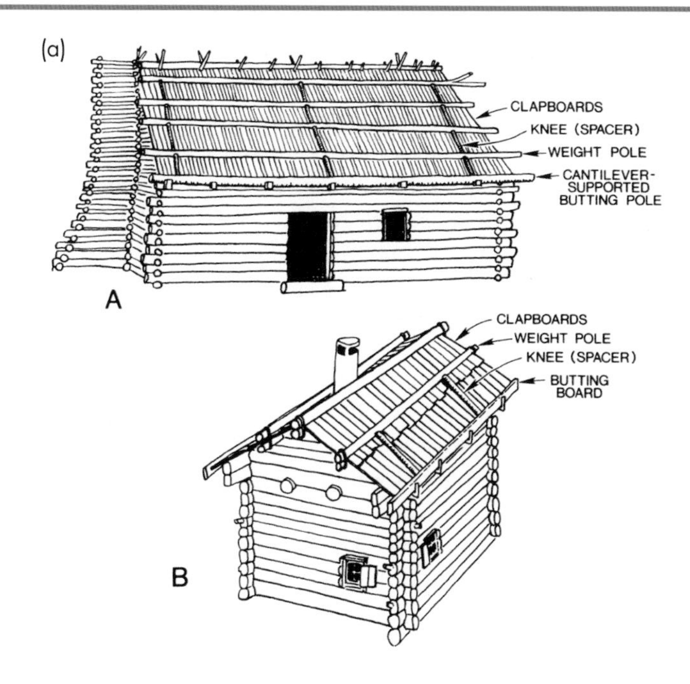

Figure 19.2 Imprints of Savo-Karelian culture on the North American frontier. a) Drawings show similarities of backwoods pioneer and Karelian house roofs. Key: A = America, B = Karelian Isthmus. (b) An American open-passage dog-trot cabin in Pennsylvania, showing also a Karelian board roof with weight pole and, in the foreground, a Karelian-style zig-zag rail fence with X-shaped supports. Credit: (a) *Re-reading Cultural Geography* (Foote et al., 1994); (b) *Old Redstone, or historical sketches of Western Presbyterianism* by J. Smith (1854)

SUMMARY

- A long-established approach to material culture in Cultural Geography is to see things as the inscriptions of a culture on the landscape.
- Reading these material inscriptions allows the Cultural Geographer to map out the spatial origins, diffusions and extents of different cultures.

The stuff of cultural reproduction

However, there are problems with seeing material geographies in terms of an unwitting imprint left by a culture on the landscape. For a start, there is a danger of assuming 'cultures' as entities that act in and on the world. This pays too little attention to the invention, construction and contestation of cultures in the modern world. Moreover, it is limiting to frame material forms as just visible evidence of the cultures that make them. This ignores how material culture is an active part of cultural life or, to put that another way, how cultures don't only make things, but things make cultures. Such concerns are responded to in work that centres the role of material culture in forming and contesting cultural identities in the modern world. Let me elaborate, first by referring back to me and my nuclear family in our VW campervan, and then through a more substantial case study, concerned with **class** and **ethnicity** in the city of Vancouver, British Columbia.

My previous mention of our family trips to continental Europe in our campervan follows an embarrassingly common pattern – one in which, by hook or by crook, I will somehow manage to mention to people that I drive a 1970s VW Type 2 Westphalia campervan (the details can be difficult to force in but they matter, so I make sure they get said somehow; just as I have now!). I do this for reasons I understand only too well: this campervan is an object that I am very pleased to own (actually, my partner owns it, but I ignore that detail); I like to be associated with it because I think it represents meanings that I value and says something about me that I like; and, in outline, these are something about discernment (it's not just the usual tin box family car), alternativeness (there's the aura of surfer culture about it; though I can't actually swim a stroke let alone surf) and youthfulness (when I know I am resoundingly middle-aged, a husband and father in the archetypal nuclear family unit). Further details of my tastes and self-regard need not detain us here, but the broader argument is worth an explicit re-statement: 'my' VW campervan is a fragment of material culture that plays a role in the fashioning of my cultural identity.

In his chapter on 'identities' (Chapter 42), Peter Jackson says more about identity and its relations to material goods. His focus is on food, but here I want to follow on from the last section and its preoccupation with material landscape forms such as houses. The case study we will be considering is the residential neighbourhood of Shaughnessy Heights in Vancouver (see Figure 19.3). In David Ley's telling (1995), in 1990 local resident Harry Liang (a pseudonym) decided to fell two large sequoia trees on the front lawn of his newly built family home. His Anglo-Canadian neighbours protested. Yellow ribbons were tied and people joined arms around the trees. The dispute became headline news in Vancouver and well beyond (for example, gaining coverage in the *South China Morning Post*, the main English-language newspaper in Hong Kong). To understand why these trees and this neighbourhood squabble generated such interest we have to reflect on the history of those sequoias and the role they played in broader cultural politics of identity.

Bordering Vancouver's central business district, Shaughnessy Heights was developed as an elite neighbourhood by the Canadian Pacific Railroad from 1907 onwards. Landscaped by Frederick Todd, its layout of rolling hills, curved roads and large plots materialized particular Anglophile aesthetics, reproducing eighteenth-century English elite landscape tastes for the picturesque countryside (Duncan, 1999). Neo-Tudor and Victorian architecture

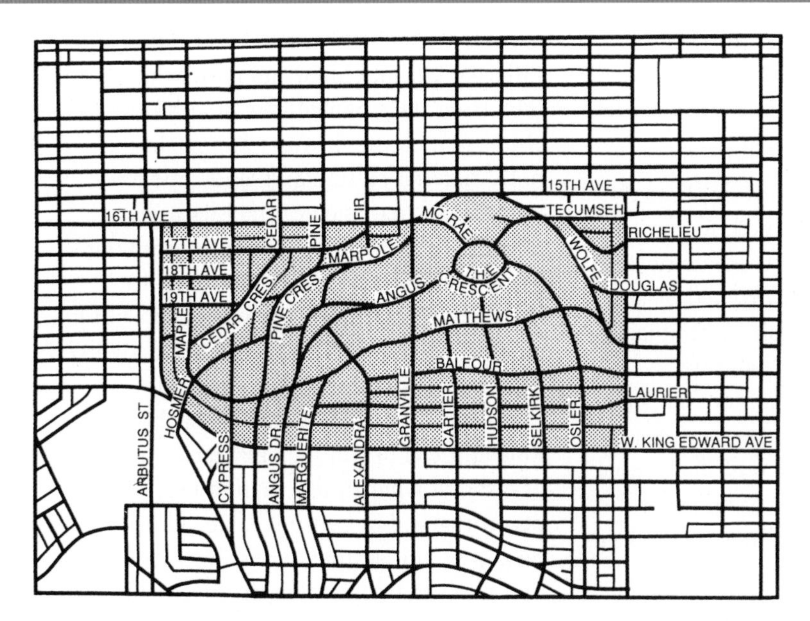

Figure 19.3 Map of Shaughnessy Heights, Vancouver (the shaded area represents Shaughnessy Heights). Source: Anderson and Gale, 1999

Figure 19.4 A Shaughnessy mansion. Credit: Joseph Lin

dominated the housing stock (see Figure 19.4). This material culture of houses, trees and streetscapes expressed the cultural tastes of the elite social class the development was being marketed to. But, as Jim Duncan argues, over time it also came to be implicated in the cultural reproduction of that class fraction. It was not just made for them, it came to make them. In consequence, defending the material landscape became a form of social reproduction. Residents formed the Shaughnessy Heights Property Owners Association (SHPOA) to lobby for planning regulations that would maintain the material fabric of the neighbourhood, in particular against encroachment by non-elite social groups (e.g. through property or plot sub-division). By the 1970s the SHPOA was becoming particularly skilled at making this case in ways that did not reveal naked self-interest – for example, through appeals to ideas of heritage or conservation.

More recently, the class relations implicated in this elite landscape were joined by a more explicit defence of British ethnicity. Shaughnessy Heights was constructed by and for a British **diaspora** in British Columbia. Harry Liang, by contrast, was an ethnic Chinese-Canadian, reflecting a more recent elite migrant population into this west-coast city. Liang's newly built home was an example of what became known locally as 'monster houses', characterized by prominent symbols of wealth such as Greek columns, spiral staircases and tall cathedral entry halls, all put on public display by the clearing of vegetation from plots (hence the sequoia felling). This elite group formed a rival resident organization, the South Shaughnessy Property Owners Rights Committee (SSPORC), which as the name suggests appealed to ideologies of Canadian democracy, free markets and property rights. The SHPOA framed the residential materiality of Shaughnessy Heights as about a (British-Canadian) heritage to be conserved, related to the defence of the social position of elite, anglophile Canadians in Vancouver. The SSPORC framed the issue as about their freedom as elite Chinese-Canadians to materialize their identities in Vancouver through property development.

If we compare Ley's and Duncan's interpretation of Shaughnessy Heights to the study of pioneer material culture by Jordan and Kaups discussed earlier, clear differences in approach are apparent. First, Ley and Duncan see material culture as the active expression of differing cultural identities rather than the unwitting trace of a culture or cultures. Second, chiming with much wider approaches to both material culture and landscape, they understand material geographies to be meaningful. The 'stuff' that makes up our worlds can be treated as signs, open to interpretation not only by researchers but by all of us within the conduct of our daily lives. Material geographies are implicated, for example, in aesthetic judgements about the look and feel of things. They raise questions of taste, of what things we find beautiful, becoming and of cultural value. Third, the meanings and aesthetics of things are caught up in cultural politics, in disagreements and struggles. The same thing may not be read or appreciated in the same way by different people. Our judgements of things are not pure but implicated in processes of social distinction (Bourdieu, 1984). Thus Mr Liang's sequoia trees are seen as *both* a symbol of heritage and a valued urban nature, *and* as symbolic of a restriction on private freedom and a block to a clear view of and from his residential presence in Vancouver. Finally, these struggles over the meanings and physical existence of things are central to the production and reproduction of distinctive cultural groups. Elite Anglo-Canadians use large neo-Tudor houses with tree-lined streets and lawns to materialize their cultural values and to maintain a privileged social position. Elite Chinese-Canadians establish their presence in the social landscape of British Columbia through large new houses and clear plots. Material culture doesn't just reflect these groups; it is central to their ongoing formations.

SUMMARY

- Cultural groups do not just exist, they have to be produced and reproduced. Material culture is a means through which this (re)production can take place.
- The (re)production of cultural groups is implicated in various forms of cultural politics. Thus, so too is material culture.
- The things that make up material culture play a role in cultural (re)production in part through their meanings and aesthetics.
- In consequence, when Cultural Geographers 'read' or interpret pieces of material culture they need to do more than culturally identify and map them – they need to tease out their multiple and contested meanings and evaluations.

The stuff of practice

Did I mention that I drive a 1970s VW campervan? . . . Actually, I do have a good reason to talk about it again. In the preceding section, I suggested that my campervan was in part an object I relate to through its symbolic character and the meanings it can evoke for both myself and others. However, it's not only that. My family and I go away in it. We sleep in it. We park up, use the cooker to make a cup of tea, and have picnics in it. I enjoy how its lack of speed makes me drive in more relaxed ways. Driving it around usually brings a smile to my face. To cut a long story short, this campervan is not just a symbolic object, it is put to practical use. It does things, we do things with it, and it does things to us. This banal observation points to a rather different approach to material geographies and material culture, one focused on our relations to stuff in practice. Cultural Geographers are especially interested in how such practical relations are involved in the production and consumption of space.

This practical orientation emerges in part from research on the consumption of material goods (Miller, 1987, 1995a). Geographers and others have argued that to consume is to do far more than simply purchase things; it is to use them (see Chapter 26 for a fuller discussion). Consumption thus becomes a process for re-appropriating objects and spaces produced by distant bodies and forces. Work on the home, for example, has looked at how practices of decoration and improvement (or 'DIY') allow people to fashion domestic environments that may be less alien to them, to make space 'homely' (Miller, 1988, 2001). Increasingly such work has gone beyond a pre-occupation with how commercial goods are 'domesticated' through their consumption, opening up a wider range of questions about how we 'live with things'. For example, Nicky Gregson examines how domestic spaces are made through the provisioning, the display, the storage, the maintenance and cleaning, and the disposal or 'ridding' of all kinds of stuff (Gregson, 2007). Elizabeth Shove and her co-authors consider how domestic consumption involves not only having but 'doing', looking at how new products and materials have co-evolved with changing forms of domestic life (Shove *et al.*, 2007).

Similar lines of analysis can be developed in public space too, focused on how people inhabit and fashion their environments through

Figure 19.5 The MP3 player is a technology used to affect everyday experiences of urban space. Credit: Startraks Photo/Rex Features

particular practices and using particular technologies and materials. Think for example of the personal stereo, developed as the portable cassette player or Walkman (du Gay *et al.*, 1997) and now predominantly the iPod or another MP3 player. How do people use personal stereos and with what effects on urban space? Those are the questions that Michael Bull pursues in his account of 'iPod culture and urban experience' (Bull, 2007). He constructs a typology of how personal stereos are used (see Case Study box) (Bull, 2000). In essence, they enable the management of interpersonal interactions with other people; the sonic mediation of relations to the urban environment; and the sensory activation of emotional states within the user. They are technologies for remaking our surroundings and our embodied perceptions of them.

The architectural theorist Iain Borden (2001) makes somewhat similar arguments about the skateboard and its use in urban spaces. He argues that skateboarding recomposes the city: functional or monumental spaces devoted to efficiency or symbolic power are transformed through improvisational practices that make urban surfaces the stage for skilful performance and bodily excitement. The Cultural Geographer Laura Price (2013) similarly reviews how wool and practices of knitting can intervene in the materialities of the city, most obviously in acts of 'yarn bombing' where knitted 'graffiti' is deployed to re-craft the city, to soften and domesticate it, at the same time as stitching a stereotypically feminine practice into the urban fabric.

In such accounts, things are understood as more than the passive recipients of cultural values, meanings and projects; they have their own capacities. Thus technological changes in material culture can also involve changes in the very nature of our geographies. In attempting to trace current trends in this regard, Nigel Thrift (2003) looks to the world of toys, and in particular to the development of interactive, performative toys such as the pioneering Furby. He argues that these toys – with their abilities to sense, to communicate, to learn – presage a much wider shift in the material culture of the wealthiest societies, associated with emergent computational and artificial intelligence technologies. This shift is towards 'intelligent environments', spaces that don't just receive our projections of meaning or allow us to act on/in them but that perform and have consciousness. Thrift's prime concern is the potential for these new kinds of environments and things either to offer new possibilities for expression and empowerment to their users, or to further

CASE STUDY

A summary of Michael Bull's typology of common usages of the personal stereo in urban space

1. To block out unwelcome noises and to reimpose control over auditory environments.
2. To aid the management of interactions with strangers in the city, allowing users to signal a lack of interest in interaction and to reimpose their own senses of personal space in public.
3. To 'personalize' the user's experience of the city, replacing the city's noise with the user's own selection of music.
4. To create a pleasurable experience of the city, in particular by matching successfully music to visual scenes, creating an aesthetically convincing, cinematic experience of the city.
5. To provide some stimulation in those urban places that are dull and boring, such as crowded commuter trains, and to reclaim these repetitive, dispiriting times and spaces.
6. To avoid feeling alone in the city, by being accompanied by familiar sounds and a companionable technology.
7. To help concentration and reflection, allowing the user to organize their thoughts and adjust their moods.
8. To gain a sense of rhythm, energy and purpose from the music being listened to.
9. To listen with others via a shared player and headphone, thereby constituting an intimate and exclusive group in public space.

(See Bull, 2000; especially pages 186–90.)

Figure 19.6 Yarn bombing. Credit: Shrewdcat (Wikimedia Creative Commons)

control them through an ever tighter designing of space and behaviour within it. He emphasizes that the balance between these two outcomes is not determined by the technology itself, but will depend on how we develop and engage with it. In this regard, this case is illustrative of wider concerns in technology studies with both the material shaping of society and geography and the social and geographical shaping of material forms.

SUMMARY

- We relate to the objects of material culture not only through appreciating and contesting their symbolic meanings but also through putting them to use.
- To understand the geographies of material culture we therefore need to study how objects are used, and in particular how they are used to 'inhabit' or re-fashion places and spaces. Research on domestic space and urban public space has been pioneering in this regard.
- The stuff of material culture is diverse and characterized by differing material and technological qualities. These qualities are important in shaping the geographical uses to which objects are put and the kinds of geographies their use produces.

The stuff of cultural traffic

In a seminal collection, Arjun Appadurai called for the study of 'the social life of things' (1986). Appadurai's particular concern was with following 'things in motion', a methodology designed to cast new light on the various social contexts through which they moved. His argument has been widely influential, helping to generate studies that trace out the biographies and geographies of things and explore the range of places and people that these things have existed in relation to. Notable examples include:

- research that links together the production, marketing and consumption of particular products (e.g. du Gay *et al.* (1997) on the Sony Walkman; Mansfield, 2003, on Euro-American 'imitation crab' or 'krab')
- related work seeking to 'follow things' and reconnect the producers and consumers of material **commodities** (Cook, 2004; Hughes and Reimer, 2004; Redclift, 2004)
- research on collections of objects, including in art galleries, private collections and museums (Geoghegan, 2010; Hill, 2006)
- research on the making, selling and consuming of things marked as coming from somewhere culturally or geographically 'different' or 'foreign' (Cook and Crang, 1996; Cook and Harrison, 2003; Dwyer and Jackson, 2003; Jackson, 1999)
- accounts of the complicated lives of things that become 'second-hand' (Gregson and Crewe, 2003)
- research on how objects end their lives, become 'rubbish', are disassembled and broken up, and with what geographical consequences (Gregson *et al.*, 2010).

An example may help to navigate this extensive list and draw out the significance of such work for our discussions of the cultural

Figure 19.7 The 'Italian' motor scooter was an important object to young Britons interested in design and the 'Continent' in the late 1950s and early 1960s. Credit: Popperfoto/Getty Images

geographies of material culture: Dick Hebdige's account of the Italian motor scooter (1988). Hebdige's interest is in the 'multiple values and meanings which accumulate around a single object over time' associated with 'different groups of users separated by geographical, temporal and cultural location' (1988: 80). His particular foci are Vespa and Lambretta motor scooters. Launched in 1946 and 1947 respectively, by the Piaggio and Innocenti engineering companies, these scooters began life intended as cheap post-war urban transport and a way of utilizing spare post-conflict production capacity. However, the scooter soon became fashioned around the ideal consumer of a 'new', youthful, emancipated urban Italian woman. Both models coalesced around a design aesthetic of 'clothed' machinery (the engine and most other mechanics being hidden from view under cover panels). Motor transport was removed from the masculine realm of mechanical competence and brute force and translated into feminine worlds of style, fashion and domestic utensils. In turn, those feminine worlds were given new geographies, the confinement of home and the passivity of being on display transformed into ideologies of active mobility and freedom. Advertising emphasized scooter-based tourism and travel, as well as the international popularity of these Italian products.

That international popularity introduced further elements to the scooter story, though. As a result, in Britain, scooters became implicated in particularly British cultural debates. At one level this was an argument between engineering and design expertise; the former (centred around the British motorcycle industry) deploring these gimmicky, unsafe and unmanly machines; the latter celebrating their aesthetics and their origins in an Italy that, for young trend-setting designers, represented 'all that was chic and modern' (Hebdige, 1988: 106). At a wider level, this polarization of reactions filtered down into the world of youth sub-cultures and wider popular tastes. On the one hand, there were the British motorbikes, American stylizations and macho masculinity of the 'Rockers', for whom scooters were nothing more than 'hair-dryers'. On the other hand, by the 1960s Italian scooters had become synonymous with the 'Mods' (short for modernist), a working-class, male-led, youth sub-culture that emerged in London in the 1950s, devoted to fashion consciousness and Continental (especially Italian) style. For the Mods, Vespas worked alongside Italian suits and coffee bars to signal an identification with foreign sophistication, discerning consumerism, youthful modernity and a different sort of masculinity. By the 1960s this Mod(ernist) culture had mutated into the stylized scooter-boy uniform of parka anorak, jeans and Hush Puppy shoes, maybe with a beret for added continental symbolism. Scooters themselves were increasingly customized, with added chrome, mirrors, flags and even unsheathed mechanics. Fights between Mods and Rockers at British seaside resorts on bank holiday weekends became huge media stories and have lived on through fictional renditions such as the film *Quadrophenia*.

The scooter had therefore shifted from being cheap transport to being an accessory of the 'new' Italian woman, to being an icon of Italian style in the UK, to being an object that was materially transformed (or customized) as part of a particularly British, working-class, male, youthful cultural opposition between Mods and Rockers. Today, rather like 'my' VW campervan, the Italian motor scooter has taken on 'retro' qualities, whether in the form of surviving original machines which are increasing in value or in new models like the Vespa 946 that are styled to reference the original prototype Vespa designs.

The broader insights of this story can now be elaborated. Things move around and inhabit multiple cultural contexts during their lives. Cultural Geographers are especially interested in the changes that happen to a thing in this process: both material changes and 'translations' in the thing's meanings. They are also interested in the knowledges that move with things, especially about their earlier life. How much do people encountering a thing in one context know about its life in other contexts? Who mediates this knowledge? What role do **imaginative geographies** of where a thing comes from (for example, of 'Italianicity' in the scooter's case) play in our encounters with objects, and, in reverse, what role does material culture play in wider imaginative geographies?

Cultural Geographers are also concerned with the forms of power associated with the movement of things. Whose senses of what an object means become dominant and most powerful? What are the consequences when a thing travels cross-culturally? To what extent does one set of people end up appropriating and taking over the (material) culture of another? Thus hooks (1992) argues that 'white people' have appropriated black American culture, with a range of negative consequences for black people, and Spooner (1986) narrates the dominance of Euro-American traders and consumers in defining what counts as an authentic Oriental rug, again posing all sorts of dilemmas for those people actually making them. Or, to look at the situation rather differently, Cultural Geographers also ask what possibilities are there for local populations to appropriate the globally distributed material culture of powerful, transnational corporations for their own ends? Thus, a range of studies has emerged that have looked at the 'indigenization' (the making local) of global products, from McDonald's (Watson, 1997) to Coca-Cola (Miller, 1998a).

SUMMARY

- Things travel, moving through space and time. In these travels, the meanings and material nature of things can change. Things themselves therefore have complex cultural geographies.
- Cultural Geographers have considered the understandings and knowledges that people have of where things come from as forms of 'imaginative geographies'.
- Cultural Geographers have also examined the power relations of cross-cultural movements of things. In this regard, attention has been paid to the problems of cultural 'appropriation' by dominant groups. However, on the other hand, consideration has also been given to the 'indigenization' and use of global things in apparently subordinate local cultures.

Conclusion

There is a lot of 'stuff' in Cultural Geography and a lot of cultural geographies in any bit of 'stuff'. This chapter has introduced four different, but not necessarily mutually exclusive, approaches to these geographies of material culture. First, we have seen how things can be viewed as inscriptions of cultural presence, influence and distribution. Second, we have seen how material culture is implicated in processes of cultural reproduction and in their associated cultural politics. Third, we looked at how people relate to things and materials in practice, and how those practical relations can shape places and spatial experience. And, fourth, we thought about the cultural geographies involved in the traffic of things across conventionally defined cultural–geographic borders. The general flow of argument has been away from viewing the geography of things as their distribution across the world and towards geographical analyses that get inside things, their meanings and uses, and tease out their role in the production of the places and spaces we inhabit.

DISCUSSION POINTS

1. Construct an outline history of Cultural Geography's interest in material culture.
2. In what ways is material culture bound up with questions of cultural politics?
3. What role does the material culture of landscape play in processes of cultural reproduction?
4. Taking the example of either urban public space or domestic space, discuss the practices, materials and objects through which we 're-make' these spaces.
5. Do you think people should consume and use the material culture of other societies? Outline the potential pros and cons of such cross-cultural exchanges.
6. What might we learn about our cultural geographies by 'following things'?

FURTHER READING

Borden, I. (2001) *Skateboarding, Space and the City: Architecture and the Body*. Oxford: Berg.

Written by an architectural theorist, this book is a celebration of 'street skating'. Borden argues that cities are largely produced to serve the interests of economic efficiency, but that skaters reclaim the city for their own purposes, using their boards as tools to re-make it as a stage for their own 'compositions' and 'performances'. You may find the academicism of some of this grating, but it is still a great book, offering a particularly powerful example of a performative account of material culture and space. Focusing on Chapters 1, 7 and 8 is a good start.

Duncan, J.S. and Duncan, N.G. (2004) *Landscapes of Privilege: The Politics of the Aesthetic in an American Suburb*. New York: Routledge.

A study of the town of Bedford in Westchester County, New York State, 'an archetypal upper class American suburb' according to the book's back-cover blurb. The focus is on the physical presentation of place and thus the material culture of landscape. The argument is that while looking natural, landscapes convey cultural codes and are implicated in the reproduction of social classes and hierarchies.

Gregson, N. (2007) *Living with Things: Ridding, Accommodation, Dwelling*. Wantage: Sean Kingston Publishing.

An analysis of how we live with things in our domestic lives, based on detailed ethnographic research in the UK.

Jackson, P. (1999) Commodity cultures: the traffic in things. In: *Transactions of the Institute of British Geographers*, NS24, 95–108.

A clear and thought-provoking review of what the notion of things having 'social lives' means for Cultural Geography. Sets out the key questions involved in so-called 'cross-cultural' material cultures and summarizes some key studies in the field concisely.

Smith, J.M. and Foote, K.E. (eds.) (1994) How the world looks. In: K.E. Foote, P.J. Hugill, K. Mathewson and J.M. Smith (eds.) *Re-reading Cultural Geography*. Austin: University of Texas Press, 27–163.

I have a special affection for this reader of classic papers in Cultural Geography, produced in the USA from a very American perspective. This section of the reader focuses on work that looks to 'read' material landscape forms and to understand the processes behind their production.

WEBSITE

www.followthethings.com

Set up by the Cultural Geographer Ian Cook, this website pools studies and views on what it means to 'follow things' so that one gains a better understanding of who makes the things that we buy.

SECTION FOUR

DEVELOPMENT GEOGRAPHIES

Introduction

However we measure it, or choose to describe it, it is clear that some spectacular inequalities exist in terms of development across the globe. A recent research report for Christian Aid (2012: 5) contained the following data:

- The richest 1 per cent today controls 40 per cent of the world's wealth, while the poorest 50 per cent own just 1 per cent.
- The poorest 20 per cent of the world population account for only 1.3 per cent of total consumption.
- 1.4 billion people live in absolute poverty, without enough to live a decent life.
- Almost 1 billion people have no access to fresh water.
- 925 million people do not have enough to eat.
- 1.4 billion people have no access to electricity.

Issues of 'development' are an inescapable part of our everyday lives. The ideas we hold of global development are regularly fuelled by the graphic images of 24/7 news channels and documentary programmes reporting on the latest famine, warfare, impoverishment or refugee populations in some far off 'third world' country. Mega charity spectacles such as Children in Need and Comic Relief annually pull at our heartstrings, and purse strings, in response to the victims of 'underdevelopment'. Our environmental concerns, for example, over the chopping down of rainforests or the erosion of biodiversity, bring us into direct contact with issues of how the commercial conduct of economic development (in this case of agricultural production) is often at odds with global ecological objectives. The more discerning of us may even allow the components of our everyday diet to remind us of the global geographies of food and of the likelihood that what we are eating is directly connected with unfair trade and production conditions that benefit big international firms but impoverish those whose labour has been directly involved in food production. Similarly, we are now more aware than ever that the clothes we wear, phones we text and talk on and computers we work on, may have been produced elsewhere in the world in conditions of extreme poverty and exploitation.

Alongside opportunities for charitable giving, popular campaigns have now emerged to enable us to 'do something about' the conditions in 'third world' or 'underdeveloped' countries'. For example, from small beginnings the fair trade movement has made significant strides. Since the 1970s, alternative trading organizations such as Oxfam and Traidcraft have established networks of fairly traded products, building positive trading relationships with groups of third world producers and their communities and selling these trademarked 'fair trade' goods in countries such as the UK. Originally selling only from catalogues, charity shops and in churches and other such organizations, the fair trade movement has now moved into the mainstream of selling in supermarkets. In 1998 the UK sales of Fairtrade-mark products in supermarkets and other large stores were worth £16.7 million. In 2004, sales topped £100 million, and in 2010 they topped £1 billion. In 2012 it is estimated that they will top £1,500 million. Sales of Fairtrade chocolate alone are now worth almost £500 million in the UK. Other campaigns have also made their mark. For example, the Jubilee 2000 campaign for the cancellation of international debt mobilized millions of people worldwide to call on governments and international institutions to wipe clean the historic slate of unpayable debts owed by the poorest countries of the world. More recently, a phalanx of campaigning organizations have come together for the Trade Justice Campaign, calling for fundamental changes to the rules that govern international trade, so that they work in the interests of the world's poorest people.

Given the energy of these campaigns and the consistent media spotlight on the plight of the world's poorest and most vulnerable people, it is unsurprising that geographies of development are a vital part of bringing the global into the local – and of helping us to understand that our wealth is directly related to others' poverty. With that in mind, some of the early writings on development geography tended to be overly tied to detached indicators of the state of development in different countries. Thus, details of production, consumption, investment, demographic characteristics, health, education, income and so on, came to dominate the agenda. However, geographers have gradually spread the news that development is diverse, complex and often contradictory, and the real-life experiences of people in the countries and regions concerned are made trivial by being reduced to a series of impersonal indicators.

Nevertheless, as Katie Willis shows us in Chapter 20, there is still more work to be done by geographers interested in development in order to unpack the in-built assumptions on which our studies are often founded. Indeed, in many ways the origins of development theory and practice – both as governmental *and academic* exercises – are inseparable from the historical process by which a *colonial* world of inter-war and post-war years has been reconfigured and transformed into a *developing* world. The unpacking of development involves both an understanding of the material realities of the development process itself and a grasp of how development works as a social imaginary; in other words, how discourses of development are produced and circulated.

Chapter 20, then, presents a discussion of some of the most popular ideas about development and reflects rather different understandings about the unequal nature of power relations in the world. It traces a line from development as modernization, reflecting how the 'North' has provided technical assistance to overcome the perceived obstacles to development in the 'South', to more recent neo-liberal ideas of development, which are anchored in market mechanisms and trade liberalization as the most efficient ways of achieving development.

These discourses of development need to be subjected to thorough critical scrutiny, not least because they are integrally associated with the continuing disparities and uneven geographies highlighted at the start of this introduction. Sarah Radcliffe lays out some of the foundations for such a critique in Chapter 21. She demonstrates how European colonialism and neo-colonialism have shaped the experience of development, pervading contemporary development practices and studies of those practices. Post-colonial theory, she argues, will therefore help to put nuanced geographies and histories back into essentialized binaries such as first world/third world, North/South, developed/underdeveloped and modern/traditional ways of life. Equally, feminist theory can shed new light on development practices. Although it is important to recognize multiple femininities within development, geographies of development need to address the gendered power relations that result in women working longer hours for less money and with less security than men in developing contexts. This is especially important given the increasing recognition of the role of women as active agents in development processes. Radcliffe's third strand of rethinking development recognizes that under the auspices of neo-liberalism, the role of the state in development has been reconfigured. States remain important, but NGOs and hybrid institutions are becoming more important, and so maps of development should not simply be confined to nation-states.

The unpacking of development will also need to involve listening to 'other voices' in developing countries and Paul Routledge takes us in this direction in Chapter 22. He shows how many development practices have resulted in the '**pauperization**' and marginalization of indigenous people, with peasants, tribal people, women and children usually being viewed as impediments to progress and thereby being excluded from participation in the development process. It is important, therefore, to hear the voices of *these* people, who will sometimes organize themselves into place-specific social movements so as to resist threats to their economic and social survival. Routledge shows us how social movements articulate resistance within society across a range of different realms, especially the economic, political, cultural and ecological. These are not simply small localized outbursts of resistance. Many movements act across multiple scales and are increasingly involved in globalizing networks of resistance.

In Chapter 23, Katherine Brickell broadens this perspective on 'other' voices, by reminding us that development geographies are bound up in a whole set of activities through which those living in the Global South build their own 'everyday' geographies. These are, of course, much richer and more complex than they have tended to be portrayed in traditional development work, and Brickell uses examples drawn from India and Africa to illustrate this. She shows how the practices of home, work and leisure offer rich sources of geographical material across both public and private spaces.

Development geographies are therefore becoming more attuned to many of the wider theoretical considerations of Human Geography. Sensitivities to resistance and to gender, to youth and to culture, are being allied with perspectives drawn from post-colonial and post-developmental theory. The key question is how such post-colonial and post-developmental geographies might be implemented in practice. As neo-liberalism becomes ever more dominant, how far is it possible to engender wider access to, and participation in, capitalist development – without that development merely being a vehicle for yet more economic dominance and exploitation?

FURTHER READING

Escobar, A. (1995) *Encountering Development – The Making and Unmaking of the Third World.* Princeton: Princeton University Press.

This may have been written a while ago, but it remains a key contribution to questioning the very ideology of development based on an industrialized and North American/European model.

Potter, R., Binns, T., Elliott, J. and Smith, D. (eds.) (2008) *Geographies of Development: An Introduction to Development Studies.* Harlow: Pearson.

This is the third edition of a comprehensive introductory textbook on Development Studies.

Williams, G., Meth, P. and Willis, K. (eds.) (2009) *Geographies of Developing Areas – The Global South in a Changing World.* Oxon: Routledge.

A wide-ranging consideration of daily life in the Global South.

Journals such as *Third World Quarterly* are also worth browsing through and regular reviews of contemporary themes in development appear in *Progress in Human Geography.*

CHAPTER 20 THEORIES OF DEVELOPMENT

Katie Willis

> . . . let's reject the cynicism that says certain countries are condemned to perpetual poverty, for the past half century has witnessed more gains in human development than at any time in history . . . From Latin America to Africa to Asia, developing nations have transformed into leaders in the global economy. Nor can anyone deny the progress that has been made toward achieving certain Millennium Development Goals. The doors of education have been opened to tens of millions of children, boys and girls . . . Access to clean drinking water is up. Around the world, hundreds of millions of people have been lifted from extreme poverty . . . Yet we must also face the fact that progress towards other goals that were set has not come nearly fast enough. Not for the hundreds of thousands of women who lose their lives every year simply giving birth . . . Not for the nearly one billion people who endure the misery of chronic hunger. This is the reality we must face – that if the international community just keeps doing the same things the same way, we may make some modest progress here and there, but we will miss many development goals. That is the truth. With 10 years down and just five years before our development targets come due, we must do better.
>
> (US President, Barack Obama, September 2010)

Introduction

Barack Obama's remarks at the 2010 Millennium Development Goals (MDGs) Summit at the United Nations in New York, encapsulate many themes associated with 'development' at the start of the twenty-first century. At the heart of the concept of **development** is the improvement of people's lives, represented in Obama's quote by reference to poverty reduction, clean water and education. However, what constitutes an 'improvement', which improvements should be addressed and how they should be achieved are highly contentious issues. In particular, it is important to consider which actors are involved in defining and achieving the progress to which Obama refers. While Obama refers to Latin America, Africa and Asia as places where development has occurred, he is also clear that development has not been rapid or extensive enough to benefit everyone. He is therefore presenting development as a dynamic process which can be spatially and socially extended to different parts of the world, if the international community adopts the right policies.

The World Bank divides the world's countries into three categories based on the Gross National Income per capita (GNI p.c.) of each country (Figure 20.1). GNI p.c. (also known as Gross National Product, GNP p.c.) is a measure of the wealth in a country divided by the population and is often used as an indicator of development. Thus, at a global level, this measure puts western Europe, North America, Japan, South Korea, Singapore, Australia, New Zealand, Saudi Arabia and the United Arab Emirates in the category of 'most developed', while most of Sub-Saharan Africa is 'least developed'. Development in this case is presented as a goal to be achieved. High levels of national income mean you can join the 'club' of developed nations.

However, what do these national income measures tell us about the lives of individuals living within these countries? Greater economic wealth at a national level does not automatically mean better standards of living for everyone, due to inequality. For example, residents of poor neighbourhoods in England die, on average, seven years earlier than those in richer areas (Marmot Review, 2010: 16), while in low-income Rwanda, the richest 10 per cent of the population have access to over 40 per cent of the country's income (World Bank, 2011). In addition, is greater economic wealth

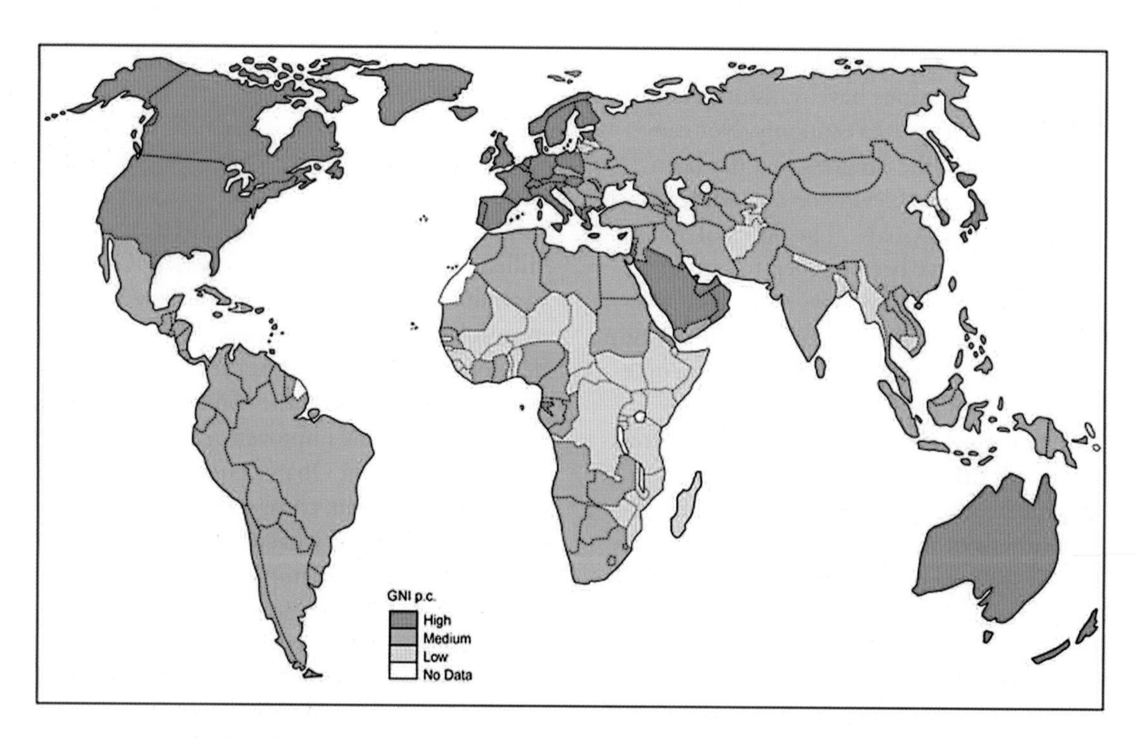

Figure 20.1 World Bank income classifications, 2010. Source: K Willis (based on data from World Bank, 2012)

what everyone desires? Does development have to include increased incomes, or do the environmental destruction and social upheaval that may be associated with this form of development mean that the focus on economic factors should be reconsidered? This questioning of development measures can be seen in the range of indicators available. These include the Human Development Index (HDI) adopted by the United Nations Development Programme (UNDP) in the early 1990s. It includes GNP p.c., but also measures of life expectancy and school enrolment.

Despite attempts to measure development, as exemplified by the World Bank GNI p.c. data and the HDI, it remains a highly complex and fluid concept. As with many concepts, we often use it without thinking how our understandings and use of the term have come about. Constructions of development vary across time and space, but particular understandings become more widely used. This reflects power relations, not only at a global scale, but also within countries, communities and households.

In this chapter, I am going to discuss some of the most popular theoretical ideas about development in the post-Second World War period. This temporal focus does not mean that development was not theorized before the Second World War, but reflects the fact that from the mid-1940s onwards there was an explicit attempt by western countries to promote development in Africa, Asia, Latin America and the Caribbean. In addition, concentrating on academic theories, as well as policies implemented by governments, multilateral organizations and development agencies, does not mean that other groups of people did not have different ideas about development. Alternative constructions of development and forms of resistance are covered in later chapters.

SUMMARY

- Definitions of development vary greatly and reflect differential understandings of the world and unequal power relations.
- It is important to consider scale in any examination of development.
- Development can be considered as both a goal and a process.

Modernization

Development is often understood as a process of becoming '**modern**'. As discussed later in the book (Chapter 34), modern can just mean 'of the moment', but in most cases there are particular understandings of what modern is and should be. In development terms, modern has usually applied to being like 'the West'. This might be in economic terms (for example, commercial rather than subsistence agriculture), social terms (such as the importance of class groups rather than family or tribal affiliations), or political terms (for example, democratically elected governments rather than tribal or religious forms of organization). Ideas of development as modernization can, therefore, often be viewed as **Eurocentric**. The idea of modern being presented as desirable is based purely on the experiences and values of countries and societies in the Global North.

A number of theorists, often termed 'modernization theorists', have attempted to describe and model this form of development. The best known of these is Walt Rostow who, in 1960, published his book *The Stages of Economic Growth: A Non-Communist Manifesto*. From examining the history of a number of countries, most notably the United States of America and some European countries, he argued that there were five stages of economic growth that could be identified, ranging from 'traditional' societies to what Rostow termed 'The Age of High Mass Consumption' (see Table 20.1). While Rostow's focus was on economic growth rather than development *per se*, it is clear that the shifts in economic, social and political organization could all be encompassed within post-Second World War ideas about development.

In Harry S. Truman's US Presidential inauguration speech in 1949, he referred specifically to the need for assistance to be provided to poorer countries of the world:

> For the first time in history humanity possesses the knowledge and the skill to relieve the suffering of these people [the world's poor] . . . I believe that we should make available to peace-loving peoples the benefits of our store of technical knowledge in order to help them realize their aspirations for a better life . . . What we envisage is a program of development based on the concepts of democratic fair dealing . . . Greater production is the key to prosperity and peace. And the key to greater production is a wider and more vigorous application of modern scientific and technical knowledge.
>
> (Truman, 1949 in Escobar, 1995: 3)

Truman's sentiment fits into the understanding of development as modernization. Truman

Stage	Characteristics
Traditional	Agriculture and hunter-gatherer societies; pre-Newtonian science and technology; social structures dominated by family, clan or tribal groupings; pre-nation-state
Preconditions for Take-Off	Savings and investment rates above population growth rates; increased importance of nation-state and national institutions; elite status not based on family or clan allegiance; changes often triggered by external intrusion
Take-Off	Triggered by internal or external stimulus e.g. political revolution, colonialism, technical innovation; higher rates of investment and saving; substantial manufacturing sector; banks and other institutions in place
Drive to Maturity	Expansion of use and range of technology; growth of new economic sectors; investment and savings 10–20 per cent of national income
Age of High Mass Consumption	Widespread consumption of durable consumer goods and services; increased spending on welfare services

Table 20.1 Rostow's stages of economic growth. Source: Adapted from Rostow (1960)

argued that the poverty and poor living conditions in many parts of the world were an indication of a lack of development. As the USA had the technical knowledge to promote development, it should help other parts of the world. This idea of sharing knowledge to achieve development became the key to aid policies for many northern countries in the 1950s onwards. Without external aid, it was argued, the poorer countries of the world would never develop. These projects met with different degrees of success (see box).

Despite external assistance, the development 'problem' was presented very much in terms of factors that were internal to the countries involved. These poorer countries did not have the skills, knowledge and capital to develop. Once they were assisted and shown how to develop, it was argued that success would follow. Rostow, for example, argued that colonialism had been crucial for the economic growth trajectory of many countries. Development obstacles were, therefore, seen as endogenous (i.e. relating to factors within countries). As I will argue in the next section, and as **post-development** and **post-colonial** theorists have argued (see Chapters 21 and 33) this focus on particular internal factors as explaining a lack of development has been greatly criticized.

CASE STUDY

The East African Groundnut Scheme

In the post-Second World War period, the British government became concerned about the reduced availability of food oils on the world market because of the potential impact on Britain's import bill. The East African Groundnut Scheme was an attempt to increase the world supply of groundnuts and oil, as well as providing opportunities for social welfare improvements and economic development in Tanganyika (now part of Tanzania), which was then part of the British Empire.

The scheme aimed to clear and cultivate approximately 1.3 million hectares, providing jobs directly and also through local multiplier effects. In addition, the project aimed to provide a demonstration effect to local people so they could see the benefits of technology and mechanization.

The UK government spent millions of pounds on equipment and seeds, but the project proved to be a disaster. By the time the project was finally ended in 1955, only about 9000 hectares had been cleared and yields were poor. The reasons for failure were multiple. Soil, vegetation and climatic conditions in Tanganyika were unsuitable for the use of heavy machinery such as tractors. In turn, when the machinery broke down, there were insufficient spare parts or trained mechanics to mend it. There was a lack of appropriate housing and service provision for workers, and attempts to export groundnuts were hampered by problems with the railways and port facilities.

In this case, and many others, attempts to modernize agriculture in this part of East Africa using northern ideas and equipment proved to be highly inappropriate.

Source: Adapted from Hogendorn and Scott (1981)

SUMMARY

- The post-war period saw the start of large-scale development planning aimed at the Global South.
- Development was understood as 'modernization' along the lines followed by the USA and western Europe.
- Northern countries were key in providing technical assistance to overcome perceived development 'obstacles' within southern countries.

Structuralism and dependency

While modernization theorists and northern policy makers tended to focus on endogenous factors, other theorists and activists argued that factors external to the poorer countries (i.e. exogenous factors) were of much greater importance. Many of these ideas and theories developed in parts of the Global South, reflecting the importance of considering how understandings and approaches to development are a product of temporally and spatially specific contexts.

Structuralism, most associated with Raúl Prebisch and the UN Economic Commission for Latin America (ECLA), was one of the earliest theoretical approaches of this type. ECLA (known as CEPAL in Spanish) was set up in the 1940s and its aim was to promote development in Latin America. Prebisch, its first director, argued that it was important to consider the current reality of Latin America when choosing development policies, rather than following northern-based models. While free trade policies had been important in the growth of northern economies, ECLA argued that Latin American economies in the post-Second World War world needed protection from northern competition. The world economy was very different in the 1940s and 1950s than in the nineteenth century when western Europe industrialized. This approach was termed 'structuralism' because it focused on the structures of the global economy and how these might help or hinder development attempts (Clarke, 2002; Preston, 1996: 181–9).

The policy implications of this were to promote state intervention and protectionist policies in Latin America. By doing this, Latin America would be able to develop (i.e. industrialize, urbanize and increase standards of living) as western Europe had done. Thus, development as a goal was the same as described by the modernization theorists, but the process was to be slightly different.

Import-substitution industrialization (ISI) was a key element of ECLA policy recommendations. ISI policies focused on building up domestic manufacturing, rather than relying on imports from elsewhere. Because of the greater experience of European and American companies, it was important to protect Latin America's infant industries from competition. This was done through imposing import tariffs. In addition, government finance and technical support were targeted at manufacturing concerns. In many cases, these policies were successful and allowed the growth of manufacturing industries in Latin America.

For example, manufacturing production increased annually by 4.0 per cent on average in Latin America in the 1950s and 6.3 per cent per annum on average in the 1960s (World Bank, 1983, in Sheahan, 1987: 85). However, success was not always achieved. Small domestic markets and the need to import technically advanced equipment often created obstacles to self-sufficiency (Sheahan, 1987).

A similar set of theories focusing on development in the context of the global economy and exogenous limits, emerged in the 1960s and 1970s. These were termed **dependency theories** and were again focused on the Latin American experience. As with structuralism, dependency theorists argued that Latin America could not follow a European development path, because the global context was very different. For most dependency theorists, however, the solution was not to adopt protectionist measures as outlined by the structuralists, rather the solution was a more dramatic withdrawal from the global economic system.

The best known dependency theorist is André Gunder Frank (1967). He used the examples of Chile and Brazil to demonstrate his arguments. Frank argued that since the arrival of the Europeans in what is now Latin America, the continent's peoples and resources had been exploited. Rather than being used to help the development of Latin America, profits had been taken out of the region. This had created what Frank termed 'the development of underdevelopment' – that is, Latin America's 'underdevelopment' was a direct consequence of the relationships of exploitation.

There is a strong spatial dimension to Frank's work. He highlighted how the chains of exploitation and dependency run not just from the periphery of Latin America to the core of Europe, but that within Latin America there are unequal relationships between urban and rural groups and between landowners and labourers, such that local-level exploitation and inequality reflects the inequalities on a global scale (see Figure 20.2). Walter Rodney in his work on

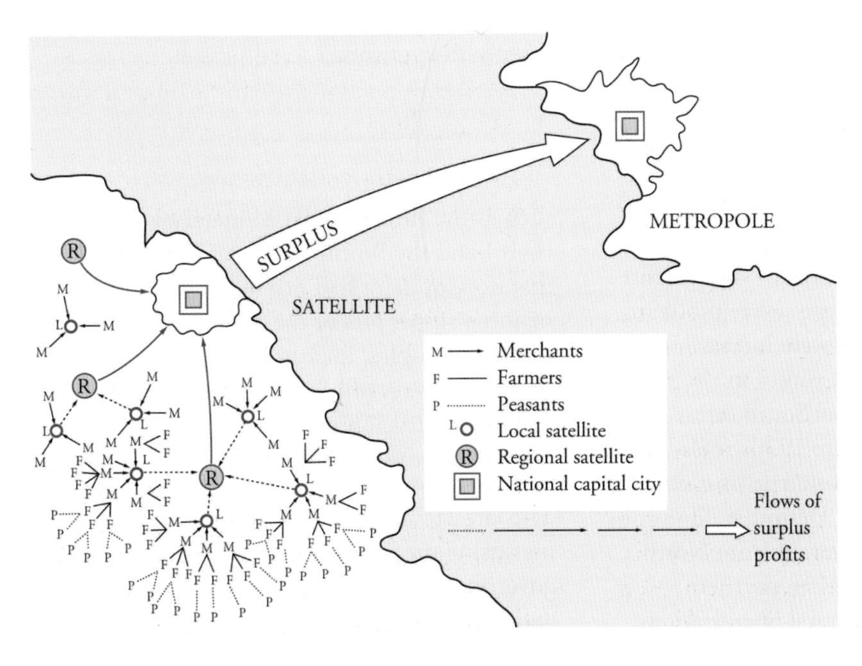

Figure 20.2 Graphical representation of dependency theory. Source: Potter *et al.* (2008: 111)

African underdevelopment also follows a dependency argument:

> Colonialism was not merely a system of exploitation, but one whose essential purpose was to repatriate the profits to the so-called 'mother country'. From an African view-point, that amounted to consistent expatriation of surplus produced by African labour out of African resources. It meant the development of Europe as a part of the same dialectical process in which African was underdeveloped.
>
> (Rodney, 1974: 162)

For dependency theorists, it is the global capitalist system dominated by the northern countries which is an obstacle to autonomous development in the Global South. Thus, the solution and route to development is through withdrawing from the global economic system and setting up alternative forms of society and economy. Without this, southern countries are destined to remain poor and exploited and development will not take place.

Dependency approaches attracted a great deal of criticism, not least because of their static construction of global relationships. Even as Frank and others were writing, some parts of the Global South, most notably the newly industrializing countries (NICs) of East Asia (see below) were experiencing rapid economic growth and improvements in standards of living without withdrawing from global capitalism.

SUMMARY

- Structuralism and dependency approaches blamed the structure of the global political economy for low levels of economic development.
- Protectionism or withdrawal from the global capitalist system were advocated as development solutions.
- Policies based on structuralist and dependency theories experienced some success, but there were problems with their long-term viability.

Neo-liberalism

As the ISI experiments based on structuralist theories began to experience severe problems, and dependency theories were increasingly unable to present realistic routes to development in the Global South, ideas began to emerge in the late 1970s regarding new ways in which development (in a western form) could be achieved. Despite millions of dollars of aid being transferred from North to South since the late 1940s, northern governments and multilateral organizations such as the International Monetary Fund (IMF) and the World Bank were concerned that development had not reached the vast majority of peoples of the South. For these southern peoples, increasingly aware of alternative ways of life elsewhere in the world, the goal of development in a western style became increasingly desirable, if unattainable.

The late 1970s and early 1980s saw the increasing prominence of **neo-liberal** ideas in government policymaking in western Europe and North America, as well as in development policies directed towards the South. At the heart of neo-liberal ideas is the belief that government intervention in the economy always leads to inefficiencies and that it is far better to let market forces determine wages, prices and what should be produced and where. The market, it is argued, is neutral and does not try and benefit one group over another (Toye, 1993).

The election of Margaret Thatcher as UK Prime Minister in 1979 and the inauguration of Ronald Reagan as US President in 1981 reflects this move to reduced state intervention in northern countries. Both leaders felt that government inefficiencies had led to poor economic performance in the 1970s and reduced choice for individuals, families and communities. They both advocated market-led reforms such as privatization and reduced state expenditure (Harvey, 2007).

These neo-liberal policies were also introduced into development policies towards the Global South, most notably through structural adjustment policies (SAPs). Just as with the modernization ideas outlined earlier, SAPs reflect the imposition of ideas and understandings of development on Southern countries from those of the Global North. In the late 1970s/early 1980s most Southern countries found themselves increasingly unable to meet the payments on their national debt because of rising oil prices, increasing interest rates and global recession. The debt crisis which resulted triggered the widespread implementation and adoption of SAPs. Unable to borrow money from private banks, southern governments were forced to agree to SAPs in order to gain further funding from the IMF and the World Bank.

SAPs aimed to stabilize a country's economy and then to restructure it in order to promote development in the future. Policies were aimed at reducing the role of the state and opening up national economies to foreign investment and competition, so ending ISI protectionist policies. While SAPs achieved their aims of stabilization, in many cases the restructuring was associated with increasing levels of social inequality as unemployment increased and welfare provision was slashed (Mohan *et al.*, 2000). Despite these problems, SAPs continued to be promoted throughout the 1980s and early 1990s and were also implemented in the transition economies of Eastern Europe and the former Soviet Union after the collapse of the Communist bloc (Bradshaw and Stenning, 2004). In the late 1990s, the rigidly designed SAPs were replaced with Poverty Reduction Strategy Papers (PRSPs), which were meant to be more context-specific and promote grassroots participation (see below). However, the neo-liberal fundamentals remain the same.

The East Asian countries have often been highlighted as examples of the successes of neo-liberal development policies. The newly-industrializing countries (NICs) of Hong Kong, Taiwan, Singapore (Figure 20.3) and South Korea, along with other regional nations such as Thailand, the Philippines, Indonesia and Malaysia were regarded as what the World Bank (1993) termed the 'Asian Miracle'. The World Bank argued that these nations had successfully achieved economic growth without increasing levels of inequality by following neo-liberal market-led policies (Figure 20.4). However, this failed to recognize the key role of government subsidies and protectionist tariffs in some countries.

The global economic crisis of 2007 onwards has been interpreted by some observers as an indication of the limits of neo-liberal policies and of capitalism more generally (Harvey, 2010). Government expenditure on bailing out financial institutions, or supporting key manufacturing industries in the USA and

Figure 20.3 Downtown Singapore. Credit: Urban Redevelopment Authority, Singapore

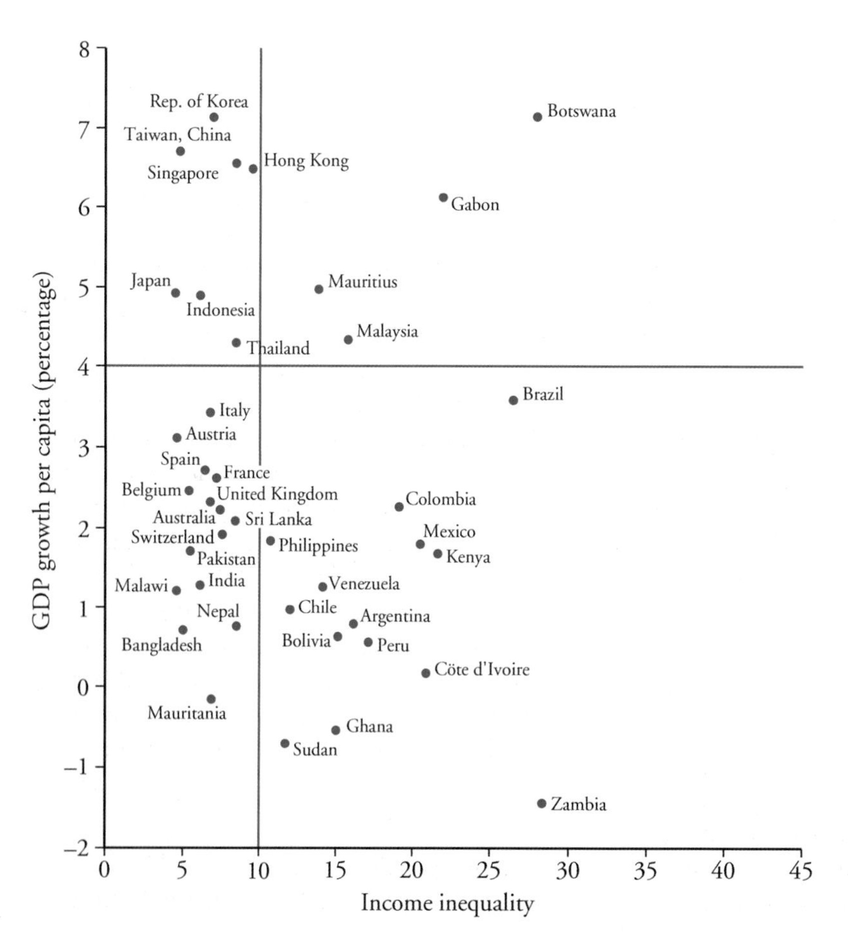

Figure 20.4 Income inequality and growth of GDP, 1965–89. Credit: J. Page (1994) *World Development*, Elsevier, p. 616

western Europe, certainly suggests that policies involving greater state intervention could become more widespread (Willis, 2011: 66–7). However, while some national governments in the Global South, most notably in Latin America (Panizza, 2009), have moderated their neo-liberal policies at the start of the twenty-first century, international development policy, as framed by the IMF, World Bank and northern governments remains committed to key neo-liberal principles.

SUMMARY

- Neo-liberal theories focus on the role of the market as the most efficient route for achieving development.
- Government intervention in production and service provision is reduced, and private and voluntary organizations are supposed to replace government activities.
- The success of East Asian development has been used as example of successful neo-liberal policies.
- Neo-liberalism remains the key theoretical grounding for government and multilateral development assistance despite global economic crisis.

Grassroots development

While all the theories outlined above aimed to improve the lives of individuals, the focus of policies has been at the level of the nation-state. The argument has been that if appropriate policies are adopted nationally, then the benefits will 'trickle-down' to those at the grassroots level. However, in far too many cases this has not been the case. Instead, national-level policies have exacerbated inequalities or have left the majority of the population without a say in what changes (if any) they would like to their lives.

The national-level focus of modernization theories, structuralism and dependency approaches and neo-liberal policies hides, however, some attempts at engaging with development at the grassroots. For example, in the late 1960s and early 1970s, many multilateral organizations and northern governments began to advocate a 'basic needs approach' in their work. This was a response to the realization that many of the aid policies that had been implemented had failed to reach the very poorest. A basic needs approach identified key needs which had to be met, including shelter, food, health, education and employment. Rather than focusing on the large-scale projects of earlier periods, it was argued that smaller-scale activities, such as concentrating on the work of informal sector traders and producers in urban areas would be much more effective in helping the economically poor (Hunt, 1989). Unfortunately, this approach was not widely adopted due to shortages of funding and fears by many southern governments that focusing on small-scale projects would hinder economic growth. For these governments, 'development' was largely economic.

'Helping people to help themselves' is a common slogan used by **non-governmental**

organizations (NGOs) working in the development field, particularly in southern countries (Figure 20.5). This focus on individuals, families and communities represents a grassroots or bottom-up approach to development and has become increasingly common since the early 1980s. Development NGOs vary greatly in size, scope and membership, and can be found in both the Global North and South (Lewis and Kanji, 2009). For example, Oxfam is northern-based, but it works with partner organizations in the South.

The growth of NGOs reflects the supposed benefits which such organizations provide to local communities. Because they work with people at a local level, NGOs are meant to provide more efficient and appropriate services, such as health care, infrastructure or agricultural assistance. In comparison with the large-scale projects described earlier, NGO projects are viewed as more appropriate and therefore more likely to succeed (Edwards and Hulme, 1995; Lewis and Kanji, 2009). In addition, NGOs are regarded as promoting participation and empowerment among marginal communities and groups. It is argued that government activities often exclude the voices of the marginal, while NGOs are ideally positioned to engage with the needs and opinions of the poor. This may be particularly the case when they work in partnership with community-based organizations (CBOs).

While many NGOs have achieved a great deal in terms of providing the services local people want in an appropriate and affordable manner, there are many cases of NGOs being unable to live up to the 'development panacea' label. While the number of NGOs has increased massively since the 1980s, there are still insufficient numbers to meet the massive demand for their services. In addition, they are often spatially and socially concentrated in particular sectors. Despite supposedly meeting the needs of the most vulnerable, NGOs often work in more physically accessible communities, so contributing to spatial inequalities (Mercer, 1999). In addition,

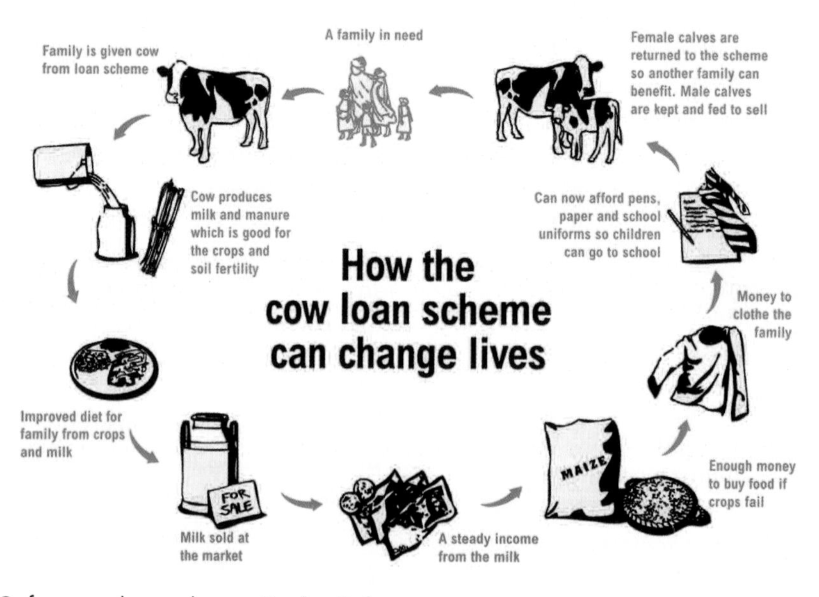

Figure 20.5 Oxfam cow loan scheme. Credit: Oxfam

Figure 20.6 Community pharmacy sponsored by international NGO, Handicap International, Sigor, NW Kenya. Credit: Katie Willis

because most NGOs rely on donations from individuals or governments, their work is often directed by the issues donors regard as important, rather than those identified by local people (Hulme and Edwards, 1997).

Finally, as discussed earlier, development policy has been dominated by neo-liberal ideologies since the early 1980s. The rise of NGOs during this period is no coincidence. As governments have reduced state expenditure and privatized many services, such as healthcare, education and infrastructure provision, poor communities have been left unable to afford these services. NGOs have, therefore, been crucial in attempts to fill the gaps left as the state has withdrawn (Figure 20.6). In addition, NGO activities often fit within neo-liberal understandings of decentralization, greater local democracy and choice. They have, therefore, been encouraged. While the majority of people would regard choice and democracy as desirable, in reality, neo-liberal policies have left large numbers of marginal people even less able to access the services required to improve their lives (Smith *et al.*, 2008).

SUMMARY

- Since the 1980s there has been an increasing focus on the sub-national level of development.
- NGOs have become key agents in development practice because they are regarded as providing effective, efficient and appropriate services.
- While NGOs can be very important service providers and channels for consciousness-raising, there are limits to what they can do.

Conclusion

Despite its widespread use, the concept of development is highly contested. Within a general understanding of improvements in individuals' lives lies a wide variety of definitions and policies. The post-war period heralded an era of development planning, where the economically richer countries of the West tried to help the poorer countries of Africa, Asia, Latin America and the Caribbean. This help took the form of technology, education and support which mirrored that used in the development of the West. Such Eurocentric theories and approaches did not remain unchallenged. Rather than seeing a lack of development as a reflection of internal deficiencies, structuralist and dependency approaches argued that global structures of inequality created and exacerbated situations of poverty and marginalization. Without addressing these inequalities, development would remain a distant dream. Despite these challenges, the 1980s saw a reassertion of development theories and practices based on northern definitions and priorities with neo-liberalism. While the details of these policies have been adapted since that period, neo-liberalism remains at the forefront of international development policy-making.

These major themes in development theorizing since the 1940s demonstrate the importance of considering not only how levels of development vary spatially, as demonstrated by the World Bank data in Figure 20.1, but also how understandings of 'development' are a product of particular geographies. Latin America, for example, proved a particular environment for the elaboration of structuralist and dependency ideas challenging Eurocentric theories. The politics of development policies and theories are highlighted in the chapters that follow.

DISCUSSION POINTS

1. What are the advantages and disadvantages of using GNI per capita as a measure of development? Look at the latest UNDP *Human Development Report* (available at www.undp.org) and evaluate the range of national-level measures used to indicate development.
2. How can external factors affect the economic development of a country?
3. Why have NGOs been regarded as appropriate agents for development and what limitations may there be to their effectiveness?

FURTHER READING

Bebbington, A. J., Hickey, S. and Mitlin, D. C. (eds.) (2008) *Can NGOs make a Difference? The Challenges of Development Alternatives*. London: Zed Books.

Useful collection of chapters on different elements of NGO activity. Provides a range of examples and highlights the ways in which NGO successes are shaped and limited.

Lawson, V. (2007) *Making Development Geography*. London: Hodder Education.

A clear overview of geographical perspectives on development.

Peet, R. and Hartwick, E. (2009) *Theories of Development* (second edition). London: The Guilford Press.

A detailed consideration of different approaches to development in its broadest sense. While a range of theories are covered, the authors make a strong case for considering radical alternatives to neo-liberal orthodoxy.

Willis, K. (2011) *Theories and Practices of Development* (2nd edn.). Abingdon: Routledge.

Useful introductory textbook covering a range of development theories and associated policies.

WEBSITES

www.developmentgateway.org

Development Gateway: links to a range of development information, including latest news and discussion groups.

www.eldis.org

Portal for development-related information run by the Institute of Development Studies, University of Sussex.

www.imf.org

International Monetary Fund: provides information on the approaches and activities of the IMF. Contains useful material on poverty reduction policies.

www.oneworld.org/home

One World: an excellent site for up-to-date development information.

www.un.org/millenniumgoals/

United Nations: provides information on UN Millennium Development Goals.

www.worldbank.org

World Bank: details the activities of all members of the World Bank Group.

CHAPTER 21
RETHINKING DEVELOPMENT

Sarah A. Radcliffe

Introduction

At the start of the twenty-first century, we – if we live in the Global North – often assume that we are living in an integrated world or a 'global village', where not everyone has an i-Phone but they do have a mobile phone. For many people living in the Global South, however, resources and integration are a mirage – you may have a mobile phone, but your settlement lacks adequate local health care services and a decent school for your children. And the contrasts between rich and poor are becoming ever more entrenched: if you live in the Global South, you may work for someone whose wealth exceeds that of a member of the European middle class, but you may well be paid less than a living wage. Development – to put it simply – is not a single condition, a common shared experience; for some, as Esteva so eloquently puts it, 'development stinks'. Cosmopolitan jet-setters in São Paulo live one kind of development while women in sub-Saharan Africa, who walk for hours to collect water, experience a completely different kind of development. How do we listen to this difference? What analysis best suits this geographical diversity, while also linking our analyses to the obvious class, ethnic and race-based inequalities we see? Rethinking development means reconsidering the categories we use in development geography, and unpacking the power relations that shape them.

Since the explosion of post-development thinking into development studies in the mid-1990s, there has been a lot of rethinking of development going on. As noted in the other chapters in this section, development has been analysed in terms of the differentiated and highly uneven impacts of capitalist relations on countries' income and life quality levels (Chapter 20); and a variety of protest movements and campaigns struggle to provide alternatives to development (Chapter 22). Although at one time overwhelmingly economic, development studies today specifies and interrogates the interrelationship of development's political, cultural and institutional facets, which are examined in this chapter (see also Chapter 23). Rethinking development is primarily concerned with thinking about the power relations bound up in development (as a process and as intervention) – its operations, its geographies, its highly uneven distribution, contradictions, and strategies for achieving it, so the analysis of power is central. In the remainder of the chapter, three windows onto development's power will be examined, namely the power of writing, the power of gender and the powers of the state.

Geography and post-colonialism

Post-colonialism has become an important framework in recent years for geographers trying to understand development, as it challenges us to write about 'developing countries' in ways that recognize the social, economic and political impacts of colonialism, while also drawing out the interconnections between 'North' and 'South' (see also Chapter 23; Gregory, Johnston *et al.*, 2009). As a critique of legacies of colonialism and its rigid thinking, post-colonialism covers a broad terrain of analysis, which has been selectively used by Human Geographers in their understanding of development. Post-colonialism is thus not simply a question of independence that was gained unevenly (for example, the Spanish American colonies gained independence between the 1820s and 1898, while British and French colonies in Africa and Asia became independent during the middle years of the twentieth century) (McClintock, 1995) (see Figure 21.1). A post-colonial country is one whose struggle for independence subsequently turned into the search for development, epitomized by a western-style modernization process (Schech and Haggis, 2000). Southern post-colonial countries often took development as *the* framework for national action and identity (Power, 2003), defining their identities and devising policies to reach a western-defined standard which, inevitably, they could not replicate (Gupta, 1998). Post-colonial frameworks critique neo-colonialism's uneven impact on the materialities of people's lives, livelihoods affected by nation-states' failure to provide services to their entire populations, income inequalities, hierarchical social relations and diverse urban structures (Power, 2003). Post-colonialism also criticizes the ways in which the West has constructed its knowledge about the South, in order to 'decolonize the mind' in Ngugi wa Thiong'o's words. By changing our 'mental maps', decolonization of the mind begins to provide an empirically-rich, geographically-informed understanding of the ways in which global processes of capitalism and development interventions intersect with local and national processes of giving meaning and living everyday lives. Moreover, post-colonialism challenges notions of a single 'path' to development. Human Geographers are in a good position to provide just such a detailed empirically rich discussion of highly differentiated post-colonial situations. By deconstructing the categories of 'poor countries' and 'third world women', the power of these categories to shape development interventions in inappropriate ways can be replaced by contextualized and empirically based knowledge.

Post-colonialism also challenges our expectations about individuals and societies in Africa, Latin America and Asia. It throws out negative stereotypes about 'basket-case African countries' or tin-pot dictators in South America (Simon, 1997; Power, 2003), while forcing us to question 'our' static view of 'tradition', seeing the third world as unchanging. While post-colonial perspectives have been criticized for their over-romanticization of 'hybrid cultures' that blend local and global elements (see Chapter 3), geographers recently have begun to document the harsh realities of everyday lives in the South. Geographers too are attuned to how seemingly disparate and distant lives are connected, through economies of production,

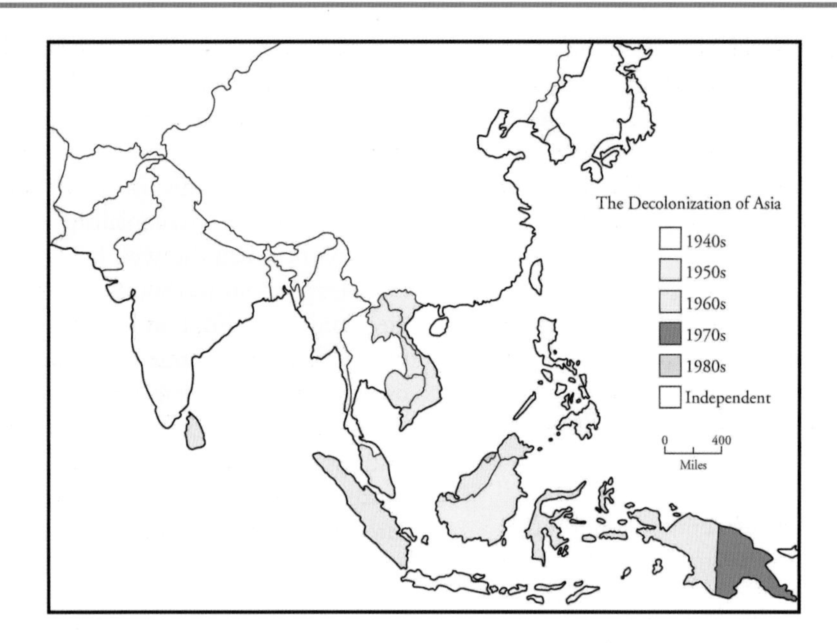

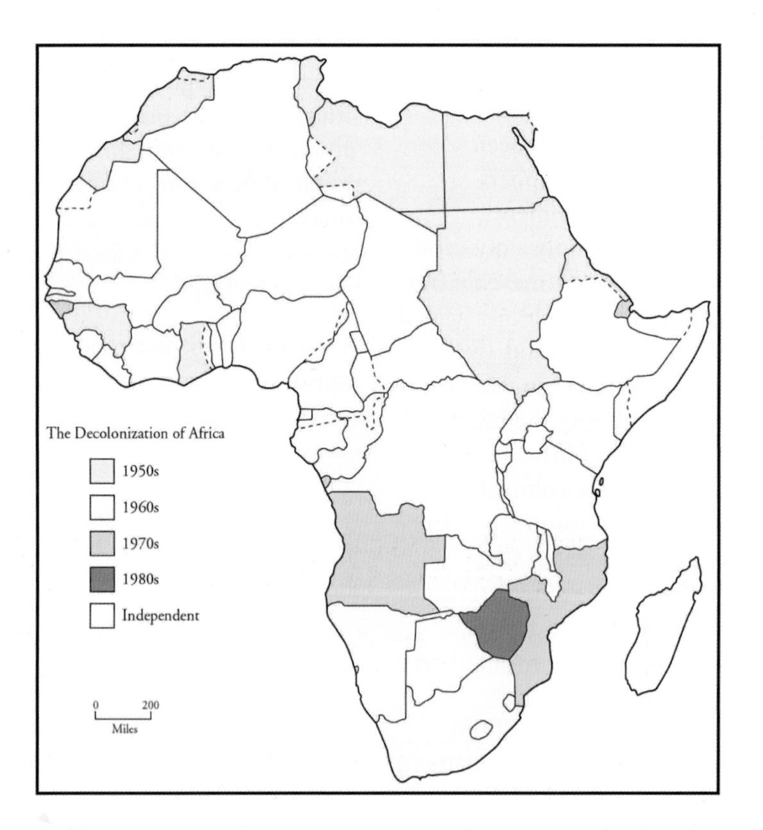

Figure 21.1 Decades of independence for former colonies in the South, by decade: (a) Asia; (b) Africa. Credit: (a) and (b) Sarah Radcliffe/Cartographic Unit, Department of Geography, University of Cambridge

exchange and financial flows and political decision making at local, regional, national and global scales. For example, shanty-town dwellers in Thailand might find their earnings buy little, due to global financial instability, that their work is outsourced to China and their environment is affected by garbage waste coming from Europe. Post-colonialism challenges us to understand these connections as the result of longstanding narratives about which countries and people 'matter', and the lack of priority given to low-income groups in certain places.

While addressing the gulfs in income and well-being that characterize North–South relations (see Figures. 20.1 and 20.4), post-colonial approaches acknowledge the diversity of priorities in development. Needs – for clean water, adequate food, education – are not pre-given but are determined by social identities and local–regional cultures within politicized and culturally specific contexts, however self-evidently beneficial they appear at first sight. The 1980s Basic Needs Approach (BNA) to development tried to face up to the needs of 1.2 billion people in absolute poverty, yet inevitably intersected with wider political and social changes occurring in the diverse countries of Africa, Asia and Latin America. Many groups and individuals in these countries demanded clean water and electricity for their neighbourhoods (i.e. basic needs), while also pursuing demands for wider 'strategic' needs, such as political rights as citizens and participants in the development process. While great strides have been made in pursuing basic goods, the politics and economics of neo-liberal development (see Chapter 20) mean that these wider strategic needs often remain at the forefront of development debates in the South. Post-colonial approaches question development experts' ability as 'outsiders' to identify groups' needs completely, and remind us that whoever claims to speak on behalf of the world's poor is always themselves located within the very structures of political economy that produce global inequality.

Post-colonialism also encourages us to consider that we – sitting in our libraries, far from women collecting water in Nigeria – have a role to play in all of this. 'Our' identities and our worlds are made in that same global economy, as it works its uneven way around the world in ever faster circles. Development is thus not only about economic relations and income differentials, but also about the politics of needs, about a political economy of racism (how colonialism and its relations underpin our attitudes to issues as varied as AIDS, famine relief and population growth), and about how tradition can be re-invented.

SUMMARY

- Post-colonialism is a critique of the ways in which colonialism and neo-colonialism shape the experiences of, and the writing about, development. It also hopes to contribute to new ways of 'knowing' about development.
- Post-colonial perspectives encourage us to map out a more complex geography than a world divided into first and third world, North and South, and to rethink what is meant by traditional versus modern.

Contextualizing development: feminism and gender

> In a world where women do two-thirds of the world's work, earn 10 per cent of the world's income and own less than 1 per cent of the world's property, the promise of 'postcolonialism' has been a history of hopes postponed.
>
> (McClintock, 1995: 19)

In her critique of development, Anne McClintock highlights the ways in which post-colonialism celebrates too much, too soon, when it comes to the position of women. National independence and the new international division of labour result in a world where women and men often experience widely differing lives, whether in terms of citizenship rights, work or kin and family relationships. To take a global view, these differences reduce women's power, visibility and access to resources.

Yet while McClintock points to women's continued marginalization, she represents women as a homogeneous group, which is defined more by its internal sameness than by heterogeneity. This quote from McClintock galvanizes our political solidarity with women and human rights, yet it also reiterates what the feminist film-maker Trinh Minh-ha calls the 'third world difference' – that is, a way of identifying all women in the South in a distinct category to western women. As Chandra Mohanty has argued, the third world difference portrays third world women as powerless, burdened, cloistered and oppressed by their (male) kin (Mohanty, 1991). Found in writings by colonial officials, development experts and some early geographers, this 'third world difference' is an example of a colonial legacy. A post-colonial perspective therefore attempts to examine third world women as agents of change and as highly differentiated. **Gender** divisions of labour and a gendered distribution of resources and income differentiate women along lines of class, location, religion, age/generation, race/ethnicity and sexuality (Visvanathan, 1997). The gulf in living standards between women of different classes often translates into highly diverse strategies. Depending on their country, urban or rural location, family forms and sources of income, poor women face difficulties in securing income, health and a voice for themselves and their communities, whereas many high-income women enjoy careers and living standards comparable with (or better than) those of their western counterparts. Such differences are graphically illustrated by a middle-class, urban woman with a university degree and career next to her live-in maid, who is most likely from the countryside with limited formal education and few career opportunities (see Figure 21.2; also Radcliffe, 1990).

Political coalitions across these gulfs are fraught with difficulties, especially for poor women who already face a 'triple burden' of production, reproduction and community management work (Moser, 1993). In contrast to their passive stereotype, women in the South have diverse strategies for dealing with a combination of forces, including capitalist economies that see them as the cheapest labour, racism, colonialism and male power in families and states. As Mohanty points out, women's 'agency is . . . figured in the minute, day-to-day practices and struggles' (Mohanty, 1991: 38; Radcliffe and Westwood, 1993). Women who particularly face an uphill struggle in guaranteeing security for themselves under conditions of growing global inequalities and political instability are female heads of households, and refugee women. Female heads of households account for an estimated

Figure 21.2
Credit: © Angela Martin

one in three of domestic units in the third world (Chant, 1997; Varley, 1996). Of the approximately 17 million refugees around the world in 1992, around 80 per cent were women and their dependants. Gender-sensitive approaches to refugee flows recognize that women must be able to gain income, become empowered, be free of sexual harassment and claim refugee status as persecuted women.

Ways of 'being female' are intimately bound up with livelihood and production relations, the form and discourse of the nation-state, cultural ideologies and political culture as well as kinship, all of which vary historically and geographically. Consequently, third world women experience their lives in the intersection of these processes; their actions are not driven 'just' by gender concerns, but by gender *in association* with struggles over class, race and ethnicity, colonial legacies and global capitalism. Third world feminism has challenged first world feminism for ethnocentric and racist views.

Acknowledgement of women's diverse engagements with development (both as a process and as an intervention) has only slowly entered development planning. During 'modernization', planners believed that third world women were outside development, and advocated their incorporation into the development process. Pioneer feminists of development, such as Esther Boserup, pointed out the fallacy of this argument, showing that women were disadvantaged precisely because of the 'male bias' embedded in development. 'Gender and Development' (GAD) policy attempts to rectify this bias by analysing male–female power relations in different settings. Repositioning gender and development issues around questions of rights, democracy and participation extends the long-term interest in women's empowerment in the development process (Rowlands, 1997). The establishment of women's ministries and female quotas in national elections are indicative of sustained change over the past decade in many countries of the South.

CASE STUDY

Masculinities and development: the case of southern Peru

Indigenous ('Indian') small farmers in the Peruvian Andes are engaged in a multiplicity of livelihood activities that include subsistence farming, cash crop production, migration and local waged work. Gender divisions of labour in agriculture are clear-cut, with male farmers ploughing, while women sow, weed and harvest. Development assistance, usually in the form of credit and technical training, largely goes to men as heads of households. Small farmer confederations are the means by which men, and some women, address their concerns to government and agencies, and coordinate local development initiatives. Leaders are overwhelmingly men, although a small number of rural women promoted female concerns (e.g. political representation, individual land titles). Male attitudes to women's struggles and demands are crucial in determining the outcome of negotiations, either in the household or the unions. If men demonstrate *comprensión* (understanding), women travel the country, carry out promotional and organizational work, and gain support in confederations for gender-aware policies. Supportive masculinities contrast with obstructive attitudes that block women's appointment to leadership positions, marginalize their concerns and prevent female mobility. Where this happens, rural women have often established parallel, gender-specific organizations in the Andes.

Source: Personal field notes (1988–90)

Development thinking has also begun to deconstruct the category 'men', as men's role in influencing long-term social change, supporting gender-sensitive initiatives, and male-on-male violence is increasingly recognized. Early development thinking tended to assume that male heads of households were fair to all family members. However, it was shown that in areas where historical and cultural conventions sanctioned uneven access to household goods, children and women particularly felt the consequences of illness, malnutrition and dis-empowerment. A new focus on masculinities highlights the diverse positions and attitudes of men, varying with generation, work, family role, location, and so on, to replace the notion of 'patriarchy' (universal male power). Whether development projects are targeted at women-only groups or at mixed groups, the attitudes of men in the community and their assistance or objections affect the project's outcome as well as women's confidence, as a case study from peasant associations in southern Peru illustrates (see box).

SUMMARY

- Gender relations in the South are diverse, as place-specific interactions of multiple femininities and masculinities offer widely diverse opportunities and create overlapping forms of disadvantage.
- Women, especially those in low-income groups and younger age groups, are – in comparison with menfolk in their society – more likely to work longer hours, earn less money for their work, and have less secure access to land or property title, credit or development assistance.
- Third world women are active agents in development, yet have been stereotyped by much writing from the North.
- Development policy needs to address gendered power relations, women's agency and the diversity of masculinities found in regional cultures.

Development, power and the state

The nation-state remains a highly significant development actor in the South. During the middle years of the twentieth century, the state was seen as *the* institution to carry out development, as it was presumed to have its population's interests in mind and the organizational capacity to administer change. States varied, however, in their institutional make-up, political commitments and development visions. In many areas, colonization created heavily militarized priorities, political hierarchies and a limited public sphere for democratic debate, as in Latin America, Pakistan and sub-Saharan Africa. In other areas, governments – sometimes unelected – prioritized benefits for all, the reform of agriculture and a universal education system (e.g. South East Asia). Where a dynamic civil society was complemented by a strong legislative structure, as in Sri Lanka, more inclusive and equitable development outcomes were possible (Martinussen, 1997). Thus, state structures and their relationship with civil society (that is, citizens) affect the negotiation and implementation of development priorities. A 'developmentalist state' is one that combines an ability to organize a productive agricultural sector (through land redistribution and taxation), together with publicly supported principles of the rule of law, guaranteed territorial sovereignty, and independent judiciaries and administrators, to bring about general well-being (Leftwich, 1994).

During the 1980s, the role of the state in development was called into question by the rise of **neo-liberal** thinking that advocated policies that would promote deregulated markets in sales, property and labour (see Chapters 20 and 28). The existence of legislative controls on markets (e.g. subsidies on fuel for public transport and poor families), labour (e.g. protection of workers' working conditions and remuneration) and land (e.g. measures to protect collective territory) had characterized the mid-twentieth century model of development in the Global North and South. These measures were 'rolled back' under

Figure 21.3 Democratic decision making and development. Credit: Eric Miller/Still Pictures

neoliberal development models, although the extent and nature of market dominance that resulted did vary from country to country. As the neo-liberal agenda drove international development policy, particularly in the Bretton Woods institutions of the World Bank and IMF, so too third world states laid off state workers, privatized state-owned resources (mines, airlines, telecommunications, and so on), and reduced subsidies on basic food items, transport and electricity (Toye, 1993). According to the 'Washington consensus', the state was to concede its **developmentalist** role to the market through 'cutting back' its administration and oversight of development.

The past two decades have seen the normalization and extension of neo-liberal mantras in development, which has had a number of implications for the provision of public services (education, health) as well as for the state's structure and function. Three key aspects have been explored in geography:

1. Many countries in the South have experienced political decentralization in recent years, which is promoted as a means to create closer and more effective ties between government and the governed. Regional authorities, municipalities or ethnically-defined areas have been created or strengthened with tax-raising powers and decision-making rights. For example, the Bolivian Law of Popular Participation in 1994 redirected power to local organizations – including Indian communities – in a programme to provide 'growth with equity', combining a neo-liberal economic environment with access to decision making for mostly rural, marginalized poor populations.
2. Although it has now generally been recognized that nation-states do provide an irreplaceable role in providing some services and political guidance, the neo-liberal revolution dramatically transformed the political landscape of development. The widespread impacts of neo-liberalism on societies, as well as the forms of government used by neo-liberal states, have meant that geographers have become interested in understanding the nature of **civil society** and its role in development. Although defined as the individuals and groups active in opposition to or 'outside' the market and

the state, geographers have shown that civil society has often been central to maintaining societies and organization. On the one hand, neo-liberal development worked closely with **non-governmental organizations (NGOs)**, seeing them as a low-cost and efficient means of providing development. Mostly staffed by citizens of the country in which they work, NGOs stepped into welfare provision, extension work and credit and job-creation programmes that would, at one time, have fallen automatically to the state. NGOs vary greatly in their remit, structures and geographies, and can provide services of housing, healthcare, waste disposal and even education, often drawing on international funds. In Latin America, it is estimated that around 11,000 NGOs are working in multiple activities, from popular education to agricultural development, through to human rights work. NGOs have often been seen as closer to grassroots beneficiaries than other development agencies, and they can be innovative proponents of appropriate technology, participation and new development ideas. However, NGOs are not elected and they often become accountable to donors than to local people, which can pressure them to work to global agendas rather than local needs.

On the other hand, civil society has been the site for protest against neo-liberal measures (see also Chapter 22), and also crucially an electoral force. As countries became formal democracies through the 1980s and 1990s ordinary people became able to vote. Geographers continue to document how exclusionary political cultures marginalize or silence certain groups, such as ethnic minorities, women, rural populations or disabled people. Yet although national level politics often remain under the control of economic and political (frequently urban) elites, ordinary people's sense of themselves as citizens and as political actors means that the political geographies of development are dynamic and highly diverse. The 1980s and 1990s were a period of re-democratization around the world, as dictators and unelected governments fell in the wake of popular protest or collapse from within. While the formal aspects of democracy are not a guarantee of citizenship rights, civilian governments have risen dramatically in number from 25 per cent globally in 1973 to 68 per cent in 1992 (Leftwich, 1993: 614). As a result, the number of governments chosen in regular elections in the South has increased, as have the number of young voters. Many commentators view the 2011 'Arab Spring' as a civil society grassroots-led process of re-democratization. Re-democratization may result in the re-writing of constitutions and legislation, and the establishment of new rights for marginal groups. Nevertheless, the post-colonial context for development remains important: powerful countries in the North seek to provide aid to 'friendly' countries or to support particular types of democracy. Full citizen involvement in local decision making and the implementation of development often remains a distant promise rather than a reality. Nevertheless, despite neo-liberalism's efforts to encourage individualized responses to change, civil groups and networks remain active interlocutors in development processes (Escobar, 2008).

3. Neo-liberal development has been challenged by voters who have elected anti-neo-liberal leaders. These 'post-neo-liberal' states make investments in public infrastructure, give the state a central role in development planning, encourage civil participation in decision making (to counterbalance financial and political elites) and promote social redistribution

(to counter the inequalities of income that resulted from neo-liberal development at a number of scales). Such development experiments are most closely associated with Latin America (Brazil, Venezuela, Bolivia and Ecuador) but South Africa and other countries in Asia have also adopted a state-centred model of development (Sader, 2009).

SUMMARY

- States have remained important development factors over the past half century, although the nature of their intervention and the kinds of government used are highly variable spatially and socially.
- Civil society has become an important part of development 'delivery'. However, civil society is highly diverse and involved in a broad range of political, development and organizational activities.
- Re-democratization, decentralization and the emergence of anti-neo-liberal politics contribute to complex and dynamic decision-making processes, at a number of interlinked scales from the local, through the national, to the international.

Conclusions

Development geographies are changing quickly. Not only are the multifaceted processes undergone by countries in the South being transformed by successive waves of political-economic change and social relations of protest and accommodation, but also the analytical frameworks we bring to bear on these varied situations have diversified in recent years. Development geographers have engaged extensively with post-colonial perspectives, in ways that challenge our understanding of a neatly bounded category of 'developing countries' and that ask us to recognize the sheer diversity and groundedness of development experiences for individuals, groups and countries.

The gendering of development has also been a key aspect of recent development thinking. Whereas in previous decades, development was assumed to be gender neutral and applicable to men and women, recent decades have highlighted the male bias of much development planning and the highly differentiated impacts of development processes on femininities and masculinities. Politically, the context for decision making and the structures for administering development outcomes have also been radically re-thought in recent years. The shifting role of civil society, the decentralization of (some) budgets and (some) decision making have characterized many countries at the start of the twenty-first century. Overall, development remains a highly contested process, and however much the priority is to reduce poverty and bring an end to exclusionary political and development cultures, the means by which development is to be achieved remains the topic of highly passionate and globally significant debate. Under the umbrella of increased global integration, the complex geographies of development remain to be mapped and discussed, by participants and students of development alike.

DISCUSSION POINTS

1. Post-colonial approaches advocate listening to different voices in developing countries in order to get different kinds of information to that in World Bank reports. In their 2008 article 'The fiction of development: literary representation as a source of authoritative knowledge', Lewis, Rodgers and Woolcock argued that literary novels are useful sources of alternative knowledge about processes of development (or indeed of the lack of development). Examine one of the novels they list (see Further reading below) and compare that account with an official report.
2. What do we mean when we talk about femininities and masculinities in development? As an exercise, try to identify the ways in which male and female identities and differences are represented and practiced in the society you are most familiar with. Is it a useful framework for analysing how processes work in your society, and does it help us think about how society might change? You might like to think about this question in relation to divisions of labour, political representation and decision making at different scales and access to services such as education, health and training.
3. How has development changed in countries that experienced the 'Arab Spring'? Can you find out how their development models and priorities have changed since 2011? Which multilateral agencies and countries are funding or influencing their development? How are countries addressing the poverty and social inequality? And under what constraints?

FURTHER READING

Sheppard, E., Porter, P.W., Faust, D.R. and Nagar, R. (2009) *A World of Difference: Encountering and Contesting Development* (second edition). New York: Guildford Press.

An imaginatively written and empirically rich textbook talking about development conceived as a series of differentiated knowledges, livelihoods and social relations.

Escobar, A. (2008) *Territories of Difference: Place, Movement, Life, Redes.* London and Durham: Duke University Press.

Through a detailed ethnography of people and places in Colombia, this anthropologist of development focuses on how development is experienced under conditions of power. Escobar documents neo-liberal effects and resistance, using concepts familiar to geographers (nature, development, place, networks).

Radcliffe, S. A. (2006) Development and geography: gendered subjects in development processes and interventions. In: *Progress in Human Geography* 39(4), 524–32.

An introductory overview to how geographers are looking at gender relations in the developing world and policy.

Memmi, A. (2003 [1957]) *The Colonizer and the Colonized.* London: Earthscan.

A North African's vivid account of colonial oppression, which was written at a time of fierce contests over how to achieve decolonization.

Young, R.J.C. (2003) *Postcolonialism: A Very Short Introduction.* Oxford: Oxford University Press.

An accessible discussion of what it's like to be colonized, to experience formal decolonization and to live in a world still affected by colonial power. The book uses narrative, case studies and individual's life histories to introduce the main points of post-colonial theory.

Mohanty, C. (1991) Cartographies of struggle: third world women and the politics of feminism. In: C. Mohanty, A. Parker and A. Russo (eds.) *Third World Women and the Politics of Feminism.* London: Routledge.

A classic post-colonial feminist discussion of development (and development 'experts'') addresses third world women.

Lewis, D., Rodgers, D. and Woolcock, M. (2008) The fiction of development: literary representation as a source of authoritative knowledge. In: *Journal of Development Studies* 44(2), 198–216.

CHAPTER 22
SURVIVAL AND RESISTANCE

Paul Routledge

Introduction: the threat to people's survival

On New Year's Day 1994, ski-masked Mayan guerrillas emerge from the Lacandon jungle, capture the town of San Cristobal de las Casas in Chiapas, Mexico and declare war on the Mexican state.

In late April 1997, on the horizon before Brasilia, the hoes and machetes of thousands of *campesinos* glint in the sunshine as a movement of Brazil's landless peasants approaches the country's capital to demand land from the government.

In July 1999, on the Narmada river, India, a fleet of fishing boats, slogans written on their white sails, protest the construction of the mega-dams along the river valley, as part of a week-long 'rally for the valley'.

In December 2009, on the snow-covered streets of Copenhagen, Denmark, a multi-pronged swarm of demonstrations converge on the UN Climate talks being held in the city, demanding a radical climate justice agenda.

While very different, these four moments are examples of conflicts that are occurring in both the developed and developing countries. These conflicts represent mobilizations of the dispossessed, the poor, the marginalised: those threatened with, or experiencing, displacement, **cultural ethnocide**, climate change and the grinding ravages of poverty. They speak of struggles over the allocation of resources, over self-determination and over rights of economic and cultural survival, and it is to a consideration of these struggles that this chapter is devoted.

Within the global economy, economic development has had a variety of results. Certain locally based development practices have improved health and education services, enhanced environmental quality and generated employment opportunities that foster equity and self-sufficiency. However, many development practices are increasingly influenced by the ideology of **neo-liberalism** which is committed to 'free market' principles of free trade which privilege privatization and deregulation, while undermining or foreclosing alternative development models based upon social redistribution, economic rights or public investment (Peck and Tickell, 2002). Within particular countries, such development has emphasized economic growth, modernization and industrialization as the panacea for poverty. Capital intensive schemes have displaced traditional and subsistence economies which are labour intensive (often resulting in unemployment) and western values (of capitalist production, economic growth) have been emphasized at the expense of indigenous and traditional systems of knowledge, economy and culture. It has facilitated both the state's and transnational corporations' securing control over natural and financial resources and consolidated the power of those directing and benefiting from the development apparatus – national ruling elites and international institutions (Nandy, 1984).

Neo-liberalism entails the centralization of control of the world economy in the hands of transnational corporations and their allies in key government agencies (particularly those of the United States and other members of the G8), large international banks and international institutions such as the International Monetary Fund (IMF), the World Bank, and the World Trade Organization (WTO). These institutions enforce the doctrine of neo-liberalism enabling unrestricted access of transnational corporations (TNCs) to a wide range of markets (including public services), while potentially more progressive institutions and agreements (such as the International Labour Organization and the Kyoto Protocols) are allowed to wither (Peck and Tickell, 2002). Neo-liberal policies have resulted in the **pauperization** and marginalization of indigenous peoples, women, peasant farmers and industrial workers, and a reduction in labour, social and environmental conditions on a global basis – what Brecher and Costello (1994) term 'the race to the bottom' or 'global pillage'. Moreover, the neo-liberal global economy is based upon the extensive use of fossil fuels (for energy, travel, industry and agribusiness) causing serious ecological consequences, including climate change which disproportionately impacts already disadvantaged groups (Dalby, 2009).

In addition to the threats posed by neo-liberal development are those posed by the practices and policies of particular states. These may be aimed at enabling development practices to take place, or they may be aimed at securing the state's political (and economic) control over resources and territory. These territories are frequently inhabited by groups who perceive such state policies as an intrusion on their political and cultural rights. The assault upon the lifeworlds of these groups – which include indigenous peoples and peasants – has led to the emergence of myriad social movements who articulate struggles for political autonomy, and cultural, ecological and economic survival. This chapter considers the four examples of contemporary popular resistance that opened this chapter: the Landless Movement of Brazil, the Zapatistas of Mexico, the Save Narmada Movement of India, and emerging climate justice networks. It investigates how geography – and geographers – can lend important insights into these struggles.

Geography and resistance

Human Geography can lend some important understandings to people's resistances, providing valuable insights into the place-specific character of struggles, explaining why these conflicts arise, and why they emerge where they do. This is because different social groups endow space (and its associated resources) with a variety of different meanings, uses and values. Such differences can give rise to various tensions and conflicts within society over the uses of space for individual and social purposes, and the control of space by the state and other forms of economic and cultural power such as transnational corporations. As a result, particular places frequently become sites of conflict between different groups within society, which reflect concerns of ecology (e.g. struggles to prevent deforestation and pollution), economy (e.g. peasant struggles to secure land on which to grow food), culture (e.g. struggles to protect the integrity of indigenous people's communities) and politics (e.g. struggles for increased local autonomy). These concerns are also associated with what Gedicks (1993) terms the 'resource wars': the struggle over the remaining natural resources between indigenous and traditional peoples, state and national governments, and transnational corporations.

In response to these different concerns, people frequently organize themselves into social movements, which are ongoing collective efforts aimed at bringing about particular changes in a social order. Many of these social movements are place-specific, actively affirming local identity, culture and systems of knowledge as an integral part of their resistance. In doing so, these movements articulate localized '**terrains of resistance**' with their own place-specific idioms of protest, which motivate and inform their struggles. However, resistances are also becoming increasingly global in character, spanning both national and international space. This is because with the impacts of globalization –where localities are increasingly influenced by non-local economic and cultural forces – many social movements feel obliged to focus their resistance both within and beyond the confines of their immediate locality, in order to attract as wide a support to their struggle as possible. Such 'global justice networks' (Routledge and Cumbers, 2009) involve different social movements and **non-government organizations** (NGOs) that coordinate their struggles across a variety of scales in response to the global economy, the actions of particular governments and the threats of climate change.

SUMMARY

- Different social groups endow space and its resources with different meanings, uses and values arising in conflicts over the uses and control of particular places.
- Certain development practices have emphasized economic growth and industrialization as the solution to poverty, and viewed indigenous and traditional systems of knowledge as impediments to progress.
- As states and transnational corporations seek control over resources and territory for development, so the inhabitants of the areas concerned are frequently displaced and economically marginalized, while the environment is degraded.
- In response to such development processes, people often organize themselves into place-specific social movements to resist threats to their economic and cultural survival. Many such movements act across multiple scales and are increasingly involved in globalizing networks of resistance.
- Geography can provide insights into the place-specific character of resistances, explaining why these conflicts arise and why they emerge where they do.

Social movement struggles: resistance for survival

Social movements operate on a number of interrelated 'levels' within society. At the level of the economy, they articulate conflicts over access to productive natural resources such as forests and water, as well as conducting struggles in the workplace. The economic demands of social movements are not only concerned with a more equitable distribution of resources between competing groups, and the integrity of local, traditional forms of economic practice. They are also involved in the creation of new services such as health and education in rural areas (Guha, 1989). Indeed, social movements have emerged in many areas, including civil liberties, women's rights, and science and health, that are themselves often related to problems caused by the neo-liberal development. At the level of culture, social movement identities and solidarities are formed, for example, around issues of class, kinship, neighbourhood and the social networks of everyday life. Movement struggles are frequently cultural struggles over material conditions and needs, and over the practices and meanings of everyday life (Escobar, 2008). Social movements frequently affirm and regenerate local (place-specific) identity, knowledge and practices, which at times are expressed in the language and character of the struggles. Local resistance may incorporate local linguistic expressions, such as songs, poems and dramas, which imbue and affirm local experiences, beliefs and cultural practices.

At the level of politics, social movements challenge the state-centred character of the political process, articulating critiques of neo-liberal development ideology and of the role of the state. Movements are frequently autonomous of political parties (although some have formed working alliances with trade unions, voluntary organizations and non-government organizations). Their goals frequently articulate alternatives to the political process, political parties, the state and the capture of state power. By articulating concerns of justice and 'quality of life', these movements enlarge the conception of politics to include issues of gender, ethnicity and the autonomy and dignity of diverse individuals and groups (Guha, 1989). At the level of the environment, social movements are involved in struggles to protect local ecological niches – such as forests, rivers and ocean shorelines – from the threats to their environmental integrity through such processes as deforestation (e.g. for logging or cattle grazing purposes) and pollution (e.g. from industrial enterprises). Many of these social movements are also multidimensional, simultaneously addressing, for example, issues of poverty, ecology, gender and culture – such as the survival of peasant and tribal populations.

Social movements are by no means homogenous. A multiplicity of groups including squatter movements, neighbourhood groups, human rights organizations, women's associations, indigenous rights groups, self-help movements among the poor and unemployed, youth groups, educational and health associations and artist's movements are involved in various types of struggle. Many of these struggles take place within the realm of civil society, i.e. those areas of society that are neither part of the processes of material production in the economy nor part of state-funded organizations, and can be either violent or nonviolent in character.

While social movements in the developed and developing countries share some of the broad characteristics mentioned above, for instance they articulate such issues as ecology, gender and ethnicity, there are also important differences between them. In the developed countries, social movements have often concentrated upon 'quality of life' issues, whereas in developing countries, movements have often focused upon the access to economic resources. An example of this difference is represented by the issues faced by ecological movements. In the developed countries, the ecology movement has taken much of the industrial economy and consumer society for granted, working to preserve nature as an item of 'consumption', as a haven from the world of work. In the developing countries, however, those affected by environmental degradation – poor and landless peasants, women and tribes – are involved in struggles for economic and cultural survival rather than the quality of life. Such groups articulate an 'environmentalism of the poor' (Martinez-Allier, 2002), whose fundamental concerns are with the defence of livelihoods and of communal access to resources threatened by commodification, state take-overs and private appropriation (e.g. by national or transnational corporations) and with emancipation from material want and domination by others.

Social movements rarely articulate their demands at only one level. As we shall see below, economic struggles may also contain political dimensions, political struggles may also contain cultural elements, and so on. Moreover, the responses of state authorities to social movements vary according to the type of movement resistance and the character of the government involved. When faced with social movement challenges, governmental responses include repression, co-option, cooperation and accommodation. Repression can range from harassment and physical beatings, to imprisonment, torture and the killing of activists.

Zapatistas

Concerning resistance to neo-liberal development, the *Ejercito Zapatista Liberacion National* (the EZLN or the Zapatistas) in Chiapas, Mexico, has articulated resistance to the North American Free Trade Agreement (NAFTA) and the Mexican state (see Figure 22.1).

Figure 22.1 Although based in the Lacandon jungle, the Zapatistas have been able to globalize their resistance through creative use of the internet. Credit: epa/Juan Carlos Rojas

The Zapatistas, a predominantly indigenous (Mayan) guerrilla movement, have emerged in Chiapas due to a number of factors:

1. The state of Chiapas is rich in petroleum and lumber resources which have been ruthlessly exploited, causing deforestation and pollution.
2. The increasing orientation of capital intensive agriculture for the international market has led to the creation of a class of elite wealthy farmers and forced Indian communities to become peasant labour for the extraction and exploitation of resources, the wealth of which accrues to others. In addition, large landowners and ranchers control private armies which are used to force peasants off their land and to terrorize those with the temerity to resist.
3. Although it is resource-rich, Chiapas is among the poorest states in Mexico with 30 per cent of the population illiterate and 75 per cent of the population malnourished.
4. The production of two of the main crops from which *campesinos* (peasants) earn a living in Chiapas – coffee and corn – have undergone severe economic problems in recent years, and will be further damaged by NAFTA.
5. Government reforms in 1991 enabled previously protected individual and communal peasant land-holdings to be put up for sale to powerful cattle ranching, logging, mining and petroleum interests.

The Zapatistas initially engaged in a guerrilla insurgency by occupying the capital of Chiapas and several other prominent towns in the state, as noted at the beginning of the chapter. However, they staged their uprising in a spectacular manner to ensure maximum media coverage, and thus gain the attention of a variety of audiences, including civil society, the state, the national and international media and international finance markets. The appearance of an armed insurgency, at a moment when the Mexican economy was entering into a free trade agreement, enabled the Zapatistas to attract national and international media attention. Through their spokesperson, Subcommandante Marcos, the Zapatistas engaged in a 'war of words' with the Mexican government, fought primarily with rebel communiques (via newspapers and the internet), rather than bullets. Through their guerrilla insurgency and their war of words the Zapatistas have attempted to raise awareness concerning the unequal distribution of land and economic and political power in Chiapas; challenge the neo-liberal economic policies of the Mexican government; articulate an indigenous worldview which promotes Indian political autonomy; and articulate a call for the democratization of civil society. The success of the Zapatista struggle has lain in its ability, with limited resources and personnel, to disrupt international financial markets and their investments within Mexico, while exposing the inequities on which development and transnational liberalism is predicated (Harvey, 1995; Ross, 1995). In addition, although Zapatista guerrilla bases were in the Lacandon jungle in Chiapas, the Zapatistas were particularly concerned to globalize their resistance, and were the catalyst for the emergence of certain international networks of grassroots social movements challenging neo-liberalism, such as the World Social Forum (WSF).

The WSF, which commenced in 2001 in Porto Alegre in Brazil, is a gathering of diverse social movements, NGOs and trade unions from across the world who oppose neo-liberal globalization and the World Economic Forum (an annual meeting of political leaders, business and financial magnates held in Davos, Switzerland). They meet annually in different locations to discuss alternatives to neo-liberal globalization and engender a process of dialogue and reflection and the transnational exchange of experiences, ideas, strategies and information between the participants regarding their multi-scalar struggles (see Sen *et al.*, 2004).

However, despite certain successes, the movement has been faced with repression from the Mexican government. Over 15,000 army personnel have been deployed in Chiapas; villages suspected of being sympathetic to the Zapatistas have been bombed and peasants suspected of being Zapatistas have been arrested and tortured. The Zapatistas increasingly concentrated on providing health and social care in the autonomous regions they controlled and on developing agricultural cooperatives, and an uneasy stand-off with the government is in place. Since its emergence in 1994, the Zapatistas have posed more than just a political challenge to the Mexican state. In their demands for equitable distribution of land, their calls for indigenous rights and ecological preservation (such as an end to logging, a programme of reforestation, an end to water contamination of the jungle, and preservation of remaining virgin forest), they also articulate an economic, ecological and cultural struggle.

Movimento Sem Terra or Landless Movement

The *Movimento Sem Terra* (MST), or Landless Movement, in Brazil provides an interesting example of the organized struggle for access to land resources (see Figure 22.2).

Figure 22.2 Sem Terra activists hold aloft machetes and hoes, which have become symbols of their struggle for land. Credit: © Sebastiao Salgado/Amazonas/nbpictures

The MST is a mass social movement of some 220,000 members which has developed during the past 25 years, and is made up of Brazil's dispossessed – the croppers, casual pickers, farm labourers and people displaced from the land by mechanization and by land clearances. Many of those involved are homeless or live in roadside tents and earn less than sixty pence a day. Of Brazil's population of 165 million people, fewer than 50,000 own most of the land, while four million peasants share less than 3 per cent of the land. While approximately 32 million people in Brazil are malnourished, over 42 per cent of all privately-owned land in Brazil lies unused. Hence the principal demand among the dispossessed has been for land. The MST targets Brazil's vast estates that lie unused. First, groups of people illegally squat the uncultivated land in 'land invasions'. After the area has been 'secured' the MST resettles massive numbers of people on the squatted sites, who then construct houses and schools and commence farming. Since 1991, the MST has occupied 518 large ranches and resettled approximately 600,000 people. The process has been far from peaceful, as the large landowners and their private armies have attacked and killed the squatters. For example, in one incident in 1996, in the state of Para, 19 MST activists were shot dead by police in the pay of local landowners (Wright and Wolford, 2003).

While the right of the government to redistribute land that is not being farmed is enshrined in the Brazilian constitution, successive regimes have failed to exercise this right, due in part to the political power wielded by the country's landowners. However, the Brazilian government has begun to tentatively initiate agrarian reforms, providing credit for new settlements and confiscating some of the ranches in the state of Para to settle some of the families of those MST members killed in 1996. This is due, in large part, to the success of the MST in mobilizing popular support for its cause and its ability to develop alliances with trade unions and other grassroots organizations, both in Brazil and internationally. This is evidenced by its participation in the *La Via Campesina* network (the peasant way), an international farmer's network established in 1993 (Desmarais, 2007). Evidence of this support was dramatically shown at the mass demonstration in Brasilia in April 1997, mentioned at the beginning of this chapter, where over 120,000 landless people lined the streets to demand land reform (Vidal, 1997). In attempting to change the government's agrarian policy, the MST is also operating within the field of political action.

Narmada river valley project

Another example of a 'resource war' is that of the resistance against the Narmada river valley project in India (see Figure 22.3).

The Narmada river, which is regarded as sacred by the Hindu and tribal populations of India, spans the states of Madhya Pradesh, Maharashtra and Gujarat, and provides water resources for thousands of communities. The long-term government project envisages the construction of 30 major dams along the Narmada and its tributaries, as well as an additional 135 medium-sized and 3,000 minor dams. When completed, the project is expected to flood 33,947 acres of forest land and submerge an estimated 248 towns and villages. According to official estimates (based on the outdated 1981 census), over 100,000 people – 60 per cent of whom are tribal – will be forcibly evicted from their homes and lands. With two of the major dams already built, opposition to the project has been focused on the Sardar Sarovar reservoir, the largest of the project's individual schemes. The resistance

Figure 22.3 During the 'rally for the valley' fishermen conduct a boat protest on the Narmada river. Credit: Paul Routledge

to the project has been coordinated by the *Narmada Bachao Andolan* (Save Narmada Movement) which consists largely of peasant farmers and indigenous people and demands the curtailment of the scheme.

The movement's repertoire of protest has included mass demonstrations, road blockades, fasts, public meetings and disruption of construction activities. In addition, the movement has deployed peasant and tribal testimonies, songs and poems to give 'voice' to the struggle. While the movement has been almost completely nonviolent, its leaders and participants have been harassed, assaulted and jailed by police. While localized protests have occurred along the entire Narmada valley, wider public attention has been drawn to spectacular events such as mass demonstrations and the 'rally for the valley' mentioned at the beginning of the chapter (Routledge, 2003a). In addition, the movement has expanded its resistance to regional, national and international scales, developing networks of solidarity and support and participating in such initiatives as the World Social Forum.

However, despite the resistance, construction of the dams continues. In representing a threat to the ecology of the area surrounding the Narmada river, the construction of the dams also threatens the economic survival of the tribal and peasant peoples who will be evicted from their homes and lands – from which they earn their livelihoods – when the land is submerged. Moreover, these inhabitants have a profound religious connection to the landscape around the Narmada river. This spiritual connection to place – which eviction threatens to sever – intimately informs their customs and practices of everyday life. Hence opposition to the dam also articulates the inhabitants' desire for cultural survival. In addition, many of the villages that border the Narmada are demanding a level of regional autonomy, seeking 'our rule in our villages', thereby

Figure 22.4 Protesting against climate change. Credit: Paul Routledge

Figure 22.5 Getting the message across. Credit: Paul Routledge

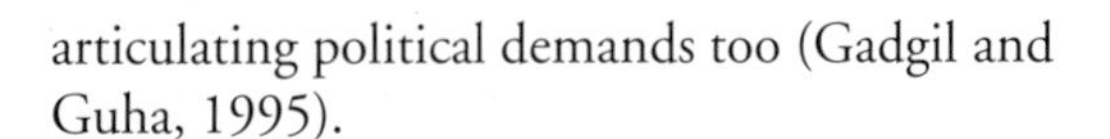

articulating political demands too (Gadgil and Guha, 1995).

More recently, as the mobilizations in Copenhagen attest, a range of differentially placed and resourced networks are forging solidarities concerned with issues of climate change and justice (e.g. see Routledge and Cumbers, 2009). Such 'climate justice networks', building upon those developed during the alter-globalization mobilizations, involve the prosecution of political action both within and beyond the scale of the State (see Figures 22.4 and 22.5).

Climate justice

The concept of 'climate justice' emerged from the Global South and refers to attempts to conceptualize the interrelationships between, and address the roots causes of, the social injustice, ecological destruction and economic domination perpetrated by the underlying logics of capitalism that has seen industrialized countries reap the benefits of fossil fuel-intensive development (e.g. see the Bali Principles of Climate Justice, 2002). Such processes are especially pertinent for the poor of the Global South who face multiple injustices. They are the victims of resource conflicts generated by capitalism's search for profits; they are located at the frontline of the effects of climate change; they possess small carbon footprints and they do not have the resources to mitigate against its effects.

Climate justice activism is currently focused around a coalition of grassroots social movements, including the US-based Indigenous Environment Network, *La Via Campesina*, and autonomous activist groups such as the UK's Camp for Climate Action and the Climate Justice Action Climate Justice Now! and Rising Tide networks (Wolf and Mueller, 2009). All of these groups were represented in the Copenhagen protests. The geographical reach of climate justice networks is exemplified by a global day of action organized on 24 October 2009 that saw 5200 protests in 181 countries unite in a call for an equitable and meaningful solution to the climate crisis (White, 2009).

An initial requirement for the construction of common ground between groups has been the construction of 'convergence spaces' (Routledge, 2003b; Cumbers *et al.*, 2008) where groups can meet one another, exchange experiences and plan collective strategies. For example, in Copenhagen, 2009, an alternative conference of activists and environmental practitioners called the *Klimaforum* lasted twelve days and featured 202 debates, seventy exhibitions and forty-three films covering a wide range of climate-related issues. It provided a space for bringing together activists from different struggles. In so doing it forged common ground exemplified by the *Klimaforum* Declaration. This declaration was signed by 466 civil society organizations (predominantly NGOs) and articulated a series of series of principles around which different campaigns concerned with issues of climate justice, located in different local and national contexts, could forge common ground as the basis for solidarity and cooperation. This declaration included a range of demands:

- leaving fossil fuels in the ground
- reasserting people's and community control over production
- re-localizing food production
- massively reducing over-consumption, particularly in the Global North
- respecting indigenous and forest people's rights
- recognizing the ecological and climate debt owed to the peoples in the Global South by the societies of the Global North, necessitating the making of reparations (see http://declaration.klimaforum.org/).

Climate justice networks articulate concerns at the political, economic, cultural and ecological levels. Moreover, such networks and each of the other struggles discussed in this chapter are examples of people's struggles that think and act both locally and globally.

SUMMARY

- Social movements articulate resistance within society in the realms of economics, politics, culture and the environment.
- At the economic level, social movements articulate conflicts over the productive resources in society, involving demands for a more equitable distribution of resources, the creation of new services and the integrity of local, traditional forms of economic practice.
- At the political level, social movements challenge the state-centred character of the political process, articulating critiques of neo-liberal development ideology and of the role of the state.

- At the cultural level, social movements frequently affirm and regenerate local (place-specific) identity, knowledge and practices, which at times are expressed in the language and character of the struggles.
- At the ecological level, social movements struggle to protect remaining environments from further destruction, and to ensure the economic (and cultural) survival of peasant and tribal populations.

Conclusion

The four examples described above are but a few of the resistances to transnational neoliberalism and repressive governments that have proliferated across the world during the past fifteen years. These have involved leftist guerrillas, social movements, non-government organizations, human rights groups, environmental organizations and indigenous people's movements. Frequently, coalitions have formed across national borders and across different political ideologies in attempts to revitalize democratic practices and public institutions, promote economic and environmental sustainability, encourage grassroots economic development and hold transnational corporations accountable to enforceable codes of conduct. Such resistances are frequently responses to local conditions that are in part the product of global forces, and resistance to these conditions has taken place at both the local and the global level. In contrast to official political discourse about the global economy, these challenges articulate a 'globalization from below' that comprises an evolving international network of groups, organizations and social movements.

Human Geography can provide valuable insights into the place-specific character of these different forms of resistance, explaining not only why conflicts emerge, but why they arise where they do. Hence, in our examples, the landless peasants of the MST have focused their struggle in those areas of Brazil that have large estates with unused land; the Narmada movement has been most active in those areas threatened by submergence by the construction of dams; and the Zapatistas have emerged in Mexico's poorest and most economically and ecologically exploited state. Geography can also provide insights into how and why movements act across multiple scales, for example by being part of international resistance networks such as those converging around issues of climate justice. As geographers, then, we can contribute to the understanding of struggles for survival in different cultural contexts. Given that we live within an increasingly interdependent world, we might also consider ways of attempting to contribute towards these struggles, to make our own contribution towards an environmentally sustainable and socially just world.

DISCUSSION POINTS

1. Why, and in what ways, do challenges to neo-liberalism articulate a 'globalization from below'?
2. What opportunities and/or constraints might social movements face when they wage struggle at local as well as global levels?
3. In what ways might geographers contribute towards struggles for a more environmentally sustainable and socially just world?

FURTHER READING

Peet, R. and Watts, M. (1996) *Liberation Ecologies: Environment, Development, Social Movements.* London: Routledge.

A fine collection of essays on the themes of development and social movements.

Routledge, P. and Cumbers, A. (2009) *Global Justice Networks: Geographies of Transnational Solidarity.* Manchester: Manchester University Press.

A discussion of globalization and resistance.

Notes from Nowhere (eds.) (2003) *We are Everywhere.* London: Verso.

Mertes, T. (ed.) (2004) *The Movement of Movements: A Reader.* London: Verso.

Juris, J.S. (2008) *Networking Futures: the Movements Against Corporate Globalization.* London: Duke University Press.

These texts offer recent overviews of grassroots social movements and initiatives around the world.

WEBSITES

http://climatecamp.org.uk

Camp for Climate Action

www.climate-justice-action.org

Climate Justice Action

www.climate-justice-now.org

Climate Justice Now

www.ienearth.org

Indigenous Environment Network

www.viacampesina.org

La Via Campesina

www.mstbrazil.org

Movimento Sem Terra

http://www.narmada.org/

Narmada Bachao Andolan

http://risingtide.org.uk

Rising Tide

www.forumsocialmundial.org.br

World Social Forum

www.actlab.utexas.edu/~zapatistas/

The Zapatistas

Alternative news sources for articles, commentaries and reports on many issues can be found at:

www.indymedia.org

Indymedia

www.zmag.org

ZNet

CHAPTER 23 HUMAN GEOGRAPHIES OF THE GLOBAL SOUTH

Katherine Brickell

Introduction

When I think back to my undergraduate degree, any lectures that mentioned countries and communities of the Global South – then referred to as the 'Third World' – usually began by pointing to a 'problem'. One of my most memorable courses, 'Poverty in Tropical Africa', did just this, and I learned about the problem of **poverty**, the problem of violence, the problem of inequality. While such problems *are* problems, and *should* be taught to and critically reflected upon by students, this should not mean that when you think about people's lives in the Global South the only human experience that you should imagine is material poverty.

A first reason why this is a mistaken overemphasis is that the Global South is a complex entity in terms of social and economic development, with entrenched poverty co-existing with a growing middle-class and emergent elite (Figure 23.1). Such contrasts, you may have noticed, re-surface in the UK media as fodder for debates over the politics of British aid, a notable example being the struggle to alleviate the plight of India's poor set against what is cast as its ruling elite's indulgent funding of a space programme. The second related point to make is that the Human Geographies of the Global South should not be limited to understanding inhabitants' responses to development challenges. While most students are introduced to the Global South through this narrow lens, there is a serious argument to be made that *all* the geography courses you take on social, cultural, economic and political geography should include scholarship from developing areas that moves beyond this poverty-only straightjacket. As this book emphasizes from its outset, a diverse range of topics make up human geographies, and this is just as true of the Global South – both internally and through its myriad connections to the Global North.

Bearing all of this in mind, this chapter examines some of the imaginative and fascinating work being produced by geographers and other social scientists engaged in research with communities of the Global South, in India and in a range of African countries. We use the 2008 Oscar-winning film *Slumdog Millionaire* as a vehicle to explore three

Figure 23.1 New wealth in Vietnam. Credit: Getty Images

related themes – on slum tourism, youth spaces and love. These themes are attracting a lively interdisciplinary scholarship which challenges the tendency to reduce studies of the Global South to the 'problems' of development, to the detriment of understanding diverse human experiences, aspirations and emotions.

Slum tourism

Slumdog Millionaire tells the story of Jamal Malik, a teenager, who grew up in a Mumbai slum and becomes a contestant on the Indian version of *Who Wants To Be A Millionaire?* Answering all the questions correctly, he is suspected of cheating and is arrested by the police. Under interrogation he explains through flashbacks to his childhood and growing up why he knows the answers. Since its phenomenal success the film has been subjected to both journalistic and academic critique for its shallow and impressionistic portrayal of poverty. It has been accused of fostering a reductive view of slum spaces and even an entire nation, fitting in with the broader idea proposed at the start of this chapter that the Global South is often subject to monolithic representation. In one of the best academic analyses of the film, Sengupta (2010) passionately contests the demeaning depictions of slums and slum-dwellers devoid of order, care and productivity. The motion picture begins, for example, with Jamal and his brother Salim being chased by aggressive guards off a private runway, running over corrugated roofs and through claustrophobic alleys before we are shown their everyday existences playing, working and even sleeping in piles of rubbish (Figure 23.2).

While commentators agree that such poverty *does* exist and its story *should* be told, what Sengupta (2010: 602) rails against is that 'it tells a misleading story of this poverty, and of resistance to this poverty, that does little service to the poor, but rather lends legitimacy to policies, practices and attitudes that further undermine their dignity and agency'. In response to the abject despair and total absence of associational life in the film (for example, no self-help group, no community organization,

Figure 23.2 Everyday existences searching for reusable scrap in Mumbai, India. Credit: FP/Getty Images

no recreational club and no charitable society), accusations have been made that it is essentially '**poverty porn**'. Indeed, the television media has not been exempt from such criticisms with the 2010 documentary *Slumming It* filmed in Dharavi (the slum used to film *Slumdog*) and shown on Channel 4 also berated by the press for making exploitative poverty porn to increase audience ratings (think also of Comic Relief's 2011 *Famous, Rich and in the Slums*). According to newspaper accounts (including sensationalist stories in the *Daily Mail*), the Indian tourism industry has denounced the show for creating an unrepresentative impression of the country in a programme they thought was actually meant to be about the architecture of Mumbai. Maybe why the Indian tourism industry felt the need to denounce this portrayal is that the camera is a powerful technology – it directly impacts upon international public perceptions of places beyond one's own. As Sengupta (2010: 600) explains, films (and other visual media) 'offer potential visitors a set of images about a country even before they arrive and, at times, compensate for visits altogether, leaving viewers with a sense of familiarity, even memories, about places they may never actually encounter' (see Chapter 53).

Such politics of representation also extends to the phenomenon of '**slum tourism**' and many commentators find it difficult to distinguish between a worthy initiative or just another example of voyeuristic 'poorism'. In fact *Slumdog* has actually initiated an increased demand for slum tours (Rolfes, 2010). From the cinema screen to reality tours, tourists seek a quest for real and authentic experiences. As *The Observer* reporter from Delhi writes in a 2006 article 'Slum tours: a day trip too far?':

> By the end of the walk, the group is beginning to feel overwhelmed by the smells

> of hot tar, urine and train oil. Have they found it interesting, Javed [the guide] asks? One person admits to feeling a little disappointed that they weren't able to see more children in action – picking up bottles, moving around in gangs. 'It's not like we want to peer at them in the zoo, like animals, but the point of the tour is to experience their lives', she says.
>
> Amelia Gentleman, The Observer, 7 May 2006

In fact I was once a 'slum tourist'. In the summer of 2000 after my first year at university I went to South Africa and was taken around by a local guide to the Nelson Mandela Camp in Soweto, Johannesburg, with the aim of experiencing 'real life' outside the lecture hall in the city I was visiting. While at this time there were very few companies running such tours, today slum tourism is big business, with a whole industry built up in metropolises of South Africa, Brazil and India. Geographers have been particularly active in researching this emergent type of tourism with the majority of studies tackling the question of why tourists choose to participate in slum tours. Some of the main motivations revealed include tourists' interest in a country's culture and its residents' living conditions, as well as the desire to experience the complexity and the diversity of the visited destination (Rolfes, 2010). This curiosity to see poverty however is also tempered in Rolfes' multi-sited study by expressed moral doubts and a certain sense of guilt in anticipation of such sightseeing. In similar research conducted by Meschkank (2011: 48), but specifically in Dharavi, the perplexing question that also arose was 'how a branch of tourism aimed at the sightseeing of poverty and squalor can be so successful?' While we might assume that tourists perceive slums as places purely of poverty, Meschkank discovered that after the tour, a change could be identified through a shifting of the negative semantics of poverty towards a more positive one. The tours transformed, at least in part, slums from places largely defined by their passivity, stagnation and desperation, to places also characterized by activity, development and entrepreneurship.

SUMMARY

- Film, television and other visual media are not neutral carriers of information. On account of financial gain or ideological bias they show a tendency to reinforce or generalize negative portrayals of the Global South as a whole, a particular country, city or group of peoples.
- The growth of slum tourism reflects many developing countries' promotion of tourism as a main strategy for achieving economic growth.
- Work in geography on the phenomenon of slum tourism tends to privilege the viewpoints of tourists over those of host communities.

Youth spaces

In the past ten years, geographers have brought to life a diverse set of experiences of urban life and have demonstrated the extraordinary perseverance, ingenuity and agency of **youth**. In this section we move beyond any general discussions of slums to consider smaller-scale spaces of the city which young people like Jamal are using and actively giving meaning to – the street corner and barbershop.

Space is an integral element in the constitution of daily life. As an everyday feature of the contemporary landscape, the street has held a longstanding fascination for geographers as a site of exclusion, resistance and social control, yet also one of work, leisure and socialization. It is especially significant in identity terms for un- or under-employed young men. The street is not just a space of marginality and crisis however, but can also be one of agency where young people forge meaningful social spaces than help them overcome the constraints they feel in their lives. According to Langevang's (2008: 227) study of the Ghanaian capital, Accra, you are immediately 'likely to be struck on the one hand by the masses of people moving around, and on the other hand by the crowds of young people – especially young men – hanging about on the street, and who never seem to move'. Langevang argues through interview research that:

> while hordes of young men hanging out at the so-called 'bases' [street meeting places] tend to be viewed by the surrounding world as either potentially dangerous or as a sign of marginalization and immobility', for the young men themselves, these places are also full of motion and serve to orient their lives socially and materially through practices of sitting, discussing and networking.
>
> Langevang (2008: 227)

This is partly because in recent years, many of these 'bases' have consolidated their presence by turning into youth clubs so 'rather than just spending their time sitting around doing nothing or fighting, they are showing the surrounding society that they are responsible citizens who are doing something useful for their community' (Langevang, 2008: 238). This involves strengthening the ties and loyalties between members and improving the relationship between base members and older members of the community through the use of membership fees to build wooden benches painted with the club's logo and signboards to construct a more positive image.

The second everyday space we are going to look at is the barbershop, another gathering or meeting place for young people. In the context of Tanzania, Weiss (2002; 2009) argues that one of the most important ways in which educated young men in Arusha have tried to negotiate the uncertainties of widespread unemployment is through establishing barbershops. These barbershops offer minimal financial return but nevertheless are centres of male friendship where information, both local and international is consumed. Rather than fulfilling a stereotype of disruptive immobile youth, Weiss contends that these spaces provide unemployed young men with opportunities to project their sense of global connectivity despite their marginality. And they do this working and hanging out in shops named after aspirational locations such as 'Brooklyn Barbers', named after a suburb of New York (Figure 23.3). So although you once might have just thought barbershops as nothing more than informal businesses filling the streets of Arusha, these actually function as hubs for young men to 'coiffure' their futures.

Weiss (2002: 95) describes, for example, young men who use the circulation of newspapers, clothing and tailored outfits, to display their transnational connections:

Figure 23.3 Brooklyn Barbershop in Arusha, Tanzania. Credit: Brad Weiss

> The staff and clientele at Classic are interested in much more than haircuts. They spend most of their time in the shop reading an assortment of daily and weekly papers, listening to music performed in French, Lingala, English, and occasionally Swahili, and watching satellite television broadcasts from the United Kingdom, South Africa, India and the United States.
>
> Weiss (2002: 95)

The barber shop therefore becomes an important space for the dissemination of varieties of media from across the world as well as a major part of youth's public presentation of the self, a self which they style through their contact with the global flow of images, objects and persons that pass through these shops. In this way, youth can be cast as transnational actors whose geographies are much more complex than simply their micro-connections with the street (van Blerk, 2005). Young men use barbershops as a means not to just get a haircut, but to transport them imaginatively, if not physically, to places of the modernity and perceived success that they crave.

CASE STUDY

The call centre

Towards the end of *Slumdog Millionaire* Jamal becomes a 'tea boy' in a call centre, an instantly recognizable marker of globalization in India. The development of information and communications technologies (ICT) has facilitated the emergence of a complex global urban system in which Mumbai, like other Indian cities, plays an integral role. Indeed, the relocation of business processes and services based upon ICT from the 'developed' Global North to the so-called 'developing' Global South is no longer in its infancy. The burgeoning call centre industry in India has been hailed as having played a major part in transforming the country's economy from one of slow growth to fast growth.

However, according to Vira and James (2011), the call centre represents an empirical focus that falls outside the domains of both traditional economic geography and development geography:

> For economic geographers, this is a space that falls outside the western core of developed capitalism. Yet the highly sophisticated technological nature of the call centre industry has also served to keep it out of the remit of traditional development research, which instead tends to focus on rural society and agrarian structures, manufacturing industry, and the

ROY NIXON
"FOR A CALL CENTRE IN BANGALORE PRESS 1 · FOR ONE IN HYDERBAD PRESS 2 · FOR ANY OTHER CALL CENTRE..."

informal sector in both rural and urban settings, usually contexts characterized by a lack of material sophistication in terms of technologies or infrastructure.

Vira and James (2011)

While it is argued that call centres are thus 'no-go zone for geographers', they have nevertheless captured the attention of cartoonists (Figure 23.4) and a supposedly 'enraptured urban middle-class India':

It has all the dramatic elements of a Bollywood movie: the struggle between instant and delayed gratification, arranged marriages and romantic love, consumerist and ascetic religious values, and the ominous prospect of generational rupture.

(Nadeem, 2009: 103)

Figure 23.4 **'For a call centre in Bangalore press 1; for Hyderabad press . . .' Credit: www.CartoonStock.com**

SUMMARY

- Youth mediate between different cultures or social worlds. They experience the social relations that are located in the place in which they are corporeally standing, but also transcend place, to experience social relations that are located in places elsewhere through media and clothing.
- By taking into account both imagined as well as physical mobility it becomes easier to disrupt the erroneous dichotomy identified between the Global North and Global South.

Love, lust and intimacy

There are 19 million people in this city Jamal. Forget about her. She is history.

Source: Salim in *Slumdog Millionaire*

While in the courses you study on developing areas you are likely to be asked to write an essay evaluating policies and initiatives designed to address reproductive health issues or HIV/AIDS, you are probably less likely to be asked to write a piece on the more emotionally charged topic of 'love' – 'the sentiments of attachment and affiliation that bind people to one another' (Cole and Thomas, 2009: 2).

Romantic love, especially, tends to be cast as a Eurocentric concept, which has been exported to other parts of the world, an argument which is premised on emotion being an indulgence in situations of poverty. In *Slumdog* however we witness the pain that Jamal experiences, and the desperate risks he takes, to be reunited with his childhood companion and 'lost love' in the mega-city of Mumbai. Indeed, Jamal's quest sits against what Sengupta (2010: 604) sees as a 'barren world' presented in the film 'without familial bonds of any depth, or even specific individuals capable of empathy and compassion'.

Powerful correctives to love as a luxury thesis are now emerging. In their edited collection *Love in Africa*, Thomas and Cole (2009: 8) argue that although love is commonly depicted as absent in Africa, lust is not, thus erroneously situating blacks as morally and spiritually inadequate in comparison to their European counterparts. There is a long history of Westerners deploying such arguments of hypersexuality to dehumanise Africans and to justify degrading policies, with love constructed as part of a civilizing and development process. The Kenyan author Binyavanga Wainaina (2005) highlighted this historical denial of love several years ago in a comedic yet cutting piece entitled 'How to write about Africa'. For the author who writes about Africa it is apparently necessary to avoid at all costs 'taboo subjects: ordinary domestic scenes, love between Africans (unless a death is involved)' (Wainaina, 2005). Indeed, Thomas and Cole (2009) concur with the mocking nature of Wainaina's analysis and argue that despite the absence of love in writing about Africa for a Western and global audience, in-continent its popularity in novels and songs is huge. This absence of emotion from discussions on Africanist scholarship on intimate relations is, according to the authors, an absence that has become ever more startling since the late 1980s with the explosion of analyses of the HIV/AIDS epidemic. As they argue: 'whereas countless studies have analysed how sexual behaviour fuels the epidemic, few have explored how that behaviour is embedded in emotional frameworks' (Cole and Thomas, 2009: 4).

An important corrective to this bias is Mark Hunter's (2010) book *Love in the Time of Aids*, an ethnography that examines why, in KwaZulu-Natal, South Africa, AIDS emerged so quickly. While an influential explanation for AIDS in Africa has always been men's long history of circular migration to the gold and diamond mines, Hunter argues that this analysis fails to capture other significant factors at work. Although policy makers tend to see low condom use in narrow terms of 'male power', it is often in affairs between 'boyfriends' and 'girlfriends' – positioned as being about love – where men and women are least likely to use condoms and in the most commodified relationships, prostitution, where condoms are used the most. Hunter (2002) describes how men often use subtle discourses of persuasion with their girlfriends to have unprotected sex. As Xolani, age twenty-three, explains:

> Men say: 'when you tell me that you want to use a condom it is because you don't trust me. I love you so much. Why use a condom when I am faithful to you . . . you must be sleeping with another guy, that is why you want me to use a condom.' They just trap you so that you feel guilty. You just say, 'OK'.
>
> (Hunter, 2002: 109)

As Xolani's experiences suggest, men will try to convince women that using a condom represents unfaithfulness and that true love is symbolized by flesh to flesh sex. As this research powerfully illustrates then, love is not too frivolous an object of enquiry for geographers (see Chapter 49). Moreover, while love is often cast as an entirely private affair, it has wide-ranging material and development implications, with the public crisis over HIV/AIDS a stark example.

SUMMARY

- In Geography to date there has been a failure to consider love in either the Global South or Global North, resulting in a serious lack of understanding about this vital element of social life.
- Meanings of love and intimacy influence sexual practices and the shape of HIV transmission patterns.

Conclusion

This chapter has introduced you to some of the Human Geographies of the Global South which lay outside the usual 'development' framework. The examples drawn on have showcased the vitality of research being produced by geographers working in a whole host of countries from India to South Africa, Tanzania to Ghana. These have revealed some of the more 'everyday' geographies experienced by some of the inhabitants of the Global South, as they seek to negotiate and make their own a range of public and private spaces. In time, work on these topics should not just be restricted to specialist development courses, but should become a component in a range of broader Geography courses. Ater all, they represent the day-to-day experiences of the majority of the world's population.

DISCUSSION POINTS

1. What do you think the motivations are for those who take part in slum tourism? How would you feel about undertaking such tours?
2. Apart from Hollywood films or pop music emanating from the USA or UK, what other international cultural flows of film and video are there?
3. What types of love, except for romantic, do you think would be important to research and why?
4. Do you think there is any point talking about the Human Geographies of the Global South, given its diversity?

FURTHER READING

Rigg. J. (2007) *An Everyday Geography of the Global South*. London: Taylor and Francis.

Williams, G, Meth, P. and Willis, K. (2009) *Geographies of Developing Areas – The Global South in a Changing World*. London: Routledge.

These two books are fantastic introductions to the Global South beyond a poverty-only focus.

Jeffrey, C. (2010) *Timepass: Youth, Class and the Politics of Waiting*. Stanford: Stanford University Press and Cambridge University Press.

Hansen, L., Dalsgaard, A.L. and Gough, K.V. (2008) *Youth and the City in the Global South*. Indiana: Indiana Press.

These two books are excellent for geographers interested in the effects of globalization and neoliberalism on the everyday experiences and future prospects of urban youth in the developing world.

Padilla, M.B, Hirsch, J.S., Munoz-Laboy, M., Sember, R.E. and Parker, R.G. (eds.) (2007) *Love and Globalization: Transformations of Intimacy in the Contemporary World*. Nashville: Vanderbilt University Press.

Valentine, G. (2008) The ties that bind: Towards geographies of intimacy. In: *Geography Compass*, 2 (6), 2097–2110.

These texts offer a good introduction to scholarship on love and intimacy.

SECTION FIVE

ECONOMIC GEOGRAPHIES

Introduction

The contemporary economy is an extraordinarily complex set of processes, operating in and around a huge variety of institutions and activities. It embraces everything from a teenager receiving and spending pocket money to the most advanced manufacturing technologies in the world being employed by global corporations. It touches almost every aspect of our daily lives and directly affects where we live, how we earn money, what we eat, how we dress and how we get around. We are surrounded and confronted by advertisers extolling us to purchase their products, and spend ages agonizing over which ones to buy. Huge swathes of our towns and countryside are devoted to the production of goods and services. Even the most pastoral rural scene of animals grazing in a field is shot through with economic relations and processes that connect the small farm in the UK to a global food industry.

The usual way in which geography has dealt with such complexity is to break it down by economic activity or sector – agricultural geography, transport geography, industrial geography, and the geographies of trade and services, for instance. This section introduces you to a different way of breaking up the economic world into manageable chunks – not by sectors of economic activity but into different parts of a single unified process, covering production, money, consumption, commodities and globalization. Within the contemporary economy, the dominant sets of relations are **capitalist** in nature, and thus the unified process that represents most of the global economy is known as the circuit of capital, as set out diagrammatically in Figure V.

Money (M) is placed into the circuit at the top of the circle by those who wish to invest. This largely takes place in and through financial centres like the City of London or the New York Stock Exchange on Wall Street. Moving clockwise around our diagrammatic circle, this money, fictitious or real, is then used to purchase **commodities** (C) in the form of labour power (LP) – say, car workers – and the means of production (MP) – say, an assembly line and bits of steel, rubber and various other metals. These are then combined in a production process (P) – say, in a car factory – which produced further commodities (C') – say, in the form of a car. This new commodity is then sold, for more money (M') than was originally invested (our initial M). The difference between M and M' is known as surplus value (S), or more usually in everyday language as profit. This amount is ready to re-invest in a further round of production. The realization of a surplus is the rationale behind the capitalist economy – those firms, organizations and individuals who do not manage to do this will quite simply go bust. Those who manage it on a regular basis will thrive and prosper. Under capitalism, therefore, production, consumption and exchange are all combined as the means to an end of making a

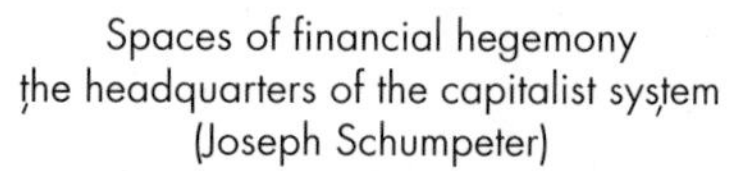

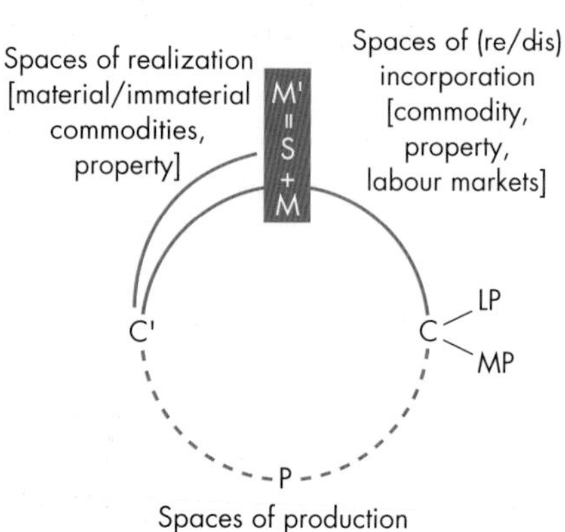

Figure V The circuit of capital. Source: Lee (2005a)

profit – or generating surplus. But as we shall see over the next five chapters, this process is never quite as simple as this, or as linear, and indeed the geographies of the various elements of the process are intrinsic to how the process works (or doesn't work).

The geographies of all this are highly dynamic and mobile since one of the defining features of capitalism is that its key component parts – capital and labour – are notionally free. The creation of value, or a surplus, takes place in and through specific economic spaces. We can break these spaces down into four main kinds: those of production, those of money and finance, those of consumption and those of commodities. As we shall see in Chapter 28, these spaces are being combined and bought together in an increasingly globalized fashion, but first, in Chapter 24, Kris Olds explores the geographies of production. He points out that production is not just confined to the formal and traditional economic spaces of the factory and the office, but instead is taking place through the production of knowledge, via networks which stretch across the globe. Moreover, as production itself has changed, so have our ideas about it, and Olds also traces how geographers have developed new ways of thinking about production. In Chapter 25, Sarah Hall looks at what finances such production in the first place by interrogating the notion of money and finance. She shows how money, although footloose and highly mobile, has a very particular geography. A small number of world financial centres, which are the major source of investment capital, form a very tight core within the global finance industry. She then shows how this global situation is mirrored at the national level, where financial services tend, again, to be concentrated and how this in turn leads to increased 'financial exclusion' when large segments of the community are denied access to these services. By contrast, in Chapter 26, Juliana Mansvelt discusses the pervasive nature of consumption and reproduction and seeks to understand the ways in which consumption connects people, places and things. She shows how consumption both requires and makes particular geographies, which may extend from the individual to the global. In Chapter 27 Michael Watts focuses more closely on one element of consumption – that entailed in the production of commodities. He looks at how commodities are taken for granted – indeed, these are the 'things' which we buy and consume everyday day – when in fact they hide a complex array of social relationships which stretch across the globe. He also shows how the environment itself is becoming increasingly commodified – and how such a system of commodification is prone to economic crisis.

Through the act of consumption we close one circuit of capital and help to contribute to the beginnings of the next one. But the chapters all show how this is not just an economic process and is instead one that is rooted in key sets of social and cultural processes. The crucial question for geographers to explore is how these processes combine, and can only work, through a whole host of very precise spatial settings. In this sense the economy only becomes realized in particular spaces. It never takes place in the abstract, as in Figure IV but different sections of the circuit are constantly being played out to different degrees in different places. Yet as all the chapters indicate, this is not a simple mapping of economic activities on to space – the spaces are themselves constitutive of the economies and are crucial to their success or failure. Thus in Chapter 28, Andrew Jones explores how such economic spaces have become increasingly stretched and globalized. Even though processes of economic globalization have a long history, he shows that we are now witnessing their intensification, which in turn is starting to produce a shift in the patterns of both economic activity and economic power. Although Europe and North America may have dominated the world economy for the past few centuries, this pattern is unlikely to prevail in the future and the emergence of Asia and South America as locations of significant economic activity make a geographical perspective on the economy more necessary than ever.

FURTHER READING

Mackinnon, D. and Cumbers, A. (2011) *Introduction to Economic Geography*. Harlow: Pearson.

This does what the title says and provides an introduction to the major topics covered in economic geography.

Leyshon, A., Lee, R., McDowell, L. and Sunley, P. (eds.) (2011) *The Sage Handbook of Economic Geography*. London: Sage.

This is a comprehensive collection of new chapters by a range of leading economic geographers.

Barnes, T.J., Peck, J., Sheppard, E. and Tickell, A. (eds.) (2008) *Reading Economic Geography*. Oxford: Blackwell.

This usefully draws together a collection of previously published chapters in one place.

For the most up-to-date writings on economic geography you should scan recent editions of the journal *Economic Geography*. This will give you a sense of the current topics being pursued in this field.

CHAPTER 24
SPACES OF PRODUCTION

Kris Olds

Introduction

This is a fascinating time for students to take a geography course. From concerns about environmental change to migration and economic globalization, the development and application of a geographical imagination can shed much light on the nature of the processes in focus, and their role in generating uneven development patterns and outcomes.

The focus of this chapter is on the economic geographies of production – the 'intentional application of labour' in the creation of transformations (from the miniscule to the systemic) that can be monetized (Lee, 2005b). As Lee pointed out, *production* needs to be related to *exchange* and *consumption*: the three moments are interrelated – they depend upon each other for they are co-constitutive (see also Chapters 26 and 27). Lee used the notion of 'circuits of value' to capture this interrelationship and suggested that it is 'social relations' that enable this circuit of value to be brought to life. Social relations is a neutral term, for such relations can be mutually supportive and reciprocal or very uneven and exploitative. As economic geographers have pointed out for decades, these social relations differ from place to place at a range of scales (from the local to the global) and they are historically specific. It is also worth noting that the creation of a transformation that can be monetized is not necessarily the goal above all else: monetization can provide the leverage to scale up **innovations**, for example, in the development of means to treat multiple sclerosis, democratize knowledge via new open-source technologies or address global challenges like climate change via innovations in carbon sequestration.

In this chapter I will focus on enhancing understandings of intentional application of labour in transformations associated with the production of space, knowledge, knowledgeable people and ideas about globalized production. This is an admittedly wide-ranging and multi-level approach to the 'spaces of production' theme (which have traditionally focused on material items), but a narrow view on production is a problem when conceiving of the fullness of life, the development process, and the diverse nature of the space economy.

This chapter is built up around the development of two disparate analytical vignettes – mini-case studies really – which progressively become more abstract and global in scale. These vignettes are not all encompassing, but they capture the full essence of how to conceptualize spaces of production from a Human Geographic perspective. Instead, the vignettes and the concluding comments are meant to illustrate the application of a geographic imagination to the phenomenon of production, broadly defined, and sensitize students to the array of issues that could be factored in when thinking about the topic.

The first vignette focuses on the early days of the development of a new space of knowledge production in a classic global city – New York. The vignette unfolds on Roosevelt Island, a 60-hectare island with a history dating back to the pre-nation state era when the Canarsie Indians controlled it. Colonial geographies (see Chapter 33) brought the island under the sway of the Dutch and then British merchants, before it became formally governed by the City of New York and known as a dumping ground for destitute, sick and criminalized people (including, for a while, the famous singer, Billy Holiday, and as well as the then bawdy entertainer Mae West). As was the case with respect to London in the seventeenth and eighteenth centuries (brilliantly captured in Miles Ogborn's *Spaces of Modernity: London's Geographies, 1680-1780*), institutions on Roosevelt Island like the New York Lunatic Asylum were designed to socialize and reform people; to produce new citizen-subjects for the 'modern' (at this time) era. It is not for nothing that the island was known as Welfare Island from 1921 to 1973.

Following a period of decline until the 1960s, and then stability as a residential space, the City of New York, under Mayor Michael Bloomberg's leadership, decided to transform a large part of Roosevelt Island into a space of production, not of materials (in the classic factory sense), but of ideas, knowledge, knowledgeable people and revenue via commercialization processes linking universities and industry. This is a different aspirational modernity, clearly, than the modernity associated with social reform and welfare on Roosevelt Island (pre-1970s). The vignette in this chapter outlines how this process was launchedand how tangible shifts in development strategy are reshaping the site's geographies in a myriad of ways.

The second vignette is an analytical one, more abstract in nature, and deals with the globalization of economic activity. The vignette focuses on the production of a now influential concept in Human Geography – Global Production Networks (GPNs) – designed to make sense of spaces of production, especially with respect to manufacturing. The GPN concept emerged in the early 2000s and has helped spur on a significant volume of research about sectors including automotive production, personal computers, cut flowers, and so on. The purpose of developing this vignette is to convey how Human Geographers think about spaces of production and how geographical concepts develop around changes in the world we are studying, while also conveying empirical material about the production process itself. By tacking back and forth – between the production of an economic geographic concept and the phenomenon of production – I hope to convey the productive tension between the political economy of production and the transformation of economic geographic ideas for making sense of this world. The chapter concludes with some caveats about what is absent from this particular take on spaces of production.

Producing knowledge, knowledgeable people and skilled workers via a new space (Applied Sciences NYC)

In late June 2012, Deborah Estrin, an esteemed professor and founder of UCLA's Center for Embedded Networked Sensing, became the first professor to be hired at a new $2 billion USD tech campus on Roosevelt Island in New York City, known as Applied Sciences NYC.

Professor Estrin, a computer scientist and engineer, is one of the world's leading thinkers regarding network and routing protocols in association with the global systems that underlie, and enable, society to coordinate and analyse the development process (at multiple levels and in multiple sectors). As her personal website states, Estrin's current work focuses on 'participatory sensing systems, leveraging the location, activity, image, and user-contributed data streams increasingly available from mobile phones'. *The Verge*, a NY-based media outlet, notes that Estrin was called by CNN one of the '50 People Who Will Change the World' as well as one of the '10 Most Powerful Women in Tech' (Carmody, 2012). On this latter point, a significant amount of noise emerged in the Twittersphere about this appointment, frequently associated with the #womenintech hashtag. The underlying cause of this interest is that the USA, and indeed most other countries around the world, is facing major challenges redressing the gender imbalance associated with the Science, Technology, Engineering and Mathematics (STEM) fields.

The hiring of the first faculty member in association with Applied Sciences NYC sounds like a mundane event: after all, faculty members are hired every day around the world, including in the institution you are enrolled in. However, from a geographical perspective, Estrin's hire is one of those formative moments, because it is the peopling of a new space of now prioritized knowledge production. After all, people need to be in place to use technologies and their networks to gather information and transform it into knowledge that can be utilized, scaled up and sometimes monetized. Estrin brings along a particular skill set and a pedagogical approach that is itself very geographical in nature for it requires students to learn outside of the classroom, in on-campus incubators, and in places of work via cooperative education programmes that are often linked up with spin-off firms:

> She'll be teaching courses on networking, Internet architecture and mobile computing – what she called her 'genre' of material. However, it's the school's apprenticeship program that has her most excited about the new position.
>
> 'Students are going to be able to be more creative and inventive, doing deeper dives into what they're studying', said Estrin.
>
> 'Having New York City – along with its thriving technology sector and startup ethos – as a background doesn't hurt either', she added.
>
> 'Why do people come to New York,' Estin asked. 'It's because of energy. And [in New York], it's allowed a startup culture to grow in an urban environment. I think there's a tremendous opportunity [with Cornell NYC Tech] to be a part of the startup community and to give them an outlet and a focus that expands it and expands the business of the city.'
>
> (Fitzpatrick, 2012)

But what is Applied Sciences NYC and how did it come to emerge as a space of production, circa 2011-2012, for Deborah Estrin and everyone who will follow after her?

Just over a year ago, in December 2010, New York City stirred up interest around the world with the issue of a Request for Expressions of Interest seeking 'Responses from the Academic World for a Partnership with the City to Create a State-of-the-Art Applied Science Campus.' As the December 2010 press release put it:

> Mayor Michael R. Bloomberg, Deputy Mayor for Economic Development, Robert K. Steel and New York City Economic Development Corporation President Seth W. Pinsky today announced that the City is seeking responses from a university, applied science organization or related institution to develop and operate an applied sciences research facility in New York City. In order to maintain a diverse and competitive economy, and capture the considerable growth occurring within the science, technology and research fields, the City is looking to strengthen its applied sciences capabilities, particularly in fields which lend themselves to commercialization. The City will make a capital contribution, in addition to possibly providing land and other considerations, commensurate with the respondent's investment.
>
> (City of New York, 16 December 2010)

In the end, 18 responses from a total of 27 institutions were received, including universities located in Finland, India, the USA, England, Canada, Israel, South Korea and Switzerland (see Olds (2012) for more details). No other jurisdiction has ever received such an expression of interest in such a concentrated period of time, highlighting the hold New York City has on imaginations worldwide.

As Mayor Bloomberg noted at the same time:

> 'We were enormously optimistic that this once-in-a-generation opportunity would draw the interest of top caliber universities from New York City, the region and the world, and the number and breadth of responses is as strong an endorsement of the idea as we could have hoped for,' said Mayor Bloomberg. 'The institutions that responded recognize the historic opportunity this initiative represents – to grow a presence in the world's most dynamic, creative and globally connected city. For New York City, it's an opportunity to increase dramatically our potential for economic growth – a game-changer for our economy.'
>
> (City of New York, 17 March 2011)

To cut a long story short, these expressions of interest were vetted and used to develop a comprehensive Applied Sciences Request for Proposals (RFP) to develop the applied sciences campus. The RFP, which was issued on 19 July 2011, had a closing date of 28 October 2011.

On 31 October 2011, the City of New York issued another press release noting that the City had received seven full proposals put together by these 17 higher education institutes:

- Amity University
- Carnegie Mellon University/Steiner Studios
- Columbia University
- Cornell University/Technion-Israel Institute of Technology
- New York University/University of Toronto/University of Warwick (UK)/The Indian Institute of Technology, Bombay/City University of New York and Carnegie Mellon
- New York Genome Center/Mount Sinai School of Medicine/Rockefeller University/SUNY Stony Brook
- Stanford University/City College of New York.

By mid-December 2011 a decision was made that one proposal (put together by Cornell University from Ithaca in upstate New York

Figure 24.1 Cornell NY Tech on Roosevelt Island. Credit: © Cornell University

state in partnership with Technion-Israel Institute of Technology) would get the formal go-ahead. The winning bid put forward a 100 per cent new campus of 185,000 square metres for approximately 2,500 students and 280 faculty members (with Deborah Estrin being the first). The City of New York provided access to City-owned land on Roosevelt Island and up to $100 million of City capital.

When completed, the City of New York notes that 70 per cent more engineering (including computer engineering) students will be based in New York's higher education landscape, with a substantial increase in the numbers of high-tech spin-off companies. Applied Sciences NYC is therefore designed to produce new labourers who will then support the emergence of new segments of the knowledge economy *not* associated with finance (on the geographies of finance, see Chapter 25). In other words, this is a state-led but university-enabled drive to refashion the nature of production in New York's post-industrial economy.

From a spaces of production perspective, Applied Sciences NYC is also noteworthy because it is generative of the following:

- The establishment of new research and teaching programs that will have significantly more latitude for configuration given that this is a purpose-built campus. In short, some distance from the main campuses of both Cornell and Technion is designed to help the two universities break free from established practices and unwritten codes of convention on their main campuses.
- The creation of physical presence in New York to enable the forging of informal

relations of trust and interdependency with city-based actors (be they US or foreign). This is, for example, one of the benefits reaped by some of the foreign universities that established a relatively deep presence in the global city of Singapore in 2001 (see Olds and Thrift, 2005). The unruly processes associated with innovation often take time: they often occur by accident and serendipity often comes into play. Serendipity is better facilitated, if indeed it can be facilitated, via proximity and territorial co-presence.

- The creation of a space of production that ties into the archipelago of network-forged nodes Cornell and Technion have established over the years.

Applied Sciences NYC is a long-term experiment in reconfiguring the university-territory relationship for the purpose of creating a novel space of production – of computer and engineering 'talent', of discoveries that can be monetized, and of innovations that will (in theory) leave a mark on the development process at local, urban, national and global scales.

SUMMARY

- As Western economies evolve, some contexts, especially cities, increasingly focus on the development and diversification of the services sector.
- University–industry linkages are posited in cities around the world to facilitate the innovation process via the production of knowledge that can be monetized.
- Applied Sciences NYC is an attempt by the state, in association with two leading international universities (Cornell and Technion), to engender structural change in the local labour market, especially with respect to the growth of the computer science and engineering sector.

People, institutions and making sense of global production networks

While the first vignette began with the peopling of Applied Sciences NYC in 2012, the second vignette focuses on the people, institutions and networks that propelled the development of a now influential concept in Human Geography known as global production networks (GPNs) that is designed to shed light on the globalization of production processes. Through this vignette, we will explore the relationship between the globalization of economic activity and the production of ideas about the changing geographies of production.

We begin in the Department of Geography, National University of Singapore in the late 1990s. During the second half of 1997 Peter Dicken, author of one of Geography's most influential books, *Global Shift* (now in its sixth edition), spent June to December in Southeast Asia on sabbatical leave from the University of Manchester. He was investigating the development of trade and investment relationships between Singaporean firms and the European Union, and was drawn to

Singapore both because of its important place as a location point for multinational firms and as a major recipient/source of foreign direct investment (FDI). Stepping back in time for a moment, the case of *Global Shift* is an interesting one, and it ties directly into the theme of spaces of production. *Global Shift*, published in multiple editions in 1986, 1992, 1998, 2003, 2007 and 2010, outlines a framework, backed by substantial empirical material (primarily of a sectoral nature), regarding the causes and consequences of the globalization economic activity.

The book's origins can be tied to Dicken's visceral reactions to industrial decline and economic change in the landscape around him, both in the Manchester region, but also in England as a whole in the 1970s and 1980s – an era of systemic industrial restructuring and employment losses – as well as inward foreign investment into the production sector in select sectors of the economy (largely based in the south of England). In *Global Shift*, Dicken attempted to interrogate the central claims, sectoral- and firm-specific dynamics and regional impacts, of the **new international division of labour** (NIDL). In doing so he analysed the spatiality of global economic transformations, including the interdependencies between firms and place, and the role of particular nations and regions in driving global economic transformations. From this, he developed an exposition of the organizational transformations taking place within firms, especially transnational

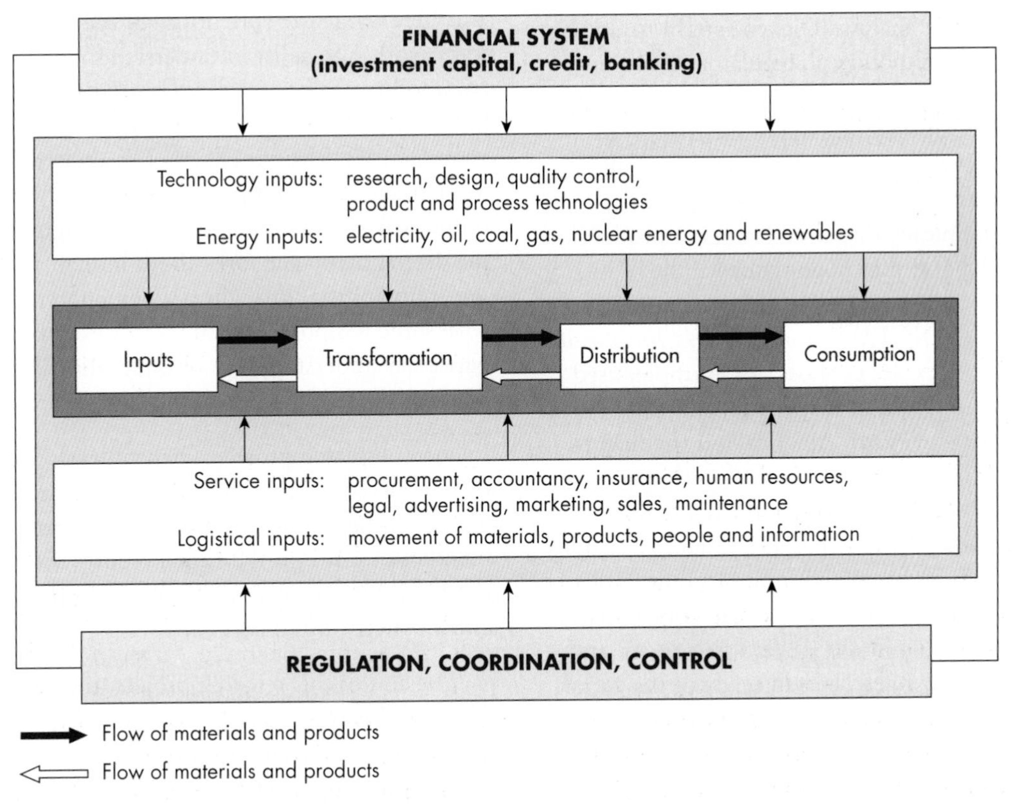

Figure 24.2 Basic components of a production circuit. Source: Dicken (2010)

corporations, as they seek to adjust to and exploit structural change at a range of scales and provided sectoral-specific analyses that highlighted the multifarious changes that shape firm strategy, relevant state policies and historically specific regional transformations (see also Olds, 2004).

Dicken's institutional political economy framework is, at its heart, centred on the core steps associated with the creation of monetizable transformations that we mentioned at the beginning of the chapter (Lee, 2005b). As noted above, this transformation is enabled by a circuit of value – a 'production circuit' – in Dicken's terms:

Processes related to the globalization of production and the variegated ways in which multinational firms related to territory demanded a global framework for economic geographic analysis. Dicken sought to do this because technological, regulatory and ideological changes in countries around the world were enabling the prime shapers of production circuits (firms) to relocate all or parts of them to take advantage of geographically dispersed factors of production (e.g. labour) that might enhance the monetization process (Knox *et al.*, 2003; also see UNCTAD's annual *World Investment Report*). Textiles or shoes, for example, used to be produced in many cities and towns in Europe and North America, but the majority of the input and transformation elements of textile and shoe production circuits were relocated to Asia (especially China). Thus the new international division of labour (NIDL) emerged as both a material and conceptual phenomenon in the 1970s, spurring on and influencing researchers in many of the social sciences, including in geography. Linked concepts and phenomenon like post-Fordism also emerged, as researchers sought to understand the organizational dynamics associated with the reshaping of production circuits and capitalism more generally. In short, geographical concepts tend to develop around changes in the world geographers are studying.

Dicken's sabbatical stay at the National University of Singapore in the second half of 1997 coincided with completion of the third edition (1998) of *Global Shift*. Intellectually, it also coincided with the emergence of vibrant debates in economic sociology and development studies about the nature of **global commodity chains** (GCCs) and **global value chains** (GVCs) and the rising influence of **Actor-Network Theory** (ANT) in the social science and humanities in many Western countries. As noted above, ANT also emerged as a major influence in Human Geography in the 1990s and 2000s, spurred on by work in nature-society geography (see Chapters 10 and 11). While the core utility of the GCC framework is to assist in understanding how the commodity chains associated with industries like textiles, clothing and automobiles are geographically structured, ANT is a relatively more micro-scale and practice-oriented framework that was designed to shed light on the diverse actor-networks (both human and non-human) that brought phenomenon to life with some temporal durability (for example, particular understandings of international finance).

In a later research project, Dicken developed these concepts further, into the idea of the global production network. As Dicken and Henderson put it in their final report (2005) to the ESRC, this project was both an empirical and a concept-building one:

> The aim of this research project, therefore, was to investigate the extent to which the EU, East Asia and Eastern Europe constitute an inter-connected nexus of economic relationships and to explore the implications

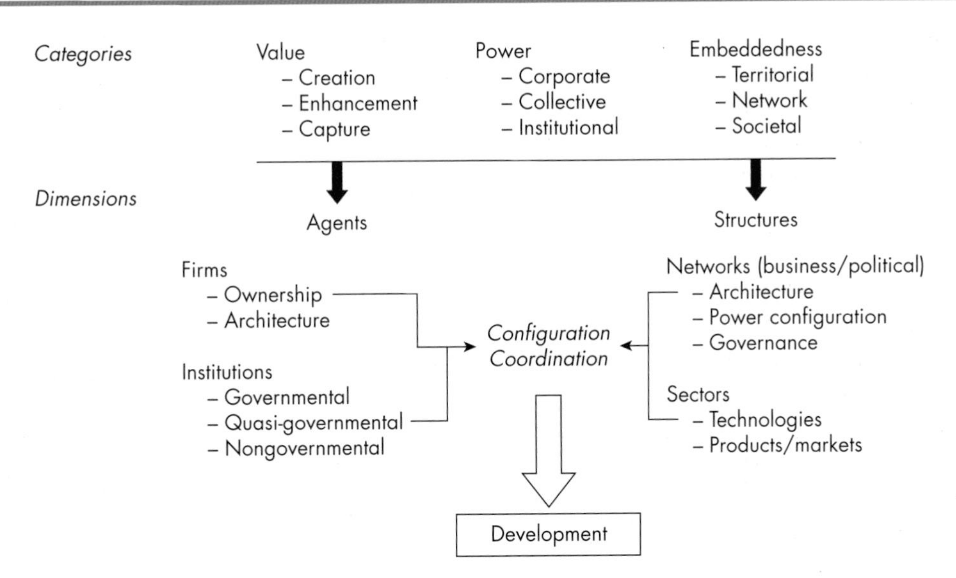

Figure 24.3 Global production network framework – a schematic representation. Source: Coe (2009)

> for the economies in which these are embedded. Although an analysis of FDI data was undertaken, *our primary approach was to develop a specific conceptual framework – the global production network* (GPN) – that allowed us to adopt a disaggregated scale of analysis. This conceptual framework extends and develops other similar frameworks, notably that of the global commodity chain. [Author's emphasis.]
>
> (Dicken and Henderson, 2005)

And the concept itself? As Neil Coe (2009) usefully puts it, the:

> GPN framework emphasizes the complex intra-, inter- and extrafirm networks involved in any economic activity, and how these are structured both organizationally and geographically. A GPN can thus be broadly defined as the globally organized nexus of interconnected functions and operations of firms and nonfirm institutions through which goods and services are produced, distributed, and consumed.
>
> (Coe, 2009: 557)

In graphic form, a GPN can be conceptualized as shown in Figure 24.3:

As alluded to above, and in Figure 24.3, the GPN framework is multi-scalar and is designed to shed light on how an assemblage of actors, networks and governance procedures articulate and become generative of outcomes that shape regional development processes. Thus there is a concern with how 'the globally organized nexus of interconnected functions and operations of firms and nonfirm institutions through which goods and services are produced, distributed, and consumed' (Coe, 2009: 577) but always with attendance to the articulation of the space of globalized production territory (hence the use of terms like **strategic coupling**).

Figure 24.4 provides a graphic representation of the take-up of these ideas, with a structural context of increasingly globalizing production. As is very evident in this series of figures, there was a concurrent spike in publications about GPNs, GCCs, GVCs and ANT, highlighting the interrelationship between the development of these concepts, but also showing how a

a) Actor-Network Theory (ANT)

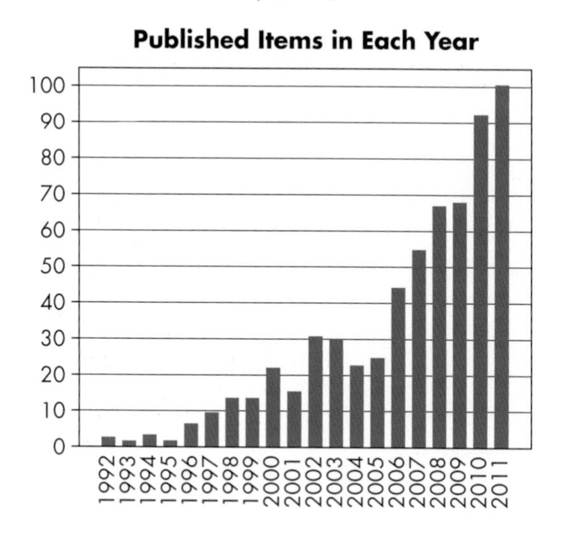

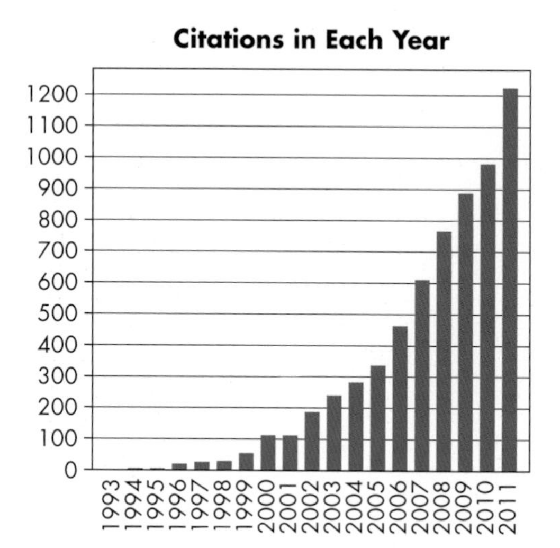

b) Global Commodity Chains (GCCs)

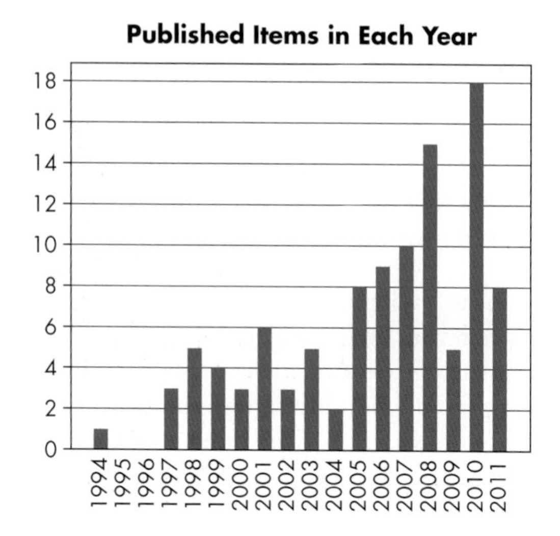

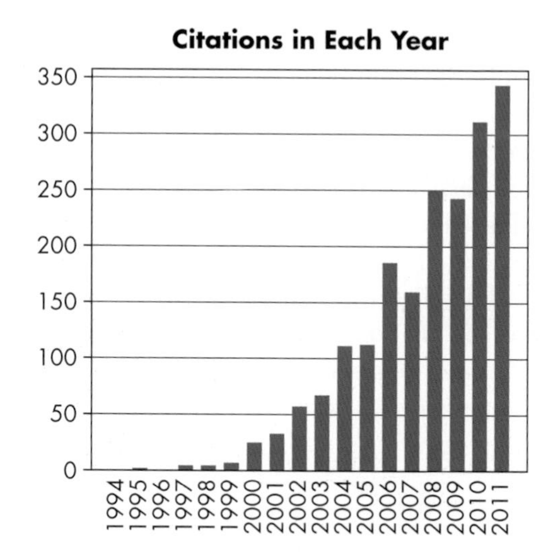

Figure 24.4 Publishing on Chains and Networks (to June 2012). Credit: Web of Knowledge Citation Reports (as at 2 July 2012)

roiling world (in economic, cultural, social, and political senses) fuels and supports the production of ideas.

Following this focus on conceptual development, Human Geographers then provided some insightful applications of the GPN framework through empirical research on the globalized spaces of production. Yank, for instance, looked at the redistribution of personal computer investment and production complexes in coastal China (e.g. Yang, 2009), and Yang *et al.*, (2009) investigated strategic coupling in the development of

c) Global Commodity Chains (GCCs)

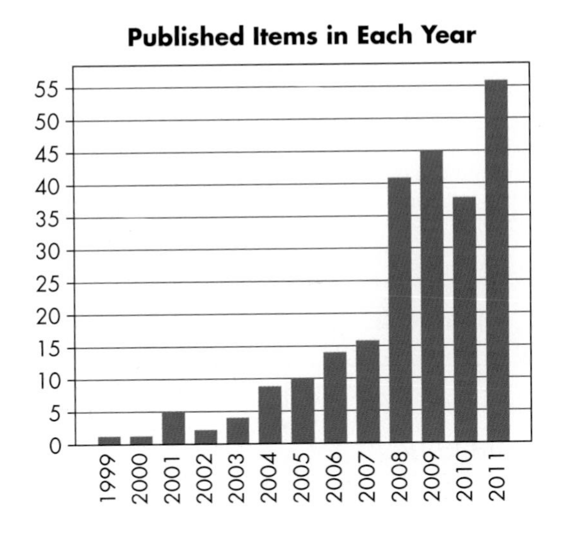

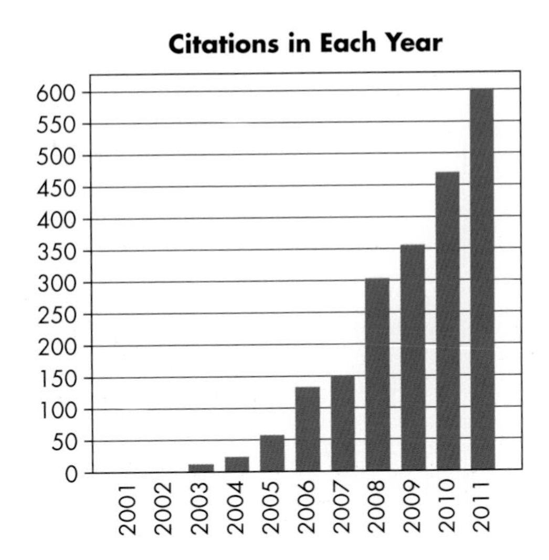

d) Global Production Networks (GPNs)

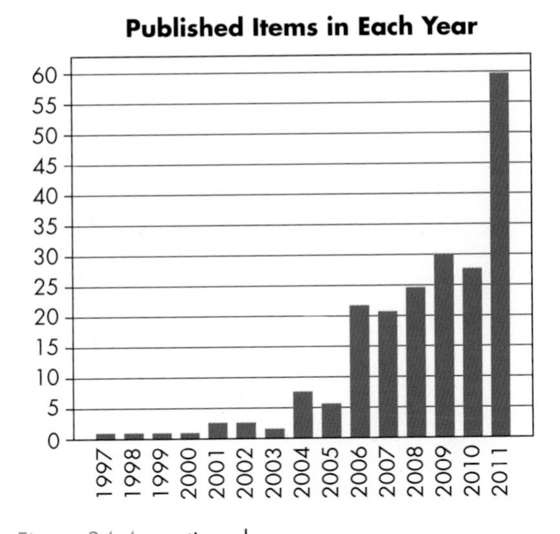

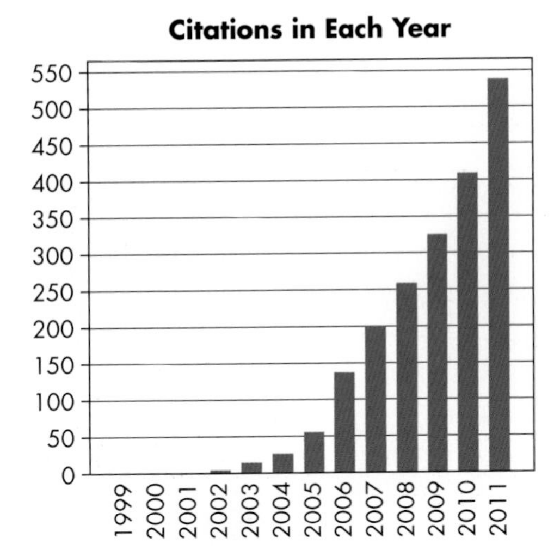

Figure 24.4 continued

high-technology parks in Taiwan. A relatively recent review of some of this work is provided by Coe (2011).

Returning to Dicken, it is worth noting that the process of driving the development of the GPN concept reshaped the fourth, fifth and sixth editions of *Global Shift*. There has been a move away from relying on GCC language and models to the inclusion of GPN-related language and models, including this key conceptual figure which has appeared in the fifth and sixth editions of *Global Shift*:

So we see the development of ideas about spaces of production being transformed as the

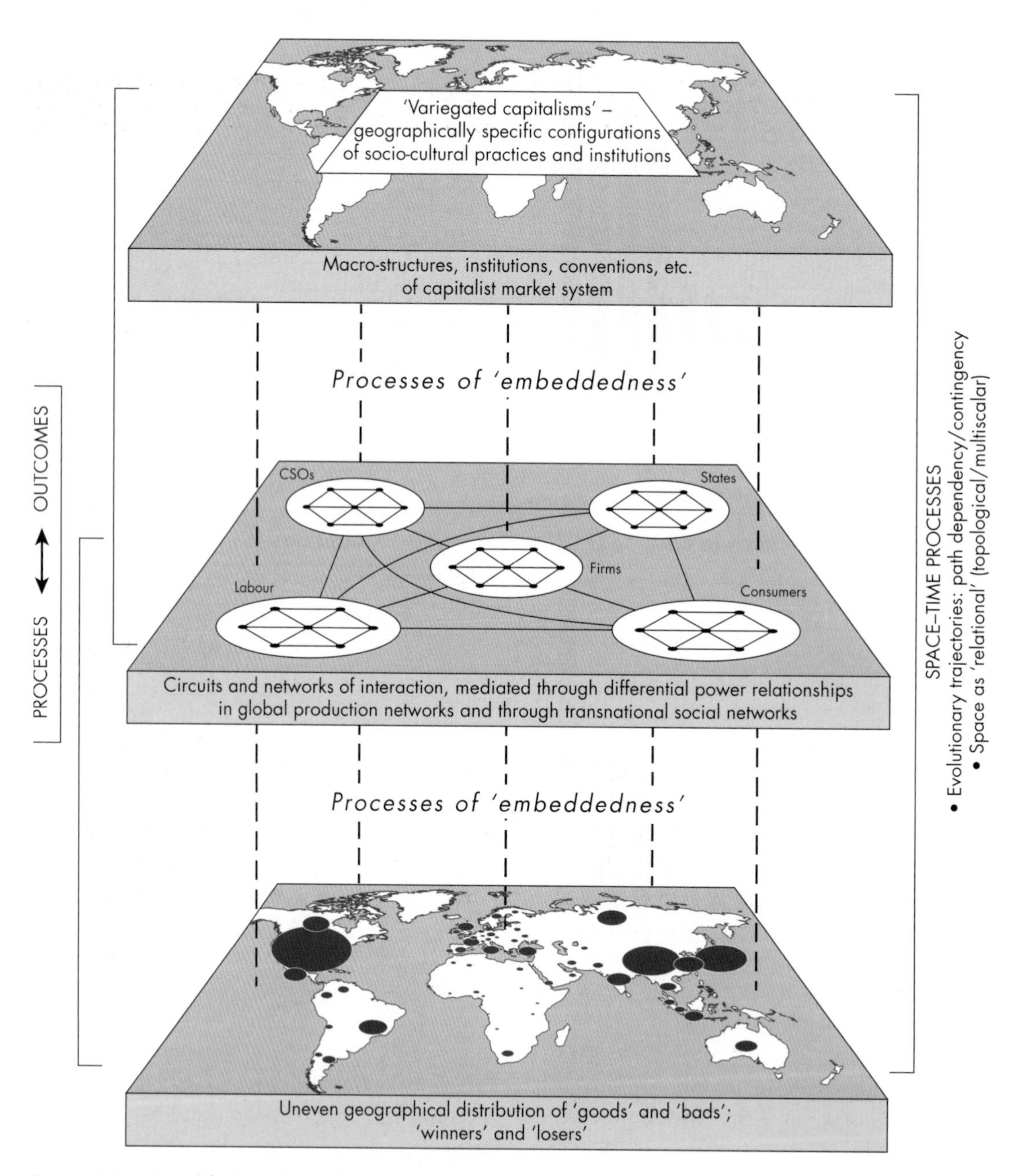

Figure 24.5 Simplified analytical framework of the global economy. Source: Dicken (2010)

'real world' is transformed. In short, there is a dialectical and symbiotic relationship between the globalization of economic activity and the production of ideas about geographies and geographical transformations. Geographers reside in a restless world, indeed.

SUMMARY

- The globalization of economic activity has transformed and integrated the spaces of production. Key actors shaping this transformation include the state, firms and labour.
- The development of conceptual thinking about this transformation is both focused upon, and inspired by, the globalization of economic activity.
- The global production networks (GPN) framework emerged in the late 1990s and early 2000s. It was designed to shed light on how the globalization of economic activity organized, governed and shaped regional development processes and outcomes.
- Professor Emeritus Peter Dicken (University of Manchester) was the key thinker behind the birth of the GPN framework.

Conclusion

Geographical research on the spaces of production has been going on for hundreds of years. Colonial era research by geographers, for example, helped European states survey vast landscapes for the purposes of identifying concentrations of natural resources (such as lumber and minerals), as well as the optimal routes through which to access them, extract them and either transform them in the 'periphery' or else immediately transport them back to the 'core' (using world-systems parlance) for consumption or as inputs into the production process. Circuits of value, binding production, exchange and consumption grew in depth and scale during the colonial era, and while they waxed and waned over time in the nineteenth and twentieth centuries, they grew increasingly global, faster-paced and wide-spread throughout the twentieth century and into the current era.

The main objective of this chapter has been to use two vignettes to introduce students to some ideas and debates associated with spaces of production. The first vignette focused on the planned construction of a new knowledge space (Applied Sciences NYC), co-constituted by the City of New York, Cornell University and Technion-Israeli Institute of Technology, on Roosevelt Island. This space of production is an explicit attempt to diversify New York's economy by educating new types of workers (for example, computer engineers), as well as providing the interdisciplinary spaces for the production of commercializable discoveries. At the end, this is a new space of production – not a factory, but still production for commercial exchange and exploitation.

The second vignette focused on the production of ideas about production – in this case, globalized production. The vignette situated the discussion of a now influential concept in the context of the globalization of production, the emergence of conceptual frameworks for making sense of networks and chains, and the critically important influence of a key figure in

Human Geography, the now retired Peter Dicken. The GPN framework is currently being utilized to shed light on how global network configurations constitute, and are constituted by, regional development processes.

It is important for you, the student, to realize that the utilization of the vignette as a pedagogical device has its pros and cons. On the positive side of the equation vignettes have the potential to bring Human Geographies to life; reflective as they are of real people, places and times. On the negative side, vignettes are geographically and historically specific by design, which means I've held off informing you about other issues, places, frameworks and debates rumbling on in economic geography. For example, I've been unable to discuss the illuminating ideas of J.K. Gibson-Graham regarding how economic geographies and markets are constituted, including via a myriad of forms of unpaid labour (Gibson-Graham, 2006). I've also failed to shed light on the critically important informal sector that dominates the economies of cities like Lagos, Manila and Dhaka (e.g. see Simone, 2004). Other important absences could also be identified.

These silences, along with many others, are inevitable in a brief introductory chapter. That said, I hope this chapter, alongside the other chapters in *Introducing Human Geographies*, will function as a vehicle to introduce you to what is a relatively heterogeneous and ever evolving discipline, tangled as it is in the existing world we are all situated in.

DISCUSSION POINTS

1. Cities including Amsterdam and London are closely following the Applied Sciences NYC development and are considering mimicking it. What factors are likely to enable or inhibit London and/or Amsterdam from achieving success with this type of development initiative?
2. To what degree is the GPN framework likely to be relevant in 10–15 years, given the likely trends in the globalization of economic activity?
3. What are the pros and cons of a discipline (Human Geography) and a subfield (economic geography) changing as the context in which society is embedded changes?
4. Is the relative power of organized labour in shaping the globalization of economic activity likely to rise or decline over the next decade? Give reasons for your answer.

FURTHER READING

Beaverstock, J.V. (2010) Peter Dicken. In: P. Hubbard, R. Kitchen and G. Valentine (eds.) *Key Thinkers on Space and Place* 2nd edn.) London: Sage, 108–12.

An insightful profile of Peter Dicken.

Castree, N., Coe, N., Ward, K. and Samers, M. (2004) *Spaces of Work: Global Capitalism and Geographies of Labour*. London: Sage.

An important and readable introduction to the role of labour in relationship to production, trade and finance.

Coe, N., Kelly, P. and Yeung, H. (2007) *Economic Geography: A Contemporary Introduction*. Oxford: Blackwell.

A long but valuable introduction to the subfield of economic geography by three scholars associated with the development of the GPN framework. The book works well when partnered with Peter Dicken's *Global Shift*.

Herod, A. (2011) *Scale*. New York: Routledge.

A focused introduction to the concept of scale by a Human gGeographer who propelled thinking about labour geographies.

Leyshon, A., Lee, R., McDowell, L. and Sunley, P. (eds) (2011) *The SAGE Handbook of Economic Geography*. London: Sage.

A helpful guide to the 'ideas and modes of thinking' in economic geography, including how and why research foci have changed over time.

Sheppard, E., Porter, P., Faust, D. and Nagar, R. (2009) *A World of Difference: Encountering and Contesting Development*. New York: Guilford Press.

A well-crafted integrated study of the 'nature and causes of global inequality and critically analyses contemporary approaches to economic development across the third world'. Section III, in particular, is an insightful introduction to geographies of production.

UNCTAD (2012) *World Investment Report*. Geneva: UNCTAD.

A yearly report that always includes useful data and analyses regarding the globalization of economic activity.

Wood, A.,and Roberts, S. (2011) *Economic Geography: Places, Networks and Flows*. New York: Routledge.

A brief yet insightful introduction to the field of economic geography and the uneven development process.

CHAPTER 25
MONEY AND FINANCE

Sarah Hall

Introduction

The international financial system plays an increasingly important role in shaping the everyday economic activities of households, firms and nation states. For example, high street banks and building societies that sell us the financial products we use on a daily basis, such as personal loans, credit cards and savings accounts, rely on the activities of financiers working in **international financial centres** such as London and New York for the production of these financial products. Firms operating in sectors seemingly far removed from the glistening towers of Canary Wharf and Wall Street, such as manufacturing and food production, are tied into the international financial system as they face growing pressures to meet the financial targets expected of them by their shareholders and rely on global financial markets to access the capital needed to invest in new machinery and technology. Finally, in several advanced Western economies, banking, accountancy and financial markets – collectively known as the financial services industry – make a highly significant contribution to national economic growth and employment. This trend is particularly marked in the UK. Here, over one million people were employed in financial services in 2011, accounting for almost four per cent of total employment in the UK. Moreover, the financial services sector contributed £124 billion to the UK national economy in 2009, making up ten per cent of total economic output (The CityUK, 2011).

Economic geographers, together with other social scientists, term the process by which these everyday economic activities of individuals, households and firms increasingly rely upon the international financial system, '**financialization**'. During the 2000s, this process went largely unnoticed by academics, politicians and the media as the global economy enjoyed a period of significant expansion and households saw their standard of living and purchasing power increase, fuelled to a significant extent by relatively easy access to cheap credit. However, what Mervyn King, Governor of the Bank of England, termed the NICE decade (no inflation, constant expansion) of the 2000s ended with the global financial crisis dating back to late summer 2008. This crisis saw the collapse of several of the financial firms that had played a vital role in facilitating the process of financialization, notably the investment bank Lehman Brothers, and the bailing-out of a number of others including the multinational insurance company AIG. These events placed the relationship between ordinary households, firms and national economies, on the one hand, and the international financial system, on the other, in the political, media and academic spotlight. Indeed, people seemingly distant from the world of high finance, such as homeowners who could no longer access mortgage finance and public sector workers facing redundancy as governments sought to fund bailouts for their banking system, found themselves experiencing the full effect of the crisis.

In addition to revealing the extent to which economic life had become financialized during the 2000s, the crisis also reveals the importance of thinking geographically about financialization and money and finance more generally. Indeed, while it is often referred to as the 'global' financial crisis, this hides the heterogeneous ways in which it has been experienced by different people in different places. For example, James Sidaway demonstrates how minority ethnic and racial groups in the USA have suffered disproportionately through the loss of their homes because of their uncompetitive mortgage finance rates (Sidaway, 2008). Building on this work, in this chapter I show how recent research on the geography of money and finance reveals how geography matters to both the operation of the international financial system, our own economic practices and the links between the two. In particular, I argue that understanding the cultural and social dimensions of money and finance is vital to revealing how these geographies are produced and their uneven effects on everyday economic life.

Placing and spacing the international financial system

Despite the importance of money and finance to the global economy, economic geographers only began to focus on money as a substantive research concern from the late 1970s onwards. David Harvey's detailed examination of the role of money and finance in shaping urban environments through investment strategies marks one of the most significant contributions to this early research on the geographies of money and finance (Harvey, 1982). This work provided the basis for the emergence of a sub-field of economic geography focusing on money and finance that developed most significantly from the early 1990s onwards. Research at this time continued to adopt the critical approach to the (il)logics of the international financial system that had been initiated by Harvey. For example, following economic geography's broader interests in macro-scale socio-economic transformation associated with the intensification of globalization processes, research focused on the changing geo-politics of the international financial system. This work examines how geographically specific financial regulations and working cultures combined to produce an international financial system that was anchored in a small number of international financial centres (notably London and New York) and offshore financial centres which have attractive regulatory environments for financial firms (Martin, 1999).

By pointing to the continued importance of a relatively small number of places within the international financial system, this research refutes claims that the deepening integration of financial markets driven by technological innovation and deregulation would herald the demise of geography as a key determinant of the location of financial services activity. For example, O'Brien (1991) in his 'end of geography' thesis, argued that greater use of technology and virtual forms of communication would mean that financial firms would no longer have to co-locate within financial districts in order to conduct business. However, echoing geographical work on money and finance, the business pages of newspapers and the financial news on television and radio clearly show how financial services activities remain concentrated in a small number of international financial centres.

This clustering of financial services activity is not new. For example, London's financial services cluster began to develop from the 1700s onwards, supporting the rise of Britain as an imperial power (Cain and Hopkins, 1986). However, the clustering of financial services activity intensified throughout the twentieth century, giving rise to a distinct hierarchy of international financial centres (see Table 25.1). The relative importance of different financial centres can be partly explained by the geographical reach of the markets they service (Beaverstock *et al.*, 1999). At the top of the hierarchy sit London and New York. These alpha world cities offer the full range of financial services (including investment banking, insurance, trading of financial products on stock exchanges and new services such as those associated with Islamic finance) to global clients. Following these two centres are secondary or beta centres such as Hong Kong and Singapore, that service pan-regional markets. Below these sit sub-regional or gamma centres, such as Chicago and Zurich that offer more geographically and organizationally specialist services. However, while London and New York have dominated this hierarchy for at least 50 years, a number of smaller centres have grown rapidly in recent years. For example, Singapore and Hong Kong have developed from regional centres into established financial centres in their own right. Meanwhile, cities such as Shanghai and Seoul are currently climbing the hierarchy rapidly, reflecting the growing importance of South East Asia in the international financial system.

Financial centre	2011 Rank	2010 Rank	Change
London	1	1	–
New York	2	2	–
Hong Kong	3	3	–
Singapore	4	4	–
Shanghai	5	6	+1
Tokyo	6	5	–1
Chicago	7	7	–
Zurich	8	8	–
Geneva	9	9	–
Sydney	10	10	–
Toronto	11	12	+1
Boston	12	13	+1
San Francisco	13	=14	+1
Frankfurt	14	11	–3
Shenzen	15	=14	–1
Seoul	16	24	+8
Beijing	17	16	–1
Washington DC	18	17	–1
Taipei	19	19	–
Paris	20	18	–2

Table 25.1 Ranking of top 20 international financial centres. Adapted from The Global Financial Centres Index 9, ZYen (2011)

Beyond the size of the markets different financial centres service, geography is also important in explaining the dominance of a small number of financial clusters in other ways. For example, the continued dominance of London and New York reflects their historic importance as trading centres more generally, as this gave them a competitive advantage over other cities as the international financial system developed. Their location in different time zones is also an important factor as this allows financial firms to trade around the clock if they have offices in both London and New York. Moreover, the nature of financial services work itself is an important factor in driving the continued concentration of financial markets in particular cities. Three agglomeration benefits are particularly significant in this respect. First, financial clusters facilitate liquid financial markets. This means that already existing financial markets attract more buyers and sellers of financial products because they are most likely to find customers for their own products there. Second, the clustering of financial firms within financial centres gives rise to 'buzz' between financiers (Bathelt *et al.,* 2004). This buzz, built around dense inter-personal and inter-firm relations, facilitates processes of innovation and the production of new financial products because financiers can learn about the specific demands of their clients and gather information on their competitors' activities. Third, once a financial centre becomes established it continues to attract the highly skilled labour force that is necessary for the production of bespoke financial products.

Figure 25.1 Financial traders on the trading floor. Credit: Getty Images

This interest in the nature of financial services work and its role in shaping the geographies of the international financial system has been the focus of much of the research conducted by economic geographers from the late 1990s onwards. This work is significant because it marks an expansion in the theoretical toolkit used by geographers interested in money and finance beyond its earlier focus on the geo-politics of finance. Instead, this more recent work examines how the social and cultural dimensions of money and finance are also important factors in explaining the continued dominance of a small number of financial centres. A focus on the financiers working within international financial centres is particularly important in this respect since by understanding their activities, it is possible to enhance our understanding of the uneven geography of the international financial system more generally.

This work focuses on the routinized, formal and informal actions of individual financiers. This includes studies of their daily working practices and their outside-of-work networking in social spaces such as bars and restaurants within financial centres, airport departure lounges and expatriate clubs, echoing the global nature of financial careers (e.g. Beaverstock, 2002). In so doing, sophisticated accounts of the ways in which working practices associated with the information-rich nature of financial services is crucial in shaping the continued importance of a small number of international financial centres have developed.

Three aspects to this work are particularly important. First, research has revealed how establishing inter-personal relations within and between financial centres is vital for financiers to produce and circulate technical knowledge about financial products and the demands of customers (Clark and O'Connor, 1997). Indeed, the importance of these networks gives rise to frequent travel between financial centres, either for specific meetings or for longer-term secondments as financial firms seek to disseminate their corporate knowledges between financial centres. Second, research has examined how financial services work also involves a more embodied and emotive set of knowledges that are played out through bodily performances. In this respect, individuals gain access to the personal networks that are important for their daily work by dressing and acting in similar ways to their peers in order that the trust necessary for building such networks might be fostered (Thrift, 1994). Moreover, McDowell (1997) has demonstrated how such bodily performances are gendered such that women working in financial services often adopt similar bodily comportment to their male counterparts in order to gain acceptance and recognition within elite financial labour markets. Third, while educational background has been well documented as an important element in securing access to these networks, recent research has demonstrated how ongoing training and education within financial workplaces through schemes such as graduate training courses within investment banks and MBA degrees from leading business schools are also important activities. It is through such activities that individuals learn how to act as an international financier and gain access into the personal networks within financial centres that are crucial for their own career success.

SUMMARY

- Despite processes of technological innovation and regulatory reform, global finance remains concentrated in a small number of international financial centres.
- There is a dynamic hierarchy of international financial centres that reflects their relative importance as command and control points within global finance.
- In order to explain the continued importance of international financial centres, research needs to consider money and finance as social and cultural practices as well as sets of economic and political relations.

Geographies of everyday financialized lives

The geographies of the international financial system and the continued importance of international financial centres within it are not only important for the working practices of financial elites. Our own everyday lives are increasingly tied into this financial system through our consumption of financial products, and, here too, geography matters. In this respect, economic geographers have well-established research interests in retail finance – that is, financial services offered to households through high street banks and building societies and internet providers. Indeed, early research on mortgage finance, negative equity (meaning home-owners who owe their mortgage provider more than the current value of their home) and financial exclusion provided an important impetus for the development of a sub-field of geographical work on money and finance in the 1990s (Leyshon, 1995).

These research topics have re-emerged in the 2000s, driven by the ways in which personal finance was central to the 'global' financial crisis of 2008. Again, issues of mortgage finance have been placed centre stage as the inability of sub-prime (higher risk) borrowers in certain parts of the USA to meet their mortgage payments was a critical factor in triggering the crisis, as mortgage lenders had to react to the fact the loans they had made might not be paid back. Moreover, in the wake of the crisis, personal finance has been affected significantly, with continued stock market falls impacting negatively on pension fund performance and home loans still being highly limited as lenders seek to minimize their exposure to risk. Echoing research on the international financial system, this more recent interest in personal and retail finance has adopted a more socially and culturally sensitive mode of analysis as compared to earlier work, focusing in particular on questions of financial inclusion, exclusion and their uneven geographies.

Financial exclusion refers to 'those processes by which individuals and households face difficulties in accessing financial services' (Leyshon *et al.,* 2008: 447). Two processes have increased the extent of financial exclusion during the 2000s. First, de-regulatory reforms of the financial services industry in advanced capitalist economies has allowed financial firms to develop new financial products that reflect the fact that individuals and households, rather than the state, are increasingly responsible for their own future financial security. This process is particularly marked in terms of pension provision as individuals are tied into the international financial system through the decline of defined benefit pension provision in which individuals were guaranteed a level of retirement income. Instead, individuals are reliant upon becoming active managers of their retirement income through using a range of investments in international financial markets (Clark, 2003). In this way, individuals are expected to act as responsible financial consumers who, through suitable education from financial literacy schemes, manage and take risks in order to manage their own financial futures (Langley, 2008). However, these products are often targeted at the most profitable individuals, and others, including those unable to engage with financial literacy, are marginalized from them. Second, technological innovation and the development of new financial products, particularly those linked to the international financial system through processes of securitization has led to a number of new channels for financial services being developed that increasingly use virtual forms of communication, particularly through activities like internet banking (Leyshon and Pollard, 2000).

Two consequences of the intensification of these financial exclusion processes are particularly significant, both of which are inherently geographical. First, research has demonstrated how the growing focus on profitable financial customers has given rise to a highly uneven geography of financial services withdrawal. This is most marked in terms of bank and building society branch closures that have shrunk by about one third since 1989 in the UK (Leyshon *et al.*, 2008). Geographers have played a vital role in demonstrating the importance of space and place to such processes. In this respect, the physical infrastructure of bank and building society branches has been conceptualized as a network, the scope and density of which can be measured, both by region, but also, and more significantly, along socio-economic lines. By adopting the latter approach, Leyshon *et al.*, (2008) have demonstrated the disproportionate impact of service withdrawal in socio-economically deprived wards in the UK. This focus on the geographies of processes of financial exclusion has been developed further through the identification of different forms of retail financial ecologies (Leyshon *et al.*, 2004). This metaphor is used to demonstrate how the working practices of financial service providers, particularly in terms of their assessment of potential customers 'at a distance' using a range of credit scoring techniques, is co-constitutive of financial landscapes. This approach has identified two contrasting idealized types of ecology: first, the middle class ecology in which highly financially literate customers use a range of distribution channels to access financial services and hence maintain a strong physical and virtual bank and building society branch network; and second 'relic' ecologies in which socio-economically deprived groups suffer both the demise of mainstream financial provision on the basis of their lack of profitability and are instead subjected to a range of more exploitative forms of financial provision such as credit offered by door-to-door lenders. This is manifested in the built environment through the greater branch closure rates in more economically deprived neighbourhoods (Figure 25.2).

The relationship between space, socio-economically deprived financial subjects and retail financial services provision has also been examined through work that examines access to credit, notably sub-prime mortgage finance (Aalbers, 2005; 2008). This work demonstrates how high street financial firms accessed credit through commercial banks within the international financial system through processes termed securitization in order to increase

Figure 25.2 Bank branches in economically deprived neighbourhoods are more likely to be subject to closure. Credit: QEDimages/Alamy

financial inclusion in the 2000s. However, access to such credit was highly uneven, with race and class acting as important factors when lenders were making decisions concerning which potential customers were deemed credit-worthy and how favourable (or otherwise) the terms of any home loan would be. In addition to revealing the different ways in which individuals and households have experienced the fallout from the financial crisis, this work also points to the relationship between elite and 'everyday' financial systems that were fostered through processes of financialization in the 2000s.

SUMMARY

- Individual households are increasingly tied into the international financial system through investment products, most notably mortgage finance and pension funds.
- This process is termed 'financialization' but has led to financial exclusion as certain groups are unable to access financial services.
- Financial exclusion disproportionately affects minority and lower socio-economic groups. This was particularly noticeable following the global financial crisis that began in 2008.

Conclusions

Economic geographical research into both the international financial system and the everyday financial geographies of households has developed rapidly since being framed as central research issues for geographers from the 1990s onwards. The rise of virtual forms of communication, technological innovation and deregulation might appear to lead to a decline in the importance of space and place to these economic activities. However, the research presented in this chapter clearly demonstrates that conceptualizing money and finance geographically is vitally important in order to understand the uneven production and consequences of both the international financial system and its links to household economies. In order to understand these geographies fully, recent research has revealed the importance of considering finance as a cultural and social practice, as well as a set of economic and political relations. In particular, the value of this approach lies in documenting how the activities of financial elites working in a select number of international financial centres are linked to the financial products that individuals and households increasingly rely upon for their own financial futures, particularly pension funds and mortgages.

However, the research discussed above on these processes of financialization also has its own, partial geography in that the vast majority of research has been conducted within the heartlands of the international financial system, particularly the USA and Western Europe. This selectivity in research sites is important because it means that the accounts generated of money and finance are themselves partial. This partiality has been revealed most clearly by what is commonly termed the global financial crisis of Summer 2008, that has actually been experienced rather differently beyond the USA and Europe. In response, it is important that future research into the geographies of money and finance expands its own geographical horizons to develop more fully accounts of financial geographies in emerging economies. Within this emerging research agenda, the cases of South East Asia and the BRIC economies (Brazil, Russia, India and China) are particularly important, since these are home to both rapidly growing international financial centres such as Singapore and Shanghai as well as vast domestic markets for everyday financial products. Understanding these emerging geographies of money and finance will be vitally important if economic geographers are to continue to understand the dynamic and uneven geographies of international finance and the consequences of this for individuals and households.

DISCUSSION POINTS

1. The majority of research into the geographies of money and finance has been conducted in the USA and Western Europe. What are the limitations of this approach and why are they important?
2. What is the value of understanding money and finance as social and cultural practices?

3. In what ways are the geographies of the international financial system linked to household economies and why do these linkages, known as processes of financialization, matter?
4. is your daily life tied into the international financial system?

FURTHER READING

Corbridge, S., Thrift, N.J. and Martin, R. (1996) *Money, Power and Space*. Oxford: Oxford University Press.

Knorr Cetina, K. and Preda, A. (eds.) (2004) *The Sociology of Financial Markets*. Oxford: Oxford University Press.

Leyshon, A. and Thrift, N.J. (1997) *Money/Space*. London: Routledge.

Martin, R. (ed.) (1998) *Money and the Space Economy*. Chichester: John Wiley.

These texts introduce the broad field of economic geographical research into money and finance and the value of a socially and cultural sensitive approach.

Langley, P. (2008) *The Everyday Life of Global Finance: Saving and Borrowing in Anglo-America*. Oxford: Oxford University Press.

McDowell, L. (1997) *Capital Culture*. Oxford: Oxford University Press.

More specialist texts that deal with the gendered nature of financial services work and the links between everyday and international finance respectively.

A number of academic journals regularly publish papers reporting research into the geographies of money and finance. The following journals are particularly important, several of which have published special issues on the topic:

Journal of Economic Geography

Economic Geography

Environment and Planning A

Geoforum

Transactions of the Institute of British Geographers

Acknowledgements

I am grateful for the support of the ESRC (RES-061-25-0071), which has allowed me to develop the ideas presented in this chapter.

CHAPTER 26
CONSUMPTION–REPRODUCTION

Juliana Mansvelt

Introduction

What comes to mind when you first think of consumption? For many of my students it is 'purchase', 'shopping' and 'leisure'. Such terms do allude to a common and partial understanding of what consumption is – that consumption is about buying and spending and that it is a non-productive activity. In this chapter I hope to challenge such assumptions. We will learn that consumption is not simply a matter of economic purchase, nor of specific sites of exchange, but that it is an important sphere of social practices in which things are valued and de-valued economically, socially and symbolically across time and in place.

Consumption, like production, involves cultural and economic processes which are integral to the creation of built environments and the maintenance of everyday life. In a globalizing world many of us are exposed to commodities through a variety of mediums and spaces – print, aural and visual media, through the internet, tourism, retail and leisure landscapes, and spaces of work and dwelling. We may even be invited to contribute to the production of consumer goods and spaces: through product and retailer loyalty schemes, as members of 'brand tribes' conveying our enthusiasm about products to others, and in the evaluation and/or design of consumer goods and services. While products, places and practices of consumption seem to penetrate almost every aspect of everyday life, practices and experiences of consuming are experienced unevenly. Access to and the ability to acquire and use consumer services and goods and the outcomes of consumption for people, non-human things and environments may be vastly different – raising questions about what is consumed, where, why, and the equity and sustainability of consumption practices.

The impacts of consumption are not limited to the economic realm. Advertising and marketing industries appeal to us as 'consumers', promoting the necessity and desirability of goods and services as items of personal and cultural value. Such appeals do not simply focus on our material wellbeing, but on the role of commodities as a means of fashioning identities. Acquisition, possession and disposal of commodities also have a social function, symbolizing both connection and difference to others who may be proximate or distant. Consumption is consequently a moral and political practice, one connected with geographies and relationships of power which may stretch across multiple countries and scales. Furthermore, the political, moral and social nature of consumption can place additional demands and responsibilities on us with regard to our consumption choices. Our friends, firms, governments and NGOs may all invoke us to make appropriate consumption choices. Seen in this light, consumption is not only a significant place-making activity but a powerful one which involves the social reproduction of cultural and economic systems.

Consumption: connections to production and social reproduction

Consumption can be defined as a set of social relations, practices and discourses which centre on the sale, purchase and use and disposal of commodities. Though this definition is premised on the notion of a commodity as a good or service which is purchased in a market (see Chapter 27), it understands consumption to be more than a momentary act of purchase or a space of economic exchange. This definition encapsulates the range of material and symbolic practices and meanings which centre around the sale, choice and selection of goods and services, their purchase, use, re-use or re-sale, and their eventual wasting. This chapter will explore some of these extended practices and meanings in more detail, but before we do so it is important to note that for a long time economic geography tended to concentrate on the geographies of production (see Chapter 24), with consumption seen primarily as a consequence of the operation of the capitalist mode of production. Nevertheless, this theorization provided valuable insight into why both consumption and production are necessary to ensure the survival of economic systems, the ways in which these spheres are inseparable and how they each involve the other in the production, maintenance and destruction of value – themes subsequently taken up by a range of social and economic geographers (see box).

CASE STUDY

A view from production: connecting consumption to production via the commodity chain

Viewed in relation to production, consumption can be seen as part of a commodity chain – a series of discrete but connected relationships and activities which are in involved in the production and consumption of commodities. A very simple commodity chain might look like that shown in Figure 26.1.

Figure 26.1 shows a linear diagram, suggesting that each discrete activity is a consequence of the previous activity. While such a simplified diagram could not hope to capture the complexity of the relationships between consumption and production – the extent of processes involved in consumption itself (purchase, use, re-use, disposal, wasting), or the ways in which

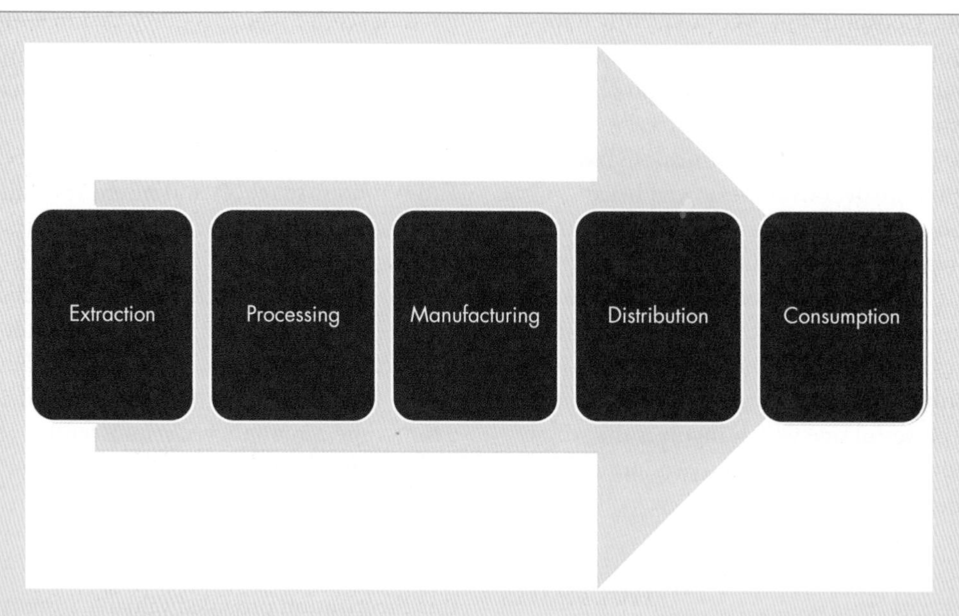

Figure 26.1 A simplified commodity chain. Credit: Juliana Mansvelt

consumption is an integral part of production, not just an end product of it – the commodity chain is a powerful metaphor.

It has provided a basis for understanding a number of issues, including:

- how chains are governed, for example by producers (suppliers) or buyers (at the consumption end by brand name firms or retailers – see Gereffi, 2001)
- how value is shaped through production and consumption relations
- considerations of why and where chains touch down in particular places (for example, with regard to how networks of production are stretched out across the globe and how the practices of consumers might be linked to the environmental, human and animal rights issues through production processes – see Hughes and Reimer, 2004).

Commodity chains have provided a starting point for scholars to follow the social and spatial lives of things through wider networks of production and consumption (see Chapter 27) and have also enabled geographers to consider how the concept has been deployed by firms, NGOs and consumer activists to invoke and morally frame the actions of producers, consumers and governments (Jackson *et al.*, 2009).

However, until recently, this production-based view of consumption meant consumption research was primarily centred upon geographies of retailing and on spaces and practices of purchase by consumers. In the latter decades of the twentieth century, geographers started to research consumption as an important set of processes in its own right. The work of social and cultural geographers emphasized how consumer talk, practice and experience are not just connected to economic relations, but to social and political relations operating at and across a variety of scales. A series of ethnographic studies of second-hand consumption spaces such as car-boot sales, charity and retro shops and households revealed the complexity of consumption beyond spaces and moments of purchase (Miller *et al.*, 1998; Clarke, 2000; Gregson and Crewe, 2003; Valentine, 1999). These studies highlighted what might be termed as 'the production of consumption' – the work, skill and knowledge which is necessary in order to acquire, **appropriate** and appreciate, maintain, keep, value and dispose of commodities (see Crewe, 2011). Consumption came to be seen less as a consequence of a relatively linear series of relations deriving from production and more as a sphere of social relationships which could only be understood in the context of multiple and changing connections between people, things and places. Accordingly, consumption isn't simply the expression of production but a critical part of how it proceeds, influencing the constitution of space, material and social life and ways places themselves are imagined and consumed (see Redclift, 2010).

Consumption has an important role in social reproduction – the means by which societies reproduce themselves over time. Social reproduction encompasses the material and bodily things which people need to reproduce from one generation to the next, such as food, shelter and relationships. It also includes the social transmission of practices, traditions, cultural values and discursive norms, thereby reproducing structures associated with consumption, production and exchange. Such reproduction inevitably includes spaces of both production and consumption and the ways in which these are reproduced through social relationships – which may be simultaneously economic, cultural, political and social. It is to how these connections are played out in the creation of spatialities, socialities and subjectivities that we now turn.

SUMMARY

- Consumption is a set of social relations, practices and discourses which centre on the sale, purchase and use and disposal of commodities.
- Consumption is important, not simply because it is necessary to ensure the continued survival of economic systems, but because it has a wider social, spatial and symbolic function.
- Consumption and production are connected. Consumption involves the production of people, commodities, places and relations and vice versa.
- The concept of social reproduction is concerned with how people live and the ways in which societies are reproduced over time. Consumption has a significant role to play in this.

Dimensions of consumption

Spatialities

Consumption produces particular spatialities (arrangements of space) which are expressed through multiple scales from the global to that of our own bodies. In contemporary society much is made of the ways in which place is produced through processes of globalization. The existence of flows of global brands (Nike, Nestlé, Shell, Toyota, etc.), consumer communities (e.g. fan clubs, brand tribes) which extend beyond national boundaries, the serial repetition of consumer spaces (fast food restaurants, retail and leisure chains, theme parks, etc.) and the colonization of landscapes by commercial advertisers and sponsors (for example sports fields, urban environments, the internet) seems to provide evidence of the significance of consumption in reproducing places in a globalizing world. Yet the role of commodities, consuming practices and sites of consumption in making places are experienced unevenly, with simultaneous tendencies towards greater diversity and fixity and fluidity of consumers, commodity flows and material landscapes (Goodman, Goodman and Redclift, 2010). For example, the same commodities may be used by individuals to create very different lifestyles and places; variations in economic livelihoods may result in diverse consumption practices and built environments (e.g. slums and wealthy gated communities); and flows of consumer goods or practices which we think of as mobile and global (such as tourism or internet shopping) depend on sites of local fixity (such as airports and websites). Such tendencies demonstrate the ways in which spatialities of consumption are relational, with variation in the constitution and expression of production, consumption and reproduction influenced by practices in other places, across other scales and over time (see box).

Space and place are not just produced through consumption; they are also consumed through both production and social reproduction (Goodman, Goodman and Redclift, 2010). Resources may be used up, land, sea or air may become commodified and existing uses of space may be reconfigured (e.g. for new factories, housing, leisure pursuits or schools). Place meanings can be altered or displaced as a consequence. The development of viticulture in New Zealand over the last thirty years for example has seen the emergence of the branding of regions and wines in terms of *terroir*. A concept traditionally associated with French place-based associations, *terroir* is being deployed to enhance the distinctiveness of the wine and its value (Overton, 2010). Wine consumption in New Zealand and overseas also validates the re-working and displacement of place associations by wine marketers (Overton, 2010) and hence through consumption, New Zealand regions come to be imagined, known and consumed differently.

The shaping of place through consumption is not a neutral process. Goss (1993a) suggests that mega malls, as spectacular retail sites, perform important ideological functions, obscuring processes of production in the imaginative associations of place and commodities (as festival or heritage,or spaces of elsewhereness). More mundane spaces of retail and commodity use are also sites for the expression of economic and social power, with ethnographic studies of consumption revealing the ways in which family, gender, ethnicity, age and class are shaped through socialities and the discourses which arise from them (e.g. Colls and Evans, 2008; Hall, 2011).

CASE STUDY

Relational geographies: Crang on globalization and ship-breaking

Moving beyond conceptions of consumption derived from the realization of value through purchase, a number of geographers are conceptualizing consumption in terms of its connection with devaluation and revaluation through depletion and wasting (Gregson *et al.*, 2010). Mike Crang's (2010) study of photographic images of ship-breaking on Asian beaches demonstrates the ways in which spatialities of consumption are relational, taking their meaning and form in the context of relations between practices, place and routes of both global production and consumption. Crang notes how containerization, a practice which exemplifies globalization (and the global extension and expansion of consumer commodities) has a counter image in the destruction of container ships on beaches in the Global South. Ironically, the 'death of big ships' and the **spatiality** associated with this (80 per cent of which occurs in South Asia) means that ships that once carried cargoes are sold for scrap according to the same logics of cheap location that actually enabled containerization routes and global flows of consumer goods (Crang, 2010: 1084). Hetherington (2004) argues with regard to disposal that wasting is a spatial and social matter of moving and 'placing' things away. Through the use of images of ship-breaking in his article, Crang makes visible the spatiality of wasting that is part of global consumption and production connections. Images depicting the materiality of waste and the appalling labour conditions and environmental degradation on Asian beaches have consequently been capitalized on by NGOs advocating bans on ship-breaking. However, as Crang (2010) suggests, such calls do not address the significance of ship-breaking in the livelihoods and economies of the countries and regions in which it occurs, nor does it leave open possibilities for altered practice with fewer negative environmental and social impacts.

Figure 26.2 Ship-breaking on Asian beaches. Credit: Naquib/Flickr, Creative Commons

Socialities

As well as creating spatialities, consumption plays an important role in reproducing socialities – the ways in which we relate to other people and things. Socialities arising from consumption practice are important in the formation, maintenance or contestation of relations of subordination and domination which may extend beyond interpersonal relationships to the ways in which citizenship, class, gender or ethnicity are produced through discourse.

Bourdieu's research has demonstrated how consuming practices can become part of a class-based system of distinction (Bourdieu, 1984). Both everyday and exceptional objects (clothes, food, appliances, furnishing, art, appliances) can symbolize our belonging and difference to others, providing a way of communicating our 'selves' (Noble, 2008). The objects of our consumption and the kinds of practices we engage in also form a basis for the moral ascriptions of others whether we intend this or not. Consider how one might interpret an individual's use of public transport: it might symbolize a level of economic status, an active commitment to a green lifestyle, or be the consequence of more pragmatic reasons connected with time and ease of access. Such interpretations are context dependent, and for the individual, work may be required to either accept or reject them.

In appropriating commodities, we don't just form relationships with others, but also with non-human things, making them part of our everyday spaces and practice through rituals such as cleaning, displaying, wearing. These acts are part of the process of appropriation through which the object is de-alienated (Miller, 1987). We can also externalize our 'selves' and feelings onto objects – hence why we might wish to discard or treasure objects, photos or clothing which are reminders of affirmative or damaging relationships to others. Finally, our social lives don't just accommodate non-human things, they also accommodate us (Hand *et al.*, 2007). Consequently the material capacities of things may cause us to act and respond differently – think how the material properties of new digital music technology (e.g. downloadable MP3 files) have changed the ways in which we purchase and listen to music. Similarly, when things break down and need repair, it reveals the taken-for-granted functioning of social-object relations (Graham and Thrift, 2007).

The consumption of things helps shape **cultural politics** and culture itself. Gregson and Crewe's (2003) research on second-hand cultures and places demonstrated that consumption is concerned less with an individual's identity production than with the maintenance of social relations. Things purchased, used and given away demonstrated relationships of love, care and friendship. They were about the ways in which things were valued and devalued economically, symbolically and socially in different contexts and through different events (e.g. death of a spouse, break-up or formation of a relationship, having a family, moving home). Consumption provides a way of shaping both intersectional relations (class, gender, sexuality, religion, etc.) but also intergenerational relationships (see box).

Socialities are important in constituting notions of spaces such as 'home' and 'work' and how these are reproduced through practices such as eating, interior design, fashion and discarding. Thus a mundane act like using a domestic appliance or disposing of waste can connect one to a series of social relations which shape politics at other scales, a politics that is shaped not just through the self but in wider social formations – families, firms, communities, schools – which may both reproduce or contest the existing power relations and forms of governance.

CASE STUDY

Shaping 'good' parents and families through consumption

In the last few years I have been involved in research projects which examine the ways in which New Zealanders in mid- and later-life interact with families, friends and organizations through consumption. For older adults who were parents, purchasing and acquiring commodities provided a means of maintaining **sociality** with adult children and grandchildren, particularly where these children lived distant. Here, Stephen reflects on how his online shopping for his adult children is a means of demonstrating love and care:

> Yeah. Well I think that, I think that's important [to look out for and buy stuff for the kids] even although we don't live particularly close to our kids, I kind of like to. I was just thinking this morning, we're not a particularly close family but I kind of like to try to encourage that cause I'm, bit of . . . see my mother died quite young.
>
> (Mansvelt, 2010)

Figure 26.3 Shopping with and for children can be significant in maintaining familial relationships. Credit: Stewart Cohen/Getty Images

For Stephen and many other parents, 'shopping for' family was also a strongly moralized practice – a means of shaping family materially and symbolically through selfless provisioning which characterizes the 'good' parent and grandparent. Where family lived nearby, the pleasure and sociality of 'shopping with' children and grandchildren was also significant in maintaining familial relationships. The capacity to gift and purchase for others differed significantly across living standards. However, it was not so much the economic value of purchases, or even the material qualities of things gifted, passed on, shared or bought which was important to establishing familial and parental identities, rather it was the symbolic meanings of practices of giving, the representational traces of things moved on and the sociality of consuming 'for' and 'with' which seemed to matter most (Mansvelt, 2012).

Subjectivities

Thinking about consumption in terms of **subjectivities** helps us consider the ways in which consumption provides a medium through which our identities are reproduced and positioned. Many of us may not question that we are consumers and yet, as historian Frank Trentmann (2006) argues, the consumer identity is one which, only over the course of the last couple of centuries, has become an important subject position as part of economic, cultural and political discourse and practice. While 'the consumer' is a means of framing an individual's subjectivity, it is a problematic concept because it is framed and reproduced materially and discursively in a variety of way by organizations such as retailers, marketers, states and media, in and across different spaces (Gabriel and Laing, 2006).

Though geographers have shown that much of our daily consumption practice is concerned with notions of sociality rather than individual, anxiety-ridden identity production, our subjectivities are shaped in part through consumption (see box). Katz and Marshall (2003: 5) suggest that across the life course we are appealed to as consumers because 'All activities from leisure to healthcare, including ironically, death and dying have become personal, consumerist, and "lifelong" experiences'. Consequently a vast range of commodities such as foods (e.g. vegan, vegetarian, 'ethnic', slow food, fast food, low-carb, organic) may provide a basis for establishing and communicating different senses of sense of self. Choices do not occur in a vacuum and are influenced by material, social and practical considerations (cost, preferences of others in household, availability of produce). In addition, through consumption our subjectivities may be mapped onto bodies. For example, we can discipline and shape bodies through the purchase and use of exercise equipment, adorn and dress bodies as a sign of distinction or similarity to others, and eat to 'become the body'. Tattooing or body piercing, like other embodied consumption practices, can be a basis for social and spatial inclusion or exclusion based on gender, socio-economic status, ethnicity, etc., in different contexts, cultures and times.

Individuals are not only encouraged to buy, but to make the right consumption choices. Responsibilities to consume 'ethically', in ways that care for human and non-human others, the environment, the future sustainability of the planet, place and for the self may place

CASE STUDY

'DIY – It's in our DNA'

The tag line above has been used in a series of advertisements for a retail chain in New Zealand which sells hardware and home renovation products. The association of the 'Kiwi' identity, and masculinity in particular, with ingenuity, productiveness and an amateur do-it-yourself culture is not new, but emerged in part from European Settler Society and the frontier culture associated with the colonization of indigenous land and peoples. Berg (1994] argues this shaped New Zealand masculinity as both rural and productive – with urban spaces and masculinities coded feminine and 'soft'. The practicality and resourcefulness required to purchase and use tools as opposed to domestic appliances (which were the domain of women) became part of the formation of the stereotypical 'good Kiwi bloke' of the 1950s to 1970s, facilitated by state policies aimed at securing and maintaining a privately owned home (Phillips, 1996). While Phillips argues that this stereotypical view of masculinity no longer has such resonance, it is still a narrative drawn on in alcohol and home renovation advertising and in the establishment of 'Menz Sheds' (workshop spaces for older men). Thus, symbolic imaginaries generated around consumer subjectivities can be powerful, shaping the ways in which people produce and project a subjective sense of self (as masculine or feminine), but also impacting on socialities which may lead to social and spatial inclusion and exclusion and the constitution of material places (for example homes, gardens, city and rural spaces).

Figure 26.4 The consumption of commodities and the commodity practices surrounding them can have an important role in the ascription of gendered identities. Credit: iStockphoto

additional demands and responsibilities on people. Numerous firms and organizations promote consumer subjectivities associated with purchase of so called ethical products, despite the fact that everyday consumption involves moral and political choice and action (Barnett *et al.*, 2005). This is exemplified in 'the celebritization of consumption' (Goodman, 2010) where famous movie and music personalities (e.g. U2 lead singer Bono, Coldplay's Chris Martin) exhort individuals to purchase or engage in consumer activism in relation to ethical causes as a form of 'cool consumption' (Andrews *et al.*, 2011). Governments have also encouraged consumers to consume appropriately and responsibility as part of connecting citizenship and consumption. Consequently, while reflections on the shaping of identities may reveal we consume as a part of 'accumulating being' (Noble, 2004), we also consume because of who we are connected to and the kinds of relations and territories we are embedded in. Spatialities, socialities and subjectivities are intimately connected!

SUMMARY

- Three dimensions of consumption that geographers have focused on are spatialities, socialities and subjectivities. While these dimensions have been separated for the purposes of this chapter, they are intimately connected and often inseparable.
- Consumption both requires and makes particular arrangements of spaces that are expressed as spatialities, and these may extend from bodies to the global. Geographers have been instrumental in examining the ways in which places are created relationally through consumption practices that involve spaces of commodity exchange/purchase, use and disposal.
- Spaces, practices and discourses of consumption play an important role in shaping socialities – the ways in which we relate to other people and things. Such relations are important in the formation and maintenance of contestation of relations of power.
- Subjectivities are also shaped through practices and discourses arising from consumption, with the commodities bought and ways these are used and disposed of contributing to the production of identity and moralized subject positions.

Conclusion

Geographical research on consumption has demonstrated that it is much more than the purchase of commodities. Consumption involves the production of spatialities, socialities and subjectivities playing a critical role in the ways in which places, people and things are connected and made meaningful and experienced. As a sphere of social processes involving purchase, use, disposal and wasting of commodities, consumption takes place in particular spatial configurations and through numerous forms, such as bodies, homes, shelters, streets, shops, workplaces, urban and rural areas. Spaces of consumption and the consumption of space are not simply expressed in changing material forms but as sites of meaning, practice, power and imagination. Socialities and subjectivities associated with consumption play an important part in the social reproduction of societies, with consumption practices significant in cultural politics and the production of morally inflected subject positions. Geographers have highlighted the ways in which consumption provides a critical lens on the world; literature which speaks to the ways in which things, places and peoples are valued and devalued economically, politically and socially. They have begun to examine how, why and where such valuations are played out in a globalizing world, providing a basis to engage in debates about the possibilities of, and limits to, consumption as we currently know it.

DISCUSSION POINTS

1. What do you think consumption is? Why do you think it is important to consider practices of commodity use and disposal, as well as their purchase as part of consumption?
2. Think of your last purchase of clothing. How easy is it to separate the cultural and economic aspects of value connected with the purchase and use of this item? What is produced through the consumption of this item, and what has been consumed in its production?
3. Can you think of a specific example of a consumption practice in your own life which connects you to places across multiple scales?
4. Consider how relations of age, gender, class and race might be produced through the consumption of food, music and/or housing.
5. Why might it be in the interests of the state to make connections between consumers and citizenship?
6. What demands and responsibilities are placed on you as a consumer? From where (and whom) do these derive?

FURTHER READING

Goodman, M.K., Goodman, D. and Redclift, M. (eds.) (2011) *Consuming Space. Placing Consumption in Perspective.* Farnham and Burlington: Ashgate.

Mansvelt, J. (2005) *Geographies of Consumption.* London: Sage.

These books provide a range of chapters exploring differing aspects of geographies of consumption.

Gregson, N. and Crang, M. (2010) Materiality and waste: inorganic vitality in a networked world. In: *Environment and Planning A*, 42(5), 1026–32.

Gregson, N. and Crewe, L. (2003) *Second-hand Cultures*. Oxford: Berg.

This book provide a view of consumption beyond purchase of commodities, extending insights into the subjectivities, spatialities and socialities associated with second-hand consumption and disposal.

Clarke, D.B., Doel, M.A. and Housinaux, K.M.L. (2003) *The Consumption Reader*. London: Routledge.

This edited collection contains a range of social science perspectives on consumption, both theoretical and empirical, and includes a section on geographies.

Jayne, M. (2006) *Cities and Consumption*. Abingdon, Oxon and New York: Routledge.

An insightful analysis of the way in which consumption shapes urban areas and vice versa.

WEBSITES

www.thewasteoftheworld.org/

Waste of the World website: introduces the work by consumption geographers in collaboration with both social scientists and scientists into the materiality of waste and the relational geographies which arise from processes of devaluation and destruction.

www.youtube.com/watch?v=OqZMTY4V7Ts&feature=relmfu

A series of YouTube videos called 'The Story of Stuff' which draw on the linear commodity chain metaphor to encourage consumers to think about the connections between production and consumption and the ways in which these might be shaped more sustainably.

CHAPTER 27
COMMODITIES

Michael Watts

> A commodity appears at first glance a self-sufficient, trivial thing. Its analysis shows that it is a bewildering thing, full of metaphysical subtleties and theological capers.
>
> (Karl Marx, *Capital*, 1867)

A commodity is a bewildering thing: the capitalist cosmos and the world of commodities

With its price tag, said the great German critic Walter Benjamin, the commodity enters the market. In the capitalist societies, that is to say the market economies that we inhabit, this appears perfectly obvious. The *Oxford English Dictionary* defines a commodity as something *useful* that can be turned to *commercial advantage* (significantly, its Middle English origins invoke profit, property and income); it is an article of trade or commerce, a thing that is expedient or convenient. A commodity, in other words, is self-evident, ubiquitous and everyday; it is something that we take for granted.

Commodities surround us and we inhabit them as much as they inhabit us. They are everywhere, and in part define who and what we are. It is as if our entire cosmos, the way we experience and understand our realities and lived existence in the world, is mediated through the base realities of sale and purchase. This cosmos, one might say, is dominated by shopping. But the commodity economy is *more* than retailing and it is this surfeit that led Marx to refer to the metaphysical subtleties of the commodity. Virtually *everything* in modern society *is* a commodity: books, babies (is not adoption now a form of negotiated purchase?), debt, sperm, ideas (intellectual property), pollution, a visit to a national park and human organs are all commodities. An Italian tourist company offered the experience of war – a two-week tour of ethnic cleansing in Yugoslavia – as a commodity for sale; a group of internationally known models ('the commodified face') put their ova up for auction ($80,000.00 and up!). The advent of the internet and the rise of eBay and electronic auctions of various sorts have, of course, vastly expanded the ability to engage in commodity exchanges – to bring buyer and seller together virtually – for a bewildering array of detritus. Even things that do not exist as such appear as commodities. For example, I can buy a 'future' on a basket of major European currencies, which reflects the average price (the exchange rate) of those national monies at some distant point in time. Other commodities do not exist in another sense;

they are illegal or 'black' (heroin, stolen organs). Others are fictional (for example, money scams and fraud). Visible or invisible, legal or illegal, real or fictious, commodities saturate our universe.

As someone once said: 'in America virtually everything is for sale . . . which means virtually everything is a commodity'. This may be of little comfort to you. But one way of thinking about contemporary capitalist societies like the USA or the United Kingdom, in which virtually everything is a commodity (i.e. for sale), is that it is a *commodity economy*. It is, in other words, a system of commodities producing commodities. So why examine commodities if they are so trivial and ubiquitous? And why might they be of interest to geographers?

Well, one issue is that commodity-producing societies – by which I mean the dominating principle is commodities producing commodities – are quite recent inventions historically speaking, and many parts of the world, while they may produce for the market, are not commodity societies in the same way as our own. Socialist societies (and perhaps parts of China and Cuba today), stood in a quite different relationship to the commodity than so-called advanced capitalist states. Low-income countries, or the so-called third world, are 'less developed' precisely because they are not mature commodity-producing economies (markets are undeveloped or incomplete, as economists might put it). In the peasant village in which I lived in northern Nigeria in the 1970s, much of what was produced by family farmers (i.e. peasants) did not pass through the market at all. It was directly consumed or entered into complex circuits of gift-giving and non-market exchange. The rural household as a unit of production was not a commodity producer; it was not fully *commoditized*.

So the full commodity form as a way of organizing social life has little historical depth; that is to say it appeared in the West within the last 200 years. It is derivatively part of what Max Weber called the 'spirit of modern **capitalism**', but it remains an unfinished project if viewed globally. Over large parts of the earth's surface the process of *commodification* – of ever greater realms of social and economic life being mediated through the market as a commodity – is far from complete. Perhaps there are parts of our existence, even in the heart of **modernity** (see Chapter 32), that never will take a commodity form. After all, I do not purchase my wife's labour power or affection . . . yet; neither do I buy the chance to go biking with my young son . . . yet. But in a commodity economy in which the logic of the market rules, the prospect of converting social intimacy into a commodity is always present. Indeed, it is happening before our eyes. Adoption is part of a market in babies; child and elderly care is sustained through market transaction. My neighbour's pet canine is walked by a paid and 'professional' dog walker.

Another peculiarity of a commodity economy is that some items are traded as commodities but are not intentionally produced as commodities. Cars and shoes are produced to be sold on the market. But labour, or more properly labour power, is also sold – I sell something of myself to my employer, the University of California – and yet it (which is to say me as a person) was not conceived with the intention of being sold. Since I am not a slave, I was not in any meaningful sense produced, like a manufactured good or a McDonald's hamburger. This curious aspect of labour as a commodity under capitalism is as much the case for land or nature. These sorts of curiosities are what Karl Polanyi in his book *The Great Transformation* (1947) called 'fictitious commodities'. Polanyi was of the opinion that market societies that do not regulate the processes by which these fictitious commodities become commodities will assuredly tear themselves apart. The unregulated, free market, commodity society would eat into the very fabric that sustains it by destroying nature and by tearing asunder the most basic social relationships. We need look no further than the booming trade in human organs. In her book *Contested Commodities*, Margaret Radin shows how the fact that a poor Indian woman sells her kidney and other organs out of material desperation 'threatens the personhood of everyone' (1996: 125).

Not least, there is the tricky matter of price, which after all is the *meaning* of the commodity in the capitalist marketplace: how it is fixed and what stems from this price fixing. For example, the running shoe that a poor inner-city child yearns for is Air Nike, which costs slightly more than the Ethiopian GNP per capita and perhaps more than his mother's weekly income; or the fact that a great work of art, Van Gogh's *Wheat Field*, is purchased for the astonishing sum of $57 million as an investment.

The problem of the determination of prices and their relations to *value* lay at the heart of nineteenth-century classical **political economy**, but it is an enormously complex problem that really has not gone away or in any sense been solved. The 'metaphysical subtleties' that Karl Marx refers to are very much about the misunderstandings that arise from the way we think about prices (doesn't it have something to do with supply and demand?) and what we might call the sociology or social life of commodities. But if there is more to commodities than their physical properties and their prices, which are derived from costs of production or supply and demand curves, there is a suggestion that commodities are not what they seem. Commodities have strange, perhaps metaphysical, effects. For example, the fact that a beautiful Carravagio painting is a commodity – and correlatively, that it is private property and only within the means of the extravagantly rich – fundamentally shapes my experience of that work and of my ability to enjoy its magnificent beauty in some unalloyed way. Its commodity status has tainted and coloured my appreciation of it. Commodities cannot escape the grip of money and money carries a particular odour (of shit, said Sigmund Freud!) and a cool, glacial quality (Simmel, 1990).

A commodity, then, appears to be a trivial thing – here's a car for sale, it has these fine qualities – but it is in fact bewildering, even theological. The commodity, said Walter Benjamin, has a phantom-like objectivity, and it leads its own life after it leaves the hands of its maker. What on earth might this mean?

SUMMARY

- A commodity is something useful that enters the market.
- Commodification refers to the process by which more and more of the material, cultural, political, biological and spiritual world is rendered as something for sale.
- A commodity-producing economy is one in which the logic of commodification is dominant.

A commodity biography: the social life of the chicken and US capitalism

> A century ago you'd eat steak and lobster when you couldn't afford chicken. Today it can cost less than the potatoes you serve it with. What happened in the years between was an extraordinary marriage of technology and the market.
>
> (John Steele Gordon)

Once in a while I will bring into my undergraduate class a freshly dressed chicken – oven-ready in poultry parlance – and ask students to identify this cold and clammy creature that I've tossed upon the lectern. After five minutes of 'it's a chicken', 'it's a dead bird', 'it's a virtual Kentucky Fried Chicken', I solemnly pronounce that it is none of the above: it is in fact a bundle of social relations.

So let's examine the humble chicken. According to the latest Agricultural Census, 7 billion chickens were sold in the USA in 1994 (roughly 30 per person). In 1991, chicken consumption per capita exceeded beef, for the first time, in a country that has something of an obsession with red meat. The fact that each American man, woman and child currently consumes roughly 1.5 pounds of chicken each week reflects a complex vectoring of social forces in post-war America. First, a change in taste driven by a heightened sensitivity to health matters and especially the heart-related illnesses associated with red meat consumption. Second, the fantastically low cost of chicken meat, which has in real terms *fallen* since the 1930s (a century ago Americans would eat steak and lobster when they could not afford chicken). And not least the growing extent to which chicken is consumed in a panoply of forms (Chicken McNuggets, say) which did not exist 20 years ago and which are now delivered to us by the massive fast-food industry: a fact that, itself, points to the reality that Americans eat more and more food outside of the home (food consumption 'away from home' is, by dollar value, 40 per cent of the *average* household food budget).

The vast majority of chickens sold and consumed are broilers (young chickens) which, it turns out, are rather extraordinary creatures. In the 1880s there were only 100 million chickens. In spite of the rise of commercial hatcheries early in the century, the industry remained a sideline business run by farmers' wives until the 1920s. Since the first commercial sales (by a Mrs Wilmer Steele in 1923 in Delmarva, who sold 357 in one batch at prices five times higher than today), the industry has been transformed by the feed companies, which began to promote integration and the careful genetic control and reproduction of bird flocks, and by the impact of big science, often with government backing. The result is what was called in the 1940s the search for the 'perfect broiler' (Boyd, 2001). Avian science has now facilitated the mind-boggling rates at which the birds add weight (almost five pounds in as many weeks!). The average live bird weight has almost *doubled* in the last 50 years; over the same period the labour input in broiler production has fallen by 80 per cent. The broiler is the product of a truly massive R&D campaign; disease control and regulation of physiological development have fully industrialized the broiler to the point where it is really a cyborg: part nature, part machine. The chicken is industrial and fully commodified: its genome is mapped, particular breeds owned through intellectual property rights, and its shape and form the product of human intervention: a featherless chicken has recently been 'invented' in Israel and Pfizer owns the patent on a rooster developed for massive combs (the red fleshy growth –

Figure 27.1
Battery farm chickens.
Credit: UNEP/Still Pictures

hyaluronan – on its head) which is used in various surgical operations and for reconstructive surgery. Our understanding of chicken nutrition now exceeds that of any other animal, *including* humans! Applied poultry science and industrial production methods have also been the key to the egg industry. A state-of-the-art hen house holds 100,000 birds in minuscule cages stretching the length of two football fields; it resembles a late twentieth-century high-tech torture chamber (see Figure 27.1). The birds are fed by robot in carefully controlled amounts every two hours around the clock. In order to reduce stress, anxiety and aggression (which increases markedly with confinement), the birds wear red contact lenses, which, for reasons that are not clear, reduce feed consumption and increase egg production. It's pretty weird.

Broilers are overwhelmingly produced by family farmers in the USA, but this turns out to be a deceptive statistic. They are grown by farmers under contract to enormous **transnational corporations** – referred to as 'integrators' in the chicken business – who provide the chicks and feed. The growers (who are not organized into unions and who have almost no bargaining power) must borrow heavily in order to build the broiler houses and the infrastructure necessary to meet contractual requirements. Growers are not independent farmers at all. They are little more than underpaid workers, what we might call 'propertied labourers', of the corporate producers who also dominate the processing industry. Work in the poultry processing industry, in which the broilers are slaughtered and dressed and packaged into literally hundreds of different products, is some of the most underpaid and dangerous in the country (in the *New York Times* of 9 February 1998, p. A12, a government report cited almost two-thirds of all poultry processing plants as in violation of overtime payment procedures). Immigrant labour – Vietnamese, Laotian and Hispanic – now represents a substantial proportion of workers in the industry. The largest ten companies account for almost two-thirds of broiler production in the USA. Tyson Foods, Inc., the largest broiler producer, accounts for 124 million pounds of chicken meat per week and controls 21 per cent of the US market with sales of over $5 billion (two-thirds of which go to the fast-food industry). According to Don Tyson, CEO of Tyson Foods, his aim is to 'control the center of the plate for the American people'. Yet it is an industry on the verge of crisis. Disease resistance to the massive chemically-based nutrition and health regimen associated with confinement and industrial production, the worldwide threats of massive avian flu and the documentation of extraordinary levels of toxins in chicken meat (most recently terrifying levels of arsenic ingestion), all speak of the rickety structure of the contemporary chicken industry.

The heart of the US chicken industries is in the ex-slave-holding and cotton-growing south. Until the Second World War the chicken industry was located primarily in the Delmarva peninsula in the mid-Atlantic states (near Washington, DC). During the 1940s and 1950s the industry moved south and with it emerged the large integrated broiler complexes – what geographers call flexible or **post-Fordist** capitalist organization. The largest producing region is Arkansas (the home state of former US President Bill Clinton) and the chicken industry has been heavily involved in presidential political finance and lobbying, including a case in which the Secretary of Agriculture was compelled to resign. The lowly chicken reaches deep into the White House.

The USA is the largest producer and exporter of broilers, with a sizeable market share in Hong Kong, Russia and Japan. Facing intense competition from Brazil, China and Thailand, the chicken industry is now global, driven by the lure of the massive Chinese market and by the newly emerging and unprotected markets of Eastern Europe and the post-Soviet states. The world chicken market is highly segmented: Americans prefer breast meat, while US exporters take advantage of foreign preference for leg quarters, feet and wings to fulfil the large demand from Asia. The chicken is a thoroughly global creature – in its own way not unlike the global car or global finance.

You start with a trivial thing – the chicken as a commodity for sale – and you end up with a history of post-war American capitalism.

The commodity

> It is in this sense that a product is a commodity; it is simultaneously a use value for the other, and a means of exchange for the producer.
>
> (M. Postone, *Time, Labour and Social Domination*, 1993)

One way to think about the commodity is derived from Karl Marx who begins his massive treatise on capitalism (Volume 1 of *Capital*) with a seemingly bizarre and arcane examination of the commodity, with what he calls the 'minutiae' of bourgeois society. The commodity, he says, is the 'economic cell form' of capitalism. It is as if he is saying that in the same way that the DNA sequence holds the secret to life, so the commodity is the economic DNA, and hence the secret of modern capitalism. For Marx, the commodity is the general form of the product – what he calls the generally necessary form of the product and the general elementary form of wealth – *only* in capitalism (Postone, 1993). A society in which the commodity is the general form of wealth – the cell form – is characterized by what Postone (1993: 148) calls 'a unique form of social interdependence': people do not consume what they produce, and produce and exchange commodities to acquire other commodities.

But the commodity itself is a queer thing because while it has physical qualities and uses, and is the product of physical processes that are perceptible to the senses, its *social* qualities (what Marx calls the social or value form) are obscured and hidden. 'Use value' is self-evident (for example: this is a chair, which I can use as a seat and has many fine attributes for the comfort of my ageing body) but 'value form', the social construction of the commodity, is not. Indeed this value relation – the ways in which commodities are constituted, now and in the past, by social relations between people – is not perceptible to the senses. Sometimes, says Marx, the social properties things acquire under particular circumstances are seen as inherent in their natural forms (that is, in the obvious physical properties of the commodity). The commodity is not what it appears. There is, then, a hidden life to commodities and understanding something of this secret life might reveal profound insights into the entire edifice – the society, the culture, the political economy – of commodity-producing systems.

Let's return to the chicken as a commodity. It has two powers: First, it can satisfy some human want (my need for a chicken curry). This is what Adam Smith called a *use value*. Second, it has the ability to command other commodities in exchange. This power of exchangeability Marx called *exchange value* or value form. Use values coincide with the *natural form* of the commodity – its chicken-ness – whereas the value form expresses its *social form*. Use values express the qualitative incommensurability of commodities – the uses of chicken can never be commensurable with the uses of a car – whereas exchange value expresses quantitative commensurability (I can exchange 20,000 chickens for one car). But commodity exchange in turn requires a universal equivalent to facilitate this quantitative commensurability: which is to say, money. The commensurability of commodities is expressed phenomenally through money, that is, in the form of a price (a chicken is $1.00 per pound and turkey is $2.00 per pound, which means that chicken exchanges for turkey at the rate of two chickens for one turkey). *Commodity circulation* refers to the process by which a commodity is exchanged for money, which in turn permits the purchase of another, different, commodity.

But the basis of comparing commodities through price is only its phenomenal form. The real basis is *value*. But what is value? For Marx it turns on what he calls abstract labour. Any labour whatsoever is viewed as a process of consumption of human energy, so that value is expended labour. Capitalism, however, is a reality not an abstraction, and rests upon commodities producing commodities in quite specific sorts of ways. More specifically, a capitalist starts with money, purchases labour power and the **means of production**, and produces commodities that are sold for money. This process generates more money for the capitalist than he began with (i.e. profit). Put differently, money and commodities circulate as expressions of the expansion of capital (not just the circulation of commodities) which rests upon the existence of a commodity-producing system and the emergence of a universal equivalent (money) (see later).

Exchange value is, to summarize, bound up with the particular form of social labour as it exists in a commodity-producing economy (capitalism). This particular form of social labour is wage labour. That is to say, there is a class of individuals (workers) who sell their labour power to another class (capitalists), which organizes production. It is this aspect of labour – as a commodity, labour power, sold to someone else at a price called the wage – that is unique to a commodity-producing economy. A commodity can thus be seen as being composed of three aspects or relations: variable capital (that paid to labour as wages), constant capital (that covering fixed capital costs such as machines) and surplus value (profits). Marx tried to explain where this surplus came from and how it emerged from the disparity between the value that the workers embody in the commodities they produce and the value they require for their own reproduction.

Whether Marx's theory of value or account of the origins of profit is right or wrong is of less relevance than the fact that the commodity allows us to analyse the forms that arise on the basis of a well-developed or full commodity economy. Capitalism is unique because it rests upon commodities that are fictitious. Labour, capital and money are all commodities (the wage, rate of interest and exchange rates determine their respective market value) but are not produced as commodities. The commodity is the way into the problem of value (its origins and its forms) and it establishes a sort of toolkit with which we can understand something of the distinctiveness of living in capitalist societies. Property, money, value in its various forms, class interests, the circulation of commodities, they are all implied in the '**dialectical** union' of use and exchange value that the commodity contains.

SUMMARY

- A commodity is a unity of exchange value and use value.
- Exchange value expresses the social and hidden form of the commodity.
- Commodity exchange requires a universal equivalent (money).

The commodity circuit

As tangible, physical things and as the embodiment of particular uses and values, commodities have lives, or *biographies*. They are made, born or fabricated; they are fashioned and differentiated in a variety of ways; they are sold, retailed, advertised and ultimately consumed or 'realized' (and perhaps even recycled!). The life of the commodity typically involves movement through space and time, during which it adds values and meanings of various forms. Commodities are therefore pre-eminently geographical objects.

Returning to the US chicken, it is possible to construct a diagrammatic 'biography' of the broiler from production to consumption, which depicts many of the actors involved in the commodity's complex movements and valuations. This is a *commodity circuit* or a *commodity chain* (in French it is referred to as a *filière*) (Friedland *et al.*, 1981). Figure 27.2 depicts the US broiler *filière* (Boyd and Watts, 1996). At the centre of this figure is the broiler complex and the large transnational integrator – Tyson Foods, Perdue Farms, ConAgra, and so on – but there are obviously a multiplicity of other actors: the public extension systems, the R&D sector, the fast-food chains, the exporters, the retailers, the service providers, the state and local government. The starting point of the commodity circuit might be the breeding units, but this itself is a collaborative effort that has involved a half-century of genetics research to produce breeding flocks. What is bred is a peculiar creature that has been 'industrialized' to maximize productivity (but with the danger of massive disease problems, which itself

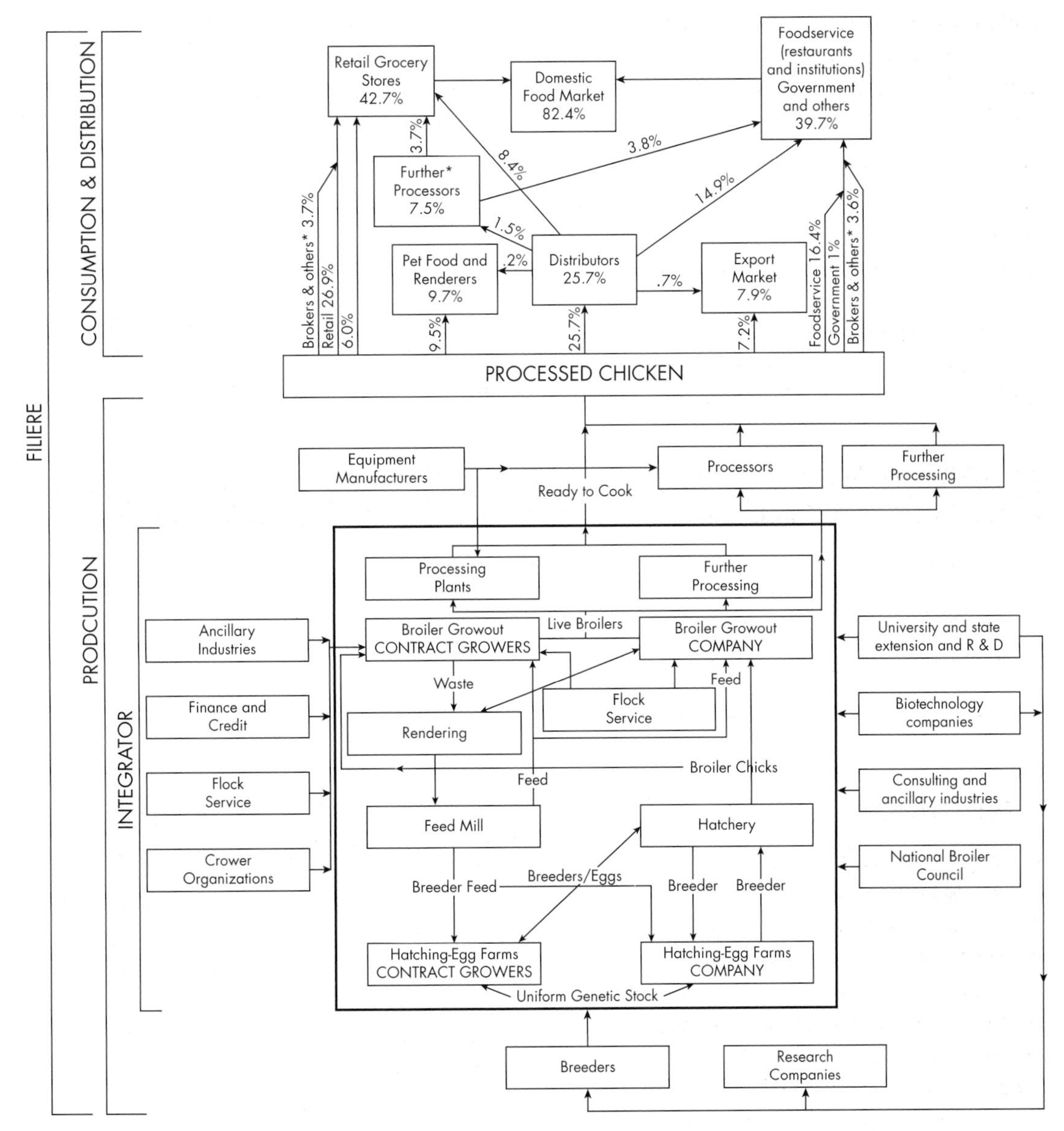

Figure 27.2 The broiler *filière c.* 1999. Source: Boyd and Watts (1996: 205)

generates one large part of the commodity circuit devoted to chicken 'health care'). The terminus is the consumption of chicken in a bewildering array of forms: as a complement to other foods in a TacoBell burrito, as an 'organic free range chicken' bought ready to cook in a yuppy store, or as 'mass' chicken parts destined for institutions like schools or hospitals. The chicken as a commodity has, in short, been further commodified. And the process is seemingly endless. At all points along the commodity circuit there are opportunities for the creation of all manner of new forms of chicken commodity.

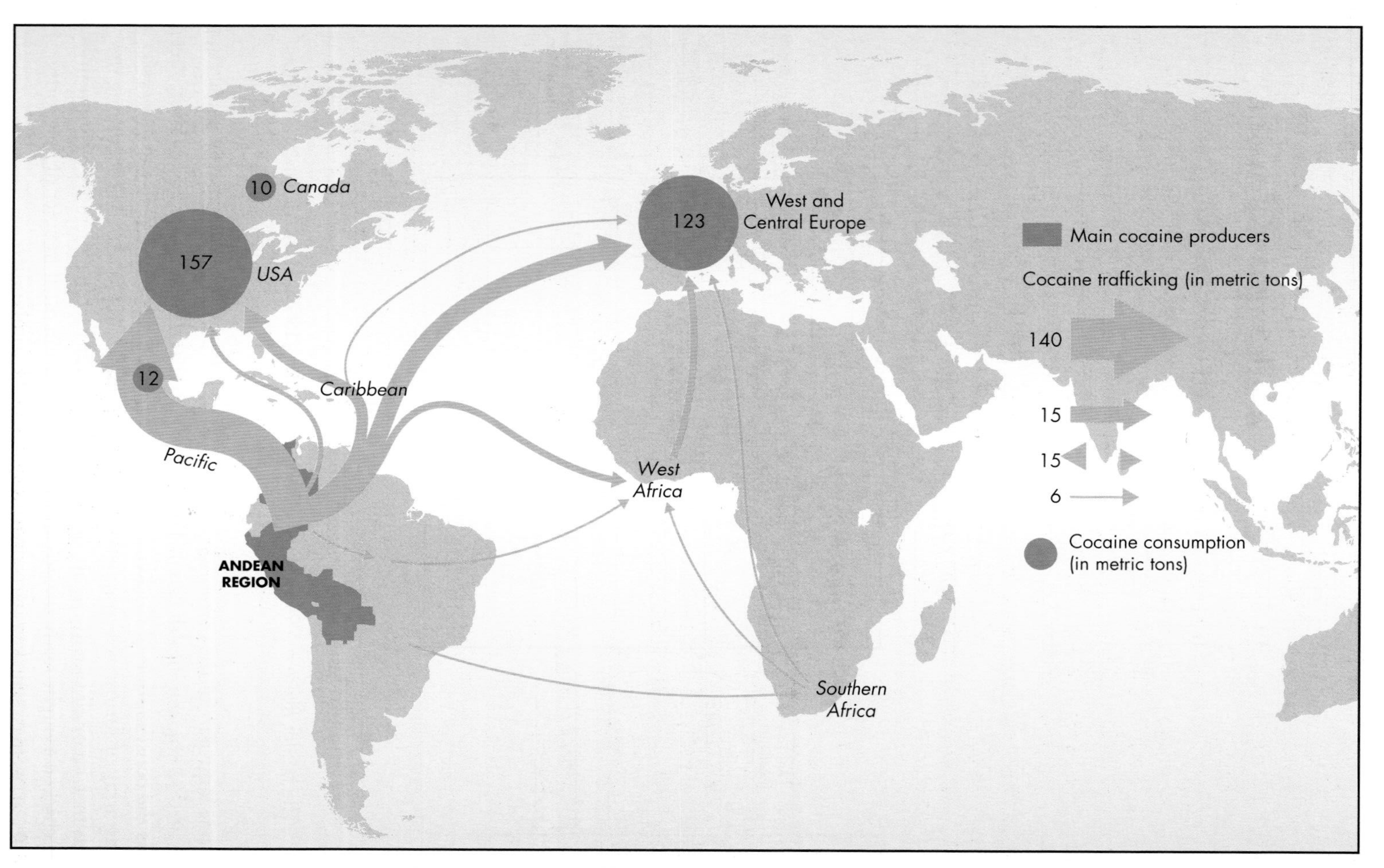

Figure 27.3 Cocaine commodity chain. Credit: Main global cocaine flows, 2009, World Drug Report 2011, United Nations Publication, Sales No. E.11.XI.10

Commodity circuits can depict different types of commodity chains and contrasting commodity dynamics. Figures 27.3 and 27.4 depict global commodity circuits for cocaine and for the apparel and automobile sectors. In the former, the peasant grower of coca leaf in Colombia is linked through a series of agents (processors, wholesalers, transporters) to the street dealer in, say, Detroit. It is an *illegal* commodity chain, which links the third world as producer to the first world as consumer. This has historically been the case for many third world drug commodity circuits (tea, coffee, sugar) which are, however, usually typically legal and dominated by rather different agents and actors (agribusiness companies rather than the Medellin Cartel). Figure 27.4 reveals different dynamics within two contrasting commodity circuits (Gereffi, 1995). The buyer circuit for which the apparel industry is the prime case is dominated by the *retailers*. The high-end fashion retailers (Armani, Donna Karan) typically produce apparel in sweatshops in the core countries or in newly industrialized countries like Hong Kong. Low-end supermarket retailers (such as WalMart) have producers in particularly poor and low-wage countries (Sri Lanka, Philippines).

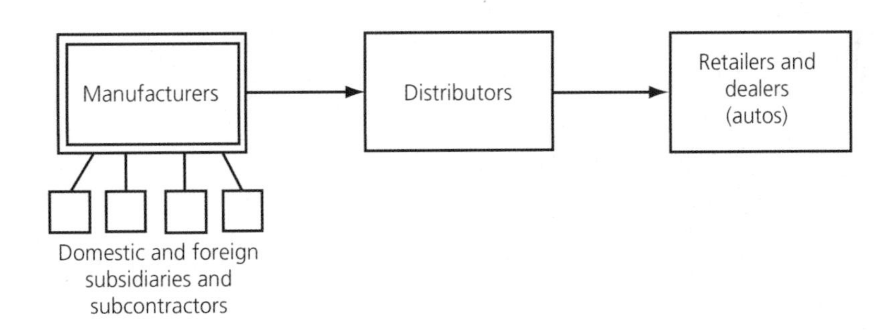

Buyer-driven commodity chains

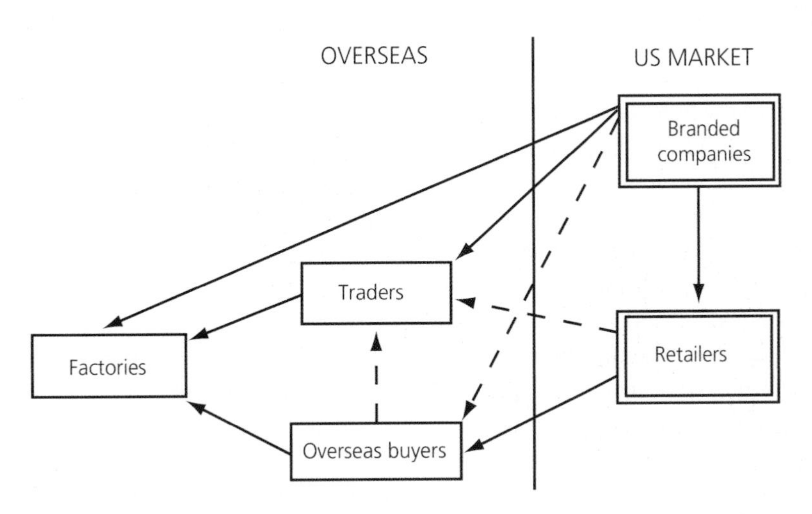

Figure 27.4 Producer-driven commodity chain. Source: Gereffi (1995)

Producers and retailers are often held together by complex subcontracting arrangements. In producer-driven chains, conversely (the automobile is an exemplar) transnational companies (TNCs), as integrated industrial enterprises, play the central role. Toyota or Ford have integrated production complexes embracing literally thousands of parent, subsidiary and subcontractor firms dotted around the world. The producer-driven commodity chain produces the 'world car' in which component parts are produced in multiple locations, although the final commodity, say, the Ford Fiesta, is assembled at one single site.

My earlier discussion of the US chicken industry highlights a number of key geographical aspects of commodity circuits. First, that different actors and agents in the circuit are linked together in complex market *and* non-market relations. At particular locations ('nodes') within the commodity circuit there are especially dense sets of social and institutional relations: contracts between growers and integrators, cooperative relations between companies and government, and so on. Second, throughout the course of the circuit the commodity itself is differentiated in enormously complicated ways into a panoply of new products and processes. According to the industry there are now literally hundreds of 'chicken products'. Third, the process of moving through the commodity chain is simultaneously a process of adding value. Growers command a low proportion of the final product price, which raises the question of who captures the value-added (in cocaine it is clearly the wholesalers and not street-level dealers or peasant growers). And finally, within each commodity chain there are particular *nodes* – which can be seen as 'sinks' of especially intense activity – in which the commodity acquires particular meanings and attributes (these are values but not necessarily economic values). In the broiler *filière*, for example, these qualities may have to do with freshness or the acquisition of a brand name (the 'Rocky Road' chicken), or the attachment of 'quality'. Quality is typically about status (think of the cachet of Michael Jordan Air Nikes), and the **semiotics** of the commodity (this is a *real* Stilton cheese produced *traditionally* in Leicestershire). A commodity circuit can, then, display both the space–time attributes of the commodity, many of which are now global, of course, and also of what Marx called the 'social' (and partially hidden) qualities of commodities: their value, their meanings and their fetish qualities.

Commoditization/ commodification

> [T]he idea of a self-adjusting market implied a stark utopia. Such an institution could not exist for any length of time without annihilating the human and natural substance of society.
>
> (Karl Polanyi, *The Great Transformation*, 1947)

The process by which everything becomes a commodity – and therefore everything comes to acquire a price and a monetary form (*commoditization/commodification*) – is not complete, even in our own societies where transactions still occur outside of the marketplace. Perhaps one might say that the process can never be complete, but history will always throw up new frontiers for the commodity process to penetrate. The reality of capitalism is that ever more of social life is mediated through and by the market. Karl Polanyi referred to this process as the *embedding* of social relations in the economy. On the one hand, he said more of social life is embedded in the logic of a commodity economy – industries are given permits to pollute, which can be bought and sold; nature is patented by private companies – and on the other, the market itself, if left to its own devices, becomes *disembedded* from social institutions.

Karl Polanyi was concerned to show that societies dominated by the self-adjusting market, in which individuals relentlessly pursued their own interests as Adam Smith suggested, would be no society at all. Rather, every person is pitted against each other in a state of quasi-war. Smith recognized the costs of unbridled accumulation and saw civil society as the necessary saviour of a market system whose powers he so admired. If the genesis of market-regulated societies carries the prospect of disembedded markets and economically embedded social relations, they remain tendencies rather than inevitabilities. The case of the chicken industry revealed, of course, that in a highly competitive and market-driven broiler industry, markets and commodities are indeed socially embedded. Economics is, as Polanyi put it, *an instituted process*. This institutedness takes the forms of social alliances, networks and studied forms of trust between actors in the commodity system. Firms build up relations of trust and cooperation between one another; the relationship between grower and integrator is contractual and not a pure market relationship. The vertically integrated, patrimonial Korean conglomerates (*chaebol*) such as Daewoo and the Taiwanese flexible, contractually linked, family firms are the socially embedded forms of market, commodity-producing behaviour associated with the so-called Asian miracle. Commodities are always fashioned in institutional and cultural ways.

In the far-flung corners of the globe, there are societies that are not commoditized at all, or at least the pursuit of 'commercial advantage' represents a minor part of their social existence. Some contemporary Indian communities in the Amazon, for example, or hunter-gatherer communities in Zaire, produce almost nothing for the market and buy little in the way of consumer goods. Some back-to-the-land communes in northern California also aspire to self-sufficiency. However, even these non-market (or non-capitalist) societies are typically commodified in some way. When I worked among isolated pastoral nomads in West Africa in the 1970s – small mobile families who depended entirely on livestock for their survival – these seemingly traditional communities did view cattle and other animals as property, and indeed would sell limited numbers of animals, particularly during the dry season when lactation rates of their animals had fallen due to the deterioration of pasture, in order to buy grain, and tea and sugar. Similarly, the communards on Albion ridge in northern California participate in a moral economy of barter and exchange with other communes (what Polanyi would have called a sort of administered trade).

Statistically speaking, one of the largest classes of people in the world is the peasantry and they are defined specifically by their *partial commoditization*. Peasants own the means of production: they directly work their land with their own family labour, which means that they do sell their labour power as a commodity. But peasants *are* involved with the market to some degree, selling part of what they grow (often export crops such as cotton or tea) to acquire money to buy clothes, pay taxes and cover school fees for their children. This partial commoditization can have some unusual consequences. Henry Bernstein (1978) pointed out that in such circumstances, a peasant family may produce commodities in order to gain cash, but should the price of this commodity fall (the price of bananas on the world market falls by, say, 40 per cent), the peasant may be forced either to produce more of a commodity whose price is falling or work harder just to meet his irreducible family needs. For a family with a small plot of land this may mean working longer and harder and exploiting the soil in order to, as it were, stay in the same place. Bernstein referred to this conundrum of price squeezes (commodity prices falling) and partial commoditization (household enterprises with irreducible consumption goals) as 'the simple reproduction squeeze'.

Polanyi's 'great transformation' was the process by which economy and society were separated during the course of the rise of the self-adjusting market: the emergence of the market economy as a *totality* in contradistinction to the patchwork market economy prior to the Industrial Revolution. The reach of the market, which is to say the extent to which everything has become a commodity, has deepened since the 1750s. What Marx called the 'vulgar commodity rabble' now encompasses much of what we do and feel. Human life has become dependent upon the market and commoditization inevitably charges inward into the refuges of social life. My colleague Nancy Scheper-Hughes, an anthropologist at the University of California, Berkeley, has been a major voice in the exposure of the so-called 'rotten trade' in human organs (Scheper-Hughes, 2002). Commercialization has entered almost every sphere of contemporary medicine and biotechnology and practices such as mortuary practices and tissue and DNA harvesting represent what she calls 'human strip farming'. Heart valves, cornea, bone fragments and other body parts are traded as a basis for research and teaching. Hughes has documented a new kind of organ trade emerging within the interstices of the global economy, what he calls 'transport tourism', which has its brokers and outlaws who short-circuit the problem of long waiting lists or donors and international codes of ethical conduct by acquiring organs from poor sellers illegally and sometimes by outright theft (kidneys stolen from Chinese prison inmates, for example). Organ brokers freely advertise on the internet, able to provide 'a living donor next week'. Jürgen Habermas (1993) calls this the 'colonization of the lifeworld'.

Commodities run amok: the financial and global climate change crises seen through the commodity lens

Credit, the universal equivalency of money and insurance are customarily hailed as the signal achievements of modernity, and specifically of capitalism's power to overcome not just a pre-modern notion of fate but to secure the very foundations of modern urban-industrial life. Credit is the lifeblood of our social and economic existence, not just simply of trade or enterprise, and insurance is a source of security (and profit) for financiers, capitalists, states and workers alike. The core of the American dream, to take one example, is the family-owned house: property, typically secured through credit (a mortgage).The great innovation of socialized insurance, for example, was to supplant the notion of individual fault by that of calculable accident: the probability accident and loss became calculable even if the specific distributions of harms (who actually suffers) were not (Johnson, 2011). Credit, insurance and risk became the touchstone of not just modern life but of modern rule and modern expertise.

Banking and insurance (and re-insurance) have a long and complex history, of course. The development of insurance and credit was central to the contribution of maritime trade to what has been called the 'commercial revolution' of early modern Europe. But this history, and the forms of insurance and finance associated with it, have vastly expanded in scale, scope and character in the period since the Industrial Revolution and indeed they have become the touchstones of globalization itself. Money continues to be lent through banks of various sorts but money itself can be abstracted and parcelled into bundles of different currencies. In other words, bundled into abstract averages of a basket of national currencies, and can be bought, sold and speculated upon in global financial markets. Money can be converted from conventional financial instruments like mortgages into different tranches of different degrees of risk and bought and sold as investment vehicles (the slicing and dicing of mortgage-backed securities). In the same way that property of various sorts can be insured, so too can life itself – so-called life insurance technologies – and virtually any sort of catastrophic risk (earthquake, hurricanes). The 'catastrophe modelling' of new emerging risks (and it goes without saying that technological innovation is typically synonymous with the creation of new risks) include terrorism, pandemic flu, longevity risk, catastrophic mortality, litigation epidemics, nuclear meltdowns and much more. As Leigh Johnson (2011) shows in her brilliant work, all of these perils were at one point considered 'uninsurable':

> . . . all have since been subjected to statistical manipulation and modeling in order to estimate their return periods (the frequency with which an event of a certain magnitude can be expected to recur) and damage curves (proprietary algorithms that model the types and extent of damages resulting from a simulated peril).
>
> (Johnson 2011: 19)

The disciplining of chance has, in short, become a huge market. Modelling firms in turn construct models and estimates for the ever-changing risk landscape in such a way that 'each new risk identified has a new cost . . . each new protection . . . makes visible a new form of insurable insecurity . . . security becomes an inexhaustible market' (Defert, 1991: 215). At the same time risk and security became an important object of modern rule. That is to say, populations are managed – both through public and private means – through a logic of security which integrates many aspects of production, psychology and human behaviour.

Why is the proliferation of complex financial transactions by massive corporate banks (often seen to be 'too big to fail') and the desire to render risk of various sorts calculable by the huge global insurance and reinsurance industry relevant to a discussion of the commodity? One answer is that the global financial crisis which exploded in 2008 with the collapse of Bear Stearns and Lehman Brothers – two major Wall Street investment banks – is an almost textbook case of Karl Polanyi's fictitious commodification run amok. Here, the so-called financialization of US capitalism assumed the form of creating, packaging mortgage and selling mortgage-backed securities through transactions that were neither monitored nor subject to any form of regulatory oversight. Credit-rating agencies whose function it is to assess independently the risks of particular tranches of high risk securities were in collusion with the investment banks and assigned ratings (from which they benefitted financially) wildly inconsistent with the riskiness of the security portfolios. Unregulated capitalism, coupled with the ever more complex commodification of credit and money and, it needs to be said, outright corruption and greed, produced a classic Polanyian crisis (in our case the great recession of the late 2000s as opposed to, in his case, the Great Depression of the 1930s).

Johnson's (2011) ground-breaking work of CAT (catastrophe) modelling and the reinsurance industry strategy to build a market in hurricane risk is another compelling illustration of commodification (and of Polanyi's spectre of the potentially disastrous consequences of fictitious commodification). She shows how the new institutional form of *reinsurance* was deliberately developed in response to the problem of urban fire, which was financially crippling for direct insurers due to the simultaneous and extreme nature of losses incurred across a large number of properties in one location (2011:18). The world's two largest reinsurers, Munich Re and Swiss Re emerged as a result of many ventures to convert disaster insurance into a hard market. Modern catastrophe reinsurance coverage for tropical storms, winter storms and earthquakes is still the primary mechanism by which property insurers survive extreme losses due to natural catastrophes (in the USA it generally includes losses due to fire and wind/snowstorm damage). In 2009, globally insurers collected US $4.1 trillion dollars in premiums for all lines of coverage, equivalent to seven percent of world GDP. Johnson's research shows how the climate science relevant to CAT modelling is deployed by the large insurers – scientific knowledge is both privatized and commodified in sum – and the financial instruments by which various climatic risks such as drought, hurricanes and floods are financialized.

Figure 27.5 Richard Branson and Al Gore. Credit: Getty Images

A similar case can be made as regards the governance of global climate change. Take for example the picture shown in Figure 27.5. It is a striking image: a global capitalist whose personal wealth is rooted in an industry – air transportation – distinguished by its massive carbon footprint, and Al Gore, a Nobel Prize-winning US politician and former vice president honoured for his contributions in placing global climate change, and the scientific work of the Intergovernmental Panel on Climate Change (IPCC) in particular, on the global political agenda. Tossing the globe into the air, British tycoon Sir Richard Branson announced to the world in 2007 that he was offering a $25 million prize for the scientist who discovers a way of extracting greenhouse gases from the atmosphere – a challenge to find the world's first viable design to capture and remove carbon dioxide from the air. Big Science meets Big Business meets Big Politics.

The United Nations Framework Convention on Climate Change (UNFCCC), which the United Nations adopted at the 2012 Rio Earth Summit, aimed to achieve the stabilization of greenhouse gas concentrations in the atmosphere at a level that would prevent dangerous anthropogenic interference with the climate system. The Kyoto Protocol, which was signed in 1997 and came into force in 2005, was established to realize these goals. CO_2 emissions are now 30 per cent higher than when the UNFCCC was signed; atmospheric concentrations of CO_2 equivalents are currently 430 parts per million. At the current rate, they could more than triple by the end of the century. In effect, this would mean a 50 per cent risk of global temperature increases of five degrees Celsius (the average global temperature now, for example, is only five degrees Celsius warmer than the last ice age). The seemingly inevitable catastrophic losses associated with global warming is, of course, central to Johnson's story of the re-insurance industry and its search for securable risks – which is to say the further financialization of the natural world. But the solution to the carbon emissions crisis has also been pursued by other market means: that is through various forms of 'cap and trade' in which permits to emit can be bought and sold (with the expectation that putting a price on carbon will produce innovations capable of substantially reducing emissions). All of which is to say carbon is converted into a commodity, and that commodity constitutes a carbon market. In turn, this market could be another avenue for financialization (carbon-based hedge funds, collateralized carbon credits) including catastrophic reinsurance (for the wealthy North but certainly not for the impoverished South).

Critics like Naomi Klein (2007) make the point that in our ascendant market order, even environmental calamity and reconstruction becomes a source of corporate profit and capitalist consolidation (so-called 'disaster capitalism'). Environmental risks, environmental catastrophes and environmental reconstruction all become sources of accumulation, not simply for an aggressively dominant finance capitalism, but for many sectors of the industrial economy (engineering, construction, extraction industries). In the current neoliberal order, the short-term future depends on a precise costing of climatic impacts, on getting the prices right, and on converting risks into both a market and a form of rule.

These illustrations of commmodification and market construction at work suggest a larger question, given the contemporary dominance of what has been called the 'neoliberal thought collective'. Namely, that the sort of hyper-commodification which is seemingly characteristic of our epoch rests upon a particular historical form of the person or subject, a form in which life is synonymous with the management of commodities. This has been put powerfully by Michel Foucault in his description, written presciently before the rise of Margaret Thatcher and Ronald Reagan, of the neoliberal order:

> I think the multiplication of the enterprise form within the social body is what is at stake in neoliberal policy . . . The stake of all neoliberal analyses is the replacement every time of *homo economicus* as partner of exchange with a *homo economicus* as entrepreneur of himself, being for himself his own capital, being for himself his own producer . . . The individual's life itself – with his relationships to his private property, for example, with his family, his household, insurance and retirement – must make him into a sort of permanent and multiple enterprise.
>
> (Foucault 2008: 148, 226, 241)

In our times, Marx's reflection upon the commodity – bewildering and metaphysical – is perhaps rendered true in ways he might never have anticipated.

Commodity fetishism and the commodity spectacle

> It is an enchanted, perverted, topsy-turvy world in which *Monsieur Le capital* and *Madame La Terre* do their ghost walking as

social characters and at the same time directly as things.

(Karl Marx, Capital, III, 1959)

In 1993, in a media stunt for her animal rights book, Rebecca Hall offered four men $2,500 each to live like battery hens for a week – in other words, barefoot in a small wire cage with a sloping floor, with 24-hour light, automated food delivery and a cacophonous noise (see Figure 27.6). They lasted 16 hours. Two years later one of the major US broiler firms hired a nationally known chef to use 'fresh' oven-ready chicken to play bowls outside of the US Congress, to demonstrate the fact that its competitors were in fact selling purportedly fresh chickens that were frozen as hard as titanium. Or take a look at the advertisement in Figure 27.7 for an 89 Male broiler: top 'livability', superior growth, excellent efficiency: chicken as machine.

These chicken tales are each speaking to quite different aspects of the commodity under capitalism, namely the spectacular and the fetishistic. A fetish is a material object invested with magical powers. Marx invoked *commodity fetishism* to describe the ways in which commodities have a phantom objectivity. The social character of their making is presented in a 'perverted' form. By this he meant a number of complex things. First, that the social character of a commodity is somehow seen as a natural attribute intrinsic to the thing itself.

Figure 27.6 Four men attempt to live like chickens in Rebecca Hall's cage. Credit: Martin Argles/Guardian

Figure 27.7 The Industrial Chicken advert. Credit: Avian Farms/Cobb Vantress Inc.

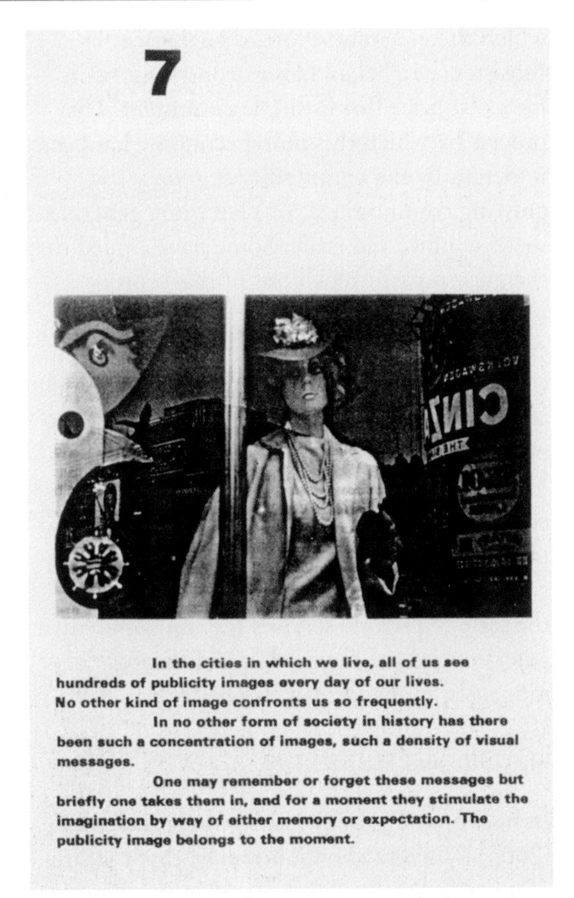
7

In the cities in which we live, all of us see hundreds of publicity images every day of our lives. No other kind of image confronts us so frequently.

In no other form of society in history has there been such a concentration of images, such a density of visual messages.

One may remember or forget these messages but briefly one takes them in, and for a moment they stimulate the imagination by way of either memory or expectation. The publicity image belongs to the moment.

Figure 27.8 Society as spectacle. Credit: Sven Blomberg

Second, that the commodities appear as an independent and uncontrolled reality, apart from the producers who fashioned them. And third, in confusing relations between people and between things, events and processes are represented as timeless or without history, they are **naturalized**. Another way to think about this is that commodity production – the unfathomable swirl of commodity life – produces particular forms of **alienation** and **reification**. Let me elaborate.

We come to accept the creature we buy as an oven-ready chicken as a natural product that stems from its use-value (as food). In fact, it is a sort of machine, something created by science to be an input (of particular proportion, colour, efficiency, and so on) into industrial manufacture; something that, far from being natural, is a social artefact containing many and complex forms of value. Moreover, the social value of the chicken appears, as it is under capitalism, in terms of relations between things (growth, efficiency, white meat). In our society virtually all of our existence appears as a thing – it is reified – but these reified things interact with us to give the impression that the social really *is* natural. The chicken futures market is 'up' this week – as though the market has a life of its own – which is confirmed to me because my shares in Tyson Foods carry increased dividends. This confusion or obfuscation of relations between people – the huge number of people involved in the community *filière* – with relations between things is central to the alienation rooted in a world in which everything is for sale, and everything is a thing.

In his book *Society of the Spectacle* (1977) Guy Debord argues that in a world of total commodification, life presents itself an as immense accumulation of spectacles. The spectacle, says Debord, is when the commodity has reached the total occupation of social life and appears as a set of relations mediated by images (see Figure 27.8). The great world exhibitions and arcades of the nineteenth century were forerunners of the spectacle, celebrating the world as a commodity. But in the contemporary epoch in which the representation of the commodity is so inextricably wrapped up with the thing itself, the commodity form appears as spectacle, or as a spectacular event, whether four men trying to be chicken or a chef playing bowls with a frozen broiler. Whatever else it may be, the terrifying events of September 11th 2001 and the collapse of the Twin Trade Towers represented an enormous spectacle in the Debordian sense; and a spectacle for which there could be no spectacular response of equal measure. Necessarily this spectacle of spectacles was a product of commodification and necessarily has become a commodity itself. Within weeks of the attacks, ground zero in New York City had become a small marketplace for September 11th T-shirts and other mementoes, just as shirts bearing the image of Osama and the falling towers were selling like hot cakes in Bangkok, Jakarta and the West Bank as icons of anti-imperialism.

Once they leave the confines of their makers, commodities take on a life of their own.

Resisting the commodity

Is commoditization an inevitable process, an implacable logic which must engulf the entire globe? As I explained earlier in this chapter, the process of commoditization is far from complete. Colin Williams (2005) shows that the contours of commodification are very uneven and often resisted, especially in societies undergoing a radical and rapid transition to a market economy. The **moral economy** – the embeddedness of economic relations – in pre-capitalist or partially capitalist societies (whether peasant communities in contemporary Borneo or in late eighteenth-century France) breaks down slowly and unevenly in the face of the onslaught of commodity production (Thompson, 1991). In my peasant community in Nigeria, for example, land was rarely sold, in large part because of the cultural and spiritual meanings attached to the land and the anti-social character of land sale (this remains an issue, for example, in Native American Indian communities in the USA). Poor families would steadfastly deny that they sold their labour because of the shame surrounding the fact of being perceived to be not self-sufficient. These are elements of a larger moral economy in which there is some effort to guarantee subsistence rights and also strong sentiments for a just price (for bread, for example). The process by which this moral economy has been undercut by the commodity economy (by growing commoditization) has often generated social conflict and strife. Some have argued that the great peasant rebellions of this century (Tonkin in the 1930s, Mexico earlier in the century) were efforts to defend the moral economy of the peasantry against the onslaught of the world market and the irrepressible logic of the commodity economy (Scott, 1976; Thompson, 1991). Of course, the moral economy is still not entirely dead; something of this remains in English villages or the crofting communities of the Hebrides. The efforts by the state to provide services (i.e. unemployment relief) outside of the market and through moneyless exchange can also be seen as an effort to *decommoditize* some realms of social life (Offe and Heinze, 1992).

In his important book *A Commodified World?* (2005) Williams shows, however, that the size of the non-commodified sector is vast. The scale of those activities – domestic work, car repair, working in community organizations, non-monetary exchanges of labor and resources, doing nothing – reveals however the contradictory ways in which non-commodified work and non-market exchange co-exist with the market in our world. On the one hand there are those millions who are as it were disposable, the casualties of a capitalism which excludes and immiserates (what one might call a planet of the wageless), and on the other a population for whom there are opportunities to escape or resist the implacable logic of the commodity. Williams recognizes that these different sorts of non-commodified work do not easily map onto developed and non-developed regions of the world, and also that they are deeply gendered (the domestic sphere is a case in point in which women conduct a disproportionate share of non-commodified work).

Williams' point – and it is important – is to emphasize that there is a need to recognize and promote what he calls 'plural economies', a world beyond the commodity which might encompass local trading schemes, time banks, mutual aid contracts, employee mutuals, basic income schemes, active community service, and so on. In other words, grassroots and state-led initiatives to move beyond a commodity economy.

SUMMARY

- A commodity circuit is a methodological device to reveal a commodity as an organized space–time system.
- Commodification always involves complex processes of economic and social embedding and disembedding.
- Commodity fetishism refers to the ways in which alienation and reification operate in commodity-producing societies.

Conclusion: commodities and the immense cosmos of capitalist accumulation

> Commodity production in its universal, absolute form [is] capitalist commodity production . . .
>
> (Karl Marx, *Capital, II*, 1978)

We began with the commodity as a trivial thing and have ended with a world of commodities that 'actually conceals, instead of disclosing, the social character of private labour, and the social relations between the individual producers' (Marx, *Capital*, I: 75). But this hidden history of the commodity allows us to expose something unimaginably vast, namely the dynamics and history of capitalism itself. The proliferation of commodities as exchange values presupposes a universal equivalent; that is, money as an expression of value. But this has its own precondition: property is implied by money-mediated exchange and correlatively the entire superstructure that sustains property relations and poses the knotty problem of the relations between price and value. In exploring the question of value we have seen that it turns in large part on the peculiarities of labour and money themselves being commodities. Whether Marx was right that labour is a special commodity that has the ability to produce greater value than it has, the relations between commodity, price and value do nonetheless lead inexorably to the centrality of class relations between labour and capital as a fundamental aspect of capitalism. This itself poses the question of how capital and wage labour come to be and how they are reproduced under conditions of contradictory interest (profit versus wage). We may, as reasonable men and women, differ in our accounts of how capitalism as a class system secures the conditions of its own reproduction, and at what cost, but the commodity as its 'cellular form' is surely one of the keys to unlocking the secrets of what Max Weber (1958) called the capitalist cosmos.

DISCUSSION POINTS

1. What does it mean to say that a product bought and sold in the marketplace can take on a life of its own, or that a commodity might have a 'social life' or a 'biography'?
2. Why is it that 'fictitious' commodities are so charged? That is to say, why is the commodification of land, labour, space and money often the object of furious social reaction?
3. As a sort of thought experiment, take the idea that the commodity is the 'cell form' of the modern capitalist world. How might you explain to someone that such a line of reasoning can shed light on the make-up or the way in which capitalism in experienced?
4. Take a global commodity (the Apple iPhone, the Nike running shoe, say) and construct a commodity chain in which you try to understand key nodal points in which 'value' and 'quality' are created.

FURTHER READING

Appadurai, A. (ed.) (1987) *The Social Life of Things*. Cambridge: Cambridge University Press.

Buck-Morss, S. (1990) *The Dialectics of Seeing*. Cambridge, MA: MIT Press.

Comaroff, J. and Comaroff, J. (1990) *Ethnography and the Historical Imagination*. Boulder, CO: Westview Press.

Harvey, D. (1982) *The Limits to Capital*. Oxford: Blackwell.

Mintz, S. (1985) *Sweetness and Power*. New York: Viking Books.

Ollman, B. (1972) *Alienation*. Cambridge: Cambridge University Press.

Postone, M. (1993) *Time, Labour and Social Domination*. Cambridge: Cambridge University Press.

Pred, A. and Watts, M. (1992) *Reworking Modernity*. New Brunswick: Rutgers University Press.

Taussig, M. (1980) *The Devil and Commodity Fetishism in Latin America*. Durham, NC: University of North Carolina Press.

Weiss, B. (1997) *The Making and Unmaking of the Haya Lived World*. London: Duke University Press.

CHAPTER 28

ECONOMIC GLOBALIZATION

Andrew Jones

Introduction

There is an enormous debate in Human Geography and other social sciences disciplines about the concept of globalization, understood as the broad integration and growing interconnectedness of all aspects of social life at the planet-wide scale. Yet within this debate a large proportion of discussion has focused on a narrow aspect – the globalization of economic activity. In fact, in popular debates in the media and politics, globalization is itself often equated with economic globalization – the power of transnational corporations, the shift of manufacturing production to different locations around the globe or the nature of globalized finance. The concept of globalization itself refers to far more than this (see Chapter 1), but undoubtedly the globalization of economic activity has and continues to play a very central role in this process. We can define economic globalization therefore as the growing integration and interconnectedness of a range of different dimensions to the world economy, and while this has been going on for many centuries, it is the more intense phase of economic globalization that has occurred since the end of the Second World War that has most concerned geographers. Since the later part of the twentieth century, economic geographers and other social scientists have argued that processes of economic globalization have made it increasingly appropriate to refer to one, integrated global economy. There are a range of factors that have led to this situation in the last forty or fifty years – the changing nature of international politics, deregulation, new information and communication technologies are just a few – but overall the degree to which economic activity in the twenty-first century is interconnected across the globe is greater than at any point in human history. For economic geographers, central to their analysis is to try to better understand and theorize how these processes have been *uneven*, with very different impacts in different parts of the globe.

The next section in this chapter examines how the globalized world economy that exists today has come about and considers how economic geographers have theorized this change as a transformation of economic activity. The third section then considers how we might think about indicators of economic globalization in general terms by examining patterns of trade and **foreign direct investment**. Sections four and five then move on to examine the key significance of transnational firms in the economic globalization of recent decades, discussing respectively both the development of globalized firms and the way in which they are leading coordinators of globalized production, distribution and consumption. Finally, the chapter ends with a conclusion which considers future trends in uneven global economic development in light of ongoing economic globalization in the twenty-first century.

The emergence of a global economy

Something that might be called world economy has existed throughout human history, but until relatively recently economic activity largely was confined to the places and localities where it was undertaken. In pre-industrial societies, economic activity entailed the production of food and various manufactured goods that were largely produced and consumed in the same local areas. However, the earliest form of what we might regard as economic globalization does have a long history in the form of trade that took place across continents, regions and, more recently, national borders. In that sense, the integration of economic activity stretches back into antiquity, and human history over the last three millennia has seen a variety of different local, regional and globally extensive trading systems. Early processes of economic globalization are evident a surprisingly long way back with, for example, the Roman Empire organising cross-continental economic activity around trade. The Chinese Empire that existed for more than a thousand years in the Middle Ages also extended currencies, trade and other limited forms of economic activity at an inter-continental scale. In that sense, in the medieval period we can talk about a world economy, but its degree of global integration is very limited even if a few global-scale interconnections did exist (Held *et al.*, 1999).

In this respect, what we mean by economic globalization in today's world is related to the nature of today's capitalist world economy that has developed since the sixteenth century, and which has been 'global in its scope' since the nineteenth century (Wallerstein, 2004). Capitalism as a form of economic organization emerged in Western Europe and spread out through the globe through European colonial expansion and then later empires. During the twentieth century, an international system of nation-states gradually replaced these empires to cover the world map (see Chapter 37). However, early economic globalization was sporadic in nature. During the nineteenth century, there was considerable integration of many new parts of the world into the capitalist system, but the two World Wars and their political consequences in the first half of the twentieth century interrupted and in fact reduced some of this economic integration (Hirst and Thompson, 1999). The world that emerged in 1945 after the Second World War was divided between the capitalist first world, the communist second world and the developing third world. This 'tri-partite' world had a range of barriers to further economic integration, with states regulating how much money and how many goods and services could be traded across national borders. A large part of the global map was communist and disconnected from the capitalist world economy altogether.

However, from the early 1970s this situation changed and the disconnected world economic system began to become more interconnected in a number of ways. During the 1970s, the degree of regulation of money exchange, flows and trade was progressively reduced as nation states and international organizations removed restrictions. In the advanced industrial first world, it became much easier to move money around the globe, for companies to invest overseas and for goods and services to be exported to new markets. This financial globalization was therefore an important basis for wider economic globalization as, when combined in the 1980s with new informational and communications technologies, it made it easier to move goods, people and services, for overseas investments to be made and for economic activity to organized at the international level (see also Chapter 25). However, the extent and pace of these economic globalization processes accelerated during the 1990s. The central reason was the collapse of the Soviet Union and the reintegration of most of the communist second world into the global capitalist economic system. Even those states that remained communist – most notably China – largely sought to open their economies to the world capitalist economy. Combined with further deregulation and liberalization of international trade associated with an increasingly dominant neoliberal ideology (see box) and ongoing advances in information technologies, economic activity became increasingly interconnected across all national borders. It is therefore in the last 25 years that it has become meaningful to refer to a globalized economy, since more or less most nation-states across the globe are integrated into the capitalist world economy.

Economic geographers have examined how different national and regional economies have been affected by this economic globalization in the last forty years in very different ways. Some regions have experienced deindustrialization as economic activities decline in the face of global competition (Hudson, 2002); examples include, many of the old industrial areas of Europe or North America, such as the north-eastern rustbelt states including cities like Detroit (see Figure 28.1).

Other regions have benefited enormously from economic globalization as new industries have emerged that are able to generate wealth by selling to increasingly globalized markets. Consider the success of regions such as **Silicon Valley** in California which has been at the

CASE STUDY

Neoliberalism

Neoliberalism is arguably the political and economic ideology that has underpinned economic globalization since the 1970s. It is based on a view that states should intervene, manage and regulate economies as little as possible, and that free markets and free trade lead to economic growth and prosperity (Peck, 2001; Harvey, 2005). In this respect, the deregulation and liberalization of the capitalist world economy was to a large extent driven by those who came from this ideological position. By the 1990s neoliberal ideas dominated the view of what policies should be used to manage the global economy by supranational organizations like the **World Bank** and **International Monetary Fund** (IMF). This broad international policy agreement on how individual nation states and international organizations should manage the global economy was known as the 'Washington Consensus'.

Figure 28.1 Industrial decline in Detroit, USA. Credit: Getty Images

centre of the computer and software industries for many decades now. An important aspect of geographical thinking has been concerned with the degree to which groups or 'clusters' of firms in different industries benefit (or not) from being close together in a regional economy, enabling them to compete more effectively in a globalized world economy (Martin and Sunley, 2003).

Yet equally at a broader level, a key concern for geographers has been how economic globalization continues to produce a changing global economic map with economic activity growing and declining across different national economies and regions. The major characteristic of this geographical change has been what Peter Dicken calls a 'global shift' in the last forty years, beginning in the 1970s and which has become more and more pronounced in recent decades (Dicken, 2010). During the 1970s and 1980s, it became evident that a number of newly-industrializing countries (NICs), mainly in south-east Asia, had rapidly growing economies that were exporting to increasingly global markets. The growth of these economies was primarily based on manufacturing industries with products exported to the existing first world economies. By the end of the twentieth century, this 'global shift' in economic activity had moved into many different industries, not just manufacturing, and many more countries were joining the list of more developed countries. Figure 28.2 shows a stylized diagram of the complexity this has produced in the global economy.

However, in the twenty-first century, this process has arguably reached a critical point. Economic globalization has brought about a re-balancing of the global economy towards previously less developed economies to a point where the old geographical dominance of the first world is now increasingly questionable. Since the global economic downturn in 2007, the older advanced industrial economies have experienced ongoing economic difficulties while Asian and other emerging economies have continued to grow at significant rates (Figure 28.3).

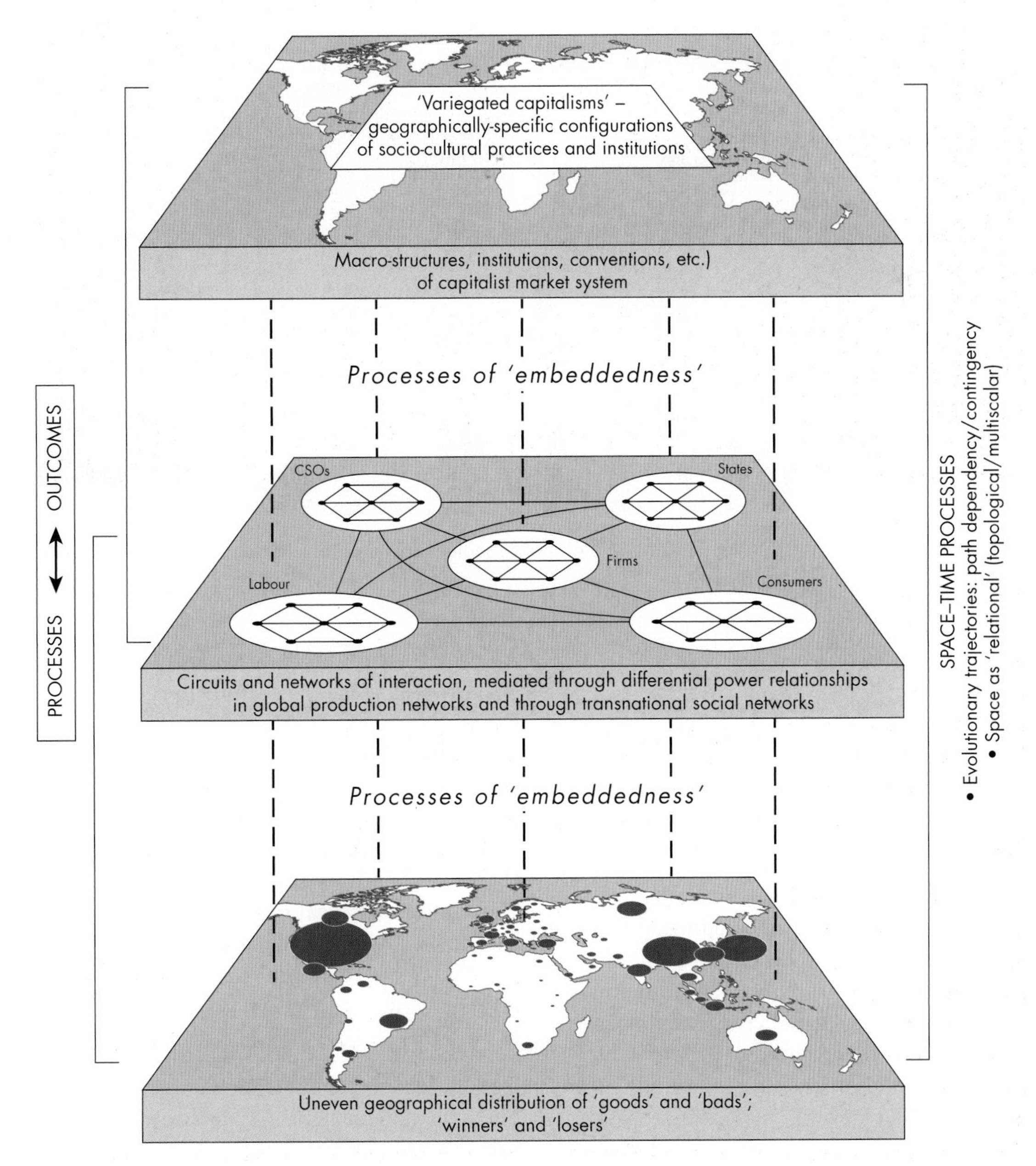

Figure 28.2 A simplified analytical framework of the global economy. Source: Dicken (2010: 53)

Figure 28.3 Victoria Harbour, Hong Kong: the global economic downturn has had little impact on growth.
Credit: iStockphotos

In effect, many of the major 'emerging economies' are rapidly catching up with the old first world – most notably the BRICs (see box). In this respect, in the second decade of the twenty-first century, what we mean by economic globalization and its nature has arguably changed significantly. Asian economies and also Latin American countries like Brazil and Venezuela are increasingly significant and represent the leading areas of economic growth on the world map. The uneven map of global economic integration is therefore a highly dynamic one, and longer-standing ideas of economic globalization corresponded to simple Americanization or Westernization look ever more inappropriate as descriptions of global economic integration in today's world.

Globalization of trade and foreign direct investment

In considering the nature of economic globalization in broad terms, two indicators are often used to understand patterns of economic integration: trade and foreign direct investment (FDI). Both indicators reveal how economic globalization in recent decades had both accelerated and produced a more complex set of interconnections in the economic activity.

Let us consider FDI first. It is a central factor in global economic integration and describes direct investment across national borders by a firm. This usually entails a firm either buying a controlling interest in a firm in another

CASE STUDY

The rise of BRICs

The acronym BRICs was termed by an economist, Jim O'Neill, in 2001 to group together four of the largest developing world economies that were experiencing significant economic growth – Brazil, Russia, India and China. From his position at the US investment bank Goldman Sachs, O'Neill argued that these four economies were at a similar stage of advancing economic development and would soon become major economic powers on the global stage. The idea is not that these four countries will be in any kind of political alliance, simply that they will dominate the global economy in terms of their size. This varies by industry, with the original argument being that China and India will dominate manufacturing and services, while Russia and Brazil will dominate energy resources and mineral extraction industries.

It is worth emphasizing that the largest of the BRICs countries by far is China. By 2010, China's ongoing rate of growth had reached a point where it is estimated that by 2040, it will have overtaken the USA to become the world's largest economy (Price Waterhouse Coopers, 2010). On current trends, India will also be the world's third largest economy by 2050. Perhaps most significant, however, is the prediction by an updated report from Goldman Sachs that by 2050 the four BRIC countries could eclipse the richest countries of the world in terms of the size of their economies. In that respect, by the middle of the twenty-first century the map of global economic power could be very different from the familiar one from the later twentieth century.

country, or setting up a subsidiary (Dicken, 2010). It is called 'direct' investment to distinguish it from 'indirect' investment, where shares might be bought in another firm. In terms of understanding economic globalization, the key thing to appreciate is the dramatic growth in the level of FDI in the world economy since the end of the Second World War (see Figure 28.4). This has not been continuous – there tend to be decline in periods of recession. However, in terms of understanding economic globalization, geographers are particularly interested in the changing patterns of FDI. In this respect, most FDI until the 1970s originated from the advanced economies of North America and Europe. However, during the 1970s, Japan became an important source, joined by other Asian 'Tiger' economies during the 1980s (for example, South Korea and Singapore). More recently, these patterns have become more complex as firms in economies formerly regarded as developing have begun to invest themselves. Asian and Latin American firms are thus becoming progressively more important in making FDI. In terms of global integration, the pattern is becoming ever more complex, reflecting an increasing level of global interconnectedness in global economic activity.

In relation to trade, both the total volume as well as the patterns of global trade are important measures that provide an overview of how economic activity has become more globally interconnected. Trade in the world

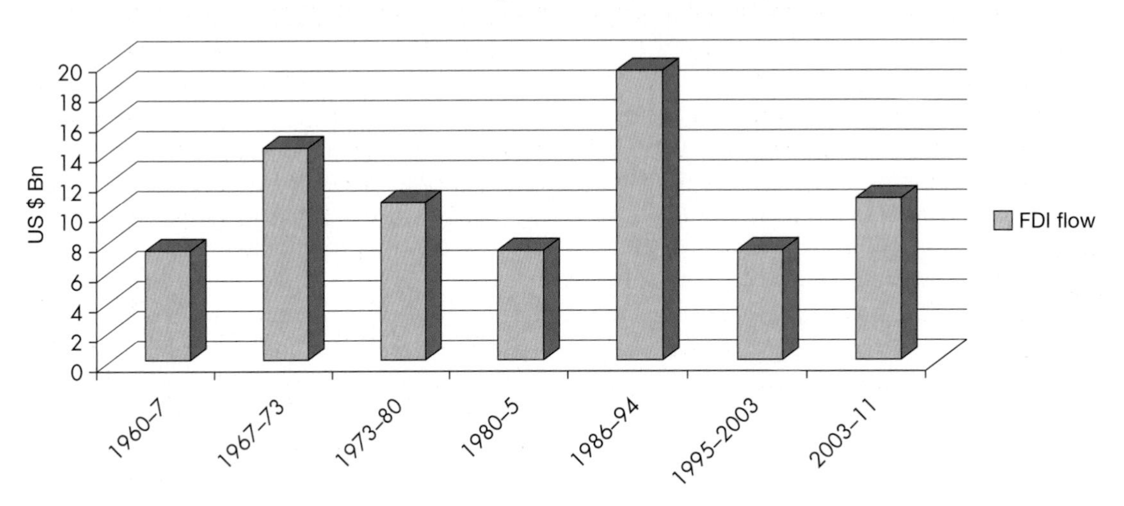

Figure 28.4 Average FDI flows in the global economy since 1960. Credit: Andrew Jones

economy refers simply to the buying and selling of goods and services between actors (individuals, firms, organizations) in different places. In the post-Second World War period, trade liberalization has been an important facilitator of increasing amounts of trade in the global economy. From the 1950s onwards, nation states negotiated trade liberalization through successive rounds of meetings between countries concerned with the General Agreement on Tariffs and Trade (GATT). During the 1990s, these trade liberalization negotiations became further formalized as the World Trade Organization (WTO) was set up to oversee ongoing reductions in barriers to trade.

As the world economy has globalized, total trade has grown enormously, but trade benefits some localities and not others, depending on the nature of their economies. While growth in total world trade stalled during the 2007–2009 economic down turn, the long-term trend has been one of enormous expansion (IMF, 2011). To get some idea of this, in 2008 total world trade measured in terms of goods exported from one country to another amounted to $15.8 trillion US dollars. In the same year exports of commercial services was worth $3.7 trillion US dollars but this remains unevenly distributed across the globe (OECD, 2009). As with FDI, much international trade is concentrated between the wealthier countries in the global economy, but, as discussed above, in today's world this is changing fairly rapidly. In the last decade countries like China and India have experienced huge trade growth with China's trade surplus becoming an increasing source of tension in international politics (IMF, 2011).

Another issue that Human Geography is concerned with is how globalization processes have made understanding the idea of trade much harder. The reason is simple: patterns of trade have become ever more complex and what we might count as 'trade' is more difficult to measure. The conventional way was to measure trade at the national level, with nation states counting how many goods and services they exported and imported. However, globalization processes have complicated this in a number of ways. For one thing, a growing proportion of world trade is different parts of the same large transnational firm 'trading' with another part – this undermines the widely held assumption that trade as an activity comes to an end with the consumption of a good or a service. Another issue is the nature of what is traded, with not only services but also new digitized products (such as software, music and film) being hard to measure because they are sold and bought in different parts of the global economy.

Development of transnational corporations

Analyses of trade and foreign direct investment in the global economy do not provide a very detailed picture of the major actors in economic globalization – transnational corporations (TNCs) – in today's global economy. It is these firms that produce the majority of products which are traded and make direct investments. While there is a long history to the internationalization of business enterprises, it is primarily the growth in numbers and the extensiveness of corporate operations in the last 50 years that is emphasized as distinctive in contemporary economic globalization.

Unfortunately, the idea of a TNC is ambiguous as it is the successor to earlier (and similar) concepts – the multinational corporation (MNC) or multinational enterprise (MNE). These acronyms are all often still used interchangeably. Essentially, the 'trans-' prefix in TNC intends to imply that large international firms now exist 'across' national economic borders rather than just operating in multiple countries (as the prefix 'multi-' denotes). Economic geographers have charted and mapped the rise of such corporations since the 1980s, but it was really in the 1990s that the term TNC came to be used for some of the largest, most globalized firms. Occasionally, TNCs may also be referred to as 'global corporations', but this concept is used lazily and any differences with the more technical terms 'TNC' or 'MNC' are unclear.

The theoretical basis for distinguishing between a 'multinational', 'transnational' or even 'global' firm rests around the degree to which these economic actors are globalized in three dimensions: how they produce goods or services, where they sell them (markets) and how the firm is set-up as an organization. Some business commentators started talking about 'global corporations' as early as the 1970s, but in reality these companies only operated in a handful of countries at that time and in many ways just repeated their operations in each country separately. Car makers like Ford or General Motors, for example, bought foreign firms like Vauxhall in the UK or Opel in Germany which made cars in their respective national markets. In other words, multinational firms became multinational either by setting up new, wholly separate operations in another country or buying up existing foreign firms that already made the same products in another country. Since the 1980s, however, this has changed in several ways:

1. There are far more firms operating in many countries and many different industries. While early multinationals tended to be in mineral extraction or manufacturing, service industries like banking, hospitality (e.g. hotel chains), retail and software are all increasingly dominated by transnational firms (Dicken, 2010).
2. Today's transnational firms are not just companies from the rich Global North but originate from many economies. Several of

the biggest transnational shipping companies originate from Singapore, for example, and large oil and mineral companies have emerged from Latin America and Australia. Nine of the top ten steel firms in 2010 were Asian (see Figure 28.5).

3. Transnational firms these days are set up very differently, with companies organizing many parts of their business at the global rather than national scale. New product research, finance and advertising are often all done at the global level where once each national operation had its own research or finance department. This is what the idea of a shift to a transnational or global corporate organizational form aims to capture (Jones, 2005). However, economic geographers point to the highly variable and uneven way in which this shift has taken place. Some companies now are very much transnational whereas others, despite being very large, are much less so. Even small companies can be globalized, and some in certain industries can be so as soon as they are founded – known as 'born global' firms (Melen and Nordman, 2009). The degree to which firms themselves are globalized in today's world economy therefore varies between firms of different sizes, from different countries of origin and in different industries.

Globalization of production and consumption

Having considered the wider significance of transnational firms in the twenty-first century global economy, we need to consider their role in organizing globalized production across many different national territories. Most people generally understand an economy to be the economic activity taking place within a nation state, but globalization has dramatically changed this. In this respect, the economic geographer, Peter Dicken, argues that the world economy today should be understood as a complex set of what he calls globalized

Figure 28.5 A Mittal Steel plant. Credit: Bloomberg via Getty Images

'geo-economies'. This argument stands in contrast to the traditional view that nation states each have an economy based and largely contained in the territory they govern. Dicken's point is that globalization has opened up serious questions about what we mean when we refer, for example, to the US, German or Australian economies. He argues that economic globalization has produced 'a new geo-economy' that is different from previous eras in terms of how the processes of production, consumption and distribution are organised (Dicken, 2010). All three of these processes no longer just happen in a small number of specific places within states, but exist as connections of many activities between places that are linked through flows of material objects (manufactured goods, components) and non-material elements (ideas, knowledge, services). The new global geo-economy is made up of many networks that span the whole globe, with different actors (TNCs, individual workers, consumers, nation states) linked into these networks as 'nodes' in different ways. However, among these actors TNCs play a leading role insofar as they are the primary organizers of production and distribution of goods and services.

In order to understand this role better, geographers have developed the concept of the global production network (GPN) (Coe *et al.*, 2008). The key issue is that national economies 'can no longer be said to contain production' inasmuch as many manufactured goods 'get made' in multiple places. A product like a car or even a laptop computer, for example, is likely to have many different components made by different firms at production facilities in many different countries around the globe. Components are shipped from one factory to another, and to make matters even more complicated, other aspects of production like design might take place in yet another set of locations. This often makes the labels 'Made in the USA' or 'Made in China' both misleading and inaccurate. It also means that it is increasingly difficult to see production as a process that occurs in one given place at a given time. The concept of the GPN therefore aims to provide a better way of understanding the multiple relationships between different firms that are involved in making something. All GPNs have to operate across a range of scales: the local places where factories are situated, nation states which have governments and global markets where they have to eventually sell products across a world with much social and cultural diversity. The important thing to realise, therefore, is that although these are production networks, the consumption of their products is also a key factor because GPNs are ultimately driven by 'the necessity, willingness and ability of customers to acquire and consume products, and to continue doing so' (after Dicken, 2010). This links back to the detailed discussion of production in Chapter 24 and consumption in the global capitalist economy in Chapter 26.

SUMMARY

- Economic globalization refers to the ongoing interconnectedness of all aspects to economic activity at the planet-wide scale.
- Processes of economic globalization have a long history but the pace of integration has increased dramatically over the last forty or fifty years.

- Human Geographers are particularly interested in understanding the uneven nature and impacts of global economic integration on economics, industries, firms, markets and labour.
- The key actors in contemporary economic globalization are transnational corporations (TNCs), which increasingly account for a growing proportion of global output of goods and services and organize globalized networks of production and distribution.
- Economic globalization has produced a global shift in the patterns of economic activity, which in the twenty-first century is characterized by a shift of economic power to emerging economies.

Conclusion: a new phase of economic globalization?

Human Geographers have been very much concerned with the nature and consequences of economic globalization, and as a process it continues to represent one of the major areas of research in relation to many different dimensions. Geographical work is not only interested in the way that regional and national economies have been affected by and responded to economic globalization, but also how it is transforming firms, industries and markets as well as reconfiguring the nature of production, distribution and consumption in global economic space. Over the last forty years, the tripartite global map of first, second and third worlds that characterized the decades after the Second World War has been dramatically altered by processes of economic globalization. A geographical perspective on this enables an understanding of the inherently uneven nature of these processes and the way in which global economic integration is producing new patterns of complexity in the spatial organization of economic activity. The last forty years has thus seen a significant opening up of national economies and a shift away from national to transnational production to a degree that it is now meaningful to refer to an increasingly integrated global economy rather than a world economy made up of many different national economic spaces.

However, early in the second decade of the twenty-first century, it also appears that processes of economic globalization have reached a critical point in shifting economic power to new regions of the globe. Ongoing low growth in the older, advanced industrial economies has only exaggerated the growing power of BRICs economies and other increasingly wealthy emerging economies in what used to be called the third world. It seems likely therefore that the patterns of global economic integration that characterize the global economic map by the mid twenty-first century will be very different from ideas of economic globalization as a spread of western and American capitalism that were proposed in the 1990s. In that sense, while economic globalization remains very much a feature of the world economy today, and appears unlikely to diminish in the coming decades, it needs to be understood as a process increasingly influenced and led by Asian and other emerging economies. A geographical perspective on economic globalization is therefore more of a necessity than ever.

DISCUSSION POINTS

1. Discuss whether or not transnational corporations really are as powerful as some critics of globalization have argued, given the ongoing significance of states in managing economic globalization.
2. What are the benefits of conceptualizing production in terms of a global network that includes not just transnational firms but also consumers, workers and other actors in different places?
3. To what extent has economic globalization led to a shift in power to Asian economies over the last decade?

FURTHER READING

Dicken, P. (2010) *Global Shift: Mapping the Changing Contours of the World Economy*. London: Sage.

Chapters 1–5 in particular are essential reading in terms of understanding the nature of production in a globalized world economy.

MacKinnon, D. and Cumbers, D. (2011) *An Introduction to Economic Geography: Globalization, Uneven Development and Place*. Harlow: Prentice Hall.

Chapters 5, 6, 9 and 10 in different ways discuss various aspects of economic globalization from a geographical perspective.

SECTION SIX

ENVIRONMENTAL GEOGRAPHIES

Introduction

In July 2012, Greenpeace activists temporarily shut down over 100 petrol stations in London and Edinburgh as a performative protest against Shell's plans to drill for oil in the Arctic and BP's environmentally damaging oil extraction in places such as the Gulf of Mexico and the Alberta Tar Sands region (see Figure VI). Banners were raised, a polar bear costume was used to connect Arctic imagery to everyday spaces of consumption and peaceful protest momentarily interrupted the everyday lives of motorists. Arrests followed and the normal operations of the petrol stations quickly resumed, but another brief but effective flare of publicity had been achieved in the ongoing international campaign to expose and confront environmental abuse and to champion environmentally responsible solutions. Such protests, and the issues underlying them, now seem part of our everyday existence, with new threats to environment and health seeming to be brought constantly into our purview. A brief glance at Greenpeace's own website (http://greenpeace.org) reveals a stark list of contemporary environmental concerns:

- Stop climate change
- Go beyond oil
- Protect forests
- Defend oceans
- Save the Arctic.

The list goes on and on, and this is only one of a wide range of organizations dedicated to defend the natural world and to present peaceful and responsible answers to pressing environmental questions.

In many ways, environmental issues are a 'natural' area of concern for geographers. Indeed traditional definitions of the subject have tended to emphasize the relationships between people and their environment and geographers have for a long time been interested in the hazardous connections between human activity and environmental degradation. For many years, geographers tended to adopt rather formal approaches to studying these interconnections, for example, a 'science' of *environmental impact assessment* has grown up that suggests methods for cataloguing the environmental consequences of a particular project or for a more structured quantitative accounting of the costs and benefits to the environment of that project. More recently, however, Human Geographers have turned their attention to environmentalism as a focus of study. Environmentalism collectively describes a wide range of ideas and practices that demonstrate a concern for nature-society relations. Many of these ideas and practices are to be found in the everyday lives of ordinary people. For example, Human Geographers have taken a keen interest in different political and ethical forms of environmentalism, from the deep ecology movement, which has explored the principle of living in harmony with nature, to the different elements of the 'green' movement, which have used popular protest and individual and household strategies to suggest new ways of envisaging sustainable politics and enacting sustainable lifestyles. Applications here range from superficial (or even cynical) 'greenwashing' of existing practices, to fundamental shifts in the philosophical, political and even spiritual understandings of how humanity and nature are mutually interconnected.

As we progress through the new millennium, the global and local consideration of different environmentalisms will become ever more important. We have already seen how long-term shifts in public opinion towards quality of life issues, while the increasingly active and effective work of environmental pressure and protest groups have begun to result in a range of policy and lifestyle changes. However, there remains a strong suspicion that these changes simply represent shades of **technocentrism** – the management of growth so as to permit the further exploitation of the environment by human beings – rather than any more **ecocentric** alternative. What we need now is a critical and sensitive Human Geographical exploration of both the technocentricity and the ecocentricy in evidence in contemporary responses to urgent environmental threats.

The three chapters in this section demonstrate how Human Geography is getting to grips with this task in among the complex and often contentious nature of environmental issues. In Chapter 29, Sally Eden explains how environmental issues – recycling, pollution, food safety, deforestation and so on – have become part of our everyday life, both through the attention of the media and in terms of everyday household activities. However, this 'everyday life' is often coloured and shaped through the presentation of environmental problems to the waiting world as particular social constructions that are strongly influenced by powerful claims-makers. Thus governments, businesses, scientists and pressure groups are all active in the process of identifying, defining and evaluating environmental issues so as to make claims about them. Moreover, the spatial and temporal context of any particular set of issues significantly influences the ways in which we make sense of our environment and find ways of 'fixing' its problems. Global issues such as stratospheric ozone depletion have attracted technological fixes, but at the same scale, biodiversity loss has shown that many such environmental problems can only be dealt with through fundamental political, economic and cultural change. Similarly, local problems such as waste management require responses involving both recycling technology and changes in individual and household behaviour.

So, environmental geographies cannot simply be approached through the straightforward application of scientific principles. We need to recognize that the ways in which environmental issues are discussed in the public arena matters enormously. Mark Whitehead takes up this theme in Chapter 30 in his discussion of how the notion of sustainability has been taken to the level of a global buzzword which works precisely because it is rarely defined with any precision. So while sustainable development has become a significant political idea, used by an increasingly diverse set of policy-makers, corporations and pressure groups, there has been considerable confusion about what it actually means to be sustainable. Sustainability can be seen in scientific terms to denote resource extraction (for example in forestry, fishing and water use) that does not exceed resource growth or availability, but more often it has involved complex geopolitical negotiations so as to reach international agreements on issues of environmental protection. Such negotiations often pit the protection of the global environment against concern for the kinds of economic development that can alleviate poverty in the 'two-thirds world' countries of the Global South. Inevitably they entertain both weak versions of sustainability in which continuing development is underpinned by technological fixes and stronger versions in which the limits of nature cannot be overcome by human ingenuity. Geographers have again argued that the spatialities of sustainability – the places in which sustainability is practiced, and the ways in which the spatial organization of life impacts on the sustainability of a society – are crucial in understanding the phenomenon. Attention to these spatialities is one way of preventing the idea of sustainability from becoming all things to all people.

The issue of climate is perhaps the archetypal arena of contemporary environmental geographies. Climate change is one of the dominant ideas of the twenty-first century; the nearly ubiquitous agreement that climate change requires urgent attention has entered the very fabric of our society. Human Geographers have made a leading contribution to research which seeks to understand both the emergence, extent and nature of this emerging consensus, and the implications of actions being undertaken to address the problem. In Chapter 31 Harriet Bulkeley explains that climate change has conventionally been viewed as a global problem requiring global solutions. The potential for carbon dioxide and other gases to contribute to the warming of the earth's atmosphere has been recognized in a series of hopeful international agreements and protocols, and in the protests that have expressed disappointment at the inefficacy of these global responses. Over the last decade, however, these international moves have been supplemented by more local scale initiatives, with the emergence of city-wide plans for carbon reduction in energy use, and more local sites of radically alternative living such as the Transition communities in the UK. There is an urgent need to understand the different ways in which climate change is being framed, understood and acted upon, and to grasp how and why it becomes constructed as a problem that requires attention. Such understanding is critical not only in realizing the possible scope for action, but also in order to address justice-related issues of whose interests are being served in our responses to climate change.

The issues discussed in this section of the book are fundamental to the future of the earth and to the task of promoting recognition and understanding of both the scientific and social scientific dynamics of how we should respond to the environmental degradation that has so often accompanied the economic development strategies of different societies. As such, these environmental geographies also present a golden opportunity for Human and Physical Geographers to work together, finding ways to overcome methodological and philosophical differences to produce new compatible academic technologies for dealing with environmental problems in ways that recognize the interconnectivity of nature and society.

Figure VI.[illegible]ce activist puts up a p[illegible]rd as they block off a BP [illegible]rol station. Credit: [illegible]peace

CHAPTER 29

GLOBAL AND LOCAL ENVIRONMENTAL PROBLEMS

Sally Eden

Introduction

It's August in Yorkshire and it's raining – more than raining, it's blustery and wet and downright chilly for the time of year. Now, is this the tail-end of a hurricane lashing across the Atlantic from its recent vandalism in Florida? Another manifestation of the disruption of our weather that we can expect from climate change? In which case, should I be reducing my impact on climate change, such as by driving to work less often? Or is it just a typical British summer, in which case should I consider flying to sunnier places for my holiday next year? In other words, how do I – and all those other people sitting and looking at the rain – make sense of the environment, and how does this influence what we, as individuals and as a society, do in response?

Environmental problems have now become part of everyday life: climate change, recycling, deforestation, air and water pollution, genetically modified food – we experience them all through media reporting and household activities. They have also become part of Human Geography, both through curricula for teaching and agendas for research. And so they should, because geographers have often taken their cue from society's concerns. But what exactly do we mean by 'environmental problems'? This may seem an obvious question, but it is not until we are clear about our terms and thinking that we can be clear about how we might attempt to 'solve' such problems.

Environmental problems and social construction

First of all, we should not see 'environmental problems' simply as reflections of 'real' environmental change. This is because we need to consider the theory of the **social construction** of reality (see Chapter 1). This idea has been around for 30 years now and seeks to understand how people make sense of the world, particularly through endowing it with meaning and, often, changing that meaning as society changes. It is sometimes criticized as meaning that there is *no* reality except in our collective imagination. This is mistaken. Talking about the social construction of environmental problems does not mean that such problems are merely imaginary, but it does mean that we can only think about them in socially defined ways – and that it is on the basis of these social definitions, rather than on the 'real' conditions, that we act. Whether I see the rain outside as climate change or a British summer, the rain continues to fall. But the meaning that rain has for me might make me act differently – how we *perceive* things matters because it directly affects what we *do* about them. So, social construction is about meaning but it is also about action.

These socially constructed meanings are often shared and can be greatly influenced by powerful groups. Because of this, environmental problems have been fought over for decades in a tug-of-war between governments, businesses, scientists and pressure groups. An added complication is that many environmental problems occur far away from people or are not measurable by them, such as an oil spill in the North Atlantic Ocean or rainforest loss along the Amazon. For social problems like crime or unemployment, I might understand them through drawing on my own experiences or those of my neighbours but, sitting here in England, I rely on other people – particularly scientists and environmental activists – to tell me what is going on in the Atlantic or the Amazon, and why it matters.

Applying social construction to environmental problems in this way focuses upon what we can think of as a 'claims-making' process (Hannigan, 1995), specifically:

- Who makes the claims that there is (or is not) an environmental problem?
- Who makes up the audience that is listening (or not listening) to these claims?
- How are these claims made and contested, using facts, rhetoric and **metaphor**?

We are influenced by various 'claims-makers', like pressure groups, scientists, industrial companies and journalists, who identify and define an environmental problem and bring it to our attention through the media (see examples in Hansen, 2010). They frequently 'encode' it in particular ways, to suit their own agenda and how they want to persuade their audience. Sometimes claims-makers tailor their information directly to their audiences, so that the coverage of environmental problems on the BBC's *Newsround* for younger viewers will be quite different in length and detail from that on the BBC's *Newsnight* programme and different again from their Twitter feeds. Another example is resource packs for schools that help teachers by providing easy-to-use information, brightly coloured briefing sheets and well organized tasks. But they are not neutral: they bring into the classroom the arguments that the pack's producers favour, often in very subtle ways. Similarly, environmental groups use newsletters, magazines, websites and online social networking to speak to their members (who are already in a sense 'converts') and also to those not yet in tune with their views, without being filtered by editors who control newspaper or TV content.

But claims-makers cannot control their audience's reaction. Some people will 'decode' the environmental meanings presented to them quite differently from how the producers intended. Geography matters: familiarity with and proximity to the environments being described may influence how people respond. People living around big factories and nuclear power plants understand the environmental impacts and risks of those sites in different ways, influenced negatively by 'folk memories' of smells, accidents and company PR, and

Figure 29.1 Photogenic wildlife make good visuals for environmental stories. Credit: Alaska Stock LLC/Alamy

positively by the employment provided, so their views may differ substantially from those of people living far away. Some studies suggest that women, younger people, those with more education and those in more socially supportive occupations (teaching, healthcare, local government) tend to be more sympathetic to environmental messages. But other studies contradict this or show that sympathy does not translate into action. Lifestyle choices also matter: some groups identify themselves as 'green consumers' whereas others are more concerned with fashion, for example, which means people will think about buying a new car in quite different ways, whatever claims the car manufacturers make. Together, the production and consumption of environmental meanings (Burgess, 1990) are critical in constructing and re-constructing environmental problems over time, as many layers of encoding and decoding accumulate.

A distinctive name or metaphor is often coined for an environmental problem, like 'the greenhouse effect' or 'Frankenstein foods', and this is a really effective shorthand way of conveying very complex ideas. Images are also hugely important. Photogenic or visually arresting images, such as furry animals or soaring birds to illustrate environmental stories, may influence TV producers in deciding whether and how to cover a story and persuade people to pay attention (Figure 29.1).

Timescale also matters. News coverage of chronic conditions, i.e. long term but low level, is difficult to keep exciting: 'pollution still happening' is not headline-grabbing, compared to the sudden drama of flooding or oil spills. Claims are often geographical: global, national and local are all ways in which environmental problems are constructed. Local protests against development in the countryside are a good example and frequently make the news. Restoration of species to 'the British landscape' is a storyline that plays on national views of wildlife. All these images are part of constructing the urgency, impact and size of environmental problems, especially problems with which the audience is not familiar.

SUMMARY

- Environmental problems do not simply reflect 'real' environmental change but are socially constructed.
- This construction is very much influenced by different 'claims-makers' – like pressure groups, scientists, industry and journalists – who identify and define an environmental problem and bring it to our attention through the media.

Environmental problems in time and space

Environmental problems, moreover, are products of specific times and places. Let us take time first. Although we tend to think of environmental problems as modern, they are very old. Ancient Greek writers considered problems of soil erosion, and early laws against pollution (especially smoke) were enacted in London in the thirteenth century. Campaigners in Victorian England were concerned about the impact of increasing urbanization and industrialization on the environment and people's health, leading to campaigns for public sanitation, personal hygiene, nutrition, exercise and access to open space. In America, George Perkins Marsh's *Man and Nature* is sometimes cited as the first 'modern' take on environmental problems and a touchstone for geographers. Published in 1864, it was before its time in *not* assuming that environmental resources were inexhaustible or that the environment was resilient to human impact (Lowenthal, 2001). But although Marsh's book helped to inspire US forest protection in the late nineteenth century, it then fell into neglect until rediscovered in the 1930s, when another US environmental problem – agricultural erosion – gained public and political attention. In the UK in the 1930s, the rise of outdoor recreation generated a drive for public access to land and for national parks (first developed in the USA in the 1860s) that came to fruition in the 1949 National Parks and Access to the Countryside Act. These debates constructed environmental problems in social terms by portraying landscapes as national assets to be protected for the public.

Even what we might think of as 'modern' environmental problems have their origins in the 1960s, in concerns about nuclear weapons and energy, industrial pollution and endangered species, concerns that we think of today as **environmentalism** and associate with the environmental movement and pressure groups like Greenpeace, Friends of the Earth, the Sierra Club and WWF. However, there is a range of environmentalist views, some being more radical and some more reformist, depending on how far they are willing to work with technology, with politicians and with companies and on the degree of change that they want to see in modern, industrialized systems. Despite this range, in most public debates environmental pressure groups tend to be simplified as collective representatives of environmentalism standing symbolically against those of transnational companies – the 'green' activists against the forces of 'brown' capitalism, one might say. Greenpeace in particular has used this storyline very successfully over the years to obtain media coverage and public support for its environmental campaigns,

Figure 29.2 'David versus Goliath' in environmental action. Credit: Argus/Still Pictures

sometimes symbolized as 'David versus Goliath' battles (Figure 29.2).

In the late 1960s and early 1970s, public interest in environmental issues rose markedly, generating considerable support for environmental groups. This also drew on science – both the growth of **ecology**, which provided scientific evidence for environmental campaigns, and space travel, which nurtured modern ideas of the global environment as 'Spaceship Earth', a small, fragile ball in space, not least through the visual power of photographs from space (Cosgrove, 1994). Scientists thus became important 'claims-makers' and evidence of this successful construction of environmental problems is seen in the series of international conferences organized by the United Nations since 1972 (see Chapter 30), which increasingly drew on scientific findings to develop policy on issues as diverse as climate change, species protection and ocean management.

Environmental problems are thus differently constructed over time. The climate is one example. In the 1970s, climate change was often constructed as a problem of global cooling, raising the prospect of 'the next ice age'. In the 1980s, as an undergraduate, I was taught about climate change through the concept of a nuclear winter – what would happen to the climate if the **Cold War** turned into a nuclear war. In the 1990s, climate change was constructed as global warming and by 2010 was increasingly connected to ideas of carbon control and a low-carbon economy (see Chapter 31). These different constructions reflect not rapid shifts in climate but rapid shifts in society and science that affected how environmental problems were seen and prioritized. In a classic paper, Downs (1972) suggested that environmental problems go through a life cycle: they rise from relative obscurity to reach a high public profile and enthusiasm for action, but then decline to a back seat in people's minds. The claims makers change too: the environmental pressure groups established in different periods have very different characters even now (Table 29.1).

As well as being products of specific times, environmental problems are products of specific places. Sometimes this is an effect

	Year established
Manchester Association for the Prevention of Smoke	1843
Commons, Open Spaces and Footpaths Preservation Society	1865
National Trust	1895
(now Royal) Society for the Protection of Birds	1889
Wildlife Trusts (originally the Society for the Promotion of Nature Reserves)	1912
Ramblers (previously known as the Ramblers Association)	1925
Campaign to Protect Rural England (originally the Council for the Preservation of Rural England)	1926
Soil Association	1946
WWF	1961
Friends of the Earth	1969
Greenpeace	1971
Woodland Trust	1986

Table 29.1 Establishment of key environmental groups in England

of the physical sort of geography, for example, the concern since the 1980s over increasing pollution in the shallow, closely bounded North Sea on the part of those countries that surround it, or the 'dust bowl' agricultural problems in the USA in the 1930s, when soil erosion contributed to unemployment, massive migration and economic loss. At other times, it is the social and political sort of geography that matters. In Florida, USA, 16 people died in the damage caused by Hurricane Charley in 2004, which was covered extensively in the British news. A few weeks earlier, 2000 people had died in flooding in Bangladesh, which received less than half as much coverage, despite being far more lethal. Newspaper editors make choices like these every day, prioritizing places for coverage and thus influencing our global consciousness.

Sometimes the politics is rooted in cultural differences: European countries have been embroiled in heated debates over genetically modified crops, whereas the USA has been steadily planting them for decades with little public debate, not least because of the huge power of agribusiness there. So, the meanings given to different environmental problems and solutions reflect cultural predispositions, political structures and, often, power and they may be shared or contested between different places.

As well as place, scale is important in constructing environmental problems. As geographers, we are of course interested in scale as an organizing principle and as a construct itself (see Chapter 1). The next section illustrates how scale is important, particularly for environmental problems, by comparing how environmental problems are constructed at global, national and local scales.

SUMMARY

- The social construction of environmental problems is influenced by specific times and places, because context strongly influences the way we make sense of our environment.

Global environmental problems

Let us start with some environmental problems that are often constructed as global: stratospheric ozone depletion and biodiversity loss. When I first began teaching environmental problems in 1991, ozone depletion was a high-profile global environmental issue. Today, it is often regarded as 'solved', and has been largely overtaken by others in the media and in curricula as debate has moved on to other global issues. By comparison, biodiversity loss continues to be problematic, despite many attempts to solve it. Let us consider why these two problems differ so much.

Constructing environmental problems

Stratospheric ozone depletion was initially identified by scientists, who became the initial claims makers for this particular problem. This is because of the atmospheric geography of ozone: its main concentration is known as the ozone layer and is up in the stratosphere about 15–22 km above our heads. This means that ordinary people cannot see or otherwise detect for themselves what is happening to ozone concentrations, but scientists can, using specialized equipment. Depletion of the ozone layer high up over Antarctica was famously 'discovered' by the British Antarctic Survey in 1984 and confirmed by NASA scientists in 1985 (Figure 29.3). These were big, very credible scientific groups and they provided a clear, international consensus that CFCs emitted by human activities were threatening stratospheric ozone. The idea of depletion was often referred to in the news as the ozone hole – a neat example of the sort of shorthand or storyline I mentioned earlier – despite being a little misleading, because depletion actually involves only thinning of the layer, not complete loss.

The chemistry involved in ozone creation and depletion was initially established in the 1970s and the scientists involved were jointly awarded the Nobel Prize in Chemistry in 1995 for their contribution to our understanding. Ozone (O_3) is created in sunlight as oxygen (O_2) molecules break down into two O atoms and then recombine unequally with others to form ozone. Ozone is broken down again when the reverse happens in sunlight, and catalysts such as chlorine and bromine encourage this destruction. Chlorine and other catalysts are largely anthropogenic, produced from the breakdown of chlorofluorocarbons (CFCs) and similar compounds emitted from human activities and products, such as air conditioners, freezers and fire extinguishers. So, the environmental problem that was identified through science was that emissions of CFCs and the like from human activities were destroying ozone in the stratosphere faster than it was being created.

In addition, the geographies of ozone depletion meant that the problem was globalised. Ozone-depleting compounds like CFCs were emitted round the world at varying levels, but because

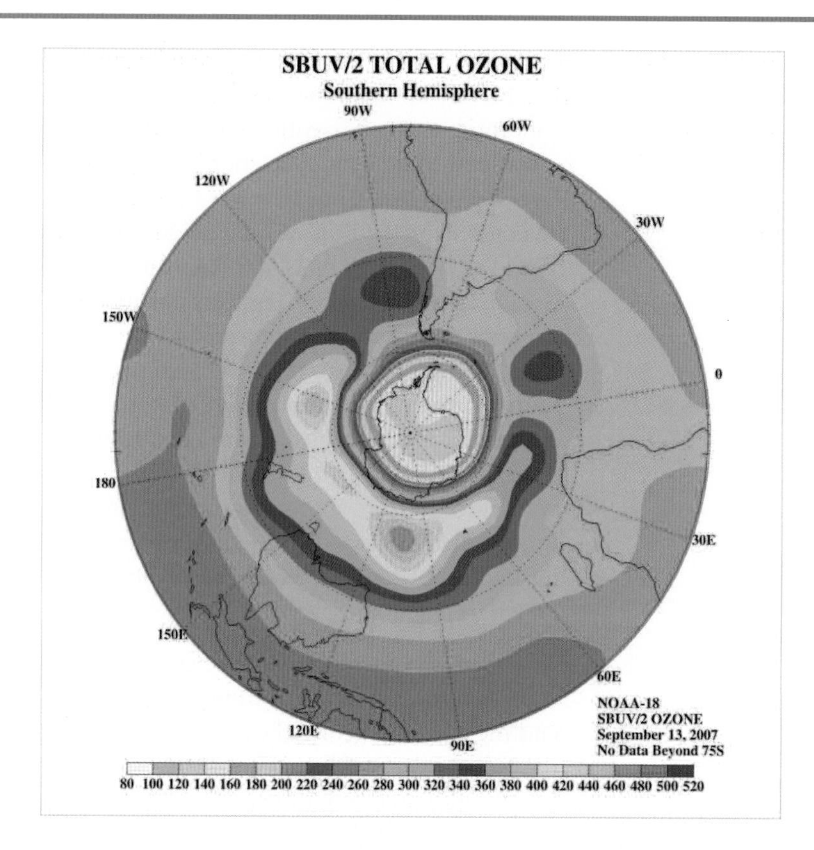

Figure 29.3
An international problem seen through science: the ozone hole in 1985. Credit: www.noaa.gov

of the way that weather patterns develop over Antarctica and the Artic, ozone loss was particularly concentrated in those areas.

The consequences of this depletion were potentially very damaging. The ozone layer absorbs and reflects ultraviolet (UV) rays from the sun, and because UV can cause skin cancer, ozone depletion has been linked with rising rates of skin cancers (approximately a one per cent rise for every one per cent ozone loss), eye cataracts and problems with the immune system in humans, as well as possibly reducing the productivity of agricultural crops and fisheries. This meant that the falling ozone concentration was rapidly constructed as a clear environmental 'problem' because of its effects on humans. In this case, humans in areas near the Arctic and Antarctic (especially Australia) were most affected, mainly because of compounds emitted into the air by other humans many thousands of miles away.

Let us now compare this with how biodiversity loss has been constructed as a global environmental problem. Biodiversity (or biological diversity) is the variability of all living organisms and stands as a symbol for wildlife and nature in general. Fossil evidence tells us that many species have become extinct in the past, but the problem that has been identified is that the rates of loss have been speeding up due to human activities. For example, the 'background' rate of loss over thousands of years is estimated to have been about one specie of bird or mammal per 100–1,000 years, but in the 50 years from 1900 to 1950, 60 species became extinct, that is, species were being lost about 100–1,000 times more quickly than in the past.

Unlike ozone depletion, biodiversity loss does not require complex atmospheric science or technology to measure, because plants and animals share spaces with humans and therefore can often be identified and counted by eye. For example, the UK's Royal Society for the Protection of Birds runs 'Big Garden Birdwatch', involving hundreds of thousands of householders reporting bird counts in their gardens, and uses the results to monitor changing bird populations nationally.

But on bigger scales, especially looking at the global level, it can be far more difficult to monitor animal and plant populations, for many reasons. Some species may be unknown to science in the first place. Although estimates vary a great deal, a common assumption is that there may be around 14 million species on the planet, but only about 1.8 million of them are known to science and for only a fraction of those do we have enough information to be able to measure them and develop protection plans.

Figure 29.4 A bittern. Source: iStockphoto

Some species may also be highly mobile and thus difficult to track, such as fish in the oceans. For example, we know that cod stocks in the North Atlantic have been greatly depleted by European and North American fishing fleets over the last couple of centuries, but no one is exactly sure how many cod are left and how long it would take them to recover

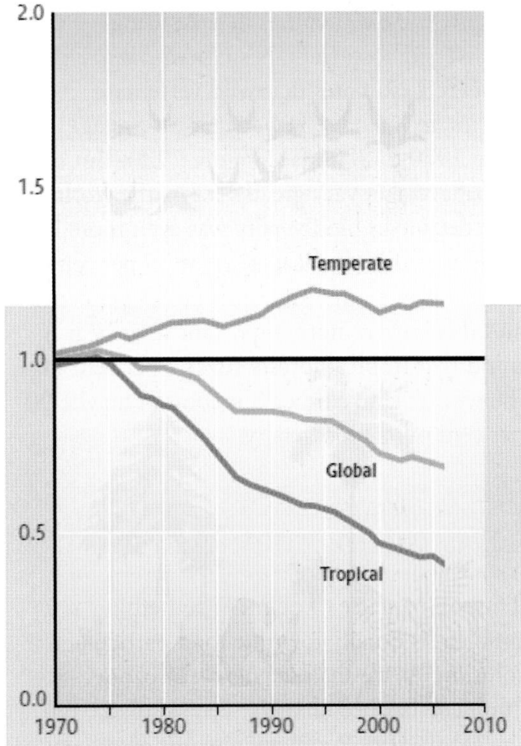

Figure 29.5 The global Living Planet Index (LPI) shown here by the middle line, has declined by more than 30 per cent since 1970, suggesting that on average, vertebrate populations fell by nearly one-third during that period. The Tropical LPI (bottom line) shows a sharper decline, of almost 60 per cent. The Temperate LPI showed an increase of 15 per cent, reflecting the recovery of some species populations in temperate regions after substantial declines in the more distant past. Credit: WWF/Zoological Society of London/ UN (2010) Global Biodiversity Outlook 3 (www.cbd.int/gbo3/)

to previous levels if left alone (which in any case is unlikely). Estimates of fish populations are therefore critical to defining how great the 'problem' in the oceans is, but they are also subject to great dispute between different nations and their fishing fleets, all of whom are interested in exploiting this important resource, so that, unlike the clear results for ozone depletion discussed above, claims are heavily contested.

Despite these difficulties in measuring biodiversity loss, the UN's *Global Biodiversity Outlook 3* estimates that wild vertebrate species declined by 31 per cent on average from 1970 to 2006 (see Figure 29.5). But such global averages vary greatly across the world. The decline in biodiversity was even more pronounced in tropical areas (at 59 per cent) and freshwater environments (at 41 per cent) and this is even more important if we bear in mind that tropical forests cover 7 per cent of the world's land surface but contain maybe 90 per cent of its biodiversity.

Constructing solutions

The consequence of defining something as a problem is usually to argue for a particular solution. As we shall see next, the ozone 'problem' has been largely seen as solved, but the biodiversity 'problem' remains extremely difficult to tackle. Why is this?

In the case of ozone depletion, one reason is that the solution was constructed as a technological one or 'technofix' – there was a clearly identified and delineated culprit in CFCs and their like, so the offending compounds could be substituted with less damaging ones by changing industrial technologies. The 1987 Montreal Protocol on Substances that Deplete the Ozone Layer was the global agreement to reduce the use and emissions of CFCs and similar compounds across the world. All the major producing and consuming countries signed up to it and the extensive public and **NGO** support led to manufacturers changing products to make them CFC-free. The Protocol is now seen as a political success, which is why I have stopped teaching ozone depletion – the 'problem' is regarded as 'solved'. But that is only in political terms – it is likely that the ozone layer will not recover sufficiently until perhaps 2040, because of the legacy of CFC emissions in the past. In the meantime, affected populations have to adapt and learn to live with the problem by changing their behaviour in the sun. Public campaigns in Australia therefore continue to tell sunbathers how to protect themselves from the higher levels of ultraviolet radiation ('slip on a shirt, slop on some sun-cream, slap on a hat') to avoid skin cancers developing.

Biodiversity loss is more problematic. There is no clear technological culprit; instead, biodiversity is depleted by a wide range of everyday human activities in vulnerable environments, from agriculture to building houses, as well as by direct killing for food or sport, and these vary greatly in terms of the speed and geography of their impacts. For example, deforestation is a continuing problem in many tropical countries, at the same time as forest cover is increasing in many countries in Europe and North America.

Unlike the problem of stratospheric ozone depletion, the problem of biodiversity loss is often caused by conflict between the interests of humans and the interests of other species. For example, protecting habitat for orang utans or whales may reduce development opportunities for local people through farming or fishing. Protecting animals may even involve damaging humans, such as when reserves are created to protect key species of value for tourism but in the process resident humans are displaced, damaging indigenous cultures. Sometimes biodiversity loss is caused by conflict between the interests of different non-human species: where species have been introduced – both intentionally and unintentionally – to new environments as humans have migrated, they have sometimes proved invasive, out-competing existing species and causing the latter to decline.

These are some reasons why, despite species loss being identified as a problem since the earliest environmental pressure groups were established, the problem is far from solved. For example, although the extent of protected areas (national parks, nature reserves, etc.) has increased globally in recent years, many species and habitats have continued to decline. The UN designated 2010 as the International Year of Biodiversity to promote action to protect species, but because none of its targets for the year were met, it has now designated a full Decade on Biodiversity for 2011–2020 to continue to push for change.

Protecting biodiversity in the oceans is even more difficult. For most of history, the oceans were open-access resources for all nations and consequently were heavily exploited. Dwindling numbers of many species, from cod to whales, has now prompted international agreements to control fishing in many areas, especially the North Sea and the North Atlantic. But often the protection of environmental resources is lost in political wrangles between nation states or in debates about the uncertain estimates of remaining populations.

To address some of these issues, the human benefits of protecting biodiversity are increasingly promoted through the idea of 'ecosystem services'. These include not only the ways in which humans make money from wildlife (e.g. eco-tourism, logging forests) but also the ways in which humans benefit without any economic exchange being involved (e.g. carbon storage by forests, reed beds that filter pollutants out of water, soils that support agriculture, insects that pollinate fruit crops and the simple beauty of a landscape). The idea of ecosystem services is a good example of claims-making designed deliberately to protect the environment, using not only natural science but also economics to promote the monetary value of environmental functions. The hope is that constructing the problem using the language of money will be more persuasive and influential in protecting biodiversity for the future.

SUMMARY

- Science is important in identifying and constructing many global environmental problems, but solutions usually require not only science but also politics, economics and cultural change.
- Environmental problems have different geographies, both physical and human, meaning that a 'global' problem is actually a complex mosaic of national and international factors.
- The techno-fix solution to stratospheric ozone depletion is largely seen as a success, whereas solutions to biodiversity loss, particularly in the oceans, continue to be practically and politically difficult.

Local environmental problems

By comparison with the examples of ozone depletion and climate change, environmental problems that are typically constructed as local often seem far less glamorous. A good example is waste. Waste management is rarely the subject of 'big science'; it does not dominate the agenda at international conferences run by the United Nations, it rarely makes the front pages of newspapers and it often, frankly, involves rather smelly and mundane questions. But it is hugely important (Figure 29.6).

Constructing the problem

The environmental problem of waste is both simpler and more complicated than ozone depletion and biodiversity loss. For one thing, it is close at hand: where I live, everybody puts their rubbish out on Wednesday mornings in black plastic bags for the local authority to collect and dispose of. So, if I wanted to measure the amount of waste produced by my neighbours and myself, I could go out and weigh the bags – much easier than measuring ozone 22 km above my head or counting whales in the ocean. But in each bag will be a great variety of materials: broken glasses wrapped in newspapers, potato peelings, abandoned plastic toys and CDs, cardboard cereal boxes. The low-tech solution to all this for decades across the world has been to dump it in holes in the ground, often old quarries, cover it over and ignore it – sometimes euphemistically referred to as 'landfilling'. Of the 400+ million tonnes of waste currently produced every year in the UK, most comes from construction, mining and other industries; only about 23 million tonnes comes from households and the vast majority of it is landfilled.

Waste disposal has environmental and health effects. Landfill risks land contamination and leakage to the water table from the buried materials, or a build-up of methane gas from decomposition, which can explode or be emitted. Methane is also a greenhouse gas and about a third of it in the UK comes from landfill, which contributes to climate change

Figure 29.6
The environmental problem of waste. Credit: Thomas Raupach/Still Pictures

(see Chapter 31). Moreover, finding new landfill sites is becoming more difficult as the UK runs out of holes to fill, particularly in the south-east. As this is also where most of the waste is produced, transport and its environmental impacts will also increase as waste is moved around the country in search of landfill space. Incinerating waste to get rid of it can have health effects and risks environmental acidification when the emissions fall on surrounding land; it also releases carbon dioxide and thus contributes to climate change.

In the past, these problems were exacerbated by the fact that waste was growing – indeed, some saw waste as a measure of human success, as more production and more wealth means more consumption and more material things and packaging that eventually end up in the bin. Product design with built-in obsolescence, such as the laptop computer on which I typed this chapter, speeds up the process of waste creation and the amount produced over time, but can be hugely profitable. Waste is especially problematic because it is involved in everything we do, from building houses to cooking dinner. It is therefore politically difficult to ask people to buy less and thus to waste less, which leaves politicians in a dilemma: they want people to buy more (to help the economy) but produce less waste (to help the environment).

Constructing the solution

Despite this, in the UK, even in the few years between the second and third editions of this book, the overall trend has turned around and the amount of waste produced has begun to (slowly) decline, year on year. How has this happened? The most permanent solution to waste disposal is to reduce the problem by producing less waste. Failing that, we can deal more effectively with the waste that *is* produced. Both strategies have been applied in the 2000s but there is dispute over the best solutions: whether recycling is better than waste-to-energy (incineration), for example, or whether the emissions from an incinerator are so damaging that they outweigh the advantages in disposing of waste and generating electricity in the process. Indeed, proposing incineration as waste-to-energy is not necessarily a solution, because it will typically run into local opposition because of smells, visibility, emissions, increased traffic and so on.

Recycling is often constructed as everyday common sense and therefore an obvious and well-established solution to the problem of waste. But infrastructure, culture and geography matter. The Netherlands and Germany, for example, have had household recycling systems for far longer than the UK has, so they recycle and recover far more material and correspondingly landfill half as much, as well as reducing the amount of waste they produce. The social construction of the solution in such countries is 'normalized' and has become part of daily routine, whereas recycling, particularly separating materials like glass, plastics and food waste inside the house, has been portrayed as difficult, inconvenient or just plain messy in Britain. However, the social construction of recycling has shifted in recent years here. Public support for recycling has increased – around 40 per cent of household waste was recycled in 2011, which is a huge increase compared to 11 per cent in 2001 – and putting out the recyclables for collection is becoming increasingly normalised (although still at a level lower than other European countries).

Unlike ozone depletion and biodiversity loss, waste management in Europe is driven by the idea (we could say a 'construct') of a 'waste hierarchy' (Figure 29.7). This is like a ladder with the 'best' options at the top – which are waste minimization and reuse of waste – and the less good ones further down – which are

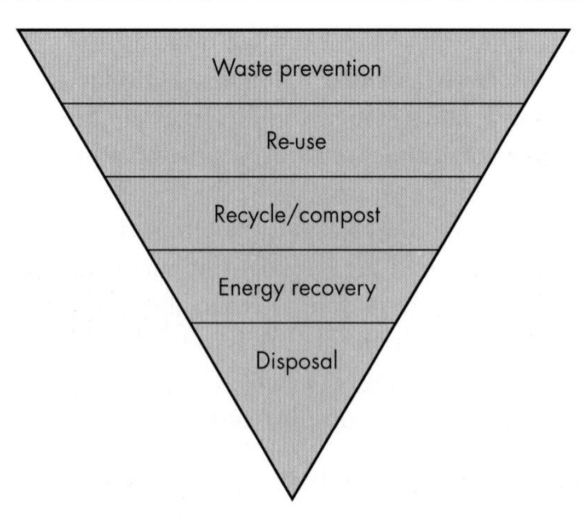

Figure 29.7 The waste hierarchy. Credit: DEFRA © Crown Copyright

recycling and recovery (usually through incineration that produces electricity) – with landfill last. Countries in the EU try to 'move up' the hierarchy.

For example, the UK government has used various strategies to move up the hierarchy and away from the country's reliance on landfilling waste. In 1996, a tax on landfill was introduced, paid by local authorities and companies, to make landfill more expensive and encourage them to consider other options. This constructed the waste problem as a waste of *money* as well as of resources and used rhetoric like 'waste minimization', 'waste-to-energy', 'de-materialization' (such as using emails or texts instead of paper letters), 'industrial ecology' and 'eco-efficiency'. Lots of books have been written arguing that changing to more efficient (less wasteful) operations is not only possible but increasingly being implemented by business (e.g. von Weizsäcker *et al.*, 1997). By constructing environmental problems as to do with efficient technology (or lack of it), this avoids discussing the structural problems of underdevelopment, unequal access to resources and wasteful consumption.

Waste management is also a good example of a problem that gets relegated down to lower scales of government. Local authorities in the UK are responsible for collecting and disposing of household rubbish but, as planning authorities, they are also responsible for approving new disposal sites, such as incinerators. They are also charged with collecting and safely disposing of unwanted electrical goods and cars under the delightfully named WEEE Directive (Waste from Electronic and Electrical Equipment) and the End of Life Vehicles Directive respectively, both passed by the European Union. Because local authorities often have to pay for waste disposal, the landfill tax has also pushed them to develop recycling systems, especially kerbside collection from local residents.

As already stated, UK attitudes to waste have changed greatly between the second and third editions of this book. The social construction of waste in the UK has shifted from being 'out of sight, out of mind' in landfills to being a significant economic drain on the budgets of companies and local government, due to the landfill tax. Coupled with national targets for recycling, this financial push has led to new systems and technologies for collecting and recycling waste. Now, the recycling 'solution' is being increasingly normalized as part of how people manage their households, making recyclable waste very visible on our streets every collection day. Rather than a 'technological fix' (as for ozone depletion), this has generated a 'behavioural fix' for waste through changing disposal practices.

This is not to say that it has been an easy or unproblematic shift. Recycling processes are increasingly globalised as rubbish is exported (with consequences for climate change in terms of transport) and can involve very complex dismantling and disposal processes (with consequences for human and environmental health in terms of releasing toxins); the market for products made from recycled materials is also not keeping pace with recycling rates. However, the case of waste does demonstrate for us that environmental problems – and what count as acceptable solutions to them – continue to change in response not only to environmental but also social and political factors.

SUMMARY

- Waste is constructed as a local environmental problem, with less scientific and international controversy than biodiversity loss and ozone depletion.
- Constructing solutions for waste involves technology, but also consumption change by way of a behavioural fix in households.

Conclusion

These three environmental problems have different geographies and histories. The same claims-makers are involved in all three, particularly environmental groups, national governments and, for waste in particular, local governments. Science is an important claims-maker, although the individual scientists speaking on each issue differ and some areas of science are more contentious than others. Media coverage of the global issues has been frequent, but waste is rarely headline news. All the issues depend upon shorthand, whether this is 'the ozone hole' or 'the waste hierarchy', to summarize a set of complicated ideas, many of which are contested.

The boundaries that we put around a 'problem' are socially constructed, because we tend to compartmentalize. I have separated out 'local' and 'global' problems here, but in reality they are connected: ozone is depleted by CFCs from waste fridges, and biodiversity loss and waste production both grow worse with development – the 'problems' are products of success – although this link seems to have been successfully decoupled in recent years in some countries, in terms of waste generation at least.

So, environmental problems are not just reflections of 'real' environmental change but reflections of how our society sees itself, its environment and the possibilities for improving both. That is not to say that environmental problems are not important or worth addressing. As I said, sometimes critics of social construction argue that it denies external reality. Not so. What it does is to show more clearly how little we know and how much our social and geographical context affects what we understand by 'environmental problems' and how we act to solve them.

DISCUSSION POINTS

1. Choose a recent environmental problem that you have seen or heard discussed in the media and make a list of the major 'claims-makers' involved. How influential is each one, in your view?
2. What do you think are the main obstacles to (a) reducing waste and (b) protecting endangered species? How are these different in different countries?
3. What do you feel is the most effective scale (global, national, local) at which to develop solutions to environmental problems? How do solutions differ between scales, in terms of how they are constructed and promoted?

FURTHER READING

Hannigan, J.A. (2006) *Environmental Sociology* (2nd edn.). Abingdon: Routledge.

Explains how social construction works and applies it to environmental issues.

Hansen, A. (2010) *Environment, Media and Communication*. Abingdon: Routledge.

Explains and gives examples of how environmental issues are constructed and given meaning through the mass media.

Harris, F. (2011) *Global Environmental Issues* (2nd edn.). Chichester: Wiley.

Covers a range of environmental issues, including biodiversity and waste.

Hulme, M. (2009) *Why We Disagree About Climate Change*. Cambridge: Cambridge University Press.

Shows the different ways in which climate change is given cultural meaning and why this becomes a problem.

WEBSITES

www.rspb.org.uk/birdwatch/

RSPB Big Garden Birdwatch

www.cbd.int/gbo3/

UN Global Biodiversity Outlook 3

www.cbd.int/2011-2020/

UN Decade on Biodiversity 2011-2020

http://ozone.unep.org/new_site/

UN Montreal Protocol

www.defra.gov.uk/statistics/environment/waste/

UK waste management statistics from the Department for Environment, Food and Rural Affairs (DEFRA)

www.wrap.org.uk/

WRAP

www.wwf.org.uk/

WWF

www.greenpeace.org/international/en/

Greenpeace International

www.wildlifetrusts.org/

Wildlife Trusts

www.iwcoffice.org

International Whaling Commission

CHAPTER 30
SUSTAINABILITY

Mark Whitehead

Introduction

In July 2009, *The Economist* organized an online debate on the topic of sustainable development. The debate centred on the statement 'the house believes that sustainable development is unsustainable'. The ensuing six-day discussion covered topics ranging from the laws of thermodynamics to international development, climate change to resource scarcity, technological innovation to urbanization, and patterns of domestic consumption to eco-terrorism. This debate and its content are important for at least three reasons:

1. They indicate the enduring significance of sustainable development as a policy goal and subject of political conjecture.
2. They suggest an emerging set of concerns about the value and future role of sustainable development thinking.
3. They illustrate the broad field of concerns that are connected to sustainable development.

At the end of the debate, 59 per cent of respondents claimed that sustainable development was sustainable and had an important role to play in shaping our collective futures. Sustainable development is the most prominent political expression of a series of a policies and philosophies that focus on the notion of sustainability (see Connelly, 2007; McManus, 1996; Myerson and Rydin, 1996; Whitehead, 2006). Although questions of sustainability have come to occupy an increasingly significant position within international and national policy-making over the last thirty years, questions of sustainability have been an ongoing issue throughout human history. From hunter-gatherers, who were concerned about the availability of their precarious food supply, to the anti-nuclear weapon campaigners of the twentieth century, who anticipated atomic catastrophe, sustainability is a field of concern that occupies the minds of any individual or society that concerns itself with threats to its future survival. This chapter provides an introduction to the history of sustainability as a concept and policy goal. It also considers the particular role that geography and geographers play in the analysis of sustainability.

Unpacking sustainability

Designed ambiguity: the scientific and geopolitical origins of sustainability

Figure 30.1 is a fictitious representation of the increasing frequency with which the word sustainable is being used. While this graph clearly exaggerates the point, it does indicate two commonly held beliefs: first, that sustainability is being used by an increasingly diverse set of user groups, including public policy-makers, corporations and environmental groups; second, that the increasing frequency with which the term is being used is leading to confusion concerning precisely what it means to be sustainable. The world sustainable has a long and complex history. The origins of the term can be traced back to the early part of the eighteenth century. According to the Merriam Webster's Collegiate Dictionary, to be sustainable means:

> a. relating to, or being a method of harvesting or using a resource so that the resource is not depleted or permanently damaged; b. of or relating to a lifestyle involving the use of sustainable methods.

This relatively simple definition points us in the direction of the scientific origins of sustainable thinking.

The scientific origins of sustainability can be traced back to the birth of modern, forest

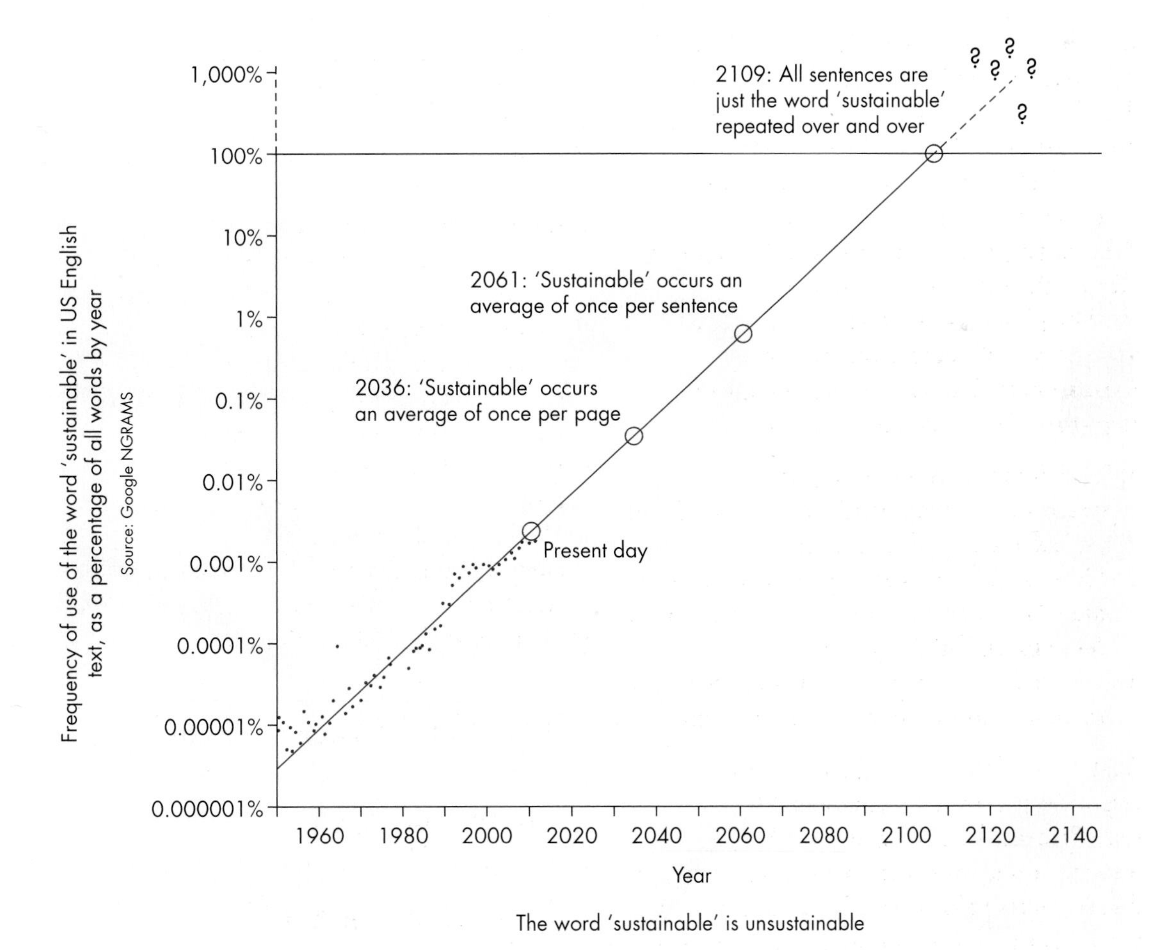

Figure 30.1 The increasing significance of the word sustainable. Credit: Randall Munroe/www.xkcd.com (Creative Commons)

management practices and the associated principle of *sustainable yield*. Sustainable yield forestry was a branch of environmental conservation that sought to find a middle path between the ecologically destructive practices of forest clearance associated with the industrial era, and the preservationist goals of the early environmental movement (Dresner, 2002: 20). Advocated by prominent conservationists such as Gifford Pinchot (who acted as Chief of the Forestry Division of the US government at the turn of the twentieth century), sustainable yield forestry was opposed to the idea of conservation for conservation's sake, and instead advocated the wise use of forest resources in order to simultaneously protect nature and to provide for human need (Dresner, 2002)). Sustainable yield forestry was based upon scientific studies that suggested that if an optimal amount of resources were taken from a forest ecosystem, but no more, that that forest could naturally regenerate itself. Modern forms of sustainable development essentially take these principles of wise use and apply them to all forms of human environmental relations.

According to the Organization for Economic Development and Cooperation (OECD), sustainable yield is defined 'as the extraction level of the resource which does not exceed the growth' (OECD, 2005). On these broad terms, sustainable yield relates not only to forestry, but also to the rate of water extraction from aquifers, fish from marine ecosystems or nutrients from soils. In the early 1990s the notion of sustainable yield became popularized within the American TBS broadcasting network's animated television programme

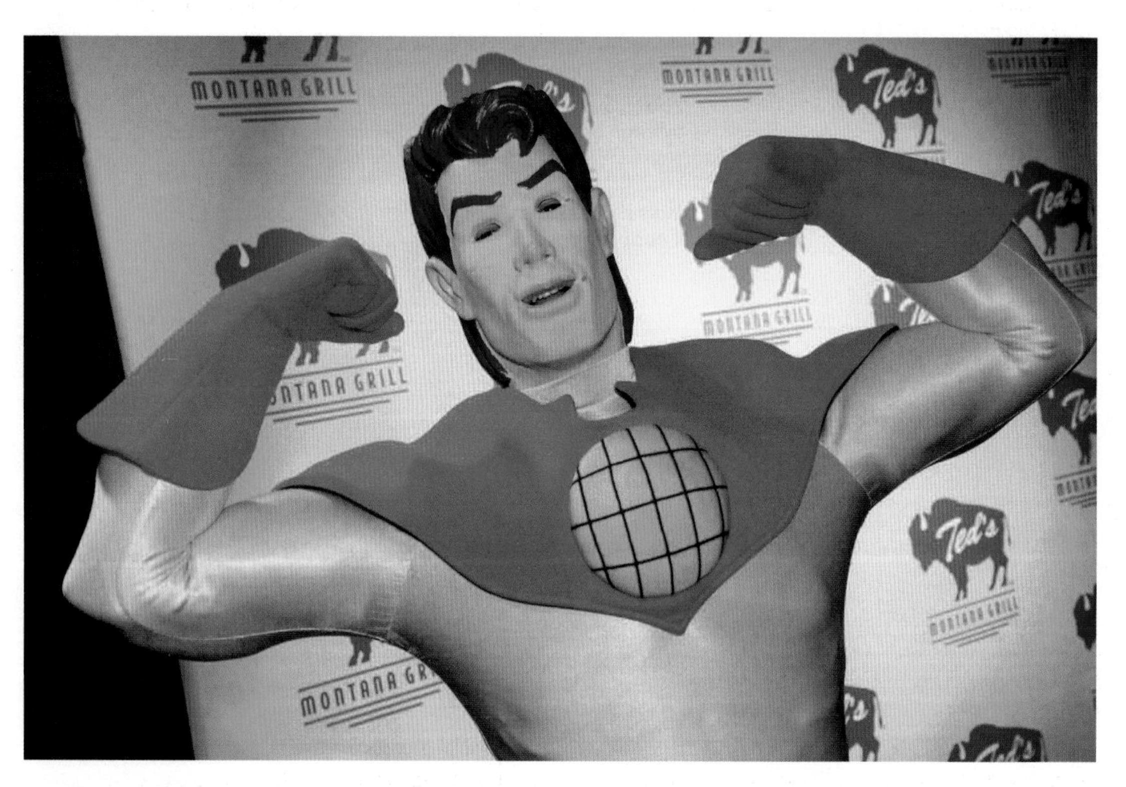

Figure 30.2 Captain Planet and the Planeteers. Credit: Rick Diamond/WireImage for Edelman Public Relations Worldwide/Getty Images

Captain Planet and the Planeteers (see Figure 30.2). In this animated series Captain Planet battles an assortment of pollution-causing villains, by drawing on his superpowers and a PhD in sustainable yields. While the assessment of sustainable yields appeared to come easy to Captain Planet, in the real world such calculations are much more difficult to achieve. The problem is that while the assessment of sustainable yields has historically been based upon the observation and management of specific resources (water, fish, soil) in specific places (forests, river catchments, organic farms), sustainability is actually the product of a much more complex set of ecological interactions operating at a variety of scales (see Redclift, 2005). Returning to the example of sustainable yield forestry, while it may be possible to assess a sustainable rate of timber extraction, it is much harder to assess the impacts that managed forest clearances have on the other forms of vegetation and wildlife that rely on the forest for their own sustainability. At one and the same time, it is also difficult to assess the impacts that sustainable forest management practices may have on the sustainability of the broader ecosystems within which forests are located. Woodland clearances can, for example, lead to soil runoff and the contamination of rivers and lakes downstream.

Although the science of sustainable yields provided the basis for thinking about sustainability, the emergence of sustainable development as a policy goal has more to do with geopolitical negotiation than science. At its core, sustainability reflects a kind of grand international compromise. This is a compromise forged between the concerns of More Economically Developed Countries (MEDCs) over the deterioration of the global environment, and Less Economically Developed Countries over persistent forms of poverty and socio-economic disadvantage (see Whitehead, 2007: 22–5). During the 1960s, the emergence of various environmental movements in MEDCs resulted in the prioritization of global environment protection as a key national policy goal. As many forms of environmental pollution are trans-boundary (from acid rain to climate change), it was recognized that the effective protection of the global ecosystem could not be achieved by the unilateral action of nation states acting on their own. It was in this context that during the late 1960s the first attempts were made to try and develop an international consensus on the protection of the global environment. Coordinated under the auspices of the United Nations, these early steps towards multilateral environmental action culminated in the 1972 UN Conference on the Human Environment, which was convened in Stockholm. This first major international meeting on the global environment brought together representatives from 113 different nation states, but it also exposed some deep geopolitical divisions. At the heart of these divisions was the concern among representatives from LEDCs that the imposition of new, international environmental laws and agreements could undermine their sovereign control over the use of natural resources and hinder their attempts to develop economically. In a famous speech delivered to the 1972 Stockholm conference, the then Prime Minister of India, Indira Gandhi, stated, 'poverty is the biggest polluter' (Figure 30.3). There are two implications of Gandhi's account of the pollution of poverty: first, that the most pressing concerns of LEDCs are not global environmental protection, but the alleviation of life threatening poverty; and second, that poverty itself is a cause of significant environmental pollution.

In light of the apparent stalemate that was reached in Stockholm, the conference developed a series of principles, which sought

Figure 30.3 Indian Prime Minister Indira Gandhi who spoke of the 'pollution of poverty' in 1972. Credit: Mondadori via Getty Images

to preserve the resource sovereignty of states, while establishing a framework for international environmental policy development.

In his reflections on the 1972 Stockholm Conference, Steven Bernstein argues that it produced a 'weak norm complex of environmental protection' within which the needs of environmental conservation were largely subservient to the needs for international trade and economic development, and within which environmental protection and economic development were seen as competing policy objectives (Bernstein, 2000: 470). It was in this context that sustainability, and more specifically sustainable developed, emerged in the 1980s as a 'breakthrough idea' (Bernstein, 2000: 470). Sustainable development was a breakthrough idea to the extent that it not only suggested that environmental protection and economic development were not mutually incompatible, but it also claimed that if they were properly orchestrated such goals could be mutually supportive. It is within this assumption that sustainable development, as we now know it, moves beyond the science of sustainable yields. While the idea of a sustainable yield suggests that it may be possible to utilize natural resources at a rate that enables the environment to replenish itself, sustainable development claims two additional things:

1. that a lack of economic development (and the persistence of poverty) actively contribute to overshooting sustainable yields
2. that economic growth and wealth creation lead to both a greater respect for the carrying-capacities of nature, and, at times, to increases in the threshold at which sustainable yields exists.

Each of these assumptions requires explanation. The first assumption is derived from something called Time Preference Theory. When applied to questions of sustainable development, Time Preference Theory concerns the extent to which environmental decision making shows a preference for long- or short-term priorities and needs. The principles of sustainability suggest the importance of decision making that is future-oriented, and which can allow for the future welfare and recovery rates of the environmental resources on which people depend. Theories that connect poverty with unsustainable behaviours suggest that the pressures of poverty force people to prioritize the short-term need for nutrition over longer-term environmental care. Porritt thus observes that:

> it is poverty that drives people to overgraze, to cut down trees, to adapt ecologically damaging shortcuts and lifestyles . . . – in short to consume the very seed corn on which the future depends, in order to stay alive today.
>
> (Porritt, 1992: 35)

Others, however, have argued that Time Preference Theory actually works in the opposition direction (see Maxwell and Frankenberger, 1992). Recent research indicates that poor families in drought-prone regions display clear signs of future-oriented decision making (for an overview of debates see Gray and Moseley, 2005). This research suggests that it is those who have the least amount of assets to fall back upon who are most likely to plan carefully for their future environmental needs. This perspective reveals possible weaknesses in the second assumption associated with sustainable development thinking. Wealth does not always result in a greater concern for the environmental future. What wealth does generate, and which constituted an important component of sustainable development theory, are opportunities to invest in new forms of technology (including high yielding varieties of crops and more energy efficiency devices), which means that it may be possible to get more from nature while remaining sustainable.

Notwithstanding such debates, sustainable development provided a blueprint in and through which it became possible to forge major international agreements between the Global South and North on issues of environmental protection. The principles of sustainable development were initially developed and formally codified by the World Commission on Environment and Development (WCED). The WCED was established by the General Assembly of the United Nations in 1983. Following significant consultation and debate, the WCED outlined the core principles and practices associated with sustainable development in its 1987 report *Our Common Future*. In essence, *Our Common Future* provided the basis upon which the principles of sustainable development could be ratified at major UN conferences held in Rio de Janeiro in 1992 (United Nations Conference on Environment and Development) and in Johannesburg in 2002 (United Nations World Summit on Sustainable Development). What ultimately appears to unite sustainable development thinking is a desire to ensure that policies reflect the complex ways in which

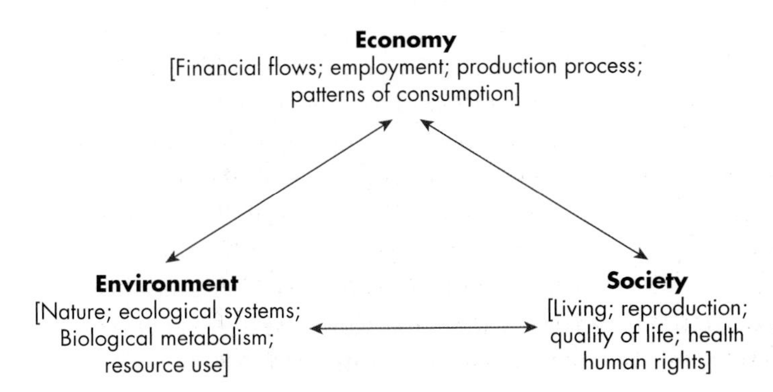

Figure 30.4 Sustainable development as integrated policy-making. Credit: Mark Whitehead

social, environmental and economic systems are interconnected (see Figure 30.4).

The spectrum of sustainabilities

While sustainable development provided a basis for geopolitical compromise and international agreements on major environmental issues ranging from forest conservation to climate change, many have been critical of sustainable development policy-making. Some have argued that it reflects a form of 'designed ambiguity,' through which it can be interpreted in very different ways by different people and organizations (Bernstein, 2000: 471). Others have argued that by tethering the goals of environmental protection and social justice to economic growth, it has resulted in a 'business as usual' approach to international policy development (Redclift, 2005). What is clear is that the long and highly political history of sustainability has resulted in the production of a wide range of associated policies and practices. To these ends, when people as diverse as the Secretary General of the United Nations, the Chief Executive Officer of Exxon, the President of Mali, or the Director of Friends of the Earth talk about sustainability, you can be sure that they are actually talking about quite different things.

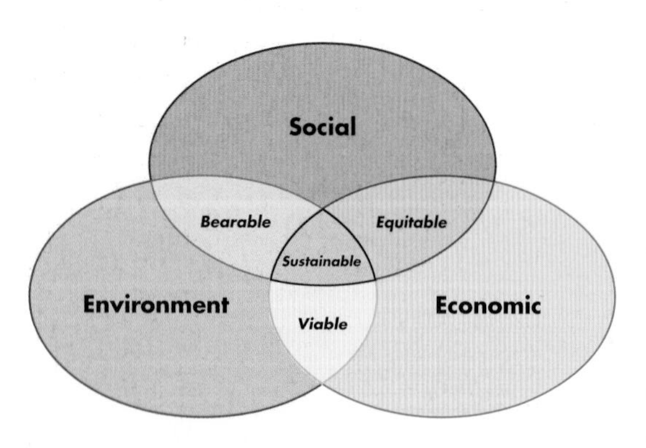

Figure 30.5 Sustainability as an elusive centre. Credit: Wikimedia Creative Commons

There are different ways in which we can think about the diverse forms of sustainability that exist (for an excellent overview see Connelly, 2007). In some ways it is helpful to think of sustainable development as a form of Venn diagram of overlapping circles (Figure 30.5). While the goal of sustainability may be to find the elusive space where social, economic and environmental priorities are achieved, in reality policies for sustainable development will tend to prioritize different socio-economic and ecological goals. Myerson and Rydin (1996) suggest that sustainability is best thought of as a spectrum of options (Figure 30.6). At one end of the spectrum are the so-called cornucopian optimists (cornucopian is taken from the Greek for 'overflowing store'), who believe that the abundance of natural resources and advances in human technology mean that there is little need for socio-economic reform in order to achieve long-term sustainability. At the other end of this continuum are those who believe that there are environmental limits to growth, and that only through strategies of degrowth (we return to this idea in the conclusion) can long-term social and ecological justice be achieved.

Underlying the diagrams and spectrums that can be used to depict the different types of sustainability that exist is a broader distinction between *weak* and *strong* sustainabilities (see Neumayer, 2004). Weaker interpretations of sustainability suggest that human development, particularly in the technological realm, can be substituted for natural capital, in order to enable socio-economic development to continue. Stronger definitions of sustainability are based on the assumption that the limits of nature should be respected and cannot be entirely overcome through human ingenuity. The distinction between strong and weak

	Cornucopian (Optimists)	Green Economics	New Economics	Limits to Growth
Development	Current Growth	Marginal Change	Substantial Change	No Growth/ Degrowth
Environmental Protection	Positive Feedback	Trade-offs	Address negative feedback	Absolute limits
Social Equity	Trickle-down	Some Redistribution	Necessary redistribution	Substantial redistribution
Key Figures	Von Hayek	Pearce	Simms	Meadows
Sustainability Spectrum	Weak	Moderate	Moderate	Strong

Figure 30.6 The spectrum of sustainabilities. Credit: Mark Whitehead (after Myerson and Rydin, 1996)

versions of sustainability is illustrated neatly in the ecological economist Herman Daly's reflections on annual fish catches (2005: 103). According to Daly, weak versions of sustainability would suggest that it is the number of available fishing boats, and the effectiveness of the human techniques and technologies of fishing, that determine how many fish can be caught in a year. Stronger versions of sustainability suggest that in the long term fish catches have little to do with the number of fishing boats and are instead determined by the natural ability of fish populations to regenerate naturally (Daly, 2005: 103). There is debate and speculation concerning the extent to which contemporary sustainable development policies are predominantly weak or strong in their orientation. In Bernstein's fascinating account of the evolution of UN-sanctioned sustainable development policies, he claims that the Organization for Economic Cooperation and Development had a crucial role in ensuring that weaker versions of sustainability are the default policy option throughout large parts of the world (Bernstein, 2000).

SUMMARY

- Theories of sustainability have their scientific origins in the practices of sustainable yield forestry.

- Modern forms of sustainable development reflect a form of geopolitical compromise between attempts to protect the global environment and concern for the alleviation of poverty in LEDCs.
- There are many contemporary forms of sustainability. Policies for sustainability range from weak, more reformist, programmes, to strong, and more radical, initiatives for socio-ecological reform.

Geographical perspectives on the sustainability society

Geographers have made important contributions to the study of sustainable development. In part, these derive from the strong connections between the discipline and notions of sustainability. As a series of chapters in this book have illustrated, geography combines a concern with the operation of human socio-economic systems with the study of physical environmental processes. To these ends, it is well positioned to analyse the forms of socio-environmental interactions with which sustainability is ultimately concerned. Geographers such as Bill Adams and Michael Redclift were among the first to produce detailed critical accounts of sustainable development policy-making (see Adams, 1990; Redclift, 1987). Prominent geographers such as Piers Blaikie also produced pioneering research into the connections that exist between socio-economic systems and environmental degradation (Blaikie, 1985; Blaikie and Brookfield, 1987). Through his empirical research into the practices of peasant farmers in Asia and Africa, Blaikie revealed that soil erosion and associated forms of environmental degradation were neither the fault of mismanagement at the farm level or the product of broad climatic shifts. Instead, Blaikie claimed that environmental degradation was often the product of unfair land ownership structures and trading regimes. In this context, Blaikie's work shifted discussion about the connections between poverty and environmental degradation from the level of the individual farmer to the broader processes that produced social poverty and disadvantage in the first place.

Beyond this early pioneering work, geographers have played an important role in exploring the spatialities of sustainability. By spatialities of sustainability I mean two things:

1. the spaces and places in which sustainability is practised
2. the way in which the spatial organization of life impacts on the sustainability of a society.

In relation to the first dimension of the spatialities of sustainability, geographers have explored the ways in which distinctive patterns of sustainable development policy have emerged in different geographical contexts. This work has gone some way to explaining why we see such diverse forms of sustainability. At large geographical scales, related work has explored the distinctions between the types of business-oriented forms of sustainability which predominate in many Western states (which tend to prioritize greater environmental efficiencies in industry and a move towards dematerialized economic sectors such as finance), and more radical forms of sustainability in the Global South (which focus on questions of social and environmental justice, and are often connected with new social movements) (see Williams and Mawdsley, 2006; Peet and Watts, 2004). Geographers have also explored the emergence of sustainable development policies in the post-socialist world (see Oldfield and Shaw, 2002; 2006; Pavlinek and Pickles, 2000). Related work has illustrated the ways in which sustainability provided an alternative paradigm of development to socialism, and how post-socialist versions of sustainability have been strongly influenced by Soviet scientific traditions and pre-socialist cultural norms (see Oldfield and Shaw, 2002). Taken together, this geographical research has illustrated that the role that political structures, culture, history and identity play in shaping which sustainable development policies and practices emerge in different places. While *et al.*, (2004) argue that varied geographical expressions of sustainability reflect specific spatial fixes for communities' social, economic and environmental needs. Through the notion of a *sustainability fix*, While *et al.* (2004), argue that sustainability is not only an international political compromise, but always reflects a series of local compromises within which the relative balance of social justice, environmental protection and economic development are worked out.

The second aspect of the spatialities of sustainability concerns the ways in which the spatial organization of human activity impacts upon sustainable development. To these ends, geographers have explored how the suburban spread of cities has been based upon the availability of cheap oil and has facilitated the property-based urban development of many cities. Such patterns of development have, however, made living sustainable lifestyles difficult as the routines of daily life (from commuting to collecting groceries) often involve the use of high levels of non-renewable resources. Geographers have subsequently become interested in the ways in which cities can be planned and designed in ways that maximize their sustainability. Smart urbanism thus seeks to achieve higher density patterns of metropolitan development that can make living lower energy lifestyles more achievable (see Krueger and Gibbs, 2008). Beyond the urban scale, geographers have also been active in considering the potential role of regional planning in facilitating more sustainable patterns of living. Regions exist at a scale above that of the city and tend to incorporate agricultural land and natural resources within their remits. Related work has considered the potential role of the region as a more localized source for a population's resource needs than the global economy, and as a spatial context within which it may be easier to measure and monitor sustainability (Whitehead, 2003).

SUMMARY

- Geographers have played an important role in exploring the complex relations that exist between poverty and environmental degradation.
- Studying existing forms of sustainability in different places has enabled geographers to explain why different forms of sustainable development policies exist in different locations.
- There are strong connections that exist between the spatial organizations of a society and its relative levels of sustainability.

Conclusion: beyond the sustainable society

In this chapter we have charted the emergence of sustainability from sustainable yield forestry and geopolitical compromise to become one of the dominant policy frameworks of the early twenty-first century. We have discovered that sustainability is essentially concerned with forging a balance between economic growth, environmental protection and social justice. In exploring the nature and origins of sustainability, we have also seen the great diversity that exists in the actually existing sustainabilities that are being practised throughout the world. While it is clear that there are strong forms of sustainability, which call for society to live within the carrying capacities of the natural environment and the effective redistribution of wealth, it appears that weaker, more conformist, styles of sustainability are more common.

This chapter has also illustrated the contribution that geographers have made to the study of sustainability. At one level, these contributions have shed valuable light on the complex connections between poverty and environmental degradation. In more general terms, geographers have consistently exposed the spatial dimensions of sustainability. By thinking about sustainability spatially, geographers have revealed the role that politics and culture play in shaping the forms of sustainability that emerge in different places. At one and the same time geographers have considered the ways in which the spatial organizations of society conditions the ability of that society to be sustainable.

Sustainable development continues to be an important national and international policy objective. Indeed, Rio+20, the United Nations Conference on Sustainable Development that will be convened in June 2012 is likely to re-endorse the core principles of sustainability. Notwithstanding this, it is clear that sustainability is now being challenged by a series of competing paradigms of social development and environmental protection. At one level, many now claim that climate change is the dominant environmental policy goal. While climate change in essence recreates the kinds of tensions between economic development and environmental reform that sustainability attempts to resolve, there is a real danger that in the rapid pursuit of greenhouse gas emission reductions, issues of social justice may become marginalized (see While *et al.*, 2010). At another level, concepts such as resilience, decoupling and degrowth are all emerging as alternative paradigms of development. Unlike the optimism that is inherent within visions of sustainability, resilience thinking suggests that society should prepare itself for socio-economic and environmental shocks to the system, and not assume that the future will be characterized by managed stability (see Folke, 2006). Emerging ideas of decoupling on the other hand argue that economic development needs to be rapidly and permanently separated from the need to use non-renewable resource utilization (Swilling (undated)). On the other hand, ideas of degrowth assert that economic growth is now environmentally unsustainable and new forms of social organization need to be instigated that do not rely on economic expansions for their survival (Martinez-Alier, 2010). The extent to which contemporary forms of sustainability are unsustainable is sure to become a crucial issue in the decades ahead of us.

DISCUSSION POINTS

1. Will the drive for continued economic growth ultimately undermine attempts to construct a more sustainable global society?
2. What can a geographical perspective offer the study of sustainability?
3. What are the key geographical obstacles that prevent people from leading more sustainable lifestyles?

FURTHER READING

Adams, W.M. (2001) *Green Development: Environment and Sustainability in the Third World* (2nd edn.). London: Routledge.

This volume provides a detailed account of the political evolution of policies for sustainable development. Focusing mainly on sustainability in LEDCs, Adams' volume helps to place sustainable development in historical context.

Bernstein, S (2000) Ideas, social structure and the compromise of liberal environmentalism. In: *European Journal of International Relations* 6, 464–512.

This is a challenging, but highly rewarding paper. It provides a detailed account of the processes in and through which stronger visions of sustainability have been marginalized by neo-liberal visions of a much weaker approach to sustainable development.

Neumayer, E. (2004) *Weak Versus Strong Sustainability: Exploring the Limits of Two Opposing Paradigms*. Cheltenham: Edward Elgar.

This text introduces the key distinctions between weak and strong forms of sustainability. It also provides a detailed analysis of natural capital.

Whitehead, M (2006) *Spaces of Sustainability: Geographical Perspectives on the Sustainable Society*. Abingdon: Routledge.

This book provides an overview of the contribution that geography has made to the study of the sustainable society. Exploring sustainability in a range of different geographical contexts, and at a series of different scales, this volume illustrates the importance of developing a spatial perspective on sustainable development.

WEBSITES

www.unep.org/

United Nations Environment Programme: provides helpful information on emerging international sustainable development policy initiatives.

www.guardian.co.uk/environment/sustainable-development

The Guardian newspaper's sustainable development site: provides links to up-to-date media stories relating to sustainable development and relevant blogs.

http://anthropocenedotcom.wordpress.com/

Placing the Anthropocene blog: provides commentary and analysis on a range of themes related to questions of sustainability.

CHAPTER 31
CLIMATE CHANGE

Harriet Bulkeley

> Climate change is everywhere. Not only are the physical climates of the world everywhere changing, but just as importantly, the *idea* of climate change is now to be found active across the full parade of human activities, institutions, practices and stories. The idea that humans are altering the physical climate of the planet through their collective actions, an idea captured in the simple linguistic compound 'climate change', is an idea as ubiquitous and powerful in today's social discourses as are the ideas of democracy, terrorism, or nationalism.
>
> Mike Hulme, 2009: 322

Introduction

Living in the UK, **climate change** is part of our daily lives. The supermarkets where we do our routine shopping inform us of their intentions to reduce the greenhouse gas (GHG) emissions that lead to climate change on our behalf. Tesco even provide 'carbon footprints' on everyday products, such as milk and orange juice, to assist consumers to take personal action on the issue. Leading corporations such as HSBC have become '**carbon neutral**'[1], while others have set ambitious targets for reducing carbon dioxide emissions. To take just two examples, B&Q, a large DIY retailer, has pledged to pursue a '90 per cent reduction in CO_2 by the end of the 2023 financial year', while BT, the telecoms company, have reduced their UK carbon footprint by 59 per cent below 1996 levels and are now aiming to 'achieve an 80 per cent reduction in CO_2 intensity world-wide by 2020'[2]. This has not been an issue championed by corporate actors alone. Civil society groups – from Oxfam to the Women's Institute – have also made addressing climate change part of the core activities of their organizations. Several hundred places in the UK have also pledged to become 'low carbon' or 'transition' communities. At the same time, universities are now charged with addressing their own carbon footprints in order to meet government targets and student groups across the country have organized debates, events and action around this issue. In this manner, climate change has entered the very fabric of society.

Coupled with growing national and international policy efforts to address the issue, this wealth of climate-related activity has led several people to point to the near ubiquity of the consensus that climate change is a problem that requires our urgent attention. As Mike Hulme argues in the quote used to introduce this chapter, the *idea* of climate change is as prevalent in contemporary societies as those of 'democracy, terrorism or nationalism'

Figure 31.1 Tesco zero-carbon store. Credit: AP Images

(Hulme, 2009). The ubiquity of the idea of climate change has been accompanied, Erik Swyngedouw (2010: 215) argues, by a growing consensus which 'is now largely shared by most political elites from a variety of positions, business leaders, activists and the scientific community' about the nature of the climate change problem and how it should be addressed. This consensus, he goes on to suggest, is part of an emerging 'post-political' condition 'structured around the perceived inevitability of capitalism and a market economy as the basic organizational structure of the social and economic order, for which there is no alternative' (Swyngedouw, 2010: 215). While acknowledging the ubiquity of the notion of climate change and the consensual nature of much of climate politics in the UK and at least some parts of Europe, the extent to which such perspectives hold true in other parts of the world is certainly debatable. Understanding the Human Geography of climate change, therefore, means interrogating the emergence, extent and nature of this consensus, and its implications for why, how and for whom action to address climate change is being undertaken.

This chapter explores these issues by examining how the idea of climate change has been constructed within different arenas. While conventionally viewed as a 'global' problem, requiring 'global' solutions, climate change is now also mobilized at multiple other spatial scales and through new forms of networked political space (Bulkeley, 2005). For example, different forms of carbon market have been established which cut across traditional global, national and local boundaries (Bumpus and Liverman, 2009; Newell and Paterson, 2010). These include the EU Emissions Trading Scheme through which large users of energy in Europe can 'trade' their emissions with one another in order to meet certain targets or 'caps', as well as voluntary schemes through which individuals and businesses seek to '**offset**' their own emissions through paying for schemes that reduce emissions elsewhere. Climate change is also often regarded as a 'business' problem, and there are many examples, as noted above, of how corporations have started to integrate climate change into their strategic goals and daily operations which also cut across traditional geographic divides. Alternatively, climate change is often framed as a problem that individuals need to address. In the early 1990s, governments across the world launched information campaigns to try to persuade people to take action on the issue, such as the UK's 'Helping the Earth Begins at Home' initiative. While the tools and techniques through which individuals are now encouraged to understand their contribution and to take action to address the issue have grown in sophistication, the basic framing of climate change as an issue with which we should all be concerned in our daily lives has remained the same (Paterson and Stripple, 2010). There are, then, multiple Human Geographies of climate change, and these are central to the ways in which the idea of climate change is understood and acted upon. In the rest of this chapter, just two such geographies are considered: climate change as a global problem and the notion that cities are central to addressing climate change. These two sets of debates show that while the idea of climate change is increasingly ubiquitous, its geography makes a significant difference to how it is, and could be, addressed.

Making climate change a global problem

It is often taken for granted that climate change is a global issue. That greenhouse gases, wherever they are emitted, contribute to the phenomenon of climate change, and that this will have differentiated impacts across the world (Figure 31.2), suggests that a global response is required to address the problem. Indeed, this is an argument that is often used by those who seek to promote international action. However, the apparently natural connection between the scale of the problem and the scale at which a solution should be sought masks the complex process through which climate change has come to be known as a global problem, and through which efforts to create an international response have been forged and contested.

The initial international response

While the potential for different gases, including most significantly carbon dioxide, to contribute to a warming of the earth's atmosphere was the subject of scientific investigation throughout the nineteenth and twentieth centuries, it was not until the late 1980s that the issue made its way onto the global stage. While the first international assessment of climate change was conducted in the 1980s, the Villach Conference of 1985, at which it reported its findings and concern about the potential for 'anthropogenic', or human-induced, climate change, failed to capture political or public imagination in a context of prevailing concern for the diminishing ozone layer. In 1988, when the Canadian government sponsored a conference in Toronto on 'The Changing Atmosphere',

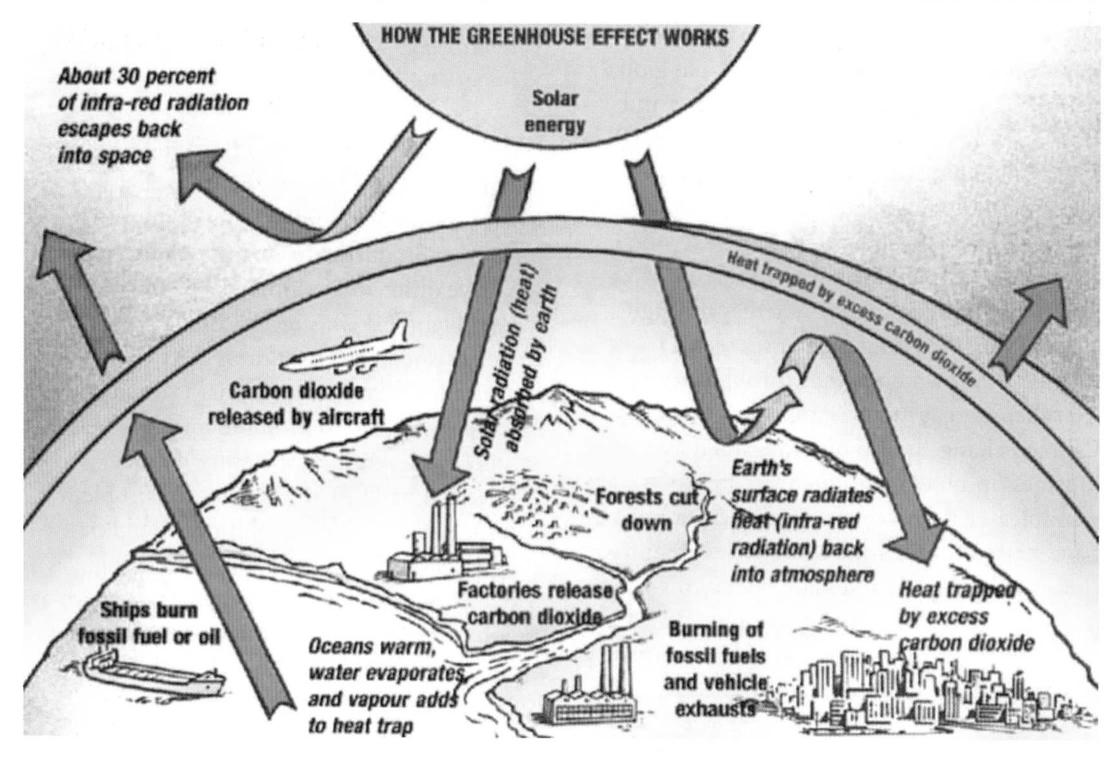

Figure 31.2 What is the greenhouse effect? Credit: www.fao.org

a series of extreme weather events coupled with a statement by NASA Chief Scientist to the US Senate on the importance of taking climate change seriously served to galvanize international action (Paterson, 1996). Likening the potential effects of climate change to that of nuclear war, participants at the Toronto Conference agreed a voluntary target of reducing their carbon dioxide emissions to 20 per cent below 1988 levels by 2005. Following this meeting, national governments established the Intergovernmental Panel on Climate Change (IPCC) to act as an authoritative body for synthesizing and assessing the state of scientific knowledge on the issue. In their first report, the IPCC concluded with certainty that increased concentrations of greenhouse gases in the atmosphere could lead to climate change. These findings were endorsed so that by the end of 1990 political momentum was such that 'negotiations towards an international convention were virtually unavoidable' (Paterson, 1996: 48).

The outcome of these negotiations was the United Nations Framework Convention on Climate Change (UNFCCC) which was agreed at the United Nations Conference on Environment and Development (sometimes referred to as the 'Earth Summit') in Rio in 1992. Reaching an international agreement was fraught with difficulties, as different national governments argued over the extent to which they should be given responsibilities for reducing **GHG emissions** and how far the international community should provide financial and technical assistance for these responses. Key differences which emerged during this period lay between the USA, who were opposed to legally binding targets, and the

EU, who argued for strong commitments. In addition, a group of developing economies known as G77 persuasively argued that their contribution to atmospheric GHG emissions were historically minimal in comparison to those of the industrialized countries so that they should not be obliged to commit to such targets. The final agreement, which was signed by 186 countries including the USA, Australia, China and India, aimed to avoid 'dangerous interference in the climate system' through stabilizing levels of GHGs in the atmosphere. This was to be achieved through individual countries reducing their GHG emissions in line with the principle of 'common but differentiated responsibility' which established that industrialized countries should take voluntary action but that any action on the part of developing countries would be subject to the receipt of additional aid and technical support (Bulkeley and Newell, 2010; Paterson, 1996).

From Rio to Kyoto, Copenhagen to Cancun

The UNFCCC was an important landmark in the process of building scientific knowledge and political commitment for an international response to climate change. However, while framing climate change as a global problem, it also established the highly differentiated nature of responsibilities and positions in relation to the idea of climate change. By the mid-1990s, signatories to the UNFCCC began further rounds of negotiation within which these differences became even more marked. In the light of a second assessment by the IPCC in 1995 which established the likelihood that climate change was already taking place, the commitments reached at Rio appeared increasingly inadequate. In 1997, the Kyoto Protocol was agreed, committing 38 industrialized countries to reduce GHG emissions by an average of 5.2 per cent below 1990 levels during the period 2008–12 and establishing a set of 'flexible mechanisms' through which individual national targets can be reached. These included the provisions for establishing 'carbon trading' and the 'Clean Development Mechanism' through which industrialized countries could undertake projects to reduce GHG emissions in developing countries as part of reaching their national targets. Despite the inclusion of these measures, in 2001 the USA announced that they would not commit to signing the Kyoto Protocol. With the withdrawal of the USA – the world's largest emitter of GHG emissions – many assumed that the international effort to address climate change was undone. However, the EU and G77 redoubled their efforts to achieve agreement and with the Russian ratification of the Kyoto Protocol in 2005 it entered into force (Bulkeley and Newell, 2010).

Despite international agreement and the existence of binding targets for those industrialized countries who are signatories to the Kyoto Protocol, negotiations have continued to address the detail of the mechanisms through which it is implemented and to consider how the international community should develop responses for the post-2012 period. Central to these negotiations have been concerns about how to account for the role of forests and 'avoided deforestation' in attenuating climate change; how **adaptation** to the effects of climate change might be funded; and whether 'transition' economies, such as China, India, Brazil, Mexico and South Africa should undertake specific international commitments to reduce GHG emissions. The 2009 Copenhagen Conference was heralded as a critical juncture for reaching agreement on these, and other, issues. The palpable sense of disappointment among business and civil society organizations, the global media and

Figure 31.3 Copenhagen protests. Credit: SIPA Press

various publics globally at the failure to reach agreement at Copenhagen is perhaps testament to just how much has come to ride on the outcomes of this continuing process of international negotiation.

The hope and the disappointment that Copenhagen has come to embody also signify something more fundamental – the reworking of the notion of climate change as a 'global' problem. Although built into the process of international negotiation over the past two decades, the experience of Copenhagen has highlighted the differentiated and fragmented nature of the global response to climate change. In its wake, a new form of global response emerged as developed, rapidly industrializing countries, including China and India, and developing countries, pledged individual commitments under the Copenhagen Accord. The subsequent 2010 meeting in Cancun, Mexico, provided UNFCCC backing for this approach, while also confirming the commitment of developed countries to commit to a billion dollar Climate Investment Fund for supporting developing economies in their efforts to address the issue. In effect, national commitments and market-based mechanisms remain central to the international response to climate change. This particular 'global' response to the issue is, therefore, constituted through a particular set of relationships between nation states, markets and scientific expertise.

Such a response appears to contain the essential ingredients of Swyngedouw's post-political consensus, dependent as it is on the perceived inevitability of capitalism and of the market economy as the critical means through which actions may be mobilized. However, the history of the climate negotiations given above shows that this consensus is underpinned by deep divisions over what such markets may be able to achieve and dissent concerning who should bear the responsibilities and who should gain the rewards of addressing climate change. At the same time, the negotiations themselves have been subject to growing and vociferous movements campaigning for climate justice, within which these fundamental roles of the state and the market are questioned. Furthermore, the growing dissatisfaction with international efforts has led Matthew Hoffman to identify new forms of 'governance experiment' emerging among government, business and civil society organizations, many of which organize 'transnationally' creating new political arenas within which the global and local dimensions of climate change may be addressed (Hoffman, 2011). Taken together, this suggests that while global responses to climate change are constitutive of a 'post political' consensus, they are also opening up new spaces for (political) response, many of which are to be found in the city.

SUMMARY

- The history of climate change as a global problem has been marked by growing scientific consensus about its significance and increasing political conflict as to what should be done.
- Despite the framing of climate change as a global issue, since the beginning of the international negotiations there have been profound geographical differences in terms of the positions taken by different countries and of their responsibilities for action.
- Although commentators often point to the failure of the climate change negotiations, over the past twenty-five years most countries have committed to the UNFCCC and its goal of stabilizing the earth's climate, and many have also signed the legally binding Kyoto Protocol.
- In the aftermath of the 2009 Copenhagen meeting, we can see a new form of 'global' climate politics emerging in which industrializing economies are playing a much stronger role and where national, rather than international, commitments dominate.
- In addition, we are also seeing the emergence of multiple climate change 'governance experiments' as cities, private companies and communities seek to respond to the issue themselves.

Climate change and the city

With estimates suggesting that 70–80% of anthropogenic emissions of global carbon dioxide emissions related to energy use may be attributed to urban areas (IEA, 2008), and significant concerns about the potential vulnerability of cities to the effects of climate change (World Bank, 2010; UN-Habitat, 2011), it is perhaps to be expected that climate change should have become an increasingly important issue for the world's cities. What is perhaps more surprising is that urban responses to the climate change issue have only recently begun to gather pace. Dominated by the global framing of climate change discussed above, and with limited capacity and resources, the opportunities and challenges for cities in this arena are only slowly being realized. This section explores how and why the idea of climate change has come to be seen as an urban issue, and considers what this means for how the issue can be addressed.

While climate change is now increasingly framed as an urban issue, it is important to remember that not all cities contribute equally to global emissions and that different forms of vulnerability will be found in different cities. Figure 31.4 shows different ways of measuring the ten cities in the world that contribute most to climate change – in terms of total GHG emissions, per capita emissions and in terms of the amount of GHG emissions generated to produce economic outputs (measured in Gross Domestic Product or GDP). The variations that are evident show that how we measure and assess such contributions is important. It also shows how different energy needs (e.g. for heating in Russia) and forms of energy supply (whether they are coal-based, as in many cities in China, or oil/diesel-based, as in some cities in Africa) makes a significant difference to GHG emissions. This suggests that we must be careful not to conclude that all cities should and can address climate change in the same way (Dodman, 2009).

Emergence of an urban response to climate change

Initial urban responses to the issue of climate change emerged in Europe and North America in the late 1980s and early 1990s with municipal governments, often through EU-funded projects and transnational municipal networks, seeking to take action to reduce GHG emissions (Bulkeley and Betsill, 2003). Given the role that municipal authorities could play in critical policy areas that affected the levels of GHG emissions, such as land use planning, transportation, waste and in some cases energy, cities became framed as part of the 'solution' to addressing the climate change problem. In addition, cities were championed as arenas within which new approaches and technologies could be tested and where public engagement with the seemingly remote global issue of climate change might effectively be fostered. Throughout the 1990s, urban responses were concentrated in cities in North America and Europe, often led by municipal authorities and frequently focused on energy efficiency. Actions were predominantly focused on those areas where such authorities had some level of direct control – such as their own buildings – though they also developed a range of tools and approaches to enable other stakeholders and members of the community to take action.

In the early 2000s, urban responses to climate change received further momentum. In the USA, the withdrawal of the federal government from the Kyoto Protocol process led some progressive municipal governments to form the US Mayors Agreement (Gore and Robinson, 2009). This engagement of locally elected politicians with the climate change agenda has been replicated globally, most recently with the launch in 2009 of the European Covenant of Mayors, which requires signatories to pledge to go beyond the EU target of reducing carbon dioxide emissions by 20 per cent by 2020 through the formation and implementation of a sustainable energy action plan (CoM, 2011a). By 2011, more than two thousand members had signed up to this commitment (CoM, 2011b). Global cities have also sought to engage with the climate change agenda,

Total GHG (MtCO2e)	Total GHG per capita (MtCO2e/cap)	GHG per GDP (ktCO2e/US$bn)
New York 196	*Moscow 15.4*	Tianjin 2316
Tokyo 174	*St Petersburg 15.4*	Beijing 1107
Moscow 167	Los Angeles 13.0	Shanghai 1063
LA 159	Chicago 12.0	St Petersburg 971
Shanghai 148	Miami 11.9	Moscow 922
Osaka 122	Philadelphia 11.9	*Lagos 893 (total = 27Mt)*
Beijing 110	Shanghai 11.7	Bangkok 799
Chicago 106	Toronto 11.6	*Riyadh 726*
Tianjin 104	*Dortmund 11.6*	*Tehran 560*
London 73	Tianjin 11.1	Wuhan 554
Bangkok 71	Bangkok 10.7	*Kinshasa 598 (total = 6Mt)*
Miami 65	Beijing 10.1	Istanbul 384

Table 31.1 Comparing the relative contributions of cities to GHG emissions. Source: Data from Cities and Climate Change: An Urgent Agenda, World Bank 2010. Note that figures in italics are estimates.

primarily through the C40 Cities Climate Leadership Group. This network was instigated in 2005 by the then Mayor of London, Ken Livingston and his deputy, Nicky Gavron, together with The Climate Group, a not-for-profit organization based in London, and has attracted the participation of 40 of the world's most politically and economically influential cities as well as the financial support of the Clinton Climate Foundation. Through such new initiatives, as well as in the activities of existing networks, such as the Climate Alliance and the Cities for Climate Protection programme, there is an increasing focus on the issues of climate change for cities in developing countries and on including concerns for adaptation and resilience alongside reducing GHG emissions. At the same time, a wider range of social and economic interests are becoming involved in urban climate responses, from HSBC and the World Bank to Action Aid and WWF.

Securing resources or addressing inequalities?

Within these efforts to position cities as a critical site for responding to climate change, several different processes can be discerned which have served to promote urban responses within which climate change is at once a more strategic concern and one which is more closely aligned to the concerns of urban growth and resource security that dominate urban agendas (Hodson and Marvin, 2010). For Aidan While and his colleagues, this is the result of a processes of eco-state restructuring focused on 'carbon control' and creating a 'distinctive political economy associated with climate **mitigation** in which discourses of climate change both open up, and necessitate an extension of, state intervention in the spheres of production and consumption' (2010: 82). As this politics of carbon control 'comes to ground' the 'calculative practices of urban management' are shifted and new forms of financial strategy and economic development are enacted. Mike Hodson and Simon Marvin (2010) make a similar argument for the growing strategic importance of climate change in terms of ensuring resilient infrastructure and securing resources for economic development, particularly in the forms of energy and water, which in turn is leading to a distinct set of urban political economies orchestrated by and through the climate change issue.

Examples of strategies and projects to achieve these kinds of carbon control and resource security can be found in cities as diverse as London, Mexico City, Cape Town and Bangalore. For example, in London the 'Decentralised Energy and Energy Masterplanning (DEMaP) Programme has been developed to assist both public and private sector to identify Decentralised Energy (DE) opportunities in London' in order to 'contribute towards the Mayor's target of providing 25% of London's energy supply from decentralised energy sources by 2025'[3]. In Bangalore, concerns over a lack of infrastructure services and depleting resources, particularly water supplies, have led to the emergence of new forms of low carbon housing. The Towards Zero Carbon Development (T-ZED) is one such project targeted at the higher income residents of the city who primarily work in the IT and related sectors (Figure 31.4). The development was initiated by a private company, Biodiversity Conservation India, and occupies a location at the outskirts of the city and, like many other developments in this area, is a gated community or compound. Its design included measures to reduce the embodied energy of the building (including the use of local materials), renewable energy generation, energy efficient technologies (including air conditioning, refrigeration and lighting), rainwater harvesting, as well as landscaping and conservation. In addition, the development seeks both to offer 'no compromise' in lifestyle while also engaging residents in a suite of low carbon practices, such as behavioural changes to reduce energy and water use, organic vegetable gardening and community-based activities.

Figure 31.4 T-ZED Bangalore. Credit: T-ZED Homes, BCIL

Innovations for addressing climate change are also emerging in the urban arena. One prominent example is the Transition Towns movement, established in the small town of Totnes, Devon, UK in 2006, and which has since spread internationally (Bailey *et al.*, 2010). As Transition Town Brixton, London, explains:

> Transition Town Brixton is a community-led initiative that seeks to raise awareness locally

of Climate Change and Peak Oil. TTB proposes that it is better to design that change, reduce impacts and make it beneficial than be surprised by it. We will vision a better low energy/carbon future for Brixton. We will design a Brixton Energy Descent Plan. And then we will make it happen.

Source: TTB, 2011

Rather than focusing on the strategic dimensions of resource security and carbon control, in Brixton the Transition Towns group focuses on issues of individual and community resilience, and is concerned with addressing wider social concerns. Such initiatives which focus on 'climate resilience' are also increasingly being established in cities in developing countries. Development challenges are being framed in relation to addressing and adapting to climate change. For example, UN-Habitat has founded a 'Cities and Climate Change Initiative' which seeks to document urban vulnerabilities to climate change in a range of cities in Africa, Asia and Latin America and to determine how resilience to the effects of climate change might be put into practice[4]. At the same time, the potential to harness climate change to address fundamental development challenges, including access to affordable energy, is slowly being realized. One example is the Kuyasa project in the Khayelitsha area of Cape Town. Led by the NGO SouthSouthNorth, the project involved retrofitting ceilings, providing energy efficient light bulbs and solar hot water heating, which together reduced energy use in households (hence yielding carbon savings) and energy poverty, providing direct financial benefits. This project was unusual in using the Clean Development Mechanism as the basis for its funding (SouthSouthNorth, 2011).

These brief examples illustrate that many urban responses to climate change, in common with their global counterparts, share the fundamental belief in the ability of market mechanisms and existing political systems to address the issue. As Erik Swyngedouw (2010) suggests, this can create a 'post-political' condition where a particular notion of what climate change is and how it should be addressed becomes dominant and there is no room for dissenting voices. However, these examples also suggest that this is not the complete picture. Cities are becoming sites within which alternative responses to climate change are emerging, some of which may use market mechanisms to their own ends and others which articulate very different forms of social relations. Rather than witnessing the straightforward emergence of a uniform idea of the urban climate change problem, these examples point to a more fractured landscape where the potential for contestation and conflict is ever present.

SUMMARY

- Cities are significant contributors to global GHG emissions and are potentially vulnerable to the effects of climate change. As well as being a global issue, climate change is therefore an urban issue, although it is important to understand the significant differences between different cities.
- Although cities have sought to address the climate change issue since the early 1990s, it has only been in the last ten years that this has gathered momentum.

- Underpinning these responses have been concerns for enhancing resilience, securing resources and for achieving 'carbon control', which align climate change as an issue primarily related to economic and urban development.
- However, alternative approaches can also be seen in cities where climate change is mobilized in order to address issues of social and environmental justice.
- Rather than witnessing the emergence of a 'post-political' consensus, it appears that climate change is an issue that remains contested.

Conclusions

Climate change has certainly come a long way in the past twenty-five years. From the initial concerns of a small group of scientists, it is, as Mike Hulme has suggested, now one of the dominant ideas of the twenty-first century. In the UK, this idea is now so familiar that it has become part of normal government policy and business operations, and a common feature of everyday life. While the UK may be unusual in terms of the depth of this engagement, the involvement of most of the world's countries in the international process of negotiating an international agreement and the growing interest in cities around the world in the issue mean that this idea now touches on the lives of millions of people.

However, as we have seen in this chapter, the ubiquity of the term climate change masks very many differences in how the term is framed, understood and acted upon. There are important geographies at work that have served to create climate change as a global issue, a business problem, a matter for government policy, an urgent agenda for cities, disaster relief, planning conflict, and an issue of individual responsibility, to name just a few. How, why and by whom climate change becomes constructed as a problem that requires attention is therefore critical in terms of the scope of possible action and whose interests are served. As climate change has come to dominate the international agenda, very real concerns have been raised that it has served to marginalize other critical environmental problems – such as soil loss, water scarcity and deforestation. There are concerns too that the worthy efforts to create new forms of climate change finance may be shaping the development agenda and skewing the provision of aid in undesirable ways. At the same time, the consensus that existing political and economic approaches can effectively deal with the problem has caused some to ask whether we are able to depart from 'business as usual' and address the underlying issues of under-development and over-consumption that lie at the heart of the issue. Understanding these geographies is critical. As Saleemul Huq and his colleagues have argued:

> [the] kinds of changes needed in urban planning and governance to 'climate proof' cities are often supportive of development goals. But . . . they could also do the opposite – as plans and investments to cope with storms and sea-level rise forcibly clear the settlements that are currently on floodplains, or the informal settlements that are close to the coast.
>
> (Huq et al., 2007: 14).

Ensuring that our responses to climate change are undertaken in a just manner means keeping open the possibilities for contestation and dissent – in making sure that climate change remains political.

DISCUSSION POINTS

1. The chapter has explained the diversity of responses being undertaken in response to climate change. a) Is such a diversity useful, or should there be a more unified approach? b) What are the possibilities and limits of trying to achieve a more uniform response?
2. Examine a popular text about climate change: this could be a book, a newspaper article, a policy document. a) How is the issue framed geographically? b) Is climate change portrayed as a global or local issue, one for collective or individual responses? c) What sorts of solutions are suggested? Consider the authors and audiences of this document. d) Why do you think climate change is framed in this way? e) What alternative framings might you suggest?
3. a) How and why are different cities vulnerable to the effects of climate change? b) What sorts of measures need to be taken to address this vulnerability? c) Can these all be taken at the urban scale?
4. If you live in a city or are studying at university in one, try and find out about its response to climate change. a) What are municipal authorities doing? b) Are other actors – business, civil society or community groups – taking action? c) What are the similarities and differences in the approaches which they are undertaking? d) Why is this the case?

FURTHER READING

There are now many journals that carry articles on climate change with a Human Geography perspective, including *Climate Policy* and *Global Environmental Politics*, while both *Local Environment* and *Environment and Urbanization* frequently have articles on urban and community-based responses to the issue of climate change. The following sources are also particularly useful:

Bulkeley, H. and Newell, P. (2010) *Governing Climate Change*. Abingdon: Routledge.

Provides a short introduction to climate politics from different perspectives, covering the history of international negotiations, conflicts between North and South and the emergence of new forms of climate governance, including by private sector and community organizations.

Giddens, A, (2009) *The Politics of Climate Change*. Cambridge: Polity Press.

Accessible introduction to climate politics in the UK and internationally from one of the UK's leading social scientists.

Hodson, M. And Marvin, S. (2010) *World Cities and Climate Change: Producing Urban Ecological Security*. Milton Keynes: Open University Press.

This book examines how world cities are responding to climate change, providing an overview of the global situation and detailed examples from several major cities, including London, New York and Shanghai.

Hoffman, M.J. (2011) *Climate Governance at the Crossroads: Experimenting with a Global Response*. New York: Oxford University Press.

Explores the failure of international negotiations and the multiple new forms of climate governance that are emerging in their place.

Hulme, M. (2009) *Why We Disagree About Climate Change*. Cambridge: Cambridge University Press.

A reflective book by one of the world's leading climate scientists, providing a novel interdisciplinary perspective on why climate change has become such a contentious scientific and political problem.

WEBSITES

www-01.ibm.com/software/solutions/soa/innov8/cityone/index.html

IBM CityOne: online game that allows players to explore the consequences of different types of energy and water system for the future of cities.

http://unfccc.int/

United Nations Framework Convention on Climate Change: comprehensive coverage of the international negotiations and annual conference of the party's meetings.

www.unhabitat.org/content.asp?typeid=19&catid=555&cid=9272

UN-Habitat (2011) *Global Report on Human Settlements: Cities and Climate Change*, UN-Habitat, Nairobi, Kenya: recent analysis of the vulnerability of cities to climate change, their contribution to GHG emissions, measures for mitigation and adaptation, and future policies. The UNGRHS website contains several background studies on individual cities as well as links to other reports on urban sustainable development.

NOTES

1. See: www.hsbc.com/1/2/sustainability/protecting-the-environment/carbon-neutrality (accessed September 2011)
2. See: www.btplc.com/Responsiblebusiness/Protectingourenvironment/Climatechange/Reducingourownfootprint/index.htm (accessed September 2011)
3. See: www.londonheatmap.org.uk/Content/home.aspx (accessed September 2011)
4. See: www.unhabitat.org/categories.asp?catid=550 (accessed September 2011)

SECTION SEVEN

HISTORICAL GEOGRAPHIES

Introduction

A simple, but overly simplistic, definition of Historical Geography is that it is concerned with the geographies of the past. If it's dead, old or a little bit musty, then it's Historical Geography. Certainly, Historical Geographers do study the past, and they cover a vast range of thematic issues, time periods and geographical areas from a variety of perspectives. Historical Geography is not confined within some sort of intellectual ghetto; in this book, for example, you will find historical research being discussed in many chapters outside of this particular section. However, although sometimes cast in such terms (for example, Mitchell (1954)), Historical Geography is neither just about human geographies in past times nor solely about applying established approaches in Human Geography to the past. Rather, Historical Geography brings together the intellectual projects of Geography and History in a richer dialogue.

Historical Geography thus has a special role to play in Human Geography, through combining Geography's focus on '**spatiality**' with History's attention to '**historicity**'. The historicist emphasis on how human life varies over time means attending both to historical specificity and processes of historical transformation. So, in this sense, Historical Geography is not just about the past, but involves understanding how and why things change or stay the same, and how this is involved in the production of places, spaces and landscapes. It is that part of the discipline where we most directly reflect on what role history plays in our human geographies. Historical Geography is also an opportunity to recognize the potential difference of other times and places from our own, in so doing widening our horizons and preventing us from making uncritical generalizations about how things can and might be. Historical Geography might also involve discovering parallels to present situations in the past, thereby qualifying a frequent sense that we live in unheralded times, and helping to sharpen our sense of what may really be 'new' and different about our present circumstances.

Picking up on such issues, Miles Ogborn's opening chapter on 'Modernity and modernization' deals directly with the dominant framing of world history in terms of a progression towards a modern present. If you thought Historical Geography was just about geography in the past this might seem a surprising topic, given how notions of the modern are associated with some sense of 'nowness'. Actually, though, the 'modern' is very much an historical subject. For a start, people in the past were modern too. What we might now think of as old and traditional was once modern (to pick somewhat random examples, think about steam railway engines or the Charleston dance). Moreover, ideas of the modern involve constructing relations between historical entities such as 'past', 'present' and 'future'. The very notion of the modern is fundamentally concerned with questions of historicity. In his chapter, Miles explores two different ways of thinking about the relations between past, present and future within modernity. In the first, modernization is cast as 'progress' – a move away from the past. We often adopt this way of thinking, for example when we distinguish between a modern 'western' world and more 'traditional' societies elsewhere that need to modernize. Focusing on the ideas of W.W. Rostow about 'stages of development' and the case of Turkey, a country long imagined as somehow bridging a modern West and a more traditional East, Miles argues that modernization in reality is not as simple as breaking from the past by importing practices from elsewhere, throwing up much more difficult experiences and complex geographies. Illustrating his argument through the famed case of the urban development of nineteenth-century Paris, Miles argues that modernization is better captured using Marx's notion of 'creative destruction', with its combined emphasis on both invention and loss. Modernization benefits some, but often at the expense of others. Throughout, the chapter shows how modernization is a profoundly geographical as well as historical process. It involves the re-making of places and their landscapes so that they can embody modernity (in modern housing, roads, countryside, and so on). Modernization also works out across space in complex ways, so that while modernity is a long-standing global presence, particular places and times experience different modernities. Modernity has, then, an historical geography.

Richard Phillips' chapter 'Colonialism and post-colonialism' offers a complementary analysis. It argues that colonialism has been fundamental to the shaping of the modern world, establishing global geographies of power that still resonate today. Cutting across some of the intellectual divisions often associated with Human Geography (for example, between Economic, Cultural and Development Geographies), so-called post-colonial studies have been an important area of scholarship for more than twenty years now, notably led by work within Historical Geography. Richard's account is divided into two parts. In the first, he explores modern European colonialism, an era that reached its peak just before the First World War, by which time, for example, more than 400 million people in Africa and India were subject to British rule. He emphasizes the different kinds of experiences that people had of colonialism, obviously in part dependent on whether they were colonizers or colonized, and also the wide-ranging impacts it had through massive population movements, resettlements and transformations of the global connections linking people and places. In the second part of the chapter, Richard considers how there is an 'ongoing colonial hangover' that shapes the supposedly post-colonial world of today. This includes relics from past empires that are live political issues (think, for example, of the Falkland Islands/Islas Malvinas, or indeed Tibet). Important too are the traces of colonial geographies in contemporary geopolitical arrangements (most obviously in the national boundaries formed during colonial rule and in the cartographic politics of formal decolonization). But also apparent are new forms of imperial power (often termed neo-colonialist), whether those be associated with military intervention (as when recent US foreign policy in Afghanistan and Iraq is described as colonial), overseas economic investment (as sometimes attributed to Chinese investments in Africa or US corporate investments in Mexico) or the institutions that help to govern the world-system today. While we might have thought that colonialism is a thing of the past, Richard sets out clearly why we should neither forget it nor romanticize it in 'pith helmet chic' (Gregory, 2004). Understanding colonialism is a central requirement for grasping both the geographies of the past and the geographies of the present.

The question of how the past is remembered in the present is developed further in the third and final chapter, where Nuala Johnson discusses the relationships between 'Space, memory and identity'. Her interest here is in how Historical Geographers and others have researched, in her words, 'the network of sites that connect the past to the geographies of identity in contemporary life'. Recent scholarship has explored how memory is not just a facet of personal life but also takes social or collective forms. Central to these are particular sites and landscapes: memorials, monuments and museums, for example. Also important is the recent growth of what we might term the 'heritage sector', through which places of memory are caught up in the growth in tourism as well as popular forms of history. These sites of memory are both politically powerful and contentious. They have had an important role, for instance, in presenting ideas of national identity through the memory work that they do. While they may thus serve the interests of powerful groups or institutions, sites of memory are open to contention, both in arguments over their production and design (over what should be remembered and how) and in the varied interpretations and experiences that people take from them (which may well escape the intentions of those making them). Nuala's discussion is illustrated through two case studies. The first, resonating with Richard Phillips's previous chapter on colonialism and post-colonialism, concerns the representation of British global trade and empire at the National Maritime Museum in Greenwich, London. The second is a newly designed monument, memorializing on its tenth anniversary the bombing in 1998 in the town of Omagh, Northern Ireland by a dissident Republican group (the Real Irish Republican Army), which killed twenty-nine people (or thirty-one if a murdered mother's unborn twins are included in the count) and seriously injured many more. In both cases, Nuala draws out both the politically contentious nature of collective memory and how those politics are worked through the material design of memory sites.

All of the chapters in this section demonstrate, then, that Historical Geography is not a dry, antiquarian pursuit. History is important to our human geographies; indeed to live in the present demands an understanding of past and future too. But, in turn, geography is important to those histories: we remember through sites such as museums and memorials; we understand the contours of our world through creating places of modernity and tradition. Our geographies are always historical; our histories always geographical. Historical Geography is that part of the discipline where those relations are most determinedly brought to the fore.

FURTHER READING

Journal of Historical Geography

This is the leading journal for the sub-discipline of Historical Geography. It might be useful to have a look through some of the recent issues to get a feel for the variety of work done by Historical Geographers, as well as some of the key debates they are engaged in. It's an excellent journal. If you want more compressed summaries of developments, the regular 'progress reports' on Historical Geography in the journal *Progress in Human Geography* are also an excellent resource, though they can be quite dense.

Baker, A.R.H. (2003) *Geography and History. Bridging the Divide.* Cambridge: Cambridge University Press.

There are many brilliant books of Historical Geography, but far fewer books about Historical Geography. Alan Baker's book on the relations between Geography and History is one of the best, developing a powerful case both for the past achievements and future potential of bringing them together.

Graham, B. and Nash, C. (eds.) (1999) *Modern Historical Geographies*. London: Longman.

A collection of essays considering the Historical Geographies of key processes shaping the modern, post-medieval world. Issues considered include those developed by the chapters in the section, namely modernization, colonialism and memory/heritage.

Philo, C. (1995) 'History, geography and the still greater mystery of historical geography'. In: R. Martin, D. Gregory and G.E. Smith (eds.) *Human Geography: Society, Space and Social Science*. London: Macmillan, 252—81.

An exploration of the contribution Historical Geography can make to both Human Geography and History.

CHAPTER 32
MODERNITY AND MODERNIZATION

Miles Ogborn

Introduction: This is the modern world

What does it mean to call something 'modern'? In part it simply means that it is new, up to date or of the moment. This might relate to (modern) technology, art or life as a whole, and these descriptions are about understanding how the world is changing. Sometimes this idea of '**modernity**' – the condition of being modern – is used to celebrate newness, perhaps to encourage people to adopt an innovation or to attract a wider range of voters to a new set of policies. Sometimes it is part of complaints that the modern world is moving on and leaving much that is valuable behind, as in some discussions of the difficulties found in understanding modern art. Whichever point of view is taken, the idea of 'modernity' situates people in time. It suggests that time is divided up into past, present and future. It gives a certain value or significance to the past (positive or negative) and it makes the present important as a time of change and of decisions about what the future should be. Calling something 'modern' makes people think about historical change.

Understanding modernity as 'newness' allows an appreciation of the modernity of any point in time. Modernity is not only 'now'. There was newness and modernity in the past too. This helps to explain the excitement that accompanied the coming of the railways, electricity, the cinema or buildings made of concrete, steel and glass. It also helps in understanding the sense of danger they brought too – the fear that the world was changing too fast. In each case the relationship between past, present and future established by modernity is experienced by people and understood in terms of how it might change their lives for better or worse. For example, Raphael Samuel has examined the ideas of modernity in Britain in the 1950s to argue that '[t]he ruling ideology of the day was forward-looking and progressive, the ruling aesthetic one of light and space. Newness was regarded as a good in itself, a guarantee of things that were practical and worked' (1994: 51).

He explores this by showing that post-war 'home improvement' meant – for those unlucky enough to live in 'ugly', 'old-fashioned' Georgian and Victorian houses – tearing out fireplaces, removing draughty sash windows, knocking down partition walls and covering over 'dust-collecting' plasterwork or banisters. In their place came central heating, fluorescent strip lighting, fixtures and fittings of easy-clean, smooth and colourful plastic, Formica and fibreglass and kitchens and broom cupboards full of labour-saving devices – washing machines, electric cookers and Hoovers (see Figure 32.1). Houses and people's lives were to be transformed through this 'appetite for modernization' (1994: 56). This example shows both the intensity with which people experience modernity (in this case as something desirable) and its role in making new geographies. Here it is new domestic geographies. These houses were made into different sorts of *places*; they looked and felt different. Indeed, the same impulse also

Figure 32.1 Home improvement as modernization

transformed the geography of many cities and their inhabitants' lives and experiences. The motorways, housing estates and civic and shopping centres so characteristic of the urban planning of the 1950s and 1960s can also be seen as an attempt to make spaces that were new, clean and easy to use. In both cases modernization involves geographical change – transforming places, spaces and landscapes – as a part of historical change (Hubbard and Lilley, 2004).

These examples also begin to suggest that modernity and modernization are about more than just 'newness'. There are particular sorts of historical and geographical change involved in the production and understanding of the new. Without setting this out too rigidly (for the many opinions about what 'modernity' is, see Ogborn, 1998), important processes can be identified: the application of scientific principles to human and natural worlds; the development of industrial economies (capitalist and non-capitalist); and the formation of states that govern many aspects of life through their bureaucracies (Giddens, 1990). These are all both historical and geographical processes. They create particular forms of modern life through specific (and often very rapid) types of historical change, and a crucial part of this is the transformation of spaces, places and landscapes. For example, Figure 32.2 shows a dramatic modern landscape: the Hoover Dam built across Black Canyon on the Arizona–Nevada border between 1931 and 1936. It is 725 feet high, incorporating 2.5 million cubic metres of concrete and creating a lake of 210 square miles. Understanding this dam in terms of modernity means seeing it as a product of the technological and scientific control of nature by geologists and engineers, funded largely from public money, and planned by one of the world's most powerful states to provide irrigation and hydro-electric power for the development of capitalist industry and agriculture. The Hoover Dam has not only transformed the physical geography of Black Canyon, but also the economic geography and environment of the western United States. It should also be understood in terms of the responses it provokes:

> Confronting this spectacle in the midst of emptiness and desolation first provokes fear, then wonderment, and finally a sense of awe and pride in man's skill in bending the forces of nature to his purpose. In the shadow of the Hoover Dam one feels that the future is limitless, that no obstacle is insurmountable, that we have in our grasp the power to achieve anything if we can but summon the will.
>
> (Stevens, 1988: 266–7)

The idea of 'man' against nature is raised again later, but for now it is important to note the sense of wonder, excitement and fear in how the Dam seems to open up the future. The Hoover Dam is a landscape that is the product of the scientific, industrial and political processes of modernization, and is experienced in terms of the challenges and dangers of modernity. It combines, therefore, the ideas of modernity as both 'newness' and as particular forms of historical and geographical change.

Figure 32.2 The Hoover Dam: a landscape of modernity. Credit: Lester Lefkowitz/Corbis

The historical geography of modernity and modernization (and the difficult question of '**postmodernity**') is a huge area. In many ways much of this whole book, and much of Human Geography, is about the geographies of the modern relationships between people, science and nature; industry, capitalism and space; and politics, power and territory. Rather than trying to cover this array of issues in miniature, in this chapter I want to concentrate instead on a more specific issue, one of particular relevance to how we think about Historical Geography and the relations between history and geography. My focus will be on (some of) the different ways of understanding the historical and geographical changes involved in processes of modernization. I want to start with an idea that is there in both the examples of the Hoover in the 1950s house and the Hoover Dam: modernization as 'progress'.

SUMMARY

- Modernity is a matter of the experience of 'newness' and the specific understanding of historical time that this experience involves.
- Modernity is also a matter of a specific set of interconnected economic, political, social and cultural changes that produce 'newness'.
- Modernization involves changes in people's lives and in geographies at all scales. These changes can be understood in a range of different ways. This chapter now turns to consider some of those ways, in particular how changes are understood as 'progress' and, more ambivalently, as 'creative destruction'.

Modernity and modernization as progress

Understanding modernization in terms of 'progress' suggests that a society makes a clean break with a problematic past and does what is necessary to move forward into a better future. An influential version of this was the application of 'modernization theories' that were applied to Latin America, Asia and Africa after 1945. In his book *The Stages of Economic Growth* (1960) W.W. Rostow suggested that the history of each society could be understood through five stages: traditional society; the preconditions for take-off; take-off; the drive to maturity; and the age of high mass consumption. Britain, with its Industrial Revolution, had been through this process first and could be followed by the other countries of the world (see Figure 32.3). Rostow argued that 'traditional society' prevented regular growth through its non-scientific attitude to nature, lack of social mobility, failure to see the potential for change and non-centralized political power. This situation was to be changed in the 'preconditions' period by removing the technological, social and political constraints on economic growth through, for example, newly centralized states (controlled by new, often nationalist, elites keen on modernization) building roads and railways and encouraging science, technology and key industries. These geographical changes in agriculture, industry, transportation and urbanization, combined with a positive attitude to modernization, would prompt 'take-off' – 'the great watershed in the life of modern societies . . . when the old blocks and resistances to steady growth are finally overcome' (1960: 7) – into a stage of continual growth and less state intervention. This economic growth would utterly transform the society into a complex modern industrial economy during the 'drive to maturity'. Eventually it would reach a final stage of 'high mass consumption' (modelled on the affluent parts of the USA in the 1950s) where production and consumption were based on consumer durables: a land of automobiles and suburban homes equipped with refrigerators and televisions.

Rostow presented modernization as progress. He argued that it was absolutely necessary for countries to make the decisions that would promote modernization or they would lose out to others. He also presented it, through the idea of 'take-off', as a dramatic transformation.

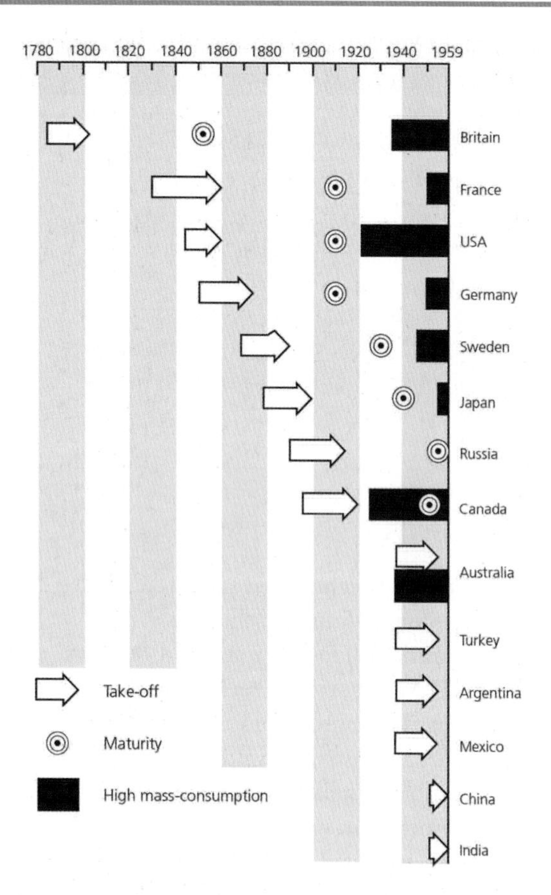

Figure 32.3 Rostow's stages of economic growth. Credit: W.W. Rostow, 1960: *The Stages of Economic Growth*, p. xii

However, by understanding it as an orderly progression of stages that others had successfully gone through he was able to claim that there was only one path to follow for successful modernization, and that this break with the past and the move into a brighter future would be relatively painless and full of benefits in the long run. This meant that this form of modernization could be offered as the way forward for the countries of what people were starting to call the 'third world'. They could become like Western Europe and the USA.

One country that modernized in this way was Turkey (although Rostow argued that the Turkish state pushed for take-off too early). It is a useful example of some of the changes in people's lives and in geographies that 'modernization as progress' involves. It also suggests some of the problems with this version of modernity. Turkey's transformation involved opening itself up to the forces of western modernization. This began during the nineteenth-century Ottoman Empire, but was dramatically accelerated with the foundation of Turkey as a **nation-state** in 1923 under Mustapha Kemal (he later called himself Atatürk, 'Father of the Turks'). The modernization process was to be a total transformation of Turkish economy, society, politics and culture, and a total break with the past. Mustapha Kemal declared that 'the new Turkey has no relationship to the old. The Ottoman government has passed into history. A new Turkey is now born' (Robins, 1996: 68). Moving into this promising new future involved economic policies that transformed parts of the country. During the 1930s the state intervened to modernize the economy and to make it serve national ends rather than being oriented to the export of raw materials (see Figure 32.4). Import controls were coupled with agricultural policies aiming at national self-sufficiency in food through scientific farming and irrigation schemes. A modern capital city was built at Ankara. Extensive railway construction produced a national network and a Soviet-inspired five-year plan (1934–38) provided large textile, sugar, paper, cement and steel plants spread across the country in order to substitute Turkish goods for imports from the West. These were combined with earlier (1920s) policies designed to remove what were seen as cultural barriers to modernization. The influence of Islam on people's lives was to be reduced by separating state and religion, abolishing religious courts and centres of religious learning and closing down shrines and religious brotherhoods. Turks were to be reoriented towards the West and its ideas by prohibiting the fez in favour of the western hat for men; the adoption of Latin rather than Arabic script; the metric system; family surnames and the Gregorian calendar; and, along with the westernization of new architecture, the playing of American jazz in public places. Finally, any identification with causes other than that of the modern Turkish nation-state were prevented by the prohibition of internationalist organizations and programmes of repression and assimilation aimed at ethnic minorities (Kurds, Georgians and Armenians). Turkey's modernizers understood the process of modernization as one of making people make a clean break with the past and encouraging them to progress into a bright (western) future (Landau, 1984; Parker and Smith, 1940; Schick and Tonak, 1987).

There are clearly some very real problems with this single, progressive and western version of modernization. First, it can never be a matter of simply following a western model. The world's countries are not separate entities ranged along a developmental path; they are connected together so that the development of one has consequences for others (Taylor, 1989). Also, within Turkey, the adoption of a model from outside has meant that 'modernization has been an arid and empty affair' (Robins, 1996: 67), unconnected to the cultures of the place and lacking dynamism. Second, this version of modernization underplays the disruptiveness of the transformation for people and the landscapes that they live in. Robins shows how Kemalist policies that sought to achieve a western modernity by suppressing the past, the specific nature of Turkish society and the cultural differences within the country had very traumatic consequences, which are only now being dealt with.

Why modernization theorists put forward such ideas is best explained by their political context. In an era of **decolonization**, the Cold

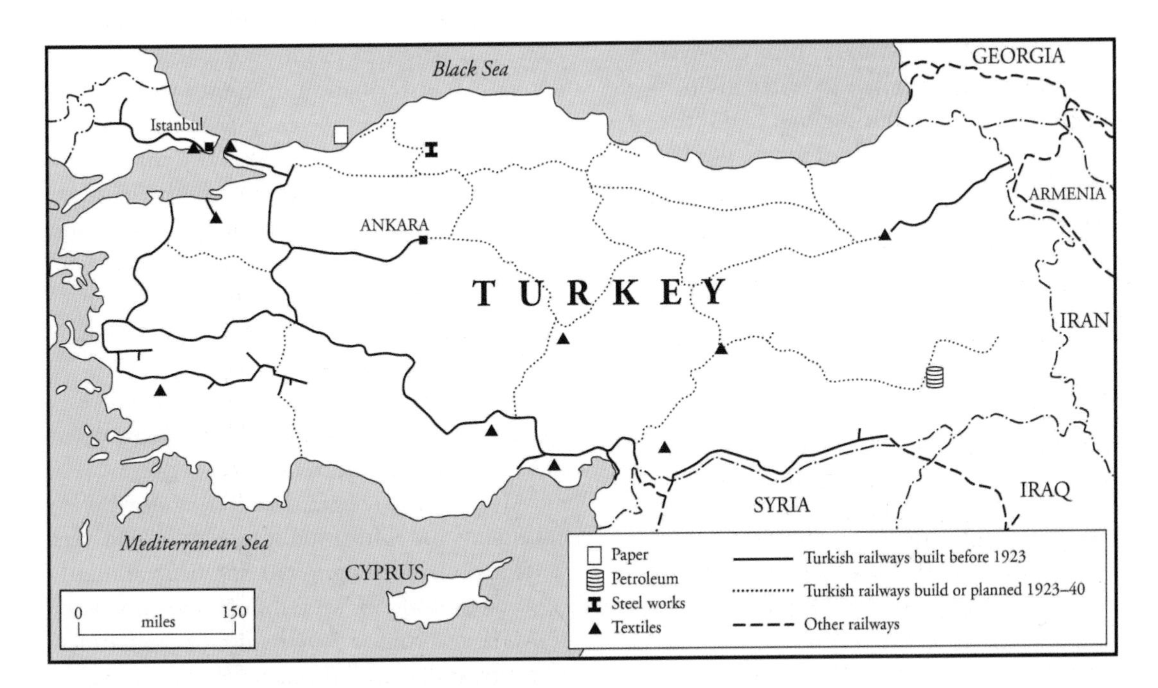

Figure 32.4 Turkish modernization: industry and transport

War and a post-war boom in Western Europe and the USA, the idea of modernization as westernization (and seeing that as orderly progress) was part of a political move to combat the appeal of communism to 'third world' countries. Rostow certainly saw himself as offering an alternative and subtitled his book 'a non-communist manifesto'. In the next section I will look at some of the ideas that Rostow was reacting against to see how modernization is understood in a different way: as creative destruction.

SUMMARY

- Rostow's theory presented modernization as progress: a single (western) path to modernity that is positive, beneficial and relatively painless.
- Turkey's experience of 'modernization as progress' shows the dramatic changes in lives and geographies that it involves, and that ideas like Rostow's simplified the difficulty, complexity and trauma of modernization.

Modernity and modernization as creative destruction

Understanding modernization as 'creative destruction' suggests that the changes involved are dramatic and unsettling ones, and that making a new future always means destroying many of the geographies and ways of life of the past and present. Marshall Berman (1982) has suggested that Karl Marx and Freidrich Engels' (1848) *Communist Manifesto* puts forward this view of historical and geographical change in its discussion of capitalism. He sums it up using a phrase (borrowed from Shakespeare) from the *Manifesto*: in the modern world 'All that is solid melts into air'. This poetic image suggests the sense of constant change and uncertainty that Marx and Engels argued is necessary for **capitalism** to be successful. This arises from the continual need to develop the 'productive forces' – labour power, raw materials, machinery, science, communications and transportation – needed to produce **commodities** and make a profit. It generates the constant search for new materials and new markets that drove global exploration and settlement and radically transformed the lives and landscapes of many in Africa, Asia and the Americas, as well as in Europe. It also meant a dramatic transformation in forms of work, most vividly seen in the coming of the factory system, where all sorts of labour, no matter how menial or prestigious, became activity that was done for wages and was often controlled by the workings of a machine. Finally, it involved radical geographical changes, the building of huge industrial cities, the making of modern nation-states and what Marx and Engels described as the 'Subjection of Nature's forces to Man, machinery, application of chemistry to industry and agriculture, steam-navigation, railways, electric telegraphs, clearing of whole continents for cultivation, canalisation of rivers, whole populations conjured out of the ground . . .' (1848: 85). So, even though they opposed it, Marx and Engels were in awe of capitalism's capacity to make 'all that is solid melt into air', destroying entire ways of life and whole landscapes as it developed the forces of production. More importantly, they saw capitalism as destroying what it had itself previously created – factories, docks, whole cities – as they became unprofitable. For them, the process of modernization is both dramatic and traumatic. It offers all sorts of possibilities for changing the world, but also destroys ways of life and places that had become known, accepted and familiar. This making and breaking of social relations and their geographies is modernization understood as creative destruction.

An example can help here. Between 1850 and 1870 Paris was transformed by this sort of 'capitalist modernization'. Many of what are now that city's best-known features – tree-lined boulevards, monumental architecture and pavement cafés – were constructed during this period. However, to create this new urban geography much of Paris had to be destroyed and many lives were seriously disrupted. How and why did this happen? David Harvey (1985) argues that the impetus was the economic crisis of the 1840s – the worst that France had experienced. There were many workers who could not find work and, at the same time, many investors who could find nowhere profitable to invest. In addition, Paris's eighteenth-century infrastructure made it more of a hindrance than a help in escaping from this crisis. The solution orchestrated by Emperor Louis Napoleon and his minister Baron Haussmann was a massive reorganization of the geography of France and its capital through a huge public works programme – building new railways, roads and telegraph systems across France and transforming Paris with new streets, water supply, sewers, monuments, public buildings, parks, schools and churches. The rebuilding programme would put people and capital to work in the short term, and, in the longer term, would provide the sort of modern city within which profitable investments in land, manufacturing or commerce could be made.

The most significant part of this new urban geography was Haussmann's boulevards. Figure 32.5 shows the streets built between 1850 and 1870. In all they were 85 miles long and, on average, three times wider than the ones they replaced (Pinckney, 1958). The extent of the remodelling of traffic flows is clear. The aim was to make Paris into a single, functioning unit rather than a series of separate neighbourhoods. Particularly significant is the cross at the heart of the city made by Rue de Rivoli, Boulevard de Strasbourg, Boulevard de Sébastopol and Boulevard Saint Michel, which linked both banks of the River Seine to an axis that ran right across Paris. Other boulevards connected new railway terminals to the city centre. These new streets brought other changes too. Building them combined modern science and engineering, as well as a lot of hard work (in the mid-1860s, 20 per cent of the working population of Paris was employed in construction). Their planning necessitated the first accurate topographic and land ownership maps for the city. They were also politically important. Haussmann ensured that the boulevards provided vistas of monuments or buildings that were symbols of France, religion and empire – like the Arc de Triomphe – so that the new city combined both tradition and modernity. It was also rumoured that these wide, straight roads made it harder for revolutionaries to put up barricades as they had done in 1830 and 1848 and easier for the army to ride their horses or fire their artillery down them. These huge streets certainly made it easier to travel faster through the city by horse and carriage and pedestrians were now confronted with huge numbers of speeding vehicles (Berman, 1982). Finally, they provided new opportunities for private investment in the plush apartment

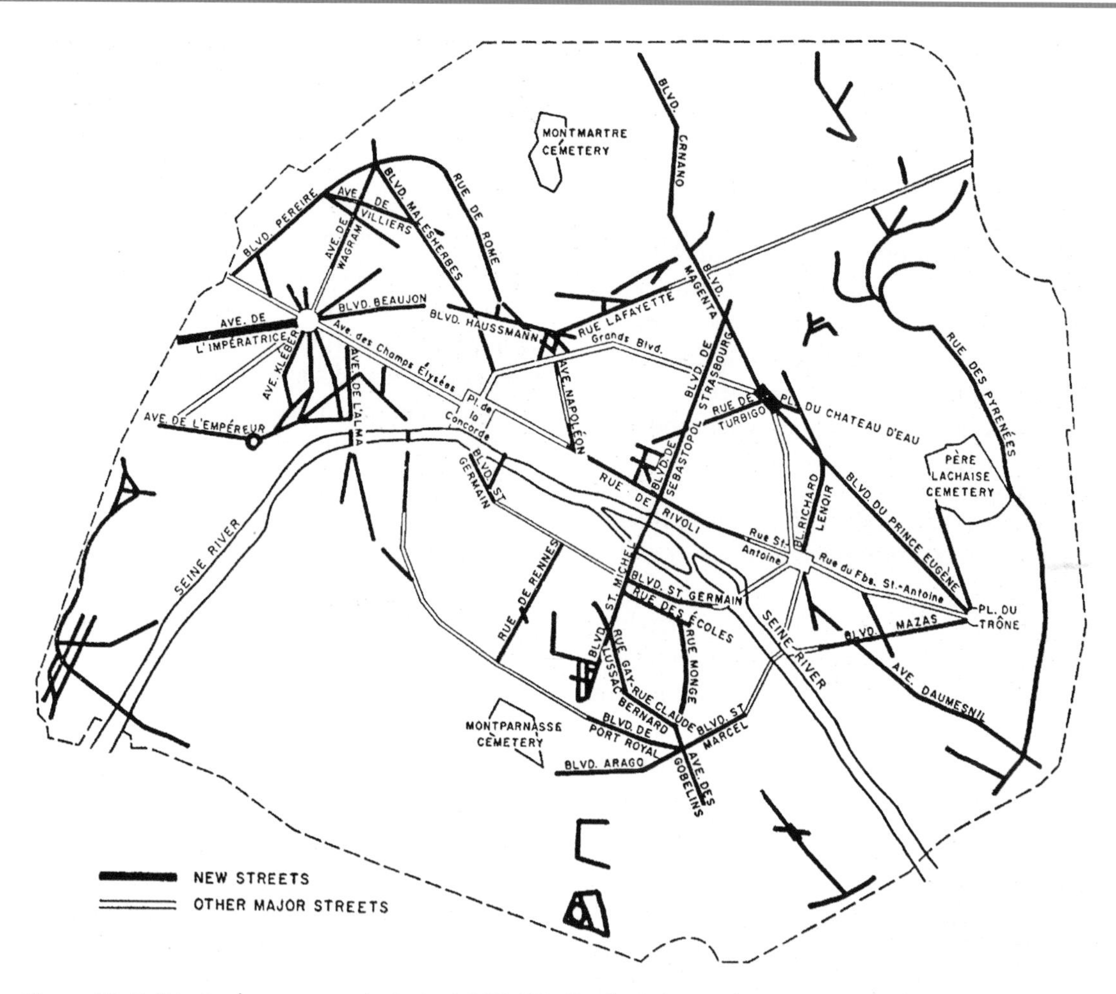

Figure 32.5 Principal new streets in Paris, 1850–70. Credit: D.H. Pinckney, 1958: *Napoleon III and the Rebuilding of Paris*, p. 73

buildings, exclusive hotels, fashionable pavement cafés and dazzling shops (including vast department stores) that lined the gas-lit boulevards. As Harvey argues, they 'became corridors of homage to the power of money and commodities, [and] play spaces for the bourgeoisie' (1985: 204). Paris was transformed – a new geography was created – and some people benefited from the opportunities for profitable investment and pleasure.

However, to build the boulevards, the old city of Paris had to be torn apart. The new streets carved their way straight through ancient districts of winding streets, hidden corners and higgledy-piggledy buildings housing vast numbers of Parisian workers. In 1850 the Ile de la Cité (the island in the Seine where Notre Dame Cathedral stands) housed 14,000 people. By 1870 only a few hundred were left to defy Haussmann's plans to devote the island to institutions of law, religion and medicine. For 20 years the centre of Paris was a building site and thousands of people were displaced. They were either forced out to the suburbs and had to walk for several hours to get to work,

Figure 32.6 'But here is where I live – and I don't even find my wife.' Credit: Honoré Daumier cartoon, 1852. In: D.H. Pinckney, 1958: *Napoleon III and the Rebuilding of Paris*

Figure 32.7 Potsdamer Platz, Berlin. Montage by Michael Pryke. Credit: www.open.ac.uk/socialsciences/berlin

or they crowded into the remaining central areas and paid exorbitant rents for appalling accommodation. Figure 32.6 shows a contemporary cartoon that, while making fun of the situation, gives a sense of the disruption felt by many ordinary Parisians as their city was destroyed. As well as the elimination of neighbourhoods it also hints at the disruption of family relationships. This man's wife may not have been in the shell of what had been their home because, as Harvey shows, the economic restructuring that accompanied Paris's transformation meant that both men and women were forced to work, often very long hours, if they and their families were to survive. In Paris, therefore, we can see an example of the wider process of 'creative destruction' whereby the forces of modernization benefit some and disadvantage others as they dramatically transform lives and landscapes. However, this is not just a historical matter. The same processes and effects can be seen in many contemporary sites. For example, Figure 32.7 is an attempt to use a montage of images to capture the dramatic and contradictory transformations of the city of Berlin in the years since 1990.

SUMMARY

- Understanding modernization as creative destruction shows that it involves the destruction of lives and geographies as well as the construction of new geographies and new ways of life.
- The example of nineteenth-century Paris shows that some people benefit and some people lose out through the processes of modernization.

Conclusion: many modernities

'Modernization as progress' and as 'creative destruction' offer very different ways of understanding these historical and geographical changes. They give different versions of the relationship between past, present and future and, as presented by Rostow and Marx, suggest very different political responses. However, there are three points that can be made about both versions that help to develop the idea of modernity. First, they both present modernity as a radical break with the past. This may be true for some parts of Turkey in the 1920s and 1930s or for many Parisians in the 1850s and 1860s. However, it is worth thinking about more subtle ways of understanding these historical geographies of modernization which recognize that change happens at different rates in different places, and that instead of being eradicated by modernity, past lives and landscapes are often remade, reinvented or reincorporated with new ways of doing things and new geographies. Second, they are both primarily concerned with economic transformations and particularly with capitalism. While this is clearly important we need to remember that historical geographies of modernity also need to be about political transformations (e.g. modern bureaucratic and territorial nation-states), technological transformations (e.g. the application of science to both nature and society), and social and cultural transformations (e.g. modernist art's attempts to deal with processes of modernization). All these processes shape the modern world together and they have done so in non-capitalist as well as capitalist economies.

These two issues lead towards the third point: that there are many different modernities. 'Modernity' is different at different times and it has many different strands. It is also different in different places. Both versions discussed here see Western Europe and North America as the most important places in these historical geographies of modernity. However, attention to the experience of modernity in other places, particularly those colonized by or drawn into new economic, political and cultural relationships with 'the West', reveals similarities to what went on in London, Paris and New York but also differences (Miller, 1994). Some have argued that there are 'alternative modernities' produced beyond 'the West' through processes of 'creative adaptation'. These show how modernity's transformative impulses are actively reshaped in particular ways in different cultural contexts as part of a constant debate about what the future should look like (Gaonkar, 2001:18; Watts, 2002). Uncovering these other modernities and their histories also demonstrates how particular Western ideas of modernity – of democratic politics or of the modern city – have come to be mistaken for universal notions (Robinson,

2006). This urges us to expand our ideas of who is modern and how. Paul Gilroy (1993a) argues that black Africans transported to the Americas and to Europe under slavery can be seen as the first modern people because of their experience of (and resistance to) globalization, industrial work on the plantations and attempts to use science to justify racism. There are also different modernities for different social groups defined in terms of class and gender. For example, this means that processes of modernization have different effects on, and are experienced differently by, men and women: modernity is often talked about in terms of dramatic confrontations between 'Man' and 'Nature' which present the whole process in gendered terms. Turkey's modernization programme meant that women were given the vote (as well as equal rights in many other areas) and in Paris, the new pleasures and hardships of the modernized city were experienced differently by men and women who had different access to the new spaces of the city and very different expectations of them when they were on the streets or in the cafés or department stores (Pollock, 1988).

What we end up with, therefore, are many different modernities. There are various processes of modernization that have been transforming lives and landscapes across the globe in different ways for hundreds of years and that are experienced differently by different sorts of people. Each one makes for a different historical geography of modernity and raises different questions about understanding the past, present and future.

DISCUSSION POINTS

1. Try and find examples of policies being put forward to change particular geographies (perhaps regenerating parts of a city, or for development in areas of the global South) which offer different versions of the future. What forms of 'modernity' are being promoted here?
2. Is modernization ever only a matter of progress or does it always involve creative destruction?
3. Is it possible to understand historical and geographical change without thinking of it in terms of modernization?

FURTHER READING

Given that it is a good idea to read original works and make up your own mind, the two books that I use for this discussion are as follows:

Rostow, W.W. (1960) *The Stages of Economic Growth: A Non-communist Manifesto*. Cambridge: Cambridge University Press.

Especially Chapters 1 to 4, which cover the stages up to 'take-off'.

Marx, K. and Engels, F. (1967, originally published in 1848) *The Communist Manifesto.* Harmondsworth: Penguin Books.

Section 1, 'Bourgeois and proletarians', includes the material discussed here.

Other, more general treatments of the issues raised are covered in the following texts:

Berman, M. (1982) *All That is Solid Melts into Air: The Experience of Modernity.* London: Verso.

Offers a discussion based on the idea of 'creative destruction' that covers Marx (Chapter 2) and Paris (Chapter 3). Quite a difficult book to read, but has many interesting insights and examples.

Ogborn, M. (1998) *Spaces of Modernity: London's Geographies 1680–1780.* New York: Guilford Press.

Chapter 1 contains an overview of theories of modernity that relates them to questions of historical geography.

For a range of work in Historical Geography which uses ideas of modernity and modernization, see:

Dennis, R. (2008) *Cities in Modernity: Representations and Productions of Metropolitan Space, 1840–1930.* Cambridge: Cambridge University Press.

Graham, B. and Nash, C. (eds.) (2000) *Modern Historical Geographies.* Harlow: Longman.

CHAPTER 33
COLONIALISM AND POST-COLONIALISM

Richard Phillips

Introduction: mapping empire

This chapter argues for the importance of colonialism in shaping not only the past but also the contemporary world. Maps represent this. The world map shown in Figure 33.1 appeared in the opening pages of an official *Atlas of Canada*, published by the Dominion of Canada's Department of the Interior in 1915. It is typical of maps that were produced and consumed around much of the world in the first half of the twentieth century. People often see maps as straightforward representations of space – facts about geography – but the map makers have made all sorts of choices: what to include, how to depict it and what to leave out (see Chapter 13).

The first choice the cartographers have made, perhaps their most powerful device, is one of colour. In the Canadian world map, originally 40 cm by 60 cm, colour is used to represent colonial power, as the key makes clear. Each empire is a different colour. Above all, this is a map of colonial empires, most of them European. The map is cleanly printed. Its even blocks of colour cover regions, nations, even whole continents uniformly. They depict a world that is tidily, uniformly colonized.

The colour scheme is not arbitrary. The map makers have chosen colours that reinforce the colonial theme of the map. The colour red seems to occupy much of the land surface of the earth. The key shows that Britain and British colonies are coloured bright red. Other colonial powers appear in more delicate or demure colours: the French Empire, for example, is a gentle mauve. Each colour has a symbolic and graphic function within the map. Red symbolizes authority (think of 'red tape'), aggression (a red sports car), confrontation (like a 'red rag' to a bull) and England (England's flag, showing the cross of St George, is white and red); it is also the colour of blood. Graphically, too, it is an aggressive colour, appearing larger mile-for-mile than its more lightly shaded neighbours, also pushing out and spilling into their territory. Looking at this map – imaginatively 'losing oneself' in it – it is easy to imagine a time, not too far into the future, when the colour red will be everywhere. So it is not surprising that the map makers chose the colour red for themselves. Other map makers, including French and Germans, often did the same in their own maps.

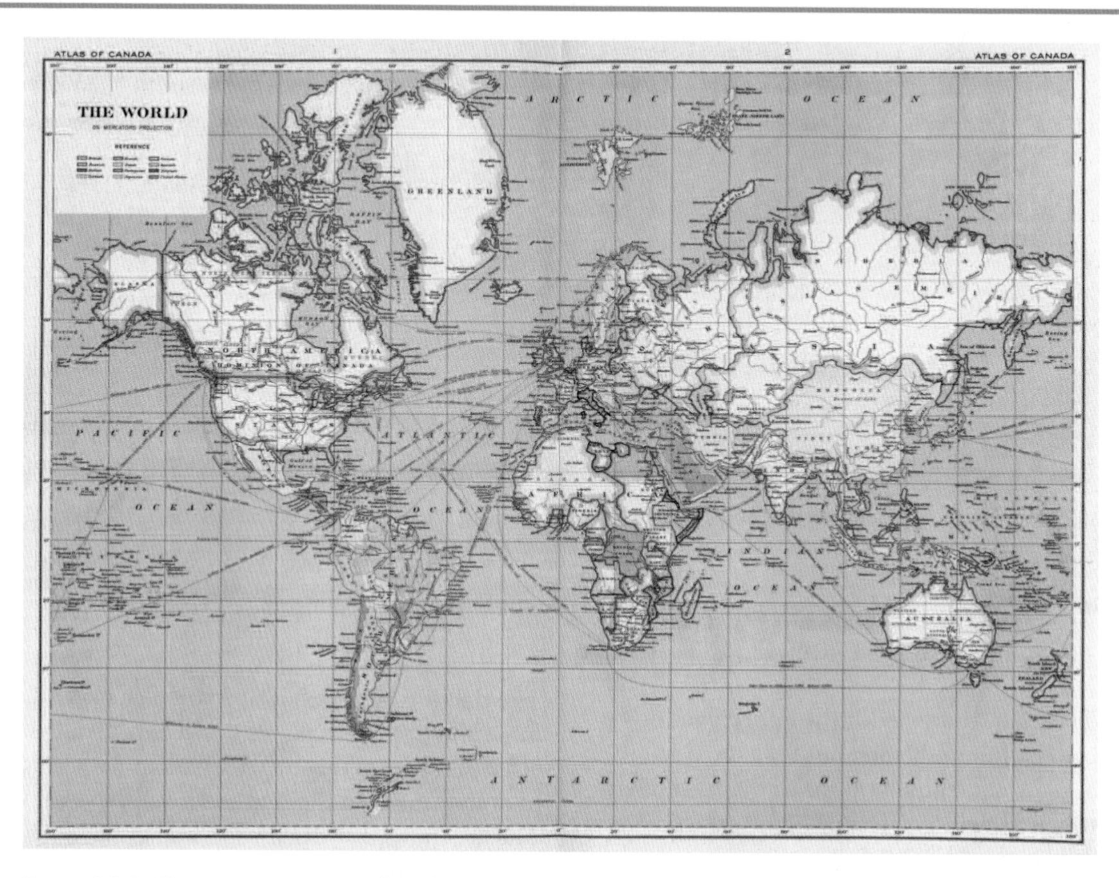

Figure 33.1 This map represents colonial power and is a vehicle of that power. World map, 1915. Source: *Atlas of Canada*, Ottawa: Government of Canada. Credit: The British Library Board/Atlas of Canada

Notice the words. They are in English. Canada is not and never has been a nation just of anglophones. When the map was made, many Canadians spoke French, as well as a host of other languages, some native North American, others imported from eastern and northern Europe, Asia and elsewhere. But the map is in English – the dominant, official language. The world is named with the place names chosen by English-speaking peoples; everywhere is what the English say it is. Thus the map asserts an English way of seeing the world.

Neither is composition merely factual, an innocent artefact of scientific cartography (see Harley, 1992). The map depicts a world with a centre and margins; it turns some places into centres and others into margins. Anywhere could be at the centre – Fiji, France, the Falkland Islands. But Europe, and specifically England, is at the centre. The centrality of England is also formally marked through lines of longitude, which as most people know centre on Greenwich, London. Before 1881 when England got other nations to accept their own prime meridian and thereby established an international convention respected by many (but not all) nations, most map makers drew lines of longitude that were centred upon their own capitals or other cities. Americans, for example, generally used Washington or Philadelphia. Persuading others to respect the

Greenwich meridian, the Europeans persuaded them to see London and Britain/Europe as the centre of the world. In case any map reader should miss the point, shipping lines were included, drawing the eye from all corners of the world into Liverpool, Bristol, Southampton and the Thames Estuary, towards a metropolitan centre. The shipping lines also suggest the importance of England as a centre of trade and commerce; every other nation seems to be measured and located according to its links with England.

The projection of the map also adds to the importance of England and other northern countries. Here, the map makers have chosen to use the Mercator projection, which was developed in the Netherlands – a leading European imperial power – in the sixteenth century. The Mercator projection distorts space, making areas further from the equator appear bigger (by land area) than they really are. This makes western European nations appear disproportionately large. England, part of a small island in the North Atlantic, is exaggerated in geographical importance, as are British dominions such as Canada, New Zealand and Australia. The projection also serves as a reminder of England's competitor in Eastern Europe and Asia, the Russian Empire, which looks very imposing on the map. Conversely, it makes India appear much smaller, perhaps more easily ruled by the English, than it might otherwise appear.

This 1915 map suggests how colonial power shaped the world one hundred years ago. Looking at it now might also provoke us to reflect on how different the world is today. The two main parts of this chapter explore those two questions in turn: considering first the colonial geographies and experiences of empire in the past, with a particular focus on the British empire; and then attending to the 'post-colonial' era of today, where in fact colonial geographies continue, in both old and new forms.

SUMMARY

- The 1915 world map represents a world dominated by colonial power.
- It also functions as an ideological vehicle of that colonial power.
- It provokes us to think about the importance of colonial geographies to both the past and present.

Experiences of empire

European colonial empires reached their peak around 1914, before the First World War broke out. Among these empires, the British was the greatest, with the French in a relatively poor if absolutely large second place. Table 33.1 shows the breakdown of British, French and other possessions, by population and area. The British controlled huge areas of land in America (mainly Canada) and Australia, although it was its African and especially its Asian colonies (mainly India) that qualified Britain as the world's largest imperial power. While British Canada and Australia accommodated just over 12 million increasingly self-governing people,

	Area in sq. km	Pop. in 000s
1. In British possession		
A. Mediterranean	10	517
B. Asia	5,199	324,114
C. Africa	9,392	50,824
D. America	10,407	10,082
E. Australia/South Seas	8,267	2,508
F. Other	18	599
Total:	33,293	388,644
2. In French possession		
A. Asia	803	14,871
B. Africa	9,499	30,514
C. Other	116	807
Total:	10,418	46,192
3. In Dutch possession	2,036	38,248
4. In Russian possession	16,153	22,605
5. In Japanese possession	288	19,200
6. In German possession	2,954	13,784
7. In American possession	388	10,299
8. In Belgian possession	2,365	10,000
9. In Portuguese possession	2,244	9,146
10. In Italian possession	1,641	1,850
11. In Spanish possession	441	640

Table 33.1 Colonial empires at the outbreak of war in 1914. Source: Veit Valentin: Kolonialgeschichte der Neuzeit, Tübingen, 1915. Source: Veit Valentin: Kolonialgeschichte der Neuzeit, Tübingen, 1915

the combined populations of British Africa and India numbered something closer to 400 million, and these colonial subjects were subjected to something closer to absolute colonial rule.

Tables such as this group land and people into columns and aggregate statistics, categorizing them under labels such as 'empires', 'imperial' and 'colonial' – terms we have already found ourselves using. There is, at the outset, a need to clarify this vocabulary. **Imperialism** refers, very broadly, to 'an unequal human and territorial relationship . . . involving the extension of authority and control of one state or people over another' (Clayton in Johnston *et al.*, 2009: 373). **Colonialism** is defined more narrowly around the 'domination' and/or 'dispossession' of 'an indigenous (or enslaved) majority' by a 'minority of interlopers' or colonizers (Clayton in Johnston *et al.*, 2009: 94).

Definitions and labels, like statistics and maps, make the world seem tidier, more ordered and more generic than experience tells us it is. Within the areas painted red, or listed in columns of British statistics, or labelled under headings such as 'colony', there remain a great variety of different experiences and perspectives. These differences – between colonizers, colonies and colonized peoples – must be remembered, because they are often greater than the similarities. As historian Ronald Hyam puts it in *Britain's Imperial Century* (1993: 1), 'When you think about it, there was no such thing as a greater Britain – India, perhaps apart. There was only a ragbag of territorial bits and pieces, some remaindered remnants, some pre-empted luxury items, some cheap samples'. Colonialism, he goes on to explain, is messier than our maps of it. Colonialism takes many different forms and is experienced by different people in different ways.

The experiences of colonists have been many and varied, as a selection of British colonists illustrates. David Livingstone, a Scotsman, spent much of his life as a missionary and explorer in Africa (see Driver, 2001). Like many colonists, he was driven by faith in Christianity, civilization and commerce. He campaigned practically against slavery and for Christianity, in both cases by exploring the continent, thereby opening Africa to trade, development and (a British idea of) progress. As an explorer, Livingstone is remembered for the epic journey to the great waterfall that the British were to name after their Queen Victoria. Most colonists were less famous and less idealistic than Livingstone. Daisy Phillips, for example, was one of many who emigrated from England in search of a new home and a better livelihood in the colonies. Along with her husband, Jack, she embarked upon the long journey from England to the interior of British Columbia, where she did her best to set up home. Frontier life proved difficult and lonely, although Daisy was to remember it fondly. The couple returned to Europe when war broke out and Jack soon died on the battlefield. Daisy's letters home, reprinted in *Letters from Windermere* (Harris and Phillips, 1984), tell a story of literate, middle-class colonial experience (see also Moodie, 1986). Other emigrants, whose colonial experiences were generally less well documented and considerably less comfortable than Daisy's, include the convicts transported to Australia in the late eighteenth and early nineteenth centuries. Most of these were Irish or English, six out of seven were male, and many were transported for petty offences. All found themselves in a brutal world of violence, hard labour and alienation, which Robert Hughes vividly describes in *The Fatal Shore* (1987). When they had served their sentences, few could afford the cost of a passage home, so they stayed as free colonists. Other British colonists

Robinson Crusoe rescues Friday.

Figure 33.2 Imaginative geographies accommodated colonial encounters, such as that between Crusoe and Friday, which inspired and legitimized real colonial acts. Frontispiece and title page of *Robinson Crusoe in Words of One Syllable* by Mary Godolphin (1868). Credit: The British Library Board

had very different experiences, which were also shaped (like Livingstone's) partly by their faith and ideals and (like Daisy Phillips's) by their class and gender, as well as by other aspects of their identities including their race and sexuality (see McClintock, 1995; Phillips, 2006).

Meanwhile, many other Britons experienced colonialism from an altogether different angle: they stayed at home. There they imagined the colonial empire from a distance, hearing or reading about it. Books such as *Robinson Crusoe* (see Figure 33.2), with exotic settings and lively storylines, narrated and mapped colonial encounters between Europeans and others (Phillips, 1997). Popular geographical narratives such as *Robinson Crusoe* presented an acceptable and exciting face of colonialism to their British, French, German and other readers. They popularized empire, persuading many to support their governments' wider imperial projects and inspiring others to seek adventures of their own, many in actual and would-be colonies. So **imaginative geographies** of empire did not just represent the empire, they helped construct it (see Said, 1993, and Chapter 16 of this book, in which Felix Driver examines imaginative geographies and their significance in the real world). European armchair and shop-floor imperialists did not just dream about empire; they also consumed it and produced goods for it. Europeans consumed colonial products such as tea from India, sugar from the Caribbean and furs from Canada. They helped to produce the manufactured goods such as railway engines, textiles and guns that Britain shipped to its colonial markets. They posted letters to friends and relatives in (what to them were) far-flung corners of the world. When they received replies, those who noticed stamps often also noticed the head of their own monarch. In their personal communications, as in their material life (as producers and consumers) and in their dreams, Europeans participated in a system that was both global and imperial.

The experiences of colonized peoples were equally varied; being colonized meant different things to different people. Again, let us consider cases from the British Empire. Chief Sechele, for example, was the only African known to have been converted by the famous British missionary, David Livingstone. Sechele followed his spiritual mentor's advice and agreed to change his polygamous ways; he annulled all but one of his marriages, and in 1848 he was baptized in front of hundreds of weeping spectators. Soon, however, he missed the three wives he had sent away, and resumed sexual relations with at least one of them. Louis Riel was also the leader of a people colonized and fundamentally changed by British imperialism, although of a very different sort. He emerged as leader of the Canadian Métis nation, a **hybrid** people born through centuries of cultural and sexual contact between Amerindians and French-Canadian *voyageurs*. Riel led two uprisings against the British North American authorities, whose plans to settle western Canada threatened the Métis people's way of life. Riel was executed and his people were displaced, as colonists swarmed west. Unlike the Métis, some colonized peoples were entirely wiped out by colonists. This was the fate of many Australian Aboriginal peoples who were displaced by penal colonies. Truganini, the last Tasmanian, saw her countrymen and women broken by English colonists (see Figure 33.3). The English hunted down some of the inhabitants and transported the others to an island reserve (Flinders Island, 40 miles to the north) that, for most, was to be their grave. Others experienced colonization in very different ways, which, like those of the colonists depended partly upon, and in turn reshaped, their class, race, gender and sexuality. Some colonized people grew rich from the colonial encounter, while others lost their language, their livelihoods, even their lives. Like the colonists, many colonized peoples consumed stories and maps of the wider world and participated in an increasingly global system. But unlike the European colonists they were

LALLA ROOKH, OR TRUGANINA, THE LAST TASMANIAN WOMAN.
(*Photographed by* MR. WOOLLEY.)

LALLA ROOKH, OR TRUGANINA.
(*Photographed by* MR. WOOLLEY.)

Figure 33.3 Truganina, the last Tasmanian. Credit: *The Lost Tasmanian Race* (1884) by James Bonwick (Fellow of the Royal Geographical Society)

not the principal architects of this world. Their colonial encounters were from positions of relative weakness.

Colonialism transformed the world, simultaneously producing a 'New World' and modernizing the 'Old World'. Changes in the non-European world are perhaps most evident. Some effects of British colonialism were superficial – statues of Queen Victoria, for example. Other colonial imprints were more fundamental and can never be reversed. Enormous population movements, both forced and free, scattered Europeans and Africans, particularly in the Americas. Some 20 million people left Britain and Ireland between 1815 and 1914, in a **diaspora** motivated variously by hunger (particularly during the Irish famine), displacement (notably in the Scottish clearances) and the desire for a better (freer and/or wealthier) way of life (by middle-class emigrants). Europeans and Euro-Americans also engineered an African diaspora, in which 12 million Africans were sold into slavery and shipped overseas, mainly to the Americas (see Chapter 45 for a more general discussion about the contemporary implications of such diasporas). Vast tracts of the world were partly or wholly resettled, as aboriginal peoples were deliberately or accidentally wiped out and replaced with free and/or forced immigrants. Movements of capital also transformed and integrated far flung territories. Railways and shipping lines, which rapidly encircled the Victorian globe, were important both as capital investment and as infrastructure through which capital and goods, people and information were moved. The result was an increasingly global society and economy, a modern **world-system** that underpinned a global geography of development and underdevelopment that we continue to grapple with today (see Chapters 20–23 for introductions to development geographies, and reflect on how these are also colonial and post-colonial geographies).

While colonialism transformed the non-European world, it also reshaped Europe. Europeans, the principal architects of the modern world-system, constructed a modern world in which they were at the centre (see Chapter 32 for more on these global geographies of modernity and modernization). Europe grew rich, and European lifestyles were enhanced by the non-European products, labour and markets that Europe controlled. Non-European economies supplied the raw materials and provided the markets that enabled Britain and other European countries to industrialize. Buoyed by prosperity, many Europeans grew arrogant, confident of their economic, political, religious and racial superiority. European confidence, expressed for example in its maps, redefined its place in the world and mapped out its future. This was not a stable or harmonious arrangement, however, as Winston Churchill, First Lord of the Admiralty and future British Prime Minister, admitted before the outbreak of the First World War:

> We are not a young people with an innocent record and a scanty inheritance. We have engrossed to ourselves . . . an altogether disproportionate share of the wealth and traffic of the world. We have got all we want in territory, and our claim to be left in the unmolested enjoyment of vast and splendid possessions, mainly acquired by violence, largely maintained by force, often seems less reasonable to others than to us.
>
> (Winston Churchill, quoted by Ponting, 1994: 132)

SUMMARY

- European colonization was experienced in many different ways by many different peoples, but when Europeans encountered the peoples of Asia, the Americas, Oceania and Africa, they generally had the upper hand.
- European colonialism established and formalized unequal territorial relationships between peoples, ensuring that modern maps and modern geographies were and are colonial maps and colonial geographies.

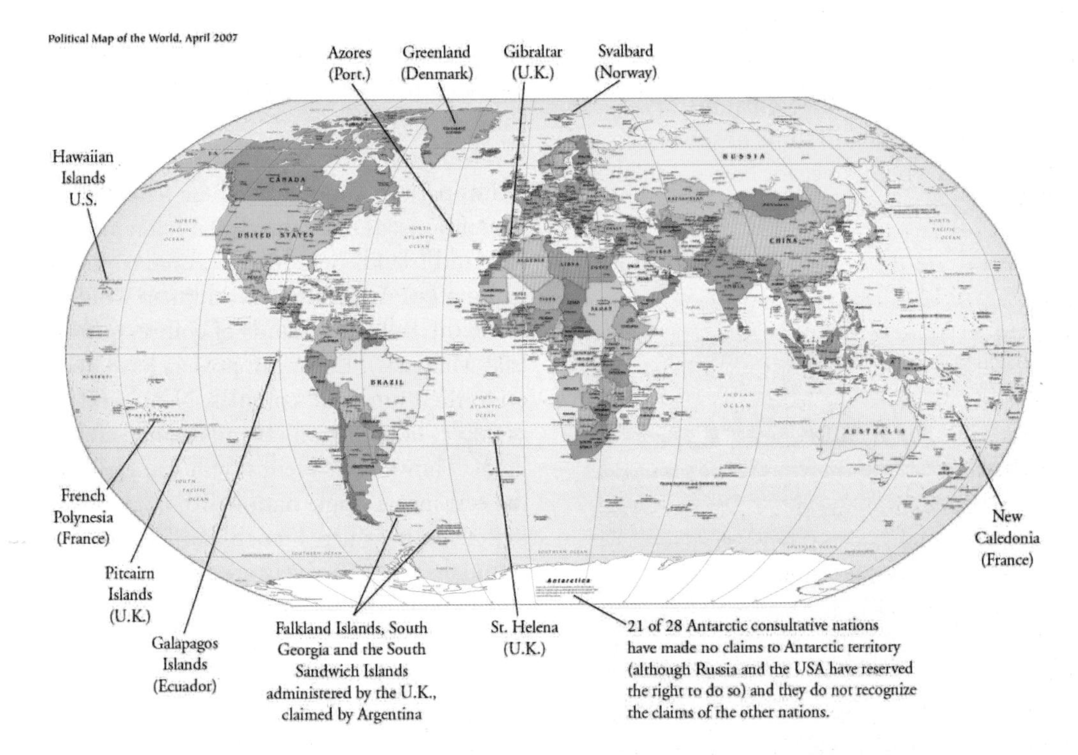

Figure 33.4 Colonialism is less obvious in 2013 than in the 1915 world map. Credit: The World Factbook

'Post-colonial' geographies

Now compare the 1915 map (Figure 33.1) with a more up-to-date map of the world (Figure 33.4). What differences do you see? What similarities? What evidence of colonialism do you see? You may need to look closely.

Perhaps the most striking difference between the two maps is the colour scheme. The domination of a few European colonial empires, symbolized in the handful of strong colours in the 1915 map that we discussed, has gone. In particular, the colour red has receded. Some patches of red (now a softer pink) remain, but you have to look hard to find them. People often say that the sun has now set on the British Empire (a response to the old boast that it never did: in an Empire that spanned the globe, it was always day time in one British territory or another). In July 1997 the British handed over control of their last major colony, Hong Kong. Britain's *The Guardian* newspaper (see Figure 33.5) described the handover of Hong Kong as the end of an era of European empire building that began five centuries ago: 'the eclipse of an empire that lasted more than 400 years'. As European colonial empires were formally dismantled, a chapter of colonial history ended, and according to some observers a **post-colonial** chapter began. Post-colonial sometimes refers to that which is after or

The real competition at Wimbledon
The battle of the sponsors
G2 with today's TV

Rachel Cusk
Terminate the abortion debate
This section, page 17

With all the best jobs
education

45p

NEWSPAPER OF THE YEAR
The Guardian

China becomes new master of Hong Kong

A last hurrah and an empire closes down

Eclipse of the colonies on a sodden, barren rock

There are no other islands to compare with Hong Kong. Alan Rusbridger witnesses the immense scale of its passing

Figure 33.5 An empire closes down but is Hong Kong decolonized? Credit: *Guardian Newspapers*

against colonialism, but neither meaning signals a complete break. There is an ongoing colonial hangover in which former colonies are plagued by relics and legacies of the defunct colonial order, as well as by some new forms of colonialism.

If you look closely at the new world map you will see relics, traces and new forms of colonialism. Let me consider each of these in turn. First the *relics*: not all of the old colonies have been handed back. The British, for example, retain dependencies around the world, mainly small islands ranging from Gibraltar (on the southern tip of Spain) to the Falkland Islands (off the coast of Argentina) and Montserrat (in the Caribbean). Britain also claims part of Antarctica. These relics can be contentious, as the cases of Gibraltar and the Falkland Islands demonstrate, even to their naming (an Argentinian world map would not name the Falkland Islands of course, but the Islas Malvinas). European powers have also held on to their 'internal' colonies. Notice, for example, how Brittany blends seamlessly into France, how Wales blends with its neighbour to the east into a single nation. To many people in Brittany and Wales, French and English colonialisms respectively are very much alive. The same is true for immigrants, such as British Asians and French Algerians, who have moved to western countries from former colonies and who continue to experience colonial and quasi-colonial relationships and inequalities (Phillips, 2009a). Internal colonialism can also be seen outside Europe. For example, some people see Tibet and Hong Kong as colonies of China and the North West Territories as a colony of Canada (Morris, 1988; 1992).

This begins to show how colonialism is not just present in the contemporary world map as a series of historical relics. It has left much more prevalent *traces*. Looking, for example, at the configurations of states and national borders, you will see traces of European colonialism. Many present-day nations are legacies of colonialism. Canada, for example, is a nation born of British hegemony over territory settled by English and French emigrants, and the borders of present-day Canada are a fossil of Empire – albeit one that Quebec separatists would like to dismantle. Other borders reflect the colonial inheritance. The border between India and Pakistan was drawn by a British civil servant, Cyril Radcliffe, a stranger to British India who was given just 36 days to complete his momentous task. Armed only with a pile of outdated maps, some crude census returns and a case or two of claret, Radcliffe produced the 'Wiggly Line' that initiated one of the world's greatest population movements (15 million Muslims and Hindus crossed the new borders) and set the stage for half a century of animosity between the two nations (Khilnani, 1997). In Africa, as in India, many borders and states are essentially colonial creations transformed into independent states. Their boundaries, shapes and sizes are part of the colonial inheritance (as Griffiths shows in *The Atlas of African Affairs*, 1994). Bob Geldof, the Irish former pop star who now campaigns to help Africans living in hunger or with AIDS, attributes many contemporary African problems to the continent's geopolitical inheritance of empire. 'No country in Africa,' he told a human rights committee in 2004, 'is free from problems of access, security, and economic stability that is directly attributable to the boundaries they inherited from the colonial era' (Geldof, 2004).

However closely you look at the contemporary world map (Figure 33.4), though, you will not learn very much about *new* forms of colonialism. For, as overt enthusiasm for the imperial project has receded, map makers and others have tended to downplay and disguise continuing and new forms of colonialism. Strong, confident, aggressive colours are succeeded by an apparently arbitrary pattern of equal and different national colours. Even though colonial powers have lowered their flags over most former colonies, ending formal colonial rule, **decolonization** has not meant an end to 'unequal economic, cultural and territorial relationships'; in other words, it has not meant an end to imperialism (as defined above). As some old imperial powers have fallen, others have risen. A transformed imperial order was revealed in 1956, when Britain and France finally retreated from Egypt, which they had invaded to regain control of the Suez Canal. They retreated because the Soviets had told them to and because the Americans chose not to intervene. Those superpowers continued where their western European imperial ancestors left off. They did not generally set up formal colonial governments, nor did they tend to found new overseas colonial settlements, both of which their predecessors frequently did, so their activities cannot always be labelled colonial. But the external influence and ventures of the USA, the USSR (until its break up in 1990) and more recently China can all be described as imperial, and they are sometimes said to be **neo-colonial**.

The term neo-colonial was coined by the first president of independent Ghana, Kwame Nkrumah (1965), who was disillusioned by the failure of decolonization to liberate Africans from the power of Europe and the USA. Colonial governments had been replaced by international monetary bodies, multinational corporations and a variety of educational and cultural organizations, reconfiguring what one US critic has called 'imperialism without colonies' (Magdoff, 2003). These elusive organizations and institutions have assumed some of the power that was previously concentrated in the hands of nation-states. Some critics argue that the political centres of gravity have shifted radically away from nations to a new world order that Michael Hardt and Antonio Negri (2000) call, tellingly, 'empire'. Others claim that the new geopolitics is defined by conflict between established and rising powers: nation-states vying for power (Callinicos, 2009).

Allegedly neo-colonial or imperial ventures involve some of the same tactics that were used by former colonial powers such as the British, with many of the same objectives (including preferential access to resources, labour and markets). These tactics sometimes include military intervention. In 2003, the USA and a few close allies invaded Iraq in what was officially presented as a defensive measure, provoked by global terrorism in the wake of the September 11th attacks, but which many saw as a war for oil and global domination, as a form of imperialism (Phillips, 2011). The reasons for going to war were complex and contested, but consistent with a broad strategy made explicit by Richard Haass, a key figure in the US government. In a paper entitled 'Imperial America', presented in 2000, Haass advocated 'an imperial foreign policy' in which 'the US role would resemble nineteenth-century Great Britain' (Foster, 2003). Other senior US politicians, speaking through an organization known as the Project for the New American Century, directly argued that more overseas bases, particularly in the Gulf, were fundamental to any strategy to 'preserve *Pax Americana*' – a phrase that reiterated the parallel with British and before that Roman imperialism: *Pax Britannica* and *Pax Romana* (Foster, 2003). But these comments were exceptional. Governments generally distance themselves from colonialism, using the term to distinguish their own actions from those of other governments, past and present. During a visit to Africa, for example, US Secretary of State Hilary Clinton complained that a 'new colonialism' was being imposed on Africa. 'We saw that during colonial times, it is easy to come in, take out natural resources, pay off leaders and leave. And when you leave, you don't leave much behind for the people who are there. We don't want to see a new colonialism in Africa.'[1] Clinton did not refer to China by name, but her remarks were clearly pointed, and reflected in particular growing Chinese investments in Africa. A Chinese Government spokesperson responded. 'We have never imposed our will on African countries . . . we hope that those concerned can view Chinese-African cooperation objectively and fairly.'[2]

In the neo-colonial order, corporations have access to global resources, markets (Coca-Cola

Figure 33.6 Call centre workers in India. Credit: Indranil Mukherjee/Getty Images

and McDonald's, for example) and labour, which often comes cheap (see Figure 33.6). Large companies tap cheaper foreign labour by setting up assembly plants or 'outsourcing' – subcontracting work to local producers – and/or 'offshoring' this work (see also Chapter 28). European countries have exported manufacturing and more recently call centre and other service jobs to former colonies, with which they retain many business contacts and share languages. The USA, with fewer former colonies as such, has invested in Asia and Latin America and also joined British companies in India. As one British newspaper put it in 2003, 'The list of companies on a passage to India reads like a Who's Who of British business' (*The Sunday Times*, 2003). If outsourcing enriches western (and other) investors, it does so at the expense of workers not only in developing countries, but also in the West. The secretary general of the British Trades Union Congress (TUC) estimated in 2004 that two million jobs could be outsourced from wealthy nations in the next five years, and that between two and three per cent of all European and American service employment would be offshored by 2015 (TUC, 2004).[3] Offshoring, a movement that began in the private sector, has become increasingly common in the public sector, particularly in western countries such as the UK (Public Finance, 2011).[4]

One response to their mutual vulnerability to international capital has been for workers and governments around the world to undercut and compete with each other. Another, however, has been to recognize shared experiences and interests and to struggle against certain forces of globalization or neo-colonialism. Western workers and consumers have campaigned on behalf of their counterparts in developing countries, for example by using their buying (and boycotting) power to pressurize large companies to improve conditions and pay for workers and suppliers. Activists around the world have also coordinated their protests on issues such as the invasion of Iraq in 2003, multiplying efforts and impacts (Phillips, 2011). Interventions such as these, not only by dissident figures but also governments, rest upon understandings of global interdependence and inequality. The historical sensitivities of governments were evident in the 2000s, when the USA, France and Britain all intervened to suppress bloody civil wars in West Africa. Each sent troops and aid to the nation with which it had a specific historical colonial connection and, it was felt, responsibility: the USA to Liberia (where African-Americans were settled in the eighteenth century), the French to the Ivory Coast (where slaves, ivory and plantation goods were once obtained) and the British to Sierra Leone (first a slave trading centre, later the base for Britain's campaign against slavery) (*The Guardian*, 2003). Interventions such as these, though welcomed by many, have also been criticized as echoes of colonialism or expressions of neo-colonialism. After the earthquake that hit Haiti in 2010, a French Government minister accused the USA of using the relief operation to extend its power in the region. Alain Joyandet said that the USA was giving priority to its own military and relief flights ahead of other nations' aid flights, adding that: 'This is about helping Haiti, not about occupying Haiti' (Huffington Post, 2010).[5]

Historical understandings of imperialism have also informed a range of other actions in the present, concerned with reconciliation between and within nations. Thus, for example, British prime minister Tony Blair apologised to the people of Ireland for centuries of oppression, and Australian prime minister John Howard apologised to Aborigines, whom the nation's school systems and adoption agencies had taken from their families and attempted to assimilate

(Gooder and Jacobs, 2002). In some cases, this dialogue with the colonial past – informed by historical and post-colonial geographers – is being followed through in financial reparations, land claims and sovereignty disputes, in which colonized and native peoples are renegotiating their relationships with the descendants of colonists and inheritors of the colonial state. This is taking place within nations – at the provincial and federal scale within Canada, for example (Harris, 2002) – and also on an international level through organizations including the United Nations. In 2007, the UN agreed a Declaration on the Rights of Indigenous Peoples, which began with an acknowledgement 'that indigenous peoples have suffered from historic injustices as a result of, inter alia, their colonization and dispossession of their lands, territories and resources' and used this as a point of departure for defining new relationships between indigenous peoples and governments. The Declaration establishes that 'states shall provide effective mechanisms for prevention of, and redress for', among other things, 'any action which has the aim or effect of depriving them of their integrity as distinct peoples, or of their cultural values or ethnic identities' and 'any action which has the aim or effect of dispossessing them of their lands, territories or resources'.[6] The UN provision illustrates how recognition of the colonial past can provide a springboard for positive change in the post-colonial present.

SUMMARY

- Empires have shaped modern world maps, dictated and policed borders and configurations of states, controlled systems of production and consumption, and interfered with patterns of culture and settlement.
- Geographies of colonialism and imperialism are not necessarily historical geographies, in the sense of being in the past, since the demise of western European colonial empires has been matched by the rise of other, mainly western and northern, imperial and/or neo-colonial powers.
- An understanding of the colonial past, in which historical geographers have an important part to play, speaks to contemporary issues ranging from relationships between former colonial powers and colonies to aboriginal land disputes and racial identity politics.

Conclusion

Geographies of colonialism and post-colonialism begin in the past, but due to the inertia of old empires and the emergence of new ones, they continue into the present. Notwithstanding some important relics, European colonial empires have largely disappeared, but they live on in territorial boundaries and relationships, and within the social, political, economic and cultural ordering of the modern world. While British, French and other European empires have receded, leaving legacies, memories and material traces, others have advanced. In many ways the USA has taken over where the British left off, but *Pax Americana* is not simply a copy of *Pax Britannica,* nor is China's rising star directly comparable with those of its American and European predecessors and rivals. The ongoing history of empire is one of continuity and change: continuity in patterns of unequal territorial relationships and (within the modern period) western domination; and change in the ways that this is organized, and in its ever increasing scope. Colonial and post-colonial perspectives draw together many traditional sub-fields of the discipline – including geographies of development, trade, economy, migration, language and culture – in an approach that is synthetic, historical and critical.

DISCUSSION POINTS

1. What are the legacies – good and bad – of European imperialism?
2. Discuss the relationships between geographical knowledge and imperial power.
3. How might it be possible to map colonial encounters from the perspectives of colonized peoples?
4. Discuss the relationships between imperialism and globalization.
5. Compare and contrast European colonial empires with their American and Chinese successors.

FURTHER READING

Butlin, R. (2009) *Geographies of Empire: European Empires and Colonies, c.1880–1960.* Cambridge: Cambridge University Press.

A historical geography of empire more than a post-colonial geography, this book is informative and accessible, and will appeal to students with an interest in Historical Geography.

Livingstone, D.N. (1992) *The Geographical Tradition.* Oxford: Blackwell.

This substantial contribution to the history and philosophy of geographical knowledge is a must for any undergraduate geographer. Livingstone pays particular attention to relationships between colonialism and the production and use of geographical knowledge.

McEwan, C. (2009) *Postcolonialism and Development.* London: Routledge.

This book is particularly useful for students who are interested in seeing how the colonial and post-colonial are implicated in international development and contemporary global economic inequalities.

Said, E. (1993) *Culture and Imperialism.* London: Chatto & Windus.

Real and imagined geographies are at the heart of Said's exploration of relationships between western culture and imperialism. The brief introductory section on 'Empire, geography, and culture' is particularly useful. This book develops ideas about 'geographical discourse' that Said first introduced in his influential if sometimes difficult *Orientalism* (1978).

Sharp, J. P. (2009) *Geographies of Postcolonialism*. London: Sage.

The first and most comprehensive textbook on post-colonial geographies, this is the most useful book for most Geography students taking courses on post-colonialism.

NOTES

1. Source: Flavia Krause-Jackson, Bloomberg, 11 June 2011: Clinton Chastises China on Internet, African American "New Colonialism"'. www.bloomberg.com/news/print/2011-06-11/clinton-chastises-china-on-internet-african-new-colonialism-.html (accessed 15 June 2011)
2. Source: Reuters, 14 June 2011 10:17am GMT, 'China dismisses US swipe on "colonial" role in Africa', http://af.reuters.com/articlePrint?articleId=AFJOE75D0BE20110614 (accessed 15 June 2011)
3. Source: TUC (2004) Submission to Department of Trade and Industry (DTI) Consultation Exercise on Global Offshoring, 8 March 2004, cited by: Huws, U., Dahlmann, S. and Flecker, J. (2004) *Status Report on Outsourcing of ICT and Related Services in the EU*. Brussels: European Commission, 13–4. Available at: www.eurofound.europa.eu/pubdocs/2004/137/en/1/ef04137en.pdf
4. Source: http://opinion.publicfinance.co.uk/2011/03/offshoring-the-bigger-picture-by-john-tizard/
5. Source: www.huffingtonpost.com/2010/01/18/french-minister-rips-us-t_n_427366.html
6. Source: United Nations Declaration on the Rights of Indigenous Peoples, Adopted by General Assembly Resolution 61/295 on 13 September 2007, www.un.org/esa/socdev/unpfii/documents/DRIPS_en.pdf (accessed 29 June 2011)

CHAPTER 34
SPACE, MEMORY AND IDENTITY

Nuala C. Johnson

Introduction

The wreaths have been removed. The ceremonies are over. As Remembrance Sunday passes once again, the statue to fallen soldiers who were members of Queen's University in Belfast returns to its slightly invisible status as a memorial icon. During the days surrounding 11th November each year, this monument, like many others around the country, assumes renewed significance in the maintenance and cultivation of a public memory dedicated to Ulster soldiers killed in the First World War and subsequent conflicts. This spectacle of remembrance amplifies Nijinsky's response to the Great War: 'Now I will dance you the war, with its suffering, with its destruction, with its death' (cited in Eksteins 1990: 273). This dance takes the form of a seasonal ritual of wreath-laying, memorial services held in the churches and the wearing of the poppy. All play a central role in the memory-work rehearsed annually in cities and towns across the United Kingdom. This memory-work, however, is not confined to the spaces of commemoration associated with the First World War, nor to the UK of course, but forms part of a larger network of sites that connect the past to the geographies of identity in contemporary life. Drawing on a corpus of scholarly work within and beyond Historical Geography, these relationships between space, memory and identity are the concern of this chapter. Specific foci will include heritage sites, museums and memorials.

Memory work

The term memory is a complex one. In this chapter, our focus is not on memory as a purely personal process. Rather, our concern is with how personal emotions relate to the collective memories that are materially made through the landscape and through the performances and rituals of remembrance. The role of the built landscape – museums, monuments, artefacts, heritage sites – in creating a sense of a common identity through memory has been the dominant research thrust among scholars from a variety of disciplines, but in particular from Human Geography.

Maurice Halbwachs' book *On Collective Memory* (1950) was the first critical attempt to give a definition to the idea of social memory. Moving away from highly individualized and psychological notions of memory which focused on the Freudian unconscious, Halbwachs proposed that personal memory was rooted and constructed through the social relations of the present. Social groups, tied by class, kinship or religion, would link individuals within them to a common shared identity. This would be cultivated through the repetition of shared cultural performances and activities that enhanced remembrance of a collective past, including commemorative rituals, story-telling, place naming and the accumulation and display of relics. All these activities would trigger a social memory which would endure best when there was a 'double focus – a physical object, a material reality such as a statue . . . and also a symbol, or something of spiritual significance, something shared by the group that adheres to and is superimposed upon this physical reality' (1992 [1941]: 204).

Halbwachs was innovative in socializing the concept of memory, although Charles Withers (1996: 382) has queried his 'rather uncritical, "superorganic" notion of culture', that is to say thinking of culture as independent of the actions of the individuals who are associated with it.

Drawing on this early tradition of work, Pierre Nora's (1996, 1997, 1998) pioneering and influential project on the construction of the French past has traced, in detail, the development of French national identity through the analysis of a variety of 'lieux de mémoires' or 'sites of memory'. These include popular festivals, public monuments, architecture, literary texts and song. At a conceptual level, Nora (1989: 9) has suggested that with the demise of peasant societies, *true memory* – which he claims 'has taken refuge in gestures and habits, in skills passed down by unspoken traditions, in the body's inherent self-knowledge, in unstudied reflexes and ingrained memories' — was replaced by *modern memory*, which was self-conscious, historical and archival. He asserts that we 'must deliberately create archives, maintain anniversaries, organize celebrations because such activities no longer naturally occur' (1989: 12). The primordial memory of peasant societies embedded in what he calls milieux de mémoires (environments of memory) has been substituted by much more self-consciously created lieux or sites of memory. Modern society with its intensified globalization, mass media and self-conscious history has been the catalyst of such change. While Nora's distinction between true and modern memory may be overwrought, his work has provided the impetus for a vast array of studies on different types of memory spaces. Geographers have been particularly interested in the manner in which place and space are central to the creation of modern memories and the cultivation of individual and social identities.

One key source of identity formation in the modern era has been the idea of a national identity (e.g. Spanish, American or Indian identity) (see also Chapter 37). Consequently the role of the **nation-state** in framing memory has been the focus of much research on public monuments, memorials and heritage sites. David Atkinson and Denis Cosgrove have demonstrated, for instance, how officials in Italy tried to define a united nation by invoking a timeless, imperial Rome through the ornate and grandiose design of the Vittorio Emanuele II monument complex (Atkinson and Cosgrove, 1998). A patriotic impulse similarly undergirded the design and representation of the Canadian past in the George Etienne Cartier monument in Montreal (Osborne, 1998). Kenneth Foote also highlights the veneration of heroes and martyrs in his analysis of the debates underlying the commemoration of four assassinated American presidents. He claims, 'the question of *whether* a victim should be commemorated is often fought out over the issue of *how* commemoration is accomplished' (Foote, 2003: 37). All these studies connect the process of memory making with the cultivation of a national imaginary and in particular the role of elite and dominant memory, mobilized by the powerful to achieve this end.

Other questions of identity cut across the memorial production of the nation. How gender is symbolically represented in landscapes of memory has been investigated in a variety of contexts. For instance, the juxtaposition of historical men and mythological women at monument complexes can indicate the different roles awarded to men and women in the representation of the 'nation' (Osborne, 1998). In war museums and battle sites, for example, the voices and roles of women can be subsumed under a trope of 'motherhood', which disguises their more active role in warfare (Johnson, 2003). Memory spaces can also be strongly racialized. The politics of race within the American South underpinned much of the debate surrounding the location of the Arthur Ashe monument in Richmond, Virginia (Leib, 2002). Derek Alderman (2003) has examined how African Americans struggled to control and determine the scale of streets in which Martin Luther King Jnr. would be remembered and thus the scale at which his memory would find public expression. He notes that the issue of scale was 'open to redefinition not only by opponents to his political/social philosophy but also people who unquestionably embraced and benefited from this philosophy' (2003: 171).

Popular memory can also be a vehicle through which dominant, official renditions of the past are resisted by mobilizing groups to create subaltern and counter-memories (Legg, 2005). Geographers have regularly noted the contested nature of meaning surrounding particular symbolic sites. Whether it be the German Historical Museum in Berlin (Till, 2005), the Peace Day Parades in Dublin and Belfast in 1919 (Johnson, 2003) or the Wolfson Gallery of Trade and Empire in Greenwich, London (Duncan, 2003), the production and consumption of these sites all involved conflict. They are neither uniformly designed nor interpreted and geographers have been particularly noteworthy at foregrounding the significance of space in the construction of that meaning (Johnson, 2005).

Recently geographers have begun to address the role of performance in the inauguration of memory spaces. Drawing on Thrift and Dewsbury's observation that performance is 'a means of carrying out a cultural practice – such as memory – thoroughly', they are taking seriously the role of bodily and non-bodily practices in the making of memorial landscapes (quoted in Hoelscher, 2003). These might include the rituals surrounding behaviour around specific memory places; for instance, the role of silence at a war memorial. By emphasizing the practices and performances involved in the making of collective memory, geographers have begun to analyse the spatiality of memory through a broader lens than the textual-visual reading of memorial landscapes. This has aided and added to the identification of the role of agency in the constitution of memory places. It has also centred questions about how people feel about sites of memory, not just in the sense of having opinions about them, but in the sense of being moved by their **affects**.

SUMMARY

- Memory is not just individual but also social and collective.
- An influential way of understanding social memory has been to focus on 'sites of memory'.
- A host of studies have explored how these sites of memory relate to the production of national identities, and to other aspects of identity such as gender and race.
- Sites of memory have been analysed in relation both to their material landscapes and the performances and feelings with which they are associated.

What should we make of heritage?

An important generic site of memory is associated with 'heritage'. Over the past 30 years there has been a huge growth in the number of heritage sites, coinciding with the expansion of tourist activity worldwide. Experiencing a growth rate of 5 to 6 per cent per annum, tourism has become one of the largest employers in the twenty-first century. According to the UN's World Tourism Organization there were 940 million international tourism arrivals in 2010, representing a 6.6 per cent increase over the previous year. Tourism involves not only commercial transactions; it is, as MacCannell (1992: 1) notes, 'an ideological framing of history, nature and tradition; a framing that has the power to reshape culture and nature to its own needs' (see Chapter 53 for further discussion). Museums, stately homes, heritage centres, folk parks, nature reserves, memorials and a myriad of other sites designed to convey historical and geographical knowledge emphasize the ways in which our efforts to represent and remember the past are mediated through complex and sometimes contradictory lenses.

Heritage usually denotes two related sets of meanings. On the one hand it refers to tourism sites with a historical theme that often have been preserved for the nation-state (e.g. the Tower of London in the UK). On the other hand, heritage is used to refer to a suite of shared cultural values and memories, inherited over time and expressed through a variety of cultural performances – for example, song and dance (Peckham, 2003). The following discussion will underline the claim that the relationships between heritage and history and between tradition and modernity continue to play a significant role in contemporary society. The heritage site itself frequently forms the epicentre upon which these issues are scrutinized. If histories are constructed and memories are mapped onto the past, the manner in which these stories and recollections of the past are related is constantly open to contestation, to alternative renderings of history and to the spaces in which histories are mediated and interpreted.

The relationship between history, heritage and memory has been subject to much debate recently among geographers, historians, cultural critics and others. Heritage, as a concept, begins with a highly individualized notion of what we either personally inherit or bequeath (e.g. through family wills and legacies). In thinking about heritage today, however, we are more concerned with collective notions of heritage that link us as a group to a shared inheritance. The basis of that group identification varies in time and in space. It can, for instance, be based on allegiance derived from a communal religious tradition or a class formation or a 'nation'. However, it is with respect to cultivating the 'imagined community' of nationhood that heritage is often most frequently linked. Three different, albeit interrelated, approaches to understanding heritage have gained currency in recent years. Briefly these comprise the view that: 1) heritage is an ideological form of inauthentic history that legitimates nation-states; 2) heritage is primarily part of a process of tourism expansion and postmodern patterns of consumption; and finally, 3) heritage is a contemporary manifestation of a longer historical process whereby human societies actively cultivate a social memory. The next few paragraphs will deal with each approach in turn.

While the nation-state's origins may be relatively recent, the national state is based on the assumption that this group identity derives from a collective cultural inheritance that spans centuries. As Benedict Anderson (1983: 15) has put it, nations are collectively 'imagined communities' because 'members of even the smallest nation will never know most of their fellow-members . . . yet in the minds of each lives the image of their communion'. And that communion is traditionally conceived as historical. National states therefore attempt to maintain this identity by highlighting the historical trajectory of the cultural group through the preservation of elements of the built environment, through spectacle and parade, through art and craft, through museum and monument. As Peckham notes (2003: 2), 'Many of the institutions through which heritage is promoted, including museums, folklore societies and other educational establishments, played a formative role in the nation-building project'. The heritage industry, then, is viewed as a mechanism for reinscribing nationalist narratives in the popular imagination (Wright, 1985). Lowenthal (1994: 43) claims 'heritage distils the past into icons of identity, bonding us with precursors and progenitors, with our own earlier selves, and with promised successors'. Heritage signifies, then, the politicization of culture where cultural forms are mobilized for **ideological** purposes. This contrasts with the work of professional historians where 'testable truth is [the] chief hallmark' and where 'historians' credibility depends on their sources being open to general scrutiny' (Lowenthal, 1996: 120).

Other writers do not locate heritage so firmly in the service of the nation-state. Many of the conventional assumptions about the nation-state, they argue, have been called into question at the beginning of the twenty-first century as globalization, multiculturalism and border change have all challenged the easy demarcation of a national 'us' and a foreign 'them'. Moreover, it is suggested, the expansion in the number of heritage sites over recent years may have less to do with nation-states than with the **postmodern** cultural forms associated with post-industrial economies and contemporary tourism. According to Urry (1990: 82), 'postmodernism involves a dissolving of boundaries, not only between high and low cultures, but also between different cultural forms, such as tourism, art, music, sport, shopping and architecture'. Consequently, the distinction between representations and reality, between genuine history and false heritage is made problematic. Baudrillard (1988) suggests that meaning has been replaced with spectacle, where historical and futuristic images coalesce. For instance, the Lascaux caves in France are now closed to the public but a replica of them can be visited 500 metres from the original. The original has become redundant as its replacement by a **simulacrum** provides a **hyper-real** representation of the caves. Consequently heritage tourism is seen as 'prefiguratively' postmodern because it has long privileged the visual, the performative and the spectacular in how it 'does history' for popular consumption.

In a third approach, the argument is that the past mediated through heritage is just one element, albeit an increasingly important one, in a whole suite of historical representations. Consequently, rather than viewing heritage as a false, distorted history imposed on the masses, we can view heritage sites as forming one link in a chain of popular memory. While some critics have queried this 'museumification' of the past, the historian Raphael Samuel (1996) has celebrated such democratization, claiming that industrial museums and heritage sites with past interiors of domestic life have been progressive developments in heritage preservation and have diminished the tendency for heritage to purvey a white, elite, European and male perspective on the past. Samuel questions whether one should accept a rigid line of demarcation between, on the one hand, the past as narrated by professional historians (history) and, on the other, the past as performed by the heritage industry (an ideological social memory).

Bearing these points in mind, the remainder of this chapter will examine in detail two particular sites of memory: the Wolfson Gallery of Trade and Empire, which challenges conventional popular representations of the British Empire, and the development of a public memorial to mark the painful memories of the bombing of Omagh, Northern Ireland.

SUMMARY

- Heritage sites are important and increasingly numerous sites of memory.
- There has been a long-standing debate about the authenticity of the historical narratives offered at heritage sites. It has frequently been suggested that heritage is merely a form of bogus history mobilized either in the service of national ideologies or the tourism industry and its postmodern consumer culture.
- Others, however, are more open to the positive potential of heritage sites, understanding them not as opposed to proper 'history' but as one link in a broader chain of popular memory.

Museums: bringing the British Empire home

Since the eighteenth century, museums have served to augment, naturalize and locate national and imperial identities. As sites that collect and display objects, the processes involved in the evolution of museum cultures provide us with an important vehicle for understanding the development of collective meanings, values and memories.

Museums involve processes of both disconnection and reconnection. Objects and narratives are shifted from the milieu in which they arise and are relocated, re-presented and made 'authentic' by their re-siting in the static context of a museum display. As Cosgrove (2003: 123) states, 'Preserving the heritage fragment inescapably involves its relocation, reconstruction and representation within the different landscape of the present'. Thus the museum acts as part of a process of de-territorialization and re-territorialization that is associated with **modernity**. In a museum display objects are removed from their original material social relations and are reintegrated into a new set of institutions and hierarchies. For instance, a weaving machine's meaning changes when it is removed from a linen factory in Belfast and relocated to a space on industrial heritage within the Ulster Museum. The reworking of objects in their new setting provides both constraints and opportunities for creative interpretations of the past. In their analysis of museum culture, Sherman and Rogoff (1994) suggest four conceptual keystones in the arch of museum politics and practices that will help to throw light on how their representation of the past works. First, museums are comprised of a series of objects, which are ordered and classified in a specific sequence to offer a coherent meaning to the display. Second, these sequences of objects are woven into an external narrative that may relate, for instance, to local history, class relations or the nation. Third, museums are designed to serve a specified public and exhibits are structured to disclose the story to that public. Finally, the audience's response to a display becomes an integral part of the design process. I will now consider these processes through the example of the Wolfson Gallery of Trade and Empire, located in the National Maritime Museum in London, UK. This

Figure 34.1 The National Maritime Museum, Greenwich, UK. Credit: Getty Images

exhibition opened in 1999 and closed in 2007, when it was reconfigured into two new parallel galleries that opened in 2007 and 2011, focused on the imperial geographies of the Atlantic and Indian and Pacific Oceans respectively.

Established in 1834, the National Maritime Museum is at the centre of the old British Empire, located as it is on the prime meridian (Longitude 0) in Greenwich, southeast London. Greenwich lay at the symbolic heart of Britain's naval empire. It is the site of the National Observatory and the Greenwich Hospital. The latter, which was designed by Sir Christopher Wren, was established to support seamen and their families. It became a Royal Naval College in 1873 and trained officers in the skills of naval science, navigation and geographical knowledge. In 1998 the Royal Navy left and the old Naval College became part of the UNESCO-designated Maritime Greenwich World Heritage Site. In the following year the Wolfson Gallery of Trade and Empire was opened. Thus the space in which this exhibition was located had changed function and meaning over time, from being the epicentre of naval training and imperial naval planning to a place where Britain's history of empire was re-displayed and relayed to popular, public audiences. The gallery's empire exhibition, however, proved controversial. Seeking to offer a radical critique of the exploitative and racist underpinning of Britain's overseas empire, it aroused protest from a variety of different quarters within UK society. The exhibition began with two large posters that highlighted the purpose and structure of the gallery. One poster claimed that:

> Although the British Empire was sometimes oppressive, its power was seldom absolute and there was always two-way traffic in wealth, ideas, goods and people along imperial trade routes.
>
> (Cited in Duncan, 2003: 20)

As Duncan observes, there was an attempt at the 'reconstruction of social memory [to] tell the old story differently' (2003: 21). This different telling involved changing the narrative structure to one that emphasized some of the negative consequences of empire and that told the story from a variety of different perspectives (i.e. not just from the point of view of those building the Empire).

The gallery was roughly organized around four sections. It avoided a chronological approach and instead rotated the narratives and the displays around themes. In the first section the arrival and impact of non-European migrants (Africans, African-Caribbeans and Asians) to Britain was the focus. Referring to these migrants as 'Imperial Travellers', the exhibit disrupted our usual expectation of imperial travellers as British, white, male explorers. Instead the very positive impact of black and Asian culture on contemporary British society was emphasized (e.g. in references to carnival). This first section, Duncan claims, 'explicitly seeks to restructure social memory by displacing the emphasis away from the British abroad and towards a celebration of the empire come home to Britain' (2003: 22).

The second part of the exhibition focused on the tea trade from China and on the West African slave trade. Using a variety of visual displays, this section linked the trade in tea, consumed by elegant eighteenth-century British society, to the development of the trade in slaves. With a 'drawing room' forming the visual centrepiece of this part of the story, the links between commerce (global trade relations), slavery and exploitation and ideas of an imperial civilizing mission were highlighted.

The third section of the gallery tacked between the eighteenth century and the present. Appropriation was perhaps the common theme. At the centre of the field of vision was an

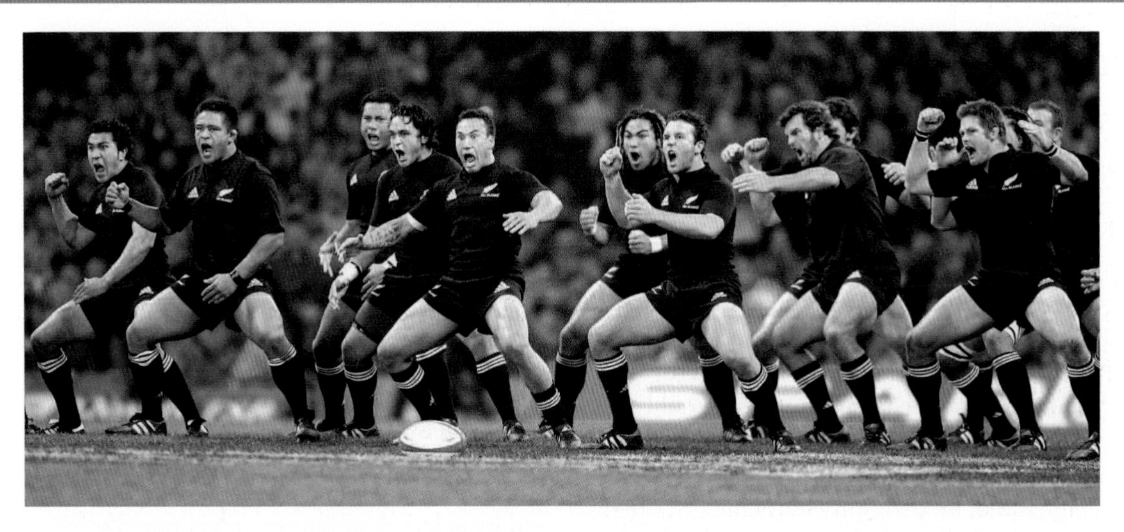

Figure 34.2 New Zealand All-Blacks performing the haka. Credit: Scott Barbour/Getty Images

embodiment of today's All-Blacks New Zealand rugby team performing the Maori ceremonial dance, the haka. There was some discussion as to whether the haka should be used at rugby matches as it 'appropriates' Maori tradition for a 'European' sport. The other element in this part of the gallery brought us back two centuries and to the British attempt to control the trade in palm oil in West Africa. Used as a lubricant in British industry and in the manufacture of margarine and soap, palm oil was important to British economic interests. The display told the story of palm oil through the actions and records of one minor British colonial official, Lt Walter Cowans. The exhibit highlighted the attacks on local African chiefs, the destruction of settlements and local economies, the killing of native people and the looting of African artefacts to later be transported and displayed in European museums and galleries. This section drew the audience's attention to the banality of greed that undergirded the quest for palm oil; it also highlighted the devastating social and cultural consequences of the competition to control the flow of this commodity.

The final part of the exhibition, relayed through a 12-minute film, documented how contemporary Britons saw their Empire and how it has been depicted through popular cultural media (e.g. movies). Narrated in a voice that seems to replicate BBC World Service English, and thus perhaps implying the authenticity of the report, this part of the story was directed towards a white British audience. It sought to ask white Britons to reflect on their image and their attitude towards their imperial past.

While the opening section of this exhibition celebrated the economic, cultural and political contribution of people of colour to British society, the remainder of the gallery highlighted 'the havoc caused by white British people abroad . . . what these exhibits do is refashion the social memory of trade and empire as a story of banality and violence' (Duncan, 2003: 24). So this particular museum's re-presentation of empire shifted in tone from celebration to repentance. It highlighted the indignities and suffering experienced by those subjected to imperial rule and it simultaneously celebrated the achievements and resilience of Britain's imperial subjects. Using a variety of visual media (e.g. maps, photos, text, film) this exhibition sought to reposition the dominant narratives of empire.

The gallery provoked considerable response, especially from white Britons. In the letters pages of the national press, some viewers objected to the museum's representation of empire, deeming it to be biased, negative and 'politically correct'. Some saw it as overtly political and argued that museums should be 'neutral' relayers of culture and history. However, as the earlier part of this chapter emphasized, modern museums emerged often as arms of the state where the nation's 'heritage' could be put on display. Hostile commentators were therefore not only concerned about an absence of neutrality, but about what the exhibition said about British identity. They saw the gallery as suppressing 'the good and exalt[ing] the bad in our colourful past' (cited by Duncan, 2003: 25) or as an attempt 'to deprive our children of their national identity' (2003: 25).

The directors of the gallery strongly defended the interpretation and representation of empire displayed at Greenwich. The gallery did, however, make some revisions. First, it introduced the Royal George figurehead signifying the abolition of slavery and alluded to other positive aspects of empire. Second, it raised the issue of the museum acting as a contact space between different cultures, and thus it actively encouraged viewers to enter into debate about the meaning of heritage and its role in the construction of contemporary social memories. While the museum may have sought to blur the boundaries between 'racial' categories and to emphasize the multicultural basis of British society, the very deployment of the black–white dualism in many parts of the exhibition gave 'authenticity' to these categories in everyday life. In that sense the very political project in which the museum was engaged and its practices of interpretation posed some contradictions. As Duncan (2003: 24) puts it, 'the contradiction is troubling, for it fails to address questions of complicity, collaboration and anti-racist dissent among the colonizers'. Nonetheless this exhibition did, at the very least, provide an image of Britain's Empire that is self-critical, thought-provoking and reflective. That ethos continued in the galleries that came to replace it: the first devoted to The Atlantic: Slavery, Trade and Empire (opened in 2007, and related to the bicentennial of the passing in 1807 of the UK's Act for the Abolition of Slavery); and the second devoted to Traders: the East India Company and Asia (opened in 2011).

SUMMARY

- Museums can be understood as sites of memory.
- At the National Maritime Museum exhibition on the history of trade and empire, the social memories were derived from an interplay of object, narrative and audience within the gallery space.
- This exhibition challenged orthodox representations of Britain's empire by focusing on the experience of those who were colonized rather than colonizing. In so doing, it made a contentious intervention into wider memories of Britain's imperial past.

Memorials: remembering the Omagh bombing and the politics of reconciliation

The Belfast Agreement, established on Good Friday 1998, brought about a series of governing frameworks and legislation which laid the foundation for a new power-sharing, devolved political administration in Northern Ireland. After thirty years of violent conflict this agreement was to put an end to the 'Troubles' and set in train a political solution to Northern Ireland's divisions. However, on 13 August 1998 a 500-pound car bomb exploded in the centre of the town of Omagh, Co Fermanagh, killing twenty-nine people and injuring scores of others. A dissident republican group, the Real Irish Republican Army (RIRA), claimed responsibility for the bomb. RIRA was formed in 1997 from a splinter group of the Provisional IRA who were opposed to the peace process and the Belfast Agreement. The bomb at Omagh stood out in

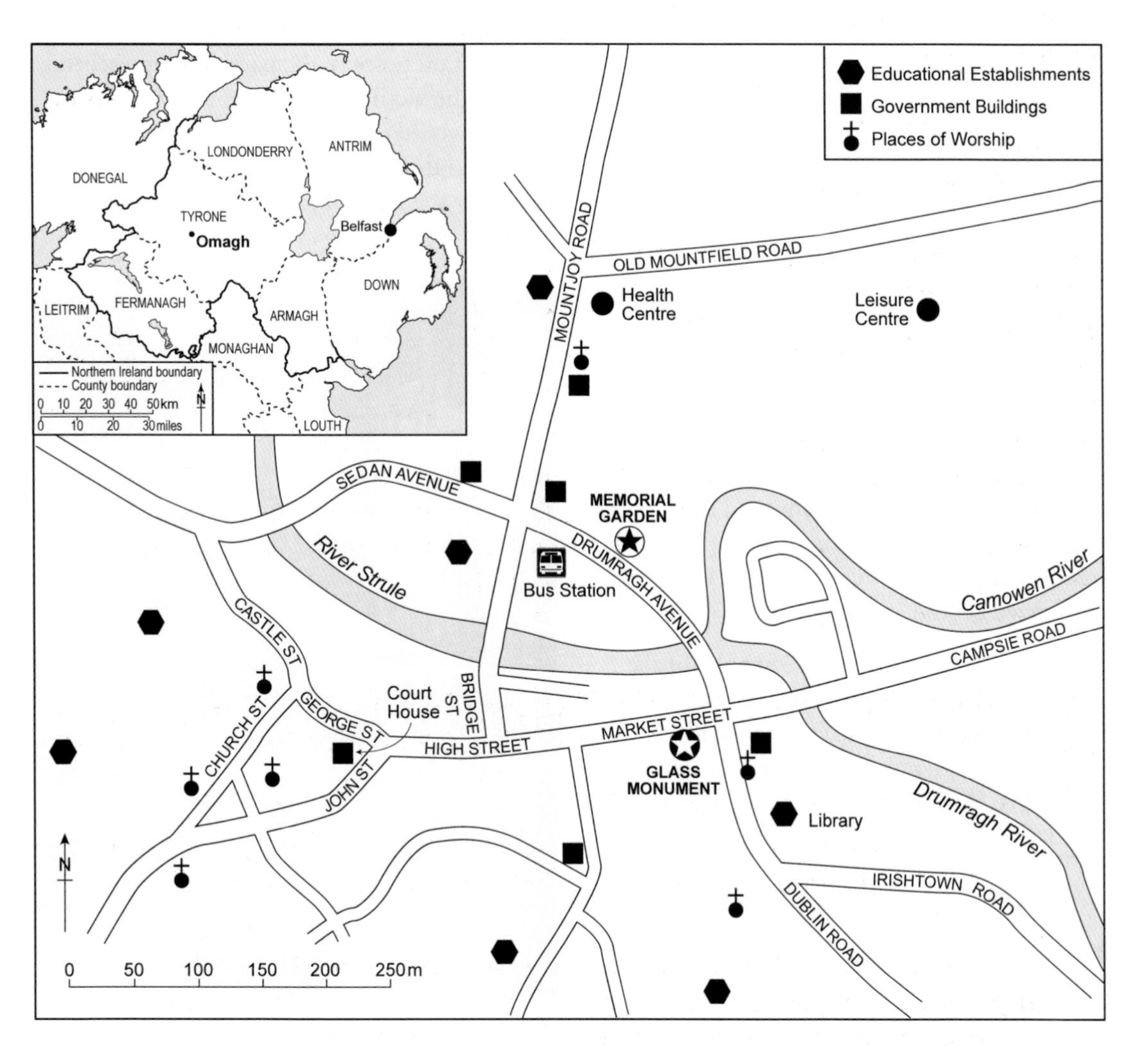

Figure 34.3 Map showing the locations of the Omagh Bomb Memorial. Credit: Maura Pringle

three ways from other attacks in Northern Ireland. First, it represented the largest loss of life in a single incident since the violent conflict began in 1969, and while it is not the only paramilitary attack to occur since the signing of the peace agreement it is the most significant with respect to casualties. Second, the bomb struck people from across the religious spectrum and over a wide geographical area. Of the twenty-nine killed, 17 were Catholic, 11 were Protestant and one was Mormon. This included a woman who was seven months pregnant with twins. Among the victims 18 were adults and 11 were children. Although most of those killed were from the local hinterland, there were three children from Buncrana, Co Donegal and two Spanish students from Madrid among the casualties. There were over 400 people injured by the blast and 165 of them were seriously harmed. Third, the attack took place in a religiously mixed community where approximately 68 per cent of the town was Catholic and 29 per cent Protestant.

After the bombing, an Omagh Fund was established by Omagh District Council and an international competition was held to design a memorial which would be unveiled to mark the tenth anniversary of the atrocity. In the design brief produced for the competition, space was considered central:

> we must be aware of the geography of the actual site of the bomb and what it is capable of accommodating physically . . . we need to be reflective and reverential but positive, optimistic and forward looking in however we mark the event.
>
> (Omagh Bomb Memorial Working Group, 2007: 2)

The memorial was to be a split site location, at the bombsite itself on Market Street and at a garden 300 metres away and around a corner on Drumragh Avenue (see Figure 34.3).

The entry chosen as the winning design, 'The Garden of Light Project', was proposed by the Northern Irish artist Seán Hillen and the Dublin-based landscape architect Desmond Fitzgerald. In their memorial plan, light was at the conceptual and material centre. The design comprised of a glass column at the site of the bomb blast with a heart-shaped, faceted crystal at its centre (see Figure 34.4). Around the corner in the memorial gardens would be a heliostat or computer controlled sun-tracking mirror that would throw light onto 29, 1m^2 gridded mirrors. They would be angled to reflect light onto a further set of mirrors mounted on poles in the park and representing each of the victims (see Figure 34.5). These mirrors would carry light along Drumragh Avenue and through the use of further mirrors transport the light onto Market Street, the site

Figure 34.4 Glass column memorial, Omagh. Credit: Edwin Aiken

Figure 34.5 Memorial garden, Omagh. Credit: Edwin Aiken

of the bomb, where they would reflect onto the glass and make it sparkle. As the bomb went off on the north-facing side of the street the designers imagined that the relative darkness and horror of the space where the bomb exploded would become a space of hope through this projection of light.

While the iconography of the memorial was well received, the inscriptions attending to it generated controversy and dissent among some of the grieving families. A group of some of the bereaved families – the Omagh Support and Self-Help Group – wished the plaque beside the glass monument to state that 'dissident republican terrorists' were responsible for the bombing, whereas the proposed script read 'A car bomb exploded at this site on Saturday 15 August 1998 at 3:10pm. This act of terror killed 31 people, injured hundreds, and changed forever the lives of many'. Derek Alderman (2010) has noted the importance of debates about inscriptions in the context of remembrance of slavery in the American south. He has claimed that conflict over wording '. . . may serve as a political strategy for raising the visibility of one vision or claim on the past while reducing or diluting the prominence of any other' (Alderman, 2010: 94). For some of those directly bereaved by the Omagh bombing, and in the absence of any criminal prosecution of the perpetrators, there was a desire to have blame directly attributed at the site of the blast itself.

As a consequence of this dispute a group of three independent facilitators was appointed by Omagh District Council to conduct interviews with all those affected directly and indirectly by the bomb. They concluded that there was no single group representing the views of the bereaved and that there was no single view on the issue of inscription. Some people supported having the perpetrators named at the obelisk, while others did not want the names of loved ones associated with those that killed them and did not want the oxygen of publicity to be given to the bombers. These conflicting views reinforce the contention that achieving reconciliation with a painful memory does not follow a single path. The recommendation of the facilitators was that the original inscription be used at the bomb site while the alternative wording suggested by the Omagh Support and Self-Help Group be placed on a plaque in the garden (see Figure 34.6). Thus the geography of the inscriptions mattered, and this recommendation was accepted by the District Council.

The memorial complex was unveiled on 15 August 2008 to mark the tenth anniversary of the atrocity. Church leaders, local councillors, political leaders from Northern Ireland, Westminster, the Irish Republic and Madrid as well as the victim's families and local community attended the ceremony. On a wet afternoon the glass monument was unveiled while local young people scattered red petals along Market Street. Ironically the rain melted the rose petals and sent rivulets of red water cascading along the street, which unintentionally reminded the audience of the red streams of blood that had run along the street on the day of the bombing when

Figure 34.6 The inscription in the memorial garden, Omagh. Credit: Edwin Aiken

the water mains burst with the impact of the blast. The crowd moved from the glass monument to the garden where the heliostat was turned on for the first time. The day was a dull, cloudy one, preventing the glass monument from sparkling at that moment. Nonetheless the design emphasized that the conditions of light would influence the affect of the monument and that the mood refracting from it would change under different light conditions. This changeability in the appearance of the monument mirrored the fact that the mood of those most bereaved by the bomb would also be changing; periods of darkness would be interspersed with periods of light. The flexibility of memory would be reflected in the changing character of the memorial itself. The designer, Seán Hillen, commented that he wanted a monument that went beyond 'simply being adequate. I wanted it to be really marvellous, not least because I wanted it to be something redeeming and positive for the victims, for the community at large and for Omagh in particular' (quoted in Johnson, 2012: 254). Not all bereaved families attended the unveiling ceremony. Members of the Omagh Support and Self-Help Group boycotted the event to express their displeasure at the inscription decision and their dissatisfaction with some of the other people attending the unveiling. This group held their own special ceremony a week later. The long-term effect of the memorial on public remembrance of the bombing remains to be determined.

SUMMARY

- Commemorating the bombing at Omagh reveals the complex processes involved in memory work and acts of public remembrance. Creating a monument that would allow a diverse community to heal from the wounds of a painful memory, as well as to provide some hope for the future, was a challenge.
- The spatial qualities of monuments are central to how they do their memory work and negotiate the challenges of remembrance. In Omagh, while the final memorial had light at its physical and metaphysical centre, the mutability of light conditions in Co Fermanagh echoed the mutability of memory for the bereaved. Just like at the glass column itself, moments of darkness and melancholy could easily impinge upon moods of light and hope among the town's population.
- Monuments are contested spaces of memory. In Omagh the dispute surrounding the content and the location of an inscription that would attribute blame for the bombing indicates that divergent views did emerge and that the politics of reconciliation in Northern Ireland is an ongoing and fragile process.

Conclusion: spaces of history or spaces of memory?

Criticisms of the heritage industry's attempts to narrate the past as little more than bogus history are often overdrawn. More generally, sites of memory need to be taken seriously, as they fashion our pasts and our senses of identity through spaces such as heritage attractions, museum galleries and memorials and monuments. These different kinds of memory places clearly have distinct purposes and forms, but they can also blur (think, for example, about tourists visiting Auschwitz). Geographers have analysed the memory work done in such sites, through attention to both their material forms and the social performances and feelings with which they are associated. In this chapter I focused on two examples to illustrate this kind of scholarship. Through the case of the Wolfson Gallery of Trade and Empire, I attempted to highlight how the past can be explored provocatively through a heritage installation. Rather than focusing on whether heritage conveys inaccurate history, the more interesting questions for geographers relate to examining the ways in which the spaces of heritage translate complex cultural, political and symbolic processes into the popular imagination. Through the example of Omagh, a provocative case certainly, I explored both the ways in which public memories are made material and performed, and how these illustrate the complexity of how individuals and communities make sense of particularly painful experiences. The relationship between light and ideas of hope was especially central to this site. The row over the inscription, however, also exemplifies the moral dilemmas that are thrown up by the seemingly contradictory desires to remember, to forget and to come to terms with the past. Such tensions ripple through the relations between space, memory and identity.

DISCUSSION POINTS

1. The heritage industry is a modern replacement for serious, scholarly and objective history. Discuss.
2. The Wolfson Gallery of Trade and Empire was an exercise in 'political correctness'. Do you agree?
3. Space is central to understanding the development of memory places. Discuss, using examples.

FURTHER READING

Foote, K.E. (2003) *Shadowed Ground: America's Landscape of Violence and Tragedy* (revised edn.). Austin: University of Texas Press.

Moving beyond the mostly European examples addressed in this chapter, Foote's book looks at how sites of tragedy and violence are marked, memorialized and performed in the American landscape. Its examples range from the site of the Battle of Gettysburg to the site of the 9/11 attack in New York.

Johnson N.C. (2012) The contours of memory in post-conflict societies: enacting public remembrance of the bomb in Omagh, Northern Ireland. In: *Cultural Geographies*, 19(2), 237–58.

This article outlines in more detail my own work on the case of the Omagh bomb memorial.

Lowenthal, D. (1996) *The Heritage Crusade and the Spoils of History*. London: Viking.

This text offers the most comprehensive overview of the distinction between history and heritage by a geographer.

Peckham, R.S. (ed.) (2003) *Rethinking Heritage: Culture and Politics in Europe*. London: IB Tauris.

This collection provides a range of studies considering the issue of heritage, with a European focus. It includes Jim Duncan's account of the Wolfson Gallery of Trade and Empire.

Till, K. (2005) *The New Berlin: Memory, Politics, Place*. Minneapolis: University of Minnesota Press.

Berlin is a fascinating city in which to consider questions of space, memory and identity. This book explores the practices and politics of place making in Berlin, and how these mediate and construct forms of social memory and identity. It is both substantively rich and conceptually innovative.

SECTION EIGHT

POLITICAL GEOGRAPHIES

Introduction

The statement that the 'personal is political' has achieved quite widespread currency since it was first used as part of the feminist movement in the 1960s. It neatly sums up the idea that politics is to be found in each and every aspect of our daily lives, and is not restricted to the more formal machinery of parliament and government. The latter concern with formal politics dominated political geography for many years and led to an emphasis on a seemingly distant and specialized sphere of activity to do with political parties, elections, governments and public policy. Although there was the notion that everyday life was affected by such processes, there was little conception that politics was actually part of our day-to-day lives. It was something that was carried out by other people (politicians and civil servants) and that went on elsewhere (in government institutions).

More recently, however, this view has changed in two key ways. First, there has been a realization that the formal politics of government and the state have much more impact on our daily lives than hitherto thought and, second, there has been a far broader examination of a whole host of informal politics – taking place in the home, in the workplace, in the street and in the community. The four chapters in this section all examine these twin developments, but do so at different levels of enquiry. Initially, in Chapter 35, Joanne Sharp takes a global perspective by concentrating on the issue of geopolitics or the relations between states. She shows how these relations are not fixed, but instead are fluid and always specific to particular historical and cultural circumstances. She then looks at how the geography of international relations is constructed through particular sets of discourses and at how even these very distant formal politics are connected to our everyday lives via elements of popular culture such as Hollywood movies and computer games. These elements in turn help to shape the 'geographical imaginations' that we all hold of the world and its political interrelationships at a global level.

In Chapter 36 Scott Kirsch examines the ways in which these geopolitical relations flare up into wars – often fought in the poorest and least developed parts of the globe. He also shows how it is becoming increasingly difficult to distinguish between war and peace, in an era where the military practices demanded by the global 'war on terror' bring a perpetual state of security alert to those countries not at war. In Chapter 37 Pyrs Gruffudd helps to show that the scale of analysis is also fluid by looking at the contested notions of the 'nation' and 'nationalism'. He shows that what for some people is only a region within a bigger entity is to others a fully-fledged 'nation'. Often the tension between these two views boils over into war and turmoil, as witnessed by events in the Middle East and in the former Yugoslavia. Pyrs looks at how the symbolic role of geography is hugely implicated in the very construction of national identity and nationalist movements, which are often centred on a struggle for land and territory.

In Chapter 38 Mark Goodwin examines the twin concepts of governance and citizenship, and in doing so moves the scale of analysis down to the more local level. He shows how the two themes are related and how, together, they cover the issues of what it means to be a citizen, of what rights and obligations one has as a citizen and of the ways in which we as citizens are governed. The chapter points out how both citizenship and governance are experienced differently by different social groups and examines how these divisions create differential spaces where different degrees of citizenship and governance are experienced and played out. Crucially, these differential geographies have become critical to the building and sustaining of new forms of participatory citizenship around localized community issues.

Taken together, these chapters provide excellent examples of the movement of political geography away from a concentration on formal politics to an exploration of the myriad spaces of informal politics. Yet they also show the continued importance of the formal political sphere. The challenge for the future, perhaps, is to examine these two spheres together, by looking at how each affects the shaping of the other, rather than continuing to see them as separate. Each chapter gives a pointer to the very exciting areas of enquiry that emerge if this is done.

FURTHER READING

Painter, J. and Jeffrey, A. (2009) *Political Geography*. London: Sage.

The second edition of a thoughtful textbook that deliberately interrogates politics and geography in order to explore the relationship between the two.

Flint, C. and Taylor, P. (2007) Political Geography: World-economy, Nation-state and Locality. Harlow: Pearson.

The fifth edition of a now classic text book, with a particular grounding in world-systems theory.

Cox, K., Low, M. and Robinson, J., (eds.) (2008) *The Sage Handbook of Political Geography*. London: Sage.

An extensive, edited collection of chapters covering most aspects of political geography.

For the most up-to-date writings on economic geography you should scan recent editions of the journal *Political Geography*. This will give you a sense of the current topics being pursued in this field.

CHAPTER 35
CRITICAL GEOPOLITICS

Joanne P. Sharp

Introduction

> Every nation, in every region, now has a decision to make. Either you are with us, or you are with the terrorists. From this day forward, any nation that continues to harbor or support terrorism will be regarded by the United States as a hostile regime.
>
> (George W. Bush, 2001)

One of my most vivid memories of watching coverage of the attack on the Twin Towers was of an interview with a passer-by (that by all accounts was repeated in conversations throughout the USA). The interviewer asked the man what Bush should do now. The man replied 'bomb them'. When asked to whom he referred, the man simply replied 'all of those people, we need to bomb all of those countries'.

At first, there was a real, but unfocused, sense of there being danger 'outside' of us-them, inside-outside, good-evil . . . a world of Manichean boundaries. Within the USA, it seemed to be impossible to critique US policy and Bush, or even to express any sense of comprehension of why the terrorists might have borne a grudge against the USA.

It became clear quite quickly that Al Qaeda was not like foes the USA had faced before (see Figure 35.1). This was not a territorially based group fighting for sovereignty but was a network of cells. Rather than emerging from the poorest of the Third World, as many western commentators had assumed in predictions of the new threats against the 'civilised world' after the end of the **Cold War**, the bombers were from educated middle class backgrounds in the Middle East, or had been recruited from mosques in the West. They raised money through credit card fraud and stock market speculation. They used mobile phones to communicate. And yet, despite all the evidence which suggests that Al Qaeda is a global network, in his rhetoric of the 'axis of evil' and his insistence that 'you are with us or you are with the terrorists', President George W Bush redrew clear lines of good and evil on to the global political map. What is it about his view of the world that led him to do this? The use of geographical imaginaries in global political models like this is called 'geopolitics'.

The *Dictionary of Human Geography* defines geopolitics as an element of the practice and analysis of statecraft that considers geography and spatial relations to play a significant role in the constitution of international politics. Certain 'laws' of geography, such as distance, proximity and location, are understood to influence the development of political

Figure 35.1 Former Al Qaeda leader Osama Bin Laden. Credit: Reuters/Corbis

situations. In geopolitical arguments, the effect of geography on politics is based upon 'common sense', rather than ideology: the 'facts' of geography are seen to have predictable influences upon political processes.

However, some authors have challenged such arguments about the political innocence of geography to suggest that rather than being a timeless concept, geographical relationships and entities are specific to historical and cultural circumstances: the nature of the influence of geography on political events can change. Given this, the meaning of geography can be *made to* change: there is a politics to the use of geographical concepts in arguments about international relations. After a brief introduction to traditional geopolitical concepts, this chapter will explore the alternative political arguments of 'critical geopolitics'. It will use various examples from the Cold War to the more recent War on Terror to explore the uses and critiques of geopolitical concepts.

The geopolitical tradition: geography as an aid to statecraft

The term 'geopolitics' was first used by the Swedish political scientist Rudolf Kjellén in 1899, but did not become popular until used in the early twentieth century by British geographer and strategist Halford Mackinder. Mackinder wanted to promote the study of geography as an 'aid to statecraft', and he believed that geopolitics offered one such way in which geographers could inform the practices of international relations. The study of geopolitics focused on the ways in which geographical factors shaped the character of international politics. These geographical factors included the spatial layout of continental masses and the distribution of physical and human resources. As a result of geography, certain spaces are seen as either easier or harder to defend, distance has effects on politics (proximity leading to susceptibility to political influence) and certain topographical features promote security or lead to vulnerability.

The concept of security is fundamental to the study of geopolitics. This refers to the maintenance of the state in the face of threats, usually from external powers. Geopoliticians argue that they can aid national security by explaining the effects of a country's geography and that of potential conquerors on future power-political relations. A student of geopolitics claims to be able to predict which areas could strengthen a state, helping it to rise to prominence, and which might leave it vulnerable. An oft-quoted line from Nicholas Spykman illustrates the necessity of a geopolitical vision by insisting that 'geography is the most important factor in international relations because it is the most permanent' (quoted in Nijman, 1994: 222). As a result, geopolitics has traditionally been considered to be a very practical and objective study: the actual *practice* of international relations has been seen to be quite separate from political theory.

One central feature of geopolitical reasoning is that it presents the world as one closed and interdependent system. It is perhaps not accidental that the rise of geopolitics as a way of understanding the world occurred at a time when global space was 'closing', the entire world was now fully explored by western colonists and imperialists so that it was now all available for state territorial and economic expansion (see Agnew, 1998). European **colonialism** had reached its height. Geopolitics offered a way states could protect territorial holdings at a time when the 'blank spaces' on the world map were finally all filled in by European powers.

Mackinder's best-known geopolitical argument is presented in his 'Heartland Thesis', which insisted upon the importance of the Asian Heartland to the unfolding history of great powers (see Figure 35.2). Mackinder believed that controlling the territory of the Heartland provided a more or less impenetrable position and could thus lead to world domination. For Mackinder, unless checked by power in the 'outer rim' of territory proximate to the Heartland, the occupying power could quite easily come to control first Europe and then the world. In 1919 Mackinder famously stated that:

> Who rules East Europe commands the Heartland;
>
> Who rules the Heartland commands the World Island;
>
> Who controls the World Island commands the World.
>
> (Quoted in Glassner, 1993: 226–7)

His conclusion was that British statesmen would need to be wary of powers occupying the Heartland, and should create a 'buffer zone' around the Heartland to prevent the further

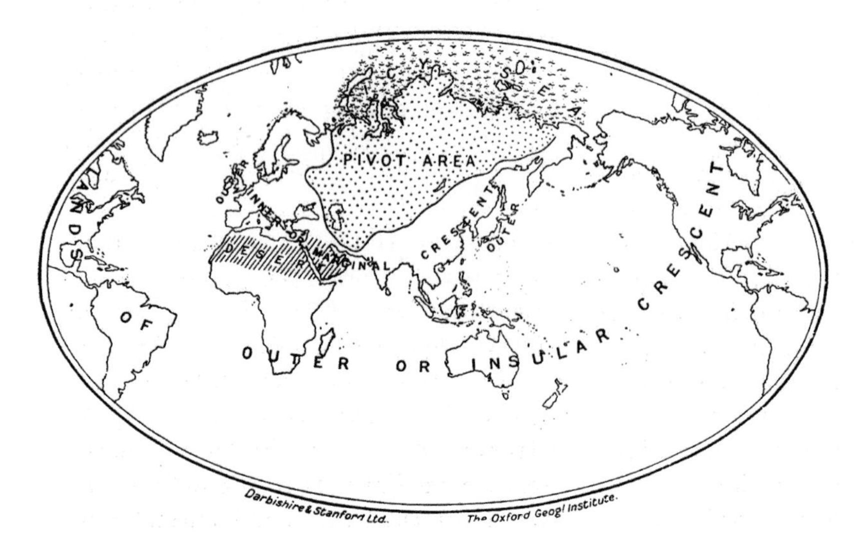

Figure 35.2 Mackinder's Heartland map. Credit: in public domain

accumulation of power that might challenge the hegemony of the British Empire.

Such geopolitical reasoning was heady stuff indeed, and there is evidence that it has both influenced foreign policy and the popular imagination. However, despite geopoliticians' insistence upon their geographical laws, their conclusions to the location of power differed. For example, whereas Mackinder promoted the power of territory, American strategist Mahan viewed control of the sea as paramount, and later others highlighted the importance of air power. Each came up with different core areas from which political dominance could be exercised.

Associations with Nazi expansionist *Geopolitik* policies (also inspired by Kjellen's work) meant that geopolitics, expressed formally in formal spatial models, fell out of use. Models still remained in textbooks, however, and were periodically updated to keep up to date with changing technology (especially the dominance of airpower and the introduction of inter-continental ballistic missiles). More significantly, a form of implicit geopolitical reasoning persisted in international relations theory and state practice throughout the Cold War.

Cold War geopolitics

One of the formative documents of the Cold War was sent as a telegram from Moscow by George Kennan – 'Mr X' – a US official in the Soviet Union at the end of the Second World War. Kennan argued that the Soviet Union was so different from the USA that there could not be compromise between the two. This image of two distinct and incompatible territorial blocs was reinforced by the political rhetoric of various political figures: in Stalin's pronouncements of the threat of capitalist expansion, Churchill's image of an Iron Curtain dividing Europe and, more recently, Reagan's depiction of the Soviet Union as an Evil Empire. A number of interrelated geopolitical concepts reiterated this binary geography in political discourse. These were, most importantly, containment, domino effects and disease metaphors.

Containment, first outlined by Kennan, referred to the military and economic sequestration of the Soviet Union. Russia's historical geography, and not simply its political and cultural difference, was invoked to give this argument scientific respectability: the USSR was seen as an inherently expansionary force that had to be kept in check. Pietz suggests that in Cold War rhetoric, the USSR was presented as nothing more than traditional Oriental despotism plus modern police-state technology (Pietz, 1988: 70).

The inevitability of Soviet expansion was also expressed in metaphors of dominoes or disease. Such metaphors saw the spread of communism or socialism not as a complex political process of adaptation and conflict but instead merely as a result of proximity to territory ruled by Soviets. The Domino Theory assumed that Soviets, communists and socialists everywhere 'were, and are, unqualifiedly evil, that they were fiendishly clever, and that any small victory by them would automatically lead to many more' (Glassner, 1993: 239). For US Admiral Arthur Redford, speaking in 1953, for example, an American nuclear strike on Vietnam was essential in order to halt a Viet Minh victory, which would set off a chain reaction of countries falling to the communists, 'like a row of falling dominoes' (in Glassner, 1993: 239). The Domino Effect can actually be seen to underlie the Vietnam War more generally. As Glassner (1993: 241) put it: 'The argument went that the United States had to fight and win in Vietnam, for if South Vietnam "went communist," then automatically, like falling dominoes, Cambodia, Laos, Thailand, Burma, and perhaps India would as well.' This process would not stop until it reached the last standing domino, the USA, and made future political action appear inevitable, unless proactive action – such as containment or pre-emptive strike – were enacted here and now.

The domino metaphor simultaneously embodied a power-political system where only two forces existed (the USSR and the USA), where only force could oppose force and where the unfolding of the process was inevitable – once started, the continuing fall of states was as unavoidable as stopping a line of dominoes from toppling once the first had been pushed. Disease metaphors were structurally very similar, relying upon notions of contagion or the malign spread of infection, again depending upon a simple notion of geographical proximity as the basis for social and political change. Even more so than with dominoes, disease metaphors illustrated the necessity for immediate action in order to prevent the further spread of the malady.

SUMMARY

- Traditional geopolitical models explain the effects of geography on international relations.
- Especially important was the question of territorial security and the danger of proximity to territory ruled by an opposing power.
- Traditional geopolitics were seen to represent a very practical application of geographical 'laws' to understanding international politics.

Critical geopolitics

More recent approaches to geopolitics, sometimes called 'critical geopolitics', refuse to accept the objectivity and timelessness of the effects of geography on political process. Critical geopolitics encompasses a range of engagements with more traditional forms of geopolitics. Some have highlighted geopoliticians' over-emphasis on the state as the main, or only, actor in international politics. Clearly other powers are involved both at the sub-state level, such as ethnic, regional and place-based groups, and at the supra-state level, such as transnational corporations and international organizations including the UN and NATO.

Critical geopolitics has been especially interested in questioning the language of geopolitics, or 'geopolitical **discourse**'. Language is not unproblematic, somehow *simply* describing what is there. Language is metaphorical, explaining through reference to other, known, concepts. Thus, there is always a choice of words and metaphors. The type of terms used – the conceptual links made – affects the meaning of what is being described. There is, as a consequence, a politics of language.

Geopolitical discourse

Critical geopolitics is influenced by postmodern concerns with the politics of representation, with 'the use of particular modes of discourse in political situations in ways that shape political practices' (Dalby, 1990: 5). To Gearoid Ó Tuathail (1996: 1), geography is not a collection of incontrovertible facts but is instead about power. What he means is that geography is not an order or facts and relationships 'out there' in the world awaiting description. Instead, geographical orders are created by key individuals and institutions and then imposed upon the world. Geography is thus the product of cultural context and political motivation.

Critical geopolitical approaches seek to examine how it is that international politics are imagined spatially or geographically and in so doing to uncover the politics involved in writing the geography of global space. Ó Tuathail (1996) calls this process 'geo-graphing' – earth-writing – to emphasize the creativity inherent in the process of using geographical reasoning in the practical service of power. Those adopting critical geopolitical approaches grant a range of power to language, from Agnew and Corbridge (1989) who see language becoming out of sync with the geopolitical reality it seeks to describe and so causing inappropriate state practice, to a figure like French philosopher Jean Baudrillard, for whom language and representation are everything (he famously suggested that the Gulf War only occurred on television; see Norris, 1992).

Rather than arguing over the true effects of geography on international relations – whether land or sea powers are strongest, as Mackinder and Mahan might have debated – critical geopolitics asks whose models of international geography are used, and whose interests these models serve. This approach owes much to the work of Michel Foucault (1980), who argued that power and knowledge are inseparable. For geopoliticians, there is great power available to those whose maps and explanations of world politics are accepted as accurate because of the influence that these have on the way the world and its workings are understood, and therefore the effects that this has on future political practice.

Critical geopolitics aims to challenge the objectivity of the geopolitician. For example, the privileging of sight (especially with the use of maps and diagrams) over other senses in geopolitical reasoning allows the geopoliticians to write as if from afar, as if somehow unconnected to the world being surveyed. This reinforces the idea of an objective account rather than one written from a position grounded within the events being discussed. It hides the fact that the geopolitician has his or her own point of view and loyalties. Although it is generally accepted that Nazi *Geopolitik* had a political agenda, this is considered to be an aberration of the 'science' of geopolitics. Yet other geopoliticians have not been innocent of interest. For example, Mackinder wanted to help maintain the British Empire and its **hegemony** over world affairs and Mahan, a naval historian, was interested in building up the US Navy at a time when other technologies seemed to make naval power less important.

Critical geopolitics looks to analyse the geography in any political description of the world. As Ó Tuathail and Agnew (1992: 194) have suggested, 'geopolitics is not a discrete and relatively contained activity confined only to a small group of "wise men" who speak in the language of classical geopolitics'. Simply to describe a foreign policy is to engage in geopolitics and so normalize particular worldviews. Any statement concerning international relations involves an implicit understanding of geographical relationships or a worldview.

Similarly, any geographical description can influence political perception. Descriptions of other places and the character of the people who inhabit them can be as significant as measurements of distance and calculations of location in constructing people's geographical imaginations. For example, the constant use of terms such as 'Evil Empire' to describe the USSR in America reinforced a binary geography of superpower stand-off that legitimated US military build-up and intervention.

Critical geopolitics and identity

Perhaps the most important claim of critical geopolitics is that traditional geopolitical arguments are in fact profoundly a-geographical. Rather than being concerned with understanding geographical process, geopolitics reduces spaces and places to concepts or **ideology**. Space is reduced to units that singularly display evidence of the characteristics that are used to define the spaces in the first place (Asia *is* exoticism, the USSR *is* communism, Iran *is* fundamentalism, the USA *is* freedom and democracy, and so on).

In the contemporary political system, dominated by the territorial state, the geography of geopolitics tends to reduce the complex workings of politics into two spheres: the domestic sphere under the control of the modern territorial state and the international realm facing anarchy without higher power to control it. Thus the state invokes discourses of security through which any different characteristics are excluded through practices of territorial control, such as patrol of borders. In security discourse, it is difference that threatens the states so that, for critic Simon Dalby (1990: 185):

> the essential moment of geopolitical discourse is the division of space into 'our' place and 'their' place; its political function being to incorporate and regulate 'us' or 'the same' by distinguishing 'us' from 'them,' 'the same' from 'the other'.

In arguing this, critical geopolitics suggests that geopolitics is not something simply linked to describing or predicting the shape of international politics, but is central to the ways in which identity is formed and maintained in modern societies. National identity is not simply defined by what binds the members of the nation together but also – perhaps even more importantly – by defining those who exist outside as different from members of the nation. Drawing borders around territory to produce 'us' and 'them' of the nation and those who are different, does not simply reflect the divisions inherent in the world but helps to create differences. Again, geopolitics does not simply reflect the facts of geography but in dividing the world into a state and the international realm helps to form geographical orders and geographical relationships.

The construction of 'otherness', and particularly the sense of danger that this presents, has implications for the practice of domestic affairs in addition to foreign policy. Thus Dalby (1990: 172) suggests that geopolitics is 'about stifling domestic dissent; the presence of external threats provides the justification for limiting political activity within the bounds of the state'. The construction of otherness simultaneously presents a normative image of identity. This happened in many ways, for instance, Elaine Tyler May explores heightened anxiety over sex at the time of heightened insecurity during the Cold War. Just as the containment of political threats outside the boundaries of the nation was planned and mapped, political leaders also sought to contain sex in early marriage and the management of individuals through the stabilizing effects of the family. May includes an illustration from a government civil defence

Figure 35.3 This illustration from a government civil defence pamphlet personifies dangerous rays as sexually flirtatious women. Credit: Defense Preparedness Agency, 1972

pamphlet that personifies alpha, beta and gamma radiation as 'sexually flirtatious women' – bombshells – only harmful if not tamed and domesticated (see Figure 35.3).

Critical geopolitics and popular culture

As a result of the influence of cultural context, different countries' geopolitical traditions draw upon specific metaphors to create images of international geography. Political elites must use stories and images that are central to their citizens' daily lives and experiences. By reducing complex processes to simple images with which their audiences would be familiar, geopoliticians could render political decisions quite natural, or could make the result of the process appear predetermined (as the domino example has demonstrated). For example, sport metaphors have been particularly prominent in the USA. Such language points to the 'essential' differences between national potentials for world-class performance and naturalizes a global arena in which the rules of the game are understood, and within which there are clear (unequivocal) winners and losers. Agnew (1998) argues that in so doing, the ambiguities of conflict are reduced to technicalities in game play. Michael Shapiro (1989: 70) points out that comparing world politics to sporting contests serves the geopolitical purpose of emptying world space of any particular content: places lose their uniqueness and world politics becomes a strategy played out on a familiar sports field.

One of the effects of broadening the scope for analysis in critical geopolitics is a consideration of a wide range of sources for analysis. More traditional approaches to geopolitics have concentrated upon the writings and pronouncements of political leaders and their academic advisers. More recently, some theorists have considered popular culture to be an important source of information.

It is possible to see the influence of popular culture on state practice directly in the central role of CNN as a source of information during the Gulf War for American leaders, or indirectly in the role of popular culture in the construction of hegemonic cultural values that shape both the actions of politicians and the expectation of societies.

Whether in films, TV programmes or in print culture, the opposition between a heroic and moralistic America and the communist Evil Empire was taken as a given. Although never using the term, the American magazine *Reader's Digest* frequently invoked geopolitical reasoning in its explanation of world affairs to its readership (see Sharp, 1993; 1996; 2000). *Reader's Digest* believed in the limitless expansive potential of communism and that only power could effectively oppose power, so that the USSR 'will not stop at international frontiers unless it is opposed' (Chennault, 1948: 121). The magazine used geopolitical arguments to warn its readers about their country's vulnerability: 'The United States is naked – incredibly naked – against a Russian atom-bomb attack' (Taylor, 1951: 85).

New technologies invite more active participation on the part of audiences. Video games, for instance, not only produce increasingly realistic landscapes, but game narratives provide a scenario through which players actively negotiate. Various authors have critiqued the Orientalist spaces produced in some games, where Middle Eastern urban landscapes are used as spaces of danger through which the player is to triumph. Equally controversial is the 2010 release of the incredibly successful *Medal of Honor* franchise, where the gameplay is set in Afghanistan 2002, and where multiplayer teams can choose to play as either the USA or the Taliban. Such images, it is argued, recreates a sense of relentless danger in the Middle East and inscribes the relationship between it and the (presumably western) player as one mediated first and foremost in violence. The increasing *impression* of realism which these games promote suggests that the boundaries between 'real' representations of different parts of the world and 'fictional' representations are blurring. For example, in response to George W Bush's War on Terror, a computer game company launched the game *America's 10 Most Wanted* in the summer of 2004 (see Figure. 35.4). The developers enthuse that the game:

Figure 35.4 Advert for the computer game *America's 10 Most Wanted*

> enlists the military expertise of CIFR's [Criminal Interdiction and Fugitive Recovery] top agent in the hunt for US fugitives listed on the FBI website, including the notorious leader of the Al 'Qaeda terrorist network Osama Bin Laden and the ousted leader of Iraq, Saddam Hussein. Commentary is provided by Dan Rathner, star news anchor for the US television network CBS over licensed news footage from CNN . . . [It] draws its inspiration from a real concern which is engulfing the world – the need to stamp out terrorism and violent crime. . . . We invite you to take the patriotic challenge and join the ranks of CIFR.
>
> (GameInfo, 2003)

The narration of territory and identity then emerges from the formal arena of politics but also through spaces of media (movies, magazines, computer games). But these spheres are not separate: the well-known news anchorman lends reality effect to games, while movie plot writers and directors are consulted by the state for suggestions on possible future scenarios.

Critiques of critical geopolitics

While critical geopolitics has had a profound influence on political geography, it has faced critique. Some have expressed concern at the over emphasis critical geopolitics gives to textual sources – political speeches, newspaper reporting, films, video games – rather than the range of other embodied practices that makes up the political world. Feminist commentators have remarked on the lack of women in the history of geopolitics and contemporary theory and practice of international relations. Cynthia Enloe suggests that women have been written out of international politics. The story of international politics has traditionally been one of the spectacular confrontation of mighty states led by powerful statesmen, of the speeches and heroic acts of the elite, and the specialist knowledge of 'intellectuals of statecraft'. Enloe (1989) refuses to accept this story as covering the full extent of the workings of international relations, and instead focuses on those elements the traditional story excludes and silences: the role of international labour migration, the availability of cheap female labour for transnational corporation investment, the availability of sex workers for the tourist industry in South East Asia, and so on. Enloe's is a very different account of international politics than the traditional story, and certainly one that lacks its glamour. She links international geopolitics to everyday geographies of **gender** relations. Her account links the personal and the political, arguing that these alternative political geographies need to be uncovered because, 'if we employ only the conventional, ungendered compass to chart international politics, we are likely to end up mapping a landscape peopled only by men, most elite men' (Enloe 1989: 1).

While critical geopolitics of the impacts of the American-led invasions of Afghanistan and Iraq have focused on the ways in which the people and landscape of the region have been visualized and reported, feminist geopolitics has also considered the impacts of this geopolitical event on the experiences of those who actually live in these places, attempting to link together their everyday lives with changing global geopolitics. Fluri (2011: 285) demonstrates the links between the everyday issue of clothing and the geopolitics of the international conflict in Afghanistan:

> Clothing choices such as military attire provide corporeal security to military bodies, while simultaneously communicating threat, distance or distrust to Afghan civilians . . . Similarly, 'other' forms of dress (such as a chadori /burqa or turban) represent security and respectability in public space from a particular perspective, while being viewed by 'other' bodies as a corporeal form of oppression or aggression, respectively.
>
> (Fluri, 2011)

SUMMARY

- Critical geopolitics understands geography to be a 'discourse', created by powerful individuals and institutions, and used as a map or script with which to make sense of the world.
- Critical geopolitics seeks to denaturalize geographical statements in international relations which appear to be so self-evident as to be 'common sense'.
- Critical geopolitics recognizes the power of geopolitical arguments not only in the context of elite debates but also in popular accounts which help to form the 'geographical imaginations' that all people hold of the world and its political interrelationships.

- Feminist geopolitics argues for the need to include understanding of other scales – especially that of the everyday – to geopolitics.

Conclusion

Geopolitics continues to be a powerful form of geographical reasoning, used in support of powerful political interests. While the content of geopolitical arguments may change (the enemy is different, the location of the heartland may move), the structure of the argument is relatively stable – geopolitics can confidently produce moral maps of the world, and locate enemies beyond the borders of the familiar.

Critical geopolitics represents an important challenge to the common-sense production of these political representations and offers the possibility for imagining alternative connections and linkages between people and groups around the world. While critical geopolitics initially focused on representations of politics, it is now engaging much more directly with a range of political practices at different scales.

DISCUSSION POINTS

1. What geographical laws are used in geopolitics? Is this how you understand 'geography'? How else might you think about geographical influences on international relations?
2. What have been the most significant changes in the shift from Cold War to post-Cold War geopolitics? What has stayed the same?
3. What are the main criticisms that critical geopolitics levels at traditional geopolitics?
4. In what places are geopolitical arguments created and communicated to others?
5. What kind of people are included in geopolitical representations? Who is excluded from them?

FURTHER READING

Ó Tuathail, G. (1996) *Critical geopolitics*. Minneapolis: Minnesota University Press.

This text offers the best examination of traditional forms of geopolitics and recent critical approaches.

Special issue, *Political Geography* (1996) 15: 6–7.

Special issue, *GeoJounal* (2010) 75

Dodds, K., Kuus, M. and Sharp, J. (eds) (2013) *The Ashgate Research Companion to Critical Geopolitics*. Farnham, Surrey: Ashgate.

Take a look at these collections to see the range of work going on under the title of 'critical geopolitics'. In what ways has critical geopolitics changed?

Campbell, D. (1992) *Writing Security: United States Foreign Policy and the Politics of Identity*. Minneapolis: University of Minnesota Press.

Dalby, S. (1990) American security discourse: the persistence of geopolitics. In: *Political Geography Quarterly*, 9(2): 171–88.

The above texts explore the interdependencies between geopolitics and American identity during the Cold War.

Der Derian, J. (2002) The war of networks. In: *Theory and Event*, 5(4).

Smith, N. (2001) Ashes and aftermath. In: *The Arab World Geographer* forum on 11 September 2001 events.

See these texts for discussions of post-September 11th geopolitics.

Enloe, C. (1989) *Bananas, Beaches and Bases: Making Feminist Sense of International Relations*. Berkeley: University of California Press.

Fluri, J. (2011) Bodies, bombs and barricades: geographies of conflict and civilian (in)security. In: *Transactions of the Institute of British Geographers*, 36, 280–96.

Special issue, *Geoforum* (2011) 42.

The above texts offer a different map of international relations that includes everyday processes and gendered images rather than just high-profile events and confrontations.

CHAPTER 36
WAR AND PEACE

Scott Kirsch

Introduction: a new type of war?

Vignette One. Speaking to airline workers at Chicago's O'Hare airport in late September 2001, US President George W. Bush assured his audience, ten days before the start of the war on the Taliban and Al Qaeda in Afghanistan, that their nation was already 'adjusting to a new type of war.' In the weeks following the September 11th 2001 attacks on the World Trade Center and Pentagon, Bush had been at pains to evoke this notion of a *different type of battlefield*, a *new type of struggle*, and a *different type of war* in speeches and photo opportunities with foreign leaders, forecasting an ambiguously 'long campaign' to be carried out, in military and intelligence operations as well as the realms of finance and law enforcement, against an enemy which 'knows no borders' and 'has no capital' (Ó Tuathail, 2003). At O'Hare, a busy hub airport suddenly located on the front lines of what would be dubbed the Global War on Terrorism, Bush continued:

> This isn't a conventional war that we're waging. Ours is a campaign that will have to reflect the new enemy. There's no longer islands to conquer or beachheads to storm . . . These are people who strike and hide, people who know no borders.

Instead, the new war was to be fought, as the President described it in a national radio address two days later, 'wherever terrorists hide, or run, or plan. Some victories will be outside of public view in tragedies avoided and threats eliminated. Other victories will be clear to all' (Bush, 2001). No borders, no capitals, no islands, no beachheads. Battles to be fought 'wherever'. Judging by Bush's statements throughout the month, the new war was going to be waged without geography in important ways, or somehow beyond it.

Vignette Two. Ten years later, we have again been promised a new kind of war. In late September 2011, a US Central Intelligence Agency (CIA) Predator Drone – part of a growing fleet of 'unmanned aerial vehicles' piloted remotely from, primarily, a US Air Force base in Nevada – attacked the convoy of the cleric and alleged terrorist Anwar al-Awlaki in Southern Yemen, killing two US citizens, Awlaki and Samir Khan, along with two other associates of Al Qaeda of the Arabian Peninsula. The event constituted 'further proof', according to US President Obama, that 'Al Qaeda and its affiliated will find no safe haven anywhere in the world. Working with

Figure 36.1 Drone wars. US CIA remotely piloted drone aircraft have become increasingly active in states such as Yemen, where the national government has limited sovereignty or effective authority over territories within its borders. The view from above, which favours a sense of precision and control over the territory below, contrasts with ground-level views of chaos and rubble, as in the second photograph, shot in the aftermath of a Hellfire Drone attack in Jaar, Yemen. Credit: (a) Rex Features; (b) Chris Cobb-Smith

Yemen and our other allies and partners, we will be determined, we will be deliberate, we will be relentless, we will be resolute in our commitment to destroy terrorist networks that aim to kill Americans, and to build a world in which people everywhere can live in greater peace, prosperity and security' (in Curtis, 2011). Two weeks' later, another drone attack in Yemen – a set of five separate strikes carried out under the same CIA–Air Force partnership – killed twenty-four alleged militants, including Awlaki's sixteen-year-old son, reflecting the escalating usage of drones in territorial settings, such as Southern Yemen, the Afghan-Pakistani borderlands, Libya and Somalia, characterized by state failure, weak states and the lack of effective territorial sovereignty.

Picking up on these trends, news agencies brandished new images to accompany reporting on the US drone attacks in Yemen and other settings, often featuring a 'view from above' that evokes a sense of precise control over the territory below (see Figure 36.1). Such perspectives appear to foreclose on even the possibility of a ground level perspective, one from which the sight of a Predator hovering at low altitudes (during the course of the hundreds of thousands of flight hours already logged under the CIA program) surely does not evoke a sense of control but one of terror from the skies (Sloterdijk, 2009) and death from above. Fragments of this perspective, based in the experience of 'non-combatants' who, because of where they live, have been caught up in the violence of war, still circulate, often as emotive images of aftermath, including photographs of dead and broken children and video documentation of bombed-out buildings, funerals and angry protests. In such images of suffering, the drone itself has long departed from the frame, but its traces are burned into the landscape; its pilot has perhaps returned home to rest in a suburb of Las Vegas, Nevada.

There are many questions that can be raised from these images and stories, and from the evolving script of the 'new war' that continues to grip the planet, albeit in complex and contested ways. The high-profile targets of drone attacks in Yemen, following the small scale 'lightning raid' that killed Al Qaeda leader Osama bin Laden in Pakistan just months earlier, has attracted attention to broader shifts in US **geopolitics**, focusing on geographical as well as technological dynamics of warfare. These shifts have included the December 2011 withdrawal of combat troops from Iraq, one of the central venues where the 'new war' declared in the first vignette had, since 2003, been territorialized (despite tenuous links to the 'war on terrorism'). These developments also include the US emphasis on small scale and flexible interventions in failed or weak states, where the lack of effective state sovereignty over parts of the country is frequently cast as *both* a safe haven for terrorist and criminal enterprises *and* an opportunity for asymmetrical warfare from the skies, one that is more easily seized in the seemingly lawless spaces of might-makes-right (see Elden, 2009). Some observers have identified a turning point in US geopolitics under the Obama Administration, animated by a recognition of limits arising in an age of financial austerity, in emphasizing the merger of military and intelligence activities (reflected in the expanded drone wars in Yemen) as a replacement for costlier territorial interventions and occupations. For others, however, this blurring of boundaries between spaces of war and spaces of (ostensible) peace, law and order, and security, has provoked thorny legal and political issues. The killing of the Awlackis, in particular, caused one observer to wonder whether summary execution of US citizens by their own government without open proceedings or published standards of evidence – 'death by drone' in sites far from the battlefield – had just been normalized (Gusterson, 2011). Or, as geographer Derek Gregory (2011) has argued, are we now absorbed in an *everywhere war*, in which no longer is anyplace truly distinct from the battlefield?

While the notions of new war, permanent war, and 'long campaign' all call attention to temporalities of the contemporary war and security terrain, a number of academics have been attuned to the emergent geographical dimensions of more than a decade of US-lead and UK-bolstered overseas invasions and military occupation in Iraq and Afghanistan (see Harvey, 2003; Gregory, 2004; Flint, 2005; Gregory and Pred, 2007; Cowan and Gilbert, 2008; Elden, 2009; and Graham, 2011). For Gregory (2011), the legal, political, and territorial apparatuses that constitute the 'extended war zone' reflect a widespread diffusion or dispersal of military, paramilitary and both state and non-state terrorist violence that can, 'in principle, occur anywhere.' And yet, while the seemingly universal nature of terrorist threats has been mobilized in efforts to garner support for a range of different interventions, blurring the boundaries between legal and political, military and intelligence activities, the effects of this conflict, and the patterns of human vulnerability that take shape around it, continue to be geographically patterned and differentiated in critical ways. Hence, Gregory's research has emphasized the emergence of historically specific geographies of war, violence and destruction in the early twenty-first century that are not located everywhere but somewhere, including in particular the emergence of new 'borderlands' such as the 'Af-Pak' military theatre and the US-Mexico border region, where both the US border patrol and drug war have been increasingly militarized in tactics and technology. If war is everywhere, it is only because the geographies of war and peace shade into one another in complex ways, defined on the ground through evolving relations of power.

Certainly, war is highly geographical, constituted through relations between places occurring across multiple scales of time and space. At the same time, war recreates places, regions and landscapes, in brutal fashion, and reshapes the possibilities for human existence and of suffering. This chapter introduces elements of a geography of war and security, while also raising questions about the spaces of peace, and the meaning of peace, in an age of dispersed but geographically differentiated warfare. What is the function of peace, in other words, in a world wherein it is only *through* war that we can hope 'to build a world in which people everywhere can live in greater peace, prosperity and security'? Thinking geographically through concepts such as spatial variation, **territory**, boundaries, scale, representation, place and landscape, as we will see in the next section, can help to make sense of war's uneven distributions, while more focused discussions of the changing nature of contemporary US **hegemony**, in particular, allow us to further explore critical questions about the interconnections between geographies of war and peace.

SUMMARY

- We are witnessing increasing references to 'new wars', with no clearly defined geography to them.
- These new wars can in principle be fought anywhere, but tend to occur in territorial settings where particular conditions, such as the lack of effective sovereignty, prevail.
- The geographies of war and peace increasingly blur into one another.

Geographies of war and security

We need not look beyond the conditions of our own everyday lives to grasp the fact that war *is* geographical, reflected in varying distributions of events and processes across the earth's surface. Most obviously, one's everyday safety from the dangers of war, terrorism and other forms of systematic political violence (for example, the 'everyday' risks of exposure to derelict landmines, rape or forced conscription of juveniles) depends to a great extent on where one happens to live. John O'Loughlin's (2005) map of the geography of conflict and development since the Second World War – in which geographical locations of battle deaths are plotted against three classes of United Nations Human Development Index (HDI) scores – illustrates this point well (Figure 36.2). Seen from this global perspective, the geographical articulation of war with poverty over the past six decades is evident. While the map obviously does not constitute a predictive model of future conflicts, its implications are clear: security from the localized effects of war, with a few notable exceptions, has become something that those living in the world's wealthier regions – the upper third of the HDI represented in dark blue on the map – are, in a sense, more likely to afford than those living in the middle and lower HDI classes, in light blue and yellow on the map.

This is not to suggest that the comparatively peaceful territories of the North Atlantic powers are therefore disconnected from conflicts occurring elsewhere, as we will discuss below. Geographic (and cartographic) perspectives – in this case simply comparing the spatial distribution of war deaths with

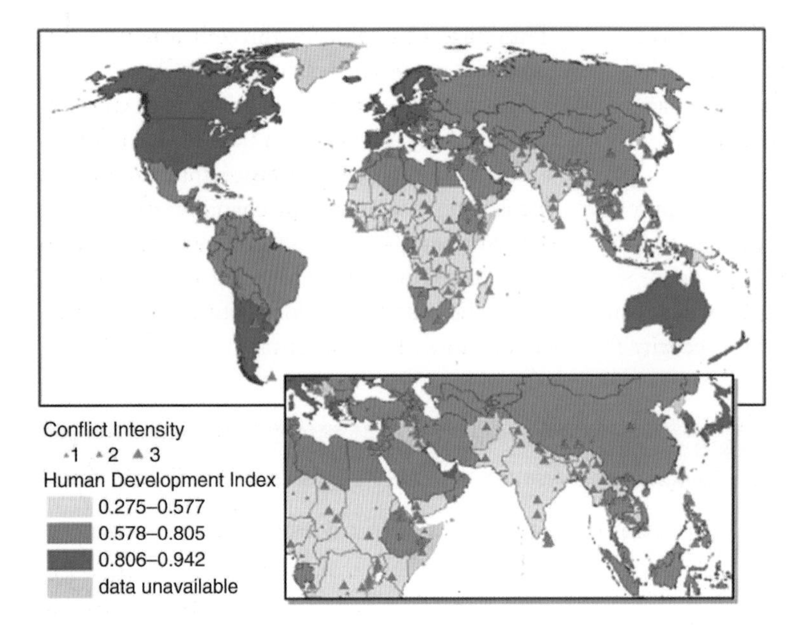

Figure 36.2 Mapping the geographic distribution of conflict against the UN Human Development Index scores. Conflict intensity is measured in three classes: 1) 25–1,000 total battle deaths; 2) over 1,000 battle deaths in the conflict but fewer than 1,000 per year; and 3) 1,000 or more battle deaths per year of conflict. Source: O'Loughlin (2005)

that of poverty or human development at an international scale – can provide useful starting points to raise new and more difficult questions. Why are wars and other forms of organized violence more likely to occur in poor regions than in wealthy ones? In what ways are geographies of violence produced by local and regional processes, and how do they reflect differing regional effects of global dynamics? We can also examine how the effects of war, including the persistence of organized violence, have often been built into the peace – in institutions, landscapes, subjectivities, bodies – through processes of reconstruction in 'post-conflict' societies, and explore the ways that war continues to structure power relations long after the treaties have been signed (Kirsch and Flint, 2011).

Clearly, the impacts of warfare are exacerbated by systemic poverty and political instability. In Congo, for example, where perhaps three million people have died as a result of persistent civil and interstate warfare in central Africa since 1998, it has been estimated that, for every violent death in the country's warzone, there are 62 'non-violent' deaths, mostly by disease and malnutrition (Lacey, 2005). Some 28 of the 62 would be children under five. These bleak statistics call our attention to the immense destructiveness of war, particularly in regions where pervasive violence has caused the disruption of precarious livelihoods and exposed the vulnerability of human life in refugee camps and other settings of displacement. In Congo, war and paramilitary violence have been instrumental in reproducing a dangerous geography of insecurity, characterized by persistent dangers to civilians from army and militia groups.

Whether insecurity has been mapped *out* of the world's wealthier regions is a different question. Returning to Figure 36.2, our eyes are drawn to the lone triangle located over New York City, reflecting the 2001 World Trade Center attack, which appears as something of a break in the pattern. And yet, if September 11th and continuing terror plots since – in Bali, Madrid,

London and Mumbai – act, albeit hopelessly and fleetingly, as 'equalizers' in the geographical redistribution of violence and fear, then such terror attacks also call attention to the affective and spectacular dimensions of contemporary war and other forms of organized violence, whether accomplished by hijacked passenger jets crashing into skyscrapers or regular drone flights over Waziristan. Organized political violence, of course, is often intended precisely to be witnessed, whether that observation occurs directly – think of the long history of public executions (see Foucault, 1995) – or increasingly, as channelled through media networks. It is, in part, by drawing on this visceral currency of witnessing that organized violence continues to impact our worlds (Thrift, 2007b). While war can be mapped and measured in body counts, this does not exhaust the range of its effects, for violence 'isn't intended to stop with the crippling of physical bodies. Violence is employed to create political acquiescence; it is intended to create terror, and thus political inertia; it is intended to create hierarchies of domination and submission based on the control of force" (Nordstrom, 2004, in Thrift, 2007b: 275). Maps and statistics can illustrate some of these relations, while they are invariably blind to others. It is clear that, initially under the rubric of the global war on terror, security for some places has become bound up elsewhere (or at least, made to *appear* bound up) in a range of new military and security practices, extending the presence of banal militarization – and war itself — in everyday life.

US political and military responses to 9/11 thus contributed to a watershed geopolitical moment, albeit for some, one likely to mark the acceleration of US hegemonic decline (Smith, 2005; Wallerstein, 2003; 2006). Here, too, questions about the **spatiality** of conflict provide a starting point for a range of critical interventions. Did Al Qaeda's intricately plotted attack on the World Trade Center (and Pentagon) serve to shatter a long standing geography of US security, in which the USA, while engaged in wars, invasions and troop deployments across the planet, had remained safe from the dangers of military conflict within its territory? Or has the attack and the hawkish US response to it served in a sense to reinforce such asymmetrical geographies of war and military power?

The latter thesis was precisely the idea among neoconservative ideologues who dominated US foreign policy under the George W. Bush Administration (2001–2008), who were intent on mobilizing military force, in the reaction to 9/11, to reshape regional power relations to US advantage in the Middle East and Central Asia, and indeed, to usher in what one influential Washington think tank had already branded a 'new American century' (Glassman, 2007). Again, we can see how the persuasive power of violence is intimately tied to – and may serve to validate – actual violence on the ground. For political geographer Gerard Toal (Ó Tuathail, 2003), a politics of *affect over intellect* thus helped to mobilize consent for the invasion of Iraq among the US polity, despite widespread international and significant internal opposition:

> Although the secular Iraqi regime of Saddam Hussein had no connection to the 9/11 attacks, was condemned by Osama bin Laden, and was deterred from using weapons of mass destruction, it was the object of a 'pre-emptive war' made possible by their channeling of the public affect unleashed by 9/11. In the geopolitical window of opportunity generated by September 11, 2001, the Bush administration interpreted the attacks in a sweeping, simplistic, and politically opportunistic manner, and after a brief war against Afghanistan, turned its 'war

> against terrorism' into a campaign against the regime of Saddam Hussein.
>
> (Ó Tuathail 2003: 857)

Consent for the war was organized around (and promoted) an affective politics of fear and risk. In the UK, participation in 'Operation Iraqi Freedom' was premised on the visceral language of imminent threat of attack from weapons of mass destruction – 45 minutes and counting! What would you do in those 45 minutes? – reflecting similar efforts to mobilize fear of violence and trauma in organizing consent for foreign wars.

Writing in the wake of the first Gulf War, the economist Samir Amin (1994) observed with concern the formation of what he saw as a new model of global warfare for the emerging post-**Cold War** world order: the North *against* the South. Although, as we have seen, the contemporary geography of war is by no means limited to North on South conflicts, it is clear that the nature of warfare itself is changing in ways that reflect this asymmetrical geographical model. In Afghanistan between October and December 2001 (the period of heaviest fighting), the so-called collateral damage of civilian deaths amounted to between 4,200 and 5,000; deaths of US allies' – including local warlords empowered to call in US air strikes – totalled in the hundreds; deaths of 'enemy combatants' numbered in the thousands or perhaps tens of thousands, while the number of US military deaths during the same period was exactly *one* (O'Loughlin 2005; Shaw, 2002). For O'Loughlin (2005), conditions of disproportionate American economic and military power should also be understood in relation to the asymmetric nature of modern technological warfare, including practices of high altitude bombing that render this asymmetry through altitudinal distances and vertical spaces: 'While there is little doubt that the US tries to avoid needless civilian loss of life, the disturbing numbers of civilians killed in "accidents" illustrates another fact of US-style modern war. In order to reduce the risk to US troops, weapons are fired from even greater distances. The advances in the electronic battlefield, combined with the use of global positioning systems, has pushed US military technology far ahead of any other country . . . These distances lead to more 'accidents' since they allow the US to fight wars at little risk to its troops. (How risky is it to drop laser-guided bombs from 29,000 feet against an enemy with weak air defenses?)'. Wars are thus increasingly fought 'at a distance', when possible, minimizing the risk to powerful countries and their soldiers, but increasing the risk to civilian populations in theatres of war.

The trend of high civilian casualties continued in the American and British invasion and occupation of Iraq throughout the decade, as has been painstakingly documented on the website www.iraqbodycount.org, though in quite different ways. Clearly, the neoconservative geopolitical moment has resulted in significant strains on US efforts to reshape power relations in the region, although the implications of these events for evolving geographies of war and security, and for the role of powerful and potentially hegemonic states in shaping these geographies, are still difficult to assess. As Immanuel Wallerstein (2006) has argued, with some irony, the loss of currency resulting from the 'project for the new American century' has served 'to accelerate the decline of US hegemony rather than reverse it'. Wallerstein, a theorist of long term historical 'world-systems', suggests that we are witnessing instead the emergence of a new moment in the metageography of war and security – portending a period dated 2001–2025 – characterized by a global deconcentration of power among states:

> The world has entered relatively unstructured, multilateral division of geopolitical power, with a number of regional centres of varying strength manoeuvring for advantage, the US, the UK, Western Europe, Russia, China, Japan, India, Iran, Brazil at the very least. There is no overwhelming superiority – economic, political, military or ideological-cultural – in any one of these centres.

After 2025, he envisages the rise to global prominence, if not hegemony, of an East Asian alliance, with the US falling in, in this geopolitical scenario, as a kind of elder partner (and political glue) among historically conflicting states of Japan, China and Korea. For this to happen, of course, is to presume that settlements are achieved for significant sources of political tension in the region, including Taiwan – for long a centrepiece in the apocalyptic gaming of potential World War III provocations.

Historical perspectives suggest that such periods of transition are rarely peaceful (but then, what periods are?). Thinking about geopolitical futures can provide intriguing and open-ended ways of imagining future territorializations of war and security, albeit in ways that are perhaps still limited by their focus on state actors and interstate relations. Meanwhile, in the geopolitical present, it must be noted that US military spending is roughly equal to that of the next nineteen countries *combined* (see Table 36.1), hence for now, asymmetrical US military power will continue to structure the world's spaces of security and insecurity in profound ways, including the persistence of long term entrenchments of military power that are themselves not easily categorized as 'war' or 'not war'.

The US policy statement known as the Bush Doctrine (Bush, 2002), which made the case for pre-emptive invasion of Iraq on the basis of self-defence against 'emerging threats before they are fully formed', may ultimately be considered a 'one-off'; the well-documented manipulation of intelligence, along with the drawn out nature of the conflict and nine-year occupation that followed, have combined to make such adventurism both less likely and less sustainable for the USA. But what might be called 'the real Bush Doctrine' embedded within this new 'right' of pre-emptive warfare – that states accused of harbouring terrorists were to be held responsible for groups operating within their territory, even if the state lacked effective sovereignty over those areas or groups (Kirsch, 2003) – has been transformative in the territorial dimensions of contemporary US warfare, particularly in redefining relations between state and non-state actors. Holding states accountable for the 'contents' of their territories in this manner provided a means of effectively locating the 'war on terror' within sovereign boundaries, tethering violence to specific state territories (Elden, 2009). This dynamic has served as a leveraging point for opportunists on all sides, permitting US military and intelligence access to many settings through drone flights and other means while allowing power blocs within states to use counter-terrorism as an advantage for strengthening their position against internal enemies, or settling old scores (Gregory 2004; Glassman 2007)[1]. Thus, while the notion of an overarching 'Global War on Terror' may be discredited – in fact, the Obama Administration summarily dropped the phrase in 2009 in favour of the nebulous, 'Overseas Contingency Operation' – its impacts, and its policies, persist in contemporary spaces of security and insecurity.

Entering what may become a new age of austerity following the 2008 economic crisis, it has become more clear that, while the USA

Rank	State	Military Expenditure (2010)	Percentage of GDP (2009)
1	United States	698,105,000,000	4.7%
2	People's Republic of China	114,300,000,000	2.2%
3	France	61,285,000,000	2.5%
4	United Kingdom	57,424,000,000	2.7%
5	Russia	52,586,000,000	4.3%%
6	Japan	51,420,000,000	1.0%
7	Germany	46,848,000,000	1.4%
8	Saudi Arabia	39,200,000,000	11.2%
9	Italy	38,303,000,000	1.8%
10	India	36,030,000,000	1.8%
11	Brazil	27,120,000,000	1.6%
12	Australia	26,900,000,000	1.9%
13	South Korea	26,550,000,000	2.9%
14	Spain	25,507,470,000	1.1%
15	Canada	21,800,000,000	1.5%
16	Israel	16,000,000,000	6.4%
17	United Arab Emirates	15,749,000,000	7.3%
18	Turkey	15,634,000,000	2.7%
19	Netherlands	11,604,000,000	1.5%
20	Poland	10,800,000,000	1.8%

Table 36.1 Military expenditure by state (2010). The USA now spends about as much on its military as the next nineteen countries combined. What changes to this 'geography of international military spending' can we expect during the twenty-first century? Source: Stockholm International Peace Research Institute (SIPRI), Military Expenditure Database for 2010 (www.sipri.org)

may have emerged from the Cold War in a position of overwhelming military superiority, its powers are limited, and as Wallerstein reminds us, they are by no means guaranteed to persist as such through another 'American century'. Our future geographies of war and security, like other human geographies, remain open questions. What will be the role of

American economic, military and geopolitical power in producing and maintaining the structures of military power, and what work do its partners and allies do in enabling, inflecting or transforming these processes? How are existing geographical arrangements of power nevertheless resisted, in a variety of ways, by actors and institutions operating at different scales and settings? What territorial forms does this resistance take, and to what effects? Addressing *these* questions would require a more sustained inquiry than can be presented here, but it should be clear from the rough sketch of geographies of war, security and military power presented in this section that our understandings of war and its outcomes are improved by – and in a sense *require* – spatial and geographical perspectives. Surely many would agree that the vast sums spent on military apparatuses (see Table 36.1) could be used to tackle more urgent human problems. But peace is more difficult to understand geographically. If war is everywhere, is peace anywhere? In the next section, we will attempt to read peace, paradoxically, *through* geographies of war, examining some ways that they are being reconfigured in spatially differentiated, hybrid landscapes of both war and peace.

SUMMARY

- Warfare is still more likely to occur in the world's poorer regions.
- Military power is increasingly asymmetrical, favouring powerful economic nations.
- The military spending of the USA is roughly equal to that of the next nineteen nations combined, enabling long-term military entrenchments overseas that are not easily categorized as 'war' or 'not war'.

Peace, *in other words*, war

It remains important not to overlook the idea of peace – war's presumed opposite – in geographical examinations of war. But, for some, it has become 'merely a deceptive illusion fostering the power of disorder and its threat – *urbi et orbi* – against the security of the world' (Alliez and Negri, 2003: 110). The notion of 'peace, *in other words*, war' thus signifies the extent to which peace and war have become 'absolutely contemporary with one another' – a hybrid condition wherein 'peace no longer appears as anything other than *the continuation of war by other means*'. Has the idea of peace been reduced to that of militarized security, or, as President Obama's comments on the Yemen drone attack, highlighted in the second vignette in this chapter, also suggest, as a justification for military strikes? Rather than opposites, war and peace seem to exist, and to circulate together, in strange unions: conditions of peace frequently serve as extensions of the militarized power relations of war, while war may be premised on the extension of a (secured) peace.

In a recent collaborative work examining the social production of 'post-conflict' geographies in a wide range of international settings, Kirsch and Flint (2011) reject the distinction between *war* and *post-war*, and that between war and peace more generally, as false binaries. Indeed, war, counter-insurgency and even genocide have all been framed, in various ways, as elements of post-war reconstruction, renewal or rebirth, while peaceful efforts to rebuild war-torn or devastated areas and to reconstruct devastated societies, often occurs *during* wars, not only in 'post-war' contexts. While some places, such as Myanmar/Burma, remain mired in an entrenched state of 'not peace, not war', reflecting complex, regionalized geographies of sovereignty, paramilitary power and governance, in other settings war persists in peacetime, in a sense, through social relations – such as agricultural labour regimes, military basing rights or gender norms in the military – that are forged under conditions of wartime or post-war crisis[2]. Of course, it is not just *peace* that has been shaped by the events and practices of *war*; conditions of war, in some contexts, have become indistinguishable from what are traditionally recognized as peace-making activities, such as democracy-building and the provision of humanitarian aid. In both Afghanistan and Iraq, US-lead wars were galvanized around a model, as Dahlman (2011) has described it, of *breaking* and *remaking* the invaded state, or of 'reconstruction as war'. Institution- and capacity-building in state and civil society have been carried out alongside and integrated with dynamics of military occupation, counter-insurgency and costly securitizations on the ground, thus articulating geographies of war- and peace-making in novel combinations; never before, it would seem, has war been so constructive.

If ideas of peace have been transformed – many would say impoverished – through the practical and philosophical intertwining of war and peace described above, then it is also true that war has been transformed in its articulation with peace, although the identification of war with notions of freedom and liberation (which are often associated with peace and 'post-war'), has been widespread. But exactly what constitutes peace – beyond *that which is not war* – remains a difficult problem of definition. Geographers, as Nick Megoran (2011) observes, have been far better at studying war than peace. So far as *this* chapter is concerned . . . *mea culpa*. This emphasis is perhaps not surprising though; in a discipline that has increasingly trained its students and practitioners to think critically about existing power relations, war and other forms of organized violence and conflict must remain key topics of inquiry. But Megoran argues that geographers need to devote more attention to understanding peace as well, including notions of positive peace that are premised on qualities such as sustainable and just relationships, equal protection and enforcement of law, and fair distributions of power and resources, rather than simply the absence of overt warfare (see also Inwood and Tyner, 2011). Here too, however, peace is very much intertwined with conflict. The work of active peace building along these lines frequently occurs, and is most effective, precisely amidst contexts of violence and insecurity. Among the settings where Megoran locates a *pacific geopolitics* is the Tent of Nations southwest of Bethlehem in the Palestinian West Bank, a non-violence centre and vineyard devoted to peace-making and environmental education among Palestinians, Israelis and other parties. It is a space surrounded by hostile Israeli settlers and situated within a complex patchwork of military and police sovereignties. We could say that active peace-making processes are structurally *drawn* to geographies of war, where their juxtaposition with conditions of war and military authority potentially garners media attention and, it is hoped, some political leverage. Of course, the need for peace is often felt most acutely in such settings as well.

Conditions of crisis – the perceived rupture of normal conditions that invites or compels response – have also been critical to the capacity of some humanitarian groups to respond actively to reshape the landscapes of war along its margins, such as Médecins Sans Frontières (MSF), an organization that has developed an interventionist approach to medical ethics under crisis conditions (Redfield, 2005; 2013). Founded in Paris in 1971 in response to the humanitarian disaster of the Nigerian civil war, MSF is currently operating in about seventy countries, creating, effectively, new transportable forms of spatial sovereignty through the mobile and rapid provision of care for those who may be in the greatest need but are most out of reach. And yet, as anthropologist Peter Redfield (2005: 344) warns, 'MSF may contribute to a migrant mode of sovereignty, administered through Toyota Land Cruisers, satellite phones, and laptop computers, but we should not forget that this sovereignty is not only mobile but is also at the same time attenuated', dedicated to an absolute defence of human life but, in the absence of adequate political authority, leaving questions of human dignity deferred. Meanwhile, and almost unnoticed, the defence of life itself has become 'a fluid and rhetorically dominant value amid contemporary secular ethics, now publicly claimed by agents of war as well as those of peace and framed by a similar justification through crisis' (Redfield 2005: 348). A richer understanding of the mechanisms of crisis and the circulation of values around the preservation of life, will also help us to better understand the articulation of geographies of war and peace in the contemporary world.

SUMMARY

- Conditions of war and peace are not binary opposites and are increasingly difficult to analytically separate.
- 'Post-war' reconstruction can be a continuation of war by other means.
- Geographies of peace have to be actively constructed and established and do not just happen when conditions of war cease.

Conclusion

While rejecting stark distinctions between war and peace, this chapter has explored some of the ways that geographies of peace and security depend, or are *made* to depend, on geographies of conflict, war and organized violence.
As we have seen, contemporary warfare is characteristically geographical in a number of important ways, from the patterned geography of conflict and human suffering to the evolving geopolitics of troop deployments and basing rights to the vertical dynamics and distant virtual control systems of the contemporary battlefield. The past two decades have been characterized by a highly disproportionate distribution of military power in terms of US military and technological capabilities, and yet, as we have seen, the nature and capacity of US power in a world of rising competitors, new practices of resistance and unpredictable economic shifts, has been called into question,

not just for the future but in the present. By attempting to understand the complex, sometimes paradoxical interrelations between spaces of war and peace – as mutually implicated, but still profoundly differentiated processes – geographical perspectives also help us to understand how different political arrangements might be possible, and what obstacles have prevented their emergence.

DISCUSSION POINTS

1. What risks do war and terrorism pose to you as an individual? How do these compare to other risks that you encounter in your everyday life (for example, from toxic substances, traffic accidents or crime)?
2. How does the question of where you live condition the risks that you as an individual face? What does this tell us about the geography of war and peace?
3. Since the end of the Cold War in 1991, how has the USA changed in its projection of military power beyond its borders? What do you expect for the future?
4. Why does Derek Gregory claim that we are now absorbed in an *everywhere war*?
5. Why is it increasingly difficult to tell conditions of war from conditions of peace?
6. Where would you expect the next major war to be fought, and why?

FURTHER READING

A number of recent works have explored the geographies of war and peace from critical perspectives, challenging distinctions between state and non-state terror, and analysing changing practices of geopolitics, territoriality, citizenship and militarism in contemporary and historical contexts.

Cowan, D. and Gilbert, E. (eds.) (2008) *War, Citizenship, Territory*. New York: Routledge.

Elden, S. (2009) *Terror and Territory: The Spatial Extent of Sovereignty*. Minneapolis: University of Minnesota Press.

Graham, S. (2011) *Cities under Siege: The New Military Urbanism*. London: Verso.

Gregory, D. and Pred, A. (eds.) (2007) *Violent Geographies: Fear, Terror, and Political Violence*. New York: Routledge.

Kirsch, S. and Flint, C. (eds.) (2011) *Reconstructing Conflict: Integrating War and Post-war Geographies*. Farnham: Ashgate.

WEBSITES

http://iwpr.net/

The Institute for War & Peace Reporting website is worth visiting for coverage of war and its consequences. It encourages local voices and attempts to build institutions of journalism in settings characterized by persistent conflict.

www.npr.org/2011/11/29/142858358/drone-pilots-the-future-of-aerial-warfare?ft=1&f=142858358

www.npr.org/2011/12/19/143926857/report-high-levels-of-burnout-in-u-s-drone-pilots

The National Public Radio website contains two audio reports that provide more on the ascendancy of the US drone program in Nevada.

http://dronewarsuk.wordpress.com/

The activist Drone Wars UK website offers an information clearinghouse and 'Drone Crash Database'.

NOTES

1. See www.newamericancentury.org/ As Glassman (2007) illustrates in the context of Southeast Asia – and as the global reach of this website's cartographically organized homepage suggest – US efforts to reshape regional and sub-national power relations, and the efforts of various local actors and groups to rework the 'global war on terrorism' to their own advantage, was not limited to the Middle East and Central Asia.
2. Chapters by Carl Grundy-Warr and Karin Dean, Don Mitchell, Takashi Yamazaki and Lorraine Dowler effectively illustrate these issues, respectively.

CHAPTER 37
NATIONALISM

Pyrs Gruffudd

Introduction

The sociologist Anthony Smith argued in 1991 that 'Nationalism provides perhaps the most compelling identity myth in the modern world' (1991: viii). The nation has arguably grown in influence since then, despite globalization, trans-national politics and cosmopolitan lifestyles. Feelings of identification with, even loyalty to, a particular nation remain powerful. In the United Kingdom the politics of nationality are shaped by challenges from two directions. On the one hand, some see threats of a pan-European state shaping economic, social and cultural life; on the other, devolution is an ongoing process, with Wales and Scotland charting sometimes radically different paths from Westminster, and a referendum on Scottish independence on the horizon. Elsewhere, small nations have re-emerged peacefully or violently after the demise of states which had hitherto suppressed their identities. After the dissolution of the Soviet Union in 1991, post-independence Kazakhstan embarked on a process of nation-building, including monumental plans for a new capital city called Astana (Koch, 2010). In 2011 the new nation of South Sudan emerged from the ruins of Africa's longest civil war, facing many legislative and practical challenges but already with some of the symbolic apparatus of national identity – an anthem, flag and designs for a capital city. In places like Northern Ireland, Corsica (see Figure 37.1) and the Basque region of Spain, sectarian and separatist tensions continue to simmer, occasionally threatening to burst into life despite peace agreements and ceasefires.

This chapter highlights some of the complexities of nationalism. The first concerns precisely which situations are commonly considered 'nationalist'. Billig (1995: 5) argues that 'In both popular and academic writing, nationalism is associated with those who struggle to create new states or with extreme right-wing politics'. So, the President of the USA would not normally be considered 'a nationalist' but a separatist politician in Quebec would, as would the leader of an extreme right-wing party such as France's *Front National*. But Billig claims that this narrow understanding of nationalism – what he calls 'hot nationalism' – always locates it on the colourful periphery while overlooking the nationalism of established states. In its broadest sense, nationalism is an **ideological** movement that draws upon national identity in order to achieve certain political goals; it therefore covers a far wider range of contexts and narratives than we might initially suppose. The usual tendency is to imagine that nationalism

Figure 37.1 A Corsican road sign amended by nationalists. Credit: Pyrs Gruffudd

is a negative phenomenon characterized by bigotry, racism, violence and aggressive social exclusion. For many others, however, nationalism refers to a positive celebration of identity in the face of oppression and the eventual attainment of social and political liberation. The conditions against which nationalist groups struggle can be extreme – racial or religious intolerance, for instance – but they can also be simple dissatisfaction with the structure of government. The struggle can, therefore, be carried through by force of arms, but also through democratic processes.

Exploring nationalism

Nationalism is a complex assemblage of phenomena and in this section I will highlight some of the ways academics have made sense of this. Perhaps the broadest distinction is between civic and ethnic forms of nationalism. Civic nationalism originated in eighteenth- and nineteenth-century Europe when modernizing states such as France and Britain sought to create homogeneous societies around the capitalist system. Civic nationalism 'envisages the nation as a community of equal, right-bearing citizens, united in patriotic attachment to a shared set of political practices and virtues' (Ignatieff, 1993: 3–4). This liberal nationalism therefore is a set of state-building practices – such as electoral systems, education and economic development – which bind a people together. Sometimes, however, its lack of conceptual regard to race, religion, colour and so forth can be problematic when the nation assumes that cultural homogeneity is desirable and refuses to acknowledge difference. In

France, for example, the government may claim that a 2011 ban on the wearing of the *burqa* in public is based on the civic ideal of a colour-blind French republican unity, but it clearly identifies Islam as a threat to certain definitions of national identity. Ethnic nationalism, on the other hand, replaces this formal, rational language of 'rights' and 'practices' with the language of historical 'belonging'. What makes the nation a place to which people pledge allegiance is 'not the cold contrivance of shared rights, but the people's pre-existing ethnic characteristics: their language, religion, customs and traditions' (Ignatieff, 1993: 4). This resilient form of nationalism can be a positive celebration of cultural heritage and diversity; an ethnic group that considers itself oppressed by the existing state of which it is a part may build its nationalism around ethnic identity and, as 'national separatists', seek political autonomy in order to protect that identity. But in other contexts ethnic nationalism can be exclusionary and can promote intolerance or even extreme violence. Nazi genocide was an extreme form of ethnic nationalism, and the definition of the nation in racial terms by many right wing nationalists, such as in the killings in Norway in 2011, contains direct echoes of that understanding. In truth, however, most nationalisms combine both 'civic' and 'ethnic' elements – at different times, in different proportions, or according to the different audiences being addressed.

The most influential work on nations has examined the internal processes through which we are socialized into feeling that we belong to a particular nation. Benedict Anderson (1991) argues that the nation is not 'real', as such, but that subtle mechanisms mould us into what he calls an 'imagined community'. The nation is an imagined, and not real, community because you will only ever meet or even be aware of a tiny proportion of your fellow nationals. However, you may nonetheless have a strong image in your mind of a shared belonging to 'our' nation and of the emotional warmth of communality – what Anderson (1991: 7) calls 'a deep, horizontal comradeship'. Famously, Anderson suggested that newspapers provided a daily, visible and shared reminder of the imagined nation. While digital media and rolling news channels may have undermined the daily newspaper, the media as a whole still has the power to shape identity, particularly at times referred to as national crises, mourning or rejoicing (see, for example McGuigan, 2000). Anderson also examined overt forms of national bonding like the singing of national anthems. Such occasions provide an experience of simultaneity: 'At precisely the same moments, people wholly unknown to each other utter the same verses to the same melody . . . Singing the Marsellaise, Waltzing Matilda, and Indonesian Raya provide occasions for unisonality, for the echoed physical realization of the imagined community' (1991: 145) (Figure 37.2). Similar moments of national unity are promoted, for example, by the ringing of bells across Britain at 8 a.m. on the first day of the Olympic and Paralympic Games in 2012 or – more poignantly – by the increasingly observed two minute silence on 11 November in commemoration of those killed in conflict (see Walter, 2001).

Michael Billig highlights the very mundaneness or banality of this process of national **socialization**. He argues that, because studies of nationalism have tended to focus on 'hot' and exotic forms, 'the routine and familiar forms of nationalism have been overlooked. In this case, "our" daily nationalism slips from attention' (Billig, 1995: 8). What he calls 'banal nationalism' is not the flag being passionately waved but the flag hanging limply and

Figure 37.2 Unisonality: the Welsh rugby team and fans sing the national anthem. Credit: Getty Images

unnoticed on public buildings like post offices in the USA. He looks also at how the contents of daily newspapers implicitly mould us into a community. He analyses weather forecasts in papers and on television and their unquestioned references to 'the country' or 'the nation' – as well as their reinforcement of 'our place' by apparently omitting the weather in neighbouring states. Together, these banalities make the homeland look 'homely, beyond question and, should the occasion arise, worth the price of sacrifice' (Billig, 1995: 175).

Similarly, Tim Edensor (2002: 2) argues that we need to understand the ways in which popular culture and everyday life can be 'expressed as national'. His analysis ranges from material objects like foods and clothing through to the very ways in which we *perform* our identities bodily through language, gesture and time-space routines (Figure 37.3). To give an example, what Edensor (2004) calls 'automobility' shapes or reflects national identity in a number of ways. As material objects, cars – ranging from the Rolls Royce to the Mini (even if now German-owned) – are seen, and advertised, as icons of Britishness, and roadscapes (such as the design of signs and petrol stations) become a familiar, even homely, part of the landscape. But national identity is also performed through automobility. Drawing on the work of the sociologist Pierre Bourdieu, Edensor suggests that we acquire the practical knowledge ('popular competencies') required to own and operate a car legally and safely through our membership of a national community. But we also perform our national identity when we carry out the unreflexive,

Figure 37.3 Performing our identity bodily through gesture. Credit: Jeremy Woodhouse/Getty Images

habitual act of driving on our roads: just consider what is deemed to be 'proper' road etiquette in the UK, particularly when contrasted with stereotypical ideas of what drivers are like in other countries.

Crucially, it has been argued by several writers that we are far from passive 'dupes' who are moulded into a top-down national identity or who uncritically absorb the messages of the media. Thompson (2001) argues that we are knowing agents who make the nation real in the course of our own deliberations and interactions and signify this in our 'common sense' understandings of the world around us. In a geographical sense, this means that most of us experience and make sense of the national at a *local* scale (see for example Jones and Desforges, 2003) or even within the home.

SUMMARY

- National identity is, arguably, becoming more significant in the modern world, despite globalization and cosmopolitanism.
- Nationalism is a complex phenomenon. It can be both a positive and a negative social force, and one found in both the geographical core and periphery.
- There is a major difference between 'civic' and 'ethnic' nationalism. The former is concerned with citizenship, laws, rights, etc., while the latter is concerned with **ethnicity**, culture and belonging.
- Recent influential studies of nationalism examine its day-to-day presence in our lives – its 'banality' or its role in creating 'imagined communities'. These studies also redress the balance towards popular culture and argue that identity is something that we perform or enact.

Geographies of nationalism

Nationalism is primarily a territorial ideology that derives its logic and inspiration from the relationship between a particular group of people and a particular parcel of land (see Williams and Smith, 1983). This process frequently draws on an awareness of history:

> 'The nation's unique history is embodied in the nation's unique piece of territory – its "homeland", the primeval land of its ancestors, older than any state, the same land which saw its greatest moments, perhaps its

mythical origins. The time has passed but the space is still there.'

(Anderson, 1988: 24)

Legends are therefore placed within the nation's space and national heroes and heroines are located through their birthplaces, graves or the site of their greatest act, thus confirming the link between a particular people and that place. Such sites of memory can often become places of pilgrimage at which explicitly or implicitly 'national' political ceremonies might be staged (see for example Withers, 1996).

Israel provides one of the most powerful examples of this mythologizing and politicization of territory. Hooson (1994) argues that:

> 'The endowment of religious symbolism upon a piece of land . . . precipitated by the establishment of Israel after the Second World War ... alongside the Moslem religious significance of the area, will make that tiny piece of land a tortured example of multiple overlapping national identities for a long time.'
>
> (Hooson, 1994: 10)

Zerubavel (1995) notes how Zionists sought to legitimate Jewish nationalist ideology through the recovery and reinvention of an ancient and potent history of settlement in the region. Masada was a mountain-top fortress above the Dead Sea whose Jewish occupants were besieged by the Romans at the end of the first century AD. According to legend, as the Romans were about to break through the defences the occupants committed mass suicide rather than be enslaved. The legend came to represent a major turning point in Jewish history and made Masada 'a locus of modern pilgrimage, a famous archaeological site, and a contemporary political metaphor' (Zerubavel, 1995: 63). Despite its grisly end, the legend became symbolic of Jewish resistance in the face of overwhelming odds. Youth trips to Masada were organized from the 1930s onwards and a pilgrimage to Masada and a pledge that it will not fall again was part of Israeli military training until the late 1980s. Elsewhere the 'science' of archaeology has played a prominent part in ideology, uncovering traces of Jewish presence in Jerusalem, for instance, thus adding to claims about the legitimacy of present-day settlement and control (Silberman, 2001). This operates alongside the extremely modern

Figure 37.4
Willy Lott's cottage – a symbol of tranquillity from John Constable's iconic *The Hay Wain* – reimagined as a branch of Tesco's. Credit: Jonathan Edwards/Guardian Newspapers

reconfiguration of the internal space of the Israeli state and the severing of the land into discontinuous three-dimensional layers by settlement and by military and strategic infrastructure – what Eyal Weizman (2007) calls the 'politics of verticality'. The division of Jewish and Palestinian areas in the West Bank by a controversial wall has further exacerbated this series of spatial divides.

Nature can also become emblematic of national identity in a process which Zimmer (1998) refers to as the 'nationalization of nature'. Certain sites, types of landscape and even indigenous species of flora and fauna (see Smith, 2011) come to be indissolubly linked with a particular nation, and that nation even derives some of its presumed character from that nature – rugged, organic, majestic, etc. As Stephen Daniels (1993: 5) puts it: 'Landscapes, whether focusing on single monuments or framing stretches of scenery, provide visible shape; they picture the nation. As exemplars of moral order and aesthetic harmony, particular landscapes achieve the status of national icons'. Daniels argues that the gentle, pastoral lowlands of southern England and paintings of them by artists like John Constable, have come to symbolize 'Englishness' and have been used at times of social tension (such as wartime) as reassuring emblems of national identity or, creatively modified, as warnings of what might come (see Figure 37.4). In the USA, a more rugged and individualistic sense of national identity coalesces around the idea of the frontier (e.g. Slotkin, 1992). But while rural landscapes tend to be imagined as the 'authentic' essence of the nation, this idealization also extends to the people, or 'folk', living within them in idealized, organic communities. Historically, many nationalist movements had close ties with groups protecting the legacy of the folk, be it their customs and way of life or simply their buildings, costume or music. Some nationalist groups went so far as to advocate moves 'back to the land' – that is, to resettle the population away from the cities and in the rural 'heart' of the country – in order to regain some essential, threatened form of national identity (e.g. Gruffudd, 1994).

SUMMARY

- Nationalist ideology is nearly always 'geographical' in that it is based around a territorial claim and it proclaims a clear sense of place.
- History and folk culture provide a nation with a long-standing imaginary bond to the land. Apparently neutral practices like archaeology can thus play a role in nationalist movements by defining the authorized national past.
- This geographical aspect can be manifested equally in material acts of nation-building (e.g. road networks) and in cultural forms like landscape painting, etc.

The matrix of nationalism

Edensor (2002) argues that national identity exists as part of a 'matrix', in a dynamic and promiscuous relationship with a series of other forms of identity and aspects of social life. But 'the historical weight of national identity means that it is hard to shift as the pre-eminent source of belonging, able to draw into its orbit other points of identification whether regional, ethnic, gendered or class-based' (2002: 35). Nationalism, therefore, is flexible and multidimensional and this is part of the reason for its success. To illustrate this I will briefly look at inflections of nationalism in different contexts, involving issues of economics, 'race' and language.

Some of the most influential early writings on nationalism tried to explain its persistence or emergence by reference to socio-economic conditions. Tom Nairn (1977) argued that the resurgence of nationalism in Scotland and Wales since the 1960s was a populist and romantic response to uneven development in the UK and a growing awareness of disadvantage. There are parallels in David Harvey's (1989: 306) argument that 'localism and nationalism have become stronger precisely because of the quest for the security that place always offers' in the face of the uncertainties of capitalist globalization. But there is no inevitable correlation between poverty, ethnic identity and nationalism. The Basque Country in Spain was historically prosperous and industrial modernization heightened, rather than diminished, its sense of identity. However, an economic recession in the 1980s *did* give Basque nationalism a new surge of energy, manifested in support for democratic parties as well as in an upturn in the activities of the terrorist group ETA (*Euskadi Ta Azkatasuna* – Basque Homeland and Liberty) (see Figure 37.5). Catalonia, on the other hand, has established itself as one of the most dynamic and prosperous regions of the European economy, but where politics is characterized by a powerful sense of Catalan (as opposed to Spanish) identity and demands for greater independence from Spain. There has been a less attractive echo of this in Italy and its long-standing disparity between the affluent north and the poorer south. There the *Lega Nord* (the Northern League), formed in 1991 and led by its charismatic leader Umberto Bossi, harnessed the grudges of the 'industrious' north against the south and its supposedly 'corrupt' politicians. In a ceremony on the banks of the River Po in 1996, Bossi declared an independent north Italian state called Padania, with its capital in Venice (see Huysseune, 2010). The League's electoral success initially lagged way behind its historically-inspired rhetoric (see Agnew, 2002) but it has been part of successive right-wing coalition governments in Italy, though ironically Bossi stood down in 2012 after a corruption scandal.

The *Lega Nord*, however, is also notable for its racist rhetoric and opposition to immigration. As I have already noted, nationalist politics are frequently articulated around the issue of ethnic 'belonging', with 'ethnic cleansing' in Bosnia – the forcible construction of ethnically homogeneous areas – perhaps the most painful recent example in Europe. While Basque nationalism was originally, and in part, based on racial distinctiveness and superiority (Conversi, 1990), very few of the 'mainstream' nationalist political parties in Europe are now overtly racist. However, that does not mean that issues of racial belonging are not a powerful part of any understanding of the nation. At the level of the state, immigration policy serves to exclude and include primarily on ethnic lines and below that level a whole range of popular discourses about race are also woven together with nationality (see Jackson

Figure 37.5 Demonstrators – including two in traditional costume – carry the Ikurriña (the Basque flag) to protest at the banning of a separatist political party by the Spanish state, 2003. Credit: Rafa Rivas/Getty Images

and Penrose, 1993). Contemporary debates about British identities in the wake of **devolution**, for instance, have highlighted the fact that many minority ethnic groups feel uncomfortable about assuming the identity of the constituent nations (Wales, Scotland, etc.) as opposed to a hyphenated sense of, say, 'Asian-British' (Bryant, 2003). Far right parties, on the other hand, consistently advance simplistic and exclusionary correlations between race and nationality.

Ethnicity more generally is central to any understanding of the vast majority of nationalist situations, but that ethnicity may be expressed as the politics of language or of religion rather than some crude notion of 'racial belonging'. Often, we need to think of these nationalisms in terms of the celebration of ethnic diversity in the face of homogenizing or oppressive forces, rather than in terms of aggressive exclusion. Recent Spanish nationalisms can be partly understood as a response to the ethnic suppression that characterized the Fascist dictatorship of General Franco from 1935 until 1975. Franco abolished the historic parliaments and legal rights of both Catalonia and the Basque Country and ruthlessly repressed their regional identities and languages, with books destroyed and place names changed to Castilian (what we call 'Spanish'). While Basque nationalism spawned violent elements, Catalan nationalism was almost entirely peaceful and based around the maintenance of folk culture. Since the restoration of Catalan autonomy in 1980 the language has regained its status in all aspects of life. According to the first leader of the new regional government 'If some issue is crucial to Catalonia, it is language and culture, because they are core elements of our identity as a people . . . Catalonia did not want autonomy for political or administrative reasons, but for reasons of identity' (quoted in Conversi, 1990: 56).

Welsh nationalism has historically also been a predominantly cultural movement. *Plaid Cymru* (The Party of Wales) was formed in 1925 in response to the perceived decline of traditional Welsh rural life and the Welsh language. The language had been marginalized and outlawed by the British state since Wales's incorporation under the Act of Union of 1536, but modernity (in the form of radio, tourism, etc.) further threatened cultural distinctiveness. Many of *Plaid Cymru's* campaigns opposed the British state's incursions (for water, military training, even tourism) into those parts of rural Wales imagined as the cultural heartlands of the nation, or sought to re-establish symbolic, if not actual, control over Welsh territory. Rhys Jones and Peter Merriman (2009), for example, use Billig's banal nationalism thesis to explore campaigns by *Plaid Cymru* and *Cymdeithas yr Iaith Gymraeg* (the Welsh Language Society) against monolingual English road signs in Wales in the 1960s and 1970s. Drawing on other studies of the politics of street- and place-names (for example Gade, 2003) they argue that such signs were viewed as symbols of linguistic and cultural injustice. While the campaign for bilingual signs involved technical and governmental negotiation, it was most notable for the many protests in which monolingual signs were defaced or symbolically removed and broken. Jones and Merriman use this story to caution against too simple a distinction between hot and banal forms of nationalism.

SUMMARY

- Nationalism is a very flexible or multidimensional ideology that can be manifested in a number of contexts. It is, perhaps, best thought of as existing within a matrix of concerns and identities.
- It has frequently been associated with groups highlighting poverty and economic and social injustice along ethnic lines, although several nationalist movements represent 'rich' areas.
- More commonly, nationalism interacts with ideas of ethnicity, race, language, religion, and so forth. In these forms it can be either culturally repressive or a reaction against cultural repression.

Conclusion

Nationalism, then, is an extremely complex – and even bewildering – phenomenon, and one that is not always reducible to simple measures of 'good' and 'bad'. It can be liberating as well as oppressive, peaceful as well as violent, progressive as well as reactionary, traditional as well as modern, even rural as well as urban. Little surprise then that it has often been characterized as 'Janus headed', after the two-headed Roman God of doorways (after whom January is named). It is also wrong to think of nationalism as something out at the margins. As Billig and Anderson show, nationalism is as much a feature of established states as it is of those nationalist groups struggling for expression and sovereignty. A thorough and sensitive analysis of nationalism, then, should consider the context within which it is being expressed, the issues that it identifies as central, the balance between ethnic and civic strategies and the way in which these factors evolve over time. It should also look at how national identity – or competing identities – come to be politicized. And, of course, for geographers, considerations of space, **territory**, landscape and scale are not only crucial but also open avenues of study that reveal in fascinating detail how powerful, creative, and often destructive, nationalism can be.

DISCUSSION POINTS

1. Do you agree that nations are, if anything, becoming more important nowadays? Who or what is challenging their supremacy?
2. Thinking of Edensor's 'matrix of national identity', how important is your national identity to you and how does it sit alongside your other forms of identity?
3. Can you identify the 'banal' ways in which national identity is manifested in popular and material cultures in everyday life?
4. What is the place of your nation? What are its iconic spaces and landmarks, and which of its landscapes might be considered the most 'authentic'?

FURTHER READING

Guibernau, M. and Hutchinson, J. (eds.) (2001) *Understanding Nationalism*. Cambridge: Polity Press.

A useful survey of the sometimes bewildering debates on how best to understand nationalism.

Harvey, D., Jones, R., Milligan, C. and McInroy, N. (eds.) (2002) *Celtic Geographies: Old Culture, New Times*.(London: Routledge.

A lively collection of essays on how the idea of 'the Celtic' underpins identities and actions in a variety of places around the world.

Kingsbury, P. (2011) The World Cup and the national thing on Commercial Drive, Vancouver. In: *Environment and Planning D: Society and Space* 29, 716–37.

Uses some challenging theory to understand fans' emotional responses to the 2006 football World Cup.

Leuenberger, C. and Schnell, I. (2010) The politics of maps: constructing national territories in Israel. In: *Social Studies of Science* 40(6), 803–42.

A discussion of how various Israeli social and political groups have used maps to legitimize their claim on national space.

Vale, L. (2008) *Architecture, Power and National Identity*. New Haven: Yale University Press.

This book examines parliamentary buildings (built ones as well as unrealized plans) in capital cities including Washington DC, New Delhi and Brasilia.

Jazeel, T. (2005) 'Nature', nationhood and the poetics of meaning in Ruhuna (Yala) National Park, Sri Lanka. In: *Cultural Geographies* 12(2), 199–227.

A study of how history, race, religion and politics can affect the meanings of nature and conservation.

WEBSITES

www.facebook.com/asenevents

Take a look at the *Facebook* page for the Association for the Study of Ethnicity and Nationalism, especially their *Newsbites* section on current affairs.

www.nationalismproject.org/

The Nationalism Project is an online resource containing information, bibliographies, reviews and a blog relating to nationalism.

CHAPTER 38

CITIZENSHIP AND GOVERNANCE

Mark Goodwin

Introduction

Issues of citizenship and governance continue to hold a prominent position in political and academic debates. Indeed, during the spring of 2012 when this chapter was being written, fierce arguments were raging, in the UK for example, over the questions of independence for Scotland, the involvement of the private sector in providing healthcare and running underperforming schools, the control of immigration and the charging of tuition fees to university students. Move outside the UK and the issues become even fiercer, ranging from the protests of the Arab spring, to the civil war being waged in Syria and the legitimacy of European financial institutions seeking to intervene in national state economies. Running through these and other similar debates are the twin themes of citizenship and governance – covering the issues of what it means to be a citizen, of what rights and obligations one has as a citizen, and of the ways in which the collective citizenry are governed.

A good example of how these twin themes come together has recently been provided by the debates over Scottish independence and Welsh **devolution**. At the heart of these debates are concerns about how different parts of the United Kingdom are governed, and about the different identities felt by those who are citizens of this supposedly 'united' nation. Recent research, for instance has shown that in Scotland, 67 per cent of people claim either exclusively or mainly Scottish identity, and only 11 per cent claim to be exclusively or mainly British. In Wales, 44 per cent claim to be exclusively or mainly Welsh, but the proportion claiming to be exclusively or mainly British is almost twice that in Scotland at 18 per cent. In England, on the other hand, only 33 per cent of people claim to be exclusively or mainly English (Jeffery et al., 2010: 17). The area of the UK where people feel most British is Northern Ireland, where 37 per cent identified themselves as British, as opposed to 28 per cent as Northern Irish and 25 per cent as Irish (Northern Ireland Life and Times Survey, 2010).

Asking these relatively simple questions about identity raises a host of issues about the complex mixtures of states, territories and areas of government within which we all live, and about how these relate to our shared understandings of what it means to be a citizen of one or other of these spaces. The difference between the percentage of people who feel British in Scotland and Northern Ireland stems from a host of political, cultural, social and economic processes, which stretch back some 500 years, but which have a very obvious contemporary manifestation. These questions promise to become more complex in the years ahead, as identities and belongings are rewoven by migration and globalization. This chapter explores such complexity by investigating the twin notions of citizenship and governance. It looks at the ways in which geography is bound up with both concepts and with how they operate in practice, and at how an appreciation of these issues helps to broaden the way we look at Human Geography.

The contested spaces of citizenship

Issues of citizenship are usually seen as the province of political science, not geography, and the usual definition of citizenship is provided in political terms as referring to 'the terms of membership of a political unit (usually the nation-state) which secure certain rights and privileges to those who fulfil particular obligations. Citizenship is a concept, rather than a theory, which formalizes the conditions for full participation in a community' (Smith, 1994: 67). A moment's thought, however, will indicate that even this narrowly political definition is bound up with geography. The political unit that one is a member of has a certain territory, thus, by definition, one is a citizen of a particular place. Also, the community or communities that one participates in, whether fully or not, are bounded too, and geographically situated. Indeed, the whole concept hinges around notions of inclusion and exclusion: to be a citizen is to be included in both a social and spatial sense, and this has offered fertile ground for geographical work. As Staeheli *et al.* put it (2012, 14) 'Citizenship varies across place, across time, and for different people. It is inseparable from the geographies of communities and the networks and communities that link them'.

In particular, geographers have questioned the basis on which such inclusion and exclusion takes place. The political definition of citizenship stresses the inclusive nature of the term – it implies that anyone within a certain territory who meets certain obligations will be included as a citizen, with corresponding rights and privileges. Yet matters are not this simple and the act of residence within a definable and bounded space does not necessarily secure citizenship. Somewhat paradoxically, the **globalization** of both capital and labour that has led to increased flows of people around the world (see Chapters 28, 41 and 53) has resulted in attempts by many governments to legislate for tighter immigration controls. This has two main implications: some people are excluded from residence altogether and others are denied full citizenship rights. In Britain, for example, there is an explicit link between immigration and those means-tested benefits that, as part of the social security system, should be available to all citizens. As Oppenheim points out (1990: 89), claiming income support, family tax credits or even access to housing under the homeless legislation can endanger the chances of bringing the rest of one's family to the UK, or create problems for the claimant themselves or their sponsor. The Immigration Act 1971 meant that the wives and children of Commonwealth citizens could only enter the country if a sponsor could support and accommodate them 'without recourse to public funds', defined clearly for the first time in 1985 to include the three major means-tested benefits referred to above. This means people who may have worked and paid taxes here for decades are only allowed to be joined by their families on condition that they do not claim benefits for them or turn to the welfare state for accommodation. More surreptitiously, perhaps, the increasing frequency of passport checks on black claimants at benefit offices, regardless of whether they were born in the United Kingdom, helps to create a climate of opinion that views welfare as the entitlement of white Britons rather than of black 'outsiders' (Oppenheim, 1990).

What we might term the bounded spaces of citizenship within the nation-state are therefore not 'straightforwardly inclusionary' (Painter and Philo, 1995: 112). Divisions along racial or ethnic lines as referred to above are fairly widespread and long established; in Germany, for instance, migrants who arrived to fill labour shortages in the 1950s and 1960s were labelled as 'guest workers' and denied social rights and freedom of movement. This labelling denied them even the status of ethnic-minority migrants – they were expected to return 'home' when no longer needed. Figure 38.1 shows the abject living conditions many such workers endured, often in spatially segregated 'ethnic enclaves'. The denial of full citizenship rights to guest workers and their geographical and social marginality in special hostels or dormitories have played no small role in the rise of racism and fascism in the newly unified Germany. This is one example of the ways in which inclusionary citizenship is actually riddled with divisions that are at once spatial and social.

Other examples of these divisions have emerged as recent debates about the term have broadened the scope for new avenues of geographical inquiry. It quickly became apparent that many groups who seem to enjoy full citizenship are actually limited in terms of the places and spaces in which this can be exercised. A highly visible example has been provided by the recent debates about **sexuality**. Geographers have increasingly been investigating the links between sexuality and citizenship by exploring what Bell (1995: 139) has called 'the spaces of sexual citizenship'. His starting point for this exploration is to ask the question, quoting Diana Fuss, 'What does it mean to be a citizen in a state which programmatically denies citizenship on the basis of sexual preference?' (1995: 139). His answer begins from the fact that such preferences can be played out in some locations and not in others. What it does mean to be a citizen in this case is partly dependent on where you are. As Bristow observes, 'In Britain, it is possible to be gay [only] in specific places and spaces: notably, the club scene and social networks often organized around campaigning organizations' (quoted in Bell, 1995: 141). It is only in these spaces that those who are gay can feel comfortable, and as Painter and Philo (1995: 115) put it, 'if citizenship is to mean anything in an everyday sense, it should mean the ability of individuals to occupy public spaces in a manner that does not compromise their self-identity, let alone obstruct, threaten or even harm them more materially'.

Valentine (1993b), in her exploration of the geographies of gay friendships, has confirmed that gay people are often forced to inhabit marginal spaces. The result is the growth of 'dense and heterogeneous networks formed around a limited geographical base [which] foster a sense of community' (Valentine, 1993b: 113). As this suggests, the more marginal spaces of sexual citizenship can be positive as well as negative. Valentine goes on to explain that 'because lesbians find it difficult to make friends and express their lifestyles outside these gay contexts ... their identities can become embedded in the networks formed in and around these places'. In some senses, then,

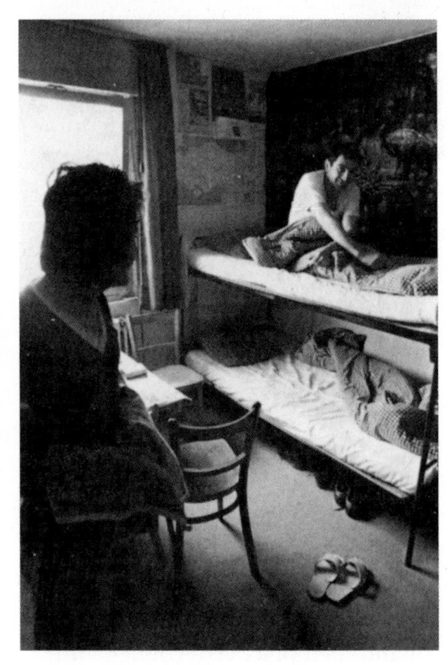

Figure 38.1 Social and spatial marginality of migrant workers denied full citizenship. Credit: Jean Mohr

we can see how these restricted geographies can offer the chance for a reconstituted citizenship to emerge around a series of 'alternative' or 'underground' spaces. In the case of gay people, whole neighbourhoods have now grown up as places of deliberate congregation, such as Castro in San Francisco (see Castells, 1983), West Hollywood in Los Angeles (Jackson, 1994), Soho in London or the City of Amsterdam (Binnie, 1995). In these areas we can find a clustering of bars, restaurants, bookshops, theatres, clothes shops and other retail outlets all catering for a gay clientele (see Figure 38.2). In these spaces we can see the flowering of an alternative culture, which can act as the basis for an alternative kind of citizenship, in which members of certain groups can establish rights and obligations to each other. Through this type of work, the notion of citizenship has been expanded beyond a somewhat narrow concern with rights and responsibilities bestowed by the nation-state. As Desforges puts it, 'rather, citizenship is conceptualized as a set of social processes in which individuals and social groups negotiate, claim and practice not only rights, responsibilities and duties but also a sense of belonging which enables full participation within a multiplicity of "communities"' (2004: 551).

This 'negotiation of belonging' within multiple communities has led geographers to look at the multiple spaces within which such negotiation takes place (Desforges, 2004; MacKian, 1995). Attention has turned away from the nation-state to examine the practice of citizenship within the home (Fyfe, 1995; McEwan, 2000), the neighbourhood (Pile, 1995), the school (Pykett, 2009), the urban (Brown, 1997) the rural (Parker, 2002) and the global (Desforges, 2004). Other geographers are looking at how the emergence of notions of sustainable development and the 'sustainable citizen' has led to the development of a stretched or 'distanciated' mode of citizenship. Such distanciation involves an 'enlargement of the public sphere within which citizenship is conceived of and then practiced' (Bullen and Whitehead, 2004: 7). This enlargement is three-fold. It covers not just a stretching beyond the nation-state to encompass an engagement with others across the globe, but also a sense of obligation towards future generations and a concern with environmental rights and responsibilities. As geographers explore these new spaces of citizenship, it has become apparent that they have been used to develop alternative forms of what we may call 'participatory citizenship'. A collection of writings from the USA documents how different marginal groups have refashioned spaces of social control into sites of resistance, and how in this process have been able to contest dominant views and assumptions about their 'place' in society (Smith, 1995). The groups include squatters resisting urban development in Michigan; the homeless campaigning for decent housing in Chicago; an anti-gentrification coalition seeking to preserve low-cost housing in New York; low-income African-American women living in a public housing project in New Orleans involved in community development; and immigrant Mexican agricultural labourers in California campaigning for employment rights. All the groups are using the appropriation of space to claim what they see as legitimate citizenship rights for their section of society. One chapter describes the setting up of 'Tranquility City', a homeless encampment of 22 plywood 'huts' in a run-down industrial district of West Chicago. The author concludes that:

> for a brief period Tranquility City became a *mini-movement area* in which a different way of living poor was experimented with: a possibility was created for the formation of a homeless community free of institutional shelter restraints. Within this mini-movement area, residents of Tranquility City were able to construct a collective identity centred around issues of social justice for other homeless individuals and collective action in helping each other acquire housing and needed services.
>
> (Wright, 1995: 39, original emphasis)

As this suggests, the spaces that such groups inhabit can be crucial in helping to form a group identity, which helps to underpin a

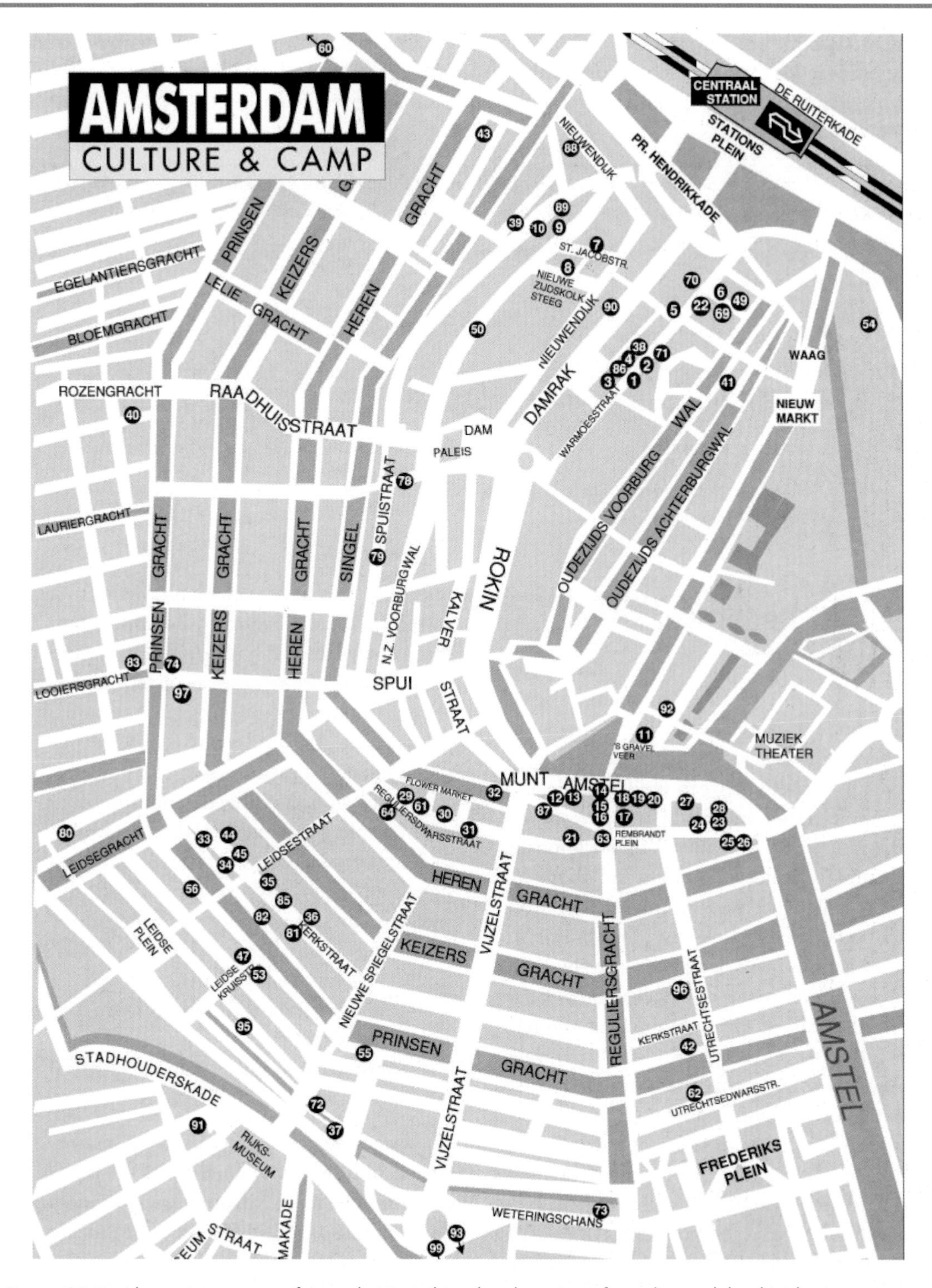

Figure 38.2 Alternative spaces of sexual citizenship: the clustering of gay bars, clubs, hotels, restaurants, shops, cinemas and fitness centres in Amsterdam. Credit: www.matchoman.com/gay/community/map

localized community that can experiment with different ways of living, which in turn is able to campaign around issues of social justice and citizens' rights. In this way geography becomes critical to the establishment of new forms of participatory citizenship.

SUMMARY

- The concept of citizenship is inherently geographical, hinging around social and spatial inclusion and exclusion.
- Such inclusion is never straightforward and is riddled with divisions.
- These divisions create differential spaces where different degrees of citizenship are experienced and played out by different groups in society.
- These spaces can be positive as well as negative, and can be used for resistance as well as control.

The changing geographies of governance

The renewed interest in issues of citizenship within both academic and policy circles can be viewed as evidence of a rethinking of the relationship between individuals, the communities they are part of and government. Citizenship lies at one pole of this relationship, concerning as it does the rights and obligations of those being governed. At the other pole are the processes and institutions of governmen, and the actions of those who do the governing. And just as new forms of citizenship have emerged as part of this rethinking, so too have new forms of government. Indeed, so prevalent are these new forms that the term governance, and not government, is now used to describe these new structures. The use of governance is widely accepted across a variety of academic and practitioner circles. It is also routinely heard in the speeches of politicians from across the political spectrum. Put simply, it 'refers to the development of governing styles in which boundaries between and within public and private sectors have become blurred' (Stoker, 1996: 2). Thus the term governance is not simply an academic synonym for government. Its increasing use signifies a concern with a change in both the meaning and the content of government. As Rhodes puts it (1996: 652–3), the term is now used to refer 'to a new process of governing, or a changed condition of ordered rule, or the new method by which society is governed'. Where government signals a concern with the formal institutions and structures of the state, the concept of governance is broader and draws attention to the ways in which governmental and non-governmental organizations work together, and to the ways in which political power is distributed, both internal and external to the state.

The way in which we are governed touches all aspects of our lives. It determines the type of education we receive, from nursery school to university. It dictates the level of healthcare we are provided with, and it concerns the provision of housing and the provision of jobs and training. It concerns planning and environmental issues, as well as transport and social services. In the UK, and indeed in many western European countries, for much of the post-war period these services were mainly provided by central, regional and local governments. These elected institutions would set the policy agenda and be largely responsible for the delivery of services. The new structures of governance that have emerged during the past three decades have transformed this system into one that now involves a wide range of agencies and institutions drawn from the public, private and voluntary sectors. They will still include the institutions of elected government, at central, regional and local level, but will also involve a range of non-elected organizations of the state, as well as institutional and individual actors from outside the formal political arena, such as voluntary organizations, private businesses and corporations and supranational institutions such as the European Union (see Goodwin and Painter, 1996: 636). The concept of governance focuses attention on the relations between these various actors, and crucially from our perspective, draws attention to the complex geographies now involved in the act of governing any particular area. The governing of localities is no longer exclusively, or even mainly, a local matter, but instead is a complex, fragmented and multi-scale process.

Geographers have investigated these issues at a variety of levels. Some have studied the international scale and looked at the emergence of multinational or global forms of governance. Others have looked at the way subnational or regional governance is developing, both in Britain and elsewhere, and many have looked at the changing nature of local governance, in both urban and rural areas. Some common threads can be discerned running through many of these studies. In particular the new emphasis on governance raises the following questions:

- The purpose of the new governing mechanisms – how and why were the particular agencies involved brought together, and what are their interests and rationales?
- The effectiveness of the various agencies involved at working together – how does each blend its particular capacities with the others?
- The links between different forms of governance and different rates of economic and social development.
- How do the new mechanisms of governance cope with uneven development and with the specific problems that might emerge in one place rather than another?
- The nature of democracy and accountability in non-elected agencies – just how does the public influence what are in the main unelected and appointed institutions, and how is the declining scope of elected state activity squared with the fashionable political notions of inclusion and empowerment?

Governance and devolution

The geographies of governance are constantly evolving, as new institutions and agencies are set up in response to a variety of social, economic and political pressures. Crucially for geographers, they are often established to cover different territories and operate at different scales than their predecessors. A good example of this can be found in the recent evolution of governing structures in the UK, initially triggered by the devolution of powers to Wales, Scotland and Northern Ireland in the late 1990s. At the broadest level devolution refers to the transfer of powers between different levels of government. Through the UK's devolution

Major devolved powers		
Scotland	Wales	Northern Ireland
Agriculture, forestry and fishing	Agriculture, forestry and fishing	Agriculture
Education	Education	Education
Environment	Environment	Environment
Health	Health and social welfare	Health
Housing	Housing	Enterprise, trade and investment
Justice, policing and courts	Local government	Social services
Local government	Fire and rescue services	Justice and policing
Fire service	Highways and transport	
Economic development	Economic development	
Some transport		

Table 38.1 Summary of the main policy areas devolved as part of the UK's asymmetrical devolution settlement

settlement new governance arrangements were put in place in England, Scotland, Wales and Northern Ireland. Crucially, these were different in each country, and as a result the UK has established what has become known as 'asymmetrical' devolution. Although new state structures were introduced across the UK, they were given varying competencies and powers (see Table 38.1 and Figure 38.3)

In addition to having authority over different policy areas, the governments of each territory in the UK have different legal competencies. The Scottish Parliament has primary legislative competence (i.e. law-making powers) over non-reserved matters, as does the Northern Ireland Assembly. However, Scotland has tax varying powers (by 3 pence in the pound up or down) while Northern Ireland does not, and in the latter there is a distinction between reserved matters which may be devolved at some point in the future and excepted matters which are kept permanently for Westminster. Wales also has a governing Assembly, but this has no primary legislative powers. Instead, Legislative Competence Orders were introduced by the Government of Wales Act 2006 and are a type of secondary (or 'subordinate') legislation which transfer powers from the UK Parliament to the Assembly in defined policy areas – though still subject to the veto of the Secretary of State or the Westminster Parliament. To make matters more complex, the Northern Ireland Assembly was suspended between October 2002 and May 2007 when formal power was transferred back to the Northern Ireland Office at Westminster (although the Northern Ireland civil service remained a strong influence throughout this period). England saw two elements of devolution: The Greater

(a)

(b)

Figure 38.3 The new institutions of governance in the UK. (a) Welsh Assembly Building; (b) Scottish Parliament; (c) Northern Ireland Assembly building, Northern Ireland Assembly Building at Stormont; (d) Greater London Assembly Building. Credit: (a) and (b) Getty Images; (c) Robert Paul Young (Creative Commons); (d) J.P. Lon (Wikimedia Creative Commons)

Figure 38.3 continued

London Authority, consisting of an elected Assembly and Mayor, was established in 2000, while eight regional development agencies (RDAs) were set up in 1999 and given power over economic development in their regions. These however were abolished on 1 July 2012.

The result was a completely new geography of governance in the UK. However, geographical reconfiguration did not stop at the level of England, Wales, Scotland and Northern Ireland. Indeed, a key feature following the initial establishment of the new Scottish Parliament and Welsh and Northern Irish Assemblies was further rounds of geographical change to the governance structures *within* each of these four nations. The result was not four territories of governance across the UK, but many more as the four nations began the task of setting up their own local and regional institutions. In the policy area of economic development for instance, this governance restructuring produced nine economic regions in England, four in Wales and two in Scotland, alongside sub-regional and local partnerships in each territory (see Goodwin *et al.,* 2012 for details). This differentiation is matched in other policy areas and the UK is now covered by a complex array of governing institutions and mechanisms, which vary from place to place and across different areas of service provision.

SUMMARY

- The concept of governance refers to new ways of governing society involving a range of participants drawn from the public, private and voluntary sectors.
- The new mechanisms of governance are operating at a variety of scales, from the local to the global, and these scales are constantly changing as new structures and institutions are put in place.
- In the UK, the devolution of power to Wales, Scotland and Northern Ireland has ushered in a host of changes to the governance structures within these countries.

Conclusion: spaces of citizenship and governance

There are very real concerns over the nature of contemporary citizenship and governance that promise to keep these issues at the forefront of geographical inquiry. Official policy statements, concerning many areas of economic and social development at all levels from the local to the European, emphasize the important role that is envisaged for new institutions and networks of governance operating beyond the formal structures of government. Yet once these new mechanisms of governance are in place they raise a number of critical questions about participation and citizenship. In many instances the new structures of governance, especially those that involve the private sector, are blurring the distinction between the public as citizen and the public as customer and consumer. Yet the citizen has rights that differ from the customer, and in the long term the legitimacy of the new structures of governance will rest on the granting of consent and support from the public. Moreover, if groups are unable to achieve their goals through these formal citizen rights, they will often set up alternative forms, and spaces, of participatory citizenship, which offer a chance to build new communities of interest based on shared and collective identities that lay beyond that of the formal citizen.

A crucial issue, however, is the territory to which the citizen feels they belong. Increasingly, as we saw at the beginning of the chapter, people are identifying not with the established nation-state but with entities that are either supranational (such as the EU) or subnational (such as the region or the locality). The debates over devolved government in Scotland and Wales, over elected mayors and regional assemblies in England, and over an increasingly integrated Europe are all evidence of this. As this uncertainty indicates, the very complexity of the new structures of governance may well ensure that all kinds of new spaces of citizenship will emerge to replace those previously linked only to the nation-state.

As we saw, many of these spaces will be created and defined by those who presently feel that they can only exercise partial citizenship. These kinds of redefinitions point to the ways in which the processes of citizenship and governance will continue to impact upon one another, and they indicate that these issues will be of continuing importance for the political geographer.

DISCUSSION POINTS

1. Which area, or country, do you identify with? To which place do you feel you belong? Of where do you consider yourself a citizen? Why do you think you have this identity?
2. Can you identify in your area of residence the formation of new forms of participatory citizenship? Who is involved and why did they seek these new spaces of engagement?
3. How is your local city governed? Make a list of the agencies and institutions that are responsible for running the city and think about who is involved in them.
4. Which groups are excluded from these agencies? What does this tell you about the structures of power in contemporary capitalism?

FURTHER READING

Smith, S. (1989) Society, space and citizenship; a Human Geography for the 'new times'? In: *Transactions of the Institute of British Geographers* 14(2), 144–56.

This is the paper that set the original agenda for the study of geography and citizenship and it is still worth reading for the suggestive links it draws between geography and social justice.

Painter, J. and Philo, C. (1995) Spaces of citizenship. In *Political Geography*, 14(2).

This is a special edition of the journal which contains nine separate articles on various aspects of geography and citizenship, and gives a strong sense of how geographers first began to engage seriously with other literatures on citizenship.

Esin, I. and Turner, B. (2002) *Handbook of Citizenship Studies*. London: Sage.

This is still the definitive collection of papers on citizenship and covers a huge range of contemporary issues and debates. It is useful in shedding light on the perspectives of other disciplines outside geography.

Goodwin, M., Jones, M. and Jones, R. (2012) *Rescaling the State: Devolution and the Geographies of Economic Governance*. Manchester: Manchester University Press.

This is an empirical study of the changing geographies of economic governance across the UK following devolution.

Peck, J. and Tickell, A. (1995) *International Journal of Urban and Regional Research*, 19(1).

Peck, J., Tickell, A. and Cochrane, A. (1996) *Urban Studies*, 33(8).

Peck, J. and Tickell, A. (1997) *Transactions of the Institute of British Geographers*, 21(4).

For excellent case studies of the emergence of new structures of local governance in the UK in the 1990s, see any one of a series of papers by Jamie Peck and Adam Tickell on urban politics and urban governance in Manchester. Articles by them on this theme appear in all of the above journals.

SECTION NINE

POPULATION GEOGRAPHIES

Introduction

The study of populations and their spatial variation has long been a concern for the Human Geographer. Indeed, it could be argued that the distribution, composition, migration and growth or decline of a population provide the basic building blocks underpinning the shape and structure of all communities. And these building blocks, of course, vary from place to place – both between countries and within them. Whatever spatial scale we take, we will find differences in the characteristics of any given population – measured by aspects such as gender, age, marital status, ethnic membership or religious affiliation. These differences themselves are the result of a whole host of social, political, cultural and economic processes. Even changes in the most 'simple' measure of population – the total number of people living in any given area – are the net result of changes in birth rates and death rates, in fertility and mortality, in patterns of health and well-being, and in the rates of in-migration and out-migration. The differences are quite widespread. Within the EU, for example, data from the Office for National Statistics shows that between 2001 and 2011 the populations of England, France, Italy and Sweden all grew by just over 5 per cent, whereas that of Germany declined by 1 per cent. At the extremes, the population of Latvia declined by 12 per cent and Bulgaria by 9 per cent, at the same time as those in Luxembourg, Ireland and Cyprus all increased by over 15 per cent. But if a population grows, it could be because elderly people are living longer, or fewer people of various ages are dying, or because more children have been born, or because more people have moved into the area – or, as is usually the case, a combination of all these. Trying to account for why such changes are happening takes us to another level of analysis – why are people living longer? Why are more babies being born? Are more people moving in to an area for economic reasons – to work, for instance, or are they moving for social and cultural reasons – say, to retire or to commute or to gain access to a school catchment area for their children? Unpacking all these elements has been at the heart of population geography.

However, although the study of populations and their variation was often considered core to Human Geography, such study has tended to be fairly descriptive, and 'population geography' became a rather specialized branch of the discipline, occupied by those with the mathematical and statistical ability to handle and manipulate large-scale population datasets. The increasing complexity of these datasets, built up through an extensive record-keeping and registration of population composition and movements, was matched from the 1960s onwards by the increasing ability of computers to store and analyse numerical information. However, in the 1980s and 1990s this stress on the descriptive analysis of quantitative information was challenged. Population geographers increasingly examined the broader sets of economic, cultural and political processes contributing to, and influenced by, the various elements of population change. As Bailey put it, 'rather than view population groups as the end-product of individual demographic events, population geography examines the two-way relationships between the acts, performances, social institutions and discourses that make up these groups and their geographic organization' (Bailey, 2005: 2). This move opened the field up to explore population change as a set of processes, rather than a set of statistics.

The three chapters in this section are all written from this new perspective. In the first, Chapter 39, Peter Kraftl looks at one of the foundational elements of any population, that of age. But rather than enumerate, categorize and describe different aspects of a population's age, he emphasizes the idea of ageing, and of ageing as a process which is shot through with social, cultural and political considerations. He uses the elderly to explore cultural issues around retirement and mobility and political issues around the provision of care and who should pay for it. He goes on to stress how age can be seen as relational and that rather than just concentrating on different age groups – children, the elderly, adolescents – population geographers should also look at how different age groups mix together and relate to each other. He concludes with a section on the politics of ageing, by exploring how various elements of age have become central to a set of political debates around education, economic development and health.

In Chapter 40, David Conradson turns his attention to health and well-being. As he notes, issues of health, fertility and mortality have long been staple to population geography, but they tended to be explored via a stress on the spatial patterning of disease and illness. Instead, Conradson emphasizes the relationship between health and place and shows how health has come to be seen as a positive indicator of well-being, rather than as the simple absence of disease. He then explores how this relationship between health and place plays out, using the examples of access to water in Bangladesh, urban neighbourhoods in the UK, Netherlands and New Zealand and therapeutic landscapes in Europe and Japan. His emphasis throughout is on the ways in which health and place are intertwined.

In Chapter 41, Khalid Koser examines migration, a topic that has long been at the centre of population geography. He explores how this has become a politically charged topic, loaded with all kinds of cultural and social meanings. Indeed terms such as 'asylum seeker', 'refugee' and 'illegal migrant' are regularly used in the popular press and broadcast media to portray migration as a negative process, yet almost always they are used in ways which alarm rather than inform. Khalid sets out to provide the information to enable you to understand and analyse migration, by looking at what we mean by migration, at what is revealed by migration statistics, and who and what are the main types of migrants. The chapter reveals the complexity behind the rather simplistic headlines which tend to dominate contemporary debates about migration. Taken together the chapters show some of the intriguing questions which emerge when we start to unpack and explore what lies behind the everyday notions of population composition and change.

FURTHER READING

Bailey, A. (2005) *Making Population Geography*. London: Hodder.

This book traces how political geography has changed and developed as a sub-discipline in response to events inside and outside academia.

Weeks, J. (2012) *Population: An Introduction to Concepts and Issues*. Belmont, CA: Thomson Wadsworth.

This is the eleventh edition of a best-selling US textbook, which covers the entire field of population studies.

Bruce Newbold, K. (2010) *Population Geography: Tools and Issues*. Lanham, MD: Rowman and Littlefield.

Another wide ranging introductory textbook, written from a more geographical perspective than Weeks' book.

The journal *Population, Space and Place* publishes articles with the latest research across the field of population geography.

CHAPTER 39
AGE

Peter Kraftl

Introduction

Ageing is something that happens to all of us in some form. Simply put, ageing describes an inescapable biological process that sees bodies grow and shrink, their capacities coming and going. In most cultures of the Minority Global North, ageing is viewed as a linear process: from birth, through childhood, into adulthood, to older age. As we age, our ways of interacting with the world around us change.

Yet it is worth stopping and thinking about the ageing process. It is important to understand the implications of 'being of a certain age' (whatever that age is), to question how ageing is experienced in different geographical contexts and to analyse how social, economic and political forces shape what it means to be a 'teenager' or 'older person'. Consider the following four (of many) ways of thinking about age and ageing:

1. Age is commonly denoted by a number. That number defines a person's legal status ('child' or 'adult'). That number is used as an instrument to control where citizens of a given country may or may not go (school, work) and what they may or may not do (smoke, drink, vote, have sex).
2. Age is one of several indicators of a person's identity. Youth cultures and subcultures – around music, fashion, sport or technology – are examples of this. But, do you have to be of a *young age* to be part of a youth subculture (see Figure 39.1)?
3. Ageing is accompanied by a series of changes in the material stuff that surrounds us. From school uniforms to wedding rings, our material belongings, however meagre, represent transitions we make throughout the lifecourse.
4. In different geographical contexts, age categories bring with them assumptions about how people should act. In several contexts, children are viewed as both angels (naturally innocent) and devils (naturally evil) (Valentine, 1996). Older people tend to be viewed quite negatively in the UK while in Japan they are viewed as 'cute'. These are, of course, generalizations, but they demonstrate that age is the product of socially-constructed values as much as natural processes (see also Chapter 43).

All of these points show that, just like other aspects of identity, ageing is a fundamentally complex process. This is one reason why age has become such an important topic for debate by Human Geographers. Yet there are two further reasons why geographers should be interested in age. First, we live in a so-called 'ageing world'. The issue of older age is

Figure 39.1 An example of a youth 'subculture'. Adrenaline Alley in Corby in the East Midlands, UK. This facility is a large indoor skateboard and BMX park. It is noteworthy because, while skateboarding and BMX-riding are viewed as 'youth' sports, many of the people using the park and taking part in its attendant sporting cultures are adults in their twenties, thirties and forties. In fact, visitors travel from all over Europe to use the facility – which rather goes against the accepted wisdom that skate parks should be built as a way of providing 'positive' activities for young people! Credit: © Sophie Hadfield-Hill

becoming a pressing concern for countries in both the Majority Global South and Minority Global North. Second, there are huge geographical variations in terms of how age matters – at global and local scales, in urban and rural contexts, and so on. Therefore, this chapter has two main objectives. It begins by exploring some of the implications (and complexities) of our 'ageing world' before going on to outline some of the main ways that geographers have analysed age and ageing in different contexts.

An ageing world?

A key way to think about geographies of age is to focus attention on the global scale of the ageing 'problem'. Certainly, national governments and international organizations have drawn upon large demographic datasets to argue for an increasingly urgent political response to our 'ageing world'. Figure 39.2 shows an example of these kinds of datasets.

Take a close look at Figure 39.2. What do you notice in terms of differences between the two continents? Are there any similarities? What do you think are the possible implications of these projected trends?

The most notable similarity is that the demographic structure of both continents is being squeezed upwards. By 2050, there will be far fewer young people and a far greater proportion over the age of 60. There are many reasons for these changes, not least improved healthcare and reduced birth rates in many countries (see Chapter 40).

There are also notable differences. The first is that despite the long-term ageing of the African population, in 2000 (and at the time of writing) nearly half of the population was aged under fifteen. A second difference is that the proportion of adults of middle age in Africa will be much greater in 2050 than in Europe. A final difference is the gender disparity in life expectancy. In Europe, by 2050, women will live on average six years longer than men (to age 84).

I also asked you to think about the potential implications of these trends – of which there are several. A key issue for many European governments has been the provision of state pensions. As the number of people of working age decreases, and the number of retirees increases, forecasters are concerned that there

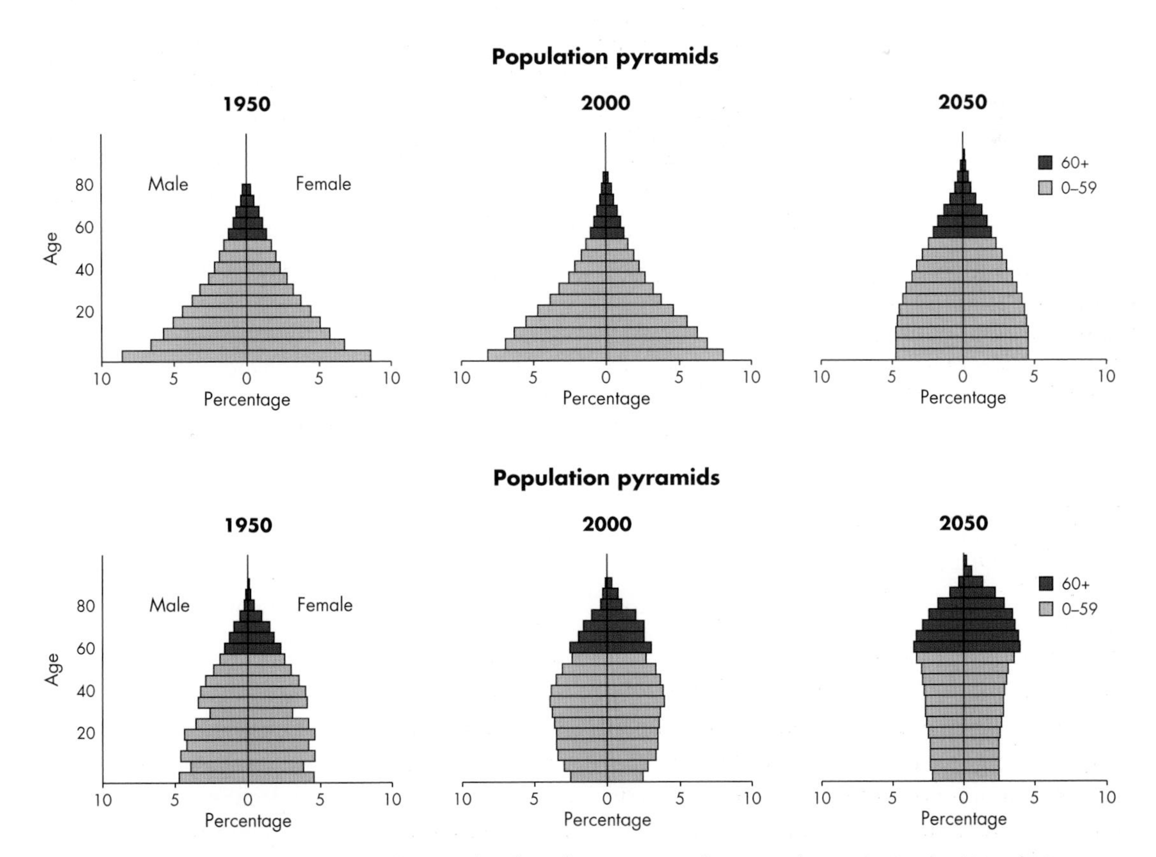

Figure 39.2 Population pyramids, by gender, for Africa (top) and Europe (bottom). Credit: United Nations Population Division (2002), *World Population Ageing* 1950–2050

will be insufficient contributions (for instance through tax) to support the pension system. Another important issue has been the provision of health and residential care for an increasingly older population – raising questions about where such care should be located, what it should look like, and who should pay for it.

Reher (2004) poses the question of whether these will become important issues as the demographic structure shifts upwards in countries outside Europe. He argues that:

> [the] breakneck pace of the demographic transition among the countries who initiated their own transformations more recently may not leave these countries with the same sort of margin for economic growth and transformation [as European countries]. For them, a situation of explosive population growth with extremely young populations may become one of rapidly ageing populations and diminishing supplies of labour in very rapid succession . . .
>
> (Reher, 2004: 34)

While the provision of pensions and healthcare are clearly pressing issues, geographers have explored other implications of ageing societies (see Andrews *et al.*, 2007). A prominent concern has been the mobility of older people. For instance, King *et al.* (1998) explore international retirement migration (IRM), arguing that it is becoming an increasingly significant component of **migration** processes in Europe. Their study focused on British retirees moving to four areas in southern Europe – the Costa del Sol, Algarve, Tuscany and Malta – and revealed several notable findings.

- Among migrants themselves, the experience of moving to a new country was varied. A significant proportion reported increased life satisfaction and improved consumption/ leisure possibilities. At the same time, many retirees reduced their (nearby) family and kinship support networks with particular implications as they moved from active old age to frail old age.
- As the number of migrants has grown in scale, and those migrants live longer, there may be increasing demands upon health and welfare provision – in addition to those of 'host' populations – demands that are concentrated in particular regions of countries like Spain and Italy.
- Residential migration has given rise to a number of spatial forms of different sizes: from converted farmhouses in Tuscany, to purpose-built *urbanizaciones* (residential estates) on the Costa del Sol.

Broadly speaking, geographers can also expose some of the contextual factors that affect the relationship between mobility and old age. For instance, Stockdale (2011) argues that while we know much about older people's mobility in urban areas, we know little about those living in rural areas. The lack of support networks experienced by many older people who can no longer drive (Schwanen and Páez, 2010) may be doubly problematic for those living in relatively remote rural places. Schwanen and Páez (2010) also make the important point that we need to move beyond the idea that mobility simply decreases with age, usually because of biological changes to the body. Instead, we need to understand the two-way relationship between older age and mobility. Schwanen and Páez (2010) discuss the effects of national schemes that evaluate the aptitude of older people to drive a car, usually once they reach a specific chronological age:

> . . . all [such schemes . . .] are rooted in a particular, biological discourse about ageing, according to which ageing is a process of inevitable incremental decline in people's [. . .] abilities. [But] by defining who is able to drive and by promoting certain driving practices in people depending on their

bodily competencies, these schemes enact old age in a certain way: they co-create aged mobile subjectivities.

Schwanen and Páez (2010: 593)

SUMMARY

- Age and ageing are the result of biological and social processes. We can understand age in relation to identities, material cultures, laws and much else besides.
- We live in an ageing world – but there are global disparities in the nature and rate of that process. These disparities have implications for labour markets, pension provision and health care.
- Older age affects (and is affected by) personal mobility. Age may restrict mobility but social rules about car driving, for instance, may also construct particular assumptions about old age.

Relational geographies of age

So far, we have concentrated primarily upon research that looks at one particular age group: the elderly. This is understandable given the trends displayed in Figure 39.2. An equally justifiable development has been the huge increase of work on children's geographies (see Chapter 43), not least given the large proportion of under-15-year-olds currently living in Africa. The problem, however, as Hopkins and Pain (2007) point out, is that these two age groups tend to be studied in isolation. The effect is that interactions between different age groups tend to be downplayed (Vanderbeck, 2007).

Hopkins and Pain (2007) make a powerful case for what they call more 'relational geographies of age'. They suggest that geographers could focus on three features of age and ageing that emphasize the relations *between* different age groups:

1. Intergenerationality: how people of different age groups mix, care for, conflict or collaborate in their everyday lives.
2. Intersectionality: how other kinds of identity-categories – like ethnicity, dis/ability, religion or socio-economic class – cut across age and may offer important commonalities between people of *different* age groups.
3. The lifecourse: how an individual does not experience 'phases' of life in isolation – one's life is a flow through childhood and adulthood, marked by complex transitional periods.

In this section, I will look at each of these ideas in turn to outline how Human Geographers have developed studies of age. I finish the section by looking at a couple of broad criticisms of these approaches.

Intergenerationality

The first idea is *intergenerationality*. While many geographers have explored intergenerational relations in outdoor spaces (such as Tucker, 2003), I want to look at a different social space: the family (Holloway, 1998; Hallman, 2010). Anna Tarrant's (2010) work, for instance, looks at the social geography of grandparenthood. She argues that grandparenthood brings with it lifestyle changes – like making time and space for new relationships with their grandchildren. She focuses on the body as a space where intimate intergenerational relationships take place. Sometimes these can be as welcome as they can be awkward to come to terms with, as two quotations from participants in her study demonstrate:

> James (age 62): she [his wife] performs some of the functions [and] I do the others, so for instance [my wife] won't change nappies so that immediately becomes my job.
>
> Mervyn (age 72): [his grandson] puts his arms round me and I kiss him, but I kiss him on the top of the head because I think it would be too embarrassing if I were to kiss him on the cheek.
>
> (Tarrant, 2010: 193)

Valentine (2008) argues that these kinds of intimacy are crucial for understanding the identities, responsibilities and forms of care that take place in sites like the family home. As Valentine argues, practices of intimacy vary over time and space and, indeed, across the lifecourse (one wonders if Mervyn was so embarrassed to kiss his own children when they were young).

Returning to the African context is helpful here. If you remember, nearly half of Africa's population is aged fourteen or under. Simultaneously, inhabitants of several African countries are affected by illnesses like HIV AIDS. This combination of factors means that many children are becoming carers for other family members. In fact, Evans, R. (2010: 1478) suggests that 'well over 4% of children in many African countries may be regularly involved in caring for parents, siblings and relatives'. She provides examples of the kinds of care in which African children are involved (see Table 39.1). The implications of such work are not just micro-spatial – in contexts with already over-burdened healthcare systems (including countries outside Africa), children can be a key, national resource for the provision of care.

Caring activity	Examples
Household chores	Cooking, washing dishes, sweeping, cleaning, fetching water or firewood
Health care	Reminding family member to take medication, giving/collecting medication, assisting with mobility
Personal care	Washing/bathing/dressing relative, assisting to eat or use the toilet
Child care	Getting siblings ready for school, supervision, help with school work
Self-care	Personal care of self, school work, training, developing livelihood strategies
Income generation	Cultivating crops for sale, rearing livestock, working in a factory, shop or bar
Community engagement	Maintaining social networks, cooperating with neighbours and NGOs, participating in community meetings and events

Table 39.1 Selection of socio-spatial and embodied dimensions of children's care work in Africa. Source: Adapted from Evans, R. (2010: 1481)

Intersectionality

The second key conceptual frame is *intersectionality*. It is particularly powerful because it reminds us that, for instance, it is impossible to place all children in the same category (Matthews and Limb, 1999). What it means to be a young person may vary substantially depending upon that child's **gender**, religion or **sexuality**. Linda McDowell's (2003) work on young working-class men was a relatively early example of this work in geography. She showed how a *combination* of factors led to affect their aspirations for future employment: from the expectations of working-class families upon young males, to their senses of masculine identity that were reinforced through the places they frequented in their daily lives (including the workplace).

In another study, Hopkins *et al.* (2010) explored the complexities of young Scottish Christians' understandings of what it meant to be 'religious'. The study group was therefore defined not just by age, but by religion and geographical location, and showed how young people adopt and transmit particular religious practices (like worship, prayer or song). Their work also speaks to intergenerational concerns. For instance, they showed that 'the places of intergenerational religious interaction [are] being shifted away from the home and the church' (Hopkins *et al.*, 2010: 325) to spaces like the car (en route to church) and youth clubs. These new spaces are also arenas in which young people can assert their own control over religious practices and even reverse traditional assumptions that religious practices are 'handed down' generations – several young people introduced practices to their parents.

Lifecourse

A third way of studying age is to look at the *lifecourse*. Bailey (2009) argues that:

> . . . interested in patterns of order and orders of patterns in the often banal practices of everyday life, lifecourse scholarship seeks to describe the structures and sequences of events and transitions through an individual's life.
>
> Bailey (2009: 407)

Thus, it is possible to discern three features of lifecourse scholarship (after Hopkins and Pain, 2007; Bailey, 2009):

1. A focus on individual biographies: on the transitions that, for instance, mark an increase (or decrease) in a person's mobility.
2. An interest in how individuals synchronize their lives in relation to others – how, for instance, adult children and elderly parents may decide where to live upon the event that the latter reach a certain age (also Pettersson and Malmberg, 2009).
3. A critical examination of the inequalities that present themselves over the lifecourse – for instance, how health inequalities between different socio-economic status groups may start in childhood but become accentuated with age.

Geographers have also attended to lifecourses that do not conform with predominant expectations in countries of the Majority Global North (Jeffrey and McDowell, 2004). Research by Butcher and Wilton (2008) shows how young people with intellectual disabilities remain, in essence, 'stuck in transition'. While those young people may take part in educational courses designed to enable them to secure paid work, a shortage of relevant jobs often means that 'youths spend considerable time in "transitional spaces", such as the vocational training centre, sheltered workshop, and supported placements' (Butcher and Wilton, 2008: 1079). In a different context, van Blerk's (2008) research in Ethiopia looked at unconventional routes to adulthood taken by 60 children living in **poverty**. Those children had moved from agricultural regions of Ethiopia to the cities of Addis Ababa and Nazareth to find work. Van Blerk showed how sex work provided a risky, although frequently successful, way for children to achieve the independence that is often symbolic of adulthood. Both studies show that geographers are steadily acquiring a more nuanced picture of the diversity of ways in which the lifecourse can be experienced in different geographical contexts. They also demonstrate some of the places (vocational training centres) and spatial processes (rural–urban migration) that form an inescapable part of lifecourse transitions.

These three inter-related ideas offer powerful ways to explain how identities and social spaces shape and are shaped by age. However, some geographers have offered words of caution, or, rather, sought to supplement these approaches (Horton and Kraftl, 2008). A key criticism of work on intergenerationality and intersectionality is that it remains too focused upon identifiable, classifiable traits of identity – like generation, **class** or gender. It has been suggested that there are additional ways to understand age that cannot be captured by such neat terms. For example, it is hard to neatly superimpose an identity onto some forms of playing – whether done by adults or children (Harker, 2005).

A second criticism is that work on the lifecourse still tends to assume a relatively linear, one-way, irreversible progression from young to old age. In response, some geographers have sought to understand how life simply 'goes on' in ways that involve more complex **temporalities** (Horton and Kraftl, 2006; Evans, B., 2010). For example, if you have left home and your parents still live in the same house, consider what happens when you 'go home'. Do you slip back into the same old habits you had as a child? Do certain smells evoke long-lost emotions? Do these habits and emotions form an implicit part of your life elsewhere? The implication is that transitions are incomplete and that we can – to certain extents – 'go back' as well as forwards as we muddle through life (Philo, 2003).

A final consideration, which I look at in the next section, is that there remain very good reasons for focusing upon particular age categories. This is because certain age-based traits (like youthfulness) become political issues in themselves. Moreover, these traits do not always map neatly onto numerical age categories, thus questioning the status of 'generations' in intergenerational relations. It is important to note, though, that none of these criticisms invalidates the three approaches outlined above; rather, as I suggested in the introduction, they are part of related attempts to *question* assumptions about age.

SUMMARY

- Studies of intergenerational relations have helped geographers to overcome the compartmentalization of different age groups.
- The notion of intersectionality is significant because it breaks down catch-all categories like 'child' and 'elderly'.

- Research about the lifecourse helps tie together different geographical scales – from the transitions that individuals make to the expectations made upon them by family, peers or wider society
- Some geographers have criticized ideas of intergenerationality, intersectionality and the lifecourse, seeking to supplement these with non-linear approaches that emphasize that life simply 'goes on'.

The politics of ageing

The previous section examined ways in which geographers can produce more balanced studies of ageing. In this section, I explore how age and ageing have become politicized. By this, I do not simply mean that ageing is now 'on the agenda' because of the kinds of data represented in Figure 39.2. Rather, I mean that age has become involved in politicized debates about, for instance, education, economic development and health (e.g. Kraftl *et al.*, 2012). Many geographers have taken a critical stance on these issues, attempting to pick apart the dominant assumptions that policy makers and the media are making about age. In this section, I highlight four features of that work, focusing on young age.

The first point relates to the use of *age categories as a justification* for a particular political or economic system (take **capitalism**, for instance). Several commentators have noted that childhood is far from a simple by-product of capitalism. Instead, as Sue Ruddick (2003: 327) argues, 'youth and childhood can be located at [the] literal and figurative core' of industrial capitalism. So, doing something 'for the good of children' has been used as a justification for all kinds of projects associated with capitalism. This argument can be made about policies initially *aimed* at children – like education for citizenship or IT skills training – but which have been central to ways in which societies define their future priorities (Hanson Thiem, 2009). Those priorities can be as diverse as social inclusion and economic competitiveness and may, actually, have very little to do with what is genuinely 'good' for *today's* children (Kraftl, 2012).

A second feature of work on the politics of ageing is its analysis of *age as an 'investment'*. Focusing on childhood, Cindi Katz (2008) argues that better-off families in the Majority Global North harbour deep-set fears about the future – about the economy, security or the environment. In the face of such insecurity, those families have increasingly used children as 'safety net'. Katz (2008: 10) argues that a major strategy for doing this has been to commodify children – to view them as a spectacular 'accumulation strategy'. This could happen in a number of ways (after Katz, 2008: 10-12):

- through the 'over-elaboration' of child-rearing in self-help guides, support groups, educational toys and all manner of other paraphernalia
- in the appearance of such over-elaboration at ever-earlier stages in the lifecourse – for instance in competition for 'prestigious' pre-school places
- through the practice of moving house to ensure that a child falls into the catchment area for a 'good' school
- in the phenomenon of the 'overscheduled' child – rushed from school, to play dates, to sport clubs, to dance classes.

These kinds of practices illustrate how the politics of *young* age are actually pervaded by some of the ideas we noted in the previous section: from specific kinds of intergenerational relations and practices within families, to a view of the lifecourse as a series of strategies to invest in an uncertain future.

A third way to understand the politics of ageing is to focus on the *emotional power of childhood*. Consider the following quotation, taken from the mission statement of a prominent charity that works on behalf of children:

> As a faith-based organization, every project we have has a goal of giving needed physical, medical and humanitarian help to children at risk and in need, with the ultimate purpose of bringing spiritual life and accompanying hope, with life-changing faith, strength and direction to all those who are served. . . . Children are our world's future and if they do not find hope, health and homes, their future – our world's future – is in jeopardy.
>
> Children's Hope International Foundation website (cited in Kraftl, 2008: 83)

Few people would disagree with the sentiment in this quotation. But that is precisely the point. Children – as a catch-all category – have been invested with a deep-seated feeling of hope that few people dare challenge. But, as Evans, B. (2010) argues, those same *feelings* can be used to manipulate whole populations – for instance, to garner support for tackling childhood obesity. It seems to be a winning combination: 'doing it for the kids' while 'doing it for the future'. Ruddick (2003) illustrates how the manipulation of adult emotions about childhood has questionable political consequences. Figure 39.3 shows an image of a child that may be familiar to you in style. It is similar to those used by charities working on behalf of destitute children. My intention is not to single out these charities, nor to claim that their work is intrinsically wrong. Rather, as Ruddick (2003) argues, it is crucial that we critically analyse how, in images like this, a child is singled out through a tightly framed, close-up photo that

> is not a portrait: it does not communicate the distinctive characteristics of a child. This 'child' comes to stand as the universal child of developing nations, disconnected from context [. . .]. What I am asked to consider is this child's aloneness – his/her absolute dependence on me as a funder. [. . .] The absence intrigues me: The images disturb me for their absence [of context . . .] they make wistful references to a (usually unspecified)

Figure 39.3 African toddler holding a cut-out of Africa. Credit: Getty Images

> larger project of modernization, the 'something more' that becomes implicated with the dispensing of mere antibiotics [or] the provision of clean water.
>
> Ruddick (2003: 341)

You may find the quote above a difficult passage to read, especially juxtaposed with the image in Figure 39.3. Certainly, it challenges many of the assumptions that donors in the Majority Global North have about their motivations for supporting a charity. To repeat, the argument is not that charitable work is necessarily wrong, nor that a donor's impulse to help is misguided. Rather, it is the emotional manipulation effected by the image itself that is under scrutiny here: we need to *question* the ways in which (young) age can be turned into an image that ends up making potentially problematic assumptions about places outside the Majority Global North.

SUMMARY

- Age can be manipulated for political and economic reasons; in many contexts, young age has been used as a justification for social policies and economic development.
- Among better-off populations in the Majority Global North, young age has been used as a kind of investment against an uncertain future.
- Powerful emotions can be associated with young age; these emotions may be manipulated to justify intervention into poorer children's lives as well as broader interventions into the societies in which they live.

Conclusion

This chapter has highlighted some of the diverse approaches that geographers have taken to studying age. Age cuts across various areas of the discipline – taking in population geography, development geography, cultural geography and social geography. I have also indicated that there are healthy debates about how geographers conceptualize age. However, it should be apparent from reading this chapter that these approaches complement one another and that geographers are making important contributions to the study of age and ageing. A central contribution has been to show that age denotes far more than a person's numerical age or their stage in the lifecourse. Ageing is not simply something that happens *to* people. Instead, age is as much a product of government policy making, family circumstances, economic conditions and individual feelings as it is a biological process. Thus, age is entwined with geographical processes in all kinds of ways. While assumptions about a person's (in)ability to drive may tell us something about how old age is understood, we have also seen that the idea of 'youth' may be a foundational concept on which entire political-economic systems are built. Moreover, concerns about age and ageing cut across multiple scales – from fears about our 'ageing world' to the built spaces (schools, hospitals, care homes, retirement communities) in which the segmentation of people into different generations is made concrete. Ultimately, geographers play an important role in questioning how societal assumptions about age are central to the organization of social spaces.

DISCUSSION POINTS

1. Can you think of other ways in which age 'matters' than those presented here? How are these expressed in spatial processes or particular places?
2. What are the advantages and disadvantages of living in a society that emphasizes 'youthfulness'?
3. In the context in which you are reading this book, can you describe some of the relationships between different generations? Are these largely positive or negative?

FURTHER READING

Hopkins, P. and Pain, R. (2007) Geographies of age: thinking relationally. In: *Area* 39: 287–94.

The authors make a clear case for a focus on intergenerational relations, intersectionality and the lifecourse, as well as the use of participatory methods in research on age. This paper spawned a debate with Horton and Kraftl (2008), available in *Area* 40: 284–92.

Reher, D. (2004) The demographic transition revisited as a global process. In: *Population, Space and Place* 10: 19–41.

The demographic transition model is a classic one for population geographers. Using large-scale quantitative analyses, Reher explores how contemporary global trends in ageing intersect with other trends, such as in healthcare and economic development.

Ruddick, S. (2003) The politics of aging: Globalization and the restructuring of youth and childhood. *Antipode* 35: 334–62.

An important example of a critical stance on the political uses (and abuses) of age. Ruddick explores how young age has been central to the development of capitalism, and looks at cultures of youthfulness among older generations.

Del Casino, Jr., V. (2009) *Social Geography: A Critical Introduction*. London: Wiley.

An accessible, thought-provoking introduction to social geography. Chapters 4–6 deal specifically with age categories throughout the lifecourse. The book also sets age into the wider context of other identity categories.

CHAPTER 40

HEALTH AND WELL-BEING

David Conradson

Introduction

For many people in the Western world, health and well-being are words that describe positive states of body and mind. When our bodies and minds allow us to move, think and feel in the ways we deem normal and desirable, we are likely to consider ourselves well and in good health. If some aspect of our body or mind ceases to work in the usual or expected way, however, or causes us difficulty or distress, then we may understand that experience as poor health or perhaps as a form of illness. In response, we may seek rest and, if we have the opportunity and means to do so, assistance from a health professional.

Bodily function is therefore an important part of health. In addition to anatomy and physiology, however, our bodies and minds are affected by the social, physical and built environments around us. The relationships we have with family members, friends and work colleagues typically influence our diet and exercise habits, for example, as well as whether we smoke or drink and how we deal with stress. Alongside their social characteristics, the places in which we live provide a certain quality of air and water, access to particular foods, opportunities for activity and rest, as well as some exposure to environmental pollutants. These aspects of our environment may support and promote health, but they also have the capacity to erode it.

As Figure 40.1 suggests, health is thus determined not only by age, sex and constitution but also by social networks and general living conditions. Paying attention to these more-than-individual factors is central

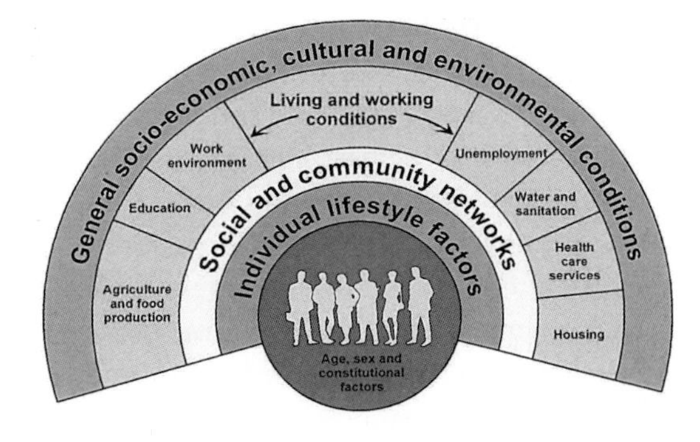

Figure 40.1 Social and community determinants of health. Credit: Dahlgreen and Whitehead (1991)

to a **socio-ecological perspective** on health (Curtis and Taket, 1996). For geographers, this perspective enables us to understand health as shaped by environments in numerous ways. This chapter considers three particular examples of health–environment interactions, looking first at disease ecology in the developing world, then at the relationship between neighbourhoods and health, and finally at therapeutic landscapes. To provide some context for these examples, the next section offers an overview of academic work in medical geography and the geographies of health.

Medical geography and the geographies of health

There have been two main strands of geographical work on health and the environment. The first, medical geography, has focused primarily on disease ecology and health services (Earickson, 2009; Mayer, 1982). Disease ecology researchers have mapped the spatial patterns of infectious and communicable diseases such as malaria, cholera and influenza. They have sought to analyse how these patterns reflect the influence of local environmental and biological factors (Learmonth, 1988). Work on health services has examined the spatial distribution of healthcare facilities, such as doctor surgeries and hospitals, and the impacts these distributions may have on community access to healthcare. The second strand of work, described as the geographies of health, emerged during the 1990s in response to the perceived limitations of medical geography. This scholarship has approached health from a more holistic biopsychosocial perspective, looking at the influence of place on health rather than analysing broader spatial patterns of disease or health services (Moon, 2009; Gesler and Kearns, 2002).

The efforts of Dr John Snow to halt an outbreak of cholera in London during 1853–54 are often considered to be an early example of disease ecology. The mechanism by which cholera spread was not fully understood at this time. After plotting the location of cholera-related deaths on a map of London, however, Snow observed a distinct clustering of deaths in a neighbourhood surrounding a particular public water pump. The local authorities were reluctant to acknowledge that cholera might be spread by drinking contaminated water, but Snow persuaded them to remove the handle from the pump, thereby preventing its further use. The pump was subsequently found to be located near a cesspit for human waste, which is likely to have contaminated its supply. A number of London water companies were also discovered to be extracting water from sewage-polluted sections of the river Thames. Although not the only factor, decommissioning the pump and publicly identifying the offending water companies coincided with a gradual reduction in the cholera outbreak. Snow's evidence-based lobbying of local authorities regarding the importance of clean water is now seen as a significant and early public health intervention. His work also underlined the analytical and political utility of disease mapping.

Contemporary disease ecology continues to map infectious diseases, although in view of the **epidemiological transition** in the west, the focus is now primarily on the developing world. Some exceptions include historical studies of cholera in the USA in the nineteenth century (Pyle, 1969) and internationally comparative work on measles (Cliff *et al.*, 1993) and polio (Smallman-Raynor *et al.*, 2006). A number of international atlases of epidemic diseases have also been produced (Cliff and Haggett, 1988; Cliff *et al.,* 2004). In terms of analytical tools, Geographical Information Systems (GIS) are widely used to examine the spatial relationships between disease incidence and environmental factors. Although early disease ecology sometimes overlooked the social and political factors that may contribute to disease outbreaks (e.g. inadequate government investment in sanitation systems), contemporary studies demonstrate a greater awareness of important political and structural conditions. Many studies of communicable diseases now reference notions of vulnerability and resilience, as well as engaging with debates in political ecology (Oppong, 2009).

Health services are the second area of medical geographical interest. Research here has examined the geographical distributions of health services within cities and regions and considered how these distributions affect community access to healthcare. Access to doctors, nurses and hospital care is important because these services have the potential to reduce the duration and impact of illness, while also promoting health (e.g. through immunization or advice and information regarding diet and activity). There are several dimensions of health service accessibility (Joseph and Phillips, 1984; Ricketts, 2010): *Geographical* accessibility reflects the distance or time a person in a particular community would incur travelling to the nearest healthcare facility (e.g. a doctor's clinic or hospital). This is often calculated using GIS models of transport networks, which are able to account for factors such as congestion as well as private and public transport options (Martin *et al.,* 2008). The *financial accessibility* of a health service reflects its affordability. When a person has to pay upfront to see the doctor, even if the cost is adjusted in view of their financial situation, some patients are likely to be dissuaded from using the service. The *cultural accessibility* of health services is a further issue. Ethnic minorities and new migrant groups are known to under-utilize primary care services (Adamson *et al.,* 2003; Smaje and Le Grand, 1997). These lower than expected utilisation rates are thought to reflect the cultural unfamiliarity which some experience – or at least anticipate – when dealing with the healthcare system in their new locales (Asanin and Wilson, 2008).

Access to health services varies not only by social class and ethnicity, but also between places. The tendency for communities with the greatest health need to be undersupplied with doctors is famously expressed in Hart's (1971) inverse care law, which asserts that:

> The availability of good medical care tends to vary inversely with the need for it in the population served. This inverse care law operates more completely where medical care is most exposed to market forces, and less so where such exposure is reduced.
> (Hart, 1971: 405)

In relatively affluent areas, where better population health typically reduces the need for healthcare, Hart noted that there was often a relative abundance of doctors (measured in terms such as the local doctor-to-patient ratio). The opposite situation tended to prevail in low-income areas, even though healthcare needs in these places were typically greater. Although factors such as diet, physical activity and working conditions are a major influence on health and illness, inability to access local healthcare services has the potential to exacerbate existing health inequalities. As Hart asserts, geographical inequalities in service provision are also likely to be greater in market-led healthcare systems, as profitability rather than population need becomes a key determinant of service provision.

These variations in the local availability of health services reflect the broader structure and nature of a healthcare system (Barnett and Barnett, 2009). As Rosenberg (1988) has noted, the organization of a healthcare system – including the roles played by state, market and not-for-profit providers in that system – is an inherently political issue. The mix of healthcare facilities in any given place typically reflects processes of negotiation, contest and compromise.

These two aspects of medical geography – disease ecology and health services – continue to be lively and productive research areas. During the 1980s and 1990s, however, they were also subject to critique and evaluation. There were calls for greater attention to issues of inequity and justice (Eyles and Woods, 1983) and to social differences in terms of ethnicity, gender and disability (Pearson, 1989; Dorn and Laws, 1994; Kearns, 1995). Instead of employing quantitative data to analyse spatial distributions of disease and illness, Kearns (1993) argued that geographers should focus on health and its relation to place and be open to qualitative as well as quantitative methodologies. Some medical geographers reacted negatively to this proposal, arguing that it went too far and risked discarding the hard-won insights of disease ecology (Mayer and Meade, 1994; Paul, 1994). Kearns (1994) responded by stating that the intent was to supplement and extend medical geography, rather than to displace or discard it.

In response to these concerns, a new geography of health emerged during the 1990s. This focused more directly on the relationship between health and specific places, rather than on patterns of illness and service provision across space. A positive understanding of health as biopsychosocial well-being was advanced, rather than equating health with the absence of disease. Qualitative research methods began to be taken seriously and there was a more proactive engagement with social theory (Kearns and Moon, 2002; Gesler and Kearns, 2002). A range of new research topics was developed, with studies examining therapeutic landscapes (Gesler, 1992; Williams, 2007), geographical gerontology (Andrews and Phillips, 2005; Andrews *et al.,* 2007), mental health (Parr, 2008; Curtis, 2010), the health experiences of women (Dyck *et al.,* 2001; Dyck, 1995), complementary and alternative medicine (Doel and Segrott, 2004; Clarke *et al.,* 2004; Andrews, 2009) and indigenous health knowledge and practices (Panelli and Tipa, 2007; Wilson and Young, 2008). Taken together, this work has made a significant contribution to our understanding of the dynamics of health and place.

Medical geography and the geographies of health are now overlapping communities of enquiry within a broader field of geographic interest in health and the environment. There are many areas of shared interest as well as points of methodological and theoretical difference. A growing number of geographers work across and between these communities, without necessarily making a strong distinction between them.

SUMMARY

- Medical geography has traditionally focused upon disease ecology, examining the links between disease incidence and local environmental characteristics, and the spatial distribution of health services. Geographic Information Systems have enabled increasingly sophisticated analyses of these issues in recent years.

- A new geography of health emerged during the 1990s, with a focus on the dynamics of health and place rather than the spatial patterning of disease and health services.
- While there remain theoretical and methodological differences between medical and health geography, these overlapping communities of enquiry share many themes and substantive foci. Their interconnectedness is signalled in phrases such as 'health and medical geography'.

Health and the environment

Having outlined the emphases of medical and health geography, we can now turn to some examples of the interactions between health and the environment. These examples – focusing on contemporary disease ecology, neighbourhoods and health and therapeutic landscapes – all draw on research undertaken by geographers.

Contemporary disease ecology: water, wells and well-being in Bangladesh

Access to clean water for drinking, washing and cooking is critical for human health. While most westerners are able to obtain clean water from a domestic tap, millions of people in the developing world have no such facility. Instead, they must gather water from wells or from streams and rivers, lakes and ponds. In densely populated regions and areas with limited sanitation infrastructure, these sources of surface water may become contaminated by human and animal waste. Those drinking from them then risk contracting diseases such as cholera, dysentery and typhoid. These diarrheal diseases are a major global health problem, with an estimated two billion cases annually that lead to the death of around 1.5 million children. Diarrheal diseases are in fact the leading cause of child mortality and morbidity worldwide, a situation which is all the more remarkable given that they are both treatable and preventable.

Diarrheal diseases have been a particular challenge in Bangladesh, as sanitation systems remain unevenly developed and surface water sources are prone to contamination by human waste. In response, around ten million shallow domestic wells – typically less than 45m in depth – have been installed across the country since the 1970s, enabling water to be drawn from underground aquifers. Over 90 per cent of Bangladeshi households now obtain their water from these shallow wells and this development has been a significant factor in the reduction of diarrheal disease in recent decades (Escamilla *et al.*, 2011).

In the 1990s, however, a number of these shallow wells were found to contain dangerous levels of naturally occurring arsenic (van Geen *et al.*, 2011). The World Health Organization guideline limit for arsenic exposure is 10μg/litre, but around a third of the Bangladeshi population were found to be drinking water that exceeded even their own national limit of 50μg/litre. Exposure to arsenic at these levels is known to cause cardiovascular disease, as well as cancers of the lung, liver and bladder (Escamilla *et al.*, 2011). There was widespread concern that the shallow well solution to diarrheal disease had inadvertently created another major health problem, albeit one likely to emerge over longer time scales. As part of the *Bangladesh Arsenic Mitigation and Water Supply Programme*:

> millions of households stopped drinking from their own shallow high-arsenic wells and switched to neighbouring wells that were low in arsenic, although often also shallow. It has recently been shown, however, that groundwater pumped from shallow low-arsenic wells is more likely to be contaminated with human waste than groundwater from shallow high-arsenic wells because of the nature of local hydrogeology combined with the high population density and poor sanitation.
>
> (Escamilla *et al.*, 2011: 521)

Those seeking to avoid arsenic contaminated water thus again risked exposure to water contaminated by human waste (van Geen *et al.*, 2011). The community response has been to install deep wells, typically of over 150m depth, so as to tap into parts of the subsurface aquifers that avoid both arsenic and faecal contamination. Around 165,000 of these wells have been installed since the mid-1990s. Because deep wells are more expensive and time-consuming to install than shallow wells, however, few individual households can afford them. Instead, these wells have typically been

Figure 40.2 Tubewells in Bangladesh. Credit: Impact Foundation (www.impact.org.uk)

located in central parts of villages, so as to maximize community access to them.

In order to determine the health impact of these deep wells in Bangladesh, medical geographers have examined how exposure to microbial pathogens (e.g. *E. Coli*, rotavirus, shingella) and arsenic varies with well type and depth (Emch, 1999; Escamilla *et al.*, 2011; van Geen *et al.*, 2011; Wu *et al.*, 2011a; Wu *et al.*, 2011b). This work has been conducted in Matlab, a rural area about 50 km southeast of the capital Dhaka, with over 100 villages and a total population of around 220,000. One aim has been to investigate how water quality is affected by well type and depth and by environmental and biological processes (e.g. water percolation and flushing within the aquifer; the capacity of bacteria such as *E. coli.* to survive in sub-surface sediments; the periodic contamination of some wells by the ingress of surface water during floods). The cultural factors which influence the use of well technologies and education around sanitation practices have also been of interest.

The researchers found that households drawing their water from deep wells had lower levels of childhood diarrhoea during 2005–06 than those using shallow wells (Escamillo *et al.*, 2011). Indeed, the *decrease* in diarrhoea rates in this period among deep-well using households was almost twice that of households using shallow wells. These findings are important, for they suggest that deep wells not only reduce exposure to arsenic but also to the microbial pathogens that can cause diarrheal disease. As mentioned earlier, however, Escamillo *et al.* (2011) note that deep wells are expensive, especially in comparison to the shallow wells that individual households have previously been able to self-fund. The ability to resource further deep wells is likely to depend upon negotiation and collaboration between Bangladeshi governmental agencies and their international development partners.

These investigations in Matlab exemplify several features of contemporary disease ecology research. The overall aim is to understand how environmental and social factors contribute to disease in a particular place or region. As in this case, the work is usually undertaken by a multidisciplinary team, drawing on both field measurements and previously gathered health and environmental data. GIS and spatial statistical techniques are commonly used in the analysis of data. When developing recommendations for future action, the researchers take account of local cultural and socio-political dynamics. In Matlab, for instance, there was recognition of the difficulties of digging further deep wells in such a resource-constrained setting.

SUMMARY

- Contemporary disease ecology considers how environmental and social factors contribute to the emergence and diffusion of disease in particular places. This work makes an important contribution to the global effort to combat infectious diseases.
- Disease ecology research is typically undertaken in multidisciplinary teams that may include health geographers, epidemiologists, public health specialists, engineers and ecologists. The data and methods are largely quantitative and GIS is commonly used to examine spatial relationships between disease and environmental factors.

Neighbourhood effects on health

In many countries there are significant health differences between people living in different places. In the affluent areas of Kensington and Chelsea in London, England, for instance, life expectancy for children born between 2008 and 2010 was 85.1 years for males and 89.8 years for females. In stark contrast, life expectancy for children born in Glasgow City in the same period was 71.6 years for males and 78.0 years for females (ONS, 2011). In other words, we expect boys and girls born in Glasgow to die more than ten years earlier than their Kensington and Chelsea counterparts. This is a striking and troubling difference, especially as it occurs within a first world country. It is not, however, uncommon. Similar degrees of health inequality have been observed in a number of developed countries (Acheson, 1998; Kawachi and Berkman, 2003; Wilkinson, 1996).

Some of these geographical variations in health outcomes reflect differences in the composition of local populations. These **compositional effects** reflect factors such as the socio-economic status, age and ethnicity profile of a community. We know that mortality and morbidity vary by age and socio-economic status, for instance, with the older and less well-off typically experiencing worse health. Place or **contextual effects** have also been observed, such that people in disadvantaged areas typically experience worse physical and mental health than those in more affluent areas, even after the influence of social factors has been taken into account (Diez Roux, 2001; Kawachi and Berkman, 2003; Macintyre *et al.*, 2002).

These place effects derive from the influence of social, built and natural environments on health. Social environments may influence health-related behaviours in a number of ways, as people typically adjust their diet, levels of physical activity, smoking and drinking in relation to local norms and culture. The density and character of built environments may also influence health, by virtue of its impact on levels of physical activity and exposure to environmental pollutants. When these factors are considered together, it is perhaps not surprising that where one lives is correlated with the likelihood of experiencing a range of illnesses, including coronary heart disease (Ellaway and Macintyre, 2010).

Researchers are often interested to disentangle place effects on health from the influence of population or compositional characteristics. Studies have also attempted to identify the causal pathways by which a particular aspect of the social, built or natural environment is able to influence health. If a clear link can be drawn between an environmental factor (e.g. relaxing age restrictions on cigarette purchases), a particular health-related behaviour (e.g. smoking) and a health outcome (e.g. lung cancer), then this provides a stronger basis for health promotion interventions.

Studies of urban greenspace and obesogenic environments highlight a number of ways in which places affect health, both positively and negatively. Urban greenspace refers to publicly accessible parks and vegetated areas in cities (Figure 40.3). These settings have been identified as a possible environmental determinant of good health. Groenwegen *et al.* (2006) suggestively refer to them as 'vitamin G', noting studies which link access and proximity to greenspace with better cardiovascular and mental health. The pathways underlying these associations are not fully understood, but proximity to greenspace may encourage physical activity, such as walking and sport, or perhaps enhance social contact and connection (Maas *et al.*, 2008; Maas *et al.*, 2009). Other research suggests that simply being able to see greenspace, irrespective

Figure 40.3 Urban greenspace. Credit: Benkid77 (Wikimedia Creative Commons)

of whether one is physically present within it, can foster mental relaxation and relieve stress (e.g. Ulrich, 1984; Hartig and Staats, 2003).

An independent health effect for access to greenspace has been observed in several places (Sugiyama *et al.,* 2008), but the strength of this effect varies socially and geographically. In a densely populated country like the Netherlands, the associations between greenspace and health are strongest for youth, the elderly and those of lower socio-economic status (Maas *et al.,* 2006). These groups may be more inclined to use greenspace or perhaps they have the greatest potential to benefit from using it. In any case, such a finding accords with Macintyre *et al.'s* (2002) observation that place effects on health are often strongest for particular social groups. In New Zealand, where residential densities are typically much lower, the positive health effect of urban greenspace is less clear (Richardson *et al.,* 2010). This may be because urban greenspace presents less of a contrast to the average residential environment in New Zealand, as many homes come with associated land and gardens.

As a less positive form of place effect, the term 'obesogenic environments' describes settings that foster obesity (Pearce and Witten, 2010). When measured in terms of Body Mass Index (BMI), the proportion of overweight people in western countries such as the USA, Canada,

Britain, Australia and New Zealand has increased significantly over the past fifty years. This is a major public health issue, as being overweight is associated with greater risk of coronary heart disease, diabetes and hypertension, as well as reduced life expectancy. Although genetic factors and diet play a role in obesity, environmental influences are also important (Lake *et al.*, 2010; Pearce and Witten, 2010). The low density nature of many North American and Australasian cities has been identified as problematic, for instance, as it encourages the use of private cars rather than more active forms of transport such as walking and cycling. The growth of relatively sedentary office jobs and the changing nature of urban food environments are also significant factors. It can be difficult to access affordable or good quality fresh food in some disadvantaged inner city neighbourhoods, particularly when supermarkets are located in the suburbs or urban periphery (Wrigley, 2002). It is also often comparatively easy to obtain energy dense fast food in many cities. In parts of New Zealand, fast food outlets have been found in clusters around schools, significantly affecting the food environment to which pupils are exposed (Day and Pearce, 2011). If governments are serious about addressing population obesity, Walton *et al.* thus argue that they should '[limit] the location of [fast] food outlets surrounding schools' (2009: 841) and find ways to encourage greater use of active modes of transport (e.g. walking, cycling, skating). These recommendations illustrate the practical implications of health geography research and its connections to policy and planning debates.

SUMMARY

- Significant variations in health and life expectancy can be observed between places. Some of these variations reflect compositional differences in local populations, but the contextual influence of physical, social and built environments is also important.
- Urban greenspace is a potentially health promoting environmental feature, though its effect appears to vary between places and social groups. Greenspace is thought to promote health because it has the potential to facilitate physical activity, social connectedness and mental relaxation.
- Work on obesogenic environments has highlighted the potentially negative health impacts of low density cities and sedentary forms of work, both of which tend to reduce physical activity. In some areas, limited access to affordable healthy food and the prevalence of fast food outlets may also be an issue.

Therapeutic landscapes

The notion of therapeutic landscapes was developed by Gesler (1992), prompted by his examination of several places with historical or contemporary reputations for healing and health enhancement. His research included the city of Bath in England, with its natural hot springs, and Epidaurus in Greece, the site of a revered healing temple in the classical world. Gesler explored how the environmental, social and symbolic dimensions of these places converged to facilitate subjective well-being for many of the people who visited them.

A number of geographers have pursued the idea of therapeutic landscapes, with studies examining mountain and forest settings, coastal and lake areas, as well as sites of spiritual pilgrimage and retreat (Williams, 2007). Attention has also been given to everyday settings such as domestic gardens, urban parks, the local beach and even home environments. In contrast to the quantitative methods and statistical analysis common in medical and some health geography, therapeutic landscape researchers have typically used qualitative methods to explore how particular places are able to foster subjective well-being.

Milligan *et al.'s* (2004) study of communal gardening among older people in the northwest of England illustrates several aspects of therapeutic landscape research. Nineteen people over the age of 65 were recruited to a study of communal gardening in two allotment sites. To obtain some pre-study baseline measures, each person was asked to assess their health and well-being and to describe their social networks. After nine months of regular allotment gardening, the participants were interviewed to explore the impact, if any, of

Figure 40.4 A Kaiyu-shiki-teien or Japanese strolling garden. Credit: Wikimedia Creative Commons

communal gardening activities on their subjectively perceived health and well-being. Those who had been gardening experienced a sense of achievement and satisfaction, as well as benefitting from the company and support of their peers. The study did not take clinical measures of physical health, such as blood pressure or capacity for independent mobility, but instead focused on subjective well-being and the embodied experiences of gardening with others.

As with the research on greenspace and obesogenic environments, work on therapeutic landscapes has potential application. For older people, for example, Milligan *et al.* (2004: 1781) note that 'communal gardening sites offer one practical way in which it may be possible to develop a more "therapeutic landscape"'. As they acknowledge, however, not everyone has the opportunity or desire to work in a garden. In addition, the inhabitants of most large cities cannot easily access forests, alpine or wilderness areas. The challenge of developing urban environments which promote mental and physical well-being nevertheless remains important. If we are able to design urban spaces that foster social connectedness and positive interaction, rather than simply maximize profitability or vehicular through-flow, then our cities might become more conducive to happiness. It is here that the research on therapeutic landscapes connects to the more applied disciplines of town planning, landscape and health architecture.

Public parks and gardens are one kind of space that has the potential to promote urban well-being. While some public parks and gardens may be considered uninviting or places of fear, perhaps especially after dark, many are also appreciated for their capacity to foster more positive thinking and feeling states. This is one reason for the prominence of these sites in many European and North American cities. In Japan there are long-standing traditions of public gardens intended for contemplation and meditation, as well as for recreation and aesthetic pleasure (see Figure 40.4). Although people inevitably experience these places in a variety of ways, they arguably represent a form of therapeutic landscape within the city.

SUMMARY

- The notion of therapeutic landscapes, developed by Gesler (1992), has become an important focus within health geography.
- A number of studies have built upon Gesler's conceptual framework to examine the physical, social and symbolic dimensions of places which people experience as health enhancing.

Conclusion

Geographers examine the connections between health and the environment from a range of perspectives. As illustrated by work on water quality in Bangladesh, disease ecology studies employ quantitative data to identify how physical and social factors contribute to disease outbreaks. When examining health services, significant geographical inequalities in access to primary and secondary care have been observed, in line with Hart's (1971) inverse care law. With the assistance of GIS and population data, recommendations for more equitable service distributions can be derived relatively quickly. In an effort to understand how neighbourhoods influence health, geographers have also studied the influence of greenspace, urban density and food retail choices. These topics all have significant policy and planning implications. With a focus on health promotion, a number of authors have then employed qualitative methods to develop Gesler's (1992) work on therapeutic landscapes. Recent engagements with the idea of well-being add a further dimension to this research (Fleuret and Atkinson, 2007; Kearns and Andrews, 2010; Atkinson *et al.*, 2012). Understanding how health and place are intertwined remains a crucial task for geographers.

DISCUSSION POINTS

1. Consider some of the different settings in your everyday life: your home, neighbourhood, places of study and (perhaps) work, and where you relax. How might these settings affect your health? Over what kinds of timescales might such health effects become visible?
2. Research on greenspace and obesogenic environments suggests that planners and policy makers should intervene in particular ways to foster healthier cities. What kinds of interventions might foster healthier cities? What factors might inhibit the development of such cities?
3. What places, if any, do you experience as therapeutic landscapes? Can you identify what makes them therapeutic for you? Is the notion of therapeutic landscapes only a luxury for affluent people in the Global North?
4. What are the strengths and limitations of quantitative and qualitative methods when examining the relationship between health and environment? Can these methods be combined in a productive fashion?

FURTHER READING

Brown, T., McClafferty, S. and Moon, G. (eds.) (2010) *A Companion to Health and Medical Geography*. Oxford: Wiley-Blackwell.

A comprehensive overview of contemporary health and medical geography, with chapters written by leading researchers in the field.

Cliff, A.D., Haggett, P. and Smallman-Raynor, M. (2004) *World Atlas of Epidemic Diseases*. London: Arnold.

A fascinating cartographic exploration of the major epidemic diseases which have impacted human life in the past and present.

Gatrell, A.C. and Elliott, S. (2009) *Geographies of Health: An Introduction*. Oxford: Blackwell.

An upper level undergraduate textbook on the geographies of health, with an emphasis on health inequalities, health services and environmental determinants of health. Attention is given to key concepts and methods in health geography.

Gesler, W.M. and Kearns, R.A. (2002) *Culture/Place/Health*. London: Routledge.

An engaging look at the intersections between culture, place and health, written by two leading health geographers.

Moon, G. (2009) Health Geography. In: R. Kitchin and N. Thrift (eds.) *International Encyclopedia of Human Geography*. Oxford: Elsevier, 35–55.

An accessible overview of health geography, describing its development and emphases, while also considering its relationship to medical geography.

Meade, M.S. and Emch, M. (2011) *Medical Geography* (3rd edn.). New York: Guilford Press.

An upper level undergraduate textbook that draws on the North American tradition of medical geography, emphasizing disease ecology, health-environment interactions and quantitative methods, including the use of GIS.

Williams, A. (ed.) (2007) *Therapeutic Landscapes*. Aldershot: Ashgate.

An edited collection on therapeutic landscapes, showing how geographers and others have used the concept to explore the healing and health-enhancing effects of a diverse range of settings.

CHAPTER 41
MIGRANTS AND REFUGEES

Khalid Koser

Introduction

Human Geography inevitably engages with issues that are controversial, emotive and politically charged. One challenge for students is to see beyond the news headlines and political 'spin' in order to understand the central issues and objectively assess the facts and figures to reach their own conclusions. Indeed these are among the key 'transferable skills' of any Geography degree.

There are probably few greater contemporary challenges in the UK in this respect than the study of migrants and refugees. In recent years **migration** has charged to the top of political agendas in the UK and elsewhere in Western Europe, and it has also become something of an obsession in the media, especially for more right-wing newspapers. Such papers uniformly portray migration negatively, and in so doing often misrepresent the realities of migration. Concepts are unclear – the terms '**asylum seeker**', '**refugee**' and '**illegal migrant**' are regularly used interchangeably. Statistics are quoted in ways that alarm rather than inform. Only a very partial picture of migration is presented – by no means, for example, are all

The large influx of refugees and asylum seekers with little grasp of English is disrupting the education of tens of thousands of our children

– the verdict of official school watchdog OFSTED

ASYLUM: THE FINAL DISASTER

A WAVE of asylum-seeker children is seriously disrupting the education of tens of thousands of pupils, a schools chief warned yesterday.

Education watchdog boss David Bell said all children suffer when there is a large influx of refugee pupils.

And the problem is so severe that some schools are facing closure unless

By **Joel Wolchover** and **Greg Swift**

exam results improve. At least 3,500 schools – mostly in urban areas – have taken in the children of asylum seekers in the past year.

And an estimated 250,000 pupils are now in classes where at least one in 10 is a refugee. The situation is more acute in London, where around one in 10 pupils in every classroom is from a refugee family.

Unveiling his annual report, Mr Bell, head of the Office for Standards in Education, said of the refugee situation: "It's disruptive for the education of all children." He added: "A number of schools do

TURN TO PAGE 9, COLUMN 1

Figure 41.1 Front page of the *Daily Express*, 6 February 2003. Credit: Express Newspapers the migrants entering the UK today asylum seekers. Overall, the real diversity and complexity of migration is ignored.

Against this background, the purpose of this chapter is to try to cut through the hype and provide students with the sort of information they require to understand and analyse migration, and hopefully to engage in reasonable debate. It focuses on four key questions:

1. What is migration?
2. What do the statistics mean?
3. How many migrants are there?
4. What are the main types of migrant?

What is migration?

The answer to this question is by no means straightforward. International migration can be considered a sub-category of a wider concept of 'movement', which embraces a whole range of forms of human mobility, from daily commuting at one end of the spectrum to permanent emigration at the other. What is defined as migration thus becomes an arbitrary choice and may be time-specific. In the UK, for example, a migrant is defined as someone who has been living abroad continuously for one year or more, whereas in the Netherlands the time period is only six months.

The concept of permanent migration has been epitomized in the idea of new lands of opportunity, perhaps typified by Australia (Castles and Miller, 2009). Today, however, what is meant by 'permanent' migration is less clear. Even people who have lived abroad for most of their lives often retain a 'dream to return' to the place of their birth and it is now relatively rare for people to migrate from one country to another and remain there for the rest of their lives. Despite this, the

CASE STUDY

Return migration

Writing in the geographical journal *Area* in 1978, Russell King lamented the lack of data, theory and analysis concerning the phenomenon of return migration (King, 1978). Almost 35 years later his comments are largely still valid – return migration remains a neglected aspect of population geography, even though millions of people return to their countries of origin each year. Return seems set to become even more important in the UK and elsewhere for some of the following reasons.

- Return is increasingly viewed as one way of reversing the so-called 'brain drain', whereby skilled people leave poorer countries to work in richer countries.
- Return is viewed as one way of combating irregular migration.
- There is growing pressure to return rejected asylum seekers.
- As conditions in their countries settle, some refugees may consider going home – in recent years it has become safer to return to the Balkans, Somaliland and Sri Lanka.
- Increasingly, work permits are only issued for short periods of time, with the expectation that migrants will return home at the end of the permit period.

phenomenon of return migration remains sorely under-studied and poorly understood (see box).

In addition to the time period concerned, there are three main ways that migrations are categorized. A common distinction is between voluntary and involuntary (or forced) migration. The Office of the United Nations High Commissioner for Refugees (UNHCR) estimates that, worldwide, there are about 15 million people (including 5 million Palestinians) who have been forced to leave their home countries, usually as a result of conflict or persecution. There are far more migrants in the world today who have moved outside their country voluntarily – over 200 million worldwide.

A second distinction that is often drawn is between people who move for economic and those who move for political reasons. The former are often described as labour migrants and the latter as refugees.

The final main distinction is between legal and illegal or 'irregular' migrants. There are at least three ways an individual might be classified as an 'irregular migrant'. First, they can enter a country illegally, without passing through an immigration checkpoint. Second, they can enter a country legally for a limited period of time, for example, with a short-term visa, but then stay illegally after the expiry date of their visa. These are also described as undocumented migrants. Third, they can be legally resident but involved in illegal activities. Worldwide it is estimated that there may be some 40 million irregular migrants.

Categorizations always simplify reality and this is true of the above migration categories in at least three main ways. First, there is some overlap *between* the different categorizations. Thus most voluntary migrants are also economic migrants, and most involuntary migrants are political migrants or refugees. Second, individuals can transform from one type of migrant to another *within* the different categorizations. A legal migrant may overstay his or her visa and become an illegal migrant or an individual might leave his or her country voluntarily but then not be able to return, as a result for example of the start of a war, and thus effectively become an involuntary migrant.

Third, the sharp distinctions drawn between migrants within each categorization are often more blurred in reality. Very few migrations, for example, are purely voluntary or involuntary. Many large corporations, for instance, consider moving staff between international offices to be part of their training. So while employees moving within, say, IBM from London to Tokyo are ostensibly moving voluntarily, they may have no option if they want to keep their job in that company (Koser and Salt, 1998). At the other end of the spectrum, even refugees have choices other than to migrate – they might, for example, stay and take a risk, or move internally to a neighbouring village or town, or join one side of a conflict or the other.

The same blurring applies to distinctions between economic and political migration. Consider the case of someone who leaves their home because they lose their job. Ostensibly they are moving for economic reasons. But what if they have lost their job because of their race or religion, or because the factory where they worked has been bombed during a conflict? In that case you might argue they are fleeing for political reasons. The analytical challenge here is to distinguish between underlying causes of migration and its immediate precipitants.

SUMMARY

- International migration can be considered a sub-category of a wider concept of 'movement'.
- There are three main ways of categorizing migration, distinguishing voluntary from involuntary, economic from political, and legal from irregular migrants.
- These categorizations oversimplify reality; there is overlap between the different categorizations and blurred boundaries within them.

What do the statistics mean?

Perhaps the most potent weapon in the ongoing debate about migration in the UK, and often elsewhere, is statistics. They are often used selectively by the press and occasionally by politicians too, in order to paint a particular, normally negative, picture of immigration.

There are three very important observations to make about statistics on migration in the UK. First, although they are the most accurate available, even official migration statistics cannot provide a complete picture of international migration in the UK. To put this rather more bluntly, not even the government can state with any certainty exactly how many people enter or leave the country each year. The most obvious reason is that official migration statistics do not include numbers on irregular migrants. Statistics on irregular migrants in the UK are no more than guesses, and really should not be used to sell newspapers or, worse, make policy (see box).

Second, there are important reservations surrounding the statistics on migration that the government does record. Most published statistics on migration into and out of the UK are based on the International Passenger Survey (IPS). This is a small sample survey conducted at sea and airports. Passengers are interviewed about their intentions of staying in the UK (or abroad, if emigrating). Those who intend to stay in or out of the UK for a year or more, having lived abroad or in the UK for a year or more, are counted as migrants. One problem is coverage: only a small fraction of the population is interviewed and the results are scaled up. Another is that people's intentions change: they may or may not stay as long as they intended. Adjustments are made to the IPS figures to try to take account of such factors.

There are two other main sources of data on migration flows. Work permits issued measure the entry of workers, but only from outside the European Economic Area (EEA). Asylum statistics show how many people apply for protection in the UK, but care is required in interpreting them, as sometimes they include dependants (spouses and children) and sometimes not. Alternative indicators of numbers of foreign migrants entering the UK include the Labour Force Survey, which records nationality and address a year ago, but again is based only on a sample of households. The national census also records address a year ago, but it does not record nationality, only country of birth, and it takes place only once every decade.

CASE STUDY

Counting illegal migrants

Counting illegal migrants is an imprecise science. One reason is conceptual – as we saw in the first part of this chapter, the term covers a range of people who can be in an illegal situation for different reasons. Another is methodological: people without legal status are likely to avoid speaking to the authorities for fear of detection, and thus go unrecorded.

Various methods have been used to try to at least estimate the size of the illegal migrant population, although it needs to be emphasized that none of these is comprehensive.

- In some countries (though not the UK) amnesties are periodically declared, whereby foreign nationals residing or working illegally can regularize their status.
- Direct surveys of illegal migrants have taken place.
- It is possible to compare different sources of recorded migration data and population data to highlight discrepancies that might be accounted for by illegal migration.
- Surveys of employers can indirectly reveal foreign workers without legal status.

A third observation, which is of course true of any statistics, is that migration statistics can be presented in different ways with different implications. In 2011 about 19,000 asylum seekers arrived in the UK. This figure can be portrayed negatively – it was an increase on the previous year in the UK and meant that the UK was the fifth largest recipient country for asylum seekers in Europe. Alternatively, it can be made to seem really rather insignificant if placed in a different context. In 2011 over 100,000 asylum seekers arrived in South Africa – almost as many as the total across all 27 European Union member states.

SUMMARY

- Statistics on migration are often used to alarm rather than inform.
- Even the UK government cannot state with any certainty exactly how many people enter or leave the country each year and recording illegal migrants is particularly problematic.
- There are reservations about the statistics even for migration that is recorded regularly.
- Depending on the way statistics are presented, migration can be portrayed positively or negatively.

How many migrants are there?

Reading the press, it is easy to run away with the idea that the UK already has been, or is about to be, overrun by migrants. But how many migrants are there in reality? As we have just seen, data problems mean that it has never been easy to say how many people come to stay in the UK or leave to live elsewhere. According to the Office for National Statistics in 2011 589,000 people arrived to live in the UK for at least a year. Once again, it is worth taking a moment to look behind this bald statistic. Media portrayals of 'floods' of immigrants rarely acknowledge, for example, that substantial numbers of people leave the UK too – it is estimated that they numbered 338,000 in 2011. This means that the net balance of migration in the UK in 2011 was 252,000. A longitudinal perspective reinforces this perspective. It was not until the mid-1980s that the UK became a net immigration country, after more than a century of losing people. Today, recorded migratory movements both in and out of the UK are greater than they have ever been.

Similarly, the media often concentrates undue attention on immigrants originating in poorer countries. In fact, many of those arriving in the UK in 2011 were British citizens, most of whom would, earlier, have been emigrants.

CASE STUDY

Key statistics relating to migration in the UK

In 2011, 589,000 people came to the UK as immigrants and 338,000 left as emigrants; the net balance was 252,000.

In 2011, 250,000 foreign students entered the UK to begin their first year of study in a higher education institution.

In 2011, about 19,000 applications were received for asylum, an 11 per cent increase on the former year. The main countries of origin of applicants in 2011 were Pakistan and Sri Lanka.

SUMMARY

- It is important to consider how many migrants leave the UK each year as well as how many arrive, in order to arrive at a figure for net immigration.
- A good proportion of those arriving in the UK each year are British citizens.
- Immigrants arrive in the UK from around the globe, and not just from poor countries.

What are the main types of migrant?

Political and media discourse in the UK tends towards a simplistic view of international migration. It focuses mainly on asylum seekers, and to a lesser extent, irregular migrants, and often confuses the two categories. This section is concerned with trying to introduce some clarity to the asylum debate.

The first point to emphasize is that most immigrants in the UK are not asylum seekers. In 2011 about 25,000 people, including their dependants, applied for asylum in the UK. That compares with about 250,000 foreign students who entered the UK to commence their first year of study and hundreds of thousands of people who came to work. The media – and more surprisingly the government – rarely publicize the immigrant doctors and nurses who help maintain the National Health Service, or the foreign students whose fees help fund universities, or the highly skilled migrants who contribute to the financial and industrial sectors in the UK.

The second point is that the category asylum seekers, just like all other migrant categories, covers a wide diversity of people. Asylum seekers are not all the same. They come to the UK for a variety of reasons, they originate in a wide range of countries, they have diverse educational backgrounds and skills, some come alone and others with family members and there are increasing numbers of unaccompanied children claiming asylum (Bloch, 2002).

We have seen that media coverage fails to place asylum seekers in the context of other immigration in the UK and underestimates the diversity of people who are asylum seekers. Perhaps most importantly, however, there is a lack of clarity over precisely who is an asylum seeker. An asylum seeker is simply a person who has applied for protection in the UK. Most do so upon arrival in the UK, although it is also possible to apply for asylum in the UK at a British Embassy in another country.

CASE STUDY

Who is a refugee?

The most widely used definition of a refugee is that employed in the 1951 Convention relating to the Status of Refugees, according to which a refugee is someone who '. . . owing to a well-founded fear of being persecuted for reasons of race, religion, nationality, membership of a particular social group or political opinion, is outside the country of his nationality . . .'.

One observation worth making is that this is an explicitly geographical definition: you need to be outside your own country in order to qualify for refugee status. The many millions of people who have fled for reasons similar to those that displace refugees, but have not left their countries, are referred to as internal displaced persons (IDPs).

Many critics argue that it is not just IDPs, but many other people of concern that are also excluded by this definition. It does not, for example, cover people who have been persecuted on the basis of their sex or sexuality.

Asylum seekers' applications are judged by the criteria of the 1951 United Nations Convention relating to the Status of Refugees (see box).

In the UK over the last decade, about 10–20 per cent of asylum seekers have been considered to satisfy the criteria in the Convention and are granted refugee status. A further 20–30 per cent do not satisfy the Convention criteria, but are granted 'Exceptional Leave to Remain' because it is accepted that it is currently unsafe for them to return to their country of origin. This means that somewhere in the region of 50–70 per cent of asylum seekers are not recognized as in need of any form of protection. Those rejected have the right to appeal and some are subsequently granted protection. The majority have their appeals rejected and are then expected to return to their countries of origin. Many, however, do not, and remain in the UK illegally.

It is simply misleading, it follows, to conflate the categories 'asylum seeker' and 'irregular migrant'. Some asylum seekers arrive in the UK illegally, others break the law once they are here, and many rejected asylum seekers stay on without authorization. But the majority of asylum seekers are not also irregular migrants. And, conversely, most irregular migrants in the UK are not asylum seekers.

SUMMARY

- Media and political discourses ignore the true complexity and diversity of immigration.
- The majority of immigrants arriving in the UK are not asylum seekers, and even within the single category 'asylum seeker' there is great diversity.
- Far clearer and more consistent terminology is needed in the asylum debate.

Conclusions

This chapter has tried to show the complexity of migration flows into and out of the UK. People come for different reasons, stay for different periods and fulfil different roles. There are inflows from all parts of the globe. Simultaneously, thousands of people each year return to their country of origin or migrate elsewhere. And British citizens are part of the movement in both directions.

The UK is not being 'flooded' with immigrants from the poorer parts of the world. Most immigrants are not asylum seekers and most asylum seekers are not irregular migrants.

If a rational debate about migration is to take place, it is essential to consider the totality of population movements and not to focus on one group, still less to demonize it. The debate needs to involve, among other things:

- considering the definition of a migrant, and using the word in a clear, consistent and non-discriminatory way
- understanding the value and limitations of statistics
- having a balanced overview of migration patterns and trends in the UK
- considering all the different strands of migration, and the circumstances and contributions of each.

DISCUSSION POINTS

1. What is migration?
2. Why is migration so hard to measure?
3. Who is a refugee?

FURTHER READING

Bloch, A. (2002) *The Migration and Settlement of Refugees in Britain*. London: Palgrave.

Focusing on refugees from Somalia, Sri Lanka and the Democratic Republic of Congo, this book demonstrates the wide diversity of the asylum-seeking experience and analyses the impact of UK policy.

Castles, S. and Miller, M. (2009) *The Age of Migration: International Population Movements in the Modern World*. Basingstoke: Macmillan.

This is probably the key text on international migration. It places contemporary migration in historical and global perspective, and analyses the causes and consequences of immigration for receiving countries.

King, R. (1978) Return migration: a neglected aspect of population geography. In: *Area*, 10, 175–82.

Although dated, this brief article by geographer Russell King is still the best overview of return migration.

Koser, K. (2007) *International Migration: A Very Short Introduction*. Oxford: Oxford University Press

This short volume explains why migration matters and provides an overview of recent trends and current debates

UNHCR (annual) *State of the World's Refugees*. Oxford: Oxford University Press.

This is an annual review of the world's refugees, providing the most authoritative data and analysis available.

SECTION TEN

SOCIAL GEOGRAPHIES

Introduction

Social Geographers investigate the relations between geography and our social relations, institutions, attitudes and experiences. Social Geography is that part of the discipline where the dimensions along which society is produced and structured are brought to the fore. Gender, age, class, ethnicity, race, sexuality, dis/ability are all issues that this section of the book will discuss.

Our route into these issues is the central question of identity: who are we and how do we differ from others? One understandable response to that question is to emphasize an individualized sense of identity. You are you, unique, not exactly like anyone else. Most of us resent being reduced to crude social determinants of class, gender, race, age, sexuality, nationality and so on. While we may often be guilty of stereotyping others, we don't want it done to us. On the other hand, with just a little reflection, it is also fairly obvious that we only possess and express our individual identities through wider social dimensions: through the relations we have to others (the three of us writing this could not be lecturers without students, for example); through the organizations and social settings that we inhabit (our families, our friendships, our work, our haunts); through our senses of collective identification. Strip all that away and most of us would fall apart as individuals. Identity is a social production.

Think about our gendering. It is quite right to question simplistic judgements about all men or all women having certain kinds of characteristics. But being women or men (or something beyond these options) is something that we all have to deal with. It is a part of what makes us who we are, both to ourselves and others. In conjunction with our class, ethnicity, age, sexuality, dis/ability, and so on, it structures our lives, too, impacting on how we are treated, the life chances we have, the expectations placed upon us. Generally, then, we cannot be ourselves outside of these social dimensions; and these social dimensions have crucial effects on who and what and where we can be.

The 'where' matters here. Social Geographers are especially interested in understanding how social processes operate through spaces, places and environments. Social Geography goes far beyond detailing the spatial distributions or environmental impacts of social groups, while leaving direct investigation of these groups to others such as sociologists and anthropologists. Instead, it emphasizes how geography is central to the social processes and relations through which social groupings and differences are made.

The section is comprised of four chapters taking different cuts through the social geographies of identity. We begin with the idea of 'identities' and its suturing of the individual and the social. Peter Jackson, one of the founding figures of the revitalized Social Geography that emerged in the 1980s, argues against over-simplifying identity into fixed (and fixing) labels. Our identities are multiple, he suggests. They are 'relational', formed in relation to others rather than stemming from innate characteristics of our own. Identities always combine self-assertion – 'this is me' – and social judgement – 'that is you'. What is more, how we 'form' our identities is shaped by wider historical currents and processes. Many writers argue that today our identities are less something we inherit through our social and geographical locations and more something that we have to fashion for ourselves within wider social fields in which we are judged and socially positioned. Peter concludes by thinking about how complex, relational identities can be portrayed and researched. Illustrating his argument through his own research on people's 'food stories', he suggests that biographical narratives or 'life stories' are one particularly promising method, allowing an expression of how our identities are given a narrative form as individual biographies embedded within wider social and material contexts.

In Chapter 43 Sarah Holloway picks up on the argument about the relational nature of identity, and homes in on the role of socially constructed differences in shaping our experiences of space. Substantively, while Peter Jackson's examples mainly concern issues of race, ethnicity, class and gender, Sarah's chapter focuses on recent work in Social Geography on age, dis/ability and sexuality. Social Geographers explore how differences in these areas are not simply biologically determined but also socially constructed. They also emphasize two ways in which identities are geographically constituted: first, identities vary over time and space (so what it means to be a woman, say, is not fixed but varies in different places, times and contexts); and second, identities are partly defined in relation to where they belong, where they are 'in and out of place'. More specifically, Sarah's chapter sets out the state of play and current agendas in Social Geography's research on age, dis/ability and sexuality. These discussions provide the reader with clear overviews of, and routes into, these important bodies of work and the kinds of arguments that are currently shaping them. Two more general suggestions also emerge. First, Sarah argues that Social Geography's attention to 'different' identities has been invaluable, shaking up a host of assumptions about the character of the 'human' in Human Geography, but also that the sub-discipline needs to avoid being silent on those identities that are in some way seen as 'the norm'. Second, while distinctive bodies of research have emerged on different facets of identity and social difference (such as age, dis/ability and sexuality), in reality these various dimensions of identity positioning 'intersect'.

In Chapter 44, Jon May deploys the relational understandings of identity discussed by Jackson and Holloway to explore the processes and politics of 'exclusion'. Exclusion here is understood as a simultaneously social and spatial process in which some members of society are placed outside 'the norm' or the mainstream. A key writer in this area is the Social Geographer David Sibley, through his work on those cast as 'outsiders' and on the psychic underpinnings of our desires to distinguish between an included 'us' and an excluded 'them'. In the second half of the chapter Jon focuses in particular on the example of how homeless people experience such geographies of exclusion. He considers research that has shown how homeless people are excluded from 'normal' society and its urban public spaces by policies that look to 'clean up' cities and remove problematic street populations. However, Jon also points to in-depth empirical research that suggests the complexity of homeless experiences and geographies. This shows, for example, how forms of exclusion and inclusion combine in spaces such as the homeless day centres that seek to combat homeless people's exclusion from society and the city. Overall, Social Geography's research on exclusion is notable for how it combines theoretical sophistication with practical commitment and applicability.

In the last chapter in this section, Claire Dwyer brings together the previous discussions of identity, difference and exclusion in the context of a burgeoning body of work on 'diasporas'. Originally referencing a 'dispersed' people or population, the notion of diaspora highlights the migrations that have moved people and identities around the world and stretched social fields across national borders. To quote Claire, 'the term diaspora is an attempt to encompass the different and complex belongings of peoples who may be dispersed across geographical boundaries and may have connections to several different places they call "home"'. Such complex belongings, associated with combined processes of migration and globalization, could be seen as increasingly prevalent within modern formations of identity. Social geographies of identity are no longer nationally contained. Other, 'transnational' spaces can be the arenas within which identities are produced, as Paul Gilroy's seminal work on 'black identities' and what he calls 'the Black Atlantic' suggests. Claire focuses in particular on research on 'diaspora cultures' that emphasizes their 'syncretic' character, their fusing and mixing of different elements, an argument exemplified in her own work on British-Asian fashion. She also emphasizes the role for geographers in understanding what she calls 'diaspora spaces', places and landscapes within which nationalized territories are disturbed and diasporic identities developed. Here she considers both commercial spaces (think for example of the many 'Chinatowns' in North American and European cities) and religious spaces (focusing on a Hindu and a Jain temple in London). Overall, Claire shows how research on diasporas includes work on both the everyday possibilities and difficulties of living transnationally and on the challenges this makes to many geographical notions, in particular the national framing of societies and cultural identities.

Finally, we would advise that you also consult the preceding section in this book on 'population geographies', which, with its foci on 'age', 'health and well-being' and 'migration' has particularly strong resonances with the field of Social Geography too.

FURTHER READING

Social and Cultural Geography

As the title suggests, this journal is not limited to Social Geography alone, representing wider confluences of Social and Cultural Geography. A survey of recent volumes will give you a feel for the breadth of Social Geography; more generally, the journal is a principal resource for reading the best contemporary research in the field.

Del Casino, V.J. (2009) *Social Geography. A Critical Introduction.* Oxford: Wiley-Blackwell.

A slightly more advanced summary of the field. In some parts it mirrors the structure of the discussion in the chapters here; for example there is an extended discussion of the social geographies of the life-course, ranging across childhood, mid-age and older-age. In other parts, it discusses issues with a slightly different emphasis; for example, in chapters on geographies of health (covered in this book in Chapter 40), communities and organizations, and social activism and social justice.

Pain, R., Barke, M., Fuller, D., Gough, J., MacFarlane, R. and Mowl, G. (2001) *Introducing Social Geographies.* London: Hodder.

An introductory textbook that will let you read further on how Social Geographers have studied various dimensions of social difference and identity, and on how these insights have been applied to social policy debates, including crime, housing and socio-economic exclusion.

Panelli, R. (2004) *Social Geographies: From Difference to Action.* London: Sage.

This textbook shares the focus on questions of identity and difference developed within this section. It also has some portraits of the research being done by some exemplary Social Geographers.

Valentine, G. (2001) *Social Geographies: Space and Society.* London: Prentice Hall.

Rather than being organized around social categories or aspects of social identity, this book foregrounds their geographies through an analysis of spatial scales ranging from the body to the nation. Discussion focuses on how social identities and relations are constructed and contested in these spaces. The book is typically well written, by one of the leading authors in the field.

CHAPTER 42
IDENTITIES

Peter Jackson

Introduction: approaching identity

I'd like to begin with an example. As part of a project called 'Food Stories' with which I was closely involved (Jackson, 2010), we interviewed Rosamund Grant, who talked about the role that food has played throughout her life. Rosamund was born in Guyana in 1946 and came to England in the 1960s. Her brother, Bernie Grant, was England's first Black MP and Rosamund's interview gives a strong sense of food's political significance in her life, during which she has run a Caribbean restaurant and worked as a teacher. Here is an extract from her life history, recorded by my colleague Polly Russell:

> I think Europeans tend to see Caribbean food in a particular way, and they would say it's ackee and salt fish, rice and peas or curried goat and mutton . . . Or people say to me 'oh, you know, I like spicy food' and by that they mean pepper really, or they don't like spicy food, and I say 'but you can have the spice without the pepper and separate'. I like people to be clear about, you know, how they think about it because they stereotype it and 'all Caribbean food is spicy' and, do you know what I mean, and 'all of it is ackee and saltfish'. You know, and so I try and broaden out people's way of thinking. So if I'm going to cook for a European crowd, or people who want to eat Caribbean food, you know, some of what I've got over the years is kind of like, oh, you know, 'don't put so much pepper in it' or 'can we have Caribbean stew?' Well I say ok well what is Caribbean stew? Caribbean stew is not a generic you know thing, it's kind of like it can be lots of different types of ways of doing things.
>
> (http://www.bl.uk/learning/resources/foodstories/index.html)

From this short extract it's immediately apparent that Rosamund's identity is closely linked to the politics of food. She wants people to be clear about the diversity of Caribbean food and not reduce it to 'ackee and salt fish, rice and peas'. I'll have more to say about the way identities are often expressed through material things like food later in the chapter. But I also want you to notice how Rosamund Grant is addressing herself in this extract to a specific audience: 'Europeans' (or, later, 'a European crowd') who tend to see Caribbean food in a particular way. This sense of our identities being crafted with respect to a real or imagined audience is another theme that runs throughout the chapter, which social scientists sometimes describe as the 'relational' nature of identity construction, where definitions of self and other are closely connected.

Immediately following the previous extract, the interviewer asks Rosamund Grant why the diversity of Caribbean food is important to her. She replies with characteristic candour:

> It's important to me because I hate stereotyping, I hate being put in a box, I hate being limited, and I hate being seen through the eyes of Europeans. I don't like defining myself through the eyes of somebody who's white or is European. I have my own definition of myself and I think that is really important for me, you know, as a black woman to make an impression in that field 'cause this is my field, you know. And because I suppose there's a lot of passionate feeling left over from slavery and the impact of slavery and migration and displacement and all that kind of stuff, and I just think, you know, ok it's time to speak for ourselves. And so I will define who I am and I will define therefore what I'm cooking, and it's not that I'm closed about that but I feel we have been put into a box and we have been closed in, and hello folks you know we're here and this is really what this is about. And so it does irritate me when people talk about 'exotic food', you know, exotic through whose perspective, is it mine or somebody else's?
>
> (http://www.bl.uk/learning/resources/foodstories/index.html)

Within a few seconds here, the narrative moves from ethnic stereotyping (being 'put in a box' and 'seen through the eyes of Europeans') to the history of slavery and the legacy of Empire ('migration and displacement and all that kind of stuff'). While Rosamund is keen to define herself and avoid being stereotyped by others, she is deeply aware of her own positionality 'as a black woman' and of the complex histories within which her personal identity is entwined. She is clearly irritated by people who over-simplify that history, referring to exotic or spicy food, for example, without considering the perspective from which some foods are defined as exotic while others are regarded as ethnically unmarked or unproblematically mainstream, a relational issue on which others have also remarked.

As you can probably tell, I think this is a really interesting interview and there is much more to be said about the potential of life history interviewing as a means of uncovering how people's personal identities are connected to wider social histories and the role of (individual and collective) memory in this process (Jackson and Russell, 2010). There is also much more to be said about constructions of the 'exotic' and particularly how this is expressed through people's relationship with food (Heldke, 2003). But I want to move on now to invite you to reflect on your own identity and how this might reflect particular material and social geographies (including what you eat and where you come from).

Personal identity

If someone stopped you in the street and asked you about your identity, you'd probably find it hard to answer their questions. 'Identity' isn't something you can easily express in words and your answer would probably depend on who was asking the question. You would probably give rather different answers to a census enumerator, a police officer or a market researcher, for example. Your answer might also vary in different times and places. When I'm on holiday abroad, for example, I'm often quite conscious of being 'British', aware of other British holidaymakers around me and of all the other nationalities who come together for a week or two in some 'foreign' place. People's attitudes to foreign food also often serve as a cultural marker of identity, whether they are keen to try new things or resistant to all but the most familiar foods. When I'm back home in England, I often describe myself as a Londoner (although I was born just south of London in Surrey and have lived in Yorkshire for nearly twenty years). In other situations, I might be more conscious of my age or sex, or how I dress, or the kind of music I enjoy. These subjective 'lifestyle' issues are just as central to my sense of identity (who I am and how I differ from other people) as more objective issues like nationality or place of birth. All this suggests that it's usually better to talk about *multiple identities* than to try and reduce complex identity issues to a single 'dimension' (such as age or class). As the cultural historian Frank Mort once remarked:

> We are not in any simple sense 'black' or 'gay' or 'upwardly mobile'. Rather we carry a bewildering range of different, and at times conflicting, identities around with us in our heads at the same time. There is a continual smudging of personas and lifestyles, depending where we are (at work, on the high street) and the spaces we are moving between.
> (Mort, 1989: 169)

As these brief examples suggest, identity is a slippery concept. On the one hand it implies a fixed and deeply personal sense of who I am. It is about individuality and what marks us out from the crowd (in Rosamund Grant's words: 'I have my own definition of myself' and 'I will define who I am'). On the other hand, our identities are forged in relation to other people, both those who we share things with ('as a black woman') and those who we distinguish ourselves from ('a European crowd'). They are not so much essences of what we are, as performed in relation to others in specific contexts.

But it should be no great surprise that identity is such a complex and contested term once we realize the political intensity associated with ideas of identity and belonging. Identity can sometimes be, quite literally, a matter of life and death as in the case of 'ethnic cleansing' in the former Yugoslavia or the history of genocide in Rwanda (to name just two recent conflicts where 'ethnic' identities became highly politicized). Identity has been a constant source of debate within the social sciences and in Human Geography and there is no single 'right way' to approach the topic. A good way of engaging with these issues is to ask questions and think of examples of your own, relating the different theories of identity that we'll discuss in this chapter to your own experience.

Conventional approaches to identity have tended to assume that we each inhabit a particular, relatively fixed, identity. So, for example, we might talk about the identity of 'single mother' or 'black youth'. Within this perspective, identities might be made up of multiple strands (such as age, gender, ethnicity or marital status), but each individual's identity is assumed to be singular and relatively stable over time and there are assumed to be a finite number of identity positions. The assumption that groups of people share a common identity position and that we can 'read off' their attitudes and characteristics from that position (that women are natural homemakers, for example, or that Welsh people are good at singing) is often described as **essentialism**. Within the social sciences, particularly from feminist scholars, there have been many criticisms of essentialist thinking (see Fuss, 1989). This chapter therefore adopts a different position, arguing that identities are plural and fluid, complex and contested. We should not assume that all single mothers share any common characteristics beyond their marital and parental status, and we should be particularly wary of categorizations like 'black youth' as such labels are often applied indiscriminately to demonize whole groups of people.

The approach that is taken in the rest of this chapter can best be described as *relational*, arguing that our identities are constructed in relation to perceived similarities and differences. Some of these differences relate to the most intimate scale of the body (to notions of sexual and racial difference, for example) while others (such as citizenship or nationality) relate to the wider scales of the nation and the world. Questions of identity are politically and emotionally charged because they are simultaneously about the most personal issues of embodiment and subjectivity, but they also relate directly to processes of inclusion and exclusion where inequalities of power often result in discrimination and injustice (Sibley, 1995). The chapter concludes that identities are rarely fixed or stable because they are always in process of formation.

KEY CONCEPT

A relational approach to identity

A *relational approach* to identity starts from the position that our identities are formed in relation to others. Identities are not defined in terms of individual characteristics that are 'innate' to particular groups of people. Thinking relationally implies much less bounded notions of the self. We become aware of who we are through a sense of shared identity with others (those who speak the same language or share our tastes and ideas, for example) and by a process of setting ourselves apart from those we consider different from ourselves. A relational sense of identity implies a sense of fluidity and flux, where identities are subject to porous boundaries and changing alliances. Relational thinking opposes fixed and antagonistic notions of identity, where people define themselves through assumptions of innate difference (such as ideas of racial superiority) or where social and cultural differences are equated with differences of power (as in most theories of social class). Relational approaches to identity challenge essentialist definitions, emphasizing the 'play of difference' between people and their (human and non-human) environments.

SUMMARY

- This chapter argues for a relational approach to identity in contrast to traditional (essentialist) approaches.
- It suggests that identities are often expressed through material things (like food) and can be researched through various methods (including the recording of life histories and other interview-based methods).
- Identities involve the most intimate aspects of our personal lives but are also related to wider notions of social inclusion and exclusion.

Formations of identity

As well as understanding identities as always in formation at any point in time, it is also important to think about how identity formations change over time. Many social theorists have argued that questions of identity became a particularly salient issue through the process of **modernization**. In pre-modern societies, people's social reach was relatively limited and their day-to-day interactions were generally conducted on a face-to-face basis with already familiar people (in the family and immediate community around the home). With the modernization of society, individuals increasingly came into contact with relative strangers, where questions of sameness and difference had to be negotiated on a daily basis. Structures and institutions rapidly emerged that enabled individuals to handle these issues, including class-based identities at work and gender-based identities at home (with clear overlap between these and other forms of identity). Authors such as Ulrich Beck have argued that in late modernity, the hold of these traditional bases of identity has weakened through a process of 'reflexive modernization' (Beck, 1992b; Beck *et al.*, 1994). Beck argues that **post-industrial** society is characterized by the decreasing constraints of social structures associated with class, gender, family and work, and that a process of increasing individualization is taking place. This process of de-traditionalization, where old structures of authority are increasingly questioned, leads to increased uncertainty and risk. This, in turn, requires new forms of self-management where individuals are (to varying degrees) able to construct their own personal narratives that enable them to understand themselves and control their future lives. Developing Beck's ideas about 'reflexive modernity', sociologists like Anthony Giddens have traced the processes through which traditional narratives of class and gender have been destabilized as self-identity has become a 'project' whereby individuals sustain their sense of identity through a constantly revisable and future-orientated 'narrative of the self' (Giddens, 1991). Unlike many of our ancestors, for example, we are now faced with a wide choice of what to eat and our selection of one type of food rather than another has the capacity to serve as a marker of our identity.

While it is easy to overstate the demise of earlier forms of work-based identity associated, for example, with notions of social class, it is clear that modern life now revolves around a host of other identity markers. Authors such as Maffesoli (1995) have talked about the present day as the 'time of tribes', referring to the proliferation of ephemeral social groupings that come together for temporary social interaction at a nightclub or other social gathering for example. Sociologists like Pierre Bourdieu (1984) have explored the way that social distinctions are based on issues of **cultural capital** as well as economic capital, with tastes in music, food and leisure serving as markers of collective identity and social differentiation. Many of these distinctions are rooted in people's consumption practices rather than in their connection to the world of work, which was the defining feature of class-based models of social identity. As 'traditional' sources of identity have declined in significance, new sources of authority have risen to guide people through the potential minefield of consumer choice (examples include the proliferation of consumer lifestyle magazines and the popularity of 'reality TV' shows such as *Big Brother* or *Come Dine With Me*).

Some authors have insisted on the continued salience of class in contemporary accounts of identity formation, particularly in association with other identity markers such as gender. A particularly good example is Beverley Skeggs' study of *Formations of Class and Gender* (1997), based on long-term ethnographic work with working-class women in north-west England. The subtitle of Skeggs' book, 'becoming respectable', provides a good indication of the subtle ways in which class continues to shape women's everyday lives, whether in terms of their childcare practices or how they dress when going out for the evening. As one of Skeggs' informants told her:

> All the time you've got to weigh everything up: is it too tarty? Will I look a right slag in it? What will people think? It drives me mad that every time you go to put your clothes on you have to think 'do I look dead common? Is it rough? Do I look like a dog?'
>
> (Skeggs, 1997: 3)

Skeggs argues that contemporary formations of class and gender involve a complex process of identification and dis-identification (including the refusal to be identified or labelled as 'working class') as social judgements are constantly being made and re-made. Here in the UK, recent media interest in the rise of 'binge drinking' among women plays into these kind of questions of class and gender (Jayne *et al.,* 2006), as well as demonstrating how the way we behave in public frequently carries moral connotations (see Figure 42.1). This kind of coverage sits within a wider context, where the demonization of the British working class has been commonplace, channelled through pejorative identity labels such as 'Chav' (Jones, 2011).

Collective identities are also being reshaped by new communications media such as mobile

Figure 42.1 Binge drinking. Source: Getty Images

phones and satellite TV and by social media such as Facebook and Twitter (see also Chapter 54). Though the geographical penetration of such media is still highly uneven (reflected in phrases such as the 'digital divide'), these technologies can help transcend the barriers of distance and enable communication across an increasingly globalized world. Such media are particularly significant in a context of increasing geographical mobility where many people (such as migrants, asylum seekers and refugees) typically conduct their lives over long distances, often involving a series of transnational connections (Hannerz, 1996). David Morley and Kevin Robins (1995) have argued that new 'spaces of identity' are emerging in response to these diasporic conditions where Europe's relations with its most significant others (America, Islam and the Orient) are being redefined. Others have argued that increasing numbers of people now inhabit this diasporic or **transnational** space. Moreover, this space is not just occupied by those who are themselves members of specific transnational migrant communities (Brah, 1996; Jackson *et al.*, 2004). Examples would include the popularity of 'Indian' food or 'Asian'-inspired music and fashion among white British consumers, many of whom participate in the symbolic and material spaces associated with other cultures without ever leaving home.

As these examples show, new conceptions of space and place, associated with ideas of **diaspora** or transnationality, are giving rise to new ideas about personal and collective identity (see also Chapter 45). Though conventional wisdom defines places as coherent and bounded entities, such ideas are easily challenged. Doreen Massey (1991), in particular, has argued for a more 'progressive' sense of place in which the emphasis is on the connections between places, looking outwards, rather than on the assertion of internal coherence, looking inwards (see also Chapter 1). This assertion of the flows and networks through which places are made and re-made is entirely consistent with relational notions of identity, seen as multiple and contingent, rather than notions of a fixed or singular identity, based on some allegedly intrinsic characteristic. As Roger Rouse has argued (in the context of Mexican migration to the USA):

> The comfortable modern imagery of nation-states and national languages, of coherent communities and consistent subjectivities, of dominant centers and distant margins no longer seems adequate . . . During the last 20 years, we have all moved irrevocably into a new kind of social space.
>
> (Rouse, 1991: 8)

A similar argument about the shifting boundaries of space, place and identity can be made in many different parts of the world, not just in those that are characterized by high levels of transnational migration as in the case discussed by Rouse. In a world that is always 'on the move', identities are increasingly complex and unstable, and our theories need to adjust to these changing circumstances.

SUMMARY

- Modern formations of identity are often said to be more complex than those in the past. Identity has become a reflexive project rather than something one simply inherits or possesses.

- As traditional sources of identity (associated with work, family and religion) have waned, new forms of identity have emerged (around consumption issues, for example). These new formations of identity involve social judgements about others and oneself.
- Place-based identities are also increasingly complex, shaped through forms of diasporic, transnational, and globalized space.

Narratives of identity

In this section we will look at the implications of taking a relational approach to identity in terms of how we might put these ideas into practice (in an undergraduate project or dissertation, for example). One implication of my earlier argument is that direct questions, such as are asked in national population censuses and social surveys, are of only limited use in relation to complex issues of identity. The National Office of Statistics has had enormous difficulties devising an appropriate question about 'ethnic identity', and questions about religion have been even more highly charged in the context of the 'war on terror' and its undercurrents of Islamophobia. As a result, over 675,000 people listed themselves as of 'Mixed' ethnicity and over 230,000 as 'Other' in the 2001 UK Census, unable to align themselves with any of the main categories such as White, Asian/British Asian, Black/Black British or Chinese. Censuses and surveys are usually better at asking for objective data (such as nationality or birthplace) with qualitative methods better suited to more subjective issues such as ethnic or religious identity. This is particularly true where such labels are politically contested, as is undoubtedly the case with religion and **ethnicity**. While many different approaches are possible (see Limb and Dwyer, 2001), I'd like to focus on just one set of methods which I shall call **narrative** approaches to identity.

Giddens' (1991) account of 'reflexive modernization' (referred to above) describes identity as a reflexive project, shaped by the institutions of late modernity and sustained through narratives of the self that are continually monitored and constantly revised. However, Giddens says very little about how these ideas might be investigated in practice in terms of empirical research. For that we need to turn to other sources, such as Margaret Somers' (1994) account of identity as a discursively constituted social relation, articulated through narratives of the self. According to Somers:

> Narrative identities are constituted by a person's temporally and spatially variable place in culturally constructed stories composed of (breakable) rules, (variable) practices, binding (and unbinding) institutions, and the multiple plots of family, nation, or economic life. Most importantly, however, narratives are not incorporated into the self in any direct way; rather, they are mediated through the enormous spectrum of social and political institutions and practices that constitute our social world.
>
> (Somers, 1994: 635)

As indicated by this dense but suggestive extract, there are a number of important features to Somers' account of the narrative construction of identity. An individual's narration must always be located within wider stories associated with family life and in relation to changing institutional structures.

From this it should be apparent that we do not simply construct our own narratives on an individual basis. We are also located within narratives that are not of our own choosing. Second, narrative identities involve a range of discursive practices that go beyond the textual construction of individual biographies. Narratives are culturally constructed, emplotted within the context of other lives and mediated by a range of external factors. Narratives are subject to social regulation through cultural norms and expectations (as in Skeggs' account of respectability, referred to above). Finally, our sense of self is socially and spatially constituted: *who* we are is related in fundamental ways to *where* we are (cf. Bell and Valentine, 1997; Keith and Pile, 1993).

As the examples at the start of this chapter also remind us, narratives of identity are not a purely *social* construction and cannot be understood entirely through textual means. Our identities are also expressed through our relationship to particular material goods, as explored in a compelling way by Daniel Miller in *The Comfort of Things* (2008) (see also Chapter 19). Based on ethnographic fieldwork in South London, Miller shows how the residents on a single street use a variety of material things to express often intense feelings of enchantment and attachment. These things can be as mundane as a photograph or a CD collection. They can involve attachments to animate or inanimate objects (a family pet or a personal computer). In each case, Miller suggests, an 'aesthetics of care' is at work which 'applies equally and indiscriminately to object and persons, since one always turns out to be the vehicle for the other' (Miller, 2008: 29). Recent work on the material culture of the home also shows how our identities involve complex processes of 'objectification', where our relationships with things are a crucial part of our sense of self and our relationships with others, helping us to define our place in the world (Gregson, 2007). Our identities are narrated not only through words, but also through things.

One way of exploring the relational construction of narrative identities and the role of material goods in shaping those identities is through the method of life-history interviewing, though other methods such as participant observation might lend themselves more readily to exploring the significance of people's relationships with things in practice. I would like to take just one example that demonstrates how our personal identities are crafted in relation to these wider relationships. Mirroring where this chapter began, the example relates to food as a material of identity. The following interview extracts were recorded by Polly Russell in the course of her PhD research on British culinary culture (Russell, 2003). The interviewee, Stephen Hallam, was born in 1956 in Nottinghamshire in the English Midlands and he is currently Managing Director of the food company Dickinson & Morris (see Figure 42.2). The company has made pork pies since 1851 and places great emphasis on the quality and authenticity of its products. Despite their commitment to tradition and authenticity, however, to satisfy growing consumer demand the production process at Dickinson & Morris has become increasingly automated and many of its pies are now mass produced. During the interview, Stephen Hallam reflected critically on this process, reconciling the inevitability of technological change with his own passionate commitment to traditional craft skills:

> The pastry case is raised by a machine, the meat is deposited by a machine, the lid is then initially secured, so a stamp comes down to make sure it is well and truly attached to the pie, then it's a pair of hands that puts the crimp on the top, it's a pair of

Figure 42.2 Stephen Hallam, pork-pie maker. Credit: Stephen Hallam, Dickinson & Morris

> hands that puts the hole in the top, it's a hand that puts the glaze on the top . . . Yes, it's a compromise but I don't think you'll ever get a machine that does it all . . . You can't substitute that technical knowledge, that knowledge with the hands, the eyes, the nose. You will smell a pie baking and your nose will tell you if something is wrong. Now you tell me a machine that can do that . . . As long as you still embody the essence of the heritage there's nothing amiss to using technology to assist you with that.
>
> Russell (2003)

The embodied nature of identity is abundantly clear from this extract as Stephen Hallam reflects on the artisanal nature of craft production and the potential compromises that are involved through increasing mechanization of the baking process. The interview also emphasizes the significance of place in the construction of personal and institutional identities. Dickinson & Morris's website proudly proclaims that it is 'the oldest pork pie bakery and the last remaining producer of authentic Melton Mowbray pork pies based

Figure 42.3 'Ye Olde Pork Pie Shoppe', Melton Mowbray, Leicestershire. Credit: Dickinson & Morris

in the town centre' (www.porkpie.co.uk). Dickinson & Morris project this image to the public through their shop in Melton, 'Ye Olde Pork Pie Shoppe' (see Figure 42.3), which was renovated following a fire in 1992. Today, however, as Stephen readily admits, most of its pies are made in a larger bakery in nearby Leicester:

> To a lot of consumers, a lot of people, Dickinson & Morris is just the pork pie shop and they believe all these packets of pork pies and sausages you see in the supermarket on shelves come from the shop but in reality they can't. We don't have the space, it's impossible. So we utilise the resources of our sister companies and divisions in the group . . . And the branded products, i.e. all those that are in supermarkets, come from the pork pie bakery facility just on the outskirts of Leicester.
>
> (Russell, 2003)

These short extracts cannot do justice to the richness of the longer interview. However, they do demonstrate how Stephen Hallam's personal identity is wrapped up in wider processes of social and technological change, including those associated with particular place identities, as well as being expressed through his relationship with particular material goods. Hopefully, too, these extracts demonstrate how interviewees are able to reflect critically on the significance of these wider changes and relationships in shaping their own identities.

KEY METHOD

Oral history interviewing

There is an extensive literature on oral history interviewing, with established guidelines on ethics and etiquette, ways of preparing for an interview and putting respondents at ease (see Perks and Thomson, 1998, for a critical introduction). As well as recording the interview, researchers usually prepare a recorded summary and/or full transcript of the interview. The challenge is then the interpretative one of making sense of the data – see Limb and Dwyer (2001) and Jackson and Russell (2010) for some useful guidelines.

SUMMARY

- Identities often assume a narrative form, in the sense that individual biographies are emplotted within wider contexts.
- Identities are often expressed through our relationships with particular material goods.
- One useful way of understanding narrative identities is through the analysis of life histories, an approach that sets personal biographies in their wider relational context.

DISCUSSION POINTS

1. What are the most important aspects of your own identity, and can you articulate this in terms of similarities to and differences from other people?
2. To what extent have identities become more complex in recent years and how would you account for these changes?
3. To what extent are identities related to particular places and how does this work at different spatial scales (from the body to the home, from the neighbourhood to the nation)?
4. Taking food as a starting point, how does your identity reflect your relationship with particular material goods?
5. Using a narrative approach, plan out how you would conduct a life history interview in order to gain a better understanding of your respondent's identity.

FURTHER READING

Jenkins, R. (2008) *Social Identity* (3rd edn.). Abingdon: Routledge.

Part of the 'key ideas' series, this is an accessible introduction to theories of identity, written from the perspective that identities are always social, involving the interplay of similarity and difference.

Woodward, K. (ed.) (2004) *Questioning Identity: Gender, Class, Ethnicity* (2nd edn.). Abingdon: Routledge.

Produced as part of an Open University course on understanding change, the book explores how new identities are forged in changing times.

DuGay, P., Evans, J. and Redman, P. (eds.) (2000) *Identity: A Reader*. London: Sage.

A wide-ranging book that contrasts cultural studies, sociological and psychoanalytic approaches to identity.

Morley, D. and Robins, K. (eds.) (2001) *British Cultural Studies: Geography, Nationality, and Identity*. Oxford: Oxford University Press.

Focuses on British identity, including the role of tradition and heritage in contemporary culture and features several essays by Human Geographers.

WEBSITES

www.bl.uk/learning/histcitizen/foodstories

The 'Food Stories' website provides access to a series of life history interviews including the extracts from Rosamund Grant with which this chapter begins. One section deals specifically with food, nation and cultural identity.

CHAPTER 43

IDENTITY AND DIFFERENCE: AGE, DIS/ABILITY AND SEXUALITY

Sarah L. Holloway

Introduction

This chapter extends Peter Jackson's discussion of how Human Geographers approach the question of identities today (Chapter 42), with a more direct focus on the relations between identity and social geographies of difference. It develops its arguments through specific foci on age, dis/ability and sexuality, showing how these are important areas of enquiry for Social Geographers and complementing the previous chapter's stronger focus on issues of class, gender, race and ethnicity.

Geographical debates about identity and difference have taken two forms in recent decades. One approach has been to dispute **essentialist** assumptions about identity, and argue instead that the differences between us are **socially constructed**. The idea is that identities are not simply grounded in our individual biology but are socially constituted through the interleaving of wider processes and the individual's biographical narratives (what Peter Jackson called a 'relational' approach to identity in Chapter 42). This social rather than natural constitution of identities means that they are not universal but vary across time and space, reflecting temporal and spatial variations in social relations. Laurie *et al.* (1999) demonstrate, for example, that what it means to be a woman varies over time and space, by tracing changing **hegemonic** forms of femininity in Britain over the past 200 years and contrasting these notions with the very different understandings of women's capabilities and responsibilities in communist East Germany and Latin America. Interestingly, as well as varying between times in history and places in the world, socio-spatial relations also shape our ideas of where particular groups are seen to be 'in' and 'out of place'. For example, women's place has, at some points in British history, been idealized as in the home, whereas men are seen to be more at home in the workplace (Laurie *et al.*, 1999) (for more on how identity and difference relate to being 'in and out of place', see Chapter 17).

A second, contrasting, but not necessarily contradictory, approach to the study of identity and difference draws on **psychoanalytic** traditions. Sibley (1995) is a prominent figure in the use of such approaches within Social Geography (see also Chapter 44 for a discussion of how Sibley approaches 'geographies of exclusion'). Drawing on Klein (1960), he argues that a sense of border developed in infancy forms the basis from which the Self seeks to distance itself from objects and people defined as 'abject', as **Other**. These ideas of Self and Other are not innate, but are culturally produced through interaction with the social milieu. One important way in which the subject mediates interaction with this social milieu is through stereotypical understandings of good and bad objects and people. These stereotypical representations of Others not only define the Self (by representing what it is not), but again they also inform social practices of inclusion and exclusion, processes through which different social groups come to be constructed as in and out of place in particular settings (Sibley, 1995). For example, romantic representations of gypsy travellers as an exotic race living in harmony with nature in gaily painted, horse-drawn caravans construct them as perfectly at home on the highways and byways of rural England; this unreal picture based somewhere in a mythical past works against contemporary city-based gypsy travellers who are often considered 'out of place' in urban space (Holloway, 2005; Sibley, 1995).

What is striking is that, despite their very different starting points, both social constructionist and psychoanalytic research on identity suggest a need to pay attention to questions of place and space. On the one hand, both approaches reject essentialist understandings of identity and emphasize that the constitution of identity and difference is a social process and is therefore specific to time and space. On the other hand, they both highlight that these processes can produce highly spatialized understandings of difference, for example as we hold spatially specific understandings of where particular social groups are in or out of place.

This chapter explores these approaches to identity and difference through a focus on age, dis/ability and sexuality. We consider each of these axes of identity in turn, but as you read on you will see how these different axes of identity intersect with one another. This issue of 'intersectionality' is something we return to in the conclusion.

SUMMARY

- Human Geography's approach to forms of identity and difference – such as age, dis/ability and sexuality – questions their determination by essential, innate or biological characteristics alone.
- On the one hand, this has involved an emphasis on the 'social construction' of differences (including age, dis/ability and sexuality). On the other hand, it has also involved an engagement with psychoanalytic understandings of identity and difference.
- In both cases, questions of place and space are recognized as central to how identity and difference are constituted.

Age

Geographical research on age emerges from a variety of traditions rather than forming a coherent body of sub-disciplinary interest (see also Chapter 39). The most common way in which Social Geographers have engaged explicitly with age-related issues is through a focus on those at the extremes of the age spectrum, with children and young people attracting somewhat more attention than older adults. However, those adults who fall between these bookend generations have rarely been studied as adults, unless in a parenting role. This has led to recent calls for studies of explicitly adult geographies, and more importantly, extra-familial intergenerational relations.

Since the 1990s, there has been a rapid increase in the attention paid to children's and young people's geographies (Aitken, 2001; Ansell, 2005; Holloway and Valentine, 2000a; Holt, 2011; Skelton and Valentine, 1998), and since 2003 this sub-disciplinary field has had its own journal, entitled *Children's Geographies*. Most of the research in this period has been allied to broader social studies of childhood. Researchers in this field argue that child, far from being a biological category, is a socially constructed identity (James *et al.*, 1998). Historical research has shown that the social category 'child' is a relatively recent invention; it did not exist in the same way in the Middle Ages, for instance, when young people were regarded as miniature adults rather than conceptually different from adults (Ariès, 1962). Moreover, the qualities supposed to be 'natural' in children have actually changed over time. Historically, some versions of Christian doctrine viewed children as 'little devils', as inherently naughty, unruly, unsocialized beings who need to be saved through strict discipline and a religious upbringing (Newson and Newson, 1974; Schnucker, 1990). In contrast, Enlightenment thinkers such as Rousseau contested this way of thinking about and relating to children, instead imagining them as 'little angels' with natural talents and virtues that could be developed though gentle coaxing by adults (Jenks, 1996). Recognition that childhood is a socially constructed phenomenon has fuelled research into its construction, contestation and consequences. It has also prompted a focus on children as competent social actors in their own right, with studies not only concentrating on forces of socialization such as school and family but also on how children themselves think about their childhoods and conduct themselves in practice (Holloway and Valentine, 2000a).

Geographers have made a distinctively spatial contribution to social studies of childhood (Holloway and Valentine, 2000b; Holloway

Figure 43.1 Many Social Geographers have focused on those at the extremes of the age spectrum. Credit: Tanya Constantine/Getty Images

and Pimlott-Wilson, 2011). One important contribution has been geographers' insistence that socially constructed ideas about childhood are not only time- but also place-specific. This means that many of the 'normal' assumptions that we hold about childhood in the West – that children are less able and less competent than adults and thus need to be educated into their future adult roles at the same time as they should be allowed a childhood of innocence and freedom from adult responsibilities – are culturally specific. For example, while childhood in the Global North is often seen as a time of dependence, in the Global South many children make essential contributions to the economic and social reproduction of their households through forms of domestic, agricultural and paid work which are often (though not always) seen as children's tasks (Bromley and Mackie, 2009; Evans, 2011; Punch, 2001). Geographical research can thus expose how many of our ideas about childhood are **Eurocentric**, forming assumptions that underpin advances such as the UN Convention on the Rights of the Child. Controversies over issues such as child labour show the contributions that geographers can make. While it is important to recognize cultural difference and avoid judging countries of the South by northern norms, we must balance this with the imperative to ask questions about the global distribution of resources, which means that some children must work to ensure household survival while others are faced with an 'epidemic' of obesity and ill health through reduced physical activity and over-consumption.

In addition to this focus on global differences in childhood, geographers have also studied the everyday spaces in and through which children's identities are made and remade, in so doing considering the ways our ideas of childhood inform / are informed by wider spatial discourses about home, school, city, rural idyll and nation (see Holloway and Valentine, 2000a, for a more extensive review). One example of this kind of thinking is research on children's use of public space. There is a relatively long history of research into children's attachment to and use of space (Hart, 1979; Ward, 1978), but in the 1990s this began to chime with growing public concern about children's presence in public space. This concern centred on the twin fears that some children ('little angels') are vulnerable to dangers in public places, while the unruly behaviour of other children ('little devils') can threaten adult control of public space (Valentine, 1996a; 1996b). Initially, geographic research considered how parents and children conceive of and negotiate these risks in socio-economically mixed urban and rural areas of the Global North, emphasizing both the importance of local parenting cultures and children's agency in the construction and contestation of family rules about use of the street (Valentine, 1997; Tucker and Matthews, 2001). More recent research in the Global North and South emphasizes the continuing importance of societal concerns about children's independent use of public space, as well as the importance of these spaces to children of diverse ages, genders, class backgrounds and locations (Benwell, 2009; Gough and Franch, 2005; Kullman, 2010; Lees, 2003; Pain, 2006). This emphasis on age, gender, class and so on, draws attention to the importance of differences between children within the Global North and South, which exist alongside those differences between North and South that were discussed earlier.

At the other end of the age spectrum, research on older people has numerous parallels with the field of children's geographies. Research into older people's geographies has a relatively long history within Human Geography (Rowles, 1978a) and witnessed an expansion of interest (though not of a comparable size to that in children's geographies) from the mid-1990s (Hardill and Baines, 2009; Harper, 1997; Harper and Laws, 1995; Laws, 1994; 1997; Tarrant, 2010). Here, too, researchers base their work on a social rather than biological understanding of identity, and consider the articulation of age with other social differences:

> Rather than defining and employing old age as a chronological descriptor, many now argue that the socially and economically constructed aspects of old age have most influence on the condition of older people's lives.
>
> (Pain *et al.*, 2000: 377)

Pain *et al.*'s (2000) study of the way discourses of old age intersect with class, bodily ability and gender in the framing of older people's leisure spaces is an interesting example of this type of work. In this, Pain and her co-authors explore the ways some older people frame their leisure activities in ways that maintain positive images of themselves through contrast with other groups of older adults. For example, some of the retired men in her study who attended a senior men's club organized around educational lectures saw themselves as maintaining a mentally active lifestyle they had been used to at work, and reinforced this positive image of themselves by contrasting their own club with one aimed at retirement-aged middle-class women, which they constructed simply as a place for tea and gossip (despite the fact that it, too, had a talk each week).

Children's and young people's geographies have now been established as a vibrant field and there is considerably more scope for research on older adults. However, a note of caution needs to be sounded as the current practice of focusing on either end of the age spectrum inadvertently suggests that those in the middle are unmarked by age. This not only establishes this middle group as a norm against which young and old are Othered; it could also mean that an analysis of the meaning of age for those in the middle years is lost to the geographical research agenda. The response to this problem has been a call for geographers to shift from studying the bookend generations and to concentrate instead on a 'relational understanding' of age (Hopkin and Pain, 2007). This current research agenda is being taken forward in two particularly interesting ways. First, increased attention is being paid to the relationships between generations. There is a reasonably long history of feminist work on intra-familial intergenerational relations through research on parents and their children (Holloway and Pimlott-Wilson, 2011) but, as Vanderbeck (2007) has been at the forefront of arguing, there is also the need for intergenerational research in the extra-familial context, exploring how different generations relate to each other or become socially and spatially segregated and separated. Second, there has been growing interest in geographical perspectives on life transitions, exploring the spatial specificity of movements between different phases of life, for example from youth into adulthood (Hörschelmann, 2011). Rather than understanding age and its social construction as defining a distinct social group and geographies, this focus on life transitions poses questions about how our identities run their course as we age. This research agenda speaks to a number of pressing questions of social policy. In the UK, for example, these include social care for the elderly and transitions from 'family' to 'residential' homes; and the transition to adulthood, in the context of property and labour markets that make leaving the parental home a different and more difficult process than it had been for the preceding generation. In both those cases, we can see how life transitions are simultaneously social and spatial (see also Chapter 2).

SUMMARY

- Childhood is a socially constructed identity: it is a recent invention and the qualities supposed natural in children have varied over time.
- Geographers have emphasized global differences in childhood as well as the sites of everyday life in and through which diverse children's identities are made and remade.
- Old age has attracted less attention, though past and present studies demonstrate the potential for future research into the social construction of old age and its articulation with other axes of identity.
- The focus on the twin extremes of the age spectrum inadvertently normalizes those in the intervening years. This has led to calls for a relational understanding of age.
- In Human Geography, this relational approach has led to current research that highlights both the importance of intergenerational relations and individuals' life transitions.

Dis/ability

Geographical research on illness, impairment and disability has a complex relation to debates about the social construction of identity and difference. To date, research in this fairly diverse field has been shaped by two contrasting models of disability and by subsequent attempts to move the latter of these two positions forward.

The first, the *medical* model of disability, has wide social currency within the Global North (Parr and Butler, 1999). In this model, disability is regarded as an 'individual medical tragedy' (Shakespeare, 1993, cited in Parr and Butler, 1999: 3). Medical problems are seen to impair some people's bodies such that they cannot undertake activities regarded as normal for a human being; for example, a baby born blind cannot see, or someone paralyzed as the result of a road traffic accident cannot walk. This is regarded as an individual tragedy and the person disabled by their medical condition is often seen as deserving of sympathy and dependent on help from able-bodied society (although social discomfort and discrimination are also common reactions to disability – see Key Issue box).

The response from the medical establishment is to provide interventions that aim to give the individual as normal a life as possible, for example through the provision of cochlea implants to some deaf people or prosthetic limbs to those missing an arm or leg. In some circumstances medical intervention can improve the quality of life for people with disabilities, but it is not universally welcomed. For example, some deaf people are against cochlea implants, arguing that the procedure – which gives some though not full hearing but involves the risk of invasive surgery, potentially destroying any residual hearing, and possible side-effects such as headaches – is an attempt by hearing society to make deaf people conform to their norms, and as such is an attack on well-developed deaf signing culture. The best-known geographer to adopt the medical model is Golledge (1993). He argues that disabled people's experience of space is fundamentally different to that of the

Figure 43.2 **Social Geographers debate how best to understand disability: as medical or social? Credit: Zigy Kaluzny/Getty Images**

KEY ISSUE

Othering people with disabilities

Notwithstanding the dominance of debates about the medical and social model of disability, there have been some attempts to understand the Othering of people with disabilities through a psychoanalytic lens (Dear *et al.*, 1997; Holt, 2003; Kitchin, 1998; Pain *et al.*, 2001). Holt (2003) provides a particularly useful review of these developments: she argues that the discourses that comprise the medical model of disability parallel the symbolism of abjection in contemporary western society. In psychoanalytic theory, abjection, part of the process through which Self defines itself as separate from Other, involves both elements of attraction and repulsion. These twin understandings can be seen in respect to disability: on the one hand we have 'positive' representations of disabled heroes overcoming adversity; on the other, 'negative' narratives in which people with impairments are seen as abnormal, dependent, frightening and unattractive beings. In different ways both these types of 'stories' harm people with disabilities, as neither extreme represents the everyday, ordinariness of living with an impairment.

Dear *et al.* (1997) argue that this construction of an abject category of 'disabled' reflects a desire on the part of the Self to impose an artificial distinction of disabled/non-disabled on a continuum of human capacity, and thereby to distance and protect itself from its Other. This distancing is also seen in broader socio-spatial relations, as Holt (2003: 19), in a review of the literature in this field, explains: 'disabled people's historical geographies of socio-spatial exclusion in asylums are [now] being replicated on a smaller scale by the new geographies of "de-institutionalisation"' (see also Dear and Wolch, 1987; Gleeson, 1997).

Source: Holt (2003)

able-bodied population; that as geographers we need to understand these different ways of knowing the world; and that having done so we will be able to suggest useful interventions, ranging from the spatial criteria to be included in the assessment of learning disabilities to tactual maps suitable for visually impaired users.

A second approach, which has had wider currency in activist and academic circles, draws on the *social* model of disability. This model recognizes bodily differences in terms of impairment, but locates disability as a product of society. This is evident in the oft-cited definitions of the terms impairment and disability by the Union of Physically Impaired Against Segregation (UPIAS 1976: 3–4; see Hall, 1995; Holt, 2003; Parr and Butler, 1999):

> Impairment – Lacking all or part of a limb, or having a defective limb, organism or mechanism of the body.
>
> Disability – The disadvantage or restriction of activity caused by a contemporary social organization which takes no or little account of people who have physical impairments and thus exclude them from the mainstream of social activities.

In this model, disability is seen to be a social construction, as the organization of society rather than an individual's medical condition is identified as the root cause of any problems. It is therefore society, rather than the individual, that is seen to be in need of change.

One popular way in which this model has been utilized in Geography is in studies that seek to show how the current social organization of the built environment disables people with impairments (Imrie, 1996). In a classic piece early in these debates, Hahn (1986) discusses a range of problems that the urban environment presents for people with impairments, ranging from insufficient provision of dropped kerbs (which enable wheelchair users to get on and off pavements/sidewalks), to a lack of accessible public transport, to poor-quality housing provision. These he locates squarely within a social model of disability:

> the major problems resulting from a disability can be traced to a disabling environment; and the solution must be found in laws and policies to change that milieu rather than in unrelenting efforts to improve the capacities of a disabled individual.
>
> (Hahn, 1986: 274)

More recent research has also sought to show how social relations in a range of spheres can either enable or disable people with impairments. Wilton and Schuer (2006), for example, focus on accommodation and the exclusion of workers with disability in the contemporary labour market. Though not universally embraced, this social model of disability has had a far greater influence on geographic studies of disability than its medical counterpart. However, as Parr and Butler (1999) make clear:

> in acknowledging the undeniable value of the social model, it is easy to forget that it too is not without fault . . . At present society plays the dominant role in constructing disability, but the role of different physical and mental impairments cannot be ignored by the social model if it is to continue to be valued and respected.

Much contemporary research on disability thus starts with an appreciation of the social model of disability but also seeks to nuance it in important ways. One way in which this is achieved is to reincorporate appreciation of the pain and physical difficulties impairments can cause into more recent work, thus treating disability both as a social construct and as an **embodied** experience. Equally important has been the move away from an overemphasis on physical impairments and the incorporation of a focus on chronic illness (Driedger, 2004; Dyck, 1999; Moss, 1999) as well as mental difference, including mental ill-health, learning disabilities and, most recently, emotional and behaviour difficulties (Hall, 2010; Holt, 2010; Lemon and Lemon, 2003; Madriaga, 2010). Underlying all of this is an appreciation that dis/ability inevitably articulates with other social differences (Parr and Butler, 1999).

SUMMARY

- The medical model locates disability as an 'individual medical tragedy'. Geographical studies based on this model aim to identify ways in which disabled people can be helped to cope with their problems.
- The social model posits disability as a socially constituted problem, stemming from the current organization of society and space. Geographical studies based on this model aim to expose the ways in which the physical and social environment currently disables people, and suggest alternative ways forward.
- Most current research favours the social over the medical model of disability, but suggests that it needs to incorporate an appreciation of disability as an embodied experience, and to broaden its focus to include chronic illness, mental health and learning disabilities.
- Dis/ability articulates with other forms of social difference.

Sexuality

Geographers focusing on sexuality have tended to pay less explicit attention to debates about **essentialism**. Those who have addressed these debates highlight the difficult issues raised for activists and academics alike (Bell and Valentine, 1995a). On the one hand, some lesbians and gay men are happy to support scientific research that seeks to uncover a genetic basis for sexual orientation. Finding a 'gay gene' would substantiate claims that homosexuality is 'natural', and so ought not to be cause for discrimination and abuse. On the other hand, these essentialist arguments are rejected both by some lesbian feminists who argue that sexuality is a choice, and more broadly by social constructionists who argue that sexuality, like all other identities, is socially constituted. For some, what is called 'strategic essentialism' is a way out of this dilemma: academics and activists employing this tactic hold a broadly constructionist approach to identity but deploy essentialist arguments when these are politically useful (Bell and Valentine, 1995a; see also Hubbard, 2002, on the possibilities of marrying social constructionist and psychoanalytic approaches). Regardless of whether opposite-sex or same-sex attraction has some basis in biology, however, the way hetero-, bi- and homosexual identities are shaped, experienced and performed is inevitably (also) shaped by socio-spatial relations, such that what it means to be a heterosexual, a bisexual, a lesbian or a gay man varies not only over time but also between places and spaces.

One of the first ways in which geographers began to look at sexuality was through an exploration of the formation and impact of 'gay neighbourhoods' in the urban landscape of the Global North (Binnie and Valentine, 1999). Paralleling early research on the segregation of racialized minorities, initial studies examined the choice and constraint factors that produced clusters of commercial gay venues and drew gay men to the city (Castells, 1983; Lyod and Rowntree, 1978; Weightman, 1981; Winchester and White, 1988). As this approach developed researchers sought to explore the links between **capitalism** and 'gay space': this has included studies of the gentrifying role some gay men have played in the urban land market (Knopp, 1990) and research into the impact business marketing has on sexualized space – for example, the implications of branding Manchester's Gay Village as a cosmopolitan spectacle, authentically gay but open to all consumers regardless of sexual orientation (Binnie and Skeggs, 2004). For the most part, this research has centred on gay villages in the Global North. Tucker's (2009) study of Cape Town is an exception to this trend and reminds us of the importance of differences between gay men, for example in terms of class and race, and its implications for their inclusion in, or exclusion from, 'gay space'.

Initially, lesbians were excluded from these debates about 'gay neighbourhoods', on the grounds that as women they had less economic power and fewer territorial aspirations than gay men (Castells, 1983). Rothenberg's (1995) study of Park Slope, Brooklyn, New York, demonstrates that this assumption is untenable and there are indeed 'lesbian neighbourhoods'. However, she questions the orientation of previous work that emphasized commercial venues, by demonstrating the importance of social networks in community formation in residential locations (see also Adler and Brenner, 1992; Peake, 1993; Valentine, 1993a). Podmore (2006) puts this contrast between commercial gay villages and networked 'lesbian neighbourhoods' in historical context, through a study of Montréal, Canada. This shows that there was a visible lesbian bar culture between the 1950s and 1980s, but that this declined from the 1990s onwards, both as a result of gentrification and as lesbian-separatism gave way to engagement with gay men under the banner of queer politics, resulting in lesbians' integration into mixed-gender sites within Montreal's gay village, and thus the deterritorialization of lesbian identity in the city.

The sense of time which is prominent in Podmore's (2006) paper is important, as there is growing recognition that 'lesbian and gay spaces' in the city do not remain static. They can change: as we saw in the case of Manchester above, where 'gay lifestyle' is being incorporated into the mainstream and marketed to a wider public; or, as in the case of Oxford Street in Sydney, where gentrification displaces gay residents, bars and shops, with the area being colonized by groups who are less tolerant to non-heterosexual others (Ruting, 2008). This fluidity in the urban landscape has prompted interest in queer-friendly neighbourhoods, which unlike gay villages are 'localities where same-sex-attracted residents, businesses, and institutions are welcomed in a dominantly heterosexual milieu, and intergroup interaction fosters dialogue' (Gorman-Murray and Waitt, 2009: 2870). However, while mixing and dialogue between different groups can be seen to promote social cohesion, urban living does not always lead to openness to others (Andersson *et al.*, 2011).

The overwhelming urban emphasis of many of these studies is increasingly being counterbalanced by research that focuses on 'the lesbian and gay experience' in suburban and rural landscapes (Kirkey and Forsyth, 2001; Smith and Holt, 2005) and indeed on the two-way migratory flows between provincial and metropolitan settings (Waitt and Gorman-Murray, 2011). More generally, there have been long-standing efforts to broaden the way geographers look at lesbian and gay landscapes. Valentine's (1993c) work was key in this respect. Through a study of lesbians in an unnamed UK town, she explored the ways in which everyday spaces such as the street, the workplace, social spaces, service environments and the home are often 'heterosexed' – that is, shaped by normative expectations of heterosexuality with 'aberrant' behaviour being policed through stares and gestures, as well as verbal and physical abuse. For example, the heterosexing of public space means that while heterosexual couples can hold hands or kiss goodbye in the street, the same behaviour from same-sex couples is likely to attract attention and sometimes verbal or physical sanctions against them. Understanding of the heterosexing of society and space, and

Figure 43.3
A heterosexual geography of the home. Credit: Lorna Ainger

challenges to this, has been taken forward through research on everyday residential, employment, shopping and social spaces (Browne, 2007; Kitchin and Lysaght, 2003; Podmore, 2001) as well as more out-of-the-ordinary events such as parades (Brickell, 2000; Marston, 2002; Johnston, 2007; Mulligan, 2008). However, as attitudes to lesbian and gay people have become more tolerant in some parts of the world, research is also emerging which suggests that spaces outside 'gay ghettoes' are not automatically 'heterosexed' (Browne and Bakshi, 2011).

Research on heterosexuality emerged later than that on sexual minorities, but has expanded rapidly during the last decade. Studies on 'immoral' forms of heterosexuality were the first to gain critical mass, with earlier work (Symanski, 1981) rigorously and critically developed in a twenty-first century focus on the importance of (mainly) heterosexual prostitution in cities across the globe (Hubbard and Whowell, 2008). This research has been diverse in nature, examining different actors important in the industry (Bailey *et al.*, 2011), from young people who use sex work to carve out a living, to residents who object to prostitution in their neighbourhood (Hubbard, 1999, 2002; Hubbard and Sanders, 2003; van Blerk, 2008). As the decade has progressed, the proliferation of 'adult entertainment' such as lap dancing clubs in mainstream rather than marginalized spaces has drawn increasing attention, as has gender diversity in the sex work industry (Hubbard *et al.*, 2008; Hubbard and Whowell, 2008).

Research on 'moral' forms of heterosexuality (i.e. forms of heterosexuality that are framed as 'normal' or 'respectable' rather than in some way 'immoral') developed much more recently, perhaps because it is harder (for heterosexuals at least) to see how everyday social institutions and spaces – homes, schools, churches, bars, supermarkets and the like – are shaped by these normative forms of heterosexuality (Hubbard, 2000). However, this is now a vibrant area of research. Historical and contemporary studies have analysed the production of moral populations (and through this, their 'immoral' Others) in relation to, for example, the family (Oswin, 2010; Phillips, 2009b) and respectable forms of feminine sexuality ('decent girls', to use the discourse analysed by Crowley and Kitchin, 2008). Ethnographic accounts have traced the more and less 'moral' making and performance of heterosexuality in different environments, from the leisure spaces of urban nightlife, to rural communities, tourist destinations and 'expatriate' socializing (Boyd, 2010; Jacobs, 2009, 2010; Little, 2003; Malam, 2008; Thomas, 2004; Waitt *et al.*, 2011; Walsh, 2007). Overall, Social Geographers are now exploring how heterosexuality is thought, regulated, practised and spatialized.

SUMMARY

- Sexual identities are shaped by socio-spatial relations, such that what it means to be a heterosexual, a bisexual, a lesbian or a gay man varies over time and space, and with other axes of social difference.
- Geographers have explored the formation of 'gay and lesbian neighbourhoods' in the urban landscape and how these change over time. This is now being matched by research on suburban and rural areas.

- The (contested) 'heterosexing' of public, work, residential and other spaces is also a key focus of concern.
- Marginal forms of heterosexuality initially attracted more attention than their 'moral' counterparts; however, the latter are now being explored in both historical and contemporary contexts.

Conclusion

Each of the axes of social difference discussed in this chapter has witnessed a blossoming of research over the last 10–20 years. Although age, dis/ability and sexuality are widely assumed in society to be biological givens, researchers in each of these fields have teased out the ways these identities are made and remade through socio-spatial relations. Most of this work has been informed by social constructionist thinking, as the use of psychoanalytic approaches is still less common in respect to age, dis/ability and sexuality than for some other forms of social difference. Regardless of which approach is adopted, it is crucial that we think through the intersections of these axes of identity. A body of work on identity and difference – centred on the concept of 'intersectionality' – has pursued that task (Valentine, 2007). All the identity positionings discussed in this and the previous chapter (where class, race, ethnicity and gender featured more strongly) are mutually articulated such that, for example, an older person's experience of age is shaped by their social class, just as a gay man's sexuality is experienced through his dis/ability, ethnicity, and so on. Place further shapes identity positionings and their intersections. The challenge for Social Geographers in the future is to explore the intersections of these diverse normalized and marginalized identity positionings, in diverse contexts spanning the Global North and South.

DISCUSSION POINTS

1. There is a bias in the geographical research agenda towards studying identity in the Global North. Why is this problematic and what areas for future research can you identify by focusing on the Global South?
2. How might a psychoanalytic approach usefully inform geographical research on age?
3. Why do social differences such as age, class, ethnicity, gender and sexuality matter in geographical research on disability?
4. Why is it important for geographers to study heterosexuality, and how might we set about studying normative versions of this particular sexual orientation?

FURTHER READING

Holt, L. (ed.) (2011) *Geographies of Children, Youth and Families*. Abingdon: Routledge.

This edited volume draws together current research on children, youth and families from around the globe.

Pain, R. (2001) Age, generation and lifecourse. In: R. Pain, M. Barke, D, Fuller, J. Gough, R. McFarlane and G. Mowl (eds.) *Introducing Social Geographies*. London: Arnold, 141–63.

A nice introductory textbook chapter specifically about the social geographies of age.

Canadian Geographer (2003) Special issue on Disability in Society and Space, 47(4).

A wide-ranging special issue on dis/ability.

Podmore J.A. (2006) Gone 'underground'? Lesbian visibility and the consolidation of queer space in Montreal. In: *Social and Cultural Geography* 7(4), 595–625.

An interesting historical analysis of urban change and lesbian and gay visibility in the city.

Boyd, J. (2010) Producing Vancouver's (hetero)normative nightscape. In: *Gender, Place and Culture* 17(2), 169–89.

An ethnographic exploration of heterosexualities (and other social differences) through a focus on urban nightlife.

Sibley, D. (1999) Creating geographies of difference. In: D. Massey, J. Allen and P. Sarre (eds.) *Human Geography Today*. Cambridge: Polity, 115–28.

A good introduction to psychoanalytical approaches to difference.

CHAPTER 44
EXCLUSION

Jon May

Introduction

'Exclusion' is a key concept for Social Geographers. One of the clearest definitions of the term is provided by Chris Philo, for whom exclusion refers to: 'A situation in which certain members of society are, or become, separated from much that comprises the normal "round" of living, and working within that society'. Exclusion, he argues, should be thought of

> as simultaneously *social and spatial.* Indeed, excluded individuals will tend to slip outside, or even become unwelcome visitors within, those spaces which come to be regarded as the loci of 'mainstream' social life (e.g. middle class suburbs, upmarket shopping malls, or prime public space)
>
> (Philo, 2000: 751)

Building on this basic definition, five further facets of exclusion can be noted:

1. Exclusion is simultaneously *material* and *symbolic*, the two mutually reinforcing each other. In its symbolic form it is closely related to the concept of **stigma** (severe social disapproval of an individual or group) with stigmatized groups often suffering exclusion.
2. Exclusion is an *active process:* referring to the deliberate distancing of an individual or group by and from another individual or group.
3. Processes of exclusion are irretrievably intertwined with processes of *inclusion*: the one necessarily producing the other.
4. Processes of exclusion operate across *multiple axes* – whether of class, gender, race and ethnicity, sexuality, age, mental or physical health and so on – and often act across more than one such axis at any one moment.
5. Processes of exclusion are *never absolute* and will always to some extent be *resisted.*

Taking these five facets as its starting point, this chapter examines the early emergence of a concern with processes of exclusion in the work of the Social Geographer, David Sibley, before exploring three contrasting examples of these processes. The first, connecting closely with some of Sibley's key themes, examines 'life at the margins' among Cornish tin miners between the mid fifteenth and late nineteenth centuries. The second brings things more up-to-date, tracing the growing exclusion of homeless people from the streets of US and UK cities. In the final section, the concept of exclusion is deliberately made more complicated in an account of the complex inclusions and exclusions articulated in homeless day centres – spaces set up to challenge the exclusions faced by homeless people on the streets.

A geography of outsiders

The first person to explore the geographies of exclusion in any detail was the Social Geographer David Sibley. In his *Outsiders in Urban Society* (1981) Sibley noted how stigmatized groups commonly inhabit the geographical margins of society. Through a detailed investigation of the life-worlds of gypsies, travellers and the North American Inuit, Sibley investigated the role that spatial boundaries play in maintaining social boundaries. As he recognized, those considered socially 'marginal' are either pushed towards, or in an attempt to avoid confrontation and abuse seek out, geographically marginal spaces. Space thus emerges as both an expression of and a means by which exclusionary practices gain purchase and meaning.

Developing these ideas further, in his later book *Geographies of Exclusion* (1995), Sibley set out to explain the dynamics of such practices. Drawing upon the ideas of Object Relations Theory (a body of work associated with the psychoanalysts Melanie Klein and Julia Kristeva), Sibley argued that western identities are structured by an innate need to differentiate between those considered to be broadly the same, and those identified as fundamentally Different – or 'Other' (see also Chapter 43). Having constructed such a distinction in their infancy, people are then engaged in a life-long struggle to maintain this distinction lest that which is Different undermines a clear and coherent sense of 'Self' (what Sibley refers to as the fear of pollution, or 'abjection'). One way of maintaining this distinction is to ensure a suitable physical distance is also maintained between 'Self' and 'Other', though as Sibley recognizes distance *per se* is rather less important than the fact that a clear and unambiguous boundary be established between the two (see Chapter 17's discussion of place and boundary making for further discussion).

Though in Object Relations Theory these ideas are mainly developed at the inter-personal level, it is not difficult to see how they can be applied at a broader scale, or why they might be attractive to Social Geographers. Certainly, it should be possible to think of numerous examples of the 'boundary marking' processes Sibley draws attention to. One of the most obvious concerns the nature and design of prisons (Martin and Mitchelson, 2008). While prisons are in one sense designed to keep people *in*, they are also understood as a way of keeping prisoners at a *safe distance from* 'decent society'. This is one reason why prisons housing the most dangerous prisoners (the 'criminally insane', for example) are often built some distance away from centres of population (in the UK, in places like Dartmoor – see below). The use of space as a way of separating the 'good' from the 'bad' continues within prisons themselves: with those who break prison rules consigned to solitary confinement.

At a more general level, inspired by Sibley's work, an increasing number of Social Geographers have begun to unpack the experiences of those positioned at the margins of society – with a growing number of studies of children, the mentally ill, physically disabled, traveller communities and so on (Holloway, 2005; Holloway and Valentine, 2000; Imrie and Edwards, 2007; Parr, 2008). Further, one of Sibley's key contributions was the recognition that processes of exclusion can be traced in the academy too. Building on his ground breaking reading of the exclusion of alternative forms of knowledge from mainstream social science in *Geographies of Exclusion*, scholars interested in the experiences of those positioned outside of mainstream society have sought for their studies of such groups to transform mainstream Geographical scholarship. However, in an important development to such debates, concern has recently been raised over whether this newfound interest in the 'margins' may have been at the expense of a more developed understanding of the 'centre'. In other words, as well as exploring the experiences of people of colour, or gay and lesbian groups, for example, geographers should be examining the power relations inherent in constructions of whiteness or heterosexuality, power relations which all too often leave these constructions as unremarked upon and understudied (Bonnett, 2005; Hubbard, 2008) (Chapter 43's discussion of sexuality provides an example of this trend to study 'centres' as well as what are designated as 'margins').

SUMMARY

- Western identities are structured by an innate need to differentiate between 'Self' and 'Other'.
- To avoid contamination of the Self by the Other, people strive to establish clear and unambiguous boundaries between the two. These boundaries often take physical form.
- Spatial boundaries therefore play an important role in maintaining social boundaries.

Life at the margins

As Sibley recognized, one of the earliest and most powerful sources of 'abjection' is the fear of bodily residues (Sibley, 1995). As such, it is not surprising that particularly reviled individuals and groups may in turn also come to be associated with 'dirt' or 'shit' and hence be shunned by mainstream society. Alternatively, a process of 'stigma by spatial association' may come into play, whereby those who live or work in 'dirty' environments may themselves come to be viewed as 'dirty' and treated accordingly.

Exactly such a process is explored by Chris Philo in his study of tin miners in the south-west of England. For Philo, the exclusions suffered by 'tinners' can be understood as a result of the associations drawn between tinners and the harsh and unusual environment in which they lived and worked.

From the mid-fifteenth to the closing decades of the nineteenth century, British tin mining was concentrated in the far south-west of the country, including on Dartmoor; a wild and remote upland area (later made famous in fiction as the setting for Conan Doyle's tale of Sherlock Holmes and *The Hound of the Baskervilles*). Because of its remoteness from surrounding towns and cities, the tinners of Dartmoor led a relatively isolated life: living and working in small settlements, cut off for long periods of time from their wives and families and from the rest of society.

This geographical isolation encouraged a perception of tinners as likewise socially remote. Such ideas were in turn fuelled by the associations that were drawn between the relatively 'wild' environment in which tinners worked and the wild nature of tinners themselves (see Case Study box).

CASE STUDY

The wild tinner

In the extract below, the 'wildness' of the miner is presented as a product of the dangers of mining (with 'rough work' producing 'rough characters'). But the extract also hints at a certain closeness of fit between the relatively wild environment in which the miners work and the character of the miners themselves: as they forego the 'civilized' pleasures of a cup to drink directly from their shovels, for example.

> . . . his apparel is course [sic], his dyet sklender, his lodginge harde, his feedynge commonly course breade and hard cheese, and his drincke is water, and for lacke of a cupe he drynketh it out of his spade or shovell . . . his lyffe most commonly is in pyttes and caves under the grounde of a greate depth and in greate danger because the earthe above his hedd is in sundry places crossed and posted over with tymber to keepe the same from falling.
>
> (Greeves, 1981: 79–80, quoted in Philo, 1998: 165)

Such apparent wildness positioned tinners in an ambiguous light. For some contemporaries, the unusually close relationship between tin miners and the environment they worked positioned tinners as 'close to nature' and thus as enjoying a lifestyle to be envied. Hence, in 1630 Thomas Westlake noted how the ability of the tinners to work according to the rhythm of 'nature's demand' enabled them to 'sleep soundly without careful thoughts' (quoted in Greeves, 1981: 80; in Philo, 1998: 166). Re-working the nature–culture binary, for others this same wildness invoked only the obvious distance between the tinners' lives and the moral order of settled society. Thus, in stark contrast to Thomas Westlake, nineteenth century commentators drew a series of disparaging comparisons between the productivity and moral fortitude of local farming communities and the idleness and immorality of those in the mining camps.

Importantly, however, the exclusions suffered by tinners did not go unchallenged. More specifically, while the isolation of the mining settlements was a major factor in the stigma that came to be attached to tin mining, it also provided mine workers with certain freedoms not enjoyed elsewhere. In contrast to colleagues in less specialized industries, for example, tinners were able to exert considerable control over their pay and working conditions. Under the remit of what was called 'Stannary Law', miners also stood beyond the reach of Common Law; being tried and sentenced by specialist 'stannary courts' instead – in front of a jury made up entirely of local tinners.

Though such freedoms should not be over-estimated (and there is good evidence to suggest that the stannary courts were anything but lenient) they are important. Not least, they remind us that while processes of exclusion may on occasions be vigorously resisted, a life at the margins may also, in some cases at least, offer those cast as 'outsiders' the chance to exercise a certain autonomy over their lives.

SUMMARY

- One of the most powerful sources of 'abjection' is the fear of bodily residues. As a result, especially reviled individuals or groups may also come to be viewed as 'dirt' or 'shit'.
- Those living or working in dirty environments may also come to be viewed as 'dirty' via a process of 'stigma by spatial association'.
- Such groups often come to inhabit marginal spaces and places – whether because pushed there, or as they seek some reprieve from confrontation and abuse.
- Processes of socio-spatial exclusion rarely go unchallenged.
- Marginal spaces may provide marginalized groups with a certain autonomy over their lives.

Contemporary geographies of exclusion

One of the most obviously excluded groups in contemporary societies is people who are homeless. In an important early essay, the American geographer Lois Takahashi (1996) outlined four main reasons that homeless people are so often stigmatized. First, because either under or unemployed, homeless people tend to be viewed as unproductive. In societies that accord a privileged status to economic productivity, homeless people are afforded at best a marginal position and at worst come to be viewed as a drain on collective resources. Second, having apparently lost contact with friends and family, homeless people are commonly perceived as 'disaffiliated' – as existing outside the comforts (and constraints) of mainstream society. Third, whether because of their own habits or because of stereotyping, homeless people often come to be associated with other stigmatized groups: drug addicts, alcoholics and the mentally ill, for example. Fourth, because life on the streets provides few opportunities to keep clean or even to use a toilet, for example, the sense of difference that homeless people convey, and the apparent threat they pose, is heightened by fears of pollution, or abjection.

Importantly, as Takahashi recognizes, the degree of stigma suffered by homeless people differs according (in part) to the extent to which a person might be understood as 'responsible' for the circumstances in which they find themselves. Hence, while a homeless child might be afforded considerable sympathy, a single homeless man is likely to engender a far less charitable response. Indeed most Western welfare systems continue to draw distinctions between the 'deserving' and 'undeserving' homeless; extending some kind of state sponsored support (however limited) to homeless families, for example, but leaving the care and accommodation of single homeless people to the voluntary or charitable sector.

Over the past twenty years or so it is clear that the stigma homeless people endure has translated into ever more sophisticated attempts to exclude them from the city's streets. In one of the earliest and still best known accounts of such exclusion, Don Mitchell (1997) outlined what he terms the 'annihilation of space by law' in the USA, charting the growing use of new legislation (for example, 'no camping

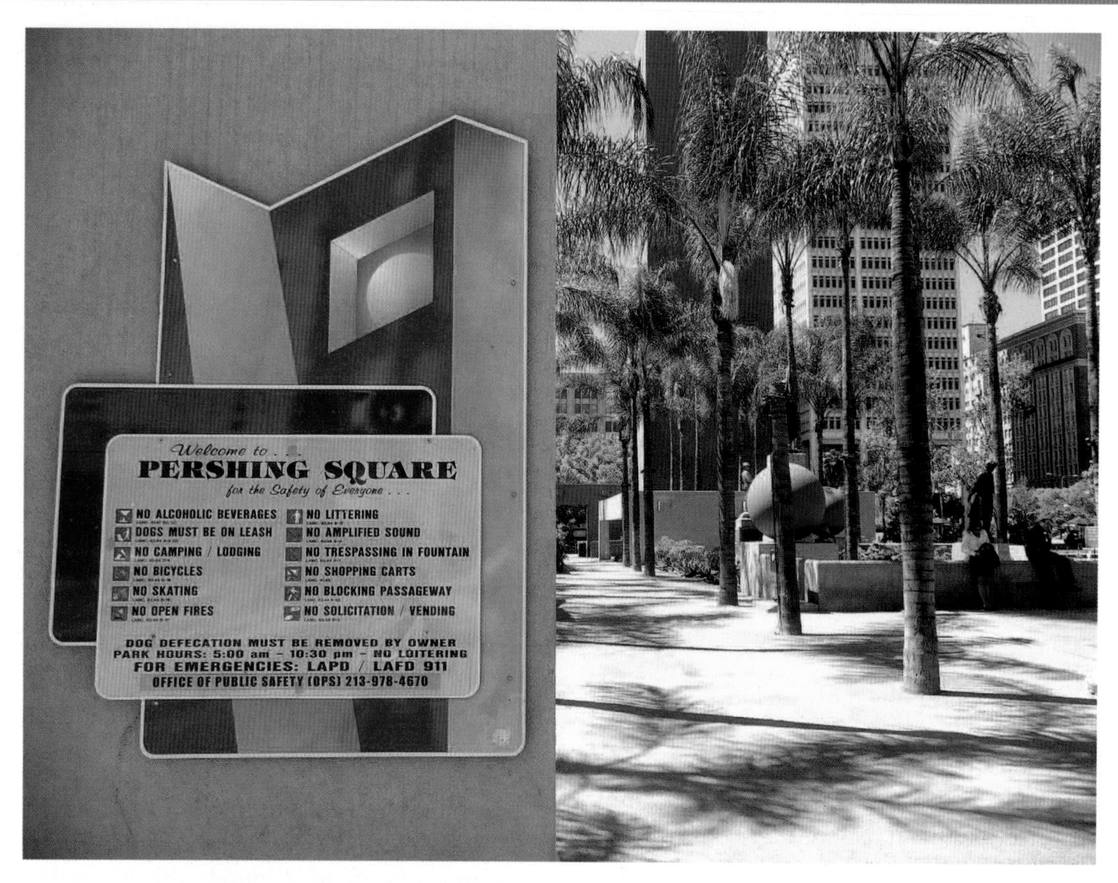

Figure 44.1 Public Space Ordinance, Pershing Square, Downtown Los Angeles. Note the careful wording which proclaims the ordinance to be for the 'safety of everyone', but which includes a number of clauses specifically designed to exclude homeless people – no 'camping' (i.e. sleeping out), no shopping carts (there are no supermarkets within a mile of this site, but shopping trolleys are often used by homeless people to move their possessions around the city) and no 'solicitation' (i.e. begging). Credit: Jon May

ordinances'), more aggressive policing (constantly moving homeless people on), and a range of new architectural developments (including the removal of public toilets and the installation of sprinkler systems and 'bum proof' benches) designed to keep homeless people out of the downtown core (see Figure 44.1 and Case Study box). Pushed out of the prime spaces of the city, street homeless people in the USA are becoming more and more concentrated in 'skid row' districts, typically located in more marginal parts of the inner/central city, though these districts too are now becoming increasingly subject to the pressures of redevelopment and gentrification.

While depressingly commonplace across the USA, with organizations like the National Coalition for the Homeless regularly publishing lists of 'America's 10 meanest cities', such developments are becoming more common elsewhere too. In the UK, the last ten years or so has seen the introduction of a range of new measures designed to help curb 'problem street culture', including the increasing imposition of Anti-Social Behaviour Orders (commonly

CASE STUDY

The rise of defensive architecture

The emergence of new forms of architecture designed to safeguard prime public spaces from the incursion of 'undesirables' was first noted by Mike Davis in his 1990 book *City of Quartz*, an account of the 'militarization of space' in Los Angeles. In an attempt to capture such changes Davis coined the phrase 'defensive architecture', a concept that includes a variety of forms: from the walls and gates that separate up-market developments from the streets beyond, to 'softer' forms of landscaping (for example, the use of spiky plants or decorative railings). Such features are often now an integral part of urban redevelopment plans and have been accompanied by a proliferation of other forms of security, from CCTV to the deployment of private security guards (paid for by local retailers) alongside the police in many towns and cities. Just as anti-homeless legislation may at first sight often appear quite reasonable (why would 'camping' be allowed in the city centre?), one of the most invidious elements of such designs is that they are hard to identify by anyone other than those they are meant to exclude. So, while the armrests used on bum proof benches may indeed make the bench more comfortable for the casual user, they also make it impossible for anyone who may need to sleep there to do so.

banning recipients from designated areas of the city, with those breaking these orders subject to arrest and imprisonment) on those who 'persistently' beg; the introduction of Designated Public Place Orders (giving police discretionary powers to confiscate alcohol from those drinking in public); and the redesign of public and commercial spaces to 'design out' 'problematic' street activities (Johnsen and Fitzpatrick, 2010).

In seeking to explain the growing exclusion of street homeless people from the central city, Mitchell points to growing concern among many local governments and businesses to bolster their city's image in an era of intense inter-urban competition: reading such developments as an attempt to secure the *local* conditions (of a prosperous and attractive cityscape) demanded by *international* capital, consumers and tourists (Mitchell, 2003). Whilst there is indeed a powerful economic logic at work here, it is also important to recognize the symbolic register through which such exclusion works. In Britain at least, recent attempts to 'secure the streets' from 'problem street culture' are also part and parcel of amoral campaign to re-instigate a culture of 'respect' (in the terms used by the Labour party/ government) and to help mend 'Broken Britain' (in terms used by the Conservative party and Coalition government). Read through this lens, it is possible to see the connections between processes of exclusion and inclusion, with attempts to exclude homeless people from the public spaces of the city – and hence from a wider 'public' itself – simultaneously signalling an attempt to secure a more coherent sense of the 'respectable' public at a time when civic culture is understood to be in crisis.

It is also important to recognize that such developments are by no means absolute or universal, and continue to be resisted – by both homeless people themselves and others. Hence, ethnographic studies have revealed the uneven geographies and variations in homeless policing regimes that unfold according to different 'contours of tolerance' across the city, for example, with some security officers and police more likely to turn a 'blind eye' to street homeless people than others, and with varying intensities of policing in different parts of the central city and according to different times of day and night (see Figure 44.2). They also show the sophisticated negotiation of such developments by street homeless people themselves who, far from being passive victims, display remarkably detailed knowledge of when and where their presence in prime public space is likely to be tolerated, and of other, more marginal spaces (side streets and alleyways, for example) within the central city where they can continue to make shelter for themselves (Cloke *et al.*, 2008).

More generally, alongside these attempts to exclude, a number of scholars have also traced the continuation of much more supportive approaches to the 'management' of street homelessness in recent decades, including increased state funding to those providing services for street homeless people, in cities across the USA and elsewhere (DeVerteuil

Figure 44.2 Homeless man, Pershing Square, Downtown Los Angeles. A homeless man catches up on some sleep in Pershing Square, feet away from the square's 'bum proof' benches and a passing police officer. Credit: Jon May

et al., 2010; Laurenson and Collins 2006, 2007). For Cloke *et al.* (2010), the continued presence of such services – the majority of which are provided by voluntary sector organizations and charities – signals a still deep rooted and powerful 'urge to care' for homeless people among many members of the housed public, and of attempts to include rather than exclude.

SUMMARY

- Homeless people are highly stigmatized.
- Over the past twenty years or so this stigma has translated into growing attempts to exclude street homeless people from the central and business districts of many cities across the USA and UK.
- The means of such exclusion are many and varied, including new legislation and new architectural designs, which together are pushing homeless people in to more marginal parts of the city.
- Such practices are neither universal nor absolute, and continue to be resisted by homeless people and others.

Inclusion/exclusion

In the previous section it was suggested that the increasing exclusion of homeless people has not gone uncontested. One of the most obvious and important challenges to such exclusion is articulated by the growing number of services catering to the needs of street homeless people in both the USA and Britain. Focusing on homeless day centres in Britain, this section examines the complex interplay of processes of inclusion and exclusion articulated within these spaces.

Day centres first emerged in Britain in the late 1970s as a response to the limited state services available to homeless people. The majority are provided by voluntary or charitable organizations, with many of the services they offer provided directly by volunteers. At a basic level, day centres offer a challenge to the spatial dynamics of homeless people's exclusion. Most obviously, perhaps, with nowhere to go but increasingly unable simply to stay still, street homeless people must 'always keep walkin', as Vanderstay puts it (1992: 39; quoted in Kawash, 1998: 328). As such, day centres offer homeless people an important place of pause and the chance to recuperate within the daily round of movement. They also cater to the immediate material needs of homeless people by providing advice and information, inexpensive or free food and drinks, clothing and bedding, bathing, laundry services and so on.

At a deeper level of analysis, day centres seek to provide ways of overcoming the stigma that many homeless people experience. In the previous section it was suggested that an important element of this stigma relates to the challenges of keeping clean on the streets (Takahashi, 1996), with the dirt and smell of the homeless body mapping onto longstanding and powerful fears of abjection among members of the housed public (Kawash, 1998).

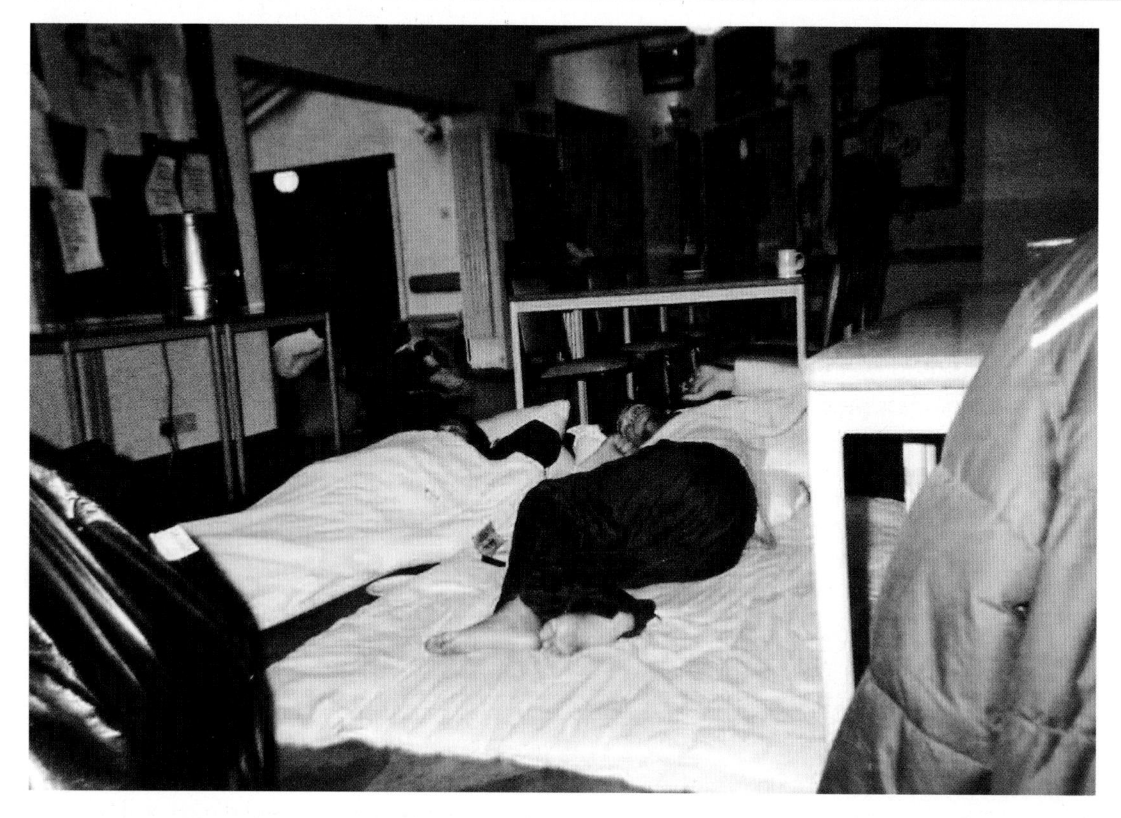

Figure 44.3 Day centre for the homeless. Credit: Jon May

Understood in these terms, the opportunity that day centres provide to use a toilet, take a shower or change one's clothes takes on very considerable significance.

Within the space of the day centre itself, staff and volunteers may also seek ways to confront such stigma more directly. Indeed, it is clear that most day centres are extremely tolerant of expressions of difference, providing a space of 'unusual norms' where bodily appearances, odours or behaviours that would likely be considered 'odd' or 'inappropriate' in other contexts are accepted without remark, as Cloke *et al.* (2010: 130) demonstrate (see Case Study box).

CASE STUDY

Geographies of license

In their examination of the geographies of homeless service provision, Cloke *et al.* (2010) undertook extended ethnographic research in a range of homeless service spaces. In the extract below, taken from their field diary, they describe the reaction of other day centre users to an elderly man suffering from problems of mental ill-health. In contrast to the responses such behaviour might more usually elicit (ranging from studied indifference to fear or hostility) it is noticeable that his increasing agitation is met with quiet acceptance and even kindness. The extract offers a good example of what Parr (2000), drawing on Goffman (1968), refers to as a 'geography of license'; the idea that different behaviours are understood as more or less acceptable in different spaces.

> The elderly man with a scraggly beard, quite emaciated and dressed in a filthy grey trenchcoat, was seated by himself at the table behind us, eyes half closed and chin on chest, incessantly mumbling to himself as he usually does. At one point the volume of his voice rose dramatically and he began to swear profusely, appearing to be very upset (although nothing inadvertent had happened). One of the other service users seated at my table called out to him 'You're alright Bob' and then re-entered our conversation, seemingly unphased. The old guy immediately calmed down, and resumed mumbling quietly in his usual manner . . . I've witnessed this happen on several different occasions. It appears that his mumbling is quietly accepted by the other service users who only intervene (and then kindly) when he gets genuinely upset by whatever is going on in his head.
>
> (Source: Cloke *et al.*, 2010: 130–1)

In a challenge to the exclusions they face on the streets, day centres therefore seek to provide homeless people with a more inclusive space of their own. They offer both material resource and, importantly, a place where the bodily aesthetics (dirt and smells) and behaviours (falling asleep during the day, because tired having spent the night moving around the city rather than risk sleeping in the open, for example) that are an inevitable and unavoidable part of the homeless experience are less likely to be considered 'out of place'.

As argued at the beginning of this chapter, however, processes of exclusion and inclusion are irretrievably connected. Just as the increasing exclusions suffered by homeless people on the streets has helped engender a more inclusive response from voluntary organizations and charities, within these service spaces themselves the same dynamics that work to include one group will often work to exclude others. Such exclusion works across a number of axes. As Cloke *et al.* (2010) argue, far from some kind of homogenous 'homeless community', those using day centres reflect the same diversity and prejudices apparent in wider society and confront the same problems of sexism, racism and homophobia. In particular, most day centres are male-dominated spaces (with Cloke *et al.* estimating some 74 per cent of users to be men) and can be intimidating places for women. Many homeless women try to avoid them altogether or, if turning to day centres for help, will attempt to keep a 'low profile' when doing so because they are wary of attracting the attention of other service users. This makes it more difficult for such women to catch the attention of staff and hence access the support they need (see May *et al.*, 2007).

In a rather different vein, Rowe and Wolch (1990) have argued that day centres expose the newly homeless to what may for them be very alien experiences of homelessness, poverty and substance abuse. As a result, what might be characterized as a 'space of licence' for those more familiar with such spaces, may be a distressing and even frightening place for others – if only because it is full of homeless people.

Elaborating on these divisions within the street homeless population, Cloke *et al.* (2010) outline divisions between what day centre users themselves refer to as 'straightheads', 'pissheads' and 'smackheads': those with no addictions, alcoholics and drug users. While there is often considerable animosity between these different groups, and between the first two and 'smackheads' in particular, it is noticeable how when describing the latter homeless people sometimes draw upon and reinforce the same constructions of stigma they themselves face from members of the housed public. In the quotation below, for example, 'Frank' complains of 'smackheads" apparent lack of respect for their own body and personal hygiene, reactivating fears of abjection and outlining the dangers (in this case, from discarded hypodermic syringes) such people apparently pose:

> They are really dirty people, most of them . . . vile . . . They've never heard of soap and water and leave their syringes all around the day centre . . . They're all scummy to me. I wouldn't trust them as far as I could throw them.
>
> ('Frank', homeless day centre user, quoted in Cloke *et al.*, 2010: 135)

Far from a simple or singular space of inclusion, which can be counterpoised with the exclusions suffered by homeless people on the streets, day centres themselves therefore articulate a complex mix of inclusions and exclusions. Day centre staff are well aware of these dynamics, of course. In practice, most recognize that in order to provide as inclusive a service as possible to as many people as possible, some exclusionary practices are necessary – whether a ban on drugs and alcohol, or the barring of individuals who are violent or abusive towards staff or other services users, for example – even if such practices necessarily exclude some of their most vulnerable clients.

As a result, day centres tend in fact to operate according to a classically libertarian conception of rights which is – ironically perhaps, given its place in the 'free market' philosophies that underpin interurban competition – no longer extended to those on the streets (Mitchell, 2005): with services available to all whose use does not damage another's ability to access those services. Inherent within this model of rights is the recognition that the parameters of inclusion and exclusion – policed by means of inclusion and exclusion from the space of the centre itself – are mutually constitutive; the inclusion of some necessarily excluding others, with this balance always the subject of intense negotiation.

SUMMARY

- Day centres offer a space of respite and material resource to homeless people.
- Within day centres, managers, staff and clients work to produce a more inclusive space of 'unusual norms' and 'license', in which bodily dispositions and behaviours considered 'out of place' and perhaps threatening elsewhere are accepted as the norm.
- Not all people are equally included in day centres, with divisions and exclusions evident across multiple axes.

- Day centres reveal processes of inclusion and exclusion to be irretrievably interconnected – the one necessarily producing the other – and hence subject to conflict and negotiation.

Conclusions

The study of exclusion is central to Social Geography. Spurred on by Sibley's call to consider the geographies of 'outsiders', Social Geographers have explored the experiences of such groups, the reasons for their exclusion and the role that space plays in maintaining the distinction between mainstream society and those constructed as Different or Other.

Though a range of 'outsider' groups have been considered, a good deal of recent work in this field has focused on the growing exclusion of street homeless people from urban public space. The best of such work recognizes that exclusion works across multiple dimensions, is inevitably contested, and traces the complex interconnections between processes of inclusion and exclusion.

DISCUSSION POINTS

1. Identify particular groups you feel might suffer from exclusion (for example, children, elderly people or homeless people). How is their exclusion articulated – in which spaces are such groups tolerated and from which are they excluded?
2. How is such exclusion resisted – and by whom? To what extent and how might an excluded group 'take ownership' of a space at the margins?
3. Identify a particular institutional space (for example, prisons, universities or homeless hostels). How is the space of this institution organized so as to promote the exclusion/inclusion of different groups?
4. What kinds of *methods* are best suited to the study of exclusion?

FURTHER READING

Sibley, D. (1995) *Geographies of Exclusion: Society and Difference in the West*. London: Routledge.

Still the most comprehensive account of the theories of exclusion and very readable.

Special Issue on 'Exclusion', *Geoforum* 1998, Volume 29, Issue 2.

Useful examples of studies of the geographies of exclusion, inspired by Sibley's work.

Mitchell, D. (2003) *The Right to the City: Social Justice and the Fight for Public Space*. New York: The Guilford Press.

A collection of landmark essays exploring the politics of public space and offering detailed accounts of the growing exclusion of homeless people from the streets of American cities.

Cloke, P., May, J. and Johnsen, S. (2010) *Swept up Lives? Re-envisioning the Homeless City*. Oxford: Wiley-Blackwell.

An important counterpoint to Mitchell, charting the complex and varied attempts to care for homeless people articulated by homeless service providers in Britain.

CHAPTER 45
DIASPORAS

Claire Dwyer

Introduction

Between 2005 and 2010 the Arts and Humanities Research Council in the UK funded their first strategic research programme called 'Diasporas, Migration and Identities'. Over the five-year period, 45 different research grants were allocated to a diverse range of projects undertaken by researchers across the humanities, including a number of geographers. Included were a study of Afghan music in London; 'Tuning-In', a project about the BBC World Service; an exploration of diasporic food cultures and their sensual experiences; a history of London's Chinatown; 'Home and Away', a study of the experiences of transnational South Asian children; an analysis of the migratory networks of Roman Britain; a study of films about Irish migrants to America in the early twentieth century; and an analysis of conversational 'code-switching' in bilingual immigrant families (see www.diasporas.ac.uk for more details). A collaborative project between geographers and curators at the Victoria & Albert Museum, 'Fashioning Diaspora Space', examined the history and geography of cultural exchange between Britain and South Asia in the fields of textiles, dress and fashion (see Figure 45.1) (Breward *et al.*, 2010). Although the projects spanned different time periods, used different methodologies, focused on very different material objects and cultural practices (including music, radio, TV, film, dance, literature, fashion and textiles) and were about a wide range of different immigrant communities and places, all of the projects were united by a particular theoretical framework – one which linked the study of migration to the idea of diasporas. In this chapter I explore the idea of diaspora as a way of thinking about migration flows and the transnational connections which they produce. I consider what such flows and connections mean for the social geographies of identity, difference and place. In particular, I want to use the concept of diaspora – and some examples of the social, economic, cultural and political practices associated with diaspora populations – to emphasize geographies of flows and connections between places. I will argue that theories of diaspora offer new ways of thinking about the connections between global–local geographies (see also Chapter 1) and in so doing, enable us to rethink the relations between society and space beyond a national frame.

I begin by emphasizing the links between migration and globalization, suggesting the ways in which these two processes are linked in the production of spaces that cross national boundaries – what I term 'transnational spaces'. I then provide a more detailed discussion of the concept of diaspora and the different kinds of

Figure 45.1 London flats and Indian textile collage by Helen Scalway (2009). An image produced as part of the Fashioning Diaspora Space project. Credit: © Helen Scalway

geographies it might suggest. In the second half of the chapter I discuss some examples of what I define as diaspora cultures and diaspora spaces to illustrate some of the different ways in which geographers and others have used this concept.

Globalization, migration and transnational space

Migration has long been of interest to geographers (see also Chapter 41). However, contemporary studies of migration have revealed new issues. In particular, there is less attention to models or theories of migration which suggest that migration is a permanent move from one place to another. Instead, geographers emphasize that many current migration flows are associated with webs of **transnational** linkages and connections. Migration movements can be seen as an integral part of **globalization** – those economic, social, cultural and political processes whereby places across the globe are increasingly interconnected (Giddens, 1990) through the compression of time and space (Harvey, 1989). Migration results in the 'stretching out' of social relations between people across space (Massey and Jess, 1995), as individuals retain links within a community that is spread out across national boundaries. One of the earliest studies to demonstrate these new transnational communities was an ethnography of migrants from Aguililla in Mexico living in Silicon Valley in California (Rouse, 1991), whose everyday social fields operated across national borders within one 'transnational space' (Jackson *et al.,* 2004).

While transnational links have long been maintained through visits or the exchange of letters, the globalization of telecommunications means that transnational space can be much more immediate, complicating older understandings of presence, absence and distance. So, for example, families may

CASE STUDY

Our Town with the Whole of India! by Daljit Nagra

Our town in England with the whole of India sundering
out of its temples, mandirs and mosques for the customised
streets. Our parade, clad in cloak-orange with banners
and tridents, chanting from station to station for Vaisakhi
over Easter. Our full-moon madness for Eidh with free
pavement tandooris and legless dancing to boostered
cars. Our Guy Fawkes' Diwali – a kingdom of rockets
for the Odysseus-trials of Rama who arrowed the jungle
foe to re-palace the Penelope-faith of his Sita.

Our Sunrise Radio with its lip sync of Bollywood lovers
pumping through the rows of emporium cubby holes
whilst bhangra beats slam where the hagglers roar
at the pulled-up-back-of-the-lorry cut-price stalls.
Sitar shimmerings drip down the furbishly columned
gold store. Askance is the peaceful Pizza Hut . . .
A Somali cab joint, been there for ever, with smiley
guitar licks where reggae played before Caribbeans
disappeared, where years before Teddy Boys jived.

Our cafes with the brickwork trays of saffron sweets,
brass woks frying flamingo-pink-syrup-tunnelled
jalebis networking crustily into their familied clumps.
Reveries of incense scent the beefless counter where
bloodied men sling out skinned legs and breasts
into thin bags topped with the proof of giblets.
Stepped road displays – chock-full of ripe karela,
okra, aubergine - sunshined with mango, pineapple,
lychee. Factory walkers prayer-toss the river of
sponging swans with chapattis. A posse brightens
on park-shots of Bacardi – waxing for the bronze
eyeful of girls. The girls slim their skirts after college
blowing dreams into pink bubble gum at neck-
descending and tight-neck sari-mannequins. Their grannies
point for poled yards of silk for own-made styles.
The mother of the runaway daughter, in the marriage
bureau, weeps over the plush-back catalogues glossed
with tuxedo-boys from the whole of our India!

(Source: *Look we have coming to Dover!* by
Daljit Nagra, published by Faber & Faber, London, 2007)

communicate frequently with each other via Skype, mobile phone and email. A study of female Filipino migrant care workers in Britain shows how migrant women are involved in 'transnational mothering': talking to their children in the Philippines on a daily basis via the internet, advising about homework and sharing problems (Madianou and Miller, 2011). The media are also important in creating transnational social fields – satellite television viewers in Germany, Turkey and Britain may all watch the same programmes on the same channel (Robins and Aksoy, 2003), while the Bollywood film industry is now orientated as much to the lucrative audience of Indians living outside India as to the domestic audience (Desai, 2004; Gillespie, 1995). Diaspora audiences may also seek news about their homeland and international events in their own language or via local networks, as Marie Gillespie's study of British families watching the television station Al-Jazeera to get information about the war in Iraq explains (Gillespie, 2002). As Gillespie's research suggests, telecommunications facilitate not only closer familial contact and shared cultural experiences, but also enable citizens living outside their country of origin to participate politically too. Khalid Koser describes the ways in which Eritreans living abroad were involved in the making of the new Eritrean state (Koser, 2003), while many studies point to the significance of remittances, in the form of both money and the transfer of goods, from overseas migrant populations to relatives and communities from their homeland (Mercer *et al.*, 2009). For the states from which migrants have left, diaspora populations abroad may represent significant sources of both political influence and economic investment and many countries actively cultivate connections with their diaspora populations.

These forms of transnational relations raise important questions about culture and place. If social, political and cultural relations are stretched out across space, then cultural identity can no longer be thought of as being bounded in one place (such as the nation). To return to the Bollywood example mentioned above, in its global popularity we see how Indian culture is neither confined nor solely oriented to India. Nor is culture simply being transported by migrants from one place to another. Instead, these transnational links suggest a more dynamic conceptualization of cultures being formed and transformed across and between national boundaries. These ideas about dynamic and transnational cultures are often expressed particularly effectively in poetry, music, literature and art. Daljit Nagra is a British poet, born in the UK to migrant parents from India. Nagra's poetry captures the dynamics of growing up 'in between' Britain and India, particularly through his expressive language, which mixes various forms of English and Punjabi. The extract in the box is from his award-winning first book of poetry *Look we have coming to Dover!* and reflects on the urban landscapes of his hometown and their shaping by migrants from India. Poetically, Nagra here evokes transformations of place and social life of great interest to geographers in a wide variety of contexts. Patricia Ehrkamp (2005), for example, has analysed the urban social geographies resultant from Turkish migration to Germany, showing how the public spaces and streets of cities (specifically the neighbourhood of Marxloh in the city of Duisburg, sometimes called 'Little Istanbul') play out positionings with regard to both German and Turkish society and space.

SUMMARY

- Geographers increasingly understand migration as a complex system of transnational movements rather than a permanent move from one country of residence to another.
- Migration movements are integral to the processes of globalization whereby different places and people across the globe are increasingly interconnected.
- Migrant communities have 'stretched out' social relations and networks that cross-cut national boundaries and borders to form 'transnational spaces'.

Diasporas

A concept that is helpful in understanding these 'stretched out' or 'extroverted' social fields is the idea of *diaspora*. The term diaspora is an attempt to encompass the different and complex belongings of peoples who may be dispersed across geographical boundaries and may have connections to several different places they call 'home' (Clifford, 1994). While diaspora has sometimes suggested a fixed point of origin from which a group of people initially dispersed, its evocation is more often understood as a challenge to the idea of rooted identities with fixed places of origin. Writings about diaspora are often an attempt to explore the ways in which cultures are created through the fusion and mixing of different elements.

Etymologically, the word diaspora derives from the Greek verb *speiro* (to sow, or to scatter) and the preposition *dia* (over, or through). In Ancient Greece the term referred to the process of migration and colonization. A dictionary definition states that the term diaspora refers to 'dispersion from', suggesting some sort of exile from an original homeland. This definition resonates with the biblical description of the dispersion of the Jewish people after their exile in Babylon, often seen as the original use of the term diaspora. While some writers have defined the Jewish diaspora as an 'ideal type' (Safran, 1991), the term now has a much wider currency to describe a variety of different kinds of diasporas. What all these diaspora communities share, however, is the memory of shared journeys or migrations and a recognition that another place, another 'homeland', has some claim on their emotions and loyalties. Since the term diaspora has been used to evoke a range of different historical experiences there has been much debate about its theoretical reach. As James Clifford (1994) reflects, there is often slippage between diaspora as a theoretical concept and distinct historical experiences of diaspora. In order to engage with this tension I want to outline briefly a few different examples of diasporas, recognizing that each has a specific history and geography. However, I then move beyond this delineation of different types of diasporas and their different geographies to explore the theoretical possibilities of the term diaspora for reconceptualizing the relationships between people and place.

In his book *Global Diasporas*, Robin Cohen suggests that the potentially common features of diasporas include: 'dispersal from an original homeland, often traumatically, to two or more foreign regions'; 'a collective myth and memory about the homeland'; 'a strong ethnic group consciousness sustained over a long time and based on a sense of distinctiveness, a common history'; 'a sense of empathy and solidarity with co-ethnic members in other countries of settlement'; 'a troubled relationship with host societies suggesting a lack of acceptance' (Cohen, 2008: 17). Within this commonality, Cohen produces a typology of different kinds of diasporas related to different causes for migration and different dynamics of connection. For instance, one of his types is 'trade diasporas', including the migration of Chinese traders and business people to other parts of Asia and North America, both historically and more recently (Ley, 2010; Ong, 1999), and Lebanese traders' migration to Europe and North America. He labels a second type 'victim diasporas'. Along with the classic Jewish diaspora, under this rubric he includes forced population movements such as the violent slave trade that transported millions of Africans to the Caribbean and North and South America to work in tropical plantations in the eighteenth and nineteenth centuries. While the term African diaspora was not really used until the 1950s, evocations of the Jewish exile recur in the musical and religious cultures that have developed within this transatlantic African diaspora, most obviously in Rastafarianism and reggae music, epitomised in Bob Marley and the Wailers' famed album *Exodus.*

Another example of a global diaspora associated with coerced population movement is the indentured Indian labour deployed in the plantation colonies of the British Empire (framed by Cohen as a 'labour diaspora'). In the nineteenth century, slavery was replaced by the movement of indentured labourers from the British Empire in India to work on colonial plantations in Africa and the Caribbean. One legacy of this population movement was another, later forced population movement – the expulsion from Uganda in 1972 of Ugandan Asians who were the descendants of indentured labourers. Many of these expelled Ugandan Asians came to Britain, where they still held citizenship, thus completing another process of migration within the Asian diaspora, as 'twice migrants' (Bhachu, 1985).

More generally, in the late twentieth and early twenty-first century processes of globalization have transformed population movements and, of course, new political and economic circumstances shape migration flows. Brazilian diasporic populations are now apparent in global cities such as New York and London (Margolis, 1998); a complex, differentiated Zimbabwean diaspora can be identified in the UK (Pasura, 2010); Filipino domestic and care workers reside in cities around the world, from Hong Kong (Law, 2001) to Vancouver (Pratt, 2012); and so on. At the same time, however, many contemporary population movements have close associations with their historical antecedents. In the post-war period of labour recruitment in Britain, new migrations took place from South Asia and the Caribbean. While this migration was voluntary rather than forced or enslaved migration, it continued the historical connections established under the labour conditions of the Empire and established new webs of diaspora connections as part of what Cohen calls a 'cultural diaspora'.

As suggested above, writers like James Clifford have argued that in addition to the mapping out of distinctive historical diasporas we can also use the term diaspora as a theoretical tool to describe a particular understanding of the relationship between people and place. For Clifford, a diasporic identity is one that encompasses both a consciousness of displacement and a recognition of multi-locational attachment. It is also a form of resistance, a cultural politics, opposed to nationalism and assimilation. Avtar Brah (1996), a British Asian writer interested in questions of identity and belonging, argues that conceptually diaspora is about holding in 'creative tension' the idea of 'home' and 'dispersion' (1996: 192). What she means by this is that while we can recognize a 'homing desire' we should also challenge discourses about fixed origins. So diaspora thinking is a challenge to **essentialist** and often even racist discourses that imagine the world in relation to fixed identities, borders and nations. Instead, it is a celebration of the possibilities of cross-border thinking and of making connections and networks. As Brah suggests: 'Diaspora identities are at once local and global. They are networks of transnational identifications' (1996: 196). While many of the writers who use the term diaspora may not be geographers, we can see here how the term is a key geographical notion. Diaspora emphasizes a way of thinking about geography in terms of connections, flows and networks – what Doreen Massey describes as an 'extroverted geography' of links and interconnections (Massey, 1991) – rather than in terms of boundaries and borders. In the much quoted words of cultural theorist Paul Gilroy (1993), diaspora thinking is about *routes* not *roots*.

Of course, there are tensions within this field of diaspora studies. As you may have gathered, the politics of diaspora work are both 'hopeful' and 'critical': hopeful that diasporic lives, cultures and concepts can help to forge relations between society and space in less bounded, exclusionary ways; but critical of the suffering caused to diasporic populations in their experiences of displacement and marginalization. Relatedly, diaspora studies combine more abstract theorizing with analysis of the lived realities of individuals and social groups. So, for example, in responding to the lives of Filipino domestic workers in a city like Vancouver, Canada, diaspora scholarship might both recognize and celebrate how their presence troubles narrow, exclusionary thinking about Canadian society and nationhood, and also highlight the profound, intimate pain caused to Filipino mothers and their children as they conduct their own family lives transnationally (Pratt, 2012).

In the next two sections I want to look at some examples of different ways of thinking about the geographies of diaspora, both hopeful and critical. First, I give two examples of 'diaspora cultures', drawing on the work of Paul Gilroy on Black Atlantic musical cultures and then considering work on British-Asian fashion cultures. I then turn to explore the idea of 'diaspora spaces', drawing on some of my own recent work on transnational suburban religious landscapes.

SUMMARY

- Diaspora describes dispersed communities that share multiple belongings to different places or 'homes' in different national spaces.
- Diaspora studies have sometimes been historically and socially focused, describing the migration histories and lived realities of specific diasporic groups.
- Diaspora studies have also had a more conceptual and political emphasis: challenging the idea of fixed 'roots' or origins, questioning the framing of society in terms of insiders and outsiders, and emphasizing instead transnational connections and linkages.

Diaspora cultures

The black Atlantic

As I outlined above, one of the most important migrations in history was the 'triangular trade' of slaves, which dispersed black people from Africa across the Americas. This forced migration formed the basis for what Paul Gilroy (1993) describes as a black Atlantic diaspora. This is a diaspora culture that links African-Americans, Caribbeans, black British people and peoples in Africa. In his exploration of the cultures of the black Atlantic diaspora, Gilroy uses metaphors of travel – recalling the ships that first took slaves to America as well as those that brought Caribbean migrants to Europe in the post-war period. He argues that the cultures of the black Atlantic are characterized not by a return to African 'roots' but by the interconnection of many interlinked 'routes' between different places.

This is emphasized through his exploration of the connections between different kinds of musical forms. He argues that the music of the black Atlantic diaspora – blues, reggae, jazz, soul, rap – have all been produced through particular fusions of influences in different places. Thus on the plantations African music was transformed, particularly through a fusion with other kinds of music such as European religious music. When slaves later migrated to the cities, new musical forms such as rhythm 'n' blues and jazz were created, and in each case these were shaped by new interconnections in particular places. New Orleans jazz was a different music from other forms of jazz, such as Afro-Cuban. Gilroy traces these connections to music produced in Britain by post-war migrants including British adaptations of Jamaican reggae, while noting reggae itself was born of fusions between Jamaican folk music and American rhythm 'n' blues. Similarly, the production of contemporary black music, such as rap, had its origins in the 1970s, in the fusion of the music of the Jamaican sound systems with the 'talk-over' of New York Bronx DJs (Hall, 1995b) (see Figure 45.2).

Gilroy's argument is that all these musical forms are the results of fusions of different cultural traditions within the black Atlantic diaspora. While all of the musical forms retain some distinctive elements of what might be deemed 'African', they have been transformed within different geographical and national contexts. Through these connections Gilroy unsettles a notion of tradition or fixed origins. The musical forms of the black Atlantic are not diluted forms of 'traditional' African music but are new **syncretic** forms produced within a

Figure 45.2 Tinchy Stryder. Credit: Getty Images

diasporic culture through cultural flows or interconnecting routes. And indeed these routes are not simply one-way – American and British black music is also played and bought in Africa, influencing the ways in which contemporary African music is produced.

Gilroy (1987) also traces these interconnections within black British youth cultures – black styles, music, dance, fashion and language – emphasizing the extent to which black British culture must be understood as an integral part of a black Atlantic diaspora. He argues that black cultures draw inspiration from those developed by populations elsewhere. In particular, the culture and politics of black America and the Caribbean have become raw materials for creative processes that redefine what it means to be black, as they are adapted to distinctively British experiences and meanings (Gilroy, 1987: 154). In this way, black British youth culture is actively made and re-made within a diasporic community of connection across the black Atlantic. At the same time, those black British youth cultures increasingly transform a British youth culture that is not confined to those with black identities as ascribed by race or **ethnicity**. British culture is transformed by diasporic currents that ripple through and beyond the social category that has come to be termed 'Black and Minority Ethnic', through place-based friendship networks and youth culture media and industries.

There are other examples of diaspora musical cultures that we could also draw on here to explore these issues further, such as bhangra and other kinds of Asian dance music (see Huq, 2006). However, I want to turn to an alternative cultural space, that of fashion, for my next example.

Transnational fashion

Fashion is a rich, but complicated, cultural space for examining the making of diaspora cultures. On the one hand, we can think of the ways in which clothes are often used as fixed markers or boundaries for 'ethnic' identities. Think of the ways in which 'traditional' costume is used at key international events like sports competitions to denote national identities, or the marketing of costumed dolls as tourist souvenirs. On the other hand, anthropological accounts demonstrate the extent to which dress codes are never fixed and are constantly open to change and redefinition (Eicher, 1995). More particularly, there has been much debate about the 'appropriation' of different forms of ethnic dress by elite fashion designers in the current vogue for cross-over or hybrid designs (Niessen *et al.*, 2003). How fashion operates as a site of cultural exchange and fusion is therefore complex. Here I want to consider some examples from a case study of the British-Asian fashion industry to demonstrate the possibility of diaspora cultures.

In a book about Asian fashion in Britain, Parminder Bhachu (2004) emphasizes the 'diaspora economies' within which the making and marketing of Asian clothing in Britain are being transformed. Bhachu traces a 'cultural narrative' of the salwaar-kameez, or Punjabi suit, from an item of negatively coded 'ethnic clothing' to a highly fashionable garment that was given particular prominence on the global fashion stage when worn by the late Princess Diana. Bhachu's analysis of these diaspora economies focuses on two groups of women entrepreneurs: those involved in design and retail, and the home-based seamstresses who produce outfits on a much smaller scale. Her argument is that the making of these suits involves complex, dialogic and diasporic processes, for example in the adaptation of patterns, fabrics and designs for different markets. The older African-Asian women she interviewed were pioneers in adapting patterns, sharing designs with relatives abroad and procuring fabrics from different places, producing unique garments that represented all these diasporic connections and networks. Younger Asian designers combine British and Asian sensibilities to produce 'fusion' designs that are both highly fashionable and ethnically inflected. Like the musical cultures described by Gilroy, these garments only make sense at this crossroads of cultural space. To illustrate this further I want to use the examples of two different designers I interviewed as part of a wider research project on South Asian commodity culture and transnationality (see Crang *et al.*, 2003).

Figure 45.3 Asian Wedding magazine, autumn 1999. Credit: Vincent Dolman/Asian Wedding magazine

Raishma is a fashionable boutique selling designer Asian fashion in East London (Dwyer, 2004, 2010). The eponymous designer, a young British-Pakistani woman, is a fashion graduate with experience in both the mainstream bridal fashion industry and the specialist Asian fashion industry. She specializes in evening and wedding wear and has high-profile clients including the comedy actress Meera Syal. Raishma describes her designs as 'East/West', emphasizing that they represent a combination of both 'western fashion', evident in the styles and shapes used, and 'eastern traditions', evident in the fabrics and the traditional hand-beaded embroidery. For example, her wedding outfits deliberately use colours like cream and white rather than the more traditional red, while also developing variations on more conventional outfits such as her separate bodices and skirts, or her backless, 'sari-dress'. In her marketing Raishma also self-consciously challenges expectations by putting Asian models in rural, 'English' landscapes (see Figure 45.3).

Raishma's clothes are bought by young British-Asian women seeking outfits that reflect their syncretic identities as simultaneously British *and* Asian, but they are also bought by a wider clientele as Raishma capitalizes on the current enthusiasm for an Asian aesthetic. For example, Raishma obtained widespread media coverage when she designed an outfit worn by one of the guests at the wedding of Prince William and Catherine Middleton in April 2011.

Raishma's clothes reflect a diasporic aesthetic in their fusion of different influences and they are also evident of a process of production that depends upon Raishma's own positioning within the Pakistani diaspora. Raishma's clothes are produced in Pakistan and design sketches are faxed back and forth between Lahore and London as the transnational process of production is completed.

Liaqat Rasul was the owner of the designer label Ghulam Sakina, which was launched in

1999 with Liaqat's degree show (for an extended discussion of this case study see Dwyer and Crang, 2002; Crang 2010). Unlike Raishma, Liaqat's clothes were not sold within the specialist Asian retail market; they were sold particularly in the London department store Liberty as well as a number of different outlets including some in Japan, thus reaching a diverse market. However, like Raishma, Liaqat's fashion designs can be understood within a diasporic cultural space. British born of Pakistani parents, Liaqat seeks through his clothes to develop a 'multicultural aesthetic'. He defines this as a juxtaposition of different elements 'jarring against each other'. His clothes are also about 'longevity and nostalgia' and are made from natural fibres that are dyed using traditional block-printing techniques and hand embroidery. The resulting garments represent complex combinations of different elements. In the dress shown in Figure 45.4, these include old packing fabric printed with Hindi script, a cutting taken from a borrowed paisley duvet, and a decoration based on the Bengali Kantha, or running stitch, traditionally used to recycle materials by sewing them together.

Figure 45.4 Liaqat Rasul Graduate Collection 1999 'Observational Composite'. Credit: Masoud Golsorkhi

If Liaqat's garments are a diasporic fabrication in their weaving together of many different elements, their production also emphasizes transnational connections. Liaqat used contacts gained through his work in India with Indian designer Ritu Kumar, and specialist handicraft producers Anokhi, to produce clothes in collaboration with Indian craftsmen as well as in more commercial workshops. Like Raishma, he was constantly involved in a transnational dialogue and travelled between London and Delhi to monitor the production process. Diasporic aesthetics are related to transnational social fields.

These two examples illustrate the emergence of a diaspora material culture. Like the musical cultures of the black Atlantic, these fashion spaces emerge out of the crossroads of diaspora history, representing both an aesthetic of fusion and syncretism and a material production process shaped by transnational networks. These diasporic fashions are also shaping the wider fashion economy within which fusion styles like those described here are increasingly part of the 'mainstream' western fashion market. Whether or not this **commodification** of difference can be straightforwardly interpreted as politically progressive and evident of a broader diasporic consciousness is, of course, complex and contested (Jackson, 2002; Hutnyk, 2000). However, it is clear that, like musical cultures, fashion can be a very important site within which individuals may carve out spaces of identity which make sense

Figure 45.5 Muslim schoolgirls dress a dummy at the Faith and Fashion workshop, London School of Fashion, Hackney, London (see Lewis, 2013). Credit: Getty Images

of diasporic histories and geographies. A good example are the young women interviewed by anthropologist Emma Tarlo in a book about Muslim women and fashion in the UK (Tarlo, 2010). These young fashionable Muslim women have adopted hybrid Islamic fashions which re-work the traditional forms of dress of their parents' homelands in Pakistan or Bangladesh to create new styles and designs which seek a distinctively 'British' and 'modern' Islamic style, challenging expectations of modesty, femininity, fashion and piety (see Figure 45.5).

SUMMARY

- Diaspora cultures are 'syncretic', characterized by interconnections and fusions that cut across or transform geographical boundaries.
- These diaspora cultures are often fashioned within transnational social fields and their movements of people, materials and ideas.
- It is in music, art, fashion and other cultural forms that diaspora cultures have been most readily recognized.

Diaspora spaces

The previous section emphasized that geographies of diaspora are characterized by networks, flows and connections which link multiple locations. Now I want to turn more directly to the implications of these diasporic connections for, and in, particular contexts and places.

Avtar Brah's work provides a very useful understanding of this in her description of what she defines as 'diaspora spaces'. For her, diaspora spaces are 'the point at which boundaries of inclusion and exclusion, of belonging and otherness, of "us" and "them" are contested' (Brah, 1996: 209). What is particularly interesting about Brah's definition is that she includes within these spaces not only those who have migrated and their descendants but also those who are constructed or represented as 'indigenous'. She argues that this concept of diaspora space (as opposed to that of diaspora) includes 'the entanglement, the intertwining of the genealogies of dispersion with those of "staying put"' (1996: 209). What she means by this is that the multiple connections characteristic of diaspora communities also affect those who have not migrated. So, for example, British youth cultures are bound up with and inflected by the transnational exchanges of black British 'black Atlantic' cultures even for those who are not part of these diasporas. Similarly, English men and women who remained 'at home' during Britain's imperial reign in India did not remain unaffected by the social, political, cultural and economic relationships that developed between Britain and India. Those who remained in the metropolitan centres of Empire were also interlinked into colonial relationships, for example through the consumption of colonial products like cotton or tea; the images of other peoples and places in education or entertainment (Ploszajska, 2000; Ryan, 1997); and the shaping of 'domestic' landscapes via imported plants, street names and architecture (Gilbert and Preston, 2003; King, 2004).

There are many different examples of places we might define as 'diaspora spaces'. For example, we can think of parts of cities that have been shaped by the activities of migrant communities, such as the 'Chinatowns' in many North American and European cities. These may have begun as segregated residential spaces (Anderson, 1991) and as key sites for the provision of specific ethnic goods and services, but have often been revitalized as commodified tourist spaces through urban regeneration programmes. As suggested by debates about the regeneration of London's Brick Lane (Jacobs, 1996), including its association with a new neighbourhood designation of 'Banglatown' to reflect Bangladeshi migration and business activity in the area, such processes can be controversial or risk accusations of 'celebratory multiculturalism' (Hutynk, 2000) which simplify or essentialize ethnic identities. The concept of diaspora spaces looks to resist such simplifications.

Another form of diaspora space visible within urban landscapes is that of religious buildings. More generally, recent studies of migration and diaspora have drawn attention to the importance of religion for migrants and the ways in which religious practices may be discontinued, revitalized or transformed through the migration process (Levitt, 2007). One of the most interesting phenomena in recent years has been the emergence of new forms of transnational religious architecture in cities and suburbs (Dwyer *et al.*, 2012). Thinking about the transnational connections that produce urban and suburban spaces is not new, as a number of examples of research in the British context illustrate. The architectural historian, Anthony King, traces the connections within suburban landscapes, such as the distinctive architectural form of the suburban bungalow – translated from its colonial formation in India (King, 2004). Similarly, geographers Simon Naylor and James Ryan (2002) relate the history of Britain's first mosque, the Shah Jehan Mosque built in Wimbledon in 1889, showing how its elaborate Orientalist architecture was echoed in other suburban spaces such as cinemas and bingo halls. The emergence of some spectacular new religious buildings in London in recent years – such as the Swaminaryan Hindu Temple in Neasden, the Sri Murgan Temple in East Ham or the Shikharbandhi Jain Temple in Potters Bar (Figure 45.6) – are good examples of new diaspora religious spaces.

The opening of the Shikharbandhi Jain Temple in 2005 was the culmination of more than three decades of fund raising and planning by the Oshwal Jain community. This community is a diaspora ethnic group which trace their origins to Gurjurat in India, although many of those in the UK are migrants, or descendants of migrants, from East Africa. The Jain Temple is both a careful recreation of Jain Temples in India, ensuring that approved religious symmetry is followed in its construction, and a 'hybrid' architectural form with accommodations to both British planning regulations and the vagaries of the English weather (Shah *et al.*, 2012). It is also an important site of religious and communal

Figure 45.6 The Shikharbandhi Jain Temple, Potters Bar, London. Credit: Claire Dwyer

activity for Jains across the UK and indeed more widely within the European and North American diaspora. At the same time it is gaining importance as a site within which non-Jain visitors may learn about the religion and simply appreciate the aesthetic and architectural qualities of this semi-pastoral site. The Swaminaryan Temple in Neasden, also a diaspora site attracting Hindu worshippers from across London and beyond, has actively promoted itself as a site for visitors and has an extensive programme welcoming tourists and school parties to learn about Hinduism.

Such sites of religious activity are rather different from the more commercial diaspora spaces such as London's Chinatown and some commentators have been critical of proselytizing activities by religious communities within them. However, these diasporic religious sites have also been incorporated into narratives of tourism and urban regeneration; for example, the Swaminarayan Temple in Neasden was promoted as an official tourist site during the 2012 London Olympics and Paralympics. Like other diaspora spaces, religious sites offer multiple possibilities for the remaking and shaping of identities. While the buildings themselves take hybrid forms as they evolve in new landscapes, so too religious practices change as they are shaped by the experiences of diaspora populations and spaces of interaction within the diaspora. For example, the ethnographer Ann David describes the transformation of worship practices for diasporic Sri Lankan Tamils in London as women take on new sacred roles (David, 2011).

SUMMARY

- The concept of 'diaspora spaces' emphasizes the relations between diaspora and particular places, landscapes and contexts.
- These relations between diaspora and place involve people both within and beyond diasporic groups.
- Examples that have been studied include commercialized ethnic neighbourhoods and religious spaces.

Conclusion: possibilities of diaspora?

In this chapter I have suggested that global migrations have produced diasporas. Diasporas are defined as peoples dispersed across geographical boundaries but linked through connections to particular places that are imagined as multiple 'homes'. Research on diasporas includes work on the transnational spaces and social fields diasporic populations inhabit and on the possibilities and difficulties of living transnationally. I have emphasized the ways in which thinking about diasporas challenges many geographical notions; in particular, the evocation of multiple 'homes' unsettles fixed geographical and national boundaries and the association of culture with particular places. Diasporic cultures are celebrated for their fusion of differences. The concept of 'diaspora spaces' shows these conjunctions shape particular places and affect people not only within but also beyond diasporic populations. While diaspora cultures are characteristic of particular groups, I have also suggested that they may be transformative of all national cultures since they challenge the very notion of fixed national cultures within geographical boundaries. Diasporas reconfigure the relations between society and space.

DISCUSSION POINTS

1. What does it mean to identify as a 'diaspora'?
2. Are diaspora cultures and spaces inherently progressive?
3. Can you think of other diaspora spaces than those suggested here? What kinds of transnational linkages underlie the making of such spaces?
4. What kinds of challenges are raised for both nation-states and individuals by transnational migration and the development of diaspora populations?

FURTHER READING

Collins, F.L. (2009) Transnationalism unbound: detailing new subjects, registers and spatialities of cross-border lives. In: *Geography Compass* 3(1), 434–58.

A review article that examines Human Geography's contribution to the interdisciplinary field of transnational studies. A useful further orientation to debates on transnationalism and transnational spaces.

Dwyer, C., Gilbert, D. and Shah, B. (2012) Faith and suburbia: secularization, modernity and the changing geographies of religion in London's suburbs. In: *Transactions of the Institute of British Geographers* (in press)

This article discusses in more depth the diaspora spaces of religious buildings, focusing on past and recent cases from the suburbs of London, UK, elaborating the research on which this chapter's discussion is based.

Hall, S. (1995) New cultures for old. In: D. Massey and P. Jess (eds.) *A Place in the World*. Milton Keynes: Open University Press, 175–214.

Stuart Hall, a sociologist, is one of the foremost writers associated with ideas about diaspora and cultural transformation. Despite its age, this chapter is still important reading, as well as being accessible. It gives further examples, such as cricket, of the notion of diaspora cultures.

Knott, K. and McLoughlin, S. (eds.) (2010) *Diasporas: Concepts, Intersections, Identities*. London: Zed Books.

Gives a sense of the range of ways in which the idea of diaspora has been developed by scholars in many different disciplines. It has short overview chapters on key concepts, on substantive themes such as religion, film, music, development and on case studies of diaspora groups spanning many different countries.

Ley, D. (2010) *Millionaire Migrants: Trans-Pacific Life Lines*. Oxford: Wiley-Blackwell.

Focused on wealthier migrants, this book explores the transnational social fields of Chinese business and professional elites, mapping out migration careers between East Asia, Canada, Australia and the USA through a mixture of quantitative and qualitative methods. It is especially useful in extending this chapter's introduction to the transnational geographies of migration. It also examines a trans-Pacific setting, complementing the trans-Atlantic and South Asian-European foci here.

Mavroudi, E. (2007) Diaspora as process: (de)constructing boundaries. In: *Geography Compass* 1(3), 467–79.

A review article that considers a range of work on diaspora. It complements this chapter though its focus on the role of diaspora in constructions and deconstructions of nation-state, community and individual identity boundaries.

WEBSITE

www.diasporas.ac.uk

This website has information about some of the 45 projects funded by the UK's Arts and Humanities Research Council (AHRC) within its 'Diasporas, Migration and Identities' research programme.

SECTION ELEVEN

URBAN AND RURAL GEOGRAPHIES

Introduction

Urban and rural are two of those taken-for-granted concepts we use all the time in an almost unthinking manner. Indeed, the everyday use of the terms betrays an almost self-evident acceptance of the distinction between them – we know what 'rural' means, we know what 'urban' means, and we know how they differ. Geography has played on this distinction and for much of its post-war history has tended to study the two as separate entities. It was able to do this partly because the division between urban and rural was premised around a physical distinction, and partly because the rural became conflated with agriculture. But this physical reading of both urban and rural was increasingly challenged and gradually replaced by an emphasis on process. As Harvey put it with reference to the urban:

> The study of urbanization is not the study of a legal political entity or of a physical artefact. It is concerned with processes of capital circulation; the shifting flows of labour power, commodities and money capital; the spatial organization of production and the transformation of space relations; movements of information and geopolitical conflicts . . . I prefer to . . . concentrate on urbanization as a process.
> (Harvey, 1995: xvii)

Freed from the moorings of studying the urban and rural as somewhat fixed and distinct physical entities, the scope emerged for all kinds of research that explored the two in terms of process. For if urbanization could be studied as a process, so too could rurality. A number of key themes soon became prevalent in work on both – the role of economic restructuring in shaping town and country and the influence of each on that restructuring; the social recomposition of each, sometimes as a response to economic restructuring and sometimes as a precursor to it; the role of the state and structures of governance, in both responding to economic and social change and trying to shape it. The cultural turn across the social sciences brought a concern with different sets of processes – those to do with the ways in which both 'urban' and 'rural' are key socio-cultural constructs. Both act as **signs**, **signifiers** and **referents**, and have been increasingly interpreted as a set of **social constructions** that convey a host of social, moral and cultural values. Through these processes both urban and rural lose their geographical anchors – socio-cultural spaces of rurality do not necessarily coincide with the countryside, and socio-cultural spaces of urbanism are not confined to the city. Instead, each reaches out to pervade wider society. Indeed, some have claimed that with ever sharpening processes of counter-urbanization, migration and commuting, and increasing flows of goods, services and people between the two, the social and cultural views which are thought to be attached to rural and urban provide clearer grounds for distinguishing between the two than do their geographic differences.

The chapters in this section take up and explore all these themes. In so doing they introduce you to some exciting contemporary themes in both urban and rural studies. We have chosen to hold the two separate, with distinctive chapters on each, although you will see there are several points at which such distinctions become blurred. Initially, in Chapter 46, Chris Hamnett explores some of the key aspects of contemporary urban form. One of these is the rise of safe, socially selective and high security gated communities. Others are the trends of ghettoization and gentrification, the latter exploiting images of loft living and urban renaissance. Underlying each is a worsening inequality that plays on, and sharpens, social and racial divides and produces distinctive sets of urban experiences. In Chapter 47 Lisa Law draws out what it means to experience the city. Using Singapore as an example she explores the way in which different groups visualize and feel the city, through a combination of senses. She reminds us that there is more to the city than its material construction. In Chapter 48 Paul Cloke also considers some of these themes, but in relation to the countryside. He, too, examines how the countryside is commodified and sold as an image and also how it is experienced. He also shows how geographers are exploring social divisions in the countryside by looking at rural poverty and marginalization. Indeed Cloke writes about an increasing blurring of the city and the country, bringing changes to both. It is interesting that these themes of inequality, experience and commodification crop up across the chapters – showing how geographers are now looking at similar sets of processes within the country and the city. Exactly how they operate in each setting will be different of course, but this only brings us back to the richness of contemporary work on both.

FURTHER READING

Williams, R. (1985 [1973]) *The Country and the City*. London: Hogarth Press.

As it says on the back, 'From the moment of its first publication this book has been hailed as a masterpiece'. For once the blurb is true. This is a classic book that sets about unpacking the shifting meanings of city and country from the sixteenth to the twentieth century.

Barnett, A. and Scruton, R. (eds.) (1998) *Town and Country*. London: Jonathan Cape.

Wilson, A. (1992) *The Culture of Nature*. Oxford: Blackwell.

Two books that blur commonly held distinctions between the city and the country.

Cloke, P., Marsden, T. and Mooney, P. (2006) *Handbook of Rural Studies*. London: Sage.

An excellent collection of original writings by leading rural scholars covering social, economic, cultural and political aspects of rural change.

Bridge, G. and Watson, S. (eds.) (2010) *The Blackwell City Reader* (2nd edn.). London: Wiley.

An excellent contemporary collection of chapters on the urban.

CHAPTER 46
URBAN FORMS

Chris Hamnett

Introduction: living in an urban world

We are living in an increasingly urbanized world. In Britain over 80 per cent of the population live in urban areas, and in most developed western countries the proportion is over 70 per cent. Although there has been rapid suburbanization and urban population decline in recent years, most people in western countries live in urban environments. Nor is the process of urbanization confined to western countries. In 2004 the UN suggested that 60 per cent of the world's population will be living in cities by the year 2030. Much of this growth will be in the third world and much of it will be in large cities such as Mexico City, São Paulo, Lagos, Jakarta and Shanghai. By 2012 China's population was 50 per cent urban for the first time, compared to just 20 per cent in 1980, and China now has over 120 cities with one million people. Both India and China are projected to see massive increases in their urban populations over the next 40 years, along with Nigeria and Indonesia. Of the urban population, almost half is forecast to live in cities of over one million by 2025, with more living in megacities of over 10 million (UN, 2012). Cities such as Delhi, Mumbai, Shanghai, Mexico City and São Paulo had populations of 20 million or over by 2011 and these will continue to grow.

Urban living will be a defining characteristic of life in the twenty-first century for the majority of the world's population. Consequently, the changing form, economic base and social structure of cities will continue to be of immense importance. We need to know how cities are changing and what the implications for urban life are now and in the future. These issues are frequently a subject of movies, including Spike Lee's *Do the Right Thing* on ethnic conflict in New York, *City of God* on gang conflict in the favela of Rio de Janeiro, *Slumdog Millionaire* on life for the poor in Mumbai (see Chapter 23) and *Blade Runner* set in a futuristic Los Angeles. Films and novels give us an insight into urban life, or representations of it, but as geographers we want to probe a little more deeply beneath the surface.

In this chapter, I want to concentrate on the changing nature of urban form in the contemporary developed western world, focusing on three very different aspects of urban living: first, counter-urbanization and the rise of ex-urban 'edge cities'; second, the experience of urban economic decline in the black ghettos of Chicago; and third, the 'back to the central city' movement seen in some major western cities where it is associated with **gentrification**. Finally, I want to look briefly at the growth of inequality in **global cities**, focusing on Jonathan Raban's journalistic view of New York.

From urbanization to counter-urbanization

Urbanization in many western countries peaked in the early decades of this century. In Britain, 70 per cent of the population lived in cities by 1900. Although urbanization has grown in many eastern and southern European countries in the post-war period, with rapid rural depopulation and urban growth in Spain, France, Portugal, Italy and other countries, much of north-west Europe has experienced a process of rapid suburbanization and counter-urbanization (Champion, 1989; Cheshire, 1995) and urban population decline. A growing share of the population and employment has moved out of what were frequently seen as dirty, declining, derelict and crime-ridden central and inner cities. First to the growing suburbs but, more recently, aided by a rapid increase in car ownership levels, to smaller towns and villages, far beyond the city boundary. This trend was accurately foreseen by H.G. Wells, the novelist and science fiction writer, almost 100 years ago in his book *Anticipations*, published in 1902. As Wells put it:

> We are – as the Census returns for 1901 quite clearly show – in the early phase of a great development of centrifugal possibilities. And since it has been shown that a city of pedestrians is inexorably limited by a radius of about four miles, and that a horse-using city may grow out to seven or eight, it follows that the available area of a city which can offer a cheap suburban journey of thirty miles an hour is a city with a radius of 30 miles . . . But 30 miles is only a very moderate estimate of speed and the available area for the social equivalent of the favoured season ticket holders of today will have a radius of over 100 miles . . . Indeed, it is not too much to say that the London citizen of the year 2000 may have a choice of nearly all England and Wales south of Nottingham and east of Exeter as his suburb . . . The country will take to itself many of the qualities of the city. The old antithesis will disappear . . . to receive the daily paper a few hours late, to wait a day or two for the goods one has ordered, will be the extreme measure of rusticity save in a few remote islands and inaccessible places.
>
> H.G. Wells, *Anticipations of the reactions of mechanical and scientific progress upon human life and thought* (1902)

What Wells anticipated has come to pass and the trend towards population deconcentration was discussed by Berry (1970) in another perceptive paper that envisaged the urban population of the USA becoming increasingly dispersed by the year 2000 into low-density, ex-urban settlements, with their own shopping malls, factories, office parks and entertainment facilities. In the USA Garreau (1991) terms these 'edge cities' and in Britain Herrington (1984) coined the term the 'outer city'.

There is no doubt that, in aggregate terms, population dispersal and counter-urbanization have been the single most important trend in urban structure in many western countries, particularly in North America, Australia and north-west Europe. There has been a major redistribution of the population in the USA from the older, industrial 'snow-belt' cities of the north-east such as Detroit and Pittsburgh, towards the 'sun-belt' cities of the south and west, such as Phoenix, Tucson, Las Vegas, Atlanta, Dallas and Houston. Some cities such as Detroit have seen massive population loss and dereliction as a result of de-industrialization and 'white flight' to the suburbs. This has been reflected by what Beauregard (1994) in his book *Voices of Decline* views as a long-standing and powerful **discourse** of urban decline concerning the nature, causes and consequences of urban change in older, industrial, cities. In Beauregard's view, this discourse has often functioned as a rationalization of, and a response to, racial change in American inner cities.

The magnitude of this change is indisputable. In the space of a few decades, the racial and ethnic composition of American cities has changed dramatically as a result of both high levels of immigration, particularly into key 'gateway' cities (Clark, 1996) and **white flight** to the suburbs (Massey and Denton, 1993). The changing ethnic composition of the city of Los Angeles has been clearly shown by Bill Clark of UCLA. The Los Angeles of the 1930s or the 1950s, depicted in films such as *Chinatown* and *Back to the Future*, has been replaced by a Los Angeles more akin to that of *Blade Runner* in ethnic composition. Los Angeles is now a 'majority minority' city, the second-biggest Spanish-speaking city outside Mexico City, and what Ed Soja (1992) terms 'the capital of the Third World'. The Anglo population has fallen from 80 per cent to 33 per cent of the total, while the Hispanic population has risen from 10 per cent to 45 per cent, and Asian-Americans make up another 12 per cent. Similar trends characterize Miami, which has the highest rate of immigration in the USA. In 1960 Latin Americans accounted for just 5 per cent of the metropolitan population but 50 per cent by 1990, while the Anglo share of the population decreased from 80 per cent to 32 per cent (Nijman, 1996), as many of the Anglo population moved north out of the city in the 1980s. Almost half of the present population was born abroad and 60 per cent do not speak English at home. Nijman states that Miami has made the transformation from a 'southern US city' to a 'northern Latin American city'. In Detroit (Deskins, 1996) the white population fell by 71 per cent in 1970–90, from 851,000 to 250,000, but rose by 240,000 in the suburbs. As a result, Detroit is now 75 per cent black against 43 per cent black in 1970. Similar trends have characterized Washington DC, and other cities (Knox, 1991). A black inner city is surrounded by a predominately white suburban ring.

One of the defining characteristics of edge cities in the USA is the emergence of what are termed 'gated communities': safe, socially selective, high security residential environments in which the predominantly white, upper-middle-class residents can turn their backs on the growing social and economic problems of the ethnically diverse central cities and retreat behind the walls, protected by security staff, electronic surveillance and 'rapid response' units. The rise of such gated communities is particularly marked in California and the south and west, though white suburbanization is common in many large American cities. This phenomenon has been documented by Mike Davis (1990) in his book *City of Quartz* and by Ed Soja (1992) who discusses the growth of Mission Viejo and other gated communities in the southern half of Orange County, south of Los Angeles. Davis suggests that we are seeing the emergence of what he terms 'Fortress LA' characterized by the withdrawal of the affluent behind defensive walls, undermining any notion of the USA as a mixed society and hardening social divisions in space as social protectionism firmly excludes minorities from suburban homes and jobs. We are now seeing the rise of gated luxury residential developments in parts of some British cities, for example, London's Docklands, where developers perceive potential demand from buyers concerned about crime and security

Figure 46.1 A gated development in London. Credit: Lorna Ainger

(see Figure 46.1). The social implications of this are arguably negative as such developments imply greater residential segregation of the well-off, the exclusion of other groups, and a reduction in overall social trust and integration. This means that the use of the term 'gated *community*' is actually problematic as such developments tend to lessen rather than increase a sense of community.

SUMMARY

- In many northern European and American cities, counter-urbanization has proceeded apace in recent years, with population decentralizing from the old urban cores to new suburbs and, most recently, to ex-urban centres.
- This out-migration has been very socially and racially selective in the USA, with large-scale ethnic immigration into the cities accompanied by a growing suburbanization of the white middle class.
- Many American cities, particularly in the southern and western states, have seen the growth of 'gated communities', which are highly socially selective by income, in an attempt to shut out the problems of crime and violence.

When work disappears: ghetto poverty in Chicago

Many large western cities have experienced dramatic transformations in their social and residential structure in recent decades as a result of a combination of a massive decline of manufacturing employment, large-scale immigration, rapid suburbanization, white flight, inner-city decay and social despair. This transformation has been particularly marked in many of the older American cities of the mid-west and north-east such as Pittsburgh, Philadelphia, New York, Detroit and Chicago. The transformation of New York has been extensively studied by Mollenkopf and Castells in their book *Dual City*, but the city I want to concentrate on here is Chicago, which has been the subject of major research by the black sociologist William Julius Wilson (1996) in his book *When Work Disappears: The World of the New Urban Poor*.

Wilson's work is important for a number of reasons, but first and foremost it addresses a crucial urban social issue: that of the new black urban poor and the social dislocation and breakdown that follow from it. In this respect, his work is committed to real social issues, not to the intellectual games of **postmodernism**. This is the social world explored in gritty and demanding films such as Spike Lee's *Do the Right Thing*, *Grand Canyon* and *Boyz 'n' the Hood*. The latter focuses on two talented young blacks who live in the ghetto of South Central Los Angeles and struggle against the forces of crime, violence and despair to get a college education. The French equivalent – *La Haine* – is the world of young immigrant males in one of the big social housing estates. They are depressing films, but worth seeing. *Trainspotting* offers a less bleak and more humorous Scottish equivalent set in Edinburgh's social housing estates.

Wilson takes on two sets of intellectual opponents: the social conservatives who attribute the problems of the black ghetto to the attitudes and behaviour of its residents, who are seen to be irresponsible, criminal and feckless welfare dependants, thinking only of the moment and happy to live by a combination of crime and welfare hand outs. The second set of opponents comprises the liberals who ignore or deny the reality of anti-social behaviour in the black ghetto and have yielded the field to conservative popularizers such as Charles Murray in his book *Losing Ground*. Wilson argues that ghetto social problems are only too real and cannot be ignored or brushed aside; he believes instead that they are a result of the changing employment, racial and demographic structure of American inner cities aided by systematic discrimination. Wilson's view is that the attitudes and behaviours found in black poverty areas are a response to the problems faced by their residents rather than reflecting an innate culture of lawlessness, criminality and immorality, as conservatives often believe.

As Wilson (1996) points out:

> It is important to understand and communicate the overwhelming obstacles that many ghetto residents have to overcome just to live up to mainstream expectations involving work, the family and the law. Such expectations are taken for granted in middle-class society. Americans in more affluent areas have jobs that offer fringe benefits; they are accustomed to health insurance that covers paid sick leave and medical care. They do not have to live in neighbourhoods where attempts at normal child rearing are constantly undermined . . . and their family's prospects for survival do not require at least some participation in the informal economy . . . I argue that the disappearance of work and the consequences

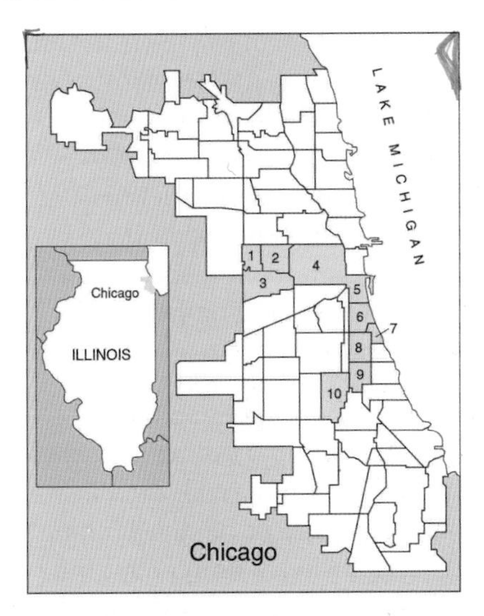

Community Areas in Chicago's Black Belt

1. West Garfield Park
2. East Garfield Park
3. North Lawndale
4. Near West Side
5. Near South Side
6. Douglas
7. Oakland
8. Grand Boulevard
9. Washington Park
10. Englewood

Figure 46.2 Chicago Black Belt. Source: Wilson (1996)

> of that disappearance for both social and cultural life are the central problems in the inner city ghetto.
>
> (Wilson, 1996: xix)

Wilson's research is set in the neighbourhoods of Chicago's Black Belt (see Figure 46.2) and incorporates a powerful mixture of quantitative and qualitative sources. He points out that less than one in three of the poor in the USA lived in metropolitan central cities in 1959 but by 1991 the figure had risen to almost half, and that much of the increase in concentrated poverty has occurred in African-American neighbourhoods. In the ten communities that constitute the historic core of the Black Belt, eight had poverty rates in 1990 that exceeded 45 per cent, three had rates over 50 per cent and three of over 60 per cent. In 1970 only two neighbourhoods had poverty rates of over 40 per cent. Overall, the poverty rate in the Black Belt rose from one-third in 1970 to half in 1990. The increase in poverty has a simple explanation according to Wilson: the rapid growth of joblessness. In 1990 only one in three adults held a job in the Black Belt compared to 60–70 per cent in the 1950s. The increase in joblessness is, in turn, a result of **de-industrialization** and the replacement of manufacturing, transportation and construction jobs, traditionally held by males, by jobs in high technology and services that hire more women. These changes, says Wilson, are related to the decline of mass production in the USA or, perhaps more accurately, the automation of mass production and the consequent job losses. He points out that in the 20-year period from 1967 to 1987, Philadelphia, Chicago, New York City and Detroit each lost between 50 and 65 per cent of their manufacturing jobs. These manufacturing job losses particularly affected black males, who disproportionately worked in the sector and had lower education levels that did not equip them for the new jobs in business services and high technology. In addition, many of the new jobs are either located in the city centre or in the expanding suburbs. He notes that, in the last two decades, 60 per cent of the new jobs created in the Chicago metropolitan areas have been located in the north-west suburbs of Cook and Du Page counties in which African-Americans comprise less than 2 per cent of the population. Consequently, blacks living in the inner city have less access to employment and they are far less likely to own a car to enable them to get to the new jobs in a country where public transport is particularly poor. As one of his respondents, a 29-year-old unemployed South Side black male noted:

> You gotta get out in the suburbs, but I can't get out there. The bus go out there but you don't want to catch the bus out there, going two hours each ways. If you have to be at work at eight that mean you have to leave for work at six, that mean you have to get up at five to be at work at eight. Then when wintertime come you be in trouble.
>
> (Wilson, 1996: 39)

Wilson notes that nearly half the housing stock in the black neighbourhood of North Lawndale has disappeared since 1960 and the remaining units are mostly run-down or dilapidated. And whereas in the past the Hawthorne plant of Western Electric employed 43,000 workers, International Harvester employed 14,000 and the world headquarters of Sears Roebuck, the mail-order firm, employed 10,000, all have now closed. The departure of the big plants, says Wilson,

> triggered the demise or exodus of the smaller stores, the banks, and other businesses that relied on the wages paid by large employers . . . In 1986, North Lawndale, with a population of over 66,000, had only one bank and one supermarket; but it was home to forty-eight state lottery agents, fifty currency exchanges, and ninety-nine licensed liquor stores and bars.
>
> (Wilson, 1996: 35)

Another of his interviewees, a 29-year-old unemployed black male, stated that:

> Jobs were plentiful in the past. You could walk out of the house and get a job. Maybe not what you want but you could get a job. Now you can't find anything. A lot of people in this neighborhood, they want to work but they can't get work. A few, but a very few, they just don't want to work: but the majority they want to work but they can't find work.
>
> (Wilson, 1996: 36)

The social consequences of mass joblessness are profound. Wilson argues that: 'Neighborhoods plagued by high levels of joblessness are more likely to experience low levels of social organization' and 'High rates of joblessness trigger other neighborhood problems that undermine social organization, ranging from crime, gang violence and drug trafficking to family breakups' (1996: 21).

The decline of job opportunities among inner-city residents has increased the incentives to sell drugs and addiction to crack cocaine has been paralleled by the rise of violent crime among young black males. Wilson points out that whereas the homicide rate for white males aged 14 to 17 increased from 8 to 14 per 100,000 between 1984 and 1991, the rate for black males more than tripled over the same period from 32 per 100,000 to 112. He argues that neighbourhoods plagued by high levels of joblessness and disorganization are unable to control the volatile drug market and the violent crimes related to it. As the informal social controls weaken, so the social processes that regulate behaviour change and Wilson instances the spread of gun culture:

> Drug dealers cause the use and spread of guns in the neighbourhood to escalate, which in turn raises the likelihood that others, particularly youngsters will come to view the possession of weapons as necessary or desirable for self-protection, settling disputes, and gaining respect from peers.
>
> (Wilson, 1996: 21)

Wilson argues that many inner-city ghetto residents clearly see the social and cultural effects of living in high jobless and impoverished neighbourhoods, particularly the effects on attitudes and behaviour. A 17-year-old black male living in a poor ghetto neighbourhood on the West Side stated that:

> Well, basically, I feel that if you are raised in a neighborhood and all you see is negative things, then you are going to be negative because you don't see anything positive . . . Guys and black males see drug dealers on the corner and they see fancy cars and flashy money and they figure: 'Hey, if I get into drugs I can be like him'.
>
> (Wilson, 1996: 55)

The problem of the black ghettos, as Wilson sees it, is one of historic racial discrimination and segregation compounded by de-industrialization, which has dramatically reduced the formal employment opportunity structure and led to destructive social behaviours that are pulling neighbourhoods down.

It is sometimes asserted that European cities are being Americanized and that some areas are becoming ethnic ghettos. But, as Loic Wacquant (1993) and Peach (1996) have argued, this is a fundamental misconception. The overall proportion of ethnic minorities in European cities is far smaller than in the USA, and there is nothing approaching the levels of ethnic concentration in American cities. In answer to the question 'Does Britain have ghettos?', Peach unequivocally says 'No'. There are areas with high minority concentrations in some British cities, such as Leicester and Bradford, but minorities comprise a majority of the population in only a relatively small number of enumeration districts. Nonetheless, the ethnic minority population of London rose by over 50 per cent between 1991 and 2001 and ethnic minorities comprised 29 per cent of the population in Greater London in 2001 and 34 per cent in Inner London. In two boroughs (Brent and Newham) the non-white population now exceeds 50 per cent and in Tower Hamlets is reaching 50 per cent, with several other boroughs such as Hackney over 40 per cent (Hamnett, 2003b). Although these figures are considerably lower than in American cities such as New York and Los Angeles, they suggest that London and some other British cities such as Leicester, Bradford and Birmingham may be moving more towards the levels of ethnic diversity found in some American cities. Leicester is projected to be the first British city with a minority ethnic majority population, and in London Department of Education data show that in inner London boroughs over 50 per cent of secondary school pupils are now from ethnic minorities – a figure which rises to 80 per cent in Brent, Tower Hamlets and Newham (Hamnett, 2012).

It is rare in most European cities, however, to find the extensive concentrations of ethnic minorities that are found in American cities. This is not to deny that some groups fare badly in labour and housing markets (Madood and Berthoud, 1997), but this is not the same as the concentration of ghetto poverty found in the USA. The scale of the problem is quantitatively and qualitatively different.

SUMMARY

- In the predominantly black inner cities of the north-eastern USA, large-scale de-industrialization has been associated with a massive increase in unemployment and poverty. These problems are found to a lesser extent in some British and European inner-city areas.

- The collapse of inner-city manufacturing jobs, particularly for males, and the growth of predominantly low-wage service sector jobs, linked to the out-migration of jobs to the white suburbs, have generated major social problems.
- The social and behavioural problems found in inner-city black areas are very real, but they should be seen as the consequence of de-industrialization and discrimination rather than innate social characteristics. They represent a response to a changed set of economic and social conditions.

Gentrification and loft living in the central city

The economic decline of older, industrial cities such as Detroit, Pittsburgh, Manchester, Liverpool, Glasgow, Lille and the Ruhr, has been paralleled by the rise of a small number of major world or global cities (Sassen, 1990) as the command and control centres of the world economy and finance system. These cities, which include London, New York, Paris, Tokyo, Toronto and others, have all experienced massive deindustrialization, but they have also seen the rapid expansion of business and financial services such as banking, legal services and management consultancy as well as the continued growth of a number of creative industries such as advertising, film and video, music, fashion and design. The cultural and creative industries are becoming increasingly important in global cities, both in terms of production and attracting visitors.

These cities have been characterized by the transformation of their industrial, occupational, income and residential structure. As Ley (1996) shows in the context of Toronto and Vancouver, the rise of a service-based economy has been paralleled by the growth of a new professional, managerial, technical and creative middle class, generally highly educated and highly paid. The rise of this group, with its cultural interests and housing market demands has, in large part, been responsible for the growth of gentrification in post-industrial inner cities. Many of them work in business or creative industries in the central city or its environs, have long or irregular hours and want to live close to work and the cultural and entertainment facilities offered by the central city. But traditional central and inner-city high-status residential areas are expensive and in short supply. Consequently, the new middle class have sought out new living opportunities in the inner city, aided by developers and estate agents who have seen the prospects for profitable transformation of these areas.

There is a large literature on traditional forms of gentrification (Butler, 1997; Hamnett, 2003b; Ley, 1996; Smith, N., 1996), which commonly involve conversion of nineteenth-century multi-occupied rental housing (much of it originally built for middle-class occupancy) back into single family houses or apartments.

In New York and London, however, there has been a trend towards conversion of older industrial buildings into spacious if expensive city centre apartments. In New York, this was first concentrated in the SoHo area of downtown Manhattan, adjacent to the financial district and characterized by elegant late nineteenth-century multi-storey industrial lofts,

Moving in and up

More of us are middle class and we're flooding back to the city, says **Chris Hamnett**

Village life: Greenwich Village was at the forefront of gentrification in New York Corbis

Gentrification has become a widespread phenomenon. The movement of the middle classes into previously working class or run-down inner city areas has occurred from New York, London, Paris, Sydney and Toronto to Budapest, Prague and even Moscow.

The term has passed effortlessly into common parlance but, although many people know intuitively what it means, few probably know where it originated. The answer lies with a London-based émigré Jewish journalist turned urban sociologist called Ruth Glass.

Glass noticed that something strange was happening in run-down areas of central London such as Chelsea and Pimlico in the late 1950s and early 1960s. Small privately rented mews cottages and larger multiple-occupied terraced houses (generally Georgian or Victorian) were being sold when their leases expired to be converted by the middle classes for owner occupation.

As readers of Jane Austen will know, in 18th and 19th century England the gentry occupied an intermediate position in the class structure between the landed aristocracy, typified by Mr Darcy at the top, and yeoman farmers below. In 1963 Glass ironically termed the process "gentrification": the coming of a new urban gentry.

From Chelsea, the process has spread rapidly over the past 40 years, first to the select areas of imposing, centrally located period housing such as Barnsbury in Islington, parts of Camden Town, Notting Hill and Primrose Hill, and then onwards and outwards. These areas are now expensive but in the late 1960s much of Notting Hill was run-down and decaying, with stucco peeling off the collapsing porticos.

In New York, Greenwich Village in downtown Manhattan became attractive to artists in the 1960s, as did Georgetown in Washington, and property prices began to creep up. By 1974, the process of converting buildings in run-down but centrally located areas had begun to spread to old iron-framed industrial buildings and warehouses in New York's SoHo and the loft apartment was born. This was followed by TriBeCa, Chelsea, and most recently the old downtown meatpacking district.

In Paris at around the same time a similar, though superficially rather different process was beginning to take place in the Marais district, east of the centre. An area of imposing aristocratic town houses, including the majestic Place des Voges, which had been in decay since the court moved westwards and was finished off in the Revolution, saw the start of a process of renovation that has led to the area becoming the Covent Garden of Paris.

Now the process has spread southwards and eastwards, out to the Bastille and beyond and into the previously working class 19th and 20th *arrondissements*.

In Sydney, the old inner city working-class areas around the Harbour such as Glebe, Redfern and Paddington have been spruced up, as have the 19th century terraced houses in Boston, Melbourne, Toronto and San Francisco. In the last 10 years gentrification has grown in eastern Europe. After the Berlin Wall came down the middle classes began to make their way back towards the old, attractive housing found in places such as Prenzlauerberg in east Berlin, or the inner city areas of Budapest and Prague.

So why has gentrification become so commonplace. Theories range from the availability of cheap domestic technologies to changes in the relationship between property prices and land prices, and the coming of age of the postwar baby boomers.

The most convincing explanation is in the relationship between changes in the structure of the economy, in class structure and changing tastes. Go back 40 years and almost all the cities mentioned above had large industrial sectors employing large numbers of manual workers. The middle classes were relatively small.

City centres are much more attractive, convenient places to live

In the interim, much of the old industrial base has disappeared and with it the manufacturing workforce. In its place is the rapidly growing financial, business services (law, advertising, management consultancy) and creative industries that employ large numbers of professionals and managers. The social class structure of big western cities has changed dramatically.

The result was predictable. Increasing numbers of the middle-class, many aged in their 20s and 30s, began to hit the housing market in the early 1970s. Many worked in or near the city centre, they were university educated and enjoyed the cultural buzz of the inner city. Last, but not least, decaying old terraced houses in inner city areas were cheap.

Will gentrification continue and if so where? If one accepts the argument that it is simply a product of postwar baby boomers looking for a cheap place to live, it should already have peaked. Most of this group are already in their 40s and 50s and are well established in the housing market. But the evidence is against this. Gentrification is thriving and spreading to some of the older industrial cities such as Manchester, Leeds and Liverpool.

Much more likely is that rising real incomes and the continuing trends towards smaller households will intensify the demand for city-centre living, aided by the fact that to many city centres and inner cities are much more attractive, convenient places to live.

The evidence points to the expansion and extension of gentrified areas.

In London, the process has spread from Islington into neighbouring Hackney, and Tower Hamlets, and south of the river into Lambeth and north Southwark. Leafy Crouch End and Finsbury Park have undergone the process. In New York, parts of Brooklyn have been gentrified and pioneers are moving into southern Harlem.

Rising demand and limited supply leads to price rises and the gradual colonisation of areas further from the centre. Meanwhile, what remains of the working class and those who cannot afford the high prices? They are either squeezed into the remaining council housing or into the outer suburbs.

Gentrification has remade the social structure of the post-industrial city. The middle classes fled the inner cities in the late 19th century and first half of the 20th centuries. Now they are back and it looks like they will remain for a while yet.

The tests will be crime and education. Young middle-class households enjoy social mix but they also require safety and the ability to ensure a good education for their children. If either requirement becomes too difficult to achieve gentrification could stall.

Chris Hamnett is professor of geography at King's College London and the author of 'Unequal City: London in the Global Arena'

Figure 46.3 Article from the *Financial Times*, 10 July 2004. Credit: © Chris Hamnett

but it has since spread into Tribeca and other areas where industrial buildings are available (Zukin, 1982). Unlike the gentrification of single family housing, conversion of such buildings generally involves property developers who have the finance and expertise to carry out the work. An insight into loft living in SoHo is seen in the 1985 film *Desperately Seeking Susan* featuring Madonna and Rosanna Arquette.

In London, the process got under way in the 1980s in Docklands with the conversion of some of the old riverside warehouses along the Thames in Wapping and elsewhere, aided by the efforts of the London Docklands Development Corporation to socially transform the area (Goodwin, 1991). But in the last few years, there has been a dramatic expansion of loft conversions in the Clerkenwell area of London, just west of the City of London and north of the Inns of Court. This area was formerly one of the industrial districts of London, as SoHo was for New York, but with the rapid decline of manufacturing in the 1960s onwards, it became increasingly derelict and empty. One or two pioneering developers, such as the aptly named Manhattan Loft Corporation, saw their potential and their

Figure 46.4 New River Head Development. Credit: The Berkeley Group Holdings plc

proximity to the City and initiated the process of conversion of old industrial and warehouse buildings into luxury residential apartments.

Some of the conversions are very dramatic, such as the old headquarters of the New River Company (Thames Water) in Rosebery Avenue, near Sadlers Wells just south of the Angel, Islington (see Figure 46.4). Manhattan Loft Corporation is currently converting the old hotel in St Pancras station into luxury apartments.

As the area has increased in desirability, aided by marketing and promotion as a fashionable place to live, prices have soared as they did in SoHo in the 1970s. Clerkenwell lofts have become home to bankers, lawyers and highly paid creative workers, and the process has spread rapidly in recent years into Shoreditch, Spitalfields and Hoxton, as well as south of the river to Borough Market. In many ways the creative centre of gravity of London has shifted sharply eastwards, with many new bars,

Figure 46.5 Price's Candle Factory, Clapham, London, has been converted into luxury apartments. Credit: Lorna Ainger

galleries and restaurants in the old city fringe. Similar developments are also taking place in a number of other cities such as Manchester, Liverpool, Leeds and Newcastle. The conversion of old industrial and office buildings into luxury apartments has been one of the defining characteristics of changing urban form in recent years. One of the most emblematic conversions has been the old Jewish Soup Kitchen for the Poor, built in Brune Street in the City of London in 1902, into luxury loft apartments.

SUMMARY

- In addition to the rise of edge cities and ex-urban development, and inner-city decline, there has a been a widespread growth in the middle classes in the central and inner areas of some major cities where economic change in the structure of employment has created new jobs in the creative industries and financial services.
- Many of the workers in these new growing industries have chosen to live in the central cities, leading to the growth of gentrification and 'loft living'. This latter trend has been associated with the conversion of industrial buildings to residential uses.
- Areas such as SoHo in New York and Clerkenwell in London have become fashionable residential areas for the new wealthy professional middle classes.

Inequality in the global city

It is clear from the discussion of the loft conversion market in London and New York, and ghetto poverty in Chicago, that the modern city is marked by sharp inequalities. Recent analyses of earnings in London (Hamnett, 2003b), New York (Mollenkopf, 1998) and Paris (Preteceille, 2001) reveal that inequalities have grown very sharply in recent years, aided by the rapid rise in earnings and bonuses in financial and legal services, where earnings of hundreds of thousands of pounds is not uncommon. Friedmann and Wolff (1982) and Sassen (1991) argue that these trends are inscribed in the global city rather than being merely incidental. These trends are very marked in London where the influx of the global rich in search of safe investment havens has led to huge increases in the price of housing in the most desirable areas such as Kensington. This is most clearly seen in the new luxury housing development at 1 Hyde Park, where, at the time of writing (2012), apartment prices now go for £6,000 per square foot and one penthouse apartment sold for £135 million. It is now estimated that for properties over £2 million, 70 per cent of buyers in central London are from overseas. Such increases at the top end of the market have had the effect of pushing up prices elsewhere in London as potential buyers are forced further afield in search of affordable property and large terraced houses in gentrified inner London can now routinely command over £1 million.

At the other end of the spectrum, the riots in London in August 2011 highlight the level of inequality. The riots and looting were mostly concentrated in poorer inner city areas such as Hackney (see Figure 46.6) and a high proportion of those appearing in court, many from ethnic minorities, had low levels of education and high levels of unemployment. Although there is strong political disagreement about the motives of the riots, with the right arguing it involved opportunistic looting of consumer goods and the left pointing to economic and social deprivation, government budget cuts and policing, it is clear that the groups involved were largely from the most deprived areas with limited economic and social opportunities. To this extent, the riots can be seen, at least in part, as an expression of growing inequality. The report by the National Centre for Social Research (2011) makes useful reading on the motivations of the rioters.

In his book on America, *Hunting Mr Heartbreak*, Jonathan Raban (1990) makes the brilliant distinction between the 'air people' and the 'street people' of Manhattan in terms of the total separation of their economic and social circumstances and their ways of life. Although his treatment is journalistic, and perhaps rather overdrawn, he puts his finger on the massive inequality that characterizes contemporary cities (see box).

Raban's picture of New York is impressionistic journalism and travel writing, not social

CASE STUDY

An extract from Jonathan Raban's *Hunting Mr Heartbreak* (1990)

The beggars slept much of the day away on benches on the subway platform. By night, they scavenged. Returning home late after dinner, I would meet them on the cross streets around East 18th, where small knots of them went tipping over trashcans in search of a bit of half-eaten pizza, or the lees of someone's can of Coors. . . .

The current term for these misfortunates was 'street people', an expression that had taken over from bag ladies, winos and bums. The Street People were seen as a tribe, like the Beaker Folk or the Bone People, and this fairly reflected the fact that there were so many more of them now than there had been a few years before. In New York one saw a people, a poor nation living on the leftovers of a rich one. They were anthropologically distinct, with their skin eruptions, their wasted figures, poor hair and bony faces. They looked like the Indians in an old Western . . .

(pp. 77–8)

There were the Street People and there were the Air People. Air People levitated like fakirs. Large portions of their day were spent waiting for, and travelling in, the elevators that were as fundamental to the middle class culture of New York as gondolas had been to Venice in the Renaissance. It was the big distinction – to be able to press a button and take wing to your apartment . . . access to the elevator was proof that your life had the buoyancy that was needed to stay afloat in a city where the ground was seen as the realm of failure and menace.

> In blocks like Alice's, where doormen kept up a 24 hour guard against the Street People, the elevator was like the village green. The moment that people were safely inside the cage, they started talking to strangers with cosy expansiveness . . .
>
> (p. 80)
>
> Everyone I knew lived like this. Their New York consisted of a series of high-altitude interiors, each one guarded, triple locked, electronically surveilled. They kept in touch by flying from one interior to the next, like sociable gulls swooping from cliff to cliff. For them, the old New York of streets, squares, neighbourhoods, was rapidly turning into a vague and distant memory. It was the place where TV thrillers were filmed. It was where the Street People lived . . .
>
> (p. 81)
>
> For Diane, places like Brooklyn and the Bronx were as remote as Beirut and Teheran. Nobody went there. The subway system was an ugly rumour – she had not set foot in it for years . . . I sometimes joined her on evenings when she was dining out uptown – evenings that had the atmosphere of a tense commando operation. At eight o'clock, the lobby of her building was full of Air people waiting for their transport. A guard would secure a cab, and we'd fly up through New York to the West 60's or the East 80's . . .
>
> (p. 84)
>
> It was a white knuckle ride. Diane sat bolt upright, wordless, clinging to the grab-rail in front of her, while the cab flew through the dismal 30s. At this level, at this hour, all of New York looked ugly, angular, fire-blackened, defaced – bad dream country. The sidewalks were empty now of everyone except Street People. This was the time when things began to happen that you'd see tomorrow on breakfast television, and read about, in tombstone headlines, in the Post and Daily News . . .
>
> (p. 85)
>
> Few of these journeys last more than ten or eleven minutes: they were just long enough to let you catch a glimpse of the world you feared. Then, suddenly, there was another guard, dressed in a new exotic livery, putting you through Customs and Immigration in another lobby.
>
> (p. 86)

science, and for a more considered approach you should look at Mollenkopf and Castells' (1991) book *Dual City? Restructuring New York*, which provides an excellent overview of trends and inequalities. But Raban's piece is one of the most powerful and vivid insights into New York today, as is Tom Wolfe's 1987 novel *Bonfire of the Vanities*, since made into a film. Getting to grips with the form and culture of the modern city is as much about film, video and novels as it is about census data, interviews and questionnaire surveys. What matters is that we try to understand what is going on, and writers, journalists and film-makers provide us with powerful visions and interpretations that can stimulate social scientific research.

Conclusion

Contemporary cities are changing in complex and often contradictory ways. Continuing suburbanization is paralleled, in some cities, by inner-city urban decline (the two are frequently causally linked) and by central city urban regeneration and gentrification. As a consequence, modern western cities are frequently characterized by growing inequality, both between rich and poor and between different ethnic groups. In some cities this is accompanied by growing social segregation between those with greater resources and choice and those with limited resources and limited choice. While some changes are clearly the result of a degree of choice and preference for different lifestyles and environments, others are often unwilling victims of economic and social processes largely outside their influence and control. While some people may be living in a postmodern urban lifestyle playground, others have to live in a post-industrial wasteland.

DISCUSSION POINTS

1. Some commentators have suggested that western cities are being turned inside-out. What do you think this means, and what are the processes contributing to this?
2. Discuss whether gated communities make positive or negative contributions to urban life.
3. Why are cities becoming increasingly unequal places to live?
4. What do you think should be done to lessen this social and economic inequality?

FURTHER READING

Wilson, W.J. (1996) *When Work Disappears: The World of the New Urban Poor*. New York: Alfred A. Knopf.

To understand what is happening in the black inner areas of some American cities you can do no better than read W.J. Wilson's *When Work Disappears*, which combines quantitative and qualitative material in an illuminating way. The urban poor are given voices and outline their situation in their own words.

O'Loughlin, J. and Friedrichs, J. (eds.) (1996) *Social Polarization in Post-industrial Metropolises*. Berlin and New York: De Gruyter.

Deskin's chapter in the above text gives a powerful picture of the changes that have affected Detroit in recent decades.

Clark, W. (1998b) Mass migration and local outcomes: is international migration to the United States creating a new urban underclass? In: *Urban Studies* 35(3), 371–84.

Frey, W. H. (1995) Immigration and internal migration 'flight' from US metropolitan areas: towards a new demographic Balkanisation. In: *Urban Studies* 32, 733–57

These texts offer an insight into the impact and implications of large-scale immigration in the USA.

Peach, C. (1996) Does Britain have ghettos? In: *Transactions of the Institute of British Geographers* 21(1), 216–35.

Hamnett, C. (2012) (forthcoming) Concentration or diffusion? The changing geography of ethnic minority pupils in English secondary schools, 1999–2009. In: *Urban Studies.*

Read these articles for an assessment of the scale of ethnic change and segregation in Britain.

Ley, D. (1996) *The New Middle Class and the Remaking of the Central City.* Oxford: Oxford University Press.

Hamnett, C. (2003) *Unequal City: London in the Global Arena.* London: Routledge.

Smith, N. (1996) *The New Urban Frontier: Gentrification and the Revanchist City.* London: Routledge.

These texts present an analysis of gentrification, its causes and effects.

Sorkin, M. (ed.) (1992) *Variations on a Theme Park: The New American City and the End of Public Space.* New York: The Noonday Press.

Davis, M. (1990) *City of Quartz.* London: Verso.

These provide an understanding of the changes taking place in the edge cities of America.

Champion, A.G. (1989) Counterurbanisation in Europe. In: *The Geographical Journal* 155, 32–59.

Cheshire, P. (1995) A new phase of urban development in western Europe: the evidence for the 1980s. In: *Urban Studies* 32(7) 1045–64.

Rigorous social scientific analyses of counter-urbanization.

Beauregard, B. (1994) *Voices of Decline: The Post-war Fate of US Cities.* Oxford: Blackwell.

A good textual analysis of the representation of urban decline in the USA.

United Nations (2012) *World Urbanization Prospects, the 2011 Revision*, United Nations: New York

National Centre for Social Research (2011) The August Riots in England: understanding the involvement of Young People. Available at: www.natcen.ac.uk/study/the-august-riots-in-england-

Provides a useful analysis of the motivations of the rioters.

CHAPTER 47
URBAN SENSES

Lisa Law

Introduction

Take a moment to think about what the word 'city' means to you. Where does your imagination begin? Do you think of what a city looks like? Do you visually map it, picturing its architecture and urban form? Do you look down at it from above or do you imagine yourself wandering through the streets? Perhaps instead of seeing the city, and conjuring it from your visual imagination, you contemplate the city's distinctive aromas or sounds. Do you smell car exhaust fumes, bakeries, coffee, perfume? Do you hear traffic, mobile phones, music, multiple languages? Or perhaps you have a disability and think of the city in a haptic register, possibly as a difficult place to get around. Do any of these senses evoke negative feelings, such as fear, or positive associations such as freedom or pleasure? How do we begin thinking about the city and the different experiences it evokes? Can we connect these seemingly routine experiences to culture and politics? This chapter encourages you to think about the senses – sight, smell, sound, touch and taste –- as one possible way to investigate these issues.

Geography is conventionally understood as a visual discipline, with an emphasis on optic methods and techniques such as observation and mapping. Our cartographic roots encouraged our reliance on sight, but this has had a profound impact on the way we contemplate the world around us. Recall various geography texts you have read (or are now reading) and consider how we describe our research and our knowledge:

> Questions are 'looked at' in a particular way, pilot studies are completed 'with a view' to developing a larger project, things are kept in 'perspective', events are 'seen' in one way or another, futures are 'envisaged', empirical research 'sheds light' on theoretical concerns, 'insights' are gained, problems are 'looked into'.
>
> (Smith, 2000: 615)

Seeing, in other words, is equated with believing; we assume vision is clear and transparent. Until recently geographers have unwittingly associated this visual world with reliable knowledge, without considering how the other senses might shape our everyday lives and geographies. But in the past few decades some geographers have suggested that looking and seeing – although dominant in the discipline – are only one of several possible pathways to knowledge and understanding.

Cultural and **feminist** geographers first brought questions of the reliability of vision to the fore, querying how vision helps produce geographic knowledge. Denis Cosgrove (1984; 1985b;

2001), a cultural geographer, has examined how 'perspective' shapes our seemingly innocent depictions of landscape. He explains how landscape descriptions – whether in written texts, paintings or photographs – have evolved through time, and how they are more akin to a 'way of seeing' than a 'true' depiction of what is actually seen. 'We *learn* to see through the communicative agency of words and pictures', he argues, 'and such ways of seeing become "natural" to us' (Cosgrove, 2003: 250). Landscape descriptions are thus not clear or transparent; they are a visual ideology that expresses what the world looks like from particular 'points of view'. No one view is omniscient, and there are multiple perspectives on any one phenomenon. Think again about your images of the city when you began reading this chapter and compare them with those of your classmates. Presumably you have different ways of representing urban spaces. Yet it is likely that at least one of your images would include a bird's eye view with abstract symbols for landforms, roads and significant buildings. Geographers suggest this depiction occurs repeatedly in western culture and is linked to the West's scientific and cultural histories (especially the cartographic project of **colonialism**). We *learn* to see the city this way, from above, like a map. But is this portrayal innocent? What is at stake when we represent the world in a particular way? Cultural geographers ask us to contemplate the relationship between vision and power: who is looking, what do they see and what conditions enable this point of view to be obtained?

These questions of **positionality** also interest feminist geographers. They sensitize us to how different people interpret and represent landscapes, and thinking about *whose* images predominate in any one culture provides clues to power. Gillian Rose (1993b; 1997b), a feminist geographer, suggests that the dominant 'way of looking' in geography has been normalized as white, male and heterosexual. That is, the way geographers look at landscapes has been informed by the vantage point of a group that has considerable economic, political and social power. It is a position that assumes its vision is clear and objective, without bearing in mind how it might exclude the viewpoints of different **genders**, **sexualities**, **races**, and so on. But let us consider this argument in relation to urban planning, as the work of geographers is often implicated in state power. Urban planners imagine our cities from above – with the aid of zoning, land use and other maps – building into designs an idealized city. This might include a tree-lined boulevard to introduce nature and aesthetic beauty to your neighbourhood, but the feelings this evokes for a range of people would likely be quite different. As Gill Valentine (1989) discusses, women can find tree-lined boulevards frightening at night as they increase shadows and reduce the possibility of being seen. She calls these 'geographies of fear' that shape women's access to public space. If women find dark boulevards frightening, we might also consider how senior citizens or minority ethnic groups experience these spaces (Pain, 2001b). Green environmental ideals are now internalized in the planning process and help this desire for green spaces appear unproblematic, objective and 'natural'. But, as we will discuss in relation to our case study below, the street *looks* and *feels* different depending on the perspectives of those inhabiting urban spaces; in other words, it depends on whether you view it from 'above' or 'below'.

Rather than tackling the question of vision on its own terms, as geographers such as Cosgrove and Rose have done, other geographers consider geographies of the broader sensorium.

After all, what our eyes see does not encapsulate the totality of our experience; we learn about and become **embodied** in the world through our whole bodies and the several senses. Perhaps the distancing, 'objective' reliance on vision has been opposed to the more embodied, 'intuitive' reliance on other forms of knowledge. But challenging the dominance of vision has been more straightforward than efforts to include other sensory experiences in our understanding of geography. We might intuitively know how the feel of the subway, the smell of food cooking, the sound of music, and so on, is crucial to the making of place, but how do we include these experiences in our research? Geographers had not developed a vocabulary to describe sensory experiences until recently, partly because of our reliance on sight but also because they are difficult to record in fieldwork. Geographers have thus generated new terms such as 'smellscapes' (Porteous, 1985), 'soundscapes' (Ingham, 1999; Leyshon *et al.*, 1998; Revill, 2000; Smith, 1997; 2000), 'visceral geographies' (Hayes-Conroy and Hayes-Conroy 2010)) and 'sensory landscapes' more generally (Law, 2001; Rodaway, 1994), in the hope that new language will aid in our comprehension of more embodied geographies (related interdisciplinary work on 'sensescapes', the 'sensory city' and the role of the senses in shaping urban experiences more generally can be found in Adams and Guy (2007) and Pink (2007)). The importance of **ethnography** and a range of in-depth qualitative research methods in addressing people's lives and experiences cannot be overstressed (Paterson, 2009; Crang, 2003); embodied senses of place are difficult to observe from a distance. Geographers working in this field thus use research methods that help define methods for a sensuous geography, such as cooking (Longhurst *et al.*, 2009), eating (Law, 2001), photo diaries (Middleton, 2010), soundwalks (Adams *et al.*, 2006) and walking more generally (Ingold, 2004; Lund, 2005) (to name only a few).

While studies of the senses have been inspired by a desire to express a multi-sensuous experience of space and place (Law, 2001; Longhurst *et al.*, 2009; Paterson, 2007; 2011), and most recently by debates in **non-representational theory** that displace symbolic/textual analyses of everyday life (Dixon and Straughan, 2010; Paterson, 2009), people with disabilities have inspired other research. The latter have asked geographers to consider how those without the capacity for sight or who are deaf, for example, might have different experiences of place (see Kitchen *et al.*, 1997; Hetherington, 2003; Macpherson, 2009). Until recently, these people have been marginalized in our discussions of the city.

In this chapter we examine how thinking about the senses can help us understand our urban experiences, using a case study of Singapore to illustrate these ideas. We begin with 'sight', and visions of Singapore as a global city, and briefly recount the history of Singapore's development using the perspectives of its urban planners and politicians. These perspectives draw on a modernist, utopian view from above, but tell us more about state power than about people's everyday lives. Views of the urban planning landscape are like a freeze-frame and often fail to appreciate the practices that make the city lively and give it meaning. We thus turn to de Certeau's (1984) ideas of 'spatial practices' that lie below planners' thresholds of visibility and consider urban living as enacted through our everyday routines. We consider a more embodied sense of place in Singapore's famous hawker centres (outdoor restaurants), with their distinctive sounds, aromas and tastes. Here we find alternative maps and views of the city and can connect a range of senses to urban life and to an unconventional conception of politics.

SUMMARY

- The way we experience urban places is not only shaped by vision; a whole range of senses embody us as subjects.
- Geography has tended to rely on vision for the acquisition of knowledge, although feminist and cultural geographers have challenged this tradition.
- Understanding the urban as a sensory landscape, which includes, but is not limited to, vision, can provide insights to the everyday politics of cities.

Seeing Singapore

Looking at Singapore from the 72nd floor of the Raffles City complex is an extraordinary sensation (see Figure 47.1). Your gaze is drawn unavoidably south, beyond the old colonial Padang and Cricket Club, to the glittering skyscrapers of Shenton Way. Here lies Singapore's stock exchange and the headquarters of many multinational institutions and corporations. Much like other **global cities**, this is where Singapore's business and financial services compete in the world economy. At the south-eastern edge of the business district stands a row of gentrified shop-houses along Boat Quay, a waterfront lined with riverboats until the 1980s but now home to cosmopolitan eating places and nightclubs. Shift your gaze north-east to see the evolving Marina Bay financial district, which should double the size of the existing district, and the Marina Bay Sands Integrated Resort (home to the world's largest public cantilevered platform). This area will also host Gardens by the Bay, an interconnected set of gardens and conservatories aimed at enhancing Singapore's Garden City image and providing top-quality urban outdoor space. Continue on to see Esplanade, a world class performing arts centre locals have nicknamed 'The Durians' (a spiked, fragrant fruit), as if to give such internationally inspired architecture a local flavour. Esplanade is part of Singapore's vision to become a 'global centre for the arts' and, together with places like Boat Quay, is imagined as transforming Singapore's erstwhile image as monotonous and uncreative (Chang, 2000; Chang and Lee, 2003; Kong and Yeoh, 2003). Global cities are now thought to maintain their competitiveness by fostering creativity and entrepreneurship, a principle embraced by Singapore's politicians. Places like Esplanade, Boat Quay and Gardens by the Bay are being promoted as sites to be creative and have fun. Continue your gaze along the coastline to see the Singapore Flyer, the world's largest observation wheel and now one of the best spots to glimpse Singapore's evolving landscape. In the distance condominiums with spectacular sea views stretch out almost as far as Singapore's first class Changi Airport. It's a city obsessed with being 'global', 'first' and 'world class'.

Participating in the global economy means building cities that can compete with others in the global capitalist economy. Thus places like Singapore – or Tokyo, London, Paris, and so on – have changing urban forms that shape and reflect new kinds of urban spaces, but also new kinds of experiences. In some cities this has meant demolishing the old to make way for the new, and this is certainly true in Singapore. In

Figure 47.1 (a) Raffles City is a massive I.M. Pei-designed complex that includes shops, restaurants, offices, conference facilities and a luxury hotel; (b) the view from the top floor of the hotel is touted as offering the best views of Singapore. Credit: (a) Kevin R. Morris/Corbis; (b) unknown

Boat Quay, for example, the redevelopment of this site all but erases its history. No longer is it a bustling anchorage where immigrants arrive or where goods are sold and exchanged, although a few riverboats still ply the river on scenic cruises and statues commemorate this history. The nearby redevelopment of Chinatown has been most controversial, however, as gentrification has brought an entirely different life to what was initially a bustling community of Chinese migrants. No longer is this where Chinese culture is *lived*, it is a place where it is *seen*. The visual *look* of the architecture has been given highest priority, in preference to the daily practices of trades and lifestyles that made this an ethnic neighbourhood (Yeoh and Kong, 1994). It is now a place for tourists, complete with hotels, museums and upmarket restaurants. This morphing of Singapore's urban form is not a new phenomenon and could be said to reflect a privileging of 'time' over 'space' (Koolhaas, 2000). Since independence, Singapore's restless landscape has reflected the state's desire to join the ranks of the world's most developed nations.

Let us continue with our Raffles City vista, but shift our focus north and west, where there are views less familiar to visitors (see Figure 47.2). They also tell stories of Singapore's developmentalism and provide immense insight into the everyday life of its inhabitants. High-rise government flats of various shapes and sizes dominate the panorama; almost 80 per cent of Singapore's inhabitants live in public housing. Unlike in places such as North America and Europe, where public housing is targeted at low-income groups, in Singapore the majority of the population aspire to own their own flats and eventually upgrade them. The state's Housing Development Board (HDB) administers the scheme, an institution originally established in 1960 to provide mass housing to Singapore's low- and middle-income households (Perry *et al.*, 1997). The HDB's early mandate was to redevelop the squalor that had become concentrated in the inner-city area – a relic of the colonial administration that neglected to address racialized inner-city neighbourhoods and urban sprawl. The goal was to transform the housing landscape into one representing a 'well-planned modern metropolis', while at the same time integrating Chinese, Indian, Malay and other ethnic groups into less segregated neighbourhoods (Goh, 2001: 1589). The landscape is thus shaped by state ideologies of modernity and multiculturalism.

While the state did rapidly transform living conditions, providing mass housing to the population, the shift to new accommodation was stressful for those relocated to these sites. It broke up communities and dramatically changed ways of life. Furthermore, because housing is shaped so strongly by state ideology, policy decisions have discernible impacts on people's everyday lives. Class and racial mixing are built into HDB neighbourhoods, for instance, and ideologies of what constitutes the 'Asian family' also shape people's access to housing. Preventing singles from buying government flats and preferential treatment for extended families are two examples found in the history of HDB policies. Chua (1995: 139) suggests that the sheer material presence of countless blocks of flats are 'powerful symbolic monuments to [the] government's efficacy', visually reinforcing the state's legitimacy and achievements. Public housing thus tends to be a

Figure 47.2 HDB housing: the majority of Singaporeans live in public housing; the architecture of these buildings is often criticized as excessively repetitive and anonymous. Credit: (b) Dean Conger/Corbis

powerful sign in the landscape that overpowers other meanings, such as the extent to which ideology regulates people's everyday sense of class, race, gender, sexuality, and so on. While there is resistance to state housing policies in popular culture, as in Eric Khoo's film *12 Stories* (see box), Chua (1995: 140) suggests that alternatives to the current system generally fail to become incorporated into people's 'everyday rationalisation of their life-world'. We will return to this idea of state planning and architecture, and people's capacity to resist dominant meanings, in our discussion of hawker centres.

Our touristic view of Singapore tells us much about how the state manipulates the landscape to accord with its own, idealized, vision of Singapore. The early housing landscape, with its minimalist and anonymous architecture, mirrored **Fordist** modes of production that were driving the economy of the time (Goh, 2001). The contemporary central business district, with its multinational headquarters and cosmopolitan sites, illustrates the state's desire to network itself into a global **space of flows**. Singapore's landscape thus tells a story of the state's ongoing desire to compete in the global economy, organize its population and

CASE STUDY

Eric Khoo's *12 Storeys*

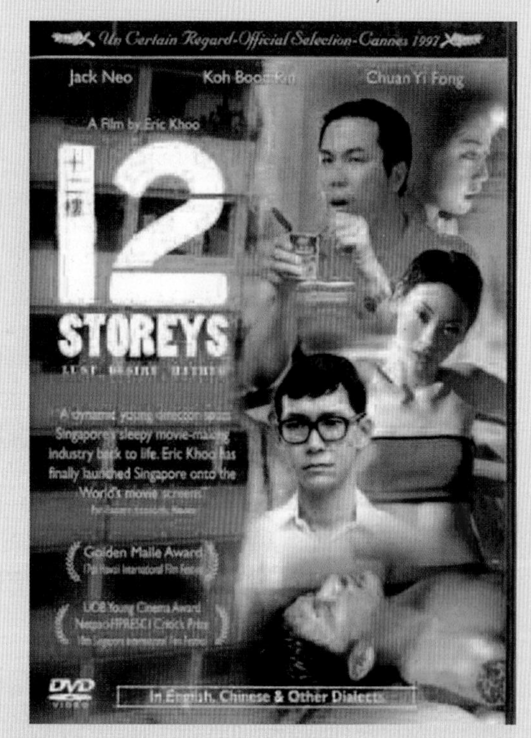

Figure 47.3 *12 Storeys* © Zhao Wei Films

Eric Khoo's film, *12 Storeys*, premiered in Singapore and at Cannes in 1997. Touted as Singapore's return to the world screen, the film offers a critique of Singapore's archetypal image as clean, efficient and upwardly mobile. *12 Storeys*, as the title implies, is set within a typical working-class HDB block (many of which have 12 storeys). Khoo portrays the worlds of three households inhabiting the block, and whose experience of HDB living is claustrophobic and full of loneliness and alienation: a dutiful but repressed brother obsessed with disciplining his siblings' work ethic and sexuality; a frustrated, sexless relationship between a bucktoothed noodle seller and his acquisitive China bride; and an overweight daughter of a malicious Cantonese immigrant woman who devoted her life to raising other people's children. One thread holding the stories together is the spirit of a young man who commits suicide at the beginning of the film;

he witnesses these people's dramas, but never speaks to them. Another thread is a group of unemployed men that collectively idle at the local coffee shop (a phenomenon discussed further in the main text of this chapter). The characters express the 'nightmarish contradictions in self-understanding and interpersonal relations' that Khoo believes haunt modern Singapore and reveal what most Singaporeans just 'know' (De, 2002: 196).

Khoo's characters are ordinary Singaporeans in a dysfunctional society; they are 'representative types of Singaporeans who have fallen out of step with the Singapore Success Story' (Chua and Yeo, 2003: 119). As disaffected HDB-dwellers, they do not symbolize the state-promoted imagery of happy families, efficiency and economic prowess. In this way, Khoo provides us with an alternative portrayal of everyday life and an oblique critique of the state. He does so from the margins of the modern metropolis, however, and this is reflected in the audience he reaches: *12 Storeys* is not mainstream cinema and has limited circulation in the art house circuit. Without engaging Singapore's mass audience, and without empathizing with (rather than merely representing) their worlds, Khoo stands accused of 'superficial ethnography' and an inability to represent the 'embodied psychological dramas' of life in Singapore (Chua and Yeo, 2003: 120). (Khoo's more recent film, *Be With Me*, has a broader appeal and reflects a multi-sensuous Singapore (see Law *et al.*, 2011)).

rationalize its spheres of production and reproduction (De, 2002). It is a vision from above, available from vantage points like Raffles City, but tells us little about the people who inhabit and make sense of the city at street level. As Michel de Certeau (1984) once observed in relation to the view of New York City from the (now destroyed) World Trade Center, to be lifted to such heights is,

> to be lifted out of the city's grasp . . . one's body is no longer clasped by the streets . . . nor is it possessed . . . by the rumble of so many differences . . . It transforms the bewitching world . . . into a text that lies before one's eyes. It allows one to read it . . . looking down like a god.
>
> (de Certeau, 1984: 383–4)

This vision from above fixes the landscape into a static image that denies its lived complexity.

Michel de Certeau (1984), a theorist of the everyday as counter-concept in western societies, has written about how urban practices can interrupt the hegemony of state ideologies of urban development. His interest lies in disrupting totalitarian systems that become incorporated into our sense-making activities and routines, and in exploring the possibilities of **human agency** that are foreclosed when we focus too narrowly on reading representational systems in the urban landscape. In Singapore, a country stereotypically known as a surveillant state, it is important to understand how state power is established through a mechanism such as urban planning. We can 'read' in the landscape how the state organizes its economy and society, finding evidence of how ideologies materialize in urban space. People live in mixed neighbourhoods, labour at cosmopolitan workplaces and have fun in places like Boat Quay. At the same time, Singaporeans do not necessarily carry out their lives in ways intended by the state. We might also consider how people resist, subvert and/or transgress state ideology. Drawing on our critiques of vision discussed earlier, we might ask ourselves: What does Singapore look like from the sidewalks of the city and suburbs? Is living in Singapore akin to living within the confines of a city planning map? De Certeau advises that we should not undervalue the capacity of the everyday pedestrian walking in the city to craft new meanings into spaces designed by planners. He terms these subversions and transgressions 'spatial practices' that make up the 'lived space' of the city.

Everyday life contains habits, or what de Certeau terms 'spatial practices', that interrupt, rather than resist, state or other ideologies. He describes these fragmentary routes through the city as 'tactics' in that they are not resistant in the conventional sense; rather, they are enacted through a myriad of footsteps that undermine state ideology. Pedestrians recreate even the most highly planned city by living in it, by weaving new stories into the urban landscape that can transgress the narratives preferred by the state. Even though pedestrians transgress ideologies in ways that are small, individual and non-confrontational, de Certeau suggests these practices do create another 'mobile' city within the highly planned one. He suggests that spatial practices are not *separate* from the administration of everyday life by the state, but they are '*foreign* to the "geometrical" or "geographical" space of the visual, panoptic, or theoretical constructions' that preoccupy urban planners (de Certeau, 1984: 93, [emphasis added]). Using de Certeau's ideas allows us to think about what the city looks like 'from below', and thus to understand the complex engagement Singaporeans have with state ideology. As we will see in our discussion of hawker centres below, de Certeau's conceptualization works particularly well in Singapore, where people find individual (and collective) ways to survive life under an interventionist state.

SUMMARY

- Seeing the city from above, as urban planners do, fixes the city's meaning and denies its lived complexity.
- State ideology is built into the city's built form and can inscribe a dominant meaning into space.
- State ideology is not the only meaning inscribed in urban landscapes. De Certeau suggests that the everyday pedestrian can craft new and sometimes subversive meanings into the urban environment. His idea of 'spatial practices' allows us to consider the possibilities of human agency.

Sensing everyday geographies

Let us now shift our attention to the lived space of Singapore and consider how state ideology and human agency are practised in an everyday environment. One space of considerable significance is the ubiquitous hawker centre, an open air food court typically attached to a wet market or HDB housing estate (see Figure

47.4). Hawker centres have several stalls selling a variety of ethnic fare, including Chinese, Indian, Malay and even western food. Customers order and pay at their chosen stall, where food/drink is served immediately or soon thereafter to a numbered table in an open seating area. Because the price of food at hawker centres is reasonable, and because there are so many of them (more than 130 centres with almost 18,000 individual stalls are regulated by the government) many families across socio-economic groups take advantage of hawker fare to supplement or even replace home cooking. Frequent patronage makes hawker centres an important social space in Singapore, whether it be for family meals, housewives stopping for a chat, couples on dates, youths having coffee or men gathering to drink beer. Hawker centres are thus densely meaningful sites, where state ideology meets the practices of everyday life.

Hawker centres have been regulated by the state since their planned construction in Singapore after independence. Their history can be traced to the state's ideology of a well-ordered modernity and to the sanitization of the city bound up in the garden city movement of the time. Mobile hawkers that once plied their trade on the streets were gradually moved into state-regulated centres complete with electricity, rubbish disposal, water and sewerage facilities. Their relocation into purpose-built centres was thought to be a solution to the problems of pollution, the spread of food-borne diseases and 'the encumbrance of land for other national development projects' (Perry *et al.*, 1997: 210). While regulations have changed over time, the state maintains a surveillance role in managing these sites. For example, each stall must be granted a permit to commence operation, which allows the state to ensure a mix of hawker fare at any one centre. As in the case of HDB estates, the state favours racial mixing and can ensure hawker centres are reflective of Singapore's multicultural population and sensitive to religious dietary restrictions (e.g. providing halal sections for Muslims). Furthermore, each food handler must take a

Figure 47.4 Hawker centres

state-run course and pass various health examinations and each stall is officially 'graded' according to state-defined sanitary conditions (ranging from 'A' to 'D', with 'A' being the cleanest). Hawker centres are important everyday sites that could be 'read' as a powerful inscription in the landscape. Their design and form corresponds to the state's ideology of class and racial mixing and orderliness. But do Singaporeans merely live out the state's ideals of order and cosmopolitanism in hawker centres?

If we return to de Certeau's ideas of spatial practices, we can rethink hawker centres as part of the everyday lived experience of the city. Because hawker centres are so ubiquitous and frequented by people across different social groups, they form an important part of the rhythm of daily life. They are spaces where people gather for many reasons, from the functional act of eating a meal to renewing social networks or simply stopping to chat and gossip. But patronage does not mean adhering to state ideology, as people's visits are harnessed to different everyday lives and routines. If we drew a map of people's everyday routes we would have a sense of how important hawker centres are to the flow of daily life, but we would not be able to tell the meanings people attach to their visits. We would not know, for instance, if they experience pleasure, apprehension or different degrees of feeling in or out of place. Hawker centres look and feel different depending on who you are.

For the purposes of our discussion, let us take one group that forms a regular feature of life in hawker centres, especially at the ever present *kopi tiam* (literally 'coffee shop'). The *kopi tiam* is an important meeting place for blue collar men in HDB neighbourhoods, where they gossip and tell stories in vernacular groups. Conversations range from idle banter to discussions of world affairs and the men 'freely mingle current historical information and rational knowledge with folk wisdom' thus rejuvenating 'a heterogeneous tradition of conversation characteristic of ethnic village life in colonial Singapore' (De, 2002: 201; see also Chua, 1997). By upholding village life in the midst of the modern metropolis, the men sustain traditions that subtly undermine (rather than overtly resist) state ideology. They do not gather in groups that imply class or racial mixing and they speak in ethnic languages or in Singlish – a dialect spoken colloquially in Singapore but officially discouraged by the government (see box). These 'traditional' practices are antithetical to the state's 'modern' ideals. The vernacular soundscapes of hawker centres thus offer us different possibilities in terms of thinking about power and resistance, but require listening to — and not just looking at — what is taking place.

While soundscapes provide important clues to the everyday world of hawker centres, this does not mean that vision remains irrelevant. What blue collar men see at the *kopi tiam* is not the glittering skyscrapers of Shenton Way, the orderly HDB landscape or the more general vision of Singapore as well-planned metropolis. Instead they watch people pass by, sometimes making disparaging comments about today's modern youth (e.g. the bizarreness of dyed hair) or they scan the landscape for lucky numbers to enter into the next 4D lottery (illegal bookies are in close proximity to take their wager). This is not what the state *wants* them to see, nor is this how it *wants* hawker centres to be used. These textures of place 'from below' are multifaceted and require a degree of ethnographic awareness.

Yao Souchou (2001), an ethnographer, argues that an important feature of a *kopi tiam* gathering is the social pleasure of poking fun at the state: a sophisticated yet informal practice of bitching about the government. He suggests that by 'eating air' and 'talking cock', the men

CASE STUDY

Singlish in Singapore

Singlish is spoken vernacularly in Singapore and can be thought of as an Asian dialect of the English language. It is a mixture of British (and increasingly American) English with some grammar and vocabulary drawn from Chinese (especially Hokkien), Malay and Tamil languages. Singlish is influenced by the grammatical and tonal structure of Chinese, making it difficult to imitate for non-Chinese or non-Singaporean speakers. Importantly, Singlish is officially denounced by the state, as it is understood as unrefined pidgin that impedes equality and Singapore's access to the outside world. On the other hand, many Singaporeans think:

'Bladi Garmen si peh kaypoh one, why always so bedek kacang hor?'

According to www.free-definition.com this sentence can be broken down as follows:

- 'bladi Garmen' – bloody government
- 'si peh' – very (from Hokkien, a Chinese dialect)
- 'kaypoh' – busybody (from Hokkien)
- 'one' – extraneous modifier
- 'why always so' – indication of harboured displeasure
- 'bedek kacang' – lit. 'aiming at peanuts' (Malay); in this sentence, can probably be taken to mean 'meddlesome' or 'annoying'
- 'hor?' – Chinese prompt for affirmation, somewhat like the French *n'est-ce pas?*

Since the late 1990s, schools have carried out Speak English Campaigns and a broader Speak Good English Movement was initiated in 2000. It is unlikely that Singlish can be eradicated by the state, however.

enact a visceral condition that tells us much about everyday life in Singapore. In local parlance 'eating air' evokes leisurely evening strolls that provide respite from the tropical heat, as well as opportunities to chat with friends and share a meal. 'Talking cock' is a Singlish term for the informal, artful verbal exchanges that embellish social facts with a mixture of humour and wisdom. Yao argues that in the cool of the evening these men 'eat air' and 'talk cock' as an everyday, subversive practice that tells us much about the pleasure and burden of life under a robust state. Crucial to these exchanges – which introduce everything from the price of cars to the enduring fortunes of the Lee family – is their seasoning with food and drink. Food arrives in bite-sized pieces small enough not to inhibit talking and alcohol lubricates the discussion. Eating air and talking cock thus represent a visceral embodiment that can be connected to discussions of politics in Singapore, but this is not a politics that understands talking cock as formal, conscious resistance. Instead, it is an embodied form of political engagement that subtly undermines the state (which, in any case, talks cock too).

Understanding these textures of place and sensory embodiments provides important clues to life in everyday Singapore, where resistance is not generally 'seen'. In this context it might be more important to understand phenomena such as idling in hawker centres, where more subtle politics unfold though the everyday habits of gathering and sharing a meal. Such communal assembly is a spatial practice (in de Certeau's sense) that helps rewrite Singapore's landscape in ways that are more meaningful to its inhabitants. These everyday geographies lie below the state's threshold of visibility and summon a whole range of senses to give them meaning: the soundscapes of vernacular language, the taste of ethnic food and the feel of the cool evening air. While these everyday practices might not alter the political world most Singaporeans live in, they do provide alternative maps and views of a city undergoing profound transformation. To access these worlds geographers must move beyond optic methods and techniques and learn to hear, taste and smell their meanings.

SUMMARY

- Hawker centres in Singapore are one example of a state narrative inscribed in the landscape, but people craft new meanings into these sites. They are thus the 'lived spaces' that tell us about urban experience.
- While spatial practices are about vision, it is also important to remember that a range of senses embody people at these sites; the senses work in tandem to produce everyday, embodied experiences.

Conclusion

Although the discipline of geography has historically relied on vision for its understanding of urban experience, there is another world of understanding that lies beyond (within?) that which is in view. We must not rely solely on our sense of sight or on the visions of powerful others, otherwise we end up investigating only 'remarkable' and 'elite' landscapes, ones that are more readily seen from above than below. As Cresswell (2003: 280) argues, our challenge is to produce 'geographies that are lived, embodied, practiced; landscapes which are never finished or complete, not easily framed or read'. The senses can provide geographers with fresh insights to the complexity of everyday urban life and new opportunities for embodied research.

DISCUSSION POINTS

1. What are some of the problems encountered when thinking about urban experience from above?
2. Why is it important to move beyond geographers' critiques of vision?
3. What methods would you use to investigate the link between the senses and urban experience?
4. Does an understanding of the senses provide us with insight to human agency?

FURTHER READING

Rodaway, P. (1994) *Sensuous Geographies: Body, Sense and Place.* London and New York: Routledge.

Pocock, D. (1993) The senses in focus. In: *Area,* 25(1), 11–6.

These texts offer a good introduction to how the senses are relevant to geography.

Ingham, J. (1999) Hearing places, making spaces: sonorous geographies, ephemeral rhythms, and the Blackburn warehouse parties. In: *Environment and Planning D: Society and Space,* 17, 283–305.

Law, L. (2001) Home cooking: Filipino women and geographies of the senses in Hong Kong. In: *Ecumene,* 8(3), 264–83.

Smith, S. (2000) Performing in the (sound)world. In: *Environment and Planning D: Society and Space,* 18, 615–37.

These three articles examine the senses and our urban experiences.

Adams, M., Cox, T., Moore, G., Croxford, B., Refaee, M. and Sharples, S. (2006) Sustainable soundscapes: Noise policy and the urban experience. In: *Urban Studies* 43, 2385–98.

Middleton, J. (2010) Sense and the city: Exploring the embodied geographies of urban walking. In: *Social and Cultural Geography* 11(6), 575–96.

These two articles give insight into how debates about the senses can be related to urban planning policy.

De Certeau, M. (1984) *The Practice of Everyday Life.* Berkeley: University of California Press.

Michel de Certeau's ideas about 'spatial practices' are to be found in this book.

Howes, D. (ed.) (2004) *Empire of the Senses: The Sensual Culture Reader.* Oxford and New York: Berg Publishers.

Outside geography, this is an excellent collection of essays on the senses.

CHAPTER 48
RURALITY

Paul Cloke

Introduction: the countryside comes to town?

In recent years the British countryside has been given an ever more strident political voice. Countryside marches and rallies in 1997 and 1998 saw over 100,000 rural people gathering in central London to protest about the encroachment of urban-based bureaucracy into country life, as epitomized by the proposed (now enacted) ban on fox hunting (see Figure 48.1). In the words of the *Daily Telegraph*,

Figure 48.1 The country comes to town. Credit: Dan Chung/Reuters/Corbis

in the annals of popular protest, there can seldom have been a noisier plea to a British government to do absolutely nothing than yesterday's Countryside Rally . . . they had come from farms, moors and fells, emptying villages and leaving nature to its own devices for a day in order to let the urban majority know that the rural minority wishes to be left alone.

(*Daily Telegraph*, 11 July 1997)

Such protests have continued, often under the political banner of the Countryside Alliance, to present a conservative and conservationist politics of 'no change' in the face of what has been perceived as a string of other threats to the essential character of the countryside. New centralized planning and housing policies risk destruction of the character of country towns and encourage house building on green belt or other protected land. The sale of state-owned forests risks the annihilation of woodland and a clampdown on public access to land. Proposals to build a high speed rail link (HS2) between London and Birmingham risks despoliation of some of the most beautiful countryside and vibrant communities in England (see Figure 48.2).

In many ways such protests are indicative of a peculiarly British collection of landscapes, traditions and cultural practices associated with the countryside. Here we are offered the view of a somewhat timeless, highly valued and all-embracing country life that needs to be preserved at all costs from the ravages of urbanism. It is, however, the view of a small but powerful minority, which can grab the imagination about what country life stands for. Our **geographical imaginations** of the country

Figure 48.2 (a) A protest against planning laws; (b) A protest about the selling off of forests; (c) A protest against the high-speed rail link HS2.
Credit: (a)–(c) Associated Press

are often produced and reproduced from 'stuff' such as this. By contrast the distinguished travel writer, Jonathan Raban, records a visit to rural Alabama in his book *Hunting Mr Heartbreak*. Here he emphasizes the shock experienced by some Europeans when they encounter some of the countrysides of America. The scale, colour and 'savagery' of nature in the American outdoors do not easily accommodate direct comparisons with more familiar European landscapes.

> It was how Europeans had always seen American nature – as shockingly bigger, more colourful, more deadly, more exotic, than anything they'd seen at home. When the urban European thought of the countryside, he imagined a version of pastoral that was akin to, if a good deal less exaggerated than, that on offer in Ralph Lauren's Rhinelander Mansion on Madison Avenue. The 'country' was an artefact – hedged, ditched, planted, well patrolled . . . The European landscape was a mixture of park, farm and garden; the nearest we come to wilderness was the keepered grouse moor and the occasional picturesque crag. We were astonished by America, its irrepressible profusion and 'savagery'.
>
> (Raban, 1990: 153)

In highlighting these differences, Raban also shows us that rural areas represent a vivid and often specific facet of the geographical imagination. Not only do we carry around with us an **idyll-ized** sense of what our rural areas look like, and therefore of what they are like to live in and visit, but we are often shocked when encountering other stereotypes of our countrysides or other countrysides.

In this chapter I want to discuss how rural areas have become exciting contexts for study in Human Geography. In recent years there has been something of a resurgence of interest in rural studies, partly as it has embraced the 'cultural turn' that is evident in the broader social sciences, but also in part because the significance of 'nature' and 'rurality' has gone beyond rural geographical space. Rural areas themselves have offered fertile ground for the study of more mainstream cultural ideas, of which four have been of particular importance:

1. *A focus on landscape*, emphasizing the meanings, myths and ideologies that are represented therein. Geographical study of landscape can range from deep countryside to urban street and from deep history to the imaginative futuristic landscapes of science fiction. However, countryside landscapes demonstrate particular power relations as well as being objects of desire that many would wish to conserve.
2. *A focus on how nature relates to space*. Again, nature is by no means confined to rural areas, but countrysides are often represented as the 'obvious' spaces of nature. Here the relations between culture and nature (see Chapter 10) are often a visible element of country life – as in the 'hunting' debate mentioned above – and consequently the country provides fertile ground for the study of how humans and non-humans interact.
3. *A focus on food,* notably the shift from mass-produced to more variegated and niched food production, in which 'local', 'sustainable', 'organic' and 'fair' foods have become entwined with powerful forces of marketing and ethics, played out in both big supermarkets and small farm shops.
4. *A focus on 'hidden others'*. Countrysides are rich in myth and they represent territories where an overriding cultural gloss on life can mask very significant socially excluded groups. Issues of **gender**, **sexuality**, poverty and alternative lifestyles are important in this context.

What links these themes together is the importance of an idyll-ized view of the rural.

Countrysides are seen as places where people can live close to nature and in harmony with surrounding landscapes. Country living is characterized by a happy, healthy and close-knit community and a problem-free existence that differs markedly from urban life. Such an idyll reflects the power of those who can afford to buy into and enjoy rural life and deflects any 'problems' that don't fit the image. And it is this idyllic cultural image that transfers itself into broader society such that the country is no longer confined to the spatial boundaries of recognizably rural areas. Through the **commodification** of nature and rurality within contemporary consumption (as indicated not only through media attention and in advertising but also in 'country' consumer goods ranging from four-wheel drive vehicles to furnishings and clothing), the importance of the country, and the meanings attached to it, have spread throughout society.

It is important to note that these cultural themes of the country are not the only geographies to be told of rural areas. Indeed, these more recent cultural geographies are being overlaid on to existing accounts of behavioural and political studies of economic change and demographic and relational studies of social change. What makes the country important in Human Geography, however, is a combination of the idyll-ized imagined geographies peddled by media and advertising and held by people as significant reference maps for spatial behaviour, and the specific material changes occurring in rural geographical areas. This mix of imagined and spatialized countrysides begs questions about how we recognize rurality when we see it, and it is important here to reflect briefly on these debates surrounding the nature of rurality.

The blurring of country and city

When we study rural geographies we have to keep hold of two rather different kinds of change. First, there are the changes that are occurring in rural areas themselves. For example, over the past 30 years there has been a hugely significant reversal of the trend of the previous century whereby population had been concentrating into urban centres in most western countries. In the USA, for example, the 1950s and 1960s saw a strong positive correlation between settlement size and population growth rate, but during the 1970s 'smallness' became associated with growth. In the UK, the 1981 census revealed that rural districts had begun to experience demographic growth over the previous decade, and this trend continued over the next 20 years (see Table 48.1). Although this broad picture of rural in-migration masks a diversity of localized patterns of growth and decline, it suggests

	1981–1991	1991–2001	1981–2001
Rural districts	+7.1%	+4.9%	+12.4%
Urban districts	+1.4%	+0.9%	+2.4%
England total	+3.0%	+2.0%	+5.0%

Table 48.1 Population change in rural and urban districts of England, 1981–2001. Source: Woods (2004), drawing on data from the Countryside Agency

important changes in rural society. In-migrant populations were at one and the same time seeking out the perceived advantages of rural lifestyles and bringing with them attributes of urban living and expectations that were likely to transform the very communities they had been attracted to. Traditional rural life had already been transformed by the near universal availability of urban-based media and universal telecommunications, and now this has been reinforced by the infusion of migrants, often from urban places, who were seeking to live out imagined geographies of rural life in particular geographical places and spaces, often leading to some turbulence with more 'indigenous' populations.

Demographic change has usually gone hand in hand with economic change. As the size of the agricultural workforce has diminished, the notion that rural areas are dominated by agriculture has in many places become more applicable to the dominance of agricultural landscapes than to the agricultural economy. Counter-urbanization was often accompanied by an urban-to-rural shift in new manufacturing growth and although the economic impact of this shift has sometimes been short-lived, new forms of service sector employment have often added to the economic potential of these non-metropolitan areas. Indeed, it is now commonly assumed that the growing importance of telecommunications and information technology will metaphorically 'shrink' the geographic distances between rural areas and major urban centres and thereby favour service sector growth in rural areas. An online personal computer allows many contemporary work tasks to be performed from the rural home, and internet shopping and service provision are in some ways transforming the economic, and even geographical, isolation of many rural residents (Commission for Rural Communities, 2009).

This general picture of change itself masks considerable variation both within and between nations, so it is useful to talk of rural geographies rather than a rural geography. It is an obvious but often forgotten fact that what we regard as western or 'developed' nations vary enormously in scale. As Figure 48.4 demonstrates, the scale of influence exerted by major metropolitan areas differs widely, such that the urban pressure on the country in Britain will be far more intense than those on certain areas of the USA and Australia. Such variation means that particular places will be located rather differently in the mosaic of change described above. For some, it would be no exaggeration to suggest that they reflect suburban characteristics, performing a dormitory role for metropolitan labour

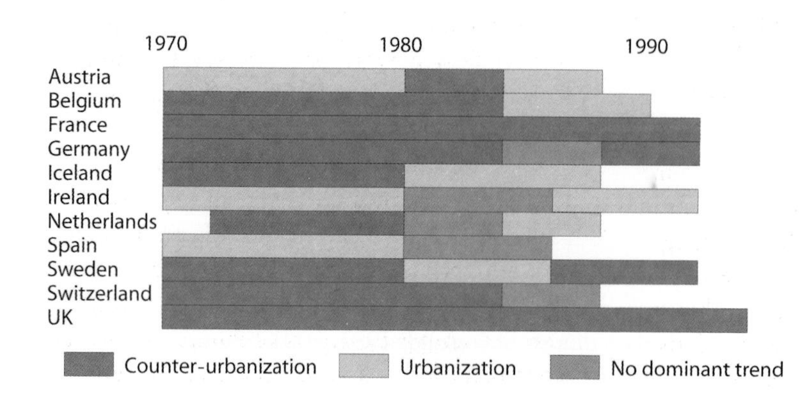

Figure 48.3 Predominance of counter-urbanization and urbanization for 11 European countries. Source: Woods (2004)

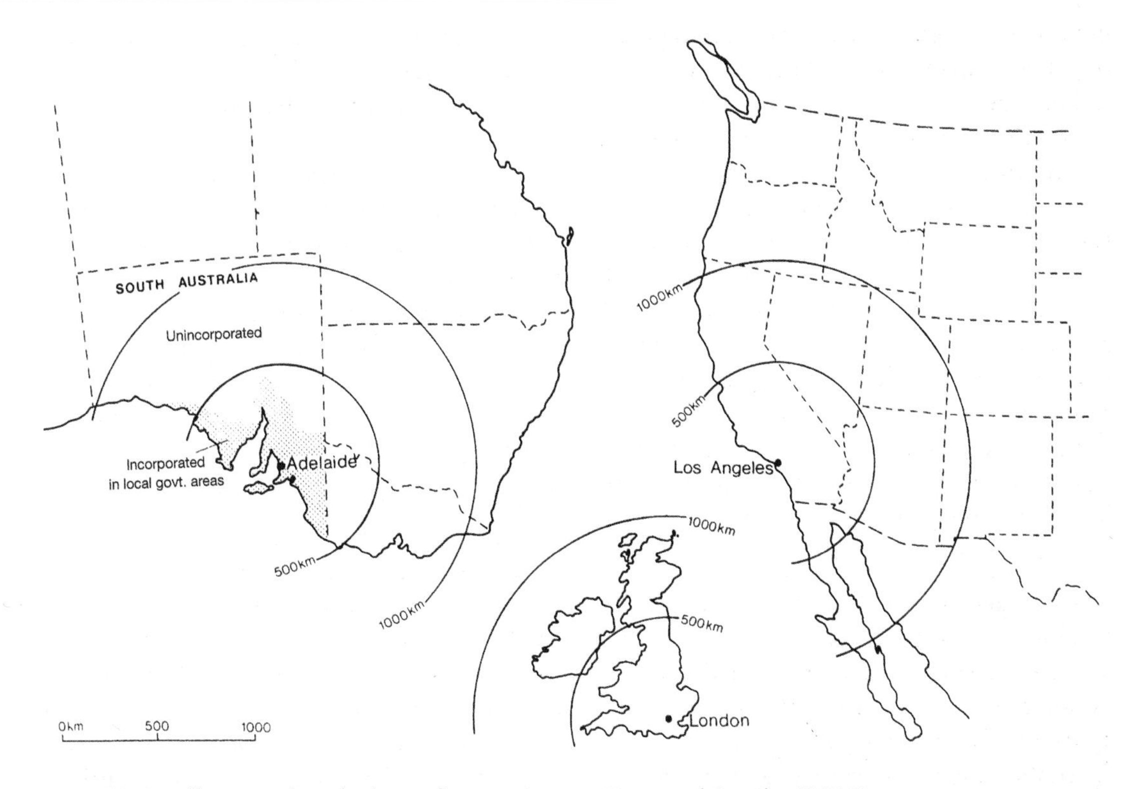

Figure 48.4 Different scales of urban influence. Source: Hugo and Smailes (1985)

markets. Elsewhere, agriculture will remain as the dominant economic as well as landscape feature. Elsewhere again, extreme geographical marginality reflects characteristics of 'outback', 'wilderness' or even desert – each posing particular questions of nature–culture relations and of the potential for commodification. So, the changes occurring in rural areas themselves are irregular and particular attention has to be given to the geographies of particular places within the overall framework of change.

Accepting the importance of these differences in nature–culture relations, Marc Mormont (1990) has suggested that another key question about rural change concerns the changing relationship between space and society, and it is increasingly clear that this relationship is no longer only about the traditional divisions between rural and urban or town and countryside. He argues that such dualisms have been completely overtaken by events and outlines a series of changes relating to personal mobility and new economic uses of the countryside that indicate the outmoded nature of any view that sees rural society and rural spaces as being welded together. Mormont's analysis of change relates to Belgium, but appears relevant to many countrysides as is his conclusion, which is to suggest that there is no longer a single rural space, but rather a multiplicity of social spaces that overlap the same geographical area. The supposed opposition between the geographic spaces of city and countryside is being broken down, but oppositions between the social significances of city and countryside remain. For Mormont, then, rurality is a category of thought – a social construction – and in contemporary society the social and cultural views that are thought to be attached to rurality provide clearer grounds for differentiating between urban and rural than do the differences manifest in geographic space.

Other commentators (see, for example, the seminal writings of Raymond Williams in the British context) have also noted the blurring of the country and the city. A very interesting contributor to this debate is Alexander Wilson (1992), who in his book *The Culture of Nature* suggests that recent land development in North America – suburbs, theme parks, shopping centres, executive estates, industrial parks, tourist developments and the like – has served to reproduce misleading ideas about city and country. He argues that the form of this development fragments geographies into those devoted to work and leisure and production and recreation, which are oppositions that obscure more than they reveal about the nature of city and country. He cites the West Edmonton Mall (Edmonton, Canada – see Figure 48.5) as an example of this jumble of country/city compromises. Its suburban location, 45 hectare size and 2.5 kilometre-long concourse suggest the monstrous urbanism of an indoor shopping centre. However, it includes a 1 hectare lake with dolphins, sharks (and four full size submarines, which give rides in the lake). Moreover, the mall also houses an 18-hole golf course, a water park with 6-foot surfing waves and hundreds of animals in aquariums and cages.

Figure 48.5 West Edmonton Mall. Credit: Simon Grosset/Alamy

In this way, the West Edmonton Mall encapsulates objects and beings from the natural world into its commodified urban space. The 'landscape' is artificial, but the control over nature is very much part of the attraction. In yet another way, the binary opposition between city and country, and indeed culture and nature is blurred in such place-making.

Another form of city-country blurring is evident in Jean-Didier Urbain's (2002) observation that the spread of the city out into the country has effectively *ruralized* a significant part of the urban. The nature of the city has been changed both by centralizing tendencies and by decentralizing practices, with the result that an important slice of contemporary urbanity can now be found in the village and that the urban form thereby now encapsulates very strong rural characteristics and influences. Equally, urban managers seem increasingly to be striving for a set of virtues in the city that are more commonly associated with the rural: seemingly fundamental values such as protection, solidarity, community spirit and identity.

Therefore the blurring of rural–urban distinctions is bringing crucial changes to urbanity as well as rurality.

These examples suggest that the assumed differences between the geographical spaces of city and country have been somewhat undermined by changes occurring in the social, economic and built environments concerned. It is again important to emphasize that such changes differ in scale and intensity in different places. Some countrysides will appear relatively untouched while others will have been visibly transformed. Nevertheless, the increasing importance of rurality as a social rather than geographical construct applies very widely. These factors constitute the first set of changes that rural geographers have to grasp.

The second kind of change relates to the way in which geographers have offered different ways of understanding rurality itself. Any given rural geographical space can appear different according to the theoretical perspectives adopted. For example, geographers have traditionally mapped rurality by equating it with particular functions: thus rural areas are dominated (currently or recently) by extensive land uses such as agriculture and forestry or large open spaces of undeveloped land. Rural areas contain small, lower-order settlements that demonstrate a strong relationship between buildings and surrounding extensive landscapes and are thought of as rural by most of their residents. Rurality engenders a way of life that is characterized by a cohesive identity based on respect for the environment and behavioural qualities of living as part of an extensive landscape.

This type of analysis has been useful in generating indicators of rural territory and remains useful, especially in those areas that are less transformed by the process of blurring described above. However, different theoretical epochs in the social sciences have produced critiques of these definitions of rural space. From political economy approaches came the insight that rural areas were increasingly linked into changing globalized economies, with the causes of rural change usually stemming from outside of the rural areas concerned. From this viewpoint 'rural' places were not particularly distinct, and for some, this realization led to a call to do away with rural as an analytical category. Recent moves to highlight the theoretical power of *actor networks* give a different perspective to these spatial linkages (see Cloke, 2005; Murdoch, 2003). Here there is a recognition that human and non-human actors are bound together relationally into *hybrid collectives* that transcend simple spatial and temporal boundaries. As a result, the 'here and now' of rural areas may not reflect the complexity of the networks concerned.

Other windows on rurality have focused on the power of the rural as a significant category of the imagination. Drawing on more *postmodern* and *poststructural* ways of thinking, rural researchers (see, for example, Marc Mormont's work discussed above) have suggested that rurality can be regarded as a **social construct**, and that the importance of the 'rural' lies in the fascinating world of social, moral and cultural values that are thought to be significant there. Far from 'doing away with' the rural, then, the idea of rurality as a social construct invites researchers to study how behaviour and decision making are influenced by the social and cultural meanings attached to rural places. In particular, there is considerable interest in how meanings of rurality are constructed, negotiated and experienced (Cloke and Milbourne, 1992). While such meanings may have much in common, there will be many different versions of rurality perceived by different individuals and organizations.

Leaning on the philosophical writings of Baudrillard, Keith Halfacree (1993) discusses these multiple meanings of rural in terms of three levels of divergence. The sign (= rurality) is increasingly being detached from the signification (= meanings of rurality) as social representations of rurality become more diverse. Equally, sign and signification are also becoming more divorced from their referent (= the rural geographical space). He points out that it is a characteristic of postmodern times that symbols are becoming more detached from their referential moorings and therefore that socially constructed rural space is becoming increasingly detached from geographically functional rural space. Indeed, Jonathan Murdoch and Andy Pratt (1997) believe that we have reached the stage of 'post-rural' studies, reflecting a willingness to believe in an idea of countryside that we realize may no longer be authentic in terms of the material reality of rural society and space.

These different approaches to 'mapping' or 'knowing' the country introduce a constructive tension to rural studies, especially when held together with the material changes occurring in what are commonly recognized as rural geographical spaces. For some, the country will be seemingly knowable and apparently atheoretical. For others, the country is best recognized through the lens either of how rurality is perceived as a lived-in and often affective landscape, or of how the rural is performed in a visceral, embodied and often mobile manner (see Cloke, 2006; Halfacree, 2006). For yet others, the complexities of power, practice and process will render the category 'rural' unknowable as any kind of geographical or social entity. The more postmodern the country seems, the more blurring seems to occur between country and city. Many residents and visitors do appear to act as if the countryside exists in some knowable form. Others, however, appear to know their 'rural' places differently, seeing them in other regional or local ways.

SUMMARY

- Rural areas themselves are changing demographically, socially and economically. A geographically 'rural' space may now be overlapped by many different social spaces, thus transforming traditional countrysides.
- Many new land developments – theme parks, shopping centres, tourist developments, etc. – also blur the difference between country and city.
- Geographers have also changed the ways in which they have sought to understand rurality itself. Defining rural space by the functions that go on there has been challenged by those who view rurality as a social construct – a category of thought – or as a place that is co-constructed by the performance of humans and non-humans.
- The cultural meanings associated with the country have become increasingly detached from rural geographical space and are now important throughout society.

Commodifying the countryside

There is evidence that the continuing importance of the country will in part lie in attempts to commodify the countryside as a particular type of attraction within postmodern consumption. Popular culture now serves us up with what Raymond Williams referred to as 'a continuing flood of sentimental and selectively nostalgic versions of country life'. Films such as *Hot Fuzz, War Horse* and *Tamara Drewe*; television series such as *Midsomer Murders, The Vicar of Dibley* and *Doc Martin*; children's favourites such as *Postman Pat* and *Sylvanian Families*; magazines such as *Country Life* and *Countryman*; all merely add to classics in art, literature and media in their focus on cosy and nostalgic aspects of countryscape.

Moreover, the advertising industry repeatedly borrows from the treasure chest of positive meanings vested in the countryside – the 'goodness' of nature to sell bread; the 'classiness' of the country house to sell cars; the 'pioneer spirit' of rural America to sell jeans or cigarettes – and in so doing reinforces these references to nature, heritage, nostalgia and so on, in popular constructions of contemporary rural life. In this way, the meanings of country are attached to products that are themselves often aspatial. The country escapes from its geographical referent and inhabits the wider world of taste and consumption. An excellent example of this escape can be found in the way in which the Laura Ashley company purposefully commodified the appeal of country tradition to create a style that is applicable to many kinds of geographical space. In this example, the past rustic traditions are sieved through the 'colourful mixture of prints and textures' and the Welsh farmhouse, Long Island house and Swiss chalet are made available to anyone, anywhere. Rustic tradition becomes contemporary commodity and nature's countryside is bought and sold as fabric and furnishings.

The country is thus being commodified within both the geographical spaces and social spaces it inhabits. Part of the background to the significant shift in the nature and pace of commodification in rural areas in many developed nations is the perceived transition from productivist agriculture, where industrialization and scientization of farming have been deployed to generate increasingly efficient production outputs, to post-productivist agriculture. Although such a transition is far from clear-cut (there are many complexities and ambiguities that defy a simple translation from an overall productivist scheme to an overall post-productivist scheme – see Evans *et al.*, 2002), there are some common underlying changes occurring in rural land use, often encouraged by state regulation, such as the reforms to the Common Agricultural Policy of the EU. One aim has been to extensify food production by using fewer chemical inputs; witness, for example, the growth in organic food production. Another has been to recognize and reward the role of farmers as guardians of landscape and environment, capable of switching to farm practices that are conducive to broader goals of sustainability and conservation. A third trend has been to diversify farming, promoting a more pluriactive countryside in which farmers are seeking both to enhance the value of their food products (for example, through the growth of local farmers' markets) and to engage in new enterprises, notably in the tourism and leisure sectors. Alongside this farm-based commodification, rurality more generally has given rise to a series of new markets for countryside commodities: the countryside as an exclusive place in which to live; rural communities as a context to be bought and sold; rural lifestyles that can

Figure 48.6 Rural attractions: (a) Morwellham Quay; (b) Dollywood. Credit: (b) Pat O'Hara/Corbis

be colonized; icons of rural culture that can be crafted, packed and marketed; rural landscapes with a new range of potential, from 'pay-as-you-enter' national parks, to sites for the theme park explosion, and so on.

As the country becomes commodified, particular meanings and characteristics are emphasized that come to represent its very essence. For example, in a study of how new and revamped rural tourist attractions were being advertised in parts of Britain, it was found that particular meanings, signs and symbols of countryside were clearly being represented (Cloke, 1993). These socially constructed ruralities reflected the perhaps predictable themes of nature, outdoor fun and history. They also reflected the slightly less predictable themes of family safety, 'hands-on' or 'up-close' experiences of nature and the specific commodity links with souvenir craft and particular foods and drinks that form integral components of the packaged day out in the countryside. The study demonstrated that many of these new countrysides were based on the production of a spectacle for visitors. For example, Morwellham Quay on the border of Devon and Cornwall recreates a Victorian copper port (see Figure 48.6a) and in so doing offers an outdoor theatre of rural history. According to the brochure,

> the quay workers, cooper, blacksmith, assayer and servant girls dressed in period costume, recreate the bustling boom years of the 1860s . . . Chat with the people of the past. Sample for yourself the life of the port where a bygone age is captured in the crafts and costumes of the 1860s . . . Try on costumes from our 1860s wardrobe.

The invitation is to spectate and participate in the history that is 'captured' by the attraction and presented to visitors in the form of spectacle.

Examples of commodification of country spaces abound. Indeed, Howard Newby (1988) has suggested that rural Britain in general has become a theatre for visiting townsfolk, with rural people, and especially farmers, being the scene-changers and bit-part actors for that theatre. In other nations, this commodification of often reconstructed ruralities is also strongly represented in the changing rural scene. Alexander Wilson (1992) discusses examples in southern Appalachia:

> Scattered along the roads of Tennessee, Kentucky, and the western part of North Carolina are restored villages that recall and reconstruct ideas about the way things once were in those mountains. Some are within nature parks, some are part of theme parks, some are simply an assemblage of buildings that evolved out of someone's backyard, some promote religion, others consumerism.
> (Wilson, 1996: 206)

His examples range from gaudy theme parks such as 'Dollywood' (The Dolly Parton story – Figure 48.6b) to 'authentic' museums, but each attempts to recreate a lost time and culture in this part of rural America. Here, too, then, the nostalgia of rural life – and in this case the pioneerism and specific culture of rural mountain folk – are commodified as attractions for visitors to the contemporary countryside. Once again, the character of the present is vested in the symbols and meanings of the past.

At this point, though, we need to appreciate that the process and practices of commodification do not always reinforce nostalgic countryside notions of English rural idylls or the pioneerism of rural America. Although rural areas will usually be trading on their past, there are interesting examples now whereby the commercialism of place-making has begun to forge new identities for the country. An example of this can be drawn from rural New Zealand, where the growth and

Figure 48.7 The 'Awesome Foursome' experience: helicopter, bungy, jet-boat and white water rafting experience near Queenstown, New Zealand

Figure 48.8 The Yucca Mountain nuclear waste dump in Mercury, Nevada. Credit: Associated Press

commodification of adventure tourist facilities, practices and subcultures have added new dimensions to the lives of many of its people and to its landscape (Cloke and Perkins, 1998). Adventure tourism has influenced the production and reproduction of new imagined geographies of the country in South Island, New Zealand. In particular, whatever the activity and place being advertised, there are repeated allusions and references to freshness:

- *A fresh look at spectacular environments* – what was previously thought to be the spectacular scenery of New Zealand can be made more spectacular by participation in (or watching others participate in) adventurous pursuits in places of natural or historical significance.
- *Fresh, youthful thrills* – adventure tourists are provided with the white-knuckle excitement of contemporary theme parks, but in a 'natural' outdoor setting.
- *The freshness of eager experimentation* – rural areas of New Zealand are being 'branded' by continual experimentation with bigger, better and more exciting thrills in the outdoors environment.

Invitations to 'crack the canyon with the Awesome Foursome' (see Figure 48.7) reflect a different form of countryside commodification, where the relationship between tourist and landscape reflects a far more active, performative and 'embodied' experience of nature relations than is provoked by conventional countryside nostalgia.

This account of commodification in the reshaping (yet often reinforcing) of countrysides has tended to emphasize the natural and positive cultural attitudes of rural areas. Against these potentially idyllic representations of rurality, however, can be set rather more dystopian narratives of rural life and landscape (see, for example, David Bell's (1997) analysis of 'horror' films set in small-town America). Commodified rurality is not unambiguous. The glorious isolation of 'wilderness' or 'outback' can also be attractive to those wishing to dump hazardous waste that is unwanted in more populated regions (Figure 48.8).

SUMMARY

- Rural commodification is linked to changes from productivism to post-productivism.
- The country is being made into a commodity in many different ways and this process contributes to the emphasis on particular meanings and symbols as being important indicators of rurality.
- Some meanings reflect a nostalgic representation of the country as being dominated by the virtuous values of the past, involving closeness to nature and close-knit community. These meanings are often presented in various kinds of spectacle.
- Other commodification has led to new 'country' meanings such as the fresh and adventurous encounters and performances with nature sponsored by the rise of adventure tourism in New Zealand.

Out of sight and out of mind: rural others?

Figure 48.8 also emphasizes that beneath the innocent idyll of the country there lie other characteristics that hit the headlines less often. The underbelly of idyll-ized countrysides is rarely exposed. Indeed, the stories used to promote the country as idyll usually serve to mask any contradictory social conditions. Unemployment or underemployment, the scarce availability of affordable housing, the rationalization of local services into larger centres and the shrinking of public transport services have all served to disadvantage low-income households in rural areas. Moreover, two significant volumes (Cloke and Little, 1997; Milbourne, 1997) have emphasized different processes and practices by which certain rural people can be marginalized. Here, the emphasis is on individuals and groups that are 'other' than the mainstream, with identities characterized by gender, race, sexuality, age, class, alternativeness and so on. I use the 'otherness' relating to rural poverty as an example here, but the story of how a rural problem is hidden discursively away in public discourses as well as geographically could just as easily refer to the rural homeless or indeed any of the identity groupings listed above.

A key study has shown that the percentages of households in or on the margins of poverty in 12 case study areas in England ranged between 39.2 per cent and 12.8 per cent, with 10 of the 12 areas having more than 20 per cent levels of poverty (see Table 48.2). Yet, in successive government policy documents for rural areas, the word poverty does not appear. Two reasons may be advanced for this. First, the political propaganda of 1980s and early 1990s Britain has effectively pronounced the end of poverty, claiming that absolute poverty has been eradicated by economic success and that relative poverty is a figment of the academic imagination and should more properly be labelled as 'inequality'. Second, despite the censorship of the word 'poverty' from government pronouncements, there is a strong sense in which it has been easier to deny the existence of poverty in rural areas than in the cities. Here, we can link the imagined geographies of idyll-ized rural lifestyles with the

Table 48.2 Percentage of households in or on the margins of poverty* in 12 case study areas in rural England. Source: P. Cloke et al., (1994) Lifestyles in Rural England. London: Rural Development Commission

Nottinghamshire	39.2	North Yorkshire	22.0
Devon	34.4	Shropshire	21.6
Essex	29.5	Northamptonshire	14.8
Northumberland	26.4	Cheshire	12.8
Suffolk	25.5	West Sussex	6.4
Wiltshire	25.4	Across 12 areas	23.4
Warwickshire	22.6		

Note: * measured as less than 140 per cent income support entitlement

idea that poverty in rural areas is being hidden or rejected in a cultural dimension both by decision-makers with power over rural policy and by rural dwellers themselves (including those who appear, normatively, to be poor). Rural people can be recognized as 'deprived' of ready access to the advantages of urban life, but are not as such impoverished by such deprivation because rural living somehow offers perceived compensations for any such disadvantage. In this way, rurality appears to signify itself as a poverty-free zone, and constructs of rural idyll at the same time exacerbate and hide poverty in rural geographic space.

Poverty in other developed nations, such as rural America, is an equally important issue and one that is also influenced by the impact of dominant cultural constructions of rurality. In contrast to Britain, the USA does have an official poverty line and so by state-defined statistics, levels of rural poverty are currently reckoned to be around 20 per cent of households. Rather than attempting to deny poverty politically, the emphasis has been to differentiate between the deserving and the undeserving poor. In this way, urban underclasses are signalled as 'undeserving' while non-metropolitan low-income workers are cast as 'deserving'; the spotlight of publicity falls on the former not the latter. As a result, cultural construction of poverty in America has a distinct spatial outworking, with impoverished rural Americans being lost in the shadows. And even when poverty is recognized as an issue in rural areas, there is a tendency to assume that it is restricted to key problem areas, notably Appalachia and the south. But as McCormick (1988) graphically indicates: 'Today, the problem has no boundaries. A tour of America's Third World can move from a country seat in Kansas to seaside Delaware, from booming Florida to seemingly idyllic Wisconsin' (1988: 22).

What has been discussed here in terms of poverty also applies to other issues, notably rural homelessness (Cloke *et al.,* 2002). The tendency to regard social problems in the country as out of sight and out of mind is directly related to the dominance of prevailing social constructions of rurality. These in turn are essentially interconnected with the relations of power at work in and beyond rural areas. It is unsurprising, then, that the most significant demonstration of 'rural' opinion in Britain in recent years should be focused on the specific issue of fox hunting and the more general issues of being left alone to exercise personal freedom in the idyllic villages, farms, moors and fells. Neither is it surprising that the major demand of demonstrators is that nothing should change. Such is the conservatism of the powerful, not the powerless.

SUMMARY

- Low-income households in rural areas are disadvantaged by changes to job markets, housing markets and the provision of services.
- 'Other' individuals and groups are marginalized in rural society on the grounds of gender, race, sexuality, age and so on.
- The example of rural poverty demonstrates how idyllic representations of the country mask the occurrence of significant social problems and nullify the perceived need for policy responses to those problems.

Conclusion: powerful geographies of countryscapes?

These complex and often ambiguous relations between nature and culture, and society and space, which underpin the mosaic of geographical space of the country make for an important and exciting territory of geographic enquiry. It should be made clear that there are many specific fascinations within these countryscapes that have hardly been touched on here – for example, the restructuring and reregulation of agricultural landscapes in a 'post-productivist' era; the cultural importance of food; the conservation of valued landscape and habitat; the evolution of community and social relations in countrysides; the power relations and governances that pertain there; and so on. Those wishing to delve into particular issues such as these might like to use the further reading suggested at the end of this chapter. None of these concerns, however, is immune from the emerging core of significance in rural studies that has been outlined in this chapter – namely, the interconnections between the socio-cultural constructs of 'country', 'rurality' and 'nature' that seem to be so important in (re)producing our geographical imaginations of rural space (geographical and social) and the actual experiences of how lives are practised within these spaces. Such practices will need to be viewed both from the outside looking in (taking full account of 'structuring' influences) and from the inside looking out (taking full account of individual difference, embodiment and identity). Chris Philo (1992) catches the mood of these significances in his review of what he regards as neglected rural geographies in Britain. His contention is that most accounts of rural life have viewed the mainstream interconnections between culture and rurality from the perspective of typically white, male, middle-class narratives. There is therefore an urgent need to look through other windows on to the rural world.

Myths of rural culture often marginalize a range of individuals and groups from a sense of belonging to, or in, the rural. We need to make our geographies of the rural more open to the circumstances and voices of other people in order to overcome the neglect of 'other' geographies. Such a conclusion is exciting, but not unproblematic. As Jonathan Murdoch and Andy Pratt (1993) have suggested, simply by 'giving voice' to others we do not necessarily uncover the power relations that lead to marginalization or neglect. A range of important questions arises here, relating in particular to the power of the researcher, the potential reinforcement of marginalized identities by labelling them as 'other', and the potential for flippant rather than politically grounded engagement with marginalized people. However, the mosaic geographies of the country will be richer for the addressing of these questions than for their neglect. After all, when the huntsmen come to town, are they really the voice of the country?

DISCUSSION POINTS

1. Evaluate the arguments for and against 'doing away with rural' (Hoggart, 1990) as an appropriate geographical category.
2. In what way does the blurring of the urban–rural divide consist of different processes of urbanizing the rural and ruralizing the urban?
3. Discuss, with examples, what it might mean to understand rurality by *performing* the rural.
4. How appropriate is the idea of a transition from productivist and post-productivist in the understanding of contemporary rural areas?
5. Why is it important to recognize the increasing range and scale of commodification in rural areas?
6. How useful is it to think about rural 'others'? Using the example of a particular social group, explore how exactly rurality is associated with rendering that group 'out of sight and out of mind'.

FURTHER READING

Cloke, P. (ed.) (2002) *Country Visions*. Harlow: Pearson.

Cloke, P., Marsden, T. and Mooney, P. (eds.) (2006) *Handbook of Rural Studies*. London: Sage.

Murdoch, K., Lowe, P., Ward, N. and Marsden, T. (2003) *The Differentiated Countryside*. London: Routledge.

Woods, M. (2004) *Rural Geography: Processes, Responses and Experiences in Rural Restructuring*. London: Sage.

Woods, M. (2011) *Rural*. Abingdon: Routledge.

These texts offer comprehensive accounts of the parallel importance of the changing nature of rural areas and the changing ways in which geographers and others have sought to interpret rurality.

Bunce, M. (1993) *The Countryside Deal*. London: Routledge.

A view of the idyll-zation of the country in North America and Britain.

Barnett, A. and Scruton, R. (eds.) (1998) *Town and Country*. London: Jonathan Cape.

Wilson, A. (1992) *The Culture of Nature*. Oxford: Blackwell.

These texts discuss the blurring of the country and the city.

Cloke, P. and Little, J. (eds.) (1997) *Contested Countryside Cultures*. London: Routledge.

Cloke, P., Milbourne, P. and Widdowfield, R. (2002) *Rural Homelessness*. Bristol: Policy Press.

Deals, among other things, with the issue of marginalized 'others' in the country.

PART THREE

HORIZONS

SECTION ONE

NON-REPRESENTATIONAL GEOGRAPHIES

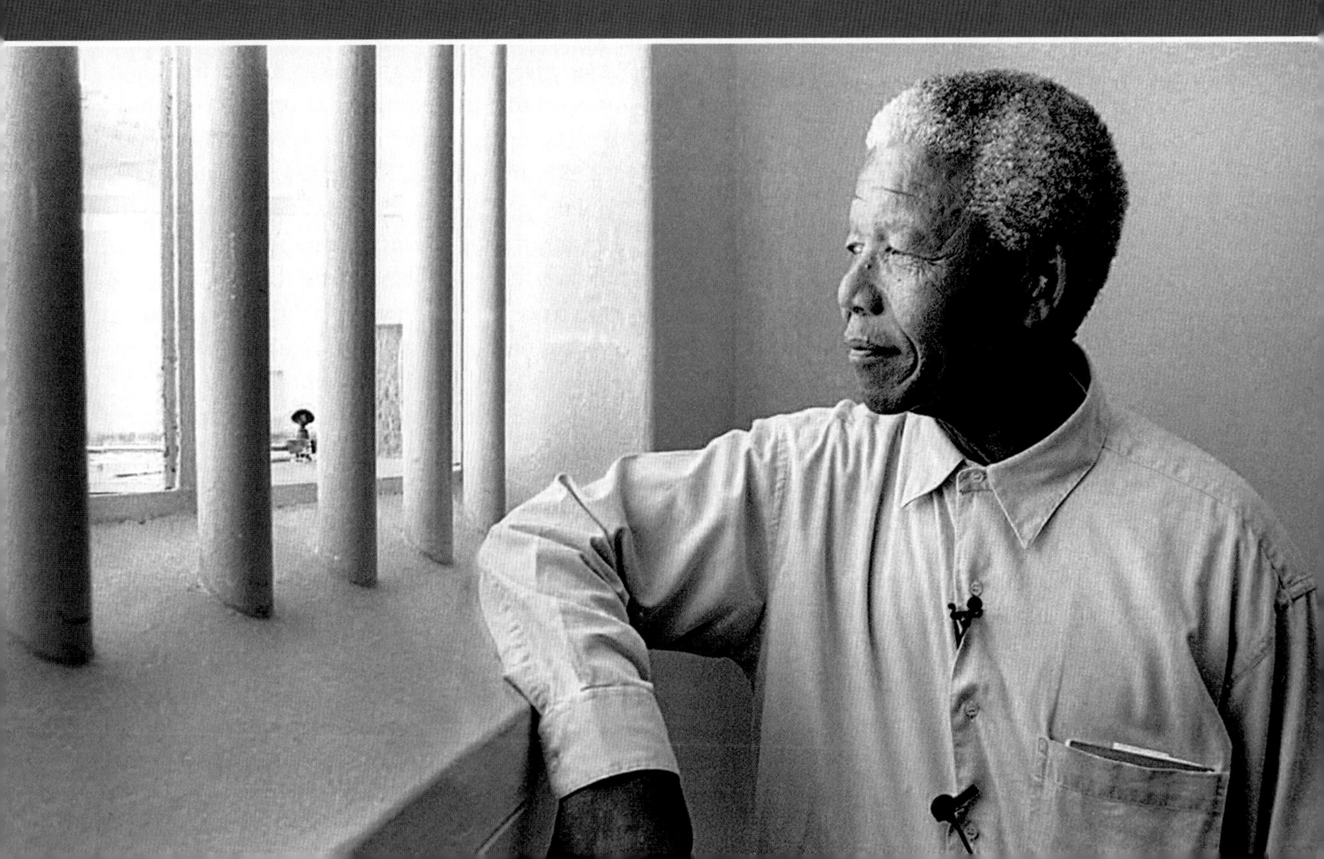

Introduction

I can remember several very special occasions when words and ideas were simply incapable of capturing the multidimensional vitality of my life as it unfolded: examples include the birth of my children, especially that first time of holding their new lives in my arms; an extraordinary half hour of swimming around and 'playing' with a sea turtle, oblivious to all else on the Great Barrier Reef; and the haunting co-presence of histories – of the terrible injustice of apartheid and of the astonishing reconciliation that followed – when visiting Robben Island, the prison in which Nelson Mandela was imprisoned (see Figures VII and VIII). For every one of these extraordinary memories, there are myriad everyday equivalents in which the subconscious or non-conscious playing out of life defies easy description and representation, and in recent years Human Geography, led by Nigel Thrift (2007a; see also Anderson and Harrison, 2010a), has sought to get to grips with a series of different *non-representational geographies* that reflect the ordinary practices by which life unfolds.

Non-representational geographies urge us to go beyond our previous obsessions with representation and meaning, which tend to emphasize that which is fixed, or dead, rather than that which is in the state of becoming, or alive. The perceived problem here is not necessarily representations or meanings themselves, but the seemingly overwhelming geographical desire to engage in representationalism – to impose fixed structures of meaning onto our world. The challenge from non-representational thinking is to respond to the worlds that are not easily fixed by representation; to refocus our attention on the bodily practices, performances and manifestations of everyday life in order to witness the vitality of the world as it unfolds. Accordingly, Thrift (1997: 124) urges us to grasp the significance of 'mundane everyday practices that shape the conduct of human beings towards others and themselves in particular sites' in order to achieve a geography of what actually happens. Such an approach involves several key requirements: that we take the body seriously, especially precognitive aspects of embodied life; that we take the nonhuman world seriously, especially in how the social relates to the material, and how the body evolves alongside and in relation to things; and that we take technologies of being seriously, especially the coming together of assembled hybrids of networks and interconnections.

This section, then, begins to engage with some of this non-representational thinking, via some of the body-practices – emotions, affects and performances – that indicate the need to push beyond orthodox representational ways of understanding. In Chapter 49, Hester Parr argues that emotions are central to the experience of spaces and places and help to open up understandings of Human Geography that champion subjective, insider, accounts of the human world. She traces the history of human geographical interest in emotions, beginning with the early attempts in humanistic geographies to use feelings and experiences to build existential meanings and to focus on human emotional attachments. These early geographies of emotion often presented a rather undifferentiated account of humans/bodies/emotions, but ensuing feminist politics of knowledge allowed for a more complex understanding of emotional subjectivity and geography. Specifically, emotions became recognized as evidence of an unstable subjectivity – one in which experience of fear, illness, joy and so on contributed to how different subjects experience the world differently. This focus on emotions not only opened out experiential worlds of difference, but also helped to situate knowledge production by interconnecting emotional subjectivity to the experience of conducting research.

Further insights came from psychoanalytic reflection connecting the unconscious and emotional self with experiences of the social and material world. The recognition of deep-seated psychoanalytical feelings about who and what are acceptable or not in a particular place helps to explain how anxieties and desires structure social life into particular relations of self and other. Thus spatial or social proximity to forms of 'unacceptable' otherness can represent a challenge to emotional stability and self-identity, and can prompt attempts to achieve separation from this otherness by establishing borders, purifying the space within those borders, and thereby achieving some kind of social exclusion. These kinds of analyses have been used, for example, to explain why different groups perceived as social others (travellers, homeless people, people of different ethnic backgrounds) have often experienced exclusion from rural communities. To do so requires that these deep-seated emotional responses can be known and made known, and the increasing popularity of non-representational geographies has therefore posed important conceptual and methodological questions about our capacity to make sense of emotions. How is it possible to represent the process by which internal mental states and emotional atmospheres move people and affect events? Indeed, how is it possible to access the meaningfulness of emotions, moods and atmospheres? Non-representational studies have explored new ways of attending to gestures and performances through testimony and witnessing, although the exact relationship between emotions and non-representational affects (see Chapter 50) has been closely contested. Parr's answer is to see affect as the 'how' of emotion – the precognitive or non-cognitive feelings and sensations that presage a more cognitive understanding of what emotions entail. In so doing she offers yet more fertile ground for geographical work in the interstitial spaces between representation and non-representation.

If affect is the 'how' of emotion, then in turn emotion may express the body's capacity to affect and be affected. This highly significant idea of affect has emerged from the introduction of non-representational theories into geography, and serves to capture the experiences of life as lived immanently. As Ben Anderson explains in Chapter 50, affect has been used to give conceptual expression to a wide range of life experiences: background moods, focused emotional involvement, immediate visceral responses, shared atmospheres, outpourings of passion, fleeting feelings, neurological body responses and many more. Essentially, however, the idea of affect helps us to understand that living in the present tense of our experience – in other words how we are in the process of *becoming* at any given moment – is partly about a non-conscious and background sense that is located under the radar of thought, deliberation and reflection. Emotions can thus be understood as some of the ways in which this non-conscious background is subsequently named and interpreted. What is more, these affects are not just an individual phenomenon, but can also take a more collective form in terms of atmospheres, vibes and structures of feeling that can be detected by multiple bodies sharing the same geographical or social space. Affects can also reflect particular sensibilities or dispositions associated with particular groups of people.

Anderson explains to us that affects are thereby essentially social in that they invariably connect us to other things, places and people beyond ourselves. In this way, affects connect us to myriad other geographies, and so affective life needs to be taken very seriously because it performs a significant role in all lived geographies. Human Geography has indeed begun to take affective life seriously, not least in seeking to grasp how affect is differentiated over space and over time – a differentiation that often emerges via particular encounters between people, or between people and sites. We can thus begin to take seriously some of those life experiences that are difficult to put into words: the wondrous or sometimes scary encounter with the vibrant matter of nature; the haunting and ghostly encounter with some kind of history that manifests itself in the present; the hard-to-put-into-words adrenaline rush of adventurous pursuits, and so on. These encounters entangle us in transpersonal geographies of connection and disconnection, yet when it comes down to it, they are simply manifestations of the mundane practices of our everyday lives.

Amanda Rogers takes up this theme in Chapter 51 in her discussion of performances, in which she explains the idea of performance as a metaphor for the practice of social life, both as an expression of identity and in ways that continually reinvent identity. In this way, performances are emotive – they use and evoke places, they blur the distinction between real and imagined, and they permit everyday encounters to take on greater meaning. Rogers suggests three ways in which Human Geography has been excited by the idea of performance. The first connects performance with identity-making, whether by deliberate impression-making, by subconsciously performing according to expectations that normalize how identity should be expressed, or by way of spontaneity, improvisation and creativity that subvert or recast more conscious and representational modes of acting out identity.

The second relates to embodied performances such as in dance or theatre. Here, scripted presentations can present people and their identities in creative ways, enabling the imagination of a resistance to or subversion of the political norms of the day. However, embodied performance can also be improvised, going beyond prescribed scripts to accentuate new modalities of experience based on intuition, feeling and sensation. Unscripted performance makes connections that are often difficult to express or make sense of, but they can open out through new understandings of the materiality of the body and its capacity to suggest different ways of dwelling in and experiencing the places in which we live.

The third set of ideas about performance draws directly on non-representational geographies to suggest that performance is the mechanism by which everyday spaces and activities are constituted and given life. Gestures, small actions, skills and subconscious ways of being all add vitality to the spaces we inhabit, helping to create their liveliness as part of the everyday relational networks between people and non-human beings, materials and ideas. Some Human Geographers have suggested that these non-representational frameworks present us with a new register of grasping the ways in which everyday life is and becomes. Accordingly, we need to wire ourselves into a non-representational style of thinking if we wish to grasp what happens before conscious thought and wilful representation perform their rationalizing and often politicizing interpretation of events, encounters and relations. Others (see for example, Nash, 2000; Lorimer, 2005) argue that a focus on both the representational and the non-representational (and indeed the interconnected spaces between these two) is required to balance our exploration of performances, relationships, materialities and texts that constitute everyday life. These three excellent chapters on emotion, affect and performance give us significant scope to think through these debates.

Figure VII Nelson Mandela revisits his prison cell on Robben Island. Credit: Getty Images

Figure VIII Nelson Mandela's 3m x 3m prison cell on Robben Island. Credit: Getty Images Europe

CHAPTER 49

EMOTIONAL GEOGRAPHIES

Hester Parr

Geography and emotions

> Geographers are not always talking about the same thing when they talk about affect and emotion . . . their conceptual, empirical, ethical and political emphases will differ, and often profoundly so'
>
> (McCormack, 2006: 330)

That Human Geography is concerned with emotions, or that we can discuss 'emotional geographies', may seem rather surprising. Certainly, geographers themselves have struggled to come to terms with the notion that emotions are central to our experience of spaces and places. This chapter offers a basic introduction to emotional geographies, charting a partial disciplinary account of the different ways in which geographers have encountered emotions as a focus of study. The quote above indicates that there are different elements at stake here – and as you will see below – very different ways of conceiving what those stakes are. As the chapter unfolds, the different terminologies and ideas that geographers have begun to use and grapple with in order to understand these human sensibilities will become clearer.

It has often been mooted that Geography is a discipline built upon so-called '**objective**' scientific truths and investigations, whereby the '**subjectivity**' of the geographer is actively ignored, dismissed and written out of accounts of the physical and human world that are produced by such knowledge. In this construction, Human Geography can be seen as a rational discipline, predicated on notions of a scientific order that is (supposedly) devoid of any human emotional content. By implication, Human Geographical work that does not simply represent the 'facts' of the world in a detached, scientific and objective manner runs the risk of being irrational and therefore not as valuable a form of knowledge. What counts as irrational knowledge? For some geographers, irrationality has been a category linked with disorder, chaos, corporeality, femininity and, crucially, emotionality. Therefore geographical research that is seen to work with notions of disorder, the body, gender and emotion could be deemed irrational knowledge that is not valuable in the canons of the discipline. In other words, Human Geography can be considered as built on the basis of a profound binary dualism. This dualism is effected, on the one hand, through constantly reproducing so-called objective, rational scientific knowledge that is considered valuable and, on the other, through ignoring and not reproducing supposed irrational and worthless knowledges. There are specific feminist readings of this dualism that we address further below.

There has hence been an historical reluctance to encounter emotions in geography, until recent times. Disciplinary knowledges are never static, however, and this silence has been addressed. Partly as a result of more women entering the academy and more geographers responding to associated debates about a 'politics' of knowledge (see the section on feminist geographies below), there has been a critique of so-called rational and objective knowledge production (Rose, 1992). Although this critique has taken many forms, one effect of this questioning of knowledge production has been an increasing interest in and validation of subjective knowledges – ones that privilege 'insider' accounts of the human world. In doing so, new research work has opened up different understandings of everyday human lives and the ways in which everyday geographies contain complicated relations between selves, bodies and spaces. A key part of this recent re-evaluation has been the analysis of subjective *emotional geographies* (Davidson and Bondi, 2004; Davidson and Milligan, 2004), as part of a disciplinary struggle to encounter emotion in different ways. Since the initial flurry of interest and accounts of emotional geographies, this is a field that has been sedimented by further diverse study (Davidson *et al.*, 2005; Smith *et al.*, 2009), but one which has also prompted controversy (Thein, 2005; Tolia-Kelly, 2006; Pile, 2010). Recent debates have questioned what we understand by 'emotion' and 'affect', and these terms and tensions relate to the philosophical basis of contemporary geographical enquiries that discuss human feelings. The reminder of this chapter characterizes some of these histories, developments and encounters.

SUMMARY

- Emotions have been neglected in the discipline of Human Geography and this is partly explainable by the construction of so-called rational and objective knowledge.
- Geographical work that focuses on emotions, gender, the body and disorder have been constructed as irrational knowledge in the past.
- New research values subjectivity and subjective emotional knowledge as the result of debates about the politics of knowledge construction.
- There are new debates about the significance and constitution of emotion and affect as part of the contemporary discipline.

CASE STUDY

Moving emotions into geography: humanistic geography

One of the first sustained attempts to incorporate emotions as part of geographers' thinking became apparent through the project of humanistic geography in the 1970s. Humanistic geography emerged as a critical reaction to the perceived *dehumanising* effects of an increasingly technical Human Geography as *spatial science*, involving the translation of the human world from a language of 'substance' (involving rich textures of people and places) to one of 'geometry' (involving mapped and graphed X, Y coordinates). Part of this critique was to incorporate versions of the philosophical base of humanism into Geography. The result was an explicit agenda that sought to validate and centre subjective understandings of the relations between humans and world. As the famous humanistic geographer Yi Fu Tuan claimed early on, 'humanistic geography achieves an understanding of the human world by studying people's relations with nature, their geographical behaviour, *as well as their feelings and ideas* as regards space and place' (Tuan, 1976: 266). Tuan and others like him produced work that investigated deeply held attachments to place. However, creating a human(istic) geography that was about 'feelings' was no simple task, and humanistic geographers articulated the need for the incorporation of feelings and emotions in geographical analysis through an appeal to a complex set of humanist philosophies known as existentialism and phenomenology. This is not the place to explain these influences in any great detail (for this see Cloke, Philo and Sadler (1991) and Peet (1998)), but suffice to say that core geographical concepts such as place were re-evaluated through a deliberate appeal to subjective human experience:

> Within this humanist perspective concepts of traditional significance in geography are given existential meaning or a focus of human emotional attachment . . . for example, place is defined as the centre of meaning or a focus of human emotional attachment.
>
> (Entrikin, 1976: 616)

> Existentialism . . . deals with the emotional life, the feelings, the moods, and affects through which people are involved in the world. Existentialism, then, differs from positivist science in its emphasis on inner experience, knowledge by participation rather than observation and its celebration of subjectivity over objectivity.
>
> (Peet, 1998: 36—7)

A clear conceptual agenda was established that centred 'feelings, emotions, moods and affects', although how these were actually differentiated and written through as part of substantive geographical study is more controversial. Humanistic geography has been criticized as a search for essences: *essential* human emotional responses to place, for example, that were largely *undifferentiated* by location, gender, age, class or race and generally structured on mythical masculine norms (Rose, 1992). Some have also criticized humanistic geographers for only really trying to consider their *own* emotions towards places, and only then partially, while others criticize these geographers for reducing human subjectivity to feelings that are somehow 'beyond rational scrutiny' (Daniels, 1985: 151). Although there were important exceptions (see Rowles, 1981), humanistic geographers' approach to emotions were often deemed as flawed and introspective.

SUMMARY

- Humanistic geographers validated subjective knowledge as a means to understanding human–world relations.
- Place, and the emotions it evokes for people, was a central focus of this work.
- However, humans, their bodies and emotions were largely undifferentiated by humanistic geographers.

Feminist geographies

As argued by many **feminist** geographers, the western academy, including the Geography discipline, is dominated by white, heterosexual, non-disabled men, and this is reflected in the form and content of the knowledge produced by this academy. Feminist geographers have critiqued how, in particular, valued geographical knowledge is inscribed by gender relations. Here, a valued rational Human Geography is also understood as a masculinist knowledge, although many feminists have tried to destabilize the binary distinctions and associations between masculinist rationality and feminine irrationality. In the context of this chapter, feminist analyses of a rational geography discipline suggest subjective emotional matters may be ignored or essentialized by masculinist knowledge productions.

CASE STUDY

Situated knowledges

This term is most associated with feminist geographers and their critiques of the process of knowledge production. Drawing inspiration from the philosopher Donna Haraway (1991b), who comments critically on the construction of powerful scientific knowledge, feminists have challenged the truth claims of detached, disembodied means of knowing the world. Haraway argues for a 'situated knowledge', which refers to the notion that knowledge can be partial, located and embodied – in other words, and put simply, knowledge always comes from someone, somewhere. Conventionally, the western academy has constructed the most valuable forms of knowledge as ones that are impartial and deeply authoritative because of 'the god trick of seeing everything from nowhere' (Haraway, 1991b: 189) and the refusal to situate claims relative to personal, social and geographical contexts. Feminist researchers argue that situating knowledge enables more critical thinking whereby 'transcendent' knowledge is replaced by 'a politics and epistemologies of location, positioning and situating where partiality and not universality is the condition of being heard to make rational knowledge claims' (Haraway, 1991b: 195).

Feminist geographical research has been particularly influenced by such critiques of knowledge production and this has been evident in debates about research methods, reflexivity and power relations in the discipline (England, 1994). These debates in turn have influenced feminist research practice, as researchers have tried to employ a range of qualitative research techniques that enable sensitive, non-exploitative and embodied encounters with a range of actors and agents in a variety of locations. Such encounters and the knowledges that they produce 'do not depend on a logic of discovery, but a social relation of conversation, [and hence] the world neither speaks for itself nor disappears in a favor of a master coder' (Peet, 1998: 269). Producing situated knowledge also entails processes of reflexivity, itself a kind of 'self conscious analytic scrutiny' (England, 1994: 82) in order to avoid the 'god trick' of the objective master gaze. However, such processes are fraught with difficulties, as Rose (2002: 257) argues when asking 'in what terms can we describe our "situation"? What is our "position"?' in a world often conceptualized as constituted by fluid and uncertain boundaries. Haraway's account of 'situating knowledge' hence challenges feminist and other researchers to destabilize taken-for-granted forms of academic authority, but without 'fixing' our positions and differences.

For the likes of Rose (1992: 60), the possible 'overlaps between **humanistic** geography and feminist geography were more apparent than real', and feminism offered a more complicated understanding of emotional subjectivity and geography. Feminist geography work, especially since the 1990s, contributes to 'offer[ing] distinctive . . . perspectives on the human subject' in ways that 'unsettle and move beyond the denial and neglect of emotion and corporeality characteristic of most social research' (Bondi, 2002: 6), including that of humanistic geography. There are many examples of the work of feminist researchers who have sought to centre emotions as part of a re-assertion of unstable subjectivity within the discipline. Studies of fear, embodied panic, feelings associated with music, childbirth, pregnancy and illness, have all contributed to the multiple ways in which both men and women's emotional lives are 'alternative sources of knowledge' for geographers (WGSG, 1997: 87). The 'turn' to write through explicit and substantive emotional geographies (Davidson *et al.*, 2005) is still an ongoing project; although one of the key ways in which a concern for emotions has been translated within and beyond feminist writings is in the realm of research practice (the 'doing' of geography).

Although, as Rebecca Widdowfield (2000: 200) argues, there is nothing 'inherently feminist about recognizing the influence of emotions in academic research', feminist geography has been enormously influential in the exposure of emotional relations in research *practice* (WGSG, 1997). Widdowfield discusses the importance of acknowledging her own emotions in research on lone parents in run-down council estates in northern England. Adopting a multi-method approach (drawing on both quantitative and qualitative materials), she relates the emotional experience of conducting interviews in places where the physical and social environment seemed like 'a desolate landscape', and where:

> Going into the LDNs ('less desirable neighbourhoods'), I was *angry* at the

injustice and inequality in society . . . I was *upset* at the unpleasant and unhappy circumstances in which many people live and *distressed* by the fact that I could not see a solution . . .

(Widdowfield, 2000: 204)

While acknowledging that it is not always appropriate that researchers should detail their feelings in each piece of research they undertake, Widdowfield argues that reflecting on her 'demoralized' and 'disillusioned' emotional state allowed her some insights into the reality of the social circumstances of her interviewees in Newcastle's deprived estates. At the same time, noting the 'positive feelings' among some of her interviewees and their strong sense of community challenged her own emotional responses charted above. Over the course of the research project, her own negative feelings changed somewhat as she became more familiar and less threatened by the environments she researched and, through this process, became aware of the danger of 'outsider perspectives in making policy prescriptions' based on particular emotive readings of place (2000: 205). In attempts to expose knowledge production (and even policy-making) as subjectively constituted, Widdowfield (2000: 205) argues that 'writing emotions into research accounts can facilitate a better understanding of the work undertaken and forms an important part of situating knowledge'. There are now other accounts of the personal and relational nature of gaining geographical knowledge, broadly addressing this call to 'emotional writing' (see examples in Limb and Dwyer, 2001).

Apart from reflecting on the researcher's experiences, discussing emotions as part of the research process is an increasingly appropriate activity given that qualitative methods are often used to 'empathize with' the relationships between people, social groups and their everyday environments. Building upon humanistic geography traditions that validate the inclusion of emotionally subjective experiences in the world, and also the feminist mantra of 'the personal is political', it has become commonplace for Human Geographers to seek out 'insider' accounts of everyday life. Key methods for eliciting these accounts have been the use of in-depth interviews, biographical life histories and ethnography. Such methods are best understood as a means to collect in-depth and intensely personal understandings of lived geographies, and it should come as no surprise that such methods produce encounters that are laden with a powerful emotional content, both for the researcher and the researched. Feminist scholarship has been at the forefront of a call to acknowledge these emotional exchanges in academic work (but see also Thomas, 2010), in ways that hopefully enlighten us as to how and whether our research methods and the emotions they evoke are implicated in the knowledges produced. Some geographers have written of the difference that 'emotionally safe' interview space can make to knowledge production (Parr, 1998), while others have problematized the emotive power relations that are always present in the management of interviews (Burgess, 1998a, 1998b; Pile, 1991), for example.

Criticisms of this 'emotional turn' have included a questioning of our abilities to be able to envision clearly and lay bare our multiple feelings and relations in research encounters (Rose, 1997). Rose also asks whether such a project has hallmarks of a rationalist geographical knowledge, in so far as emotional subjectivity is ordered and made intelligible. Similarly, others have worried that a concern with emotionality is being increasingly ghettoized within certain parts of Human Geography, as 'what little talk of emotion there is occurs squarely in the cultural (and often feminist) corners of the discipline' (Anderson and Smith, 2001: 7), and that this runs the risk of reinforcing a wider divide between a perceived rational, relevant and a 'frothy', irrelevant geographical knowledge (Hamnett, 2003). This latter worry is arguably less relevant in 2013, as much has been done to establish the field and it has resonated within economic and geopolitical debates in the context of powerful meta-narratives about globalization and fear, for example (see Pain 2009).

Since these debates, a distinctive field of 'emotional geographies' has emerged, much of it inspired from and by feminist writings. Davidson, Bondi and Smith (2005) have initially led this field (and see Smith *et al.*, 2009), which has now become populated by many studies of the relationship between emotion and space (and see the journal *Emotion, Space and Society* for examples of this trend). The relational constitution of emotion is a key feature of these studies and a key task is to evoke the vitality of lives and worlds that geographers study (Smith *et al.*, 2009: 15).

SUMMARY

- Some feminist geographers read the dualism of rational and irrational knowledge as gendered and seek to destabilize this construct.
- Acknowledging the emotional subjectivity of the researcher can help situate geographical knowledge.
- Qualitative methods can evoke powerful emotions for respondents.
- Writing 'emotional geographies' is an established tradition and there are fewer worries about the 'ghettoization' of this approach within feminist and cultural geography.

Psychoanalytic geographies

There are still other ways in which geographers have recently encountered emotions. In work influenced by **psychoanalytic** thinking, they have shown how reflecting on the psyche and theories of the self can serve to connect the unconscious, emotions, and the social and material world (Callard, 2003; Kingsbury, 2009; Sibley, 1995; Pile 2005, 2010). Although it has been said that psychoanalysis is often theoretically 'tamed' in geography (Callard, 2003), psychoanalytic theory has proved attractive to geographers who want to understand more about the ways in which anxieties and desires structure social life. One of the more accessible ways to introduce this literature is to look at work on psycho-social geographies (Parr and Davidson, 2011). As David Sibley (1998) has argued,

> In mapping the topography of the mind, the psychoanalyst is necessarily concerned with feelings. Feelings of repulsion and desire, of nervousness, elation and so on, contribute to . . . the avoidance of certain places and people, or conversely, attraction to particular place and social milieu.
>
> Sibley (1998: 116)

Using insights from object relations theory, Sibley (1998: 117) argues that certain people and places can be constituted as 'good' or 'bad'

Figure 49.1 Dale Farm traveller site. Credit: Getty Images

objects in the world, and that they can then be depositories for feelings of pain and anxiety, pleasure and excitement. It has hence been possible to appreciate more about how negative emotional states are provoked and sustained in relation to certain human groupings in certain places (gypsies and travellers often being identified as such a group – see Figure 49.1).

In general, psychoanalytic geographers would argue that adult emotional behaviour cannot be understood outside of the ways in which early psychological conflicts are experienced. In other words, how we develop as infants and children has a massive impact on our adult social and spatial relationships. Reflecting on this process helps us to see how humans relate to their world through the formation of strong beliefs about who and what is acceptable and who and what is not ('the Self' and 'the Other'). The latter represents a category that is rejected and distasteful. The significance of infant development in this theory is as follows: In the western world our infant development has common features, in that we usually experience the world in clean, safe, warm environments, and the way in which we are socialized into the world by our parents in infanthood is deemed important. In particular, we are often socialized into keeping things clean from an early age. We learn to distance ourselves from dirty things all the time when we are young, such as bodily waste, nasty things in the garden and people designated as dirty. This has a huge significance for psychoanalytic geographers, who maintain that how we learn about purity and cleanliness when young holds meanings for later adult life and the creation and maintenance of social and spatial boundaries between people and places. Strong feelings and fears about the possible pollution of the self and body (ideas learned in childhood) come to the fore when adult people are presented with people and places that they deem to be 'dirty'. Who and what is deemed dirty is often the result of different social stereotypes in western society, but the results are often the same. Feelings of fear and anxiety about pollution and dirt result in social and spatial distancing between groups:

> The sense of border in the infant in Western society [e.g. border between pure/dirty] becomes the basis for distances from 'others' . . . [this feeling] assumes a much wider cultural significance [in adulthood].
>
> (Sibley, 1995: 7)

So Sibley argues that the stereotypes or categories that societies assign to people constructed as 'ugly, dirty or imperfect' are associated with the border between Self and Other, reflecting deep, ultimately unconsciously rooted feelings that make some people distance themselves from others. Extreme forms of spatial purification based on such strong feelings include examples of 'ethnic cleansing' whereby one ethnic group actively seeks the exclusion of another from particular spaces (usually nation-states).

This therefore becomes a very geographical argument about the constitution of human identity and emotional relations as these relate to the constitution of space. Geographers hence argue that there is an explicit link between the inner workings of the human mind (the psyche) and the social landscape and how it is organized spatially. In particular, proximity to 'otherness' (other groups or people deemed 'different' by social stereotypes) can constitute a challenge to the security of emotional stability and self-identity, and so individuals strive to separate themselves from that they deem different, in order to try to sustain a pure identity:

> Separation is a large part of the process of purification – it is the means by which defilement or pollution is avoided . . .
>
> (Sibley, 1995: 37)

Psychoanalysis has provided geographers with a distinctive theoretical basis for understanding the emotional and psychic processes that fuel geographies of social exclusion, such as those hinted at in the above quotations. This work deals directly with embodied irrationality in ways that complement but also differ from humanistic and feminist geographies. Since the mid-1990s psychoanalytic geography has diversified and engaged with a range of Freudian and post-Freudian theory (Kingsbury, 2009; Pile, 2005, 2010, 2011), albeit in particular ways:

> While psychoanalytic geographies have tended to be suspicious of expressed emotions, this does not mean that they do not take them seriously. Instead, psychoanalytic geographies have tended to focus on desires and anxieties, phobias and pleasures – in the middle ground between inexpressible affects and expressed emotions . . .
>
> (Pile, 2010:14)

Where psychoanalytic geography is most distinctive is in its dealings with unconscious human life, but in ways that allow expansive conceptualizations of the city (Pile, 2005) and nation (Nast, 2000) among other geographies. It has, however, been claimed that psychoanalysis has been neglected by geographers (Pile, 2011), partly because it can be 'politically unpalatable' with disciplinary trends (Callard, 2003) and deals with a range of emotional states that still render geographers rather 'squeamish'.

SUMMARY

- Psychoanalytic geography explores the connections between unconsciousness, selves and spaces.
- Object relations theory has provided an accessible way to understand social and spatial exclusion.
- It is claimed psychoanalytic theory is 'tamed' in human geography writing.
- Psychoanalytic geography is now a diverse field which complements and challenges feminist, emotional and affectual geographies.

Emotion and affect: the 'non-representational turn'

In the range of work discussed above, geographers have sought to represent emotional processes and subjective experiences in various ways, normally as meaningful entities that hold data about human–environment relations. Yet emotions are such complex phenomena that some researchers are beginning to problematize whether they can be represented. An emerging body of challenging work linked to '**non-representational theory**' concerns itself with 'mundane everyday practices that shape the conduct of human beings towards others and themselves in particular sites' (Thrift, 1997: 126–7). Developing a non-representational theory involves 'not a project concerned with representation and meaning, but rather the performative "presentations", "showings" and manifestations of everyday life' (Thrift, 1997: 126–7). The emphasis in this style of work, as Catherine Nash (2002: 655) argues, is on practices 'that cannot adequately be spoken of, that words cannot capture, that texts cannot convey'. Emotional geographies, and the bodily, sensual and instinctive moments and practices that constitute them, seem to be a possible case in point here.

Central to this literature is a specific proposition that the words of human language cannot adequately do the work of representing many interior mental and emotional states and how emotional atmospheres move people and events. Here the assumption is that emotions can never be reconvened in words that somehow represent their interiority, nor their intangible threads of connection with everyday spaces, thus pointing to the 'ineffability' (or unspeakability) of some aspects of what we understand to be emotions (but see Laurier and Philo, 2006a (abstract)), as a counter to this notion, as they

> argue that this impossibility should not become a warrant for withdrawing from the world . . . instead, [they] propose that close descriptions can still be offered of particular encounters, attending in the process to the situated, embodied sense-making work being (unavoidably) undertaken by the peoples involved that makes those encounters what they are.

Such ideas have promoted some geographers to try and encounter emotions not through exhaustive self-reflection or interview data, but rather through exploring new ways to 'attend to' smiles, movements, gestures, moods, feelings, types of talk, and 'atmospheres' in order to witness different forms of emotional practice and their orderings or ''becomings'. To be more specific here, a new academic language and way of thinking has accompanied these developments, and this has made some distinctions (and connections) between 'emotions' and 'affect'. This latter term is often distinguished, although not uncontroversially, from emotion. For example, Derek McCormack (2003: 495) argued that emotion 'works in an already established field of discursively constituted categories in relation to which the felt intensity of experience is articulated'. In other words, we all more or less know how to behave and represent our feelings in relation to what we perceive after the event as sadness, happiness or anxiousness, for example. However, when working with the notion of 'affect', a notion that is not reducible to emotion, often the focus becomes more the sensational properties of embodied experience *before* they are registered by conscious thought (and thus represented as readily identifiable emotions). There are numerous related 'definitions' of affect (see Pile, 2010), that orientate around affect as 'a transpersonal capacity which a body has to be affected (through an affection) and to affect (as the result of modifications)' (Anderson, 2006: 735). The basic argument is that thinking in terms of affect, researchers are more likely to be attuned to practices or 'thoughts in motion' and that this effectively opens up exciting possibilities for new understandings of human being-in-the world and their 'doings'. For Thein (2005: 451) 'affect is the *how* of emotion. That is, affect is used to describe . . . the motion of emotion'. Already it is possible to see both the distinctions and overlaps which might be suggested by an orientation to emotion and affect.

The basic premise articulated above inevitably obscures much detail, theoretical nuance and debate about the relationship between emotions and affects and the role geographical analysis can play in understanding how such human sensibilities are spatialized and indeed, how they shape human geographies. For our purposes, I want to be selective and highlight particular aspects of this debate in order to try and simply convey something about how geographers are conceptualizing the significance of human feelings, as well as how best to study them.

In 2005, Thein wrote one of the first critiques of the 'non-representational turn' with reference to emotion, and she highlights Thrift's attempt to formulate affectual analyses that 'goes beyond the simple romanticism of somehow maximising individual emotions' (Thrift, 2004: 68). The result of such a stance for Thein is that '"affect" [is] a term and a concept [which] is employed here in masculinist, technocratic and distancing ways', and this then deliberately unsettles a feminist project of the personal-as-political:

> The jettisoning of the term "emotion" in favour of the term "affect" seems compelled by an underlying revisiting, if in a more theoretically sophisticated register, of the binary trope of emotion as negatively positioned in opposition to reason, as objectionably soft and implicitly feminized. In this conceptual positioning, these transhuman geographies re-draw yet again not only the demarcation between masculinist reason and feminized emotion, but also the false distinction between "personal" and "political" which feminist scholars have extensively critiqued.
>
> (Thein, 2005: 452)

What Thein is also arguing here is that non-representational writings seem to 'hold to one side' or ignore, expressed 'emotional geographies' precisely because they are overtly 'emotional' and too loaded with social significance to be of use to a non-representational project, and therefore of limited interest to those intent on theorizing 'unmarked' beings, doings and atmospheres and intensities between people and things in the world. Thein reads this as another example of a masculinist construction of knowledge in the geography discipline.

Responses to this position, and other critical commentaries (for example, Anderson and Harrison, 2006; McCormack, 2006; Pile, 2010; Tolia-Kelly, 2006), are still reverberating in the discipline, orientated in part by a dilemma concerned with how best to 'witness', understand and then write geographies of emotion and affect; and considerations about what is lost or limited by an orientation to either. Tolia-Kelly (2006: 214), for example, argues against the universalizing tendencies of affectual analysis in the context of race and gender and a need for 'historical contextualizing towards a non-universalist understanding of emotional registers'. Anderson and Harrison (2006: 334) provocatively ask in response, 'What does it mean to ground emotions in the figure of the human?', as they critique what they see as a caricaturing of writing on affect. Instead they offer an expansive potential agenda for study including: 1) the materiality of affect and emotion; 2) the nature(s) of subjectivity; 3) new vocabularies (including silence, contagious affects, poetics); and 4) political practices and interventions. This exciting list contributes to a sense of an emergent 'lively' scholarship, although it is sometimes presented in complicated forms (and see Lorimer's 2005 review of 'more than representational' endeavours for an accessible over-view of work in this vein).

More recent extensions of these arguments have highlighted the limitations of the project of 'recovering' emotional geographies (of love, hate, anxiety, denial, repression, etc.), risking

CASE STUDY

Steve Pile and the place of emotions in geography

In a recent article which has ignited new debate about the place of emotions and affect in human geography, Pile (2010) has critiqued the ways in which geographers have conceptualized, encountered and researched and written about emotions. He acknowledges that geography has recently become saturated with references to emotion:

> geographers have described a wide range of emotions in various contexts, including: ambivalence, anger, anxiety, awe, betrayal, caring, closeness, comfort and discomfort, demoralisation, depression, desire, despair, desperation, disgust, disillusionment, distance, dread, embarrassment, envy, exclusion, familiarity, fear (including phobias), fragility, grief, guilt, happiness and unhappiness, hardship, hatred, homeliness, horror, hostility, illness, injustice, joy, loneliness, longing, love, oppression, pain (emotional), panic, powerlessness, pride, relaxation, repression, reserve, romance, shame, stress and distress, suffering, violence, vulnerability, worry. Even those who declare themselves suspicious of the language of emotions – such as 'hatred, shame, envy, jealousy, fear, disgust, anger, embarrassment, sorrow, grief, anguish, pride, love, happiness, joy, hope, wonder' (Thrift, 2004: 59) – have nonetheless attended to anger, boredom, comfort and discomfort, despair, distress, enchantment, energy, enjoyment, euphoria, excitement, fear, frustration, grace, happiness, hope, joy, laughing, liveliness, pain, playing, rage, relaxation, rhythm, sadness, shame, smiling, sorrowfulness, 'Star Wars affects', surprise, tears (crying), touching, violence, vitality.

Pile's point in producing such a list, is to question how and whether these terms (and the feelings they 'represent') are part of a thorough-going 'emotional epistemology'. For Pile, there is a risk that geographers are simply 'adding emotions' to their analyses and presenting these in ever-more intimate, caring and poignant ways, and he argues that, instead, 'emotional geography must know *why* emotions are important and interesting' (Pile, 2010: 17, author's italics).

'an ever expending shopping list of expressed emotions' (Pile, 2010: 17) (and see box). Indeed, Pile (2010, 2011) questions the similarities and differences between 'emotional geography' and 'affectual geography', as well as a contemporary neglect of psychoanalytical theory in this field. Robust detailed debate about the purpose of this scholarship (Bondi and Davidson, 2011; Curti *et al.*, 2011; Dawney, 2011) shows tensions in the discipline in relation to how best to conceptualize and research these often intangible, but fundamentally important, aspects of our earth-life. Part of what is at stake for Pile, for example, is how emotional and affectual geographies can seem to be advancing certain kinds of 'performative' conceptual orthodoxies – or quite formulaic and limited ways in which emotion and affect might be discussed and thought through (a claim disputed in the exchange cited above). Above all, there is a sense in which work in this area – even though rather tense and intense (to attribute metaphors of feeling) – seeks to develop 'new and better questions' (Pile 2011: 606) about how emotion and affect configure the human and non-human world.

SUMMARY

- Non-representational geographies focus on what people do as opposed to what they say they do.
- In research work attention is drawn to practices, becomings, moods and atmospheres rather than representations.
- Pre- or non-cognitive feelings and sensations are of more interest than emotions (understood as socio-cultural constructions).
- There is controversy about how to research the non-representational and what exactly 'affect' is.

Conclusion

Despite being an academic discipline forged on oppositions of rationality and irrationality, and objectivity and subjectivity, geographers have seemingly encountered emotions in exciting and different ways throughout the recent past. Although there are important differences in terms of how different geographers have understood emotions and how they are relevant to geographical enquiry, collectively, humanistic, feminist, psychoanalytic and non-representational approaches have all sought to validate the place of emotions in the discipline.

DISCUSSION POINTS

1. Why have emotions been ignored by geographers in the past?
2. In what ways does the study of emotion differ in human geography?
3. What emotions has your own research work evoked for you?

FURTHER READING

Bondi, L., Davidson, J. and Smith, M. (2005) *Emotional Geographies*. Aldershot: Ashgate

This was the first edited book of chapters relating to a range of work on emotions and geography.

Widdowfield, R. (2000) The place of emotions in academic research. In: *Area*, 32(2), 199–208.

This article is clear and straightforward. It was discussed in the main text of this chapter and highlights how the doing of research is an emotional activity and that reflecting on such matters can be useful.

Pile, S. (2010) Emotion and affect in recent human geography. In: *Transactions of the Institute of British Geographers* 35, 1, 5–20.

This is a recent and provocative article which has sparked a string of responses about how emotion and affect are understood in Human Geography.

Bondi, L. and Davidson, J. (2011) Lost in translation: a response to Steve Pile. In: *Transactions of the Institute of British Geographers*, 36, 595–8.

This is one of the responses to Pile's paper (see Dawney, 2011; Curti *et al.*, 2011; Pile, 2011) which argues about the risks of presenting a dualistic framing of emotion and affect.

Tolia-Kelly, D.P. (2006) Affect – an ethnocentric encounter? Exploring the 'universalist' imperative of emotional/affectual geographies. In: *Area* 38, 213–17.

A key article which argues that affectual analysis can be devoid of attention to ethnocentricism, power, history and culture and thus produces universalizing accounts of affects.

CHAPTER 50
AFFECTS

Ben Anderson

Introduction

Scene one: Becoming terrified

The 2003 Iraq war began with a US-led bombing campaign based on a military doctrine known as 'shock and awe', or 'rapid dominance'. Creating the twin affects of 'shock' and 'awe' through airpower would destroy or damage the ability of an enemy to fight, or so claimed the authors of the doctrine of overwhelming force, information supremacy and battlefield dominance. For those people living in Baghdad, what might it have felt like to be bombed? What might it have felt like to wait and hear bombs explode and buildings fall?

Figure 50.1 'Shock and awe'. Credit: Associated Press

Scene two: Becoming energized

We are at a protest in London against cuts to the Education Maintenance Allowance and the introduction of university tuition fees of £9,000. Walking next to one another, we repeat a phrase: 'no ifs, no buts, no education cuts'. Sometimes sung, sometimes shouted, sometimes screamed, the repetition of the demand connects us. Producing a kind of felt solidarity, participation in the chant energizes the school children, university students and other people participating in the protest. Momentarily, repetition of the phrase gives hope. Later, the protest will be 'kettled' by the Metropolitan Police: contained in a specific site for a period of hours. One effect of the action

Figure 50.2 Kettling. Credit: Padmayogini/Shutterstock.com

of kettling is to contain the energies of the crowd. Protest becomes boring, exhausting, draining and uncomfortable. At the same time, images of the sporadic violence that emerges in the kettle are transmitted to a media audience who may themselves be energized by the spectacle of protestors and police clashing.

Scene three: Becoming bored

Figure 50.3 Being bored. Credit: Phil Boorman/ Getty Images

You are sat trying to listen to a first year lecture in a compulsory module. Interest begins to wane. A lecturer attempts to hold your attention as you sit in lines in a tiered lecture theatre with other students, some making notes, some not. Before you notice it you start to draw something, doodling. Boredom overtakes you. You switch off. Something like boredom seems to infuse the room: a stilling and slowing of life that interrupts the demand to always be interested and results in various escape attempts. Distractedly, you look around the room. You daydream. Occasionally, your interest returns, piqued by a word, an image.

Shock and awe in bombing, the energies of protest, boredom in a lecture: this chapter discusses why geographers should consider these and other experiences. It does so by introducing recent geographical work on the topic of affect. Unlike many of the other ideas you have encountered in the book, the term 'affect' is likely to be new to you. Affect is not generally a part of our ordinary vocabularies. While we will look in more detail at definitions below, to start with we can define affect in relatively straightforward terms: affect refers to the experience of life as it is lived. We can get a sense of what is meant by this by looking at the example of protest. One of the things that might be shared at a protest is an *affect of participation*. And the space of the protest is in part constituted through that shared affect; an affect that changes as the protest marches and chants and is kettled and an affect that mixes and merges with the joy, anxiety, excitement and other affects of protest. The starting point of recent work on affect is that the role of affect is not unique to intensified spaces such as protests: any and all spaces are spaces of affect.

Beginning from this insight, a wide range of work has developed over the last ten years on what Pile (2010) has named 'affectual geographies'. The first section introduces and summarizes some of the different ways in which the term 'affect' is being used in contemporary Human Geography. While Thrift (2004: 59) is correct to stress that there is no 'stable definition of affect', there are nevertheless a number of commonalities between different understandings. The second section considers some of the different 'spaces of affect' that research has focused on to develop the general claim that understanding affect matters to geography. By way of a conclusion, I will reflect on some of the issues around affect-based research.

Understanding affect

The term affect has been used by geographers to describe a wide range of quite different phenomena: background moods such as depression, moments of intense and focused involvement such as euphoria, immediate visceral responses of shame or hate, shared **atmospheres** of hope or panic, eruptions of passion, lifelong dedications of love, fleeting feelings of boredom, societal moods such as anxiety or fear, neurological bodily transitions such as a feeling of aliveness, waves of feeling . . . among much else. In Human Geography alone, the term has been used to understand a wonderfully diverse range of geographies: fathering (Aitken, 2009), popular geopolitics (Carter and McCormack, 2006), landscape relations (Wylie, 2009), new forms of work (Woodward and Lea, 2010); race and racism (Lim, 2010; Swanton, 2010), alcohol (Jayne *et al.,* 2010), obesity (Evans, 2009), dance (McCormack, 2003), war and violences (Anderson, 2010), therapeutic landscapes (Conradson, 2005; Lea, 2008), and animals and other non-humans (Greenhough and Roe, 2011) to name just some.

Across this varied work, you will find that the term affect is used in multiple ways. Sometimes this can be confusing. Not only is there no consensus about what is or is not included within the category, but something about the type of experience being described appears to be difficult to fully capture in a definition. Nevertheless, we can point to three main uses of the term (see Gregg and Seigworth, 2010; Thrift, 2004 for others). These uses are not mutually exclusive, but they do have different consequences for how geographers understand and make sense of affective life. In the main, the uses emerge from the introduction of non-representational theories (see box) into Human Geography (Thrift, 2007a; Anderson and Harrison 2010b). However, the emergence of affect as a topic of interest is indebted to the work of feminist and humanistic geographers. Both argue that geographers should engage with the emotional dimensions of life and, in different ways, both practice styles of research practice and theory that value emotional knowledges.

CASE STUDY

Non-representational theories

'Non-representational theories' is an umbrella term for a group of predominantly post-structural theories that attempt to attend to life as it happens. What is shared between them are three questions:

1. How do sense and significance emerge from ongoing practical action?
2. How, given the contingency of orders, is practical action organized?
3. How to attend to events and the chance of something different that they open up.

(After Anderson and Harrison, 2010b: 23–4).

The plural – theories, rather than theory – is important: for there are now a range of non-representational theories that respond to the three starting problematics. While a concern with the dynamics of life crosses between them, the interest in affect has emerged most strongly in work influenced by the French philosopher Gilles Deleuze and the cultural theorist Brian Massumi (see McCormack, 2003; Lim, 2010 and outside of geography, Clough, 2007).

Deleuze outlines a philosophy of becoming that attempts to be adequate to the 'singularity' of events: 'bottlenecks, knots, foyers, and centers; points of fusion, condensation, and boiling; points of tears and joy, sickness and health, hope and anxiety, "sensitive" points"' (Deleuze 1990: 52). For Deleuze, events open up the chance of something different within processes of becoming (see Dewsbury, 2000). As such, events are folded into the dynamics of 'a life': the ever changing, ever renewed, assembling of heterogeneous things into some kind of order.

As part of this attempt to understand a world of events, Deleuze (and his long-time collaborator Félix Guattari) draws a definition of affect from an encounter with the philosopher Baruch Spinoza's *Ethics*. Brian Massumi (1987: xvi), the translator of *A Thousand Plateaus*, summarizes the concept in the following terms: '

> [Affect] is a prepersonal intensity corresponding to the passage from one experiential state of the body to another and implying an augmentation or diminution of that body's capacity to act.
>
> (Massumi, 1987)

In his own work at the interface between cultural theory and political theory, Massumi (2002: 36) equates affect with one's 'sense of aliveness' and stresses that this 'sense' expresses our participation in relations with others, may come to be perceived, but also exceeds conscious thought. For this reason, affect offers a 'point of view' into what Massumi after Deleuze and Guattari calls 'virtuality': the potentialities that are folded into life as it happens.

Uses of the term 'affect

First, the term affect is used to describe a body's **capacities to affect and be affected** that may be expressed in emotions or feelings. Consider what it feels like to play a video game

James Ash's (2010) research shows how the careful designing of an 'environment' in games conditions a specific feeling of immersion in game play. This immersion is not necessarily consciously reflected upon by the game player, it acts more as a taken-for-granted background that enables him to play the game. It is this sense of the taken-for-granted background of life that research on affect has focused on. For example, in their work on the role of alcohol in urban places, Jayne *et al.*, (2010) show how understanding drunkenness as an affective experience is bound up, in the UK, with fun and shame offers a new perspective on 'problem drinking' as a policy issue. On a different register, Emma Roe (2006) thinks that our affective and for the most part taken-for-granted relationship with food conditions how we engage with ethical issues around animal welfare and food production. Often this work has involved a claim that the lived present of experience is *partly* **non-conscious**, in the sense that something about experiences such as playing games, drinking alcohol or eating meat are at least partially below the threshold of conscious reflection and deliberation. Normally this use involves invoking a distinction between affect and emotion: emotion being used to refer to the ways in which that non-conscious background is named, interpreted and reflected on by people. By contrast, affect refers to the felt quality of an experience: a quality that

Figure 50.4 Video games and affective attunement. Credit: James Ash

provides something close to the background sense of an event or practice or space.

Debate rages about how geographers should understand non-conscious capacities to affect and be affected. Should we, for example, learn from the insights of recent types of neuroscience and try to integrate an understanding of neurochemical processes into our understanding of affective life, even though this may risk a form of biological reductionism (Thrift, 2007a; Papoulias and Callard, 2010)? Alternatively, perhaps we should be attuned to the specificity of bodily experience through forms of phenomenology and attempts to describe experience as it takes place and constitutes subjects (Wylie, 2009). Might we think with the insights of recent corporeal feminism and begin with the forces that form and are formed by bodily differences (Colls, 2012; see also Hayes-Conroy and Hayes-Conroy (2008) on viscerality)? No matter which approach is used, the assumption is that the background feeling of what happens matters to how geographies are lived.

Less commonly, but as importantly, affect is used to refer to explicitly collective experiences that are shared between individuals. One example would be the idea that a space has a characteristic **atmosphere**. Whether it is a 'cosy' room, 'uplifting' building or 'romantic' setting, atmospheres are strange things for geographers to try and research (Anderson,

2009; Bissell, 2010a; McCormack, 2008). Even though the quality of an atmosphere can be named, they are still vague. It is unclear where the atmosphere is located or what its boundaries are. Like a gas or air, an atmosphere might appear to infuse a whole space without being definitely locatable anywhere. Conradson (2005) captures this in his careful description of the atmospheres of care that can come to envelope particular places and make them therapeutic. He shows how atmospheres emanate from the particular array of things and people that make up spaces such as retreats or drop-in centres. Collective affects are not only linked to specific sites, however. We might talk, for example, of the characteristic mood of an event like a protest or a carnival. A protest might be 'good natured' or 'threaten to turn violent' or a carnival might be 'fun' or 'exciting'. Stepping back from specific sites, we might try and apprehend the distinctive **structure of feeling** that characterizes a particular society or culture. One example would be the claim that in the midst of the 'war on terror' contemporary liberal democracies have been marked by an 'age of anxiety' or a 'culture of fear'; another example might be the periodic panics about crime or other forms of urban disorder that seem to erupt and spread across the media.

Closely linked, and third, affect has been used to identify particular sensibilities or dispositions associated with distinctive groups of people. Very often this involves carefully mapping how affective lives are lived in distinctive ways. For example, work on the geographies of employment in post-Fordist economies has detailed how an experience of perpetual insecurity has become the norm for some segments of the population. Against this taken-for-granted background, Lauren Berlant (2011: 1) describes the 'cruel optimism' that infuses the lived experience of particular segments of the lower middle class in the United States. She describes a situation in which 'something you desire is actually an obstacle to your flourishing'. Her examples range widely, including the promise of particular types of family relations or participation in forms of waged work. On a similar register, Eliasoph (1998) shows how a disposition of apathy conditions how particular segments of American society disengage from the formal realm of politics. What this work emphasizes, perhaps more than the other two uses of affect, is that affective life (re)produces the ways in which society is patterned.

We can identify some initial commonalities between these three uses of the term affect. To begin, they all assume that affects such as immersion or cruel optimism or anxiety are fundamentally social in that they connect us to other people and things. We tend to think of feelings as our own and, as such, as private and idiosyncratic. This is a view of emotional life with its origins in a distinction between the rational and emotional and an assumption of individuality. Feminist research on emotions has done much to critique this divide, showing the negative consequences of implicitly or explicitly identifying reason as masculine and emotion as feminine (see Rose, 1993; Bondi, 2005; Pain, 1997). One of the main claims of research on affective life is that affects are not simply personal. We find that our experience always connects us to people or things or places beyond ourselves. There is something of a paradox here: what appears to be most personal, most unquestionably ours, is an index of our participation in wider processes. For one example of how affective experience connects us to processes beyond ourselves consider a recent literature on the peculiar persistence of the past in the present, or the ways in which places are made through pasts that live on – what Edensor (2008: 331) nicely terms a 'vaguely recognizable but elusive form'. Thinking with the vocabulary of haunting and through a case of the remnants of working-class life in a regenerated area of Manchester, Edensor shows us how what might appear to be distant in time or space is folded into the affective experience of urban life. Manchester becomes a site of 'ghosts' as remains and traces of past industry live on in the material fabric of the city: '. . . illegible traces, memories and forms of hearsay from the past that continue to make their mark' (Edensor, 2008: 331).

All three uses therefore presume that affects are part of other geographies. There is no such thing as a circumscribed realm of 'affective geographies' that sits separately from the other more familiar geographies that you will have encountered. Rather, affects such as fear or boredom can be part of any and all geographies. We might, for example, think of how the economic geographies of the current financial crisis are inseparable from affects such as panic, greed, exuberance and confidence. Cities are a 'roiling maelstrom' (Thrift, 2004) of affects: fear, longing, indifference, and so on. Consequently, all research on affective life presumes that affects are a vital part of any and all geographies and as such are integral to how life is lived. The next section sets out how geographers have researched such spaces of affect.

SUMMARY

- There are multiple definitions of 'affect', and understanding affect involves also considering the meaning and use of terms such as emotion and feeling.
- What is common across these definitions is that affect is used to refer to the taken-for-granted 'background' of life and thought: the feeling of what happens.
- Claims are often made that affects are non-conscious or only partly conscious.
- Although the term affect is used in different ways, work in the field of affective geographies begins from a presumption that affects emerge from relations between people and between people and things.

Spaces of affect

David Bissell (2010b) has recently argued that geographers are particularly well placed to describe 'affective topologies', or 'how different affects are intensified or quiesced at different times and in different spaces' (2010b: 82). His point is that boredom, fear and other affects are not timeless, ageographical, experiences. Rather, affect and affects are formed and emerge through particular geographies. Let us look, then, at some examples of how recent affect-based research has attempted to attune to and understand what Bissell terms the 'temporally and spatially differentiated characteristics of affect' (2010b: 82).

Work on affective life often starts by paying close attention to how particular affects emerge through **encounters** between people or between people and sites. We can get a sense of what is meant by this through an example of ethnographic research on what it feels like to live in a multicultural town (Swanton, 2010; also Wilson, 2011). What Swanton does is focus on specific, situated occasions where different people and groups encounter one another in Keighley, a former mill town in Northern England. What he shows is that affects such as suspicion or fear form in, and give shape to, mundane moments of intercultural contact in spaces such as cafes, pubs, bookmakers and back streets (see Figure 50.5).

Alongside the tolerances and indifferences that conditions many ordinary encounters, occasionally hatreds will intensify. In the event of a terror alert, for example, ordinary encounters become charged with different affects as particular Muslim men become the object of suspicion and subjected to forms of everyday racism from some white residents. Ordinary affective life in Keighley is constantly being made and remade through very mundane encounters through which differently racialized bodies are enacted.

Encounters might be between two or more individuals or groups, but they also might be between an individual and a particular set of things. In her ethnography of children's use of objects that are or become toys, Tara Woodyer (2008) shows how encounters with things create the affective worlds of childhood. Drawing on Jane Bennett's (2010) work on the

Figure 50.5 Ordinary affective life in Keighley. Credit: Dan Swanton

'vibrancy' of matter, Woodyer pays attention to how some toys have a liveliness – the capacity to enchant, delight and move. Encounters might also be between an individual and what we might call the spiritual: the '. . . immanent sensation of something more going on' (Dewsbury and Cloke 2009: 708). Taking as their example Christian spiritual practices such as prayer, baptism and contemplation, Dewsbury and Cloke (2009) describe how 'something more' is brought into being through practices and performances and comes to be encountered as a vague, amorphous, absent-presence.

It is not enough, however, to only focus on spaces of encounter. One criticism of work on affect has been that it is concerned only with small-scale interactions, ignoring or forgetting wider geographies of connection and disconnection. To address this criticism, recent work has focused on the extended, translocal geographies of (dis)connection through which ordinary affective lives are constituted. So we should note that encounters always involve connections to other spaces and times (as we saw in the example of haunting described earlier). For example, Swanton's example of the terror alert shows how something that might seem far away – terrorist events in New York – conditions an encounter in a place like Keighley. What seems to be far away becomes part of encounters.

Let's develop this point with another example – this time of hope and transnationalism in the context of Australian national culture. In an insightful book, Ghassen Hage (2003) wonders about how hope is kept alive among peoples who have been forced to move between countries. His concern is with 'the availability, circulation, and the exchange of hope'. So he interprets migration not only as a phenomenon involving the movement of people between places, but also, at times, the production and distribution of hope. However, he argues that various policies he associates with neo-liberalism have diminished the capacity Australian society has to hope for a better world. The result, for Hage, is the emergence of a form of 'paranoid nationalism' infused with worry about 'immigrants' and others felt to be a threat to Australian national culture. Leaving aside the details of Hage's argument, why this kind of approach matters is that it helps us understand that while affects emerge from encounters they are also determined by the ways in which those encounters are entangled with economic, political, social and other forces that extend beyond the space of an encounter. This means that the geographies of affect are always **transpersonal**: they extend beyond the personal expression of affect in a specific feeling or emotion. Bringing together an emphasis on (dis)connections with an attunement to encounters, Kathleen Stewart (2007) gives a wonderful account of such transpersonal geographies in her ethnography of 'ordinary affects' in American culture in the early twenty-first century. Ordinary affects are:

> the varied, surging capacities to affect and be affected that give everyday life the quality of a continual motion of relations, scenes, contingencies, and emergences. They're things that happen.
>
> (Stewart, 2007: 1)

Stewart's work is exemplary for how it relates the affects of everyday life in America to spatial processes such as neoliberalism or capitalism without either privileging those processes as dominating forces that predetermine the feeling of what happens or wishing them away by romanticizing everyday life.

Straightaway, we can see that the affective topologies that Bissell writes about are complicated. They extend beyond the individual who feels to include situated

Figure 50.6 Airports and affective design.
Credit: Peter Adey

encounters and they occur in the midst of wider (dis)connections. At the same time, we should remember that specific spaces provide what Thrift (2008: 236) terms 'a series of conditioning environments that both prime and "cook" affect'. Adey (2008), for example, tracks how airports are now designed to create a smooth experience of movement through the intentional deployment of light and colour, one increasingly orientated to the passenger as consumer.

Likewise, the 'atmosphere' of a retail space such as a mall or shopping centre might be carefully designed to appeal to unconscious consumer experience. Lighting, shop window display and music might all be crafted to generate particular atmospherics designed to facilitate the activity of shopping. Of course, as Rose *et al.*, (2010) have shown, people actually experience these designed environments in ways that do not correspond to the designer's intentions. The same is true for other attempts to manipulate the affective dynamics of sites. Ash (2010) makes this point in his account of how the 'environment' of video games is designed to occupy attention and create a specific form of absorption. It is not that affective life is simply manipulated, that would be to presume that people respond automatically to efforts to shape conduct. Instead, efforts are made to shape the conditions of affective life. In the case of video games, this is by organizing the interplay between scripted and unscripted events. In the case of shopping malls, it is by producing a specific atmosphere that becomes part of consumer conduct. Developing this emphasis on how affective life is shaped, recent work has argued that the modulation of affect is a key site for the operation of contemporary forms of power (Anderson, 2012). Examples include the use of new behavioural knowledges in UK public policy in order to shape the 'predictably irrational' dimensions of human decision making (Whitehead *et al.*, 2011) or the manipulation of fear by media and political elites to justify war and other forms of violence (Ó Tuathail, 2003).

It would be a mistake, then, to think that researching ordinary affects is only a way of talking about individual people and what and how they feel. Instead, the point of research on affect is to offer careful, nuanced descriptions of life as it is lived that connect affective life to the range of forces that shape and condition it.

SUMMARY

- Affects such as hope, fear, anxiety, cruel optimism, apathy and so on are not 'natural' phenomena but the point at which individual lives are connected to spatial processes and practices.

- Affect are formed in encounters that are entangled in transpersonal geographies of connection and disconnection.
- There are multiple relations between affective life and forms of power, including but not restricted to intentional manipulation.

Conclusion

> Affect remains hazy, atmospheric, and nevertheless perfectly apprehensible.
> (Guattari, 1996: 158)

Engaging with spaces of affect is necessary if Human Geography is to understand how ordinary life is lived and connected to wider processes. For this reason, the concept has opened up new and novel directions for thinking and research, chiefly a focus on how life happens in specific encounters and on the ordinary affects of 'big', seemingly abstract, socio-spatial processes.

Understanding how affective life takes place poses particular methodological challenges (Dewsbury, 2009), for how do you research experiences such as participation in protest or the terror of being bombed that seem to be fleeting and ephemeral? How do you then analyse and represent that experience, if that is the aim? Returning to the examples of terror, participation and boredom at the start of the chapter, did my combination of words and images evoke something of the experience? Perhaps not, and perhaps that failure is endemic to any attempt to represent affective life.

Responding to the problem of how to research and represent affect, some geographers have recast qualitative methods such as interviews, using them as a means of eliciting reflection on background affect (Latham, 2003). Other geographers have begun, somewhat tentatively, to experiment with other methods. These **performative methods** aim to participate in affective life, create encounters and then evoke specific affects (Dewsbury, 2009; Simpson, 2011a). However, should the task of a geographical analysis of affect be to evoke experience? Perhaps geographical work on affect should, instead, aim to explain the conditions of affective life or critique the ways in which ordinary affects are structured? Perhaps, slightly differently, affect-based research should learn from the insights of **participatory geographies** and work with people to produce knowledge that makes a difference to their lives?

Many of the insights of recent work on affect are still to be fully developed. While promising much, research on affect has been subject to a number of criticisms and responses concerning the politics of affect-based research and the implications of researching what might appear to be 'non-representational' (Barnett, 2008; Pile, 2010; and responses in Vol. 36, Issue 4 of *Transaction of the Institute of British Geographers*). Although variegated, the debates turn around two problems or questions, in addition to the methodological challenges touched on above. First, how should we understand the processes whereby the non- or pre-conscious dimensions of thought and life are connected to the ongoing dynamics of social life? Second, if affects occur below, alongside and above the subject (Dawney, 2011), then how to understand the relation between affective life and multiple, intersecting, social differences? Debates about the political promise of affect-based research follow on: how should we understand the relation between affective life and forms of power and how, if at all, does the concept of affect allow us to get a purchase on the contemporary condition? These questions matter because affect-based research promises a way of understanding how ordinary life is lived in the midst of a range of spatial processes and practices. Or, put differently, it offers a way of understanding how spaces are made through affects such as terror on being bombed, boredom while being lectured or joy in protest.

DISCUSSION POINTS

1. Evaluate what a focus on affect contributes to geographical research and thought.
2. Consider the reasons why affect has emerged as part of Human Geography research, and how the emphasis on affect in geography relates to a renewed interest in affect and emotion across the social sciences and humanities.
3. What are the main points of difference between theories of affect?
4. Using an example of a particular site such as an airport or shopping mall, explore how different approaches to affect help us understand specific 'affective topologies'.
5. What methodological challenges does researching affect pose to Human Geography and how have geographers responded to those challenges?
6. How have geographers considered the relation between forms of power and affective life?
7. What does a focus on affect have to offer analyses of spatial processes and forms such as neoliberalism or capitalism?

FURTHER READING

McCormack, D. (2003) An event of geographical ethics in spaces of affect. In: *Transactions of the Institute of British Geographers*, 28(4), 488–507.

Thrift, N. (2004) Intensities of feeling: Towards a spatial politics of affect. In: *Geografiska Annaler B*. 86(1), 57–78.

These texts provide some of the starting points for work on affect in geography; both are connected to the emergence of non-representational theories.

Stewart, K. (2007) *Ordinary Affects*. Durham and London: Duke University Press.

Ó Tuathail, G. (2003) 'Just out looking for a fight': American affect and the invasion of Iraq'. In: *Antipode*, 856–70.

Excellent case studies of specific affective geographies that bring empirical material and theoretical claims into dialogue with one another. The papers engage with different theories of affect.

Swanton, D. (2010) Sorting bodies: race, affect, and everyday multiculture in a mill town in northern England. In: *Environment and Planning A* 42(10), 2332–50.

Lim, J. (2010) Immanent politics: thinking race and ethnicity through affect and machinism. In: *Environment and Planning A*, 42(10), 2393–2409.

A couple of inventive empirical studies that show the importance of work on affect for understanding the geographies of race and sexuality.

Pile, S. (2010) Emotions and affect in recent human geography. In: *Transactions of the Institute of British Geographers*, 35(1), 5–20.

A critical overview of recent work on affect and emotion in Human Geography. Responses to Pile's paper can be found in Volume 36, Issue 4 of *Transactions of the Institute of British Geographers*. You should read the responses alongside Pile's paper.

Barnett, C. (2008) Political affects in public space: normative blind-spots in non-representational ontologies. In: *Transactions of the Institute of British Geographers*, 33(2), 186–200.

Anderson, B. (2012) Affect and biopower: Towards a politics of life. In: *Transactions of the Institute of British Geographers*, 37(1), 28–43.

Debates about the relation between affect and politics are explored in depth in these texts.

WEBSITES

www.journals.elsevier.com/emotion-space-and-society/

The post-disciplinary journal *Emotion, Society and Space* was set up in 2008 and contains articles on affect and emotion from a wide range of theoretical perspectives in relation to a variety of empirical topics.

http://psimpsongeography.wordpress.com/

http://jamesash.co.uk/blog/

http://supervalentthought.com/

http://homecookedtheory.com/

Four blogs by geographers and others that contain regular postings or comment on affect-based research.

CHAPTER 51 PERFORMANCES

Amanda Rogers

Introduction

Port Talbot in South Wales is a town known for its heavy industry. Driving down the M4, the motorway that historically cut through the town's heartland, the skyline is dominated by steel works chimneys and dockyard cranes. This is perhaps not a place one would immediately associate with art and performance, despite being the hometown of Anthony Hopkins, Rob Brydon, Richard Burton and Michael Sheen. Yet in Easter 2011, the National Theatre of Wales, together with Michael Sheen and the theatre company WildWorks, staged *The Passion*, a three-day outdoor theatre event held in different sites around Port Talbot that dynamically re-told the history of the town and its people. In this introduction, I will briefly discuss *The Passion* in order to draw out three themes that characterize geographical literatures on performance before discussing each theme in the rest of the chapter.

The Passion was an updated passion play that used the Christian story of Jesus' trial and crucifixion to tell a part fictional, part real, history of Port Talbot and its people. The story centred on a missing schoolteacher known as 'The Teacher' (Sheen) who reappeared on the beach at Port Talbot having lost all his memory at a time of crisis. In this semi-fictional world, the town has been taken over by a global corporation that is planning a 'Passover Project' – a new road to take Port Talbot into the future. Activists protest against the development, only to be confronted by armed guards, taken away, jailed or shot dead. The Teacher becomes the town's saviour, comforting those who are outcast by listening to their stories. Without a memory, he is the vehicle for the town to express its history and to find its voice and identity once again. The Teacher stands against injustice and thus becomes a threat to the corporation that owns the town. They capture The Teacher, put him on trial and after a long procession, he is crucified. As he dies, the Teacher remembers who he is and calls out a list of places, experiences and memories associated with Port Talbot before he, and the town, are resurrected.

The Passion was therefore about Port Talbot, its residents and their stories. The relationship between performance, identity and place is the first theme of interest to geographers. The creators of *The Passion* were keen to include stories about marginalized or disadvantaged groups and to celebrate the work of local communities. There was therefore an awareness of the power relationships surrounding whose stories to tell, and extensive attempts were made to gather different oral histories from people living in Port Talbot. The theatre companies spent six months sourcing local stories, amateur films and old photographs to use in the performance. *The Passion* used a professional cast of 15 local actors, but over 1,000 people from community and support groups also performed, and hundreds of others made props and costumes. *The Passion* thus encouraged people to express themselves by becoming involved in the performance's creation and staging. This ownership continued during the show's reception as audiences joined in with local resistance chants or laughed as they recognised certain references. During the final procession, crowds spontaneously sang the Welsh hymn *Calon Lân*, signalling the expression of Welsh national pride and identity. However, this **atmosphere** was not necessarily exclusive, as the procession also included locals and visitors singing other songs together, creating a lively environment that allowed everyone to participate.

The second feature of *The Passion* that resonates with geographers is that the performance used spaces in creative ways that were politically and emotionally powerful. For instance, Llewellyn Street was partly destroyed by the construction of the M4, leaving behind one side of the street that now faces the concrete road supports. One scene staged here involved a character describing the street's history as the ghosts of old neighbours appeared around the supports, wanting to be remembered. The scene highlighted the politics of **development** in Port Talbot and its impact on the community, but many residents found this scene emotionally moving because they imagined their older family members as the ghosts and they felt a sense of personal loss. The performance also animated spaces by using them in new ways. A patch of ground with a skip on a local council estate became The Garden of Gethsemane, the place where The Teacher questioned whether he could continue on his journey. He discussed his dilemma with his father, a roofer on a nearby house. In such moments, people were drawn together through mourning, laughter and joy. *The Passion* therefore unfolded in different places and audiences recorded and shared these scenes on social networking sites. As a result, people were able to watch performances they had missed and *The Passion* generated a huge online following – so much so, that by the final day around 15,000 people from within and beyond the UK visited Port Talbot to watch the play[1].

The Passion was therefore very emotive in how it used and evoked different places, blurring the line between the real and the imagined, between everyday life and performance. In the final procession, audiences carried personal memorabilia associated with Port Talbot and they discussed these objects with other members of the crowd. Everyday actions and encounters began to take on greater meaning. Online comments detail many small acts of kindness and describe how strangers from different places shared food and drink with each other, reflecting the sense of togetherness created by the play. In the BBC Wales documentary about *The Passion*, and the subsequent film about the event, *The Gospel of Us*, the residents of Port Talbot described how they felt the performance had reawakened the town's feeling of community, with people of different ages and social classes working together without violence or resentment. The performance created a positive sense of being together, working as an expression of hope and possibility. The future of Port Talbot seemed open, with locals commenting that *The Passion* had made them think about how the town had been, how it was now, and what it might be in the future.

The Passion therefore illuminates three areas of performance that have interested geographers: performance and identity; the embodied or experiential qualities of performance; and the relationship between performance and the everyday. Although we can initially approach these themes by viewing performance as a piece of art, as we examine these three areas in more detail performance becomes a broader term. Performance is about creative skill but it can also include embodied actions that express our identities and our relationships (e.g. the sharing of food and drink with people reflects friendliness). As a result, performance becomes the different practices through which we make, reinforce, challenge and perhaps change, our everyday world. It is to the geographical understanding of these practices that we will now turn.

Performance and identity

In *The Passion* we saw how a theatrical performance explored and expressed the lives of people in Port Talbot. It is possible to take this quality of theatre and apply it to the social world more generally in order to suggest that our daily actions are like theatrical performances. This is what the sociologist Erving Goffman analysed in his book *The Presentation of Self in Everyday Life*. Goffman suggested that theatre was a metaphor for social life and that in any social setting 'we not only act, but put on an act' (Tuan, 1990: 159). We act, or perform, depending on the situation we are in and the nature of the interaction we are having. Importantly for geographers, Goffman draws attention to how these performances are spatialized – they alter according to where we are. I behave very differently when I am in my own home compared to my parents' house or my in-laws' (best behaviour!). Goffman suggests that we interact by performing socially defined roles according to our location and who we are with, but also that we manage our behaviour to maintain these role performances. Our social interactions occur in a 'front stage' where we manage the impression we give to others but there is also a 'backstage' where we slip out of socially defined expectations.

Geographers have used and critiqued Goffman's ideas to examine behaviour, particularly in the workplace, where performance is integral to selling services and products (Crang, 1994; see also McDowell and Court, 1994). Workplaces may expect that we behave in certain ways, that we wear particular types of clothes, or talk to people using particular language. Philip Crang's (1994) account of working in Smoky Joe's restaurant highlights how waiting staff reproduce a scripted form of behaviour that conforms to management expectations of 'service' – but he also demonstrates how this behaviour is altered, improvised and elaborated upon in order to obtain tips. This may mean giving the impression of being busy, looking anxious when orders are late or showing one's own personality in order to individualize the service provided. Such work highlights how our actions are continually negotiated according to management or co-worker surveillance, as well as in relation to specific encounters; performances are not simply scripted according to a single context because interaction does not remain static. There are specific micro-geographies to these interactions and there is often not such a fixed distinction between front and back stage. People often have their own ways of expressing themselves and it can be difficult to completely 'switch' from one mode of behaviour to another.

Although Goffman helps us to think about how our everyday actions are similar to performance (more about this theme later), in his account, performance is often quite a predetermined and restrictive set of practices. We always seem to know how we should behave and dutifully do so; there is less sense of spontaneity or creativity, subversion even. This is because although Goffman is interested in the social world and the interactions that comprise it, he is much less concerned with power relationships. Goffman doesn't consider how power disparities between people may affect their interactions or examine how identities such as race, ethnicity, age, gender or sexuality might be expressed through our behaviour. Other accounts of performance in geography pay much greater attention to such politics of performance, particularly regarding identity. Goffman also suggests that our actions simply express a pre-existing self. However, other accounts of performance in geography suggest that through our actions we can create and re-create our subjectivity. As a result identities are seen as much more constructed and fluid.

This second approach to performance is most evident in geographical work that draws upon the work of the feminist theorist, Judith Butler (1993, 1990), to describe the emergence of identity in spaces of power. Butler suggests that identities, particularly gender, are not something natural that we all inherently 'have'. Rather, identities are constructed social relationships that are continually brought into existence through our actions. Action, or how we 'do' identity, is viewed as performance. However, we do not necessarily have a choice over how identity performances occur because social **discourses** and expectations normalize our understanding of identity and regulate how it is performed. As a result, the construction of identity is often invisible, especially as we learn to enact these discourses about identity from an early age. Underlying these ideas is the suggestion that performing identity is not a one-off activity but an ongoing process of constructing the self. We continually repeat actions that represent normative expectations, often unconsciously, and this link between action and social discourse suggests that identity is **performative**. However, because identity is continually being enacted, it is possible to rework how it materializes. No act is ever identical in its repetition, and this suggests that performances create opportunities for subversion as identity is enacted in new ways. Geographers have been drawn to this disruptive quality of performance and have developed Butler's work by examining how identity performances are inherently situated.

Much of the geographical work on performativity examines the relationship between sexuality and space. For instance, Bell *et al.,* (1994) analyse the identity performances of gay skinheads and feminine (or 'lipstick') lesbians to demonstrate how everyday space is dominated by heterosexual identities. Skinheads are conventionally depicted as masculine, heterosexual identities, and lipstick lesbians similarly reflect social expectations of femininity. Yet to see skinheads hold hands or kiss in public can challenge these expectations, disrupting and revealing the normative production of heterosexual space. Simultaneously, however, these disruptive performances can reproduce dominant understandings of identity. Basically, how can you tell which skinheads are gay? Gay skinheads or very feminine lesbians may pass as straight. Do transgressive performances have to be recognized as disruptive or is their presence alone enough? In general, geographers view recognition as central to subverting or re-performing identity. For instance, Bell *et al.,* (1994) and Bell and Valentine (1995c) examine how small actions can perform homosexual identity, producing queer space within dominant heterosexual space, particularly when others understand the meaning of these practices. Similarly, Mahtani (2002) describes how mixed race individuals perform racial identity in ways that challenge normative expectations. They may elaborate or invent stories about their backgrounds in order to force people to reflect on their assumptions about racial identity. However, mixed race individuals may also enact the practices associated with just one part of their racial identity, such as speaking another language, performing certain customs, or discussing different foods. Such performances are contextually dependent, but they often produce a sense of belonging by repeating established expectations. Like sexuality or gender, racial identity therefore appears as a construction, a performance that is variable and constantly in formation.

SUMMARY

- Performance is a metaphor for the practice of social life. Our actions can be viewed as performances that express our social roles and identities, and these actions vary according to context.
- However, performances may not only express identity but can actively help to constitute and re-enact it.
- The performance of identity can reproduce but also subvert discursive expectations and the norms of social space.

Geographies of the performing arts

The previous discussion emphasized how we perform or 'do' identity, but there was less sense of how these practices were **embodied** or how they might be viewed as skilful, playful or experimental. As geographers have become more interested in performance, they have turned not only to sociology and gender studies, but also to the performing arts in order to focus on creative modes of being and expression. The performing arts draw attention to the embodied experience of space and place but also highlight how bodies are implicated in issues of politics and economy. Music, dance and theatre performances can therefore illuminate geographies of power and identity but through creative, bodily action.

As we have already seen with *The Passion*, the performing arts can represent people, places and identities. Geographers have analysed both what artistic performances represent but also *how* these representations are enacted through smaller geographies of practice and embodied action. Building on the previous section, artistic spaces of performance can offer opportunities to imagine people and their identities in creative ways. In my own work on theatre (Rogers, 2010), I have analysed how using and acting out scripts can help ethnic minorities challenge stereotypes by creating more complex portrayals of identity. Similarly, Pratt and Kirby (2003) examine how a theatre project in Vancouver represented the experiences of nurses in the context of union struggles over pay. Theatre was a public space where the nurses' stories could be heard but the playful qualities of theatrical performance also created a space of safety where they could discuss their experiences and the politics of their profession. The ambiguity between the real and the fictional often allows artistic performances to 'play around with norms and conventions', enabling situations of resistance to be imagined and literally acted out onstage (Pratt and Kirby, 2003: 19). This political potential of theatre has also been explored by Houston and Pulido (2002) who detail how performances ranging from street theatre to different forms of protest (including banner waving and fasting) were used by workers at the University of California to highlight their unfair pay and working conditions. Here, artistic performances used collective, rather than individual, action to help create social and economic change.

These accounts highlight the creative possibilities of the performing arts and how they can rework socio-economic relationships or change public perceptions. However, there are limits to performance's revolutionary potential, particularly as decisions have to be made over whose stories to tell. Mattingly's (2001) work on a community theatre project in San Diego discusses these issues of 'narrative authority' (2001: 447). She describes how choices were made to represent the neighbourhood of City Heights through a safe image of multicultural harmony, rather than by addressing the real problems that teenagers face. This representation was promoted because the theatre performance was part of a wider strategy of urban regeneration that attempted to improve the image of City Heights in order to make it more attractive for investment. However, residents contested this sanitized image of their neighbourhood and teenagers challenged expectations around why they should be involved. Despite the protests around representation, the agenda of local authorities and the need to stimulate economic growth ultimately determined how the performance was created and the representations it promoted. Artistic performances can therefore reproduce power relationships and such politics play out in particular spaces through live bodily action.

Much of the discussion so far has focused on how the performing arts represent bodies through action and practice. However, not everything about artistic performances is linked so overtly to representation. If we think back to *The Passion* there was also a sense of energy and wonder, of bringing spaces to life by using them differently. Artistic performances are highly imaginative and open up alternative ways of being and interacting. In geography, **non-representational theory** has taken inspiration from the performing arts, particularly dance, to emphasize the poetic qualities of bodily practice. From a non-representational perspective, representation limits the possibilities inherent within practice because it fixes and reads our actions according to established ways of understanding the social world (Lorimer, 2005). Instead, the performing arts demonstrate how embodied practices can be more open ended, improvisatory and innovative, opening up new modalities of experience. By attending to the materiality of the body, by valuing intuition, sensation and feeling, we can begin to explore these alternative forms of knowledge and understanding. The performing arts access and use these physical experiences and although they are more difficult to express and represent, they are suggestive of the different ways we inhabit, create and experience the places in which we live.

Non-representational theory has therefore encouraged geographers to attend to the practices of the performing arts in order to open up the investigation of alternative ways of knowing and communicating more generally. By analyzing embodied action we can gain insights into how our spaces of experience are constructed and interpreted, and this in turn may provide insights into how power operates through space. For instance, Wood and Smith's (2004) work on the emotional experience of music highlights the practical (and spatial) techniques that performers use to influence and create emotional relations between performers and audiences, such as the design of programmes, the creation of sets, rehearsal and improvisation. Through these actions, musical performances can create spaces of intimacy and emotional power that can be used to political effect by mobilizing feelings of belonging, struggle or resistance. The Last Night of the Proms, for instance, uses a selection of music that sounds highly emotive and people are encouraged to wave flags and dress up in order to help mobilize a very specific, and exclusive, construction of British nationality (see Figure 51.1). By attending to the practical techniques and skills of the performing arts, non-representational theory has encouraged examination of how our everyday worlds and spaces are organized to elicit particular reactions or behaviours.

The performing arts therefore illuminate the social organization of practice but they are also a crucible for experimenting with alternative ways of being. Merriman (2010) analyses how the dancer Anna Halprin and her architect husband Lawrence explored environments through dance and play in order to create a more participatory and equitable approach to

Figure 51.1 The Proms. Credit: Getty Images

urban planning. This was a radical way of trying to design living spaces that directly responded to people's needs and embodied experiences. Here, dance helped create place through its animated bodily movements, and dance is often viewed in this lively way owing to its improvisatory and rhythmic qualities. These qualities value experimentation and can be harnessed to produce new sensory worlds. If you are clubbing, for instance, there may be intense moments where you get into a groove and are transported, where how you feel, think and move is transformed in relation to how you were before (see Figure 51.2). There is a sense of possibility and difference, if only for a short time, but this experience may take time to achieve and can rely upon a whole series of other practices or factors (see Malbon, 1999). Geographers have taken inspiration from dance's dynamism to suggest that movement produces and transforms space. Rather than view space as static and abstract, dance – and performance generally – suggests the myriad possibilities through which bodily action varies, creatively producing different configurations of space in the process (McCormack, 2008).

Non-representational understandings of performance have thus encouraged greater attention to creative practice, to embodied ways of being, doing and knowing, and to the imaginative ways in which space and place are made. However, it is worth highlighting that some geographers are unsatisfied with non-representational theory's lack of grounding in specific bodies (Nash, 2000). Non-representational theory suggests that

Figure 51.2 Dancing can transform how people feel. Credit: iStockphotos

embodied feelings and practices do not necessarily relate to identity, but the problem is that bodies often appear in such writings as universal and asocial, creating debates over the constitution of politics. Many critical applications of non-representational theory therefore balance the representational and the non-representational, exploring 'practices, performance, texts, objects and images together' (Nash, 2000: 661). For instance, Cresswell (2006b) highlights how bodily movements and dance steps were regulated in early twentieth century ballroom dancing to produce an appropriate embodiment of Englishness, one that was often highly formal and white. This type of work, like much of that discussed earlier in this section (see also Saldanha (2005) on the racial dynamics of raves) therefore focuses on creative embodied performances and links these to the production and representation of social relations.

SUMMARY

- The performing arts demonstrate that performance is a series of embodied, creative and lively practices.
- Non-representational theory has used this understanding of performance to analyse how we experience space and place through our bodily senses and feelings.
- Some geographers find that non-representational approaches to performance lack an explicit grounding in social politics. Their work therefore emphasizes creative bodily practices, but highlights how the poetic and experimental qualities of performance are linked to the production and reproduction of social relations or identities.

Performance and everyday life

As we have seen, non-representational theory draws upon the performing arts in order to highlight how we know and understand the world through our bodies. As a result, artistic practices and expressions are seen as exaggerations of our everyday actions. This blurring between art and life suggests that first, performance is everywhere in our everyday lives. We are familiar with performance-driven events and their spaces, from demonstrations on the streets, to the grounds of sports matches, to the buzzy atmosphere of a city centre with its restaurants, bars and clubs. We are also creators and audiences of performance, recording all manner of activities on our smart phones and uploading or sharing these with other people. However, this blurring between art and life suggests not only that performance is ubiquitously present or that it is a metaphor for everyday life, as Goffman suggested. Rather, non-representational geographers have suggested that performance is the mechanism for *constituting* everyday spaces and activities; it is *the* primary foundation of our contemporary society (see also McKenzie, 2001). The vocabulary of performance filters into our relationships at work, into how we use and understand technology, and how we conceptualize cultural or social interaction.

For non-representational geographers, performance is all those small actions, skills and ways of going on through which we continually create everyday spaces. As such, the world is always ongoing and never finished, it is 'not a reflection but a continuous composition' (Thrift, 2003b: 2021). Performance is vital and full of life, and as such it is productive. This means that through our actions we create our everyday realities and that the world is not something 'out there' that we are separate from. Non-representational theory views our everyday world as being formed from all sorts of material elements (including not only physical matter, but also imaginative ideas) of which we are only one part. All of these elements are interwoven together, meaning that our actions can change their configuration, potentially creating new social formations or ways of being. The excessive and experimental quality of performance reflects the fact that our actions have variable meanings and their effects cannot always be anticipated. In turn, we may also unexpectedly feel the reciprocal impact of performances occurring elsewhere. This is one reason why non-representational geographers are less concerned with 'traditional' issues of identity, because performance can complicate established social categories. As a result, the social world is seen as a 'weaving of material bodies that can never be cleanly or clearly cleaved into a set of named, known and represented identities' (Anderson and Harrison, 2010b: 13).

Simpson (2011b; 2008) illustrates this intermeshing in his work on how street performance relates to the everyday rhythms of cities. Street performers do not simply come out and perform: they improvize in response to the weather, flows of city life (people, traffic, the crowd, other activities happening around them) and byelaws governing the duration and location of their act. However, even as street performers are affected by the rhythms of city life, they also affect those rhythms in turn. People may slow down as they pass by in the flow, or stop and start to gather into a crowd (see Figure 51.3). Simpson (2011b) describes how street performers momentarily change the course of people's actions and attention, crowds may start to join in with the performance by dancing or singing along, people can laugh together, sympathize with the volunteers being laughed at, and start talking to one another. In this way, the performer intervenes in the city by producing convivial public spaces through emotional and affective responses – spaces that may not otherwise be produced. As a result, all these actions and encounters combined temporarily influence the ongoing flow of everyday life.

As non-representational theory has paid greater attention to this type of everyday encounter between people, there has been a shift away from simply focusing on 'obvious' examples of performance to examining how practices construct spaces. For instance, Jones (2005) describes how his daily commute on a bicycle creates a different experience of the city compared to driving or walking. Cycling provides an embodied and affective understanding of the urban environment, and this knowledge can help develop transport policy in ways that are sensitive to how people use and relate to the cities in which they live. In a different direction, Ash and Gallagher (2011) have attended to the experiences and skills involved in playing videogames, analysing how these technologies impact on our bodies and social relations. The practice of playing, for instance, can begin to reorganize household spaces and relations, particularly when gaming becomes the focal point of the living room. Technologies therefore affect the environments in which they are played and can construct new 'virtual' worlds that become integral to our

Figure 51.3 A street performance. Credit: Getty Images

everyday experiences (see also Ash, 2009). While gaming is a particular type of practice, many embodied actions are similarly skilled and habitual as we routinely do things that constitute or reorder particular environments, often without thinking about it. Examining the conduct of the everyday means attending to these routine performances and valuing our daily actions as creative forces. This focus shares similarities with an **ethnomethodological** approach that analyses how our actions reproduce social worlds. For instance, Laurier and Philo (2006b) examine how different encounters and gestures between people in cafes create friendly, quiet or alienating places. Similarly Laurier *et al.,* (2008) analyse what people do and discuss when they share cars in order to explore how relationships are enacted in these spaces. By attending to such spaces of everyday performance, we can learn how to establish relationships of care and responsibility. Think back to the very beginning, to how strangers watching *The Passion* began to enact their own performances by giving each other bottles of water or food for nothing in return, purely just to help other people. These are the kinds of worlds that performances can help evoke, elucidate and create.

SUMMARY

- Non-representational theory views performance as everyday embodied actions and skills.
- These everyday actions are valued as creative, lively and productive of space and place.
- Performance is thus not only a means of expression, an artistic form or a form of action present in our social life, but the mechanism through which we construct (and reconstruct) the worlds in which we live.

Conclusion

In this chapter we have followed the different ways in which geographers have engaged with ideas around performance. First we saw how performance was viewed as a mode of social behaviour that varied according to context, or a way of creating identity in relation to established social norms and conventions. Second, geographers have examined the performing arts in order to draw attention to the more sensate, embodied and creative ways in which identity can be enacted or reworked. The performing arts also draw attention to the emotional and affective experiences of space and place. These ideas have been applied to everyday life in order to suggest that performance is the means by which we produce and understand the world in which we live. So while often associated with the performing arts, performance is also a range of activities that can express social relationships, produce convivial spaces and atmospheres, or just be the way we do things on a daily basis. These bodily actions provide a mechanism for thinking about how we negotiate power and how we might be able to produce a more ethical and equitable world.

DISCUSSION POINTS

1. What might we mean by performance?
2. What possibilities does an analysis of performance offer geographers? What are the drawbacks? Is performance always progressive?
3. Can you think of other examples of artistic, identity or everyday performances besides those presented here? What spaces might these performances help to create?

FURTHER READING

Gregson, N. and Rose, G. (2000) Taking Butler elsewhere: performativities, spatialities and subjectivities. In: *Environment and Planning D: Society and Space* 18, 433–52.

This article outlines the difference between applications of Goffman and Butler within geography and takes inspiration from work on performance generally in order to think about how performance produces space.

Thrift, N. (2000) Afterwords. In: Environment and Planning D: Society and Space 18, 213–55.

Thrift, N. and Dewsbury, J.D. (2000) Dead geographies – and how to make them live. In: Environment and Planning D: Society and Space 18, 411–32.

These articles describe the non-representational approach to performance and highlight the different intellectual lineages behind the turn to performance within geography.

Smith, S.J. (2000) Performing the (sound)world. In: *Environment and Planning D: Society and Space* 18, 615–37.

Takes a non-representational approach to music, examining the poetics, spaces and politics of performing sound.

Pratt, G. and Johnston, C. (2007) Turning theatre into law, and other spaces of politics. In: *Cultural Geographies* 14, 92–113.

An explicitly geographical analysis of a Marxist theatre piece staged in Vancouver, one that implemented the work of legendary theatre maker Augusto Boal. This article is significant for developing work in the performing arts and moving on from debates around non-representational theory.

There are also writings on practice-led engagements with performance. For instance:

Blunt, A., Bonnerjee, J., Lipman, C., Long, J. and Paynter, F. (2007) My home: text, space and performance. In: *Cultural Geographies* 14, 309–18.

Analyses a performance for how it may illuminate geographical work on the concept of 'home.'

Johnston, C. and Pratt, G. (2010) Nanay (Mother): a testimonial play. In: *Cultural Geographies* 17, 123–33.

Describes how a theatre performance was created to help disseminate academic research to the public, while also attempting to engage audiences in the wider issues behind the research.

NOTES

1. There are many video clips on Youtube that document different parts of *The Passion*, where it is possible to see not only the event, but people recording and sharing it. For instance: **http://www.youtube.com/watch?v=qWWw8QLG44E** and **http://www.youtube.com/watch?v=h5FZJy_2Uzo&feature=related**

SECTION TWO

MOBILITIES

Introduction

Every so often particular ways of thinking about the world come to the fore, generating new approaches and discussion across academic disciplines. It is now common for protagonists to label these as intellectual 'turns' of various sorts. The 'mobilities turn' has been a notable feature of the last decade across both the social sciences and humanities. It posits, to quote Peter Adey from his contribution to this section, 'that the social world is constituted by mobilities of people and objects, flows of information and materials, all entangled together'. It highlights, then, geographies of movement and flow. In Human Geography, it has involved a host of substantive studies: on everyday human forms of mobility (walking, driving, train journeying, using a mobile phone, to name but some); on the kinds of places that seemingly emphasize mobilities and flows of various sorts (for example airports or ships, seas and oceans); and on more obviously contentious sites of mobility (such as national borders, with their policing of mobile people and things). Human Geographers have been at the forefront in developing this approach both conceptually and empirically. The 'mobilities turn' has invigorated and brought into dialogue a number of areas of study within the discipline: including the more established, such as migration, transport geography, tourism geography and communication studies; and somewhat neglected or 'forgotten' topics, ranging from logistics and infrastructures (with their flows of energy, information, water and waste) to the humanistic interest in the choreographies of embodied movement. The three chapters in this section are designed to give you a sense of the key conceptual concerns of work on geographies of mobilities, as well as illustrating something of their substantive range.

In Chapter 52 Peter Adey sets out an argument for understanding mobilities in relation to 'three P's': politics, practices and places. Mobilities are political, he suggests, because they are both shaped by and help to shape power relations. Mobilities are not innate; they are socially constructed and differentiated. Think, for example, of the very different experiences of international mobility of, on the one hand, the so-called 'kinetic elite' (those who traverse the globe in capsular spaces of comparatively easy movement) and, on the other, of refugees or those caught up in the flows of human smuggling and trafficking. Mobilities are uneven. Those differences are apparent too in the practices of mobility. Mobility is not just a matter of abstract flows or forces; it is experienced, felt. Human Geographies of mobility attend to how mobility is done and how it feels. Finally, Pete emphasizes the relations between mobility and place. He argues against an inevitable opposition, where place stands for one set of things, mobility for another, or where mobilities always threaten places, and places constrain mobility. Rather, he suggests that mobilities shape places, and places shape mobilities. Throughout the chapter Pete develops his arguments by interweaving two very different case studies: the representation of mobilities in the film *Up in the Air* and the mobilities that shaped the catastrophic effects of Hurricane Katrina on the city and people of New Orleans.

In their chapter on 'touring mobilities', Claudio Minca and Lauren Wagner focus on one of the prime forms of contemporary mobility, tourism. Experientially, journeying away from home to stay temporarily somewhere 'different' has become one of the main ways that people get to know the world and its geographies. Economically, not only is tourism estimated to generate 9 per cent of total global GDP but it has also transformed various scales of spatial organization, from international transport networks, to regional and urban systems, to local planning and property dynamics. Touring is thus central to both our **imaginative geographies** of the world and to many of its material dynamics. It needs to be taken seriously. Claudio and Lauren pull out three issues in particular. First, they consider how touring came to be a popular form of modern mobility and the kinds of understandings of places and travel that it enacts. We may take being a tourist somewhat for granted, but tourism has a rich and complex cultural history bound up with wider ways of imagining travel across familiar and unfamiliar worlds. Second, they identify how tourism tends to infiltrate more and more kinds of spaces and, in that light, outline some of the key developments in tourism today, including cultural tourism, nature tourism and tourist sites focused on their therapeutic or regenerating qualities. Finally, Claudio and Lauren think about how tourist mobilities are not just of consequence for tourists themselves but, in their words, 'remake spaces and places at multiple scales'. They ask us to think, for example, about the regional economies and environmental transformations produced by coastal developments in the Mediterranean, or the trend towards enclave tourist developments where locals are excluded unless servicing tourists, or the impacts on local residents when cities become renowned tourist destinations. In all these cases, we see again the politics, multiple practices and transformations of place produced by mobilities.

In Chapter 54 Julia Verne introduces debates over 'virtual mobilities', looking at how we can be mobile without actually moving. Contemporary information and communication technologies (ICTs) mean that we don't have to travel physically to somewhere else to be connected to people or activities happening there. Digital flows can predominate. We chat or exchange texts on our mobiles, Skype with friends, family and colleagues far away, and visit virtual sites and environments. Such virtual mobilities are an increasingly mundane facet of everyday life. Julia argues that we need to explore them in two ways. First, we should reflect on their implications for how we think about our geographies. She looks, for example, at what happens to ideas of 'presence' and 'absence' when such virtual mobilities allow one to be both of these things simultaneously. She outlines too how they feed into debates over how to conceptualize place and distance, distinguishing between what are called 'topographic' and 'topological' theorizations of space (these labels are not as scary as they sound!). Second, though, Julia also identifies that this kind of theoretical reflection needs to go hand in hand with detailed empirical research on how virtual mobilities are experienced and practiced. Considering fascinating studies ranging from the political activism of 'smart mobs', to transnational family communications, to children's usage of mobile phones, Julia explores how virtual mobilities shape experiences of social relatedness, emotional connection and personal freedom. Throughout, her call is to get beyond over-hyped distinctions between real geographies and virtual geographies, in order to recognize virtual mobilities as commonplace but textured parts of our lives today.

Of course, while giving a strong sense of key issues these three chapters are not an exhaustive survey of all the work in Human Geography that might associate itself with the 'mobilities turn'. As a wider influence on current debates, it crops up in many places throughout this book. In particular, it is worth highlighting the discussion of human mobility and migration in

Chapter 41 and the account of diasporic social relations and cultures in Chapter 45 as bodies of work that are very directly centred on issues of mobility.

FURTHER READING

Mobilities

Launched in 2006, this is an interdisciplinary journal devoted to research developing the 'mobilities turn'. The journal's statement on its scope sets out how it aims to examine 'both the large-scale movements of people, objects, capital, and information across the world, as well as more local processes of daily transportation, movement through public and private spaces, and the travel of material things in everyday life'. The journal is a great resource for finding the latest and best scholarship in this area.

Adey, P. (2010) *Mobility*. Abingdon: Routledge.

This textbook presents mobility as a key idea in Geography. It addresses mobility as both a ubiquitous facet of the 'world out there' and as a way of engaging with that world analytically.

Cresswell, T. and Merriman, P. (eds.) (2011) *Geographies of Mobilities: Practices, Spaces and Subjects*. Aldershot: Ashgate.

This edited collection rethinks how Geographers have approached mobilities in the light of the 'mobilities turn'. It suggests that focusing on mobilities allows connections to be made between previously distinct sub-disciplinary fields focused on issues such as migration, transport and tourism.

CHAPTER 52

MOBILITIES: POLITICS, PRACTICES, PLACES

Peter Adey

Figure 52.1 Everywhere but no place to go. Credit: Getty Images

Introduction: up in the air

The airport terminal is an exemplar of our increasingly mobile world. It is truly a global place, and it has become a common focus of geographical and social scientific research which focuses on the study of mobilities. A perspective known as the 'new mobilities paradigm' – postured by sociologists such as Mimi Sheller and John Urry (2006) and geographers such as Tim Cresswell (2006a; 2010) and myself (Adey, 2010) – presupposes that the social world is constituted by mobilities of people and objects, flows of information and materials, all entangled together. Contemporary relations and obligations, such as employment, services or structures like the family, all depend upon mobilities in order to constitute and produce them. Somewhere like the airport resembles the flux of this sort of mobile global condition. The airport sees all manner of mobilities for all different sorts of reasons: a package holiday to Florida, a short break away to Paris, a business trip to Vancouver, the start of a distant migration to China, returning to Poland to see one's family for a while. From commuting, to tourism, to migration, the airport encompasses not only different transport modes but a variety of different kinds of mobility, from the more fleeting to the more permanent. Mobile societies give these kinds of movement certain sorts of values and significance. For instance, the late writer J.G. Ballard (1997) famously wrote how the airport expresses a sense of opportunity. In the airport all options are open to us – the world is your oyster, any destination is possible.

Perhaps this is the kind of air-world expressed in Jason Reitman's film *Up in the Air* (2009), based on the novel by Walter Kirn. It is all about smoothness, a quality George Clooney radiates as Ryan Bingham, an executive for Career Transitions Corporation (CTC). Companies outsource their downsizing, redundancies and firing to CTC, distancing difficult decisions and encounters. Personifying the dislocation of economic power from the places and lives of ordinary people, Bingham flies into cities and offices to handle redundancy announcements. He lives his life on the move, inhabiting a world of surfaces and distance. Swipe, click, click, swipe, swivel. Nod, smile, greet and glide. Hovering through the airport terminal, Clooney's performance evokes a sense of a smooth, cushioned transition and practiced habits of mobility. He is the embodiment of his character's effortless and frictionless travel, his easy smile a metaphor for the velvety laid-back passage that sees him reclining in first class lounges and zipping through priority queues and boarding channels. He leaves no friends and most of his connections are superficial. Passage is undoubtedly shared, but the kinds of propinquities which occupy his mobility are formed by the briefest handshake or the passing of a business card.

Bingham is also a motivational speaker. Through the device of asking 'What's in your backpack?' he argues against being weighed down by the ties of relationships, friends, family, hangers-on. No commitments, no discomfort, no worries. A smooth mobility. His new and disillusioned colleague, Natalie, who was attempting to replace his profession with teleconference style technology and auto-prompts, tells him that he is simply 'nothing'. Airy or free, his mobility is productive. He takes pleasure in the motion of being on the move – this is mobility for its own sake. Scrap that rucksack, he argues, empty it almost completely of everything but the things you really need to facilitate your ease of movement.

Tempting perhaps, except this rendering of Bingham's mobility is deeply problematic, mainly because this kind of experience of the airport (and mobility more generally) is as real as the smart and snappy editing and the soft-focus tilts and zooms that profligate director Reitman's portrayal of the airport passage. This is movement without pathos or soul. It does not matter. This is the thin veneer of Bingham's surfaces. The film begins with aerial footage of countless cities and countryside, sound tracked by a 'hippy-jazz' version of Woody Guthrie's *This Land is Your Land* (by Sharon Jones and the Dap-Kings). We don't see the aircraft, not even its wheels as the camera descends on landing. Weightless travel is accentuated here. But as the film immerses us more firmly in the reality of Bingham's life, we find that it is not actually all that fun or fulfilling. As his life, with all its flaws, is gradually peeled away in the film, so too there is a kind of 'touching down' of the movie's vizual style. It becomes less slick, less snappy. Bingham's passage is not nearly as comfortable or easy as it seems at first glance. It is disorientating, tiring, and as we focus in on the bodily practices of mobility, it is telling of how life on the move is deeply unequal. His experience of mobility is highly westernized, middle-class and privileged, as well as gendered

– it is a kind of mobility shot through with a politics we should not take for granted.

Of course, as viewers of *Up in the Air* we may already have a sense of this. In practice, our own experiences of airport mobility are likely to be much less smooth than Bingham's initially seem to be: perhaps chopped through by a struggle with overtired and hungry children, or by delays, queues at check in or passport control, jockeying for position to grab unassigned seats, arrival at/transit via an airport that seems remarkably distant from our desired destination, marked by mixtures of excitement, stress and so on. Nor is Bingham's smile, echoed in the check-in stewardesses and first class lounge attendants, very well mirrored in many people's social experiences of transit. Wider tensions often intrude on and shape spaces of mobility. To take an extreme example, in the UK in the autumn of 2011, a YouTube video called 'My tram experience' was posted, viewed by millions and trended on Twitter. It was mobile phone footage of racist abuse being levelled by a woman at other tram users (on an outer London line) for being black or Asian or Polish and, therefore, 'not British'. At the time of writing the prosecution and trial of the woman in is process, but clearly what we have here is a film clip that sits at the opposite end of the spectrum to *Up in the Air*, illustrating in extremis the kinds of tensions and indeed violence which can occupy the spaces of passenger mobility.

The kind of approaches developed in the 'new mobilities paradigm' can help us unpack the airport and other places of mobility, in order to understand their complex human geographies, their social and cultural relations and their politics. Attending to the geographies of mobilities takes the question of life on the move seriously, opening movement up to critical purchase. The vibrancy of the paradigm has been such that studies within Human Geography have addressed a number of kinds of everyday mobility, in different contexts: including not only airports but cars and auto-mobility (Merriman, 2009), cycling (Spinney, 2006; 2009), public rail transport (Bissell, 2009; 2010), budget airlines and migration (Burrell, 2011) and urban walking (Middleton, 2011), among others.

In this chapter, rather than trying to cover that substantive breadth, the aim is to focus on some of the key questions that have animated these more specific studies. To that end, we will explore mobilities in three main ways. We start by exploring how mobilities intersect with *politics*. Second, the chapter looks at how mobility is *practiced*, or in other words how it is done and experienced. Third, we will explore how mobility relates to *places*, how it might contest or actually compose them. Alongside the airport terminal and the lofty spaces of *Up in the Air*, these three foci will be developed through a very different case study: New Orleans under the flood waters of Hurricane Katrina in 2005. The chapter will elucidate the politics, practices and places of mobility of the city's evacuation and partial resettlement, including in its remit the people who could not move and who were left to fend for themselves in the storm's wake.

SUMMARY

- The 'new mobilities paradigm' argues that social worlds are constituted through flows of people, objects and information.

- We often imagine these flows to be smooth, creating worlds of easy movement and lives that float above the constraints of place. Images of airports and air travel often represent such ideas.
- In reality, mobilities are varied, shaped by and shaping forms of power and privilege.

Politics

There is a politics to mobility. Mobilities are not innate. They are both socially constructed – shaped by social relations of various sorts – and socially constitutive – shaping our social worlds. In a recent essay on the politics of mobility, Tim Cresswell suggests that mobilities 'involve the production and distribution of power' (Cresswell, 2010: 21) and gives three ways in which we might think about this. First, we might consider how mobilities are given meaning in **representations**. What do mobilities mean as they are expressed in popular culture, stories and language? How do we imagine mobility? What qualities do we associate various forms of mobility with? It is just these sorts of questions we were posing of *Up in the Air* earlier. The film, we saw, looks to represent mobility as a smooth space of movement and then to undercut that portrait, showing its partiality and emotional costs.

Second, we might focus on the practices of mobility. Rather than assuming its qualities or making general associations (for example, between mobility and freedom), we can investigate specific acts of mobility, asking questions such as 'How is mobility embodied? How comfortable is it? Is it forced or free?' (Cresswell, 2010: 22).

Third, we can consider how there are multiple mobilities, keyed into forms of social difference and inequality. Looking at the mobilities of an airport, for example, we might ask, does everyone move in the same way within it? 'Who moves furthest? Who moves fastest? Who moves most often?' (Cresswell, 2010: 21). Even in this one sort of place, there are many mobilities differing in speed, rhythm, routes and motive forces. There are the business elites like Bingham, passing through the airport within an itinerary made up of places of transit and temporary residence (airports, aeroplanes, hotels, conference centres and so on). But there are also the mobilities of the workforces who commute into the airport each day, from homes that may well be under the flight paths of the planes that the airport serves. And there are the homeless, who temporarily take shelter in airport terminals, using them as a safer and more comfortable place to 'sleep rough' than the streets. At London's Heathrow airport, for example, a 2008 survey by a homeless outreach organization suggested perhaps 100 homeless people sleeping regularly in its departure lounges, bus station and so on. Here, Bingham's advocacy of living out of a single bag takes on a rather different quality, helping the homeless to blend in with travellers sleeping overnight before flying on, functioning within lives where the airport is a home of sorts, but still leaving one vulnerable to being moved on by security guards or the police if one is identified as homeless, as not the sort of mobile person airports are meant to be for.

The politics of different mobilities are to be found well beyond the airport. On 29th August 2005 the eye of Hurricane Katrina passed to the south east of New Orleans, USA. The associated storm surge caused fifty breaches in the canal levees and by 31st August 80 per cent of the city had been flooded, with damage particularly severe in the lowest lying areas. Estimates suggest that perhaps 80 per cent of the city's population evacuated in advance, but many did not, in particular in the poorest inner urban neighbourhoods, areas with predominantly African American populations. In total, it is estimated that over 1800 people died during Hurricane Katrina across New Orleans and its environs. Television coverage painfully included both immobilized, stranded residents desperately fighting for their lives and appealing for help to leave and a presidential fly-over by Air Force One, with President George W Bush surveying the scene from above.

Writing about Hurricane Katrina, Neil Smith claims that there is 'no such thing as a natural disaster' (2006). The effects of the hurricane and storm surge on New Orleans residents were fundamentally shaped by geographies of social vulnerability (Cutter, 2006). Central to these were both immediate and more long-term geographies of mobility. Let me elaborate. Most immediately, crucial to whether people were able and willing to evacuate or not was their relationship to private car ownership. Those without cars, or without the funds to fill them with gas and to fund their departure (the hurricane hit just before the end of the month, in advance of salary and benefit payments), struggled to leave the city. The differences in car access keyed into wider social differences. There was a remarkable racial polarization; for example, 27 per cent of black people did not have access to a car, contrasted with just 5 per cent of non-Hispanic white people. Evacuation plans centred on the use of private cars, establishing highly effective contra-flow arrangements on freeways to enable traffic to leave. Public transport was not deployed in a systematic way to evacuate people from the city. Many buses were left in situ, stranded by the floods themselves (Figure 52.2).

Tasked with investigating the failures and successes of the emergency management of the disaster, the US House of Representatives Select Bipartisan Committee produced mixed conclusions when it reported in 2008. In terms of the evacuation plans for the city, it found that while there had been a recognition of issues around private car ownership these had

Figure 52.2 School buses sit stranded in a New Orlean's bus depot. Credit: Associated Press

not been acted on. In other words, despite the acknowledgement 'that there are people who cannot evacuate by themselves, the city did not make arrangements for their evacuation' (2008: 113). What seems to have mattered most was the way the identification of vulnerable people was actually fed into contingency plans; how they were deployed in reality, and the expediency of those plans. The report would praise city and emergency leaders for their management of the highway contra-flow system, while that success actually highlighted the main contradiction: namely, that the contra-flow 'only facilitated the evacuation of those who had the means to evacuate the city' (2008: 111), and not those who did not. What is more, although the city's emergency plans had advocated evacuation from the city, in actuality those unable to access private car transport were taken or directed to places of shelter within the city, to sites like the Superdome football stadium, where numbers would swell to around 25,000 people. Made immobile in the structure, the dome became a place of lawlessness and inadequate provision of food, water and sanitation. The most vulnerable of the city would take the full brunt not only of the hurricane and the floods but also of the most intense forms of social disorder in their wake. Differential mobilities constituted the disaster of New Orleans, an event turned catastrophic because of the way people could not leave or were made to leave.

The kinds of mobility available to the city's black, elderly and other vulnerable populations were not just the product of the latest city or state transport policy, but the legacy of a much longer process of transport provision in some places and neglect for it in others. The politics of mobility is rarely a new thing, but often builds on the back of much longer historical trends and of **ideologies** sedimented into plans and processes, concretized into large-scale building projects and infrastructure. Here these included a general privileging of private automobility as the norm. Elsewhere, geographers have shown how particular imaginations about mobility and the mobile individuals, identities and bodies doing all this moving are embedded within transport models, plans and the eventual roads and trains that we use (Imrie, 2000). The mobile person is often cast as universal; a 'one size fits all' policy, assuming an apparently de-sexed, de-gendered, de-classed and de-raced, but often male, white and middle class subject. Presuming this subject as universal can have dire implications for those who do not fit these profiles, categories or typologies of the mobile subject (see also Chapter 43 on dis/ability).

The failures and inequalities with respect to evacuation amplified much longer standing politics of mobility in New Orleans. Cities are fundamentally shaped by their mobilities. New Orleans came in to being as a place of traffic, located at the confluence of three navigable water bodies (the Gulf of Mexico, the Mississippi River and Lake Pontchatrain). Like other cities, it developed through these and other 'flows': of people (from daily transportation to longer term migration and residence processes); of capital and money (investments made in some areas and not in others); and of water (in New Orleans including not only water supply but flood defences and pumping systems). Urban Geographer Steven Graham encourages us to turn our focus to the urban structures and infrastructures that support the flows of mobility. If we do so, we find that these flows are never simply 'natural'. As Graham suggests, 'it is impossible to separate the natural world from the man-made one of cities, infrastructures, and technologies: they are made and function together, as a whole' (2006: 67). The flooding of New Orleans was not a purely natural disaster. The failure to shore up the levees or mobilize the stranded buses in Figure 52.2 cannot simply be naturalized as an inundation by floodwaters, somehow apart from our interventions and actions.

The urban structures and infrastructures that shaped the differential flows of the Hurricane Katrina catastrophe had been long in the making. For example, earlier trends in population displacement would come to bear on the event. Migration from rural impoverishment had seen many African American residents living in the poorest (and lowest lying) neighbourhoods, where public housing had been built in the 1950s and 1960s. These moves into the city were joined with other moves out of it, to its edges. Like other American cities, New Orleans had gone through a process of urban metropolitan development and suburbanization in the second half of the twentieth century. These are processes that Hugh Bartling (2006) places at the heart of the city's calamitous flooding. The middle class **'white flight'** to the suburbs, Bartling shows, was premised upon large-scale infrastructural projects designed to enable commuter and consumer mobilities into and out of the city. As decentralization and dispersal became the watchwords in US urban development, residence in the inner city was cast as a problem, as abnormal. For Bartling, the design of mobility infrastructures like those of New Orleans exhibit a 'cultural logic of racism' built into the provision of transport policy. City centre cores were still accessible but predominantly by automobile. Cars would make use of the newly built highways or expressways, such as the Lake Pontchartrain Causeway. Levee maintenance and protection would give way to middle class marinas and casinos. This amplified existing social and economic inequalities – the poor got poorer and the rich got richer. *Inter alia*, the white got more mobility and the black, poorer and ethnic minorities of the city got a lot less personal and flexible mobility.

The reconstruction of New Orleans has continued to be shaped by differential mobilities. Many poorer black residents have not returned. To widespread criticism, some politicians and commentators speak of Hurricane Katrina as having cleansed the city of problematic communities and neighbourhoods. Grass roots organizations are active and there have been governmental commitments to rebuilding the city, but areas such as the Lower Ninth Ward still lie partially abandoned, with only some rebuilt dwellings and an estimated 25 per cent population return. In a painful echo, coach tours now bring in tourists to look at the dereliction; they may come, but not much of the money from this tourism finds its way into the Lower Ninth and there are frequently awkward encounters with the residents who are there (Rich, 2012). The politics of mobility in New Orleans continue.

SUMMARY

- There is a politics to mobility. Mobilities are implicated in the production and distribution of power.
- One way to see the politics of mobility is to focus on questions of 'difference': how mobilities differ and how these differences key into wider forms of social difference such as class, race, dis/ability and gender.

- In the case of New Orleans and the ongoing impacts of Hurricane Katrina, differential mobilities were crucial both to the more immediate politics of the city's incomplete evacuation and to the longer term development of unequal geographies of vulnerability.

Practices

At one moment in *Up in the Air* we see Clooney's Bingham teaching Natalie Keener the ropes of airport mobility. Bingham's version of racial profiling involves him considering the best passengers to follow through the security queue. Asians, he tells her, are swift and efficient movers through security. They travel light, pack well and are prepared. This is his mantra of mobility. His 'carry-on', in contrast to Natalie's, is perfectly suited to air travel; hers is bulky and meant for the hold. Faced by airport security, he has perfected the rhythm of slipping off his shoes quickly, walking with his ticket facing the security agent through the metal arch detector. His mobility is done through honed bodily movements which are designed to save him time and discomfort. Clearly not all mobility is like this.

Back in 2005 in New Orleans, Abdulrahman Zeitoun (pronounced Zey-toon) had planned to leave the city. The focus of Dave Eggers' book about him, *Zeitoun* (2009), he had stayed behind after his wife and kids had left in late August to the shelter of relatives. The owner of a family property decoration/maintenance business, and hence with responsibilities to job sites and properties which needed securing, Zeitoun was confident his house would be safe as he could easily retreat upstairs to the second floor. Once the extent of flooding became clear, he stays, exploring the neighbourhood in a canoe and helping other stranded people with a neighbour. A muffled female voice from somewhere down the street encourages him to explore further. They reach a house and hear, 'help me'. Zeitoun jumps into the water, holding his breath and swims to the porch where he hits his knee against the masonry and lets out a gasp of pain. The front door is stuck so he breaks in. A heavy elderly woman is stuck, clinging on for dear life. Zeitoun and a friend get help from another boat and they hoist her into a boat. Zeitoun, taking most of the weight, was 'soaked and exhausted'. This was a hard but invigorating mobility. Tired and wet he had found a purpose, helping the stranded victims of the floods. 'He was needed' (Eggers, 2009: 116).

Soon, Zeitoun's mobility within the city would become far more discomforting and painful. As his wife Kathy began to worry about his lack of contact, and family started phoning in from Syria, she heard a news broadcast from President Bush's weekly radio address. Comparing the floods to 9/11 and the 'war on terror', the President's line was firm: 'America is confronting another disaster . . . America will overcome this ordeal, and we will be stronger for it', he said (cited in Eggers, 2009: 198–9). As the National Guard flooded into New Orleans in order to enforce the evacuation and sustain law and order, Zeitoun's mobility came head to head with the city's atmosphere of fear and insecurity.

Both state and private security firms, police and military personnel would reinforce class and racial divides, effectively criminalizing desperate people without food and water. Sarah Kaufman (2006) picks up on a story from the *New York*

Figure 52.3 Cover of *Zeitoun* (2009) by Dave Eggers. Credit: McSweeneys, New York

Times which compared news reports over two images which contained people wading through shoulder high waters laden with foodstuff taken from a supermarket. The first image was a picture of an African American; reports described his actions as 'looting a grocery' store. The second image was of a white couple; stories described their movements after 'finding bread and soda from a local grocery store'. The operative word here is of course 'looting', as opposed to 'finding'. Looting became admonished as the worst kind of behaviour, animalistic even. Victims of the flooding became the problem. However, describing the store as 'local' to the white couple implies some forms of belonging and ownership not afforded to the 'black looter'. Desperate times call for desperate measures and why wouldn't it be right for a couple to take essentials from their corner store? This is but another way that the actions of the mobile body were racialized in New Orleans.

The suspicion directed at (many of) those left in the city took on the overtones of a terrorist emergency, not a surprising development given the way emergency planning had been absorbed into the 'all-hazards' approach of the Department for Homeland Security, a department which would fight disasters as it would 'fight' terrorists (Lakoff, 2006). There was a militarization of the city that would have real effects on mobile practices. For Zeitoun, six armed people met him in the foyer to one of his rented properties, without having had permission to enter. Like at an airport, they asked for his ID, which they did not check. As he tried to go back for a phone number written on a piece of paper on the table, he was grabbed, turned and shoved onto a boat. He was led off the boat at an intersection where two men in bulletproof vests jumped on him. His face was pushed into the ground. A knee placed on his back, hands on his legs. He was handcuffed with plastic cable ties, bundled into a van and driven to a make-shift jail that had been established in the city's main bus 'station-turned-military base' (it would be nicknamed 'Camp Greyhound'). When Zeitoun's friend, Nasser, shifted in his seat, the guards shouted 'Sit still. Go back to your position'. When Nasser resisted, they responded, 'Stop moving!' Subjected to swearing, racial abuse and the accusation of being al Qaeda and a member of the Taliban, Zeitoun is led into a room where he is frisked, questioned, and eventually stripped, rectal cavity-searched and taken into a cage which was locked with a padlock, a temporary jail cell. He was to be kept in Camp Greyhound for three days and in a correctional facility for a further twenty. He was not allowed a mobile phone to contact his family.

Clearly, this incarcerating (im)mobility was a remarkably uncomforting, painful and scary experience for Zeitoun. His mobility as well as his rights were taken away. His very being became an object of intense scrutiny and suspicion. The response was to immobilize him: both physically and through denying the virtual mobility of communication technologies. Many in New Orleans experienced similar or worse, but there are wider implications here too. Experiences of mobility differ and those differences are part of the politics of mobility. In particular, geographies of mobility are often twinned with geographies of security. While some forms of mobility are sanctioned and eased others are subjected to intense surveillance, scrutiny and regulation. Security becomes defined as separating out 'good' and 'bad' flows. A number of geographers have considered how such security is devised and experienced, for example, across national borders and at sites such as airports (Adey, 2004; Adey, 2009; Amoore, 2006; Curry, 2004). Given what we know about forms of surveillance as well as 'profiling' in the aviation industry (including the pre-emptive identification of those in need of tighter control based upon a generic categorization of their identity), Zeitoun's experience is not a million miles apart from those of many others in our mobile world.

SUMMARY

- Human mobilities are exercised and experienced in embodied ways.
- Geographies of mobility are often paired with geographies of security.
- In the case of New Orleans, many who remained after Hurricane Katrina struck found their mobilities taken away by authorities that framed them primarily as security threats. More generally, we should reflect on how mobilities are securitized and on the different embodied experiences of this.

Places

The relations between the geographies of mobility and place have been an important theme within the 'new mobilities paradigm'. Generally, research on mobilities has sought to move beyond opposing the two, instead exploring how places shape mobilities and mobilities shape places. It has also sought to question simplistic perspectives that frame mobility as progressive and place as reactionary or, in reverse, place as social/nurturing and mobility as functional/destructive.

Let's go back to *Up in the Air*. The film's narrative works through relationships between mobility and place. As we saw earlier, the character Bingham's public persona is of someone who has escaped from attachments and the rootedness of place because of his hypermobility (he reaches his 10 million frequent flyer miles near the end of the film). As the film progresses, this escape is less positively presented. Bingham is in fact lonely, seeking out what the film's trailers described as a 'connection', an encounter of relative permanence and significance. The whole film sees him recovering those connections. He finds warmth in his family, reconnects with the place where he grew up (his high school) and starts to love a woman, whom he falls for while on the road. Later in the film and smitten with the woman he has finally connected with, he visits her only to find out that she is married with a family life of her own in her home place – his face says it all. The moment when he stands in front of the massive airport departures screen is not J.G. Ballard's moment of opportunity and freedom, but one of real imprisonment. He has nowhere to go but the airworld.

That airworld is presented as a particular kind of space. Marc Augé (1995) posited the term 'non-place' to describe the spaces of transit that Bingham inhabits: airports, hotels, freeways and so on. Non-places, in Augé's account, are 'spaces where people do not meet, where they communicate only through signs and images, and where interactions are structured by rules not defined by the people in them' (Spinney, 2007: 25). The film's director Jason Reitman described Bingham as 'living hub to hub, with nothing, with nobody' (quoted in the *Wall Street Journal*, 3 September 2009). The notion of the 'non-place' has been an important one in mobility studies, though often as a reference point for critique. The non-place is an ideal type against which real places and experiences of mobility can be compared. As I argued earlier, spaces of mobility are actually imbued with many place-like qualities, including within them a much greater range of meaningful experiences. In *Up in the Air*, the meaningful 'connections' that Bingham seeks do not have to imply fixity. His company soon decides to reject the teleconferencing and software-prompted methods of firing employees in favour of Bingham's more personable approach which necessitates his co-presence with the worker, his wealth of knowledge, experience and skill, and, of course, all his mobility in order to get there. Being on the move, in other words, might be entirely necessary in order to support meaningful connections. As a wedding present to his younger sister and brother-in-law, Bingham gifts them his air miles, allowing them to take a round the world trip for their honeymoon, cementing their commitment with mobility, as 'co-pilots' of their lives.

Work focused on place has also sought to question simplistic oppositions between it and mobility. While being attuned to how such an opposition may be mobilized in representations of places, geographers are used to thinking more critically about notions of place which are articulated as somehow authentic, pure, immobile and marked by stasis. Influentially, Doreen Massey (2005) suggests that our notions of place should become more extroverted, progressive and global, viewing them as comings-together, coagulations of flows (see also Chapters 1 and 45). Politically, Massey advocates a more porous, open sense of place rather than a defensive one where mobilities are seen as a threat to a place's integrity.

These ideas play out in complex ways in the case of New Orleans and Hurricane Katrina. In previous sections of this chapter we have already seen how New Orleans was a place long shaped by uneven geographies of mobility. We have also considered how experiences of the hurricane and flooding were characterized by varied sorts of mobile practices: driving away from the city, paddling canoes around it, being imprisoned in emergency sites of securitization, being forcibly put on to buses and so on. In summary, after Katrina New Orleans can be seen as expressing particularly sharply the tensions 'between the ways in which a city fosters attachment and the ways in which it facilitates mobility' (Steinberg, 2008: 4). Mobilities have both made New Orleans and been a threat to it. How those tensions are reconciled has been central to the ongoing debates over both conduct during the emergency and in the reconstruction of the city and its region.

During the emergency, the dominant emphasis was on how flows of various sorts threatened the region and its places. As Abdulrahman Zeitoun's experience showed, this way of seeing things was not only focused on the floodwaters but also on the people caught up in them. Militaristic policing and practices organized at keeping social order appeared prejudiced towards the mobile and homeless population, disposed to seeing them as potentially violent, insurgent and dangerous. An article in the *Army Times* described the military operation as intending to 'take the city back'. Brig. Gen. Gary Jones, commander of the Louisiana National Guard's Joint Task Force, would articulate this in more concerning terms, suggesting that New Orleans had become someplace else, someplace foreign, other. The 'place is going to look like Little Somalia', he said, 'We're going to go out and take this city back. This will be a combat operation to get this city under control' (Chennelly, 2005). Once the problem was framed as a city losing its spatial order, then it was all too easy for mobile people to become part of the problem, to be a threat to places because they themselves had been cast out-of-place (see also Chapter 17). This became especially clear as some of the displaced population left stranded in the city were turned away from the places of shelter (such as the Superdome and Convention Centre) which were full and advised to travel to adjoining areas. On the Crescent City Connection Bridge, between New Orleans and the city of Gretna on the other side of the Mississippi River, armed police formed a barrier reinforced by warning shots fired over refugees' heads, forcibly turning them away. Seeking to protect their place from becoming 'another New Orleans', the officers saw themselves as safeguarding Gretna from social disorder in the context of failed power, water and food supplies. For Bartling and others, their defensive sense of place helped demonstrate 'the impermeability and intransigence of the "color line"' (Bartling, 2006), as refugees who were mostly black were turned away from a neighbourhood with a majority white population and police force.

The disagreements have remained as profound in considerations of Katrina's aftermath. In the dialogue between Naomi Klein and Neil Smith (2008) following the publication of Klein's book *Shock Doctrine*, Klein would describe the situation in New Orleans as a kind of 'cultural genocide', 'an unmaking of a community . . . with the deepest cultural political roots in this country' (Klein and Smith, 2008: 588). Her claim is that the forced relocation of New Orleans' stranded population across the USA produced an erasure of place. In the short term, once evacuation did occur, people were put on buses and helicopters, often separated from children, siblings, parents and grandparents, let alone friends or neighbours. Locating them became a nightmare of bureaucracy through a tangled web of associations and responsibilities of private and public organizations operating at the local, state and federal levels.

More generally, in the years since 2005, putting people and places back together again has not

Figure 52.4 The swamped Superdome, New Orleans. Credit: National Oceanic and Atmospheric Administration (NOAA)

been a smooth process. Klein argues the focus of reconstruction has been much more on opportunities for private industry to profit – what she calls 'disaster capitalism' – than on the rebuilding of place and community. Many of those forcibly evacuated have not returned to New Orleans. As Tiessen puts it, 'The immobile were made mobile after the storm but in exile seem to have become, again, immobilized' (2008: 117). In the Lower Ninth Ward, the area that suffered the most extreme destruction, about 25 per cent of its residents have returned in order to salvage their homes and begin anew. Those still exiled have articulated how highly they valued the qualities of the community and neighbourhood that they left; prior residents relate how they yearn for the distinctive foods, environment and relations of their home place (Chamlee-Wright and Storr, 2009). However, New Orleans' reconstruction has so far articulated mobility and attachment according to other emphases; vast areas of the city have not yet been rebuilt, and rents have gone up in many areas by over 40 per cent since pre-Katrina. The mobilities required to remake an area like the Lower Ninth have yet to be established.

SUMMARY

- Studying the relations between mobility and place has been an important theme within the 'new mobilities paradigm'.
- The spaces of mobility have been described as 'non-places', but scholarship has also shown that being on the move can be supportive of meaningful connections to people and places.
- Places can be conceived as bounded spaces, threatened by mobilities and/or as confluences, made by mobilities. These competing conceptions of place and mobility were apparent both during the emergency of Hurricane Katrina and in debates over New Orleans' reconstruction.

Conclusion

Perhaps it is because mobility is too self-evident that it had escaped serious analysis until recently. We might overlook it because of its ordinariness. It is the flights high in the air above us, the underground subway line beneath our feet, the journey that brings a visitor or a loved one to our door, the habits and routines we take for granted as we move around places. However, Human Geographers and other social scientists are now opening up multiple mundane mobilities to scrutiny. In this chapter, we focused on more extraordinary geographies of mobility – as represented in the Hollywood film *Up in the Air* and the city of New Orleans after Hurricane Katrina – to throw into sharper relief three key concerns for research in this area. First, there is a politics of mobility, including how various people are differently mobile. Second, mobility does not just happen, it is practised, and there is a need to understand the different sorts of experiences and doings of mobility. Third, mobilities are shaped by distinctive kinds of relationships and reactions to places.

DISCUSSION POINTS

1. In what ways is mobility 'political'?
2. Compare and contrast the mobilities of Bingham and Zeitoun.
3. In what ways is your own everyday life shaped by practices and experiences of mobility?
4. How should we understand the relations between mobility and place?

FURTHER READING

Adey, P. (2010) *Mobility*. Abingdon: Routledge.

This is a textbook that presents mobility as a key idea in geography. It addresses mobility as both a ubiquitous facet of the 'world out there' and as a way of engaging with that world analytically. It provides deeper and wider consideration of the meanings, practices and politics of mobility highlighted in this chapter.

Cresswell, T. (2006) *On the Move: Mobility in the Modern Western World*. New York: Routledge.

One of the first book-length treatments of mobility in Human Geography. Cresswell understands mobility as central to what it is to be human, a fundamental of geographical existence. The first two chapters in the book discuss how to define and approach mobility. The other chapters are case studies that apply this approach in a range of contexts from the workplace, to immigration, to the dance hall. Chapter 9 examines the mobilities at Schiphol Airport, Amsterdam, arguing against its reduction to a 'non-place'. The Epilogue gives an early response to Hurricane Katrina's impact on New Orleans.

Cresswell, T. and Merriman, P. (eds.) (2011) *Geographies of Mobilities: Practices, Spaces and Subjects*. Aldershot: Ashgate.

This edited collection rethinks how geographers have approached mobilities in the light of the 'new mobilities paradigm'. It suggests that focusing on mobilities allows connections to be made between previously distinct sub-disciplinary fields such as migration, transport and tourism. The book is divided into three sections. The first focuses on 'mobile practices': the experience and performance of mobility as something that is done, ranging from walking to flying. The second section examines 'mobile spaces': relatively fixed locales which enable mobility to occur such as roads, bridges and airports. The final section deals with mobile subjects, which include the figures of the tourist, the refugee, the commuter and the migrant.

Merriman, P. (2009) Automobility and the geographies of the car. In: *Geography Compass* 3, 586–99.

Middleton, J. (2011) Walking in the city: the geographies of everyday pedestrian practices. In: *Geography Compass* 5, 90–105.

Spinney, J. (2009) Cycling the city: movement, meaning and method. In: *Geography Compass* 3, 317–35.

These three review articles give overviews of scholarship on everyday mobilities of the car, cycling and walking.

WEBSITE

http://understandingkatrina.ssrc.org/

For more on Hurricane Katrina and New Orleans, see the excellent resource *Understanding Katrina: Perspectives from the Social Sciences,* featured on this website.

CHAPTER 53
TOURING MOBILITIES

Claudio Minca and Lauren Wagner

Introduction

Shortly after 11.00 p.m. on Saturday, 12 October 2002, a series of bomb blasts shook Bali, putting an end to the peace and prosperity the island had so jealously guarded over the years. The first bomb exploded inside Paddy's Bar, followed a few seconds later by the explosion of a vehicle parked in front of the nearby Sari Club, two popular night spots in Kuta Beach, the most famous resort on the island. The bombs in Kuta killed 212 people from 22 countries, including 88 Australians and 35 Indonesians. They also injured 324 people and damaged 418 buildings.

Traumatized by the horror of the carnage, the Balinese reacted at first with an outburst of anger towards the alleged perpetrators who, in their eyes, could only be outsiders. Then came a time of 'introspection' (*mulat sarira*), of intense soul-searching. Even though the attack was the work of outsiders, it led the Balinese to wonder if they brought this curse upon themselves and to ask what they must do to set things right. The bombing was seen as a warning that something must be out of balance in Bali, that all was not well on the island of the gods. Indeed, many regarded the strike as a punishment from the gods, a divine retribution for the sins of Kuta, where drugs and prostitution were allowed to flourish shamelessly. Too busy chasing tourist dollars, the Balinese had disregarded their moral values and neglected their religious duties. Thus, in the words of one distinguished 'cultural observer' (*pengamat budaya*), Balinese psychiatrist and medium Luh Ketut Suryani: 'The destruction happened because the Balinese people have already forgotten their Balineseness'.

(Picard 2009: 99–100)

This chapter considers 'touring' as a prominent kind of mobility; one in which a person, the 'tourist', journeys away from home to somewhere else before returning. The response to the tragic events in Kuta described above speaks to how contemporary touring is at the intersection of many facets of the late modern world: our cultural mappings; our political and economic relations and differences; our practices and our moral framings of them; and our emotional expectations. Tourism is often presented as a deliberately **liminal**, spatio-temporal condition in which the need for escape from the everyday – the 'need for a break' – is perceived as something that everyone should be able to access. Yet 'taking a break' often means adopting a somewhat paradoxical subjective posture as a 'tourist' – becoming temporarily 'out of place' as the consumer of the cultural and environmental resources of a distant place. Tourism, at the same time, is also the condition in which our desires for adventure, escape, radical otherness and 'cultural enhancement' should be satisfied, or, at least, this is what the industry often promises. This transformation of an individual into a 'tourist' is not neutral: it etches coordinates and relationships in shifting world geographies and shapes the way we understand them. Contemporary touring, of which tourism is a fundamental element, is often perceived as a practice in which consumption patterns *and* understandings of late modernity, sense of place *and* geographical imaginations tend to merge and produce an ever-changing culture of mobility. This tourist mobility is based on new subjectivities and identities, and produces new feelings of *depaysement* and (lack of) belonging.

Academic work exploring contemporary forms of touring is normally presented under the interdisciplinary heading of *tourism studies* (see journals like *The Annals of Tourism Research*, *Tourist Geographies* and *Tourist Studies* for a sense of this field). Tourism studies is a relatively new field of investigation, because until recently practices of touring were not considered an 'object' worthy of academic analysis (Tribe 2005; 2006). One of the first researchers to 'take touring seriously' was American anthropologist Dean MacCannell in his book *The Tourist* (1976), where he reflects, from a Marxist perspective, on the relationship between authenticity, subjectivity and the tourist experience. That intervention, along with other shifts in anthropological perspectives on the dynamics of 'hosts and guests' (Smith, 1977), sparked an entirely new interdisciplinary field. In contrast to tourism management studies, mainly focused on tourist marketing, development and issues of logistics and hospitality, tourism studies is a meeting ground for contributions coming from sociology, anthropology, politics, history, literary studies, cultural studies and, of course, Human Geography. As part of this academic discussion, geographers are studying, among many other things, the spatial implications of the relationship between tourism and culture, politics, the environment, and concepts like identity, authenticity, subjectivity, place, landscape, community and power (Franklin, 2003; Minca and Oakes, 2006; 2011). This geographical branch considers the workings and development of tourism as an 'engine' of spatial formations or **spatialities** that affect our everyday lives, as well as considering how tourism shapes the ways in which we think of the world and of 'our own place in the world'.

The relationship between 'tourism and space' may be identified with three major manifestations which correspond, by and large, to three major areas of investigation for geographers interested in tourism. First, historically, the need and desire to 'tour' has been the result of individual and collective 'cognitive maps' of the familiar and unfamiliar world. These cognitive maps, part of wider **imaginative geographies**, allow potential travellers to imagine 'discoverable' places able to satisfy their expectations in terms of the experience of being away from home.

Second, touring is by definition a form of mobility, involving departure from one place called home in order to visit another place away from home. Motivations for this mobility are linked not only to perceptions of places but also to fascinations with travel. Geographers therefore approach tourism in relation to both destinations and forms of journeying.

Third, tourism produces actual spaces, sometimes with revolutionary consequences for the places involved and for those who inhabit them. Geographical tourism studies are thus often concerned with the spatial dimension of these transformations, together with the models that tourist planning explicitly and implicitly tends to implement in the development of a destination or a region. All of these aspects of study engage the fluxes, trajectories, density and patterns of distribution of tourist 'presence'. They place the infrastructural, environmental, social and cultural impacts of tourism at the core of geographical investigations of contemporary touring.

SUMMARY

- 'Touring' is a major of form of mobility in the modern world.
- 'Tourism studies' has emerged as an interdisciplinary field that treats tourist mobilities as a serious object of study. It considers what touring means both for tourists and for others, such as those living and working in tourist destinations.
- Human Geographers investigate both the imaginative geographies associated with tourism and the material spatiality of the transformations that it enacts.
- Human Geographers analyse tourism as about both mobility (most obviously, forms of travel and touring) and place (most obviously, the destinations through which touring is done).

Tourist imaginations

The ways in which we travel, and their genealogies, are an important element in the geographical understanding of tourism. The history of modern traveling shows how the language and the perspectives of contemporary tourism reflect in many ways models and figures that took shape in the nineteenth century European cultural and political context. While other forms of travel, such as religious pilgrimage, predate this period, the notion of touring as a leisure activity emerged more concretely through developments in this century.

The Grand Tour of Europe became a common practice among the British elite in the eighteenth and nineteenth centuries, to educate young men in classical art and literature by visiting specific sites construed as a source of knowledge (Withey, 1997). This construal of places as historical sites rather than lived environments was not limited to those with the means to travel around Europe; it was a pattern of perception matched by the growing popularity of Great Exhibitions that attracted millions of curious and admiring visitors from 1851 onwards (Morton, 2000). This 'tour-istic' mode of viewing and observing was also related to the figure of the *flâneur* – a late nineteenth century notion of a (male) external observer, a wandering subject who walked the city in order to savour its inner spirit, simply for the pure pleasure of doing so with no other declared purpose or objective (Pile, 1996).

Much like Grand Tour travellers sought the monuments of Italy while tolerating the discomfort of actually being there, the conception and layout of exhibition pavilions structured an approach towards 'other cultures' and places as essences on display, to be gazed at by decontextualized 'observers', and to be read as part of a larger patchwork composing the 'cultural geography of the world'. This 'world-as-exhibition' (Gregory, 1994; Mitchell 1991) developed into an intellectual framework and a discursive formation that was incorporated into the mass touring practices of the nineteenth and twentieth centuries for viewing cities, landscapes, countries and peoples. It is still visible in the contemporary tourist promotion of faraway and not so faraway destinations as distillable into 'must-see' essences purported to represent the complexity of a 'culture' (Kirshenblatt-Gimblett, 1998).

As practices of touring stretched further afield, notable 'tourists' of the nineteenth century were also 'explorers', whose travels contributed to the development of geographical thought. One key example from the early nineteenth century, who pushed the boundaries of the traveller-explorer-geographer, is Alexander von Humboldt:

> By combining a love of travel and a flair for danger with his passion for discovery, Humboldt became the very prototype of the scientific adventurer ... [his] expedition through Latin America was one of the great journeys of history, and it was his spirit of adventure as much as his love of science that made the young Prussian such a compelling figure among his contemporaries. His exploits, reported in the American and European newspapers (based on letters sent to friends and family en route) enthralled readers in the same way that the adventures of Robert Scott, David Livingstone, and Charles Lindbergh would captivate future generations.
>
> (Helferich, 2004: xviii)

For Humboldt – often considered the quintessential geographic explorer through his efforts to conjugate science and poetics, the 'impressions' provoked by nature and the geometry of morphology (Livingstone and Withers, 1999) – travelling was at once a methodology for scientific advancement and the pleasure of discovery in unknown places. Following his example, others who embarked on touring the world engaged in a sort of unaware political practice, attached on the one hand to scientific enterprise and the discovery of new worlds, and on the other to a new aesthetics of a European Self, seeking fulfilment *en route* (see Driver, 2001). The age of state-sponsored explorations that accompanies and follows Humboldt's leisure/scientific travels helped consolidate a culture of touring that would soon assume new tonalities in an expanding geography of empire (Mitchell, 1991; Pratt, 1992) (see also Chapter 33).

Figure 53.1 'Off the beaten track'

The figure of the explorer, and later the anthropologist, were raised to an almost mythical rank, as European travellers and scientists who acted as a formidable instrument of legitimacy for colonial domination by endowing conquest with a scientific credibility (Asad, 1973; Cooper and Stoler, 1997). Generally, exploration illustrates the dense connections between travel and geographical knowledge. More specifically, it speaks to how our ideas about geography shape our touring practices. Today, tourist brochures and travel guidebooks echo paths of 'exploration' by promoting the promise of authentic experiences of the cultural Other and 'real' knowledge of faraway places 'off the beaten path'.

Touring and exploration are no longer purely elite pursuits. As European nation-states consolidated over the latter nineteenth and early twentieth centuries, touring practices progressively extended to all social classes. New regulations on the labour market, implemented especially from the 1920s onwards, enabled leisure time. As part of this leisure, an emerging touring culture, strongly tainted by colonial, nationalist and bourgeois influences, encouraged 'pilgrimage' to the celebrated monumental spots of the nation. Visiting natural, cultural and historical wonders through the narrative of 'the nation' became a powerful exercise in the confirmation and the reproduction of a set of political and historical geographies that governments were keen to identify and support. Domestic tourism, from the 1920s onwards, represented not only a phenomenon concerning increasing portions of populations – the Fascist regime in Italy, for example, was very supportive of all forms of vacationing that would enhance the national sentiment (Michaud 2004) – but was explicitly transformed into a political project. This was often driven by nationalistic ideologies and aimed at a selective recognition of some key symbolic sites, while excluding or marginalizing others.

The persuasive language employed by the tourist industry today reveals how these models of the world-as-exhibition, the discovery of new places and knowledge through exploration, and the nation as constituted in symbolic sites are still in place and successful. Indeed, they form a taken-for-granted language through which to describe the experience of touring (Minca and Oakes, 2006; 2011). A world to see; cultures and places to be encountered; explorations to be done; discoveries to be made; iconic and historical sites to be visited: these remain powerful elements of tourist mobilities today.

SUMMARY

- The desire to tour is shaped by our imaginative geographies of the familiar and unfamiliar world.
- The language and perspectives of contemporary tourism have been forged by historical cultures of travel, exhibition, 'flanerie' and exploration.
- Touring emerged historically as a way of seeing both the wider world and its other peoples and places, and the nation to which one 'belongs'. These roles continue today.

Touring today

In *Tourism and Development in Tropical Islands*, Stefan Gössling gathers a cohort of papers discussing the potentially beneficial and potentially devastating economic and environmental impacts of tourism on some of the most sought-after destinations – namely, tropical islands around the world. Many of these islands have come to rely on tourism as a principle source of income, as their importance in other global economic patterns has declined. Ironically, inasmuch as such exotic islands are sought after because of their remoteness and isolation, their tourism economy causes disproportionate environmental destruction for that very reason:

> . . . during a typical vacation in the Seychelles ... more than 1.8 [hectares] of biologically productive land is appropriated. This can be compared to roughly two hectares of land annually available to each human being for all-encompassing living space. The major part of the tourist footprint (more than 90 per cent) resulted from transportation to the destination, rendering prominent both the environmental importance of air travel and the fact that tourism to tropical islands contributes over-proportionally to global warming. These are insights of importance given the chronic dependence of islands on aviation and their susceptibility to global environmental change, including global warming and sea level rise.
>
> (Gössling, 2003: 8)

This telling example introduces another important topic concerning contemporary touring: the relative distances marked out by changing patterns of transportation networks and accessibility to tourist destinations. The tourist frontier is endlessly infiltrating new spaces, selectively incorporating or marginalizing towns, cities, regions and countries on the basis of ever-changing political, cultural, logistical and climatic elements, often with little regard for the economic or environmental impacts left behind. These shifts in spaces and times of tourism reflect a political economy of mobilities, in which the intangible but desirable experiences offered by tourism keep moving to new locations in new forms, motivating new pathways of tourist mobilities to discover ever newer destinations.

If the roots of modern global tourism may be found in those nineteenth century models described earlier, contemporary tours are equally marked by more recent grand cultural trends, related to practices of 'being modern' that create a propensity towards non-essential forms of consumption like tourism (Flusty, 2011). These regimes have produced a literal revolution in how work and free time are regulated, creating a global political economy that includes tourism as a common form of leisure practice (Urry, 1990). 'Being a tourist' is a practice which more and more people are ready to learn, as demonstrated by the growing number of international travellers, and the emergence of the tourist industry as one of the most important economic sectors globally. The World Travel and Tourism Council, an industry body that promotes the tourism industry globally, published research by the consultancy Oxford Economics that travel and tourism's global economic contribution in 2011 was 2.8 per cent of global GDP directly, 9.1 per cent in total (taking into account investments in supply chains, wider consumer spending, etc.) (WTTC, 2011: 14). This mass-produced practice of tourism instigates collective movement – tourist mobilities – that blaze paths for individuals to travel to, and most importantly *consume,* new destinations.

The emergence of a mass tourism industry has been paralleled by more specific developments and differentiations related to these dynamics of destination. We will highlight three recent evolutions, focused on sites of 'culture', 'nature' and 'self-regeneration'. Cultural tourism signals a growing interest for touring 'culture', in all its forms. It is, by and large, marked by two different, but usually coexistent, ways of understanding touring and its 'objects'. On the one hand there is a 'patrimonialist' approach, which places particular emphasis on museums, monuments and all the forms of material cultural capital that can be clearly mapped out and catalogued (Coleman and Crang, 2002). On the other hand, there is also a growing propensity towards the observation – sometimes even the 'invasion' – of communities ('cultures') that live in more or less remote places, including indigenous cultures (Jacobs, 1996) and tribal peoples (Cohen, 1989). Both of these modes of understanding culture evoke a nostalgia for the irrevocable loss of the past, but often this is an imagined past portrayed by present-day interpreters (van den Berghe, 1994).

The elements included or excluded by these historical readings are in fact often produced by actors in the global financial and political machine of the heritage industry (Graham *et al.*, 2000; Winter *et al.*, 2008). The politics of who gets to produce culture for touristic consumption, and to what ends, is an important area of study in that context (see also Chapter 34 for more on the politics of heritage).

A similar sentiment of loss (real or potential) marks many of the tourism options somehow linked to nature and its protection (Holden, 2000; Mowforth and Munt, 2003). From the most conservationalist products, branded as ecotourism or sustainable tourism, to the more banal forms like hiking or literary walks in specific landscapes, these tours are often tainted by a vein of nostalgia for a lost or endangered nature. The perceived disappearance of 'the natural' from our everyday lives is one of the reasons behind a growing interest towards all possible forms of touring a temporary but meaningful 'return to nature' (Ingold, 2000).

The 'regenerating' holiday break is a related form, emphasizing the therapeutic role of natural environments for the tourist. This is a mode of vacationing associated with the pleasure (or even need) of fun and rest, often in highly specialized and protected environments (Pons *et al.,* 2009). The beach is the most significant example, with its long and fascinating history of a space dedicated to healthfulness, relaxation and leisure (Urbain, 1994) (Figure 53.2). Tourist villages, mega-hotels and marinas have populated the coast of many tourist 'basins' around the world – with the Mediterranean as a leading example. Earlier, we referenced the impacts of beach tourism on tropical islands. The movement to places of rest also impacts, of course, many mountain and rural areas, as green tourism in the countryside is currently the new trend for European travellers seeking regeneration. The effects in these landscapes may be somewhat less spectacularly visible, but not necessarily less profound and revolutionary for both the natural environment and the local communities that are directly and indirectly involved. Recently, Tom O'Dell has written about the spa industry (O'Dell, 2010). Focusing on Sweden, he shows how long-standing relations between touring and concerns for health and well-being are being recast in the twenty-first century as a booming spa industry fashions places to escape the stresses of everyday life, to be pampered, to relate sensuously with one's environment, and thereby to get back in touch with oneself.

As a result of these expanding political economies of tourist mobilities, coastal and mountain regions, cities, rural areas and villages, forests, mangroves – even factories

Figure 53.2 Moroccans now living in Europe flock to beaches near their Moroccan hometowns during the summer holidays. Their seasonal presence puts pressure on local infrastructure and natural resources. Credit: Wagner (2011)

and laboratories – are visited by an increasing number of travellers armed with cameras and curiosity for whatever may appear 'different' from the familiar. Touring has infiltrated and impacted places well beyond the beach. For instance, 'dark tourism' is a label applied to the touring of sites of death, disaster and atrocity (Lennon and Foley, 2000); 'slum tourism' is the growing industry of touring neighbourhoods of urban informal housing in cities of the Global South (Dyson, 2012; Frenzel and Koens, 2012). The motivation to choose these destinations is not random; it interlocks with geographies of **cultural capital** and systems of mobilities that continue to make new destinations accessible for the market force that wishes to reach them. A common concern is with finding some quality absent from other parts of everyday life, travelling in order to find, for example, culture, nature and indeed oneself. Tourism is a source of **affect** and emotions for many. The places that we tour become part of the rhythm of our social life (Edensor, 2011) and are therefore invested with expectations and meaning that are well beyond simple consumption of a service. The locales where we spend our holidays are often important points of reference for the ways in which we conceive 'our place in the world', and are therefore necessarily implicated in the production of meaning as much as our 'normal' life.

SUMMARY

- Contemporary tourism illustrates a political economy of mobility, in which new destinations are marketed in new forms in order to appeal to various sorts of consumer desires and tastes.
- Recent evolutions include cultural tourism, nature tourism and touring for self-regeneration. In various ways these forms of mobility respond to sentiments of loss – searching out culture, nature and oneself.
- The places that we tour form important points of reference for how we conceive 'our place in the world'.

The spaces of tourism

Touring is marketed as transforming the tourist, renewing them in some way. It is, however, also profoundly transformative of the spaces through which it produces its mobilities. For example, the now commonplace summer migrations of millions of continental Europeans towards the Mediterranean has contributed dramatically to transforming the geography of this basin and of its coasts, while making some of its regions entirely dependent on tourism for their economic survival. The frontier of international tourism tends to reach out and incorporate the most diverse spaces, as a result of a persistent need for novelty and on-going 'improvements' to hospitality, making new sites more accessible to distant tourists. At present, we are witnessing the consolidation of a set of touring geographies that contribute in a decisive way to declining and growing cities, the re-arrangement of functional spatial hierarchies, processes of regionalization and de-regionalization, and changing political, natural and cultural frontiers through the capacity to shift mass mobilities in new directions.

When tourism is at the heart of radical transformations of places and spaces, always already invested with meaning, the objectives and motivations that guide such transformations are necessarily contentious. Tourism, as a powerful engine of territorial organization, must therefore be studied with a critical eye in order to identify the difference between how the spatial 'product' is presented and its effective territorial impact.

Although tourism is a key element in the planning of many cities and regions around the world, there are endless cases of real or potential conflict between the tourist use of a specific area and the concurrent existence of industrial, agricultural or other activities (Gössling and Hall, 2006). Too often, in fact, forms of aggressive speculation associated with the development of tourism have colonized entire regions while producing irreversible forms of pollution and overcrowding – as is the case of the semi-anarchic developments on most of the Spanish coastline, including the phenomenon of intense cementification of the waterfront (also known as 'Marbellization', from the city of Marbella; see Lozato-Giotart, 1990). In many parts of the world, as slices of coast are converted into expensive tourist beaches, the surrounding real estate market revolutionizes the social fabric of small towns, often marginalizing the former residents to the

CASE STUDY

The 'Essaouira Spirit'

Figure 53.3 A street in Essaouira, Morocco. Credit: Claudio Minca

Essaouira, Morocco, is not a name one would necessarily recognize, but like many other places it is enmeshed in global tourism flows as a 'cultural product'. From its origins as a port town, controlled by a Mauretanian king and several Moroccan sultans, then becoming a Portuguese trading port, and later part of the French Protectorate, it was mostly forgotten when steamships overtook sailboats in maritime commerce, its port being inadequate. Essaouira's revival began through its 'discovery' by alternative tourists, who valued it as a source for artistic handicrafts, for its picturesque old city (*medina*) and for its breezy, temperate climate. Hippie crowds in the 1960s led to temporary residence by Cat Stevens and Jimi Hendrix, among other artists and musicians. In the 1980s, surfers started coming for the constant winds. In 2001, Essaouira was named a UNESCO World Heritage site.

Contemporary production of the city aims to embody the 'Essaouira Spirit'. Whether through handicrafts or the annual Gnaoua Festival that draws a worldwide audience to listen to Gnaoua and other world music, the appeal is directed to an 'alternative' tourist. This appeal

Figure 53.4 Foreign-owned residences in Essaouira. Credit: Bauer *et al.* (2006)

has a global reach, as more and more visitors come to the city and leave with a memory of the Essaouira Spirit. Along with this growth in popularity, the population of the city, particularly in the *medina*, has shifted. Bauer *et al.* (2006: 33) counted 298 non-Moroccan property owners in the tiny old city, whose homes vary from second homes, to retirement homes, to guesthouses that they operate as resident expats. To some extent, these primarily European homeowners simply mark a new phase of the historical interaction between Morocco and Europe. The importance, however, of their impact on Essaouira's identity and economic livelihood is becoming clear as local resident Moroccans are forced to move further out of the city as house prices within it rise beyond their means. Yet this importance also has global reach, as more and more visitors come to the city and leave with a memory of the Essaouira Spirit.

peripheries or colonizing the area with the greatest tourist potential. Elsewhere, forests are sometimes destroyed, some other times 'rescued' and protected – both processes in the name of 'tourism'. Chunks of mountains are converted into ski slopes; pieces of deserts into golf courses. These dynamics of destruction and conservation are not exclusively applied to 'natural' landscapes: new historical districts are gentrified and depopulated, then repopulated by new tourism related activities (souvenir shops, bars, etc.). Once popular destinations are rejuvenated or left to decline and disappear from the map of tourism as they fall out of the global flows that sustain it. New destinations, attractions, specialized enclaves, typologies, and itineraries appear and become popular in a glance and enter these ever-changing geographies of touring.

In the light of this transformative power of tourism, geographers have developed a number of lines of critical scholarship. We will outline four here. First, there have been efforts to theorize tourist destinations as if they were products with a life cycle. The 'Tourism Area Life Cycle' model (Butler, 1980) emphasized the factors that allow for a destination to attract the first tourists and subsequently 'take off', until a stage of consolidation is reached or, sometimes, stagnation or even decline. A vast geographical literature has adopted this model to study small and large destinations, and in more than one case entire countries, especially where tourism represents the main source of revenue. One line of development undertaken by geographers has been an emphasis on how tourist places, as products, may have to reinvent themselves or may be reinvented by new forms of tourism. Jon Goss (1993b) considered the tourism marketing of Hawai'i in the rest of the USA, showing how its imagination as a 'different' place was adjusted to mesh with the tastes and desires of varying target markets. John Connell (1993) traces out the history of touristic interest in Kuta, Bali, showing how this one place has been moulded by the overlapping of different 'resort cycles', from elite Western travellers in search of the exotic East, to world travellers seeking out alternative spaces of authenticity, to youthful tourists seeking places of action and hedonistic fun.

Second, geographers often focus their attention on the infrastructural dimension of the spatial organization of tourist facilities (Lozato-Giotart, 1987; Lozato-Giotart and Balfet, 2004). It is common knowledge that the public services (hospitals, police, etc.) provided by a tourist destination should be proportionate to the peaks of demand. For tourist destinations characterized by a strong seasonality this represents a huge problem in both logistical and financial terms. Can a small tourist town pay for a set of infrastructures (roads, hospitals, sewage system, etc.) that are adequate to respond to the needs of a population ten times higher, but present for only a month per year or so? In addition, sometimes the existence of an intangible critical maximum threshold of crowding, traffic, noise or other pollution can be crucial in the appreciation of a vacation or of a visit to a tourist site. Ignoring these factors could be seriously damaging for a tourist destination – the noisy enjoyment of one jet skier can spoil the relaxation of hundreds of beach relaxation seekers. All of these features – from hospital beds to typologies of tourist – may become key aspects in the spatial success or failure of a destination.

Third, inasmuch as tourism operates spatially at different scales, geographers are equally interested in how the development or the decline of tourism affects regions and regionalization (Miossec, 1976). Coastal developments, for example, may produce the virtual desertification of interior areas inland. The workings of tourist zoning or the urban hierarchies generated or modified by tourism, can create *ex nihilo* ('from nothing') many specialized spaces (think of Cancun in Mexico, for example, founded as a tourist city). Sometimes these are extraterritorial enclaves, as is the case with tourist villages (Club Med) or theme parks (Disney World and the like). Tourist enclaves are indeed proliferating everywhere, especially in lower income countries, where holidaymakers are often shielded from the 'perils' of the hosting territories (poverty, crime, disease, terrorism, etc.) (Minca, 2009). The by-products of these preoccupations are the increased securitization of many tourist spaces and restricted circulation for residents in areas that have been colonized by the tourist machinery (a typical case is Varadero, Cuba, where the only 'local people' admitted to many areas are those working with the tourists).

This brings us to a final area of interest for geographers, as they study the ways in which all these elements are related to the presence and the life of residents in these destinations, both from a social and cultural point of view and in terms of spatial practices (Meethan, 2001; Sheller and Urry, 2004). Quite often residents move to different parts of the city, if not to another city altogether, in order to avoid or reduce their contact with tourists. In other cases, they are literally forced out by the inflation produced locally by the tourist economy and the marginalization of services and shops that are essential for their everyday life. Taking Venice as an example, residents often avoid some itineraries notoriously jammed by wandering visitors. Everyday shops and facilities, like bakeries and cinemas, are disappearing from most parts of the city. Residents are thus induced to move out by lack of space or resources, which renders their everyday activity difficult, sometimes even impossible. More generally, geographers are keen to understand not only how tourists see and engage with the places they tour, but also how 'hosts' view and critically respond to tourism development and tourists (Brickell, 2012).

With this cursory review of how geographers deal with the 'hard' spatial arrangements of tourism, we have tried to illustrate how both the material geographies and geographical imaginations of tourism are constantly changing and re-articulating. Geographers try to penetrate the deeper logic of these changes in order to better understand them. In so doing they seek to help the industry and the authorities – too often keen to open their territories to speculative moves of the tourist kind – to appreciate the complexity of a phenomenon in which spaces and places, in their specificity and difference, are key resources for the mobilization of millions of tour seekers.

SUMMARY

- Tourism is a powerful engine of territorial organization at multiple scales: globally, regionally and locally. Tourist mobilities remake spaces and places.
- In studying the power of tourism, geographers have contrasted how the spatial 'product' is presented and its effective territorial impacts, paying attention to the impacts of tourism upon the lives of residents.

Conclusion: taking tourism seriously

Tourism, not only international tourism but also domestic, has become a political and economic force, described as a potential remedy to both poverty and marginalization for the deprived population of some countries and regions. The now vast and rich literature on the topic, however, shows that the hoped-for improvements in the living conditions of those very populations are rarely obtained, while many times the results are quite the opposite, with an increased pressure on communities, regions, governments and international agencies to manage its impacts (Butcher, 2007; Timothy and Nyaupane, 2009). Tourism, at the same time, is often presented as a vehicle to recover, or possibly re-invent, traditions and identity narratives, as well as to empower individual and collective subjectivities to 're-connect' with cultural and social histories presented as complete and nostalgically whole. Tourism, in its real and imagined geographies, seems to offer for many a space of temporary 'order' for subject formation and meaning. From the cultural tours of remote communities to the plastic worlds of Disneyland; from the forests translated into 'reserves' to the Mediterranean beaches; from the factories converted into museums to the literary landscapes; what we witness is a spectacle of thousands, sometimes millions of travelling people, who dream, spend, consume, socialize, take pictures, experience 'difference', learn, have fun, relax and gaze. In so doing, they activate a vast political economy of touring which creates enormous calculated and incalculable impacts on spatial formations and practices.

In western societies today, tourism seems to be virtually everywhere and everyone seems to be, in one way or another, a tourist. The geographies of contemporary touring, perhaps the most extraordinary global cultural phenomenon of our age, discover and colonize ever more new territories, producing novel images, experiences, practices, spaces and places. Tourist mobilities should be historically contextualized; there would not be tourism today without the geographical imaginations and forms of touring produced during the early stages of modernity. But at the same time, we need to recognize the contemporary significance of tourism as truly remarkable. The experience of late modernity would be very different without the penetrating presence of the powerful spatial machine of touring. We need to take tourism seriously.

DISCUSSION POINTS

1. This chapter has argued that we need to take tourism seriously. Do you agree, and what might this entail?
2. In what ways have the language and perspectives of contemporary tourism been forged by historical cultures of travel, exhibition, flanerie and exploration?
3. Why do people become tourists? What sorts of geographical imaginations animate contemporary forms of tourism that seek out culture, nature and regeneration?
4. 'Tourism is a powerful engine of territorial organization at multiple scales'. Discuss, using relevant examples.
5. 'Tourism shows how mobility happens in places, and places are produced through mobilities'. Discuss.

FURTHER READING

Hannam, K. (2008) Tourism geographies, tourist studies and the turn towards mobilities. In: *Geography Compass* 2, 127–39.

This review article surveys the field in the context of three (comparatively) recently launched academic journals: *Tourism Geographies* (launched in 1999 and emphasizing the fundamentally geographical aspects to tourism), *Tourist Studies* (launched in 2001 and providing a platform for developing critical perspectives on tourism as a social phenomenon) and *Mobilities* (launched in 2006 and focused on bringing together a range of research on mobilities). It provides a valuable overview in its own right but you should also treat these journals as important resources for your own further reading in this area.

Minca, C. and Oakes, T. (eds.) (2006) *Travels in Paradox: Remapping Tourism*. Lanham: Rowman & Littlefield.

This edited collection explores the relations between mobility and place that are created by tourism. Its approach is to see how travel happens in places rather than between them. Rather than opposing contemporary touristic mobility with a past rootedness in place, its contributors explore how tourists inflect places and places inflect tourists.

Sheller, M. and Urry, J. (eds.) (2004) *Tourism Mobilities: Places to Play, Places in Play*. London: Routledge.

This edited collection looks to engage the interdisciplinary field of tourism studies with the emergence of what has been labelled a 'new mobilities paradigm'. A range of case studies – focused on places ranging from the Caribbean to Dubai, Machu Picchu to the Taj Mahal, Rio de Janeiro's favelas to Harrogate, England – consider how tourism is implicated in multiple flows (of people, ideas, money, information, images and so on) and in the production of places (of paradise, heritage, play/pleasure and un/reality).

Williams, S. (2009) *Tourism Geography. A New Synthesis* (2nd edn.). Abingdon: Routledge.

Updated and extended from a first edition published in 1998, this textbook provides a comprehensive and accessible overview of the field of tourism geography.

CHAPTER 54
VIRTUAL MOBILITIES

Julia Verne

Introduction: going somewhere without moving

Imagine someone sitting in her office staring at the screen of her computer. Outside, the sun is shining and it's a beautiful day. While the text on the screen starts to blur, she remembers the last day she spent at the beach, wishing she could go there again this afternoon. But she's also very much aware of the fact that she had better continue working if she wants to finish her essay in time. So she suppresses the urge to leave her room and forces herself to concentrate on her work again. However, before outlining the next paragraph, she allows herself to once more check her email inbox. She decides that a new message from one of her course mates deserves an immediate reply; a notification from Facebook makes her browse through a couple of her friends' pages; and while on the internet, she then can't resist also quickly looking up some possible locations for a trip she plans to make next weekend. Almost one hour has passed when she finally gets back to her text. Slowly, she becomes more and more immersed in her subject, and it is only when her phone rings a couple of hours later that she realizes how much time has passed. At the end of the afternoon, when she finally decides to switch off the computer, her eyes are burning – a clear sign that she has stared at the screen for far too long. When getting up from her chair, she can also feel it in her back that she hasn't moved much for the last couple of hours. And still, this is what we've come to call mobility – virtual mobility!

So, what kind of mobility is virtual mobility? What does 'virtual' imply if it makes us call something mobility that at first sight seems to be a clear illustration of the immobility most people experience when spending their day in front of a computer? In everyday language, 'virtually' is often used in opposition to 'really' or to suggest that something is almost or nearly the case. And indeed, most people would probably say that this woman has *not really* been mobile. As Doel and Clarke (1999: 263) point out, according to common sense, 'the virtual is often to the real as the copy is to the original'. In this respect, 'the virtual can never be anything more than a pale imitation of the real: a mere simulation'. But why should exchanging news with friends on Facebook be any less real than talking to them in person? And is it not real when the woman we imagined looks at pictures of hotels and explores their location in Google Maps? Maybe the real and the virtual are not so simply opposed.

Discussions of the complex relationship between the virtual and the real often quote a statement from the philosopher Gilles Deleuze: 'We opposed the virtual and the real: although it could not have been more precise before now, this terminology must be corrected. The virtual is opposed not to the real but to the actual. *The virtual is fully real in so far as it is virtual*' (Deleuze, 1994: 208). From this point of view, the main characteristic of *virtual* mobility would not be that it is not 'real', but that it is not *actual* mobility – our imagined character is not actually leaving her desk meeting friends or visiting different places. But if there is no 'actual' mobility involved, why then do we still call this mobility?

Actual mobility is usually thought of as physical movement that takes people, things and ideas from one place to another. On the other hand, virtual mobility might only consist of the actual mobility of electronic particles, while the people involved do not necessarily have to be physically mobile at all. But, and this is the crucial point, it still appears as if they are moving to different places. This aspect is also stressed in the Oxford English Dictionary, in which the virtual is defined as (you may need to read this a few times!): 'anything, that is so in essence or effect, although not formally or actually, admitting of being called by the name so far as the effect or the result is concerned'.

This understanding of virtuality emphasizes the fact that, even though she is sitting at her desk all afternoon, this woman still communicates with colleagues and friends in faraway sites, visits friends' Facebook pages and learns much about their current activities and thoughts, and by watching images on Google Maps gets a vivid impression of possible holiday locations. Thus, she is 'virtually' mobile, when virtual mobility feels like going somewhere even though not 'actually' going anywhere.

In thinking about how this kind of mobility becomes possible, probably the most prominent association with the virtual today comes to mind: information and communication technologies (ICT). Any discussion of virtual mobility today is usually closely related to computer-mediated communication, the internet or mobile phones. It is also particularly in relation to the extremely rapid increase of ICT-use that virtual mobility has emerged as a topic in the social sciences and humanities. In this chapter, we will try to engage with these virtual mobilities from a geographic perspective. Developing an idea of the main challenges virtual mobilities pose to geographers' conceptions of place and space will serve as a starting point from which to look more closely at the different ways in which virtual mobilities have been addressed so far.

SUMMARY

- Virtual mobilities are real mobilities.
- Virtual mobilities involve going somewhere without *actually* having to go anywhere.
- The widespread use of information and communication technologies (ICT) has been an important prompt for studies of virtual mobility.

Virtual mobilities and conceptualizations of space

People 'meet' online, become 'friends' and 'like' each other's news on Facebook, see each other on Skype, join online discussion groups and take part in a video conference. Even though most of these practices have only been possible for a couple of years, many of them have already become part of mundane everyday life. But, while to most practitioners the use of ICT has become more or less banal, its social and cultural implications still seem a lot less clear and heat the discussions among academics. It was in 1997 that Michael Batty, in one of the earliest geographical essays on this subject, argued that computer-mediated communication technologies are changing geography and its study in subtle but dramatic ways, generating a 'new geography, which we call "virtual Geography"' (Batty, 1997: 337; see also Taylor, 1997; Graham, 1998; and Dodge and Kitchin, 2000 for similar early statements on the effects of information technologies on geography). Today, fifteen years and numerous publications later, we can identify two main aspects in respect to virtual mobilities that have caught geographers' attention, both directly addressing conceptualizations of place and space: first, the challenge of a clear distinction between absence and presence; and second, the turn away from topographical towards topological conceptions of space. Let us consider these two foci in turn.

Challenging the binary of absence and presence

Information and communication technologies are often proclaimed to change radically or even overcome space by 'unwiring and de-locating us, [. . .] taking us out of fixed spaces [and freeing us] from the confines of location' (Sanders, 2008: 181). The use of ICT devices during commuting time spent in a car, train or on the bus has become a prime example of how mobile communication technologies enable people to be more individually mobile through space, while still being able to form connections 'on the go' (e.g. Comer and Wikle, 2008). On the one hand, this means that one can, for example, do office work while commuting and therefore somehow be present in the office though actually being absent from it. On the other hand, these devices offer a possibility to 'uncouple oneself from the immediate environment' (Callon and Law, 2004: 7) by being virtually absent, for example with the help of the laptop's DVD player or its iTunes library taking oneself 'out of' a boring space of transit. In a similar vein, through Skype people can take part in family celebrations in far-away places, being able to join their conversations and watch them eat and dance while sitting on their sofa, possibly a couple of thousand miles away.

Allowing us in a certain sense to be present in more than one place at once, and to be present and absent at the same time, virtual mobilities thus seem to blur the common distinction between presence and absence, provoking us to play with the idea that these two categories might not have to be opposed to each other. New phrases and terms have been coined in response. Expressions such as 'mediated presence' or 'connected presence' are used to point at the moments in which 'the (physically) absent party gains presence through the multiplication of mediated communication gestures on both sides, up to the point where co-present interactions and mediated distant exchanges seem woven into a single, seamless web' (Licoppe, 2004: 135). Conversely, the term 'absent presence' has been coined to describe the state in which someone is 'physically present yet at the same time absorbed by a technologically mediated world elsewhere' (Villi & Stocchetti, 2011: 105; see also Gergen, 2002: 227).

Underlying these reflections on absence and presence, we can note a separation of mind and body. While the body rests on the seat in a commuter train, a person can be considered absent because the mind is fully immersed in a video. Moreover, despite not having their body physically attending a cousin's birthday party, a person is seen as present because they are all together in the same video conversation. It therefore does not depend on the body if someone is considered present or absent. Instead, body and mind, actual and virtual presence, are separated from each other and given equal weight. And it is this new weight given to non-physical forms of presence and absence that poses the main challenge when we relate virtual mobilities to ideas of place and space. If places and spaces are considered as bounded entities, a part of the earth's surface or, to use a common geographic image, as containers, the main emphasis is generally put on material and physical dimensions: it would be the location of the body that decided if someone is present in a certain place or not, denying any kind of virtual presence or absence. It is thus especially in relation to increasing ICT-use and resulting virtual mobilities that the appropriateness of classical, topographical conceptions of space have been questioned and possible alternatives have become more prominent.

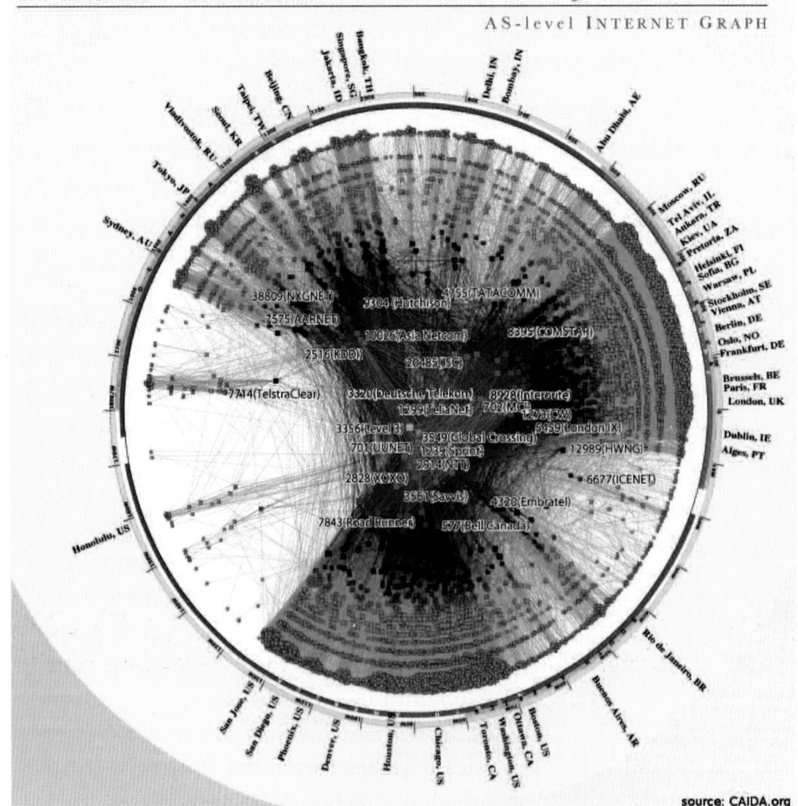

Figure 54.1 An internet topology map. This visualization represents macroscopic snapshots of IPv4 and IPv6 Internet topology samples captured in 2010. Source: © 2012 The Regents of the University of California. All rights reserved

From topographical to topological conceptualizations of space

> 'The virtual' is imagined as a 'space' between participants, a computer-generated common ground which is neither actual in its location or coordinates, nor is it merely a conceptual abstraction . . . The virtual involves a modification of understandings of locatedness and the relations between distinct places.
>
> (Shields, 2003: 49)

Thinking about mobility, most people are likely to imagine actual movements from one point to another, both situated somewhere on the earth's surface with a measurable distance between them: from A to B. Thus, mobility is commonly understood in relation to 'topographical' conceptions of space (see **topographical space**). Virtual mobility, however, does not imply any movement in topographic space. Getting somewhere by establishing close connections – generally through the use of communication technologies – instead fits the notion of relational or '**topological space**', understood as non-metric and constituted by object relations (see Figure 54.1). From this perspective, it is not physical proximity but somebody's multiple connections to other people, things and ideas that determines if he or she is 'in' a certain space or not. And since the same significance is assigned to material as well as immaterial and virtual flows, such relational conceptions of space indeed allow someone to be present in more than one place. As Amin has put it, topological thinking assumes the 'subtle folding together of the distant and proximate, the virtual and the material, presence and absence' (2007: 103).

Compared to topographical space that can be easily measured and demarcated, topological conceptions of space are generally a lot less tangible – they appear as having a more fluid, 'thread-like, wiry, stringy, ropy, capillary character' without any clear borders, their stability depending on the durability and intensity of the connections that constitute them and thus 'never captured by the notions of levels, layers [or] territories . . .' (Latour, 1996: 370). As Shields (2003: 50) has formulated it: 'Virtual spaces have an elusive quality which comes from their status as being both no-place and yet present via the technologies which enable them.'

These issues are not purely theoretical. A central question that arises from these changing conceptualizations of space is how such spaces are lived. How is it to move around in spaces made up of actual and virtual relations? What senses of place and space, what particular geographies do we encounter through virtual mobilities? Is it really, as the so-called technological utopianists imply, that the concepts of material space, distance and time, as well as the body will be rendered completely obsolete? It is to these questions that we turn in the next section.

SUMMARY

- Geography's major concern is how virtual mobilities inform and challenge our conceptions of space and place.

- Virtual mobilities are seen to enable people to be both absent and present at the same time, an idea that demands going beyond material, physical concepts of presence and absence.
- Consequently, topographical space does not seem sufficient to grasp the spatial dimensions of virtual mobilities, opening up discussions about alternative, topological conceptions of space and place.

Experiencing virtual mobilities

As Shields has so powerfully put it:

> The hype around digital virtuality over the past decade has been more about myth and less about actual cyberspaces . . . Portrayed as enabling a human virtuosity beyond the limits of the body or gravity, the legacy of the baroque echoes through the claims of Silicon Valley entrepreneurs.
>
> (Shields, 2003: 15)

On the other hand, sceptics foresee the deterioration of social life and communities due to a further retreat into virtual worlds. Large claims are made both for and against virtual mobilities and their implications for human experience. Out of these debates, however, the urge has grown to investigate more carefully to what extent virtual presence might replace actual presence, what differences this would make, and where the limits of these mixtures and minglings of absence and presence lie. How do we experience the spaces and places made up of virtual mobilities? And where do we currently encounter the limits of virtual experiences?

A sense of relatedness despite dispersion

A first theme dealing with the spatial effects of virtual mobilities is the creation of a sense of togetherness and common presence between widely dispersed individuals. Mobile phone calls and text messages, emails, instant messages and video calls are all being explored in respect to the ways in which they allow us to experience a certain proximity despite physical distance. Villi and Stochetti (2011), for example, have recently looked at the ways in which a sense of sharing, participation and fellowship is conveyed through the communication of photographs to distant others via mobile phones. But apart from virtual mobilities being able to enhance the feeling of connectedness between individuals, what seems to have attracted the attention of geographers even more is how virtual mobilities enable the formation of virtual communities. Social networking services, online forums, Twitter and blogs link huge numbers of people and create a sense of communality, even though some (or maybe even most) of them might have never met in person.

Facebook, probably the most impressive example in this respect, counted more than 750 million active users only seven years after it was founded, and thus serves as the most popular and widely spread service that allows its users to reconnect with old friends, make new ones and stay in touch with them by exchanging news in real time, looking and commenting on each other's photographs, sharing interests or playing games together. Moreover, these virtual spaces – by providing a means to let people know what they are thinking, doing and feeling, even when

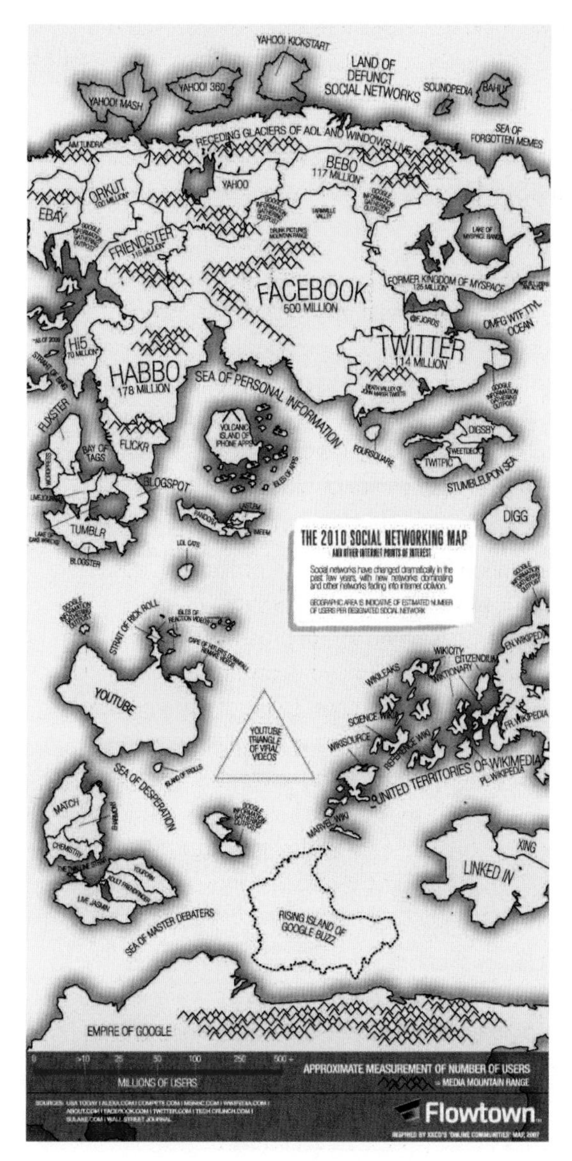

Figure 54.2 Creating an alternative topography based on virtual spaces: The 2011 Social Network Map. Credit: Ethan Bloch/www.flowtown.com/blog/the-2010-social-networking-map

being physically absent – have also proved a powerful tool in the fostering of social and political movements (e.g. Bach and Stark, 2005; Papadimitriou, 2006; Rodgers, 2003). As Rheingold (2002: xviii) has pointed out, wireless networks and mobile internet 'enable people to act together in new ways and in situations where collective action was not possible before'. In doing so, they facilitate the formation of what he has called 'smart mobs': groups of people who, by the use of communication devices, are able to act in concert even if they do not know each other ('flashmobs' are a form of smart mob, usually consisting only of a very brief encounter followed by a quick dispersion).

A sense of loneliness despite connection

As many of us probably know from our own experiences, virtual mobilities through social networking services and mobile phones indeed help to maintain connections and facilitate exchange regardless of physical distance. A great deal of research has thus concentrated on exactly that, especially in the context of **transnational** migration (Panakagos and Horst, 2006) (see also Chapter 45). Here, practices of ICT use are seen to bring life to the more abstract idea of 'transnational spaces' by creating 'a sense of involvement in each other's everyday lives' across different national territories (Horst, 2006: 156; see also Maintz, 2008; Rettie, 2008). However, a more thorough engagement with virtual mobilities among so-called 'transnational families' also shows that even though, as Larsen *et al.* (2006: 262) have put it, digital communication may 'resemble physical meetings' and can be the 'best substitutes when these cannot take place', they are still not the same. Mobility, virtual as well as physical, always includes leaving someone behind: at some point someone has to hang up, end the video conversation or log off, and in doing so leaves the parties involved alone in their respective locations. Seeing your loved ones on the screen and talking to them without being able to follow your desire to hug

CASE STUDY

Smart mobs, Twitter revolutions and the close connection between virtual spaces and physical location

What became known as People Power II in the Philippines often serves as the first example of how a head of state, in this case President Estrada, lost power to a 'smart mob', as Manila residents were mobilized and coordinated themselves by using text messages. The use of Twitter has gained considerable attention in this respect, with some even speaking of Twitter revolutions (e.g. Morozov, 2009a, 2009b). Famous examples of this kind of digital activism include the overthrow of the Egyptian government in early 2011 and the political resistance to the government of Iran: in the 2009–2010 election protest Twitter was used to spread information about upcoming protests in Teheran and to post pictures of clashes between protestors and government forces.

In the English riots in August 2011, mobile communication technologies – in particular the private, encrypted BlackBerry Messenger Service – were given a decisive role in coordinating the riots and looting. However, in order to go beyond the impression that these technologies play an important role in the formation of social movements and political resistance, there still seems more empirical research needed that attempts to get a deeper insight into how influential these virtual spaces actually are, how they work and what difference they make. After all, revolutions took place before the invention of these kinds of communication technologies.

What is also important to note here is that in all these cases the virtual coming-together of people is often closely related to actual movements, as we have most impressively seen in the images of Tahrir Square in Cairo in early 2011. The effectiveness of virtual movements depends on actual mobility. Neither protests and demonstrations, nor phenomena such as flashmobs, would be possible if the people linked to each other in virtual spaces were not also located with such physical proximity that they can actually meet in person. Thus, there are important questions about the interconnectedness of virtual and actual mobility and the continuing relevance of physical distance.

them serves as a constant reminder of the physical distance separating you from them (e.g. Verne, 2012). Embodied presence matters to people and mediated presence is not identical to embodied presence.

Overall, virtual mobilities, despite facilitating connections and exchange, also involve, and might sometimes even reinforce, feelings of separation, loneliness and longing. Although these aspects have so far been rather neglected in academic research – the work done on loneliness and virtuality has mostly been addressed to issues of social isolation and a retreat into virtual worlds (e.g. Keith *et al.*, 2011; Olds and Schwarz, 2009) – they point to the ambivalent experiences of virtual mobilities and the fact that in some respects, to some people, virtual presence might indeed only feel like a very deficient copy of 'really' being there.

CASE STUDY

The death of distance and the loss of location?

In recent years an impressive number of writers have expressed their conviction that virtual mobilities diminish the constraints of physical distance as we increasingly have the ability to do everything from everywhere. This conclusion has been reached not only in respect to specific settings, such as virtual communication within transnational families, but also with regard to social, economic and cultural domains more generally. As Paul Virilio (2000: 20) has stated, 'everything arrives without any need to depart'. Social interaction, economic transactions and political relations, all may proceed unimpeded by the need for physical proximity, so that distance simply seems to die (Cairncross, 1997). And it does not end there. This kind of thinking also tends to suggest that because of that, 'people will gradually become more decoupled from place, and that the significance of physical proximity, local community, and the home will weaken' (Vilhelmson and Thulin, 2008: 602)

Being more and more immersed in virtual spaces, the physical spaces people pass through in their daily lives are supposed to lose significance (e.g. Ling and Campbell, 2009). But, to use a common example, do people really not care about their city centre anymore because they can meet people and shop online? Even though e-shopping has increased tremendously within the last decade and more and more products have become available, as shoppers we recognize differences between browsing through the shops online and actually wandering through small boutiques or huge department stores, moving along rails of clothes, touching the material, trying them on, etc. (e.g. Leyshon *et al.*, 2005; Ren and Kwan, 2009). At times, we may prefer virtual shopping, if we want to avoid fitting rooms and prefer to try things on at home, before deciding whether to keep or return them. At other times, we may prefer to actually go shopping, so we can see and touch products 'in the flesh'. In either case, important questions remain concerning the specific experience of virtual mobilities – what does it replace, what does it not replace, and why?

A sense of freedom despite control?

Probably the biggest field of geographic research with regard to virtual mobilities has emerged around the question of increased freedom and independence. Central here is how communication technologies not only allow for a certain withdrawal from one's immediate environment but also make it possible for someone to be physically mobile while always appearing to stay within reach. Of particular interest in this respect has been the role of mobile communication technologies for children's mobility. Adolescents especially are seen to use mobile phones to free themselves from parental control, for example by promising to call regularly or at least if they encounter any trouble and thus being allowed to go further away or stay out longer (Mikkelsen and Christiansen, 2009; Nasar *et al.* 2007; Valentine and Holloway 2002).

Figure 54.3 Virtual mobility replacing actual mobility? Credit: Chris Madden

However, empirical findings suggest a more complex picture. Looking at the itineraries and sojourn maps of children and young adults, as well as analysing the negotiations between them and their parents, Pain *et al.* (2005: 826) show that, instead of bringing fundamental changes to the lives of young people, mobile phones have rather 'taken on a role in the existing ties, struggles and surveillance between young people and adults [and] in the already contradictory nature of young people's relationships with public places'. Moreover, they point out that, apart from possibly increasing a sense of freedom, 'mobile phones have brought surveillance, traditionally associated only with the state and corporate bodies, into the realm of personal relationships' (Pain *et al.* 2005: 815). Not only mobile phones, but virtual mobilities in general always seem to be connected to a so-called 'digital leash' that facilitates widespread 'mutual monitoring' by making individual activities more transparent (Green, 2002: 32; Green and Smith, 2004). Not only the mobility gained by the use of mobile phones but also one's movements through the internet can be tracked, allowing for almost complete control over freedom. Thus, as Sanders (2008: 181) has emphasized, at the same time that virtual mobilities seems to 'de-locate' us in time and space, they also 'reinforce the most traditional conventions regarding location'.

SUMMARY

- Virtual mobilities allow for a strong sense of community despite dispersion; they create a sense of space independent of any physical entity.
- The experience of virtual mobilities still differs from actual mobility and meetings, showing that physical location and distance do still matter.
- While contributing to a sense of freedom, virtual mobilities are at the same time accompanied by increasing surveillance and mutual monitoring; virtual mobilities have become trackable as well as mappable in both topographic and topological terms.

Conclusion

> Visions of new technologies revolutionising the way we live are often bold, sweeping, and millenarian. They are exciting to hear; they sell books . . . But their shelf life is roughly equivalent to that of a Big Mac.
>
> (Fischer, 1997: 113)

It was in April 1997 that the first special issue of a geographical journal was devoted to cyberspace (in *Geographical Review*; see Adams and Warf, 1997), and only a month later that Batty published his article 'Virtual Geography' (Batty, 1997). Even though there had been a few articles and conference presentations on the impacts of communication technologies on geography before (e.g. Batty, 1993), 1997 marks a central moment in the establishment of 'virtual geographies' as a research area. Since then, it has developed into a rich and diverse field of geographic inquiry and, with the many and rapid technological changes that have taken place within the last 15 years, it still possesses the special aura of the new and exciting. While this sense of novelty can be a positive thing, by attracting and indeed demanding new scholarship on the subject, it might also be a reason why so much of the discussion about virtual geographies seems to be so deeply informed by rather generalizing public discourses. As Woolgar has put it, 'much initial research on the social impacts of electronic technologies was characterised by a polarisation between narrow suspicion and uncritical enthusiasm ... Both sets of views tended to assume that the effects of these technologies would be predictable and universal' (Woolgar, 2002: 3–4). A closer look at the more recent literature suggests that these tropes have still not completely lost their power: the themes dealt with in this chapter are still framed by rather radical assumptions about the effects of virtual mobilities and have only slowly started to take more ambivalent and subtle experiences seriously. Thus, it seems as important today as it was a decade ago to move beyond these kind of 'fast food accounts' towards something 'a bit more nourishing' (Bingham, 1999: 260, referring to Fischer's quotation with which I started this conclusion). We still need 'to counteract or at least question the associations with novelty and epochal transition that cling to the subject of virtual geographies' (Crang *et al.* 1999: 3).

One way of doing this is to recognize virtual geographies generally and virtual mobilities more specifically, as mundane and everyday rather than exceptional and extraordinary. As indicated in the image of the woman sitting at her desk at the beginning of this chapter – an image that could probably be adapted to most of us without any fundamental changes – virtual mobilities are closely related to our everyday lives. Most of us might spend a couple of hours each day being virtually on the move, making it all the more astonishing that our academic debates on the subject sometimes still sound so far removed from our daily experiences. All of us, both professional researchers and students, can therefore sometimes use our experiences to engage critically with some of the broader discourses about virtual geographies and virtual mobility.

Of course, not all our experiences of virtual mobility will be the same, in extent or character. Across the world there are many differences in our everyday lives and these relate to differences in our virtual mobilities. In part this means we need to recognize the uneven geographies of adoption and connectivity that may be associated with ICTs, the multiple 'digital divides' that shape virtual mobilities (see for example Cheneau-Loquay (2007) for views from Africa). But we should also be wary of overly simplistic accounts of such differences. For example, virtual mobilities certainly do not map into stereotypical divisions between a mobile, connected western world and a more static, less connected Global South (Castells *et al.* 2007; Pfaff, 2010b). In-depth studies show much more nuanced distinctions in technology adoption and in the character of virtual and other forms of mobility (for one example, based around the biography of a single mobile phone in East Africa, see Pfaff, 2010a).

This chapter has argued that one task posed by virtual mobilities is to theorize some very basic ideas of geography, such as place, space and distance, in new ways. Topological thinking is required as well as more traditional topographic understandings. On the other hand, a second task is to move beyond abstract theorization and to ground often overly generalized discourses in a range of everyday experiences. How does it feel for different people in different places to be in a world in which it is generally possible virtually to reach any place by typing the name of it into Google? From a geographic perspective we are especially keen to examine what virtual mobilities bind together, what kind of spaces they create and how these are experienced. And indeed, there is still a lot to be explored. Virtual mobilities definitely change something, but what and how exactly?

DISCUSSION POINTS

1. What is the difference between 'topographical' and 'topological' conceptions of space? How do virtual mobilities impact on topographical ideas of space? What relationship between topographical and topological space is produced by virtual mobilities? What do you think about the argument proclaiming the 'death of distance'?
2. Even though some qualities of face-to-face interaction can be conveyed in virtual encounters, especially vision and sound, others such as touch or scent still cannot. How can we be more sensitive to the bodily and sensual experiences of virtual mobilities?
3. Why is virtual mobility today mostly related to the technological world? What about dreams and imaginations? Are we not also virtually mobile by imagining the world while looking at a holiday picture or maps?
4. How should we understand the relationship between the virtual and the 'real' world? How clear-cut are the boundaries and where are their limits?

FURTHER READING

Crang, M., Crang, P. and May, J. (eds.) (1999) *Virtual Geographies. Bodies, Space and Relations.* London: Routledge.

This is an early volume on virtual geographies, edited by Social and Cultural Geographers. It provides a sensitive engagement with a diverse range of topics that still are at the core of today's discussions.

Ling, R. and Campbell, S. (eds.) (2009) *The Reconstruction of Space and Time. Mobile Communication Practices.* New Brunswick, NJ: Transaction Publishers.

A recent collection of articles on the virtual mobilities produced through the use of mobile phones, with a focus on their effects on space and time.

Panakagos, A.N. and Horst, H.A. (eds.) (2006) Return to Cyberia: technology and the social worlds of transnational migrants. In: *Global Networks*, Special Issue, 6(2), 109–220.

This is a collection of articles dealing with virtual mobilities in transnational migrant spaces. It provides excellent examples of how ICTs are used in that context.

Woolgar, S. (ed.) (2002) *Virtual Society? Technology, Cyberbole, Reality*. Oxford: Oxford University Press

This edited volume, the outcome of a major research project on 'Virtual Society? The social science of electronic sociologies' (1997–2001), deals with the questions addressed in this chapter in a very reflexive way.

SECTION THREE

SECURITIES

Introduction

The term 'securities' is used in this book in a deliberately broad sense, and this section is designed to introduce you to a range of work within Human Geography concerned with the idea of security. The use of the plural in the section title is deliberate, and reflects the fact that we are not just talking about security in a conventional military or international relations sense – as in the security of the nation state from external threats. We are also concerned with individual security as well as security of resources at a global scale. Security can therefore be examined at all levels – from the individual, through the nation-state, to the global level. This increased spread of topics related to security can be directly traced to the emergence of what Ulrich Beck has termed 'Risk Society'. Beck defines risk as 'a systematic way of dealing with hazards and insecurities induced and introduced by modernization itself. Risks, as opposed to older dangers, are consequences which relate to the threatening force of modernization and to its globalization of doubt' (1992b: 21). For Beck then, modern society inevitably involves the proliferation of risks, not risks which are naturally produced, such as floods and epidemics, but those which are economically and socially produced as part of society's modernization. Beck calls these 'manufactured risks', and they include radiation, pollution, terrorism, toxins in food, climate disruption, germ warfare, biotechnology and toxic waste. In March 2011 we witnessed the devastating combination of natural and manufactured risk when the combined effect of an earthquake and tsunami in Japan triggered critical equipment failure and nuclear meltdowns at the Fukushima Nuclear Power Plant, resulting in the release of radioactive material. Added to these 'global risks', which are difficult to measure and assess, let alone control, may be a set of more localized risks around employment, health and crime, as family structures, employment patterns and welfare provision all undergo significant change. The result is a world full of uncertainties and insecurities.

In such a world, the search for security becomes all the more important. As we have noted, this takes on several guises – from an individual seeking employment and security for their family, to nations seeking to protect themselves from terrorism, to continents seeking to keep open supplies of raw materials and other natural resources. However, one of the lessons from the Fukushima nuclear disaster is that the insecurities and uncertainties of the contemporary world are the result of a complex mix of social, political, technical, scientific and natural factors, all operating at a variety of scales. Their complexity means that addressing them will perhaps require new sets of institutions and agencies able to operate across all these various realms. If such disasters point to the difficulty of separating 'natural' and 'man-made' risks, they also show how responses are ultimately located within a set of social and political frameworks. All the chapters in this section are designed to illustrate the social and political tensions which operate around the notions of risk and security in the modern world – and indeed all also show how the very notions of risk and security are themselves socially constructed and politically and culturally understood.

In Chapter 55, David Murakami Wood explores our responses to the idea that we are living in a risk society. David argues that the collective fears engendered within risk society are used to justify a set of responses that can be categorized under the term 'surveillance'. He shows how such surveillance takes different forms, ranging from more 'traditional' political espionage and CCTV, to contemporary uses of unmanned aerial vehicles (or drones) and information gathered by Google and social networking sites such as Facebook. He points out that although such surveillance is 'everywhere', different surveillance techniques have an uneven geography to them – and indeed are used to maintain and construct a set of social divisions between people, objects and places.

In Chapter 56, Klaus Dodds turns his attention to the securities which surround the exploitation of the globe's natural resources. As he notes, minerals, oil, gas, fishing and timber, among others, all have their own particular resource geographies and all have different concerns of risks and securities as a result. His chapter focuses on the geopolitics of resource exploitation, using examples drawn from Russia, Europe and the Arctic. He shows how competition for natural resources is represented in particular ways in the media, with some areas spoken of as resource frontiers, while other nations are seen to be involved in a 'war' for scarce resources. Given that states and societies remain wedded to the exploitation and use of carbon resources in particular, insecurities around the exploitation of these resources threaten to shape geopolitical calculations for a good while into the future.

In Chapter 57, Steve Hinchliffe shifts the focus away from national and resource security to look at the security of life at an individual level. He looks at how new risks and insecurities are arising across the globe from threats to the biological aspects of life. Issues of what he labels as 'bio-security' are now becoming pervasive around the constant threat of emergent pandemics and epidemics, triggered by infectious diseases that are often carried and spread by animals or in animal products. These new diseases will have a new geography and a different politics of security as a result, and Steve examines potential responses to them. He brings us back to a fundamental tenet of Beck's risk society – how we learn to live with the insecurities and risks which we are now generating, and in this particular case how we make human and nonhuman lives compatible with each other.

FURTHER READING

Beck, U. (1992) *Risk Society: Towards a New Modernity.* London: Sage.

Not an easy read in places, but this is the book which kick-started social science's recent engagement with issues of risk and security.

Mythen, G. (2004) *Ulrich Beck: A Critical Introduction to the Risk Society.* London: Pluto Press.

Provides a good readable summary of the key issues which emerged both in Beck's *Risk Society* book and in subsequent debates.

Graham, S. (2010) *Cities Under Seige.* New York: Verso.

An interesting exploration of the link between surveillance, warfare and urban management.

Lane, S., Klauser, F. and Kearnes, M. (eds.) (2012) *Critical Risk Research: Practices, Politics and Ethics.* Chichester: Wiley.

A collection of chapters exploring the politics and ethics of risk and risk research.

CHAPTER 55
RISK/FEAR/ SURVEILLANCE

David Murakami Wood

Introduction: risk-surveillance societies

In the latter part of the twentieth century, the processes of **globalization**, diminishing of nation-states' power and the increasing recognition of the centrality of what were previously considered the 'side effects' or 'externalities' of industrial processes, has meant a growing concern with risk. Indeed, German sociologist, Ulrich Beck (1999) described the world we are living in as a 'risk society', in which the distribution of social and environmental 'bads' – the worst possibilities – has become as important as, or even more important than, the distribution of social and environmental 'goods', such as income, housing or clean air (see box). Such developments are connected to the increased emphasis on security (see Chapter 57). The argument made here is that this risk society is both a reaction to, and a generator of, a shifting and historically and spatially contingent collection of fears, which are used to justify a response that involves the generation of knowledge that can be used either to mitigate those risks or to manage those fears. The response is surveillance.

What is surveillance? Put simply, surveillance is the purposeful monitoring or gathering of information, with the aim of either inducing or preventing change. Such inducement or prevention can include the use of new data produced through collation and manipulation of information gathered, so-called dataveillance, but it can also involve no data at all – in many cases, surveillance itself can be used simply as a form of moral suasion or intimidation. Many more authoritarian states and institutions rely on this kind of suasive surveillance, and indeed it was argued by the French historian of ideas, Michel Foucault, through his analysis of transformations in military camps, schools, hospitals and prisons, that such surveillance underpins the simultaneous construction of modern space and **subjectivity** (Foucault, 1976). Surveillance is not the same as simply looking or seeing, although it often involves both visibility and visualization – as reflected in the subheadings for this chapter – and not all monitoring is surveillance, if the monitor makes no direct or indirect attempt to induce or prevent any change in the monitored subject.

CASE STUDY

Risk, actuarialism and the culture of fear

The rise of 'risk thinking' can be seen as only the latest aspect of a move towards the more accurate description and quantification of the world that began with the enlightenment and eventually led to an array of developments including modern cartography, the invention of statistics, and the creation of academic disciplines, including geography. Judgement of risk relies on the calculation of probabilities, a process known as 'actuarialism', because it arose among actuaries, who calculate the probability of risk for the insurance industry. Risk is also fracturing into different forms in contemporary **neoliberal capitalism**, many of which increasingly emphasize not the potential harm that can be done to people, other living things or environments, but to corporations and institutions. 'Organisational risk' has become one of the most important factors in corporate decision-making, and 'reputational risk' a significant factor in policy-making, for example in competition between global or world cities.

Actuarialism is one of the major drivers of surveillance, as its underpinning assumptions creates a need for more and more data to drive increasingly refined calculations of risk and to anticipate or predict potential events. At the same time, in order for people and companies to purchase products or support particular policy directions, the actual risk of events is downplayed in favour of a concentration on the *consequences* of disaster events, in order to generate fear. This had led some to argue that we live in a 'culture of fear'.

Surveillance may be divided up according to the object of surveillance, for example between targeted and mass surveillance. Targeted surveillance specifies the object of surveillance in advance and concentrates only on that object, whereas mass surveillance provides an overview without specifying any particular object in advance. In practice, these forms are a spectrum and often interconnected. For example, an open-street video surveillance system may be used for the mass surveillance of crowds in urban streets, but if one individual arouses the suspicion of operators, or the attention of biometric or behavioural recognition software, the same system can then be used to target and track that person.

Surveillance has now become so important as a means of governance that this has led some commentators to argue that we live in a 'surveillance society' (Lyon, 2007). In this kind of society, surveillance becomes the primary mode of social ordering. One of the key questions for geographers is to what extent this ideal typical sociological statement holds across different spatial categories, such as regions nation-states, cities or locales. Would it be more true to argue that there are different kinds of surveillance societies, depending on where we might live or work, or even that there remains some places where surveillance is not the dominant mode of ordering at all?

SUMMARY

- 'Risk society' means a move from the distribution of social 'goods' to the distribution of 'bads' or risks.
- These risks are assessed using actuarial methods and these methods are spreading to more and more sectors of government and business.
- Avoiding or mitigating these risks has led to the increased use of surveillance.
- Surveillance is the purposeful monitoring or gathering of information, with the aim of either inducing or preventing change.

Picturing surveillance

When we think of surveillance, we generally imagine things that are culturally familiar. This chapter will examine surveillance by considering, in a rough chronology, several recent forms of surveillance as routes into a whole variety of wider geographical issues.

Global spies and private eyes

For those who grew up in the period of the **Cold War** from 1945 to 1990, surveillance was all about spies and undercover investigation: a world glamorized in successive generations in the Bond and Bourne films, but captured rather more accurately in the John le Carré novel, *Tinker, Tailor, Soldier, Spy*, which was based in British foreign intelligence in the 1960s. The methods included human informers, telephone-tapping and increasingly high-tech systems of satellite imaging. In this imaginary, the locus is international, covert and at once personal and human, but far removed from everyday life.

Yet at the very same time, similar techniques were in fact being used by the very same governments to gather information on their own populations. There was (and remains) a multi-scalar hidden geography of espionage and political policing. In the former East Germany, for example up to one in six of the population were enlisted as informers by the state internal security agency, the Stasi, and numerous technologies of monitoring were deployed from the opening of mail, through listening devices that worked through water pipes, to hidden cameras in everyday objects. In the USA, meanwhile, the Federal Bureau of Investigation (FBI) and the National Security Agency (NSA) collaborated illegally to monitor radical student groups, indigenous, black and environmental groups and civil rights activists. Of course, the use of informers and infiltrators was in many ways nothing new and dates back to the oldest civilisations, but is particularly connected to modern policing since the beginning of the nineteenth century.

It is also the case that such espionage was never simply the prerogative of the state. As the 1974 movie, *The Conversation*, illustrates, the corporate world was already involved in using undercover techniques to try to gain an economic or other advantage over both competing companies and workers. Surveillance within the workplace remains one of the most important ways in which people will encounter surveillance. From the invention of the modern production line, industrial management methods have always included the combination of spatial concentration and close monitoring of worker activities and even bodily movements, with the aim of increasing 'efficiency' and maximizing the profit that capital could produce from the exploitation of human labour.

The use of surveillance as an economic tool operates at many scales, and it was during the Cold War that the underpinning architecture of what we now understand as a global **neoliberal** capitalist economic order was created. This set of institutional arrangements is less familiar to most, or certainly less often considered as 'surveillance'. Many supra- and inter-national governmental institutions were created after the Second World War, designed to enforce the order that had emerged from the war: legal-political institutions like the United Nations (UN), the international police agency, Interpol, and, more recently, the International Criminal Court (ICC). However, several supranational economic regulation bodies were also created, including the World Bank and the International Monetary Fund (IMF) and much later, the World Trade Organization (WTO). The IMF developed systems of economic surveillance of national economies and imposed 'structural adjustment' programmes on countries that failed to perform to expectations – until recently these were mainly countries of the Global South, but the United Kingdom was structurally adjusted in the 1970s and the global financial crisis of 2008 onwards has seen several European states, including Greece, Spain and Italy facing similar measures. However, surveillance at the global economic level is increasingly private; global credit ratings agencies, in particular Standard & Poor's, Moody's and Fitch Ratings, operate perhaps the most important type of global dataveillance because they can, in effect, reduce whole national populations and collective life chances to economic marginality.

Eyes on the street

For those growing up in the post-Cold War era of the 1990s, surveillance was something different. Surveillance meant video cameras (closed-circuit television or CCTV). CCTV systems are essentially one or more video cameras connected via either cable or wireless to a set of recording devices (either tape or digital storage) and usually also a live feed to a control room where operatives can watch the images on monitors. In most cases, operatives have the ability to pan, tilt and zoom cameras to focus on particular people or situations of interest.

The contemporary spread of video surveillance is not unconnected to the Cold War world of espionage and military surveillance. The contemporary **political economy** of surveillance could be argued to derive from the changes that occurred in the military industrial complex towards the end and immediately after the Cold War with corporations involved in military supply seeking to find new civilian markets by exploiting the growing fear of crime in the risk society. This helps to explain the rise of CCTV with the Automated Number (or License) Plate Recognition (ANPR/ALPR) systems put into place in London after IRA terrorist attacks in the early 1990s using technologies tested in the Gulf War of 1991 (Coaffee, *et al.*, 2009) (see Figure 55.1). This has only been highlighted further by 9/11 and after, driven not just by terrorism directly, but by vaguely connected fears of immigration and infiltration.

However, the major driving forces in the expansion of video surveillance were not

Figure 55.1 CCTV camera at the site of an IRA bombing in the City of London. Credit: David Murakami Wood

directly connected to international geopolitics. CCTV has multiple origins in both state and private sector (e.g. casinos, theme parks and shops) but really took off as a method of providing reassurance and preventing stock losses (through shoplifting) in large American malls. The growth of malls in both North America, and slightly later in the UK, coincided with, and contributed to, the economic decline of urban centres. In relaxing planning legislation and encouraging private investment, the response of urban authorities was essentially to mimic the 'theme park' environments of malls – including video surveillance. And it was the UK where public video surveillance spread most rapidly (Norris and Armstrong, 1999) and where the USA and other nations looked after 9/11. The British 'example' has now helped spread open-space CCTV throughout the world through the channels opened by the increasingly networked **global cities** (Murakami Wood, 2012).

Surveillance became used as a symbol of downtown urban regeneration and 'success stories' of the effectiveness of CCTV were further built into this imaginary. The reality remains more mixed. As CCTV has proliferated, particularly in the UK, years of research has shown little evidence of its preventative effect on crime in public places, and perhaps more surprisingly, there is also evidence from police forces themselves that only a small percentage of street crimes are solved using video surveillance.

So what is CCTV used for? Essentially, it is a significant part of the process of urban sociospatial control. It is used to direct help to those who are seen to need it, but also to target those 'flawed consumers' who are perceived as out-of-place in regenerated urban centres, such as beggars, the homeless, unlicensed buskers and informal traders, youths – and particularly minority ethnic youths – 'hanging around', skateboarders and so on. As such, CCTV can be seen as a marker of what Neil Smith (1996) calls 'revanchist urbanism', the reclaiming of urban space by dominant groups from those who are marginal and of lower social status. In many places, where video surveillance is operated as part of urban regeneration schemes or other private–public partnerships, the exclusion of certain activities (and by implication, certain people) is made clear through specific local rules or bylaws. In other places, it becomes normalized through practice. Video surveillance is therefore also part of the contemporary 'splintering' of urban landscapes (Graham and Marvin, 2000) and 'enclavism' – the breaking of the broadly public urban realm into microzones covered by different rules and regulations, separated from each other symbolically, practically and in some cases physically – such as gated communities and private shopping malls (see Chapter 46).

This trend began in the Global North but is now found worldwide, and has reached its most extreme extent in some of the economically fast-growing states of the Global South (Davis and Monk, 2007), such as the Mexican gated community portrayed in the film, *La Zona*, or in Brazil where forms of fear-driven enclavism has led to whole private cities such as the Alphaville developments, protected by private security guards, 'intelligent' entry systems and digital surveillance cameras. More recently, moves have been made to wall off marginal informal settlements, the favelas, in Rio de Janeiro, combined with intensified policing through special 'pacification units', the installation of surveillance cameras and the creation of a city-wide control room that will combine monitoring, emergency dispatch and other security and policing services. However, the main difference in the case of the Rio favelas is that the inhabitants are being socially spatially segregated and monitored at least partly for the risk they are believed to post to wealthy urban residents – and potentially to the major global mega-events being hosted by the city: the FIFA World Cup in 2014 and the Olympic Games in 2016 (see box).

CASE STUDY

Surveillance and mega-events

Surveillance has been a growing aspect of mega-events – events of global significance that take place in particular places. Such mega-events include popular sports events such as the Olympic Games or the FIFA World Cup, International Expos and major international political meetings like the G8, G20 and World Summits. These events bring together the growing competition between global or world cities, with the globalization of surveillance. In many cases, local and national laws are temporarily set aside and distinct zones of exception are created where special rules apply and particular sanctions can be implemented. The transnational organizations behind these events have specific requirements for security and surveillance, driven by the perceived risk to high-level delegates and to the prestige of the organizations, as well as the fear of catastrophic risk, such as a major terrorist attack or natural disaster.

This began after the fatal attack on the Israeli team in the Olympic village by Palestinian militants in 1972, but has been further accelerated by several subsequent events including

9/11. The International Olympic Committee (IOC), for example, now has a permanent office for security and is constantly updating its best practice based on previous games. The organization can also rent out surveillance equipment to host cities. Host cities increasingly base decisions around surveillance on judgements about risk, not just to the event itself, but to the reputation of the city. Companies providing security and surveillance equipment and services have thus increasingly successfully targeted mega-events for experimental roll-outs of surveillance practices and technologies which might be unacceptable or controversial in normal circumstances, including surveillance airships (at the 2007 Pan-American Games in Rio de Janeiro), chemical-sniffing robots (at the 2008 FIFA World Cup) and unmanned aerial vehicles (at the London 2012 Olympics). In many cases, systems are often bought and redeployed in the host city after the event, as with video surveillance in Athens after the 2000 Olympics or Toronto after the 2010 G8 meeting.

SUMMARY

- In recent history, surveillance can be seen in a wide range of places and at all spatial scales from local to global.
- The main aim of surveillance in practice is to maintain order in the interests of dominant classes, facilitate 'normal' behaviour and prevent or challenge 'abnormal' activities.
- Surveillance is increasingly implicated in the sociospatial fracturing of space from the reconstruction of national borders to the urban enclavism of gated communities and privately-operated public open spaces.
- Video surveillance (or closed circuit television, CCTV) is the most commonly encountered form of surveillance in urban space.

Beyond CCTV: new forms of surveillance

Contemporary exemplars of surveillance are no longer restricted to CCTV. Even as video surveillance continues to expand and intensify globally, moving from global cities to smaller towns and even into rural settings, other forms of surveillance have come to typify the current age. Three very different forms represent both the globalization of surveillance and its movement to the domain of the virtual as well as the material. The first is the proliferation of drones or unmanned aerial vehicles (UAVs) (see Figure 55.2). We saw in Chapter 36 how UAVs are increasingly used to fight wars at a distance. Perhaps more than any other form of surveillance they illustrate the connection between the gathering of information and action or response, and also demonstrate the complex scalar geographies of such sociotechnical systems. Operators in office buildings in Florida and Nevada use video screens and wireless telemetry to fly surveillance and combat missions on the borders of Pakistan and Afghanistan, over the Gulf States or even along the borders of the USA itself. In the early days of drone warfare, UAVs were only the eyes for other more active military functions, acting as reconnaissance and targeting devices. However, increasing numbers of drones are now equipped with air-to-ground missiles and can themselves be used to launch strikes against targets that they have also been used to identify. For forces using drones, this combination of remote control surveillance and strike capacity appears to offer a bloodless and risk-free form of warfare which appeals to both funding agencies and the domestic public, but to those targeted, including many non-combatants and innocent civilians, drones constitute a non-negotiable, pitiless form of summary justice entirely outside the bounds of international law or customary forms of warfare.

But UAVs are no longer restricted to the battlefields of global 'wild zones'. Just as the geopolitical vision of security forces has shifted to encompass the internal disorder of the domestic and the urban, so drones have also spread into civil use. Internal state organizations in countries as diverse as India, Israel and the UK are now using UAVs for multiple tasks as varied as crowd control, border patrol and fisheries monitoring.

The second contemporary exemplar is the border crossing. As nation-states continue to decline in power, ironically efforts to strengthen the outward symbolism and materiality of state power have grown, particularly when it comes to border control. As with CCTV, the actual

Figure 55.2 Unmanned aerial vehicle (UAV). Credit: US Navy/in public domain

effectiveness of procedures and systems in place may be open to question – Bruce Schneier (2003) calls such measures 'security theater' – however, border surveillance exemplifies another important aspect of contemporary surveillance, that is the practices of social and spatial sorting that result from information gathering and its application to diverse populations. Border controls are no longer, if they ever were, simple checkpoints, and borders in many ways are no longer the line on the map. 'The border is everywhere' has become almost a cliché but the content of this statement is of vital concern to those forced to live with the consequences of transformations in border practices. Borders have become semi-permeable zones that no longer mark an inside/outside distinction, but have their own internal geographies of human and baggage scanners, interrogation rooms and holding and deportation areas. Such zones may leave their marks on those who pass through, meaning that for some, particularly those considered risky because of national or ethnic origin, travel patterns and even business transactions, associations and friendships, the border can never really be left behind. The border crossing now marks a nexus of state and private sector surveillance, as information taken from multiple public and private sources from intelligence to social networking activity is combined together with biometrics (such as fingerprints, iris recognition and full body scans) and increasingly behavioural observation. Such information is used to build profiles of individuals and groups and, more importantly, the actual and potential associations between them in virtual and material space that might give rise to risks such as terrorism, illegal migration or drug-smuggling and other illicit trading.

The third contemporary imaginary of surveillance is that around social networking and Web 2.0. With over 350 million users, Facebook has rapidly become perceived as an essential component of social relationships. Yet, Facebook is not a charitable service; it is a major transnational corporation with a value rated at over $1bn when its stock market flotation was announced in 2012. Where does this value come from? Essentially it comes from the information generated through the surveillance of Facebook users, much of it voluntary in the sense that this information is self-published by the user. Every time a Facebook user 'likes' something (a movie, song, brand, etc.), this data is stored and connected to other 'likes' and the 'likes' of friends to build up not just personal profiles but more importantly social graphs of users (Figure 55.3). This data is vital for contemporary marketing, in which corporations attempts to build 'brandscapes' around consumers, in order to anticipate purchasing decisions or insinuate themselves into decision making. Such brandscaping is not limited to the virtual, however. Increasingly social networking is adding locational components – through applications like Foursquare, Facebook Places or Google Latitude, which allow the user to link their social data with spatial location, so that friends (or more) can see where the user is or has been and what they are doing. Individuals can choose to have their mobile location tracked by selected friends and to track their friends' mobiles. Google has expanded its locational services in other aspects too. Google Maps has become the default GIS system for both business and individuals, and this has now been supplemented by Google StreetView, which has almost total coverage of most industrialized countries. Two major concerns have arisen over StreetView: the first is the ability to identify individuals or vehicles in embarrassing or undignified contexts through their chance exposure at the time that the Google StreetView photographic collection

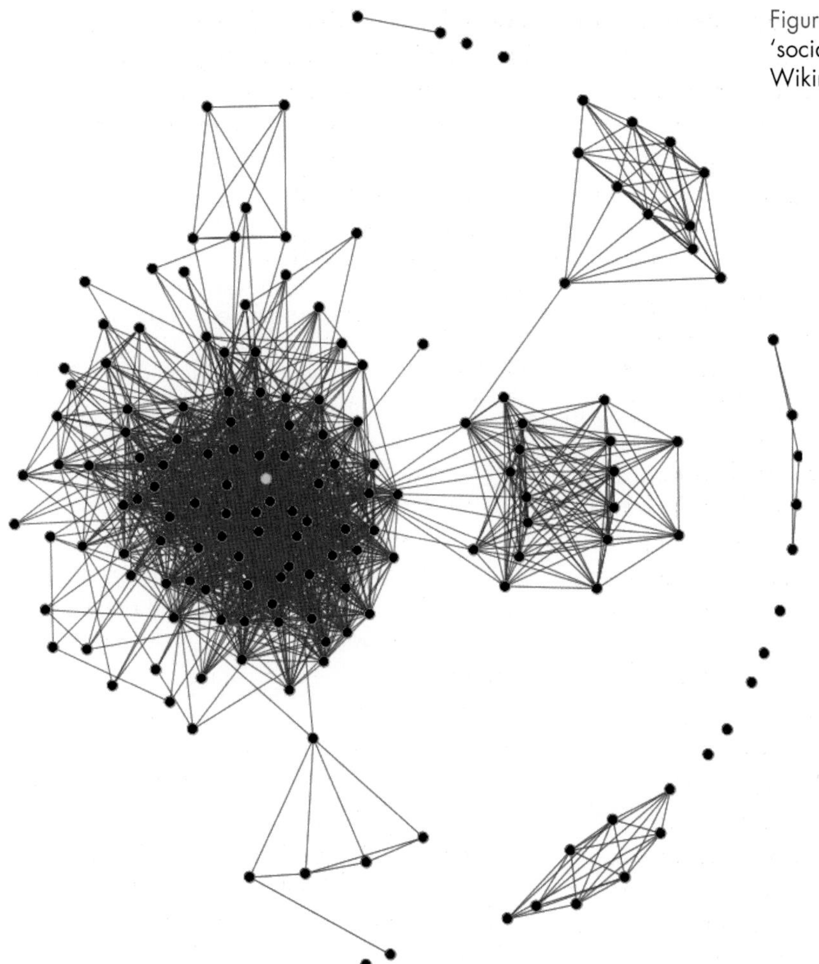

Figure 55.3 A Facebook 'social graph'. Credit: Wikimedia Creative Commons

vans were passing. The second is that Google has also been discovered to have been using its photographic vehicles to systematically 'sniff' private wireless systems and gather detailed personal information.

Locative systems themselves are mostly voluntary and consensual. However, it isn't just friends who can see this data. It gets added to the already existing mass of personal preferences and connections by social networking companies, the profiles of which are used to market to other users, and sometimes made available to developers of applications for the different social networking platforms (see box). It is mined by intelligence agencies and by other private companies who specialize in social media surveillance. It is also used by employers when assessing prospective and current employees and not just passively: interviewees for jobs are increasingly being asked to handover logins and passwords for social networking sites so that they can check on whether the candidate represents an organisational risk, i.e. whether they mesh with the corporate culture, have any poor personal habits, or have any record of making disparaging remarks about companies, etc.

CASE STUDY

Bad applets

Commercial software applets which mine social networking data, much of which is made available to app developers by social networking companies, and combine them with locational data and mapping technologies, usually Google Maps, have sometimes been controversial. One American app, Offender Locator, used public US states' data on those listed on sex offender registers, including photographs and registered addresses, and combined them with Google Maps. While it could be argued that this allows families to make choices that might protect their children's safety, such apps also provide a false sense of reassurance, a very limited form of spatial knowledge (e.g. where an offender officially lives) and could encourage paranoia, mistaken identification and vigilantism.

Another notorious example was the 'Girls Around You' app, which combined Google Maps, Foursquare and Facebook Places location data and public profile information on women, allowing the user to locate 'available' women in their immediate vicinity. This was criticized as being a gift to stalkers, sexual harassers and rapists. The social networking companies involved have since claimed to have stopped supplying data for this kind of app.

SUMMARY

- The spread and use of new technologies of surveillance are changing the spatiality of surveillance.
- Unmanned aerial vehicles (UAVs or drones) are moving from battle space to civilian use, leading to an increasingly vertical geography of everyday surveillance.
- Surveillance techniques allow borders and boundaries to be more than simple lines on a map, but to be flexible and differentially permeable.
- The use of internet and social media has produced new virtual spaces of surveillance.

Emerging trends in surveillance

Computing continues to increase in sophistication and a new era of ubiquitous or pervasive surveillance is emerging. This is as a result of the combination of computing first with communications technologies and second with the embedding of communicative computing in an increasing variety of everyday objects, buildings and even living creatures and people – so called ubiquitous or pervasive computing. This era is underpinned by the decreasing costs and size of sensors of all types, 'big data' – that is the generation of exponentially vaster quantities of information, and more open circuits of surveillance and analysis including 'crowdsourcing' – the use of the input of multiple networked individuals to build up a larger picture.

French philosopher, Gilles Deleuze, regarded the precursors of these technological changes combined with the **post-Fordist** restructuring of the economy and the crisis of the institutions as the basis of a 'control society'. While ongoing transformations and distribution of information technologies undoubtedly mean a spatial spread and an intensification of surveillance, they also offer a wider distribution of the means of surveillance and an increasing variety of state and private actants involved in carrying out surveillance: rather than one 'Big Brother', there are many little brothers – including most of us.

At the same time we should critically assess claims of the democratization of surveillance. Surveillance practices and systems, based on the underpinning actuarial logic or the relative profitability and/or risk posed by individuals, groups, places etc., threaten to reinforce and deepen existing social and spatial divisions and create new forms of recombined (virtual and material) territoriality. For example, research on the use of geodemographic software like Acorn MOSAIC has shown that they can play a key role in gentrification processes and enclavism and encourage the spatial reinforcement of class divisions (see Chapter 46).

The same kinds of problems of mediation and distanciation observed with the use of drones in war is now also visible in crowdsourced video surveillance. The short-lived Shoreditch Digital Bridge project (see Figure 55.4) allowed residents of the Haberdasher's Estate in London to see live feed from local video surveillance cameras and connected this to a 'rogues gallery' of local criminals. More recently, the Internet Eyes start-up has been broadcasting surveillance footage from paying private business customers on the internet, allowing anyone in the world to sign up as volunteer monitors. The incentive is that the system is a game in which volunteers gain points for spotting suspicious activity and even more points if this contributes to an arrest or conviction. The volunteers do not know where the actual footage is being broadcast nor can they contact the actual business or law-enforcement; all communications are mediated through the Internet Eyes system.

Figure 55.4 Shoreditch Digital Bridge project. Credit: Press Association

SUMMARY

- Virtual and physical space surveillance are in the process of recombining with the advent of ubiquitous or pervasive computing.
- New economies of surveillance are emerging in which performance, participation and consent are involved in a complex mixture with control.
- However, while more people may be involved in surveillance, we should be wary of claims of the democratization of surveillance.

Conclusions

Surveillance is often said to be everywhere these days, but geographers have important contributions to make on multiple levels. First, surveillance is one of many things currently being globalized, but this globalization is not even or smooth. The development of surveillance, like all development, is uneven and contested, and always in process. Second, surveillance has an important role – increasingly the most crucial role – in constructing space and place and in maintaining, deepening and generating sociospatial order and differentiation and division between people, groups, objects and places. Third, surveillance is increasingly embedded in our virtual and material space and therefore ever more deeply implicated in relationships between humans, nonhuman beings, technologies and environments.

DISCUSSION POINTS

1. What are the reasons for the increase in surveillance, and what are the relationships between surveillance and globalization?
2. How does surveillance construct space? You may like to think about the transformation of urban space as an example.
3. Do we all live in surveillance societies? Consider the ways in which surveillance might be similar or different in different parts of the world.

FURTHER READING

Ball, K., Haggerty, K.D. and Lyon, D. (eds.) (2012) *Routledge Handbook of Surveillance Studies.* Abingdon/New York: Routledge.

Provides a comprehensive overview of the evolving transdisciplinary field of surveillance studies.

Beck, U. (1999) *World Risk Society.* Cambridge: Polity.

The most accessible statement of Beck's ideas on risk.

Coleman, R. and McCahill, M. (2011) *Surveillance and Crime.* London: Sage.

Undergraduate-level introduction to issues around surveillance, policing and crime.

Fuchs, C., Boersma,, K. Albrechtslund, A. and Sandoval, M. (eds.) (2012) *Internet and Surveillance: the Challenges of Web 2.0 and Social Media.* London/New York: Routledge/ESF COST.

Coming out of a major pan-European research network, Living in Surveillance Societies, the first collection dedicated to this particular area of surveillance.

Graham, S. (2010) *Cities Under Siege: the New Military Urbanism.* New York: Verso.

Accessible but critical work on the links between new forms of warfare, surveillance and urban management.

Kitchen, R. and Dodge, M. (2011) *Code/Space: Software and Everyday Life.* Cambridge MA: MIT Press.

A rich and nuanced critical geographic account of the relationships between software, space and subjects.

Lyon, D. (2007) *Surveillance Studies: An Overview.* Cambridge: Polity Press.

A good, simple but not simplistic, introduction to surveillance studies.

Films mentioned in the chapter:

The Conversation (1974). Dir. Francis Ford Coppola.

La Zona (2007). Dir. Rodrigo Plá.

Minority Report (2002). Dir. Steven Spielberg.

Tinker, Tailor, Soldier, Spy (2011). Dir. Tomas Alfredson.

WEBSITE

www.surveillance-studies.net/?p=310

The Surveillance Studies Network has annotated reading lists on surveillance, including books, academic articles and reports, as well as an annotated list of surveillance feature films.

CHAPTER 56
INTERNATIONAL RESOURCES

Klaus Dodds

Introduction

A British spy is sent by his superiors in London to Baku in Azerbaijan to investigate a possible plot to disrupt a trans-national pipeline project. European energy security appears to be at stake, as an evil terrorist mastermind is threatening not only to disrupt the gas pipeline distribution system, but also assault Istanbul. Standing by a computer-generated map of the Eastern Mediterranean, the industrialist Electra King explains that the presence of rival energy projects could spell danger. Her prediction proves correct and James Bond's travels take him around the Caspian Sea region and finally to Istanbul. Bond, in *The World is Not Enough* (1999), once again navigates the eddies and shoals of global politics. Energy security is represented, in this film, as the business of states seeking to secure vital gas supplies vulnerable to trans-border and pipeline-based infrastructural disruption (Dodds, 2003).

Geographers have shown considerable interest in both the popular and policy-orientated representations and understandings of **resources** (Bradshaw, 2010; Bridge, 2010, Kuik *et al.*, 2011). Hollywood films, for example, are hugely popular and they play an important part in helping to visualize the legitimization and acquiescence as well as the resistance to resource production, circulation and consumption (Dodds, 2007, Le Billon, 2012). Resource-based stories, whether narrated through film or other popular media, are frequently presented in a sensationalistic manner. The capacity of resources to invoke and provoke a range of reactions from anger and excitement to fear and even expressions of hope is noteworthy (Sparke, 2005).

Resources are always embedded within a series of narratives, involving curses, wars and bonanzas (Bridge, 2010; Bridge and Wood, 2010; Le Billon, 2012). There are two distinct ways in which resources are often understood. The first is to view resources as caught up in the strategies of states and governments. States are assumed to be constantly on the lookout for resource advantage. Energy security becomes a zero-sum game in which resource dependencies are considered to be 'security risks'. Attention is thus given to supply, demand, transit, infrastructure and the like.

The second approach highlights how states, markets and companies are bound up with

Figure 56.1 Caspian Sea with oil derricks on the tip of the Absheron Peninsula. Credit: Amos Chapple/ Getty Images

one another in relational ways that are always dynamic and shaped by networks and multi-scale governance. As Gavin Bridge (2010) reminds us, resources do not simply exist, they become – and they become in part because substances such as coal, oil and gas are regarded as essential sources for industrialization and development more generally (Bradshaw, 2009; Bridge, 2010). These assembled networks and dependencies produce security risks rather than resource-rich regions themselves.

Resources are also intimately related to geographies of investment, extraction, production, exchange and consumption (Bridge, 2010; Le Billon, 2012). Minerals, oil, gas, fishing and timber to mention but a few have their own particular resource geographies. The contemporary Arctic region is used briefly as a case study in order to illustrate how resource exploitation is legitimized, organized, regulated and resisted by a range of actors, operating across a series of spatial scales, including indigenous organizations, energy companies, environmental groups and regional organizations such as the Arctic Council (Dodds, 2010). Resource exploitation and extraction is, more often than not, capital-intensive and may involve substantial investment in infrastructure such as pipelines, transport and/or storage facilities. So securing rights to mine, for example, not only relies on the existence of some kind of legal system but also the perceived risk/reward ratio. In the Arctic region, for instance, while the threat of conflict/unrest may appear remote compared to other parts of the world, corporations are particular mindful of the land rights of indigenous peoples in places like Arctic Canada (Nuttall, 2010).

By way of conclusion, we return to the etymological origins of resources, from the Latin word meaning 'to rise again' in order to consider the double-edged nature of what we have discussed. Resources can suggest empowerment and revival but they also invoke dependence, fear and vulnerability. While we tend to recognize and indeed praise things and people that are considered resourceful, we also need to recognize the way in which people and places/environments are embedded within dynamic resource-related geographies. Demand for resources is not only driven by human needs such as providing heat and shelter but is related to other kinds of social and cultural practices ranging from driving cars to the wearing of precious stones in the form of jewellery. But, as places like Greenland might imply, resource potential for indigenous peoples might also offer the prospect of a distinctly post-colonial future, free from the Kingdom of Denmark.

SUMMARY

- Resources are frequently the subject of news stories and film scripts.
- Resources are embedded in bonanza, curse and war narratives.
- Resources are embedded in geographies of extraction, production, exchange and consumption.

Resources: geopolitical storylines and resource bonanzas, curses and wars

Critical geopolitics and allied areas such as critical resource geographies (see Chapter 35) have contributed, over the last twenty years, to a body of literature that considers how space and place are represented within formal, practical and popular discourses – in other words, the realms of think tanks, universities, governments, policy makers and popular culture (Dodds *et al.*, 2012). As subject matter, resources and their availability routinely feature in official reports (BP Statistical Review of World Energy, for example), government speeches and, as noted earlier, popular media including newspapers, social media, television and radio programmes and films. When linked to debates about energy security and/or climate change, resource-related discussions frequently invoke a series of geographical imaginations and representations, usually aided and abetted by maps, in which supplies, circulating networks and demand are displayed (for one earlier analysis, see Ebinger, 1977). For example, in the winter of 2006– 2007 European media outlets were releasing multiple stories about Russian gas supplies and how, using the example of Ukraine, western European states were potentially vulnerable to the Russian gas corporation Gazprom potentially turning off supplies of gas in the midst of winter (see box). Ukraine's gas supplies were temporarily halted by Russian

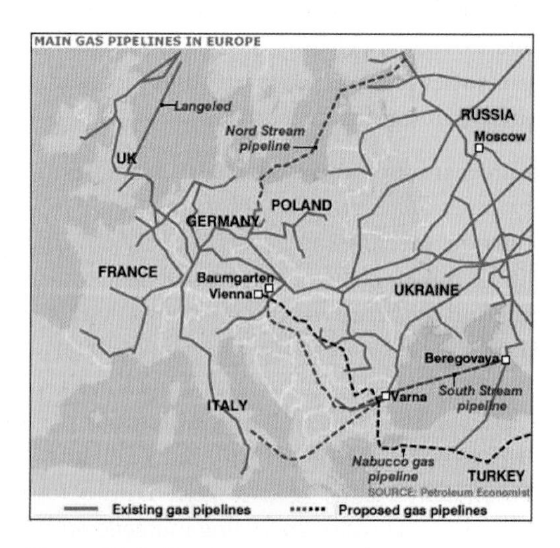

Figure 56.2 European gas pipelines

suppliers, however, because of a dispute over an unpaid gas bill and allegations that Ukraine had 'stolen' Russian gas en route to western European states. As part of that discussion, maps showing pipeline networks between Russia and Western Europe were used to highlight the connections, in a very literal sense, between the supplier and customers (see Figure 56.2).

Moreover, for reasons of location, former Soviet states such as Ukraine and Belarus were important transit countries, as piped gas had to cross their territories in order to reach western European consumers. At the height of the gas crisis, one particular important geopolitical storyline was that of Russia as an 'energy superpower' 'bullying' and 'harassing' smaller neighbours and former Soviet Republics in the Cold War era, as well as intimidating customers within the European Union. Under such a rendition, resources quickly become represented as emblematic of unequal power relations between states and regional organizations. State-sponsored organizations such as GAZPROM are then invested with the capability to manipulate demand and potentially unsettle customers, especially in the midst of a cold winter where fuel for heating is at a premium. As Marshall Goldman, a Russian expert, wrote of the senior management of GAZPROM and the state-owned oil company Rosneft, '. . . [they] had become a real life Dr No – an archetypal James Bond villain' (Goldman, 2008: 2). Why? Part of the answer lies in certain simple statistics – 40 per cent of Germany's gas supplies and 100 per cent of the Baltic States' collective gas consumption comes from Russian sources. Further afield, and thus not just restricted to Europe, countries such as the USA were also importing Russian gas and mindful of any interruption to supplies.

CASE STUDY

'Ukraine accused of stealing Europe's gas'

Russia has accused Ukraine of stealing $25m of gas exports destined for Europe after it cut off supplies to the country on Sunday.

Countries as far west as France say supplies from a pipeline running via Ukraine have fallen by up to 40%.

Ukraine denied taking the gas, but said it would siphon off a share if temperatures fell much below freezing.

The row erupted after Russia raised the price of 1,000 cubic metres of gas from $50 to $230 and Ukraine refused to pay.

Russia's gas monopoly Gazprom is still charging the lower price to some former Soviet countries, though the average price in the EU is $240 (£140).

Moldovan President Vladimir Voronin said his country had also been cut off, after refusing to pay $160 per 1,000 cubic metres, according to the Itar-Tass news agency

(Source: BBC News, 2 January 2006: http://news.bbc.co.uk/1/hi/world/europe/4574630.stm)

The so-called 'gas crisis' we have outlined represents just one example of a renewed concern for resources and their availability. In the immediate post-Cold War era, globalization debates tended to posit the idea that the geographies of supply and demand were largely stable and characterized by pacific trading relations and indeed falling commodity prices. While citizens of the Global South might be seen as in the grip of resource politics (e.g. blood diamonds), western consumers were asked to make informed choices rather than stopping purchasing. Such resource-based representations help to consolidate an **Orientalising** view of the Global South, which emphasized resource availability rather than seeing places as part of a mosaic of relationships involving the human and non-human worlds, drawing together different places and spaces (Bridge, 2009).

The financial crisis of 2008 brought into sharper relief interest in commodity prices and anxieties over resource availability, whether it be linked to 'peak oil' and/or the accessibility of particular minerals such as indium. The availability of resources, whether oil, gas, grain, timber and rare earths, revealed three fundamental themes. First, what we consider to be natural resources are implicated in the production of a wide range of commodities and waste. Second, the interaction between resource production, transformation and consumption of resources is characterized by a whole series of contradictions and paradoxes. Third, resources continue, as a consequence of the above, to generate considerable controversy and interest precisely because of the social and cultural values invested in their availability and allocation, across a wide range of geographical scales. And yet, despite all of that, the vast majority of consumers, especially in the more affluent parts of the world, would have little to no awareness, for example, of where resources such as oil, gas and grain are sourced unless there is a specific resource crisis.

When such crises do unfold, the role of geopolitical storylines is important because they help to conceptualize, albeit often implicitly, understandings of resources. A popular geopolitical storyline involves the idea of 'bonanza'. Bonanza, in this case, refers either to a rich seam of ore and/or a situation that leads to a sudden source of wealth. The discovery of resource potential in a place where none has been discovered before would be one popular adoption of the term. Alternatively, it might involve other substances, already discovered, that were previously not considered to be commercially valuable. A sudden technological innovation or shift in market demand might then be invoked as a critical factor in generating a new resource bonanza. Either way, there is a sense here that resources create opportunities (and security dilemmas) and that the value of substances like oil and gas lies in their capacity to empower those who possess such resources – so as with films such as *Mad Max* (1979) the discovery of oil (or gasoline) becomes an immediate source of power (see box).

Fundamentally, though, resources are associated with particular places and this matters because it helps to cement the idea that the location and availability of resources is largely fixed. In other words, they are rooted in place. But as resource geographers have argued, this underestimates the role that other factors play in transforming material objects into resources. The latter, therefore, tell us a great deal about what human society values and thus requires knowledge and labour in order to extract, transport and process into particular products (Bridge, 2009; Le Billon, 2012). So resources depend on this interaction between the human and non-human worlds and are thus fundamental, relational and mobile. It is not

CASE STUDY

Hollywood and resource films

Resources, and the battle over their use and value, are represented in a number of Hollywood films from the dystopian thriller *Mad Max* (1979) to more contemporary productions such as *Three Kings* (1999), *Lord of War* (2005), *Syriana* (2005), *Blood Diamond* (2006), *There will be Blood* (2007) and *Avatar* (2010). In *Three Kings*, for example, the American-led intervention in Kuwait in 1991 is depicted as deeply duplicitous and motivated more by the desire to secure access to oil supplies then it is to save lives either in Kuwait and/or in Saddam Hussein's brutalized Iraq. In *Lord of War*, on the other hand, a Ukrainian-American arms dealer travels the world circumventing UN restrictions on arms sales in order to sell ex-Soviet hardware, including AK-47s and helicopters, to military regimes and cabals using resource-related revenues as a funding stream. In an entertaining and insightful way, *Lord of War* helps the viewer trace, via a series of networks and geopolitical encounters, how the ending of the Cold War offered new opportunities for the exchange of arms, money and commodities such as 'blood diamonds' around the world.

difficult then to understand why the supply and transport of resources is thus considered critical to the national security agendas of states, regional organizations and corporations. A resource bonanza of sorts might be considered to facilitate power projection, because resources are equated with capabilities (Bridge, 2004). In the case of gas supplies, for example, countries such as Russia, Norway and Qatar are considered by many observers to possess superpower-like qualities. Qatar, the Arabian Gulf microstate, has proven oil reserves of 15 billion barrels and 26 trillion cubic metres of natural gas – about 15 per cent of the world's total – and one of the largest in terms of proven reserves. On the one hand, this has enabled rapid economic growth for a country once focused on fishing and pearl hunting but on the other hand, it means that the country is also dependent on international markets and changing patterns of demand, albeit ones conditioned by long-term contracts with customers.

The idea that resources might provoke 'resource wars' (Le Billon, 2005) is particularly popular among policy orientated commentators who warn that countries such as the USA need to remain vigilant when it comes to securing access to supplies and preventing other countries from consuming vital energy sources. One prominent advocate is Michael Klare (2002, 2005), who over a period of twenty-plus years has warned his American readers that countries such as the USA, and in particular rivals such as China, are locked into successive generations of struggles to secure access to energy resources around the world. In his article 'The new geopolitics of energy', Klare (2008a) notes that there is a geopolitical 'perfect storm' in the making as diminishing reserves of oil, gas and industrial minerals are being pursued by major industrial powers. States, in pursuit of their national security strategies, are attempting to monopolize supplies as well as using arms transfers and financial aid to leverage access to resource-rich countries (Klare, 2008b). This in turn has been cited as part of the 'resource curse' of many such countries concerned. Sometimes called the 'paradox of plenty', it has been frequently noted that countries possessing an apparent abundance of resources such as hydrocarbons and minerals tend to have lower rates of economic growth and developmental progress. While there are many possible reasons for such a set of circumstances, emphasis is often given to the role of corruption and unstable government that either steals or squanders resource wealth. In media reports involving oil exploitation and development in Nigeria, for example, it is not uncommon to read about rampant government corruption and plutocracy as opposed to the role of international markets, energy corporations, technology, geopolitical strategies of powerful states and changing patterns of world demand.

Klare calls for reducing exposure to energy insecurities by rapidly investing in alternatives to carbon sources. If the USA fails to develop alternative sources,

> A new cold war with China, with an accompanying arms race, will require trillions in additional military expenditures over the next few decades. This is sheer lunacy: it will not guarantee access to more sources of energy, lower the cost of gasoline at home or discourage China from seeking new energy resources. What it will do is sop up all the money we need to develop alternative energy resources and avert the worst effects of global climate change.
>
> (Klare, 2008a)

But achieving such profound change will not be straightforward given the dependence of the USA on access to the 'strategic triangle' shaped by oil reserves and transit routes encompassing the Arabian Peninsula, Central Asia and the Eastern Mediterranean. Under the former Bush administration, especially in the context of the War on Terror, US military bases and intervention were linked, in no small measure, to guarding against possible terrorist and/or rival state-led interference with current and future oil and gas supplies. Indeed, ever since the so-called Carter Doctrine of the late 1970s this has been a long-standing feature of US energy security policy.

These **realist**-based analyses tend to marginalize the experiences of local contexts, especially in

CASE STUDY

Oil shock wave

In July 2011, a crisis war game was organized by a campaign group called Securing America's Future Energy (SAFE) involving former Bush administration employees. It is November 2011. Al-Qaeda attack a major crude oil processing facility in Saudi Arabia, disrupting supplies and unsettling international energy markets. Iran and Venezuela threaten to 'choke-off' oil supplies raising the price of oil to over $200 per barrel. How did the US react to this fictional scenario? By advocating more aggressive oil and gas expansion in Alaska and the Gulf of Mexico, and thus expanding domestic supplies. There is little reflection on alternative energy supplies and/or climate change during the crisis war game by the participants.

(Source: The Guardian, 14 July 2011. Available at: www.guardian.co.uk/environment/2011/jul/14/us-oil-crisis-simulation-oil-shockwave)

the Global South where people may well face routine shortages, especially when it comes to clothing, food and/or heating sources such as charcoal. As the forthcoming discussion of the Arctic region will show, we should not underestimate the role of indigenous communities and their use and valuation of local and regional eco-systems, while being mindful of how global geographies of resource availability and consumption shape connections between livelihoods and landscapes (Bridge, 2009). Challenging those kinds of realist geopolitical representations of resources is difficult though, because much of the last hundred years has witnessed states and governments militarizing and securitizing resource supplies and access therein, often at the expense of local communities and places.

SUMMARY

- The 'gas crises' involving Russia and its European neighbours concerned competing representations of resources.
- Resource politics needs to be understood at a variety of interacting scales including the local, regional and global.

Multi-scale governance and resource politics in the Arctic region

The Arctic, as in earlier incarnations involving different kinds of resources such as sealing and whaling, is represented widely as a '**resource frontier**' for mineral exploration and exploitation. While onshore oil and gas exploitation is comparatively well established, a great deal of interest is being directed towards the potential resources to be found within the Arctic Ocean. The so-called Arctic Ocean coastal states (Canada, Denmark/Greenland, Norway, Russia and the USA) have all, as a consequence, released national security strategies for the Arctic region extolling both the need for environmental stewardship while stressing the strategic importance of their 'sectors' (see Figure 56.3)

In resource terms, the Arctic, if represented as a 'resource frontier' encourages a particular realist-inspired analysis. Interest tends to gather around the activities of states and their militaries alongside relationships with major corporations, whether they are state-owned, partially state owned or privately owned, such as GAZPROM, Shell, BP, Statoil and smaller companies such as the UK-based Cairn Energy. As United States Geological Survey Arctic maps illustrate, the undiscovered potential of oil and gas reside, overwhelmingly, in the exclusive economic zones of those five coastal states, so the idea of a 'resource scramble' is unlikely in the sense that the broad patterns of ownership of those resources is not disputed.

For an island such as Greenland, which enjoys an autonomous relationship with the colonial power Denmark, some indigenous peoples have argued that possible independence in the future might be realizable if there are sufficient exploitable reserves of oil and gas. So the seemingly unstoppable demands of carbon-fuelled capitalism, on the face of it, appear to offer the possibility of financial

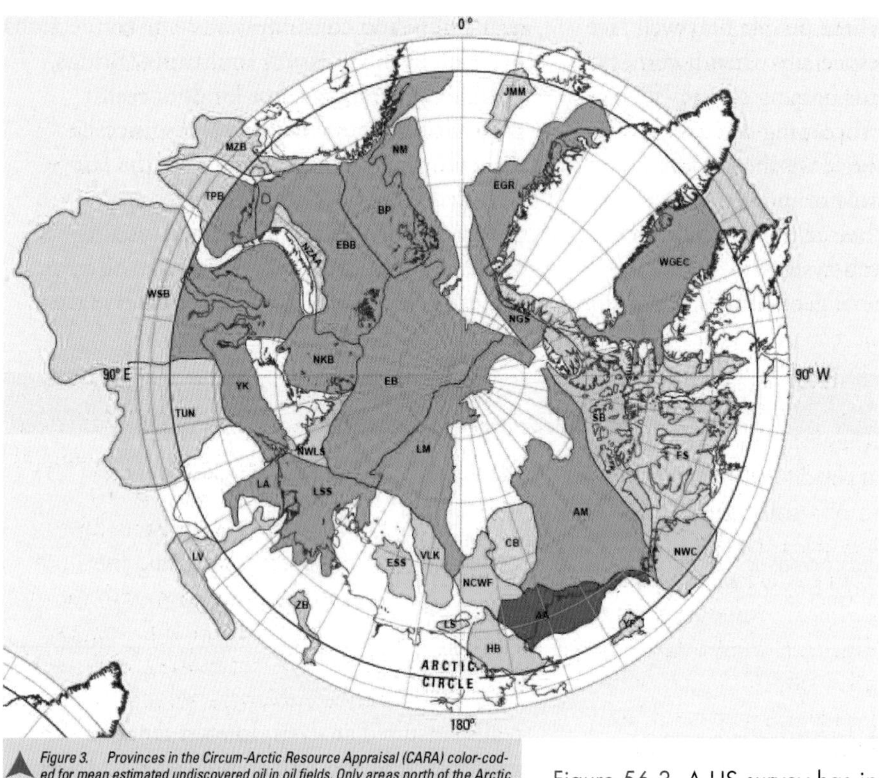

Figure 56.3 A US survey has indicated that there are still vast amounts of untapped oil and gas beneath the Arctic Circle. Credit: USGS

autonomy from Denmark, which currently provides a block grant to Greenland worth 500 million Euros per year. The Greenland government created a state-owned oil company called NUNAOIL and established a Department of Resources which is charged with facilitating international investment and production. The hope for many Greenlanders (total population 56,000) in the future is for the country's modest economic base (largely derived from fishing-related exports) to expand to include oil, gas, gold and rubies. The end result may well be the emergence of a truly indigenous independent nation-state, no longer dependent on Danish grant monies, albeit one committed to developing carbon resources widely implicated in regional and global climate change. But Greenland is not the first, and almost certainly will not be the last, to believe that mineral extraction should be the cornerstone of a post-colonial national development policy.

In the case of the continental Arctic, the role of resources is complex and varied. Indigenous, first nation and Northern communities have played an important role in shaping the geographies of investment and production. This has often co-existed uneasily with Canada acting as an energy superpower (supplying Albertan tar sands oil to Chinese/East Asian markets), while committing itself to a 'Northern Strategy' which prioritizes the sustainable development of the Canadian Arctic region. In the case of the long-running Mackenzie Gas Project and others such as Northern Gateway Pipeline Project, Aboriginal groups objected to the presence of pipelines and extraction facilities being based in their tribal territories. For the Mackenzie region, one effect of these objections was to raise projected costs of gas extraction and transportation considerably, leaving operators to complain that it was easier to work in other areas of the world such as the Niger Delta. Aboriginal communities have also been at the forefront of reminding southern constituencies of the consequences of hydrocarbon exploitation, in particular, remembering the 1989 Exxon Valdez disaster affecting southern Alaska or warning of a Gulf of Mexico-type spillage (April 2010) being replicated in the Arctic Ocean. As Mark Nuttall (2010) notes in his work on pipeline politics in Northern Canada, indigenous peoples are important actors in Canadian resource politics and ever vigilant over the evolving geographies of investment, production and consumption and the associated waste and by-product generated by oil and gas exploitation. As such, it might explain why corporations and states may be drawn to potential areas of production in offshore locations precisely because there are no indigenous treaty rights to be concerned with.

The challenge for fledging independent states and indigenous/Aboriginal communities throughout the Arctic region is to capitalize on the revenue opportunities afforded by resources while at the same time ensuring that any investment ends up facilitating sustainable development. As the 'resource curse' literature warns us, resources, whether rubies, oil, gas or gold, do not guarantee such an outcome, as resources are exploited, transported and consumed elsewhere. New geographies of inclusion/exclusion frequently accompany resource extraction and exploitation, as states and corporations alongside say indigenous communities enclose and exclude others in terms of property rights and tangible benefits. The experiences of communities in the Niger Delta provide a salutary reminder that resources can be an apparent 'curse' (Watts, 2008).

SUMMARY

- The Arctic continues to be represented as a 'resource frontier'.
- Indigenous communities in the Arctic are mindful of the possibilities resources offer to shape new political futures.
- The Arctic is seen as a region of low political risk but socially complex in places like Canada.

Conclusion

According to the BP Statistical Review of World Energy (2010), finite fossil fuels such as coal, gas and oil are being consumed at the rate equivalent to the burning of 10,000 million tonnes of oil per year. Worse still, for those concerned about global climate change, there are no signs that this level of consumption is declining globally (Barnett, 2007). In fact it is predicted that by about 2050, something in the order of 20,000 million tonnes of oil per year will be consumed per annum. If global economic growth continues to be fuelled by carbon sources then the implications for the global climate are dire – and perhaps the most profound global security challenge is to consign the carbon age to historical memory. But this will not be straightforward as we continue, as a global community, to draw upon the earth's resources in a substantial manner. As Bridge (2009) notes, more generally, apart from the exploitation of oil and gas, something in the order of 50,000 million tonnes of raw materials are appropriated in some form or other. And this extraction, for much of the last century, has intensified and magnified over time, especially as either lower quality or more inaccessible resources are now being targeted for extraction and production – this applies particularly to hydrocarbons but also to other minerals such as copper, nickel and zinc.

States and societies remain addicted to these resources, and as such it is not difficult to imagine that emerging geographies of resource extraction, production, exchange and consumption will fundamentally shape future global political developments. Our assembled dependencies and networks create risks and insecurities, and thus we should think carefully about how we represent resource geographies. If oil- and gas-rich regions such as the Arctic and the Middle East matter, they do so because of our 'need' for those substances. Carbon resources remain an integral element of the world economy and thus will continue to exercise a powerful role in contemporary and future global politics

DISCUSSION POINTS

1. Should we be worried that the next generation of wars will be fought over access to resources?
2. In what ways could we represent resources that did not reproduce storylines about bonanzas, curses and wars?
3. Should we seek to impose a mining moratorium in parts of the Arctic region, where national sovereignty is not clear-cut?
4. Why does geography matter to the movement of resources?
5. What are the political prospects of the movement to promote low carbon futures?

FURTHER READING

Klare, M. (2008) *Rising Powers, Shrinking Planet: How Scarce Energy Is Creating a New World Order*. Oxford: One World Publications.

A top analyst of resources considers the current state of energy supply and demand. He considers four major resource pools – the Middle East, Caspian Sea, Africa and Russia – and concludes by noting that energy security is a top issue for major powers.

Le Billon, P. (ed.) (2005) *The Geopolitics of Resource Wars*. Abingdon: Routledge.

This is a major collection of geographical essays on the role of violence in shaping the exploitation, transportation, processing and consumption of resources.

WEBSITES

http://secureenergy.org/

Securing America's Future Energy – this US group is described as a non-partisan campaign group which seeks to highlight the USA's dependencies on oil and to promote alternative energy resources.

www.worldenergyoutlook.org/

International Energy Agency *World Energy Outlook 2011*. This annual report provides essential information and analysis of global energy projections. It explores energy-related scenarios and contemplates future patterns of exploitation, development, investment and consumption.

CHAPTER 57
SECURING LIFE: NEW HAZARDS AND BIOSECURITY

Stephen Hinchliffe

Introduction: three securities

Today, the word security is most readily associated with nation states and armed conflict. It is also a word that seems to be used more frequently as international uncertainties grow. Some would even say there is a new age of insecurity. A war on terror and a state of continuous alert or never ending emergency is how many commentators characterize the last few decades. And yet, this martial or war-like sense hasn't always been the first thing you would think of when you hear the term security. Indeed, in the late nineteenth century and for much of the twentieth century, the word was more often associated with the health and welfare of people or populations. The term social security, associated with a welfare state, communicates a sense that some kind of safety net should be provided by a society in order to make sure that the most vulnerable are cared for and that the rest can afford to take risks in the knowledge that if things go wrong there is a collective capacity to pick up the pieces. The idea is captured in conventional insurance whereby probability is used to protect the majority from the misfortune of a minority (car insurance is the easiest to imagine – with a known incidence of accidents for a given population, it is easy to calculate a premium whereby those who are involved in accidents are financially covered). More recently, and related to both these uses, security has become associated with life, or more particularly the biological aspects of life. This chapter will explore this rise in biological security, consider its characteristics and critically engage with the kinds of issues that it raises (as well as those that it tends to ignore). A key aim is to use a **geographical imagination** to engage with the issue of how life is currently understood *as matter that requires security*. Table 57.1 summarizes the three forms of security already mentioned: nation-state, welfare or social security and biological security. An aim of this chapter will be to think about what kinds of things you could write in the final cell in the table. Is biosecurity simply a mix of previous securities, or are there other modes of operation

	Nation state security	Social security	Biosecurity
Object	Nation or territory	Population	Life
Mode of operation	Conflict or deterrent	Welfare and insurance	?

Table 57.1 Some characteristics of three securities. Source: Substantially redrawn and simplified from Lakoff (2008)

that are more fitting? To think about this, we'll first look at what the threats to life are, before looking into the making of biosecurity and then finishing by discussing what, if anything, is new about securing life itself.

Threats to life

What are the threats to life in a modern Western democracy with a high average life expectancy? It is possible to find out the answer relatively easily using statistics that are often widely available. In the USA, for example, heart disease and cancer account for the vast majority of deaths (see Table 57.2). Indeed, without going into detail (deaths are often related to multiple causes), it is fairly straightforward to suggest that the majority of threats to life in the USA are related to an ageing population and to lifestyle (i.e. what people eat, what kinds of exercise they do, what they do for a job and so on). Few, in contrast, relate to infectious diseases. Influenza is eighth on the list and diseases that would have accounted for many more deaths a century ago such as cholera, tuberculosis and typhoid are no longer major killers. Many of those infectious conditions have been reduced through improved public health – sanitation, better-quality food and water supply, better pharmaceuticals and immunization programmes, and improved overall health care.

Cause of death	Number of deaths
1. Heart disease	616,067
2. Cancer	562,875
3. Stroke	135,952
4. Chronic lower respiratory disease	127,924
5. Accidents	123,706
6. Alzheimer's disease	74,632
7. Diabetes	71,382
8. Influenza and pneumonia	52,717
9. Nephritis	46,448
10. Septicaemia	34,828

Table 57.2 Leading causes of death in the USA (2007). Source: Centers for Disease Control and Prevention (National Vital Statistics Report, 2011)

To be absolutely clear, a majority of the world's population have not been not so lucky – malaria, cholera and typhoid remain among the many communicable and infectious diseases that continue to pose a daily threat to life. But that didn't stop the French philosopher, Rene Girard, famously stating in 1974 that we live in a 'world where the plague and epidemics in general have disappeared almost altogether' (1978: 138). He was not alone. In 1978 the United Nations issued an accord predicting that even the poorest nations would undergo an '**epidemiological transition**' before 2000.

The transition is the moment at which diseases of old age overtake those relating to infection as the main causes of death. Pharmaceuticals, better health care, sanitation and improved diets would, the prediction had it, lead to health improvements for all. The optimism was short-lived. Only a year or two later, a new infectious disease which came to be called AIDS (Auto Immune Deficiency Syndrome) put paid to that confidence. In countries like South Africa, HIV/AIDS now displaces heart disease and other conditions associated with ageing and lifestyle as the foremost cause of death, and of course leads to a great deal of suffering, social stigma and hardship (Table 57.3).

Looking back at the numbers for the USA, and recognizing the need for much more detail, it might be inferred that the epidemiological transition has held firm in the wealthiest countries. One might also argue that such a transition just hasn't happened *yet* in countries with greater overall poverty. To be sure, within most if not all nations there will be areas where lifestyle diseases outweigh infectious diseases and vice versa. Indeed, we shouldn't lose sight of the huge disparities within and between nation-states where some are threatened by lack of adequate sanitation, basic care and necessary drugs and others are more prone to life style conditions. Progress, you could argue, is uneven and delayed by new diseases like HIV/AIDS, but continues nonetheless.

Cause of death	Number of deaths
1. HIV/AIDS	132,990
2. Heart disease	34,402
3. Stroke	33,866
4. Tuberculosis	28,907
5. Interpersonal Violence	27,563

Table 57.3 Leading causes of death in South Africa (2000). Source: South African Medical Research Council

However, there is an issue that doesn't show up in these headline figures. For some commentators, diseases like HIV/AIDS are not so much a temporary setback to the onward march of medical progress but mark a new age of infectivity, one that has potential to affect all corners of globe. Rather than stalling an epidemiological transition, HIV/AIDS and other newly emerging infectious diseases demonstrate the failures of the old public health models and the fragility of **modernization**. For some, we live in insecure times, constantly threatened by pandemics (diseases which affect wide portions of populations, possibly across continents). In many senses it is these *potential* diseases that are the concern of biosecurity, so we will now look at these in more detail.

SUMMARY

- Lifestyle diseases were assumed to have largely replaced infectious diseases as the biggest threats to human life.
- There is an argument that this epidemiological transition hasn't happened in most parts of the world, or where it has, it is under threat from a new set of diseases.
- We live in an era when pandemics and plagues are feared by national governments and health institutions.

The coming plague

In 2009 a swine flu pandemic was declared by the World Health Organization (WHO – the United Nations body which takes global leadership on matters of health). Seemingly emerging from a pig farm in Mexico (or what is referred to in the business as a concentrated animal feeding operation or CAFO, see Chapter 27), the flu virus quickly spread to the USA and then took hold in Europe and Australasia. Within a few months, the pandemic took lives and caused both mild and debilitating illness in Latin America, North America, Europe, Australasia, Asia and Africa. Previous to this, in 2005 and 2006, there were fears that a highly pathogenic form of avian influenza could potentially become contagious and deadly in people. Such fears have somewhat subsided, but the disease's mutability (ability to change) and the endemic status of the disease in China, Bangladesh, India, Vietnam and Indonesia, means that the threat remains. In 2003 a disease called Severe Acute Respiratory Syndrome (SARS) spread rapidly across continents, infecting people and leading to nearly 1,000 deaths. Compared to the confidence expressed in the 1970s concerning the epidemiological transition, diseases like SARS generated bio-insecurity, expressed neatly in a scientific paper entitled 'The world is teetering on the edge of a pandemic that could kill a large fraction of the human population' (Webster and Walker, 2003) as well as popular

Figure 57.1 Fear of highly infectious diseases leads some to take drastic measures. Credit: AFP/Getty Images

science books with titles like *The Coming Plague* (Garrett, 1994).

Many argue that these pandemics, epidemics and emerging infectious diseases have the potential to be disastrous, affecting large numbers of people the world over. In the next section, we will look in more detail at some of the roots of these fears, asking what is causing this potential re-emergence and/or fear of infectious disease. What creates bio-insecurity? And, given the insufficiencies of nation-state and population security, what modes of operation are being installed or employed around the matter of life?

Making bio-insecurity

When interrogating the growing fear of infectious disease there are two elements to consider: first, an investigation into the rise of disease threats and second, an investigation into how governments, publics and others are acting on these fears at this moment. We don't have to assume that new disease threats are a fiction to question why, among all the other issues that people have to deal with, these concerns come to the fore at certain moments. Both investigations are inherently geographical, as we will see. This section looks at the rise of disease threats, while the following section discusses how such threats are understood and acted upon.

What is it about contemporary society that makes disease more mobile and therefore adds to biological insecurity?

You may have listed globalization or included rapid travel between continents and increased transport and trade. A person incubating a new kind of deadly flu may not know they have a disease for a few weeks, despite being infectious, but in that time they could have flown around the world several times and interacted with many potential recipients. Richard Krause, once head of the US National Institute of Health, remarked how viruses know no country, there being 'no barriers to prevent their migration across international boundaries or around the 24 time zones' (Krause, 1993: xvii). A key term here is a connected society, sometimes referred to as **time and space compression**, or a shrinking world (something experienced by many if not even a majority of the world's inhabitants). Speed and linkages are the issue, aided by what is increasingly regarded as a borderless world. Arguably travel and trade have become less restricted by barriers (though again, the effects are uneven, with the wealthy enjoying more freedom). **Neo-liberal** approaches to trade in particular have sought to lower restrictions and encourage the free movement of goods around the world.
As Krause suggests, even if there were firm borders it is hard to imagine a way to stop viruses travelling – for they do so as invisible co-travellers.

Travel and mobility is one thing, but there is also the issue of the conditions necessary for a disease to flourish or expand. Again, there are core geographical concerns that possibly start to explain some of the new diseases. Poverty, climate and environmental change, war, urbanization, economic recession and reduced public spending on health care can all contribute to the increased incidences of new infectious disease. You might want to think why each of these plays a role in what is undoubtedly a multi-faceted issue. But there is one more contribution to bio-insecurity that is important to explore – the role of animals.

It was estimated that in the last two decades of the twentieth century there were 177 new or re-emergent human diseases. Three-quarters of these diseases originated from animals and animal products (Taylor *et al.*, 2001). Some of these diseases pass between people and animals in the same way that we catch colds from each other; others are vector borne and others circulate in animal products including food. These diseases are called zoonotic, meaning that they affect animals and people. Influenzas, for example, are zoonatic, as they tend to either originate or pass through birds, and then can infect pigs who are excellent at making them more suitable for human infection. SARS is often linked to domestic civet cats and wild bats and circulates like flu. Bovine Spongiform Encephalopathy (BSE or mad cow disease) emerged in the 1980s and crossed to people as a result of eating infected beef, while food-borne bacterial diseases like E coli and campylobacter are responsible for most cases of food poisoning in so called advanced economies. It is also important to register the diseases that don't directly cross to people but can devastate livestock, livelihoods and economies. These are epidemics of animals and are called epizootics. In the UK in 2001, a foot and mouth disease epizootic cost the nation £8 billion, with huge losses of animal life and damage done to rural society. Another epizootic, bovine tuberculosis, is a disease of cattle that costs the UK Treasury roughly £90 million per year and results in the culling of 25,000 cattle per year. These diseases may not directly threaten human health, but they can lead to potentially catastrophic collapses in food systems and threaten food security. The latter relates to the ability of a nation or group of nations to provide enough food for its population, and there are fears that fragile food systems, which are highly networked and globalized, are increasingly exposed to disease or even biological attack.

So a major element in the new security around life is the nature of the relations between people and wild and domestic animals. Protecting a population involves attending to all those species, companionable or otherwise, with whom people live, that they depend upon, eat, may threaten and/or attempt to conserve. Geographers can therefore not only use their knowledge of a globalizing planet to analyse biological insecurity, they can also draw on the research that has investigated changing relations between people, animals and other non-humans (see Chapters 3, 10, 11 and 12).

SUMMARY

- A suite of new, emerging infectious diseases have hit the headlines in the last few decades.
- Many of these diseases seem to have taken advantage of rapid environmental change, global connections and a shrinking and speeded-up planet.
- Diseases that affect human as well as non-human animals are called zoonotic.

Responding to bio-insecurity

How societies respond to biological insecurity depends on many issues, but, simplifying somewhat, the degree to which differing emphasis is placed on the two main concerns that we raised above – a shrinking planet and human-animal relations – has a large bearing on the kinds of action taken. For two accounts of emerging infectious disease, each with its own proponents, geographical imagination and set of characteristic responses, see Case Study box.

CASE STUDY

Two accounts of emerging infectious disease

1. Disease attacks from outside 'Western' society

In this version of the rise of new diseases, the latter emerge from settings where people live cheek by jowl with their animals, or from environments where people or domestic animals become infected by wildlife, either through inadvertent contact or diet. A shrinking planet means that these diseases can soon break out and spread quickly. There is a geographical imagination here whereby 'the "primitive" farms of Guangzou, like the "primordial" spaces of African rainforests, temporalize the threat of emerging infections, proclaiming the danger of putting the past in (geographical) proximity with the present' (Wald, 2008: 7). Evidence seems to come from diseases like SARS, Ebola, possibly HIV/AIDS and perhaps avian influenza. The security threat is then from the Global South and East, and the responses to this threat from the wealthier North and West are accordingly geopolitical, cordoning off disease zones and fighting disease through a biosecurity intent on keeping things out, or re-colonizing countries through large capital investment in order to foster modernization.

Figure 57.2 A backyard chicken farm in Vietnam. Credit: AFP/Getty Images

Figure 57.3
A biosecure chicken shed in England. Credit: Author

2. Disease arises from within 'Western' society

In this version, disease threats are from *within* modern settings, notably within the farms and industrialized feeding operations where human-animal relations have been radically altered in the last decades. For geographers like Mike Davis, the 'monster at our door' is not so much the threat of disease from elsewhere, but from within. Insecurity resides in the proclivities of viruses and other microbes to mutate and flourish in settings where tens of thousands of animals, all with similar genetic make-up and eating the same food, live in close proximity (Davis, 2005). Evidence here comes from BSE and also perhaps avian influenza. The security threat is from a profit-driven effort to develop secure and cheap food supplies, and the response is less a matter for geopolitics and more a matter of challenging the politics of food, land use changes and the political ecologies and economies of production.

After reading these case studies, it might be clear where my sympathies lie (with the second account), but as geographers we need to account for the prevalence of the first story. After all, it tends to inform most state and super state actions on infectious diseases. This may be a matter of politics (the notion of disease existing outside Western society seems to make the issue more tractable, controllable and of course shifts the blame elsewhere). But it is probably more than that. It also relates to cultural understandings of disease as well as the history of the term security with which we started.

In terms of cultural understandings, biological security fears are the subject of any number of novels and Hollywood films, including most recently *Outbreak* and *Contagion*. The narrative or story in these films and novels is often

Figure 57.4 (a) Testing ducks for disease on a farm, and (b) biohazard clothing worn after a suspected outbreak. Credit: (a) Stephen Hinchliffe; (b) AFP/Getty Images

similar. Indeed, the literary scholar Priscilla Wald calls it the *outbreak narrative*, where a sudden and rapidly spreading disease emerges (often from a primordial setting like a rainforest, or sometimes from outer space) and the heroes fight irrational state structures in order to secure a future (Wald, 2008). This is disease from elsewhere and the heroic attempt of **modernity** to re-install order.

These films draw on and amplify a scientific convention which understands disease through 'germ theory'. The latter tends to view disease as caused by specific microbes or pathogens which invade bodies from outside, rather than as a condition relating to multiple stresses on an organism. In this sense, germ theory and germ warfare feed directly into a common understanding of disease and of security (i.e. keeping things out). This common sense though scientifically-supported understanding of disease therefore tends to feed a return to martial, conflict-based understandings of security. And it is this dominant sense of security that may account for the prevalence of a response to cleansing and colonizing so-called disease spaces.

The other main element to the security response to emerging disease is the emergency. While efforts are made to keep things out and to sanitize disease spaces, there is the constant warning that the next pandemic will eventually arrive. It is often said it is a matter of when, not if, and when it comes it will be so intense that old methods of insurance won't cover the damage. So a significant mode of operation highlighted by social scientists for all kinds of new emergency, from hurricanes to floods, from terrorist attack to disease, is *preparedness* (Collier and Lakoff, 2008). In short this refers to the need to be prepared for events that haven't been experienced directly before. Some of this is technical – having enough emergency beds in hospitals, stockpiling or having the option to buy necessary drugs (in the case of pandemic influenzas this means antiviral drugs and capacity for vaccine development and production). Some of it is about how responsive services and people can be to an uncertain environment (are the police ready, are networks in place for people to work at home, collect prescriptions without infecting each other and so on). How will a panic be policed? What emergency measures will be needed? As events that followed Hurricane Katrina in New Orleans showed all too well, these questions raise all manner of issue about the unevenness of societies, different access to health care, food stocks and mobility (especially in that particular case, private automobiles). While some of the concerns here might return us to a welfare version of security, and certainly the swine flu pandemic of 2009 and other events in the last few years have painfully revealed the lack of basic capacity in the US health care system and have raised questions about how issues are dealt with in the UK, the tendency has been to direct funds to disease containment and crisis management. Again, the insurance model that underpins health care systems no longer seems adequate for these emergencies, so we seem to return to a more martial, nation-state security as governments aim to be prepared and private firms offer financial and other services aiming to secure the future.

If we stopped there we might return to the table with which we started and note that the final cell relating the mode of operation for biosecurity would be somewhat similar to that of nation-state security. Indeed, many commentators have likened the response to emerging infectious diseases and other bio-hazards as a new arena of conflict and use of emergency powers. Pandemic disease requires states to retain emergency powers. Linking this to similar security concerns surrounding climate and environment generally, the social scientist Melinda Cooper has this to say:

> What is being articulated here is a profoundly new strategic agenda where war is no longer being waged in the defense of the state . . . or even human life . . . but rather in the name of life in its biospheric dimension, incorporating meteorology, epidemiology, and the evolution of all forms of life, from the microbe up.
>
> (Cooper, 2008: 98)

So as we move along that table, we note how the old modes of operation based on state war and insurance are no longer enough, as diseases breach the territories associated with the state and pose such catastrophic effects that insurance, based on populations, predictions and manageable risk, no longer works. Instead, Cooper argues, we have a situation where state governments sanction, in the name of life, far reaching methods like pre-emption and militarized responses to disasters, and increasingly private companies and corporations seek to profit from the heightened levels of insecurity.

There is a lot to recommend in this analysis of biological security – it can certainly help geographers to account for and critique current policy and the movements of, for example, investment banks and drug companies into new futures markets. But there is also a tendency to strike an air of finality, making all this seem inevitable with no room for states and other actors to look for alternatives. It may be that attending again to the competing geographies of disease affords a different biological politics. In the final section we look briefly at a case disease (avian influenza) and possible responses, in order to do just that.

SUMMARY

- Emerging infectious diseases tend to be understood as 'outbreaks', arising from outside of modern, Western society.
- Responses to new biological threats tend to be through emergency planning, leading some commentators to fear a new round of militarized state and corporate power.

Biological politics

You may have noticed that when we talked about the two geographies of emerging diseases (attack from outside and arising from within), avian influenza was used as evidence for both. Indeed, highly pathogenic avian influenza (HPAI) is often considered to be a disease of wild birds that is passed to domestic poultry on farms with poor biosecurity. That is, those farms where birds can mix with wildlife are, according to the Food and Agriculture Organization (of the United Nations) and the World Health Organization, more likely to develop avian influenza. Moreover, as those farms tend also to have unregulated contact between people, especially children, other animals and poultry, these are the spaces that are most likely to see the necessary zoonotic crossings for a really dangerous form of the disease to develop. Such a narrative or outbreak story has its own answer – modernize farms, make sure poultry is raised in enclosed spaces with limited contact with other organisms, ensure slaughter is sanitary and institute disease surveillance in order to forewarn of any possible infringements. Many states have started to follow this path – in Egypt for example, after the first cases of HPAI in 2006, the authoritarian government attempted to eradicate 'backyard poultry', especially in or close to cities, and supported a massive investment in factory farming (Hinchliffe and Bingham, 2008). Similarly, large capital investment, funded by large banks seeking new revenues after the collapse of property markets in 2008, is being directed to CAFOs in Asia and Africa. These facilities are designed to exclude disease and as a result can trade their animal products internationally.

And yet, avian influenza is becoming a concern in the very settings where chickens and other poultry are enclosed and concentrated in so-called bio-secure settings. Indeed, there are those who would argue that the problem is made worse by having so many chicken bodies together in one place and growing at a rate that makes their immune systems highly compromised. Moreover, an outbreak on a concentrated feeding operation with tens of thousands of birds is harder to contain than one on a small farm with a few hundred birds. The waste outputs from the CAFO, including faeces, amount to a huge logistical and environmental problem even before there is a disease issue, but once the wastes are contaminated these problems are compounded. Moreover, avian flu is not simply something that comes in from the outside. To be sure, avian influenzas circulate in wild birds, but it is the tendency for the influenza viruses to change (i.e. to mutate or reassort) that is the biggest fear. While this can happen in the wild, it tends to do so sporadically and strains soon die out. But when this happens on a large poultry farm, the effects can infect thousands of birds overnight and circulate within the industry rapidly. Indeed, it turns out that the very measures being used to counter the bird flu threat – enclosure and industrialization – may be the cause of greater insecurity in the future. So much so that we may in fact be breeding influenza (Wallace, 2009).

The point for our purposes is this: while it may be true that new biological threats have led to emergency measures and new kinds of action wherein, for example, small farmers are being displaced by large sheds on a global scale, there is also a politics to be engaged with that questions the ways in which human and non-human relations are being changed in the name of security. Geographers can use their critical skills to question current models of threat (attacks from outside) and can use work on human-animal relations to challenge the notion that enclosure of animals is the answer. Above all, a geographical imagination – one that is alive to contingencies and openness of the world – can cast a critical eye on any claim to be able to control the world. When we believe we are in control we probably have the most to fear.

Conclusion

We may live in what the director of the World Organisation for Animal Health (OiE) called a 'global biological cauldron' (cited in Braun, 2008: 260) wherein new biological threats are being created on a daily basis. But that cauldron may have its hotspots, its points of fermentation and amplification, in places that are rather different to the conventional understandings of disease as coming from outside modern Western society. Rather than the primitive farms and viral hotspots in rainforests, they are present in the heart of modern society. There is then a different geography to disease, and possibly a different security politics to engage with once we have traced this geography. Biological security may not simply be a revision of previous, conflict ridden nation-state security, or even a rush to speculate on uncertain futures, but a political engagement with how we make human and non-human lives, and how we live together with others.

SUMMARY

- The emergency generated by new biological threats like emerging infections may not be so much one in which we should be mobilizing military and corporate resources or power to handle crises or outbreaks, but used to open up debates on how we raise animals and feed the planet.

DISCUSSION POINTS

1. What would you write in the final cell of Table 57.1?
2. Why aren't the other modes of operation listed in that table sufficient to life?
3. The current global human population is around 7 billion. Around 52 billion chickens are killed every year to give this population a high protein diet. To what extent do you think the need for abundant cheap food leads to new biological threats?
4. Look again at Figures 57.2 and 57.3 – which is the more bio-secure system of farming? Explain your answer.
5. For each of the geographical terms listed below, demonstrate how and why they explain new biological threats:
   - Globalization
   - Trade
   - Reduced species diversity
   - Human power over nature
   - Uneven development
6. Given the evidence contained in this chapter and in your wider reading, to what extent do you think humans can control life and its processes?

FURTHER READING

Davis, M. (2005) *The Monster at Our Door: The Global Threat of Avian Flu*. New York: The New Press.

A key intervention in shifting the debate away from outbreaks and distant threats to 'Western' lifestyles and towards a concern with modernization itself.

Leach, M., Scoonies, I. and Stirling, A. (2010) Governing epidemics in an age of complexity: Narratives, politics and pathways to sustainability. In: *Global Environmental Change* 20, 369–77.

A detailed critique of the outbreak narrative calling for a more sustainable approach to world livestock farming.

Hinchliffe, S., Allen, J., Lavau, S, Bingham, N. and Carter, S. (2013). Biosecurity and the topologies of infected life: from borderlines to borderlands. In: *Transactions of the Institute of British Geographers* 38:4.

More detail on the geographical tools used in understanding contemporary disease threats and the requirement to find new approaches to making life safe.

Wald, P. (2008) *Contagious: Cultures, Carriers, and the Outbreak Narrative*. Durham, NH: Duke University Press.

A fascinating cultural history of contagions and the stories we tell about disease, its origins and solutions to infectious conditions.

Wallace, R.G. (2009) Breeding influenza: the political virology of offshore farming. In: *Antipode* 41, 916–51.

A detailed and compelling argument linking the expansion of capital investment in food and chicken production to the rise in emerging infectious diseases.

Film mentioned in the chapter:

Contagion (2012). Dir. Steven Soderbergh: provides a detailed engagement with contemporary biological threat.

Game/ App: The app entitled Plague Inc. in which you play as a microbe seeking to infect the world is a good lesson in the dynamics of global mobility and microbial effectiveness.

SECTION FOUR

PUBLICS

Introduction

Accounts of the geographies of public life have tended in recent years to reflect a rather pessimistic narrative of the triumph of the individualistic over the collective. Political moves towards privatization and deregulation have dismantled some aspects of public welfare and service provision. Instead, the making of the neoliberal subject has emphasized the capacity of individuals to make their way in the world, to exercise consumer choice and to take responsibility for their own welfare and well-being. Despite spirited defence of cherished public collectives – such as the valorization of the National Health Service in Danny Boyle's opening ceremony for the 2012 London Olympics (see Figure IX) – there is often a sense of inevitability that the fragmentation and individualization of public life will mean that looking after number one (and number one's family) will transcend more corporate and collective ideas of concern and intervention. Such is the perceived withering of, and disengagement from, public life, that we have begun to assume a post-political phase in which we can't see how anything can be changed much in the formal political arena. The main sites of hope or optimism for many people lie in the new forms of public participation: in the re-birth of activism and voluntary and charitable care that occur largely outside of formal public policy, despite attempts to incorporate it into the political card-trick of the so-called 'Big Society'.

Against this background of widespread pessimism and flickering optimism, Human Geographers have been turning their critical focus onto concerns about ethics and responsibility. However, the pursuit of an appropriately critical voice with which to deal with these concerns has been tricky. Some argue that the rise of sectoral identity politics (relating, for example to gender, ecology and sexuality) has rendered ethics and responsibility increasingly complex, fragmented and easily absorbed by the surrounding blanket of supposed tolerance that seems to characterize contemporary liberal society. In so doing, it is possible that the postmodern remapping of the 'political' in Human Geography has risked emptying that 'political' of its fundamental meaning. Others have therefore called for more normative critical thought, to include more stable sets of political and ethical ideas about human good and flourishing, and to restore what some see as a lost radicalism to critical Human Geographies. Perhaps the most interesting Human Geographies of ethics and responsibility, however, are those that seek to find new ways to work with this complexity rather than to reduce it to either-or strategies that oversimplify the ways in which we shape, and are shaped by the responsibilities and irresponsibilities that surround us – responsibilities and irresponsibilities that we affect, and that affect us. Prominent among such approaches are Human Geographies that assess collective responsibilities and values in the public sphere, and those that trace the emergence of ethics in the responsibilities that are materialized in our embodied actions and dispositions towards others.

In Chapter 58, Clive Barnett introduces a discussion of the public sphere, and in particular poses the question of how and why things in the public sphere should be, and are valued. Part of the significance of this question lies in the contemporary climate of political change, in which there has been a strong sense that the public sphere has declined and that public cultures have been severely challenged. There has thus been a rather pessimistic picture painted of the deliberate shrinkage of the state and the consequent dilution of collective values due to the political championing of individualism. Such pessimism has only partly been counterbalanced by more hopeful stories of the collective innovation and originality (for example, in new forms of service, welfare and care) that have emerged under these circumstances. In among these changes, he asks what is 'public' and how is it articulated in the everyday grammars and practices of daily life.

The answers to these questions are fundamental to contemporary Human Geography. The idea of the public is sometimes associated with a significant actor, sponsoring concerted action both in specific spheres, for example, public health and public services, and in the name of elusive political and ethical values such as the public good and the public interest. Crucially, these ideas of 'the public' are often constituted by a series of specific values: equality, fairness, openness and legitimacy. Publicness can thereby be regarded as an emergent quality – informed by these values – that takes shape in a number of important but different ways. Barnett discusses three such spheres of emergence: First, he discusses public *spaces* in which new forms of citizenly engagement are developed. These may be open expressions of opinion-forming, such as the Occupy activist congregations in places such as St Paul's Cathedral, or they may take place in less famous locations where transformative narratives of public virtue, for example dealing with inequalities relating to gender and sexuality, are given prominence. Second, he points to changes in public *culture*, in which new forms of exchanging, communicating and mediating opinions serve to open out ideas about public value across what might previously have been regarded as separate and different socio-economic cohorts of the population. The post-Olympic discussion of the role of sport in education represents an example of this emergent formation of new public cultures. Third, he points to public *politics* and *ethics* by which the public will is translated into new programmes of service, welfare and care, as evidenced, for example, by the recent growth in third sector organizations dealing with homelessness and human trafficking.

What characterizes these public geographies is that public value is based on sharing, on the binding together of strangers, and that new combinations of sharing and dividing are always in formation as public values emerge and re-emerge. Increasingly, Human Geographers have looked to the idea of ethics to underpin these public values of sharing. In Chapter 59, Keith Woodward explores a range of Human Geographies that assess both our responsibilities to human and non-human others and the degree to which these responsibilities are materialized in our actions and dispositions towards others. He argues that ethical commitments are produced in and through the social relations in which we are situated. This may be partly about conscious thoughtful action, about normative moral judgements, but it is also brought about by our wider involvement in the much bigger process of making spaces and places and of participating in the ecological transformation of the environments in which we live.

Woodward's account explores two registers of ethical geographies. First he identifies *the body* as a key ethical space, not simply in terms of the capacity of the mind to make seemingly rational decisions, but also in the sense that our bodies become enrolled in a wide range of interactions with humans and other entities that are often beyond our control. Such interactions can bring about unintended consequences elsewhere, but can also affect our practices, thoughts and feelings. We thus shape, and are shaped by, the social and ecological world in which our lives unfold. Ethics are not concerned simply with our capacity to make judgements; ethics are also about how we become enlisted in myriad ways in the wider transformation of the world. These understandings are further illustrated in Woodward's second ethical register – *the space of the other*. Here he assesses our ethical responsibilities to those whose lives are detrimentally affected by processes and practices that naturalize and perpetuate unevenness, injustice and prejudice towards difference. Yet again here we encounter discourses of pessimism and optimism. The example of Union Carbide's treatment of Bhopal victims demonstrates how local people were transformed by the power of multinational capital into a set of distant others, somehow unconnected to the ethical identity of the (in this case, corporate) self. By contrast, the growth in fairly traded commodities has succeeded in some measure in bringing the care of others directly into the sphere of caring for the self. Ethical optimism arises not so much from altering consumption patterns, as from the rising up of new ethical movements in order to mobilize ethical devices (such as fair trade) that help to govern the ethical self and to interconnect that self to the process and practice of caring for others.

These ethical geographies suggest implications for Human Geographical practice. The rising influence of participatory geographies highlights the need to be part of the sharing process that underpins both new forms of publicness and particular community projects. However, even with the best will in the world, it is often difficult to grasp the complexity of ethical, cultural and political interconnections that occur when we encounter lives in other spaces. We need to infuse our geographical practices with an ethics of solidarity in order to empower a sense of collective and shared living, with an ethics of care that help to co-produce shared responsibility for communities and spaces, and with an ethics of power relations that emphasizes 'power-with' rather than 'power-over' so as to reinforce the value of collective life.

Figure IX: A celebration of the National Health Service in Danny Boyle's opening ceremony for the 2012 London Olympics. Credit: [illegible] Images

CHAPTER 58
HOW TO THINK ABOUT PUBLIC SPACE

Clive Barnett

Introduction: emergent publics

This chapter examines why geographers and other social scientists sometimes worry that public space is being enclosed, why they are concerned that public culture might be becoming less serious or less rational and why they think that public services shouldn't be privatized. Why, in short, they think that 'public' things are to be valued.

Since the translation into English of Jürgen Habermas' *The Structural Transformation of the Public Sphere* (Habermas, 1993) the concept of the public sphere has become a central reference point across a range of fields for trying to evaluate rapid changes in the institutional configurations, economic foundations, technological mediums and socio-cultural formation of contemporary public life. Habermas told a tragic story, in which the conditions which laid the ground for the emergence of a classic model of a liberal public sphere in the eighteenth century (not least, a vibrant, commercialized press) end up in the twentieth century undermining the norms of that public sphere (because people end up watching too much TV). Subjected to an enormous critical literature, what Habermas' account retains is a basic challenge for any consideration of the value of publicness: the challenge to how we approach concepts such as the public sphere, which are at once descriptive *and* evaluative.

Capturing the dynamic transformation of public life both descriptively and evaluatively is a challenge. Existing ways of discussing publicness generally don't succeed in doing both of these things well at the same time. There is a tendency to tailor descriptions of publicness to evaluative judgements.

Contemporary evaluations of new public cultures remain caught between pessimistic discourses of *decline* and optimistic discourses of *originality*. The framing of debate about publics and publicness within these two discourses has produced an impasse that blocks conceptual, empirical and normative analysis. The pessimistic perspective discerns consumerism, individualism, marketization and privatization as leading inevitably to disengagement and withering of public life. In contrast, the optimistic perspective sees new technologies as panaceas for past injustices and exclusions, leading to the restoration of direct participation, the proliferation of opportunities for personal expression and the re-birth of movement activism.

Both pessimistic and optimistic discourses of the fall and rise of the public sphere leave in place settled criteria inherited from the past. In both cases, empirical changes in the material, institutional and social configurations of public practices are judged against static criteria of what public action should look like, where it should take place, who should participate, towards what ends it should be directed and in what registers it should be articulated. It is settled images of these matters that allows the proclamation of new subjects of public life, or the proliferation of new objects of public concern, and the celebration of new mediums of public expression. But the same settled criteria also underwrite laments that public life has been coarsened, fragmented or individualized. Different narratives alight on different aspects of publicness to either celebrate new forms – the self-organizing dimensions of flash-mobs, the exuberant deliberations of online forums – or to bemoan them – the replacement of properly informed citizenly media by celebrity-dominated, emotive registers of public culture or the fragmentation of diverse publics rendering impossible any unified, concerted opinion-formation to take place.

It is between the optimistic and pessimistic narratives about public life that Habermas' challenge lies: How are we to derive criteria of evaluation of emergent trends from historically specific models which might themselves be in the process of being left behind? This chapter responds to this challenge by trying to define just what sort of values are invested in the concept of 'the public' and its various derivatives.

Figure 58.1 Online publics. Credit: AFP/Getty Images

Figure 58.2 Simon Cowell, judge and producer of television show 'The X Factor'. Credit: Getty Images

SUMMARY

- Evaluations of contemporary transformations of public life tend to oscillate between optimism and pessimism.
- The conceptual challenge of traditions of public sphere theory is to negotiate a path between these two poles.
- They key challenge in thinking about the value of ideas of public space and the public sphere is to think more carefully about how criteria of evaluation derived from one context can be applied to new contexts.

The grammar of public value

One problem that any discussion of the public, the public sphere, public life – publicness in general – immediately faces is that of definition. Just what is a public? Or is that even the right question? Is a better question 'What is public?' Consider the following questions:

- Is public a name given to particular spaces, by virtue of their openness?
- Is public a name given to certain institutions, by virtue of their function or degree of accessibility?
- Is 'the public' a collective subject of some sort, and if so, who is it composed of?
- Is public a name we give to certain sorts of action done from particular motivations – in the public interest or publicly spirited?
- Or is public a name given to actions undertaken in particular patterns of interaction, collectively, as a public, as distinct from privately?
- Or is it better to think of publicity as more like a medium into which and out of which one can move – by going public, making things known, exposing oneself or others to scrutiny of an indeterminate yet attentive audience?

By raising this cascade of questions, my point is to suggest that if we attend to the *grammar* of 'public talk', we can begin to see some of the difficulty in trying to nail down a clear and concise definition. So, what do we notice if we attend to the grammar of public talk in this way?

We notice that 'public' is at once a noun and an adjective, something one can be *in* as well as something you can *move* into (by going public). In this latter sense, public is also used as a verb, something one does, for example, as in publicizing, to publish. We also notice that 'public' is a name given to certain sorts of agents (the public, the public sector, public universities), as well as the name for certain types of action (ones distinguished perhaps by their location and/or their motivation). And we notice too that public actions are not necessarily restricted to public agents. All sorts of private agents can undertake actions that individually or collectively serve the public interest – and some people think the best way they can do so is by acting out of self-interest (see LeGrand, 2006).

Maybe to find a way through this thicket of grammatical complexity we just consult a dictionary? Well, if we consult a dictionary for the meaning of 'public', we enter into an enormous terrain of multiple meaning[1]. We find that not just is the word public, as both a noun and an adjective, variously defined, but that there are a host of designations of things, sites, activities as public, for instance public lavatory, public house, public eye, public nuisance, public opinion, public transport, public holiday, and so on and so on.

Nevertheless, I think it is possible to identify a family of recurring themes across this variety, broadly divided three ways into thinking of 'public' as an adjective, a noun and a verb.

1. There is an adjectival sense of public, defined against things private, which signifies a sense of openness. This might be a spatial sense, where public is related to exposure, to being on show or available to others, but it is a sense that resonates, too, across a political-economic terrain of definitions of the market and the public sector, without wholly capturing what is most at stake in this field.
2. There is also the nominal sense where the public is a name for a certain type of collective, a synonym perhaps for the

Figure 58.3 'Public' has multiple meanings. Credit: AFP/Getty Images

community, or the nation, or sometimes offset against these more **embodied**, substantive collectives. This second sense is crucial to different understandings of the value of publicness, because depending on which field of analysis one looks at, one finds markedly different senses of what sort of existence this collective view of the public can and should have. In certain strands of strongly republican political theory, for example, the public as a collective entity exists only in and through the reflexive medium of its own openness. This tradition sees publics as self-organising collectives, gathered together in a 'space of appearances' to consider matters of shared concern. It is a tradition that is intensely suspicious of any institutionalization or instrumentalization of the public in organizational form – for example, the idea that the state can embody the public is anathema from this perspective.

3. Yet there is a third sense of public which does open itself up to this more institutional sense of the public as a concentrated, sovereign actor. Here, public refers to certain functions that authorize some actors to act on behalf of others in a particular way, in the name of a quite abstract sense of the public, such as the public good, public health or public interest. It is this sense that is captured by the ideas of public service and the work of public servants, who act for, represent, act on behalf of and care for members of the public, sometimes up close and personal, sometimes in the most general of senses. What makes these sorts of delegated agency and trusteeship qualify as 'public' is that they are enacted in the name of values of equality and

impartiality that loop back to the first sense of openness.

Thinking of these three senses of public – with reference to the value of openness, of collectivity and of serving or representing – as the core values around which publicness is constituted, contested, and transformed might be a way of starting to get a handle on the variety of values associated with publicness. Bringing these three senses together throws into relief the importance of attending closely to the question of what sort of collective actor a public is – to the question of what distinguishes a public from society, the nation, community, the people, audiences, civil society, the state. Above all, though, and this takes us back to the lesson to be drawn from Habermas, it should alert us to the importance of attending to the variable formation of publicness, through practices, and in relation to problems and issues.

SUMMARY

- 'Public' is a name attributed to all sorts of actors, actions, objects, processes and subjects.
- Attending to the grammar of public talk – to how the value of publicness is invoked in specific contexts – might help in understanding the overlapping family of values indicated by such invocations.
- It is possible to identify three overlapping values which public talk is often used to articulate: the intrinsic value of openness, the intrinsic values of collectivity and instrumental values of concerted action.

Imagining the public

So far, I have suggested that publicness needs to be thought of as an emergent quality, which is best investigated by attending to the variable combination of practices, meanings and, crucially, values (Mahony *et al.*, 2009). The analysis of the transformation of public life should be guided by the question of 'what are publics good for?' The question is meant to attune analysis to the ways in which different repertoires of evaluation and different claims of value are used to sustain, challenge and transform settled public formations.

Approaching the analysis of publics with this question in mind means more than simply looking at what 'public' or 'the public' means in different contexts; the sense guiding us here is that 'public' is not best thought of as merely an empty signifier. Rather, there is a cluster of values that overlap around the vocabulary of publicness: openness, sharing, living together, authority, legitimacy. Attention should focus on how these are assembled and re-assembled in contested claims about what sort of thing should or should not be a matter of public debate, an object of public management or a benefactor of public financing. We should also learn to attend to how the meanings of the public emerge, not least in contrast to other values, such as those associated with 'privacy' or 'the market', for example. Finally, we need to negotiate between the idea that there are some activities which just are public by their very nature and the idea that public things are things contingently named public – if we are to do so, then we need to keep in view the importance of communicative action (e.g. deliberating, engaging, encountering) in delineating the scope and shape of public.

In this section, I want to develop a little further the idea of looking at the 'grammar' of publicness as suggested above. I want to do so in order to illustrate how the analysis of the formation of public life might proceed empirically as well as conceptually, without losing sight of the inherently evaluative quality of any investigation into public life. Below, I identify three institutional configurations which have been considered paradigms of public life in different academic fields. The reason for distinguishing these paradigms of public value is to draw out the values that are embedded in these institutional models, values which are at stake in critical and diagnostic accounts of the contemporary transformation of these models.

Public space

The first paradigm of public value is **public space**, the focus of attention in spatial disciplines such as Human Geography, urban studies, architecture and urban sociology. Public spaces are, though, also a focus of attention in political philosophical accounts of the public sphere, as either figures of the public or as empirical scenes for certain sorts of practices. This field of literature focuses on particular sorts of spaces – public parks, streets,

Figure 58.4 Women demonstrating in Tahrir Square, Egypt. Credit: AFP/Getty Images

shopping malls, cafes – as exemplary of certain values of publicness. The first thing to say about this field is that it is primarily concerned with a particular function of the public: the background conditions of a certain sort of sociality that is taken to be crucial to more formally public, citizenly forms of engagement. These are, then, in the vocabulary of public sphere theory, relatively weak publics, not strongly articulated with concentrated sites of authority, but rather scenes where the virtues and capacities of public encounter are learnt and put into practice (e.g. Watson, 2006).

One of the crucial contributions of **feminist** scholars to understandings of public space is to point out that the sorts of virtues and activities often associated with these paradigmatic public spaces might be found in more privatized, secluded or secreted locations. This work begins to unsettle the assumption that *public action* is necessarily action that takes place in *public space* (see Barnett, 2008b). This assumption is further unsettled by work on queer publics which draws into view the ways in which different configurations of public exposure might support emancipatory or oppressive relationships. Related to this functional concern with sociality is the primary concern in these fields with a particular set of public values, primarily those associated with openness, accessibility and inclusion in spaces of interaction.

Specifying the function and values associated with public space in this way allows us, in turn, to notice the ways in which transformations of public space revolve attempts to redefine these functions and values. So, for example, concerns with public order and public nuisance lead to regulatory adjustments in the management of public space which give less credence to the citizenly value of unanticipated encounters with difference than to norms of security and safety or which put a premium on not being offended. In crucial respects, this logic might amount to a privatization of public spaces in so far as it shifts the balance from being open to uninitiated contact with others towards allowing subjects in public exercise more control of the shape and content of their interactions.

Public culture

The second paradigm of public value has a different function from the first one, and focuses primarily on the opinion-forming aspect of a vibrant *public culture*. This is, of course, one aspect of the public spaces discussed above as well, but a crucial dimension of the notion of the public sphere is the way in which such 'real' spaces of face-to-face interaction are embedded within mediated circuits of communication. So the key institutions of public opinion-forming would include broadcasters (public and private), newspapers, print cultures more generally (publishing, public libraries), museums, churches, schools and universities and, just to keep up to date, social media and other internet-based spaces of interaction. These form the circulatory infrastructure through which opinions, information, science and religion is made available to dispersed populations. It is important to acknowledge that this range of institutions illustrates the degree to which vibrant public cultures are often dependent on and sustained by private institutions of different sorts, and not least, by the operations of markets. If the key public function of these institutions is that of keeping citizens informed and allowing them opportunities for free and unfettered expression, then in turn there is a primary value at stake in thinking of these sorts of institutions as public, irrespective of the source of funding or degree of selectivity of particular examples. This is the contribution such institutions make to the development and

Figure 58.5 BBC Broadcasting House, London. Credit: Getty Images

circulation of a *shared culture*, of a world held in common by all citizens and available to all to engage with and appropriate as their own (Wessler, 2009).

The institutionalized, mediated qualities of public culture in this expansive sense has always been haunted by a worry about the **paternalism** involved in presuming who knows what is good for audiences, listeners, readers and viewers by the way of information, entertainment and education. This paternalist worry has underwritten a strong trend towards market populism across various fields of cultural policy and cultural economy. But markets in cultural goods are hardly 'perfect' in their responsiveness to the needs and preferences of members of the public. The combined impact of new technologies and privatized media lead some to worry about the fragmentation of a once unified public culture into myriad enclosures that means we end up only ever being exposed to our dose of 'The Daily Me'. In contrast to this worry about *fragmentation* of the public sphere, there is a concern that market-dominated logics in cultural provisioning lead to a decline of *pluralism* in public culture. This concern with pluralism is more interesting because it does not suppose that the value of shared culture needs to be modelled on the ideal of a unified, single culture. Rather, it focuses attention on the quality and quantity of opportunities for sharing, where sharing is understood as a process of exchange and communication across difference.

These two paradigms of public value – public spaces and institutions of public culture – both emphasize the importance of recognizing that the public sphere is about much more than politics or **citizenship** narrowly conceived. They remind us that the relationship between a wide, dispersed public *culture* and the political functions ascribed to public *deliberation* are complex ones and as such, these two fields should not be thought of as identical or directly and causally connected (Wessler, 2009). The question of defining their relationship is, of course, a staple of public sphere theory from Habermas' original tragic narrative onwards.

The state as public agent

The third paradigm of public value cleaves more closely to what one might call the public function of 'will-formation' than the first two, and would include various political formations of the state, including both welfare agencies but also procedures of election, legislation and policy making. Here, the public is understood to be a collective subject whose will is embodied in and whose interests are protected by the institutions of the state – the agencies of the public. The function of institutions of will-formation is to filter dispersed opinion-formation into *actionable decisions* and to implement these through programmes of service provision and distribution. And the key value underwriting these configurations, in their idealized liberal democratic form at least, is that of equality – whether the equality of participation through

Figure 58.6 Houses of Parliament, London. Credit: Arpingstone (Wikimedia Commons)

electoral enfranchisement or the equality of impartiality embodied in expansive systems of welfare provision (see Newman and Clarke, 2009).

These three different paradigms of public value embody and enact particular values of publicness – interaction, common culture, equality, transparency – and with different emphasis on the relation between 'weak' public actions such as chatting and strolling around parks and 'strong' public actions such as electing a government or distributing the revenues from taxation.

All three paradigms illustrate the following issues that any consideration of public value should always keep in mind:

1. The value of publicness is enacted in various *practices* – in voting, in being counted, in deliberating, in shared rituals.
2. Public values are expressed through various forms of *communication* – rational ones, reasonable ones, passionate ones.
3. Invocation of public value is always with various modalities of *power* – publics can be weak or strong, they can influence or exercise power, they can act as sieges against concentrations of power or as sluices enabling its more democratic regulation.

The question remains, however, of what this way of thinking about public value implies for the ways in which geographers approach issues of public space. We consider that topic in the next section.

SUMMARY

- Three paradigms of public value can be identified across different academic and institutional fields.
- Public value is sometimes thought to be embodied in spatial configurations of action; sometimes in institutional configurations of public culture; sometimes in institutional configurations of state or 'state-like' power.
- Thinking of public value as variable across different institutional fields helps us see that public value is always emergent in relation to particular issues and problems.

Spaces of public action

To illustrate this sort of analysis of the variable formation of institutional configurations of public value, it is useful to consider the example of the recent work by the social theorist Craig Calhoun. Calhoun's work crosses a number of the fields in which the theme of public is currently at stake – social theories of multiculturalism and national integration, political theories of cosmopolitanism, public engagement of university institutions and the privatization of welfare systems. Calhoun (2009) identifies four features that make a public:

1. The collective creation of institutions and the sharing of collective life.
2. Some goods are inherently public goods, in the sense derived from economics, where we can only enjoy some things if we share them.

3. A sense of a public as joining together strangers;
4. A sense of active participation in discussing and deciding what is held to be good.

What emerges from these four features is a strong impression that the value of publicness has to do with forms of *sharing*. Now, sharing sounds like a nice value. But of course, to share something, even to share in something (a pastime or a meal), also involves dividing, appropriating, making use of. Patrick Chabal (2009) talks of the idea of partaking as a term that captures the ambivalence of collective life – the idea of *taking part* in activities with others and also of *making use* of shared resources. Chabal is interested in rethinking the value of participation, which isn't quite the same sort of value as publicness, though they are related. Publicness, in fact, seems to incorporate a certain style or mode of participation, of partaking with others. It is worth thinking about this idea that the value of public life involves certain styles of participation a little further. This is because it begins to bring out something about the peculiar spatiality and temporality of publicness that geographers should take more notice of than they are sometimes prone to do, particularly in their eagerness to affirm the importance of public space. There are two aspects to the idea of public participation as sharing-and-dividing worth mentioning:

First, for social thinkers concerned with understanding publics as a distinctive feature of **modernity**, such as John Dewey (1927) for example, a key feature of publics was their emergence through processes of 'indefinite extension', associated with the development of communications innovations such as print, railways and telecommunications. The significance of this extension of human communication lies, for thinkers like Dewey, not just in the stretching out over time and space of social interaction, but in the associated development of new forms of subjectivity shaped by the indefinite qualities of these mediums of extension. Publics are characterized as collectives in which social bonds work through rather than just in spite of anonymity and difference.

Second, in both classical literary accounts of the modern public, as well as recent literary theoretical and philosophical accounts of the public sphere, a key feature of a public is this idea of a community of strangers. Both reading publics and the modern city are recurrent figures for this type of collective, in which the value of openness of a public space or medium is specifically related to being exposed to 'initiation of communication by others'.

Now let us combine these two emphases: the quality of *indefinite extension* characteristic of the mediums of public communication, and the binding together of *strangers*. By doing so, you can begin to glimpse a sense of publicness as an emergent quality which oscillates between dependence on a pre-given background of social relations, communications mediums and urban infrastructures; and the transfiguration of this background in an excessive movement of reflexive self-organization (see Warner 2002). Keeping in mind this sense of the emergence of publicness allows one to see 'public' as a name given to a place you can go to, or an institutional configuration; and that public values can be expressed in action, performed in practice.

So here is the lesson for how to think about public space: what makes a space public, or not, is how it is used, how it can be used and to what use it can be put. Or, if you prefer, public action determines the definition, shape and extent of public spaces. We need to stop thinking of publicness primarily as a type of *space* and instead focus on the type of *action*

Figure 58.7 Occupy Wall Street movement camping out in Zuccotti Park, New York. Credit: New York Daily News/Getty Images

that is attributed the status 'public'. And public action, by definition, is not actually or conceptually contained within particular configurations of place, space or territory (see Barnett, 2008b).

SUMMARY

- A key feature of definitions of public value is an emphasis on the value of sharing.
- Sharing is understood to be foundational to public life in so far as it implies modes-of-being in common which depend upon division, diversity and anonymity.
- Public value is dependent on the contexts in which actions are undertaken and this means that the public status of spatial configurations cannot be established in advance of an analysis of the mode of action under which they are constructed and made use of.

Conclusion: investigating emergent publics

I have presented a rather abstract analytical framework for investigating the variable formation of public life. This framework is guided by a conviction that public value is an emergent quality, by which I mean that issues of public value 'break out' around problems, issues and processes that are not easily anticipated in advance. There are, I would suggest, three dimensions to the formation of any new configuration of public life; three areas to consider when investigating the contemporary transformation of public life, as described below.

- *The emergence of new objects of public action*: The concerns over which public debate and decisive action are demanded, and around which communities of affected interest are formed, have multiplied. For example, the proliferation of environmental concerns transforms the most mundane of everyday domestic practices into activities with public significance.
- *The emergence of new subjects of public action:* The identities around which collective, participatory agency is mobilized have likewise been transformed. For example, the restructuring of welfare systems generates new forms of rights-based mobilization by patients groups, while 'living wage' campaigns engage multiple identities around contingent, issue-specific campaigns.
- *The emergence of new mediums of public action:* The means through which issues emerge as public concerns, through which demands for attention are addressed, and through which action in response to these concerns is enabled have been reconfigured. For example, new communications technologies restructure the rhythms and norms of public media cultures, while the potential for markets to serve as mediums for public action is being explored by a variety of activist campaigns.

The account of public value developed in this chapter suggests that these three different dimensions of *public emergence* might be combined in different ways in specific situations. When investigating transformations of public life, it is always best to avoid idealizations in which public is offset against private, state against market, communal virtues against self-interest. Start instead by asking how existing public values are invoked and reconfigured in specific situations.

So, what is a public? Or what counts as public? The answer is: it depends.

DISCUSSION POINTS

There are no right answers to the following questions. Instead, they are intended to encourage you to think about when and where you might undertake actions which count as more or less public in their motivations, mediums or consequences.

1. What places do you think of as 'public spaces'? What is it that makes these places 'public' – is it because they are open and accessible, is it because they are free to enter or is it because they cater for wide ranging collective needs? Perhaps you can think of places that might be considered public for different reasons.

2. What activities do you undertake that might be considered public actions? What is it that makes them 'public' – is it because of where you do them or because of who you do them with? Or is it because of why you undertake them – the spirit in which you undertake these activities, or perhaps is it because of their likely consequences? Maybe you can think of activities that count as 'public' for different reasons.
3. Can you think of something you do in public space which is, nevertheless, a private activity? And can you think of something you do in private spaces which might, nevertheless, count as a public action?

FURTHER READING

Barnett, C. (2008) Convening publics: the parasitical spaces of public action. In: K.R. Cox, M. Low and J. Robinson (eds.) *The Sage Handbook of Political Geography*. London: Sage, 403–17.

A conceptual review, critique and reconstruction of how public space is theorized in Human Geography.

Ivesen, K. (2007) *Publics and the City*. Oxford: Blackwell.

This book develops an innovative account of the relationship between media, urban space and public life.

Latour, B. and Weibel, P. (eds.) (2005) *Making Things Public: Atmospheres of Democracy*. Cambridge MA: MIT Press.

A wide-ranging collection of short essays on various mediums and spaces of public life.

Parkinson, J. (2012) *Democracy and Public Space*. Cambridge: Cambridge University Press. An important reconceptualization about how and why physical spaces of public interaction matter to the performance of democratic institutions.

Schudson, M. (2009) A family of public spheres. http://publicsphere.ssrc.org/schudson-a-family-of-public-spheres/ Accessed 23 October 2009.

An accessible introduction to the different issues raised by theories of the public sphere.

Warner, M. (2002) *Publics and Counterpublics*. New York: Zone Books.

An important contribution to theorizing the plurality of subjects, spaces and mediums of public life – ranging from historical and literary publics, to contemporary publics of queer activism.

WEBSITES

http://publicsphere.ssrc.org/

The Social Science Research Council (SSRC) hosts a web resource dedicated to academic and policy debates about the public sphere from across a range of disciplines.

www.ssrc.org/

The SSRC has a number of ongoing programmes of research on aspects of contemporary public life around the world.

www.nyu.edu/ipk/

The SSRC's programmes of research are closely related to work at the Institute of Public Knowledge at New York University.

www8.open.ac.uk/ccig/programmes/publics

The Centre for Citizenship, Identities and Governance at the Open University has a Publics Research Programme, which also contains information and links to various ongoing projects of work on public life.

www.cbc.ca/ideas/episodes/features/2010/04/26/the-origins-of-the-modern-public/

The Origins of the Modern Public: A series of audio programmes from the Canadian Broadcasting Corporation in which academics from a variety of disciplines discuss the history and contemporary formulations of public life.

NOTE

1. The printable version of the *Oxford English Dictionary*'s entry on 'public, *adj.* and *n.*' runs to 40 A4 sheets. Thanks to Engin Isin for pointing the way to this excess of meaning. <http://dictionary.oed.com/cgi/entry/50191807?query_type=word&queryword=public&first=1&max_to_show=10&sort_type=alpha&search_id=bHVF-LGivue-14447&result_place=2> Accessed 27 October 2009.

CHAPTER 59
ETHICAL SPACES

Keith Woodward

Introduction

Wherever we are, we inhabit and produce ethical spaces. The task of spatial ethics is to unmask and study the geographical and practical characteristics of those sites, so as to better understand the nature of our relationships and responsibilities to others. This chapter considers some ways of accounting for ethical practice and spatial life by addressing three areas that have mattered greatly in recent Human Geography: the space of the body, the responsibility to the Other and the ethics of collective life. Along the way, it explores the diversity and reach of the field by introducing several examples of ethical spaces ranging from bodies to factories, insect drones to industrial disasters, and research sites to revolutions.

Few topics within the social sciences have had as broad a reach and as abiding an influence as ethics. Indeed, Popke has estimated that 'something on the order of 1500 articles dealing with ethics are published in academic journals each year' (Popke, 2006: 504). While ethics have a long and rich intellectual history, contemporary approaches tend to ask two kinds of questions: what *can* be done? and what *ought* to be done? These are sometimes answered in 'normative' or moral terms – that is, according to what is considered 'good', 'virtuous', or 'just'. While normative ethics remain a concern for

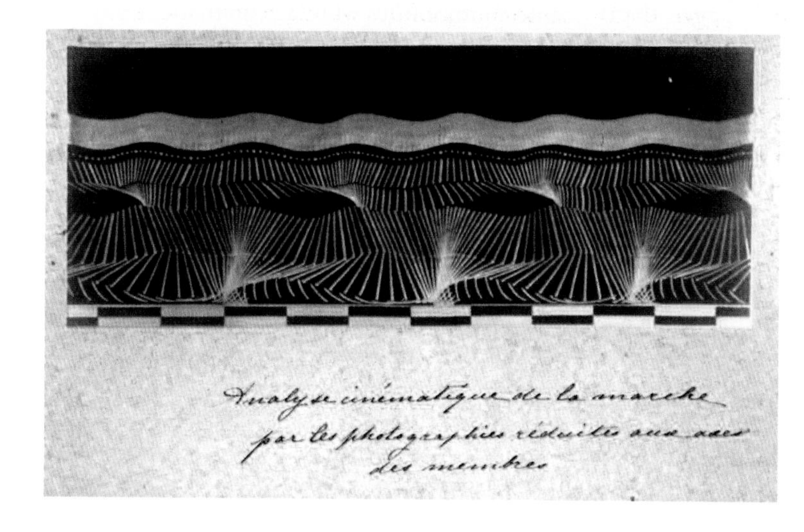

Figure 59.1 Marey's movement chronophotographs. Joinville soldier walking, 1883. Credit: public domain

Human Geographers, *spatial ethics* allow us to paint with a much broader brush and to examine:

> the nature of our interactions with, and responsibilities toward, both human and non-human others. To speak of ethical geographies, then, is to consider the nature and extent of these responsibilities, both empirically and theoretically, as well as the ways in which our actions and dispositions toward others tend to fulfill or abrogate them with particular contexts or institutional arrangements.
>
> (Popke, 2008b: 436)

Popke's language of 'responsibility' suggests that our ethical commitments are *produced* through our immersion and participation in social relations. Studying such ethical situations allows us to better understand the ways that our social relations are also *spatial* relations.

Lives that are lived together – from individual interactions to large social formations – create all sorts familiar and novel spatialities, such as homes, work and play spaces. While these are often maintained via thoughtful activities, ethical spaces are by no means produced *only* through conscious acts. The 'social production of space', for instance, can just as easily involve processes and events that are much 'larger' than our individual choices and deeds. These, too, are a matter of ethics. Parks, schools, streets, sporting venues and laboratories, for instance, all depend upon specific, *care*ful practices by their users, though we rarely enter these spaces with the explicit purpose of 'producing' them. More often, familiar practices and even unconscious habits form the ethical 'glue' that holds such social sites together. And yet, despite their seeming innocuousness, these spaces can simultaneously be locations of dramatic difference, equality and inequality, justice and injustice and liberation and oppression. Whether or not we remain vigilant over our practices in the busyness and distraction of daily life, coming to recognize how we remain socially responsible for them is at the heart of spatial ethics.

This responsibility extends beyond our local encounters to our very relation to the earth. Throughout the past century, geographers have studied the deleterious effects of modernity on 'natural' environments and biological life. Ongoing destructive industrial production practices, the privatization and plunder of natural resources and the proliferation of industrial and consumer waste have brought us to a critical point of unprecedented crisis in the environmental history of our planet. Today, some scientists have even begun to suggest that it may be too late to reverse the damage done by human-made climate change. Now, they tell us, we must prepare to adapt ourselves – and other forms of life – to potentially massive environmental transformations. Here, the scope of ethical spaces grows to encompass the globe, engendering responsibilities to others living at a great distance whose lives are nevertheless touched daily by our **consumption** and **production** practices. Many such individuals are members of vulnerable communities that have limited, if any, access to political power and financial resources. Yet it is often these same communities who face dramatic health crises and resource scarcity as direct results of global environmental and climatic changes. The growing ubiquity of this kind of environmental injustice indicates that, just as ethics contribute to the making of spaces, there is also, in turn, an *ecology of ethics* conditioning the activities that spaces can make available to us. Thus, the environment in which our acts *take place* both affects and is affected by what we do: our practices transform spaces and those transformed spaces, in turn, bring forth changes in our practices. This back-and-forth between space and practice confronts us with a bevy of *new ethical problems*.

Despite their seeming differences, much the same holds for our relation to non-human lives, whose local habitats are also being transformed or destroyed as a direct result of rising temperatures. Some creatures are successfully migrating and adapting to new habitats, but their moves are complicated and precarious. Because these corridors often run through sub/urban spaces, many other species will be less successful and face extinction. Here, environmental transformations and their urgent spatial injustices and inequalities provide a common language for describing our responsibilities to the human and non-human lives of our shared planet. Speaking ethically in terms of a shared Earth also means recognizing our responsibilities to *future life* that will inherit it. All of these ethical responsibilities are fundamentally geographical, for they take the earth as the *shared context* (or, the 'ground') upon which our collective activities take – and make – place.

SUMMARY

- Ethics asks two questions: 'What can be done?' and 'What ought to be done?' Normative ethics concerns the morality or rightness and wrongness of actions.
- There are many ethical dimensions to the social production of space, even when we are unaware that we are playing roles in its production.
- There is also an ecology of ethics whereby our activities both affect and are affected by their environments.
- Because the effects of human actions can be both spatially and temporally far-reaching, the Earth is the shared ethical context for humans and non-humans, both in the present and in the future.

What can a body do?

As the site of movement and force, desire and violence, individuality and identity, the body is an ethical space. By way of illustration, consider the early efforts to record the physiology of bodily movement by the French physiologist Etienne-Jules Marey (see Figures 59.1 and 59.2). Aided by specially constructed 'chronophotographic' cameras, Marey sought to capture the complex motions and interacting parts of human and non-human bodies in the acts of jumping, running, walking, pole vaulting, flying and so on. Through images that presented his subjects in a series of sequential movements, Marey explored how the body's moving, interacting components –bones, joints, muscles – worked dynamically and in concert to expend and conserve energy. From the viewpoint of its collaborating parts, Marey recognized – and his images illustrate – that bodily mechanism is something that is fundamentally mobile *and spatial* in nature. Even a cursory glance at his chronophotography reveals that 'the notion of space, not time, was the important one.' As Braun (1992) observes, Marey's images:

> left just the twisting arabesques of the invisible body as white lines on a black background . . . Marey made wire models of the trajectories. Both the models and the

Figure 59.2 High jump, 1890. Chronophotograph of human movement by Etienne-Jules Marey. Credit: Getty Images

> photographs made quite an impact, no doubt because they offered the first examples of how to isolate one segment of the body in movement from the intricate and almost invisible combinations it was a part of; they showed exactly how one part reacted to the movement of another. The ability of Marey's lens[es] . . . to separate out each of the moving parts of the human machine and bring them under scrutiny of the trained eye marked an important step in both science and medicine; for one thing, it meant that the precise site of certain locomotive pathologies would now be pinpointed and described more easily than ever before.
>
> (Braun, 1992: 100, 102)

Though not a geographer, Marey's influential mappings of mobile bodily components is certainly a precursor to the ways geographers today are exploring the body as an interactive, ethical system capable of functioning as register of landscape (Wylie, 2005) or as a site of the 'bio-political production' of subjectivity (Fannin, 2007).

Marey's research invites us to ask 'what can a body do?' and we find one answer in the acute and concerted movements of body parts. With this spatial ethics of the body, we see that a body does a great deal more than we may expect and that it does this in ways that are potentially enabling (by training its movements and caring for its parts) or disabling (through harm or restraint). The seventeenth-century Dutch philosopher Baruch Spinoza (Figure 59.3) is perhaps the first to conceive bodily ethics in terms of doings. For many thinkers before Spinoza (and many after), ethics concerned only the mind and its capacity for rational deliberation – the body was merely the vehicle through which the will of the individual was enacted. Refusing the separation of the mind from body, Spinoza contended that ethics was a matter of discovering and *thinking* about our bodies in the act of doing. Much as we cannot know how to swim before we start swimming, we cannot know what our bodies can do before we start doing. Likewise, for Spinoza, the question of ethics could only be answered through bodily activity. This perspective has recently become very popular with geographers concerned with **poststructuralist** and **non-representational** theories (see, for example, Chapter 50 on 'affect'). For these geographers, ethics considers the ways that our interactions with humans and other entities (from animals to advertisements) enroll our bodies and affect our behaviours, practices, thoughts and feelings, often in ways that hold heavy sway over who and what we are. Thus, the ethics of the body suggests that we are much less the legislators of our actions than we might like to presume. Consequently, Spinoza's question What can a body do? is much more complicated than it first appears. Having considered the complexity of movement for the space of the body, we can now further complicate the body by exploring how it 'works'.

Figure 59.3 Baruch Spinoza. Credit: Getty Images

Working life

Studies of the interacting segments of the body have greatly aided our understanding of its systematic workings, and yet this same segmented perspective on bodily spaces has also contributed to the development of nuanced and widespread approaches to their exploitation. Consider the spatial model of the Fordist factory (see Figure 59.4). This is a space of interacting bodily and 'machinic' movements, where specific, localized parts are employed in the processes of production. There, workers are obliged to spend long hours in awkward postures performing repetitive exertions – turning a wrench, forcing material through a machine, sewing a specific pattern – for hours on end. As a consequence, factory production over-utilizes certain parts of the body until they eventually wear themselves out. Because this process is the result of specific doings, work on different bits of material with different machinery employing different parts of the body results in a different set of stresses and strains. For this reason, philosophers such as Gilbert Simondon have observed that this space of repetitive and fragmented exertion creates an '"anatomy of work" that provokes different effects of industrial fatigue' (Simondon, 2009: 21). The factory is an ethical space not merely because it concerns bodily doings, but because such doings are purchased and manipulated for the purpose of making a profit. That is, an 'anatomy of work' merges questions about bodily doings with vital concerns about whether it is *right* or *just* to use the body of another for economic gain. How do we go about placing a *value* on these exertions, particularly when repetitive labour irreparably wears that body out?

Neither do the ethical dimensions of the factory exhaust themselves in the use of bodies. 'The workplace', Simondon continues, 'is a *human environment*' (2009: 21). Throughout

Figure 59.4 Workers wind transformer coils at the Atwater Kent radio factory in Philadelphia, 1925. Credit: National Photo Company Collection

the working day, the factory is not only the site of work; it is the local *situation* within which workers' lives are lived. As we will see in the last section of this chapter, a local situation can engender a positive ethics of community among its occupants that allows the collective to exert a strength that the individual cannot. However, in the factory, dangerous machinery, toxic chemicals and waste create a threatening environment for workers, who fall victim to chronic disease and physical injury. Aside from its obvious immorality, an ethics of 'doing' helps us understand the ways that elements of the environment can greatly improve or radically delimit the capacities of a body to act. While learning about bodily doings can be empowering, threatening or exhausting, spaces can have the opposite effect. Asking what a body can do, then, concerns both its internal workings and its external, environmental relations.

Non-human life

But the spatial ethics of working bodies are in no way exclusive to *human* bodies. It is certainly possible to inquire about what a *non-human* body can do, or, more to our purposes, what a human can do with a non-human body. As its raw material or its raw power, animals and their bodies have long played roles in industrial production (Roe, 2010). Today, corporations and governments are finding new and terrifying ways to extend their reach through the bodies of insects. As Jake Kosek's recent research on the militarization of bees illustrates, private and governmental research on the appropriation of insect life for security and warfare purposes has exploded throughout the so-called 'war on terror era'. Some of this work had been relatively benevolent, such as the 'DARPA-funded research to train free-flying bees to detect certain scents – of landmines, for example – by placing traces of the explosive chemicals near food sources' (Kosek, 2010: 657). Here, the flight patterns of trained bees disclose the hidden spaces of the minefield, allowing maps to be generated for the purposes of disarming the site (Figure 59.5).

Figure 59.5 Training bees to become 'chemical detection devices'. Courtesy of Los Alamos National Laboratories

However, other projects are much more sinister. For example, the abilities of killing machines such as US military drones are currently being modified according to principles derived from mathematical models of bee behaviour. Another DARPA (Defense Advanced Research Projects Agency) project 'aims at developing tightly coupled machine-insect interfaces by placing micromechanical systems inside insects during early stages of metamorphosis, with the aim of controlling insect locomotion' (Kosek, 2010: 661). The hope is that the 'interface will allow humans to control insect behaviors and motion trajectories via specialized GPS units with optical or ultrasonic signals' (Kosek, 2010: 661–2). Though Kosek expresses doubts about the realizability of the latter project, its very operation raises important questions about human uses and abuses of non-human life. For, as one of Kosek's interviewees observed: 'it turns out bees have minds of their own, and they can be delinquent from their training, for while they are easily reigned in some respects, they do not always do as they are told' (Kosek, 2010: 659). It is unlikely that the interviewee was suggesting that a bee has anything resembling a human brain. Yet the fact that it has *a* brain should force us to inquire about *how* we might think in an ethical relation to non-human life.

The animal rights activist and ethicist Peter Singer has spent much time troubling just such questions. Ethics of the non-human does not mean that we must read the bee, say, as an ethical or moral being (at least not in the way that we define these terms for ourselves). Nor, as Singer explains, is it a claim:

> that all lives are of equal worth or that all interests of humans and other animals are to

> be given equal weight, no matter what those interests may be. It is saying that where animals and humans have similar interests – we might take the interest in avoiding physical pain as an example, for it is an interest that humans clearly share with other animals – those interests are to be counted equally, with no automatic discount just because one of the beings is not human.
>
> (Singer, 1985: 9)

Notably, the shared interests of the DARPA projects are the result of manipulation – chemical detection training is accomplished through food association and efforts at bee mind control speak for themselves. However, what is notable is that, while bees' 'wills' are being bent to the human, our own endeavours are becoming more bee-like. Landmine maps and drone flight patterns, for instance, both access a bee-oriented spatiality – drawing upon the bees' sense of space – that would otherwise be unavailable to us. Learning to see through the senses of the bee is but one of the many ways that the non-human is a frequent component of our ethical lives.

SUMMARY

- A key dimension of ethics is bodies in relation to each other and themselves. Spinoza's famous inquiry into what bodies can do meant that ethics concerned not only moral judgements, but, further, an exploration of the capacities of our corporeal being and a caution against structures that control, delimit and exploit them.
- The body is a complex system whose inter-workings constitute their own ethical spatiality for geography.
- The complexity of the body can be enlisted in the transformation of the world in all sorts of ways. Because of this, it has also been subject to forms of exploitation and abuse and control.
- The bodies of non-humans, such as Kosek's bees, are also subject to ethical questioning, and, further, illustrate the ways in which human actions are intertwined with the lives – and even the bodily capacities – of other non-human beings.

In the space of the 'Other'

In 1949 Simone de Beauvoir (Figure 59.6) published *The Second Sex,* her germinal meditation on the politics of womanhood in the Western world. Rejecting the 'naturalization' of gender difference on the basis of biology, she famously argued that 'One is not born, but rather becomes, a woman' (Beauvoir, 1989: 267). Her point was that, while biology doubtless contributes important components of our sexual being, the notion of gendered identity is socially produced. Accordingly, in addition to understanding the physical and biological dimensions of the body, ethical perspectives also guard against the projection of difference *onto* the body for the sake of rationalizing social unevenness and

'Otherness': 'Only the intervention of someone else,' Beauvoir observes, 'can establish an individual as someone *Other?'* (267). Feminist Geography has explored the many ways that the 'feminized other' is also an uneven and unethical socio-*spatial* production. Reproductive or 'domestic' labour, for example, has long been a gendered form of work and accordingly devalued relative to the 'productive' work of men. Accordingly, we can recognize its naturalization in the spatial effect of the home space, where certain areas are identified with 'women's work', thus 'reifying' invented differences. The same holds for the gendering of behaviour: it is not uncommon to read interviews with maquiladora and sweatshop owners and managers who express a preference for hiring women on the basis of prejudicial beliefs that they are somehow more docile and easier to manage and control (Wright, 2006).

The invention of the woman as an 'Other' gains a foothold through the social production of spaces – the gendered home space or sweatshop – that make such prejudicial beliefs *appear* as though they were realistic. Thus, beyond the question of the body, Ethical Geographies concern 'spaces of difference'. Here we examine our ethical responsibilities to those whose lives are subject to and/or victimized by social and economic constructions that produce, naturalize and perpetuate unevenness and difference. Domestic spaces are a familiar example, but how to locate spatial ethics within a 'globalized' world where multinational corporations take advantage of human and material resources in distant and unfamiliar places? This 'distance effect' often makes it difficult for well-meaning individuals to imagine ways in which they might help stop – or at least avoid contributing to – the corporate exploitation of distant Others. One popular approach is to alter one's consumption patterns. Commodity chains map out their own ethical spatialities from the consumer through many points of increasingly exploitative global production practices. However, many well-meaning consumers despair that this awkward situation comes to an impasse when they realize that their practice is but a drop in the ocean of globalized consumption. Responding to this distance effect, Barnett *et al.* (2005) question the model in which 'people are implicated in their actions by reference to a linear chain of relations between free will, knowledge, voluntary action, causality, responsibility and blame' (Barnet *et al.*, 2005: 25; *see also* Popke, 2006: 508). Viewing ethics as a mediated social relation, they examine the organizational strategies of a fair trade network that 'make available policies, campaigns and especially repertoires of ethical consumption to consumers in such a way that the resultant opportunities help to govern the consuming self' (Barnett *et al.*, 2005: 36). Thus, in ethical consumption, care for the Other is wedded to

Figure 59.6 Simone de Beauvoir. Credit: Corbis Images

care for the self. To be sure, such a model of marrying ethics and consumption has become a major component of fairly traded goods, but, relatively speaking, this model is still far from the norm of globalized consumption, where less ethical corporations tend to mask rather than eradicate exploitation.

Environmental injustice and other spaces

Just after midnight on 3 December 1984, a pesticide plant owned by Union Carbide India Limited leaked 'approximately 27 tons' of lethal methyl isocyanate gas into the night sky over the city of Bhopal, Madhya Predesh, India (Dhara and Dhara, 2002: 392). As the deadly cloud descended, it first reached the poorer areas, the slums sitting just downwind from the factory, and then quickly spread across the city (Figure 59.7). Witness reports recall residents running blinded through the streets, screaming and vomiting greenish bile from chemicals that burned their eyes and lungs. 'Within hours', Broughton (2005) explains, 'the streets of Bhopal were littered with human corpses and the carcasses of buffaloes, cows, dogs and birds. An estimated 3,800 people died immediately, mostly in the poor slum colony adjacent to the UCC plant.' Another 11,000 would die in the ensuing weeks and years, and a staggering 558,000 people would suffer disabling injuries as a result of exposure. Following investigation, it was discovered that the factory was in a state of severe disrepair with a long list of faulty equipment and safety violations. Yet the parent corporation, the transnational conglomerate Union Carbide Corporation (UCC), although it had sought to capitalize on local cheap labour by siting the factory in Bhopal, refused to take responsibility for the gruesome, deadly social costs of the toxic spill.

In the *Union Carbide Report* (March 1985) delivered months after the accident, Warren Anderson, the then chief executive officer of UCC, coldly observed:

> Non-compliance with safety procedures is a local issue. UCC can't be there, day in and day out. You have to rely on the people you have in place. My board does not know what valve is on in Texas City. Safety is the responsibility of the people who operate the plant. You can't run a billion dollar company out of Danbury.
>
> (Mukherjee, 2010: 19)

Note that Anderson's report uses the very same 'distance effect' logic to absolve UCC of responsibility that Barnett *et al.* (2005) criticize above. Turning Union Carbide India's employees into spatial Others, Anderson invokes the distance between the UCC boardroom and Bhopal to invent a bizarre individualist ethics that supposedly stretches the chain of causality beyond reasonable responsibility. 'My board does not know what valve is on in Texas City', he declares, as though

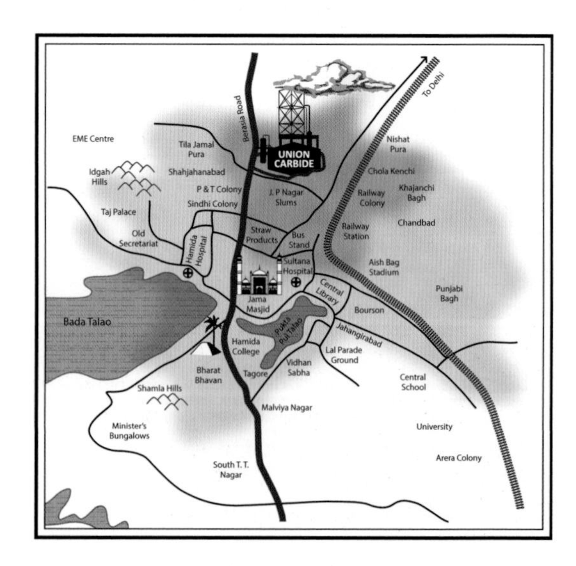

Figure 59.7 Map of Bhopal disaster. Credit: Briony Morrow-Cribbs, after *Sri Lanka Guardian*

the physical presence of board members at global chemical plants was the otherwise standard measure for assessing corporate responsibility. Conveniently for Anderson, this line of reasoning would make it impossible to *ever* find major corporations responsible for the work- and production-related accidents at its subsidiaries and subcontractors. What Anderson hopes to ignore in his slippery rationale is the fact that *all* ethical relationships (just like UCC's sources of corporate profit) are complex and mediated.

Mukherjee (2010: 19) agrees that disentangling responsibility inside the complex system of a global corporation such as UCC is indeed a difficult task. To be sure, lack of personnel training, poorly assessed and managed risks, absentee capitalism and many other factors played a role in the event. However, the distance effect in Anderson's surprisingly cold and indifferent 'report' seeks to absolve his corporation of *any* responsibility to local workers and residents. And yet, a key spatial component of global capitalism is precisely its ability to expand its reach, close distances and thus increase profit margins by exploiting spaces of economic unevenness (Harvey, 1989). Despite Bhopal's position at the time as an emerging major Indian city, the relative poverty of the region made it a lucrative prospect for a corporation seeking to capitalize on the global unevenness of labour value: 'In 1985, the monthly income of 80% of the population was below 6 US dollars/mo, and only 1.25% earned *more* than 18 US dollars/mo' (Dhara and Dhara, 2002: 392). By contrast, the net worth of Union Carbide in 1985 was several billions of dollars. For corporations, ethics intertwine with the production and exploitation of spatial and labour value in sites such as Bhopal.

Just as troubling is the uneven power that, after the accident, allows Anderson to speedily absolve UCC of responsibility – and to extract itself from the situation – while Bhopal's victims and families were forced to struggle for justice for the next several decades. It was long after one of the worst industrial disasters in history that UCC finally paid victims and survivors a pittance before finally being bought by Dow Chemical. The new owner assumed responsibility for all of UCC's holdings in the purchase, but to this day it has refused to accept responsibility for Bhopal. Just that easily, corporate responsibility disappears. Meanwhile, in Bhopal, the factory continues to stand, and though it was never subject to a proper industrial cleanup, it is now left abandoned and slowly decaying (see Figure 59.8). According to many reports, the plant's chemicals were never removed and now leak from rusting containers, bleeding into the soil and local groundwater (Eckerman, 2005: 141–6). Slums still populate the spaces surrounding the plant and, when it rains, the area occasionally floods, causing contaminated groundwater to spill into homes of the slum dwellers and the watering holes where children play.

Such stark differences between the fates of Union Carbide and the Bhopal victims lay bare the power of global corporate reach and the ability to mobilize capital over the workings of social and environmental justice. Bhopal illustrates for us the ethical crises that emerge as a result of the exploitation of spatial difference. Industrial sites are spaces of increased health risks for the populations residing in the surrounding areas. Where these get sited is never innocent: it is the result of much legal wrangling and, oftentimes, cronyism. Thus, for example, it comes as little surprise that highly pollutant industries end up being located upstream from disenfranchised and marginalized communities rather than those of the wealthy and powerful (Bullard, 2000; Pulido, 2000). Industrial accidents and covert

Figure 59.8 The Union Carbide pesticide plant (background) in Bhopal today. Credit: Getty Images

waste dumping in local waters is obviously unethical, yet the corporate responsibility to workers and local populations often gets buried or masked behind legal wrangling and the ability of corporations to mobilize more social and political power. Thus, in contrast to the fairly traded commodities of Barnett *et al.* (2005), local victims and exploited workers are transformed into Others set at a distance.

Participatory approaches

Finally, by way of closure, we should recognize that we geographers, as social scientists, also have long had an interest in engaging those who live differently and at a distance from us. The long history of the discipline has been bloodied by the unethical complicity of cartographers and explorers in colonialism, the slave trade and global resource exploitation. Today, it is the responsibility of each geographer to be vigilant against perpetuating such injustices by remaining conscientious about *what* kind of research we perform with vulnerable subjects in other places and thinking critically about *who* stands to benefit from it. The ethical geographer seeks ways to help empower communities by following such collaborative research models and more interactive, immersive methodologies such as Participatory Action Research, where researchers offer their specialized skills to *contribute to* a community project rather than simply *gathering data from* it (Pain, 2003; Cahill *et al.*, 2007). An early precursor to this approach is William Bunge's (1971/2011) *Fitzgerald* (Figure 59.9). First published in 1971 and written in collaboration with the African-American residents of the Fitzgerald neighbourhood in Detroit, Michigan, *Fitzgerald* is at once an exploration of racial difference and spatial segregation in the USA and a record of community organizing during the long civil rights era. Bunge famously moved to Fitzgerald and participated with neighbourhood members in community building and organizing. The book is a truly singular example of a work of collective scholarship and an effort 'to strengthen the science of geography so that we can write more books of service to society' (Bunge, 1971/2011).

However, the ethics of participatory approaches are complicated and precarious. As Bryan (2010) and Wainwright (forthcoming) have shown, military and corporate interests often threaten to ensnare researchers in projects that can put subjects' lives and livelihoods at risk. One such controversy arose with the México Indígena project in Oaxaca, Mexico (Figure 59.10). Envisioned as a participatory project

Figure 59.9 Bunge at a block club meeting in his Fitzgerald home. Credit: University of Georgia Press

devoted to mapping indigenous spaces, geographers associated with the project failed to mention to participating subjects 'that this research prototype was financed by the Foreign Military Studies Office (FMSO) of the United States Army and that reports on [this] work would be handed directly to this Office' (Union of Organizations of the Sierra Juarez of Oaxaca, 2009). Data containing maps, 'community member names and the associated geographic location of their plot(s) of land, formal and informal use of the land and other data that cannot be accessed via the Internet' were made available – without participants' foreknowledge – to a foreign military office with demonstrated interests in indigenous activities and property. Even if such work is done with the best of intentions, the problem for the geographer is that it fails to recognize the ethical, cultural and geopolitical complexities that come with encountering lives in other spaces.

> 'It is one thing for the military to search scholarly sources for intelligence. It is quite another to gather intelligence for the military on "vulnerable populations" such as indigenous peoples – a practice explicitly

Figure 59.10 Oaxaca and *México Indígena* controversy. Credit: © Gregory Labold

banned by the AAG's [Association of American Geographers] Statement of Ethics . . . The *México Indígena* project appears to have done just that, taking advantage of indigenous peoples' desires for land rights to gather intelligence that will let policymakers more effectively intervene in indigenous affairs.'

(Bryan, 2010: 415)

SUMMARY

- Beyond what it 'does', the body is also the site of difference and the production of Otherness.
- Responsibility in matters of consumption works by way of mediated relations.
- Sites of environmental injustice, such as the industrial disaster at the Union Carbide pesticide plant in Bhopal, remind us that corporate profits often depend on socio-spatial unevenness.
- Geographers have a history of complicity in the production of distant Others and the perpetuation of exploitation. Today, we must strive to enrich our ethical dimensions of our research.

Whose space is it anyway?

For those studying spatial ethics, 2011 – and much of 2010 – was an exciting time. In city squares and university campuses throughout the world, workers, citizens, students and teachers organized themselves collectively and 'occupied' spaces in protest against local and global social, political and economic disparities. From Chile and Puerto Rico to Greece, Spain and the UK, massive groups assembled to protest governmental austerity measures introduced in the wake of 2008's global financial crisis. At roughly the same time, during what has been called the Arab Spring, enormous viral movements broke out across northern Africa that called for sweeping democratic reforms. Challenging the unequal power held by the most wealthy, the 'Occupy' Movement's denunciation of the '1%' closed out 2011 with the occupation of spaces in hundreds of cities throughout the USA. Earlier that same year in Madison, Wisconsin, where I live and work, the newly-elected Governor and state senate introduced legislation that stripped collective bargaining rights from State employees and slashed State-funded health care for many of Wisconsin's poorer communities. Workers, students and citizens responded by also participating in a massive occupation, holding a continuous protest in the State Capitol building from late February to mid-March 2011 (Figure 59.11).

While there may have been a bad apple here or there, none of these events were driven by unruly mobs (though politicians and media pundits sometimes like to paint them that way). Rather, these events were organized by everyday people and for many, they were the largest protests they had ever seen; for some, they were the first. The glue holding such a complicated event together is not anger or violence; it is an ethic of care that grows within the protest site, a product of the being-together

Figure 59.11 A protest at the Capitol Building in Madison, Wisconsin, 2011. Credit: Getty Images

over time of many different people who are bound together by a certain struggle. By way of closure, I turn to a more positive perspective: the power of collective spatial ethics. Social practices are a collective affair and it makes good sense that we should cultivate them together for the improvement of social life. Perhaps nowhere else is this as clear as it is in popular protest.

Though it began by happenstance, occupying the Wisconsin State Capitol building quickly became a critical component in building the momentum of the protest. It provided a central, highly visible location for protesters to gather from around the state (and, eventually, the country), for keeping pressure on politicians and for motivating the media to let citizens' voices be heard. Further, the duration and enormity of the event (an estimated 100,000+ people filled the Capitol and the surrounding square on one Saturday morning alone) had a massive impact on the sense of community within Madison and inspired others in the Capitol cities of Iowa and Ohio to undertake similar efforts. These practices of community enact the first ethical dimension of collective living, the practice of solidarity, expressed both locally among group members who are involved in a struggle together, and at a distance, between different groups. Indeed, as we experienced in the Capitol occupation, support from a distance can have a very strong effect on local practices. Take for example the widely circulated photograph of an Egyptian student in Tahrir Square whose sign expressed solidarity with Wisconsin. The image of this simple statement had an energizing effect upon Capitol protesters, who circulated it and spoke of it frequently, and the image led many Wisconsinites to develop strong feelings of connectedness with the recent popular victory in Cairo.

Much like all seven of the occupied University of Puerto Rico campuses, Tahrir Square in Cairo, Zuccotti Park in lower Manhattan, and the many other occupations, the space of the Capitol building was transformed as our practices and needs changed throughout the weeks of protest. No one could have predicted that the occupation would take place, let alone that it would last as long as it did. Consequently, protesters daily lives inside the Capitol were lived 'on the fly'. In the meantime, activists with crucial skills and knowledges (medical, legal and so on) quickly formed collectives and created a medical space, a family care space, food and information spaces, meeting spaces, exercise and storage spaces, among others. Here, the transformation of Capitol space arose in concert with an emerging 'ethic of care' (Tronto, 1993; Smith, 2005; Popke, 2006: 506–8) among the community of protesters and occupiers. Indeed, the Capitol was perhaps the exemplar of a dynamic ethical space, as changing spatial and practical conditions worked off of one another to regularly present new ethical challenges to be addressed collectively by the protesters.

In such a massive body of people, decision making presents its own unique ethical challenges. As Pope has noted, 'event-spaces' such as these 'are a collective accomplishment. To speak of an ethics of enactment, then, is also to call into question the nature of the social, and the ways in which it is, or is not, figured as a site of collective responsibility and regard' (2008a: 4).

Nowhere is this truer than in the effort to create and maintain a democratic ethic that attempts to respect the decisions and interests of all of the members of the community. Here, the model of consensus-based decision making employed by the 'Occupy' Movement represents one such socio-spatial reconfiguration that invites us to rethink the political and ethical subject. Ideally, consensual democratic structures are part of a larger 'political movement that aims to bring about a genuinely free society – that is, one where humans only enter those kinds of relations with one another that would not have to be enforced by the constant threat of violence' (Graeber, 2012: 52).

Thus, participants in a consensus model come to a decision based upon what they feel they can all live with (rather than fighting for what they most want). Rejecting an ethics of 'power-over' others in favour of one that promotes 'power-with' each other, the model is an ethical system designed to maintain the strength of a collective endeavour. However, even this system requires constant, locally-focused rejigging. 'Consensus,' as Kauffman explains, 'has considerable virtues, but it also has flaws. It favors those with lots of time to spend in meetings. Unless practiced with unusual skill, it can lavish excessive attention on the stubborn or disruptive' (2011: 49).

Just as consensus requires constant care and readjustment of practices to suit the situation, so too does an ethics of collective living. While some of the struggles of 2011 were clearly victories for the masses, others were not, while still others ended ambiguously or without resolution. The intensity of struggling together is a rare kind of joy and over time, or after a loss, it can lead to a certain kind of collective

fatigue – my friends in Wisconsin referred to it as a 'protest hangover'. To this, too, there is a collective ethics: how do we continue to struggle together to make the world a better place in the face of what often seem to be insurmountable odds? For thinkers such as Badiou, ethics is a matter of discovering truths about the world through our encounters. Collective encounters, such as those in Wisconsin's Capitol, change held beliefs, challenge bureaucratic representations of people and prompt us to question social difference – in short, they make the world appear new. For Badiou, such 'truths' ask us for many different kinds of commitments – political, artistic, amorous – but for all this, their ethics are simple: 'Keep going!' (Badiou, 2001: 52). There is perhaps no better ethics for collective living.

Figure 59.12 An Occupy Wall Street General Assembly meeting, employing the spokes council model. Credit: Getty Images

SUMMARY

- Collective living is empowered, locally or at a distance, by an ethics of solidarity that draws different struggles into relation with one another.
- Central to collective living is an ethics of care that, in Wisconsin's context, helped reproduce the community and the space it transformed.
- Event-spaces are a 'collective accomplishment' that operate according to relations of power. While both sorts are usually in play to some degree, an ethics of 'power-with' versus 'power-over' tends to better reinforce the life of a collective.

DISCUSSION POINTS

1. Unfortunately, today's corporate industrial disasters are far too frequent and too large to blame on a single individual. Drawing upon a current example, such as British Petroleum's massive 2010 oil spill in the Gulf of Mexico, discuss what you see as your own personal responsibilities in relation to such 'larger scale' disasters. How do you square the limits of your own local choices and actions with processes and events that are much more global?
2. Consider a familiar social space – the classroom, your local High Street, the pub, etc. What sorts of practices go on in these spaces that produce their specific character? Doubtless, there are unwritten rules, things that are done and not done. What kinds of ethical 'know-how' are necessary for negotiating these spaces? Are these conscious or unconscious, learned or intuitive? Is access and use of these spaces easier for some and more difficult for others? How do these spaces (re)produce that inequality?
3. Try the following experiment: next time you're walking around campus or a public space, try catching the gaze of each passer-by. Or: next time you're on a train, try striking up casual conversations with those around you. How do they react? What does this tell you about the relationship between different kinds of spaces and what our bodies do in them? What do these behaviours tell you about our 'social production of space'?
4. Design a study that explores the injustices and inequalities in the local spaces surrounding your campus. In what ways would you like to involve communities and marginalized identities in your study? As a researcher, should you employ your own knowledge, skill and experience to help change spaces of inequality and unevenness? If so, how? What are your long-term commitments to such spaces and the people who live there?
5. Imagine that the cost of tuition at your university increased suddenly and dramatically. How would this transform your campus into a space of economic difference? In what ways? How might you organize with fellow students to let administrators on campus know that you weren't happy about the new costs? If you are worried about reprisals from grumpy lecturers, administrators and the vice-chancellor, how might students organize so that their voices are clearly heard while still avoiding being singled out? Where might you research how students on your campus and elsewhere have organized university spaces in the past? How might students organize a space of equality in which they gain a stronger voice in administrative conversations and decision making?

FURTHER READING

Anderson, B. and Harrison, P. (eds.) (2010) *Taking Place: Non-Representational Theories and Geography*. Burlington: Ashgate Press.

A collection of essays describing new directions in 'non-representational' approaches in Human Geography. Ethics holds a central place in this movement and is explored throughout many of these chapters.

Bunge, W. (2011) *Fitzgerald: Geography of a Revolution*. Athens, GA: University of Georgia Press.

A classic text and an early example of collaboration between a researcher and a community that addresses questions of social justice and local identity. It remains a fascinating work that synthesizes many areas of the discipline – from Historical and Social Geography to planning and cartography – to explore the multifaceted dimensions of race and racialized difference in a late twentieth century American neighbourhood.

Featherstone, D. (2008) *Resistance, Space and Political Identities*. Malden, MA: Blackwell.

A study of the changing ethics and practices of contemporary transnational movements resisting neo-liberal globalization. Looking closely at networked forms of resistance, Featherstone sets the movements in the broader context of the past two centuries of transnational resistance.

Harvey, D. (1973) *Social Justice and the City*. Baltimore, MD: Johns Hopkins University Press.

One of the definitive texts devoted to social justice in Geography. The trajectory of the book details both liberal and radical approaches to thinking about issues of social justice in the urban context (as well as illustrating Harvey's own transformation into a radical geographer).

Harvey, D. (1994) *Justice, Nature, and the Geography of Difference*. Malden, MA: Blackwell.

A geographical masterwork that provides a deep theoretical exploration of the intersections between Marxist Geography and studies of nature, identity and difference. Part IV provides several case studies that explore the implications of class, identity and environmentalism for practising social justice.

Jones, J.P. III, Nast, H.J. and Roberts, S.M. (eds.) (1997) *Thresholds in Feminist Geography*. New York: Rowman & Littlefield.

Employing a variety of case studies and theoretical approaches, these essays explore the place of feminist thought in a variety of disciplinary areas and themes. The guiding focus throughout is grounded, embodied practice and the character of identity.

Kropotkin, A. (1902) *Mutual Aid: A Factor of Evolution* (multiple editions), Project Gutenberg. Available at: **www.gutenberg.org/ebooks/4341**

This classic text devoted to Ethical Geography challenges 'social Darwinism' and the notion of the 'struggle of all against all' by discussing many forms of life for whom evolution depends upon mutual assistance. Kropotkin draws upon an enormous number of examples from ants and bees, to birds and large mammals, and finally to a variety of historic and contemporary socio-spatial formations among humans, such as villages, cities, states and worker guilds. A 'must read' for every Human Geographer.

Massey, D. (1994) *Space, Place, and Gender*. Minneapolis: University of Minnesota Press.

Massey's text challenges the notions that issues of gender and race are 'only local' and that the global is the domain of class. This work illustrates that even simply moving through one's town or consuming at the local shop involves gendered, raced and classed relations that are *simultaneously* local and global.

Mitchell, D. (2003) *The Right to the City*. New York: Guilford Press.

Looking at sites such as the People's Park in Berkeley, California, this text investigates the struggles over public space and in the marginalization of the homeless in the wake of widespread urban privatization and gentrification.

Pulido, L. (1996). *Environmentalism and Economic Justice: Two Chicano Struggles in the Southwest.* Tucson: The University of Arizona Press.

This groundbreaking text explores environmental injustice in the context of poor and minority communities, groups who are frequently over-exposed to environmental hazards yet under-represented in political decisions. Through a study of the struggles of a pesticide campaign and a grazing collective, Pulido shows how racism and economic difference combine to marginalize communities.

Smith, D.M. (1994) *Geography and Social Justice*. Cambridge, MA: Blackwell.

A key introduction to the social justice that provides a detailed review of several approaches to equality and injustice. In particular, it focuses on the intersections between ethics, social justice and Geography and provides many case studies in spaces throughout the world.

Thrift, N. (2008). *Non-representational Theory*. New York: Routledge.

While this is an often theoretically challenging work, Thrift's focus on practice, embodiment and new ways of conceptualizing ethics runs throughout. This edition collects the germinal articles on non-representational theory by the thinker who introduced the notion to our discipline.

Whatmore, S. (2002) *Hybrid Geographies: Natures, Cultures and Spaces*. London: Sage

Whatmore theorizes and examines the relational ethics of hybridity between the human and non-human and the social and the natural. This important work redefined all of these notions through studies of everything from wildlife to boundary disputes to genetically modified foods.

Glossary

This glossary provides brief definitions of key terms used but not explained fully in the course of this collection's chapters. The terms listed in the glossary are marked in bold when first used in a particular chapter. Cross-referencing between entries is facilitated by capitalizing terms that are separate glossary entries in their own right.

Actor-Network Theory (ANT): sometimes also termed 'Actant-Network Theory', ANT is an approach that emerged from the sociological study of science and technology and has become widely influential within Human Geography. Two key features can be identified here. a) First, ANT questions traditional foci on the 'social' in the social sciences and the 'human' in Human Geography, arguing for the inclusion of non-human entities (animals, plants, technologies, objects and so on). b) Second, ANT views the world as comprised of networks of 'associations' involving these entities. For ANT, it is these associations that produce forms of POWER and AGENCY.

adaptation: measures to reduce current and future vulnerabilities to the impacts of climate change.

affect: the study of affect has been promoted by a recent 'affective turn' across the social sciences and humanities, and within Human Geography by the influence of NON-REPRESENTATIONAL THEORY. In that context, affect is associated with scholarship that emphasizes EMBODIMENT and the experience of life as it is lived. More specifically, the notion of affect refers to: a) a change in bodily state, which can increase or decrease that body's capacity to act; b) how this change comes from ENCOUNTERS between various sorts of bodies and things; and c) the existence of affecting forces beyond any individual human subject/body. As this indicates, much writing on affect has also contributed to a range of more substantial studies: from work on how POST-INDUSTRIAL economies involve the engineering of affect into the products and spaces we consume to that on the ATMOSPHERES of particular sites and places.

agency: the capacity to 'act' or, in effect, to have an impact on your surroundings. Human Geographers have widely debated the character, sources and locations of agency. These debates have included the issue of whether exclusively human or whether it emerges from networks of relations between people, other forms of life and non-living things (see also HUMAN AGENCY, NON-HUMAN AGENCY and ACTOR-NETWORK THEORY). E.g., for scholars in the field of animal and plant geographies, part of the value of their work involves being open to the ways in which various humans and nonhumans do things to each other in different contexts.

alienation: a term with two interrelated meanings, the second being a more specific formulation of the first. First, then, alienation refers to a sense of estrangement or lack of power felt by people living in the modern world. In this respect it is often used to describe the experience of modern urban living, in which traditional forms of social cohesion and belonging supposedly break down. Second, and drawn from MARXIST social thought, alienation refers more particularly to the separation of labour from the means of production under CAPITALISM. To explain, since under capitalism it is capitalists who own the resources required for economic production (land, machinery, money, etc.) workers have no control over their productive lives: how production is organized, what is produced, what the product is used for and how they relate to other workers. They are therefore alienated from their work and its products.

animality: a quality or nature associated with animals.

appropriation: in relation to COMMODITIES, appropriation involves valuing/using them in such a way that they lose their status as commodities purchased in a market place and become non-commodified as possessions (such as taking off price labels, displaying items in one's dwelling, washing or altering things).

asylum seeker: someone seeking protection under the terms of the 1951 United Nations Convention relating to the Status of Refugees.

atmosphere: a concept used to understand how specific sites have a characteristic affective quality. We might say, for example, that a room is welcoming or a park is relaxing. Atmospheres are ambiguous: they surround and envelop particular sites, cannot be definitively located, but are present in that people may 'feel' an atmosphere. Atmospheres emanate from the collections of people and things that make up particular sites, without being reducible to any one element.

autoethnography: the processes by which the Human Geographer chooses to make explicit use of her or his own positionality, involvements and experiences as an integral part of ethnographic research.

biodiversity: describes the variability in nature, including of genes, individuals, species and ecosystems. Biodiversity exists both in particular living things (such as individual rare species) and also in biological processes and dynamic ecological systems.

capacities to affect and be affected: a technical term that follows from Human Geography's encounter with Gilles Deleuze's thought. It orientates inquiry to two things. First, emphasis is placed on what a body can do, i.e. on a body's capacities rather than its properties. Second, capacities to AFFECT and affected are two sided. As bodies do things, as they act, they open themselves up to being affected by things outside of themselves.

capital accumulation: the prime goal of capitalism as a mode of production. It is the deployment of capital to convert it into new (and more) capital. This process is often designated by the simple formula MCM', where M is money or liquid capital, which is invested in C which is commodity capital (land, labour, machines) with a view to profit, realized as M' which is more money (or more technically 'expanded liquidity').

capitalism: an economic system in which the production and distribution of goods is organized around the profit motive (*see* CAPITAL ACCUMULATION) and characterized by marked inequalities in the social division of work and wealth between private owners of the materials and tools of production (capital) and those who work for them to make a living (labour) (*see* CLASS).

carbon neutral: contributing zero net GREENHOUSE GAS EMISSIONS to the atmosphere: usually achieved through a combination of energy efficiency, the use of renewable energy and OFFSETTING emissions.

civil society: a concept with a long and changing history of meanings, civil society has been used in the last decade to emphasize a realm of social life and a range of social institutions that are separate from the NATION-STATE.

class: a collection of people sharing the same economic position within society, and/or sharing the same social status and cultural tastes. The precise ways in which one's economic position — for example, as a worker, a capitalist or a member of the land-owning aristocracy — is related to one's social status or cultural tastes has been much debated. However, Human Geographers have studied class and its geographies from all these perspectives: as an economic, social and cultural structuring of society.

climate change: In a general sense this term is used to refer to past, present or future changes in the earth's climate. In common use, it is used to refer to changes in the climate resulting from the introduction of greenhouse gases from anthropogenic (human) sources.

co-construction of society and space: a term that recognizes that social characteristics have a spatial dimension: they do not simply vary from place to place but are constructed differently across space. Space and place thus become integral to the ways in which social characteristics are formed.

Cold War: a period conventionally defined as running from the end of the Second World War until the fall of the Berlin Wall, during which the globe was structured around a binary political geography that opposed US capitalism to Soviet communism. Although never reaching all-out military confrontation, this period did witness intense military, economic, political and ideological rivalry between the superpowers and their allies.

colonialism: the domination and/or dispossession of an indigenous (or enslaved) majority by a minority of interlopers or colonizers. The rule of a NATION-STATE or other political power over another, subordinated, people and place. This domination is established and maintained through political structures, but may also shape economic and cultural relations. *See also* NEO-COLONIALISM and IMPERIALISM.

colonization: the physical settling of people from a colonial power within that power's subordinated colonies (*see* COLONIALISM). *See also* decolonization.

commodification: this term is used in two interrelated ways: (a) as the conversion of a thing, idea or person into a COMMODITY (the term 'commoditization' is often preferred for this sense); and (b) a wider societal process whereby an ever increasing number of things, human relationships, ideas and people are turned into commodities. Both meanings see the process of commodification as symptomatic of the penetration of CAPITALISM into the everyday lives of people and things.

commodity fetishism: the process whereby the material origins of COMMODITIES are obscured and they are presented 'innocent' of the social and geographical relations of production that produced them.

commodity: something that can be bought and sold through the market. A commodity can be an object (a car, for example), but can also be a person (the car production worker who sells their labour for a wage) or an idea (the design or marketing concepts of the car). Those who live in capitalist societies are used to most things being commodities, though there are still taboos (the buying and selling of sexual intercourse or grandmothers, for example) (*see* CAPITALISM and COMMODIFICATION). This should not disguise the fact that the 'commodity state' is a very particular way of framing objects, people and ideas.

community-based organizations (CBO): non-profit organizations with the primary goal of providing social services at the neighbourhood level, often situated in the context of urban revitalization processes.

compositional effects: the influence of the compositional characteristics of a local population, including its age, sex and socio-economic status, upon neighbourhood health. A neighbourhood with a relatively young, well-educated and largely white population is likely to have better health outcomes than a materially disadvantaged community with a significant ethnic minority population.

contextual effects: the influence of the social and physical environment upon neighbourhood health. The mechanisms which underlie contextual effects are the subject of research and debate, but possibilities include the operation of local social norms around health-related behaviours (e.g. peer expectations around smoking and drinking) and the impact of the physical environment on health conditions (e.g. the link between damp housing and child asthma). Some apparently contextual effects may reflect unmeasured individual variables, but research to date suggests that places can have significant impacts on health outcomes.

cultural capital: a term invented by the French sociologist Pierre Bourdieu, who argued that distinctions based on the judgement of taste were as important in marking social differences as access to economic capital. People with high cultural capital have tastes that are highly valued and recognized as being 'cultured'. Some people may be high in cultural capital but low in economic capital (and vice versa).

cultural ethnocide: the permanent disappearance of a particular ethnic group, associated with the eradication of their cultural practices, brought about by the effects of economic and/or political policies. *See also* ETHNICITY.

cultural landscape: traditionally this phrase has meant the impact of cultural groups in fashioning and transforming the natural landscape. More recently it has been suggested that landscape itself is a cultural image, a way of symbolizing, representing and structuring our surroundings.

cultural politics: a phrase that emphasizes the implications of culture in questions of power. On the one hand, this means reflecting on how cultural forms and judgements involve relations of power. On the other hand, it means recognizing that politics are in part undertaken through cultural relations and realms (resulting in what in the USA, especially, have been termed 'culture wars'). More specifically, a focus on cultural politics centres questions about whose culture is dominant, the role of culture in practices of resistance, whose REPRESENTATIONS portray the world, and about the IDEOLOGY of cultural forms. It is concerned with the shaping of power through the social, economic and political relations that are created through the cultural practices of everyday life. It involves how meanings are constructed and negotiated through culture and how relationships of dominance and subordination are produced, maintained and resisted.

cyberspace: the forms of space produced and experienced through new computational and communication technologies. Exemplary are the spatialities of phenomena such as the internet, virtual reality programs, hypertext and hypermedia, or the science-fictional accounts that draw on these for inspiration. Human Geographers have approached these spatialities in at least three distinct ways: in terms of the kinds of places users encounter within them; by analysing the kinds of communicational networks they facilitate; and in terms of the geographical location of their infrastructures.

de-industrialization: this term usually refers to a decline in manufacturing industry. It can designate either a fall in manufacturing output or, more commonly, a fall in the number and share of employees in manufacturing industry as a result of plant closures, lay-offs, etc. It is possible for de-industrialization to occur at the same time as manufacturing output, and exports rise if automation is associated with jobless growth. Areas that have suffered from de-industrialization are characterized by large numbers of vacant derelict factories. *See also* POST-INDUSTRIAL.

decolonization: this term has two meanings: (a) the ending of formal colonial rule by one power over another (*see* COLONIALISM); (b) the departure of a settler population from a colonized territory (*see* COLONIZATION). In both cases, however, processes of decolonization are in fact likely to be far less of a 'clean break' than these definitions suggest. Legacies from, and new forms of, colonialism are still central to post-colonial experiences.

dependency theories: theories that explain levels of economic and social DEVELOPMENT with reference to global economic structures. In particular, the idea that southern countries are exploited by those of the North and will remain in subservient or dependent positions until the global economic system is changed.

de-territorialization: the uncoupling of political and economic processes from particular national spaces, as when banks move offshore or when economic decisions are entrusted to the World Bank or the G7 powers. *See also* TERRITORY.

development: a highly contested term that, in its most general sense, means change (usually positive change) over time. It has often been used to describe processes of becoming MODERN or ideas of progress. More specifically, development has often been defined from a particular EUROCENTRIC viewpoint, such that progress is assessed in relation to the experiences of western European societies and economies. Thus, development encompasses INDUSTRIALIZATION, urbanization and increasing standards of living in relation to health, education and housing, for example. Constructions of development are not value-free. POST-DEVELOPMENT theorists, among others, have highlighted the way definitions and DISCOURSES of development reflect and reinforce existing power relations.

developmentalism: a set of propositions or policies that demand (or provide for) the transformation of pre-modern societies into MODERN societies. Industrialization is often considered to play a key role in the process of development. *See also* POST-DEVELOPMENTALISM.

devolution: the process of devolving some political power to more local levels of government, often associated with the formation of local and regional assemblies. This can take place at many scales: at a supranational scale, for example within the European Union, power can be devolved to individual nations or to regions; and within the nation-state power can be devolved to regions and localities.

dialectic, dialectical: a dialectic is a process through which two opposites are generated, interrelated and eventually transcended. This process can be purely intellectual, in so far as it is a procedure of thought. Examples in this vein would include the opposites used as the starting points for many of the chapters in Part 1 of this book (local–global, society–space, and so on), which are then worked through and beyond in the course of each chapter. Dialectical processes can also be identified in the wider world, though. An example would be the combination in a COMMODITY of both use and exchange values, in which opposites become interpenetrated in the commodity form.

diaspora: the dispersal or scattering of people from their original home. As a noun it can be used to refer to a dispersed 'people' (hence the Jewish diaspora or the Black diaspora). However, it also refers to the actual processes of dispersal and connection that produce any scattered, but still in some way identifiable, population. In this light it also can be used as an adjective – diasporic – to refer to the senses of home, belonging and cultural identity held by a dispersed population.

discourse, discursive: drawing on the work of the French philosopher Michel Foucault, Human Geographers define discourses as ways of talking about, writing or otherwise representing the world and its geographies (*see also* REPRESENTATION). Discursive approaches to Human Geography emphasize the importance of these ways of representing. They are seen as shaping the realities of the worlds in which we live, rather than just being ways of portraying a reality that exists outside of language and thought. They are also seen as connected to questions of practice – that is, what people actually do – rather than being confined to a separate realm of images or ideas. More specifically, Human Geographers have stressed the different ways in which people have discursively constructed the world in different times and places, and examined how it is that particular ways of talking about, conceptualizing and acting on people and places come to be seen as natural and common-sensical in particular contexts.

ecocentrism: a perspective favouring a humble and cautious outlook towards the scope for interfering with the planet, and arguing for a smaller-scale, more communitarian, style of living.

ecology: a way of studying living things (plants, animals or people) that emphasizes their complex and dynamic interrelationships with each other and the environment. As well as their use in Biogeography, ecological theories about competition for resources, invasion and distribution have also been applied in urban geography.

ecosystem: an ecosystem comprises: a set of plants, animals and micro-organisms among which energy and matter are exchanged; the physical environment with which, and within which, they interact. *See also* ECOLOGY.

embodied, embodiment: this concept suggests that the self and the body are not separate, but rather that the experiences of any individual are, invariably, shaped by the active and reactive entity that is their body – irrespective of whether this is conscious or not. The argument, then, runs that the uniqueness of human experience is due, at least in part, to the unique nature of individual bodies.

empowerment: a political transformation whereby marginal individuals and groups attempt to increase their own power by struggling against injustice. This process may include increased opportunities of participation, greater legitimacy of contributions, and enhanced capacity building from the less privileged individuals and social groups.

encounters: the coming together of two or more things in an event that might result in something new being produced. What is produced in an encounter cannot be accounted for by reference to any of the participants in the encounter on their own. For this reason, the concept of the encounter privileges the relation between entities. Societies are made up of innumerable encounters: between two friends who unexpectedly happen across one another in a street, between an individual and an idea, between a person and a building, between one crowd and another crowd, etc.

Enlightenment: a philosophical and intellectual movement usually dated to the seventeenth and eighteenth centuries and centred in Europe, which advanced the view that the world could be rendered knowable and explained systematically by the application of rational thought (science). Revolutionary in its challenge to the religious beliefs and superstitions that then held sway, it has since been much criticized for projecting rationality as a universal, rather than situating reasoning processes in particular social and material contexts.

environmental audit: the attempt to calculate, and list for comparison, all the likely effects of any proposed action on the surrounding ecosystems and populations.

environmental determinism: a school of thought which holds that human activities are controlled by the environment in which they take place. Especially influential within Human Geography at the end of the nineteenth and beginning of the twentieth centuries, when the approach was used in particular to draw a link between climatic conditions and human development. Some authors used this link to argue that climate stamps an indelible mark on the moral and physiological constitution of different races (*see* RACE). This in turn was used to assert the superiority of western civilization, and hence to justify and support the imperial drive of nineteenth-century Europe (*see* COLONIALSIM and IMPERIALISM).

environmental risk: real and imagined threats to existing human, social and ecological systems posed by physical events, such as earthquakes; or from the unforeseen consequences of human activity, such as releasing genetically modified organisms into the environment.

environmentalism, environmentalist: a social and political movement aimed at harmonizing the relationship between human endeavour and the presumed limits to interference of planetary life support systems.

epidemiological transition: a shift from a situation in which infectious diseases and famine are a country's leading causes of death through to a situation in which chronic illnesses such as cancer, heart disease and strokes now occupy this position. This transition, which is typically associated with processes of MODERNIZATION and DEVELOPMENT, normally reflects improvements in sanitation, diet, water quality, immunization, access to drugs and improved medical care. On the basis of their most common causes of death, nations may be characterized as pre- to early transition (e.g. Somalia), mid-transition (e.g. India) or late transition (e.g. Canada).

epistemology, epistemological: epistemology is the study of knowledge, particularly with regard to its methods, scope and validity. This technical term from philosophy refers to differing ideas about what it is possible to know about the world and how it is possible to express that knowledge. Different academic disciplines and different general approaches in Geography are marked by distinctively different epistemologies. Human Geographers are interested in the epistemological questions raised by the geographical knowledges held both by academics and by ordinary people. In studying these epistemological questions Human Geographers seek to connect up questions of content (what kinds of things people know) with structures of belief (how and why they claim to know) and issues of authority (how and why these knowledges are valued and justified).

essentialist/ists/m: the belief that identities, such as class, gender and race, as well as age, dis/ability and sexuality, are directly determined by biology. This view is opposed by social constructionists, who argue that these differences between us are shaped through the interweaving of wider socio-spatial processes and individual biographies. The term is often used negatively to emphasize the stereotyping that it can produce.

ethnicity: a criterion of social categorization that distinguishes different groups of people on the basis of inherited cultural differences. Ethnicity is a very complex idea that needs careful consideration. For instance, in popular usage ethnicity often becomes a synonym for race, but in fact there is a crucial distinction in so far as race differentiates people on the grounds of physical characteristics, and ethnicity on the grounds of learned cultural differences. Moreover, while everyday understandings of ethnicity often treat it as a quality only possessed by some people and cultures (for instance, 'ethnic minorities' and their 'ethnic foods' or 'ethnic fashions') in fact these differential recognitions of ethnicity themselves need explanation. The complexities of the concept are further emphasized by recent debates within and beyond Human Geography over the extent to which new forms of ethnicity are emerging through the cultural mixing associated with processes of GLOBALIZATION (*see also* HYBRID and SYNCRETIC).

ethnocentric: an adjective describing the tendency for people to think about other cultures, societies and places through the assumptions of their own culture, society or place. An example of ethnocentricism is the production of theories about the whole world based on the specific model of western development (*see also* DEVELOPMENTALISM).

ethnography, ethnographic: the research processes that use qualitative methods to provide in-depth explorations and accounts of the lives, interactions and 'textures' associated with particular people and places.

ethnomethodology: an approach to studying social life that focuses on how we produce everyday spaces and situations through our actions. Ethnomethodology emphasizes practical conduct but it can also include conversational analysis. This focus on talking as a mode of representational action and interaction distinguishes ethnomethodology from NON-REPRESENTATIONAL THEORY.

Eurocentric: adjective that describes the characteristic of believing that the western European experience is the only correct way to progress. This may be because there is no awareness of alternatives, or it may reflect a belief in European superiority. Is often used now to refer to theories and perspectives coming more generally from the Global North.

everyday life: a notoriously difficult term to define, as it has meant many different things to a great many different philosophers, theorists and practitioners. We can generalize that it is an arena of social life that includes repetitive daily cycles and routines that we learn but eventually take for granted.

feminism: a series of perspectives that together draw on theoretical and political accounts of the oppression of women in society to suggest how gender relations and Human Geography are interconnected (*see also* PATRIARCHY).

financialization: the growing importance of money and finance within the economy. It is a multi-faceted phenomenon that can be measured through deeper links between the international financial system and household economies; a growing reliance on the financial services sector for economic growth; and increased pressure on firms to respond to, and meet, the demands of the international financial system and their shareholders in particular.

Fordism: a form of industrial CAPITALISM dominating the economies of the USA and western Europe from the end of the Second World War to the early 1970s. It was characterized by mass production and mass consumption, where high levels of productivity (often promoted by new assembly line production) sustained high wages, which in turn led to high levels of demand for industrial products. This demand fed into higher production, which supported higher productivity, and the 'virtuous circle' began all over again. It was an economic system that was often underpinned by a political compromise between capital and labour, and by subsequent state policies towards wages, taxes and welfare provision, which helped to sustain mass production and consumption. *See also* POST-FORDISM.

foreign direct investment (FDI): an investment made in a foreign country (usually by a firm) in productive facilities in a foreign country (usually by a firm in productive facilities). This can include a direct investment in an existing firm and its production (classically, for example, a factory or mine). However, it is distinct from indirect investment such as buying shares in a foreign company.

gender: a criterion of social organization that distinguishes different groups of people on the basis of femininity or masculinity. In any one location, many masculinities and femininities interact. As a concept, gender is usually used in Human Geography in distinction to that of sex (i.e. femaleness and maleness) in order to emphasize the SOCIAL CONSTRUCTION of women's and men's roles, relations and identities. Human Geographers' accounts of the world have always been shaped through understandings of gender (*see* MASCULINISM) but explicit analyses of the geographies of gender and the gendering of geographies are comparatively recent, and associated with the growth of feminist geography (*see* FEMINISM).

gentrification: an urban geographical process commonly taken to have two main attributes. The first is the invasion (or replacement) of traditional inner-city working-class residential areas by middle-class in-migrants; the second is the upgrading, improvement and renovation of the existing housing, whether done by the new residents or by developers. Commercial gentrification refers more specifically to the replacement of older, traditional, low-rent, retail and other uses by new, stylish, fashionable boutiques, cafés, bars and other retail outlets.

geographical imagination: an awareness of the role of space, place and environment in human life. This phrase is sometimes used in the definite singular – the geographical imagination – to refer to the distinctive intellectual concerns and contributions of Geography. It is also used in the plural – geographical imaginations – to emphasize the many different ways in which academics, students and lay publics alike can develop their sensitivities to human geographies.

geopolitics: an approach to the theory and practice of statecraft, which considers certain laws of geography (e.g. distance, proximity and location) to play a central part in the formation of international politics. Although the term was originally coined by Swede Rudolf Kjellen in 1899, it was popularized in the early twentieth century by British geographer Halford Mackinder.

GHG emissions: greenhouse gases (GHGs), such as carbon dioxide and methane, have the effect of trapping heat within the earth's atmosphere and therefore contribute to climate change.

global city: a term used by Saskia Sassen in the early 1990s to denote those cities that play a key role in the operation of the capitalist world economy, particularly in terms of financial and business services. The term builds upon the concept of 'world city', put forward by Freidmann and Wolff, to capture the command and control centres of late CAPITALISM, particularly in terms of the concentration of headquarters of multi- and transnational corporations.

global commodity chain (GCC): the transnational chains or networks constructed to produce particular goods and services. A global commodity chain is collaboratively constructed by an array of actors including the state, firms and workers. This said, there are uneven outcomes with respect to the risks and rewards associated with the development process.

globalization: the economic, political, social and cultural processes whereby: (a) places across the globe are increasingly interconnected; (b) social relations and economic transactions increasingly occur at the intercontinental scale (*see* TRANSNATIONAL); and (c) the globe itself comes to be a recognizable geographical entity.

As such, globalization does not mean everywhere in the world becomes the same. Nor is it an entirely even process; different places are differently connected into the world and view that world from different perspectives. Globalization has been occurring for several hundred years, but in the contemporary world the scale and extent of social, political and economic interpenetration appears to be qualitatively different to international networks in the past.

global value chain (GVC): the distributed activities associated with the production of a good or service that can be monetized. These typically include invention, design, production, marketing and distribution. A global value chain can be constructed and managed by one or more firms.

gravity model: a mathematical model based on a rather crude analogy with Newton's gravitational equation, which posited a constant relationship between the gravitational force operating between two masses, their size and the distance between them. Geographers have used this model to predict and account for a range of flows between two or more points, especially those to do with migration and transport (*see also* SPATIAL SCIENCE).

hegemony, hegemonic, hegemon: hegemony is an opaque power relation relying more on leadership through consensus than coercion through force or its threat, so that domination is by the permeation of ideas. For instance, concepts of hegemony have been used to explain how, when 'the ruling ideas are the ideas of the ruling class', other classes will willingly accept their inferior position as right and proper (*see* CLASS). Hegemonic is the adjective attached to the institution that possesses hegemony. For instance, under capitalism the bourgeoisie are the hegemonic class. Hegemon is a term used when the concept of hegemony is applied to the competition between NATION-STATES: a hegemon is a hegemonic state. For instance, the USA has been described as the hegemon of the world economy in the mid-twentieth century.

heteronormativity: a social regulatory framework that produces binary sex division, normalizes desire between men and women, and marginalizes other sexualities as different and deviant.

historicity: to recognize the historicity of Human Geographies is to recognize their historical variability. This involves both an emphasis on historical specificity – on how historical periods differ from each other – and on historical change – on how Human Geographies are re-made and transformed over time.

homogenization: a term used to describe the process whereby places and social characteristics become more similar to each other, so that they eventually become indistinguishable. For example, the spread of global brands such as McDonald's (McDonaldization) through the world is thought to cause places to look and feel the same, thus reducing local difference.

human agency: usually indicates self-conscious, purposeful action. Geographers have used this term to stress the freedom and capacity of people to generate, modify and influence the course of everyday and world events (see also AGENCY).

human exceptionalism: the idea that humans should be accorded a special status in our understanding of the world or, in effect, that humans were in some way 'exceptional.' The idea is rooted in religious world views but has come under attack by geographers who prefer to explore the contextual interplay of various factors, instead of buying into particular world views at the outset.

humanist: an outlook or system of thought that emphasizes human, rather than divine or supranatural, powers in understanding the world. Associated with the ENLIGHTENMENT, humanism marks human beings off from other animals and living things by virtue of supposedly distinctive capacities for language and reasoning. While underscoring progressive social changes, like the idea of human rights, it is criticized for making universal claims about human nature; privileging the individual over the social relations of human being; and licensing human abuse of the natural world.

Humanistic Geography: a theoretical approach to Human Geography that concentrates on studying the conscious, creative and meaningful activities and experiences of human beings. Coming to prominence in the 1970s, Humanistic Geography was in part a rebuttal of attempts during the 1960s to create a law-based, scientific Human Geography founded on statistical data and analytical techniques (*see* SPATIAL SCIENCE). In contrast, it emphasized the subjectivities of those being studied and, indeed, the Human Geographers studying them. Human meanings, emotions and ideas with regard to place, space and nature thus became central.

hybrid, hybridity, hybridization: hybrids are the products of the combination of usually distinct things. These terms are often used to describe new plant types but are used in Human Geography to emphasize the equal and positive mixing of cultures rather than negative ideas of cultural assimilation, dilution, pollution or corruption. *See also* SYNCRETIC.

hyper-reality: a phrase most associated with the Italian semiotician (*see* SEMIOTICS) and novelist Umberto Eco, which suggests the development of simulated events and representations that out-do the 'real' events they are meant to be depicting. Thus in hyper-reality the REPRESENTATION exceeds the original, being more sensational, more exciting or so forth. *See also* SIMULACRUM.

icon: a visual image, landscape feature or other material form that comes to symbolize or stand for a wider set of meanings or phenomena.

iconography: this term means both: (a) the study of the symbolic meanings of a picture, visual image or landscape; and, less often, (b) the system of visual meaning thereby being studied. Developed in particular within disciplines such as art history, Human Geographers have extended iconography to the analysis of landscape symbolism and meaning. This analysis combines examination of the symbolic elements of a landscape with consideration of the social contexts in which a landscape is produced and viewed (*see also* SEMIOTICS).

ideological: a meaning, idea or thing is ideological in so far as it helps to constitute and maintain relations of domination and subordination between two or more social groups (classes, genders, age groups, etc.).

ideology: a meaning or set of meanings that serves to create and/or maintain relationships of domination and subordination, through symbolic forms such as texts, landscapes and spaces.

idyll-ized: the process by which dominant myths about places and spaces come to reflect circumstances of picturesque beauty, tranquillity and harmonious living conditions. The term is often used in relation to rural spaces, where social problems can be hidden by the impression of idyllic life in close-knit communities and close to nature.

illegal migrant: there are at least three ways an individual might be classified as an illegal migrant. First, they can enter a country illegally, without passing through an immigration checkpoint. Such entrants are often also referred to as 'irregular' migrants. Second, they can enter a country legally for a limited period of time, for example with a short-term visa, but then stay illegally after the expiry date of their visa. These are also described as 'undocumented' migrants.

Third, they can be legally resident but involved in illegal activities.

imaginative geographies: representations of place, space and landscape that structure people's understandings of the world, and in turn help to shape their actions. In the work of Edward Said, the term refers to the projection of images of identity and difference on to geographical space in a way that sustains unequal relationships of power.

imperialism: a relationship of political, and/or economic, and/or cultural domination and subordination between geographical areas. This relationship may be based on explicit political rule (*see* COLONIALISM), but need not be. Imperialism is an unequal human and territorial relationship involving the extension of authority and control of one state or people over another.

industrialization: the process through which societies develop an economy based on the mass and mechanized production of goods. *See also* DE-INDUSTRIALIZATION and POST-INDUSTRIAL.

innovation: the strategic attempt to develop or improve processes, products and organizations. The state, firms and non-state actors (including universities) typically seek to foster the development of 'innovation systems' via the creation of networks, regulations, institutions and systemic practices that might engender regularized forms of innovation. The capacity to be innovative varies enormously across time and space.

international financial centre: an area, often a district within a city, in which financial services are concentrated in order to take advantage of the agglomeration benefits that stem from being co-located with other financial services firms. These agglomeration benefits include: being able to share technological innovations; learn about competitor activities; the ability to work with other firms on transactions; and the opportunity to share and develop new financial knowledge through personal networks.

International Monetary Fund: founded in 1945, an intergovernmental organization that aims to foster economic cooperation with a particular focus on policies that affect currency exchange rate and balance of payments. Its stated aim is to foster economic stability and growth.

inter-generational equity: this term refers to the concept of equality in access to resources or wealth between one generation of people and another (usually a future generation, or people as yet unborn).

intra-generational equity: this term refers to the concept of equality in access to resources or wealth within a single generation (usually between different groups of people today, for example, between rich and poor classes in one country, or between rich and poor countries such as between 'first world ' and 'third world' countries).

liminal: initially developed in anthropological theorizations of ritual transformation (to describe the state within a ritual process after pre-ritual status has been lost but before post-ritual status has been conferred); in Human Geography the notion of the liminal has been extended in application to a variety of spaces where 'normal' social orders and senses of self-identity are in some way suspended.

linguistic (or cultural) turn: this phrase is used to describe changes in the social sciences and Human Geography over the last 30 years. It refers to the adoption of interpretative (qualitative) approaches to explore the ways in which meanings, values and knowledges are constructed through language and other forms of communication.

love: an emotion of strong affection and personal attachment, yet the word is difficult to consistently define as it can have distinct meanings in different contexts. Love, among other uses and meanings, can refer specifically to the passionate desire and intimacy of romantic love, to sexual love, the emotional closeness of familial love or to the platonic love that defines friendship.

Marxist, Marxism: social and economic theories influenced by the legacy of the leading nineteenth-century political philosopher, Karl Marx. Highly influential in the framing of critical geography, these theories focus on the organization of capitalist society and the social and environmental injustices that can be traced to it. *See also* CAPITALISM, MODE OF PRODUCTION, ALIENATION and COMMODIFICATION for examples of the influence of Marxist thinking in Human Geography.

masculinism, masculinist: refers to the connections between cultures of masculinity, knowledge and power, in particular, dualisms between mind and body, subject and object and the presumption that scientific knowledge can and should be objective and context free. A form of thought or knowledge that, while often claiming to be impartial, comprehends the world in ways that are derived from men's experiences and concerns. Many feminist geographers have argued that Human Geography has traditionally been masculinist (*see* FEMINISM).

means of production: the resources that are indispensable for any form of production to occur. Typically this would include land, labour, machines, money capital and knowledge/information. In MARXIST thinking the means of production and labour power together constitute the 'forces of production'.

mental map: a mental map describes our everyday notions about our spatial location. People rarely picture their spatial location to themselves through the images of a formal map. But of course we are all spatially located and aware. A mental map thus relates the elements we see as important and misses places we do not visit. Studies have shown that instead of a bird's-eye view mental maps are organized around paths and landmarks that help us find our way in daily life. *See also* PERCEPTION.

metaphor: the use of a word or symbol from one domain of meaning to apply to another. Thus a rose (a botanical term) is translated into the domain of human relationships to symbolize love. Metaphors involve this movement of concepts from their normal realm to a new realm.

migration: migration can be considered a sub-category of a wider concept of 'movement', which embraces a whole range of forms of human mobility, from daily commuting at one end of the spectrum to permanent emigration at the other.

mitigation: the reduction of greenhouse gas emissions and/or their capture and storage in order to limit the extent of CLIMATE CHANGE.

mode of production: a term taken from Marxist social thought, referring to the distinctive ways in which production has taken place in different types of society and in different historical epochs. Thus CAPITALISM is seen as one 'mode of production' distinguishable, for example, from feudalism and communism. In Marxist geography, the mode of production is seen as the fundamental determinant of the kind of society and the kind of human geographies a person has to live with and through.

modern, modernity, modernism: ideas of the modern are most commonly defined through their opposition to the old and the traditional. In this light, the adjective 'modern' is synonymous with 'newness'; 'modernity' refers both to the 'post-traditional' historical epoch within which 'newness' is produced and valued, as well as to the economic, social, political and cultural formations characteristic of that period; and 'modernism' applies more narrowly to artistic, architectural and intellectual movements that centrally explore ideas of 'newness' and develop 'new' aesthetics and ways of thinking to express these. Modernity has been most commonly located in Euro-American societies from the eighteenth century onwards, and thereby associated with their characteristic combination of capitalist economies (*see* CAPITALISM), political organization through NATION-STATES, and cultural values of secularity, rationality and progress (*see* ENLIGHTENMENT). However, increasingly, Human Geographers are recognizing that modernity is a global phenomenon that has taken many different forms in different times and places. See also MODERNIZATION.

modernization: a process of social transformation in which technological change plays a leading role and in which scientific rationality becomes widely accepted. Often associated with the rise of industrial CAPITALISM, modernization has far-reaching implications for processes of identity formation and change.

moral economy: a term used by the historian Edward Thompson to describe the fact that economic relations have a normative aspect. For instance, the moral economy in the peasant economy would refer to the idea of a subsistence ethic in which the social relations of the poor attempt to provide forms of local security and support to prevent starvation. The concept of the moral economy has connections to Karl Polanyi's idea that the economy is an instituted process embedded in social relations.

multicultural, multiculturalism: multicultural is an adjective used to describe a place, society or person comprised of a number of different cultures. Multiculturalism is a body of thought that values this plurality. As the Human Geographer Audrey Kobayashi has noted, both the multicultural and multiculturalism can be conceived of in a number of ways: as 'demographic', i.e. simply reflecting a diversity of population; as 'symbolic', i.e. as about the presence or absence of the symbols associated with particular cultural groups within wider societal or national culture (for example, in the media, or in museums, or in school curricula); as 'structural', in so far as institutions are established to reflect a multicultural society and pursue multiculturalism; and 'critical', to the extent that multiculturalism itself critiques the assumptions of distinct, separate cultural groups sometimes attached to notions of the multicultural. Within Human Geography, increasing emphasis is being placed on developing this 'critical multiculturalism' through examining the HYBRID or SYNCRETIC cultural forms emerging within multicultural societies.

narrative: narratives are particular kinds of stories that are subject to cultural conventions about authorship, plot, style and audience. They are also a form of argument that contain assumptions, beliefs and IDEOLOGIES, which may be examined or unexamined. Narrative approaches in the social sciences attempt to recover the way that social life is 'storied' in terms of a series of events and how we 'make sense' of those events. Borrowed from historical and literary sources, narrative methods often involve the collection of life-history interviews that use personal testimony to relate the particularities of individual biography to their wider social context. The purpose of a narrative is to describe the world and offer an explanation of it. Like discourses, they shape the realities of the world we live in and are deeply connected to practice (*see also* DISCOURSE).

nation-state: a form of political organization that involves (a) a set of institutions that govern the people within a particular territory (the state), and (b) that claims allegiance and legitimacy from those governed, and from other states, on the basis that they represent a group of people defined in cultural and political terms as a nation.

naturalization, naturalized: this term has two distinct and different meanings: (a) the way in which social relations, cultural norms or institutions are made to seem 'natural' rather than SOCIAL CONSTRUCTIONS; and (b) the process through which species become established in new environments, sometimes also applied to human life to refer to the formal integration of immigrants in new societies.

neo-colonialism: economic and political ties, continuing after formal independence, between metropolitan countries and the South, that work to the benefit of the North. *See also* COLONIALISM.

neo-liberalism, neo-liberal: pertaining to an economic doctrine that favours free markets, the deregulation of national economies, decentralization and the privatization of previously state-owned enterprises (e.g. education, health). A doctrine that, many Geographers argue, favours the interests of the powerful (e.g. transnational corporations) against the less powerful (e.g. peasants) within societies.

new international division of labour (NIDL): the reshaping of the division of labour such that various components of the production process are located in multiple countries. This concept emerged in the context of debates about the impact of GLOBALIZATION, especially in the 1970s when production was being relocated from European and North American contexts to Asia and Latin America to take advantage of cheaper labour costs. NIDL is also facilitated by technological change (e.g. container shipping).

night time economy: generally used to refer to the entertainment industry and the broad range of clubs, pubs, restaurants etc., that are part of the leisure and tourism activities associated with the contemporary urban environment.

non-conscious: a broad term to designate a range of phenomena that occur below the threshold of conscious awareness. The negative – non – does not designate a lack, nor does it designate a dimension of experience that is completely separate from conscious acts of thinking. Rather, conscious and non-conscious processes are intertwined with one another. The nature of the relation between the two is a key concern of the research that has emerged in relation to NON-REPRESENTATIONAL THEORIES.

non-governmental organization (NGO): non-state and non-profit organizations whose primary goals may include economic development, environmental protection and human rights advocacy.

non-human agency: AGENCY usually indicates self-conscious, purposeful action. Whereas traditionally, geographers have used this term to stress the freedom and capacity of *people* to generate, modify and influence the course of everyday and world events (see HUMAN AGENCY, increasingly they are affording comparable levels of agency to non-human animals, organisms, technologies, machines and hybrid kinds of assemblage or apparatus.

non-representational theory: introduced into Human Geography by the Geographer Nigel Thrift, this phrase – which he has since accepted may not have been 'the best-chosen of phrases' – seeks to highlight various understandings that emphasize the practical, active and EMBODIED character of the world. Rather than denying the importance of REPRESENTATIONS, non-representational theory has sought to resist an undue obsession with them (see LINGUISTIC (OR CULTURAL) TURN) through a direct focus on actions, events, moments, things.

offset, offsetting: the creation of 'sinks' for GREENHOUSE GAS (GHG) EMISSIONS, e.g. through planting trees, or the provision of low carbon energy services in another place to compensate for GHG emissions

ontology: a classic philosophical consideration that asks fundamental questions and offers contested propositions about the very nature of existence. Geographers have traditionally been concerned with these questions of 'being' in terms of human existence, though recently have extended their enquiries to wonder at all manner of living organisms, as well as objects, things and technologies, in the changing make-up of life. A new frontier of ontological politics has resulted, represented in a shift in philosophical thought, away from what is (beings) towards what emerges (becomings).

Orientalizing: derived from Edward Said's pioneering work *Orientalism* (1981), which considered how regions such as the Middle East were consistently imagined and represented by Western authors and texts as exotic, mysterious and/or dangerous (see IMAGINATIVE GEOGRAPHIES).

Other, Otherness: an Other is that person or entity supposed to be opposite or different to oneself. Otherness is the quality of difference which that Other possesses. Potentially applicable at various scales, these terms have been used in Human Geography to emphasize how ideas of difference are structured through the opposition of self (same) and other (different). In cultural terms, they also highlight how the Other is often defined in relation to the Self – as everything it is not – rather than in its own terms.

paradigm: a stable consensus about the aims, shared assumptions and practices of a particular academic discipline. The term can be used in a narrow technical sense to fit the model of change in scientific disciplines such as physics and chemistry proposed by Thomas Kuhn in *The Structure of Scientific Revolutions* (1962). Kuhn argued that most scientific activity works within a stable consensus (that he called 'normal science') and does not challenge these fundamental assumptions. Eventually normal science becomes disrupted by anomalies or issues that cannot be explained within the existing framework, and new innovative work may trigger a 'paradigm shift' to a new set of shared assumptions and practices. The term 'paradigm' has entered popular usage in a looser sense and is now used regularly to refer to general approaches, theoretical frameworks and methodologies held by significant groups both within disciplines and, particularly in the social sciences and humanities, across disciplines. It is difficult to fit the development of Human Geography to a strict model of paradigms and paradigm shifts, but the terminology is often used to draw attention both to communities of geographers who share general approaches, theoretical frameworks and methodologies, and to the episodic nature of change in the discipline.

participatory geographies: a way of doing academic research and theory based on the co-production of knowledge with participants. Although there are different forms and levels of participation, participatory research is in general animated by the principle that research should be developed with and for participants, and by the desire that it should be beneficial to those participants.

partnerships: an arrangement between a number of agencies and institutions in which objectives are shared and a common agenda is developed in pursuit of a common purpose. The partnership approach encourages collaboration and integration, and the aim is that by blending and pooling their resources the different agencies will be able to produce a capacity for action that is more than simply the sum of their individual parts.

patriarchy: a social system in which men oppress and exploit women. The term was first coined in analyses of households headed by men and organized to the benefit of those 'patriarchs' (for example, through an unequal division of domestic work, or through women's marriage vows 'to obey', or through the legality of rape by husband of wife). However, the term is now used in a wider sense to think about how unequal power relations between men and women are established through realms stretching from the social organization of reproduction and childcare, to the organization of paid work, the operations of the state, cultural understandings of gender differences, the regulation of human sexuality, and men's violence towards women.

pauperization: the progressive impoverishment of people owing to the impacts of certain development programmes. For example, the displacement of peasant and tribal peoples from their sources of livelihood (e.g. land) to make way for the construction of hydro-electric dams.

perception: the process through which people form mental images of the world. Often assumed to be both one-directional (from the world to us) and biological (neurologically controlled), many academic studies have emphasized the role of cultural filters or frames in altering how we form pictures of the world.

performative methods: ways of engaging with and encountering the world such as street theatre or walking tours that are based on what Dewsbury called a 'restless experimentation' and draw inspiration from the performing arts. Performative methods attempt to create research events that might foster new relations between people and between people and spaces rather than represent an existing world.

performativity: the process by which bodies enact (i.e. perform) identities and bring them into being. These acts constitute rather than reflect identity. However, these performances emerge from the workings of power; they are the product of DISCURSIVE constraints or socially sanctioned codes that enable identity and necessitate that it is continually repeated. Performativity thus links bodily action to the re-enactment of social conventions. Although these repeated actions consolidate the norms of identity practice, they also provide possibilities for slippage and subversion because an act is never identical in its repetition.

political-economic, political economy: the study of how economic activities are socially and politically structured and have social and political consequences. Political-economic approaches in Human Geography have paid particular attention to understanding capitalist economies and their geographical organization and impact (*see* CAPITALISM and also MARXIST). Central to such analyses have been questions concerning the class-based nature of the human geographies of capitalist societies (*see* CLASS).

positionality: the personal experiences, beliefs, identities and motives of the Human Geographer, which influence her or his work and the way in which her or his knowledge is situated.

post-colonial, post-colonialism: sometimes hyphenated, sometimes not, this term has two distinct meanings: (a) the post-colonial era, i.e. the historical period following a period of colonialism (*see also* DECOLONIZATION); (b) post-colonial political, cultural and intellectual movements, and their perspectives, which are critical of the past and ongoing effects of European and other COLONIALISMS.

post-developmentalism: a radical critique of DEVELOPMENTALISM, which demands the self-empowerment of poor or marginalized people in opposition to the powers of the state or capital.

post-Fordism: refers to the forms of production, work, consumption and regulation that emerged out of the crisis of mass, standardized forms of capitalist production (FORDISM) during the 1970s. In terms of production and work, post-Fordism turns on the importance of flexibility in work and other institutional forms of productive organization. Economic Geographers have analysed how this flexibility has been driven both by versatile and programmable machines, and by forms of 'vertical disintegration' of some firms in some sectors that make greater use of strategic alliances and subcontracting. Accompanying these changes in production are changes in consumer demand (the centrality of quality over standardization), in labour markets, in finance and legal structures, and in the broad social contract that characterized post-war Fordism.

post-industrial: a description applied to the new economic, social and cultural structures emerging in late capitalist societies in the late twentieth century, highlighting in particular, trends away from manufacturing, manual work and the mass production of physical goods (*see* INDUSTRIALIZATION) and towards the tertiary sector, forms of service employment and the production of experiences, images, ideas and relationships.

post-materialist: a philosophy relating the quality of existence, and peaceful relations between people and between people and the planet, to the manageable acquisition of material goals.

postmodern, postmodernity, postmodernism: the British national newspaper, *The Independent*, sarcastically defined postmodernism thus: 'This word has no meaning. Use it as often as possible.' In fact, the main problem for a glossary entry such as this is that ideas of the postmodern have been used so often with so many different meanings! Nonetheless, one can generalize to say that notions of the postmodern (sometimes hyphenated, sometimes not) are used to suggest a move beyond 'modern' society and culture (*see* MODERN, MODERNITY, MODERNISM). More specifically: (a) postmodern is an adjective used to describe social and cultural forms that eschew 'modern' qualities of order, rationality and progress in favour of 'postmodern' qualities of difference, ephemerality, superficiality and pastiche; (b) 'postmodernity' is the contemporary epoch, after a period of 'modernity', in which such postmodern forms supposedly predominate, an epoch characterized both by the loss of an overall sense of social direction and order and by the triumph of the media image over reality (*see* HYPER-REALITY and SIMULACRUM); and (c) 'postmodernism' refers more narrowly to a collection of artistic, architectural and intellectual movements that promote postmodernist values, aesthetics and ways of thinking. If that is not complicated enough, while some view all things postmodern as signs of a radically new historical era of postmodernity, others see them more as recent twists to the history of modernity. So perhaps a revised version of *The Independent*'s definition might be: 'This word has a host of meanings. This makes it interesting, but also means it should be used with care'!

poverty: has been traditionally characterized by the identification of the poor on the basis of monetary indicators. However, a growing number of development agencies such as the World Bank now insist on the necessity of defining poverty as a multidimensional concept rather than relying on income or consumption expenditures per capita. Poverty also encompasses low levels of health and education, poor access to clean water and sanitation, inadequate physical security, lack of voice and insufficient capacity and opportunity to better one's life.

poverty porn: (also known as development porn or even famine porn) is any type of media, be it written, photographed or filmed, which exploits the poor's condition in order to generate the necessary sympathy for selling newspapers or increasing charitable donations or support for a given cause.

power: the capacity to get something done. It involves actively assessing existing knowledges and creating new knowledges. It may also involve an intervention in the world.

psychoanalysis, psychoanalytic: largely associated with the work of Freud, psychoanalytic theory concerns itself with the mental life of individuals rather than with any overt observable behaviour, and argues that the most important elements of such mental lives are the unconscious ones. It posits that the unconscious parts of the mind (the 'id') are in perpetual conflict with both the more rational and conscious elements (the 'ego') and with those parts of the mind concerned with conscience (the 'superego'). Psychological disturbances can then be traced to these conflicts, and can be remedied through psychoanalytical therapy, which is able to give an individual insight into their unconscious mental life.

qualitative methods: research methods that involve the production and analysis of data that do not take a numerical form.

quantitative methods: research methods that involve the production and analysis of numerical data.

queer: refers to both personal and political identifications that challenge normative categories and definitions. Often used in relation to gendered or sexed identities it signifies the questioning of traditional male/female and hetero/homo binaries. 'Queering' presents an ongoing critique of normalcy and seeks to re-imagine space/place/time in new and alternative ways.

race: a criterion of social categorization that distinguishes different groups of people on the basis of particular secondary physical differences (such as skin colour). Human Geographers have studied questions of race in a number of ways including: (a) the extent, causes and implications of the spatial segregation of different racial groups within cities, regions or nations; (b) the role played by geographical understandings of place and environment in the construction both of ideas of race per se and of ideas about particular races; and (c) the forms of racism and inequality that operate through these geographical patterns, processes and ideas. Increasingly, Human Geographers have emphasized how racial categories, while having very real consequences for people's lives, cannot simply be assumed as biological realities, having instead to be recognized as SOCIAL CONSTRUCTIONS. *See also* ETHNICITY.

realist: a term used in geopolitics and the discipline of International Relations (IR) to describe an approach to world politics that prioritizes the NATION-STATE and national security priorities on the basis that the international arena is judged to be anarchic.

reflexivity: a process through which we are able to reflect on what we know, how we come to know it, and how we interact with others. The key point is that we are able to change aspects of ourselves and the structures that make up society in the light of these reflections.

refugee: a refugee is someone who, 'owing to a well-founded fear of being persecuted for reasons of race, religion, nationality, membership of a particular social group or political opinion, is outside the country of his [or her] nationality'.

regulation: the arrangement of an economy into more or less cohesive forms of production, consumption, monetary circulation and social reproduction. Derived initially from a disparate group of French political-economists (*see* POLITICAL-ECONOMY) a 'regulationist approach' therefore attends to the different ways of socially, politically and culturally organizing economic activity, especially capitalist economic activity (*see* CAPITALISM). These different ways of regulating capitalist economies are identified both historically and geographically (*see also* FORDISM and POST-FORDISM).

reification: the act of transforming human properties, relations and actions into properties, relations and actions that are seen to be independent of human endeavour and that then come actually to govern human life. The term can also refer to the transformations of human beings into 'thing-like' beings. Reification is therefore a form of ALIENATION.

relational identity: the belief that the status of any given thing should not be assumed, but rather examined according to how various factors combine to shape the way that thing is understood in specific contexts.

representation: the cultural practices and forms by which human societies interpret and portray the world around them and present themselves to others. In the case of the natural world, for example, these representations range from prehistoric cave paintings of the creatures that figured in the lives of early human groups to the televisual images and scientific models that shape our imaginations today. *See also* DISCOURSE.

resource frontier: a term used to describe the exploration and exploitation of resources such as oil, gas, diamonds in areas that might be considered remote and or beyond the immediate reach of a nation-state such as the central Arctic Ocean. This in itself can encourage particular 'framings' of places and processes such as 'scramble for resources'.

resources: The Organization for Economic Co-operation and Development defines resources as: 'Natural assets (raw materials) occurring in nature that can be used for economic production or consumption' (OECD, 2011).

semiotics: this term has two interrelated meanings: (a) the study of forms of human communication and the ways they produce and convey meaning; and (b) those forms of human communication and systems of meaning themselves. While including spoken and written language, semioticians deliberately also analyse how other social phenomena – including dress, architecture and the built environment, visual art, social gatherings and events, landscapes – are communication systems whose 'languages' can and should be analysed. Human Geographers have drawn on ideas from semiotics (*see* SIGN, SIGNIFICATION, REFERENT) but parallel work in art history on the language of painting has been more directly influential (*see* ICONOGRAPHY).

sexuality: sexual attitudes, preferences, desires and behaviours. Human Geographers have emphasized how our sexualities are not simply a biological given but complex socio-cultural constructs (*see also* SOCIALLY CONSTRUCTED and SOCIAL CONSTRUCT). They have examined how, on the one hand, these constructs sexualize our encounters with places and environments in personally and socially significant ways and, on the other hand, how our sexualities themselves are shaped through experiences and understandings of the geographies of the body, the home, the city, the nation, travel, etc.

sign, signification, referent: these are ideas relating to the DISCURSIVE construction of geographical worlds. The sign is a concept or word that is significant in the understanding of everyday meanings and places and people (for example, 'rurality'). Signification is the process by which significant meanings are attached to signs (for example, social representations and interpretations of 'rurality'). The referent is the geographical phenomenon that is being signalled (for example, rural localities).

Silicon Valley: the southern part of the San Francisco bay area in California, USA, which has an historic concentration of many of the world's leading technology companies.

simulacrum: refers to the notion of a copy without an original. If we say at the first point we have an original object, then any image represents it, or replicas are copies of it. We thus have a pattern of original and copy. However, often it is the copies themselves that are copied. The products may begin to diverge from the original. The idea of a simulacrum, or of many simulacra, takes this a step further by emphasizing the presence in contemporary postmodern landscapes of many copies with no original. An example might be the shopping-mall 'Parisian café' that clearly imitates a Parisian café, but for which there is no single original it imitates; or Disneyland's main street, which is again a representation for which there was no original. *See also* HYPER-REALITY.

slum tourism: a rising phenomenon that involves mainly international tourists paying to take guided tours of the most disadvantaged parts of a city. This tourism niche emerged during the mid-1990s and is operating in almost all major slums around the world.

social capital: the notion of social capital is used to suggest that people's social networks have wider value through the relations of trust, reciprocity, co-operation and information exchange that they engender. Models of social organization and social space that involve high levels of social interaction and hence social capital – whether that is neighbourhood communities or internet-based support groups – are counterposed positively to examples of individualistic and passive existence.

social construct: a set of specific meanings that become attributed to the characteristics and identities of people and places by common social or cultural usage. Social constructs will often represent a 'loaded' view of the subject, according to the sources from which, and the channels through which, ideas are circulated in society.

social exclusion: a state of being that is perceived as being outside of, or marginalized from, mainstream social relations and the attendant resources and opportunities that this involves.

socialization: the relations through which human beings learn about acceptable social norms and moral standards in a given society. Socialization can occur through different institutions and spaces such as the family, schools or the media.

socially constructed/ion/ists: a catch-all term that emphasizes how both human geographies in the world 'out there' and the knowledges Human Geographers have of these geographies are the outcome of socio-spatial relations. For example, in debates about identity and difference social constructionists have opposed ESSENTIALIST conceptions which suggest that the differences between us are grounded in nature and emphasize instead how they are made and remade through the interweaving of wider socio-spatial processes and individual biographies. While it runs the risk of reproducing a crude society/nature dichotomy, an emphasis on social construction has been widely drawn upon in Human Geography in order to emphasize how: (a) the things, situations and ideas that surround us are not innate but the products of social forces and practices that require explanation; and (b) nor are they inevitable, instead being open to the possibility of critique and change.

sociality: socialities can be thought of as the connections, relationships and interactions between people. These can be interpersonal relations (as in work colleagues, families, Facebook friends) but also relations in which unknown people are connected across spaces (as in ethical imaginings of distant others connected through one's consumption practices). In recent years geographers have demonstrated the importance of non-human things in the establishment, maintenance and dissolution of socialities (e.g. animals, plants, photos, food, furnishing, money and mobile technologies).

socio-ecological perspective: an understanding of health which recognizes the influence of social and environmental factors. The biomedical view which underpins western medicine tends to focus on individual bodies as biological systems. The socio-ecological perspective is more holistic, encompassing the biological, psychological, social and environmental dimensions of health. As well as treating individual illness, those working with a socio-ecological perspective recognize that health may be promoted through household, city and even national scale interventions.

space of flows: a term first coined by the urban theorist Manuel Castells in distinction to the 'space of places', this self-confessed 'cumbersome expression' emphasizes how the character and dynamics of a bounded place are reliant upon a host of connections and flows that go beyond its boundaries. These include flows of people (through many forms of travel and migration), of capital and money (think of the impacts of the global networks of the international financial system, for example), of ideas, of media imagery and of objects, among many others. The notion of the 'space of flows' is therefore a complement and corrective to Human Geographers' long-standing interest in bounded places and territories (*see* TERRITORY), perhaps particularly important in an age of intensified GLOBALIZATION.

spatial co-presence: a term used to describe a situation when several people or objects are found in the same location at the same time. For example, a holidaymaker and a tour guide are spatially co-present during a guided tour, but a farmer and the consumer who finally eats his/her product are usually spatially distant.

spatial differentiation: refers to the variation of characteristics over space. Spatial differentiation is most commonly applied to the ways in which social characteristics such as CLASS or RACE can vary from one place to the next and may be configured differently in different places.

spatiality: the spatial arrangements of relations, both between people and non-human things. The term emphasizes the production of space, i.e. how places are socially and materially created, reconfigured and experienced in the context of the changing economic, political and cultural relations between other places, people and things. A consideration of spatiality also includes the effects spaces have on these relations and the power relations associated with this.

spatial science: an approach to Human Geography that became influential in the 1960s by arguing that geographers should be concerned with formulating and testing theories of spatial organization, interaction and distribution. The theories were often expressed in the form of models – of, for instance, land use, settlement hierarchy, industrial location and city sizes. If validated, these theories were then accorded the status of universal 'laws'. Through this manoeuvre, the advocates of spatial science claimed that Human Geography had been shifted from an essentially descriptive enterprise concerned with the study of regional differences to a predictive and explanatory science. Critics claim that in its attempts to formulate universally applicable laws, spatial science ignored the social and economic contexts within which its spatial variables were located.

stigma: a social process leading to the devaluation of an individual or group(s) who over time come to be identified as embodying negative traits such as dirt, disorder, laziness, criminality or mental illness. Human Geographers are especially interested in the spatial practices by which stigmatized groups come to be separated from mainstream society, thus re-enforcing the boundaries between the 'normal' and the 'deviant'.

strategic coupling: the articulation or convergence of interests between two or more actors (e.g. firms, the state) such that it (re)shapes the territorial development process. This concept, heavily utilized in global production network-related research, is linked to the notion of 'translation' in ACTOR-NETWORK THEORY where different actors work up a collaborative process that leads to joint mobilization and action with respect to a broadly agreed upon problem.

structuration: an approach to social theory that stresses the interconnection between knowledgeable human agents and the wider social structures within which they operate. Although used across the social sciences, the notion of structuration is most closely associated with the British sociologist Anthony Giddens, who developed structuration theory in a number of writings in the late 1970s and early 1980s. A key point of such theory is that the wider structural properties of social systems are both the medium and the outcome of the social practices that constitute the systems.

structure of feeling: a concept invented by the cultural theorist Raymond Williams (1961) to describe the 'felt sense of the quality of life at a particular place and time'. A structure of feeling conditions what is and can be felt by individuals or groups and reminds us that what might appear to be most ephemeral – feelings – are organized and patterned.

subjectivity: a complex term, in general usage subjectivity refers to human individuality. Sometimes this relates to our knowledge of the world (as when we speak of subjective knowledge based on individual experience); sometimes to the very status of what makes us a human individual (as when the philosopher Descartes associated human subjectivity with thought: 'I think, therefore I am'). Human Geographers draw on a range of theories of subjectivity, but generally these emphasize how our individual subjectivity is fashioned through wider social forces. For example, a post-structuralist view of subjectivity sees it arising from the ways in which bodies and identities relate to others separate from the self. Subject formation is also often understood as an expression of power relations as they operate through practices and DISCOURSES. One area of research that considers subjectivity is that on geographies of consumption. Through consumption, individuals may actively resist or embody multiple subjectivities such as being a 'consumer' 'old', 'fat', 'wealthy', etc. Geographical work on consumption has highlighted how consuming subjects are made and performed across space and over time through practice, embodied experiences, movements and imaginings.

syncretic: an adjective applied to a culture or cultural phenomenon that is composed of elements from different sources, and that combines them in such a way as to create something new and different from those sources. Drawn from anthropological literatures, notions of the syncretic are very similar to those of HYBRIDITY, but are often preferred for their less biological undertones.

technocentrism: an outlook of optimism towards any environmental or other geographical challenge on the basis of the adaptiveness and innovation of human endeavour and technological know-how.

temporality: socially produced time. This term is used by Human Geographers to emphasize how time is socially constructed and experienced, rather than being an innate backdrop to social life (*see also* SOCIAL CONSTRUCTION).

terrains of resistance: the material and/or symbolic ground upon which collective action (e.g. by social movements) takes place. This can involve the economic, political, cultural and ecological practices of resistance movements, as well as the physical places where their resistance occurs.

territory: a more or less bounded area over which an animal, person, social group or institution claims and attempts to enforce control.

thirdspace: a term coined to encourage thinking beyond binary oppositions and to reflect the ways in which dominant relationships between space and society can be contested. Thirdspace challenges conventional understandings of the world and disrupts the taken-for-granted assumptions that privilege particular identities and their dominance in particular spaces.

time–space compression: coined by the Geographer David Harvey, this phrase has been more widely used to express (a) the transformations in TEMPORALITY and SPATIALITY produced in a world of ever more rapid turnover and ever quicker forms of communication; and (b) the subjective experience of these changes.

topographical space: literally, topography means 'place writing'. Historically, the term has been applied to a number of specific kinds of geography, ranging from topographic mapping that shows surface shape and features to local studies focused on documenting the many material features of a place (a topography). Influenced by various strands of post-structural philosophy and social theory, and their promotions of notions of TOPOLOGICAL SPACE, in contemporary Human Geography topographical concepts of space are seen as prioritizing absolute and fixed coordinates: e.g. Object/Person/Place A is located on the earth's surface here; Object/Person/Place B is located on the earth's surface there; they are 100km distant from each other. This way of thinking is seen as overly static and poorly addressed to how objects/people/places are formed through their relations.

topological space: the term 'topology' has specific uses in designating areas of mathematics and of GIS, but in contemporary Human Geography it has also taken on a more general set of associations in relation to 'post-structural' philosophies of space. Here, topological thinking complicates the ideas of place, distance and boundaries that we are used to from conventional maps (e.g. where Place A is not Place B, and it is 100 km between A and B) (see TOPOGRAPHICAL SPACE). Topological space is not based on absolute and fixed coordinates; it emphasizes how spatial objects are both constituted and displaced by networks of relations; it argues that things can be multiply located at any one time; it conceives of space as 'folded', 'pleated' and 'ruptured'

transaction costs: the costs involved in engaging in transactions, normally between productive firms. Such costs (which normally do not include the costs of COMMODITIES being exchanged or the costs of transport between points of supply and points of demand) may be associated with factors like distance. Distance may serve to increase costs as delays or misinformation may be more common when transactions are taking place between partners located some way away from each other. By contrast, geographical and social proximity may help to reduce transaction costs.

transnational: an adjective used to describe Human Geographical processes that have escaped the bounded confines of the NATION-STATE. These have been identified in the realms of the economy (*see* TRANSNATIONAL CORPORATIONS), in politics (for instance, through the political agency of groups in relation to a nation-state they do not reside in; e.g. Kurdish exiles campaigning for Kurdish nationalism) and in culture (for example, through the identification of 'transnational communities' that have dispersed from an original homeland into a number of other countries but that also have strong linkages across this DIASPORA).

transnational corporations: very large companies with offices or plants in several countries and/or companies that make decisions and accrue profits on a global basis (sometimes called multinational or global corporations).

transpersonal: a term used in work on AFFECT inspired by Gilles Deleuze (and his reading of the philosopher Baruch Spinoza) to highlight how affects move between people and therefore cannot be considered to be the exclusive property of one person existing in separation from a range of non-human or other-than-human forces.

venture capital: the provision of finance by an investor to businesses not quoted on stock markets and in the form of shares. The aim of venture capitalists is to make a very high return on their investment by selling their shares at a later date. Venture capital is a highly risky way to invest and, partly because of this, investors typically expect returns of over 30 per cent.

volunteered geographic information (VGI): a term coined by Michael Goodchild (2007) to refer to information volunteered or asserted, rather than authoritative geospatial data from institutions such as the OS or USGS.

white flight: this term refers to the ethnically specific nature of out-migration from urban areas, particularly in the USA. In many of the major metropolitan areas, large-scale ethnic immigration into the inner cities from the 1950s onwards was followed by the out-migration of affluent whites. The result has been a growing ethnic differentiation of urban space between poor minority urban areas and white suburbs. *See also* ETHNICITY.

World Bank, The: created in 1944, an international financial institution charged with providing loans to developing countries to finance capital investment programmes.

world-system: an integrated international economic system, founded upon mercantile then industrial capitalism, which originated in Europe around 1450 and spread to cover most of the world by 1900. World-systems analysis, which examines this system, treats the world as a single economic and social entity, the CAPITALIST world economy.

youth: the United Nations defines youth as persons between the ages of 15 and 24. It is acknowledged that the experience of 'being young' varies enormously across regions and within countries, however. As Ansell (2005) writes in *Children, Youth and Development* there is therefore no universal definition of 'child' or 'youth', let alone agreement about what it means to be a child or a youth.

References

Aalbers, M. (2005) Place-based social exclusion: Redlining in the Netherlands. In: *Area* 37, 100–9.

Aalbers, M. (2008) The financialization of home and the mortgage market crisis. In: *Competition and Change* 12, 148–66.

Abolafia, M. (1996) *Making Markets: Opportunism and Restraint on Wall Street.* MA: Harvard University Press.

ACME (2012) Homepage for ACME: An international e-journal for critical geographies. Available at: http://www.acme-journal.org/Home.html (last accessed 1 June 2012).

Adams, P.C. and Warf, B. (eds.) (1997) Cyberspace. Special Issue, *Geographical Review* 87(2), 139–308.

Adams, M., Cox, T., Moore, G., Croxford, B., Refaee, M. and Sharples, S. (2006) Sustainable soundscapes: Noise policy and the urban experience. In: *Urban Studies* 43, 2385–98.

Adams, M. and Guy, S. (2007) Senses and the city. In: *Senses and Society* 2(2), 133–36.

Adams, V. (1992) Tourism and Sherpas, Nepal: Reconstruction of reciprocity. In: *Annals of Tourism Research* 19, 534–54.

Adams, W. (1990) *Green Development: Environment and Sustainability in the Third World.* London: Routledge.

Adams, W. (1996) *Future Nature.* London: Earthscan.

Adams, W.M. (2004) *Against Extinction: The Story of Conservation.* London: Earthscan.

Adamson, J., Ben-Shlomo, Y., Chaturvedi, N. and Donovan, J. (2003) Ethnicity, socio-economic position and gender – Do they affect reported health-care seeking behaviour? In: *Social Science & Medicine* 57, 895–904.

Adey, P. (2004) Surveillance at the airport: Surveilling mobility/mobilising surveillance. In: *Environment and Planning A* 36, 1365–80.

Acheson, D. (1998) *Independent Inquiry into Inequalities in Health.* London: The Stationery Office.

Adey, P. (2008) Airports, mobility and the calculative architecture of affective control. In: *Geoforum* 39(1), 438–51.

Adey, P. (2009) Facing airport security: Affect, biopolitics and the preemptive securitisation of the mobile body. In: *Environment and Planning D: Society and Space* 27, 274–95.

Adey, P. (2010) *Mobility.* Abingdon: Routledge.

Adler, J. (1989) Origins of sightseeing. In: *Annals of Tourism Research* 16, 7–29.

Adler, S. and Brenner, J. (1992) Gender and space: Lesbians and gay men in the city. In: *International Journal of Urban and Regional Research* 16, 24–34.

African Biodiversity Network (ABN) 2007 *Agrofuels in Africa – the impact on land, food and forests: case studies from Benin, Tanzania, Uganda and Zambia.* Available at: http://www.africanbiodiversity.org/sites/default/files/PDFs/Agrofuels%20in%20Africa-Impacts%20on%20land%2C%20foods%20%26%20forests.pdf

Agarwal, S., Ball, R., Shaw, G. and Williams, A. (2000) The geography of tourism production: Uneven disciplinary development. In: *Tourism Geographies* 2, 241–63.

Agnew, J. (1987a) *Place and Politics.* Boston: Allen & Unwin.

Agnew, J. (1987b) *The United States in the World Economy: A Regional Geography.* Cambridge: Cambridge University Press.

Agnew, J. (1995) The rhetoric of regionalism: The northern league in Italian politics, 1983–94. In: *Transactions of the Institute of British Geographers* 20(2), 156–72.

Agnew, J. (1998) *Geopolitics.* London: Routledge.

Agnew, J. (2002) *Place and Politics in Modern Italy.* Chicago: Chicago University Press.

Agnew, J. and Corbridge, S. (1989) The new

geopolitics: the dynamics of geopolitical disorder. In: R. Johnston and P. Taylor (eds.) *A World in Crisis?* (2nd edn.). Oxford: Blackwell.

Agnew, J. and Corbridge, S. (1995) *Mastering Space: Hegemony, Territory and International Economy*. London: Routledge.

Agnew, J.A., Livingstone, D. and Rogers, A. (eds.) (1996) *Human Geography: An Essential Anthology*. Oxford: Blackwell.

Aitken, S. (2001) *Geographies of Young People: The Morally Contested Spaces of Identity*. London: Routledge.

Aitken, S. (2009) *The Awkward Spaces of Fathering*. Aldershot: Ashgate.

Aitken, S. and Zonn, L. (1993) Weir(d) sex: representations of gender–environment relations in Peter Weir's *Picnic at Hanging Rock* and *Gallipoli*. In: *Environment and Planning D: Society and Space* 11, 191–212.

Aitken, S. and Michel, S.M. (1995) Who contrives the 'real' in GIS? Geographic information, planning and critical theory. In: *Cartography and Geographic Information Systems* 22(1), 17–29.

Aitken, S. and Crane, J. (2009) Affective visual geographies and GIScience. In: M. Cope and S. Elwood (eds.) *Qualitative GIS: A Mixed Methods Approach*. London: Sage.

Alderman, D.H. (2000) A street fit for a king: Naming places and commemoration in the American South. In: *Professional Geographer* 52, 672–84.

Alderman, D.H. (2010) Surrogation and the politics of remembering slavery in Savannah, Georgia (USA). In: *Journal of Historical Geography* 36, 90–101.

Allan, S., Adam, B. and Carter, C. (eds.) (2000) *Environmental Risks and the Media*. London: Routledge.

Allen, J. and Massey, D. (eds.) (1995) *Geographical Worlds*. Oxford: Oxford University Press.

Alleyne-Dettmers, P. (1997) Tribal arts: A case study of global compression in the Notting Hill Carnival. In: J. Eade (ed.) *Living the Global City*. London: Routledge.

Alliez, E. and Negri, A. (2003) Peace and war. In: *Theory, Culture and Society* 20(2), 109–18B.

Amin, A. (2007) Rethinking the urban social. In: *City* 11, 100–14.

Amin, S. (1990) *Maldevelopment: Anatomy of a Global Failure*. London: Zed Books.

Amin, S. (1994) About the Gulf War: Reflections on the new world order, capitalist utopia, militarism and US hegemony, and the role of the United States. In: J. O'Loughlin, J. (2004) T. Mayer and E. Greenberg (eds.) *War and Its Consequences: Lessons from the Persian Gulf Conflict*. New York: HarperCollins College Publishers.

Amoore, L. (2006) Biometric borders: governing mobilities in the war on terror. In: *Political Geography* 25, 336–51.

Amsden, A. (1989) *Asia's Next Giant: South Korea and Late-Industrialisation*. New York: Oxford University Press.

Anderson, A. (1997) *Culture, Media and Environmental Issues*. London: UCL Press.

Anderson, B. (1991) *Imagined Communities: Reflections on the Origins and Spread of Nationalism*. London: Verso.

Anderson, B. (2003) Porto Allegre: A worm's eye view. In: *Global Networks* 3, 197–200.

Anderson, B. (2006) Becoming and being hopeful: Towards a theory of affect. In: *Environment and Planning D: Society and Space* 24, 733–52.

Anderson, B. (2009) Affective atmospheres. In: *Emotion, Society and Space* 2, 77–81.

Anderson, B. (2010) Morale and the affective geographies of the 'war on terror'. In: *Cultural Geographies* 17, 219–36.

Anderson, B. (2012) Affect and biopower: Towards a politics of life. In: *Transactions of the Institute of British Geographers* 37(1), 28–43.

Anderson, B. and Harrison, P. (2006) Questioning affect and emotion. In: *Area* 38, 333–5

Anderson, B. and Harrison, P. (eds.) (2010a) *Taking-Place: Non-Representational Theories and Geography*. Farnham: Ashgate.

Anderson, B. and Harrison, P. (2010b) The promise of non-representational theories. In: B. Anderson and P. Harrison (eds.) *Taking-Place: Non-representational Theories and Geography*. Farnham: Ashgate, 1–34.

Anderson, B. and Wylie, J. (2009) On geography

and materiality. In: *Environment and Planning A* 41, 318–35.

Anderson, J. (1988) Nationalist ideology and territory. In: R.J. Johnston, D.B. Knight and E. Kaufman (eds.) *Nationalism, Self-determination and Political Geography*. London: Croom Helm.

Anderson, J. (1996) The shifting stage of politics: New medieval and postmodern territorialities. In: *Environment and Planning D: Society and Space* 14(2), 133–53.

Anderson, K. (1991) *Vancouver's Chinatown: Racial Discourse in Canada 1875–1980*. Montreal: McGill-Queen's University Press.

Anderson, K. (1995) Culture and nature at the Adelaide Zoo: At the frontiers of 'human' geography. In: *Transactions of the Institute of British Geographers* 20, 275–94.

Anderson, K. (1997) A walk on the wildside. In: *Progress in Human Geography* 21(4), 463–85.

Anderson, K. and Domosh, M. (eds.) (2002) North American spaces/postcolonial stories. In: *Cultural Geographies* 9(2), themed issue.

Anderson, K., Domosh, M., Pile, S. and Thrift, N.J. (eds.) (2002) *Handbook of Cultural Geography*. London: Sage.

Anderson, K. and Gale, F. (eds.) (1992) *Inventing Places: Studies in Cultural Geography*. London: Longman.

Anderson, K. and Gale, F. (eds.) (1999) *Cultural Geographies* (2nd edn.). London: Longman.

Anderson, K. and Smith, S. (2001) Emotional geographies. In: *Transactions of the Institute of British Geographers* NS, 26, 7–10.

Andersson, J., Vanderbeck, R.M., Valentine, G., Ward, K. and Sadgrove, J. (2011) New York encounters: Religion, sexuality, and the city. In: *Environment and Planning A* 43(3), 618–33.

Andrews, G. (2009) Complementary and alternative medicine. In: R. Kitchin and N. Thrift (eds.) *International Encyclopedia of Human Geography*. Oxford: Elsevier, 234–38.

Andrews, G. and Phillips, D.R. (eds.) (2005) *Ageing and Place: Perspectives, Policy, Practice*. Abingdon: Routledge.

Andrews, G., Cutchin, M., McCracken, K., Phillips, D. and Wiles, J. (2007) Geographical gerontology: The constitution of a discipline. In: *Social Science and Medicine* 65, 151–68.

Andrews, G.J., Kearns, R.A., Kingsbury, P. and Carr, E.R. (2011). Cool aid? Health, wellbeing and place in the work of Bono and U2. In: *Health & Place* 17(1), 185–94.

Andsager, J.L. and Drzewiecka, J.A. (2002) Desirability of differences in destinations. In: *Annals of Tourism Research* 29(2), 401–21.

Ang, I. (1985) *Watching Dallas*. London and New York: Methuen.

Ang, I. (1991) *Desperately Seeking the Audience*. London and New York: Routledge.

Anger, K., Foot, J., Joy, C., Massey, D. and Rutherford, J. (2004) Thinking the globally locally: A roundtable. In: *Soundings* 26, 46–58.

Ansell, N. (2005) *Children, Youth, and Development*. Abingdon: Routledge.

Ansprenger, F. (1989) *The Dissolution of the Colonial Empires*. London: Routledge.

Anzaldúa, G. (1987) *Borderlands/La Frontera: The New Mestiza*. San Francisco: Aunt Lute.

Appadurai, A. (ed.) (1986) *The Social Life of Things. Commodities in Cultural Perspective*. Cambridge: Cambridge University Press.

Appadurai, A. (1990) Disjuncture and difference in the global cultural economy. In: *Theory, Culture and Society* 7, 295–310.

Arblaster, A. (1984) *The Rise and Fall of Western Liberalism*. Oxford: Blackwell.

Ariès, P. (1962) *Centuries of Childhood*. New York: Vintage Press.

Armstrong, J. (2004) The body within, the body without. In: *Globe and Mail*, 12 June 2004, F1, F6.

Arreola, D.D. (1988) Mexican American housescapes. In: *Geographical Review* 78, 299–315.

Arrighi, G. (1990) The three hegemonies of historical capitalism. In: *Review* 13, 365–408.

Arrighi, G. (1994) *The Long Twentieth Century*. London: Verso.

Asad, T. (ed.) (1973) *Anthropology and the Colonial Encounter*. London: Ithaca Press.

Asanin, J. and Wilson, K. (2008) 'I spent nine years looking for a doctor': Exploring access to health care among immigrants in Mississauga, Ontario, Canada. In: *Social Science & Medicine* 66(6), 1271–83.

Ash, J. (2009) Emerging spatialities of the screen: video games and the reconfiguration of spatial awareness. In: *Environment and Planning A* 41, 2105–24.

Ash, J. (2010) Architectures of affect: Anticipating and manipulating the event in practices of videogame design and testing. In: *Environment and Planning D: Society and Space* 28(4), 653–71.

Ash, J. and Gallacher, L.A. (2011) Cultural geography and videogames. In: *Geography Compass* 5, 351–68.

Association of American Geographers (2002) The geographical dimensions of terrorism: Action items and research priorities. www.aag.org/news/godt.html, accessed 22 May 2002.

Atkinson, D. and Cosgrove, D. (1998) Urban rhetoric and embodied identities: City, nation and empire at the Vittorio Emanuele II monument in Rome 1870–1945. In: *Annals, Association of American Geographers* 88, 28–49.

Atkinson, S., Fuller, S. and Painter, J. (eds.) (2012) *Well-Being and Place*. Aldershot: Ashgate.

Augé, M. (1998) *A Sense for the Other*. A. Jacobs (trans.) Stanford, CA: Stanford University Press.

Avery, S. (1989) *Up from Washington. William Pickens and the Negro Struggle for Equality, 1900–1954*. Newark: University of Delaware Press.

Bach, J. and Stark, D. (2005) Recombinant technology and new geographies of association. In: R. Latham and S. Sassen (eds.) *Digital Formations: IT and New Architectures in the Global Realm*. Princeton: Princeton University Press, 37–53.

Back, L. (1995) X amount of Sat Siri Akal! Apache Indian, reggae music and the cultural intermezzo. *New Formations* 27, 128–47.

Badiou, A. (2001) *Ethics*. P. Hallward (trans.). New York: Verso.

Baggini, J. (2005) *The Pig That Wants to be Eaten*. London: Granta.

Baier, L. (1991/92) Farewell to regionalism. In: *Telos* 90 (winter), 82–8.

Bailes, A.J.K. (2004) Lessons of Iraq. Moscow, 20 April 2004, lecture. http://www.sipri.org/people/alyson_bailes/2004042001.html. Accessed 21 May 2004.

Bailey, A. (2005) *Making Population Geography*. London: Hodder.

Bailey, A. (2009) Population geography: Lifecourse matters. In: *Progress in Human Geography* 33, 407–18.

Bailey, I., Hopkins, R. and Wilson, G. (2010) Some things old, some things new: The spatial representations and politics of change of the peak oil relocalisation movement. In: *Geoforum* 41(4), 595–605.

Bailey, A., Hutter, I. and Huigen, P. (2011) The spatial-cultural configuration of sex work in Goa, India. In: *Tijdschrift voor Economische en Sociale Geografie* 102(2), 162–75.

Baker, N.C. (1984) *The Beauty Trap*. London: Piatkus.

Baker, S. (1993) *Picturing the Beast: Animals, Identity and Representation*. Manchester: Manchester University Press.

Baldwin, E., Longhurst, B., McCracken, S., Ogborn, M. and Smith, G. (1999) *Introducing Cultural Studies*. Hemel Hempstead: Prentice Hall.

Bali Principles of Climate Justice (2002). http://www.ejnet.org/ej/bali.pdf Accessed 18 January 2010.

Ballard, J.G. (1997) Airports. In: *The Observer*, 14 September 1997.

Barbier, E.B., Burgess, J.C. and Folke, C. (1994) *Paradise Lost? The Ecological Economics of Biodiversity*. London: Earthscan.

Barker, C. (1999) *Television, Globalization and Cultural Identities*. Buckingham: Open University Press.

Barker, K. (2008) Garden terrorists and the war on weeds: Interrogating New Zealand's biosecurity regime. In: K. Dodds and A. Ingram (eds.) *Geographies of Security and Insecurity*. Hampshire: Ashgate, 165–84.

Barndt, M. (1998) Public participation GIS – Barriers to implementation. In: *Cartography and Geographic Information Systems* 25(2), 105–12.

Barnes, C. (1991) *Disabled People in Britain and Discrimination*. London: Hurst and Company.

Barnes, M. (1997) *Care, Communities and Citizens*. Harlow: Longman.

Barnes, T.J. and Duncan, J.S. (eds.) (1992) *Writing Worlds: Discourse, Text and Metaphor in the Representation of Landscapes*. London: Routledge.

Barnett, A. and Scruton, R. (eds.) (1998) *Town and Country*. London: Jonathan Cape.

Barnett, C. (2008a) Political affects in public space: Normative blind-spots in non-representational ontologies. In: *Transactions of the Institute of British Geographers* 33(2), 186–200.

Barnett, C. (2008b) Convening publics: The parasitical spaces of public action. In: K.R. Cox, M. Low and J. Robinson (eds.) *The Sage Handbook of Political Geography*. London: Sage, 403–17.

Barnett, C., Cloke, P., Clarke, N., and Malpass, A. (2005) Consuming ethics: Articulating the subjects and spaces of ethical consumption. In: *Antipode* 37(1), 23–45.

Barnett, J. (2007) The geopolitics of climate change. In: *Geography Compass* 1, 1361–75.

Barnett, R. and Barnett, P. (2009) Health systems and health services. In: R. Kitchin and N. Thrift (eds.) *International Encyclopedia of Human Geography*. Oxford: Elsevier, 58–70.

Barrell, J. (1980) *The Dark Side of the Landscape: The Rural Poor in English Painting 1730–1840*. Cambridge: Cambridge University Press, 1–33.

Barrell, J. (1990) The public prospect and the private view: The politics of taste in eighteenth-century Britain. In: S. Pugh (ed.) *Reading Landscape: Country – City – Capital*. Manchester: Manchester University Press, 19–40.

Barry, J. (1999) *Environment and Social Theory*. London: Routledge.

Bartling, H. (2006) Suburbia, mobility, and urban calamities. In: *Space and Culture* 9, 60–2.

Barton, H. (1996) The Isles of Harris superquarry: Concepts of environment and sustainability. In: *Environmental Values* 5, 97–122.

Bassett, T.J. and Zuéli, K.B. (2000) Environmental discourses and the Ivorian Savanna. In: *Annals of the Association of American Geographers* 90, 67–95.

Bathelt, H., Malmberg, A. and Maskell, P. (2004) Clusters and knowledge: Local buzz, global pipelines and the process of knowledge creation. In: *Progress in Human Geography* 28, 31–56.

Batty, M. (1993) The geography of cyberspace. In: *Environment and Planning B: Planning and Design* 20(6), 615–6.

Batty, M. (1997) Virtual geography. In: *Futures* 29(4/5), 337–52.

Baudrillard, J. (1988a) *America*. London: Verso.

Baudrillard, J. (1988b) Consumer society. In: M. Poster (ed.) *Selected Writings*. Cambridge: Polity Press.

Bauer, S., Escher, A. and Knieper, S. (2006) Essaouira: 'The Wind City' as a cultural product. In: *Erdkunde* 60(1), 25–39.

Bauman, Z. (1990) *Thinking Sociologically*. Cambridge, MA: Blackwell.

Bauman, Z. (1996) On communitarians and human freedom. In: *Theory, Culture and Society* 13(2), 79–90.

Baumann, G. (1990) The re-invention of bhangra. In: *World Music* 32(2), 81–95.

BBC News (2006) Ukraine accused of stealing Europe's gas, 2 January 2006. Available at: http://news.bbc.co.uk/1/hi/world/europe/4574630.stm

Bear, C. (2011) Being Angelica? Exploring individual animal geographies. In: *Area 43(3),* 297–304.

Beauregard, B. (1994) *Voices of Decline: The Post-war Fate of US Cities*. Oxford: Blackwell.

Beauregard, R. (1999) Break dancing on Santa Monica Boulevard. In: *Urban Geography* 20(5), 396–9.

Beauvoir, S. de (1989) *The Second Sex*. H.M. Parchley (trans.). New York: Vintage.

Beaverstock, J.V. (2002) Transnational elites in global cities: British expatriates in Singapore's financial district. In: *Geoforum* 33(4), 525–38.

Beaverstock, J.V., Smith, R.G. and Taylor, P.J. (1999) A roster of world cities. In: *Cities* 16(6), 445–58.

Beck, U. (1992a) From industrial society to the risk society: Questions of survival, social structure and ecological enlightenment. In: *Theory, Culture and Society* 9, 97–123.

Beck, U. (1992b) *Risk Society: Towards a New Modernity*. London: Sage.

Beck, U. (1999) *World Risk Society*. Cambridge: Polity Press.

Beck, U. (2002) The silence of words and political dynamics in the world risk society. In: *Logos*, 1 (4) (Fall 2002), 1–18.

Beck, U., Giddens, A. and Lash, S. 1994. *Reflexive Modernization: Politics, Tradition and Aesthetics in the Modern Social Order*. Cambridge: Polity Press.

Begg, B., Pickles, J. and Smith, A. (2003) Cutting it: European integration, trade regimes, and the reconfiguration of east-central European apparel production. In: *Environment and Planning A*, 35, 2205.

Bell, D. (1995) Pleasure and danger: The paradoxical spaces of sexual citizenship. In: *Political Geography* 14(2), 139–54.

Bell, D. (1997) Anti-idyll: Rural horror. In: P. Cloke and J. Little (eds.) *Contested Countryside Cultures*. London: Routledge.

Bell, D. (2007) The hospitable city. Social relations in commercial areas. In: *Progress in Human Geography* 3(1), 7–22.

Bell, D., Binnie, J., Cream, J. and Valentine, G. (1994) All hyped up and no place to go. In: *Gender, Place and Culture* 1, 31–47.

Bell, D. and Valentine, G. (1995a) *Mapping Desire: Geographies of Sexualities*. London: Routledge.

Bell, D. and Valentine, G. (1995b) Queer country: Rural lesbian and gay lives. In: *Journal of Rural Studies* 11, 113–22.

Bell, D. and Valentine, G. (1995c) The sexed self: Strategies of performance, sites of resistance. In: S. Pile and N. Thrift (eds.) *Mapping the Subject: Geographies of Cultural Transformation*. London: Routledge.

Bell, D. and Valentine, G. (1997) *Consuming Geographies: You Are Where You Eat*. London: Routledge.

Bell, M. (1998) *An Invitation to Environmental Sociology*. Thousand Oaks: Pine Forge Press.

Bell, M., Butlin, R. and Heffernan, M. (eds.) (1994) *Geography and Imperialism, 1820–1940*. Manchester: Manchester University Press.

Bell, M.M. (1994) *Childerley: Nature and Morality in an English Village*. Chicago: University of Chicago Press.

Benjamin, W. (1978) *Reflections*. New York: Harcourt Brace Jovanovitch.

Bennett, B. and Routledge, P. (1997) Tibetan resistance 1950–present. In: R.S. Powers, W.B. Vogele, C. Kruegler and R.M. McCarthy (eds.) *Protest, Power, and Change*. New York: Garland Publishing.

Bennett, J. (2007) Edible material. In: *New Left Review* 45, 133–45.

Bennett, J. (2010) *Vibrant Matter: A Political Ecology of Things*. Durham, NC: Duke University Press.

Benwell, M.C. (2009) Challenging minority world privilege: Children's outdoor mobilities in post-apartheid South Africa. In: *Mobilities* 4(1), 77–101.

Berg, L.D. (1994) Masculinity, place and a binary discourse of theory and empirical investigation in the human geography of Aotearoa/New Zealand. In: *Gender, Place and Culture* 1(2), 245–60.

Berger, J. (1972) *Ways of Seeing*. Harmondsworth: Penguin.

Berlant, L. (2011) *Cruel Optimism*. London: Duke University Press.

Berman, M. (2010) *All That is Solid Melts into Air*. London: Verso.

Berndes, G., Hoogwijk, M. and van den Broeck, R. (2003) 'The contribution of biomass in the future global energy supply: A review of 17 studies'. In: *Biomass and Bioenergy* 25, 1–28.

Bernstein, H. (1978) Notes on capital and peasantry. In: *Review of African Political Economy* 10, 53–9.

Bernstein, R. (1992) *The New Constellation: The Ethical – Political Horizons of Modernity/ Postmodernity*. Cambridge, MA: MIT Press.

Bernstein, S. (2000) Ideas, social structure and the compromise of liberal environmentalism. In: *European Journal of International Relations* 6, 464–512.

Berry, B.J.L. (1970) The geography of the United States in the year 2000. In: *Transactions of the Institute of British Geographers* 51, 21–54.

Bhabha, H.K. (1994) *The Location of Culture*. London: Routledge.

Bhachu, P. (1985) *Twice Migrants*. London: Tavistock Publications.

Bhachu, P. (2003) *Dangerous Designs: Asian Women Fashion the Diaspora Economies*. London: Routledge.

Bhattacharyya, D. (1997) Mediating India: An analysis of a guidebook. In: *Annals of Tourism Research* 24(2), 371–89.

Billig, M. (1995) *Banal Nationalism*. London: Sage.

Biltcliffe, P. (2005) Walter Crane and the Imperial Federation Map showing the extent of the British Empire (1886). In: *Imago Mundi* 57, 63–9.

Bingham, N. (1999) Unthinkable complexity? Cyberspace otherwise. In: M. Crang, P. Crang and J. May (eds.) *Virtual Geographies. Bodies, Space and Relations*. London: Routledge, 244–60.

Binnie, J. (1995) Trading places: consumption, sexuality and the production of queer space. In: Bell, D. and Valentine, G. (eds.) *Mapping Desire*. London: Routledge.

Binnie, J. and Valentine, G. (1999) Geographies of sexuality – A review of progress. In: *Progress in Human Geography* 23, 175–87.

Bissell, D. (2009) Visualising everyday geographies: Practices of vision through travel-time. In: *Transactions of the Institute of British Geographers* 34, 42–60.

Bissell, D. (2010a) Passenger mobilities: Affective atmospheres and the sociality of public transport. In: *Environment and Planning D: Society and Space* 28(2), 270–89.

Bissell, D. (2010b) Placing affective relations: Uncertain geographies of pain. In: B. Anderson and P. Harrison (eds.) *Taking Place: Non-Representational Theories and Geography*. Aldershot: Ashgate, 79–98.

Burrell, K. (2011) Going steerage on Ryanair: Cultures of migrant air travel between Poland and the UK. In: *Journal of Transport Geography* 19, 1023–30.

Blaikie P.M. (1985) *The Political Economy of Soil Erosion in Developing Countries*. London: Longman.

Blaikie, P.M. and Brookfield, H. (1987) *Land Degradation and Society*. London: Methuen.

Blair, T. (2004) Message to the Commissioners from the Rt Hon. Tony Blair, Prime Minister of the United Kingdom, 24 May 2004. http://213.225.140.43/commission/message_to_comm.htm, accessed 30 June 2004.

Blaut, J. (1993) *The Colonizer's Model of the World*. New York: Guilford.

Bloch, A. (2002) *The Migration and Settlement of Refugees in Britain*. London: Palgrave.

Blunt, A. (1994) *Travel, Gender, and Imperialism: Mary Kingsley and West Africa*. New York: Guildford Press.

Blunt, A. and Rose, G. (1994) (eds.) *Writing Women and Space: Colonial and Postcolonial Geographies*. London: Routledge.

Blunt, A. and McEwan, C. (eds.) (2002) *Postcolonial Geographies*. London: Continuum.

Blunt, A. and Wills, J. (2000) *Dissident Geographies: An Introduction to Radical Ideas and Practice*. London: Longman.

Bondi, L. (1992) Gender symbols and urban landscapes. In: *Progress in Human Geography* 16(2), 157–70.

Bondi, L. (2005) Making connections and thinking through emotions: Between geography and psychotherapy. In: *Transactions of the Institute of British Geographers* 30(4), 433–48.

Bondi, L. and Davidson, J. (2011) Lost in translation: A response to Steve Pile. In: *Transactions of the Institute of British Geographers* 36, 595–8.

Bondi, L., Davidson, J. and Smith, M. (2004) *Emotional geographies*. Aldershot: Ashgate.

Bondi, L., Avis, H., Bankey, R., Bingley, A., Davidson, J., Duffy, R., Einagel, V.I., Green, A-M., Johnston, L., Lilley, S., Listerborn, C., Marshy, M., McEwan, S., O'Connor, N., Rose, G., Vivat, B. and Wood, N. (2002) *Subjectivities, Knowledge and Feminist Geographies*. Oxford: Rowman & Littlefield.

Bonnett, A. (1997) Geography, 'race' and whiteness: Invisible traditions and current challenges. In: *Area* 29, 193–9.

Bonnett A. (2005) Whiteness. In: D. Atkinson, P. Jackson, D. Sibley and N. Washbourne (eds.) *Cultural Geography: A Critical Dictionary of Key Concepts*. London: I.B.Tauris, 109–14.

Bonnett, A. (2008) *What is Geography?* London: Sage.

Bonwick, J. (1884) *The Lost Tasmanian Race*. London: Sampson Low, Marston, Searle and Rivington.

Boorstin, D. (1992) *The Image: A Guide to Pseudo-events in America* (2nd edn.) New York: Vintage Books.

Booth, D. (ed.) (1994) *Rethinking Social Development: Theory, Research and Practice*. Harlow: Longman.

Borden, I. (2001) *Skateboarding, Space and the City. Architecture and the Body*. Oxford: Berg.

Borrow, G. (1989) *Wild Wales: The People, Language and Scenery*. London: Century Hutchinson.

Bouquet, M. (1987) Bed, breakfast and an evening meal: Commensality in the nineteenth and twentieth century farm household. In: M. Bouquet, and M. Winter (eds.) *Who From Their Labour's Rest? Conflict and Practice in Rural Tourism*. Aldershot: Avebury.

Bourdieu, P. (1984) *Distinction: A Social Critique of the Judgement of Taste*. London: Routledge and Kegan Paul.

Bourdieu, P. (1990) *The Logic of Practice*. Stanford, CA: Stanford University Press.

Bourdieu, P. (2010) [1984] *Distinction*. Routledge, Abingdon and New York.

Bowlby, S., Lewis, J., McDowell, L. and Foord, J. (1989) The geography of gender. In: R. Peet and N. Thrift (eds.) *New Models in Geography* 2. London: Routledge.

Boyce, J.K. (1992) Of coconuts and kings: The political economy of an export crop. In: *Development and Change* 23(4), 1–25.

Boyd, J. (2010) Producing Vancouver's (hetero)normative nightscape. In: *Gender, Place and Culture* 17(2), 169–89.

Boyd, W. (2001) Industrial meat. In: *Technology and Culture* 42, 153–78.

Boyd, W. and Watts, M. (1996) Agro-industrial just-in-time: The chicken industry and postwar American capitalism. In: D. Goodman and M. Watts (eds.) *Globalizing Food*. London: Routledge, 205

Boyer, P. (1994) *By the Bomb's Early Light: American Thought and Culture at the Dawn of the Atomic Age* (2nd edn.). Chapel Hill and London: University of North Carolina Press.

BP (2010) *Statistical Review of World Energy*. Available at: www.bp.com/statisticalreview

Bradshaw, M. (2009) The geopolitics of global energy security. In: *Geography Compass* 3, 1920–37.

Bradshaw, M. (2010) Global energy dilemmas: A geographical perspective. In: *The Geographical Journal* 175, 275–90.

Bradshaw, M. and A. Stenning (eds.) (2004) *East Central Europe and the former Soviet Union*. London: Pearson Prentice Hall.

Brah, A. (1996) *Cartographies of Diaspora*. London: Routledge.

Brantlinger, P. (1985) Victorians and Africans: The genealogy of the myth of the Dark Continent. In: *Critical Inquiry* 12, 166–203.

Braun, B. (2005) Environmental issues: Writing a more-than-human urban geography. In: *Progress in Human Geography* 29 (5) 635–50.

Braun, B. (2008) Thinking the city through SARS: Bodies, topologies, politics. In: H. Ali and R. Keil (eds.) *Networked Disease*. Oxford: Blackwell.

Braun, M. (1992) *Picturing Time: The Work of Etienne-Jules Marey (1830–1904)*. Chicago: University of Chicago Press.

Brecher, J. and Costello, T. (1994) *Global Village or Global Pillage*. Boston: South End Press.

Brecher, J., Costello, T. and Smith, B. (2000) *Globalization From Below: The Power of Solidarity*. Cambridge, MA: South End Press.

Breward, C., Crill, R. and Crang, P. (eds.) (2010) *British Asian Style. Fashion and Textiles, Past and Present*. London: V&A Publishing.

Brickell, C. (2000) Heroes and invaders: Gay and lesbian pride parades and the public/private distinction in New Zealand media accounts. In: *Gender, Place and Culture* 7, 163–78.

Brickell, K. (2012) Visual critiques of tourist development: Host-employed photography in Vietnam. In: *Tourism Geographies* 14, 98–116.

Bridge, G. (2004) Mapping the bonanza: Geographies of mining investment in an era of neo-liberal reform. In: *The Professional Geographer* 56, 406–21.

Bridge, G. (2009) Material worlds: Natural resources, resource geography, and the material economy. In: *Geography Compass* 3, 1217–44.

Bridge, G. (2010) Beyond peak oil: Political economy of energy crises. In: R. Peet, K. Robbins and M. Watts (eds.) *Global Political Ecology*. Abingdon: Routledge.

Bridge, G. and Wood, A. (2010) Less is more: Spectres of scarcity, and the politics of resource access in the upstream oil sector. In: *Geoforum* 41, 565–76.

Brohman, J. (1996) *Popular Development: Rethinking the Theory and Practice of Development*. Oxford: Blackwell.

Bromley, R.D.F. and Mackie, P.K. (2009) Child experiences as street traders in Peru: Contributing to a reappraisal for working children. In: *Children's Geographies* 7(2), 141–58.

Brook, C. and Gimpo (1997) *K Foundation Burn a Million Quid*. London: Ellipsis.

Broughton, E. (2005) The Bhopal disaster and its aftermath: A review. In: *Environmental Health* 4(6).

Brown, K., Turner, R.K., Hameed, H. and Bateman, I. (1997) Environmental carrying capacity and tourism development in the Maldives and Nepal. In: *Environmental Conservation* 24, 316–19.

Brown, M. (1997) *Replacing Citizenship: AIDS Activism and Radical Democracy*. New York: Guilford Press.

Browne, K. (2007) (Re)making the other: Heterosexualising everyday space. In: *Environment and Planning A* 39(4), 996–1014.

Browne, K. and Bakshi, L. (2011) We are here to party? Lesbian, gay, bisexual and trans leisurescapes beyond commercial gay scenes. In: *Leisure Studies* 30(2), 179–96.

Browne, K. Bakshi, L. and Law, A. (2010) Positionalities: It's not about them and us, it's about us. In: S. Smith, R. Pain, S. Marston and J.-P. Jones (eds.) *Handbook of Social Geographies*. Sage, London, 586–604.

Brownlow, A. (2000) A wolf in the garden: Ideology and change in the Adirondack landscape. In: C. Philo and W. Wilbert (eds.) *Animal Spaces, Beastly Places: New Geographies of Human–Animal Relations*. London: Routledge.

Brundtland, H. (1987) *Our Common Future*. Oxford: Oxford University Press.

Bryan, J. (2010) Force multipliers: geography, militarism, and the Bowman expeditions. In: *Political Geography* 29, 414–6.

Bryant, C.G.A. (2003) These Englands, or where does devolution leave the English? In: *Nations and Nationalism* 8(3), 393–412.

Bryson, J., Henry, N., Keeble, D. and Martin, R. (eds.) (1999) *The Economic Geography Reader*. Chichester: Wiley.

BT plc (2011) Reducing our own footprint. www.btplc.com/Responsiblebusiness/Protectingourenvironment/Climatechange/Reducingourownfootprint/index.htm Accessed September 2011)

Buck-Morss, S. (1990) *The Dialectics of Seeing*. Cambridge, MA: MIT Press.

Bulkeley, H. (2005) Reconfiguring environmental governance: Towards a politics of scales and networks. In: *Political Geography* 24: 875–902.

Bulkeley, H. and Betsill, M.M. (2003) *Cities and Climate Change: Urban Sustainability and Global Environmental Governance*. New York: Routledge.

Bulkeley, H. and Newell, P. (2010) *Governing Climate Change*. Abingdon: Routledge.

Bull, M. (2000) *Sounding Out the City. Personal Stereos and the Management of Everyday Life*. Oxford: Berg.

Bull, M. (2007) *Sound Moves: iPod Culture and Urban Experience*. Abingdon: Routledge: Abingdon.

Bullard, R.D. (2000) *Dumping in Dixie: Race, Class, and Environmental Quality*. Boulder: Westview Press.

Bullen, A. and Whitehead, M. (2004) Negotiating the networks of space, time and substance: a geographical perspective on the sustainable citizen. Unpublished paper available from authors at Institute of Geography and Earth Sciences, University of Wales, Aberystwyth.

Bumpus, A. and Liverman, D. (2008) Accumulation by decarbonization and the governance of carbon offsets. In: *Economic Geography* 84: 127–55.

Bunce, M. (1993) *The Countryside Deal*. London: Routledge.

Bunge, W. (1971/2011) *Fitzgerald: Geography of a Revolution*. Cambridge, MA: Schenkman Publishing/Athens, GA: University of Georgia Press..

Bunge, W. (1973) The geographer. In: *Professional Geographer* 25, 331–7.

Bunn, D. (1994) 'Our wattled cot': Mercantile and domestic space in Thomas Pringle's African landscapes. In: W.J.T. Mitchell (ed.) *Landscape and Power*. Chicago, IL: University of Chicago Press, 127–74.

Burgdorf, M.P. and Burgdorf, R. (1975) A history of unequal treatment: The qualifications of handicapped persons as a suspect class under the equal protection clause. In: *Santa Clara Lawyer* 15, 855–910.

Burgess, J. (1985) News from nowhere: The press, the riot and the myth of the inner city. In: J. Burgess and J. Gold (eds.) *Geography, the Media and Popular Culture*. London and Sydney: Croom Helm.

Burgess, J. (1990) The production and consumption of environmental meanings in the mass media. In: *Transactions of the Institute of British Geographers* 15, 139–61.

Burgess, J. (1993) Representing nature: Conservation and the mass media. In: B. Goldsmith and A. Warren (eds.) *Conservation in Progress*. Chichester: John Wiley.

Burgess, J. and Gold, J. (eds.) (1985) *Geography, the Media and Popular Culture*. London and Sydney: Croom Helm.

Burgess, J. and Unwin, D. (1984) Exploring the living planet with David Attenborough. In: *Journal of Geography in Higher Education* 8(2), 93–113.

Burgess, J., Bedford, T., Hobson, K., Davies, G. and Harrison, C.M. (2003) (Un) sustainable consumption. In: F. Berkhout, M. Leach and I. Scoones (eds.) *Negotiating Environmental Change: New Perspectives from Social Science*. Cheltenham: Edward Elgar, 261–92.

Burgess, J., Harrison, C.M. and Filius, P. (1998) Environmental communication and the cultural politics of environmental citizenship. In: *Environment and Planning A*, 30, 1445–60.

Burgess, J., Limb, M. Harrison, C.M. (1988a) Exploring environmental values through the medium of small groups 1: Theory and practice. In: *Environment and Planning A*, 20, 309–26.

Burgess, J., Limb, M. Harrison, C.M. (1988b) Exploring environmental values through the medium of small groups 2: Illustrations of a group at work. In: *Environment and Planning A*, 20, 457–76.

Burton, D., Knights, D., Leyshon, A., Alferoff, C. and Signoretta, P. (2004) Making a market: The UK retail financial services industry and the rise of the complex sub-prime credit market. In: *Competition and Change* 8, 3–25.

Bush, G.W. (2001) News releases for September 2001. http://georgewbush-whitehouse.archives.gov/news/releases/2001/09/

Bush, G.W. (2002) *The National Security Strategy of the United States*. 20 September 2002. Washington, DC: The White House.

Bush, G.W. (2004) President outlines steps to help Iraq achieve democracy and freedom. Remarks by the President on Iraq and the War on Terror. United States Army War College, Carlisle, Pennsylvania, 24 May 2004. www.whitehouse.gov/news/releases/2004/05/20040524-10.html.

Butcher, J. (2007) *Ecotourism, NGOs and Development: A Critical Analysis*. Abingdon: Routledge.

Butcher, S. and Wilton, R. (2008) Stuck in transition? Exploring the spaces of employment training for youth with intellectual disability. In: *Geoforum* 39, 1079–92.

Butler, D. (2006) Mashups mix data into global service. In: *Nature* 439, 6–7.

Butler, J. (1990) *Gender Trouble: Feminism and the Subversion of Identity*. London: Routledge.

Butler, J. (1993) *Bodies that Matter: On the Discursive Limits of 'Sex'*. London: Routledge.

Butler, R. (1980) The concept of a tourist area cycle of evolution: Implications for management of resources. In: *Canadian Geographer* 24, 5–12.

Butler, R. (1998) Rehabilitating the images of disabled youths. In: T. Skelton and G. Valentine (eds.) *Cool Places: Geographies of Youth Culture*. London: Routledge.

Butler, R. and Bowlby, S. (1997) Bodies and spaces: an exploration of disabled people's use of public space. In: *Environment and Planning D: Society and Space*, 15(4), 411–33.

Butler, T. (1997) *Gentrification and the Middle Classes*. Aldershot: Ashgate.

Butz, D. and Besio, K. (2009) Autoethnography. In: *Geography Compass* 3, 1660–47.

Cai, X., Xiao, Z. and Dingbao, W. (2011) Land availability for biofuel production. In: *Environmental Science & Technology* 45(1), 334.

Cain, P.J. and Hopkins, A.G. (1986) Gentlemanly capitalism and British expansion overseas: The colonial system 1688–1850. In: *Economic History Review* 39(4), 501–25.

Cairncross, F. (1997) *The Death of Distance: How the Communications Revolution will Change our Lives.* Boston, MA: Harvard Business School Press.

Calhoun, C. (2009) Remaking America: public institutions and the public good. http://publicsphere.ssrc.org/calhoun-remaking-america-public-institutions-and-the-public-good/ (accessed: 23 October 2009).

Callard, F. (2003) The taming of psychoanalysis in geography. In: *Social and Cultural Geography* 4, 295–312.

Callinicos, A. (2003) *An Anti-capitalist Manifesto.* Cambridge: Polity.

Callinicos, A. (2009) *Imperialism and Global Political Economy.* Cambridge: Polity.

Callon, M. and Law, J. (2004) Introduction: absence – Presence, circulation, and encountering in complex space. In: *Environment and Planning D: Society and Space* 22, 3–11.

Cambrensis, G. (1978) *The Journey through Wales and the Description of Wales.* Harmondsworth: Penguin.

Campbell, C. (1987) *The Romantic Ethic and the Spirit of Modern Consumption.* Oxford: Blackwell.

Campbell, D. (1992) *Writing Security: United States Foreign Policy and the Politics of Identity.* Minneapolis: University of Minnesota Press.

Canadian Geographer (2003) Special Issue on Disability. In: *Society and Space*, 47(4).

Carey, J. and Quirk, J. (1970a) The mythos of the electronic revolution. In: *American Scholar*, Winter, 219–41.

Carey, J. and Quirk, J. (1970b) The mythos of the electronic revolution. In: *American Scholar*, Summer, 395–424.

Carmody, T. (2012) Cornell NYC's first professor Deborah Estrin talks networked sensors, climate change, and mobile health. *The Verge*, http://vrge.co/LuOHaw (accessed 3 July 2012).

Carney, G. (1998) *Baseball, Barns and Bluegrass: A Geography of American Folklife.* Oxford: Rowman & Littlefield.

Carney, J. (1996) Converting the wetlands, engendering the environment. In M. Watts and R. Peet (eds.) *Liberation Ecologies. Environment, Development and Social Movements.* New York: Routledge, 165–87.

Carr, E.H. (1961) *What is History?* London: Macmillan.

Carson, R. (1964) *Silent Spring.* Boston: Little, Brown.

Carter, S. and McCormack, D. (2006) Film, geopolitics and the affective logic of intervention. In: *Political Geography* 25(2), 228–45.

Carver, S. (2001) Public participation using web-based GIS. Guest editorial. In: *Environment and Planning B: Planning and Design* 28(6), 803–4.

Castells, M. (1983) *The City and the Grassroots.* London: Edward Arnold.

Castells, M. (1989) *The Informational City.* Oxford: Blackwell.

Castells, M. (1996) *The Rise of the Network Society. Volume I: The Information Age: Economy, Society and Culture.* Cambridge, MA and Oxford: Blackwell.

Castells, M. (1997) *End of Millennium: Volume III: The Information Age: Economy, Society and Culture.* Oxford: Blackwell.

Castells, M., Fernandez-Ardevol, M., Linchuan Qiu, J. and Sey, A. (eds.) (2007). In: *Mobile Communication and Society: A Global Perspective.* Cambridge, MA: MIT Press.

Castles, S. and Miller, M. (2009) *The Age of Migration.* Basingstoke: Macmillan.

Castree, N. (1995) The nature of produced nature. In: *Antipode* 27(1), 12–48.

Castree, N. and Braun B. (eds.) (2001) *Social Nature: Theory, Practice and Politics.* Oxford: Blackwell.

Castree, N. and MacMillan, T. (2004) Old news: Representation and academic novelty. In: *Environment and Planning A* 36, 469–80.

Castree, N. and Sparke, M. (2000) Professional geography and the corporatization of the university: Experiences, evaluations and engagements. In: *Antipode* 32(3), 222–9.

Ceci, S. and Williams, W. (2011) Understanding current causes of women's underrepresentation in science. In: *Proceedings of the National*

Academy of Sciences of the United States of America, 108(8), 3157–62.
Cell, J. (1982) *The Highest Stage of White Supremacy.* Cambridge: Cambridge University Press.
Chabal, P. (2009) *Africa: The Politics of Suffering and Smiling.* London: Zed Books.
Chakrabarty, D. (1999) Adda, Calcutta: Dwelling in modernity. In: *Public Culture* 11, 109–45.
Chambers, R. (1994) Participatory rural appraisal (PRA): Challenges, potentials and paradigm. In: *World Development* 22(10), 1437–54.
Chambers, R. (2006) Participatory mapping and geographic information systems: Whose map? Who is empowered and who disempowered? Who gains and who loses? In: *The Electronic Journal on Information Systems in Developing Countries,* 25(2), 1–11.
Chamlee-Wright, E. and Storr, V.H. (2009) 'There's no place like New Orleans': Sense of place and community, recovery in the Ninth Ward after Hurricane Katrina. In: *Journal of Urban Affairs* 31(5), 615–34.
Champion, A.G. (1989) Counterurbanisation in Europe. In: *The Geographical Journal* 155, 52–9.
Chang, T.C. (2000) Renaissance revisited: Singapore as a 'global city for the arts'. In: *International Journal of Urban and Regional Research* 24(4), 818–31.
Chang, T.C. and Lee, W.K. (2003) Renaissance city Singapore: A study of arts spaces. In: *Area* 35(2), 128–41.
Chant, S. (1997) *Women-headed Households: Diversity and Dynamics in the Developing World.* Basingstoke: Macmillan.
Chaplin, M. (1998) Authenticity and otherness: The Japanese theme park. In: *Architectural Design* 131, 76–9.
Charlesworth, A. (1994) Contesting places of memory: The case of Auschwitz. In: *Environment and Planning D: Society and Space* 12, 579–93.
Chartier, R. (1989) The practical impact of writing. In: R. Chartier (ed.) *Passions of the Renaissance.* Cambridge, MA: Belknap Press.
Chartier, R. (1994) *The Order of Books: Readers, Authors, and Libraries in Europe Between the Fourteenth and Fifteenth Centuries.* Cambridge: Polity.
Chatterjee, P. (1986) *Nationalist Thought and the Colonial World.* London: Zed Books.
Chatterjee, P. and Finger, M. (1994) *The Earth Brokers: Power, Politics and World Development.* Routledge: London.
Chatterton, P. and Hollands, R. (2002) Theorising urban playscapes: Producing, regulating and consuming youthful nightlife city spaces. In: *Urban Studies* 39(1), 95–116.
Cheneau-Loquay, A. (2007) From networks to use patterns: The digital divide as seen from Africa. In: *Geojournal* 68, 55–70.
Chenelly, J. (2005): Troops begin combat operations in New Orleans. *Army Times,* 2 September 2005 http://www.armytimes.com/legacy/new/1-292925-1077495.php
Chennault, C. (1948) Why we must help China now. In: *Reader's Digest,* April, 121–2.
Chertow, M. and Esty, D. (eds.) (1997) *Thinking Ecologically: The New Guardians of Environmental Policy.* New Haven, CT: Yale University Press.
Cheshire, P (1995) A new phase of urban development in western Europe: The evidence for the 1980s. In: *Urban Studies* 32(7), 1045–64.
Choi, S.R., Park, D. and Tschoegl, A.E. (1996) Banks and the world's major banking centres 1990. In: *Weltwirtschaftliches Archiv* 132, 774–93.
Choi, S.R., Park, D. and Tschoegl, A. (2003) Banks and the world's major banking centers, 2000. In: *Review of World Economics* 139(3), 550–68.
Christian Aid (2012) *The Rich, the Poor and the Future of the Earth.* London: Christian Aid.
Christophers, B. (2009) *Envisioning Media Power: On Capital and Geographies of Television.* Lanham, MD: Rowman & Littlefield.
Chua, B.H. (1997) *Political Legitimacy and Housing: Stakeholding in Singapore.* New York: Routledge.
Chua, B.H. and Yeo, W.W (2003) Singapore cinema: Eric Khoo and Jack Neo – Critique from the margins and mainstream. In: *Inter-Asia Cultural Studies* 4(1), 117–25.
Churchill, W. (1993) *Struggle for the Land.* Monroe, ME: Common Courage Press.
City of New York (2010) Mayor Bloomberg announces initiative to develop a new

engineering and applied sciences research campus to bolster City's innovation economy, http://on.nyc.gov/NutgVG (accessed 18 May 2012).

City of New York (2011) Mayor Bloomberg announces City received 18 submissions from 27 academic institutions for new applied sciences campus, http://on.nyc.gov/LPEx2u (accessed 18 May 2012).

CityUK, The (2011) *Key Facts About UK Financial and Professional Services*. London: The CityUK.

Clark, G. and O'Connor, K. (1997) The informational content of financial products and the spatial structure of the global finance industry. In: K. Cox (ed.) *Spaces of Globalization: Reasserting the Power of the Local.* New York: Guildford Press, 89–114.

Clark, G., Gertler, M. and Feldman, M. (2000) *The Oxford Handbook of Economic Geography*. Oxford: Oxford University Press.

Clark, G.L. (2000) *Pension Fund Capitalism*. Oxford: Oxford University Press.

Clark, G.L. (2003) *European Pensions and Global Finance.* Oxford: Oxford University Press.

Clark, G.L., Thrift, N. and Tickell, A. (2004) Performing finance: The industry, the media and its image. In: *Review of International Political Economy* 11(2), 289–310.

Clark, G.L. and Hebb , T. (2004) Pension fund corporate engagement – The fifth stage of capitalism. In: *Relations Industrielles-Industrial Relations* 59(1), 142–71.

Clark, W.A.V. (1998a) *The California Cauldron: Immigration and the Fortunes of Local Communities*. New York: Guilford Press.

Clark, W.A.V. (1998b) Mass migration and local outcomes: Is international migration to the United States creating a new urban underclass. In: *Urban Studies* 35(3), 371–84.

Clarke, A.J. (2000) Mother swapping: The trafficking of nearly new children's wear. In: P. Jackson, M. Lowe, D. Miller and F. Mort (eds.) *Commercial Cultures*, 85–100. Oxford and New York: Berg.

Clarke, C. (2002) The Latin American structuralists. In: V. Desai and R.B. Potter (eds.) *The Companion to Development Studies*. London: Arnold, 92–6.

Clarke, D.B. (2003) *The Consumer Society and the Postmodern City*. London and New York: Routledge.

Clarke, D.B., Doel, M.A. and Segrott, J. (2004) No alternative? The regulation and professionalization of complementary and alternative medicine in the United Kingdom. In: *Health and Place* 10(4), 329–38.

Clastres, P. (1998) *Chronicle of the Guayaki Indians*. (Trans. by P. Auster.) London: Faber and Faber.

Clean Technica (2011) Global Biofuels Market Could Double To $185.3 Billion By 2021. Available at: http://s.tt/15M0S (last accessed 21 March 2012).

Cliff, A.D. and Haggett, P. (1988) *Atlas of Disease Distributions: Analytical Approaches to Epidemiological Data.* Oxford: Blackwell.

Cliff, A.D., Haggett, P. and Smallman-Raynor, M. (1993) *Measles: An Historical Geography of a Major Human Viral Disease. From Global Expansion to Local Retreat, 1840–1990.* Oxford: Blackwell.

Cliff, A.D., Haggett, P. and Smallman-Raynor, M. (2004) *World Atlas of Epidemic Diseases.* London: Arnold.

Clifford, J. (1994) Diasporas. In: *Cultural Anthropology* 9(3), 302–38.

Cloke, P. (1993) The countryside as commodity: New spaces for rural leisure. In: S. Glyptis (ed.) *Leisure and the Environment*. London: Belhaven.

Cloke, P. (1997) Poor country: Marginalization, poverty and rurality. In: P. Cloke and J. Little (eds.) *Contested Countryside Cultures*. London: Routledge.

Cloke, P. (ed.) (2003) *Country Visions.* London: Pearson.

Cloke, P. (2004) Exploring boundaries of professional/personal practice and action: Being and becoming in Khayelitsha Township, Cape Town. In: D. Fuller and R. Kitchin (eds.) *Radical Theory/Critical Praxis: Making a Difference Beyond the Academy*. Praxis (e) Press, 92–102.

Cloke, P. (2005) Conceptualising rurality. In: P. Cloke, T. Marsden and P. Mooney (eds.) *Handbook of Rural Studies*. London: Sage, 18–28.

Cloke, P., Philo, C. and Sadler, D. (1991)

Approaching Human Geography. London: Paul Chapman Publishing.

Cloke, P. and Milbourne, P. (1992) Deprivation and lifestyles in rural Wales II: Rurality and the cultural dimension. In: *Journal of Rural Studies* 8, 360–74.

Cloke, P. and Little, J. (eds.) (1997) *Contested Countryside Cultures*. London: Routledge.

Cloke, P. and Perkins, H. (1998) 'Cracking the canyon with the Awesome Foursome': Representations of adventure tourism in New Zealand. In: *Environment and Planning D: Society and Space* 16, 185–218.

Cloke, P., Crang, P., Goodwin, M., Painter, J. and Philo, C. (2000) *Practising Human Geography*. London: Sage.

Cloke, P., Milbourne, P. and Widdowfield, R. (2002) *Rural Homelessness*. Bristol: Policy Press.

Cloke, P., Cook, I., Crang, P., Goodwin, M., Painter, J. and Philo, C. (2004) *Practising Human Geography*. London: Sage.

Cloke, P., Crang, P. and Goodwin, M. (eds.) (2004) *Envisioning Human Geographies*. London: Arnold.

Cloke, P., Johnsen, S. and May, J. (2004) Journeys and pauses: Tactical and performative spaces in the homeless city. Unpublished manuscript available from authors at the University of Bristol.

Cloke, P. and Johnston, R.J. (eds.) (2005) *Spaces of Geographical Thought*. London: Sage.

Cloke, P., Marsden, T. and Mooney, P. (eds.) (2005) *Handbook of Rural Studies*. London: Sage.

Cloke, P., May, J. and Johnsen, S. (2008) Performativity and affect in the homeless city. In: *Environment and Planning: Society and Space* 26(2), 241–63.

Cloke, P., May, J. and Johnsen, S. (2010) *Swept up Lives? Re-envisioning the Homeless City*. Oxford: Wiley-Blackwell.

Coaffee, J., Murakami Wood, D. and Rogers, P. (2009) *The Everyday Resilience of the City*. Basingstoke: Palgrave Macmillan.

Cockburn, C. (1997) Domestic technologies: Cinderella and the Engineers. In: *Women's Studies International Forum* 20, 361–71.

Coe, N. (2009) Global production networks. In: R. Kitchin and N. Thrift (eds.) *The International Encyclopedia of Human Geography*, Oxford: Elsevier, 4, 556–62.

Coe, N. (2011) Geographies of production II: A global production network A-Z. In: *Progress in Human Geography* 36(3), 389–402.

Coe, N., Dicken, P. and Hess, M. (2008) Global production networks: Realising the potential. In: *Journal of Economic Geography* 8, 271–95.

Coe, N., Hess, M., Wai-chung Yeung, H., Dicken, P. and Henderson, J. (2004) Globalizing regional development: A global production networks perspective. In: *Transactions of the Institute of British Geographers 2004*, 7.

Cohen, A. (1982) A polyethnic London carnival as a contested cultural performance. In: *Ethnic and Racial Studies* 5, 23–41.

Cohen, E. (1989) Primitive and remote: Hill tribe trekking in Thailand. In: *Annals of Tourism Research* 16, 30–61.

Cohen, L. (2003) *A Consumer's Republic: The Politics of Mass Consumption in Post-war America*. New York: Knopf.

Cohen, R. (1997 [2008]) *Global Diasporas: An Introduction*. London: Routledge

Cohen, S. (1980) *Folk Devils and Moral Panics: The Creations of the Mods and Rockers* (2nd edn.). Oxford: Martin Robertson.

Cole, J. and Thomas, L.M. (2009) Thinking through love in Africa. In: J. Cole and L.M. Thomas (eds.) *Love in Africa*. Chicago: University of Chicago Press, 1–30.

Coleman, S. and Crang, M. (eds.) (2002) *Tourism: Between Place and Performance*. New York: Berghan Books.

Collier, S.J. and Lakoff, A. (2008) The problem of securing health. In: S.J. Collier and A. Lakoff (eds.) *Biosecurity Interventions: Global Health and Security in Question*. New York: Columbia University Press/SSRC.

Colls, R. (2012) Feminism, bodily difference and non-representational geographies. In: *Transactions of the Institute of British Geographers* (in press).

Colls, R. and Evans, B. (2008) Embodying responsibility: Children's health and supermarket initiatives. In: *Environment and Planning A* 40, 615–31.

Colyer, R. (1976) *The Welsh Cattle Drovers: Agriculture and the Welsh Cattle Trade Before and During the Nineteenth Century*. Cardiff: University of Wales Press.

Comaroff, J. and Comaroff, J. (1990) *Ethnography and the Historical Imagination*. Boulder, CO: Westview Press.

Comer, J.C. and Wikle, Th. A. (2008) Worldwide diffusion of the cellular telephone 1995–2005. In: *Professional Geographer* 60(2), 252–69.

Commission of the European Communities (1986) Television and the audiovisual sector: Towards a European policy. In: *European File* 14(86), August–September.

Commission for Rural Communities (2009) *Mind the Gap: Digital England – A Rural Perspective*. Cheltenham: CRC.

Cone, C. (1995) Crafting Selves: The lives of two Mayan women. In: *Annals of Tourism Research* 22(2), 314–27.

Connell, J. (1993) Bali revisited: Death, rejuvenation, and the tourist cycle. In: *Environment and Planning D: Society and Space* 11, 641–61.

Connelly, S. (2007) Mapping sustainable development as a contested concept. In: *Local Environment* 12: 259–78.

Conradson, D. (2003) Geographies of care: Spaces, practices, experiences. In: *Social and Cultural Geography* 4(4), 451.

Conradson, D. (2005) Freedom, space and perspective: Moving encounters with other ecologies. In: J. Davidson, L. Bondi and M. Smith (eds.) *Emotional Geographies*. Aldershot: Ashgate, 103–16.

Conversi, D. (1990) Language or race? The choice of core values in the development of Catalan and Basque nationalisms. In: *Ethnic and Racial Studies* 13(1), 50–70.

Cook, I. (2000) Social sculpture and Shelley Sacks' 'exchange values'. In: *Ecumene* 7, 338–44.

Cook, I. (2004) Follow the thing: papaya. In: *Antipode* 36, 642–64.

Cook, I. and Crang, P. (1996) The world on a plate: Culinary culture, displacement and geographical knowledges. In: *Journal of Material Culture* 1, 131–53.

Cook, I. and Harrison, M. (2003) Cross over food: Re-materializing postcolonial geographies. In: *Transactions of the Institute of British Geographers* NS28, 296–317.

Cooke, F. (2003) Maps and counter-maps: Globalized imaginings and local realities of Sarawak's plantation agriculture. In: *Journal of Southeast Asian Studies* 34, 265–84.

Cooper, A., Law, R., Malthus, J. and Wood, P. (2000) Rooms of their own: public toilets and gendered citizens in a New Zealand city, 1860–1940. In: *Gender, Place and Culture* 7(4), 417–35.

Cooper, F. and Stoler, A.L. (1997) *Tensions of Empire: Colonial Cultures in a Bourgeois World*. Berkeley: University of California Press.

Cooper, M. (2008) *Life as Surplus: Biotechnology and Capitalism in the Neoliberal Order*. Seattle: University of Washington Press.

Cope, M. and Elwood, S. (eds.) (2009) *Qualitative GIS: A Mixed Methods Approach*. London: Sage.

Corbett, J. (1994) A proud label: Exploring the relationship between disability politics and gay pride. In: *Disability and Society* 9(3), 343–57.

Corbett, J.M. and Keller, C.P. (2005) An analytical framework to examine empowerment associated with participatory geographic information systems (PGIS). In: *Cartographica* 40(4), 91–102.

Corbett, J.M., Rambaldi, G., Kyem, P., Weiner, D., Olson, R., Muchemi, J., McCall, M. and Chambers, R. (2006) Overview: Mapping for change – The emergence of a new practice. In: G. Rambaldi, J. Corbett, M. McCall, R. Olson, J. Muchemi, P. Kyem, D. Weiner and R. Chambers (eds.) *Mapping for Change: Practice, Technologies and Communication*, Participatory Learning and Action, 54.

Corbridge, S. (1993) Colonialism, post-colonialism and the political geography of the Third World. In: Taylor, P. (ed.) *Political Geography of the Twentieth Century*. London: Belhaven Press.

Corbridge, S. (ed.) (1995) *Development Studies – A Reader*. London: Arnold.

Corbridge, S., Thrift, N.J. and Martin, R. (1996) *Money, Power and Space*. Oxford: Blackwell.

Corrigan, P. (1997) *The Sociology of Consumption: An Introduction*. Thousand Oaks: Sage.

Cosgrove, D. (1985a) Prospect, perspective and the evolution of the landscape idea. In: *Transactions of the Institute of British Geographers* 10, 45–62.

Cosgrove, D. (1985b) *Social Formation and Symbolic Landscape*. London: Croom Helm.

Cosgrove, D. (1989) Geography is everywhere: Culture and symbolism in human landscapes. In: D. Gregory and R. Walford (eds.) *Horizons in Human Geography*. Basingstoke: Macmillan.

Cosgrove, D. (1994) Contested global visions: One-world, whole-earth and the Apollo space photographs. In: *Annals of the Association of American Geographers* 84, 270–94.

Cosgrove, D. (1997) Prospect, perspective and the evolution of the landscape idea. In: T. Barnes and D. Gregory (eds.) *Reading Human Geography*. London: Edward Arnold, 324–41.

Cosgrove, D. (2001) *Apollo's Eye: A Cartographic Genealogy of the Earth in the Western Imagination*. Baltimore: Johns Hopkins University Press.

Cosgrove, D. (2003) Heritage and history: A Venetian geography lesson. In: R.S. Peckham (ed.) *Rethinking Heritage: Culture and Politics in Europe*. London: I.B. Taurus, 113–23.

Cosgrope, D. (2006) Art and mapping: An introduction. In: D. Wood and J. Krygier (eds.) *Cartographic Perspectives* 53, 1–83.

Cosgrove, D. (2008) *Geography and Vision: Seeing, Imagining and Representing the World*. London: I.B. Tauris.

Cosgrove, D. and Daniels, S. (eds.) (1988) *The Iconography of Landscape: Essays on the Symbolic Design and Use of Past Environments*. Cambridge: Cambridge University Press.

Cotgrove, S. and Duff, A. (1980) Environmentalism, middle class radicalism and politics. In: *Sociology Review* 28, 335–51.

Couloubaritsis, L., De Leeuw, M., Noel, E., and Sterckx, E. (1993) *The Origins of European Identity*. Brussels: European Interuniversity Press.

Covenant of Mayors (2011a) About the Covenant. http://www.eumayors.eu/about_the_covenant/index_en.htm . Accessed March 2011.

Covenant of Mayors (2011b) Welcome http://www.eumayors.eu/home_en.htm. Accessed March 2011.

Cowan, D. and Gilbert, E. (eds.) (2008) *War, Citizenship, Territory*. New York: Routledge.

Cowen, T. (1998) *In Praise of Commercial Culture*. Cambridge, MA: Harvard University Press.

Cox, K. (2002) *Political geography: Territory State and Nation*. Oxford: Blackwell.

Craig, W.J., Harris, T.M. and Weiner, D. (eds.) (2002) *Community Participation and Geographical Information Systems*. London: Taylor & Francis.

Crampton, J. (2001) Maps as social constructions: Power, communication and visualization. In: *Progress in Human Geography* 25, 235–52.

Crampton, J. (2009) Cartography: Performative, participatory, political. In: *Progress in Human Geography* 33(6), 840–48.

Crampton, J. and Krygier, J. (2006) An introduction to critical cartography. In: *ACME: An International E-Journal for Critical Geographies* 4(1), 11–33.

Crang, M. (1998a) *Cultural Geography*. London: Routledge.

Crang, M. (1998b) Places of practice, and the practice of science. In: *Environment and Planning A* 30, 1971–4.

Crang, M. (2003) Qualitative methods: Touchy, feely, look-see? In: *Progress in Human Geography* 274, 494–504.

Crang, M. and Cook, I. (2007) *Doing Ethnographies*, London: Sage.

Crang, M. and Thrift, N. (2000) *Thinking Space*. London: Routledge.

Crang, M., Crang, P. and May, J. (1999) Introduction. In: M. Crang, P. Crang, P. and J. May (eds.) *Virtual Geographies. Bodies, Space and Relations*. London: Routledge, 1–20.

Crang, P. (1994) It's showtime: On the workplace geographies of display in southeast England. In: *Environment and Planning D: Society and Space* 12, 675–704.

Crang, P. (1997) Performing the tourist product. In: C. Rojek and J. Urry (eds.) *Touring Cultures: Transformations of Travel and Theory*. London: Routledge.

Crang, P. (ed.) (2003) Field cultures. In: *Cultural Geographies* 10, special issue, 251–378.

Crang, P. (2004) The geographies of consumption.

In: P. Daniels, M. Bradshaw, J. Bryson and J. and Sidaway (eds.) *Introduction to Human Geography* (2nd edn.). Harlow: Prentice Hall.

Crang, P., Dwyer, C. and Jackson, P. (2003) Transnationalism and the spaces of commodity culture. In: *Progress in Human Geography* 27, 438–56.

Cresswell, T. (1994) Putting women in their place: The carnival at Greenham Common. In: *Antipode* 26(1), 35–58.

Cresswell, T. (1996) *In Place/Out of Place: Geography, Ideology and Transgression.* Minneapolis: University of Minnesota Press.

Cresswell, T. (2003) Landscape and the obliteration of practice. In K. Anderson, M. Domosh, S. Pile and N. Thrift (eds.) *Handbook of Cultural Geography*. London: Sage, 269–81.

Cresswell, T. (2004) *Place: A Short Introduction.* Oxford: Blackwell.

Cresswell, T. (2006a) *On the Move: Mobility in the Modern Western World.* New York: Routledge.

Cresswell, T. (2006b) 'You cannot shake that shimmie here': Producing mobility on the dancefloor. In: *Cultural Geographies,* 13, 55–77.

Cresswell, T. (2010): Towards a politics of mobility. In: *Environment and Planning D: Society and Space* 28(1), 17–31.

Cresswell, T. (2013) *Geographic Thought: A Critical Introduction*. Chichester: Wiley-Blackwell.

Crewe, L. (2011) Life itemised: lists, loss, unexpected significance, and the enduring geographies of discard. In: *Environment and Planning D: Society and Space,* 29(1), 27–46.

Cromley, E. and McLafferty, S. (2012) *GIS and Public Health* (2nd edn.) New York: Guilford Press.

Cronon, W. (1995) The trouble with wilderness: Or getting back to the wrong nature. In: W. Cronon (ed.) *Uncommon Ground: Towards Reinventing Nature*. New York: W.W. Norton.

Cross, G. (2000) *An All-consuming Century: Why Commercialism Won in Modern America.* New York: Columbia University Press.

Crouch, D. and Ward, C. (1997) *The Allotment. Its Landscape and Culture.* Nottingham: Five Leaves Publications.

Crowley, U. and Kitchin, R. (2008) Producing 'decent girls': Governmentality and the moral geographies of sexual conduct in Ireland (1922–1937). In: *Gender, Place and Culture* 15(4), 355–72.

Cugoano, O. (1787) *Thoughts and Sentiments on the Evil and Wicked Traffic of Slavery and Commerce of the Human Species*. London: T. Becket.

Culler, J. (1988) *Framing the Sign: Criticism and its Institutions*. Oxford: Basil Blackwell.

Cumbers, A., Routledge, P. and Nativel, C. (2008) The entangled geographies of global justice networks. In: *Progress in Human Geography* 32(2) 183–201.

Cunningham, H. (1992) *Children of the Poor: Representations of Childhood Since the Seventeenth Century*. Oxford: Polity Press.

Curry, M. (2004) The profiler's question and the treacherous traveller: Narratives of belonging in commercial aviation. In: *Surveillance and Society* 1, 475–99.

Curry, M.R. (1995) Rethinking rights and responsibilities in geographic information systems: Beyond the power of image. In: *Cartography and Geographic Information Systems* 22(1), 58–69.

Curti, G.H., Aitken, S.C., Bosco, F.J. and Goerisch, D.D. (2011) For not limiting emotional and affectual geographies: A collective critique of Steve Pile's 'Emotions and affect in recent human geography'. In: *Transactions of the Institute of British Geographers* 36, 590–4.

Curtis, C. (2011) *President Obama Welcomes New Chairman of the Joint Chiefs of Staff.* The White House Blog, 30 September, 2011.

Curtis, S. (2010) *Space, Place and Mental Health.* Aldershot: Ashgate.

Curtis, S. and Taket, A.R. (1996) *Health and Societies: Changing Perspectives*. London: Arnold.

Cutter, S. (2000) Why didn't geographers map the human genome? In: *AAG Newsletter* 35(9), 3–4.

Cutter, S. (2006) The geography of social vulnerability: Race, class and catastrophe. In: *Understanding Katrina: Perspectives from the Social Sciences, http://understandingkatrina.ssrc.org/Cutter/*

Dahlgreen, G. and Whitehead, M. (1991) What can we do about inequalities in health? In: *Lancet* 338, 1059–63.

Dahlman, C. (2011) Breaking Iraq: Reconstruction as war. In: S. Kirsch and C. Flint (eds.) *Reconstructing Conflict: Integrating War and Post-war Geographies*. Farnham: Ashgate.

Daily, G.C. (1997) (ed.) *Nature's Services: Societal Dependence on Natural Ecosystems*. Washington: Island Press.

Dalby, S. (1988) Geopolitical discourse: The Soviet Union as Other. In: *Alternatives*, 13(1), 415–22.

Dalby, S. (1990) American security discourse: The persistence of geopolitics. In: *Political Geography Quarterly* 9(2), 171–88.

Dalby, S. (1994) Gender and geopolitics: Reading security discourse in the new world order. In: *Environment and Planning D: Society and Space* 12(5), 525–46.

Dalby, S. (2009) *Security and Environmental Change*. Cambridge: Polity.

Daly, H.E. (1990) Toward some operational principles of sustainable development. In: *Ecological Economics* 2, 1–6.

Daly, H. (2005) Economics in a full world. In: *Scientific American*, September, 100–7.

Daniels, S. (1982) Humphry Repton and the morality of landscape. In J.R. Gold and J. Burgess (eds.) *Valued Environments*. London: Allen & Unwin, 124–44.

Daniels, S. (1985) Arguments for a humanistic geography. In: R.J. Johnston (ed.) *The Future of Geography*, 143–58.

Daniels, S. (1993) *Fields of Vision: Landscape Imagery and National Identity in England and the United States*. Cambridge: Polity Press.

Daniels, S. and Cosgrove, D. (1993) Spectacle and text: Landscape metaphors in Cultural Geography. In: J.S. Duncan and D. Ley (eds.) *Place/Culture/Representation*. London: Routledge, 57–77.

Daniels, S. and Lee, R. (eds.) (1996) *Exploring Human Geography: A Reader*. London: Arnold.

Daniels, S. and Seymour, S. (1990) Landscape design and the idea of improvement 1730–1900. In: R.A. Dodgson and R.A. Butlin (eds.) *An Historical Geography of England and Wales*. London: Academic Press, 187–520.

Dann, G. (1996) The people of tourist brochures. In: T. Selwyn (ed.) *The Tourist Image: Myths and Myth Making in Tourism*. London: Wiley, 61–82.

Dant, T. (1999) *Material Culture in the Social World: Values, Activities, Lifestyles*. Buckingham: Open University Press.

Darby, H. (1962) The problem of geographical description. In: *Transactions of the Institute of British Geographers* 30, 1–14.

Darden, J. (2004) *The Significance of White Supremacy in the Canadian Metropolis of Toronto*. Lewiston, NY: The Edwin Mellen Press.

Darnton, R. (1979) *The Business of the Enlightenment: A Publishing History of the Encyclopaedia*. Cambridge, MA: Harvard University Press.

David, A. (2011) Embodied migration: Performance practices of diasporic Sri Lankan Tamil communities in London. In: *Journal of Intercultural Studies* (in press).

Davidson, B. (1992) *The Black Man's Burden*. New York: Times Books.

Davidson, J. and Bondi, L. (2004) Spatialising affect, affecting space: Introducing emotional geographies. In: *Gender, Place and Culture*, 11(3), 373–74.

Davidson, J. and Milligan, C. (2004) Embodying emotion, sensing space: Introducing emotional geographies. In: *Social and Cultural Geography* 5(4), 523–32.

Davidson, J., Bondi, L and Smith, M. (eds.) (2005) *Emotional Geographies*. Burlington VT and Aldershot: Ashgate Press.

Davies, G. (1999) Science, observation and entertainment: Competing visions of postwar British natural history television, 1946–1967. In: *Ecumene. A Journal of Cultural Geographies* 7,

Davies, G. (2003) A geography of monsters? In: *Geoforum*, 34(4), 409–12.

Davies, G. (2011) Writing biology with mutant mice: The monstrous potential of post genomic life. In: *Geoforum* (in press).

Davis, M. (1990) *City of Quartz: Excavating the Future in Los Angeles*. London: Verso.

Davis, M. (2005) *The Monster at Our Door: The Global Threat of Avian Flu*. New York: The New Press.

Davis, M. and Monk, D.B. (2007) *Evil Paradises:*

Dreamworlds of Neoliberalism. New York: New Press.

Dawney, L. (2011) The motor of being: Clarifying and defending the concept of affect. A response to Steve Pile's 'Emotions and affect in recent human geography'. In: *Transactions of the Institute of British Geographers* 36, 599–602.

Day, K. (2001) Constructing masculinity and women's fear in public space in Irvine, California. *Gender, Place and Culture* 8(2), 109–27.

Day, P. and Pearce, J. (2011) Obesity-promoting food environments and the spatial clustering of food outlets around schools. In: *American Journal of Preventive Medicine* 40(2), 113–21.

De Angelis, M. (2000) New internationalism and the Zapatistas. In: *Capital and Class* 70, 9–36.

De, Esha Niyogi (2002) The city between the global state: architecture and the people in Singapore's gendered imaginations. In: S. Sarker and E.N. De (eds.) *Trans-status Subjects: Gender in the Globalisation of South and Southeast Asia*. Durham and London: Duke University Press, 189–210.

De Certeau, M. (1984) *The Practice of Everyday Life*. Berkeley: University of California Press.

De Goede, M. (2001) Discourses of scientific finance and the failure of long-term capital management. In: *New Political Economy* 6, 149–70.

Dear, M. (1994) Postmodern Human Geography: A preliminary assessment. In: *Erdkunde*, Band 48, March, 2–13.

Dear, M. (1999) The relevance of postmodernism. In: *Scottish Geographical Magazine*, 115(2).

Dear, M. and Flusty, S. (1998) Postmodern urbanism. In: *Annals of the Association of American Geographers* 88, 50–72.

Dear, M. and Wolch, J. (1987) *Landscapes of Despair: From Deinstitutionalization to Homelessness*. Princeton, Princeton University Press.

Debord, G. (1977) *Society of the Spectacle*. Detroit: Black and Red Books.

Defert, D. (1991) Popular life and insurance technology. In: *The Foucault Effect: Studies in Governmentality* by G. Burchell, D. Gordon and P. Miller (eds.) Chicago: University of Chicago Press.

DEFRA (2002) The environment in your pocket: key facts and figures on the environment of the United Kingdom. Available free from www.defra.gov.uk.

Delanty, G. (1995) *Inventing Europe: idea, identity, reality*. Basingstoke: Macmillan.

De Leeuw, S. Kobayashi, A. and E. Cameron (2011) Difference. In: V. Del Casino, M. Thomas, P. Cloke and R. Panelli (eds.) *A Companion to Social Geography*, Chichester: Wiley-Blackwell.

Deleuze, G. (1990) *The Logic of Sense*. New York: Columbia University Press.

Deleuze, G. (1994) *Difference and Repetition*. London: Athlone.

DeLyser, D. and Starrs, P. (2001) Doing fieldwork. In: *The Geographical Review* 91, iv–viii.

DeLyser, D., Herbert, S., Aitken, S., Crang, M. and McDowell, L. (2010) *The Sage Handbook of Qualitative Geography*. London: Sage.

Dennis, S.F. (2006) Prospects for qualitative GIS at the intersection of youth development and participatory urban planning. In: *Environment and Planning A* 38, 2039–54.

Department for International Development (1997) *Eliminating World Poverty: A Challenge for the 21st Century*. London: UK Government Stationery Office.

Der Derian, J. (1992) *Anti-diplomacy: Spies, Terror, Speed and War*. Oxford: Blackwell.

Der Derian, J. (2002) The war of networks. In: *Theory and Event* 5(4).

Dery, M. (1993) Flame wars. In: M. Dery (ed.) *Flame Wars: The Discourse of Cyberculture*. Durham, NC: Duke University Press.

Desai, J. (2004) *Beyond Bollywood: The Cultural Politics of South Asian Diasporic Film*. Abingdon: Routledge.

Desforges, L. (2004) The formation of global citizenship: International non-governmental organizations in Britain. In: *Political Geography* 23, 549–69.

DeSilvey, C. (2003) Cultivated histories in a Scottish allotment garden. In: *Cultural Geographies* 10, 443–68.

Deskins, D.R. (1996) Economic restructuring, job opportunities and black social dislocation in

Detroit. In: J. O'Loughlin and J. Friedrichs (eds.) *Social Polarization in Post-industrial Metropolises*. Berlin and New York: de Gruyter.

Desmarais, A.A. (2007) *La Via Campesina*. London: Pluto Press.

Desmond, J. (1999) Picturing Hawai'i: The 'ideal' native and the origins of tourism, 1880–1915. In: *Positions: East Asia Cultures Critique* 7(2), 459–501.

DeVerteuil, G. (2003) Homeless mobility, institutional settings, and the new poverty management. In: *Environment and Planning A* 35, 361–79.

DeVerteuil, G., May, J. and Von Mahs, J. (2009) Complexity not collapse: re-casting the geographies of homelessness in a 'punitive age'. In: *Progress in Human Geography* 33(5), 646–66.

Dewey, J. (1927). *The Publics and its Problems*. Athens OH: Ohio University Press.

Dewsbury, J-D. (2000) Performativity and the event: Enacting a philosophy of difference. In: *Environment and Planning D: Society and Space* 18(4), 473–96.

Dewsbury, J.D. (2009) Performative, non-representational and affect-based research: seven injunctions. In: D. DeLyser, S. Aitken, S. Herbert, M. Crang and L. McDowell (eds.) *Handbook of Qualitative Geography*, London: Sage, 321–34.

Dewsbury, J.D. and Cloke, P. (2009) Spiritual landscapes: Existence, performance and immanence. In: *Social and Cultural Geography* 10(6), 695–711.

Dhara, V.R. and Dhara, R. (2002) 'The Union Carbide disaster in Bhopal: a review of health effects'. In: *Archives of Environmental Health* 57(5), 391–404.

d'Hautessere, A.-M. (1999) The French mode of social regulation and sustainable tourism development: the case of Disneyland Paris. In: *Tourism Geography* 1, 86–107.

Dicken, P. (1986) *Global Shift: Industrial Change in a Turbulent World*. London: Harper & Row.

Dicken, P. (1992) *Global Shift: The Internationalization of Economic Activity* (2nd edn.). London: Paul Chapman Publishing.

Dicken, P. (1998) *Global Shift: Transforming the World Economy* (3rd edn.). London: Paul Chapman Publishing.

Dicken, P. (2003) Global Shift: Reshaping the Global Economic Map in the 21st Century (4th edn.). London: Sage.

Dicken, P. (2004a) Geographers and globalisation: (Yet) another missed boat? In: *Transactions of the Institute of British Geographers* 29, 5–26.

Dicken, P. (2004b) Globalization, production and the (im)morality of uneven development. In R. Lee and D. Smith (eds.) *Geographies and Moralities: International Perspectives on Development, Justice and Place*. Oxford: Blackwell.

Dicken, P. (2007) *Global Shift: Mapping the Changing Contours of the World Economy* (5th edn.). London: Guilford Press.

Dicken, P. (2010) *Global Shift: Mapping the Changing Contours of the World Economy*. London: Sage.

Dicken, P., Kelly, P.F., Olds, K. and Yeung, H. (2001) 'Chains and networks, territories and scales: Towards an analytical framework for the global economy'. In: *Global Networks* 1(2), 89–112.

Dicken, P. and Thrift, N. (1992) The organization of production and the production of organization: why business enterprises matter in the study of geographical industrialization. In: *Transactions of the Institute of British Geographers* NS 17(3), 279–91.

Dickinson and Morris (2011) (http://www.porkpie.co.uk, accessed 9 September 2011).

Diez Roux, A.V. (2001) Investigating neighbourhood and area effects on health. In: *American Journal of Public Health* 91(11), 1783–9.

D'Ignazio, C. (2009) Art and cartography. In: R. Kitchin and N. Thrift (eds.) *The International Encyclopedia of Human Geography*. Elsevier, Oxford, 190–206.

Dixon, D. and Jones, J.P. (2004) Guest editorial: What next? In: *Environment and Planning A* 36, 381–90.

Dixon, D. and Straughan, E. (2010) Geographies of touch/touched by geography. In: *Geography Compass* 4(5), 449–59.

Doan, P. (2010) The tyranny of gendered spaces – Reflections from beyond the gender dichotomy. In: *Gender, Place and Culture* 17(5), 635–54.

Dobson, A. (2000) *Green Political Thought.* London: Routledge.

Dodd, L. (1992) Heritage and the 'Big House': Whitewash for rural history. In: *Irish Reporter* 6, 9–11.

Dodd, L. (1993) Interview by Nuala Johnson, Strokestown Park House, 2 September.

Dodds, K. (2003) Licensed to stereotype: Popular geopolitics, James Bond and the spectre of Balkanism. In: *Geopolitics* 8, 125–56.

Dodds, K. (2007) *Geopolitics: A Very Short Introduction.* Oxford: Oxford University Press.

Dodds, K. (2010) A polar Mediterranean? Accessibility, resources and sovereignty in the Arctic Ocean. In: *Global Policy* 1, 303–11.

Dodds, K., Kuus, M. and Sharp, J. (eds.) (2012) *The Ashgate Companion to Critical Geopolitics.* Farnam: Ashgate.

Doel, M.A and Segrott, J. (2004) Materialising complementary and alternative medicine: Aromatherapy, chiropractic, and Chinese herbal medicine in the UK. In: *Geoforum* 35(6), 727–38.

Dodge, M. and Kitchin, R. (2000) *Mapping Cyberspace.* London: Routledge.

Dodman D. (2009) Blaming cities for climate change? An analysis of urban greenhouse gas emissions inventories. In: *Environment and Urbanization* 21(1), 185–201.

Doel, M. (1994) Deconstruction on the move: from libidinal economy to liminal materialism. In: *Environment and Planning A* 26, 1041–59.

Doel, M. and Clarke, D. (1999) Virtual Worlds. Simulation, suppletion, s(ed)uction and simulacra. In: M. Crang, P. Crang and J. May (eds.) *Virtual Geographies. Bodies, Space and Relations.* London: Routledge, 261–83.

Domosh, M. and Seager, J. (2001) *Putting Women in Place: Feminist Geographers Make Sense of the World.* New York: Guildford Press.

Dorling, D. (2010) *Injustice: Why Social Inequality Persists.* Bristol: Policy Press.

Dorn, M. and Laws, G. (1994) Social theory, body politics, and medical geography: Extending Kearns' invitation. In: *Professional Geographer* 46, 106–10.

Dorra, M. (1996) La traversée des apparences, *Le Monde Diplomatique*, June, 32.

Douglas, M. and Isherwood, B. (1979) *The World of Goods.* New York: Basic Books.

Downs, A. (1972) Up and down with ecology: The issue-attention cycle. In: *Public Interest* 28, 38–50.

Drakulic, S. (1996) *Café Europa.* London: Abacus.

Dresner, S. (2002) *The Principles of Sustainability.* London: Earthscan.

Driedger, S.M., Crooks, V.A. and Bennett, D. (2004) Engaging in the disablement process over space and time: Narratives of persons with multiple sclerosis in Ottawa, Canada. In: *Canadian Geographer* 48(2), 119–36.

Driver, F. (2001) *Geography Militant: Cultures of Exploration and Empire.* Oxford: Blackwell.

Driver, F. (2003) The geopolitics of knowledge and ignorance. In: *Transactions of the Institute of British Geographers* 27,131–2.

Driver, F. (2010) In search of the imperial map: Walter Crane and the image of empire. In: *History Workshop Journal* 69, 146–57.

Driver, F. and Jones, L. (2009) *Hidden Histories of Exploration: Researching the RGS-IBG Collections.* London: RGS-IBG.

Driver, F. and Martins, L. (2002) John Septimus Roe and the art of navigation, *c.*1815–1830. In: *History Workshop Journal* 54, 144–61.

Driver, F. and Martins, L. (eds.) (2005) *Tropical Visions in an Age of Empire.* Chicago: University of Chicago Press.

Driver, F., Nash, C., Prendergast, K. and Swenson, I. (2002) *Landing: Eight Collaborative Projects Between Artists and Geographers.* London: Royal Holloway.

Dryzek, J. (1997) *The Politics of the Earth.* Oxford: Oxford University Press.

Du Gay, P., Hall, S., Jones, L., Mackay, H. and Negus, K. (1997) *Doing Cultural Studies: The Story of the Sony Walkman.* London: Sage.

Dunbar, G. (1974) Geographical personality. In: *Geoscience and Man* 5, 25–33.

Duncan, J. (1990) *The City as Text: The Politics of*

Landscape Interpretation in the Kandyan Kingdom. Cambridge: Cambridge University Press.

Duncan, J. (2003) Representing Empire at the National Maritime Museum. In: R.S. Peckham (ed.) *Rethinking Heritage: Culture and Politics in Europe*. London: I.B. Taurus, 17–28.

Duncan, J. (1995) Landscape geography 1993–94. In: *Progress in Human Geography* 19, 414–22.

Duncan, J.S. (1999) Elite landscapes as cultural (re)productions: The case of Shaughnessy Heights. In: K. Anderson and F. Gale (eds.) In: *Cultural Geographies*. London: Longman, 53–70.

Duncan, J.S. and Duncan, N.G. (2004) *Landscapes of Privilege: The Politics of the Aesthetic in an American Suburb*. New York: Routledge.

Dwyer, C. (2004) Tracing transnationalities through commodity culture: A case study of British-South Asian fashion. In: P. Jackson, P. Crang and C. Dwyer (eds.) *Transnational Spaces*. Abingdon: Routledge.

Dwyer, C. (2010) From suitcase to showroom: British Asian retail spaces. In: C. Breward, P. Crang and R. Crill (eds.) *British Asian Style. Fashion and Textiles, Past and Present*. London: V&A Publishing.

Dwyer, C. and Crang, P. (2002) Fashioning ethnicities: The commercial spaces of multiculture. In: *Ethnicities* 2(3), 410–30.

Dwyer, C. and Jackson, P. (2003) Commodifying difference: Selling EASTern fashion. In: *Environment and Planning D: Society and Space* 21, 269–91.

Dwyer, C., Gilbert, D. and Shah, B. (2012) Faith and suburbia: Secularization, modernity and the changing geographies of religion in London's suburbs. In: *Transactions of the Institute of British Geographers* (in press).

Dyck, I. (1995) Hidden geographies: The changing lifeworlds of women with multiple sclerosis. In: *Social Science & Medicine* 40(3), 307–20.

Dyck, I. (1999) Body troubles: Women, the workplace and negotiations of a disabled identity. In: R. Butler and H. Parr (eds.) *Mind and Body Spaces: Geographies of Illness, Impairment and Disability*. London: Routledge, 119–37.

Dyck, I., Lewis, N. and McLafferty, S. (2001) *Geographies of Women's Health*. London: Routledge.

Dyer, R. (1997) *White*. London: Routledge.

Dymski, G.A. and Veitch, J.M. (1996) Financial transformation and the metropolis: Booms, busts and banking in Los Angeles. In: *Environment and Planning A* 28, 1233–60.

Dyson, P. (2012) Slum tourism: Representing and interpreting reality in Dharavi, Mumbai. In: *Tourism Geographies* 14, 254–74.

Earickson, R. (2009) Medical Geography. In: R. Kitchin and N. Thrift, N. (eds.) *International Encyclopedia of Human Geography*. Oxford: Elsevier, 9–20.

Ebinger, C. (1977) *Great Power Rivalry in the Far East: The Geopolitics of Energy*. Washington, DC: Centre for Strategic and International Studies.

Eboda, M. (1997) Rum do as Reggae Boyz blow hot. *The Observer*, 18 November 1997, 12.

Eckerman, I. (2005). *The Bhopal Saga*. Hyderabad: Universities Press (India) Private Limited.

Eco, U. (1985) How culture conditions the colours we see. In: M. Blonsky (ed.) *On Signs*. Oxford: Blackwell, 157–83.

Eco, U. (1987) *Travels in Hyper-reality*. London: Picador.

Ecologist, The (1972) *A Blueprint for Survival*. Harmondsworth: Penguin.

Edensor, T. (2000) Staging tourism: Tourists as performers. In: *Annals of Tourism Research* 27, 322–44.

Edensor, T. (2002) *National Identity, Popular Culture and Everyday Life*. Oxford: Berg.

Edensor, T. (2004) Automobility and national identity: Representation, geography and driving practice. In: *Theory, Culture and Society* 21(4/5), 101–20.

Edensor, T. (2008) Mundane hauntings: commuting through the phantasmagoric working-class spaces of Manchester, England. In: *Cultural Geographies* 15, 313–33

Edensor, T. (2011) The rhythms of tourism. In: C. Minca and T. Oakes (eds.) *Real Tourism*. Abingdon: Routledge.

Edwards, E. (1996) Postcards: greetings from another world. In Selwyn, T. (ed.) *The Tourist*

Image: Myths and Myth Making in Tourism. London: Wiley, 197–222.

Edwards, M. and Hulme, D. (eds.) (1995) *Non-Governmental Organisations – Performance and Accountability. Beyond the Magic Bullet.* London: Earthscan.

Eggers, D. (2009) *Zeitoun.* New York: McSweeneys.

Ehrkamp, P. (2005) Placing identities. Transnational practices and local attachments of Turkish immigrants in Germany. In: *Journal of Ethnic and Migration Studies* 31(2), 345–64.

Ehrlich, P.R. (1972) *The Population Bomb.* London: Ballantine.

Eicher, J. (ed.) (1995) *Dress and Ethnicity: Change Across Time and Space.* Oxford: Berg.

Eksteins, M. (1990) *Rites of Spring, The Great War and the Birth of the Modern Age.* New York: Bantum Books.

Elden, S. (2009) *Terror and Territory: The Spatial Extent of Sovereignty.* Minneapolis: University of Minnesota Press.

Eliasoph, N. (1998) *Avoiding Politics. How Americans Produce Apathy in Everyday Life.* Cambridge: Cambridge University Press.

Ellaway, A. and MacIntyre, S. (2010) *Neighbourhoods and health.* In: T. Brown, S. McLafferty and G. Moon (eds.) *A Companion to Health and Medical Geography.* Oxford: Wiley-Blackwell, 399–417.

Ellwood, W. (2001) *The No-nonsense Guide to Globalization.* London: Verso.

Elsaesser, T. (1994) European television and national identity: or 'what's there to touch when the dust has settled'. Paper presented to the European Film and Television Studies Conference, London, July.

Elwood, S. (2002) GIS use in community planning: A multi-dimensional analysis of empowerment. In: *Environment and Planning A* 34(5), 905–22.

Elwood, S. (2004) Partnership and participation: Reconfiguring urban governance in different state contexts. In: *Urban Geography* 25(8), 755–70.

Elwood, S. (2009a) GIS, public participation. In: R. Kitchin and N. Thrift N (eds.) *International Encyclopedia of Human Geography* 1, 520–5. Oxford: Elsevier.

Elwood, S. (2009b) Multiple representations, significations and epistemologies in community-based GIS. In: M. Cope and S. Elwood (eds.) *Qualitative GIS: A Mixed Methods Approach.* Thousand Oaks, CA: Sage Publications, 57–74.

Elwood, S. (2010) Geographic information science: Emerging research on the societal implications of the geospatial web. In: *Progress in Human Geography* 34(3), 349–57.

Elwood, S. and Ghose, R. (2001) PPGIS in community development planning: Framing the organizational context. In: *Cartographica* 38(3-4), 19–33.

Elwood, S. and Leitner, H. (2003) GIS and Spatial knowledge production for neighborhood revitalization: Negotiating state priorities and neighborhood visions. In: *Journal of Urban Affairs* 25(2), 139–57.

Emch, M. (1999) Diarrheal disease risk in Matlab, Bangladesh. In: *Social Science & Medicine* 49, 519–30.

Emel, J., Wilbert, C. and Wolch, J. (2002) Animal geographies. In: *Society and Animals* 10, 407–12.

Emmer, P. (1993) Intercontinental migration as a world historical process. In: *European Review* 1(1), 67–74.

England, K. (1994) Getting personal: Reflexivity, positionality and feminist research. In: *Professional Geographer* 46, 80–9.

English Nature (1993) *Position statement on sustainable development.* November.

Enloe, C. (1989) *Bananas, Beaches and Bases: Making Feminist Sense of International Relations.* Berkeley, CA: University of California Press.

Entrikin, J.N. (1976) Contemporary humanism in geography. In: *Annals of the Association of American Geographers* 66, 615–32.

Entrikin, J.N. (1991) *The Betweeness of Place: Towards a Geography of Modernity.* London: Macmillan.

Epstein, B. (2003) Notes on the antiwar movement. In: *Monthly Review* 55(3), 109–16.

Escamilla, V., Wagner, B., Yunus. M., Streatfield, P.K., van Geen, A. and Emch, M. (2011) Effect of deep tube well use on childhood diarrhoea in Bangladesh. In: *Bulletin of the World Health Organization* 89(7), 521–7.

Escobar, A. (1992) Culture, economics, and politics in Latin American social movements theory and research. In: A. Escobar and S.E. Alvarez (eds.) *The Making of Social Movements in Latin America*. Boulder, CO: Westview Press, 62–85.

Escobar, A. (1995) *Encountering Development: The Making and Unmaking of the Third World*. Princeton: Princeton University Press.

Escobar, A. (2008) *Territories of Difference: Place, Movement, Life, Redes*. London and Durham: Duke University Press.

Esin, I. and Turner, B. (2002) *Handbook of Citizenship Studies*. London: Sage.

Esteva, G. (1985) Beware of participation and development: Metaphor, myth, threat. In: *Development: Seeds of Change* 3, 75–86.

ETC Group (2010) The new biomasters: Synthetic biology and the next assault on biodiversity and livelihoods. Available at: http://www.etcgroup.org/upload/publication/pdf_file/biomassters_27feb2011.pdf (last accessed 21 March 2012).

Evans, B. (2010) Anticipating fatness: Childhood, affect and the pre-emptive 'war on obesity'. In: *Transactions of the Institute of British Geographers* 35, 21–38.

Evans, N., Morris, C. and Winter, M. (2002) Conceptualising agriculture: A critique of post-productivism as the new orthodoxy. In: *Progress in Human Geography* 26, 313–23.

Evans, R. (2010) Children's caring roles and responsibilities within the family in Africa. In: *Geography Compass* 4, 1477–96.

Evans, R. (2011) We are managing our own lives . . .: Life transitions and care in sibling-headed households affected by AIDS in Tanzania and Uganda. In: *Area* 43(4), 384–96.

Eyles, J. and Woods, K. (1983) *The Social Geography of Medicine and Health*. London: Croom Helm.

Falah, G. (1989) Israeli 'Judaization' policy in Galilee and its impact on local Arab urbanization. In: *Political Geography* 8(3), 229–54.

Fannin, M. (2007) The 'midwifery question' in Quebec: New problematics of birth, body, self. In: *BioSocieties* 2, 171–91.

FAO, IFAD, IMF, OECD, UNCTAD, WFP, the World Bank, the WTO, IFPRI and the UN HLTF (2011) *Price Volatility in Food and Agricultural Markets: Policy Responses*. Available at: http://www.wto.org/english/news_e/news11_e/igo_10jun11_report_e.pdf (last accessed 21 March 2012).

Fargione, J., Hill, J. Tilman, D., Polasky, S. and Hawthorne, P. (2008). Land clearing and the biofuel carbon debt. In: *Science* 319, 1235–8.

Featherstone, D. (2003) Spatialities of transnational resistance to globalisation: The maps of grievances of the Inter-Continental Caravan. In: *Transactions of the Institute of British Geographers* 28, 404–21.

Fischer, C. (1997) Technology and community: Historical complexities. In: *Sociological Inquiry* 67(1), 113–18.

Fisher, A. (1996) Deutsche Bank in Asia facing a rough ride in the East. In: *The Financial Times*, 13 November, 25.

Fisher, A. (1997) A reluctant departure. In: *The Financial Times*, 13 May, 21.

Fisher, A. (1998) Deutsche Bank warns of lower profits. In: *The Financial Times*, 29 January, 36.

Fisher, B. and Tronto, J. (1990) Towards a feminist theory of care. In: E. Abel and M. Nelson (eds.) *Circles of Care*. Albany: State University of New York Press.

Fisher, B., Turner, R.K. and Morling, P. (2009) Defining and classifying ecosystem services for decision making. In: *Ecological Economics* 68, 643–53.

Fisher, W.F. and Ponniah, T. (2003) *Another World is Possible: Popular Alternatives to Globalisation at the World Social Forum*. London: Zed Books.

Fitzpatrick, A. (2012) Class in session: Cornell NYC Tech Campus gets first professor, Mashable, http://mashable.com/2012/06/28/cornell-deborah-estrin/ (accessed 3 March 2012).

Flint, C. (ed.) (2004) *The Geography of War and Peace: From Death Camps to Diplomats*. Oxford and New York: Oxford University Press.

Flint, C. (2004) Dynamic meta-geographies of terrorism: The spatial challenges of religious terrorism and the 'war on terrorism'. In: C. Flint (ed.) *The Geography of War and Peace: From Death Camps to Diplomats*. Oxford and New York: Oxford University Press.

Flint, C. (ed.) (2005) *The Geography of War and Peace*. New York: Oxford University Press.

Florida, R. and Smith, D.F. (1993) Venture capital formation, investment and regional industrialisation. In: *Annals of the Association of American Geographers* 83, 434–51.

Fluri, J. (2011) Bodies, bombs and barricades: Geographies of conflict and civilian (in)security. In: *Transactions of the Institute of British Geographers* 36, 280–96.

Flusty, S. (2011) The rime of the frequent flyer: Or what the elephant has got in his trunk. In: C. Minca and T. Oakes, (eds.) *Real Tourism*. Abingdon: Routledge.

Foer, J.S. (2011) *Eating Animals*, London: Penguin.

Folke, C. (2006) Resilience: The emergence of a perspective for social–ecological systems analyses. In: *Global Environmental Change* 16, 253–67.

Foote, K.E. (2003) *Shadowed Ground: America's Landscape of Violence and Tragedy*. Austin: University of Texas Press.

Foote, K.E., Hugill, P.J., Mathewson, K. and Smith, J.M. (eds.) (1994) *Re-reading Cultural Geography*. Austin: University of Texas Press.

Forsyth, T. (2008) *Critical Political Ecology*. New York: Routledge.

Fortmann, L. (1996) 'Gendered knowledge: rights and space in two Zimbabwe villages: Reflections on methods and findings'. In: *Feminist Political Ecology: Global Issues and Local Experiences*. New York: Routledge, 211–23

Foster, J.B. (2003) The new age of imperialism. In: *Monthly Review* 55(3), 1–14.

Fotheringham, A.S., Brunsdon, C. and Charlton, M. (2000) *Quantitative Geography*. London: Sage.

Foucault, M. (1980) *Power/Knowledge*. London: Harvester Wheatsheaf.

Foucault, M. (1976/1995) *Discipline and Punish: The Birth of the Prison*. A. Sheridan (trans.) (1977). New York: Vintage.

Foucault, M. (2008) *The Birth of Biopolitics*. London: Palgrave.

Fox, J. (1996) How does civil society thicken? The political construction of social capital in rural Mexico. In: *World Development* 24, 1089–103.

Frank, A.G. (1967) *Capitalism and Underdevelopment in Latin America*. London: Monthly Review Press.

Frank, T. and Weiland, M. (1997) *Commodify Your Dissent: Salvos from the Baffler*. New York: W.W. Norton and Company.

Franklin, A. (2003) *Tourism: An Introduction*. London: Sage.

Freire, P. (1972) *Pedagogy of the Oppressed*. London: Sheed and Ward Ltd.

Frenzel, F. and Koens, K. (2012) Slum tourism: Developments in a growing field of interdisciplinary tourism research. In: *Tourism Geographies* 14, 195–212.

Friedland, W., Barton, A. and Thomas, R. (1981) *Manufacturing Green Gold*. Cambridge: Cambridge University Press.

Friedmann, J. and Wolff, G. (1982) World city formation: An agenda for research and action. In: *International Journal of Urban and Regional Research* 6(3), 309–44.

Frow, J. (1991) Tourism and the semiotics of nostalgia. In: *October* 57, 123–51.

Fuss, D. (1989) *Essentially Speaking: Feminism, Nature and Difference*. London: Routledge.

Fyfe, N. (1993) The police, space, and society: The political geography of policing. In: *Progress in Human Geography* 15, 249–67.

Fyfe, N. (1995) Law and order policy and the spaces of citizenship in contemporary Britain. In: *Political Geography* 14, 177–89.

Gabel, M. and Bruner, H. (2003) *Global Inc.: An Atlas of the Multinational Corporation*. New York: The New Press.

Gabriel, Y. and Lang, T. (1996) *The Unmanageable Consumer: Contemporary Consumption and its Fragmentations*. Thousand Oaks: Sage.

Gabriel, Y. and Lang, T. (2006) *The Unmanageable Consumer*. London: Sage.

Gade, D.W. (2003) Language, identity and the scriptorial landscape in Québec and Catalonia. In: *Geographical Review* 93(4), 429–48.

Gadgil, M. and Guha, R. (1995): *Ecology and Equity*. London: Routledge.

Gaia Foundation, Biofuelwatch, African Biodiversity Network, Salva La Selva, Watch Indonesia and EcoNexus (2008). *Agrofuels and the Myth of the Marginal Lands*. Available at:

http://www.cbd.int/doc/biofuel/Econexus%20 Briefing%20AgrofuelsMarginalMyth.pdf (last accessed 25 April 2012).

Game Info (2004) 'America's 10 Most Wanted' coming to consoles in 2004. Posted on 5 December 2003 @ 23:28:18 EST by xbox, http://www.xboxsolution.com/article1083.html, viewed 2 February 2004.

Gaonkar, D.P. (2001) (ed.) *Alternative Modernities.* Durham: Duke University Press.

Garitaonandía, G. (1993) Regional television in Europe. In: *European Journal of Communication* 9(3), 277–94.

Garreau, J. (1991) *Edge City: Life on the New Frontier*. New York: Doubleday.

Garrett, B.L. (2010) Urban explorers: quests for myth, mystery and meaning. In: *Geography Compass* 4, 1448–61.

Garrett, B.L. (2011) Assaying history: Creating temporal junctures through urban exploration. In: *Environment and Planning D: Society and Space* 29, 1048–67.

Garrett, L. (1994) *The Coming Plague: Newly Emerging Diseases in a World Out of Balance.* New York: Penguin.

Garton Ash, T. (1998) Europe's endangered liberal order. In: *Foreign Affairs* 77(2), 51–65.

Gauntlett, D. and Hill, A. (1999) *TV Living: Television, Culture, and Everyday Life*. London: Routledge.

Gedicks, A. (1993): *The New Resource Wars.* Boston: South End Press.

Geldof, B. (2004) The bitter legacy colonialism left to Africa. *The Independent*, 21 April 2004, 29.

Gentleman, A. (2006) Slum tours: A day trip too far? In: *The Observer*, 7 May 2006.

Geoforum (1998) Special Issue on Exclusion, 29(2).

Geoghegan, H. (2010) Museum geography: Exploring museums, collections and museum practice in the UK. In: *Geography Compass* 4, 1462–76.

Gereffi G. (1994) The organisation of buyer-driven commodity chains: How US retailers shape overseas production networks. In: G. Gereffi and M. Korzeniewicz (eds.) *Commodity Chains and Global Capitalism.* Westport, CT: Praeger, 95–122.

Gereffi, G. (1995) Global production systems and Third World development. In: B. Stallings (ed.) *Global Change, Regional Response*. Cambridge: Cambridge University Press.

Gereffi, G. (2001) Beyond the producer-driven/buyer-driven dichotomy. In: *IDS Bulletin* 32(3), 30.

Gergen, K. (2002) The challenge of absent presence. In: J. Katz and M. Aakhus (eds.) *Perpetual Contact: Mobile Communication, Private Talk, Public Transformance*. New York: Cambridge University Press, 227–41.

Gertler, M. (1997) The invention of regional culture. In: R. Lee and J. Wills (eds.) *Geographies of Economies*. London: Arnold.

Gesler, W.M. (1992) Therapeutic landscapes: Medical issues in light of the new cultural geography. In: *Social Science & Medicine* 34, 735–46.

Gesler, W.M. and Kearns, R.A. (2002) *Culture/Place/Health*. London: Routledge.

Ghose, R. (2005) The complexities of citizen participation through collaborative governance. In: *Space and Polity* 9(1), 61–75.

Ghose, R. (2007) Politics of scale and networks of association in public participation GIS. In: *Environment and Planning A* 39, 1961–80.

Gibson, W. (1984) *Neuromancer*. New York: Ace.

Gibson-Graham, J.K. (1996) *The End of Capitalism (As We Knew It)*. Cambridge, MA: Blackwell.

Gibson-Graham, J.K. (2002) Beyond global vs. local: Economic politics outside the binary frame. In: A. Herod and M. Wright (eds.) *Geographies of Power: Placing Scale*. Oxford: Blackwell, 25–60.

Gibson-Graham, J.K. (2006) *A Postcapitalist Politics*. Minneapolis: University of Minnesota Press.

Giddens, A. (1990) *The Consequences of Modernity*. Cambridge: Polity Press.

Giddens, A. (1991) *Modernity and Self-identity: Self and Society in the Late Modern Age*. Cambridge: Polity Press.

Giddens, A. (1993) *Sociology*. Cambridge: Polity.

Giddens, A. (1999) *Runaway World: How Globalisation is Reshaping Our Lives*. London: Profile Books.

Gilbert, D. (1999) London in all its glory – Or how to enjoy London: Guidebook representations of imperial London. In: *Journal of Historical Geography* 25(3), 279–97.

Gilbert, D. and Preston, R. (2003) Stop being so English. Suburbia and national identity. In: D. Gilbert, D. Matless and B. Short (eds.) *Geographies of British Modernity: Space and Society in the Twentieth Century*. Oxford: Blackwell, 187–203.

Gilbert, E.W. (1951) Geography and regionalism. In: G. Taylor (ed.) *Geography in the Twentieth Century*. London: Methuen.

Gilbert, E.W. (1960) The idea of the region. In: *Geography* 45(3), 157–75.

Gilderbloom, J.I. and Rosentraub, M.S. (1990) Creating the accessible city: Proposals for providing housing and transportation for low income, elderly and disabled people. In: *American Journal of Economics and Sociology* 49(3), 271–82.

Gillespie, M. (1995) *Television, Ethnicity and Cultural Change*. London: Routledge.

Gillespie, M. (2002) Our ground zeros: Diasporas, meaning and memory. In: B. Zelzier and S. Allen (eds.) *Journalism after 9/11*. London: Routledge, 252–70.

Gilligan, C. (1982) *In a Different Voice: Psychological Theory and Women's Development*. Cambridge MA: Harvard University Press.

Gilligan, C. (1986) Remapping the moral domain: New images of the self in relationship. In T. Heller, D.E. Wellbery and M. Sosna (eds.) *Reconstructing Individualism*. Stanford: Stanford University Press, 237–52.

Gilroy, P. (1987) *There Ain't No Black in the Union Jack*. London: Routledge.

Gilroy, P. (1993a) *The Black Atlantic: Modernity and Double Consciousness*. London: Verso.

Gilroy, P. (1993b) *Small Acts*. London and New York: Serpent's Tail.

Girard, R. (1978) *To Double Business Bound: Essays on Literature, Mimesis and Anthropology*. Baltimore: Johns Hopkins University Press.

GIS World Interview – Ian McHarg Reflects on the Past, Present and Future of GIS (1995) In: *GIS World* 8, 46–9.

Gladwell, M. (1997) The Coolhunt. *The New Yorker*, March 17, 78–88.

Glassman, J. (2007) Imperialism imposed and invited. In: D. Gregory and A. Pred (eds.) *Violent Geographies: Fear, Terror and Political Violence*. New York: Routledge.

Glassner, M.I. (1993) *Political geography*. Chichester: John Wiley.

Globe and Mail (2003) Merit found amid video-game mayhem: violence aside researchers argue skills can be gained by playing regularly, 29 May, A3.

Godlewska, A. and Smith, N. (eds.) (1994) *Geography and Empire*. London: Blackwell/IBG.

Godolphin, M. (1868) *Robinson Crusoe in Words of One Syllable*. London: George Routledge.

Goffman, E. (1956) *The Presentation of Self in Everyday Life*. London: Penguin.

Goffman, E. (1963) *Stigma*. Englewood Cliffs, NJ: Prentice Hall.

Goffman, E. (1968) *Stigma: Notes on the Management of Spoiled Identity*. Harmondsworth: Penguin.

Goldman, M. (2008) *Petro-state: Putin, Power, and the New Russia*. Oxford: Oxford University Press.

Goodchild, M.F. (1992) Geographical information science. In: *International Journal of Geographical Information Systems* 6(1), 31–45.

Goodchild, M. (2007) Citizens as sensors: The world of volunteered geography. In: *GeoJournal* 69, 211–21.

Gooder, H. and Jacobs, J.M. (2002) Belonging and non-belonging: The apology in a reconciling nation. In: A. Blunt and C. McEwan (eds.) *Postcolonial Geographies*. London: Continuum, 200–13.

Goodman, M.K. (2010) The mirror of consumption: Celebritization, developmental consumption and the shifting cultural politics of fair trade. In: *Geoforum*, 41(1), 104–16.

Goodwin, M. (1991) Replacing a surplus population: The policies of London Docklands Development Corporation. In: J. Allen and C. Hamnett (eds.) *Housing and Labour Markets*. London: Unwin and Hyman.

Goodwin, M. and Painter, J. (1996) Local governance, Fordism and the changing geographies of regulation. In: *Transactions of the Institute of British Geographers* 21(4), 635–49.

Goodwin, M., Jones, M. and Jones, R., (2013) *Rescaling the State: Devolution and the Geographies of Economic Governance*. Manchester: Manchester University Press.

Gopinath, G. (1995) Bombay, UK, Yuba City: Bhangra music and the engendering of diaspora. In: *Diaspora* 4(3), 303–22.

Gore, C. and Robinson, P. (2009) Local government response to climate change: our last, best hope? In: H. Selin, S.D. VanDeveer (eds.) *Changing Climates in North American Politics: Institutions, Policymaking and Multilevel Governance*, Cambridge, Mass.: MIT Press, 138–58.

Gorman-Murray, A. and Waitt, G. (2009) Queer-friendly neighbourhoods: Interrogating social cohesion across sexual difference in two Australian neighbourhoods. In: *Environment and Planning A* 41(12), 2855–73.

Goss, J. (1993a) The magic of the mall: An analysis of form, function, and meaning in the contemporary retail built environment. In: *Annals of the Association of American Geographers,* 83(1), 18–47.

Goss, J. (1993b) Placing the market and marketing place: Tourist advertising of the Hawaiian Islands, 1972–92. In: *Environment and Planning D: Society and Space* 11, 663–88.

Goss, J. (1999) Once upon a time in the commodity world: An unofficial guide to Mall of America. In: *Annals of the Association of American Geographers* 89(1), 45–75.

Gössling, S. (2003) Tourism and development in tropical islands: Political ecology perspectives. In: S. Gössling (ed.) *Tourism and Development in Tropical Islands: Political Ecology Perspectives.* Cheltenham: Edward Elgar, 1–37.

Gössling, S. and Hall, C.M. (2006) *Tourism and Global Environmental Change: Ecological, Social, Economic and Political Interrelationships.* Abingdon: Routledge.

Gough, K.V. and Franch, M. (2005) Spaces of the street: Socio-spatial mobility and exclusion of youth in Recife. In: *Children's Geographies* 3(2), 149–66.

Gould, P. (1985) *The Geographer at Work.* New York: Routledge.

Gould, S.J. (2004) *The Hedgehog, the Fox, and the Magister's Pox: Mending and Minding the Misconceived Gap Between Science and the Humanities.* London: Vintage.

Gowan, P. and Anderson, P. (eds.) (1997) *The Question of Europe*. London: Verso.

Graeber, D. (2012). Occupy Wall Street's anarchist roots. In: *Occupy Everything: Anarchists in the Occupy Movement, 2009–2012.* Aragorn! (ed.), LBC Books, Berkeley, CA.

Graham, B. and Nash, C. (eds.) (1999) *Modern Historical Geographies.* London: Longman.

Graham, B., Ashworth, G. and Tunbridge, J. (2000) *A Geography of Heritage: Power, Culture and Economy.* Oxford: Oxford University Press.

Graham, S. (2006) 'Homeland' insecurities? Katrina and the politics of 'security' in metropolitan America. In: *Space and Culture* 9(1), 63–7.

Graham, S. (2011) *Cities Under Siege: The New Military Urbanism.* London: Verso.

Graham, R. (1978) *Prisoners of Space? Exploring the Geographical Experiences of Older People.* Boulder, CO.: Westview Press.

Graham, S. (1998) The end of geography or the explosion of place? Conceptualizing space, place and information technology. In: *Progress in Human Geography* 22(2), 165–85.

Graham S. and Marvin, S. (2000) *Splintering Urbanism.* London: Routledge.

Graham, S. and Thrift, N. (2007) Out of order. In: *Theory, Culture and Society* 24(3), 1–25.

Gray, A. (1992) *Video Playtime: The Gendering of a Leisure Technology.* London: Routledge.

Gray, L.C. and Moseley, W.C. (2005) A geographical perspective on poverty–environmental interactions. In: *The Geographical Journal* 171: 9–23.

Green, D. (1991) *Faces of Latin America.* London: Latin America Bureau.

Green, E., Hebron, S. and Woodward, W. (1990) *Women's Leisure, What Leisure?* Basingstoke: Macmillan.

Green, N. (2002) Who's watching whom? Monitoring and accountability in mobile relations. In: B. Brown, N. Green and R. Harper (eds.) *Wireless World: Social, Cultural and Interactional Issues in Mobile Technologies.* London: Springer, 32–45.

Green, N. and Smith, S. (2004) A spy in your pocket? The regulation of mobile data in the UK. In: *Surveillance and Society* 4, 573–87.

Greenhough, B. and Roe, E. (2011) Ethics, space and somatic sensibilities: Comparing relationships between scientific researchers and their human and animal experimental subjects. In: *Environment and Planning D: Society and Space* 29(1), 47–66.

Greenwood, D. (1989) Culture by the pound: An anthropological perspective on tourism as cultural commoditization. In: V. Smith (ed.) *Hosts and Guests*. Philadelphia: University of Pennsylvania Press.

Gregg, M. and Seigworth, G. (eds.) (2010) *The Affect and Cultural Theory Reader*. Durham and London: Duke University Press.

Gregory, D. (1994) *Geographical Imaginations*. Oxford: Blackwell.

Gregory, D. (1995) Imaginative geographies. In: *Progress in Human Geography* 9, 447–85.

Gregory, D. (2003) Emperors of the gaze: Photographic practices and productions of space in Egypt, 1839–1914. In: J. Ryan and J. Schwartz (eds.), *Picturing Place: Photography and the Geographical Imagination*. London: I.B. Tauris, 195–225.

Gregory, D. (2004) *The Colonial Present: Afghanistan, Palestine and Iraq*. Oxford: Blackwell.

Gregory, D. (2011) The everywhere war. In: *The Geographical Journal* 177(3), 238–50.

Gregory, D., Johnston, R.J., Pratt, G., Watts, M. and Whatmore, S. (eds.) (2009) *The Dictionary of Human Geography*. Chichester: Wiley-Blackwell.

Gregory, D. and Pred, A. (eds.) (2007) *Violent Geographies: Fear, Terror and Political Violence*. New York: Routledge.

Gregson, N. (2007) *Living with Things: Ridding, Accommodation, Dwelling*. Oxford: Sean Kingston Publishing.

Gregson, N. and Crewe, L. (2003) *Second-hand Cultures*. Oxford and New York: Berg.

Gregson, N., Crang, M., Ahamed, F., Akhter, N. and Ferdous, R. (2010). Following things of rubbish value: end-of-life ships, 'chock-chocky' furniture and the Bangladeshi middle-class consumer. In: *Geoforum*, 41(6), 846–54. doi: 10.1016/j.geoforum.2010.05.007

Greif, M. (2012) Occupy London: What did the St Paul's camp represent? In: *The Guardian*, 29 February 2012.

Groenewegen, P., van den Berg, A., de Vries, S. and Veheij, R. (2006) Vitamin G: Effects of green space on health, well-being, and social safety. In: *BMC Public Health*, 6(149).

Griffiths, H., Poulter, I. and Sibley, D. (2000) Feral cats in the city. In: C. Philo and W. Wilbert (eds.) *Animal Spaces, Beastly Places: New Geographies of Human–Animal Relations*. London: Routledge.

Griffiths, I. (ed.) (1993) *The Atlas of African Affairs*. London: Routledge.

Griffiths, J. (1997) F1 probe says threat to relocate racing was groundless. *Financial Times*, 15 December, 1.

Gruffudd, P. (1994) Back to the land: Historiography, rurality and the nation in inter-war Wales. In: *Transactions of the Institute of British Geographers* 19(1), 61–77.

Guardian, The (2003) Did we make it better? *The Guardian*, London, 29 May, G2, 10.

Guardian, The (2011) 'Drill, drill, drill', says Bush era officials in oil crisis war game. 14 July 2011. Available at: http://www.guardian.co.uk/environment/2011/jul/14/us-oil-crisis-simulation-oil-shockwave

Guattari, F. (1996) Ritornellos and existential affects. In: G. Genosko (ed.) *The Guattari Reader*. Oxford: Blackwell, 158–71.

Guback, T. (1974) Cultural identity and film in the European Economic Community. In: *Cinema Journal* 13(1), 2–17.

Guerrera, F. (2003) Wall Street's drive to scale the Great Wall. In: *Financial Times*, 10 December, 14.

Guha, R. (1989) The problem. In: *Seminar*, March, 12–15.

Guibernau, M. and Hutchinson, J. (eds.) (2001) *Understanding Nationalism*. Cambridge: Polity Press.

Gupta, A. (1998) *Postcolonial Developments: Agriculture in the Making of Modern India*. London and Durham: Duke University Press.

Gusterson, H. (2011) Death by drone. In: *Bulletin of the Atomic Scientists*, 13 October.

Habermas, J. (1970) *Toward a Rational Society: Student Protest, Science, and Politics*. Boston: Beacon Press.

Habermas, J. (1993) *The Structural Transformation of the Public Sphere*. Cambridge, MA: MIT Press.

Hage, G. (2003) *Against Paranoid Nationalism*. London: Pluto Press.

Hahn, H. (1986) Disability and the urban environment: A perspective on Los Angeles. In: *Environment and Planning D: Society and Space* 4, 273–88.

Haklay, M. (2010) *Interacting with Geospatial Technologies*. Chichester: Wiley.

Halberstam, J. (2005) *In a Queer Time and Place: Transgender Bodies, Subcultural Lives*. New York and London: New York University Press.

Halbwachs, M. (1950) *On Collective Memory*. New York: Harper Row [reprinted by University of Chicago Press, 1992].

Hale, A. and Wills, J. (eds.) (2005) *Threads of Labour: A Workers' View of the Global Garment Industry*. Oxford: Blackwell.

Halfacree, K. (1993) Locality and social representation: Space, discourse and alternative definitions of the rural. In: *Journal of Rural Studies* 9, 23–38.

Halfacree, K. (1996) Out of place in the country: Travellers and the 'rural idyll'. In: *Antipode* 28 (1), 42–72.

Halfacree, K. (2006) Rural space: Constructing a three-fold architecture. In: P. Cloke, T. Marsden and P. Mooney (eds.) *Handbook of Rural Studies*. London: Sage, 44–62.

Halfacree, K. and Rivera, M.J. (2012) Moving to the countryside and staying: Lives beyond representations. In: *Sociologia Ruralis* 52(1), 92–114.

Hall, E. (1995) Contested (dis)abled identities in the urban labour market. Paper presented to the Tenth Urban Change and Conflict Conference, Royal Holloway, University of London, UK.

Hall, E. (2010) Spaces of wellbeing for people with learning disabilities. In: *Scottish Geographical Journal* 126(4), 275–84.

Hall, G.B., Chipeniuk, R., Feick, R., Leahy, M. and Deparday, V. (2010) Community-based production of geographic information using open source software and Web 2.0. In: *International Journal of Geographical Information Science* 24(5), 761–81.

Hall, S. (1992) The question of cultural identity. In: S. Hall, D. Held and T. McGrew (eds.) *Modernity and its Futures*. Oxford: Polity.

Hall, S. (1992) The West and the rest. In: S. Hall and B. Gieben (eds.) *Formations of Modernity*. Oxford: Polity.

Hall, S. (1995) New cultures for old. In: D. Massey and P. Jess (eds.) *A Place in the World?* Milton Keynes: Open University Press.

Hall, S.M. (2011) Exploring the 'ethical everyday': An ethnography of the ethics of family consumption. In: *Geoforum*, 42(6), 627–37.

Hallman, B. (2010) *Family Geographies: The Spatiality of Families and Family Life*. Oxford: OUP.

Hamnett, C. (2003a) Contemporary Human Geography: Fiddling while Rome burns? In: *Geoforum* 34, 1–3.

Hamnett, C (2003b) *Unequal City: London in the Global Arena*. London: Routledge.

Hamnett, C. (2012) (forthcoming) Concentration or diffusion? The changing geography of ethnic minority pupils in English secondary schools, 1999–2009. In: *Urban Studies*.

Hamnett, C. and Cross, D. (1998) Social polarisation and inequality in London: Earnings evidence, 1979–95. In: *Environment and Planning C, Government and Policy* 16, 659–80.

Hand, M., Shove, E. and Southerton, D. (2007) Home extensions in the United Kingdom: Space, time, and practice. In: *Environment and Planning D: Society and Space*, 25(4), 668–81.

Hannerz, U. (1996) *Transnational Connections: Culture People Places*. London: Routledge.

Hannigan, J.A. (1995) *Environmental Sociology: A Social Constructionist Perspective*. London: Routledge.

Hansen, A. (ed.) (1993) *The Mass Media and Environmental Issues*. Leicester: Leicester University Press.

Hansen, A. (2010) *Environment, Media and Communication*. Abingdon: Routledge.

Hanson, S. (1999) Is feminist geography relevant? In: *Scottish Geographical Journal* 115(20), 133–41.

Hanson, S. and Pratt, G. (1995) *Gender, Work and Space*. New York and London: Routledge.

Hanson Thiem, C. (2009) Thinking through education: The geographies of contemporary educational restructuring. In: *Progress in Human Geography* 33, 154–73.

Haraway, D.J. (1991a) *Simians, Cyborgs and Women: The Reinvention of Nature*. London: Routledge.

Haraway, D.J. (1991b) Situated knowledges: the science question in feminism as a site of discourse of the privilege of partial perspective. In: D.J. Haraway, *Simians, Cyborgs and Women: The Reinvention of Nature*. London and New York: Routledge, 183–201.

Haraway, D.J. (1996) Situated knowledges: The science question and the privilege of portal perspective. In J.A. Agnew, D.J. Livingstone and A. Rogers (eds.) *Human Geography: An Essential Anthology*. Oxford: Blackwell, 108–28.

Haraway, D.J. (2008) *When Species Meet*. Minneapolis: University of Minnesota Press.

Hardill, I. and Baines, S. (2009) Active citizenship in later life: Older volunteers in a deprived community in England. In: *Professional Geographer* 61(1), 36–45.

Hardin, G. (1968) The tragedy of the commons. In: *Science* 162, 1243–8.

Harding, S. (1991) *Whose Science? Whose Knowledge? Thinking from Women's Lives*. Ithaca, NY: Cornell University Press.

Hardt, M. and Negri, A. (2000) *Empire*. Cambridge, MA: Harvard University Press.

Harker, C. (2005) Playing and affective time-spaces. In: *Children's Geographies* 3, 47–62.

Harley, B. (2001) *The New Nature of Maps: Essays in the History of Cartography*. Baltimore: Johns Hopkins University Press.

Harley, J.B. (1989) Deconstructing the map. In: *Cartographica* 26, 1–20.

Harley, J.B. (1992) Deconstructing the map. In: T. Barnes and J. Duncan (eds.) *Writing Worlds: Discourse, Text and Metaphor in the Representation of Landscape*. London: Routledge.

Harmon, K. (ed.) (2004) *You Are Here: Personal Geographies and Other Maps of the Imagination*. New York: Princeton Architectural Press.

Harmon, K. (ed.) (2010) *The Map as Art: Contemporary Artists Explore Cartography*. New York: Princeton Architectural Press.

Harper, S. (1997) Contesting later life. In: P. Cloke and J. Little (eds.) *Contested Countryside Cultures*. London: Routledge, 189–96.

Harper, S. and Laws, G. (1995) Rethinking the geography of ageing. In: *Progress in Human Geography* 19, 199–221.

Harris, C. (2002) *Making Native Space: Colonialism, Resistance, and Reserves in British Columbia*. Vancouver: UBC Press.

Harris, J., Hunter, J. and Lewis, C. (eds.) (1995) *The New Institutional Economics and Third World Development*. London: Routledge.

Harris, L. and Hazen, H. (2006) Power of maps: (Counter) mapping for conservation. In: ACME: *An International E-Journal for Critical Geographies* 4(1), 99–130.

Harris, N. (1995) *The New Untouchables: Immigration and the New World Order*. Harmondsworth: Penguin.

Harris, R.C. and Phillips, E. (eds.) (1984) *Letters from Windermere 1912–1914*. Vancouver: UBC Press.

Harris, R., Sleight, P. and Webber, R. (2005) *Geodemographics, GIS and Neighbourhood Targeting*. Chichester: Wiley.

Harris, T. and Weiner, D. (1998) Empowerment, marginalization, and community-oriented GIS. In: *Cartography and Geographic Information Systems* 25(2), 67–76.

Harrison, C.M., Burgess, J. and Filius, P. (1996) Rationalising environmental responsibilities: A comparison of lay publics in the UK and the Netherlands. *Global Environmental Change* 6(3), 215–34.

Harrison, P. (2003) 'How shall I say it . . .?' Emotions, exposure and compassion. Paper given at the Emotional Geographies Conference, Lancaster 2002, typescript provided by author.

Hart, J.T. (1971) The inverse care law. In: *The Lancet*, 405–12.

Hart, R. (1979) *Children's Experience of Place*. New York: Irvington.

Hartig, T. and Staats, H. (2003). Guest editors'

introduction: Restorative environments. In: *Journal of Environmental Psychology* 23, 103–7.

Harvey, D. (1969) *Explanation in Geography*. London: Edward Arnold.

Harvey, D. (1973) *Social Justice and the City*. London: Edward Arnold.

Harvey, D. (1982) *The Limits to Capital*. Oxford: Blackwell.

Harvey, D. (1985) Paris, 1850–1870. In: D. Harvey, *Consciousness and the Urban Experience*. Oxford: Blackwell, 63–220.

Harvey, D. (1989) *The Condition of Postmodernity: An Enquiry into the Origins of Cultural Change*. Oxford: Blackwell.

Harvey, D. (1996) *Justice, Nature and the Geography of Difference*. Oxford: Blackwell.

Harvey, D. (2003) *The New Imperialism*. Oxford: Oxford University Press.

Harvey, D. (2005) *A Brief History of Neoliberalism*. Oxford: Oxford University Press.

Harvey, D. (2007) *A Brief History of Neoliberalism*, Oxford: Oxford University Press.

Harvey, D. (2010) *The Enigma of Capital and the Crisis of Capitalism*. London: Profile Books.

Harvey, D., Jones, R., Milligan, C. and McInroy, N. (eds.) (2002) *Celtic Geographies: Old Culture, New Times*. London: Routledge.

Harvey, F., Kwan, M. and Pavlovskaya, M. (2005) Introduction: Critical GIS. In: *Cartographica* 40(4), 1–4.

Harvey, N. (1995) Rebellion in Chiapas: Rural reforms and popular struggle. In: *Third World Quarterly* 16(1), 39–72.

Hawkins, H. (2010) Placing art at the Royal Geographical Society. In: V. Patel and T.C. Ledda (eds.) *Creative Compass: New Mappings by International Artists*. London: RGS-IBG, 7–16.

Hawkins, H. (2011) Dialogues and doings: Geography and art, landscape, critical spatialities and participation. In: *Geography Compass*, 5(7), 448–530.

Hay, I. (2010) *Qualitative Research Methods in Human Geography*. Oxford: Oxford University Press.

Hayes-Conroy, A. and Hayes-Conroy, J. (2008) Taking back taste: Feminism, food, and visceral politics. In: *Gender, Place and Culture* 15, 461–73.

Hayes-Conroy, J. and Hayes-Conroy, A. (2010) Visceral geographies: Mattering, relating and defying. In: *Geography Compass* 4(9),1273–83.

Hayter, T. and Harvey, D. (eds.) (1993) *The Factory and the City: The story of Cowley Automobile Workers in Oxford*. London and New York: Mansell.

Head, L. and Muir, P. (2006) Suburban life and the boundaries of nature: Resilience and rupture in Australian backyard gardens. In: *Transactions of the Institute of British Geographers* 31, 505–24.

Healy, P. (1992) A planner's day: Knowledge and action in communicative practice. In: *Journal of the American Planning Association* 58, 9–20.

Heath, S. (1990) Representing television. In: P. Mellencamp (ed.) *Logics of Television*. Bloomington: Indiana University Press.

Hebdige, D (1988) Object as image: The Italian scooter cycle. In: *Hiding in the Light. On Images and Things*. London: Comedia, 77–115.

Hechter, M. (1975) *Internal Colonialism: The Celtic Fringe in British National Development, 1536–1966*. London: RKP.

Heffernan M. (2003) Histories of geography. In: S. Holloway, S. Price and G. Valentine (eds.) *Key Concepts in Geography*. London: Sage.

Held, D. and McGrew, A. (2002) *Globalisation/Anti-globalisation*. Cambridge: Polity.

Held, D., McGrew, A., Goldblatt, D. and Perraton, J. (1999) *Global Transformations: Politics, Economics and Culture*. Cambridge: Polity.

Heldke, L. (2003) *Exotic Appetites: Ruminations of a Food Adventurer*. London: Routledge.

Helferich, G. (2004) *Humboldt's Cosmos: Alexander von Humboldt and the Latin American Journey That Changed the Way We See the World*. New York: Gotham.

Henry, N. and Pinch, S. (1997) *A Regional Formula for Success? The Innovative Region of Motor Sport Valley*. Edgbaston: University of Birmingham.

Henry, N. and Pinch, S. (2000) Spatialising knowledge: Placing the knowledge community of motor sport valley. In: *Geoforum* 31, 191–208.

Henry, N., Pinch, S. and Russell, S. (1996) In pole position? Untraded interdependencies, new industrial spaces and the British Motor Sport Industry. In: *Area* 28.1, 25–36.

Herrington, J. (1984) *The Outer City*. London: Harper & Row.

Hetherington, K. (2003) Spatial textures: Place, touch and praesentia. In: *Environment and Planning A* 3511, 1933–44.

Highmore, B. (2011) Out of the strong came forth sweetness – Sugar on the move. In: *New Formations*, 74.

Hill, J. (2006) Travelling objects: The Wellcome collection in Los Angeles, London and beyond. In: *Cultural Geographies* 13, 340–66.

Hinchliffe, S. and Bingham, N. (2008) Securing life – The emerging practices of biosecurity. In: *Environment and Planning A*, 40, 1534–51.

Hinchliffe, S., Kearnes, M.B., Degen, M. and Whatmore, S. (2005) Urban wild things: A cosmopolitical experiment. In: *Environment and Planning D: Society and Space* 23(5), 643–58.

Hinchliffe, S., Allen, J., Lavau, S, Bingham, N. and Carter, S. (2013) (forthcoming). Biosecurity and the topologies of infected life: From borderlines to borderlands. In: *Transactions of the Institute of British Geographers*.

Hine, T. (2002) *I Want That: How We All Became Shoppers*. New York: Harper Collins.

Hirst, P. and Thompson, G. (1999) *Globalization in Question*. Cambridge: Polity.

Hitchings, R. (2007) Approaching life in the London garden centre: Providing products and acquiring entities. In: *Environment and Planning A* 39, 242–59.

Hobson, D. (1989) Soap operas at work. In: E. Seiter, H. Borcher, G. Kreutzner and E-M. Warth (eds.) *Remote control. Television, Audiences and Cultural Power*. London and New York: Routledge.

Hochschild, A. (1983) *The Managed Heart: Commercialisation of Human Feeling*. Berkeley, CA: University of California Press.

Hodge, D. (ed.) (1995) Should women count? The role of quantitative methodology in feminist geographic research. In: *The Professional Geographer* 47, 426–66.

Hodgson, D. and Schroeder, R. (2002) Dilemmas of counter-mapping resources in Tanzania. In: *Development and Change* 33, 79–100.

Hodson, M. and S. Marvin (2009) Urban ecological security: A new urban paradigm? In: *International Journal of Urban and Regional Research* 33(1): 193–215.

Hodson, M. And Marvin, S. (2010) *World Cities and Climate Change: Producing Urban Ecological Security*, Milton Keynes: Open University Press.

Hoelscher, S. (2003) Making place, making race: Performances of whiteness in the Jim Crow South. In: *Annals of the Association of American Geographers* 93, 657–86.

Hoffman, M.J. (2011) *Climate Governance at the Crossroads: Experimenting with a Global Response.* New York: Oxford University Press.

Hogendorn, J.S. and Scot, K.M. (1981) The East African Groundnut Scheme: Lessons of a large-scale agricultural failure. In: *African Economic History* 10, 81–115.

Hoggart, K. (1990) Let's do away with rural. In: *Journal of Rural Studies* 6, 245–57.

Holden, A. (2000) *Environment and Tourism*. London: Routledge.

Holdgate, M. (1996) *From Care to Action: Making a Sustainable World*. London: Earthscan.

Holloway, L. (2001) Pets and protein: Placing domestic livestock on hobby-farms in England and Wales. In: *Journal of Rural Studies* 17, 293–307.

Holloway, L. (2007) Subjecting cows to robots: Farming technologies and the making of animal subjects. In: *Environment and Planning D: Society and Space* 23, 1041–60.

Holloway, L., Morris, C., Gilna, B. and Gibbs, D. (2009) Biopower, genetics and livestock breeding: (Re)constituting animal populations and heterogenous biosocial collectivities. In: *Transactions of the Institute of British Geographers* 34, 394–407.

Holloway, S.L. (1998) Geographies of justice: preschool-childcare provision and the conceptualisation of social justice. In: *Environment and Planning C: Government and Policy* 15, 85–104.

Holloway, S.L. (2005) Articulating Otherness? White rural residents talk about Gypsy-Travellers. In: *Transactions of the Institute of British Geographers* 30, 351–67.

Holloway, S.L. and Valentine, G. (2000a) *Children's*

Geographies: Playing, Living, Learning. London: Routledge.

Holloway, S.L. and Valentine, G. (2000b) Spatiality and the new social studies of childhood. In: *Sociology* 34, 763–83.

Holloway, S.L., Price, S. and Valentine, G. (2003) *Key Concepts in Geography*. London: Sage.

Holloway, S.L. and Pimlott-Wilson, H. (2011) Geographies of children, youth and families: Defining achievements, debating the agenda. In: L. Holt (ed.) *Geographies of Children, Youth and Families: an International Perspective*. Abingdon: Routledge, 9–24.

Holloway, S.L., Valentine, G. and Bingham, N. (2000) Institutionalising technologies: Masculinities, femininities and the heterosexual economy of the IT classroom. In: *Environment and Planning A* 32, 617–33.

Holt, L. (2003) (Dis)abling children in primary school spaces. Unpublished PhD thesis, Loughborough University, UK.

Holt, L. (2010) Young people's embodied social capital and performing disability. In: *Children's Geographies* 8(1), 25–37.

Holt, L. (ed.) (2011) *Geographies of Children, Youth and Families: An International Perspective.* Abingdon: Routledge.

Homewood, K., Lambin, E.F., Coast, E., Kariuki, A., Kikula, I., Kivelia, J., Said, M., Serneels, S. and Thompson (2001) 'Long-term changes in Serengeti-Mara wildebeest and land cover: Pastoralism, population or policies?' In: *Proceedings of the National Academy of Sciences* 98, 12544–9.

hooks, b. (1990) *Yearning: Race Gender and Cultural Politics*. London: Turnaround.

hooks, b. (1992) Eating the other. In b. hooks (ed.) *Black Looks: Race and Representation*. Boston: South End Press.

Hooson, I.D. (ed.) (1994) *Geography and National Identity*. Oxford: Blackwell.

Hopkins, P., Olson, E., Pain, R. and Vincett, G. (2010) Mapping intergenerationalities: The formation of youthful religiosities. In: *Transactions of the Institute of British Geographers* 36, 314–27.

Hopkins, P. and Pain, R. (2007) Geographies of age: Thinking relationally. In: *Area* 39, 287–94.

Hörschelmann, K. (2011) Theorising life transitions: Geographical perspectives. In: *Area* 43(3), 378–83.

Horst, H. A. (2006) The blessings and burdens of communication: Cell phones in Jamaican transnational social fields. In: *Global Networks* 6(2), 143–59.

Horton, J. and Kraftl, P. (2006) Not just growing up, but *going on*: Children's geographies as becomings, materials, spacings, bodies, situations. In: *Children's Geographies* 4, 259–76.

Horton, J. and Kraftl, P. (2008) Reflections on geographies of age: A response to Hopkins and Pain. In: *Area* 40: 284–88.

Houghton, J. (1997) *Global Warming: The Complete Briefing*. Cambridge: Cambridge University Press.

Houston, D. and Pulido, L. (2002) The work of performativity: Staging social justice at the University of Southern California. In: *Environment and Planning D: Society and Space,* 20, 401–24.

Howell, P. (2000) Flush and the *banditti*: Dog-stealing in Victorian London. In: C. Philo and W. Wilbert (eds.) *Animal Spaces, Beastly Places: New Geographies of Human–animal Relations*. London: Routledge.

Howes, D. (ed.) (2004) *Empire of the Senses: The Sensual Culture Reader*. Oxford and New York: Berg Publishers.

Howley, K. (2001) Envision television: Charting the Cultural Geography of homelessness. In: *Ecumene* 8, 345–50.

HSBC (2011) *Carbon neutrality.* http://www.hsbc.com/1/2/sustainability/protecting-the-environment/carbon-neutrality (accessed September 2011).

Hubbard, P. (2000) Desire/disgust: Mapping the moral contours of heterosexuality. In: *Progress in Human Geography* 24, 191–217.

Hubbard, P. (2002) Sexing the self: Geographies of engagement and encounter. In: *Social and Cultural Geography* 3, 365–81.

Hubbard, P. (2004) Inappropriate and Incongruous: Opposition to asylum centers in the English Countryside. In: *Journal of Rural Studies* 21, 3–17.

Hubbard, P. (2008) Here, there, everywhere: The ubiquitous geographies of hetereonormativity. In: *Geography Compass*, March 2008.

Hubbard, P. (2009) Opposing striptopia: the embattled spaces of adult entertainment. In: *Sexualities* 12(9), 721–45.

Hubbard, P., Kitchin, R., Bartley, B. and Fuller, D. (eds.) (2002) *Thinking Geographically: Space, Theory and Contemporary Human Geography*. London: Continuum.

Hubbard, P. and Lilley, K. (2004) Pacemaking the modern city: The urban politics of speed and slowness. In: *Environment and Planning D: Society and Space* 22, 273–94.

Hubbard, P. and Sanders, T. (2003) Making space for sex work: Female street prostitution and the production of urban space. In: *International Journal of Urban and Regional Research* 27, 75–89.

Hubbard, P. and Whowell, M. (2008) Revisiting the red light district: Still neglected, immoral and marginal? In: *Geoforum* 39(5), 1743–55.

Hubbard, P., Matthews, R., Scoular, J. and Agustín, L. (2008) Away from prying eyes? The urban geographies of 'adult entertainment'. In: *Progress in Human Geography* 32(3), 363–81.

Hudson, R. (2001) *Producing Places*. New York: Guilford Press.

Hudson, R. (2002) Changing industrial production systems and regional development in the New Europe. In: *Transactions of the Institute of British Geographers* 25, 262–81.

Hudson, R. (2004) Conceptualising economies and their geographies: Spaces, flows and circuits. In: *Progress in Human Geography* 28(4), 447–71.

Huggins, J., Huggins, R. and Jacobs, J.M. (1995) Kooramindanjie: Place and the postcolonial. In: *History Workshop Journal* 39, 164–81.

Hughes, A., and Reimer, S. (eds.) (2004) *Geographies of Commodity Chains*. Abingdon and New York: Routledge.

Hughes, R. (1987) *The Fatal Shore: History of the Transportation of Convicts to Australia*. London: Collins Harvill.

Hugo, G. and Smailes, P. (1985) Urban rural migration in Australia. In: *Journal of Rural Studies* 1(19).

Hulme, D. and Edwards, M. (eds.) (1997) *NGOs, States and Donors: Too Close for Comfort?* Basingstoke: Macmillan in association with Save the Children.

Hulme, M. (2009) *Why We Disagree About Climate Change*, Cambridge: Cambridge University Press

Hunt, D. (1989) *Economic Theories of Development: An Analysis of Competing Paradigms*. London: Harvester Wheatsheaf.

Hunter, M. (2002) The materiality of everyday sex: Thinking beyond 'prostitution'. In: *African Studies*, 61(1), 99–120.

Hunter, M. (2010) *Love in the Time of AIDS: Inequality, Gender, and Rights in South Africa*. Bloomington: Indiana University Press.

Huq, R. (2006) *Beyond Subculture: Pop, Youth and Identity in a Postcolonial World*. Abingdon: Routledge.

Huq, S., Kovats, S., Reid, H. and Satterthwaite, D. (2007) Editorial: Reducing risks to cities from disasters and climate change. In: *Environment and Urbanization*, 19(3), 3–15.

Hurrell, N., Kerry, E. and Parsons, H. (2010) 'By day, but not by night': Counter-mapping Platte Fields Park. In: *Bulletin of the Society of University Cartographers* 44(1–2), 17–24.

Hutnyk, J. (2000) *Critique of Exotica: Music, Politics and the Culture Industry*. London: Pluto Press.

Huyssen, A. (1984) Mapping the postmodern. In: *New German Critique* 33, 5–52.

Huysseune, M. (2010) Landscapes as a symbol of nationhood: The Alps in the rhetoric of the Lega Nord. In: *Nations and Nationalism* 16(2), 354–73.

Hyam, R. (1990) *Empire and Sexuality*. Manchester: Manchester University Press.

Hyam, R. (1993) *Britain's Imperial Century, 1815–1914*. Basingstoke: Macmillan.

Hyndman, J. (2003) Beyond either/or: A Feminist analysis of September 11th. *ACME*, 2(1).

Ignatieff, M. (1993) *Blood and Belonging: Journeys into the New Nationalism*. London: BBC Books/Chatto & Windus.

Imrie, R. (1996) *Disability and the City: International Perspectives*. London: Paul Chapman Publishing.

Imrie, R. (2000) Disability and discourses of mobility and movement. In: *Environment and Planning A* 32(9), 1641–956.

Imrie, R. and Edwards, C. (2007) The geographies of disability: Reflections on the development of a subdiscipline. In: *Geography Compass* 1(3), 623–40.

Ingham, J. (1999) Hearing places, making spaces: Sonorous geographies, ephemeral rhythms, and the Blackburn warehouse parties. In: *Environment and Planning D: Society and Space* 17, 283–305.

Inglehart, R. (1977) *The Silent Revolution.* Princeton, NJ: Princeton University Press.

Ingold, T. (2000) *The Perception of the Environment: Essays on Dwelling, Livelihood and Skill.* London: Routledge.

Ingold, T. (2004) Culture on the ground: The world perceived through the feet. In: *Journal of Material Culture* 93, 315–40.

Ingold, T. (2011) *Being Alive: Essays on Movement, Knowledge and Description.* Abingdon: Routledge.

Inkeles, A. and Smith, D.H. (1974) *Becoming Modern.* Cambridge, MA: Harvard University Press.

International Centre for Integrated Mountain Development (1998) Environment, culture, economy, and tourism: Dilemmas in the Hindu Kush-Himalayas. *Issues in Mountain Development,* 3, http://www.icimod.org/publications/imd/imd983.htm (accessed 12 July 2004).

International Energy Agency (2008) *World Energy Outlook 2008.* Paris, International Energy Agency.

International Monetary Fund (IMF) (2011) *World Economic Outlook,* April. Washington: IMF.

Inwood, J., and Tyner, J. (2011) Geography's pro-peace agenda: An unfinished project. In: *ACME: An International E-Journal for Critical Geographies,* 10, 442–57.

Irwin, A. (1995) *Citizen Science: A Study of People, Expertise and Sustainable Development.* London: Routledge.

Irwin, A. (2000) *Sociology and the Environment.* Cambridge: Polity Press.

IUCN (1980) *The world conservation strategy, international union for the conservation of nature and natural resources.* Gland: World Wildlife Fund and United Nations Environment Programme.

Jackson, P. (1989) *Maps of Meaning.* London: Routledge.

Jackson, P. (1999) Commodity cultures: The traffic in things. In: *Transactions of the Institute of British Geographers* NS24, 95–108.

Jackson, P. (1999) Postmodern urbanism and the ethnographic void. In: *Urban Geography* 20(5), 400–2.

Jackson, P. (2000) Rematerialising social and cultural geography. In: *Social and Cultural Geography* 1, 9–14.

Jackson, P. (2010) Food stories: Consumption in an age of anxiety. In: *Cultural Geographies* 17, 147–65.

Jackson, P. and Penrose, J. (eds.) (1993) *Constructions of Race, Place and Nation.* London: UCL Press.

Jackson, P. and Russell, P. (2010) Life history interviewing. In: D. DeLyser, S. Herbert, S. Aitken, M. Crang and L. McDowell (eds.) *The Sage Handbook of Qualitative Geography.* London: Sage, 172–92.

Jackson, P., Crang, P. and Dwyer. C. (eds.) (2004) *Transnational spaces.* Abingdon: Routledge.

Jackson, P., Ward, N. and Russell, P. (2009) Moral economies of food and geographies of responsibility. In: *Transactions of the Institute of British Geographers* 34(1), 12–24.

Jackson, T. (1996) *Material Concerns.* London: Routledge.

Jacobs, J. (1994) Earth honoring: Western desires and indigenous knowledges. In: A. Blunt and G. Rose (eds.) *Writing Women and Space: Colonial and Postcolonial Geographies.* New York and London: Guildford, 169–96.

Jacobs, J. (1996) *Edge of Empire.* London: Routledge.

Jacobs, J.M. (1988) Politics and the cultural landscape: The case of Aboriginal land rights. In: *Australian Geographical Studies* 26, 249–63.

Jacobs, J.M. (1996) Authentically yours: De-touring the map. In: *Edge of Empire: Postcolonialism and the City.* London: Routledge, 132–56.

Jacobs, J. (2009) Have sex will travel: Romantic 'sex tourism' and women negotiating modernity in Sinai. In: *Gender, Place and Culture* 16, 43–61.

Jacobs, J. (2010) *Sex, Tourism and the Postcolonial Encounter. Landscapes of Longing in Egypt.* Farnham: Ashgate.

Jacobs, M. (1993) *Sense and Sustainability: Land Use Planning and Environmentally Sustainable Development.* London: Council for the Protection of Rural England.

Jacobs, M. (1995) Sustainable development, capital substitution and economic humility: A response to Beckerman. In: *Environmental Values* 4, 57–68.

James, A., Jenks, C. and Prout, A. (1998) *Theorizing Childhood.* Cambridge: Polity.

Jay, E. (1992) *'Keep them in Birmingham': Challenging Racism in SW England.* London: Commission for Racial Equality.

Jayne, M., Holloway, S. and Valentine, G. (2006) Drunk and disorderly: Alcohol, urban life and public space. In: *Progress in Human Geography* 30(4), 451–68.

Jayne, M. Valentine, G. and Holloway, S. (2010) Emotional, embodied and affective geographies of alcohol, drinking and drunkenness. In: *Transactions of the Institute of British Geographers* 35(4), 540–54.

Jeffrey, C. and McDowell, L. (2004) Youth in a comparative perspective: Global change, local lives. In: *Youth and Society* 36, 131–42.

Jeffery, C., Lodge, G. and Schmuecker, K. (2010) The devolution paradox. In: G. Lodge and K. Schmuecker (eds.) *Devolution in Practice.* London: IPPR.

Jenkins, H. (1992) *Textual Poachers: Television Fans and Participatory Culture.* London and New York: Routledge.

Jenkins, J.G. (1982) *Maritime Heritage: The Ships and Seamen of Southern Ceredigion.* Llandysul: Gomer Press.

Jenkins, R. (1996) *Social Identities.* London: Routledge.

Jenks, C. (1996) *Childhood.* London: Routledge.

Johnsen., S., Cloke, P. and May, J. (2005) Transitory spaces of care: Serving homeless people on the street. In: *Health and Place* 11(4), 323–36.

Johnsen, S. and Fitzpatrick, S. (2010) Revanchist sanitisation or coercive care? The use of enforcement to combat begging, street drinking and rough sleeping in England. In: *Urban Studies* 47(8), 1703–23.

Johnson, J. and Salt, J. (1992) *Population Migration.* Walton-on-Thames: Thomas Nelson.

Johnson, J., Louis, R. and Pramono, A. (2006) Facing the future: Encouraging critical cartographic literacies in indigenous communities. In: *ACME: An International E-Journal for Critical Geographies* 4 (1), 80–98.

Johnson, L. (2011) *Insuring climate change?* PhD dissertation, University of California, Berkeley.

Johnson, N.C. (1996) Where geography and history meet: Heritage tourism and the big house in Ireland. In: *Annals of the Association of American Geographers* 86, 551–66.

Johnson, N.C. (2003) *Ireland, the Great War and the Geography of Remembrance.* Cambridge: Cambridge University Press.

Johnson, N.C. (2005) Locating memory, tracing the trajectories of remembrance. In: *Historical Geography* 33, 165–79.

Johnson N.C. (2012) The contours of memory in post-conflict societies: Enacting public remembrance of the bomb in Omagh, Northern Ireland. In: *Cultural Geographies,* 19(2), 237–58.

Johnson, R. (1986) The story so far: And further transformations? In: D. Punter (ed.) *Introduction to Contemporary Cultural Studies.* London and New York: Longman.

Johnston, C. and Pratt, G. (2010) Nanay (Mother): A testimonial play. In: *Cultural Geographies* 17(1), 123–33.

Johnston, L. (2007) Mobilizing pride/shame: Lesbians, tourism and parades. In: *Social and Cultural Geography* 8(1), 29–45.

Johnston, R. (1991/1997) *Geography and Geographers: Anglo-American Geography Since 1945.* London: Edward Arnold.

Johnston, R. (2003) Geography and the social science tradition. In: S. Holloway, S. Rice and G. Valentine (eds.) *Key Concepts in Geography.* London: Sage.

Johnston, R.J., Gregory, D., Pratt, G., Smith, D.M. and Watts, M.J. (eds.) (2000) *The Dictionary of Human Geography.* Oxford: Blackwell.

Johnston, R.J. and Sidaway, J.D. (2012) *Geography and Geographers: Anglo-American Geography Since 1945*. London: Hodder.

Johnstone, C. and Whitehead, M. (2004) *New Horizons in British Urban Policy*. Aldershot: Ashgate.

Jones, A. (2005) Truly global corporations? The politics of organizational globalization in business service firms. In: *Journal of Economic Geography* 5, 177–200

Jones, J.-P. (2003) Reading geography through binary oppositions. In: K. Anderson, M. Domosh, S. Pile and N. Thrift, *Handbook of Cultural Geography*. London: Sage.

Jones, J.-P. and Moss, P. (1995) Democracy, identity, space. In: *Environment and Planning D: Society and Space* 13, 253–7.

Jones III, J-P. (2010) Introduction: Social geographies of difference. In: S. Smith, R. Pain, S. Marston and J-P. Jones (eds.) *Handbook of Social Geographies*. London: Sage, 43–53.

Jones, M., Jones, R. and Woods, M. (2004) *An Introduction to Political Geography: Space, Place and Politics*. Abingdon: Routledge.

Jones, O. (2011) *Chavs: the Demonization of the Working Class*. London: Verso.

Jones, O. and Cloke, P. (2002) *Tree Cultures: The Place of Trees and Trees in Their Place*. London: Berg.

Jones, P. (2005) Performing the city: A body and a bicycle take on Birmingham, UK. In: *Social and Cultural Geography* 6, 813–30.

Jones, R. and Desforges, L. (2003) Localities and the reproduction of Welsh nationalism. In: *Political Geography* 22(3), 271–92.

Jones, R. and Merriman, P. (2009) Hot, banal and everyday nationalism: Bilingual road signs in Wales. In: *Political Geography* 28(3) 164–73.

Jordan, T. (1994) Cultural preadaptation and the American forest frontier: The role of New Sweden. In: K.E. Foote, P.J. Hugill, K. Mathewson and J.M. Smith (eds.) *Re-reading Cultural Geography*. Austin: University of Texas Press, 215–36.

Jordan, T. and Kaups, M. (1989) *The American Backwoods Frontier: An Ethnic and Ecological Interpretation*. Baltimore: John Hopkins University Press.

Joseph, A.E. and Phillips, D.R. (1984) *Accessibility and Utilization: Geographical Perspectives on Health Care Delivery*. New York: Harper & Row.

Karp, I. and Lavine, S. (1991) (eds.) *Exhibiting Cultures: The Poetics and Politics of Museum Display*. Washington, DC: Smithsonian Institute.

Katz, C. (2001) On the grounds of globalization: A topography for feminist political engagement. In: *Signs* 26(4), 1213–34.

Katz, C. (2008) Cultural Geographies lecture: childhood as spectacle: Relays of anxiety and the reconfiguration of the child. In: *Cultural Geographies* 15, 5–17.

Katz, C. and Monk, J. (1993) (eds.) *Full Circles: Geographies of Women Over the Life Course*. London: Routledge.

Katz, S. and Marshall, B. (2003) New sex for old: Lifestyle, consumerism, and the ethics of aging well. In: *Journal of Aging Studies,* 17, 3–16.

Kauffman, L.A. (2011) The theology of consensus. In: *Occupy! Scenes from Occupied America*. C. Blumenkranz, K. Gessen, M. Greif, S. Leonard, S. Resnick, N. Saval, E. Schmitt and A. Taylor (eds.) New York: Verso, 46–9.

Kaufman, S. (2006) The criminalization of New Orleanians in Katrina's wake. *Understanding Katrina: Perspectives from the Social Sciences,* http://understandingkatrina.ssrc.org/Kaufman/

Kawachi, I. and Berkman, L. (2003) *Neighbourhoods and Health*. Oxford: Oxford University Press.

Kawash, S. (1998) The homeless body. In: *Public Culture* 10(2), 319–39.

Kearns, R.A. (1993) Place and health: towards a reformed medical geography. In: *Professional Geographer* 45(2), 139–47.

Kearns, R.A. (1994) To reform is not to discard: A reply to Paul. In: *Professional Geographer* 46, 711–18.

Kearns, R.A. (1995) Medical geography: Making space for difference. In: *Progress in Human Geography* 19(2), 251–59.

Kearns, R.A. and Moon, G. (2002) From medical to health geography: Novelty, place and theory after a decade of change. In: *Progress in Human Geography* 26(5), 605–25.

Kearns, R. and Andrews, G. (2010) Geographies of well-being. In: S.J. Smith, R. Pain, S. A. Marston and J.P. Jones (eds.) *The Sage Handbook of Social Geographies.* London: Sage, 309–28.

Keeley, J. and Scoones, I. (2003) *Understanding Environmental Policy Processes: Cases from Africa.* London: Earthscan.

Keith, M. and Pile, S. (eds.) (1993) *Place and the Politics of Identity.* London: Routledge.

Keith, N., Lauren, F. and Eun Ja, H. (2011) Core networks, social isolation, and new media. In: *Information, Communication and Society* 14(1), 130–55.

Kenrick, D. and Clark, C. (1999) *Moving On: The Gypsies and Travellers of Britain.* Hertfordshire: University of Hertfordshire Press.

Kern, L. (2005) In place and at home in the city: Connecting privilege, safety and belonging for women in Toronto. In: *Gender, Place and Culture* 12(3), 357–77.

Kern, S. (1983) *The Culture of Time and Space, 1880–1918.* Cambridge, MA: Harvard University Press.

Khilnani, S. (1997) *The Idea of India.* London: Hamish Hamilton.

Kimmerman, M. (2011) In Protest, The Power of Place. In: *New York Times*, 15 October 2011.

Kindon, S. (2010) Participation. In: S. Smith, R. Pain, S. Marston and J.P. Jones (eds.) *Handbook of Social Geographies.* London: Sage, 517–45.

Kindon, S., Pain, R. and Kesby, M. (2007) *Participatory Action Research Approaches and Methods: Connecting People, Participation and Place.* Abingdon: Routledge.

King, A. (2004) *Spaces of Global Cultures.* Abingdon: Routledge.

King, R. (1978) Return migration: A neglected aspect of population geography. In: *Area* 10, 175–82.

King, R. (1995) Migrations, globalisation and place. In: D. Massey and P. Jess. *A Place in the World?* Oxford: Oxford University Press.

King, R., Warnes, A. and Williams, A. (1998) International retirement migration in Europe. In: *International Journal of Population Geography* 4, 91–111.

Kingsbury, P. (2009) Psychoanalytic theory/ psychoanalytic geographies. In: R. Kitchin and N. Thrift (eds.) *International Encyclopedia of Human Geography.* Oxford: Elsevier 8, 487–94.

Kingston, R., Carver, S., Evans, A. and Turton, I. (1999) *Virtual Decision Making in Spatial Planning: Web-Based Geographical Information Systems for Public Participation in Environmental Decision-Making.* School of Geography, University of Leeds UK http://www.geog.leeds.ac.uk/papers/99-9/index.html

Kinnaird, V., Morris, M., Nash, C. and Rose, G. (1997) Feminist geographies of environment, nature and landscape. In: Women and Geography Research Group, *Feminism and Geography: Diversity and Difference.* London: Longman, 146–89.

Kinsman, P. (1995) Landscape, race and national identity: The photography of Ingrid Pollard. In: *Area* 27, 300–31.

Kirkey, K. and Forsyth, A. (2001) Men in the valley: Gay male life on the suburban–rural fringe. In: *Journal of Rural Studies* 17, 421–41.

Kirsch. S. (2003) Guest editorial: Empire and the Bush Doctrine. In: *Society and Space* 21, 1–6.

Kirsch, S., and Flint, C. (eds.) (2011) *Reconstructing Conflict: Integrating War and Post-war Geographies.* Farnham: Ashgate.

Kirshenblatt-Gimblett, B. (1998) *Destination Culture: Tourism, Museums, and Heritage.* Berkeley: University of California Press.

Kitchen, R.M., Blades, M. and R.G. Golledge (1997) Understanding spatial concepts at the geographic scale without the use of vision. In: *Progress in Human Geography* 21(2), 225–42.

Kitchin, R. (2006) Positivistic geography and spatial science. In: S. Aitken and G. Valentine (eds.) *Approaches in Human Geography.* London: Sage, 20–29.

Kitchin, R. and Lysaght, K. (2003) Heterosexism and the geographies of everyday life in Belfast, Northern Ireland. In: *Environment and Planning A* 35, 489–510.

Kitchin, R. and Tate, N. (1999) *Conducting Research in Human Geography: Theory, Methodology and Practice.* Harlow: Prentice Hall.

Klare, M. (2002) *Resource Wars: The New Landscape of Global Conflict.* New York: Holt.

Klare, M. (2005) *Blood and Oil.* Harmondsworth: Penguin.

Klare, M. (2008a) The new geopolitics of energy. In: *The Nation*, 1 May 2008. Available at www.thenation.com

Klare, M. (2008b) *Rising Powers, Shrinking Planet.* New York: Metropolitan Books.

Klein, M. (1960) *Our Adult World and Its Roots in Infancy.* London: Tavistock Pamphlet 2.

Klein, N. (2000) *No Logo: Taking Aim at the Brand Bullies.* New York: Picador.

Klein, N. (2001) The unknown icon. *The Guardian*, 3 March, 9–16.

Klein, N. (2002) *Fences and Windows.* London: Flamingo.

Klein, N. (2004) How Bush told his lie. *The Nation*, 23 February.

Klein, N. (2007) *Shock Doctrine.* New York: Picador.

Klein, N. and Smith, N. (2008) The Shock Doctrine: A discussion. In: *Environment and Planning D: Society and Space* 26(4), 582–95.

Kneale, P. (2003) *Study Skills for Geographers.* London: Arnold.

Kniffen, F.B. (1965) Folk housing: Key to diffusion. In: *Annals of the Association of American Geographers* 55, 549–77.

Kniffen, F.B. (1990) Cultural diffusion and landscapes: Selections by Fred B. Kniffen. In: J.H. Walker and R.A. Detro (eds.) *Geoscience and Man.* Baton Rouge: Louisiana State University, Department of Geography and Anthropology 27.

Knopp, L. (1990) Some theoretical implications of gay involvement in an urban land market. In: *Political Geography Quarterly* 9, 337–52.

Knorr Cetina, K. and Preda, A. (eds.) (2004) *The Sociology of Financial Markets.* Oxford: Oxford University Press.

Knott, K. and McLoughlin, S. (eds.) (2010) *Diasporas: Concepts, Intersections, Identities.* London: Zed Books.

Knowles, A. (1997) *Calvinists incorporated: Welsh immigrants on Ohio's industrial frontier.* Chicago: University of Chicago Press.

Knox, P. (1991) The restless urban landscape: Economic and sociocultural change and the transformation of Metropolitan Washington, DC. In: *Annals of the Association of American Geographers* 81(2), 181–209.

Knox, P. and Pinch, S. (1998) *Urban Social Geography: An Introduction.* Harlow: Pearson.

Knox, P., Agnew, J. and McCarthy, J. (2003) *The Geography of the World Economy* (4th edn.). London: Arnold.

Koch, N. (2010) The monumental and the miniature: imagining 'modernity' in Astana. In: *Social & Cultural Geography* 11(8), 769–87.

Kollewe, J. (2012) 'China buys our bacon, tripe, trotters . . .' *The Guardian*, 19 May 2012, 38.

Kong, L. and B. Yeoh (2003) *The Politics of Landscapes in Singapore: Constructions of 'Nation'.* Syracuse: Syracuse University Press.

Kosek, J. (2010) 'Ecologies of empire: On the new uses of the honeybee'. In: *Cultural Anthropology* 25, 650–78.

Koser, K. (2003) Long-distance nationalism and the responsible state: The case of Eritrea. In: E. Ostsergaard-Nielsen (ed.) *International Migration and Sending Countries.* London: Palgrave Macmillan, 171–84.

Koser, K. and Salt, J. (1997) The geography of highly-skilled international migration. In: *International Journal of Population Geography* 3, 285–303.

Koshar, R. (1998) What ought to be seen: Tourists' guidebooks and national identities in modern Germany and Europe. In: *Journal of Contemporary History* 33(3), 323–40.

Koskela, H. (1997) Bold walk and breakings: Women's spatial confidence versus fear of violence. In: *Gender, Place and Culture* 4(3), 301–19.

Kraftl, P. (2008) Young people, hope and childhood-hope. In: *Space and Culture*, 11, 81–92.

Kraftl, P. (2012) Utopian promise or burdensome responsibility? A critical analysis of the UK Government's *Building Schools for the Future Policy.* In: *Antipode* 44, 847–70.

Kraftl, P., Horton, J. and Tucker, F. (2012) *Critical Geographies of Childhood and Youth: Policy and Practice.* Bristol: Policy Press.

Kramer, J.L. (1995) Bachelor farmers and spinsters:

Gay and lesbian identities and communities in rural North Dakota. In: D. Bell and G. Valentine (eds.) *Mapping Desire: Geographies of Sexualities*. London: Routledge, 200–13.

Krause, R.M. (1993) Foreword. In: S.S. Morse (ed.) *Emerging Viruses*. Oxford and New York: Oxford University Press.

Krause-Jackson, F. Clinton Chastises China on Internet, African American 'New Colonialism', 11 June 2011. *Bloomberg*. http://www.bloomberg.com/news/print/2011-06-11/clinton-chastises-china-on-internet-african-new-colonialism-.html (accessed 15 June 2011)

Kristeva, J. (1980) *Powers of Horror: An Essay in Abjection*. New York: Colombia University Press.

Kristeva, J. (1992) Le temps de la dépression. In: *Le Monde des Débats*, October.

Kristeva, J. (1993) *Nations Without Nationalism*. New York: Columbia University Press.

Krueger, R. and Gibbs, D. (2008) Third wave sustainability: Smart growth and regional development in USA. In: *Regional Studies* 49: 1263–74.

Kuhn T. (1962) *The Structure of Scientific Revolutions*. Chicago: University of Chicago Press.

Kuhn, T. (1970) *The Structure of Scientific Revolutions*. Chicago: University of Chicago Press.

Kuhn T. (1977) Second thoughts on paradigms. In F. Suppe (ed.) *The Structure of Scientific Theories*. Urbana: University of Illinois Press.

Kuik J, Lima, M. and Gupta, J. (2011) Energy security in a developing world. In: *Wiley Interdisciplinary Reviews* 2, 627–34.

Kullman, K. (2010) Transitional geographies: Making mobile children. In: *Social and Cultural Geography* 11(8), 829–46.

Kusch, C.E. (2000) Award-winning professor addresses urban poverty. In: *Penn State Outreach*, Spring/Summer, 3–6.

Kwan, M. (2002a) Feminist visualization: Re-envisioning GIS as a method in feminist geography research. In: *Annuals of the Association of American Geographers* 92(4), 645–61.

Kwan, M. (ed.) (2002b) Special Issue: Feminist geography and GIS. In: *Gender, Place and Culture* 9, 3.

Kyem, P. (2001) Power, participation and inflexible institutions: An examination of the challenges to community empowerment in participatory GIS applications. In: *Cartographica* 38(3–4), 5–18.

Kyem, P. (2004) Of intractable conflicts and participatory GIS applications: The search for consensus amidst competing claims and institutional demands. In: *Annals of the Association of American Geographers* 94(1): 37–57.

Lacey, C. and Longman, D. (1993) The press and public access to the environmental debate. In: *Sociological Review* 41, 207–43.

Lacey, M. (2005) Beyond the bullets and blades. *New York Times*, 20 March 2005.

Laituri, M. (2003) The issue of access: An assessment guide for evaluating Public Participation Geographic Information Science case studies. In: *URISA Journal* 15 (APA2), 25–32.

Lakatos, I. and Musgrave, A. (eds.) (1970) *Criticism and the Growth of Knowledge*. Cambridge: Cambridge University Press.

Lake, A., Townshed, T.G. and Alvanides, S. (eds.) (2010) *Obesogenic Environments: Complexities, Perceptions and Objective Measures*. Oxford: Wiley-Blackwell.

Lakoff, A. (2006) From disaster to catastrophe: The limits of preparedness. *Understanding Katrina: Perspectives from the Social Sciences, http://understandingkatrina.ssrc.org/Lakoff/*

Lakoff, A. (2008) From population to vital system: National security and the changing object of public health. In: S.J. Collier and A. Lakoff (eds.) *Biosecurity Interventions: Global Health and Security in Question*. New York: Continuum.

Landau, J.M. (ed.) (1984) *Atatürk and the Modernization of Turkey*. Boulder, CO: Westview Press.

Langevang, T. (2008) Claiming place: The production of young men's street meeting places in Accra, Ghana. In: *Geografiska Annaler: Series B, Human Geography* 90(3), 227–42.

Langley, P. (2008) *The Everyday Life of Global Finance: Saving and Borrowing in Anglo-America*. Oxford: Oxford University Press.

Lapola, D.M., Schaldach, R., Alcamo, J., Bondeau, A., Koch, J., Koelking, C. and Priess, J.A. (2010) Indirect land-use changes can overcome carbon savings from biofuels in Brazil. In: *PNAS* 107(8), 3388–93.

Larsen, J., Axhausen, K.W. and Urry, J. (2006) Geographies of social networks: Meetings, travel and communications. In: Mobilities 1(2), 261–83.

Lash, S. and Urry, J. (1994) *Economies of Signs and Space*. London: Sage.

Latham, A. (2003) Research, performance, and doing human geography: Some reflections on the diary-photo diary-interview method. In: *Environment and Planning A* 35(11), 1993–2017.

Latour, B. (1987) *Science in Action: How to Follow Scientists and Engineers Through Society*. Cambridge, MA: MIT Press.

Latour, B. (1993) *We Have Never Been Modern*. Harvard: Harvard University Press.

Latour, B. (1996a) *Aramis, or The Love of Technology*. Harvard: Harvard University Press.

Latour, B. (1996b) On actor-network theory. A few clarifications. In: *Soziale Welt* 47(4), 396–81.

Latour, B. (2005) *Reassembling the Social. An Introduction to Actor Network Theory*. Oxford: Oxford University Press.

Latour, B. and Woolgar, S. (1979) *Laboratory Life: The Construction of Scientific Facts*. Princeton: Princeton University Press.

Laurenson, P. and Collins, D. (2006) Towards inclusion: Local government, public space and homelessness in New Zealand. In: *New Zealand Geographer* 62(3), 185–95.

Laurenson, P. and Collins, D. (2007) Beyond punitive regulation? New Zealand Local governments' responses to homelessness. In: *Antipode* 39(4), 649–67.

Laurie, N., Dwyer, C., Holloway, S.L. and Smith, F.M. (1999) *Geographies of New Femininities*. London: Routledge.

Laurier, E. and Philo, C. (1999) X-morphising: review essay of Bruno Latour's *Aramis, or the Love of Technology*. In: *Environment and Planning A* 31(6) 1047–71.

Laurier, E. and Philo, C. (2006a) Possible geographies: A passing encounter in a cafe. In: *Area* 38, 353–64.

Laurier, E. and Philo, C. (2006b) Cold shoulders and napkins handed: Gestures of responsibility. In: *Transactions of the Institute of British Geographers* 31, 193–207.

Laurier, E., Hayden, L., Brown, B., Jones, O., Juhlin, O., Noble, A., Perry, M., Pica, D., Sormani, P., Strebel, I., Swan, L., Taylor, A.S., Watts, L. and Weilenmann, A. (2008) Driving and 'passengering': Notes on the ordinary organisation of car travel. In: *Mobilities* 3, 1–23.

Law, J. (2009) Actor network theory and material semiotics. In: B.S. Turner (ed.) *The New Blackwell Companion to Social Theory*. Oxford: Blackwell, 141–58.

Law, J. and Mol, A. (2008) Globalisation in practice: On the politics of boiling pigswill. In: *Geoforum* 39, 133–43.

Law, L. (2001) Home cooking: Filipino women and geographies of the senses in Hong Kong. In: *Ecumene* 8(3), 264–283.

Law, L., Wan-ling Wee, C.J. and McMullan, F. (2011) The cinematic landscape of Eric Khoo's *Be With Me*. In: *Geographical Research* 49(4), 363–74.

Laws, G. (1994) Ageing, contested meanings, and the built environment. In: *Environment and Planning A* 26, 1787–1802.

Laws, G. (1997) Spatiality and age relations. In: A. Jamieson, S. Harper and C. Victor (eds.) *Critical Approaches to Ageing and Later Life*. Milton Keynes: Open University Press, 90–100.

Lawson, F.H. (2004) Political economy, geopolitics and the expanding US military presence in the Persian Gulf and Central Asia. In: *Critique: Critical Middle Eastern Studies* 13(1), 7–31.

Lawton, C. (2010) Gender, spatial abilities, and wayfinding. In: *Handbook of Gender Research in Psychology* 4, 317–41.

Lea, J. (2008) Retreating to nature: rethinking 'therapeutic landscapes'. In: *Area* 40(1), 90–8.

Leach, M., Scoones, I. and Stirling, A. (2010) Governing epidemics in an age of complexity: Narratives, politics and pathways to sustainability. In: *Global Environmental Change* 20, 369–77.

Learmonth, A. (1988) *Disease Ecology: An Introduction*. Oxford: Wiley-Blackwell.

Le Billon, P. (ed.) (2005) *The Geopolitics of the Resource Wars*. London: Frank Cass

Le Billon, P. (2007) Geographies of war: Perspectives on 'resource wars'. In: *Geography Compass* 1, 163–82.

Le Billon, P. (2012) Critical geopolitics and resources. In: K. Dodds, M. Kuus and J. Sharp (eds.) *The Ashgate Companion to Critical Geopolitics*. Farnham: Ashgate.

Lee, E. (2012) American Afterlife. Unpublished PhD dissertation, Department of Geography, University of British Columbia.

Lee, J.Y. and Kwan, M.-P. (2011) Visualisation of socio-spatial isolation based on human activity patterns and social networks in space–time. In: *Tijdschrift voor Economische en Sociale Geografie* 102, 468–85.

Lee, R. (1989) Social relations and the geography of material life. In: D. Gregory and R. Walford (eds.) *Horizons in Human Geography*. London: Macmillan, 152—69.

Lee, R. (1998) Shelter from the storm? Mutual knowledge and geographies of regard (or legendary economic geographies). Paper presented to the RGS-IBG Annual Conference. University of Surrey, Guildford, 6 January.

Lee, R. (1999) *Access to the Gods? Social Relations and Geographies of Material Life*. Routledge: London.

Lee, R. (2000) Shelter from the storm? Geographies of regard in the worlds of horticultural consumption and production. In: *Geoforum* 31, 137–57.

Lee, R. (2002) Nice maps, shame about the theory? Thinking geographically about the economic. In: *Progress in Human Geography* 26(3), 333–55.

Lee, R. (2005a) The old economy. In: P.W. Daniels, J.W. Beaverstock, M.J. Bradshaw and A. Leyshon (eds.) *Geographies of the New Economy*. Abingdon: Routledge, Chapter 2.

Lee, R. (2005b) 'Production'. In: P. Cloke, P. Crang and M. Goodwin (eds.) *Introducing Human Geographies* (2nd edn.). London: Hodder Arnold.

Lees, L. (2003) The ambivalence of diversity and the politics of urban renaissance: The case of youth in downtown Portland, Maine, USA. In: *International Journal of Urban and Regional Research* 27(3), 613–34.

Lefebvre, H. (1971) *Everyday Life in the Modern World*. New York: Harper & Row.

Lefebvre, H. (1991) *The Production of Space*. Oxford: Blackwell.

Leftwich, A. (1993) Governance, democracy and development in the Third World. In: *Third World Quarterly* 14(3), 605–24.

Leftwich, A. (1994) Governance, the state and the politics of development. In: *Development and Change* 25, 363–86.

Legg, S. (2005) Sites of counter-memory: The refusal to forget and the nationalist struggle in colonial Delhi. In: *Historical Geography* 33, 180–201.

Legrain, P. (2002) *Open World: The Truth About Globalisation*. London: Abacus.

LeGrand, J. (2006). *Motivation, Agency and Public Policy*. Oxford: Oxford University Press.

Lehmann, D. (1997) An opportunity lost: Escobar's deconstruction of development. In: *Journal of Development Studies* 33(4), 568–78.

Leib, J.L. (2002) Separate times, shared spaces: Arthur Ashe, Monument Avenue and the politics of Richmond, Virginia's symbolic landscape. In: *Cultural Geographies* 9, 286–312.

Leighly J (ed.) (1963) *Land and Life: A Selection of Writings from Carl Ortwin Sauer*. Berkeley, CA: University of California Press.

Lélé, S.M. (1991) Sustainable development: A critical review. In: *World Development* 19, 607–21.

Lemon, C. and Lemon, J. (2003) Community-based cooperative ventures for adults with intellectual disabilities. In: *The Canadian Geographer* 47, 414–28.

Lennon, J.J. and Foley, M. (2000) *Dark Tourism: The Attraction of Death and Disaster*. London: Continuum.

Leonard, M. (1998) Paper planes: Travelling the new grrrl geographies. In: T. Skelton and G. Valentine (eds.) *Cool Places: Geographies of Youth Cultures*. London and New York: Routledge.

Levi-Strauss, C. (1969) *The Raw and the Cooked*. Chicago: University of Chicago Press.

Levitt, P. (2007) *God Needs No Passport: Immigrants and the Changing American Religious Landscape*. New York: New Press.

Levy, R. (1995) Finding a place in the world economy. Party strategy and party vote: The regionalization of SNP and Plaid Cymru support, 1979–92. In: *Political Geography* 14(3), 295–308.

Lewis, D. and Kanji, N. (2009) *Non-Governmental Organizations and Development*. Abingdon: Routledge.

Lewis, D., Rodgers, D. and Woolcock, M. (2008) The fiction of development: Literary representation as a source of authoritative knowledge. In: *Journal of Development Studies* 44(2), 198–216.

Lewis, M. and Wigen, K. (1997) *The Myth of Continents: A Critique of Metageography*. Berkeley: University of California Press.

Lewis, P. (1979) Axioms for reading the landscape. In: D.W. Meinig (ed.) *The Interpretation of Ordinary Landscapes: Geographical Essays*. New York: Oxford University Press, 11–32.

Lewis, R. (ed.) (2013) *Modest Fashion: Styling Bodies, Mediating Fashion*. London: I.B. Tauris.

Lewis, W.A. (1955) *The Theory of Economic Growth*. London: George Allen and Unwin.

Ley, D. (1974) The black inner city as frontier outpost: Images and behaviour of a Philadelphia neighbourhood. In: *Association of American Geographers, Monograph Series 7*, Washington DC.

Ley, D. (1977) Social geography and the taken-for-granted world. In: *Transactions of the Institute of British Geographers* NS 2, 498–512.

Ley, D. (1995) Between Europe and Asia: The case of the missing sequoias. In: *Ecumene* 2, 185–210.

Ley, D. (1996) *The New Middle Class and the Remaking of the Central City*. Oxford: Oxford University Press.

Ley, D. (2010) *Millionaire Migrants. Trans-Pacific Life Lines*. Oxford: Wiley-Blackwell.

Ley, D. and Cybriwsky, R. (1974) Urban graffiti as territorial markers. In: *Annals of the Association of American Geographers* 64, 491–505.

Ley, D. and Mountz, A. (2001) Interpretation, representation, positionality: Issues in field research in Human Geography. In: M. Limb and C. Dwyer (eds.) *Qualitative Methodologies For Geographers: Issues and Debates*. London: Arnold, 234–47.

Ley, D. and Samuels, H. (eds.) (1978) *Humanistic Geography: Prospects and Problems*. London: Croom Helm.

Leyshon, A. (1995) Geographies of money and finance. In: *Progress in Human Geography* 19, 531–43.

Leyshon, A. (1996) Dissolving difference? Money, disembedding and the creation of global financial space. In: P. Daniels and W.F. Lever (eds.) *The Global Economy in Transition*. London: Longman.

Leyshon, A. and Thrift, N.J. (1997) *Money/Space*. London: Routledge.

Leyshon, A. and Pollard, J. (2000) Geographies of industrial convergence: The case of retail banking. In: *Transactions of the Institute of British Geographers* 25, 203–20.

Leyshon, A., Lee, R. and Williams, C. (2003) *Alternative Economic Spaces*. London: Sage.

Leyshon, A., French, S. and Signoretta, P. (2008) Financial exclusion and the geography of bank and building society closure in Britain. In: *Transactions of the Institute of British Geographers* 33, 447–65.

Leyshon, A., Burton, D., Knights, D., Alferoff, C. and Signoretta, P. (2004) Towards an ecology of retail financial services: Understanding the persistence of door-to-door credit and insurance providers. In: *Environment and Planning A* 36, 625–46.

Leyshon, A., French, S., Thrift, N., Crewe, L. and Webb, P. (2005) Accounting for e-commerce: Abstractions, virtualism and the cultural circuit of capital. In: *Economy and Society* 34, 428–50.

Licoppe, C. (2004). 'Connected' presence: The emergence of a new repertoire for managing social relationships in a changing communication technoscape. In: *Environment and Planning D: Society and Space* 22, 135–56.

Liebes, T. and Katz, E. (1990) *The Export of Meaning: Cross-cultural Readings of Dallas*. Oxford: Oxford University Press.

Lilley III, W. and De Franco, L.J. (1997) No guarantees for F1's 'Sport Valley'. *Financial Times*, 31 December.

Lim, J. (2010) Immanent politics: Thinking race and ethnicity through affect and machinism. In: *Environment and Planning A* 42, 2393–409.

Limb, M. and Dwyer, C. (eds.) (2001) *Qualitative Methodologies for Geographers*. London: Arnold.

Ling, R. and Campbell, S. (2009) *The Reconstruction of Space and Time: Mobile Communication Practices*. New Brunswick, NJ: Transaction Publishers.

Lipsitz, G. (1994) Kalfou Dangere. In: *Dangerous Crossroads: Popular Music, Postmodernism and the Poetics of Place*. London: Verso.

Little, J. (1994) *Gender, Planning and the Policy Process*. Oxford: Pergamon.

Little, J. (2003) 'Riding the rural love train': Heterosexuality and the rural community. In: *Sociologia Ruralis* 43, 401–17.

Little, J., Peake, L. and Richardson, P. (1988) *Women in Cities*. London: Macmillan.

Little, J. and Austin, P. (1996) Women and the rural idyll. In: *Journal of Rural Studies* 12(2), 101–11.

Livingstone, D. (2000) Putting Geography in its place. In: *Australian Geographical Studies* 38, 1–9.

Livingstone, D.N. (1992) *The Geographical Tradition*. Oxford: Basil Blackwell.

Livingstone, D.N. and Harrison, R.T. (1981) Hunting the snark: Perspectives on geographical investigation. In: *Geografiska Annaler* 63B, 69–72.

Livingstone, D.N. and Withers, C.W.J. (eds.) (1999) *Geography and Enlightenment*. Chicago: University Of Chicago Press.

Logan, J. and Molotch, H. (1987) *Urban Fortunes*. Beverley, CA: University of California Press.

London Heat Map http://www.londonheatmap.org.uk/Content/home.aspx Accessed September 2011.

Longhurst, R. (1995) The body and geography. In: *Gender, Place and Culture* 2(1), 97–105.

Longhurst, R. (2006) Plots, plants and paradoxes: Contemporary domestic gardens in Aotearoa/New Zealand. In: *Social and Cultural Geography* 7, 581–93.

Longhurst, R., Johnston, L. and Ho, E. (2009) A visceral approach: Cooking 'at home' with migrant women in Hamilton, New Zealand. In: *Transactions of the Institute of British Geographers* 34, 333–45.

Longley, P.A., Goodchild, M.F., Maguire, D.J. and Rhind, D.W. (2010) *Geographic Information Systems and Science*. Chichester: Wiley.

Longman (1991) *Dictionary of the English language*. London: Longman.

Lorimer, H. (2005) Cultural geography: The busyness of being 'more than representational'. In: *Progress in Human Geography* 29(1), 83–94.

Lorimer, H. (2006) Herding memories of humans and animals. In: *Environment and Planning D: Society and Space* 24(4), 497–518.

Lorimer, J. (2007) Nonhuman charisma. *Environment and Planning D* 25, 911–32.

Lorimer, J. and Driessen, C. (2011) Bovine biopolitics and the promise of monsters in the rewilding of Heck cattle. In: *Geoforum* (in press)

Lorimer, J. and Whatmore, S. (2009) After 'the king of beasts': Samuel Baker and the embodied historical geographies of elephant hunting in mid-19th century Ceylon. In: *Journal of Historical Geography* 35, 668–89.

Lowe, P. and Rudig, R. (1986) Political ecology and the social sciences: The state of the art. In: *British Journal of Sociology* 16, 513–50.

Lowenthal, D. (1991) British national identity and the English landscape. In: *Rural History* 2, 205–30.

Lowenthal, D. (1994) Identity, heritage and history. In: J.R. Gillis (ed.) *Commemorations: The Politics of National Identity*. Princeton, NJ: Princeton University Press.

Lowenthal, D (1996/1998) *The Heritage Crusade and the Spoils of History*. London: Viking.

Lowenthal, D. (1997) *Geographical Journal* 163, 355.

Lowenthal, D. (2000) *George Perkins Marsh: Prophet of Conservation*. Seattle: University of Washington Press

Lozato-Giotart, J.-P. (1987) *Géographie du Tourisme, de l'espace regardé à l'espace consommé*. Paris: Masson.

Lozato-Giotart, J.-P. (1990) *Méditerranée et Tourisme*. Paris: Masson.

Lozato-Giotart, J.-P. and Balfet, M. (2004) *Management du Tourisme: Territoires, Systèmes de Production et Stratégies*. Paris: Pearson Education France.

Luke, T. (1997) At the end of nature: Cyborgs, 'humachines' and environments in postmodernity. In: *Environment and Planning A* 29, 1367–80.

Lund, K. (2005) Seeing in motion and the touching eye: Walking over Scotland's mountains. In: *Etnofoor* 18(1), 27–42.

Lutz, C and Collins, J. (1993) *Reading National Geographic*. Chicago, IL: University of Chicago Press.

Lynn, W. (1998) Animals, ethics and geography. In: J. Wolch and J. Emel (eds.) *Animal Geographies: Place, Politics, and Identity in the Nature–Culture Borderlands*. London: Verso, 280–97.

Lyod, B. and Rowntree, L. (1978) Radical feminists and gay men in San Francisco: A social space in dispersed communities. In D. Lanegran and R. Palm (eds.) *An Invitation to Geography*. New York: McGraw-Hill.

Lyon, D. (2007) *Surveillance Studies: An Overview*. Cambridge: Polity Press.

Maas, J., Verheij, R., Spreeuwenberg, P. and Groenewegen, P. (2008) Physical activity as a possible mechanism behind the relationship between green space and health: A multi-level analysis. In: *BMC Public Health* 8(1), 206.

Maas, J., van Dillern, S., Verheij, R., Groenewegen, P. (2009) Social contacts as a possible mechanism behind the relationship between green space and health. In: *Health and Place* 15(2), 586–95.

Maas, J., Verheij, R., Groenewegen, P., de Vries, S., Spreeuwenberg, P. (2006) Green space, urbanity and health: How strong is the relation? In: *Journal of Epidemiology and Community Health* 60(7), 587–92.

MacCannell, D. (1976/1989) *The Tourist: A New Theory of the Leisure Classes*. New York: Schocken Books.

MacCannell, D. (1992) *Empty Meeting Grounds: The Tourist Papers*. Routledge: London.

MacDonald, G.M. (2003) *Biogeography. Space, Time and Life*. New York: John Wiley and Sons.

Macintyre, S. Ellaway, A. and Cummins, S. (2002) Place effects on health: How can we conceptualise, operationalise and measure them? In: *Social Science & Medicine* 55(1),125–39.

MacKenzie, D. (2000) Fear in the markets. In: *London Review of Books*, 13 April, 1–5 (http://www.lrb.co.uk/v22/n08/mack01_.html).

MacKenzie, J. (1995) *Orientalism: History, Theory and the Arts*. Manchester: Manchester University Press.

MacKian, S. (1995) That great dust-heap called history: recovering the multiple spaces of citizenship. In: *Political Geography* 14, 209–16.

MacLeod, G. and Jones, M. (2001) Renewing the geography of regions. In: *Environment and Planning D: Society and Space* 1(6), 669–95.

Macpherson, H. (2009) Articulating blind touch: Thinking through the feet. In: *The Senses and society* 4(2), 179–92.

Maddison, A. (2001) *The World Economy: A Millennial Perspective*. OECD.

Maddison, A. (2003) *The World Economy: Historical Statistics*. OECD.

Madianou N. and Miller, D. (2011) *Migration and New Media: Transnational Families and Polymedia*. Abingdon: Routledge.

Madriaga, M. (2010) 'I avoid pubs and the student union like the plague': Students with Asperger Syndrome and their negotiation of university spaces. In: *Children's Geographies* 8(1), 39–50.

Maffesoli, M. (1995) *The Time of the Tribes: The Decline of Individualism in Mass Society*. London: Sage.

Magdoff, H. (2003) *Imperialism Without Colonies*. New York: Monthly Review Press.

Mahony, N., Newman, J. and Barnett, C. (eds.) (2009). *Rethinking the Public*. Bristol: Policy Press.

Mahtani, M. (2002) Tricking the border guards: Performing race. In: *Environment and Planning D: Society and Space* 20, 425–40.

Maintz, J. (2008). Synthesizing the face-to-face experience: e-learning practices and the constitution of place online. In: *Social Geography* 3, 1–10.

Malam, L. (2008) Bodies, beaches and bars: negotiating heterosexual masculinity in southern Thailand's tourism industry. In: *Gender, Place and Culture* 15(6), 581–94.

Malbon, B. (1999) *Clubbing: Dancing, Ecstasy and Vitality*. London: Routledge.

Manguel, A. (1996) *A History of Reading*. London: HarperCollins.

Manning, R.D. (2000) *Credit Card Nation: The Consequences of America's Addiction to Credit.* New York: Basic Books.

Mansfield, B. (2003) 'Imitation crab' and the material culture of commodity production. In: *Cultural Geographies* 10, 176–95.

Mansvelt, J.R. (2012). Consumption, ageing identity: New Zealand's narratives of gifting, ridding and passing on. *New Zealand Geographies*, 68(3), 187–200. doi:10.111/j.1745-7939.2012.01233.x

Margolis, M.L. (1998) *An Invisible Minority: Brazilians in New York City*. New York: Allyn & Bacon.

Marmot Review, (2010) *Fair Society, Healthy Lives.* London: Marmot Review.

Marsh, G.P. (1965 [originally 1864]) *Man and Nature*. Introduction by David Lowenthal. Cambridge, MA: Belknap Press.

Marsh, P. (2004) Where partners fight a little war everyday. *Financial Times*, 22 June, 13.

Marston, S. (2000) The social construction of scale. In: *Progress in Human Geography* 24, 219–42.

Marston, S.A. (2002) Making difference: Conflict over Irish identity in the New York City St Patrick's Day parade. In: *Political Geography* 21, 373–92.

Martin, D.-C. (1992) Le choix d'identité. *Revue Française de Science Politique*, 42(4) 582—93.

Martin, D. (1996) *Geographic Information Systems: Socioeconomic Applications*. London: Routledge.

Martin, D. (2004) Neighbourhoods and area statistics in the post-2001 census era. In: *Area* 36, 136–45.

Martin, D., Jordan, H. and Roderick, P. (2008) Taking the bus: Incorporating public transport timetable data into health care accessibility modelling. In: *Environment and Planning A* 40(10), 2510–25.

Martin, H.-J. (1994) *The History and Power of Writing*. Chicago: University of Chicago Press.

Martin, L. and Mitchelson, L. (2008) Geographies of detention and imprisonment: Interrogating spatial practices of confinement, discipline, law, and state power. In: *Geography Compass* 3(1), 459–77.

Martin, R. (ed.) (1999) *Money and the Space Economy*. Chichester: Wiley.

Martin, R. (2001) Geography and public policy: The case of the missing agenda. *Progress in Human Geography*, 25(2), 189—201.

Martin, R. and Minns, R. (1995) Undermining the financial basis of regions: The spatial structure and implications of the UK pension fund system. In: *Regional Studies*, 29, 125–44.

Martin, R. and Sunley, P. (2003) Deconstructing clusters: Chaotic concept or policy panacea? In: *Journal of Economic Geography* 3, 5–36.

Martinez-Allier, J. (1990) Ecology and the poor: A neglected dimension of Latin American history. In: *Journal of Latin American Studies* 23, 621–39.

Martinez-Allier, J. (2002) *The Environmentalism of the Poor: A Study of Ecological Conflicts and Valuation*. London: Elgar.

Martinez-Alier, J., Pascual, U., Vivien, F. and Zaccai, E. (2010) Sustainable de-growth: Mapping the context, criticisms and future prospects of an emergent paradigm. In: *Ecological Economics* 69: 1741–7.

Martinussen, J. (1997) *Society, State and the Market: A Guide to Competing Theories of Development.* London: Zed Books.

Mattingly, D. (2001) Place, teenagers and representations: Lessons from a community theatre project. In: *Social and Cultural Geography* 2, 445–59.

Marx, K. (1976 [1867]) *Capital.* Volume 1. Harmondsworth: Penguin Books.

Marx, K. (1981) *Surveys from Exile.* Harmondsworth: Penguin.

Marx, K. and Engels, F. (1967 [1848]) *The Communist Manifesto.* Harmondsworth: Penguin Books.

Maslow, H. (1970) *Motivation and Personality.* New York: Harper and Row.

Mason, C. and Harrison, R. (1998) Financing entrepreneurship: Venture capital and regional development. In: R.L. Martin (ed.) *Money and the Space Economy*. Chichester: John Wiley.

Massey, D. (1991) A global sense of place. In: *Marxism Today* (June), 24–9; reprinted in D.

Massey (1994) *Space, Place and Gender.* Cambridge: Polity Press, 146–56.

Massey, D. (1993) Power-geometry and a progressive sense of place. In: J. Bird, B. Curtis, T. Putnam, G. Robertson and L. Tickner (eds.) *Mapping the Futures: Local Cultures, Global Change*. London: Routledge.

Massey, D. (1994) *Space, Place and Gender.* Cambridge: Polity Press.

Massey, D. (1995) The conceptualization of place. In: D. Massey and P. Jess (eds.) *A Place in the World?* Oxford: Oxford University Press.

Massey, D. (1995) *Spatial Divisions of Labour: Social Structures and the Geography of Production.* Basingstoke and London: Macmillan.

Massey, D. (2004) Geographies of responsibility. In: *Geografiska Annaler* 86B, 5–18.

Massey, D. (2005) *For Space*. London: Sage.

Massey, D. and Jess, P. (1995) Places and cultures in an uneven world. In: D. Massey and P. Jess (eds.) *A Place in the World?* Oxford: Oxford University Press.

Massey, D. and Jess, P. (eds.) (1995) *A Place in the World?* Oxford: Oxford University Press.

Massey, D., Allen, J. and Sarre, P. (eds.) (1999) *Human Geography Today*. Cambridge: Polity Press.

Massey, D.S. and Denton, N.A. (1993) *American Apartheid*, Cambridge, MA: Harvard.

Massumi, B. (1987) Foreword. In: G. Deleuze and F. Guattari. (trans. B. Massumi) *A Thousand Plateaus: Capitalism and* Schizophrenia. Minneapolis: University of Minnesota Press.

Massumi, B. (2002) *Parables for the Virtual: Movement, Affect, Sensation*. Durham and London: Duke University Press.

Matless, D. (1992) An occasion for geography: landscape representation and Foucault's corpus. In: *Environment and Planning D: Society and Space* 10, 41–56.

Matless, D. (1994) Moral geography in Broadland. In: *Ecumene* 1(2) 127–56.

Matless, D. (1995) The art of right living: landscape and citizenship, 1918–39. In S. Pile and N. Thrift (eds.) *Mapping the Subject: Geographies of Cultural Transformation*. London and New York: Routledge, 93–122.

Matless, D. (2000) Action and noise over a hundred years: the making of a nature region. In: *Body and Society* 6, 141–65.

Matthews, H. and Limb, M. (1999) Defining *an* agenda for the geography of children. In: *Progress in Human Geography* 23, 61–90.

Matthews, M.H. (1987) Gender, home range and environmental cognition. In: *Transactions of the Institute of British Geographers NS* 12, 43–56.

Maxwell, S. and Frankenberger, T. (1992) *Household Food Security: Concepts, Indicators, Measurements: A Technical Review*. UNICEF and IFAD, New York and Rome.

May, E.T. (1989) Explosive issues: sex, women, and the bomb. In: May, L. (ed.) *Recasting America: Culture and Politics in the Age of the Cold War.* Minneapolis: University of Minnesota Press, 154–70.

May, J. (1996) A little taste of something more exotic: the imaginative geographies of everyday life. In: *Geography* 81, 57–64.

May, J., Cloke, P. and Johnsen, S. (2005) Re-phasing neo-liberalism: New labour and Britain's crisis of street homelessness. In: *Antipode* 37(4).

May, J. Johnsen, S. and Cloke, P. (2007) Alternative cartographies of homelessness: rendering visible British women's experiences of 'visible' homelessness. In: *Gender, Place and Culture* 14(2), 121–40.

Mayer, J. D. (1982) Relations between two traditions of medical geography: health systems planning and geographical epidemiology. In: *Progress in Human Geography* 6, 216–30.

Mayer, J. D. and Meade, M.S. (1994) A reformed medical geography reconsidered. In: *Professional Geographer* 46, 103–6.

Mayer, T. (2004) Nation, gender, and boundaries: feminist political geography and the study of nationalism. In: L. Staeheli, E. Kofman and L. Peake (eds.) *Mapping Women, Making Politics,* Abingdon and New York: Routledge, 153–68.

McClintock, A. (1995) *Imperial Leather: Race, Gender and Sexuality in the Colonial Contest.* New York: Routledge.

McCormack, D. (2003) An event of geographical ethics in spaces of affect. In: *Transactions of the Institute of British Geographers* 28(4), 488–507.

McCormack, D.P. (2006) For the love of pipes and cables: A response to Deborah Thien. In: *Area*, 38(3), 330–32.

McCormack, D. (2008) Engineering affective atmospheres: On the moving geographies of the 1897 Andree expedition. In: *Cultural Geographies* 15(4), 413–30.

McCormick, J. (1988) America's Third World. *Newsweek*, 8 August, 20–24.

McCormick, J. (1991) *British Politics and the Environment*. London: Earthscan.

McCormick, J.S. (1992) *The Global Environmental Movement: Reclaiming Paradise*. London: Belhaven.

McCracken, G. (1988) *Culture and Consumption: New Approaches to the Symbolic Character of Consumer Goods and Activities*. Bloomington: Indiana University Press.

McDowell, L. (1983) Towards an understanding of the gender division of urban space. In: *Environment and Planning D, Society & Space* 1, 59–72.

McDowell, L. and Massey, D. (1984) A woman's place. In: D. Massey and J. Allen (eds.) *Geography Matters! A Reader*. Cambridge: Cambridge University Press, 128–47.

McDowell, L. and Court, G. (1994) Performing work: Bodily representations in merchant banks. In: *Environment and Planning D: Society and Space* 12, 727–50.

McDowell, L. (1995) Body work: Heterosexual gender performances in city workplaces. In: D. Bell and G. Valentine (eds.) *Mapping Desire: Geographies of Sexualities*. London: Routledge, 75–95.

McDowell, L. (1997) *Capital Culture: Gender at Work in the City*. Oxford: Blackwell.

McDowell, L. (2003) *Redundant Masculinities: Employment Change and White Working Class Youth*. Oxford: Blackwell.

McEwan, C. (2000) Engendering citizenship; Gendered spaces of democracy in South Africa. In: *Political Geography* 19, 627–51.

McGregor, A. (2000) Dynamic texts and the tourist gaze: Death, bones and buffalo. In: *Annals of Tourism Research* 27(1), 27–50.

McGuigan (2000) British identity and the 'people's princess'. In: *Sociological Review* 48(1), 1–18.

McKay, G. (2011) *Radical Gardening. Politics, Idealism and Rebellion in the Garden*. London: Frances Lincoln.

McKendrick, N., Brewer, J. and Plumb, J.H. (1982) *The Birth of a Consumer Society: The Commercialization of Eighteenth-century England*. London: Europa Publications.

McKenzie, J. (2001) *Perform or Else: From Discipline to Performance*. London: Routledge.

McKibben, B. (1990) *The End of Nature*. Oxford: Oxford University Press.

McKie, L., Gregory, S. and Bowlby, S. (2002) Shadow times: The temporal and spatial frameworks and experiences of caring and working. In: *Sociology* 36, 897–924.

McLafferty, S. (2002) Mapping women's worlds: Knowledge, power and the bounds of GIS. In: *Gender, Place and Culture* 9, 263–9.

McLuhan, M. (1964) *Understanding Media*. London: Routledge and Kegan Paul.

McManus, P. (1996) Contested terrains: Politics, stories and discourses of sustainability. In: *Environmental Politics* 5: 48–71.

McNamee, S. (1998) Youth, gender and video games: Power and control in the home. In: T. Skelton and G. Valentine (eds.) *Cool Places: Geographies of Youth Cultures*. London and New York: Routledge, 195–206.

McNay, L. (1994) *Foucault: A Critical Introduction*. Cambridge: Polity Press.

McVeigh, T. (2011) 'Biofuels land grab in Kenya's l Delta fuels talk of war'. In: *The Guardian,* 2 July 2011. Available at: http://www.guardian.co.uk/world/2011/jul/02/biofuels-land-grab-kenya-delta (last accessed 25 April 2012).

McVeigh, T. (2012) On Cardiff's party streets they'll put up with the price of drinks by saving on the taxi home. In: *The Observer*, 25 March 2012.

Meadows, D.H., Meadows, D.L., Randers, J. and Behrens III, W.W. (1972) *The Limits to Growth: A Report for the Club of Rome's Project on the Predicament of Mankind*. London: Pan.

Mearsheimer, J. (1990) Why we will soon miss the Cold War. In: *The Atlantic* 266(2), 35–50.

Meethan, K. (2001) *Tourism in Global Society: Place, Culture, Consumption.* New York: Palgrave.

Megoran, N. (2011) War and peace? An agenda for peace research and practice in geography. In: *Political Geography*, 30, 178–89.

Meinig, D. (ed.) (1979) *The Interpretation of Ordinary Landscapes.* Oxford and New York: Oxford University Press.

Melen, S. and Nordman, E. (2009) The internationalization modes of Born Globals: A longitudinal study. In: *European Management Journal* 27, 243–54

Mellinger, W. (1994) Toward a critical analysis of tourism representations. In: *Annals of Tourism Research* 21, 756–79.

Mercer, C. (1999) Reconceptualizing state–society relations in Tanzania. In: *Area* 31(3), 247–58.

Mercer, C., Page, B. and M. Evans (2009) *Development and the African Diaspora.* London: Zed Books.

Merriman, N. (ed.) (1993) *The Peopling of London: Fifteen Thousand Years of Settlement from Overseas.* London: Museum of London.

Merriman, P. (2009) Automobility and the geographies of the car. In: *Geography Compass* 3, 586–99.

Merriman, P. (2010) Architecture/dance: Choreographing and inhabiting spaces with Anna and Lawrence Halprin. In: *Cultural Geographies* 17, 427–49.

Mertes, T. (ed.) (2004) *The Movement of Movements: A Reader.* London: Verso.

Meschkank, J. (2011) Investigations into slum tourism in Mumbai: Poverty tourism and the tensions between different constructions of reality. In: *GeoJournal* 76, 47–62.

Meyer, D.R. (2003) The challenges of research on the global network of cities. In: *Urban Geography* 24(4), 301–13.

Meyrowitz, J. (1985) *No Sense of Place: The Impact of Electronic Media on Social Behaviour.* Oxford: Oxford University Press.

Michaud, E. (2004) *The Cult of Art in Nazi Germany.* Stanford, CA: Stanford University Press.

Middleton, J. (2010) Sense and the city: Exploring the embodied geographies of urban walking. In: *Social and Cultural Geography* 11(6), 575–96.

Middleton, J. (2011) Walking in the city: The geographies of everyday pedestrian practices. In: *Geography Compass* 5, 90–105.

Mikkelsen, M. R. and Christiansen, P. (2009) Is children's independent mobility really independent? A study of children's mobility combining ethnography and GPS/Mobile Phone Technologies. In: *Mobilities* 4(1), 37–58.

Milbourne, P. (ed.) (1997) *Revealing Rural Others.* London: Pinter.

Miller, C.C. (2006) A beast in the field: The Google Maps mashup as GIS/2. In: *Cartographica* 41(3), 187–99.

Miller, D. (1987) *Material Culture and Mass Consumption.* Oxford: Blackwell.

Miller, D. (1988) Appropriation of the state on the council estate. In: *Man* 23, 353–72.

Miller, D. (1992) The young and the restless in Trinidad. A case study of the local and the global in mass consumption. In: R. Silverstone and E. Hirsch (eds.) *Consuming Technologies.* London: Routledge.

Miller, D. (1994) *Modernity: An Ethnographic Approach.* Oxford: Berg.

Miller, D. (1995a) Consumption and commodities. In: *Annual Review of Anthropology* 24, 141–61.

Miller, D. (1995b) Consumption as the vanguard of history. In: D. Miller (ed.) *Acknowledging Consumption: A Review of New Studies.* London: Routledge.

Miller, D. (1997) *Capitalism: An Ethnographic Approach.* Oxford: Berg.

Miller, D (1998) Coca-Cola: A black sweet drink from Trinidad. In: D. Miller (ed.) *Material Cultures. Why Some Things Matter.* London: UCL Press, 169–87.

Miller, D. (ed.) (2001) *Home Possessions. Material Culture Behind Closed Doors.* Oxford: Berg.

Miller, D. (2008) *The Comfort of Things.* Cambridge: Polity Press.

Miller, D. (2010) *Stuff.* Cambridge: Polity Press.

Miller, D. (2011) *Tales from Facebook.* Cambridge: Polity Press.

Miller, D., Jackson, P., Thrift, N., Holbrook, B.

and Rowlands, M. (eds.) (1998) *Shopping, Place and Identity*. London and New York: Routledge.

Milligan, C., Gatrell, A.C. and Bingley, A.F. (2004) Cultivating health: Therapeutic landscapes and older people in Northern England. In: *Social Science & Medicine* 58, 1781–93.

Milligan, C. and Miles, J. (2010) Landscapes of care. In: *Progress in Human Geography* 34(6), 736–54.

Mills, S. (1989) Tourism and leisure – Setting the scene. In: *Tourism Today* 6, 18–21.

Milward, A. (1992) *The European Rescue of the Nation State*. London: Routledge.

Minca, C. (2007) The tourist landscape paradox. In: *Social and Cultural Geography* 8(3), 433–53.

Minca, C. (2009) The island: Work, tourism and the biopolitical. In: *Tourist Studies* 9(2), 88–108.

Minca, C. and Oakes, T. (eds.) (2006) *Travels in Paradox: Remapping Tourism*. Lanham: Rowman & Littlefield.

Minca, C. and Oakes, T. (eds.) (2011) *Real Tourism*. Abingdon: Routledge.

Mintz, S. (1985) *Sweetness and Power*. New York: Viking Books.

Miossec, J. (1976) *Elements Pour une Théorie de l'Espace Touristique*. Aix-en-Provence: CHET.

Mishan, E. (1969) *The Costs of Economic Growth*. Harmondsworth: Penguin Books.

Mitchell, B. and Draper, D. (1982) *Relevance and Ethics in Geography*. London: Longman.

Mitchell, D. (1997) The annihilation of space by law: The roots and implications of anti-homeless laws in the United States. In: *Antipode* 29, 303–36.

Mitchell, D. (2000) *Cultural Geography: A Critical Introduction*. Oxford: Blackwell.

Mitchell, D. (2003) *The Right to the City: Social Justice and the Fight for Public Space*. New York: The Guilford Press.

Mitchell, D. (2005) The SUV model of citizenship, floating bubbles, buffer zones, and the rise of the 'purely atomic' individual. In: *Political Geography* 24(1), 77–100.

Mitchell, D. and Staeheli, L. (2006) Clean and safe? Property redevelopment, public space and homelessness in downtown San Diego. In: S. Lowe and N. Smith (eds.) *The Politics of Public Space*. New York: Routledge.

Mitchell, J.B. (1954) *Historical Geography*. London: The English Universities Press.

Mitchell, T. (1991) *Colonising Egypt*. Berkeley: University of California Press.

Mitchell, W.J.T. (1994a) Introduction. In: W.J.T. Mitchell (ed.) *Landscape and Power*. Chicago and London: University of Chicago Press.

Mitchell, W.J.T. (1994b) Imperial landscape. In: W.J.T. Mitchell (ed.) *Landscape and Power*. Chicago and London: University of Chicago Press.

Mizelle, B. (2011) *Pig*. London: Reaktion.

Mohan, G., Brown, E., Milward, B. and Zack-Williams, A.B. (2000) *Structural Adjustment: Theory, Practice and Impacts*. London: Routledge.

Mohanty, C. (1991) Cartographies of struggle, Third World women and the politics of feminism. In: C. Mohanty, A. Parker and A. Russo (eds.) *Cartographies of Struggle, Third World Women and the Politics of Feminism*. London: Routledge.

Mohanty, C. (2003) *Feminism Without Borders: Decolonizing Theory, Practicising Solidarity*. Durham: Duke University Press.

Mohen, J. (2000) Geographies of welfare and social exclusion. In: *Progress in Human Geography* 24(2), 291–300.

Mohen, J. (2002) Geographies of welfare and social exclusion: dimensions, consequences and methods. In: *Progress in Human Geography* 26(1), 65–75.

Mollenkopf, J. and Castells, M. (1991) *Dual City? Restructuring New York*. New York, NY: Russell Sage Foundation.

Monbiot, G. (2003) *The Age of Consent: A Manifesto for a New World Order*. London: Flamingo.

Monk, J. and Hanson, S. (1982) On not excluding half the human in geography. In: *The Professional Geographer* 34, 11–23.

Monk, J. and Hanson, S. (2008) On not excluding half the human in geography. In: P. Moss and K. Falconer Al-Hindi (eds.) *Feminisms in Geography: Rethinking Space, Place and Knowledges*, Maryland: Rowman & Littlefield, 60–7.

Moodie, S. (1986) *Roughing It in the Bush* (with Introduction by Margaret Atwood). London: Virago.

Moody, K. (1997) *Workers in a Lean World.* London: Verso.

Moody, R. (ed.) (1988) *The Indigenous Voice.* London: Zed Books.

Moon, G. (2009) Health geography. In: R. Kitchin and N. Thrift (eds.) *International Encyclopedia of Human Geography.* Oxford: Elsevier, 35–55.

Morales, E. (1989) *Cocaine.* Tucson: University of Arizona Press.

Morley, D. (1986) *Family Television: Cultural Power and Domestic Leisure.* London: Comedia.

Morley, D. (1991) Where the global meets the local: Notes from the sitting room. In: *Screen* 32, 1–15.

Morley, D. (1992) *Television, Audiences and Cultural Studies.* London: Routledge.

Morley, D. (1992) Where the global meets the local: Notes from the sitting room. In: D. Morley *Television Audiences and Cultural Studies.* London: Routledge.

Morley, D. and Robbins, K. (1995) *Spaces of Identity: Global Media, Electronic Landscapes and Cultural Boundaries.* London: Routledge.

Mormont, M. (1990) Who is rural? Or how to be rural: Towards a sociology of the rural. In: T. Marsden, P. Lowe and S. Whatmore (eds.) *Rural Restructuring.* London: David Fulton.

Morozov, E. (2009a) Iran elections: A Twitter revolution? In: *Washington Post,* June 17 2009

Morozov, E. (2009b) Iran: Downside to the 'Twitter Revolution'. In: *Dissent* 56(4).

Morris, C. and Holloway, L. (2009) Genetic technologies and the transformation of the geographies of UK livestock agriculture. In: *Progress in Human Geography* 33(3), 313–33.

Morris, J. (1986) *The Matter of Wales: Epic Views of a Small Country.* Harmondsworth: Penguin Books.

Morris, J. (1988) *Hong Kong, Xiang Gang,* London: Viking.

Morris, J. (1990) *Hong Kong: Epilogue to an Empire.* London: Penguin Books.

Morris, J. (1991) *Pride Against Prejudice.* London: The Women's Press.

Morris, J. (1992) *O! Canada.* London: Hale.

Mort, F. (1989) The politics of consumption. In: S. Hall and M. Jacques (eds.) *New Times: The Changing Face of Politics in the 1990s.* London: Lawrence and Wishart, 160–72.

Mortimore, M. (1998) *Roots in the African Dust: Sustaining the Drylands.* Cambridge: Cambridge University Press.

Morton, P.A. (2000) *Hybrid Modernities: Architecture and Representation at the 1931 Colonial Exposition, Paris.* Cambridge, MA: MIT Press.

Moser, C. (1993) *Gender Planning and Development.* London: Routledge.

Moss, P. (1999) Autobiographical notes on chronic illness. In: R. Butler and H. Parr (eds.) *Mind and Body Spaces: Geographies of Illness, Impairment and Disability.* London: Routledge, 155–66.

Moss, P. (2001) *Placing Autobiography in Geography.* Syracuse: Syracuse University Press.

Moss, P. (ed.) (2002) *Feminist Geography in Practice: Research and Methods.* Oxford: Blackwell.

Mountz, A. (2003) Human smuggling, the transnational imaginary, and everyday geographies of the nation-state. In: *Antipode,* 622–44.

Mountz, A. (2010) *Seeking Asylum: Human Smuggling and Bureaucracy at the Border.* Minneapolis: University of Minnesota Press.

Mowforth, M. and Munt, I. (2003) *Tourism and Sustainability: Development and New Tourism in the Third World.* London: Routledge.

Mulligan, A.N. (2008) Countering exclusion: the 'St. Pats for all' parade. In: *Gender, Place and Culture* 15(2), 153–67.

Murakami Wood, D. (2012) *The Watched World: Globalization and Surveillance.* Lanham, MD: Rowman and Littlefield.

Murdoch, J. (2003a) Co-constructing the countryside: hybrid networks and the extensive self. In P. Cloke (ed.) *Country Visions.* London: Pearson, 263–82.

Murdoch, J. (2003b) Geography's circle of concern. In: *Geoforum* 34, 287–89.

Murdoch, J. and Pratt, A. (1993) Rural studies: Modernism, postmodernism and the 'post-rural'. In: *Journal of Rural Studies* 9, 411–28.

Murdoch, J. and Pratt, A. (1997) From the power of topography to the topography of power: A discourse in strange ruralities. In: P. Cloke and J. Little (eds.) *Contested Countryside Cultures.* London: Routledge.

Murdoch, K., Lowe, P., Ward, N. and Marsden, T. (2003) *The Differentiated Countryside*. London: Routledge.

Murgatroyd, L. and Neuburger, H. (1997) A household satellite account for the UK. In: *Economic Trends* 527, October, 63–71.

Murray, W.E. (2006) *Geographies of Globalization.* Abingdon: Routledge.

Myerson, G. and Rydin, Y. (1996) *The Language of Environment: A New Rhetoric*. London: UCL Press.

MyTravelGuide.com (2004) Kathmandu Valley problems, http://www.mytravelguide.com/city-guide/Asia/Nepal/Kathmandu-Valley-problems, accessed 27 July 2004.

Nadeem, S. (2009) Macaulay's (cyber) children: The cultural politics of outsourcing in India. In: *Cultural Sociology* 3(1), 102–22.

Nairn, T. (1977) *The Break-up of Britain*. London: New Left Books.

Nairn, T. (1995) Breakwaters of 2000: From ethnic to civic nationalism. In: *New Left Review* 214, 91–103.

Nandy, A. (1984) Culture, state and rediscovery of Indian politics. In: *Economic and Political Weekly* 19(49), 2078–83.

Nasar, J., Hecht, P. and Wener, R. (2007) 'Call if you have trouble': Mobile phones and safety among college students. In: *International Journal of Urban and Regional Research* 31(4), 863–73.

Nash, C. (1996) Reclaiming vision: Looking at landscape and the body. In: *Gender, Place and Culture: A Journal of Feminist Geography*, 3 149–69.

Nash, C. (2000) Performativity in practice: Some recent work in Cultural Geography. In: *Progress in Human Geography* 24(4), 653–64.

Nast, H.J. (2000) Mapping the 'unconscious': Racism and the oedipal family. In: *Annals of the Association of American Geographers* 90, 215–55.

National Portrait Gallery (1996) *David Livingstone and the Victorian Encounter with Africa*. London: National Portrait Gallery.

Nayak, A. and Jeffrey, A. (2011) *Geographical Thought: An Introduction to Ideas in Human Geography*. Harlow: Pearson.

Naylor S. and Ryan J. (2002) The mosque in the suburbs: Negotiating religion and ethnicity in South London. In: *Social and Cultural Geography* 3(1), 39–59.

Neal, S. and Agyman, J. (2006) (eds.) *The New Countryside? Ethnicity, Nation and Exclusion in Contemporary Rural Britain.* Bristol: Polity Press.

Nepal, S. (1997) Sustainable tourism, protected areas and livelihood needs of local communities in developing countries. In: *The International Journal of Sustainable Development and World Ecology* 4, 123–35.

Neumayer, E. (2004) *Weak Versus Strong Sustainability: Exploring The Limits of Two Opposing Paradigms.* Cheltenham: Edward Elgar.

Newby, H. (1988) *The Countryside in Question.* London: Hutchinson.

Newell, P. and Paterson, M. (2010) *Climate Capitalism: Global Warming and the Transformation of the Global Economy.* Cambridge: Cambridge University Press.

Newhouse, J. (1997) Europe's rising regionalism. In: *Foreign Affairs* 76(1), 67–84.

Newman, D. (1989) Civilian and military presence as strategies of territorial control: The Arab–Israeli conflict. In: *Political Geography* 8(3), 215–28.

Newman, J. and Clarke, J. (2009) *Publics, Politics and Power: Remaking the Public in Public Services.* London: Sage.

Newson, J. and Newson, E. (1974) Cultural aspects of childrearing in the English-speaking world. In: M.P.M. Richards (ed.) *The Integration of a Child into a Social World.* Cambridge: Cambridge University Press.

Niessen, S., Leshkowich, A.M. and Jones, C. (2003) *Re-orienting Fashion: The Globalisation of Asian Dress.* Oxford, Berg.

Nietschmann, B. (1995) Defending the Miskito Reefs with maps and GPS: Mapping with sail, scuba, and satellite. In: *Cultural Survival Quarterly* 18(4), 34–7.

Nightingale, A. (2003) A feminist in the forest: Situated knowledges and mixing methods in natural resource management. In: *ACME: An*

International E-Journal for Critical Geographies 2(1), 77–90.

Nijman, J. (1994) Nicholas Spykman. In: J. O'Loughlin (ed.) *Dictionary of Geopolitics*. Westport, CT: Greenwood Press.

Nijman, J. (1996) Ethnicity, class and the economic internationalization of Miami. In: J. O'Loughlin and J. Friedrichs (eds.) *Social Polarization in Post-industrial Metropolises*. Berlin and New York: de Gruyter.

Nkrumah, K. (1965) Neo-colonialism: The Last Stage of Imperialism. London: Nelson.

Noble, G. (2004) Accumulating being. In: International Journal of Cultural Studies, 7(2), 233–56.

Noble, G. (2008) Living with things: consumption, material culture and everyday life. In: N. Anderson and K. Schlunke (eds.) Cultural Theory in Everyday Practice, 98–113. Melbourne: Oxford University Press.

Noddings, N. (1984) Caring: A Feminine Approach to Ethics and Moral Education. Berkeley: University of California Press.

Nora, P. (1989) Between memory and history: *Les Lieux de Mémoire*. In: *Representations* 26, 7–25.

Nora, P. (ed.) (1996) *Realms of Memory: The Construction of the French Past. Volume I: Conflicts and Divisions*. New York: Columbia University Press.

Nora, P. (ed.) (1997) *Realms of Memory: The Construction of the French Past. Volume II: Traditions*. New York: Columbia University Press.

Nora, P. (ed.) (1998) *Realms of Memory: The Construction of the French Past. Volume III: Symbols*. New York: Columbia University Press.

Nordstrom, C. (2004) *Shadows of War*. Berkeley: University of California Press.

Norris, C. (1992) *Uncritical Theory: Postmodernism, Intellectuals and the Gulf War*. Amherst, MA: University of Massachusetts Press.

Norris, C. and Armstrong, M. (1999) *The Maximum Surveillance Society: The Rise of CCTV*. Oxford: Berg.

Northern Ireland Life and Times Survey, (2010) ARK. Northern Ireland Life and Times Survey: ARK. www.ark.ac.uk/nilt

Notes from Nowhere (eds.) (2003) *We are everywhere. The Irresistable Rise of Global Anticapitalism*. London: Verso.

Nuttall, M. (2010) *Pipeline Dreams: People, Environment, and the Arctic Energy Frontier*. Copenhagen: International Work Group for Indigenous Affairs.

Nye, D. (1991) The emergence of photographic discourse: images and consumption. In: D. Nye and C. Pedersen (eds.) *Consumption and American Culture*. Amsterdam: VU University Press.

Nyerges, T., Couclelis, H. and McMaster, R. (2011) *The Sage Handbook of GIS and Society*. London: Sage.

Oakley, A. (1981) *Subject Women*. London: Martin Robertson.

Obama, B. (2010) US President Barak Obama's remarks at Millennium Development Goals Summit, 22 September 2010. http://www.america.gov/st/texttrans-english/2010/September/20100922172556su0.2969934.html, accessed 15 September 2011.

Obermeyer, N. (1998) The evolution of public participation GIS. In: *Cartography and Geographic Information Systems* 25, 65–6.

O'Brien, R. (1991) *Global Financial Integration: The End of Geography?* London: Pinter.

O'Dell, T. (2010) *Spas. The Cultural Economy of Hospitality, Magic and the Senses*. Lund: Nordic Academic Press.

ODT Inc. (2008) *Arno Peters: Radical Map, Remarkable Man*. DVD. 30 mins. USA: ODT Maps.

Offe, K. and Heinze, R. (1992) *Beyond Employment*. London: Polity Press.

Office for National Statistics (2000) *Social Trends 30*. London: The Stationery Office.

Office for National Statistics (2004) *Social Trends 34*. London: The Stationery Office.

Office for National Statistics (2011) *Life expectancy at birth and at age 65 by local areas in the United Kingdom, 2004–06 to 2008–10*. Statistical Bulletin, 19 October 2011. Accessed: http://www.ons.gov.uk/ons/dcp171778_238743.pdf

Ogborn, M. (1998) *Spaces of Modernity: London's Geographies 1680–1780*. New York: Guilford Press.

Oldfield, J.D and Shaw, D.J.B. (2002) Revisiting sustainable development: Russian cultural and scientific traditions and the concept of sustainable development. In: *Area* 34: 391–400.

Oldfield, J.D. and Shaw, D.J.B. (2006) V.I. Vernadsky and the noosphere concept: Russian understandings of society-nature interaction. In: *Geoforum* 37: 145–54.

Olds, J. and Schwartz, R.S. (2009) *The Lonely American: Drifting Apart in the Twenty-first Century*. Boston, MA: Beacon.

Olds, K. (2004) 'Classics in Human Geography: Peter Dicken's *Global Shift* (1986)'. In: *Progress in Human Geography* 28(4), 507–11.

Olds, K. and Thrift, N. (2005) 'Cultures on the brink: reengineering the soul of capitalism – on a global scale'. In: A. Ong and S. Collier (eds.) *Global Assemblages: Technology, Politics and Ethics as Anthropological Problems*. Oxford: Blackwell, 270–90.

Olds, K. (2012) 'Unsettling the university-territory relationship via Applied Sciences NYC'. In: *Inside Higher Ed*, http://bit.ly/yhuCIj (accessed 1 July 2012).

Ollman, B. (1972) *Alienation*. Cambridge: Cambridge University Press.

O'Loughlin, J. (2004) The political geography of conflict: civil wars in the hegemonic shadow. In C. Flint (ed.) *The Geography of War and Peace*. New York: Oxford University Press.

O'Loughlin, J. and Friedrichs, J. (eds.) (1996) *Social Polarization in Post-industrial Metropolises*. Berlin and New York: De Gruyter.

O'Loughlin, J. and van der Wusten, H. (1993) Political geography of war and peace. In: P.J. Taylor (ed.) *Political Geography of the Twentieth Century*. London: Belhaven Press.

O'Loughlin, J. (2005) The political geography of conflict: civil wars in the hegemonic shadow. In: C. Flint (ed.) *The Geography of War and Peace*. New York: Oxford University Press.

Olwig, K. (2002) *Landscape, Nature and the Body Politic*. Madison: University of Wisconsin Press.

Omagh Bomb Memorial Working Group (2007) *Design Competition: Project Brief*. Omagh: Omagh District Council.

O'Neill, J. (2001) *Building Better Global Economic BRICs*. New York: Goldman Sachs.

Ong, A. (1999) *Flexible Citizenship: The Cultural Logics of Transnationality*. Durham, NC: Duke University Press.

Openshaw, S. (1991) A view on the GIS crisis in geography, or, using GIS to put Humpty-Dumpty together again. In: *Environment and Planning A* 23, 621–8.

Openshaw, S. (1992) Further thoughts on geography and GIS: a reply. In: *Environment and Planning A* 24, 463–6.

Oppenheim, C. (1990) *Poverty: The Facts*. London: Child Poverty Action Group.

Oppong, J. (2009) Communicable diseases, Globalization of. In: R. Kitchin and N. Thrift (eds.) *International Encyclopedia of Human Geography*. Oxford: Elsevier, 209–13.

Organization for Economic Cooperation and Development (OECD) (2005) *Glossary of Statistical Terms* http://stats.oecd.org/glossary/detail.asp?ID=6588) Accessed 27 April 2012.

Organization for Economic Cooperation and Development (OECD) (2009) *International Trade in ICT Goods and Services*. Geneva: OECD.

Organization for Economic Cooperation and Development (OECD) (2011) *Glossary of Statistical Terms: Natural Resources*. Available at: http://stats.oecd.org/glossary/detail.asp?ID=1740

O'Riordan, T. (1976) *Environmentalism*. London: Pion.

O'Riordan, T. (1988) The politics of sustainability. In: R.K. Turner (ed.) *Sustainable Environmental Management*. London: Belhaven Press.

O'Riordan, T. (1999) *Environmental Science for Environmental Management*. Harlow: Longman.

O'Riordan, T. and Jordan, A, (1998) Kyoto in Perspective. *ECOS* 18(314), 38–42.

O'Riordan, T. and Voisey, H. (eds) (1998) *The Transition to Sustainability*. London: Earthscan.

Osborne, B.S. (1998) Constructing landscapes of power: the George Etienne Cartier monument, Montreal. In: *Journal of Historical Geography* 24, 431–58.

Osborne, P. (1996) Modernity. In: M. Payne (ed.) *A Dictionary of Cultural and Critical Theory*. Oxford: Blackwell.

O'Sullivan, D. (2006) Geographical information science: Critical GIS In: *Progress in Human Geography* 30, 783–91.

Oswin, N. (2008) Critical geographies and the uses of sexuality: Deconstructing queer space. *Progress in Human Geography* 32(1), 89–103.

Oswin, N. (2010) The modern model family at home in Singapore: A queer geography. In: *Transactions of the Institute of British Geographers* 35(2), 256–68.

Ó Tuathail, G. (1996) *Critical Geopolitics*. Minneapolis: Minnesota University Press.

Ó Tuathail, G. (2003) 'Just out looking for a fight': American affect and the invasion of Iraq. In: *Antipode* 35(5), 856–70.

Ó Tuathail, G. and Agnew, J. (1992) Geopolitics and discourse: Practical geopolitical reasoning in American foreign policy. In: *Political Geography* 11(2), 190–204.

Ou-fan Lee, L. (1999) *Shanghai Modern: The Flowering of a New Urban Culture in China, 1930–1945*. Cambridge, MA: Harvard University Press.

Overton, J. (2010) The consumption of space: Land, capital and place in the New Zealand wine industry. In: *Geoforum* 41(5), 752–62.

Overton, M. (1994) Historical geography. In: R.J. Johnston, D. Gregory and D.M. Smith (eds.) *The Dictionary of Human Geography* (3rd edn.). Oxford: Blackwell.

Owens, S. (1997) Negotiated environments: Needs, demands, and values in the age of sustainability. In: *Environment and Planning A* 29, 571–80.

Oxfam (2004) *Trading Away Our Rights: Women Working in Global Supply Chains*. London: Oxfam.

Pacione, M. (1999) Relevance in Human Geography: Special collection of invited papers. In: *Scottish Geographical Journal*, 115(2).

Pain, R. (1997) Social geographies of women's fear of crime. In: *Transactions of the Institute of British Geographers* 22, 231–44.

Pain, R (2001a) Age, generation and lifecourse. In: R. Pain, M. Barke, D, Fuller, J. Gough, R. McFarlane and G. Mowl (eds.) *Introducing Social Geographies*. London: Arnold, 141–63.

Pain, R. (2001b) Gender, race, age and fear in the city. In: *Urban Studies* 38, 899–913.

Pain, R. (2003) 'Social geography: On action-oriented research'. In: *Progress in Human Geography* 27, 649–57.

Pain, R. (2004) Social geography: Participatory research. In: *Progress in Human Geography* 28(5), 652–63.

Pain, R. (2006) Paranoid parenting? Rematerializing risk and fear for children. In: *Social and Cultural Geography* 7(2), 221–43.

Pain, R. (2009) Globalized fear? Towards an emotional geopolitics. In: *Progress in Human Geography* 33(4), 466–86.

Pain, R., Mowl, G. and Talbot, C. (2000) Difference and the negotiation of 'old age'. *Environment and Planning D: Society and Space*, 18(3), 377–94.

Pain, R., Barke, M., Fuller, D., Gough, J., Macfarlane, R. and Mowl, G. (2001) *Introducing Social Geographies*. London: Hodder.

Pain, R., Grundy, S., Gill, S., Towner, E., Sparks, G. and Hughes, K (2005) 'So long as I take my mobile': Mobile phones, urban life and geographies of young people's safety. In: *International Journal of Urban and Regional Research* 29(4), 814–30.

Painter, J. and Philo, C. (1995) Spaces of citizenship: An introduction. In: *Political Geography* 14(2), 107–20.

Palumbi, S. (2001) Humans as the world's greatest evolutionary force. In: *Science* 293 (7 September), 1786–90.

Panakagos, A.N. and Horst, H.A. (eds.) (2006) Return to cyberia: Technology and the social worlds of transnational migrants. In: *Global Networks*, Special Issue, 6(2), 109–220.

Pan-American Health Organization (PAHO) (2002) *Health in the Americas. Volume II, 2002 Edition*. Washington, DC: PAHO.

Panell, R. and Tipa, G. (2007) Placing well-being: A Maori case study of cultural and environmental specificity. In: *EcoHealth*, 4, 445–60.

Panelli, R. (2004) *Social Geographies*. London: Sage.

Panizza, F. (2009) *Contemporary Latin America: Development and Democracy Beyond the Washington Consensus*. London: Zed Press.

Papadimitriou, F. (2006) A geography of 'Notopia'. In: *City* 10(3), 317–26.

Papatheodorou, A. (2003) Corporate strategies of British tour-operators in the Mediterranean region: An economic geography approach. In: *Tourism Geographies* 5, 280–304.

Papoulias, C. and Callard, F. (2010) Biology's gift: Interrogating the turn to affect. In: *Body and Society* 16(1), 29–56.

Park, C. (1997) *The Environment: Principles and Applications*. London: Routledge.

Park, R. (1926) The urban community as a spatial pattern and a moral order. In E.W. Burgess (ed.) *The Urban Community*. Chicago, IL: University of Chicago Press, 3–18.

Parker, G. (2002) *Citizenships, Contingency and the Countryside*. London: Routledge.

Parker, J. and Smith, C. (1940) *Modern Turkey*. London: George Routledge & Sons.

Parker, K., Parker, A. and Vale, T. (2001) Vertebrate feeding guilds in California's Sierra Nevada: Relations to environmental condition and change in spatial scale. In: *Annals of the Association of American Geographers* 91(2), 245–62.

Parr, H. (1998) The politics of methodology in 'post-medical geography': mental health research and the interview. In: *Health and Place* 4(4), 341–53.

Parr, H. (1999) Delusional geographies: the experiential worlds of people during madness/illness. *Environment and Planning D: Society and Space* 17, 673–90.

Parr, H. (2000) Interpreting the 'hidden social geographies' of mental health: ethnographies of inclusion and exclusion in semi-institutional places. In: *Health and Place* 6, 225–37.

Parr, H. (2007) Mental health, nature work, and social inclusion. In: *Environment and Planning D* 25, 537–61.

Parr, H. (2008) *Mental Health and Social Space: Towards Inclusionary Geographies?* Oxford: Wiley-Blackwell.

Parr, H. and Butler, R. (1999) New geographies of illness, impairment and disability. In R. Butler and H. Parr (eds) *Mind and Body Spaces: Geographies of Illness, Impairment and Disability*. London: Routledge, 1–24.

Parr, H. and Davidson, J. (2011) Psychic life. In: V. Del Casino, M. Thomas, R. Panelli and P. Cloke (eds.) *A Companion to Social Geography*. Oxford: Blackwell Press.

Parr, H., Philo, C. and Burns, N. (2004) Social geographies of rural mental health: Experiencing inclusion and exclusion. In *Transactions of the IBG* 29(4), 401–19.

Pasura, D. (2010) Competing meanings of the diaspora: The case of Zimbabweans in Britain. In: *Journal of Ethnic and Migration Studies* 36(9), 1445–61.

Paterson, M. (1996) *Global Warming and Global Politics*. London: Routledge.

Paterson, M. (2007) *The Senses of Touch: Haptics, Affects and Technologies*, Oxford: Berg.

Paterson, M. (2009) Haptic geographies: Ethnography, haptic knowledges and sensuous dispositions. In: *Progress in Human Geography* 33(6), 766–88.

Paterson, M. (2011) More-than-visual approaches to architecture: Vision, touch, technique. In: *Social and Cultural Geography* 12(03), 263–81.

Paterson, M. and Stripple, J. (2010) My space: Governing individuals' carbon emissions. In: *Environment and Planning D: Society and Space* 28: 341–62.

Paul, B.K. (1994) Commentary on Kearns's 'Place and health: Toward a reformed medical geography'. In: *Professional Geographer* 46, 504–5.

Paulson, S. and Gezon, L. (2004) *Political Ecology Across Spaces, Scales, and Social Groups*. New Brunswick, NJ: Rutgers University Press.

Pavlinek, P. and Pickles, J. (2000) *Environmental Transitions: Transformations and Ecological Defence in Central and Eastern Europe*. London: Routledge.

Peach, C. (ed.) (1975) *Urban Social Segregation*. London: Longman.

Peach, C. (1996) Does Britain have ghettoes? In: *Transactions of the Institute of British Geographers* 21(1), 216–35.

Peake, L. (1993) Race and sexuality: Challenging the patriarchal structuring of urban social space. In: *Environment and Planning D: Society and Space* 11, 415–32.

Peake, L. (2008) Moving on up. In: *Gender, Place and Culture* 15(1), 7–10.

Pearce, D., Markandya, A. and Barbier, E. (1989) *Blueprint for a Green Economy*. London: Earthscan.

Pearce, D. (1993) *Blueprint Three: Measuring Sustainable Development*. Earthscan: London.

Pearce, J. and Witten, K. (eds.) (2010) *Geographies of Obesity: Environmental Understandings of the Obesity Epidemic*. Aldershot: Ashgate.

Pearson, M. (1989) Medical geography: Genderless and colorblind. In: *Contemporary Issues in Geography and Education* 3, 9–17.

Peck, J. (2001) Neoliberalising states: thin policies/hard outcomes. In: *Progress in Human Geography* 25, 445–55.

Peck, J. and Tickell, A. (2002) Neoliberalizing space. In: *Antipode* 34, 380–404.

Peck, J., Barnes, T.J., Sheppard, E. and Tickell, A. (eds.) (2003) *Reading Economic Geography*. Oxford: Blackwell.

Peckham, R.S. (2003) *Rethinking Heritage: Culture and Politics in Europe*. London: I.B. Taurus.

Peet, R. (1998) *Modern Geographical Thought*. Oxford: Blackwell.

Peet, R. (1989) World capitalism and the destruction of regional cultures. In: R.J. Johnston and P. Taylor (eds.) *The World in Crisis?* Oxford: Blackwell.

Peet, R. (with E. Hartwick) (1999) *Theories of Development*. London: Guilford Press.

Peet, R. and Thrift, N. (1989) Political economy and Human Geography. In: R. Peet and N. Thrift (eds.) *New Models in Geography: Volume 1. The Political-economy Perspective*. London: Unwin Hyman.

Peet, R. and Watts, M. (eds.) (1996/2004) *Liberation Ecologies: Environment, Development, Social Movements*. London: Routledge.

Peluso, N. (1995) Whose woods are these? Counter-mapping forest territories in Kalimantan, Indonesia. In: *Antipode* 27, 383–406.

Pepper, D. (1984) *The Roots of Modern Environmentalism*. London: Routledge.

Pepper, D. (1996) *Modern Environmentalism: An Introduction*. London: Routledge.

Perks, R. and Thomson, A. (eds.) (1998). *The Oral History Reader*. London: Routledge.

Perrons, D. (2004) *Globalization and Social Change: People and Places in a Divided World*. Abingdon: Routledge.

Pettersson, A. and Malmberg, G. (2009) Adult children and elderly parents as mobility attractions in Sweden. In: *Population, Space and Society* 15, 343–57.

Pfaff, J. (2010a) A mobile phone: mobility, materiality and everyday Swahili trading practices. In: *Cultural Geographies* 17, 341–57.

Pfaff, J. (2010b) Mobile phone geographies. In: *Geography Compass* 4, 1433–47

Phillips, J. (1996) *A Man's Country? The Image of the Pakeha Male, a History*. Auckland: Penguin.

Phillips, R. (2006) *Sex, Politics and Empire: A Postcolonial Geography*. Manchester: Manchester University Press.

Phillips, R. (2009a) 'Bridging East and West: Muslim-identified activists and organizations in the UK anti-war movements'. In: *Transactions of the Institute of British Geographers* 34, 506–20.

Phillips, R. (2009b) Settler colonialism and the nuclear family. In: *Canadian Geographer* 53(2), 239–53.

Phillips, R. (2011) Vernacular anti-imperialism. In: *Annals of the Association of American Geographers* 101, 1109–25.

Phillips, R.S. (1997) *Mapping Men and Empire: A Geography of Empire*. London: Routledge.

Philo, C. (1992) Neglected rural geographies: A review. In: *Journal of Rural Studies* 8, 193–207.

Philo, C. (1995) Animals, geography, and the city: Notes on inclusions and exclusions. In: *Environment and Planning D: Society and Space* 13(6), 644–81.

Philo, C. (1997) Of other rurals. In: P. Cloke and J. Little (eds.) *Contested Countryside Cultures: Otherness, Marginality and Rurality*. London: Routledge.

Philo, C. (1998) A 'lyffe in pyttes and caves': Exclusionary geographies of the west country tinners. In: *Geoforum* 29(2), 159–72.

Philo, C. (2000) Social exclusion. In: R.J. Johnston, D. Gregory, G., Pratt and M. Watts (eds.) *The Dictionary of Human Geography*. Oxford: Blackwell, 751–2.

Philo, C. (2003) To go back up the side hill:

Memories, imaginations and reveries of childhood. In: *Children's Geographies* 1, 3–24.
Philo, C and Wilbert, C. (eds.) (2000) *Animal Spaces, Beastly Places: New Geographies of Human-animal Relations*. London: Routledge.
Philo, G. (1993) From Buerk to Band Aid: the media and the 1984 Ethiopian famine. In: J. Eldridge (ed.) *Getting the Message: News, Truth and Power*. London: Routledge.
Picard, M. (2009) From 'Kebalian' to 'Ajeg Bali': Tourism and Balinese identity in the aftermath of the Kuta bombing. In: M. Hitchcock, V.T. King and M. Parnwell (eds.) *Tourism in Southeast Asia: Challenges and New Directions*. Copenhagen: NIAS Press.
Pickens, W. (1991/1923) *Bursting Bonds*. Bloomington, IN: Indiana University Press.
Pickles, J. (ed.) (1995) *Ground Truth: The Social Implications of Geographical Information Systems*. New York: Guilford.
Pickles, J. (2003) *A History of Spaces: Cartographic Reason, Mapping and the Geo-coded World*. London and New York: Routledge.
Pickles, J. (ed.) (1995) *Ground Truth*. New York: Guildford Press.
Pietz, W. (1988) The 'post-colonialism' of Cold War discourse. In: *Social Text* 19/20(Fall), 55–75.
Pike Research (2011) *Biofuels Markets and Technologies*. Pike Research: Washington.
Pile, S. (1991) Practising interpretative human geography. In: *Transactions of the Institute of British Geographers* 16, 458–69.
Pile, S. (1995) 'What we are asking for is decent human life.' SPLASH, neighbourhood demands and citizenship in London's docklands. In: *Political Geography* 14, 199–208.
Pile, S. (1996) *The Body and the City: Psychoanalysis, Space, and Subjectivity*. London: Routledge.
Pile, S. (2005) *Real Cities: Modernity, Space and the Phantasmagoras of City Life*. London: Sage.
Pile, S. (2010) Emotions and affect in recent human geography. In: *Transactions of the Institute of British Geographers* 35(1), 5–20.
Pile, S. (2011) Reply: For a geographical understanding of affect and emotions. In: *Transactions of the Institute of British Geographers*, 4, 603–6.
Pile, S. and Thrift, N. (1995) *Mapping the Subject: Geographies of Cultural Transformation*. London: Routledge.
Pinckney, D.H. (1958) *Napoleon III and the Rebuilding of Paris*. Princeton, NJ: Princeton University Press.
Pinder, D. (1996) Subverting cartography: the situationists and maps of the city. In: *Environment and Planning A* 28, 405–27.
Pinder, D. (2007) Cartographies unbound. In: *Cultural Geographies* 14, 453–62.
Pink, S. (2007) Sensing Cittàslow: Slow living and the constitution of the sensory city. In: *Senses and society* 2(1), 59–77.
Pivar, M., Fredkin, E. and Stommel, H. (1963) Computer-compiled oceanographic atlas: An experiment in man-machine interaction. In: *Proceedings of the National Academy of Sciences of the United States of America*, 50(2), 396–98.
Platteau, J.-P. (1994) Behind the market stage where real societies exist: Parts I and II. In: *Journal of Development Studies* 30, 533–77 and 753–817.
Ploszajska, T. (2000) Historiographies of geography and empire. In: B. Graham and C. Nash (eds.) *Modern Historical Geographies*. London: Prentice Hall.
Pocock, D. (1993) The senses in focus. In: *Area* 25(1), 11–16.
Podmore, J.A. (2001) Lesbians in the crowd: Gender, sexuality and visibility along Montréal's Boul. St-Laurent. In: *Gender, Place and Culture* 8, 333–55.
Podmore J.A. (2006) Gone 'underground'? Lesbian visibility and the consolidation of queer space in Montreal. In: *Social and Cultural Geography* 7(4), 595–625.
Polanyi, K, (1947) *The Great Transformation*. Boston: Beacon Books.
Pollard, J.S. (1996) Banking at the margins: A geography of financial exclusion in Los Angeles. In: *Environment and Planning A* 28, 1209–32.
Pollock, G. (1988) Modernity and the spaces of femininity. In: G. Pollock (ed.) *Vision and Difference: Femininity, Feminism and the Histories of Art*. London: Routledge, 50–90.
Ponniah, T. (2004) Democracy vs empire:

Alternatives to globalisation presented at the World Social Forum. In: *Antipode* 36, 130–3.

Pons, P.O., Crang, M. and Travlou, P. (eds.) (2009) *Cultures of Mass Tourism: Doing the Mediterranean in the Age of Banal Mobilities.* Farnham: Ashgate.

Pontalis, J.-B. (1990) *La Force d'Attraction.* Paris: Seuil.

Ponting, C. (1994) *Churchill.* London: Sinclair-Stevenson.

Poole, P. (1995) Indigenous peoples, mapping and biodiversity conservation: An analysis of current activities and opportunities for applying geomatics technologies. In: *Biodiversity Support Program Discussion Paper Series.* Washington, DC: WWF, The Nature Conservancy, World Resources Institute.

Popke, E.J. (2006) Geography and ethics: everyday mediations through care and consumption. In: *Progress in Human Geography* 30, 504–12.

Popke, E.J. (2008a) Geography and ethics: non-representational encounters, collective responsibility and economic difference. In: *Progress in Human Geography* 32, 1–10.

Popke, E.J. (2008b) Spaces of Being in-common: ethics and Social Geography. In: *The Handbook of Social Geography.* S. Smith, R. Pain, S.A. Marson and J.P. Jones III (eds.) London: Sage, 435–54.

Porritt, J. (ed.) (1991) *Save the Earth.* London: Dorling Kindersley.

Porteous, J.D. (1985) Smellscape. *Progress in Human Geography* 9(3), 358–9.

Postone, M. (1993) *Time, Labour and Social Domination.* Cambridge: Cambridge University Press.

Potter, R., Binns, T., Elliott, J. and Smith, D. (2008) *Geographies of Development.* Harlow: Pearson Education.

Power, E. (2008) Furry families: Making a human-dog family through home. In: *Social and Cultural Geography* 9, 535–55.

Power, E. (2009) Border-processes and homemaking: Encounters with possums in suburban Australian homes. In: *Cultural Geographies* 16(1), 29–54.

Power, M. (2003) *Rethinking Development Geographies.* London: Routledge.

Pratt, G. (2004) *Working Feminism.* Edinburgh: Edinburgh University Press; and Philadelphia: Temple University Press.

Pratt, G. (2012) *Families Apart: Migrant Mothers and the Conflicts of Labor and Love.* Minneapolis: University of Minnesota Press.

Pratt, G. and Hanson, S. (1994) Geography and the construction of difference. In: *Gender, Place and Culture* 1, 5–29.

Pratt, G. and Kirby, E. (2003) Performing nursing: BC Nurses' Union Theatre Project. In: *ACME* 2, 14–32.

Pratt, J., Leyshon, A. and Thrift, N.J. (1996) Financial exclusion in the 1990s II: Geographies of financial inclusion and exclusion. In: *Working Papers on Producer Services* 38.

Pratt, M.L. (1986) Scratches in the face of the country; Or, what Mr Barrow saw in the lands of the Bushmen. In: H.L. Gates Jr. (ed.) *'Race', Writing and Difference.* Chicago, IL: Chicago University Press, 138–62.

Pratt, M.L. (1992) *Imperial Eyes: Travel Writing and Transculturation.* London: Routledge.

Pred, A. (1990) *Lost Words and Lost Worlds: Modernity and the Language of Everyday Life in Late Nineteenth-century Stockholm.* Cambridge: Cambridge University Press.

Pred, A. and Watts, M. (1992) *Reworking Modernity.* New Brunswick: Rutgers University Press.

Prendergast, K. and Nash, C. (2002) Mapping emotion again. In: F. Driver, C. Nash, K. Prendergast and P. Swenson (eds.) *Landing: Eight Collaborative Projects Between Artists and Geographers.* Egham: Royal Holloway, University of London.

Preston, P.W. (1996) *Development Theory: An Introduction.* Oxford: Blackwell.

Price, L. (2013) (forthcoming) Knitting the city. In: *Geography Compass.*

Price, M. (1996) Taming the tourists. In: *People and the Planet*, 5, http://www.oneworld.org/patp/pap_info.html, accessed 4 April 2004.

Price Waterhouse Coopers (2010) *The World in 2050.* London: Price Waterhouse Coopers.

Pringle, R. (1999) Emotions. In L. McDowell and

J. Sharp (eds.) *A Feminist Glossary of Human Geography*. London: Arnold, 68–9.
Probyn, E. (1993) *Sexing the Self: Gendered Positions in Cultural Studies*. London: Routledge.
Probyn, E. (2011) Eating roo: Of things that become food. In: *New Formations* 74, 33–45.
Pulido, L. (2000) 'Rethinking environmental racism: White privilege and urban development in southern California'. In: *Annals of the Association of American Geographers* 90, 12–40.
Punch, S. (2001) Household division of labour: Generation, gender, age, birth order and sibling composition. In: *Work, Employment and Society* 15, 803–23.
Putnam, R. (2000) *Bowling Alone: The Collapse and Revival of American Community*. New York: Simon & Schuster.
Pykett, J. (2009) Making citizens in the classroom: An urban geography of citizenship education? In: *Urban Studies* 46: 803–23.
Pyle, G.F. (1969) The diffusion of cholera in the United States in the nineteenth century. In: *Geographical Analysis* 1, 59–75.
Raban, J. (1990) *Hunting Mr Heartbreak*. London: Pan Books.
Radcliffe, S. and Westwood, S. (eds.) (1993) *Viva: Women and Popular Protest in Latin America*. London: Routledge.
Radcliffe, S.A. (1990) Ethnicity, patriarchy and incorporation into the nation: Female migrants as domestic servants in Peru. In: *Environment and Planning D: Society and Space* 8, 379–93.
Radin, M. (1996) *Contested Commodities*. Cambridge, MA: Harvard University Press.
Radway, J. (1984) *Reading the Romance: Women, Patriarchy, and Popular Literature*. Chapel Hill, NC: University of North Carolina Press.
Rambaldi, G., Muchemi, J. Crawhall, N. and Monaci, L. (2007) Through the eyes of hunter-gatherers: Participatory 3D modelling among Ogiek indigenous peoples in Kenya. In: *Information Development*, 23.
Ratliff, E. (2007) The whole earth, catalogued. How Google Maps is changing the way we see the world. In: *Wired* 15, 154–9.
Rawcliffe, P. (1998) *Swimming with the Tide: Environmental Groups in Transition*. Manchester: Manchester University Press.
Razack, S. (2000) Gendered racial violence and spatialized justice: The murder of Pamela George. In: *Canadian Journal of Law and Society* 15, 91–130.
Redclift, M. (1984) *Development and the Environmental Crisis: Red or Green Alternatives?* London: Methuen.
Redclift, M. (1987) *Sustainable Development: Exploring the Contradictions*. London: Methuen.
Redclift, M. (2004) *Chewing Gum. The Fortunes of Taste*. New York: Routledge.
Redclift, M. (2005) Sustainable development (1987–2005): An oxymoron comes of age. In: *Sustainable Development* 13: 212–27.
Redclift, M. (2010). Frontier spaces of production and consumption: Surfaces, appearances and representations on the 'Mayan Riviera'. In: M.K. Goodman, D. Goodman and M. Redclift (eds.) *Consuming Space. Placing Consumption in Perspective*, 81–95. Farnham and Burlington: Ashgate Press.
Redfield, P. (2005) Doctors, borders, and life in crisis. In: *Cultural Anthropology*, 328–61.
Redfield, P. (2013) *Life in Crisis: The Ethical Journey of Doctors Without Borders*. Berkeley: University of California Press.
Rediscovering Geography Committee, National Research Council (1997) *Rediscovering Geography: New Relevance for Science and Society*. Washington: National Academy Press.
Reed, H.C. (1981) *The Pre-eminence of International Financial Centers*. New York: Praeger.
Reher, D. (2004) The demographic transition revisited as a global process. In: *Population, Space and Place* 10, 19–41.
Ren, F. and Kwan, M.-P. (2009) The impact of geographic context on e-shopping behavior. In: *Environment and Planning B: Planning and Design* 36, 262–78.
Rengert, G., Mattson, M. and Kenderson, K. (2001) *Campus Security: Situational Crime Prevention in High-density Environments*. Monsey, NY: Criminal Justice Press.
Rettie, R. (2008) Mobile phones as network capital: Facilitating connections. In: *Mobilities* 3(2), 291–311.

Reuters, 14 June 2011 10:17am GMT. China dismisses US swipe on 'colonial' role in Africa. http://af.reuters.com/articlePrint?articleId=AFJOE75D0BE20110614 (accessed 15 June 2011)

Revill, G. (2000) Music and the politics of sound: Nationalism, citizenship and auditory space. In: *Environment and Planning D: Society and Space* 18, 597–613.

Rheingold, H. (2002) *Smart Mobs. The Next Social Revolution*. Cambridge, MA: Perseus Publishing.

Rhoads, B. and Thorn, C. (eds.) (1996) *The Scientific Nature of Geomorphology*. London: Routledge.

Rhodes, R. (1996) The new governance: Governing without government. In: *Political Studies*, 652–67.

Rich, N. (2012): Jungleland. The Lower Ninth Ward in New Orleans gives new meaning to 'urban growth'. In: *New York Times Magazine*, 25 March.

Richardson, E., Pearce, J., Mitchell, R., Day, P. and Kingham, S. (2010) The association between green space and cause-specific mortality in urban New Zealand: An ecological analysis of green space utility. In: *BioMed Central Public Health* 10(1), 240.

Ricketts, T.C. (2010) Accessing Health Care. In: T. Brown, S. McLafferty and G. Moon (eds.) *A Companion to Health and Medical Geography*. Oxford: Wiley-Blackwell, 521–39.

Rieff, D. (1993) Notes on the Ottoman legacy written in a time of war. In: *Salmagundi* 100, 3–15.

Rigg, J. (2003) *Southeast Asia*. London: Routledge.

Ritzer, G. (2004) *The Globalization of Nothing*, Thousand Oaks, CA; and London: Pine Forge Press.

Rival, L. (1998) *The Social Life of Trees: Anthropological Perspectives on Tree Symbolism*. Oxford: Berg.

Robbins, P. (2007) *Lawn People: How Grass, Weeds and Chemicals Make Us Who We Are*. Philadelphia: Temple University Press.

Robbins, P. (2012) *Political Ecology: A Critical Introduction* (2nd edn.). Oxford and Malden MA: Wiley-Blackwell (First edition: 2004).

Robins, A.G. (2002) Living the simple life: George Clausen at Childwick Green, St. Albans. In: D. Corbett, Y. Holt and F. Russell (eds.) *The Geographies of Englishness: Landscape and the National Past 1880–1940*. New Haven: Yale University Press, 1–28.

Robins, K. (1996) Interrupting identities: Turkey/Europe. In: S. Hall and P. du Gay (eds.) *Questions of Cultural Identity*. London: Sage, 61–86.

Robins, K. and Aksoy, A. (2003) Banal transnationalism: The difference that television makes. In: K.H. Karim (ed.) *The Media of Diaspora: Mapping the Global*. London: Routledge, 89—104.

Robinson, J. (2006) *Ordinary Cities: Between Modernity and Development*. Abingdon: Routledge.

Robinson, N. (ed.) (1993) *Agenda 21: Earth's Action Plan*. New York: Ocean Publications (IUCN Environmental Policy and Law Paper No. 27).

Robson, E. (1996) Working girls and boys: Children's contribution to household survival in West Africa. In: *Geography* 81, 403–7.

Rocheleau, D. (1995) Maps, numbers, text and context: mixing methods in feminist political ecology. In: *Professional Geographer* 47(4), 458–66.

Rocheleau, D., Thomas-Slayter, B. and Edmunds, D. (1995) Gendered resource mapping: Focusing on women's spaces in the landscape. In: *Cultural Survival Quarterly* 18(4), 62–8.

Rocheleau, D., Thomas-Slayter, B and Wangari, E. (1996) *Feminist Political Ecology: Global Issues and Local Experiences*. New York: Routledge.

Rodaway, P. (1994) *Sensuous Geographies: Body, Sense and Place*. London and New York: Routledge.

Rodgers, J. (2003) *Spatializing International Politics. Analysing Activism on the Internet*. London: Routledge.

Rodney, W. (1974) *How Europe Underdeveloped Africa*. Washington, DC: Howard University Press.

Roe, E.J. (2010) Ethics and the non-human: The matterings of sentience in the meat industry. In: B. Anderson and P. Harrison (eds.) *Taking-place: Non-Representational Theories and Geographies*. Farnham: Ashgate, 261–80.

Rogers, A. (2010) The geographies of performing scripted language. In: *Cultural Geographies* 17, 353–75.

Rogoff, I. (2000) *Terra Infirma: Geography's Visual Culture*. London: Routledge.

Rolfe, M. (2010) Poverty tourism: Theoretical reflections and empirical findings regarding an extraordinary form of tourism. In: *GeoJournal* 75, 421–42.

Roller, D.W. (ed.) (2010) *Eratosthenes' Geography*. Princeton, NJ: Princeton University Press.

Rose, D. and Carrasco, P. (2000) The 1996 census as a tool for measuring unpaid household labour: reflections on a controversy and preliminary explorations for the case of the Montreal region. Paper presented at the annual meetings of the Association of American Geographers, Pittsburgh.

Rose, G. (1992) *Feminism and Geography: The Limits of Geographical knowledge*. London: Polity Press.

Rose, G. (1993a) *Feminism and Geography: The Limits of Geographical knowledge*. Cambridge: Polity Press; and Minneapolis: University of Minnesota Press.

Rose, G. (1993b) Looking at landscape: The uneasy pleasures of power. In: G. Rose (ed.) *Feminism and Geography: The Limits of Geographical knowledge*. Cambridge: Polity Press; and Minneapolis: University of Minnesota Press, 86–112.

Rose, G. (1995) Place and identity: A sense of place. In: D. Massey and P. Jess (eds.) *A Place in the World?* Oxford: Oxford University Press.

Rose, G. (1997a) Looking at landscape: The uneasy pleasures of power. In: T. Barnes and D. Gregory (eds.) *Reading Human Geography*. London: Arnold.

Rose, G. (1997b) Situating knowledges: Positionality, reflexivities and other tactics. In: *Progress in Human Geography* 21(3), 305–20.

Rose, G (2002) Conclusion. In: L. Bondi, H. Avis, R. Bankey, A. Bingley, J. Davidson, R. Duffy, V.I. Einagel, A-M. Green, L. Johnston, S. Lilley, C. Listerborn, M. Marshy, S. McEwan, N. O'Connor, G. Rose, B. Vivat and N. Wood (eds.) *Subjectivities, Knowledge and Feminist Geographies*. Oxford: Rowman & Littlefield, 253–8.

Rose, G., Degen, M. and Basdas, B. (2010) More on 'big things': Building events and feelings. In: *Transactions of the Institute of British Geographers* 35, 334–49.

Rose. G. (2011) *Visual Methodologies: An Introduction to Interpreting Visual Materials*. London: Sage.

Rosenberg, M.W. (1988) Linking the geographical, the medical and the political in analyzing health care delivery systems. In: *Social Science & Medicine* 26, 179–86.

Ross, J. (1995) *Rebellion from the Roots*. Monroe, ME: Common Courage Press.

Rostow, W.W. (1960) *The Stages of Economic Growth: A Non-communist Manifesto*. Cambridge: Cambridge University Press.

Rotfeld, A.D. (2002) Introduction: global security after 11 September 2001. In: *SIPRI Yearbook 2002: Armaments, Disarmament, and International Security*. Oxford: Oxford University Press.

Rothenberg, T. (1995) 'And she told two friends': Lesbians creating urban social space. In: D. Bell and G. Valentine (eds.) *Mapping Desire: Geographies of Sexualities*. London: Routledge, 165—81.

Rouse, R. (1991) Mexican migration and the social space of postmodernism. In: *Diaspora* 1, 8–23.

Routledge, P. (2003a) Voices of the dammed: Discursive resistance amidst erasure in the Narmada Valley, India. In: *Political Geography* 22, 3, 243–70.

Routledge, P. (2003b) Convergence space: Process geographies of grassroots globalisation networks. In: *Transactions of the Institute of British Geographers* 28(3), 333–49.

Routledge, P. and Cumbers, A. (2009) *Global Justice Networks: Geographies of Transnational Solidarity*. Manchester: Manchester University Press.

Rowe, S. and Wolch, J. (1990) Social networks in time and space: Homeless women in Skid Row, Los Angeles. In: *Annals of the Association of American Geographers* 80, 184–205.

Rowlands, J. (1997) *Questioning Empowerment: Working with Women in Honduras*. Oxford: Oxfam.

Rowles, G. (1978a) *Prisoners of Space? Exploring the*

Geographic Experience of Older People. Boulder, CO: Westview Press.

Rowles, G. (1978b) Reflections on experiential fieldwork. In: D. Ley and M. Samuels (eds.) (1978) *Humanistic Geography: Prospects and Problems*. London: Croom Helm.

Royal Geographical Society (2012) *Consumption Controversies: Alcohol Policies in the UK*. London: Royal Geographical Society.

RTI International (2004) *The Local Governance Project in Iraq*. Downloaded from www.rti.org, 19 May 2004.

Ruddick, S. (1996) *Young and Homeless in Hollywood*. New York: Routledge.

Ruddick, S. (2003) The politics of aging: Globalization and the restructuring of youth and childhood. In: *Antipode* 35: 334–62.

Ruddick, S. (2004) Activist geographies: Building possible worlds. In P. Cloke, P. Crang and M. Goodwin (eds.) *Envisioning Human Geographies*. London: Arnold, 229–41.

Rundstrom, R. (1995) GIS, indigenous people, and epistemological diversity. In: *Cartography and Geographic Information Systems* 22(1), 45–57.

Rushkoff, D. (1994) *Cyberia: Life in the Trenches of Hyperspace*. New York: HarperSanFrancisco.

Russell, P. (2003) Narrative constructions of British culinary culture. Unpublished PhD dissertation, University of Sheffield.

Rutherford, T. (2004) Convergence, the institutional turn and workplace regimes: The case of lean production. In: *Progress in Human Geography* 28(4), 425–46.

Ruting, B. (2008) Economic transformations of gay urban spaces: Revisiting Collins' evolutionary gay district model. In: *Australian Geographer* 39(3), 259–69.

Ryan, J. (1997) *Picturing Empire: Photography and the Visualization of the British Empire*. London: Reaktion Books.

Sack, R. (1992) *Place, Modernity and the Consumer's World: A Relational Framework for Geographical Analysis*. Baltimore: Johns Hopkins University Press.

Sack, R. (1997) *Homo Geographicus*. Baltimore: Johns Hopkins University Press.

Sader, E. (2009). Post-neoliberalism in Latin America. In: *Development Dialogue* 51, 171–80.

Safran, W. (1991) Diasporas in modern societies: Myths of homeland and return. In: *Diaspora* 1(1), 83–99.

Safran Foer, J. (2009) *Eating Animals*. London: Penguin.

Said, E. (1993) *Culture and Imperialism*. London: Chatto & Windus.

Said, E. (1995) [1978] *Orientalism*. Harmondsworth: Penguin.

Saldanha, A. (2005) Trance and visibility at dawn: Racial dynamics in Goa's rave scene. In: *Social and Cultural Geography* 6, 707–21.

Samuel, R. (1994) *Theatres of Memory. Volume 1: Past and Present in Contemporary Culture*. London: Verso.

Sanders, R. (2008). The triumph of geography. In: *Progress in Human Geography* 32(2), 179–82.

Sassen, S. (1988) *The Mobility of Capital and Labour*. Cambridge: Cambridge University Press.

Sassen, S. (1990) Finance and business services in New York City: International linkages and domestic effects. In: *International Social Science Journal* 42, 287–306.

Sassen, S. (1991) *The Global City: New York, London, Tokyo*. Princeton, NJ: Princeton University.

Sauer, C.O. (1925) The morphology of landscape. In: *University of California Publications in Geography* 2, 19–54.

Sauer, C.O. (1952) *Agricultural Origins and Dispersals*. New York: American Geographical Society.

Saville, S. (2008) Playing with fear: Parkour and the mobility of emotion. In: *Social and Cultural Geography* 9(8), 891–914.

Sawicki, D.S. and Craig, W.J. (1996) The democratization of data: Bridging the gap for community groups. In: *Journal of the American Planning Association* 62(4), 512–23.

Sayer, A. (2000) *Realism and Social Science*. London: Sage.

Schama, S. (1995) *Landscape and Memory*. London: Harper Collins.

Seamon, D. and Mugereur, R. (eds.) (1985) *Dwelling, Place, Environment: Toward a*

Phenomenology of Person and World. Amsterdam: Nijhoff.

Shillington, L. (2008) Being(s) in relation at home: Corporealities, aesthetics, and socialnatures in Managua, Nicaragua. In: *Social and Cultural Geography* 9, 755–76.

Shiva, V. (1991) *Ecology and the Politics of Survival: Conflict Over Natural Resources in India*. London: Sage.

Scannell, P. (1988) Radio times: The temporal arrangements of broadcasting in the modern world. In: P. Drummond and R. Paterson (eds.) *Television and its Audience: International Research Perspectives*. London: BFI, 15–31.

Scannell, P. (1995) *Radio, Television and Modern Life*. Oxford: Blackwell.

Schaefer, F.K. (1953) Exceptionalism in geography: A methodological examination. In: *Annals, Association of American Geographers* 43, 226–49.

Schama, S. (1987) The Enlightenment in the Netherlands. In: R. Porter and M. Teich (eds.) *The Enlightenment in National Context*. Cambridge: Cambridge University Press.

Schech, S. and Haggis, J. (2000) *Culture and Development: A Critical Introduction*. Oxford: Blackwell.

Scheper-Hughes, N. (2002) The end of the body. *SAIS Review*, 22(1), 61–80.

Schick, I.C. and Tonak, E.A. (eds.) (1987) *Turkey in Transition: New Perspectives*. New York: Oxford University Press.

Schirmer, J. (1994) The claiming of space and the body politic within national-security states. In: J. Boyarin (ed.) *Remapping Memory: The Politics of Timespace*. Minneapolis: University of Minnesota Press.

Schlesinger, P. (1994) Europe's contradictory communicative space. In: *Daedalus* 123(2), 25–52.

Schneier, B. (2003) *Beyond Fear: Thinking Sensibly About Security in an Uncertain World*. New York: Copernicus Books.

Schnucker, R.V. (1990) Puritan attitudes towards childhood discipline, 1560–1634. In: V. Fildes (ed.) *Women as Mothers in Pre-industrial England*. London: Routledge.

Schoenberger, E. (1997) *The Cultural Crisis of the Firm*. Cambridge, MA, and Oxford: Blackwell.

Schor, J.B. (1998) *The Overspent American: Upscaling, Downshifting and the New Consumer*. New York: Basic Books.

Schor, J.B. (2000) *Do Americans Shop Too Much?* Boston: Beacon Press.

Schroeder, R. (1996) *Possible Worlds: The Social Dynamic of Virtual Reality Technology*. Boulder, CO: Westview Press.

Schroeder, R. (1999) *Shady Practices: Agroforestry and Gender Politics in The Gambia*. Berkeley, CA: University of California Press.

Schumacher, E.F. (1973) *Small is Beautiful: A Study of Economics As If People Mattered*. London: Blond and Briggs.

Schuurman N. (2000) Trouble in the heartland: GIS and its critics in the 1990s. In: *Progress in Human Geography* 24(4), 569–90.

Schwanen, T. and Páez, A. (2010) The mobility of older people – An introduction. In: *Journal of Transport Geography* 18, 591–95.

Schwartz, J. and J. Ryan (eds.) (2003) *Picturing Place: Photography and the Geographical Imagination*. London: I.B. Tauris.

Schwartz, J.M. (1996) 'The Geography Lesson': Photographs and the construction of imaginative geographies. In: *Journal of Historical Geography* 22, 16–45.

Scott, J. (1976) *The Moral Economy of the Peasantry*. New Haven, CN: Yale University Press.

Scott, R.A. (1969) *The Making of Blind Men: A Study of Adult Socialization*. London: Transaction Books.

Scottish Natural Heritage (1993) *Sustainable Development and the Natural Heritage: The SNH approach*. Edinburgh: Scottish Natural Heritage.

Seager, J. (1994) *Earth Follies*. London: Routledge.

Seager, J. (2000) 'And a Charming Wife': Gender, marriage, and manhood in the job search process. In: *Professional Geographer* 52, 709–21.

Seager, J. (2003) *The State of Women in the World Atlas*. New York: Penguin; London: Women's Press; Paris: Autremont Editions.

Seager, J. and Nelson, L. (eds.) (2004) *Companion to Feminist Geography*. Oxford: Blackwell Publishers.

Searle, A. (1996) *Antony Gormley: Field for the*

British Isles. London: Arts Council of Great Britain (Spotlight Series).

Secord, J. (2000) *Victorian Sensation: The Extraordinary Publication, Reception and Secret Authorship of Vestiges of the Natural History of Creation.* Chicago and London: University of Chicago Press.

Sehgal, R. (1995) *The Black Diaspora*. New York: Farrar, Straus and Giroux.

Seiter, E., Borchers, H., Kreutzner, G. and Warth, E.-M. (1989) Don't treat us like we're so stupid and naïve: Towards an ethnography of soap opera viewers. In: E. Seiter, H. Borchers, G. Kreutzner and E.-M. Warth (eds.) *Remote Control*. London and New York: Routledge.

Select Bipartisan Committee to Investigate the Preparation for and Response to Hurricane Katrina (2008) Final Report. Available at: http://katrina.house.gov/full_katrina_report.htm

Self, C. and Gollege, R. (1994) Sex-related differences in spatial ability: What every geography educator should know. In: *Journal of Geography* 93, 234–43.

Sen, G. and Grown, C. (1987) *Development Crises and Alternative Visions. Third World Women's Perspectives*. New York: Monthly Review Press.

Sen, J., Anand, A., Escobar, A. and Waterman, P. (eds.) (2004) *World Social Forum: Challenging Empires*. The Viveka Foundation: New Delhi.

Sengupta, M. (2010) Million dollar exit from the anarchic slum-world: *Slumdog Millionaire*'s hollow idioms of social justice. In: *Third World Quarterly* 31(4), 599–616.

Seward, D. French Minister Slams US Role in Haiti. *Huffington Post*, 19 January 2010, http://www.huffingtonpost.com/2010/01/18/french-minister-rips-us-t_n_427366.html

Seymour, S., Daniels, S. and Watkins, C. (1994) Estate and empire: Sir George Cornewall's management of Moccas, Herefordshire and La Taste, Grenada, 1771–1819. Working Paper 28, University of Nottingham, Department of Geography and *Journal of Historical Geography* 24, 313—51.

Shah, B., Dwyer, C. and Gilbert, D. (2012) Landscapes of diasporic religious belonging in the edge-city: The Jain temple at Potters Bar, Outer London. In: *South Asian Diaspora* 4(1), 77–94.

Shakespeare, T. (1993) Disabled people's self-organisation: A new social movement? In: *Disability, Handicap and Society* 8, 249–64.

Shakespeare, T. (1994) Cultural representations of disabled people: Dustbins for disavowal? In: *Disability and Society* 9(3), 283–99.

Shapiro, M. (1989) Representing world politics: The sport/war intertext. In: J. Der Derian and M.J. Shapiro (eds.) *International/intertextual Relations: Postmodern Readings of World Politics*. Lexington MA: Lexington Books.

Sharp, J. (1993) Publishing American identity: Popular geopolitics, myth and the *Reader's Digest*. In: *Political Geography* 12(6), 491–503.

Sharp, J. (1996) Hegemony, popular culture and geopolitics: The *Reader's Digest* and the construction of danger. In: *Political Geography* 15(6/7), 557–70.

Sharp, J. (2000) *Condensing the Cold War: Reader's Digest and American identity*. Minneapolis: University of Minnesota Press.

Sharp, J. (2009) Geography and gender: What belongs to feminist geography? Emotion, power and change. In: *Progress in Human Geography* 33, 74–80.

Shaw, G., Agarwal. S. and Bull, P. (2000) Tourism consumption and tourist behaviour: A British perspective. In: *Tourism Geographies* 2, 264–89.

Shaw, M. (2002) Risk-transfer militarism, small massacres and the historic legitimacy of war. In: *International Relations* 16, 343–59.

Sheahan, J. (1987) *Patterns of Development in Latin America*. Princeton: Princeton University Press.

Sheller, M. and Urry, J. (eds.) (2004) *Tourism Mobilities: Places to Play, Places in Play*. Abingdon: Routledge.

Sheller, M. and Urry, J. (2006) The new mobilities paradigm. In: *Environment and Planning A* 38, 207–26.

Sheppard, E. (1995) GIS and society: Towards a research agenda. In: *Cartography and Geographic Information Systems* 22(1), 5–16.

Sheppard, E. and Barnes, T. (2003) *A Companion to Economic Geography*. Oxford: Blackwell.

Sherman, D.J. and Rogoff, I. (1994) *Museum*

Culture: Histories, Discourses and Spectacles. Minneapolis: University of Minnesota Press.

Shields, R. (1991) *Places on the Margin: Alternative Geographies of Modernity.* London: Routledge.

Shields, R. (2003) *The Virtual.* London: Routledge.

Shilling, C. (1993) *The Body and Social Theory.* London: Sage.

Shore, C. (1996) Transcending the nation-state? The European Commission and the (re)-discovery of Europe. In: *Journal of Historical Sociology* 9(4), 473–96.

Short, J. (1991) *Imagined Country: Society, Culture and Environment.* London: Routledge.

Shortridge, J.R. (1991) The concept of the place-defining novel in American popular culture. In: *The Professional Geographer* 43, 280–91.

Shove, E., Watson, M., Hand, M. and Ingram, J. (2007) *The Design of Everyday Life.* Oxford: Berg.

Shurmer-Smith, P. and Hannam, K. (1994) *Worlds of Desire, Realms of Power: A Cultural Geography.* London: Edward Arnold.

Sibley, D. (1981) *Outsiders in Urban Societies.* Oxford: Blackwell.

Sibley, D. (1995) *Geographies of Exclusion: Society and Difference in the West.* London: Routledge.

Sibley, D. (1998) The problematic nature of exclusion. In: *Geoforum* 29(2), 119–21.

Sibley, D. (1999) Creating geographies of difference. In D. Massey, J. Allen and P. Sarre (eds.) *Human Geography Today.* Cambridge: Polity, 115–28.

Sidaway, J. (2008) Subprime crisis: American crisis or human crisis? In: *Environment and Planning D: Society and Space,* 26, 195–98.

Sidaway, J.D. (2003) Banal geopolitics resumed. In: *Antipode* 35(4), 645–51.

Sidaway, J.D., Bunnell, T., Yeoh, B.S.A. (2003) Geography and postcolonialism. In: Theme issue of *Singapore Journal of Tropical Geography* 24(3).

Sieber, R. (1997) *Computers in the Grassroots: Environmentalists' Use of Information Technology and GIS.* Working Paper # 121. New Brunswick, NJ, Center for Urban Policy Research.

Sieber, R. (2006) Public participation geographic information systems: A literature review and framework. In: *Annals of the Association of American Geographers* 96(3), 491–507.

Silberman, N.A. (2001) 'If I forget thee, O Jerusalem': Archaeology, religious commemoration and nationalism in a disputed city, 1801–2001. In: *Nations and Nationalism* 7(4), 487–504.

Silverstone, R. (1994) *Television and Everyday Life.* London: Routledge.

Simmel, G. (1990) *The Philosophy of Money.* London: Routledge.

Simmons, I. (1996). *Changing the Face of the Earth. Culture, Environment, History.* Oxford: Basil Blackwell.

Simon, D. (1997) Development reconsidered: New directions in development thinking. In: *Geografiska Annaler* 79B(4), 183–201.

Simondon, G. (2009) 'Technical mentality'. A. DeBoever (trans.) In: *Parrhesia* 7, 17–27.

Simone, A. (2004) *For the City Yet to Come: Changing African Life in Four Cities.* Durham, NC: Duke University Press.

Simpson, P. (2008) Chronic everyday life: Rhythm analysing street performance. In: *Social and Cultural Geography* 9, 807–29.

Simpson, P. (2011a) So, as you can see . . .: Some reflections on the utility of video methodologies in the study of embodied practices. In: *Area* 43(3), 343–52.

Simpson, P. (2011b forthcoming) Street performance and the city: Public space, sociality and intervening in the everyday. In: *Space and Culture.*

Sinclair, U. (1906) *The Jungle.* New York: Doubleday, Page and Co.

Singer, P. (1985). Prologue: Ethics and the new Animal Liberation Movement. In: *In Defense of Animals.* P. Singer (ed.) New York: Harper and Row, 1–10.

Skeggs, B. (1997) *Formations of Class and Gender.* London: Sage.

Skelton, T. (2000) 'Nothing to do, nowhere to go?': Teenage girls and 'public' space in the Rhondda Valleys, South Wales. In: S.L. Holloway and G. Valentine (eds.) *Children's Geographies: Playing, Living, Learning.* London: Routledge, 80–99.

Skelton, T. and Valentine, G. (1998) *Cool Places: Geographies of Youth Cultures.* London: Routledge.

Sloterdijk, P. (1995) World markets and secluded spots: On the position of the European regions in the world-experiment of capital. In: P. Büchler and N. Papastergiadis (eds.) *Random Access: On Crisis and its Metaphors*. London: Rivers Oram Press.

Sloterdijk, P. (2009) *Terror from the Air*. Cambridge, Mass: MIT Press.

Sloterdijk, P. and Mueller van der Haegen, G. (2005) Instant democracy: the pneumatic parliament. In: B. Latour and P. Weibel (eds.) Making Things Public: Atmospheres of Democracy. Cambridge, Mass: MIT Press.

Slotkin, R. (1992) *Gunfighter Nation: The Myth of the Frontier in 20th Century America*. New York: Macmillan.

Smaje, C. and Le Grand, J. (1997) Ethnicity, equity and the use of health services in the British NHS. In: *Social Science & Medicine* 45, 485–96.

Smallman-Raynor, M.R., Cliff, A.D., Trevelyan, B., Nettleton, C. and Sneddon, S. (2006). *Poliomyelitis. A World Geography: Emergence to Eradication*. Oxford: Oxford University Press.

Smith, A. (2000) Employment restructuring and household survival in 'postcommunist transition': Rethinking economic practices in Eastern Europe. In: *Environment and Planning A* 32, 1759–80.

Smith, A. (2002) Culture/economy and spaces of economic practice: positioning households in post-communism. In: *Transactions of the Institute of British Geographers* NS(27), 232–50.

Smith, A. (2003) Power relations, industrial clusters and regional transformations: Pan-Europe integration and outward processing in the Slovak clothing industry. In: *Economic Geography* 79, 17–40.

Smith, A., Rainnie, A., Dunford, M., Hardy, J., Hudson, R. and Sadler, D. (2002) Networks of value, commodities and regions: Reworking divisions of labour in macro-regional economies. In: *Progress in Human Geography* 26(1), 41–63.

Smith, A., Stenning, A. and Willis, K. (eds.) (2008) *Social Justice and Neoliberalism*, London: Zed Press.

Smith, A.D. (1991) *National Identity*. Harmondsworth: Penguin Books.

Smith, B. (1985) *European Vision and the South Pacific*. New Haven and London: Yale University Press.

Smith, D. (2004) Morality ethics and social justice. In P. Cloke, P. Crang and M. Goodwin (eds.) *Envisioning Human Geographies*. London: Arnold, 195–209.

Smith, D.M. (2000) *Moral Geographies: Ethics in a World of Difference*. Edinburgh: Edinburgh University Press.

Smith, D.P. and Holt, L. (2004) Lesbian migrants and the cultural consumption of greentrified rurality. Paper available from the authors. University of Brighton, Cockcroft Building, Lewes Road, Brighton BN2 4GJ.

Smith, D.P. and Holt, L. (2005) Lesbian migrants in the gentrified valley and 'other' geographies of rural gentrification. In: *Journal of Rural Studies* 21(3), 313–22.

Smith, J. (2000) *'The Daily Globe': Environmental Change, the Public and the Media*. London: Earthscan.

Smith, J.M. and Foote, K.E. (eds.) (1994) How the world looks. In: K.E. Foote, P.J. Hugill, K. Mathewson and J.M. Smith (eds.) *Re-reading Cultural Geography*. Austin: University of Texas Press, 27–163.

Smith, M.P. (ed.) (1995) *Marginal Spaces*. New Brunswick, NJ: Transaction Publishers.

Smith, M., Davidson, J., Cameron, L. and Bondi, L. (eds.) (2009) *Emotion, Place and Culture*. Burlington VT and Aldershot: Ashgate Press.

Smith, N. (1987) Academic war over the field of geography: The elimination of Geography at Harvard, 1947–1951. In: *Annals of the Association of American Geographers* 77(2), 155–72.

Smith, N. (1990) *Uneven Development*. Oxford: Basil Blackwell.

Smith, N. (1994) Geography, empire and social theory. In: *Progress in Human Geography* 18(4), 491–500.

Smith, N (1996) *The New Urban Frontier: Gentrification and the Revanchist City*. London: Routledge.

Smith, N. (2001) Ashes and aftermath. In: *The Arab World Geographer Forum on 11 September Events*.

Smith, N. (2003) Scales of terror and the resort to geography: September 11, October 7. In: *Society and Space* 19, 631–7.

Smith, N. (2005). *The Endgame of Globalization.* New York: Routledge.

Smith, N. (2006): There's no such thing as a natural disaster. *Understanding Katrina: Perspectives from the Social Sciences,* http://understandingkatrina.ssrc.org/Smith/

Smith, N. (2011) Blood and soil: Nature, native and nation in the Australian imaginary. In: *Journal of Australian Studies* 35(1), 1–18.

Smith, S. (2000) Performing in the (sound)world. In: *Environment and Planning D: Society and Space* 18, 615–37.

Smith, S.J. (1989) *The Politics of 'Race' and Residence*. Cambridge: Polity Press.

Smith, S.J. (1993) Residential segregation and the politics of racialisation. In: M. Cross and M. Keith (eds.) *Racism, the City and the State.* London: Routledge.

Smith, S.J. (1994) Citizenship. In: R. Johnston, D. Gregory and D.M. Smith (eds.) *The Dictionary of Human Geography*. Oxford: Blackwell.

Smith, S.J. (1997) Beyond geography's visible worlds: A cultural politics of music. In: *Progress in Human Geography* 21, 502–29.

Smith, S.J. (2005) States, markets and an ethic of care. In: *Political Geography* 24, 1–20.

Smith, S. J. and Easterlow, D. (2005) The strange geography of health inequalities. In: *Transactions, Institute of British Geographers* 30, 173–90.

Smith, S.J. and Mallinson, S. (1996) The problem with social housing: Discretion, accountability and the welfare ideal. In: *Policy and Politics* 24, 339–58.

Smith, V. (ed.) (1977) *Hosts and Guests: The Anthropology of Tourism*. Philadelphia: University of Pennsylvania Press.

Smith, W.D. (1984) The function of commercial centres in the modernisation of European capitalism: Amsterdam as an information exchange in the seventeenth century. In: *Journal of Economic History* 44, 985–1005.

Snow, C.P. (1959) *The Two Cultures and the Scientific Revolution.* Cambridge: Cambridge University Press.

Soja, E. (1989) *Postmodern Geographies,* London: Verso.

Soja, E. (1992) Inside exopolis: scenes from Orange County. In: M. Sorkin (ed.) *Variations on a Theme Park: The New American City and the End of Public Space.* New York: Noonday.

Soja, E.W. (1996) *Thirdspace*. Oxford: Blackwell.

Somers, M.R. (1994) The narrative constitution of identity: A relational and network approach. In: *Theory and Society* 23, 605–49.

Sorkin, M. (ed.) (1992) *Variations on a Theme Park: The New American City and the End of Public Space.* New York: The Noonday Press.

SouthSouthNorth (2011) *Project Portfolio and Reports* http://www.southsouthnorth.org/ Accessed March 2011.

Sparke, M. (2005) *In the Space of Theory.* Minneapolis: University of Minnesota Press.

Spillman, L. (1997) *Nation and Commemoration: Creating National Identities in the United States and Australia.* Cambridge: Cambridge University Press.

Spinney, J. (2006): A place of sense: A kinaesthetic ethnography of cyclists on Mont Ventoux. In: *Environment and Planning D: Society and Space* 24, 709–32.

Spinney, J. (2007) Cycling the city: Non-place and the sensory construction of meaning in a mobile practice. In: D. Horton, P. Rosen and P. Cox (eds.) *Cycling and Society.* Aldershot: Ashgate, 25–46.

Spinney, J. (2009) Cycling the city: Movement, meaning and method. In: *Geography Compass* 3, 317–35.

Spooner, B. (1986) Weavers and dealers: The authenticity of an oriental carpet. In: A. Appadurai (ed.) *The Social Life of Things. Commodities in Cultural Perspective.* Cambridge: Cambridge University Press, 195–235.

Staeheli, L. and Brown, M. (2003) Where has welfare gone? Introductory remarks on the geographies of care and welfare. In: *Environment and Planning A* 35, 771–7.

Staeheli, L. and Mitchell, D. (2005), The complex politics of relevance in geography. In: *Annals of the Association of American Geographers* 95, 357–72.

Staeheli, L., Ehrkamp, P., Leitner, H. and Nagel, C. (2012) Dreaming the ordinary: Daily life and the complex geographies of citizenship. In: *Progress in Human Geography*, in press.

Stallman, R. (1999) The GNU Operating System and the Free Software Movement. In: C. DiBona, S. Ockman and M. Stone (eds.) *Open Sources: Voices from the Open Source Revolution*. Sebastopol, CA: O'Reilly Media Inc., 53–70.

Steinberg, P. (2008) What is a city? Katrina's answers. In: P. Steinberg and R. Shields (eds.) *What is a City? Rethinking the Urban after Hurricane Katrina*. Athens GA: University of Georgia Press, 3–29.

Steiner, A. (2003) Trouble in paradise. In: *New Scientist*, 18 October, 21.

Stevens, J.E. (1988) *Hoover Dam: An American Adventure*. Norman, OK: University of Oklahoma Press.

Stevens, S. (1993) Tourism, change and continuity in the Mount Everest region. In: *Geographical Review* 83, 410–27.

Stevenson, N. (2002).*Understanding Media Cultures*. London: Thousand Oaks; New Delhi: Sage.

Stewart, K. (2007) *Ordinary Affects*. Durham and London: Duke University Press.

Stockdale, A. (2011) A review of demographic ageing in the UK: Opportunities for rural research. In: *Population, Space and Place* 17, 204–21.

Stoddart, D. (1986) Geography, exploration and discovery. In: *On Geography and Its History*. Oxford: Blackwell.

Stoker, G. (1996) Governance as theory: Five propositions. Mimeo (available from the author at the Department of Government, University of Strathclyde).

Storper, M. and Walker, R. (1989) *The Capitalist Imperative*. New York and Oxford: Basil Blackwell.

Sugiyama, T., Leslie, E., Giles-Corti, B. and Owen, N. (2008) Associations of neighborhood greenness with physical and mental health: Do walking, social coherence and local social interaction explain the relationships? In: *Journal of Epidemiology and Community Health* 62(5), e9.

Sui, D. (1999) Postmodern urbanism disrobed: Or why postmodern urbanism is a dead end for urban geography. In: *Urban Geography* 20(5), 403–11.

Sui, D. (2008) The wikification of GIS and its consequences: Or Angelina Jolie's new tattoo and the future of GIS. In: *Computers, Environment and Urban Systems* 32, 1–5.

Sunday Times (2003) The great Indian takeaway. London: *Sunday Times*, 8 June, 3.5.

Sundberg, J. (2004) 'Identities-in-the-making: Conservation, gender, and race in the Maya Biosphere Reserve, Guatemala'. In: *Gender, Place and Culture* 11(1), 44–66.

Sundberg, J. (2005) Looking for the critical geographer, or why bodies and geographies matter to the emergence of critical geographies of Latin America. In: *Geoforum* 36(1), 17–28.

Sundberg, J. (2011) Diabolic Caminos in the desert and cat fights on the Rio: A posthumanist political ecology of boundary enforcement in the United States-Mexico borderlands. In: *Annals of the Association of American Geographers* 101(2), 318–36.

Swanton, D. (2010) Sorting bodies: Race, affect, and everyday multiculture in a mill town in northern England. In: Environment and Planning A 42(10), 2332–50.

Swilling, M. (undated) *Africa 2050 – Growth, Resource Productivity and Decoupling* (UNEP) http://www.learndev.org/dl/BtSM2011/Africa%20Policy%20Brief.pdf) Accessed 27 April 2012.

Swyngedouw, E. (2000) The Marxian alternative: historical-geographical materialism and the political economy of capitalism. In: E. Sheppard and T. Barnes (eds.) *A Companion to Economic Geography*. Oxford: Blackwell.

Swyngedouw, E. (2010) Apocalypse forever? Post-political populism and the spectre of climate change. In: *Theory, Culture & Society*, 27: 213–32.

Symanski, R. (1981) *The Immoral landscape: Female Prostitution in Western Societies*. Toronto: Butterworths.

Synnott, A. (1993) *The Body Social: Symbolism, Self and Society*. London: Routledge.

Takahashi, L. (1996) A decade of understanding homelessness in the USA: From characterization to representation. In: *Progress in Human Geography* 20(3), 291–310.

Tarlo, E. (2010) *Visibly Muslim: Fashion, Politics, Faith*. Oxford: Berg.

Tarrant, A. (2010) Constructing a social geography of grandparenthood: A new focus for intergenerationality. In: Area 42: 190–7.

Taussig, M. (1980) *The Devil and Commodity Fetishism in Latin America*. Durham, NC: University of North Carolina Press.

Tauxe, C.S. (1996) Mystics, Modernists and Constructions of Brasília. In: *Ecumene* 3, 43–61.

Taylor, G. (1949) *Urban Geography*. London: Methuen.

Taylor, H. (1951) No watchdog for America. In: *Reader's Digest*, February, 85–7.

Taylor, J. (1997) The emerging geographies of virtual worlds. In: *Geographical Review* 87(2), 172–92.

Taylor, L.H., Latham, S.M. and Woolhouse, M.E. (2001) Risk factors for human disease emergence. In: *Philosophical Transactions of the Royal Society B: Biological Sciences*, 356.

Taylor, P. (1989) 'The error of developmentalism in Human Geography. In: D. Gregory and R. Walford (eds.) *Horizons in Human Geography*. Basingstoke: Macmillan, 303–19.

Taylor, P. (1990) Editorial comment: Geographical knowledge systems. In: *Political Geography Quarterly* 9, 211–12.

Taylor, P. and Flint, C. (2000) *Political Geography*. Harlow: Pearson.

Taylor, P. and Overton, M. (1991) Further thoughts on geography and GIS. In: *Environment and Planning A* 23, 1087–9.

Taylor, P.J. (1996) *The Way the Modern World Works: World Hegemony to World Impasse*. Chichester: John Wiley.

Taylor, P.J. (2004) *World City Network: A Global Urban Analysis*. Abingdon and New York: Routledge.

Taylor, P.J., Catalano, G. and Gane N (2003) A geography of global change: Cities and services, 2000–2001. In: *Urban Geography* 24(5), 431–41.

Tendler, J. (1997) *Good Government in the Tropics*. Baltimore: Johns Hopkins University Press.

Terborgh, J. (1999) *Requiem for Nature*. Washington, DC: Island Press.

Terlecki, M. and N. Newcombe (2005) How important is the digital divide? The relation of computer and videogame usage to gender differences in mental rotation ability. In: *Sex Roles* 53(5–6), 433–41.

Thede, N. and Ambrosi, A. (eds.) (1991) *Video the Changing World*. Montreal and New York: Black Rose Books.

thee data base (1996) The K Foundation: Why we burnt a million pounds. *thee data base* [online], 7, Available from http://members.xoom.com/databass/KFound.htm.

Thien, D. (2005) After or beyond feeling?: A consideration of affect and emotion in geography. In: *Area* 37(4), 450–6.

Thobani, S. (2007) White wars: Western feminisms and the 'War on Terror'. In: *Feminist Theory* 8, 169–85.

Thomas, H. (1997) *The Slave Trade*. London: Macmillan.

Thomas, M. (2004) Pleasure and propriety: Teen girls and the practice of straight space. In: *Environment and Planning D: Society and Space* 22(5), 773–89.

Thomas, M. (2011) Introduction: Psychoanalytic methodologies in geography. In: *Professional Geographer* 62(4), 478–82.

Thomas, N. (2010) *Islanders: The Pacific in the Age of Empire*. Yale University Press: New Haven.

Thomas, W., Sauer, C., Bates, M. and Mumford, L. (1956) *Man's Role in Changing the Face of the Earth*. Chicago, IL: University of Chicago Press.

Thompson, A. (2001) Nations, national identities and human agency: Putting people back into nations. In: *Sociological Review* 49(1), 18–32.

Thompson, E. (1991) *Customs in Common*. London: Penguin Books.

Thompson, J. (1995) *The Media and Modernity: A Social Theory of the Media*. Oxford: Polity Press.

Thompson, K.A. (2007) *An Eye for the Tropics: Tourism, Photography, and Framing the Caribbean Picturesque*. Durham, NC and London: Duke University Press.

Thrift, N. (1983) On the determination of social action in space and time. In: *Environment and Planning D: Society and Space* 1, 23–57.

Thrift, N. (1994) On the social and cultural determinants of international financial centres. In: S. Corbridge, N.J. Thrift and R. Martin (eds.) *Money, Power and Space*. Oxford: Blackwell, 327–55.

Thrift, N. (1996, orig. 1985) Flies and Germs: A geography of knowledge. In: *Spatial Formations*. London: Sage.

Thrift, N. (1997) The still point: Resistance, expressive embodiment and dance. In: S. Pile and M. Keith (eds.) *Geographies of Resistance*. London: Routledge, 124–51.

Thrift, N. (1999) Steps to an ecology of place. In: D. Massey, J. Allen and P. Sarre (eds.) *Human Geography Today*. Cambridge: Polity Press.

Thrift, N. (2001) Chasing capitalism. In: *New Political Economy* 6(3), 375–80.

Thrift, N. (2003a) Closer to the machine? Intelligent environments, new forms of possession and the rise of the supertoy. In: *Cultural Geographies* 10, 389–407.

Thrift, N. (2003b) Performance and . . . In: *Environment and Planning A* 35, 2019–24.

Thrift, N. (2004) Intensities of feeling: Towards a spatial politics of affect. In: *Geografiska Annaler* 86B(1), 55–76.

Thrift, N. (2007a) *Non-representational Theory: Space, Politics, Affect*. Routledge, Abingdon.

Thrift, N. (2007b) Immaculate warfare? The spatial politics of extreme violence. In: D. Gregory and A. Pred (eds.) *Violent Geographies: Fear, Terror, and Political Violence*. New York: Routledge.

Tickell, A. and Peck, J. (1996) The return of the Manchester men: Men's words and men's deeds in the remaking of the local state. In: *Transactions of the Institute of British Geographers* 21(4), 595–616.

Tiessen, M. (2008) Uneven mobilities and urban theory: The power of fast and slow. In: P. Steinberg and R. Shields (eds.) *What is a City? Rethinking the Urban after Hurricane Katrina*. Athens GA: University of Georgia Press, 112–24.

Tiffen, M., Mortimore, M.J. and Gichugi, F. (1994) *More People, Less Erosion: Environmental Recovery in Kenya*. Chichester: John Wiley.

Till, K. (2005) *The New Berlin: Memory, Politics, Place*. Minneapolis: University of Minnesota Press.

Timothy, D.J. and Nyaupane, G.P. (eds.) (2009) *Cultural Heritage and Tourism in the Developing World: A Regional Perspective*. Abingdon: Routledge.

Tizard, J. (2011) Offshoring: the bigger picture. In: *Public Finance* 17 March 2011 http://opinion.publicfinance.co.uk/2011/03/offshoring-the-bigger-picture-by-john-tizard/

Tolia-Kelly, D.P. (2006) Affect – An ethnocentric encounter? Exploring the 'universalist' imperative of emotional/affectual geographies. In: *Area* 38, 213–7.

Tolia-Kelly, D.P. (2011) Narrating the postcolonial landscape: Archaeologies of race at Hadrian's Wall. In: *Transactions of the Institute of British Geographers* 36, 71–88.

Toye, J. (1993) *Dilemmas of Development: The Counter-revolution in Development Theory and Policy* (2nd edn.). Oxford: Blackwell.

Transition Town Brixton (2011) About 'Transition'? http://www.transitiontownbrixton.org Accessed March 2011

Traynor, C. and Williams, M.G. (1995) Why are geographic information systems hard to use? In: I. Katz, R. Mack and L. Marks (eds.) *Conference Companion on Human Factors in Computing Systems – CHI '95*. Denver, CO.: ACM Press, 288–9.

Trelluyer, M. (1990) La télévision régionale en Europe. In: *Dossiers de l'Audiovisuel* 33, 10–55.

Trentmann, F. (2006) The evolution of the consumer: Meanings, identities, and political synapses before the age of affluence. In: S. Garon and P.L. Maclachlan (eds.) *The Ambivalent Consumer: Questioning Consumption in East Asia and the West*. Ithaca and London: Cornell University Press, 21–44.

Tribe, J. (2005) New tourism research. In: *Tourism Recreation Research* 30(2), 5–8.

Tribe, J. (2006) The truth about tourism. In: *Annals of Tourism Research* 33(2), 360–81.

Tronto, J. (1987) Beyond gender differences in a theory of care. In: *Signs* 12, 644—63.

Tronto, J. (1993) *Moral Boundaries: A Political Argument For an Ethic of Care*. London: Routledge.

Tronto, J. (1998) An ethic of care. *Generations* 22(3), 15–20.

Tuan, Yi-Fu (1976) Humanistic geography. In: *Annals of the Association of American Geographers* 66, 266–76.

Tuan, Yi-Fu (1977) *Space and Place: The Perspective of Experience*. Minneapolis: University of Minnesota Press.

Tuan, Y.-F. (1990) Space and context. In: R. Schechner and W. Appel (eds.) *By Means of Performance: Intercultural Studies of Theatre and Ritual*. Cambridge: Cambridge University Press, 236–44.

Tuan, Y-F. (2004) *Dominance and Affection: The Making of Pets*. Yale: Yale University Press.

TUC (2004) Submission to Department of Trade and Industry (DTI) Consultation Exercise on Global Offshoring, 8 March 2004 (2004). In: U. Huws, S. Dahlmann and J. Flecker *Status Report on Outsourcing of ICT and Related Services in the EU*. Brussels: European Commission, 13–4. http://www.eurofound.europa.eu/pubdocs/2004/137/en/1/ef04137en.pdf

Tucker, A. (2009) Framing exclusion in Cape Town's gay village: The discursive and material perpetration of inequitable queer subjects. In: *Area* 41(2), 186–97.

Tucker, F. (2003) Sameness or difference? Exploring girls' use of recreational spaces. In: *Children's Geographies* 1, 111–24.

Tucker, F. and Matthews, H. (2001) They don't like girls hanging around there: Conflicts over recreational space in rural Northamptonshire. In: *Area* 33, 161–8.

Tulloch, D. (2008) Is VGI participation? From vernal pools to video games. In: *GeoJournal* 72, 161–71.

UK Government (1999) *A Better Quality of Life: A Strategy for Sustainable Development for the UK*, Cm 4345, The Stationery Office, London: ISBN 0-10-143452-9.

UK National Ecosystem Assessment (2011) *The UK National Ecosystem Assessment: Synthesis of the Key Findings*. Cambridge: UNEP-WCMC.

Ulrich, R.S. (1984) View through a window may influence recovery from surgery. In: *Science* 224 (4647), 420–1.

UN Centre for Human Settlements (1996) *An Urbanizing World: Global Report on Human Settlements 1996*. Oxford: Oxford University Press.

UN-Habitat (2011) *Global Report on Human Settlements: Cities and Climate Change*. Nairobi: UN-Habitat.

UN-Habitat (2011) *Cities and Climate Change Initiative*. http://www.unhabitat.org/categories.asp?catid=550 Accessed September 2011.

United Nations (2007) *Declaration on the Rights of Indigenous Peoples*, Adopted by General Assembly Resolution 61/295 on 13 September 2007, http://www.un.org/esa/socdev/unpfii/documents/DRIPS_en.pdf (accessed 29 June 2011)

United Nations (2012) *World Urbanization Prospects, the 2011 Revision*. United Nations: New York.

UNHCR (annual) *State of the World's Refugees*. Oxford: Oxford University Press.

United Nations Conference on Trade and Development (UNCTAD) (2010) *World Investment Report*. New York: United Nations.

United Nations Conference on Trade and Development (UNCTAD) (2011) *World Investment Report*. Geneva: UNCTAD.

United Nations Development Programme (UNDP) (1997) *Human Development Report 1997*. Oxford: OUP-UNDP.

United Nations Development Programme (UNDP) (1999) *Human Development Report: Globalisation with a Human Face*. New York: Oxford University Press.

United Nations Development Programme (UNDP) (2002) *Human Development Report 2002*, Oxford: Oxford University Press.

United Nations Environment Programme World Conservation Monitoring Centre (UNEP-WCMC) (2009) *The impacts of biofuel production on biodiversity: A review of the current literature*. Available at: http://www.cbd.int/agriculture/2011-121/UNEP-WCMC3-sep11-en.pdf (last accessed 21 March 2012).

United Nations Population Division (2002) *World Population Ageing 1950–2050.* New York: UNPD.

UPIAS (1976) *Fundamental Principles of Disability.* London: Union of the Physically Impaired Against Segregation.

Upretty, R. (2002) Impact of tourist industry in Nepal. In: *Nepal Weekly Telegraph*, 14 August, http://www.nepalnews.com.np, accessed 25 April 2004.

Urbain, J.-D. (1994) *Sur la plage: mœurs et coutumes balnéaires (XIXe-XXe siècles).* Paris: Payot.

Urbain, J-D. (2002) Paradis verts: desirs de campagne et passions residentielles. Paris: Payot.

Urry, J. (1990) *The Tourist Gaze: Leisure and Travel in Contemporary Societies.* London: Sage.

Urry, J. (1995) *Consuming Places.* London: Routledge.

United States Government Accountability Office (GAO) (2001) INS Southwest Border Strategy: Resourced and Impact Issues Remain After Seven Years. http://www.gao.gov/new.items/d01842.pdf (last accessed 26 February 2013)

Vakil, A.C. (1997) Confronting the classification problem: Toward a taxonomy of NGOs. In: *World Development* 25(12), 2057–70.

Valentine, G. (1989) The geography of women's fear. In: *Area* 21(4), 385–90.

Valentine, G. (1993a) Desperately seeking Susan: A geography of lesbian friendships. In: *Area* 25(2), 109–16.

Valentine, G. (1993b) Negotiating and managing multiple sexual identities: Lesbian time–space strategies. In: *Transactions of the Institute of British Geographers* 18, 237–48.

Valentine, G. (1993c) (Hetero)sexing space: Lesbian perceptions and experiences of everyday spaces. In: *Environment and Planning D: Society and Space* 11, 395–413.

Valentine, G. (1996a) Angels and devils: Moral landscapes of childhood. In: *Environment and Planning D: Society and Space* 14, 581–99.

Valentine, G. (1996b) Children should be seen and not heard: The production and transgression of adult's public space. In: *Urban Geography* 17(3), 205–20.

Valentine, G. (1997a) 'Oh Yes I Can.' 'Oh No You Can't': Children and parents' understandings of kids' competence to negotiate public space safely. In: *Antipode* 29, 65–89.

Valentine, G. (1997b) 'My Son's a Bit Dizzy' 'My Wife's a Bit Soft': Gender, children and cultures of parenting. In: *Gender, Place and Culture* 4, 37–62.

Valentine, G. (1997c) A safe place to grow up? Parenting, perceptions of children's safety and the rural idyll. In: *Journal of Rural Studies* 13(2), 137–48.

Valentine, G. (1998) 'Sticks and stones may break my bones': A personal geography of harassment. In: *Antipode* 30(4), 305–32.

Valentine, G. (1999) Eating in: Home, consumption and identity. In: *Sociological Review* 47(3), 491–524.

Valentine, G. (1999) Imaginative geographies. In: D. Massey, J. Allen and P. Sarre (eds.) *Human Geography Today.* Cambridge: Polity Press.

Valentine, G. (2001) *Social Geographies. Space and Society.* London: Pearson.

Valentine, G. (2007) Theorizing and researching intersectionality: A challenge for feminist geography. In: *Professional Geographer* 59(1), 10–21.

Valentine, G. (2008) The ties that bind: Towards geographies of intimacy. In: *Geography Compass* 2, 2097–110.

Valentine, G. and Holloway, S.L. (2002) Cyberkids? Exploring children's identities and social networks in on-line and off-line worlds. In: *Annals of the Association of American Geographers* 92, 302–19.

van Blerk, L. (2005) Negotiating spatial identities: Mobile perspectives on street life in Uganda. In: *Children's Geographies* 3(1), 5–21.

van Blerk, L. (2008) Poverty, migration and sex work: Youth transitions in Ethiopia. In: *Area* 40: 245–53.

van den Berghe, P.L. (1994) *The Quest for the Other: Ethnic Tourism in San Cristobal, Mexico.* Seattle: University of Washington Press.

Vanderbeck, R. (2007) Intergenerational geographies: Age relations, segregation and re-engagements. In: *Geography Compass* 1, 200–21.

van Geen, A., Ahmed, K., Akita, Y., Alam, M., Culligan, P., Emch, M., Escamilla, V., Feighery, J., Ferguson, A., Kappett, P., Layton, A., Mailloux, B., McKay, L., Mey, J., Serre, M., Streatfield, P., Wu, J. and Yunus, M. (2011) Fecal contamination of shallow tubewells in Bangladesh inversely related to arsenic. In: *Environment, Science and Technology* 45, 1199–205.

van Rooy, A. (1998) *Civil Society and the Aid Industry*. London: Earthscan.

Varley, A. (1996) Women-headed households: Some more equal than others? In: *World Development* 24, 505–20.

Veblen, T. (1899) *The Theory of the Leisure Class: An Economic Study in the Evolution of Institutions*. New York: Macmillan.

Veijola, S. and Jokinen, E. (1994) The body in tourism. In: *Theory, Culture and Society* 11, 125–51.

Verne, J. (2012) *Living Translocality: Space, Culture and Economy in Contemporary Swahili Trading Connections*. Stuttgart: Franz Steiner Verlag.

Vertovec, S. (1996) Berlin Multikulti: Germany, 'foreigners' and 'world-openness'. In: *New Community*, 22(3), 381—99.

Vidal, J. (1997) The long march home. In: *The Guardian Weekend*, April 26, 14–20.

Vilhelmson, B. and Thulin, E. (2008) Virtual mobility, time use and the place of the home. In: *Tijdschrift voor Economische en Sociale Geografie* 99(5), 602–18.

Villi, M. and Stocchetti, M. (2011) Visual mobile communication, mediated presence and the politics of space. In: *Visual Studies* 26(2), 102–12.

Vira, B. and James, A. (2011) Researching hybrid 'economic'/'development' geographies in practice: Methodological reflections from a collaborative project on India's New Service Economy. In: *Progress in Human Geography* 35(5), 627–51.

Virilio, P. (2000) *Polar Inertia*. London: Sage.

Visser, G. (2003) Gay men, leisure space and South African cities: The case of Cape Town. In: *Geoforum* 34, 123–37.

Visvanathan, N., Duggan, L, Nisonoff, L and Wiegersma, N. (eds.) (1987) *The Women, Gender and Development Reader*. London: Zed.

Von Weizsäcker, E., Amory, B., and Lovins, L. Hunter (1997) *Factor Four: Doubling Wealth – Halving Resource Use the New Report to the Club of Rome*. London: Earthscan.

Wacquant, L. (1993) Urban outcasts: Stigma and division in the Black American ghetto and the French urban periphery. In: *International Journal of Urban and Regional Research* 17(3), 366–83.

Wade, R. (1990) *Covering the Market: Economic Theory and the Role of Government in East Asian Industrialisation*. Princeton: Princeton University Press.

Wagner, L. (2011) Negotiating diasporic mobilities and becomings: Interactions and practices of Europeans of Moroccan descent on holiday in Morocco. Doctoral thesis. London: Geography, University College London.

Wainaina, B. (2005) How to write about Africa. In: Granta: *Magazine of New Writing* 92. www.granta.com/Archive/92/How-to-Write-about-Africa/Page-1

Wainwright, H. (2003) *Reclaim the State: Experiments in Popular Democracy*. London: Verso.

Wainwright, J. (2008) *Decolonizing Development: Colonial power and the Maya*. Oxford: Blackwell Press.

Wainwright, J. (in press) *Geopiracy: Oaxaca, Militant Empiricism, and Geographical Thought*. New York: Palgrave MacMillan.

Wainwright, J. and Bryan, J. (2009) Cartography, territory, property: Postcolonial reflections on indigenous counter-mapping in Nicaragua and Belize. In: *Cultural Geographies* 16, 153–78.

Waitt, G. and Head, L. (2002) Postcards and frontier mythologies: Sustaining views of the Kimberley as timeless. In: *Environment & Planning D: Society and Space* 20(3), 19–344.

Waitt, G. and Gorman-Murray, A. (2011) Journeys and returns: Home, life narratives and remapping sexuality in a regional city. In: *International Journal of Urban and Regional Research* 35(6), 1239–55.

Waitt G., Jessop L. and Gorman-Murray, A. (2011) 'The guys in there just expect to be laid': Embodied and gendered socio-spatial practices of a 'night out' in Wollongong, Australia. In: *Gender, Place and Culture* 18(2), 255–75.

Wall Street Journal (2004) Outsourcing concerns some globalisation backers. In: *Wall Street Journal Europe*, 26 January, A6.

Wald, P. (2008) *Contagious: Cultures, Carriers, and the Outbreak Narrative*. Durham, NH: Duke University Press.

Wallace, R.G. (2009) Breeding influenza: The political virology of offshore farming. In: *Antipode* 41, 916–51.

Wallerstein, I. (1979) *The Capitalist World Economy*. Cambridge: Cambridge University Press.

Wallerstein, I. (1984) *The Politics of the World Economy*. Cambridge: Cambridge University Press.

Wallerstein, I. (2003) US weakness and the struggle for hegemony. In: *Monthly Review*, 55(3).

Wallerstein, I. (2004) *World Systems Analysis: An Introduction*. Durham, NC: Duke University Press.

Wallerstein, I. (2006) The curve of American power. In: *New Left Review*, 40, 77–94.

Walsh, K. (2007) 'It got very debauched, very Dubai!' Heterosexual intimacy amongst single British expatriates. In: *Social and Cultural Geography* 8(4), 507–33.

Walter, T. (2001) From cathedral to supermarket: Mourning, silence and solidarity. In: *Sociological Review* 49(4), 494–511.

Walton, M., Pearce, J. and Day, P. (2009) Examining the interaction between food outlets and outdoor food advertisements with primary school food environments. In: *Health and Place* 15(3), 841–48.

Walvin, J. (1996) *Fruits of Empire: Exotic Produce and British Trade, 1660–1800*. London: Macmillan.

Wang, N. (1999) Rethinking authenticity in tourism experience. *Annals of Tourism Research* 26, 349–70.

Ward, C. (1978) *The Child in the City*. London: Architectural Press.

Warf, B. (1988) Regional transformation, everyday life, and Pacific Northwest lumber production. In: *Annals of the Association of American Geographers* 78, 326–46.

Wark, M. (1994) *Virtual Geography*. Bloomington and Indianapolis: Indiana University Press.

Warner, M. (2002) *Publics and Counterpublics*. New York: Zone Books.

Waterman, P. (2002) The 'call of social movements' of the Second World Social Forum, Porto Allegre, Brazil, 31 January–5 February 2002. In: *Antipode* 34, 625–32.

Waterman, P. and Wills, J. (2002) (eds.) *Place, Space and the New Labour Internationalisms*. Oxford: Blackwell.

Watson, J.L. (ed.) (1997) *Golden Arches East: McDonald's in East Asia*. Stanford: Stanford University Press.

Watson, S. (2006) *City Publics*. Abingdon: Routledge.

Watts, M. (2002) Alternative modern: Development as cultural geography. In: K. Anderson, M. Domosh, S. Pile and N. Thrift (eds.) *Handbook of Cultural Geography*. London: Sage, 433–53.

Watts, M. (2008) *The Curse of Black Gold*. New York: Powerhouse Press.

Weber, M. (1958) *The Protestant Ethic and the Spirit of Capitalism*. New York: Scribners.

Webster, R.G. and Walker, E.J. (2003) The world is teetering on the edge of a pandemic that could kill a large fraction of the human population. In: *American Scientist* 91, 122.

Weightman, B. (1980) Gay bars as private places. In: *Landscape* 23, 9–16.

Weiss, B. (1997) *The Making and Unmaking of the Haya Lived World*. London: Duke University Press.

Weiss, B. (2002) Thug realism: Inhabiting fantasy in urban Tanzania. In: *Cultural Anthropology* 17(1), 93–124.

Weiss, B (2009). *Street Dreams and Hip Hop Barbershops: Global Fantasy and Popular Practice in Urban Tanzania*. Bloomington: Indiana University Press.

Weizman, E. (2007) *Hollow Land: Israel's Architecture of Occupation*. London: Verso.

Wells, H.G. (1902) *Anticipations of the Reaction of Mechanical and Scientific Progress Upon Human*

Life and Thought. London: Chapman and Hall.
Wells, H.G. (1920) *The Outline of History*. London: Cassell.
Wesely, J. and Gaarder, E. (2004) The gendered 'nature' of the urban outdoors: Women negotiating fear of violence. In: *Gender and Society* 18(5), 645–63.
Wessler, H. (2009) *Public Deliberation and Public Culture: The Writings of Bernhard Peters, 1993–2005*. London: Palgrave Macmillan.
Wex, M. (1979) *'Let's Take Back Our Space': 'Female' and 'Male' Body Language as a Result of Patriarchal Structures*. Berlin: Frauenliteratureverlag Hermine Fees.
Whatmore, S. (2002) *Hybrid Geographies: Natures, Cultures and Spaces*. London: Sage.
Whatmore, S. and Thorne, L. (2000) Elephants on the move: Spatial formations of wildlife exchange. In: Environment and Planning D 18, 185–203.
Wheeler, J.O. (1998) Mapphobia in geography? 1980–1996. In: *Urban Geography* 19(1), 1–5
Which? (1989) No entry. October 1989, 498–501.
While, A., Jonas A.E.G. and Gibbs, D.C. (2004) The environment and the entrepreneurial city: Searching for the urban 'sustainability fix' in Leeds and Manchester. In: *International Journal of Urban and Regional Research* 28: 549–69.
While, A. Jonas, A.E.G. and Gibbs, D.C. (2010) From sustainable development to carbon control: Eco-state restructuring and the politics of regional development. In: *Transactions of the Institute of British Geographers* 35: 76–93.
White, A. (2009) *The Movement of Movements: From Resistance to Climate Justice*. www.stwr.org/the-un-people-politics/copenhagen-the-global-justice-movement-comes-of-age.html. (accessed 4 December 2009).
Whitehead, M. (2003) From moral space to the morality of scale. The case of the sustainable region. In: *Ethics, Place and Environment* 6(3): 235–57.
Whitehead, M. (2006) *Spaces of Sustainability: Geographical perspectives on the Sustainable Society*. Abingdon: Routledge.
Whitehead, M., Jones, R, and Pykett, J. (2011) Governing irrationality, or a more than rational government? Reflections on the rescientisation of decision making in British public policy. In: *Environment and Planning A* 43(12), 2819–37.
Widdowfield, R. (2000) The place of emotions in academic research. In: *Area* 32(2), 199–208.
Wilkinson, R. (1996) *Unhealthy Societies: The Afflictions of Inequality*. London: Routledge.
Williams, A. (ed.) (2007) *Therapeutic Landscapes*. Aldershot: Ashgate.
Williams, A. and Shaw, G. (eds.) (1988) *Tourism and Economic Development*. London: Belhaven.
Williams, B. (1981) *Moral Luck*. Cambridge: Cambridge University Press.
Williams, C. (2005) *A Commodified World?* London: Zed Books.
Williams, C.H. and Smith, A.D. (1983) The national construction of social space. *Progress in Human Geography* 7(4), 502–18.
Williams, G. and Mawdsley, E. (2006) Postcolonial environmental justice: Government and governance in India. In: *Geoforum* 37: 660–70.
Williams, J. (1982) *Dream Worlds*. Berkeley: University of California Press.
Williams, R. (1961) *The Long Revolution*. Harmondsworth: Penguin Books.
Williams, R. (1983) *Keywords. A Vocabulary of Culture and Society*. New York: Oxford University Press.
Williams, R. (1985 [1973]) *The Country and the City*. London: Hogarth Press.
Williams S (2001) *Emotions and Social Theory*. London: Sage Publications.
Williams Paris, J. and Anderson, R.E. (2001) Faith-based queer space in Washington, DC: The Metropolitan Community Church-DC and Mount Vernon Square. In: *Gender, Place and Culture*, 8, 149–68.
Willis, K. (2011[2005]) *Theories and Practices of Development*. Abingdon: Routledge.
Wilson, A. (1992) *The Culture of Nature*. London: Routledge.
Wilson, E (1991) *The Sphinx in the City*. Berkeley, CA: University of California Press.
Wilson, H. (2011) Passing propinquities in the multicultural city: The everyday encounters of bus passengering. In: *Environment and Planning A* 43, 634–49.

Wilson, K. and Young. T. (2008) An overview of Aboriginal health research in the social sciences: Current trends and future directions. In: *International Journal of Circumpolar Health* 67(2–3), 179–89.

Wilson, W.J. (1987) *The Truly Disadvantaged, the Inner City, the Underclass and Public Policy.* Chicago: University of Chicago Press.

Wilson, W.J (1996) *When Work Disappears: The World of the New Urban Poor.* New York: Alfred A. Knopf.

Wilton, R. and Schuer, S. (2006) Towards socio-spatial inclusion? Disabled people, neoliberalism and the contemporary labour market. In: *Area* 38(2), 186–95.

Winchester, H. and White, P. (1988) The location of marginalised groups in the inner city. In: *Environment and Planning D: Society and Space* 6, 37–54.

Winter, T., Teo, P. and Chang, T.C. (eds.) (2008) *Asia on Tour: Exploring the Rise of Asian Tourism.* New York and Abingdon: Routledge.

Withers, C.W.J. (1996) Place, memory, monument: Memorializing the past in contemporary Highland Scotland. In: *Ecumene* 3(3), 325–44.

Withey, L. (1997) *Grand Tours and Cooks' Tours: A History of Leisure Travel, 1750–1915.* New York: W. Morrow.

Wolch, J. and Dear, M. (1993) *Malign Neglect: Homeless in an American City.* San Francisco, CA: Jesse-Bars.

Wolch, J., West, K. and Gaines, T. (1995) Trans-species urban theory. In: *Environment and Planning D* 13, 735–60.

Wolch, J.R. and Emel, J. (1998) *Animal Geographies: Place, Politics, and Identity in the Nature–culture Borderlands.* London: Verso.

Wolch, J. and DeVerteuil, G. (2001) Landscapes of the new poverty management. In J. May and N. Thrift (eds.) *TimeSpace.* London: Routledge, 149–67.

Wolf, E. (1982) *Europe and the People Without History.* University of California Press: Berkeley.

Wolf, F.O. and Mueller, T. (2009) Green new deal: Dead end or pathway beyond capitalism? In: *Turbulence* 5, 12–6.

Wolf, M. (2000) Why this hatred of the market? In F.J. Lechner and J. Boli (eds.) *The Globalization Reader.* Oxford: Blackwell, 9–11.

Women and Geography Study Group (WGSG) (1997) *Feminist Geographies: Explorations in Diversity and Difference.* London: Longman.

Wood, D. (1992) *The Power of Maps.* New York: Guilford Press.

Wood, D., Fels, J. and Krygier, J. (2010) *Rethinking the Power of Maps.* New York and London: Guilford Press.

Wood, N. and Smith, S. (2004) Instrumental routes to emotional geographies. In: *Social and Cultural Geographies* 5, 533–43.

Woods, M. (2000) Fantastic Mr Fox? Representing animals in the hunting debate. In: C. Philo and W. Wilbert (eds.) *Animal Spaces, Beastly Places: New Geographies of Human–animal Relations.* London: Routledge.

Woods, M. (2004) *Rural Geography: Processes, Responses and Experiences in Rural Restructuring.* London: Sage.

Woods, M. (2011) *Rural.* Abingdon: Routledge.

Woodward, K. (ed.) (1997) *Identity and Difference.* London: Sage.

Woodward, K. and Lea, J. (2010) Geographies of affect. In: S. Smith, R. Pain, S. Marston and J.P. Jones (eds.) *The Sage Handbook of Social Geographies.* London: Sage, 154–75.

Woodyer, T (2008) The body as research tool: Embodied practice and children's geographies. In: *Children's Geographies* 6(4), 349–62.

Woolgar, S. (2002) Five rules of virtuality. In: S. Woolgar (ed.) *Virtual Society? Technology, Cyberbole, Reality.* Oxford: Oxford University Press, 1–22.

Worcester, R.M. (1993) Public and elite attitudes to environmental issues In: *International Journal of Public Opinion Research* 5, 315–34.

World Bank (1992) *World Development Report, 1992.* Oxford: Oxford University Press/World Bank.

World Bank (1993) *The East Asian Miracle,* Oxford: Oxford University Press.

World Bank (1994) *Adjustment in Africa: Reforms, Results and the Road Ahead.* Oxford: OUP-World Bank.

World Bank (2010) *Cities and Climate Change: An Urgent Agenda*. World Bank, Washington, DC.

World Bank (2011) *Income Share Held by Highest 10%*. http://data.worldbank.org/indicator/SI.DST.10TH.10, accessed 15 September 2011.

World Bank (2012) *How We Classify Countries*. http://data.worldbank.org/about/country-classifications, accessed 15 January 2012.

World Tourism Organization (2002) *Compendium of Tourism Statistics*. Madrid: World Tourism Organization.

World Travel and Tourism Council (2004) *Global Travel and Tourism Poised for Robust Growth in 2004*. www.wttc.org/News31.htm, accessed 29 July 2004.

World Travel and Tourism Council (WTTC) (2011) *The Review 2011*. Available at: www.wttc.org/site_media/uploads/downloads/WTTC_Review_2011.pdf

Wright, A. and Wolford, W. (2003) *To Inherit the Earth: The Landless Movement and the Struggle for a New Brazil*. Oakland, CA: Food First! Publications.

Wright, M. (2006) *Disposable Women and Other Myths of Global Capitalism*. New York: Routledge.

Wright, M. (2011) Necropolitics, narcopolitics, and femicide: Gendered violence on the US-Mexico Border. In: *Signs: Journal of Women in Culture and Society* 36(3), 707–31.

Wright, P. (1985) *On Living in an Old Country*. London: Verso.

Wright, T. (1995) Tranquility city: Self-organisation, protest and collective gains within a Chicago homeless encampment. In: M.P. Smith (ed.) *Marginal spaces*. New Brunswick, NJ: Transaction Publishers.

Wrigley, N. (1998) Leveraged restructuring and the economic landscape: The LBO wave in US food retailing. In: R.L. Martin (ed.) *Money and the Space Economy*. Chichester: John Wiley.

Wrigley, N. (2002) 'Food deserts' in British cities: Policy context and research priorities. In: *Urban Studies* 39(11), 2029–40.

Wu, J., van Geen, A., Matin, K., Akita, Y., Alam, M.J., Culligan, P.J., Escamilla, V., Feighery, J., Ferguson, A.S., Knappett, P., Mailloux, B.J., McKay, L., Serre, M., Streatfield, P.K., Yunus, M. and Emch, M. (2011a) Increase in diarrheal disease associated with arsenic mitigation in Bangladesh. In: *PLoS One* 6(12), e29593.

Wu, J., Yunus, M., Streatfield, P.K., van Geen, A., Escamilla, V., Akita, J., Serre, M., and Emch, M. (2011b) Impact of tubewell access and depth on childhood diarrhea in Matlab, Bangladesh. In: *Environmental Health* 10, 109.

Wylie, J. (2005) A single day's walking: Narrating self and landscape on the South West Coast Path. In: *Transactions of the Institute of British Geographers* 30, 234–47.

Wylie, J. (2007) *Landscape*. Abingdon: Routledge.

Wylie, J. (2009) Landscape, absence and the geographies of love. In: *Transactions of the Institute of British Geographers* 34, 275–89.

Wylie, J. (2010) Landscape. In: J. Agnew and J. Duncan (eds.) *Handbook of Geographical Knowledge*. London: Sage.

Wynne, B. (1993) Public uptake of science: A case for institutional reflexivity. In: *Public Understanding of Science* 2, 321–37.

Yang, C. (2009) Strategic coupling of regional development in global production networks: Redistribution of Taiwanese personal computer investment from the Pearl River Delta to the Yangtze River Delta, China. In: *Regional Studies* 43(3), 385–407.

Yang, Y.-R.D., Hsu, J.-Y. and Ching, C.-H. (2009) Revisiting the Silicon Island? The geographically varied 'strategic coupling' in the development of high-technology parks in Taiwan. In: *Regional Studies* 43(3), 369–84.

Yao, S. (2001) 'Eating air', 'talking cock': Food, pleasure and the art of lying in Singapore. Paper presented to the Foodscapes: the cultural politics of food in Asia conference, National University of Singapore, 13–15 June 2001.

Yapa, L. (1996) What causes poverty? A postmodern view. In: *Annals of the Association of American Geographers* 86, 707–28.

Yeoh, B. and L. Kong (1994) Reading landscape meanings: State constructions and lived experiences in Singapore's Chinatown. In: *Habitat International* 184, 17–35.

Young, I.M. (1990a) The ideal of community and the politics of difference. In: L. Nicholson (ed.) *Feminism/Postmodernism*. London: Routledge.

Young, I.M. (1990b) Throwing like a girl: A phenomenology of feminine body comportment, motility and spatiality. In: *Throwing Like a Girl and Other Essays in Feminist Philosophy and Social Theory*. Bloomington: University of Indiana Press.

Young, I. M. (1990c) *Throwing Like a Girl and Other Essays in Feminist Philosophy*. Bloomington: Indiana University Press.

Young, J. (1993) *The Texture of Memory: Holocaust Memorials and Meaning*. London: Yale University Press.

Zaloom, C. (2003) Ambiguous numbers: Trading technologies and interpretation in financial markets. In: *American Ethnologist* 31, 258–72.

Zerubavel, Y. (1995) *Recovered Roots: Collective Memory and the Making of Israeli National Tradition*. Chicago, IL: Chicago University Press.

Ziegler, J. (2007) Press conference, United Nations Special Rapporteur on the Right to Food. 26 October. Available at: http://www.un.org/News/briefings/docs/2007/071026_Ziegler.doc.htm (last accessed 25 April 2012).

Zimmer, O. (1998) In search of natural identity: Alpine landscape and the reconstruction of the Swiss nation. In: *Comparative Studies in Society and History* 40(4), 637–65.

Zukin, S. (1982) *Loft Living: Culture and Capital in Urban Change*. Baltimore: Johns Hopkins University Press.

Zukin, S. and DiMaggio, P. (1990) Introduction. In: S. Zukin and P. DiMaggio (eds.) *Structures of Capital. The Social Organization of the Economy*. Cambridge, New York and Sydney: Cambridge University Press.

Zurick, D. (1992) Adventure travel and sustainable tourism in the peripheral economy of Nepal. In: *Annals of the Association of American Geographers* 82(4), 608–28.

Z/Yen (2011) *The Global Financial Centres Index 9*, London: Z/Yen Group and Qatar Financial Centre Authority.

Index

Note: page numbers in **bold** refer to tables; page numbers in *italics* refer to figures.

Section opener image sources

PART 2 Themes

SECTION 1: BIOGEOGRAPHIES

SECTION 2: CARTOGRAPHIES

SECTION 3: CULTURAL GEOGRAPHIES

SECTION 4: DEVELOPMENT GEOGRAPHIES

SECTION 5: ECONOMIC GEOGRAPHIES

SECTION 6: ENVIRONMENTAL GEOGRAPHIES

SECTION 7: HISTORICAL GEOGRAPHIES

SECTION 8: POLITICAL GEOGRAPHIES

SECTION 9: POPULATION GEOGRAPHIES

SECTION 10: SOCIAL GEOGRAPHIES

SECTION 11: URBAN AND RURAL GEOGRAPHIES

PART 3 Horizons

SECTION 1: NON-REPRESENTATIONAL GEOGRAPHIES

© Getty Images

SECTION 2: MOBILITIES

© Kent Miller

SECTION 3: SECURITIES

© Michael Dalder/Reuters/Corbis

SECTION 4: PUBLICS

© Getty Images